Synthetic Strategies in Carbohydrate Chemistry

Edited by

Vinod K. Tiwari
Department of Chemistry, Institute of Science, Banaras Hindu University, Varanasi, Uttar Pradesh, India

ELSEVIER

Elsevier
Radarweg 29, PO Box 211, 1000 AE Amsterdam, Netherlands
The Boulevard, Langford Lane, Kidlington, Oxford OX5 1GB, United Kingdom
50 Hampshire Street, 5th Floor, Cambridge, MA 02139, United States

ISBN: 978-0-323-91729-2

For Information on all Elsevier publications
visit our website at https://www.elsevier.com/books-and-journals

Publisher: Candice Janco
Acquisitions Editor: Gabriela Capille
Editorial Project Manager: Catherine Costello
Production Project Manager: Paul Prasad Chandramohan
Cover Designer: Christian Bilbow

Typeset by MPS Limited, Chennai, India

Contents

List of contributors xiii
Foreword xv
Preface xvii
List of abbreviations xxi

1. Glycosidic bond formation methodology: challenges and impact in oligosaccharide synthesis 1
Rituparna Das and Balaram Mukhopadhyay
   - 1.1 Introduction 1
   - 1.2 Complexity of carbohydrates 2
   - 1.3 General aspects of oligosaccharide synthesis 3
     - 1.3.1 Anomeric effect 4
     - 1.3.2 Background on classical approaches 6
     - 1.3.3 Mechanistic studies 7
   - 1.4 Glycosylation reactions 15
     - 1.4.1 Glycosyl donors 15
     - 1.4.2 Role of acceptors 40
   - 1.5 Conclusions and future outlook 41
   - References 42

2. Recent advances in stereoselective intramolecular *O*-glycosylation 53
Manoj K. Jaiswal, Anindra Sharma, Vinod K. Tiwari and Richard R. Schmidt
   - 2.1 Introduction 53
   - 2.2 Molecular clamping approach 55
   - 2.3 Intramolecular aglycon delivery concept 75
   - 2.4 Leaving group approach 85
   - 2.5 Conclusions and future outlook 90
   - Acknowledgments 91
   - References 91

3. Thioglycoside-based glycosylations in oligosaccharide synthesis 95
Nitish Verma, Maude Cloutier and Charles Gauthier
   - 3.1 Introduction 95

3.2 **Preparation of thioglycosides** 96
3.2.1 One-pot synthesis of thioglycosides from unprotected sugars 99
3.3 **Activation of thioglycoside donors with various promoters** 100
3.4 **Thioglycosides for chemoselective glycosylations** 104
3.4.1 Type A chemoselective glycosylations 105
3.4.2 Type B chemoselective glycosylations 106
3.4.3 Type C chemoselective glycosylations 107
3.4.4 One-pot chemoselective glycosylations 108
3.5 **Orthogonal glycosylations** 111
3.5.1 Orthogonal glycosylations of thioglycoside acceptors with halide glycosyl donors 111
3.5.2 Orthogonal glycosylations between trichloroacetimidate donors and thioglycoside acceptors 113
3.5.3 Orthogonal coupling of thioglycoside acceptors with *N*-phenyltrifluoroacetimidate donors 116
3.5.4 Orthogonal glycosylations of thioglycoside acceptors with other glycosyl donors 116
3.5.5 Thioglycoside-based one-pot orthogonal glycosylations 119
3.6 **Preactivation-based glycosylation strategies** 123
3.6.1 Preactivation glycosylations based on the AgOTf/p-ArSCl promoter system 123
3.6.2 Preactivation-based one-pot glycosylations 128
3.7 **Summary and future outlook** 131
References 132

4. **General strategy for the synthesis of *N*-glycosides** 139
Ariza Khanam and Pintu Kumar Mandal

4.1 **Introduction** 139
4.2 **Nucleosides and their derivatives found in nature** 140
4.3 **General strategy for the chemical synthesis of nucleosides and other *N*-glycosides** 141
4.3.1 Chemical synthesis of nucleosides via *N*-glycosylation 142
4.3.2 Chemical synthesis of *N*-glycosides excluding nucleosides 157
4.3.3 Replacing a hydroxyl group of sugar by *N*-containing groups 165
4.3.4 Direct installation of amine group into sugar moiety 168
4.4 **Conclusions and future outlook** 181
Acknowledgments 183
References 183

5. **The emerging polyhydroxyazepane-based *C*-glycosides for glycosidase inhibition** 187
Sumit and Indrapal Singh Aidhen

5.1 **Introduction** 187

5.2 **Piperidine-based iminosugars as glycosidase inhibitors: background** 190
 5.2.1 Piperidine-based iminosugar *C*-glycosides inspiring entry to azepane-sugars 192
5.3 **The emergence of polyhydroxyazepanes** 194
 5.3.1 Nomenclature and Numbering 194
 5.3.2 Various 1,6-dideoxy-1,6-imino-D/L-heptitols: synthesis and biological activities 195
5.4 **Emerging *C*-glycosides of polyhydroxyazepanes** 211
 5.4.1 *C*-2 functionalized polyhydroxyazepane 220
5.5 **Conclusions and future outlook** 222
References 223

6. **Sialic acid donors: stereoselective chemical and enzymatic *O*-glycosylations** 227
Manoj K. Jaiswal, Vinay K. Pandey and Vinod K. Tiwari

6.1 **Introduction** 227
6.2 **Some notable characteristics of sialic acids** 228
6.3 **Common approaches in chemical sialylation** 234
 6.3.1 Chemical synthesis and glycosylation 234
 6.3.2 Glycodiversification through enzymatic glycosylation 248
 6.3.3 Some glycodiversification through modular triazole forming coupling reactions 259
6.4 **Conclusions and future outlook** 264
Acknowledgments 264
References 265

7. **Common protection–deprotection strategies in carbohydrate synthesis** 273
Bhaswati Ghosh and Suvarn S. Kulkarni

7.1 **Introduction** 273
7.2 **Common protection–deprotection strategies** 273
 7.2.1 Possible protection at the primary position of saccharides 273
 7.2.2 Protecting groups for the secondary hydroxyl groups of carbohydrates 276
 7.2.3 Protecting group at the anomeric position of carbohydrates 277
 7.2.4 Cyclic protecting group of carbohydrates 279
 7.2.5 *N*-protecting groups for 2-amino-2-deoxy-glycosides 282
 7.2.6 One-pot multistep regioselective protection of carbohydrates catalyzed by acids 283
 7.2.7 Acyl migrations in carbohydrate chemistry 286

7.2.8 *De novo* asymmetric synthesis of oligosaccharides using atom-less protecting groups 286
7.2.9 Protecting group strategies for sialic acid derivatives 288
7.2.10 Protecting group strategies toward glycofuranoses 288
7.2.11 Protecting group strategies toward sulfated glycosaminoglycans 289
7.2.12 Orthogonally protected building blocks for automated glycan assembly 289
7.3 Conclusions and future outlook 291
References 292

8. Synthetic chemistry of D-fructose 297
Tanmaya Pathak

8.1 Introduction 297
8.2 D-Fructose in synthetic chemistry 298
8.2.1 Di-D-fructose dianhydrides 298
8.2.2 Partial protection of D-fructose 299
8.2.3 Epoxides derived from D-fructose 301
8.2.4 Chiral ketones from D-fructose 304
8.2.5 1,5-Anhydro-D-fructose 306
8.2.6 Aza-heterocycles from D-fructose 308
8.2.7 Vinyl sulfone-modified fructose derivatives 311
8.2.8 Vinyl selenone-modified fructose derivatives 313
8.2.9 Isonucleosides from D-fructose 317
8.2.10 Biomass and fructose 320
8.3 L-Sorbose in synthetic chemistry 322
8.4 Conclusion and future prospective 328
Acknowledgments 329
References 330

9. Recent synthetic strategies to access diverse iminosugars 335
Camilla Matassini and Francesca Cardona

9.1 Introduction 335
9.2 Monocyclic iminosugars 336
9.2.1 Four-membered iminosugars (azetidines) 336
9.2.2 Five-membered iminosugars (pyrrolidines) 339
9.2.3 Six-membered iminosugars (piperidines) 344
9.2.4 Seven-membered iminosugars (azepanes) and eight-membered iminosugars (azocanes) 349
9.3 Bicyclic iminosugars 350
9.3.1 Pyrrolizidines 350
9.3.2 Indolizidines 354
9.3.3 Quinolizidines 357
9.3.4 Nortropanes 360

9.4 Conclusions and future outlook 361
References 362

10. Vinyl sugar enol ethers in organic synthesis 365
Junaid Shafi Banday, Irshad Ahmad Zargar, Nandagopal Hudait and Debaraj Mukherjee

10.1 Introduction 365
10.2 Synthesis and reactivity of pyranoid glycals 366
10.2.1 Synthesis of pyranoid glycals 366
10.2.2 Synthesis of aryl *C*-glycosides 367
10.2.3 Synthesis of *C*-glycosides 373
10.2.4 Synthesis of nucleosides 377
10.2.5 Synthesis of *N*-glycosides from pyranoid glycals 379
10.2.6 Synthesis of *O*-glycosides from pyranoid glycals 381
10.2.7 C2 functionalization of pyranoid glycals 386
10.2.8 One-pot C1 and C2 functionalization of pyranoid glycals 389
10.3 Synthesis and reactivity of furanoid glycals 393
10.3.1 Synthesis of furanoid glycals 393
10.3.2 Synthesis of natural products from furanoid glycals 396
10.3.3 Synthesis of nucleosides from furanoid glycals 397
10.3.4 Synthesis of *C*-aryl glycosides from furanoid glycals 401
10.4 Conclusions and future outlook 403
References 403

11. Synthesis of carbohydrate-containing macrocycles 407
Anji Chen and Guijun Wang

11.1 Introduction 407
11.2 Natural product-based glycomacrocycles 408
11.2.1 Major classes of sugar-based naturally occurring macrocycles 408
11.2.2 Synthesis of Clemoarmanoside A and Clemahexapetoside A and their analogs 408
11.2.3 Synthesis of Forsythenethoside A 412
11.2.4 Synthesis of Ananatoside A and analogs 413
11.2.5 Synthesis of Ipomoeassin F and analogs 415
11.2.6 Synthesis of resin glycoside Calysolin IX 425
11.2.7 Synthesis of macrocycles by interrupted Pummerer reaction-mediated glycosylation 428
11.2.8 Synthesis of Batatin VI 434
11.2.9 Synthesis of ellagitannins and other related strained glycomacrocycles 437
11.2.10 Synthesis of glycoterpenoid and steroid derived glycomacrocyles 445

11.3 Sugar-based crown ether, azacrown ether, and other crown ether-like macrocycles 446
11.3.1 Monosaccharide-based crownether derivatives formed by vicinal diol 447
11.3.2 Synthesis of other crown ether-like macrocycles 450
11.3.3 Sucrose-based crown ether and other macrocycles 457
11.4 Synthesis of macrocycles by C–C bond forming reactions 464
11.4.1 Synthesis of macrocycles by RCM strategy 464
11.4.2 Synthesis of sugar-fused macrocyclic lactones via Sonogashira and Heck reaction 469
11.4.3 Synthesis of macrocycles via acetylene coupling reaction 470
11.4.4 Macrocyclization via aldol condensation reactions 472
11.5 Synthesis of sugar-embedded macrocycles by reductive amination, macrolactamization, and macrolactonization 473
11.5.1 Synthesis of aza-linked sugar-based macrocycles 473
11.5.2 Synthesis of macrocycles by macrolactamization 476
11.5.3 Synthesis of macrocycles by macrolactonization 479
11.6 Synthesis of macrocycles via click macrocyclizations 482
11.6.1 Pentose sugar-embedded macrocycles via click macrocyclizations 482
11.6.2 Hexose sugar-embedded macrocycles via click macrocyclization 489
11.6.3 Synthesis of sucrose-based macrocycle via click chemistry 497
11.6.4 Macrocyclization by glycosylation 497
11.6.5 Synthesis of cyclodextrin analogs by glycosylation 498
11.6.6 Synthesis of azobenzene-containing glycomacrocycles by glycosylation 507
11.6.7 Azobenzene-containing sugar-based macrocycles 509
11.6.8 Synthesis of pyranose-based macrocycle libraries by intramolecular S_NAr reaction 512
11.6.9 Synthesis of cyclooligosaccharide derivatives for gene delivery applications 512
11.7 Conclusion and future perspective 515
Acknowledgement 516
References 516

12. Recent advances in synthesis of diverse glycopeptides and glycohybrids 523
Rajdeep Tyagi, Kavita Singh, Vinay Kumar Mishra and Ram Sagar

12.1 Introduction 523
12.2 Synthesis of O-linked glycopeptides 524
12.3 Synthesis of S-linked glycopeptides 545

12.4	**Synthesis of C-linked glycopeptides**	551
12.5	**Synthesis of N-linked glycopeptides**	558
	12.5.1 Glycopeptides containing asparagine	558
	12.5.2 Glycopeptides containing arginine	568
	12.5.3 Click chemistry glycosylation	571
	12.5.4 Hybrid molecule synthesis	577
12.6	**Recent developments for the synthesis of glycohybrids**	589
	12.6.1 Anticancer glycohybrids	590
	12.6.2 Antibacterial glycohybrids	598
	12.6.3 Antimalarial glycohybrids	599
	12.6.4 Antiviral glycohybrids	600
	12.6.5 Acetylcholinesterase inhibitor glycohybrids	600
	12.6.6 Glycosidase inhibitor glycohybrids	601
	12.6.7 Carbonic anhydrase inhibitor glycohybrids	602
	12.6.8 Antifungal glycohybrids	603
12.7	**Conclusions and future outlook**	604
References		604

13. Carbohydrate-based chiral organocatalysts 611

Vinod K. Tiwari and Hélène Pellissier

13.1	**Introduction**	611
13.2	**Organocatalysts: historical background and basic aspects**	612
	13.2.1 Historical background	612
	13.2.2 Basic aspects of organocatalysts	615
13.3	**Carbohydrates as potent scaffolds for organocatalysis**	619
13.4	**Carbohydrate-based ketones: organocatalysts for enantioselective epoxidation**	620
13.5	**Carbohydrate-based prolinamides, pyrrolidine amides and pyrrolidines: organocatalysts for aldol and Michael reactions**	629
13.6	**Carbohydrate-derived amino alcohols: organocatalysts for asymmetric aldol and Mannich reactions**	640
13.7	**Carbohydrate-based thiourea amines: bifunctional organocatalysts for Michael, Biginelli, aza-Henry, and Mannich reactions**	646
	13.7.1 Carbohydrate-based thiourea-primary amine bifunctional organocatalysts	646
	13.7.2 Carbohydrate-based thiourea-secondary amine bifunctional organocatalysts	653
	13.7.3 Carbohydrate-based thiourea-tertiary amines bifunctional organocatalysts	655
13.8	**Other organocatalysts derived from carbohydrates**	665
	13.8.1 Organocatalysts with amine functionality	665
	13.8.2 Bifunctional thiourea/urea organocatalysts	668
13.9	**Conclusions and future outlook**	672
References		672

14. Emerging aspects of domino reactions in carbohydrate Chemistry 679

Mangal S. Yadav, Abhishek Gupta, Sunil Kumar and Vinod K. Tiwari

14.1 Introduction 679
14.2 Asymmetric tandem reactions using chiral carbohydrate derivatives 680
14.3 Organocatalyzed tandem reaction 684
14.4 Glycals in domino reaction 691
14.5 Reductive ring opening in tandem reactions of carbohydrates 698
14.6 Miscellaneous examples 701
14.6.1 Butenolides and spiro compound 701
14.6.2 Sugar ring contraction 708
14.6.3 Levoglucosenone 709
14.6.4 Macrocyclic synthesis 713
14.6.5 Sugar ligands 714
14.6.6 Glycosylation 716
14.7 Conclusions and future outlooks 723
Acknowledgments 724
References 724

Index 731

List of contributors

Indrapal Singh Aidhen Department of Chemistry, Indian Institute of Technology Madras, Chennai, Tamil Nadu, India

Junaid Shafi Banday Natural Products & Medicinal Chemistry Division, CSIR-Indian Institute of Integrative Medicine (IIIM), Jammu, Jammu and Kashmir, India

Francesca Cardona Department of Chemistry "Ugo Schiff" (DICUS), Sesto Fiorentino, Italy

Anji Chen Department of Chemistry and Biochemistry, Old Dominion University, Norfolk, VA, United States

Maude Cloutier Unité mixte de recherche (UMR) INRS-UQAC, Centre Armand-Frappier Santé Biotechnologie, Institut national de la recherche scientifique (INRS), Laval & Chicoutimi (Qc), Canada

Rituparna Das Sweet Lab, Department of Chemical Sciences, Indian Institute of Science Education and Research Kolkata, Nadia, West Bengal, India

Charles Gauthier Unité mixte de recherche (UMR) INRS-UQAC, Centre Armand-Frappier Santé Biotechnologie, Institut national de la recherche scientifique (INRS), Laval & Chicoutimi (Qc), Canada

Bhaswati Ghosh Department of Chemistry, Indian Institute of Technology Bombay, Powai, Mumbai, Maharashtra, India

Abhishek Gupta Department of Chemistry, Institute of Science, Banaras Hindu University, Varanasi, Uttar Pradesh, India

Nandagopal Hudait Natural Products & Medicinal Chemistry Division, CSIR-Indian Institute of Integrative Medicine (IIIM), Jammu, Jammu and Kashmir, India; Chemical Sciences, Bose Institute, Salt Lake City, Kolkata, West Bengal, India

Manoj K. Jaiswal Department of Chemistry, Institute of Science, Banaras Hindu University, Varanasi, Uttar Pradesh, India

Ariza Khanam Medicinal and Process Chemistry Division, CSIR-Central Drug Research Institute, Lucknow, Uttar Pradesh, India

Suvarn S. Kulkarni Department of Chemistry, Indian Institute of Technology Bombay, Powai, Mumbai, Maharashtra, India

Sunil Kumar Department of Chemistry, Institute of Science, Banaras Hindu University, Varanasi, Uttar Pradesh, India

Pintu Kumar Mandal Medicinal and Process Chemistry Division, CSIR-Central Drug Research Institute, Lucknow, Uttar Pradesh, India; Academy of Scientific and Innovative Research (AcSIR), Ghaziabad, Uttar Pradesh, India

Camilla Matassini Department of Chemistry "Ugo Schiff" (DICUS), Sesto Fiorentino, Italy

Vinay Kumar Mishra Department of Chemistry, Institute of Science, Banaras Hindu University (BHU), Varanasi, Uttar Pradesh, India

Debaraj Mukherjee Natural Products & Medicinal Chemistry Division, CSIR-Indian Institute of Integrative Medicine (IIIM), Jammu, Jammu and Kashmir, India; Chemical Sciences, Bose Institute, Salt Lake City, Kolkata, West Bengal, India

Balaram Mukhopadhyay Sweet Lab, Department of Chemical Sciences, Indian Institute of Science Education and Research Kolkata, Nadia, West Bengal, India

Vinay K. Pandey Department of Chemistry, Institute of Science, Banaras Hindu University, Varanasi, Uttar Pradesh, India

Tanmaya Pathak Department of Chemistry, Indian Institute of Technology Kharagpur, Kharagpur, West Bengal, India

Hélène Pellissier Aix-Marseille Univ, CNRS, Centrale Marseille, iSm2, Marseille, France

Ram Sagar Glycochemistry Laboratory, School of Physical Sciences, Jawaharlal Nehru University, New Delhi, India; Department of Chemistry, Institute of Science, Banaras Hindu University (BHU), Varanasi, Uttar Pradesh, India

Richard R. Schmidt Department of Chemistry, University of Konstanz, Konstanz, Baden-Württemberg, Germany

Anindra Sharma Department of Chemistry, Institute of Science, Banaras Hindu University, Varanasi, Uttar Pradesh, India

Kavita Singh Glycochemistry Laboratory, School of Physical Sciences, Jawaharlal Nehru University, New Delhi, India

Sumit Department of Chemistry, Indian Institute of Technology Madras, Chennai, Tamil Nadu, India

Vinod K. Tiwari Department of Chemistry, Institute of Science, Banaras Hindu University, Varanasi, Uttar Pradesh, India

Rajdeep Tyagi Glycochemistry Laboratory, School of Physical Sciences, Jawaharlal Nehru University, New Delhi, India

Nitish Verma Unité mixte de recherche (UMR) INRS-UQAC, Centre Armand-Frappier Santé Biotechnologie, Institut national de la recherche scientifique (INRS), Laval & Chicoutimi (Qc), Canada

Guijun Wang Department of Chemistry and Biochemistry, Old Dominion University, Norfolk, VA, United States

Mangal S. Yadav Department of Chemistry, Institute of Science, Banaras Hindu University, Varanasi, Uttar Pradesh, India

Irshad Ahmad Zargar Natural Products & Medicinal Chemistry Division, CSIR-Indian Institute of Integrative Medicine (IIIM), Jammu, Jammu and Kashmir, India

Foreword

Carbohydrates are fascinating molecules. They are abundant in nature, structurally diverse, and complex and play many important roles in life, materials, energy, health, disease, etc. The love for carbohydrates brought us together to a joyful journey exploring carbohydrate chemistry with a focus on their structures, synthesis, properties, functions, and applications.

I got to know Prof. Vinod Kumar Tiwari back in 2007 when he was a bright young organic chemistry lecturer at the Banaras Hindu University (BHU) in Varanasi, India and I was in the pretenure years of my independent academic career in the Department of Chemistry at the University of California, Davis. Vinod spent a year working with us on developing synthetic strategies of introducing modification to a fascinating sugar, sialic acid, by chemically modifying its six-carbon precursor that could be tolerated by carbohydrate biosynthetic enzymes to obtain complex biologically important structures containing modified sialic acids. We have now further developed strategies and efficient processes for accessing this class of synthetically challenging glycans and glycoconjugates containing sialic acids with naturally occurring or nonnatural modifications. We enjoyed spending time on research and involving in stimulating discussions, a habit continues throughout our careers.

In his independent career, Vinod has been a great mentor and a leader in the carbohydrate field. The *Synthetic Strategies in Carbohydrate Chemistry* that he edited and the chapters that he contributed as the corresponding author reflect his passion and research interest in developing synthetic methodologies that can be applied to advancing carbohydrate-based and carbohydrate-inspired materials and therapeutics. The book provides an up-to-date progress and advances on several important topics, including chemical glycosylation strategy developments and considerations for donor selection and preparation, protection, and deprotection, as well as regio- and stereoselectivity control; iminosugar and polyhydroxyazepane glycomimetics; innovative molecules synthesized from sugars; and the synthesis and application of diverse sugar-containing materials. It is an informative guide for readers who are interested in getting to know recent advances in carbohydrate chemistry and a handy reference book for practitioners in the field.

Xi Chen
Department of Chemistry, University of California-Davis, Davis, CA, United States

Preface

Due to the number of fascinating features displayed by carbohydrates, such as their structural diversity, hydrophilicity, biocompatibility, bioavailability, transport, minimal toxicity, superior ADME properties, and energy storage, there is an increased demand for exploring them as versatile scaffolds in organic synthesis for their complete chemical, biochemical, and medicinal investigations. Furthermore, carbohydrates are the key components of many important biological processes in and between all kingdoms of life that play pivotal roles notably, modulation of protein function, intercellular adhesion, malignant transformation, signal transduction, and viral and bacterial cell surface recognition. These features essentially formulate them to be quite valuable scaffolds of pharmacological importance. Moreover, high functionality, chiral-pool character, and the readily availability of carbohydrates as one of the most abundant natural products are some additional structural features that further enhance their utilities in synthetic organic chemistry of diverse applications. In this respect, we present a judicious endeavor to communicate extensive information about the growing importance of carbohydrates in organic synthesis, including, glycosylations, catalysis, molecular scaffolding, chemical diversity, macrocyclization, and tandem synthesis, to the readers.

The main objective of this book is to introduce a concept of multidisciplinary opportunity with a critical discussion about the diverse emerging relevance of carbohydrates in chemistry, biology, and medicine. This book covers 14 different topics that illustrate the scope of synthetic carbohydrate chemistry in various fields. Several leading experts who are very active in this field have contributed to this Elsevier venture and I am really grateful to them for their excellent contributions.

The glycosidic bond formation with desired selectivity in glycosidic linkages both stereo (anomeric) and regioselectivity is the key objective for the chemical synthesis of oligosaccharides. In Chapter 1, Mukhopadhyay and Das described the general consideration with a growing impact on the desired selectivities in conventional glycosylation methodologies. Furthermore, Schmidt and Tiwari in Chapter 2 briefly discuss some new insight into the three major variants in intramolecular glycosylations such as leaving group–based method, functional substituent–based method, and rigid spacer–based method through IAD for the stereoselective synthesis of bioactive *O*-glycosides. Charles Gauthier in Chapter 3 presents an interesting perspective of thioglycoside-based donors in the chemical glycosylation strategies for easy access to biologically relevant oligosaccharides. In Chapter 4, Pintu Mandal highlights interesting aspects of chemical *N*-glycosylation to access a diverse range of structurally significant *N*-glycosides, the *N*-linked glycoproteins, antibiotics, and nucleosides.

Iminosugars offer to mimic the biological action of carbohydrates and represent the most important class of glycomimetics. Indrapal Singh Aidhen in Chapter 5 briefly covers the emerging aspects of the development of diverse polyhydroxyazepanes as a seven-membered analogue of iminosugar-based *C*-glycosides as potential glycosidase inhibitors. In Chapter 6, Tiwari et al. presented a concise overview of the stereoselective chemical and enzymatic *O*-glycosylations using various sialic acid donors.

In Chapter 7, Ghosh and Kulkarni present some fascinating aspects of common protection—deprotection strategies at the different positions of various carbohydrates with future perspectives in glycoscience. With the advent of research on biomass and its application in generating "petroleum" products, D-fructose is an inexpensive ketose afforded commercially available 5-hydroxymethylfurfural and a promising biofuel dimethylfuran. L-sorbose, the 5-epimer of D-fructose, is equally useful as an inexpensive raw material. Therefore Tanmaya Pathak in Chapter 8 strives to highlight the growing synthetic applications of D-fructose and L-sorbose.

In Chapter 9, Cardona and Matassini presented a concise overview of recent synthetic strategies to access diverse classes of iminosugars. Mukherjee et al. in Chapter 10 explore the reactivity of sugar-based vinyl ethers in diversity-oriented syntheses of biological importance. In Chapter 11, Anji Chen and Guijun Wang highlight the synthesis of a diverse range of carbohydrate-based macrocycles having fascinating features, including their drug-likeness potency, used as ligand in catalysis, and also ion complexation capacity. Further, Ramsagar et al. in Chapter 12 present a comprehensive overview of recent advances in the synthesis of diverse glycopeptides and glycohybrids. The growing impact of sugar-based organocatalysts in a diverse range of asymmetric reactions with high stereoselectivities is highlighted by Tiwari and Pellissier in Chapter 13. At the end, a brief discussion about the emerging aspects of highly efficient Domino reactions explored in carbohydrate chemistry with the involvement of chiral sugar in various transformations, including chiral catalysis, ring fusion, and ring fragmentation, is briefly covered by Tiwari et al. in Chapter 14.

The chapters are devoted to diverse glycosides that have played a pivotal function in shaping the field of glycoscience and continue to pave the way toward our understanding of stereoselective glycosylations, macrocyclization, asymmetric catalysis, and Tandem synthesis of bioactive candidates. Further, topics such as basic and advanced aspects of carbohydrates, including iminosugars and sialic acids, are briefly covered in this book.

I sincerely thank all the contributors who are eminent scientists actively working on various aspects of synthetic carbohydrate chemistry. In addition to Dr. Anindra Sharma, Dr. Abhijeet Kumar, and Dr. Sanchayita Rajkhowa, I also thank all my PhDs, namely, Dr. Anand K. Agrahari, Mangal S Yadav, and Manoj K. Jaiswal who help me for the referencing and proofread of this project. My special thanks go to Prof. Richard R. Schmidt, Dr. R. P. Tripathi, Prof. Xi Chen, Prof. Peter Seeberger, Prof. S. Chandrasekaran, Prof. Robert Field, Prof. Carmen Galan, Prof. Alexei Demchenko, Prof. D. Basavaiah, Prof. Ganesh Pandey, Dr. P. K. Tripathi, Prof. G. S. Singh, and Dr. Hai Yu for their valuable suggestions and all the learned experts who provide constructive comments to make our book more interesting and useful for the readers and investigators working on a broad aspect of synthetic carbohydrate chemistry, glycosylation methodology, organic synthesis, catalysis, macrocyclization, glycopeptides, medicinal chemistry, and drug discovery and development. I would like to express my heartiest thanks and deep sense of appreciation to Prof. Xi Chen, Professor at the University of California, Davis for her consistent support and moreover for writing the venerated foreword to the Elsevier book.

Last but not least, I would like to express my deep sense of appreciation to Dr. Joseph Hayton, publishing director, and the entire editorial team associated with Elsevier Inc. notably Ms. Catherine Costello, Ms. Emily McCloskey, and Mr. Paul P. Chandramohan for their keen interest and throughout timely support in publishing this theme with the high standard of publication maintained in bringing out this volume on *Synthetic Strategies in Carbohydrate Chemistry*' with Elsevier Inc.

Vinod K. Tiwari

List of abbreviations

Ac	acetyl
Ac_2O	acetic anhydride
AcCN	acetonitrile
AcOH	acetic acid
AF	1,5-anhydro-D-fructose
AI	artificial intelligence
AIBN	azobisisobutyronitrile
All	allyl
AllTMS	allyl trimethyl silane
Aloc	allyloxycarbonyl
aq.	aqueous
Asp	asparagine
ASSPC	active-site-specific pharmacological chaperone
ATP	adenosine triphosphate
AZMB	2-(azidomethyl)benzoyl
$B(C_6F_5)_3$	pentafluorophenyl borane
BDSB	bromo(diethyl)sulfonium bromo(pentachloro)antimonate
$BF_3.OEt_2$	Boron trifluoride diethyl etherate
BHQ	2-*tert*-butylhydroquinone
BHT	dibutylhydroxytoluene
BIAB	bis (imidazolylazo)benzene
Bn	Benzyl
Boc	*tert*-butoxycarbonyl
$(Boc)_2O$	Di-*tert*-butyl dicarbonate
BODIPY	boron dipyrromethene
BOP	(bnzotriazol-1-yloxy)tris(dimethylamino)phosphonium hexafluorophosphate
BTM	benzotetramisole
BtOH	hydroxybenzotriazole
Bu	butyl
Bz	benzoyl
CAN	ceric ammonium nitrate
cat.	catalytic
CbzCl	benzyl chloroformate
CD_3OD	deuterated methanol
CDs	cyclodextrins
$CHCl_3$	chloroform
ClAc	chloroacetyl
CMF	5-chloromethylfurfural

CMPI	2-chloro-1-methylpyridinium iodide
CMP-Sia	cytidine 5′-monophosphate-sialic acid
CS/DS	chondroitin sulfate/dermatan sulfate
CSA	camphorsulfonic acid
CuAAC	copper- catalyzed azide alkyne cycloaddition
DAST	diethyl aminosulphur trifluoride
DBDMH	1,3-dibromo-5,5-dimethylhydantoin
DBTO	dibenzothiophene-5-oxide
DBU	1,8-diazabicyclo[5.4.0]undec-7-ene
DCC	*N,N'*-dicyclohexylcarbodiimide
DCE	dichloroethane
DDQ	2,3-dichloro-5,6-dicyano-1,4-benzoquinone
DE	diastereomeric excess
DEAD	diethyl azodicarboxylate
DFAs	di-D-fructose dianhydrides
DFT	density functional theory
DIAD	diethylazodicarboxylate
DIBALH	diisobutylaluminum hydride
DIC	diisopropylcarbodiimide
DIPEA	diisopropyl ethyl amine
DISAL	methyl 3,5-dinitrosalicylate
DMA	dimethylacetamide
DMAP	4-dimethylamino-pyridine
DMB	4,4-dimethoxybenzophenone
DMC	2-chloro-1,3-dimethylimidazolinium chloride
DMDO	dimethyl dioxirane
DME	dimethyl ether
DMF	*N,N*-dimethylformamide
DMJ	deoxymannojirimycin
DMM	dimethoxymethane
DMNPA	2,2-dimethyl-2-(ortho-nitrophenyl)acetyl
DMSO	dimethyl sulphoxide
DMTST	dimethyl(methylthio)sulfonium trifluoro methanesulfonate
DNJ	1-deoxynojirimycin
DR	diastereomeric ratio
DPPB	1,4-bis(diphenyl phosphine)butane
DTBMP	2,6-di-*tert*-butyl-4-methylpyridine
DTT	DL-1,4-dithiothreitol
EDC	1-ethyl-3-(3-dimethylaminopropyl)carbodiimide
EDG	electron donating group
EDTA	ethylenediamine tetra acetic acid
ee	enantiomeric excess
EEDQ	*N*-ethoxycarbonyl-2-ethoxy-1,2-dihydroquinoline
equiv	equivalent
EtOAc	ethyl acetate
EtOH	ethanol
$EtSSnBu_3$	ethyl tributylstannylsulfide

EWG	electron withdrawing group
FDA	Food and Drug Administration
Fmoc	fluorenyl-9-methoxycarbonyl
Fuc*p*	L-fucopyranose
GAGs	glycosaminoglycans
Gal	galactose
Gal*f*	D-galactofuranose
GalNAc	*N*-acetylgalactosamine
Galp	D-galactopyranose
GalpNAc	*N*-acetyl D-galactosamine
GD3	disialyl ganglioside
GlcpNAc	*N*-acetyl D-glucosamine
Glu	glucose
GM3	monosialylated glycan
GN-Tf	N,N′-di-Boc-*N*-triflyl-guanidine
HAD	hydrogen bond-mediated aglycone delivery
HATU	hexafluorophosphate azabenzotriazole tetramethyl uronium
HBTU	2-(1*H*-benzotriazol-1-yl)-1,1,3,3-tetramethyluronium Hexafluorophosphate
$HClO_4$-SiO_2	silica-supported perchloric acid
Hep/HS	heparin/heparin sulfate
$Hf(OTf)_4$	hafnium trifluoromethanesulfonate
HHDP	2,4-*O*-hexahydroxydiphenoyl
HIV	human immunodeficiency virus
HMDS	hexamethyldisilane
HMF	5-hydroxy-methylfurfural
HMPA	hexamethylphosphoramide
HOAt	1-hydroxy-7-azabenzotriazole
HOBt	1-hydroxybenzotriazole
HPLC	high-performance liquid chromatography
HyLys	hydroxy-lysine
HyP	hydroxyproline
HyPro	hydroxy-proline
IAD	intramolecular aglycon delivery
IBr	iodine monobromide
ICl	iodine monochloride
IDCP	iodonium dicollidine perchlorate
IPRm	interrupted Pummerer reaction-mediated glycosylation
IUPAC	International Union of Pure and Applied Chemistry
K_2CO_3	potassium carbonate
KDN	deaminoneuraminc acid
L.g	leaving group
LC-MS	liquid chromatography-mass spectrometry
LDA	lithium diisopropylamide
LED	light-emitting diode
Lev	levulinoyl
ManNAc	*N*-acetylmannosamine
***m*-CPBA**	meta-chloroperbenzoic acid

MDR	multidrug resistant
MeOH	methanol
MMPP	magnesium *bis*(monoperoxyphthalate) hexahydrate
MOM	methoxy methyl ether
MOM-Cl	chloromethyl methyl ether
Ms	mesyl
MS	molecular sieves
MsCl	methanesulfonyl chloride
MsCl	methansulfonyl chloride
MTBE	methyl-terz-butyl.ether
NAP	2-naphthylmethyl
Naph	naphthyl
NBD	nitrobenzoxadiazole
NBS	*N*-bromosuccinimide
***N*-Cbz**	*N*-carboxybenzyl
N-DMP	*N*-(2,4-dimethoxyphenyl)
Neu5Ac	*N*-acetyl- neuraminic acid
Neu5Gc	*N*-glycolylneuraminic acid
NFSI	*N*-fluorobenzenesulfonimide
NHC	*N*-heterocyclic Carbene
$NiCl_2dppe$	dichloro (1,2-bis(diphenylphosphino)ethane)nickel
NIS	N-iodosuccinimide
NMO	*N*-methyl morfoline *N*-Oxide
NMP	*N*-methyl-2-pyrrolidone
NMR	nuclear Maganetic Resonance
NOE	nuclear overhauser effect
Nosyl-Cl	*p*-nitrobenzenesulfonyl chloride
***N*-Phth**	*N*-trichloroacetyl
***N*-TFA**	*N*-trifluoroacetyl
***N*-Troc**	*N*-trichloroethoxycarbonyl
OPME	One-pot multi-enzyme
OPSB	*O*-2-[(propan-2-yl)sulfinyl]benzyl
OPTB	*O*-2-[(propan-2-yl)thio]benzyl
PC	pharmacological chaperones
PCC	pyridinium chlorochromate
$Pd(dba)_2$	palladium(0) *bis*(dibenzylideneacetone)
PDC	pyridinium dichromate
PEG-600	polyethylene glycol
PG	protecting group
Pico	picoloyl
Piv	pivaloyl
PMB	*p*-methoxybenzyl
PMB-Cl	*p*-methoxybenzyl chloride
PPG	phenylpropiolate glycosyl
PPh_3	triphenyl phosphine
PPTS	pyridinium *p*-toluenefulfonate
PTBD	1,5,7-triazabicyclo[4.4.0]dec-5-ene

PTC	phase transfer catalyst
PTFAI	*N*-phenyl-2,2,2-trifluoroacetimidate
***p*-TSA**	*p*-toluene sulfonic acid
Py	pyridine
r.t.	room temperature
RCM	ring-closing olefin metathesis
RRV	relative reactivity values
RTIL's	room temperature ionic liquids
RuAAC	ruthenium-catalyzed azide alkyne cycloaddition
SA	sulfuric acid
SBox	*S*-benzoxazolyl
sCO_2	superctitical carbon dioxide
SDMP	thioaglycons [*S*-dimethyl phenyl]
Ser	serine
SET	single electron transfer
SGLT2	sodium-dependent glucose transporter 2
SGM	sialylated glycan microarray
S_N1	substitution nucleophilic unimolecular
S_N2	substitution nucleophilic bimolecular
S*o*EP	*O*-ethylphenylsulfanyl
SPOS	solid-phase oligosaccharide synthesis
SPPS	solid phase peptide synthesis
SPSB	S-2-[(propan-2-yl)sulfinyl]benzyl
SRT	substrate reduction therapy
ST	sialyltransferase
SZ	sulphated zirconia
T_3P	proplylphosphonic acid cyclic anhydride
TBAB	tetrabutylammonium bromide
TBAF	tetrabutylammonium fluoride
TBAHS	tetrabutylammonium hydrogen sulfate
TBAI	tetrabutylammonium iodide
TBDMS	*tert*-butyldimethylsilyl
TBDMSCl	*tert*-butyldimethylsilylchloride
TBDPS	*tert*-butyldiphenylsilyl
TBS	*tert*-butyldimethylsilyl
TBSCl	*tert*-butyldimethylsilyl chloride
TBTA	Tris(benzyltriazoylmethyl)amine
TBTU	(1*H*-benzotriazol-1- yl)1,1,3,3-tetramethyl uranium
***t*-Bu**	*tert*-Butyl
***t*-BuONa**	sodium *tert*-butoxide
TCAI	trichloroacetimidate
TCBA	2,4,6-trichlorobenzoic acid
TCT	2,4,6-Tricholoro-1,3,5-triazine
TcTS	trypanosoma cruzi trans-sialidase
TEA	triethylamine
TEMPO	2,2,6,6-tetramethylpiperidine 1-oxyl radical
TES	triethylsilyl

Tf	trifluoromethanesulfonyl
Tf_2O	trifluoromethansulfonic anhydride
TFA	trifluoroacetic acid
THF	tetrahydrofurane
Thr	threonine
TIPDS	tetraisopropyl disiloxanylidene
TIPS	triisopropylsilyl
TMP	2,4,6-timethylpyridine
TMS	trimethyl silane
$TMS)_2S$	bis(trimethylsilyl)sulfide
TMSI	trimethylsilyl iodide
$TMSN_3$	trimethyl silyl azide
TMSOTf	trimethylsilyl trifluoromethanesulfonate
Tol	*p*-methylbenzene
TPPT-Cl	triaryl phosphorothionate chloride
TrCl	trityl chloride
Troc	2,2,2-trichloroethoxycarbonyl
Ts	Tosyl
TsCl	*p*-toluensulfonyl chloride
TTMPP	tris(2,4,6-trimethoxyphenyl)phosphine
Tyr	tyrosine
UV	ultraviolate
$Yb(OTf)_3$	ytterbium(III) trifluoromethanesulfonate
α	alpha
β	beta
μM	micro molar
[BMIM][Cl]	Butylmethylimidazolium chloride
°C	degree Celsius
2-FPyr	2-fluoropyridine

Chapter 1

Glycosidic bond formation methodology: challenges and impact in oligosaccharide synthesis

Rituparna Das and Balaram Mukhopadhyay
Sweet Lab, Department of Chemical Sciences, Indian Institute of Science Education and Research Kolkata, Nadia, West Bengal, India

1.1 Introduction

In our daily life, we often hear carbohydrates being randomly mentioned as "Pass the sugar please!"; "You won't have energy if you don't have sweets"; "He's suffering from sugar (Diabetes)" and so on. But the sugar cubes on our tea table every morning are much more than a few crystals of sweetness. They are carbohydrates, the most abundant biomolecule on earth.[1] Carbohydrates are ubiquitous in nature, which remains widely distributed in nature in various living organisms and microbes.[2] They are one of man's most versatile weapons to solve and battle the various mysteries of nature. For decades, carbohydrates were considered simply as the powerhouse of energy. Carbohydrates establish a direct link between inert solar energy and biologically usable chemical energy required by living organisms. The solar energy is trapped within the carbohydrate molecules by the green plants and blue-green algae through the process of photosynthesis. These produced carbohydrate molecules that are further harvested by nonphotosynthesizing organisms to generate huge amounts of chemical energy (adenosine triphosphate, ATP) by the process of aerobic glycolysis. This energy was utilized by living organisms to perform various life processes. One rough estimate suggests that 4×10^{11} tonnes of carbohydrates are biosynthesized each year on the earth by plants and photosynthesizing bacteria. Apart from their energy storage functions, carbohydrates were assumed to be structural and protective components of the living cells like chitins in arthropodal species and cellulose in plants.

Synthetic Strategies in Carbohydrate Chemistry. DOI: https://doi.org/10.1016/B978-0-323-91729-2.00014-8

Carbohydrate-based studies took a backseat in the backdrop of the versatility of another important class of biomolecules, protein, and nucleic acids. With the discovery of new genomes, DNA and proteins were considered as the only driving force of life. But with subsequent research, slowly but gradually, another unsolved code essential for life came into the limelight. This code proved to be a more subtle, yet much more complex alphabet in the language of life. This code unveiled itself as the most trivial item in our daily life, the sugars or rather carbohydrates. The later part of the 20th century revealed the extensive role of carbohydrates in various biological processes.[3] Carbohydrate conjugates like glycoproteins and glycolipids[4] are responsible for controlling diverse phenomena like protein folding, blood group specificity, cell growth and proliferation, embryogenesis, fertilization, and neuronal development. The carbohydrates also interact with the proteins in the cells and act as information carriers, thereby participating in central biochemical processes like cell–cell recognition, communication, differentiation and adhesion,[5] tumor and metastasis development,[6] and viral and bacterial pathogenicity with consecutive immunological recognition.[7] The unique projection of the glycoconjugates present on the external cell surface membrane is the stepping stone for various vital life processes. Thus over the past few decades in the "postgenomic era," there is a huge surge of interest in the study of glycoscience and glycochemistry, which is more traditionally known as carbohydrate chemistry.[8] Thus there was the origin of the new branch of science known as "Glycomics" attributing this class of molecules as the "third vital alphabet of life" complementing the alphabets of nucleic acids and proteins. The breakthrough studies of the past decade to decipher the "sugar code" have in turn attracted the interest of the scientific community across countries.

1.2 Complexity of carbohydrates

Polysaccharides adorn the outer cell surface of the living cells. These layers of polysaccharides are made up of repetitive oligosaccharide units. These dense arrays of polysaccharide motifs encapsulate the entire cell surface and remain protruded from the surfaces thereby providing the unique character of each living organism. These exposed carbohydrate molecules are responsible for the varied interactions in the living body, viz. interactions with adjoining protein or nucleic acids, lipids, hormones, and enzymes or with pathogenic foreign particles like viruses, bacteria, and fungi. Thus any possible communication or interaction of specific cells of a living organism occurs by knocking on the door of the exposed carbohydrate epitopes of the living cells. The repeating cell surface oligosaccharide units thus become the blueprint for possible biological transmission processes.

A proper systematic understanding of these cell surface oligosaccharides is essential to understand and solve the various unsolved riddles pertaining to vital life processes. However, oligosaccharides portray an extraordinary structural complexity owing to their varied monosaccharide counterparts, position and configuration of glycosidic linkages, and all the possible combinations of the linear and branched variants. Pertaining to the importance of carbohydrates and their structural complexity, it becomes essential to obtain homogenous carbohydrate samples and their derivatives for the proper scientific study of "Glycomes" and "Glycomics." Although carbohydrates are abundantly distributed in nature, but isolation and extraction of the required specific carbohydrate epitopes from natural sources is a highly cumbersome and laborious process and portrays as a major hurdle in the advanced study of carbohydrate and carbohydrate-based therapeutics.[9] As a major difference with other biopolymers, carbohydrates are isolated as complex mixtures in extremely low yields with batch-to-batch variation creating problems in reproducibility. Thus in such cases and in the absence of easy accessibility of specific enzymes, the importance of chemical synthesis to obtain pure oligosaccharide samples becomes highly pertinent and the need of the hour. But due to the complexity of the structural carbohydrate motifs and the abundance of free hydroxyl groups, the chemical synthesis of oligosaccharides poses as a challenging task for the scientific community. A study by the National Research Council of the National Academy of Sciences and Engineering of the United States entitled "Transforming Glycoscience: A Roadmap for the Future" speaks about the need for glycan chemistry to be brought to "the mainstream of synthetic organic chemistry rather than kept as a specialized field practiced by only a handful of glycochemists."

1.3 General aspects of oligosaccharide synthesis

In the synthesis of complex oligosaccharides, the main objective is to build the glycosidic linkages that connect the various monosaccharide units with optimum regioselectivity and stereoselectivity. This process of formation of glycosidic bonds is known as glycosylation. Over the past few decades, extensive work has been done to optimize and standardize the process of glycosylation to simplify the process of oligosaccharide synthesis. But the complexity and the selectivity involved in glycosidic bond formation have led to the conclusion that no single method can be separately identified and demarcated as the best suited for the process of glycosylation. It depends on varied and diverse set of factors ranging from regiospecificity, stereoselectivity to purification. The process becomes even more challenging due to the possibility of the formation of a pair of diastereomers during glycosidic bond formation leading to a decrease in the required yield and difficulty of purification of the required stereoisomer.[10] Thus Hans Paulsen rightly said, "*Although we have now learned to synthesize oligosaccharides, it should be emphasized that each oligosaccharide synthesis remains an independent problem, whose resolution requires considerable systematic research and a good deal of know-how. There is no universal reaction condition for oligosaccharide synthesis.*" Thus the chemical synthesis of complex oligosaccharide units for the study of intricate details of "Glycomics" requires meticulous planning and is often time consuming. Although there has been tremendous advancement in the field of the synthesis of oligosaccharides and more and more intriguing glycosylation reactions are being standardized in recent times, it still suffers from the reputation of being irreproducible and unpredictable making the field a highly eluding area of research.

1.3.1 Anomeric effect

Although the carbohydrates bear much similarity with carbocycle analogs,[11] the presence of the oxygen heteroatom in the cycle renders significant alteration in their conformational characteristics.[12] The ring oxygen in the glycosides was seen to exert an influence on the polar substituent of the anomeric carbon to predominantly occupy a particular conformation. The tendency of the anomeric substituent in the glycoside pyranose rings to occupy the axial position is commonly termed as "anomeric effect," which was originally called "Edward-Lemieux effect" in recognition of their versatile work on the anomeric position of heterocycles and glycosides.[13] It was later renamed as "anomeric effect" by Lemieux and Chu[14] in 1958 to explain the unnatural equilibrium prevalence of axial anomers in the case of glucose and its derivatives. Claire M. Filloux has concisely the definition of anomeric effect as "an equilibrium preference for the axial isomer of a 2-substituted heterocycle that exceeds expectations based on a steric model of conformation."[15] A generalized anomeric effect is observed in a system where two heteroatoms are connected to the same central carbon atom (X–C–Y). Acyclic models studied by electron diffraction also showed that preference of alkoxy, acyloxy, and halogen substituents to occupy the gauche conformation.[16] Rigorous studies on this topic for the last 50 years have revealed fascinating details about this intrinsic phenomenon of heterocycles.[17]

Numerous elaborate studies to illustrate the anomeric effect on glycosides and analogs have revealed various causes and hypotheses behind its existence. While each comes with its own set of merits and demerits, a few stand above the rest and has become the most widely accepted hypotheses. The electrostatic theory includes explanations based on classic dipole–dipole interactions, coulombic repulsion, and nonclassical hydrogen bonds, while the stereo-electronic theory includes the hyperconjugation phenomenon.

In the equatorial configuration, the lone pairs of the heteroatom and the substituent on the anomer remain partially aligned, which repels each other by a coulombic electrostatic force[18] by a phenomenon known as "rabbit ear effect," thereby destabilizing the system. But in the axial isomer, the lone pairs are oppositely oriented, which, in turn, relieves the destabilizing factor owing to the low energy barrier factor. This attractive coulombic interaction was also renamed as a nonclassical hydrogen bond between the heteroatoms and the *syn*-axial hydrogen atoms. Thus the dipole effect and the nonclassical hydrogen bond explain the prevalence of axial conformation of the glycosides. This anomeric effect was also found to be particularly solvent dependent[19] exhibiting a gradual decline in polar solvents.[20] This observation, in turn, also proves the higher ionic percentage of the equatorial isomer in comparison to the more stable axial isomer. This behavior also helps us to analyze the dipole moment theory pertaining to the anomeric effect (Fig. 1.1).[21] It illustrates that, in the equatorial conformation, dipole moments of the heteroatoms reiterate each other, resulting in a nonzero resultant dipole moment, whereas the axial conformation shows zero net dipole moment owing to their opposite configurations.

The solvent effect on the anomeric effect has also been reinforced by the hyperconjugative model, which was first proposed by Altona et al.[22] X-ray studies reveal the lengthening of the axial C–X anomeric bond in chlorinated dioxanes. It was later attributed to the stabilizing effect rendered by the interaction between the antibonding σ^*_{C-X} orbital and the vicinal oxygen

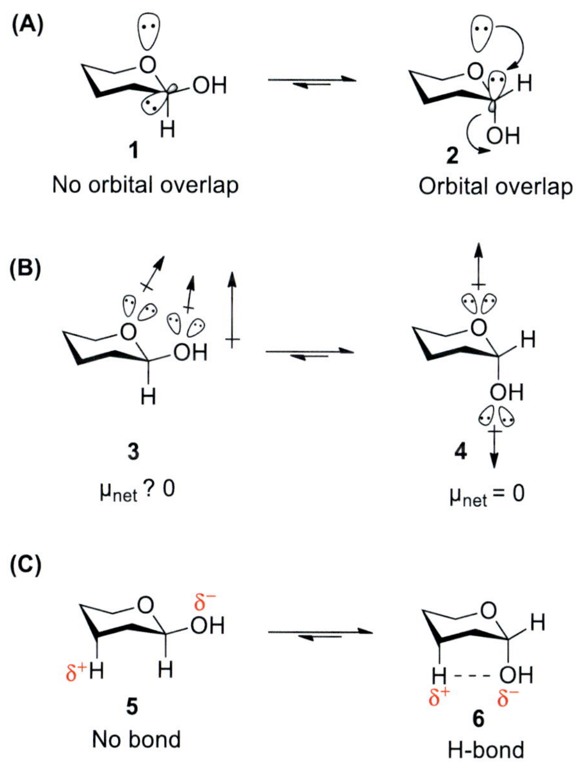

FIGURE 1.1 (A) Hyperconjugative orbital overlap; (B) dipole interactions; (C) nonclassical hydrogen bond.

nonbonding electron pair.[23] In the axial conformation, there is hyperconjugation between the antibonding exocyclic C–X bond, which remains aligned with the antiperiplanar nonbonding lone pair of the endocyclic oxygen atom, maximizing the stabilizing interaction between the two. But this activity is deactivated in the equatorial configuration where the orbitals remain misaligned, thereby destabilizing the system.

Thus the anomeric effect may be summarized as a cumulative factor in carbohydrate chemistry based on the combination of molecular orbital, electrostatic, and hyperconjugation effects. In lieu of these explanations, in most cases, the axial conformation represents the more stable thermodynamically controlled product, whereas the equatorial conformation represents the less stable kinetically controlled product. However, D-mannose and D-rhamnose exhibit completely opposite characters due to the axial 2-OH group, which renders the axial product as kinetically controlled and the equatorial product as thermodynamically controlled.

1.3.2 Background on classical approaches

The most important and essential aspect of carbohydrate synthesis lies in the phenomena of glycosylation. There has been a huge volume of work to establish and modify the concept of glycosylations. Fischer[24] and Michael[25] pioneered the formation of the first alkyl and aryl glycosidic bond by synthesizing *p*-methoxy phenyl β-D-glucopyranoside and methyl α-D-glucopyranoside, respectively. It was followed by an advanced, controlled glycosylation protocol devised by Koenigs and Knorr.[26] It involves the coupling of glycopyranosyl halides, **7**, as donors with primary alcohols as highly reactive acceptor molecules in the presence of Ag_2CO_3 or AgOTf and some promoters such as silver zeolites to produce the required glycoside **8** (Scheme 1.1). Later, the glycosidic bond formation by removal of leaving group by metal (II) salt-induced salt was developed and successfully established by Zémplén and Gerecs,[27] and subsequently by Helferich and Wedermeyer.[28]

The last couple of decades have had a humungous volume of work pertaining to glycosidic bond formation. Scientists across the world have been constantly working on methods to standardize, improve, and modify the conditions to get the optimum results from the process of glycosylation. But owing to the complexity of the carbohydrate molecules, their stereo and

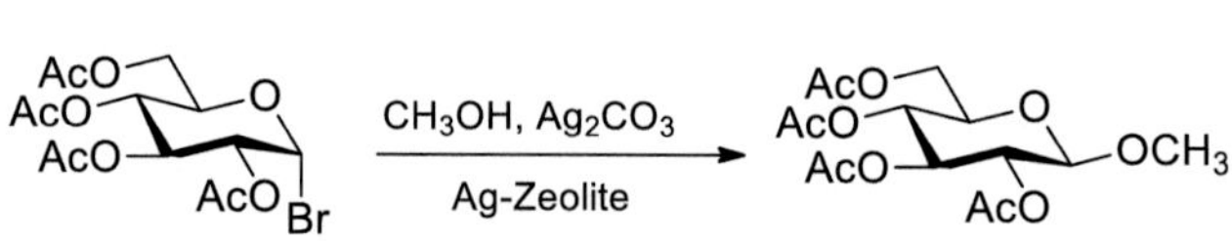

7 **8**

SCHEME 1.1 Koenigs–Knorr glycosylation with activation of glycosyl bromide, **7** with Ag_2CO_3.

regioselectivity, and their structural versatility, glycosylation remains an active domain of research. With the advent of computational chemistry, the world has seen the rise of modern techniques to modify the process. The domain of enzymatic glycosylation has also been a perfect methodology to assess, mimic, and predict biological life processes.

1.3.3 Mechanistic studies

Chemical glycosylation is the coupling of two glycosyl units with each other or with other aglycons by the formation of a glycosidic bond. The formation of an *O*-glycosidic bond is governed by a complex set of reaction mechanisms. A widespread study has been done and is also being conducted in the present times to successfully categorize the process of glycosylation in any particular subdomain of substitution reaction, but it has been proved that complete categorization of glycosylation reaction in either S_N1 or S_N2 reaction is tedious, challenging, and often quite confusing.[29] Glycosylation requires a reducing unit that remains connected by the anomeric carbon. This sugar is known as the donor, and the sugar or aglycon or a nonglycosyl unit to which the donor remains connected is known as the acceptor. The anomeric hydroxyl unit of the donor is activated by the addition of a good leaving group, and the acceptor contains a reactive hydroxyl group. As a generalized mechanistic pathway (Scheme 1.2), initially, the preactivation of donor **9** is done by attaching a good leaving group, **L**, to the hydroxyl group in the anomeric position. Further, the leaving group is activated with the aid of an

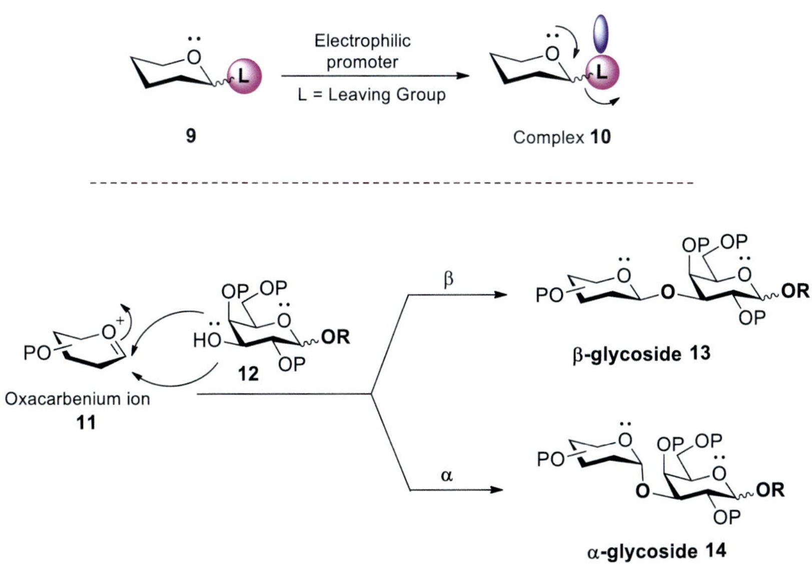

SCHEME 1.2 Generalized mechanistic pathway of glycosylation reactions.

electrophilic promoter to form an activated complex **10**. Subsequently, the removal of the leaving group leads to the formation of the oxacarbenium ion **11** in the flattened half-chair conformation. Nucleophilic attack of the acceptor **12** to this oxacarbenium ion occurs to form the glycoside **13** or **14**. Attack of the acceptor occurring from the upper face of the sugar ring leads to the formation of the 1, 2-trans or beta (β) glycoside, **13** while the attack from the bottom of the sugar ring leads to the formation of 1,2-*cis* or alpha (α) glycoside **14**.

Thus the reaction may occur via a concerted S_N2 mechanism or via a step-wise S_N1 mechanism. There has also been experimental evidence of the probability of glycosidic mechanism in the domain of S_Ni pathway involving ion pairs (Scheme 1.3). There is also plenty of evidence that suggests the extent of glycosylation as a continuum between the extremes of S_N1 and S_N2 mechanisms. Detailed studies have been done by Crich and coworkers in the past two decades, which has significantly contributed to the modern scientific idea of glycosylation mechanism.[30]

Despite this, there is a lot of controversy, speculation, and analysis on the effect of various parameters on the anomeric selectivity and stereospecificity of the required glycosidic bond. Based on the various observed outcomes and

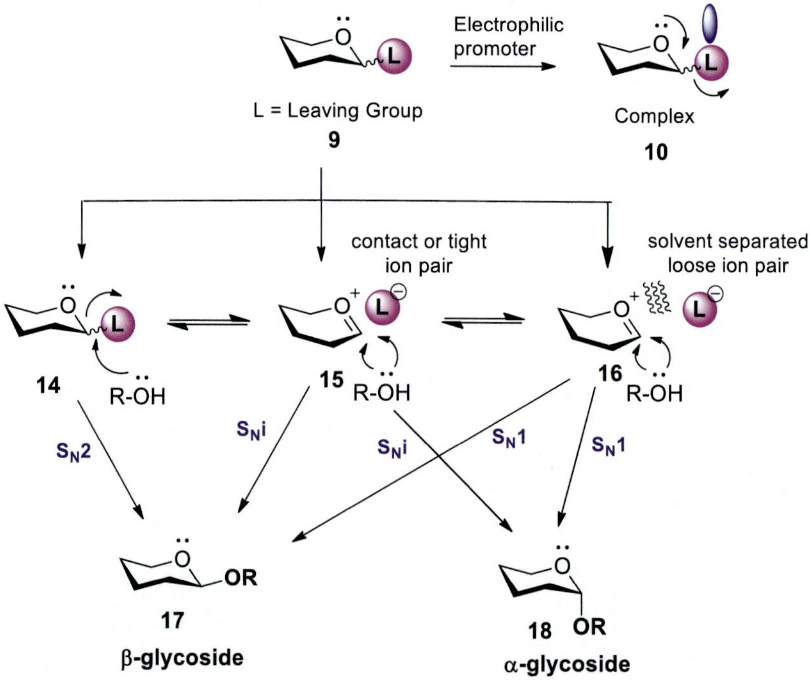

SCHEME 1.3 Continuum in the mechanistic pathway of glycosylation reactions ranging between S_N2, S_Ni, and S_N1.

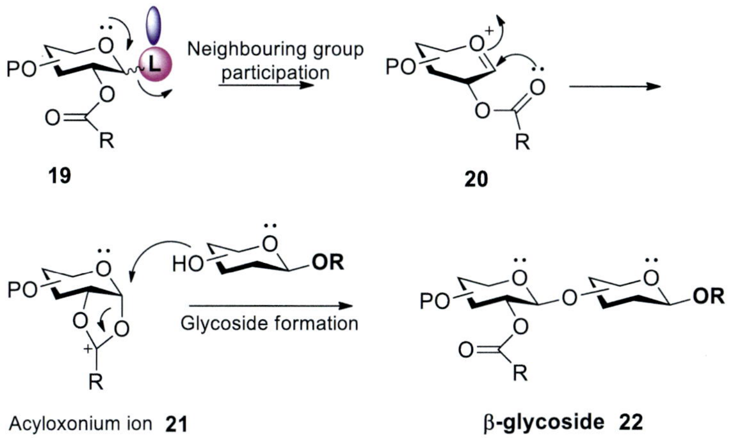

SCHEME 1.4 Formation of 1,2-*trans* glycoside by neighboring group participation of acyl protection in C-2 position.

trends, it has been observed that the presence of a participating acyl group in the C-2 position contributes significantly to the formation of β-glycosidic bonds (Scheme 1.4). The oxacarbenium ion, **20**, generated due to the cleavage of the activated leaving group, interacts with the acyl group of the C-2 position by neighboring group participation to form an acyloxonium ion complex **21**. Thus the lower face of the sugar ring gets blocked, thus facilitating the nucleophilic attack of the acceptor from the top face only forming 1,2-*trans* glycoside **22** primarily. On the other hand, to obtain alpha or 1,2-*cis* glycoside, it has been observed that the presence of nonparticipating group in the C-2 position is required in the donor molecule. Formation and identification of a stereoinverted product indicate the reaction following a probable S_N2 pathway, while S_N1 pathway leads to the formation of a mixture of compounds. This prediction can further be extrapolated to identify the stability of the glycosyl oxocarbenium intermediate formed in the process.

Various protocols have been developed to improve the yield of the desired isomer and to increase the overall efficiency of the reaction. Thus if the reaction may be designed in such a way that the donor gets activation without significant charge separation and the intermediate remains tightly bound, thereby destabilizing the oxocarbenium intermediate, then the reaction is expected to proceed via a classical S_N2 mechanism yielding an inversion product.[31] In this case, the stereochemistry of the product is mainly governed by the conformation of the intermediate ion species. However, if the oxocarbenium intermediate is loosely bound and the charge separation becomes significant, the intermediate gets stabilized causing the reaction to proceed through a two-step S_N1 type reaction protocol facilitating attack of the acceptor from both ends and thereby forming a racemic mixture or an α-β mixture as per glycoside chemistry terms.

In addition to this, there remain various other factors that directly influence the protocol of glycosylation and govern its mechanism. The significance of these factors is the solvent effects, the concentration of the approaching acceptor, temperature, duration of reaction, and so on. In sync with general organic chemistry, the oxocarbenium intermediate is stabilized in polar solvents. Thus more polar solvents shift the mechanism of glycosylation more toward S_N1 type of reaction, whereas less polar solvents often destabilize the oxocarbenium intermediate, directing the reaction to proceed through S_N2 inversion method. Thus in chemical glycosylation, the planning and design of the reactions require detailed consideration of various factors to achieve the required chemoselectivity and regioselectivity of the desired product.[32] Although umpteen scientists across the world have accomplished the synthesis and mimic various biologically important glycosides, still a common universal method with maximum reproducibility is yet to be discovered, which makes the field of carbohydrate synthesis a novel area of research.

1.3.3.1 Armed–disarmed approach

Bert Fraser-Reid was a pioneer in devising and interpreting the concept of armed and disarmed donors in glycosylation reactions.[33] It primarily describes the relative ease or difficulty of activating a glycoside as the donor molecule. Fraser-Reid first pioneered the concept with respect to *n-pentenyl-activated* donors and later extended it to various other donors.[34] He described the donor with an adjacent electron-withdrawing group like esters or azides as the disarmed donor while the donor with an adjacent alkoxy group or benzyl or silyl ethers as the armed donor.[35]

The armed donor has been shown to exhibit faster reactivity, while the disarmed donor destabilizes the oxocarbenium ion, thereby reducing the chemical reactivity. With respect to this concept, reactivity difference is obtained between the reacting glycosyl units in the process of chemoselective glycosylation. Thus Fraser-Reid's path-breaking observation[36] that protecting groups not only protect but also contribute significantly toward the reactivity of the glycosyl units has helped glycochemists all around the world to use glycosides with the same class of anomeric leaving group for both the donor as well as the acceptor. Donor **23** is armed or activated, while acceptor **24** is disarmed or deactivated by virtue of their respective protecting groups. The disaccharide **25** thus obtained can thus be directly utilized as the donor glycoside for further coupling reactions with different acceptor **26** in the presence of a more powerful promoter (Scheme 1.5). This differential protecting group also helps to activate an otherwise inactive donor and vice versa. Thus armed–disarmed concept not only helps to create differences in reactivity but also facilitates the phenomenon of sequential glycosylation reactions. In modern times, these sequential

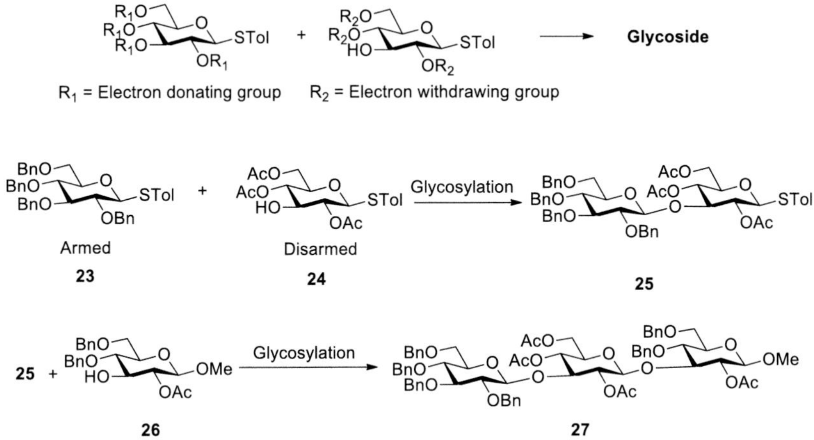

SCHEME 1.5 Armed–disarmed approach in iterative glycosylation.

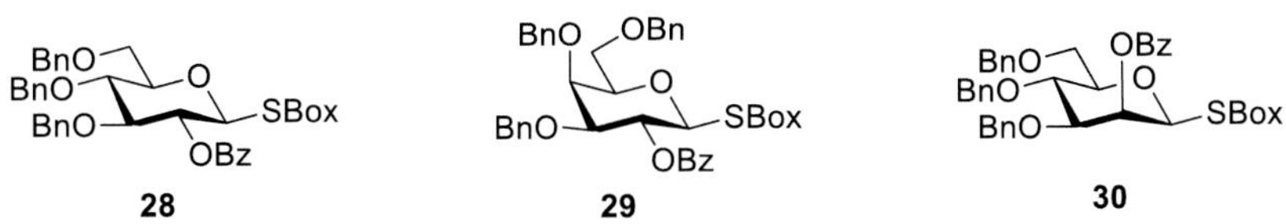

FIGURE 1.2 Superarming of thioglycosides.

glycosylation reactions are highly significant as they not only help in the reduction of the reaction steps contributing to atom economy but also in the development of modern techniques like solid-phase synthesis, automated glycoside synthesis, and high performance liquid chromatography (HPLC)-assisted glycosylation reactions. Armed–disarmed concept has now been extensively used to develop more versatile activating groups and to perform complex oligosaccharide syntheses. The concept has been clearly delineated by Boons et al.[37] and later by Kamat and Demchenko by a comparative study of benzylated and benzoylated *S*-benzoxazoyl (SBox) thioglycosides[38] that showed perbenzylated derivative to be much more reactive than their perbenzoylated counterparts. It was further superimposed to conceptualize the phenomenon of superarmed donors. It was discovered that strategic protection by common armed protecting groups leads to increased reactivity of the donor, thereby "superarming" the reacting thioglycosides. Thus if thioglycoside donors **28, 29**, and **30** have a participating group in the C-2 position and the other hydroxyl groups are perbenzylated (Fig. 1.2), then the donor is reported to portray a dual effect of β-directing as well as a superreactive thioglycoside donor.[39] This concept of "superarmed" donor can also be extrapolated to achieve the electronic effect with an established *O*-2/*O*-5 cooperative effect.[40]

1.3.3.2 Active-latent glycosylation

While the armed–disarmed approach mainly dealt with the concept of activating or deactivating the glycosyl moieties with respect to the protecting groups, the active-latent method of activation establishes a selective activation strategy of two glycosyl donors. This concept was first introduced by R. Roy and group.[41] This can be accomplished in two ways. One way involves the use of different thiophilic promoters for selective activation of the thioglycosides, and the other involves the use of the differential nucleophilicity of the thio-protecting group to modulate the reactivity of the donor molecules (Scheme 1.6). The active donor **31** is the one that has an electron donating group (EDG) in the aryl substituent of the thioglycoside. On the other hand, the thioglycoside **32** with an electron-withdrawing group (EWG) as the aryl substituent behaves as the "temporary latent group" or "temporary inactive group," thereby acting as the acceptor moiety. Thus with this "active" thioglycoside donor with the EDG group and "latent" thioglycoside acceptor with EWG, selective chemoselective glycosylations can be performed. The interesting point in this method is that the "latent" acceptor can further be made into an "active" donor easily by reactivating the sulfur heterogroup toward electrophilic promoters employed in the successive reactions. Thus "latent-active" method of glycosylation can also be implemented toward the design of successive glycosylation protocols, one-pot synthesis, and solid-phase oligosaccharide synthesis.

This method of the latent-active strategy of glycosylation was also independently introduced by Boons et al.[42] in the 1990s, where they emphasized the difference in reactivity of the anomeric group only contributing to the chemoselectivity in the glycosylation reactions. This method proved to be quite versatile as it involved the modulation of the anomeric group only without any differential protecting group manipulation. This method can progressively help chemists around the world to intuitively understand the different reactivities of the wide varieties of glycosyl donors.[43] This, in turn, has helped in increased understanding and better designing of the glycosylation protocols in oligosaccharide syntheses.

1.3.3.3 Remote participation

While neighboring group participation or anchimeric assistance by the protecting groups of the C-2 position of glycosides contributes largely toward

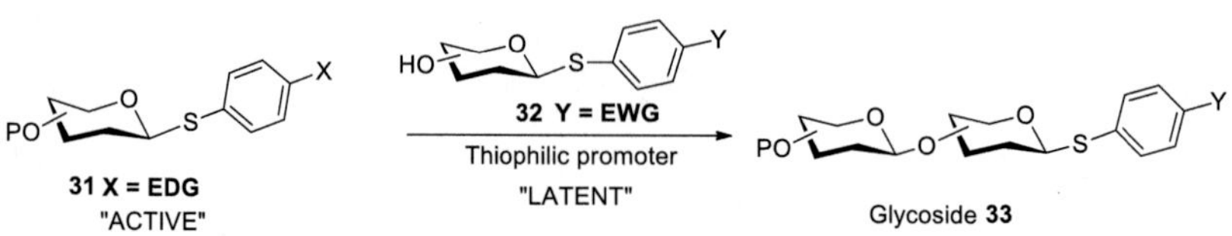

SCHEME 1.6 Active-latent approach in iterative glycosylation.

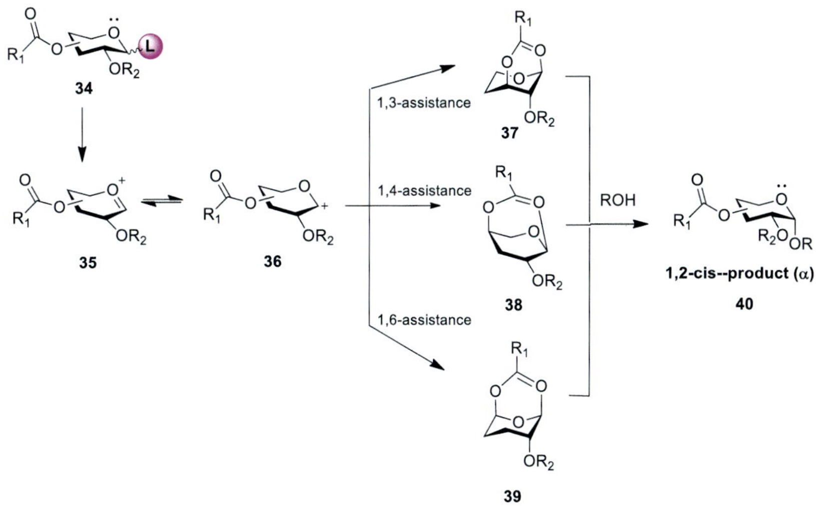

SCHEME 1.7 Remote participation of **1,3**; **1,4**; and **1,6** acyl groups on glycosylation.

determining the stereoselectivity of the desired product, remote anchimeric assistance is also much more effective by using protecting groups at C-3, C-4, or C-6 positions.[44] Acyl-protecting groups at C-2 or N-2 position, vicinal to anomeric position stabilizes the positive charge on the glycosyl carbocation intermediate and helps in the formation of almost complete 1,2-*trans* selective glycosides (Scheme 1.7).

Keeping a nonparticipating protecting group in the C-2 position if acyl protection is present in any other position (C-3, C-4, or C-5) (**34**), then it is assumed to render a remote anchimeric assistance toward the formation of desired stereoselective glycoside (Scheme 1.8).[45] While C-2 neighboring group participation mechanism proceeds through the formation of a fused bicyclic system, remote anchimeric assistance corresponds to the formation of bridged bicycles (**37, 38**, and **39**). Protecting groups at C-3, C-4, and C-6 position helps in the stabilization of the intermediate bicycles.[46] Thus a particular face of the glycosyl moiety gets hindered directing the nucleophilic attack from the other face. In this respect, it can be summarized that acyl-protecting groups in C-3 and C-6 in gluco- and galacto-glycoside moieties, along with an acyl group in C-4 of galacto series, aid in the formation of α-glycosides, whereas an acyl group in the C-4 position in mannosylpyranosides help in the formation of β-glycosides.[47] The implementation of ether as the coordinating solvent also contributes much toward the formation of 1,2-*cis* glycosides.[48] Crich et al. have developed the role of side chains in C-6 carbon by Nuclear Magnetic Resonance (NMR) spectroscopy and done a detailed analysis of the various Fisher conformational studies on stereochemical output in glycosylations.[49]

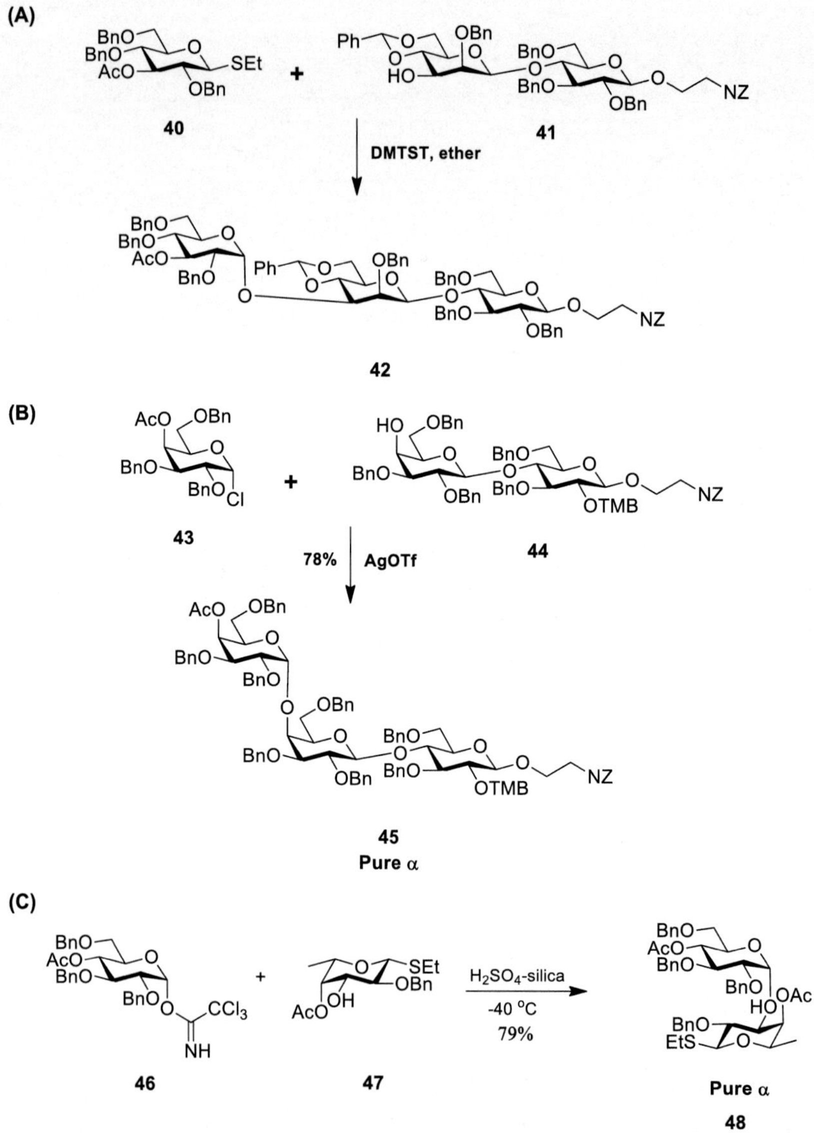

SCHEME 1.8 (A) Glycoside formation by 1,3-remote assistance; (B) glycoside formation by 1,4-remote assistance; (C) glycoside formation by 1,6-remote assistance.

Remote participation of acyl groups is, however, supported by a wide number of additional factors, and all the factors combine and act cohesively to help in the formation of our required 1,2-*cis* stereo-controlled product.[50] Thus acyl groups are more widely used more as a temporary protecting group rather than a remote assisting group. However, some illustrations

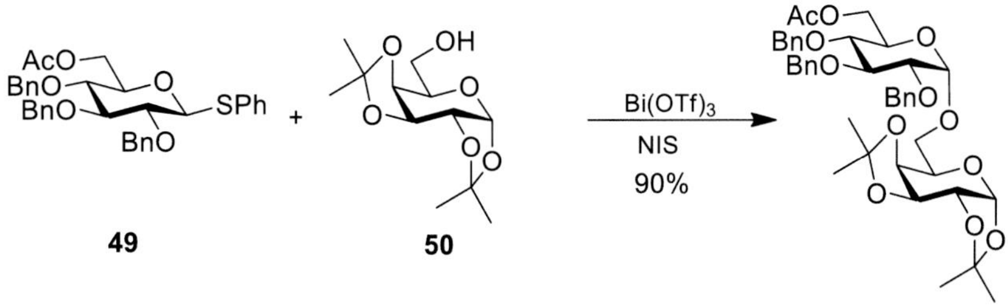

α:β = 1.6:1
51

SCHEME 1.9 Contradiction of remote participation of C6-acyl group giving a mixture of anomers.

depicting the remote participation of C-3,[51] C-4,[52] and C-6[53] acyl groups are shown in Scheme 1.8. The remote long-distance anchimeric assistance has been verified by several spectroscopic and trapping methods.[54] The mechanism of the stereodirecting properties of the ester groups by various computational methods has been explained in a detailed review by Reko Leino, Demchenko, and coworkers.[55]

It should, however, be reinforced that remote participation is not a universal method, and various studies have shown results clearly contradicting the theory of remote participation.[56] (Scheme 1.9) Sometimes, an additional remote assisting group has also been seen to contribute remote anchimeric assistance. In a recent study, Codee and group have also established the role of 2,2-dimethyl-2-(*ortho*-nitrophenyl)acetyl (DMNPA) group in remotely assisting in the formation of 1,2-*cis* glycosides.[57] Presence of the group C-6 position causes the glycosylation to proceed via ionic intermediates, thus facilitating the formation of 1,2-*cis* glycosides.

1.4 Glycosylation reactions

1.4.1 Glycosyl donors

Different glycosyl donors have been tried and tested with time for successful glycosylation reactions in terms of stereo- and regiochemical output and yield of the products. The prominent glycosyl donors extensively used in synthetic carbohydrate chemistry are subsequently discussed.

1.4.1.1 Glycosyl chlorides and bromides

In 1879 Arthur Michael first reported the procedure of coupling of an aromatic acceptor to a peracetylated glycosyl chloride.[25] Koenigs–Knorr, in 1901, first introduced the concept of glycosyl halide donors for glycosylation reactions where the bromide or chloride donors were activated with the aid of silver salts (Scheme 1.10).

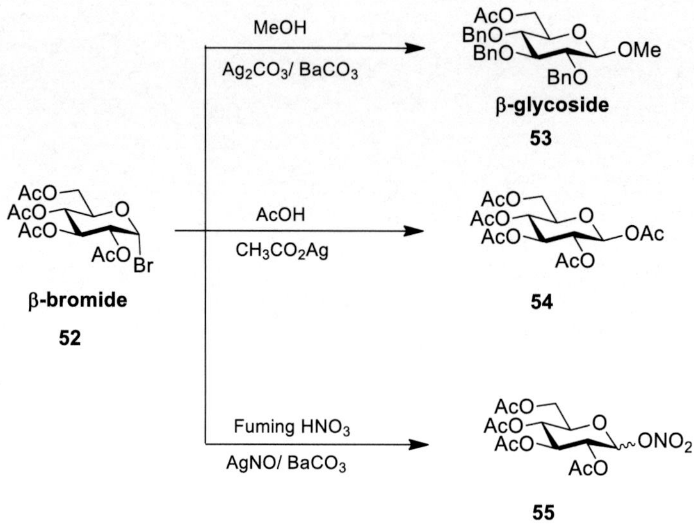

SCHEME 1.10 Koenigs–Knorr method for the activation of halides with silver salts.

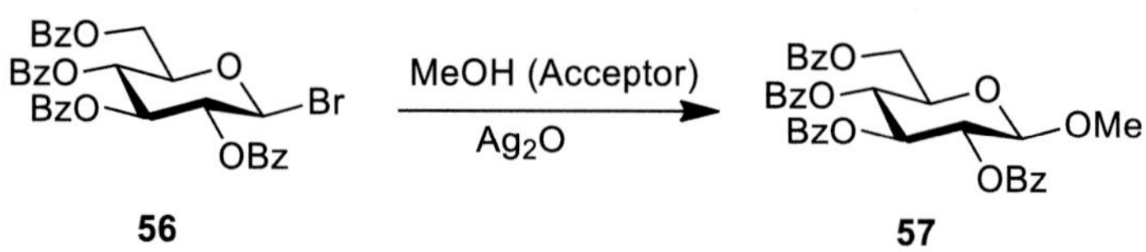

SCHEME 1.11 Use of silver oxide (Ag_2O) as the promoter in glycosylation reactions.

Similarly, Fischer and Armstrong[58] also independently worked on unprotected glycosyl chlorides, where he reported the activation of protected glycosyl chlorides with the help of silver carbonates and the further formation of glycosyl chlorides. However, working with unprotected glycosyl donors proved to be quite difficult to control and suffered from the problem of inadequate reproducibility. Thus Knoenigs–Knorr's analysis of glycosylation with protected sugars proved to be an important milestone in the history of chemical glycosylations. Their work remains seminal and versatile, which opened the avenue for a broad scientific domain of glycosylations.

The silver salt activator also played the role of acid scavenger to neutralize the effect of HBr formed during the process. Later, Fischer also replaced the use of $AgCO_3$ as the acid scavenger with the use of an organic base like quinoline.[59] The use of silver oxide (Ag_2O) not only acted as a probable acid scavenger but also introduced perbenzoylated glycosyl bromide for possible glycosylation reactions (Scheme 1.11).[60] In the following times, a lot of extensive studies were performed, which tried to standardize the use of glycosyl halides. Starting from simple sugars, complex glycosides were synthesized, and attempts were made to achieve significant stereocontrol.[61]

Glycosyl chlorides and bromides could be successfully synthesized by treating unprotected free sugars directly with acetyl chloride or bromide, respectively. This was also equally applicable for the synthesis of acetylated glycosyl bromides. This method proved effective and could be easily crystallized in pure form.[62] Hunsen et al. devised the use of the HBr–AcOH combination for the in situ preparation of HBr, which helped to establish a facile and clean protocol for the synthesis of glycosyl bromides.[63] To prevent the toxic HBr fumes produced in the process, Koto and group devised the use of dilute acetic acid and acetyl bromide for the formation of glycosyl bromides from free sugars.[64] This helped to significantly reduce the toxic adversities associated with the synthesis. This procedure was further modified by Jennings and group into a one-pot synthesis of glycosyl bromides wherein the free sugars were first acetylated with acetic anhydride followed by treatment with HBr in AcOH in the same reaction pot.[65] Similar bromination reaction where the initial acetylation step involved the addition of $LiClO_4$ along with acetic anhydride[66] followed by 33% HBr–AcOH also provided a significant yield of glycosyl bromides. The use of other reagents like ($BiBr_3$–Me_3SiBr)[67] is also well established. Sugar hemiacetals were also utilized where the primary hydroxyl groups were directly converted to the halide groups under modified Mitsonobu conditions using triphenylphosphine and appropriate *N*-halosuccinimide combination[68] in dimethyl formamide. Use of Appel reagents, that is, triphenylphosphine/carbon tetrabromide (PPh_3/CBr_4), is also reported to form highly reactive 2-*O*-benzylated glycosyl bromides.[69] The crude side product could easily be separated by precipitation using diethyl ether. To significantly reduce the use of heavy and corrosive metals and the generation of excess HBr, Jin-zhong et al. established a convenient reactor using ultraviolet irradiation for the synthesis of glycosyl bromides.[70] It involved the treatment of free sugars with acetic anhydride and $HClO_4$ in acetic acid and thereby subjected to ultrasound sonication followed by bromination by AcBr–MeOH. This process in environment-friendly reactor not only helped to reduce toxic HBr fumes but also significantly increased the yield of the reaction.

On the other hand, Ibutallin's method[71] involves the formation of β-glycosyl chlorides from their corresponding acetates by the utilization of PCl_5 in the presence of the catalytic amount of $BF_3.Et_2O$ (Scheme 1.12A). High yields of glycosyl chlorides have been reported by the same method by using $AlCl_3$ as the catalyst.[72] However, the use of polar solvents like acetonitrile could generate the required halide even in the absence of catalysts. Similarly, unprotected sugars can be directly converted to α-glycosyl chlorides exclusively under comparatively milder conditions using acetyl chloride in the presence of basic Al_2O_3 as the catalyst[73] (Scheme 1.12B). Hemiacetals can also be conveniently converted to corresponding chlorides using oxalyl chlorides in dimethyl formamide (DMF) as the solvent.[74] Protic acids like HCl or acetic acid in the presence of $SOCl_2$ produced 1,2-*cis* glycosyl chlorides, while the use of Lewis acids like $ZnCl_2$ or $BiCl_3$ produced the opposite 1,2-*trans*

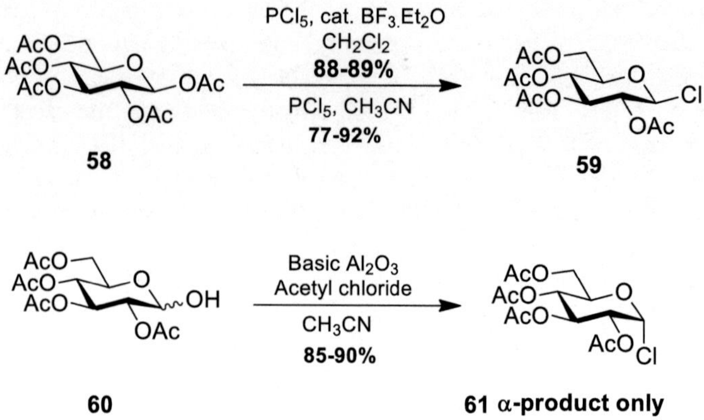

SCHEME 1.12 Glycoside formation by (A) use of PCl5 in the presence of $BF_3.Et_2O$; (B) use of basic Al_2O_3 and acetyl chloride as the promoter.

TABLE 1.1 Comparative study for the use of $SOCl_2$ and PCl_5 promoter toward the stereoselectivity of the glycosidic product.

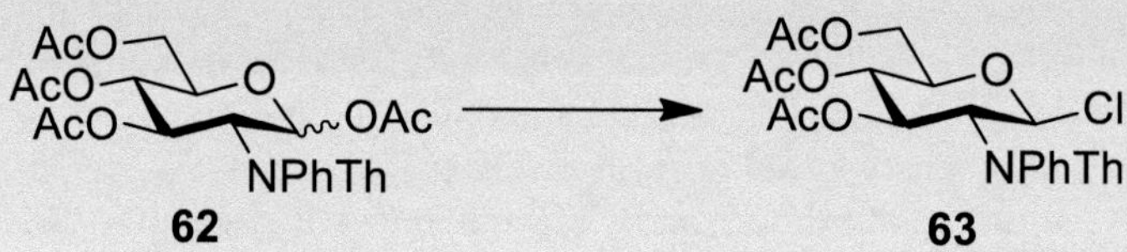

entry	Promoter	Yield (%)	α/β ratio
1	$SOCl_2/BiCl_3$	86	41:59
2	$SOCl_2/ZnCl_2$	81	44:56
3	PCl_5/BF_3	95	Β only

stereoisomers. Treatment of PCl_5/BF_3 was also seen to afford β-chlorides when applied to peracetylated phallimido glucosides,[75] but the use of $SOCl_2$ with Lewis acid afforded an α/β mixture of glycosyl chlorides. Here the neighboring group participation of the phallimido group significantly contributed to the formation of pure β-glycoside (Table 1.1).

Preparation of glycosyl chlorides and bromides from the corresponding glycosyl thioglycosides by using iodine monochloride (ICl) or iodine monobromide (IBr) under mild conditions also provided high yields.[76] Glycosyl bromides have also been produced from per benzoylated *p*-methoxy phenyl derivatives of glycosides by using zinc triflate and benzoyl bromide.[77] The method by Polat and Lidhart also proved to convert the armed benzyl protection to electron-withdrawing benzoyl protection (Scheme 1.13).

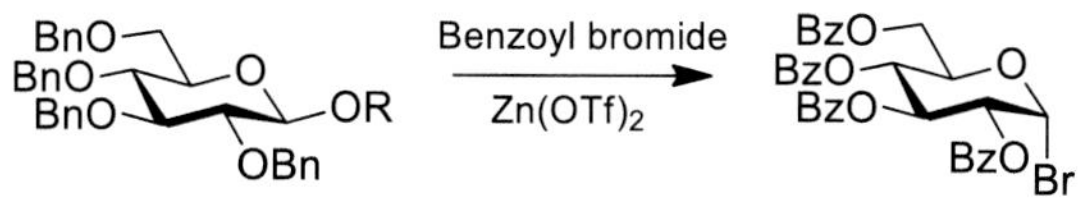

64 65

SCHEME 1.13 Conversion of armed donor **64** to disarmed donor **65**.

All the previously reported procedures for the synthesis of glycoside chlorides and bromides, however, involved harsh reaction conditions. Mong et al. reported the synthesis of lactosylchlorides from hemiacetals and orthoesters using a milder and greener combination of 2,4,6-trichloro-1,3,5-triazine (TCT)-DMF[78] via a Vilsmeier–Haack adduct. It was also found to be highly compatible with acid-labile groups such as isopropylidene and benzylidene groups. Further, in the search for more environmentally friendly and milder reaction conditions, Alfonso Iadonisi and group reported a possible synthesis of glycosyl chlorides from benzyl hemiacetals in solvent-free conditions[79] using a premixed combination of an excess of triphenylphosphine (PPh_3) and hexachloroacetone or trichloroacetonitrile at 70°C. This method is not only cheap and easy to handle, but also can be implemented in the presence of a wide variety of protecting groups. With the aim to develop greener techniques, the photo-irradiative phase-vanishing method was established for the synthesis of α-glycosyl bromide with UV light irradiation.[80] This method was further modified by Xue and coworkers[81] by using visible light in the presence of tris(2,2′-bipyridyl)ruthenium(II) chloride [$Ru(bpy)_3Cl_2$] as the photocatalyst. Recently, there has also been a report of a synthesis of glycosyl bromide using trimethylsilane and bromine (TMS/Br_2) and BHQ/Br_2.[82] Polymer-assisted synthesis of glycosyl chlorides and bromides have been accomplished by Encinas and Chiara[83] using a polymer-supported *N,N*-dialkylformamide as a catalyst. In a recent development, Bennett, Pohl, and coworkers reported the synthesis of orthogonally protected monosaccharide units in a continuous flow reactor involving the efficient use of acetyl bromide, which served a dual purpose of acetylation as well as bromination.[84] HBr generated was neutralized in the reactor itself by using saturated aqueous $NaHCO_3$. It opens a door for further effective methods for the synthesis of glycosyl bromides and chlorides. Recently, Padungros and group[85] have reported the transformation of thioglycosides to corresponding glycosyl chlorides mediated by bromodiethylsulfoniumbromopentachloroantimonate (BDSB) as a mild oxidant with Bu_4NCl as an additive at significantly low temperature.

Thus we see that although glycosylation of glycosyl halides as the donors was one of the earliest evolved methodologies in synthetic carbohydrate chemistry, still there exists a lot of interest in devising new and improved methods for the formation of glycosyl chlorides and bromides.

The moisture-lability and lower shelf life of the glycosyl bromides in comparison to other glycosyl donors often pose as a hindrance to their handling and use in glycoside syntheses. However, there has been extensive and varied research dealing with the activation of the halide donors to obtain the desired stereoselective glycoside product. It has been observed that glycosyl bromides have lower reactivity than glycosyl iodides[86] and lesser stability than glycosyl chlorides. However, glycosyl bromides presented themselves as a best-suited glycoside halide donor because of their intermediacy between glycosyl chloride and glycosyl iodides in terms of reactivity.

Classical Koenigs–Knorr glycosylation deals with glycosylation with glycosyl halides using heavy metals like mercury or silver.[87] Paulsen et al. reported the formation of highly stereoselective β-glycosides from α-glycopyranosyl halides by using heterogenous silver salts.[88] The reactions were estimated to follow S_N2 mechanistic pathway, rendering an inversion of the configuration of the glycosyl halides to afford the required stereospecific glycoside product. However, the reaction with insoluble silver salts like $AgCO_3$ or Ag_2O was found to be slow. $AgCO_3$ was also used as an acid scavenger when Hg(II) salts AgOTf was used as the promoters. Thus there was the emergence of soluble silver salts like silver acetates[26] and silver nitrates for glycosylation reactions. This simultaneous displacement of the anomeric halides by the incoming nucleophile without providing scope for further anomerization has also been successfully applied toward the synthesis of synthetically challenging glycosides, viz., β-mannosides.[89] Other soluble silver salts include silver perchlorate,[90] silver trifluoromethanesulphonate,[91] or silver triflate. Kronzer also showed glycosylation reactions with glycosyl halides in the absence of silver salts. The use of silver silicates as promoters for the activation of 2-deoxy and 2,6-dideoxy glycosyl bromides was reported by Kaneko and Herzon,[92] which showed high β-selectivity on most of the substrates (Scheme 1.14).

Demchenko also reported the activation of electronically deactivated, benzoyl-protected glycosyl chlorides with the help of iron(III) catalysts. Very recently, Demchenko and coworkers reported the use of green and light-stable, bismuth triflate, $Bi(OTf)_3$ as promoters for glycosyl halides in the absence of silver salts.[93] These reactions were seen to proceed quite fast (Scheme 1.15). The same group also reported the activation of benzoyl and benzyl-protected chlorides using silver oxide and triflic acid.[94] The advantage of the procedure was the visual identification of the progression of the reaction, identified by the disappearance of the characteristic black color of Ag_2O. They have also established a cooperative Ag_2SO_4 and $Bi(OTf)_3$ promoter system, which provided satisfactory yields for the synthesis of glycosides from corresponding anomeric chlorides.[95]

Boronic acid has also effectively been used as the catalyst for the activation of peracetylated glycosyl bromides and chlorides in the presence of Ag_2O.[96] The use of diphenylborinate as the boronate salt led to an S_N2-type displacement due to the increase in the nucleophilic parameter of boronate esters (Fukui index calculator), leading to the exclusive β-selective product. However, by

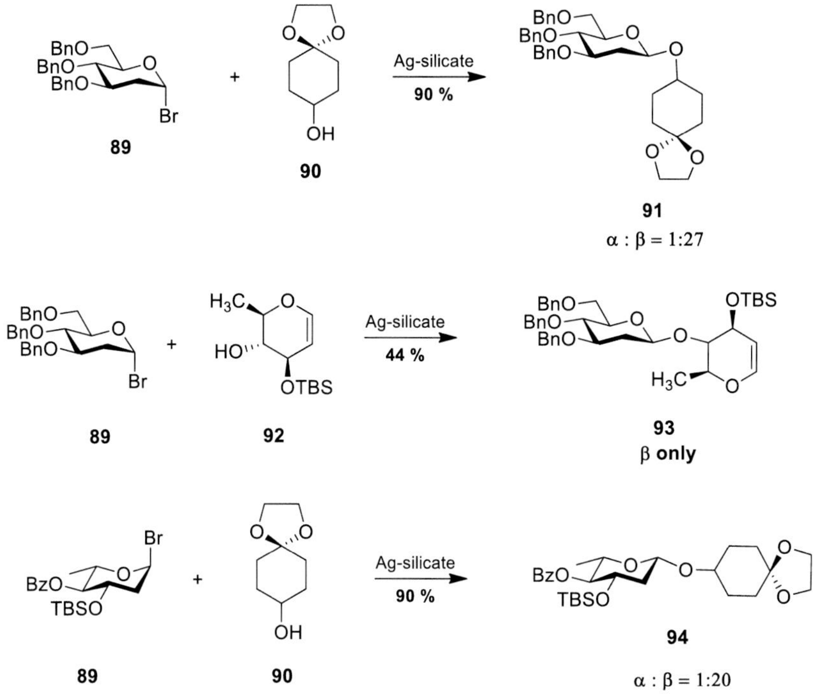

SCHEME 1.14 Activation of deoxy glycosyl bromides to form highly β-stereoselective product.

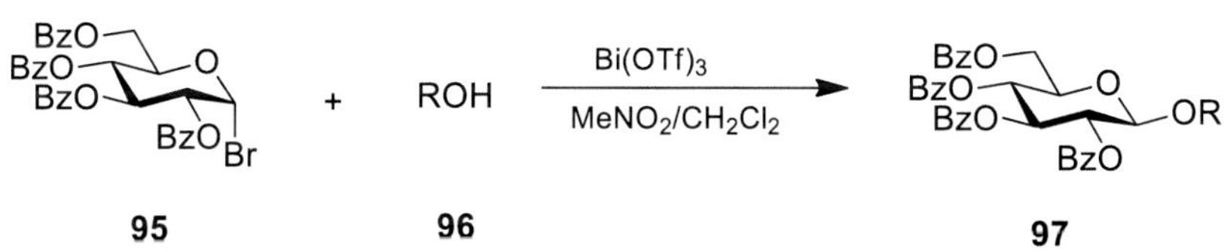

SCHEME 1.15 Activation of benzoylated glycosyl bromides (**95**) using bismuth triflate.

changing the acetyl-protecting groups to benzyl-protecting groups, both the yield and stereoselectivity of the product were affected badly.

Kuszmann and Boros reported the use of $Hg(CN)_2$ as the promoter for glycosyl bromides.[97] But when acetonitrile was used as the solvent as a promoter for glycosyl bromides, multiple decomposition products were obtained with the isolation of β-product in minimum yields. With further study, the optimized reaction conditions were reported while using 1:1 benzene-nitromethane as the solvent with increased reaction yield. However, the reaction proceeded slowly, requiring an elevation of reaction temperature. Recently, Bauer and group reported the activation protocol of glycosyl chlorides and fluorides using a new, tunable iron(III) ferrocenium salts,[98] [FcB

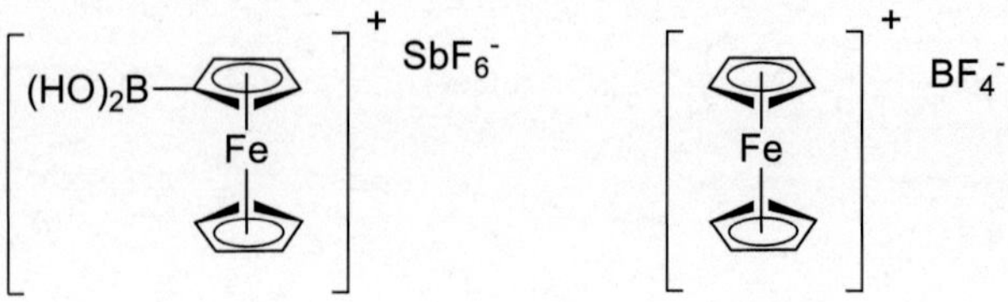

98 **99**

FIGURE 1.3 Iron(III) ferrocenium salts: $[FcB(OH)_2]SbF_6$ (**98**) and $FcBF_4$ (**99**).

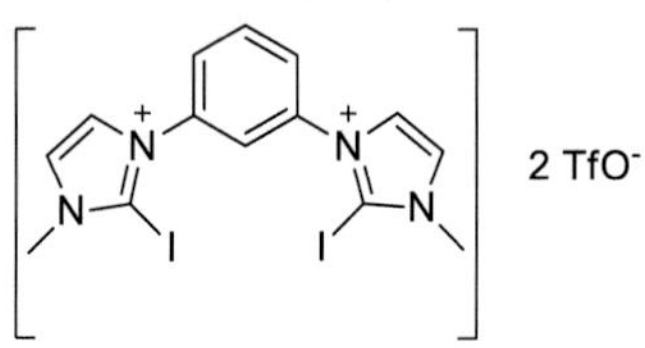

100

FIGURE 1.4 Bis(iodoimidazolium) complex, **100**.

$(OH)_2]SbF_6$ (**98**) and $FcBF_4$ (**99**) (Fig. 1.3) as promoters, by implementing the affinity of iron toward oxygen. Glycosyl bromides have also been successfully activated with the aid of silver tetrafluoroborate salts.[99]

Gervay-Hague and group also applied glycosyl iodides as donor in a hexasaccharide synthesis[100] where they activated the glycosyl iodides with tetrabutyl silyl iodide in the presence of diisopropyl ethyl amine, DIPEA, as the base to get high yields. Miyairi and group reported the activation of glycosyl iodides with the help of $LiBH_4$ in the presence of tetrahydrofuran (THF) as the solvent.[101] In an effort of its kind, Codee and coworkers tried to activate glycosyl chlorides with the help of halogen bond promoters like bis(iodoimidazolium) complex, **100** (Fig. 1.4).[102]

But fully oxygenated pyranosyl donors produced a too-low yield. So, the same group modified the glycosyl donor by removing two alkoxy substituents, which gave a successful high-yielding glycosylation reaction. On the same protocol, hydrogen bond organic catalyst like a series of substituted urea and thiourea has also been utilized[103] for Knoenigs–Knorr type activation of glycosyl chlorides. This reaction protocol provided high yield as well as high stereoselectivity without using any toxic heavy metals. More greener reaction protocol to activate pivaloyl-protected glycosyl chlorides and bromides have been reported by Isabel Matheu and coworkers where the group used supercritical carbon dioxide (sCO_2) as the solvent[104] to be highly β-selective. The reaction conditions involved pressure of 1500 psi and 60°C (Scheme 1.16).

With the emergence of room temperature ionic liquids (RTILs),[105] Toshima et al. reported the glycosylation with glycosyl halide donors in the presence of a combination of a protic acid possessing a common anion with

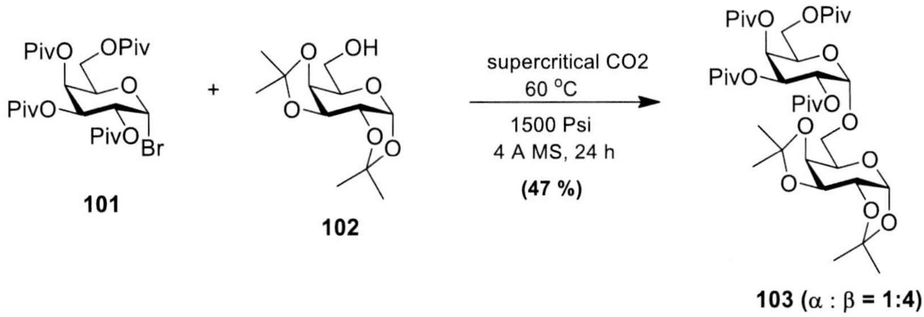

SCHEME 1.16 Activation of glycosyl bromides (**101**) using supercritical CO_2 (sCO_2).

the ionic liquid.[106] Some of the solvent system that were explored were 1-*n*-hexyl-3-methylimidazolium tetrafluoroborate (C_6mim[BF_4]) with HBF_4, 1-*n*-hexyl3-methylimidazolium trifluoromethanesulfonimidide (C_6mim[NTf_2]) with $HNTf_2$, 1-*n*-hexyl-3-methylimidazolium perchlorate (C_6mim[ClO_4]) with $HClO_4$ and 1-*n*-hexyl-3-methylimidazolium trifluoromethanesulfonate (C6mim[OTf]) with HOTf. Building on their reports, Malhotra et al. reported the activation of acetobromo glycosides with *p*-nitro phenol using a series of RTILs as the promoter,[107] of which the imidazolium halides 1-butyl-3-methylimidazolium chloride (BMIm · Cl) performed the best. With the further addition of tetrabutyl ammonium halides in the reaction medium, the yield could also be significantly improved.

1.4.1.2 Glycosyl iodides

Glycosyl iodides were first developed as an extension of Knoenigs–Knorr and Michael glycosidation processes in 1901. It was later observed by Fischer the extraordinarily high reactivity of glycosyl iodides as glycosyl donors. Kronzer and Schuerch prepared glycosyl iodides via Finkelstein exchange using sodium iodide.[108] He also observed later that by conversion of glycosyl bromides and chlorides to corresponding iodides, there was a signification increase in the reactivity, decreasing the reaction time by almost 75%. Corresponding to this observation, Thiem and Meyer reported the successful synthesis of a series of glycosyl iodides[109] from peracetates, anhydro sugars, and so on, by using trimethylsilyl iodide (TMSI). However, the high reactivity and subsequent low shelf life of the glycosyl iodides proved to be a major limitation behind its nonuse as a glycosyl donor. Despite the high degree of instability of per-benzylated glycosyl iodides, Gervay-Hague and group were the first to successfully synthesize and characterize by its in situ preparation with TMSI in CD_2Cl_2 in the NMR Spectrometer in low temperature.[110] Stachulski and coworkers in 2003 also prepared identified the crystal structure of stable 2,3,4-tri-*O*-pivaloyl-α-D-glucopyranuronate iodide.[111] Following it, there has been extensive research on the preparation and implementation of glycosyl iodides.[112]

In the preparation of glycosyl iodides, TMSI was an important reagent. However, the instability of TMSI caused a major blockage in further studies pertaining to glycosyl iodides. Field et al. addressed this problem[113] significantly whenthey synthesized TMSI in situ by reacting hexamethyldisilane (HMDS) with molecular iodine. The same group has also provided explanation about the stability, reactivity, and characterization of the glycosyl iodides.[114] Murakami and coworkers modified the same reaction by the addition of zinc iodide (ZnI_2) as an additive[115] that significantly improved the obtained yield of the corresponding glycosyl iodides.

Corresponding to the use of HBr/AcOH for the formation of glycosyl bromides, the use of HI/AcOH for the formation of glycosyl iodides was not much in vogue due to difficulty in the synthesis of pure anhydrous hydrogen iodide. Back in 1950, Ness et al. synthesized glycosyl iodides by reacting hydroiodic acid with acetic anhydride leading to the in situ formation of HI.[116] Gervay-Hague also used this method for the synthesis of iodides of *N*-acetyl neuraminic acid.[117] Recently, there has been a report of the synthesis of anhydrous HI directly from molecular hydrogen and iodine by using a rhodium catalyst.[118] However, Koreeda et al. provided a solution to the problem by the oxidation of solid molecular iodine with a suitable thiol compound (Scheme 1.17).[119]

The HI generated further reacted with the glycosyl substrate to produce the corresponding glycosyl iodides in high stereoselectivity. This solvent-free reaction with inexpensive reagents proved to be highly effective for the generation of glycosyl iodides from the corresponding acetates. In an attempt to prepare β-selective glycosyl iodides, polymer-supported reaction has also been performed under smooth and mild conditions by Caputo and coworkers using a complex of polystyryldiphenylphosphane and iodine prepared in situ in anhydrous dichloromethane at room temperature.[120] They reported that adding the protected glycosyl acetates to the polymer-bound complex and 2-equivalents of imidazole (**70**) successfully produced the corresponding glycosyl iodides in high yields and high stereoselectivity (Scheme 1.18).

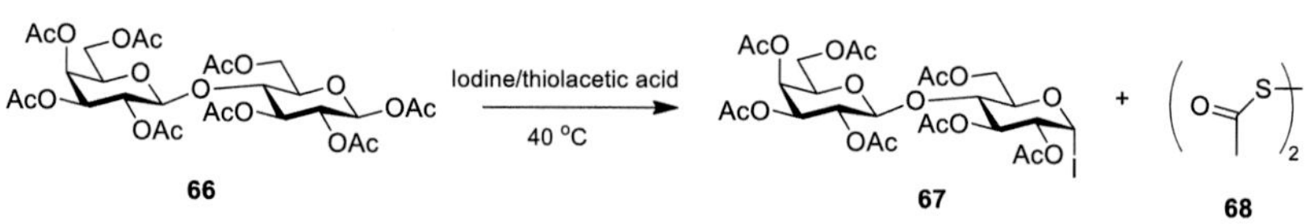

SCHEME 1.17 Use of molecular iodine for the formation of glycosyl iodides.

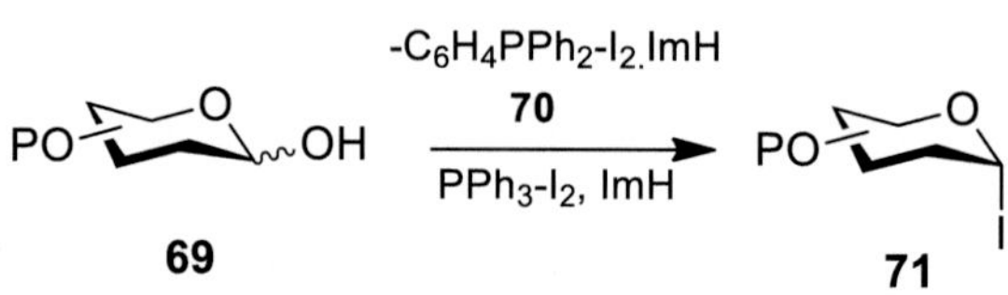

SCHEME 1.18 Use of polymer-bound imidazole, **70** for the formation of glycosyl iodides.

1.4.1.3 Glycosyl fluorides

Glycosyl fluorides come with the added advantage of higher chemical and thermal stability over other glycosyl halide counterparts. Thus glycosyl fluorides have attracted the attention of many synthetic glycochemists across the years.[121] The first glycosyl fluoride synthesis was achieved[122] by D. Braus, where he prepared anhydrous hydroflurous acid (HF) (Scheme 1.19A) by heating anhydrous potassium hydrogen fluoride. With the HF thus produced, peracetylated glycosyl acetates were converted to their corresponding glycosyl fluorides. Based on a report by Olah et al.,[123] the use of a 50%−70% mixture of hydrogen fluoride-pyridine as the possible source of HF in glycoside synthesis was first demonstrated by Noyoriet al.[124] This HF produced was used for the fluorination of glycosyl acetates. The product was found to be highly 1,2-cis stereoselective. The use of a more acidic reaction medium and higher temperature (0°C) yielded α-pyranosyl fluorides, while reactions at lower temperatures (−30°C) and with weaker 3HF-NEt_3 systems[125] yielded β-pyranosyl fluorides. This observation can easily be attributed to the anomeric effect and the thermodynamic stability of α-glycosides. The same HF-py adduct was widely used for the synthesis of various derivatives of protected glycosyl fluorides.[126] The method was also expanded to fluorinate peracetylated *N*-acetyl neuraminic acid[127] to their corresponding fluoride counterparts (Scheme 1.19B).

However, this method was best applicable in the case of peracetylated glycosides, but it did not provide satisfactory yields when applied to differentially protected hemiacetals. Mukaiyama and coworkers, however, solved the problem of conversion of hemiacetals to corresponding fluorides by treating perbenzylatedglycofuranosyl hemiacetals with 2-fluoro-L-methyl pyridinium tosylate in the presence of triethylamine.[128] A modified Mitsonobu-type reaction with PPh_3-DEAD was also applied for fluoride synthesis with glycosyl substrate, **77** with acid-labile groups[129] (Scheme 1.20A).

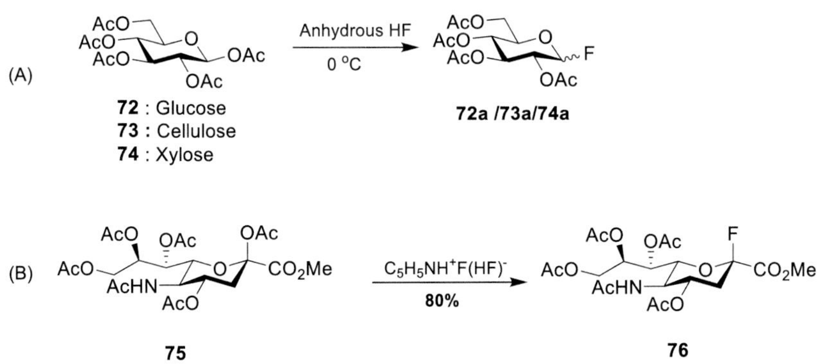

SCHEME 1.19 Conversion of glycosyl acetates to their corresponding glycosyl fluorides.

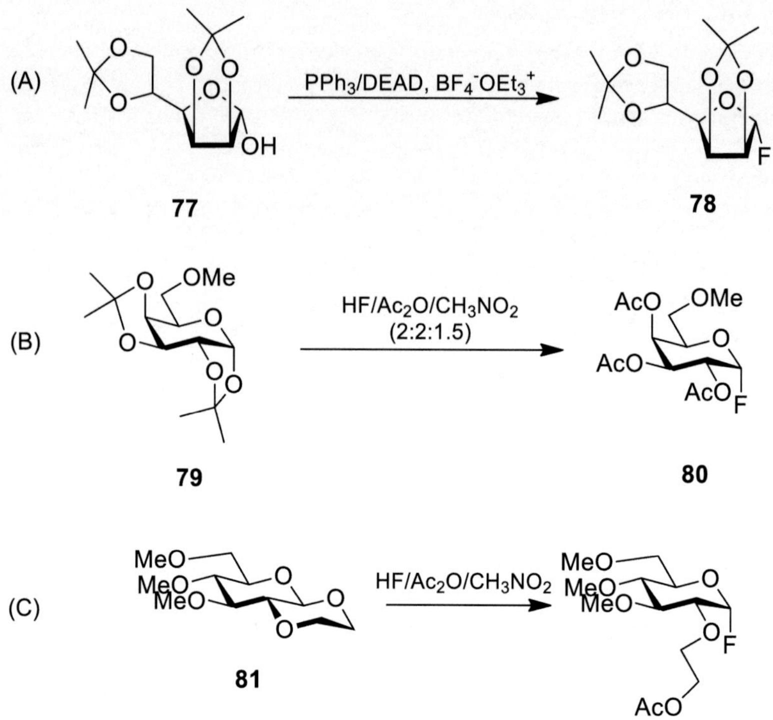

77 78 79 80 81 82

SCHEME 1.20 Synthesis of glycosyl fluorides from acetal-protected glycoside donors by (A) Mitsonobu reaction using PPh_3-DEAD; (B) HF–carboxylic acid–anhydride–CH_3NO_2 complex; and (C) by cleavage of decalin protection.

Glycosyl fluorides have also been synthesized from acid-labile iso-propylidene protected monosaccharide, **79** using HF–carboxylic acid–anhydride–CH_3NO_2 complex (Scheme 1.20B).[130] This played a dual role of the conversion of anomeric protection to corresponding fluoride and subsequent acetylation of the other free hydroxyl groups. This reagent system also easily cleaved decalin-type protected glycosides, **81**,[131] into their corresponding fluorides (Scheme 1.20C).

Preparation of glycosyl fluorides from corresponding bromides and chlorides by halide exchange mechanism has also gained much interest. Treatment of glycosyl bromides with silver fluoride (AgF) in the presence of acetonitrile initiated the halide transfer, and pure glycosyl fluorides were obtained after recrystallization.[132] In the same protocol, a differentially modified 2-benzoyloxyiminoglycosyl bromide was also converted to its corresponding β-fluoride counterpart without disturbing the benzoyloxyimino group.[133] Similarly, trifluoromethylzinc bromide-acetonitrile complex also helps in the conversion of bromides to corresponding fluorides.[134] This reagent also

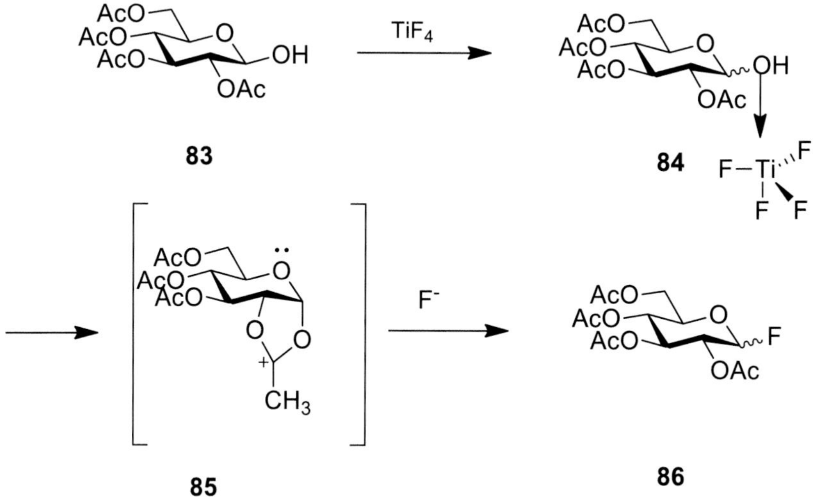

SCHEME 1.21 Synthesis of glycosyl fluorides from hemiacetal using $CF_3ZnBr \cdot 2CH_3CN$ and TiF_4 in dichloromethane.

provided evidence of the existence of the oxonium group intermediate mediated by the neighboring group participation of the acetyl group in the C-2 position. The same reagent system, however, could not provide good yield when treated with hemiacetals. But it provided excellent yields when the reagent complex was modified and replaced by $CF_3ZnBr \cdot 2CH_3CN$ and TiF_4 in dichloromethane (Scheme 1.21). The bromine−fluorine exchange has also been achieved in the presence of Et_3N-3HF/CCl_4 two-phase system.[125]

Diethyl aminosulphur trifluoride (DAST), the widely used reagent for the conversion of alcohols to fluorides, has also been used in glycoside chemistry for the conversion of hemiacetals to their corresponding fluorides. When glycosyl hemiacetals were subjected to DAST at 0°C, the fluoride products[135] were obtained in good yield. The reaction time was significantly reduced by using THF as the reaction solvent.[136] Simultaneous use of DAST and *N*-bromo succinimide (NBS) also helped in the direct conversion of thioglycosides to their fluoride counterparts.[137] However, it led to the generation of some unwanted by-products. To avoid the generation of these by-products, Kanie and coworkers used DAST in the absence of NBS, wherein dimethyl (methylthio)sulfoniumtrifluoromethanesulfonate (DMTS) substituted the use of NBS.[138] In addition to this, DAST, though highly reactive, was extremely toxic and had the potential for explosiveness. Thus to perform the reactions in a safer manner, various other fluorinating agents have been developed. Pavel Nagorny and group thus used sulfur hexafluoride, SF_6, as the fluorinating agent[139] and successfully and efficiently synthesized a library of protected glycosyl fluorides using commercially available UV-A LED source (λ_{max} = 365 nm) in the presence of 4,4-dimethoxybenzophenone (DMB) as the

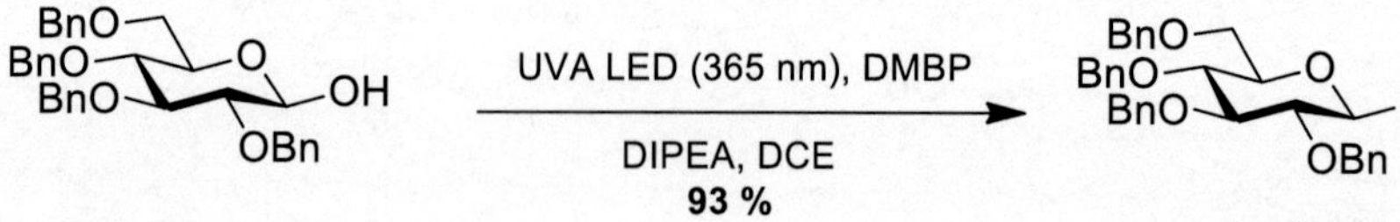

87

$\alpha:\beta$ = 1:3.7

88

SCHEME 1.22 Synthesis of glycosyl fluorides from hemiacetal using UVA-LED source and dimethyl benzophenone (DMBP) in dichloroethane in high yield (in gram scale).

photocatalyst (Scheme 1.22). The protocol proved to be highly mild, employing readily available, safe, and inexpensive reagents, and has been reported to build up the synthesis to a multigram scale in a continuous flow setup. Electrochemical synthesis of glycosyl fluorides using SF_6[140] as the fluorinating agent has been recently reported by the same group by oxidizing with Zn(II) anode. This method could easily scavenge the by-products produced to generate harmless Zn salts. Wong et al. also devised a new method of using an electrophilic fluorinating agent, readily available D-1-chloromethyl-4-fluoro1,4-diazoniabicyclo [2.2.2] octane bis(tetrafluoroborate), commercially known as Selectfluor,[141] in mild one-pot reaction conditions in the presence of SMe_2 for the synthesis of glycosyl fluorides from the corresponding glycals. In 2007 there was a report describing the conversion of thioglycosides to corresponding glycosyl fluorides catalyzed by IPy_2BF_4.[142] Hara group developed $IF_5 \cdot C_5H_5N \cdot HF$ as a new air and moisture-stable fluorinating agent[143] used to convert (phenylthiol) glycosides to their corresponding fluorides.

Vito Ferro and coworkers have also very recently developed an improved, milder, and safer protocol for β-glycosyl fluoride synthesis, which involves a combination of NBS, XtalFluor-M, and $Et_3N \cdot 3HF$.[144] Glycals have also been used for subsequent conversion to glycosyl fluorides via an epoxide intermediate that reacted with tetrabutyl ammonium fluoride (TBAF) to produce the required halide.

Glycosyl fluorides have been first used in glycosylation reactions by Mukaiyama et al. by using $SnCl_2{-}AgClO_4$ as the activator.[145] Since then, many modifications have been made by replacing $AgClO_4$ with $TrClO_4$, AgOTf, and so on. Nicolaou and coworkers reported the use of Lewis acids like $BF_3.Et_2O$[146] for the activation of glycosyl fluorides. Halide salts like SnF_4, TiF_4,[147] and TMSOtf[148] were also found to be highly efficient for the activation of glycosyl fluorides as donors in glycosylation reactions with suitable acceptor moieties. Suzuki and group devised a new class of metallocene promoters for glycosyl fluorides like $Cp_2ZrCl_2{-}AgClO_4$ and $Cp_2HfCl_2{-}AgClO_4$ for the synthesis of highly stereoselective β-glycosides (Scheme 1.23).[149] The promoters have been successfully implemented for the total synthesis of Mycinamicin IV and VII.[150] They also reported the use of $Cp_2ZrCl_2{-}AgBF_4$ and $Bu_2SnCl_2{-}AgClO_4$ combination systems for the effective synthesis

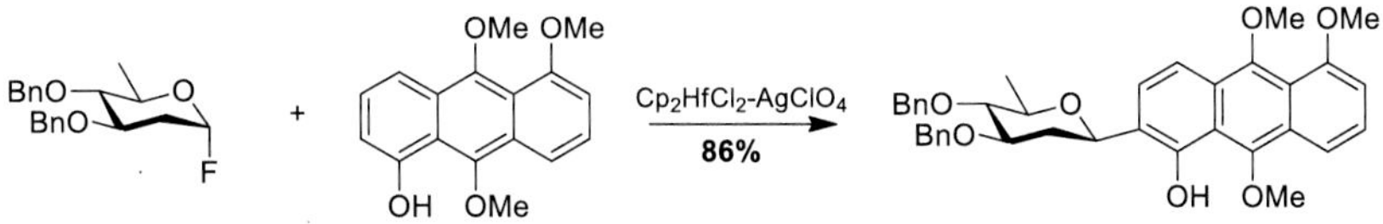

104 105 106

SCHEME 1.23 Activation of glycosyl fluorides (**104**) using $Cp_2HfCl_2{-}AgClO_4$.

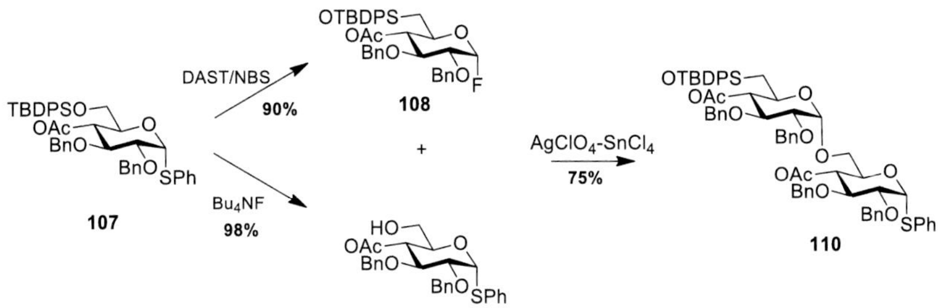

109

SCHEME 1.24 Two-step orthogonal activation of glycosyl fluorides.

of α-mannopyranosylpyranosides.[151] Dimethylgallium chloride catalyzed glycosylation[152] reactions using glycosyl fluorides have been reported to achieve completion in the presence of stoichiometric amounts of trimethylsilyl chloride (TMSCl) as the additive. Zirconocene bis(perfluorooctanesulfonate) has also been implemented for the activation of glycosyl fluorides rendering high yields but low stereoselectivity.[153] Similarly, a combination of 20 wt.% of a zirconium-based, solid acid, SO_4/ZrO_2 has also been used for the activation of mannopyranosyl fluorides.[154] Miller and group have also been successful in activating unprotected glycosyl fluorides as donors in aqueous triethylamine solution and calcium ion[155] opening a new scope of activation of glycosyl halides in aqueous medium. This also enunciates the stability of the fluoride donors even in aqueous medium.

In a report by Yang and Yu, there has been a report illustrating the use of glycosyl fluorides in selective activation, thereby helping in sequential glycosylation reaction. The fluoride glycoside was initially used as an acceptor[156] to couple with an armed thioglycoside donor. The fluoride disaccharide, thus obtained, was further used as a donor, which can be activated further for the formation of longer oligosaccharide moieties. Similarly, for a two-step activation and subsequent oligosaccharide synthesis using the same glycoside as synthon, sometimes differential activation has proved to simplify the process. Nicoloau et al., in a report, have reported a tetrasaccharide, avermectin, sequentially by converting the same molecule to its corresponding fluoride and thio derivatives (Scheme 1.24).[137] Unlike the previous application, here, the fluoride donor was activated first, followed by the activation of the thioglycoside donor.

Ogawa et al. reported the activation of glycosyl halides in the total synthesis of a lipohexasaccharide, isolated from *Bradyrhizobium japonicum.*[157] Mori et al. also reported total synthesis of a cerebroside, isolated from *Penicillium funiculosum.*[158]

1.4.1.4 Thioglycosides

Thioglycosides are the most versatile and most popular glycosyl donor employed for glycosylation reactions over the years. They are highly effective in glycosylation reactions owing to their stability toward a wide range of protecting groups[159] and their ability of being activated even in mild, environmentally friendly conditions. The soft sulfur atom of the thiophilic group has a special affinity toward soft electrophiles, contributing to the versatility of the anomeric thio-functionality.

Over the decades, thioglycosides have been prepared in a variety of ways from various glycosyl precursors. Thioglycosides preparation from peracetylated glycosides and alkyl or aryl mercaptans using a Lewis acid, $BF_3.Et_2O$ is much known in the literature on carbohydrate chemistry. But due to its repelling smell and high toxicity, mercaptans are not considered as a good nucleophile for the synthesis of thioglycosides. So, in many cases, peracetylated sugars are converted to their respective halides. Common and conventional methods involve the conversion of glycosyl halides into thiocompounds by utilizing sulfur nucleophiles like thiourea or thioacetates followed by hydrolysis.[160]

Dong and group reported the substitution of smelly alkyl mercaptans with sodium alkane thiolates in the presence of excess (8–10 equivalent) of $BF_3.Et_2O$.[161] The reaction was also seen to provide a higher yield when performed under solvent-free conditions, while the use of CH_2Cl_2 as the solvent gave high α-stereoselectivity.

Benjamin Davies and group devised a method for direct preparation of thioglycosides by concerted Lewis acid activation and displacement.[162] With that goal in mind, they demonstrated the use of Lawesson's reagent even with unprotected sugar substrates. This procedure was also expected to aid in strategic one-pot synthesis of oligosaccharides. Preparation of thio-analogs of Immunostimulant KRN7000 has also been reported by Zhu et al.[163] Murphy and group also reported the synthesis of α-S-Glycosphingolipids (Scheme 1.25).[164]

There has been a report of the preparation of thio-trehalose derivatives by the bubbling of excess hydrogen sulfide in the reaction mixture in a closed vessel.[165] However, the reaction gave a low yield owing to the formation of a significant number of side products as detected by NMR spectroscopy. Zhu et al. in 2008 also reported the direct synthesis of glycosyl thiols from anhydro sugars by following an S_N2-type mechanism using bis(trimethylsilyl)sulfide, $(TMS)_2S$, as the required sulfur nucleophile, catalyzed by trimethylsilyl

SCHEME 1.25 Ring-opening mechanism for the formation of thioglycosides from hemiacetal.

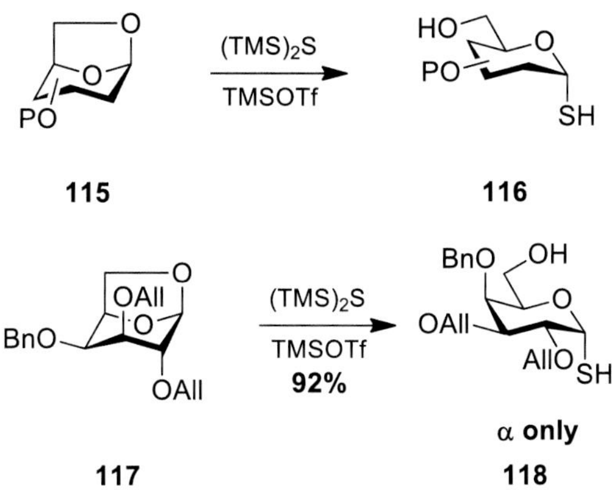

SCHEME 1.26 Ring-opening mechanism of anhydro sugars, 115 for the formation of thioglycosides.

trifluoromethanesulfonate, TMSOTf (Scheme 1.24).[166] This ring-opening reaction gave high yields affording only α-glycoside confirming the possible S_N2 mechanistic pathway. R.R. Schmidth and group also reported the ring-opening mechanism of anhydro sugars with bis(trimethylsilyl) sulfide (BTMSS) in the presence of TMSOTf as a catalyst (Scheme 1.26).[167] Demchenko et al. also reported the synthesis of thioglycosides from peracetylated sugars catalyzed by triflic acid (TfOH).[168]

With a target to find new sulfur nucleophiles, very recently, a new method has been reported by Siyu Li et al. for the conversion of glycosyl bromides to thioglycosides. They followed the principle of reductive deoxygenation coupling reaction by in situ generation of sulfenyl electrophiles from sulphonyl chlorides.[169] This was further converted to thiol nucleophiles in the presence of boron salts, B_2pin_2 or PPh_3. Manas Jana and Anup Misra devised an improved method for the synthesis of thioglycosides using sodium carbonotrithioate solution,[170] which is prepared in situ by the reaction of CS_2 with $Na_2S.9H_2O$. Use of DMF as the solvent successfully afforded only β-isomer.

Glycals have also been widely utilized for the successful synthesis of thioglycosides. Walczak and Hong successfully synthesized thioglycosides with a high degree of β-stereoselectivity from glucopyrannostannane with the aid of a Cu(I) catalyst.[171] Conversion of glycals using Lewis acids by Ferrier-type glycosylations. In this context, commercially available zeolite

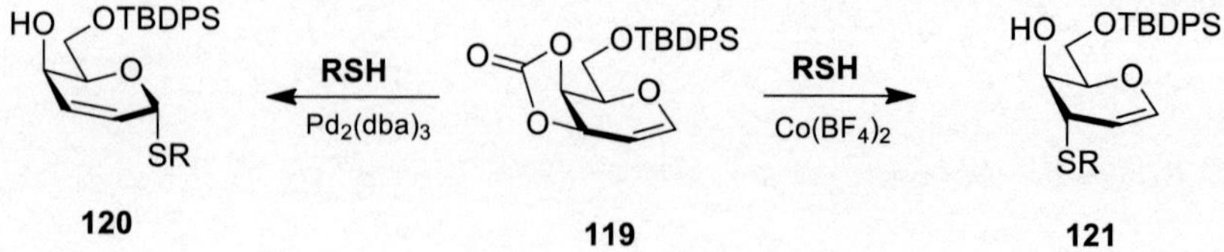

SCHEME 1.27 Synthesis of 1-thiosugar (**120**) by Pd-catalyst and 3-thiosugar (**121**) with Co catalyst.

materials, like ultrastable Y-zeolites have been utilized for the conversion of glycals to corresponding thioglycosides[172] using thiophenol as the attacking nucleophile. Hui Yao and group recently reported the synthesis of 1- and 3-thiosugars from 3,4-*O*-carbonate glycals in the presence of $Pd_2(dba)_3$ and Co $(BF_4)_2$ respectively, in high yield and high stereoselectivity[173](Scheme 1.27). Jiang et al. also established the divergent synthesis of thioglycosides via a palladium-catalyzed allylic rearrangement protocol.[174]

Activation of thioglycosides with heavy metals started the journey of using them as the probable glycoside donor in glycosylation reactions. The first activation of the thioglycoside donors was performed using Hg(II) salts like mercuric sulfate,[175] mercuric nitrate,[176] and mercuric benzoate.[177] Mukaiyama first implemented copper salts, copper triflate for the activation.[178] Following its lead, reports showed a combination of $CuBr_4-Bu_4NBr$, along with an additional promoter like AgOTf and mercuric bromide, was capable of effectively activating thioglycosides[179] to afford a high yield. Kihlberg and group also devised a protocol by using a combination of bromine with AgOTf or Hg $(CN)_2$[180] to convert the thioglycosides to corresponding bromides in situ followed by its activation by silver triflate. Kusumutu et al. introduced a wide range of promoter systems for activation of thioglycosides like NBS/Ph_2IOTf, Bu_4NOTf, Bu_4NClO_4, $LiClO_4$, or $LiNO_3$.

One of the most widely applied protocols was introduced by van Boom and coworkers who used active iodonium ion generated from halonium-based species like *N*-iodosuccinimide (NIS), NBS[181] in association with a catalytic amount of Bronsted or Lewis acid for the activation of thioglycosides in glycosylation reaction. The year 2000 saw the use of the $NIS/LiClO_4$ combination by Hirama et al. and $NIS/TrB(C_6F_5)_4$ by Mukaiyana et al. The use of all the combinations of NIS or NBS with appropriate additives is given in Table 1.2.

Supported acid catalysts appeared to be a good substitute for traditional Lewis acid promoters. Field and coworkers introduced $HClO_4$-silica for activation of disarmed thioglycosides (**entry 12**, Table 1.2) providing milder reaction conditions. On similar grounds, Mukhopadhyay and coworkers successfully developed H_2SO_4-silica for the activation of thioglycosides (**entry 14**, Table 1.2).[195] The group has also applied the promoter system for the successful synthesis of several plant and animal oligosaccharides.[196]

TABLE 1.2 Some NIS/NBS-based promoters for the activation of thioglycosides.

Promoter used	Best yield (%)	Year	Reference
NBS-4 Å MS	78	1983	Nicolaou et al.[182]
NIS/HOTf	98	1990	Van Boom et al.[183]
NIS/AgOTf	90	1990	Fraser-Reid et al.[184]
NBS/TfOH	74	1991	Sasaki et al.[181]
NBS/Ph_2IOTf	98	1995	Kusumutu et al.[185]
NIS or NBS/$TrB(C_6F_5)_4$	95	2000	Mukaiyama et al.[186]
NIS/$LiClO_4$	73	2000	Hirama et al.[187]
NIS/$Yb(OTf)_3$	80	2001	Chung and Park[188]
NBS/Me_3SiOTf	78	2002	Li et al.[189]
NIS/TrOTf	75	2003	Mukaiyama et al.[190]
NIS/$Sn(OTf)_2$ or $Cu(OTf)_2$	80: for $Sn(OTf)_2$ and 98: for $Cu(OTf)_2$	2003	Ferriéres et al.[191]
NIS/$HClO_4$-silica	87	2005	Field et al.[192]
NBS/$Bi(OTf)_3$	91	2007	Iadonisi et al.[56]
NIS/H_2SO_4-silica		2008	Mukhopadhyay et al.[193]
NIS/$La(OTf)_3$		2010	Mukhopadhyay et al.[194]

Jensen and coworkers have provided a systematic study of the effect of thioglycoside stereochemistry on reaction selectivity,[197] which showed that the α-thio donors showed higher reactivity in comparison to β-thiglycoside donors. Groups of Kita and Kajimoto et al. worked on the oxidation of sulfur with the help of hypervalent iodine reagents. A combination of phenyliodine (III) bis(trifluoroacetate) (PIFA) (**124**) with trifluoromethane sulfonic acid produced the best yield (Table 1.3).[198] The method was applicable for both armed and disarmed donors and particularly effective against glycosides having phthalimido group in the C-2 position. The low toxicity, easy handling, and easy availability of the hypervalent iodine atom make it a very effective reagent for glycosylation reactions. Similar other reagents, like phenyliodine diacetate (PIDA), p-$MeC_6H_4I(OCOCF_3)_2$, and p-$F_3CC_6H_4I(OCOCF_3)_2$, have also been reported.[199]

Nicola Pohl et al. demonstrated the use of bismuth catalysts for the activation of thioglycosides. They reported that promoters containing Bi(V) salts could be successfully used to activate thioglycosides, but Bi(II) could not.[200] They demonstrated $Ph_3Bi(OTf)_2$-mediated thiopropylglycoside activation.[201]

TABLE 1.3 Comparative study of the use of different Lewis acids on the yield of glycosylation promoted by PIFA (124).

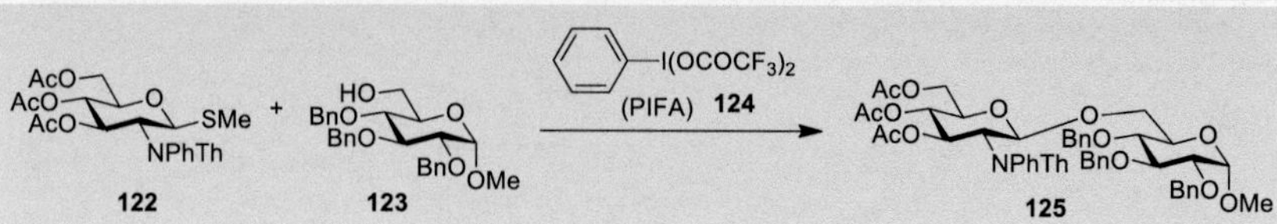

Entry	Lewis acid	Yield (%)
1.	TfOH	77
2.	TMSOTf	73
3.	(Cyclo-hex)$_2$BOTf	54

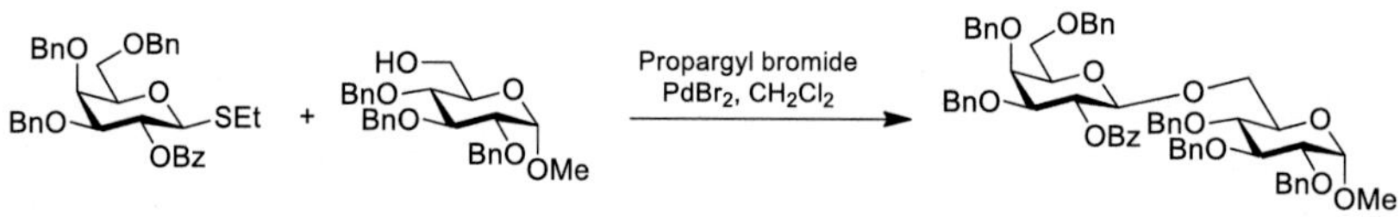

126 127 128

SCHEME 1.28 Activation of thioglycoside **126** using palladium bromide and propargyl bromide additive for glycosylation.

This method not only gave good yields but was also able to demonstrate efficient anomerization from β-isomer to corresponding α-isomer. Sureshan and group reported the activation of thioglycosides by Au(III)chloride as the catalyst[202] and the glycosylation yielded satisfactory yields even at room temperature with both armed and disarmed donors having a wide range of functional groups. Zhu et al. also demonstrated gold-catalyzed activation of thioglycosides to form deoxy glycosides.[203] The same group also used *N*-trifluoromethylthiosaccharin along with TMSOTf for glycosylation reactions[204] with thiosugars as donors. Demchenko et al. first reported the activation of thioglycosides with palladium bromide ($PdBr_2$) in the presence of propargyl bromide as an additive (Scheme 1.28).[205]

Light-induced activation of thioglycosides has also seen much development. Bowers and coworkers reported the visible light-activated glycosylation of thio-glycosides being catalyzed by Ir[dF(CF_3)ppy]$_2$(dtbbpy)PF_6 and in the presence of bromo-trichloromethane and a nonnucleophilic and protic solvent, hexafluoroisopropanol (HFIP).[206] A photo-sensitizer, $[Ru(bpy)_3]^{2+}$ was also used along with 2,6-di-tert-butyl4-methylpyridine (DTBMP)[207] for glycosylation reactions. The catalyst was used to oxidize the electron-rich aryl thioglycoside. Till Opatz et al. reported an aerobic photoredox protocol for

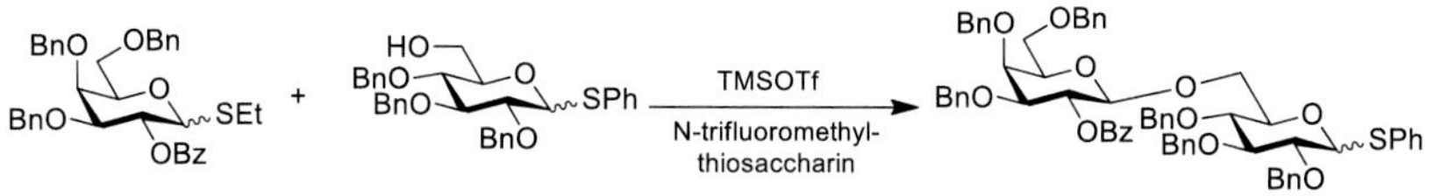

126 129 130

SCHEME 1.29 Chemoselective activation of thioethyl glycoside **126** over thiopneyl glycosides using *N*-trifluoromethylthiosaccharin in the presence of Lewis acid.

glycosylations with thioglycosides where the thio-donor was irradiated under an oxygen environment with a daylight compact fluorescence lamp (CFL) bulb using iodine as the cocatalyst.[208] The reaction showed no visible change in the iodine color concentration of the reaction mixture. Unprotected deoxythioglycosides have also been activated to synthesize O-glycosides using 2,3-dichloro-5,6-dicyano-1,4-benzoquinone (DDQ) under long wavelength UV-radiation by employing single electron transfer (SET) mechanism.[209]

Thioglycosides have also been used by preactivation step-wise one-pot procedure.[210] In 1998 Crich and group first described the formation of β-mannosides based on preactivation protocol.[211] The thioglycosides are first precooled to −78°C in the presence of DTBMP and further treated with a solution of PhSOTf. Demchenko et al. also reported a streamlined regenerative glycosylation by direct, acid-free activation of the thioglycosides using NIS and catalytic amount of HOFox.[212] Electrochemical *O*-glycosylation of primary alcohols with thioglycoside donors has been reported in the presence of sodium trifluoromethanesulphonare as the supporting electrolyte.[213]

Seeberger also reported the use of the combination of 1,3-dibromo-5,5-dimethylhydantoin (DBDMH) and TMSOTf for the activation of glycosides[214] in both normal and automated glycan synthesis setup. In another approach toward machine-assisted synthesis of glycosides using thiosugars as the donor, Nicola Pohl and group used hypervalent iodonium species, aryl(trifluoroethyl) iodonium triflimide, as the activator.[215] Thioglycosides have also been used for the synthesis of hexasaccharides in automated oligosaccharide synthesis protocol.[216] Zhu et al. also showed a competitive activation protocol of ethyl and phenyl thioglycosides[217] and demonstrated the use of *N*-trifluoromethylthiosaccharin as the catalyst in the chemoselective activation of ethyl thioglycosides **126** over phenyl thioglycosides **129** in high yields (Scheme 1.29).

1.4.1.5 Glycosyl trichloroacetimidates

Glycosyl imidates are a versatile leaving group alternative for glycosyl donors, which has a wide range of applicability in glycosylation reactions. Trichloroacetimidates and trifluoroacetimidates were first introduced by Schmidt and group in 1980,[218] and since then, it is one of the most widely used donors apart from thioglycosides. Glycosyl trichloroacetimidates can be readily formed by means of a direct, reversible, and base-catalyzed addition of alcohols with the electron deficient nitrile group of the readily available trichloroacetonitrile using inorganic (NaH, K_2CO_3) or organic (DBU) bases. The imidates formed are stable under room temperatures and on acid catalysis exhibit excellent donor characteristics. The trichloroacetamide produced as a by-product can also be converted to trochloroacetonitrile leading to the cost-effectiveness of the procedure.

The versatility of glycosyl trichloroacetimidates lies in the fact that the stereocontrol of the produced imidates can be accomplished effectively. A weak base like K_2CO_3 is applied for the exclusive formation of kinetically controlled produce, β-trichloroimidate while a strong base like NaH is applied for the exclusive formation of thermodynamically controlled, α-trichloroacetimidate. Dowex 1-X8 OH^- has also been used as a good catalyst[219] for the formation of trichloroacetimidates. Polymer-supported 1,5,7-triazabicyclo[4.4.0]dec-5-ene (PTBD) has also been used a powerful catalyst, which has also been used along with Nafion-SAC (solid acid catalyst) resin for iterative glycosylation reactions.[220] The formed imidates undergo S_N2 type reaction with the incoming acceptor alcohol molecules to yield the product with opposite stereoselectivity (Scheme 1.30). It is to be remembered, however, that the stereocontrol of the product also depends on various other factors like neighboring group participation, solvent, and temperature. Solvents of low polarity contribute toward S_N2-type reactions, while those of high polarity contribute differently. Solvents like ether facilitate the formation of thermodynamically stable axial conformation, but in the case of nitriles it facilitates the formation of the equatorial product.[221] Addition of $LiClO_4$ to the solvent system also increases its polarity, thereby increasing the probability of trichloroacetimidate activation even under neutral conditions.[222] Use of ionic liquids (ILs) also helps in glycosylation, even in the absence of acid catalysts.[223]

$BF_3.Et_2O$ is an effective Lewis acid that is used for the activation of glycosyl trichloroacetimidates in glycosylation reactions. Hyaluronic acid has

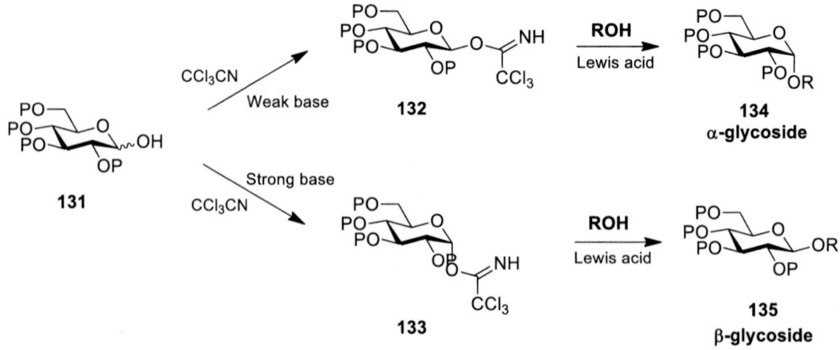

SCHEME 1.30 Role of the base on the stereochemistry of the trichloroacetimidates, **132** and **133** from hemiacetal **131** followed by glycosylation with acceptor, **ROH**.

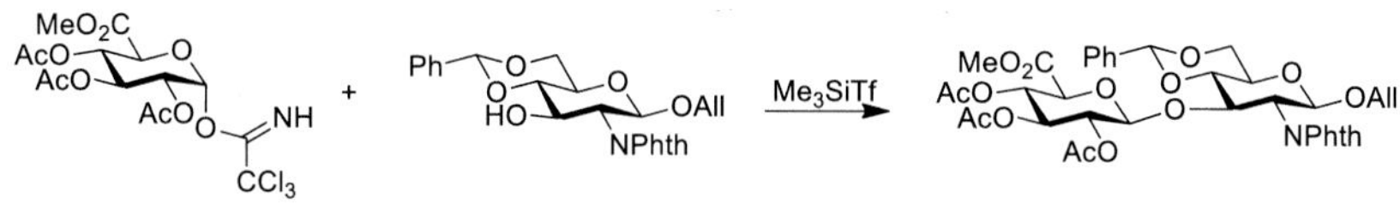

136 **137** **138**

SCHEME 1.31 Formation of hyaluronic acid **138** by activating glucuronic acid trichloroacetimidate **136**.

become a leading name in the cosmetic industry today, a part of which (**138**) is synthesized successfully by activating glucuronic acid trichloroacetimidate (**136**) using trimethylsilyl triflate as the promoter (Scheme 1.31).[224]

In 2012 Cloninger and coworkers established In(III) salts like $InCl_3$, $InBr_3$, and $In(OTf)_3$ as a more versatile promoter for the activation of trichloroacetimidates with a suitable alcohol molecule as the acceptor. Nguyen et al. established the use of stoichiometric amounts of Ni(II) catalyst for the activation of difficult imidates. In-house reagent, $Ni(4\text{-}F\text{-}PhCN)_4(OTf)_2$ catalyst, has been used for the activation of *C*-2-*N*-substituted benzylidenamino-trihaloacetimidate donors to form 1,2-*cis*-2-amino glycosidic linkages.[225] However, due to the unavailability of the reagent, the same group established stable and easily available $Ni(OTf)_2$ catalyst for the activation of trichloroacetimidates,[226] enabling the synthesis of various bioactive oligosaccharides like blood type A-type V and VI antigens. A report also illustrated the doping effect of calcium ions on sulfated zirconia (SZ) promoted glycosylation of per-*O*-benzylated galactose trichloroacetimidates.[227] This explained that the yield of the reaction was the increased Lewis acid sites on the SZ due to doping of calcium cation. This also opened a new avenue in the building of superacids. Alfonso Iadonisi and coworkers established a convenient method for the activation of both glycosyl trichloroacetimates and *N*-phenyl trifluoroacetimidates using iodine/triethylsilane as the promoter system.[228] Rina Ghosh and group established the role of iron(III) chloride for its use for orthogonal glycosylations using glycosyl trichloroacetimidates.[229] The group also applied the method for the synthesis of the O-antigen of *Escherichia coli120* in a one-pot iterative process.[230]

In instances where trichloroacetimidate formation becomes challenging, the use of *N*-phenyl trifluoroacetimidates has also evolved as a useful substitute. However, trifluoroacetonitriles being a gas are not as convenient to use as trichloroacetonitriles. So, there has been a report on the use of (*N*-phenyl) trifluoroacetimidoyl chloride for the generation of corresponding trifluoroacetimidates. Triflates of various salts of lanthanides and actinide series have also been utilized for the activation of trichloroacetimidates. Salts like $Sm(OTf)_3$,[231] $Sc(OTf)_3$, $Tb(OTf)_3$, and $Yb(OTf)_3$ were capable of activating these imidates even at room temperature.[232] $Yb(OTf)_3$ was also efficient in selectively activating armed glycosyl trichloroacetimidates over similarly armed *N*-phenyl trifluoroacetimidate glycosides (Scheme 1.32).[233]

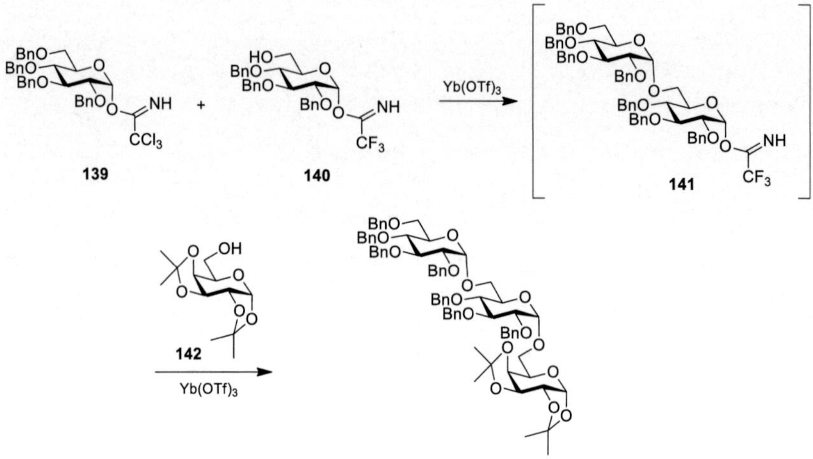

SCHEME 1.32 Selective activation of armed trichloroacetimidate **139** over armed trifluoroacetimidate **140** with the help of lanthanide triflate promoters.

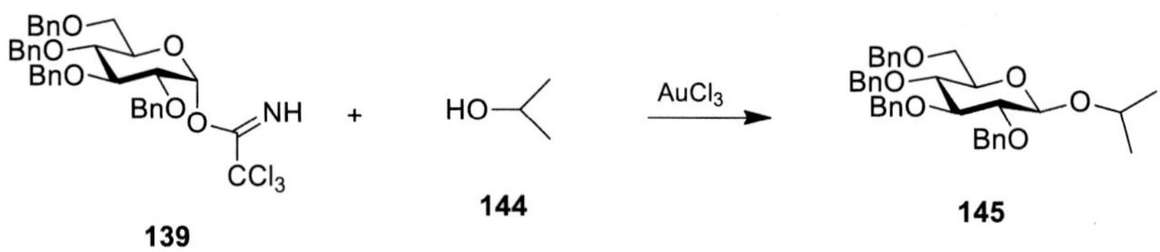

SCHEME 1.33 Activation of armed trichloroacetimidate **139** in the presence of gold chloride ($AuCl_3$).

Peng and Schmidt have also successfully devised a gold-catalyzed activation of trichloroacetimidates via an intramolecular dual catalysis pathway in the presence of $AuCl_3$ as the initiator[234] to yield high β-selective products even in the absence of participating groups in the C-2 position (Scheme 1.33).

The process mainly deals with the formation of catalyst-acceptor adduct. The same adduct formation is utilized for the activation of trichloroacetimidates using a platinum (IV) catalyst.[235] Both the procedures employ Lewis acids having less affinity toward the glycosyl donor leaving the group but having a strong affinity toward the acceptor. In the search for new catalyst systems, Schmidt and group also devised the use of phenylboron difluoride ($PhBF_2$) or diphenylboron fluoride (Ph_2BF) for the activation of glycosyl trichloroacetimidates[236] in glycosylation reactions by forming a hydrogen-bonded adduct with the acceptor, which in turn acts as the Lewis acid. But, owing to the high Lewis acidity of the reagents involved, the reactions were preferentially performed in low temperatures. So, Kandasamy et al. established a weaker and milder Lewis acid, Tris-(pentafluorophenyl)borane

($B(C_6F_5)_3$ or BCF) as the catalyst for the activation of trichloroacetimidates under milder conditions[237] resulting in high yields with effective stereoselectivity. Reusable organic photoacids are also capable of activating glycosyl imidates upon photo-irradiation,[238] thereby aiding in effective glycosylation reactions. The use of bis(2-naphthyl)disulfide as the organo-Lewis photoacid (LPA) both with and without NEt_3 upon irradiation for glycosylating trichloroacetimidate donors[239] also witnessed high yields with high stereoselectivity. Another visible light-induced photoacid-catalyzed glycosylation was reported by Wang et al. by using eosin Y[240] and PhSSPh catalyst system for activating trichloroacetimidate donors.[241] Very recently, Pederson and group have developed a method of glycosylation with trichloroacetimidate donors with a wide range of alkyl and aryl nucleophilic acceptors using commercially available pyrilium catalysts (Fig. 1.6).[242] This method produced a high yield with high β-selectivity. They established a library of compounds showing the role of protecting groups on the chemical reactivity of the donor molecules (**146**, **147**, and **148**) (Fig. 1.5). This study promises to aid the glycochemists in the further establishment of glycosylation with glycosyl trichloroacetimidates. There has also been a report of ZnI_2 mediated 1,2-*cis* glycosylation with trichloroacetimidate donors emphasizing the role of zinc cations on glycosylation.[243]

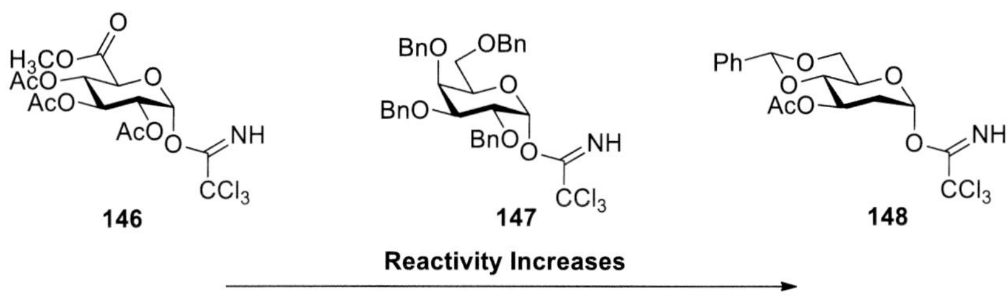

FIGURE 1.5 Reactivity difference of various protected trichloroacetimidates, **146, 147**, and **148**.

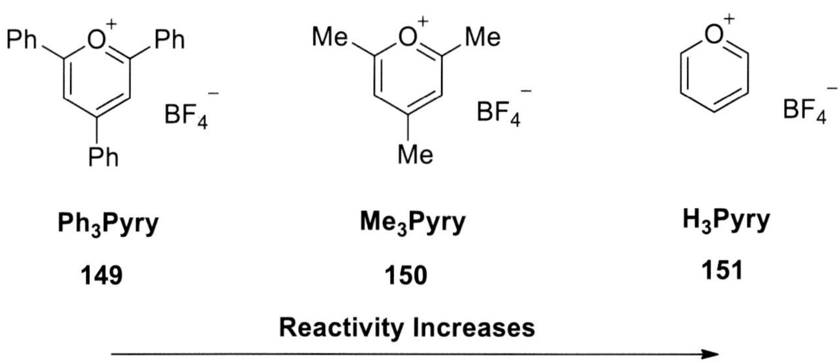

FIGURE 1.6 Reactivity difference of various pyrilium catalysts, **149**, **150**, and **151**.

1.4.2 Role of acceptors

Over the years, much study has been done, and progress has been made in modifying and tuning the activity of the reacting donors and the development of a range of effective promoter systems for chemoselective and regioselective glycosylation reactions. However, similar studies and the importance of the role of the acceptors are very convincingly missing in scientific literature. But on the contrary, the acceptors play a very important role in determining the route of glycosylation reactions. The nucleophilicity of the acceptor is highly important for the effectiveness of the glycosylation reactions. The nucleophilicity is found to be dependent on various factors, particularly electronic effects like inductive effect, steric factors, and resonance effects. Acceptors with less hindered hydroxyl groups like the primary C-6 OH is always considered to render better nucleophilicity in glycoside reactions in comparison to other crowded hydroxyl groups. In that respect, the C-3 OH position of galacto-pyranoside moiety has always been considered as a better nucleophile in comparison to the C-4 OH group of the same galactose moiety. Groups have utilized this difference in nucleophilicity in performing various selective glycosylation procedures.[244] Similarly, C-4 OH of *N*-acetyl glucosamine has been established as very poor nucleophiles, and Crich et al. have provided a detailed report explaining the reasons behind the same.[245]

Paulsen and Lebuhn illustrated the role of different glycoside acceptors on the stereoselectivity when coupled with the same α-bromo mannoside donor (**152**).[246] It was seen that conformationally restricted anhydro acceptors **154** and **155** were more prone to yield 1,2-*trans* glycosides in comparison to differentially protected, *N*-phthalimido 2-deoxy glucopyranoside moiety **153** (Scheme 1.34).

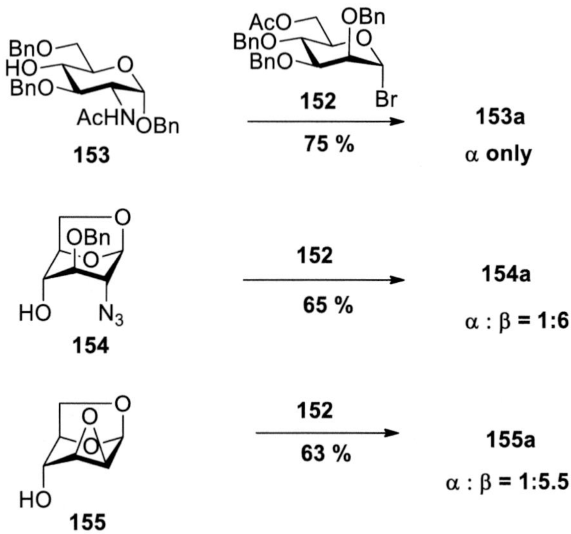

SCHEME 1.34 Comparative study of the contribution of conformationally restricted acceptors **153, 154**, and **155** on the stereochemistry of the product.

Paulsen also explained the concept of "match-mismatch" regioselectivity of a combination of donor/acceptor.[247] Ruvedaet al. illustrated this phenomenon by depicting β-D-galactosylation versus α-L-fucosylation with a set of different acceptors.[248] The same group also established the relative reactivities of different dimethylmaleimide (DMM) protected glucosamine acceptors and investigated the comparative reactivities of the same with respect to *N*-acetyl glucosamine acceptors.[249] *N*-Dodecyl mercaptan (NDDM)-protected acceptors were in turn found to increase the reactivity of the *N*-acetyl acceptors.

Kinetic isotope effects for various competitive donor–acceptor combinations have also been studied by Crich and group.[250] A detailed review of the analysis of the reactivity of acceptors in glycosylation reactions is illustrated by Codee and coworkers.[251]

1.5 Conclusions and future outlook

Chemical glycosylations are an important tool in the development of complex oligosaccharide scaffolds. This, in turn, helps in the study of carbohydrates and glycochemistry and helps in deciphering the various intricate details pertaining to life processes. Modern times have seen spurge in the development of new methodologies and new procedures for conveniently synthesizing naturally occurring oligosaccharides. Enzymatic glycosylations have also gained a lot of interest,[252] and it has gained momentum especially in the last decade. However, glycan synthesis is usually complex and no-templated, involving the use of glycosyl transferases and glycosidases. To the reduce the complexity of the procedure, the use of artificial intelligence (AI) in the analysis of data was introduced and developed by Thoedoratouet et al.[253] An algorithm for prediction of stereochemistry in glycosylation has been developed by Moon et al.[254] Stoichiometric metabolic models have also been developed to predict glycoforms.[255] All these developments have helped in the strategic planning of the oligosaccharide synthesis with greater control and precision. This enzymatic glycosylation coupled with chemical glycosylation[256] has the potential to solve the various unsolved synthetic routes of complex and long oligosaccharide units.

The development of a one-pot iterative process has contributed toward the ease of the glycosylation pathway by reducing the number of reaction steps, thereby also helping to maintain the atom economy. This primarily involves the use of armed–disarmed concept of reactivity difference for the orthogonal activation of two similarly protected donors. One-pot synthetic protocol has also helped in the development of polymer support synthetic routes in solid-phase synthesis of complex oligosaccharides, which was first developed with Merrifield resin by Fréchet and Schuerch in 1971.[257] Solid-phase synthesis promises to solve the various limitations of solution phase synthesis like purification of products from unreacted reagents.[258] Development of various instrumental facilities like fluorine labeled and decoupled ^{13}C NMR spectroscopy,[259] real bead Fourier Transform InfraRed

(FTIR), and real-time monitoring[260] has helped the glycochemists in developing and further modifying the method of solid-phase oligosaccharide synthesis (SPOS).

SPOS has eventually led to immense development in the new domain of automated oligosaccharide synthetic protocols[261] employing HPLC methods[262] and continuous flow methods. These methods significantly reduce the various manual routine procedures by using a programmed sequence of events comprising selective protection, iterative glycosylations, removal of the polymer support in situ, and subsequent deprotection of the final glycoside. The continuous flow reactor named as Glyconeer by Seeberger and Group[263] is equipped with a controlled syringe–pump driver reagent delivery system and is capable of being programmed in various temperatures and with a wide variety of temporary and permanent protecting groups. Various groups are now trying to adapt to this automated machine-driven process of oligosaccharide synthesis.[264] All these modern methods are, in turn, opening newer avenues in the synthesis of oligosaccharides aiding the glycochemists to decipher the various unknown aspects of nature.

References

1. Robyt, J. E. In Glycos*cience*: *Chemi*stry and Chemical Biology*;* Fraser-Reid, B., Tatsuta, K., Thiem, J., Eds.; Springer: Berlin-Heidelberg-New York, 2001.
2. (a) Liu, J.; Willför, S.; Xu, C. *Bioact. Carbohydr. Diet. Fiber* **2015,** *5*, 31–61.
   (b) Qin, J.; Wang, H. Y.; Zhuang, D.; Meng, F. C.; Zhang, X.; Huang, H.; Lv, G. P. *Int. J. Biol. Macromol.* **2019,** *136*, 341–351.
   (c) Stryer, L. *Biochemistry;* Freeman: W. H. New York, 1988.
3. (a) Bertozzi, C. R.; Kiessling, L. L. *Science* **2001,** *291*, 2357–2364.
   (b) Varki, A. *Glycobiology* **2016,** *27*, 3.
   (c) Varki, A. *Glycobiology* **1993,** *3*, 97–130.
   (d) Varki, A.; Cummings, R.; Esko, J.; Freeze, H.; Hart, G.; Marth, J. *New York; Essentials of Glycobiology* Cold Springer Harbor Laboratory Press, 1999.
   (e) Davis, B. G. *Chem. Rev.* **2002,** *102*, 579–601.
   (f) Linhardt, R. J. *J. Med. Chem.* **2003,** *46*, 2551–2564.
4. Horlacher, T.; Seeberger, P. H. *Chem. Soc. Rev.* **2008,** *37*, 1414.
5. (a) Reily, C.; Stewart, T. J.; Renfrow, M. B.; Novak, J. *Nat. Rev. Nephrol.* **2019,** *15*, 346.
   (b) Seeberger, P. H. *Chem. Rev* **2021,** *121*, 3598.
6. Chen, S.; Fukuda, M. *Methods Enzymol.* **2006,** *416*, 371–380.
7. Rudd, P. M.; Elliott, T.; Cresswell, P.; Wilson, I. A.; Dwek, R. A. *Science* **2001,** *291*, 2370–2376.
8. West, C. M.; Malzl, D.; Hykollari, A.; Wilson, I. B. H. *Mol. Cell Proteom.* **2021,** *20*, 100024.
9. (a) Dwek, R. A. *Chem. Rev.* **1996,** *96*, 683–720.
   (b) Cipolla, L.; Araújo, A. C.; Bini, D.; Gabrielli, L.; Shaikh, N. *Expet. Opin. Drug. Discov.* **2010,** *5*, 721–737.
10. Crich, D. *J. Am. Chem. Soc.* **2021,** *143*, 17–34.
11. Capon, B. *Chem. Rev.* **1969,** *69*, 407–498.

12. Wang, C.; Chen, Z.; Wu, W.; Mo, Y. *Chem. Eur. J.* **2013,** *19*, 1436–1444.
13. Edward, J. T. *Chem. Ind. (Lond.)* **1955,** 1102–1104.
14. (a) Lemieux, R. U.; Chu, N. J. San Francisco, CA, April *Abstracts of Papers; 133rd National Meeting of the American Chemical Society*, 1958. American Chemical Society: Washington, DC, 195831N.
    (b) Lemieux, R. U. *Pure Appl. Chem* **1971,** *25*, 527–548N.
15. Filloux, C. M. *Angew. Chem. Int. Ed.* **2015,** *54*, 8880–8894.
16. (a) Akishin, P. A.; Vilkov, L. V.; Sokolova, N. P. *Izv. Sibirsk. Otd. Akad. Nauk. SSSr* **1960,** *5*, 59.
    (b) Planje, M. C.; Toneman.; Dallinga, G. *Rec. Trav. Chim.* **1965,** *84*, 232.
17. (a) Wiberg, K. B.; Bailey, W. F.; Lambert, K. M.; Stempel, Z. D. *J. Org. Chem.* **2018,** *83*, 5242–5255.
    (b) Filloux, C. M. *Angew. Chem. Int. Ed.* **2015,** *54*, 8880–8894.
    (c) Romers, C.; Altona, C.; Buys, H. R.; Havinga, E. *Top. Stereochem.* **1969,** *4*, 39–97.
18. (a) Takahashi, O.; Yamasaki, K.; Kohno, Y.; Ohtaki, R.; Ueda, K.; Suezawa, H.; Umezawa, Y.; Nishio, M. *Carbohydr. Res.* **2007,** *342*, 1202–1209.
    (b) Takahashi, O.; Yamasaki, K.; Kohno, Y.; Ueda, K.; Suezawa, H.; Nishio, M. *Carbohydr. Res.* **2009,** *344*, 1225–1229.
    (c) Wang, C.; Ying, F.; Wu, W.; Mo, Y. *J. Org. Chem.* **2014,** *79*, 1571–1581.
19. Fuchs, B.; Ellencweig, A.; Tartakovsky, E.; Aped, P. *Angew. Chem., Int. Ed. Engl.* **1986,** *25*, 287–289.
20. (a) Lemieux, R. U.; Pavia, A. A.; Martin, J. C.; Watanabe, K. A. *Can. J. Chem.* **1969,** *47*, 4427–4439.
    (b) Vila, A.; Estévez, L.; Mosquera, R. A. *J. Phys. Chem. A* **2011,** *115*, 1964–1970.
21. Edward, J. T. *Chem. Ind.* **1955,** 1102–1104.
22. Altona, C.; Knobler, C.; Romers, C. *Acta Crystallogr.* **1963,** *16*, 1217–1225.
23. (a) Petillo, P. A.; Lerner, L. E. *American Chemical Society* **1993,** *539*, 156–175.
    (b) Romers, C.; Altona, C.; Buys, H. R.; Havinga, E. *Top. Stereochem.* **1969,** *4*, 39–97.
    (c) Ardalan, Z.; Lucken, E. A. C. *Helv. Chim. Acta* **1973,** *56*, 1715–1719.
24. Fischer, E. *Ber. Dtsch. Chem. Ges.* **1893,** *26*, 2400–2412.
25. Michael, A. *Am. Chem. J.* **1879,** *1*, 305–312.
26. Koenigs, W.; Knorr, E. *Ber. Dtsch. Chem. Ges.* **1901,** *34*, 957–981.
27. Zémplén, G.; Gerecs, A. *Ber. Dtsch. Chem. Ges.* **1930,** *63*, 2720–2729.
28. Helferich, B.; Wedermeyer, K. F. *Justus Liebigs Ann. Chem.* **1949,** *563*, 139–145.
29. (a) Nukada, T.; Berces, A.; Whitfield, D. M. *Carbohydr. Res.* **2002,** *337*, 765–774.
    (b) Fu, Y.; Bernasconi, L.; Liu, P. *J. Am. Chem. Soc* **2021,** *143*, 1577–1589.
    (c) Adero, P. O.; Amarasekara, H.; Wen, P.; Bohé, L.; Crich, D. *Chem. Rev* **2018,** *118*, 8242–8284.
    (d) Andreana, P. R.; Crich, D. *ACS Cent. Sci* **2021,** *7*, 1454–1462.
30. (a) Huang, M.; Garrett, G. E.; Birlirakis, N.; Bohe, L.; Pratt, D. A.; Crich, D. *Nat. Chem.* **2012,** *4*, 663–667.
    (b) Crich, D. *Acc. Chem. Res.* **2010,** *43*, 1144–1153.
    (c) Bohe, L.; Crich, D. *Carbohydr. Res.* **2015,** *403*, 48–59.
31. Zhuo, M.-H.; Wilbur, D. J.; Kwan, E. E.; Bennett, C. S. *J. Am. Chem. Soc.* **2019,** *141*, 16743–16754.
32. (a) Paulsen, H. *Angew. Chem. Int. Ed. Engl* **1982,** *21*, 155.
    (b) Schmidt, R. R. *Angew. Chem. Int. Ed.Engl.* **1986,** *25*, 212.

33. (a) Fraser-Reid, B.; Wu, Z.; Udodong, U. E.; Ottosson, H. *J. Org. Chem.* **1990,** *55*, 6068–6070.
(b) Mootoo, D. R.; Date, V.; Fraser-Reid, B. *J. Am. Chem. Sac.* **1988,** *110*, 2662.
(c) Mootoo, D. R.; Konradsson, P.; Udodong, U.; Fraser-Reid, B. *J. Am. Chem. Soc.* **1988,** *110*, 5583.
34. (a) Veeneman, G. H.; van Leeuwen, S. H.; van Boom, J. H. *Tetrahedron Lett.* **1990,** *31*, 1331.
(b) Friesen, R. W.; Danishefsky, S. J. *J. Am. Chem. Soc.* **1989,** *111*, 6656.
(c) Veeneman, G. H.; Van Boom, J. H. *Tetrahedron Lett* **1990,** *31*, 275.
35. Li, Z.; Gildersleeve, J. C. *Tetrahedron Lett.* **2007,** *48*, 559–562.
36. Fraser-Reid, B.; Jayaprakash, K. N.; López, J. C.; Gómez, A. M.; Uriel, C. In *ACS Symp. Ser. (Frontiers in Modern Carbohydrate Chemistry);* Demchenko, A. V., Ed.; Oxford Univ. Press, 2007; pp 91–117.
37. Zhu, T.; Boons, G. J. *Org. Lett.* **2001,** *3*, 4201–4203.
38. Kamat, M. N.; Demchenko, A. V. *Org. Lett.* **2005,** *7*, 3215–3218.
39. (a) Mydock, L. K.; Demchenko, A. V. *Org. Lett.* **2008,** *10*, 2107–2110.
(b) Mydock, L. K.; Demchenko, A. V. *Org. Lett.* **2008,** *10*, 2103–2106.
40. Premathilake, H. D.; Mydock, L. K.; Demchenko, A. V. *J. Org. Chem.* **2010,** *75*, 1095–1100.
41. Roy, R.; Andersson, F. O.; Letellier, M. *Tetrahedron Lett.* **1992,** *33*, 6053–6056.
42. Boons, G. J.; Isles, S. *Tetrahedron Lett.* **1994,** *35*, 3593–3596.
43. Das, A.; Jayaraman, N. *Carbohydr. Res.* **2021,** *508*, 108404.
44. (a) Komarova, B. S.; Ustyuzhanina, N. E.; Tsvetkov, Y. E.; Nifantiev, N. E. In *Modern Synthetic Methods in Carbohydrate Chemistry;* Werz, D. B., Vidal, S., Eds.; Wiley-VCH Verlag GmbH & Co. KgaA: Weinheim, 2013; pp 125–159.
(b) Yang, B.; Yang, W.-B.; Ramadan, S.; Huang, X. *Eur. J. Org. Chem.* **2018,** 1075–1096.
45. Hamala, V.; Šťastná, L. C.; Kurfiřt, M.; Cuřínová, P.; Dračínský, M.; Karban, J. *Org. Biomol. Chem.* **2020,** *18*, 5427–5434.
46. Komarova, B. S.; Tsvetkov, Y. E.; Nifantiev, N. E. *Chem. Rec.* **2016,** *16*, 488–506.
47. De Meo, C.; Kamat, M. N.; Demchenko, A. V. *Eur. J. Org. Chem.* **2005,** 706–711.
48. Koshiba, M.; Suzuki, N.; Arihara, R.; Tsuda, T.; Nambu, H.; Nakamura, S.; Hashimoto, S. *Chem. Asian J.* **2008,** *3*, 1664–1677.
49. Upadhyaya, K.; Bagul, R. S.; Crich, D. *J. Org. Chem.* **2021,** *86*, 12199–12225.
50. Yang, W.; Zhang, J.; Yang, C.-W.; Ramadan, S.; Staples, R.; Huang, X. *Org. Lett.* **2021,** *23*, 1153–1156.
51. Alpe, M.; Oscarson, S. *Carbohydr. Res.* **2002,** *337*, 1715–1722.
52. Nunomura, S.; Ogawa, T. *Tetrahedron Lett.* **1988,** *29*, 5681–5684.
53. Mandal, P. K.; Chheda, P. R. *Tetrahedron Lett.* **2015,** *56*, 900–902.
54. (a) Marianski, M.; Mucha, E.; Greis, K.; Moon, S.; Pardo, A.; Kirschbaum, C.; Thomas, D. A.; Meijer, G.; von Helden, G.; Gilmore, K. *Angew. Chem., Int. Ed.* **2020,** *59*, 6166–6171.
(b) Hansen, T.; Elferink, H.; Von Hengst, J. M. A.; Houthuijs, K. J.; Remmerswaal, W. A.; Kromm, A.; Berden, G.; van der Vorm, S.; Rijs, A. M.; Overkleeft, H. S. *Nat. Commun.* **2020,** *11*, 2664.
(c) Nicolaou, K. C.; Daines, R. A.; Ogawa, Y.; Chakraborty, T. K. *J. Am. Chem. Soc* **1988,** *110*, 4696–4705.

55. Hettikankanamalage, A. A.; Lassfolk, R.; Ekholm, F. S.; Leino, R.; Crich, D. *Chem. Rev.* **2020,** *120*, 7104–7151.
56. Valerio, S.; Iadonisi, A.; Adinolfi, M.; Ravid, A. *J. Org. Chem.* **2007,** *72*, 6097–6106.
57. Remmerswaal, W. A.; Houthuijs, K. J.; van de Ven, R.; Elferink, H.; Hansen, T.; Berden, G.; Overkleeft, H. S.; van der Marel, G. A.; Rutjes, F. P. J. T.; Filippov, D. V. *J. Org. Chem.* **2022,** *87*, 9139–9147.
58. Fischer, E.; Armstrong, E. F. *Ber. Dtsch. Chem. Ges.* **1901,** *34*, 2885–2900.
59. Fischer, E.; von Mechel, L. *Ber. Dtsch. Chem. Ges.* **1916,** *49*, 2813–2820.
60. Fischer, E.; Helferich, B. *Ann. Chem.* **1911,** *383*, 68–91.
61. (a) Fischer, E.; Bergmann, M.; Rabe, A. *Ber. Dtsch. Chem. Ges. B* **1920,** *53B*, 2362–2388.
    (b) Zemplen, G. Gerecs. *Ber. Dtsch. Chem. Ges.* **1930,** *63B*, 2720–2729.
    (c) Hickinbottom, W. J. *J. Chem. Soc.* **1929,** 1676–1687.
    (d) Sun, L.; Wu, X.; Xiong, D. C.; Ye, X. S. *Angew. Chem., Int. Ed.* **2016,** *55*, 8041–8044.
62. Fischer, E.; Fischer, H. *Ber. Dtsch. Chem. Ges.* **1910,** *43*, 2521–2536.
63. Hunsen, M.; Long, D. A.; D'Ardenne, C. R.; Smith, A. L. *Carbohydr. Res.* **2005,** *340*, 2670–2674.
64. Koto, S.; Yoshida, T.; Takenaka, K.; Zen, S. *Bull. Chem. Soc. Jpn.* **1982,** *55*, 3667–3668.
65. Kartha, K. P. R.; Jennings, H. J. *J. Carbohydr. Chem.* **1990,** *9*, 777–781.
66. Haworth, W. N.; Hirst, E. L.; Miller, E. J. *J. Chem. Soc.* **1929,** 2469–2479.
67. Montero, J. L.; Winum, J. Y.; Leydet, A.; Kamal, M.; Pavia, A. A.; Roque, J. P. *Carbohydr. Res.* **1997,** *297*, 175–180.
68. Hanessian, S.; Ponpipom, M. M.; Lavalle, P. *Carbohydr. Res.* **1972,** *24*, 45–56.
69. Shingu, Y.; Nishida, Y.; Dohi, H.; Matsuda, K.; Kobayashi, K. *J. Carbohydr. Chem.* **2002,** *21*, 605–611.
70. Zhao, J.-Z.; Zhang, X.-Q.; Wu, X.; Xing, Z.-B.; Yue, A.-Q.; Shao, H.-W. *Chem. Res. Chin. Univ.* **2013,** *29*, 71–75.
71. Ibatullin, F. M.; Selivanov, S. I. *Tetrahedron Lett.* **2002,** *43*, 9577–9580.
72. Nanami, M.; Ando, H.; Kawai, Y.; Koketsu, M.; Ishihara, H. *Tetrahedron Lett.* **2007,** *48*, 1113–1116.
73. Tiwari, P.; Misra, A. K. *Carbohydr. Res.* **2006,** *341*, 339–350.
74. Oberthuer, M.; Leimkuhler, C.; Kahne, D. *Org. Lett.* **2004,** *6*, 2873–2876.
75. Caoa, Z.; Liua, W.; Qua, Y.; Yaoc, G.; Gaoa, D.; Liu, W. *J. Chem. Res.* **2013,** 467–469.
76. (a) Kartha, K. P. R.; Field, R. A. *Tetrahedron Lett.* **1997,** *38*, 8233–8236.
    (b) Kartha, K. P. R.; Cura, P.; Aloui, M.; Readman, K.; Rutherford, T. J.; Field, R. A. *Tetrahedron Asymm.* **2000,** *11*, 581–593.
77. Polat, T.; Linhardt, R. J. *Carbohydr. Res.* **2003,** *338*, 447–449.
78. Chih-Wei Chang, C.-W.; Shih-Sheng Chang, S.-S.; Chao, C.-S.; Mong, K.-K. T. *Tetrahedron Lett.* **2009,** *50*, 4536–4540.
79. Traboni, S.; Liccardo, F.; Bedini, E.; Giordano, M.; Iadonisi, A. *Tetrahedron Lett.* **2017,** *58*, 1762–1764.
80. Tojino, M.; Hirose, Y.; Mizuno, M. *Tetrahedron Lett.* **2013,** *54*, 7124–7126.
81. Yuan, X.; Cheng, S.; Shi, Y.; Xue, W. *Synthesis* **2014,** *46*, 331–335.
82. Xu, L.; Luo, C.-H.; Chen, C.-S. *J. Chin. Chem. Soc.* **2020,** 1–7.
83. Encinas, L.; Chiara, J. L. *J. Comb. Chem.* **2008,** *10*, 361–363.
84. Yalamanchili, S.; Nguyen, T. A.; Zsikla, A.; Stamper, G.; DeYong, A. E.; Florek, J.; Vasquez, O.; Pohl, N. L. B.; Bennett, C. S. *Angew. Chem., Int. Ed. Engl.* **2021,** *60*, 23171–23175.

85. Chooppawa, T.; Janprasert, P.; Padungros, P. *Synlett* **2022,** *33*, 1391–1398.
86. Adinolfi, M.; Iadonisi, A.; Pezzella, A.; Ravida, A. *Synlett* **2005,** 1848–1852.
87. Nitz, M.; Bundle, D. R. Glycoscience: Chemistry and Biology. In ; Fraser-Reid, B., Tastuta, K., Thiem, J., Eds.; Vol. 2; Springer: Heildelberg, 2001; pp 1497–1542.
88. (a) Paulsen, H. *Angew. Chem. Int. Ed.* **1982,** *21*, 155.
    (b) Paulsen, H.; Kutschker, W.; Lockhoff, O. *Chem. Ber.* **1981,** *114*, 3233.
89. Paulsen, H.; Lockhoff, O. *Chem. Ber.* **1981,** *114*, 3102.
90. Khorlin, A. Y.; Nesmeyanov, V. A.; Zurabyan, S. E. *Carbohydr. Res.* **1975,** *43*, 69–77.
91. Kronzer, F. J.; Schuerch, C. *Carbohydr. Res.* **1973,** *27*, 379–390.
92. Kaneko, M.; Herzon, S. B. *Org. Lett.* **2014,** *16*, 2776–2779.
93. Steber, H. B.; Singh, Y.; Demchenko, A. V. *Org. Biomol. Chem.* **2021,** *19*, 3220–3223.
94. Geringer, S. A.; Singh, Y.; Hoard, D. J.; Demchenko, A. V. *Chem. Eur. J.* **2020,** *26*, 8053–8063.
95. Shadrick, M.; Stine, K. J.; Demchenko, A. V. *Bioorg. Med. Chem.* **2022,** *73*, 117031.
96. (a) Lee, D.; Williamson, C. L.; Chan, L.; Taylor, M. S. *J. Am. Chem. Soc.* **2012,** *134*, 8260–8267.
    (b) Beale, T. M.; Moon, P. J.; Taylor, M. S. *Org. Lett.* **2014,** *16*, 3604–3607.
97. Kuszmann, J.; Medgyes, G.; Boros, S. *Carbohydr. Res.* **2004,** *339*, 2407–2414.
98. Talasila, D. S.; Bauer, E. B. *RSC Adv.* **2021,** *11*, 36814–36820.
99. Kaeothip, S.; Pornsuriyasak, P.; Demchenko, A. V. *Tetrahedron Lett.* **2008,** *49*, 1542–1545.
100. Lam, S. N.; Gervay-Hague, J. *Org. Lett.* **2002,** *4*, 2029–2042.
101. Uchiyama, T.; Shishikura, K.; Ogawa, K.; Ohshima, Y.; Miyairi, S. *Tetrahedron Lett.* **2016,** *57*, 5294–5296.
102. Castelli, R.; Schindler, S.; Walter, S. M.; Kniep, F.; Overkleeft, H. S.; Van der Marel, G. A.; Huber, S. M.; Codee, J. D. C. *Chem. Asian J.* **2014,** *9*, 2095–2098.
103. Sun, L.; Wu, X.; Xiong, D.-C.; Ye, X.-S. *Angew. Chem. Int. Ed. Engl.* **2016,** *55*, 8041–8044.
104. Cardona, A.; Boutureira, O.; Castillon, S.; Diaz, Y.; Matheu, M. I. *Green. Chem.* **2017,** *19*, 2687–2694.
105. Galan, M. C.; Jones, R. A.; Tran, A.-T. *Carbohydr. Res.* **2013,** *375*, 35–46.
106. Sasaki, K.; Matsumura, S.; Toshima, K. *Tetrahedron Lett.* **2004,** *45*, 7043–7047.
107. (a) Kumar, V.; Talisman, I. J.; Malhotra, S. V. *Eur. J. Org. Chem.* **2010,** 3377–3381.
    (b) Talisman, I. J.; Kumar, V.; Razzaghy, J.; Malhotra, S. V. *Carbohydr. Res.* **2011,** *346*, 883–890.
    (c) Kumar, V.; Talisman, I. J.; Bukhari, O.; Razzaghy, J.; Malhotra, S. V. *RSC Adv.* **1721,** *2011*, 1.
108. Kronzer, F. J.; Schuerch, C. *Carbohydr. Res.* **1974,** *34*, 71–78.
109. Thiem, J.; Meyer, B. *Chem. Ber.* **1980,** *113*, 3075–3085.
110. Gervay, J.; Nguyen, T. N.; Hadd, M. J. *Carbohydr. Res.* **1997,** *300*, 119–125.
111. Bickley, J.; Cottrell, J. A.; Ferguson, J. R.; Field, R. A.; Harding, J. R.; Hughes, D. L.; Kartha, K. P. R.; Law, J. L.; Schienmann, F.; Stachulski, A. V. *Chem. Commun.* **2003,** 1266–1267.
112. Meloncelli, P. J.; Martin, A. D.; Lowary, T. L. *Carbohydr. Res.* **2009,** *344*, 1110–1122.
113. (a) van Well, R. M.; Kartha, K. P. R.; Field, R. A. *J. Carbohydr. Chem.* **2005,** *24*, 463–474.
    (b) Mukhopadhyay, B.; Kartha, K. P. R.; Russell, D. A.; Field, R. A. *J. Org. Chem.* **2004,** *69*, 7758–7760.

114. Bickley, J.; Cottrel, J. A.; Ferguson, J. R.; Field, R. A.; Harding, J. R.; Hughes, D. L.; Kartha, K. P. R.; Law, J. L.; Scheinmann, F.; Stachulski, A. V. *Chem. Commun.* **2003,** 1266–1267.
115. Murakami, T.; Sato, Y.; Shibakami, M. *Carbohydr. Res.* **2008,** *34*, 1297–1308.
116. Ness, R. K.; Fletcher, H. G.; Hudson, C. S. *J. Am. Chem. Soc.* **1950,** *72*, 2200–2205.
117. Gervay, J.; Gergar, T. Q. *Tetrahedron Lett.* **1997,** *38*, 5921–5924.
118. Zeng, C.; Shen, G.; Yang, F.; Chen, J.; Zhang, X.; Gu, C.; Zhou, Y.; Fan, B. *Org. Lett.* **2018,** *20*, 6859–6862.
119. Chervin, S. M.; Abada, P.; Koreeda, M. *Org. Lett.* **2000,** *2*, 369–372.
120. Caputo, R.; Kunz, H.; Mastroianni, D.; Palumbo, G.; Pedatella, S.; Solla, F. *Eur. J. Org. Chem.* **1999,** 3147–3150.
121. (a) Yokoyama, M. *Carbohydr. Res.* **2000,** *327*, 5–14.
(b) Toshima, K. *Carbohydr. Res.* **2000,** *327*, 15–26.
(c) Huonnic, K.; Linclau, B. *Chem. Rev.* **2022,** *122*, 15503–15602.
122. Brauns, D. I. *J. Am. Chem. Soc.* **1923,** *45*, 833–835.
123. Olah, G. A.; Nojima, M.; Kerekes, I. *Synthesis* **1973,** *1973*, 779–780.
124. Hayashi, M.; Hashimoto, S.; Noyori, R. *Chem. Lett.* **1984,** 1747–1750.
125. Miethchen, R.; Kolp, G. *J. Fluor. Chem.* **1993,** *60*, 49–55.
126. (a) Buchanan, J. G.; Hill, D. G.; Wightman, R. H.; Boddy, I. K.; Hewitt, B. D. *Tetrahedron* **1995,** *51*, 6033–6050.
(b) Noecker, L. A.; Edwards, J. R. *Tetrahedron Lett* **1997,** *38*, 5779–5780.
(c) Palme, M.; Vasella, A. *Helv. Chim. Acta* **1995,** *78*, 959–969.
(d) Miethchen, R.; Gabriel, T. *Chem. Ber.* **1993,** *126*, 2309–2316.
127. Sharma, M. N.; Eby, R. *Carbohydr. Res.* **1984,** *127*, 201–210.
128. Mukaiyama, T.; Hashimoto, Y.; Shoda, S. *Chem. Lett.* **1983,** *12*, 935–938.
129. Kunz, H.; Sager, W. *Helv. Chim. Acta* **1985,** *68*, 283–287.
130. Miethchen, R.; Gabriel, T.; Kolp, G. *Synthesis* **1991,** 885–888.
131. Miethchen, R.; Gabriel, T. *J. Fluor. Chem.* **1994,** *67*, 11–15.
132. (a) Helferich, B.; Gootz, R. *Ber. Dtsch. Chem. Ges.* **1929,** *62*, 2505–2507.
(b) Micheel, F.; Klemer, A. *Adv. Carbohydr. Chem.* **1962,** *16*, 85–103.
133. Lichtenthaler, F. W.; Kläres, U.; Lergenmüller, M.; Schwidetzky, S. *Synthesis* **1992,** 179–184.
134. Miethchen, R.; Hager, C.; Hein, M. *Synthesis* **1997,** 159–161.
135. (a) Rosenbrook, W., Jr; Riley, D. A.; Lartey, P. A. *Tetrahedron Lett* **1985,** *26*, 3–4.
(b) Crich, D.; Lim, L. B. L. *J. Chem. Soc., Perkin Trans* **1991,** *1*, 2209–2214.
136. Posner, G. H.; Haines, S. R. *Tetrahedron Lett.* **1985,** *26*, 5–8.
137. Nicolaou, K. C.; Dolle, R. E.; Papahatjis, D. P.; Randall, J. L. *J. Am. Chem. Soc.* **1984,** *106*, 4189–4192.
138. (a) Ohtsuka, I.; Ako, T.; Kato, R.; Daikoku, S.; Koroghi, S.; Kanemitsu, T.; Kanie, O. *Carbohydr. Res.* **2006,** *341*, 1476–1487.
(b) Suzuki, K.; Ito, Y.; Kanie, O. *Carbohydr. Res.* **2012,** *359*, 81–91.
139. Kim, S.; Khomutnyk, Y.; Bannykh, A.; Nagorny, P. *Org. Lett.* **2021,** *23*, 190–194.
140. Kim, S.; Nagorny, P. *Org. Lett.* **2022,** *24*, 2294–2298.
141. Burkart, M. D.; Zhang, Z.; Hung, S.-C.; Wong, C.-H. *J. Am. Chem. Soc.* **1997,** *119*, 11743–11746.
142. Huang, K. T.; Winssinger, N. *Eur. J. Org. Chem.* **2007,** 1887–1890.
143. (a) Kunigami, M.; Hara, S. *Carbohydr. Res.* **2015,** *417*, 78–80.
(b) Ukigai, H.; Hara, S. *Tetrahedron Lett.* **2016,** *57*, 1379–1381.

144. See, N. W.; Xu, X.; Ferro, V. *J. Org. Chem.* **2022,** *87*, 14230–14240.
145. (a) Mukaiyama, T.; Hashimoto, Y.; Shoda, S. *Chem. Lett.* **1983,** 935–938.
(b) Mukaiyama, T.; Murai, Y.; Shoda, S. *Chem. Lett* **1981,** 431–432.
146. Nicolaou, K. C.; Chucholowski, A.; Dolle, R. E.; Randall, J. L. *Chem. Commun.* **1984,** 1155–1156.
147. Kreuzer, M.; Thiem, J. *Carbohydr. Res.* **1986,** *149*, 347–361.
148. Hashimoto, S.; Hayashi, M.; Noyori, R. *Tetrahedron Lett.* **1984,** *25*, 1379–1382.
149. Matsumoto, T.; Maeta, H.; Suzuki, K.; Tsuchihashi, G. *Tetrahedron Lett.* **1988,** *29*, 3567.
150. Matsumoto, T.; Maeta, H.; Suzuki, K.; Tsuchihashi, G.-i *Tetrahedron Lett.* **1988,** *29*, 3575–3578.
151. (a) Suzuki, K.; Maeta, H.; Suzuki, T.; Matsumoto, T. *Tetrahedron Lett.* **1989,** *30*, 6879–6882.
(b) Maeta, H.; Matsumoto, T.; Suzuki, K. *Carbohydr. Res* **1993,** *249*, 49–56.
152. Koide, K.; Ohno, M.; Kobayashi, S. *Synthesis* **1996,** *1996*, 1175–1176.
153. Qiu, R.; Zhang, G.; Xu, X.; Zou, K.; Shao, L.; Fang, D.; Li, Y.; Orita, A.; Saijo, R.; Mineyama, H. *J. Organomet. Chem.* **2009,** *694*, 1524–1528.
154. Toshima, K.; Kasumi, K.; Matsumura, S. *Synlett* **1998,** *1998*, 643–645.
155. Pelletier, G.; Zwicker, A.; Allen, C. L.; Schepartz, A.; Miller, S. J. *J. Am. Chem. Soc.* **2016,** *138*, 3175–3182.
156. Yang, Y.; Yu, B. *Tetrahedron* **2014,** *70*, 1023–1046.
157. Ikeshita, S.; Nakahara, Y.; Ogawa, T. *Glycoconj. J.* **1994,** *11*, 257–261.
158. Abe, T.; Mori, K. *Biosci. Biotech. Biochem.* **1994,** *58*, 1671–1674.
159. Ghosh, B.; Kulkarni, S. S. *Chem. Asian J.* **2020,** *15*, 450–462.
160. (a) Lee, Y. C.; Stowell, C. P. *Methods Enzymol* **1982,** *83*, 278–288.
(b) MacDougall, J. M.; Zhang, X. D.; Wolgar, W. E.; Khroyan, T. V.; Toll, L.; Cashman, J. R. *J. Med. Chem.* **2004,** *47*, 5809–5815.
161. Feng, G.-J.; Luo, T.; Guo, Y.-F.; Liu, C.-Y.; Dong, H. *J. Org. Chem.* **2022,** *87*, 3638–3646.
162. Bernardes, G. J. L.; Gamblin, D. P.; Davis, B. G. *Angew. Chem. Int. Ed. Engl.* **2006,** *45*, 4007–4011.
163. Dere, R. T.; Zhu, X. *Org. Lett.* **2008,** *10*, 4641–4644.
164. O'Reilly, C.; Murphy, P. V. *Org. Lett.* **2011,** *13*, 5168–5171.
165. Defaye, J.; Gadelle, A. *Carbohydr. Res.* **1991,** *217*, 51–58.
166. Dere, R. T.; Wang, Y.; Zhu, X. *Org. Biomol. Chem.* **2008,** *6*, 2061–2206.
167. Dere, R. T.; Kumar, A.; Kumar, V.; Zhu, X.; Schmidt, R. R. *J. Org. Chem.* **2011,** *76*, 7539–7545.
168. Escopy, S.; Singh, Y.; Demchenko, A. V. *Org. Biomol. Chem.* **2019,** *17*, 8379–8383.
169. Li, S.; Wang, Y.; Zhong, L.; Wang, S.; Liu, Z.; Dai, Y.; He, Y.; Feng, Z. *Org. Lett.* **2022,** *24*, 2463–2468.
170. Jana, M.; Misra, A. K. *J. Org. Chem.* **2013,** *78*, 268–2686.
171. Zhu, F.; Miller, E.; Zhang, S. Q.; Yi, D.; O'Neill, S.; Hong, X.; Walczak, M. A. *J. Am. Chem. Soc.* **2018,** *140*, 18140–18150.
172. Levecque, P.; Gammon, D. W.; Jacobs, P.; De Vos, D.; Sels, B. *Green. Chem.* **2010,** *12*, 828–835.
173. Liu, Y.; Jiao, Y.; Luo, H.; Huang, N.; Lai, M.; Zou, K.; Yao, H. *ACS Catal.* **2021,** *11*, 5287–5293.
174. Li, J.; Wang, M.; Jiang, X. *Org. Lett.* **2021,** *23*, 9053–9057.
175. Ferrier, R. J.; Hay, R. W.; Vethaviyasar, N. *Carbohydr. Res.* **1973,** *27*, 55–61.
176. Hanessian, S.; Baquet, C.; Lehong, N. *Carbohydr. Res.* **1980,** *80*, C17–C22.
177. van Cleve, J. W. *Carbohydr. Res.* **1979,** *70*, 161–164.

178. Mukaiyama, T.; Nakatsuka, T.; Shoda, S. *Chem. Lett.* **1979,** 487–490.
179. Sato, S.; Mori, M.; Ito, Y.; Ogawa, T. *Carbohydr. Res.* **1986,** *155*, C6–C10.
180. Kihlberg, J. O.; Leigh, D. A.; Bundle, D. R. *J. Org. Chem.* **1990,** *55*, 2860–2863.
181. Sasaki, M.; Tachibana, K.; Nakanishi, H. *Tetrahedron Lett.* **1991,** *32*, 6873–6876.
182. Nicolaou, K. C.; Seitz, S. P.; Papahatjis, D. P. *J. Am. Chem. Soc.* **1983,** *105*, 2430–2434.
183. (a) Veeneman, G. H.; van Leeuwen, S. H.; van Boom, J. H. *Tetrahedron Lett* **1990,** *31*, 1331–1334.
(b) Konradsson, P.; Mootoo, D. R.; McDevitt, R. E.; Fraser-Reid, B. *J. Chem. Soc., Chem. Commun* **1990,** 270–272.
184. Konradsson, P.; Udodong, U. E.; Fraser-Reid, B. *Tetrahedron Lett.* **1990,** *31*, 4313–4316.
185. Fukase, K.; Hasuoka, A.; Kinoshita, L.; Aoki, Y.; Kusumoto, S. *Tetrahedron* **1995,** *51*, 4923–4932.
186. Takeuchi, K.; Tamura, T.; Mukaiyama, T. *Chem. Lett.* **2000,** *29*, 124–125.
187. Imai, H.; Oishi, T.; Kikuchi, T.; Hirama, M. *Tetrahedron* **2000,** *56*, 8451–8459.
188. Chung, S.-K.; Park, K.-H. *Tetrahedron Lett.* **2001,** *42*, 4005–4007.
189. Qin, Z. H.; Li, H.; Cai, M. S.; Li, Z. J. *Carbohydr. Res.* **2002,** *337*, 31–36.
190. Hashihayata, T.; Ikegai, K.; Takeuchi, K.; Jona, H.; Mukaiyama, T. *Bull. Chem. Soc. Jpn.* **1829,** *2003*, 76.
191. Périon, R.; Lemée, L.; Ferriéres, V.; Duval, R.; Plusquellec, D. *Carbohydr. Res.* **2003,** *338*, 2779–2792.
192. Mukhopadhyay, B.; Collet, B.; Field, R. A. *Tetrahedron Lett.* **2005,** *46*, 5923–5925.
193. (a) Roy, B.; Pramanik, K.; Mukhopadhyay, B. *Glycoconjugate J* **2008,** *25*, 157–166.
(b) Dasgupta, S.; Pramanik, K.; Mukhopadhyay, B. *Tetrahedron* **2007,** *63*, 12310–12316.
194. Mukherjee, S.; Mukhopadhyay, B. *Synlett* **2010,** 2853–2856.
195. (a) Dasgupta, S.; Mukhopadhyay, B. *Eur. J. Org. Chem.* **2008,** 5770–5777.
(b) Roy, B.; Field, R. A.; Mukhopadhyay, B. *Carbohydr. Res.* **2009,** *344*, 2311–2316.
196. (a) Das, R.; Mukhopadhyay, B. *Carbohydr. Res.* **2013,** *376*, 1–6.
(b) Pal, K. B.; Mukhopadhyay, B. *Carbohydr. Res.* **2013,** *379*, 26–29.
197. Trinderup, H. H.; Sandgaard, T. L. P.; Juul-Madsen, L.; Jensen, H. H. *J. Org. Chem.* **2022,** *87*, 4154–4167.
198. Kajimoto, T.; Morimoto, K.; Ogawa, R.; Dohi, T.; Kita, Y. *Chem. Pharm. Bull.* **2016,** *64*, 838–844.
199. Kajimoto, T.; Morimoto, K.; Ogawa, R.; Dohi, T.; Kita, Y. *Eur. J. Org. Chem.* **2015,** 2138–2142.
200. Kabotso, D. E. K.; Pohl, N. L. B. *Org. Lett.* **2017,** *19*, 4516–4519.
201. Goswami, M.; Ashley, D. C.; Baik, M. H.; Pohl, N. L. *J. Org. Chem.* **2016,** *81*, 5949–5962.
202. Vibhute, A. M.; Dhaka, A.; Athiyarath, V.; Sureshan, K. M. *Chem. Sci.* **2016,** *7*, 4259–4263.
203. Adhikari, S.; Baryal, K. N.; Zhu, D.; Li, X.; Zhu, J. *ACS Catal.* **2013,** *3*, 57–60.
204. Mc Carthy, C.; Tacke, M.; Zhu, X. *Eur. J. Org. Chem.* **2019,** *16*, 2729–2734.
205. Escopy, S.; Singh, Y.; Demchenko, A. V. *Org. Biomol. Chem.* **2021,** *19*, 2044–2054.
206. Wever, W. J.; Cinelli, M. A.; Bowers, A. A. *Org. Lett.* **2013,** *15*, 30–33.
207. Spell, M. L.; Deveaux, K.; Bresnahan, C. G.; Bernard, B. L.; Sheffield, W.; Kumar, R.; Ragains, J. R. *Angew. Chem. Int. Ed.* **2016,** *55*, 6515–6519 *Angew. Chem.* **2016**, *128*, 6625–6629.
208. (a) Krumb, M.; Lucas, T.; Opatz, T. *European J. Org. Chem.* **2019,** *2019*, 4517–4521.
(b) Krumb, M.; Kammer, L. M.; Forster, R.; Grundke, C.; Opatz, T. *ChemPhotoChem* **2020,** *4*, 100 100.
209. Nakanishi, M.; Takahashi, D.; Toshima, K. *Org. Biomol. Chem.* **2013,** *11*, 5079–5082.

210. Qin, X.; Ye, X.-S. *Chin. J. Chem.* **2020,** *39*, 531–542.
211. Crich, D.; Sun, S. *J. Am. Chem. Soc.* **1998,** *120*, 435–436.
212. Escopy, S.; Singh, Y.; Stine, K.; Demchenko, A. V. *Chem. Eur. J.* **2021,** *27*, 354–361.
213. Tanaka, N.; Ohnishi, F.; Uchihata, D.; Toriib, S.; Nokamia, J. *Tetrahedron Lett.* **2007,** *48*, 7383–7387.
214. Xu, F.-F.; Pereira, C. L.; Seeberger, P. H. *Beil. J. Org. Chem.* **1994,** *2017*, 13.
215. (a) Saliba, R. C.; Wooke, Z. J.; Nieves, G. A.; Chu, A.-H. A.; Bennett, C. S.; Pohl, N. L. B. *Org. Lett.* **2018,** *20*, 800–803.
(b) Kohout, V. R.; Pirinelli, A. L.; Pohl, N. L. B. *Pure Appl. Chem* **2019,** *91*, 1243–1255.
216. Manmode, S.; Kato, M.; Ichiyanagi, T.; Nokami, T.; Itoh, T. *Asian J. Org. Chem.* **2018,** *7*, 1802–1805.
217. Mc Carthy, C.; Zhu, X. *Org. Biomol. Chem.* **2020,** *18*, 9029–9034.
218. Schmidt, R. R.; Michel, J. *Angew. Chem. Int. Ed. Engl.* **1980,** *19*, 731–732 *Angew. Chem.***1980**, *92*, 763–764.
219. Yoshizaki, H.; Fukuda, N.; Sato, K.; Oikawa, M.; Fukase, K.; Suda, Y.; Kusumoto, S. *Angew. Chem. Int. Ed.* **2001,** *40*, 1475.
220. Oikawa, M.; Tanaka, T.; Fukuda, N.; Kusumoto, S. *Tetrahedron Lett.* **2004,** *45*, 4039.
221. Schmidt, R. R. New Aspects of Glycosylation Reactions. In *Carbohydrates—Synthetic Methods and Applications in Medicinal Chemistry;* Ogura, H., Hasegawa, A., Suami, T., Eds.; 68; Kodanasha Ltd.: Tokyo, 1992.
222. Böhm, G.; Waldmann, H. *Tetrahedron Lett.* **1995,** *36*, 3843.
223. Rencurosi, A.; Lay, L.; Russo, G.; Caneva, E.; Poletti, L. *J. Org. Chem.* **2007,** *70*, 7765.
224. Soliman, S.; Bassily, R. W.; El-Sokkary, R. I.; Nashed, M. A. *Carbohydr. Res.* **2003,** *338*, 2337.
225. (a) Mensah, E. A.; Nguyen, H. M. *J. Am. Chem. Soc.* **2009,** *131*, 8778–8780.
(b) Mensah, E. A.; Yu, F.; Nguyen, H. M. *J. Am. Chem. Soc.* **2010,** *132*, 14288–14302.
(c) McConnell, M. S.; Mensah, E. A.; Nguyen, H. M. *Carbohydr. Res.* **2013,** *381*, 146–152.
226. Sletten, E. T.; Ramadugu, S. K.; Nguyen, H. M. *Carbohydr. Res.* **2016,** *435*, 195–207.
227. Hinou, H.; Saito, N.; Maeda, T.; Matsuda, M.; Kamiya, Y.; Nishimura, S.-I. *J. Carbohydr. Chem.* **2011,** *30*, 575–586.
228. Adinolfi, M.; Barone, G.; Iadonisi, A.; Schiattarella, M. *Synlett* **2002,** 269–270.
229. Mukherjee, M. M.; Basu, N.; Ghosh, R. *RSC Adv.* **2016,** *6*, 105589–105606.
230. Mukherjee, M. M.; Ghosh, R. *J. Org. Chem.* **2017,** *82*, 5751–5760.
231. Adinolfi, M.; Barone, G.; Guariniello, L.; Iadonisi, A. *Tetrahedron Lett.* **2000,** *41*, 9005–9008.
232. Adinolfi, M.; Barone, G.; Iadonisi, A.; Mangoni, L.; Schiattarella, M. *Tetrahedron Lett.* **2001,** *42*, 5967–5969.
233. (a) Adinolfi, M.; Iadonisi, A.; Ravid, A. *Synlett* **2006,** *4*, 583–586.
(b) Adinolfi, M.; Barone, G.; Iadonisi, A.; Schiattarella, M. *Tetrahedron Lett* **2002,** *43*, 5573–5577.
234. Peng, P.; Schmidt, R. R. *J. Am. Chem. Soc.* **2015,** *137*, 12653–12659.
235. Li, T.; Li, T.; Zhuang, H.; Wang, F.; Schmidt, R. R.; Peng, P. *ACS Catal.* **2021,** *11*, 10279–10287.
236. Kumar, A.; Kumar, V.; Dere, R. T.; Schmidt, R. R. *Org. Lett.* **2011,** *13*, 3612–3615.
237. Mishra, K. B.; Singh, A. K.; Kandasamy, J. *J. Org. Chem.* **2018,** *83*, 4204–4212.
238. (a) Iwata, R.; Uda, K.; Takahashi, D.; Toshima, K. *Chem. Commun.* **2014,** *50*, 10695.
(b) Sangwan, R.; Mandal, P. K. *RSC Adv* **2017,** *7*, 26256.
(c) Saway, J.; Salem, Z. M.; Badillo, J. J. *Synthesis* **2021,** *53*, 489–497.

239. Iibuchi, N.; Eto, T.; Aoyagi, M.; Kurinami, R.; Sakai, H.; Hasobe, T.; Takahashi, D.; Toshima, K. *Org. Biomol. Chem.* **2020,** *18*, 851–855.
240. Liu, J.; Yin, S.; Wang, H.; Li, H.; Ni, G. *Carbohydr. Res.* **2020,** *490*, 107963.
241. Zhao, G.; Li, J.; Wang, T. *Chem. Commun.* **2021,** *57*, 12659–12662.
242. Nielsen, M. M.; Holmstrøm, T.; Pedersen, C. M. *Angew. Chem. Int. Ed.* **2022,** *61*, e202115394.
243. Zhou, S.; Zhong, X.; Guo, A.; Xiao, Q.; Ao, J.; Zhu, W.; Cai, H.; Ishiwata, A.; Ito, Y.; Liu, X.-W. *Org. Lett.* **2021,** *23*, 6841–6845.
244. Das, R.; Mukhopadhyay, B. *J. Carbohydr. Chem.* **2015,** *00*, 1–16.
245. (a) Crich, D.; Dudkin, V. *J. Am. Chem. Soc.* **2001,** *123*, 6819–6825.
(b) Paulsen, H. *Angew. Chem. Int. Ed. Engl.* **1982,** *21*, 155–173.
246. H. Paulsen, H.; Lebuhn, R. *Liebigs Ann. Chem.* **1983,** 1047–1072.
247. Paulsen, H. *Angew. Chem., Int. Ed. Engl.* **1982,** *21*, 155–173.
248. Bohn, M. L.; Colombo, M. I.; Pisano, P. L.; Stortz, C. A.; Rûveda, E. A. *Carbohydr. Res.* **2007,** *342*, 2522–2536.
249. Bohn, M. L.; Colombo, M. I.; Stortz, C. A.; Rûveda, E. A. *Carbohydr. Res.* **2006,** *341*, 1096–1104.
250. (a) Crich, D.; Chandrasekera, N. S. *Angew. Chem., Int. Ed.* **2004,** *43*, 5386–5389.
(b) Huang, M.; Retailleau, P.; Bohe, L.; Crich, D. *J. Am. Chem. Soc.* **2012,** *134*, 14746–14749.
251. van der Vorm, S.; Hansen, T.; van Hengst, J. M. A.; Overkleeft, H. S.; van der Marel, G. A.; Codeé, J. D. C. *Chem. Soc. Rev.* **2019,** *48*, 4688–4706.
252. Adlercreutz, D.; Weadge, J. T.; Petersen, B. O.; Duus, J. O.; Dovichi, N. J.; Palcic, M. M. *Carbohydr. Res.* **2010,** *345*, 1384–1388.
253. Theodoratou, E.; Thaçi, K.; Agakov, F.; Timofeeva, M. N.; Štambuk, J.; Pûcíc-Bakovíc, M.; Vûckovíc, F.; Orchard, P.; Agakova, A.; Din, F. V. N. Glycosylation of Plasma IgG in Colorectal Cancer Prognosis. *Sci. Rep.* **2016,** *6*, 28098.
254. Moon, S.; Chatterjee, S.; Seeberger, P. H.; Gilmore, K. *Chem. Sci.* **2020,** *12*, 2931–2939.
255. Antonakoudis, A.; Strain, B.; Barbosa, R.; Jimenez del Val, I.; Kontoravdi, C. *Comput. Chem. Eng.* **2021,** *154*, 107471.
256. Bennett, C. S.; Wong, C.-H. *Chem. Soc. Rev.* **2007,** *36*, 1227–1238.
257. Fréchet, J. M. J.; Schuerch, C. *J. Am. Chem. Soc.* **1971,** *93*, 492–496.
258. Kanie, O.; Ohtsuka, I.; Ako, T.; Daikoku, S., Kanie, Y.; R. Kato, R. *Angew. Chem. Int. Ed.* **2006,** *45*, 3851–3854 *Angew. Chem.* **2006**, *118*, 3935–3938.
259. (a) Mogemark, M.; Elofsson, M.; Kihlberg, J. *J. Org. Chem.* **2003,** *68*, 7281–7288.
(b) Kanemitsu, T.; Wong, C. H.; Kanie, O. *J. Am. Chem. Soc.* **2002,** *124*, 3591–3599.
260. Manabe, S.; Ito, Y. *J. Am. Chem. Soc.* **2002,** *124*, 12638–12639.
261. Seeberger, P. H.; Werz, D. B. *Nat. Rev. Drug. Disc. Nat. Rev.* **2005,** *4*, 751–763.
262. Ganesh, N. V.; Fujikawa, K.; Tan, Y. H.; Stine, K. J.; Demchenko, A. V. *Org. Lett.* **2012,** *14*, 3036–3039.
263. Krçck, L.; Esposito, D.; Castagner, B.; Wang, C.-C.; Bindsch-dler, P.; Seeberger, P. H. *Chem. Sci.* **2012,** *3*, 1617–1622.
264. Jaipuri, F. A.; Pohl, N. L. B. *Org. Biomol. Chem.* **2008,** *6*, 2686–2691.

Chapter 2

Recent advances in stereoselective intramolecular *O*-glycosylation

Manoj K. Jaiswal[1], Anindra Sharma[1], Vinod K. Tiwari[1] and Richard R. Schmidt[2]

[1]*Department of Chemistry, Institute of Science, Banaras Hindu University, Varanasi, Uttar Pradesh, India,* [2]*Department of Chemistry, University of Konstanz, Konstanz, Baden-Württemberg, Germany*

2.1 Introduction

Glycoscience and in particular glycobiology has received much attention in the past decades due to the oligosaccharide component of glycolipids, glycoproteins, and other glycoconjugates play important roles in various biological processes, such as cellular recognition, cell-growth regulation, cell differentiation, immunological response, tumor metastasis, and bacterial and viral infection.[1–7] Thus there is always an increased demand to develop a simple and facile method for easy access to diverse glycosides for their complete biological, pharmaceutical, and medicinal investigation.[8–10] Although a large number of synthetic methods for glycoside bond formation have been developed, the regio- and stereocontrol are not always well understood.[11,12] The glycosylation reaction between a glycosyl donor and acceptor bearing a free hydroxyl group leads often to undesired products and anomeric outcomes because of the lack of regioselectivity and stereoselectivity.[13] The construction of biologically relevant glycans is increasingly achieved through automated synthesis and a one-pot, simplified glycosylation approach.[14] Chemoenzymatic glycosylation reactions are highly stereoselective and offer a suitable alternative for the construction of complex and challenging glycoconjugates.[15,16] In the absence of a neighboring group participating in auxiliary, typical intermolecular glycosylation reactions usually progress with poor stereoselectivity. Early efforts to achieve some stereocontrol of glycosylations focused primarily on developing participating groups and optimized reaction conditions.[17,18] The research emphasis on chemical glycosylation is

Synthetic Strategies in Carbohydrate Chemistry. DOI: https://doi.org/10.1016/B978-0-323-91729-2.00020-3

shifted toward elucidating the additional important factors and aspects that trigger the ease of glycosylation.[19,20] A number of investigations in this regard focused on important reaction intermediate, configuration, conformation, and stereoelectronics of the starting material have been performed.[21–25] In addition to these efforts, the efficacy of glycosylation reaction in the area of intramolecular glycosylation has been established by placing both the counterparts near each other to control stereo- and regioselectivity.[26–31] The concept of intramolecular glycosylation[26] strategy may be summarized into three major classes (Scheme 2.1): (1) the accepting atom is attached via a bifunctional group to a functional group of the glycosyl donor (*Molecular Clamp concept*),[32] (2) where the glycosyl acceptor is attached with a spacer at any functional substituent at any position to the glycosyl donor (*Spacer-mediated linkage* via *nonreacting centers*, commonly name as *Intramolecular Aglycon Delivery*, IAD),[33,34] and (3) the glycosyl acceptor is attached via a spacer to the leaving group of the glycosyl donor (*Leaving Group Mediated Intramolecular Glycosylation*).[35] In many variations of the general concept of glycosylation; generally, the intramolecular approach allows for achieving better stereocontrol in comparison to that of reaction with an intermolecular fashion.[26–29] This chapter addresses the recent advancements in the area of intramolecular *O*-glycosylations, with a focus on development over the past two decades. We authors outline the techniques that have been devised to increase efficiency or improve the stereoselection of intramolecular glycosylations reactions by tethering the donor, and an acceptor counterpart is provided herein.

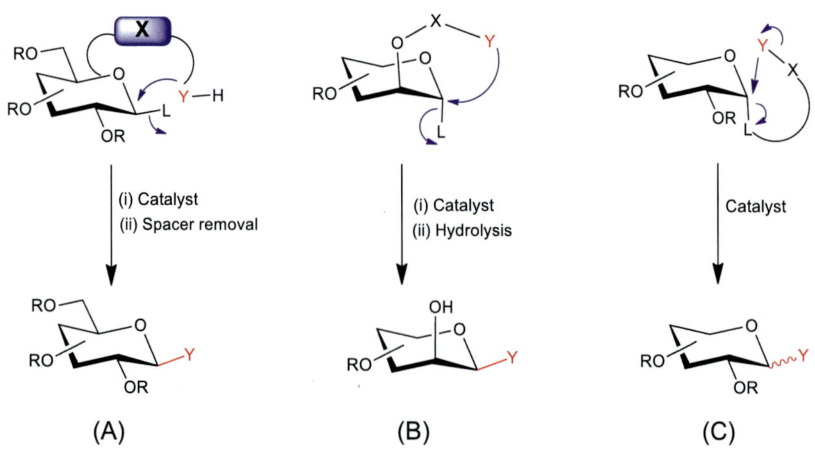

SCHEME 2.1 Common approaches for the intramolecular glycoside bond formation methodologies (L = Leaving group; R = Protecting group; X = Spacer): (A) molecular clamping approach, (B) intramolecular aglycan delivery, and (C) leaving group approach.

2.2 Molecular clamping approach

The "molecular clamp" method (Scheme 2.1A) introduced by Kusumoto et al. is one of the most common strategies for intramolecular glycosylation. The method involves the attachment of glycosyl donor and acceptor via a tether that takes place away from the reactive sites.[36] The main motto of this tethering is to attain an effective facial selectivity due to steric or geometric constraints and forces. In the literature, this concept has been also termed template-directed cyclo-glycosylation, intramolecular glycosylation of prearranged glycosides, remote glycosidation, or templated oligosaccharide synthesis.[37–40]

Kusumoto and coworkers have tried to synthesize β(1→4)-linked disaccharide having alternating glycan of glucosamine and muramic acid framework, which corresponds to the fractional structure of peptide-glycan cell wall.[36] An initial attempt to synthesize the targeted molecule by the direct glycosylation between the peracetylated oxazoline **1** and muramic acid derivative **2** could not be feasible due to the steric hindrance caused by the lactic acid moiety (Scheme 2.2).[36] The author claimed the necessity of a novel device, which facilitates the coupling and connected the two components via an ester linkage before the glycosylation. Thus the glycosylation reaction has become an intramolecular process and could be feasible. Keeping this idea in mind, the authors have synthesized the targeted molecule **3**, which was linked to the muramic acid moiety via an ester linkage to the C-6 position of the donor molecule, sterically the most favorable for β(1→4) glycosidation. Finally, the required molecule **4** was synthesized in 20% yield by successive glycosylation in the presence of *p*-toluenesulfonic acid (TsOH) at 50°C, methanolysis followed by the peracetylation. Therefore the presence of

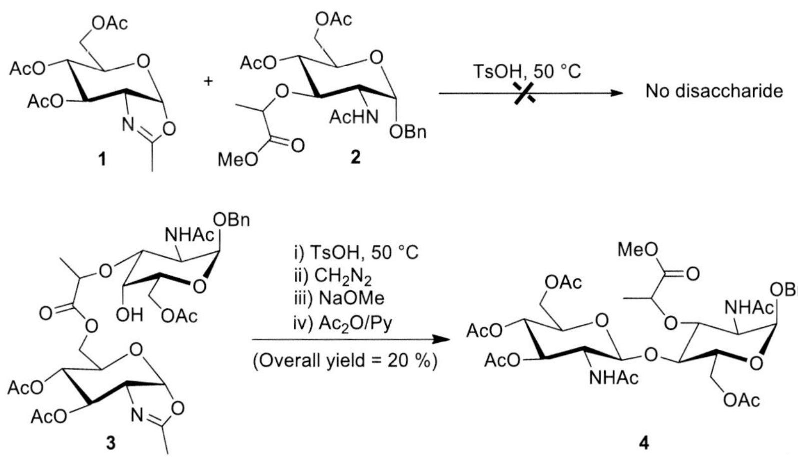

SCHEME 2.2 Synthesis of β(1→4)-linked disaccharide **4** using molecular clamp method.

desired ester linkage kept the two sugar moieties close to each other and favored the formation of the anticipated glycoside bond. This reaction is also termed entropic activation in a glycosidation reaction. In the literature, several tethers have been reported, which are used to connect the donor and the acceptor moieties away from the nonreacting centers and bring them nearby for effective glycosylation.

Ziegler and coworkers have investigated flexible succinyl linkers as a tether to obtain the glycosyl donor and acceptor in a prearranged position and thus carry out highly stereoselective glycosylation (Scheme 2.3).[37] Regioselective condensation of glycosyl donor 6 having succinyl group with 3-*O* and 4-*O*-unprotected galactose acceptor 7 resulted in 2′-3 succinyl tethered glycoside, which upon intramolecular glycosylation followed by deacetylation and perbenzoylation afford disaccharide 10 in 74% yield as a pure 1,2-*trans* isomer. Further studies of this approach to the other positions of the glycosyl donor and acceptor as well as other sugar derivatives have been also explored. It has been observed that when the succinyl tether was attached at 6′-OH of the galactosyl acceptor, an equal amount of both anomers were obtained and with glucosyl acceptor exclusively 1,2-*cis*-linked product was formed. Therefore we can say that the stereoselectivity mainly depends upon the donor–acceptor match–mismatch. Other linkers such as carbonates, oxalic, malonic, and glutaric dicarboxylic acids were also explored for intramolecular glycosylation to attain the required stereoselectivity.[41–43] This has been well accepted that the outcome of intramolecular glycosylation through the "molecular clamp method" can be unpredictable in most cases and therefore every approach that falls under this category should be studied extensively.

Initially, the "phthaloyl tether" has been investigated for regioselective glycosylation reaction by Zeigler et al.[44] followed by Valverde and

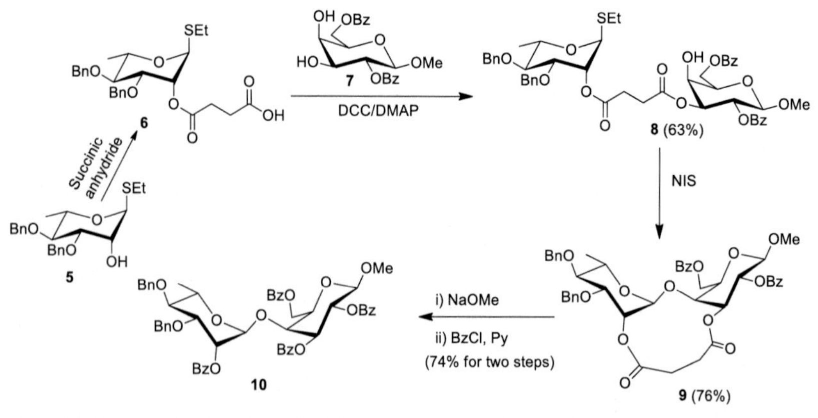

SCHEME 2.3 Succinyl as a flexible tether for intramolecular glycosylation.

coworkers.[45] The authors have prepared glycosyl donor by the reaction of glycosyl donor precursor **11** with phthalic anhydride to afford corresponding aromatic ester **12**, which upon activation with thionyl chloride coupled with *O*-2 position of methyl glycoside acceptor **13** in a regioselective manner by employing dibutylstannylidene acetals under microwave condition to yield the mixed phthalic ester **14**. Macrocyclic glycosylation of tethered compound was achieved by *N*-iodosuccinimide (NIS)/triflic acid (TfOH) to afford the 13-membered macrocyclic glycoside **15** (Scheme 2.4). In the final step, the tether group was easily removed by NaOMe and the product was completely acetylated to afford α-(1→3)-linked disaccharide **16**. During the reaction, regioselectivity was controlled by the phthaloyl linker attached to the neighboring C-2 position. In contrast to 2,6′-linked donor–acceptor pair **14**, 6,6′-donor–acceptor pair **17** led to the formation of (1→4)-linked regio-isomeric disaccharide **19**.[46] Therefore regioselectivity in these types of reactions is due to the topographic direction of the hydroxyl groups of phthaloyl tether having an effect on the transition state for the glycosylation. Furthermore, on decreasing the temperature from room temperature to −78°C, the β:α ratio changes from 1: 1 to 5.5:1. In comparison to the phthaloyl ester disaccharides, isophthaloyl ester disaccharides exhibited improved α:β stereoselectivity in several examples.[47]

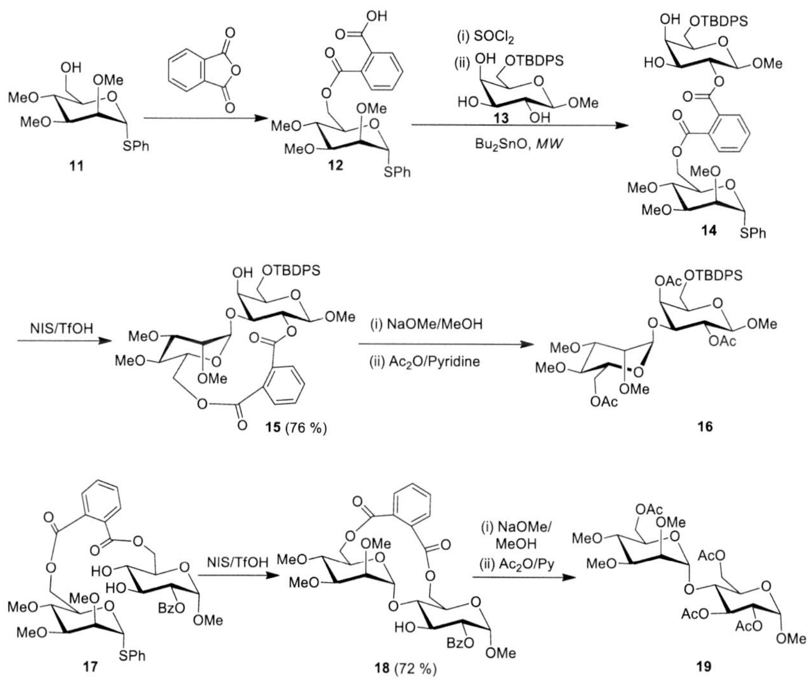

SCHEME 2.4 Phthaloyl spacer in intramolecular glycosylation reactions.

Takahashi et al. considered both the flexible succinyl tether and rigid phthaloyl tether for the synthesis of 4,6-branched trisaccharide **24** by using intramolecular glycosylation, but computation studies of the relative conformation and energies confirmed that the phthaloyl group serves as an optimal spacer in remote glycosylation.[48] Therefore the synthesis of 4,6-branched trisaccharide is achieved from thioglycoside **20** and 6-hydroxy di-glycoside **21** (Scheme 2.5). Initially, thioglycoside **20** was coupled with 6-hydroxy di-glycoside **21** in presence of dicyclohexylcarbodiimide (DCC)/dimethyl aminopyridine (DMAP), followed by treatment with diethyl aminosulfur trifluoride (DAST) and *N*-bromo succinimide (NBS) in dry dichloromethane (DCM) to convert anomeric thiophenyl group to fluoro group and then selective removal of *p*-methyl benzyl ethers was accomplished by using $Pd(OH)_2/C$ hydrogenation to afford the respective trisaccharide **22**. Subsequent remote glycosidation of compound **22** was achieved by using Cp_2HfC1_2 and AgOTf in anhydrous CH_2C1_2 under reflux conditions. Base treatment of the cyclized product **23** with NaOMe followed by acid hydrolysis of the acetal group and then acetylation using acetic anhydride/pyridine provided the required 4,6-branched trisaccharide **24**.

Chemical synthesis of cyclodextrins having α-gluco stereoselectivity is very challenging. The first total synthesis of α-cyclodextrin was achieved by Ogawa et al. via the intramolecular cyclization of a properly protected glucohexaose derivative, but these reactions were nonstereoselective and the yield of the final cyclization was very low (21%).[49] Fukase et al. have demonstrated the utility of the molecular clamp method for the synthesis of α-cyclodextrin **31** (1→4-linked α-glucoside) via a block condensation of

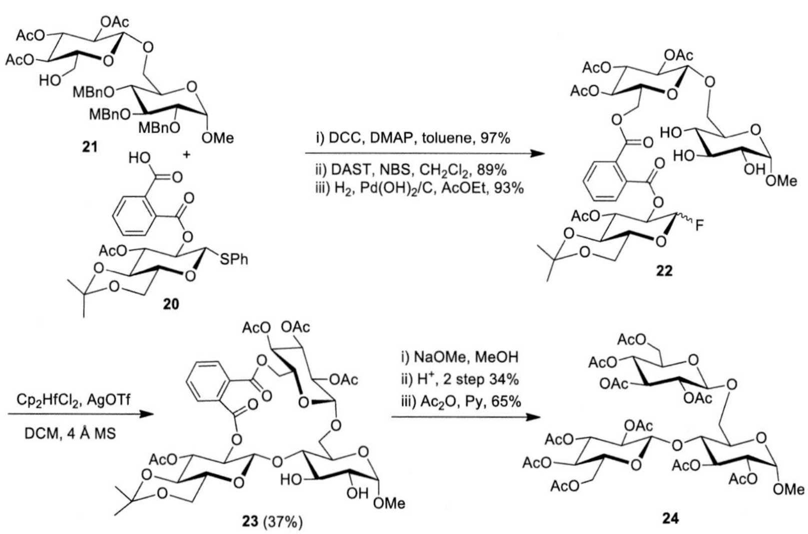

SCHEME 2.5 Synthesis of 4,6-branched trisaccharide **24** via remote glycosidation.

three maltose units (Scheme 2.6).[50] The phthaloyl spacer was used to improve glycosylation selectivity during the stepwise chain elongation process via the coupling of maltose building blocks **25** and **26**, as well as the efficiency of macrocyclization. The required macrocyclization of compound **27** was performed in a regioselective manner by using DCC, DMAP, and DMAP · HCl in refluxing CH_2Cl_2 under high dilution. For the ultimate cyclo-glycosylation, 1-*O*-allyl group of **28** was removed by cationic iridium complex followed by the treatment with iodine to give the cyclic ester **29** in 93% yield, which upon glycosylation via imidate activated by TMSOTf gave **30** only in the α-form in 66% yield. The stereochemistry of cyclo-glycosylation was controlled by the mutual effect of the phthaloyl linker and

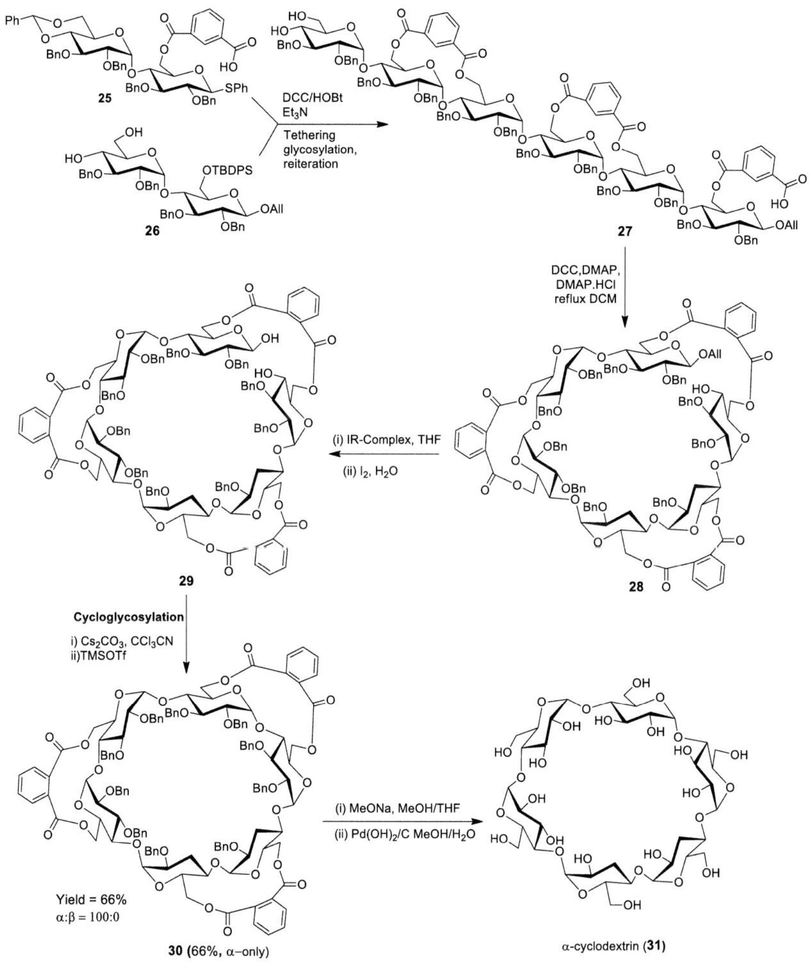

SCHEME 2.6 Molecular clamp method for the stereoselective synthesis α-cyclodextrin **31**.

the substrate. Finally, deacetylation using NaOMe followed by the hydrogenolysis of all the benzyl groups provides the desired α-cyclodextrin **31** in good yield.

It has been observed that the rigidity of the spacer plays an important role in determining the ring size in glycosylation and stereoselectivity.[51,52] Schmidt et al. have introduced a rigid spacer m-xylene for efficient intramolecular glycosylation at room temperature with high facial selectivity (Scheme 2.7).[52] Therefore ethyl β-thioglucoside **32** was first alkylated at the *C*-3 position by using NaH and 15-crown-5 as supporting reagents. The resulting intermediate **33** is then reacted with 4,6-*O*-unprotected glucoside **34** with dibutyltin oxide in toluene to afford the desired xylene-tethered pair intermediate **35**. The xylene-tethered intermediate **35** is then intramolecularly glycosylated in the presence of NIS/TMSOTf in 93% yield cyclic compound **36** with complete β-selectivity. It has been observed that the diastereofacial selectivity in intramolecular glycosylation is independent of the configuration of the glycosyl donor and exclusively only one anomer is generated. The resulting cyclic compound **36** is then subjected to hydrogenolytic *O*-debenzylation followed by comprehensive deacetylation to afford product **37** with great ease.

This concept is then further extended for the synthesis of a trisaccharide and tetrasaccharide (Scheme 2.8).[53] Therefore maltose disaccharides were selected as a glycosyl donor and linked with galactose or lactose by the rigid xylene tether to afford molecules **38** and **39**. Intramolecular glycosylation of **38** and **39** in the presence of NIS/TfOH in DCM resulted in the trisaccharide **40** and tetrasaccharide **41** with only β(1→3) linkage between the maltosyl and the galactoside or lactoside residues. Further, the usefulness of the xylene tether in the application of iterative synthesis of maltotriose has been investigated. In this request, the rigid xylene spacer was used to link the two

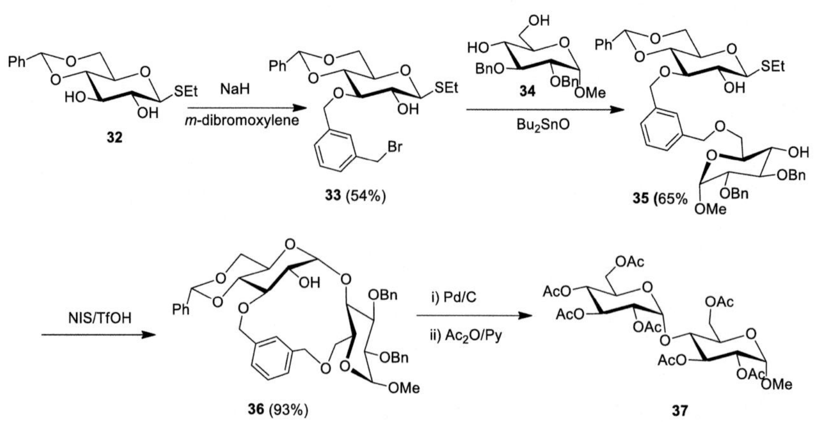

SCHEME 2.7 *m*-Xylene as a rigid tether for stereoselective synthesis of disaccharide **37**.

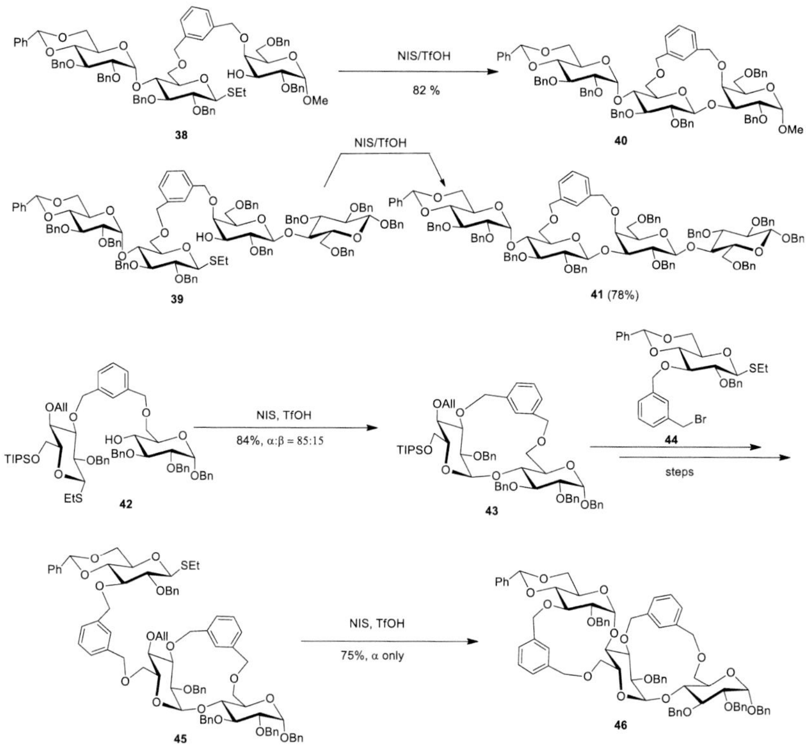

SCHEME 2.8 Synthesis of trisaccharides and tetrasaccharides by using *m*-xylene as a rigid tether for intramolecular glycosylation.

glucose units via 3- and 6-positions to produce a tether molecule **42**, which upon intramolecular glycosylation by using NIS/TfOH gives disaccharide **43** in 84% yield (α/β = 7:1). Subsequent selective desilylation at the 6′-position followed by the introduction of xylene spacer linked donor moiety **44** under standard condition afterward the removal of the 4′-allyl group gave tethered donor−acceptor combination **45**. Upon NIS/TfOH-mediated intramolecular glycosylation, compound **45** gave pure diastereoselective α-linkage exclusively, providing protected trisaccharide **46** in 75% yield. Hydrogenolysis with Pd/C, eliminating the *O*-benzyl, *O*-benzylidene, and *O*-xylylene moieties followed by *O*-acetylation with acetic anhydride in pyridine, afforded *O*-acetyl-protected maltotriosides.

Unsymmetrical rigid spacers with differentially cleavable groups have also been developed for reiterative glycosylation, because these tethers have allowed regioselective ring opening after the glycosylation step and therefore offer more flexibility in the synthesis of longer oligosaccharides.[54] Schmidt et al. have explored this strategy for the reiterative synthesis of trisaccharide (Scheme 2.9).[55] Intramolecular glycosylation of donor−acceptor-tethered

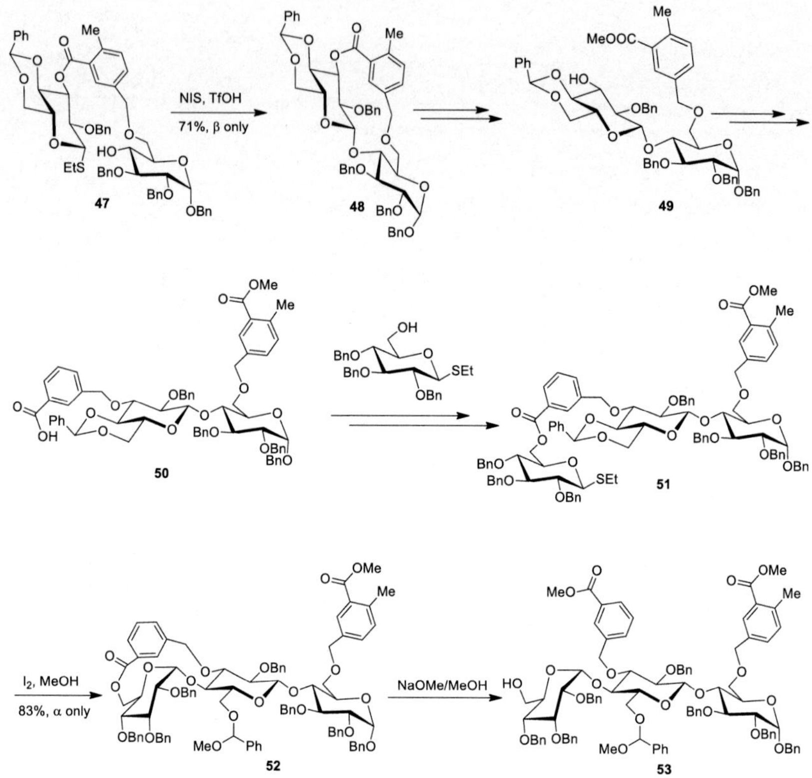

SCHEME 2.9 Reiterative synthesis of trisaccharides using a rigid nonsymmetrical spacer.

molecule **47** under NIS/TfOH facilitated condition afforded disaccharide **48** in 71% with complete β-selectivity. For further glycosylation, **48** was untethered by *trans*esterification by using NaOMe/MeOH and this led to the formation of **49** having 3′-OH group, which could serve as a center for the next tether linkage. Esterification of **50** with the next glycosyl donor under standard reaction conditions produced donor–acceptor-tethered intermediate **51** in 70% yield. The second glycosylation of **51** was performed without cleavage of benzylidene acetal ring rection by using I_2/MeOH. Under the standard condition, I_2/MeOH cleaved the benzylidene acetal ring, and being a soft atom, iodine coordinated with S-atom in contrast to the *O*-atom of acetal and therefore the carbocation would be engendered at anomeric carbon, which invites C4′-OH to furnish 14-membered macrocycles **52** in 83% yield with complete α-selectivity. Furthermore, transesterification of **52** by NaOMe/MeOH yielded the desired trisaccharide and 6″-OH of **53** will further act as an extension point in reiterative synthesis.

Short peptide chains have also been explored as a tether for intramolecular glycosylation. The reason behind this approach is the indirect use of existing automated technology for solid-phase or solution-phase peptide synthesis for the synthesis of oligosaccharides.[56–58] This way, Fairbanks et al. explored a number of peptide chains with various amino acids as templates (Scheme 2.10).[57] Aspartate-linked glycosyl donor and acceptor were evaluated for oligosaccharide synthesis by using standard *N*-ethoxycarbonyl-2-ethoxy-1,2-dihydroquinoline mediated peptide coupling protocol. NIS/TfOH-mediated intramolecular glycosylation of **54** followed by release from peptide template and subsequent acetylation resulted in a mixture of disaccharides **55a–d**. It has been observed that the variation in peptide amino acid sequence influences the stereo- and regioselectivity of glycosylation (Table 2.1). It has been observed that the stereoselectivity of these glycoside linkages varies due to the fairly low rigidity of the tethers.

Warriner and coworkers have further extended this methodology by using solid-supported peptide templates (Scheme 2.11).[59] The solid-phase peptide template sequence was started by the addition of a glycine unit to the amino methyl polystyrene-based resin, which minimizes the interaction between the

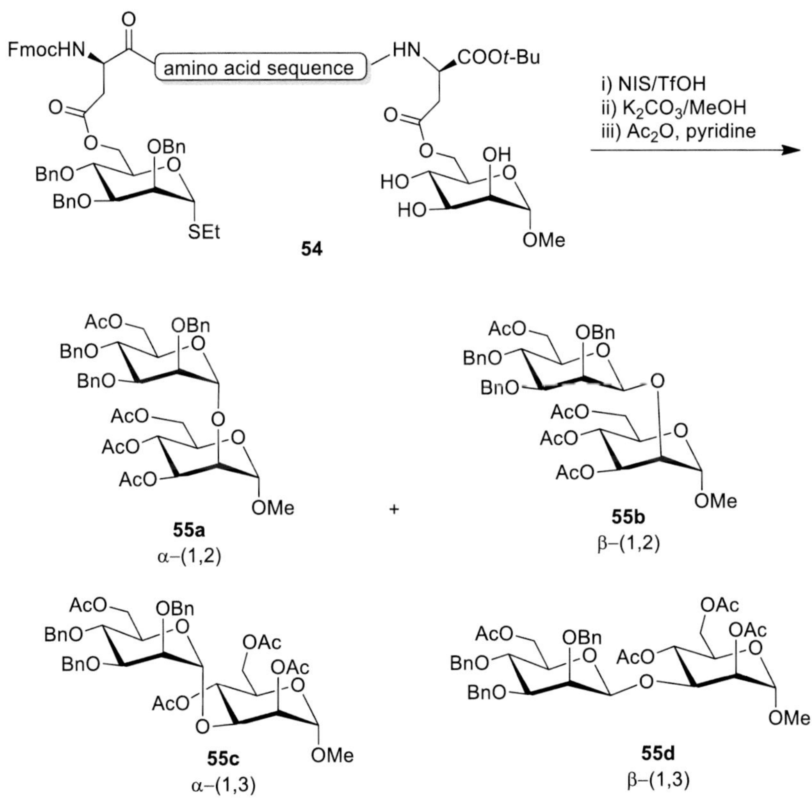

SCHEME 2.10 Oligosaccharide synthesis by using a peptide-based linker.

TABLE 2.1 Stereo- and regiochemical outcome of peptide template-based glycosidation.

Amino acid	55a α (1→2)	55b β (1→2)	55c α (1→3)	55d β (1→3)
None	–	54%	60%	78%
Gly	–	–	58%	77%
Ala	–	–	56%	59%
Phe	–	–	58%	71%
Pro	79%	73%	–	62%
GlyGly	–	–	30%	42%

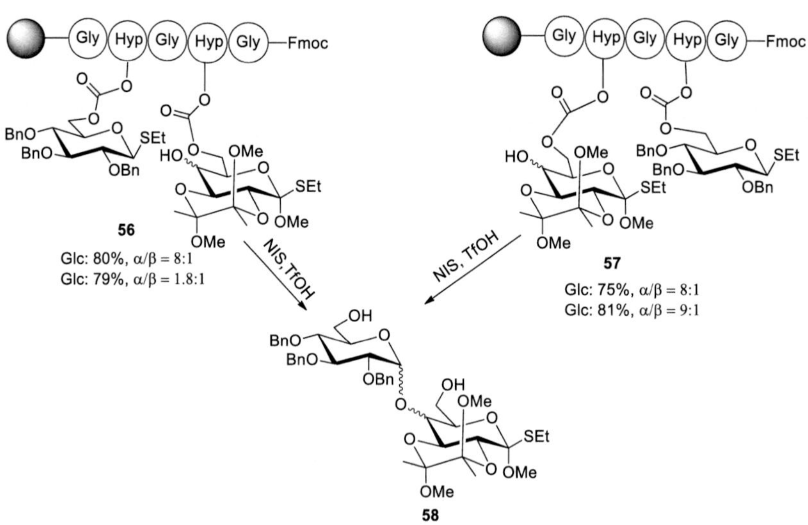

SCHEME 2.11 Effect of the relative position of monosaccharide units in solid-phase peptide-based templet mediated glycosylation [Hyp = (2*S*,4*R*)-4-hydroxypyrrolidine-2-carboxylic acid].

sugar moiety and the resin. Hydroxyproline (HyP) unit bearing the glycosyl donor was coupled to the resin-bound glycine, which was unable to couple with the acceptor having a hydroxyproline unit due to the steric hindrance caused by the bulky sugars. Therefore the insertion of an intermediate glycine unit is necessary to allow the subsequent addition of the acceptor unit and a terminal-protected glycine. Therefore glycosylation of final conjugate **56** by using NIS/TfOH afforded (1→4)-linked disaccharide **58** in 80% yields with exclusive α-selectivity. Further, it has been observed that when the position of donor and acceptor attached to the template were reversed, and then the conjugate **57** was allowed for the intramolecular glycosylation under the previous reaction condition, complete loss in stereoselectivity was observed with anomeric products in 1:1 ratio. Therefore we can say that the conformational preference of the peptide template has a significant influence on the stereoselectivity during glycosylation. The method was widely explored with a number of common saccharide acceptors including glucose, mannose, and galactose-based acceptors. Galactosyl acceptors exhibited a melodramatic effect of the relative position of the donor and acceptor on the peptide template. Interestingly, the stereoselectivity consequence was reversed (1.8:1 and 9:1) in contrast to glucosyl acceptors. After that, a similar perception was applied to the mannosyl acceptor, and a low 2:1 stereoselectivity was exhibited regardless of the comparative position of the reaction counterparts. Therefore peptide-based templates were explored for the practical synthesis of a small library of disaccharides.

Demchenko et al. have introduced templated oligosaccharide synthesis, which also comes under the category of molecular clamp method, but there is a slight difference in comparison to the general molecular clamp method.[60,61] It allows not only the glycosylation of different hydroxyl groups which are adjacent to the tether but also allows for connecting multiple building blocks. The authors have utilized bisphenol A (BPA) as the template and succinoyl, glutaryl, or phthaloyl linkers to tether glycosyl donors and acceptors together (Scheme 2.12). When the linker is L1 is shorter than L2 (succinoyl vs. glutaryl) or equal to L2 (succinoyl vs succinoyl), glycosylation of tethered 4-OH acceptor **59** yielded disaccharide **60** with complete α-selectivity (bottom-face delivery of acceptor). However, when the delivery of the acceptor is from the top face, that is, the linker L1 is longer than L2 (glutaryl vs succinoyl), α-selectivity decreases and an anomeric mixture with α:β ration 2.8:1 have been obtained. Additionally, complete α-selectivity was attained in the synthesis of disaccharides resulting from secondary glycosyl acceptors (L1 = succinoyl; L2 = glutaryl); however, the (1→6) linked disaccharide exhibited lower selectivity, which may be due to the more

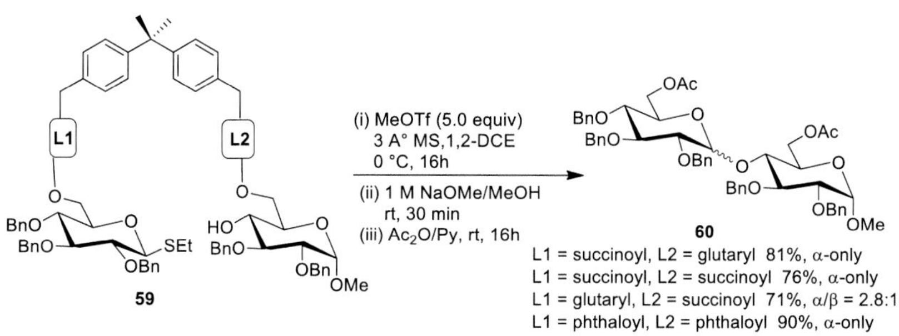

SCHEME 2.12 Rigid BPA template with various linkers and their effect on the stereoselectivity of glycosylation.

flexibility of primary 6-OH group and therefore acceptor became able to reach from both the faces of activated donor via formation of oxocarbenium ion. A more rigid spacer such as a phthaloyl tether allows the glycosyl donor and acceptor to stay near complete α-selectivity and sufficient flexibility to glycosylate the remote hydroxyl groups. Changing the solvent from 1,2-dichloroethane to acetonitrile (a solvent that is known to enhance β selectivity) could not change the anomeric stereoselectivity, indicating that the template effect is stronger than the solvent effect.[62] Alternatively, the synthesis of β-linked disaccharides has been achieved by using a benzoylated glycosyl donor tethered to 3-OH of the acceptor due to the participatory effect of the neighboring ester group. The mechanistic studies revealed that phthaloyl linker allows enhanced stereoselectivity and better yields in glycosylation reactions.

Furthermore, this concept has been utilized for the synthesis of glycoazobenzene macrocycles with high α-selectivity[63] and longer trisaccharides by using trimellitic anhydride as a precursor for the bridging linker (Scheme 2.13).[60] Initially, BPA-linked glycosyl donor **61** was connected with trimellitic anhydride to afford the intermediate **62**, which reacted with BPA-conjugated glycosyl acceptor **63** to afford tethered donor–acceptor pair **64** in 95% yield. Tethered molecule **64** could be used for further template elongation via the carboxyl group. Therefore BPA-conjugated glycosyl acceptor **65** linked with BPA-conjugated glycosyl donor **64** by using DCC/DMAP mediated coupling reaction to afford templated conjugate **66** in 82% yields. The selective activation of *the* *S*-ethyl leaving group by using MeOTf, and the glycosylation of the middle building block took place with the associated removal of the *p*-methoxybenzyl (PMB) group. The *O*-allyl phenyl leaving group was activated with NIS/TfOH, and again the PMB group of the acceptor was removed during the glycosylation step. The subsequent tethered trisaccharide was released from the template by reaction with NaOMe in MeOH to afford the resultant trisaccharide **67** with an overall 58% yield with complete α-stereoselectivity for both glycosidic linkages.

In addition to the above-mentioned suitable system as rigid spacer, including malonyl, succinyl, glutaryl, and peptide, and their arrangement with rigid spacers such as *o*- or *m*-xylene, phthaloyl, or isophthaloyl residue, a modular "click chemistry" inspired regioselective 1,4-disubstituted 1,2,3-triazole can also serve as a similar rigid spacer for the desired intramolecular aglycan delivery.[64] Since the last 21 years, this highly reliable "Click Chemistry" is widely explored in almost all fields notably in carbohydrate chemistry for the convenient synthesis of triazole-appended carbohydrate-containing molecular architectures such as glycohybrids, glycoconjugates, neoglycoconjugates (e.g., glycopeptides, glycoproteins, glycopolymers, glycoclusters, glycodendrimers, etc.), which served diverse applications in medicinal chemistry, macrocyclization, bioorthogonal coupling,

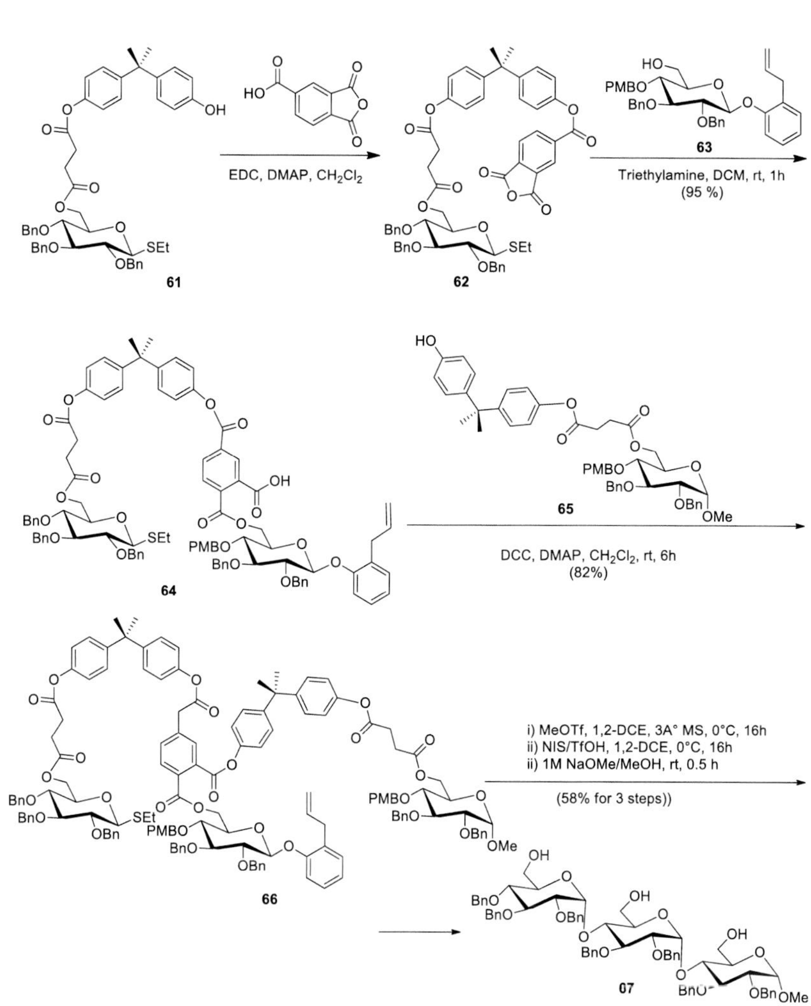

SCHEME 2.13 BPA-templated convenient synthesis of α,α′-linked trisaccharide **67**.

glyconanotechnology, sensing, gelation, chelation, catalysis, and glycosylation reactions.[65,66]

A common route for the successful macrocyclization with high selectivity from corresponding azide-alkyne scaffolds can be achieved by a number of well-established routes including Ru-catalyzed ring-closing metathesis (RCM), intramolecular ruthenium-catalyzed azide alkyne cycloaddition, and intramolecular Cu(I)-catalyzed azide alkyne cycloaddition (CuAAC)-mediated intramolecular glycosylation as part of intramolecular aglycan delivery (IAD). Toward this end, Schmidt and Tiwari in 2012 reported a "Click Chemistry" mediated regioselective 1,2,3-triazole heterocycle as the rigid spacer in intramolecular glycosylation reaction (Scheme 2.14).[64] The authors

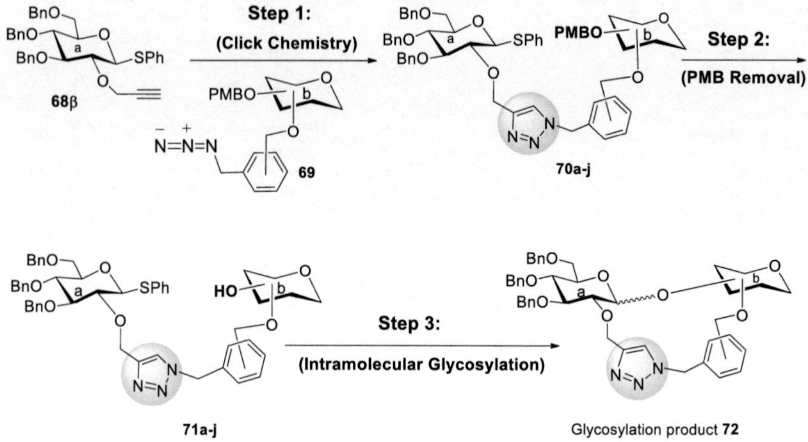

SCHEME 2.14 Triazole as a rigid spacer in intramolecular aglycan delivery (IAD).[64]

explored triazolylmethyl residue in combination with *o*- or *m*-xylylene moiety as a rationally rigid spacer system in intramolecular glycosylation. Thus phenyl-3,4,6-tri-*O*-benzyl-2-*O*-propargyl-1-thio-D-glucopyranoside **68** (a orthogonally protected glycosyl alkyne as part of the donor) was selected for the regioselective CuAAC click coupling with an *O*-(2-or 3-azidomethyl-benzyl)-protected glycosyl azide **69a–h** (as part of acceptor) to give the corresponding triazole-linked disaccharide **70a–h** (**Step 1**). 2,3-dichloro-5,6-dicyano-1,4-benzoquinone (DDQ)-mediated removal of *p*-methoxybenzyl group from the individual disaccharides **70a–h** resulted in the required triazole-linked donor–acceptor systems **71a–h** (**Step 2**), which at the end underwent intramolecular glycosylation (IAD) using NIS/TfTMS in DCM and furnish good yields of the desired disaccharide-containing triazole-linked macrocycles **72a–h** with more or less an anticipated selectivity (**Step 3**) (Scheme 2.14).[64]

The result of intramolecular glycosylation of triazole-linked donor–acceptor disaccharide **71a–h** having triazolylmethyl residue in combination with *o*- or *m*-xylene moiety (as part of acceptor) as rationally rigid spacer system to respective disaccharide-containing triazole-linked macrocycles is depicted in Table 2.2. The intramolecular glycosylation of **71a** using TMSOTf/NIS in dry DCM for 2 hours under argon condition resulted in the desired disaccharide-containing macrocycle **72a** with >98% β-selectivity, while similar intramolecular glycosylation of disaccharide **71e** afforded respective macrocycle **72e** in although poor yield but exclusively with the α-selectivity.

As 1.1 equiv. of NIS resulted in the moderately low yield of macrocycle **72a** in comparison to the IAD performed in the presence of 2 equiv. of NIS, the glycosylation result is concentration-dependent. Although the stereo (or

TABLE 2.2 IAD through click derived triazole as the rigid spacer.

Entry	Triazole-linked donor−acceptor disaccharides (71a−h)	IAD-mediated disaccharide-containing macrocycle (72a−h)	Yield and selectivity
1	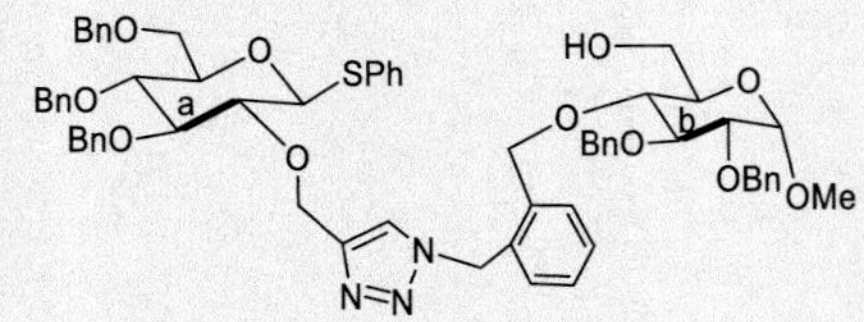 **71a**	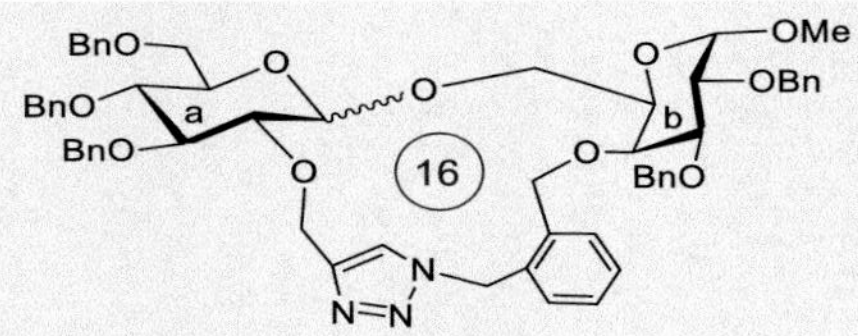 **72a**	Yield: 95%, Selectivity: α/β = 1:2
2	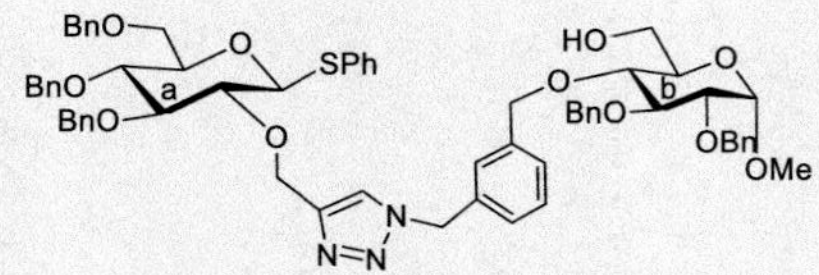 **71b**	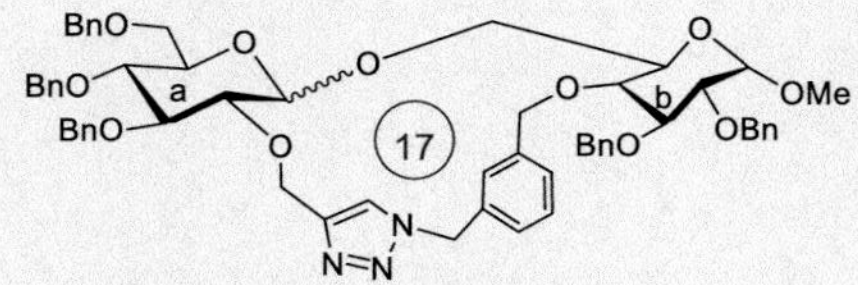 **72b**	Yield: 80%, Selectivity: α/β = 3:2
3	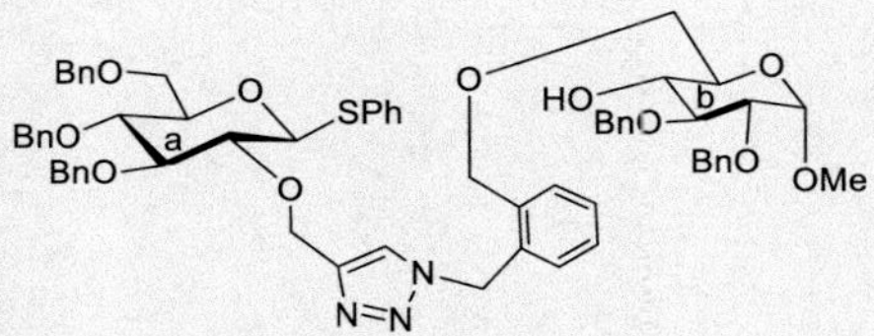 **71c**	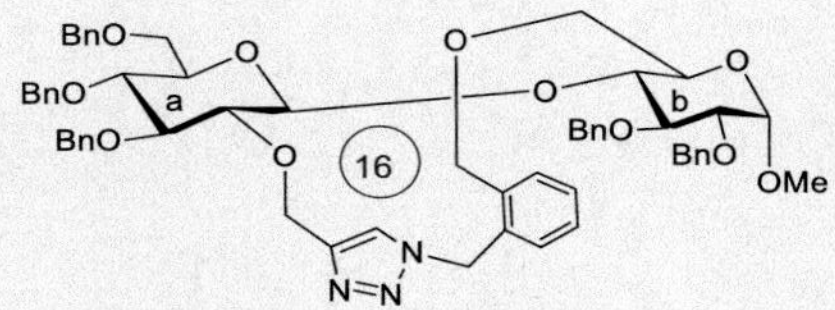 **72c**	Yield: 90%, selectivity: α/β = 3:1

(Continued)

TABLE 2.2 (Continued)

Entry	Triazole-linked donor–acceptor disaccharides (71a–h)	IAD-mediated disaccharide-containing macrocycle (72a–h)	Yield and selectivity
4	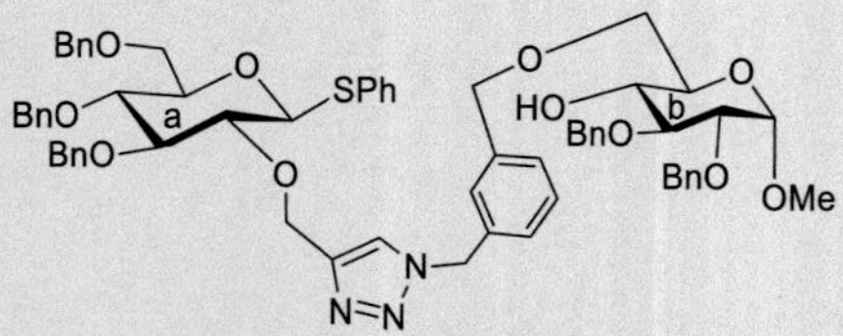 **71d**	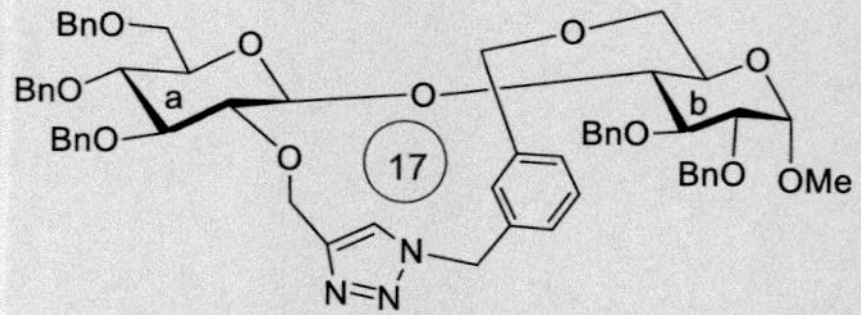 **72d**	Yield: 82%, selectivity: $\alpha/\beta = 1:2$
5	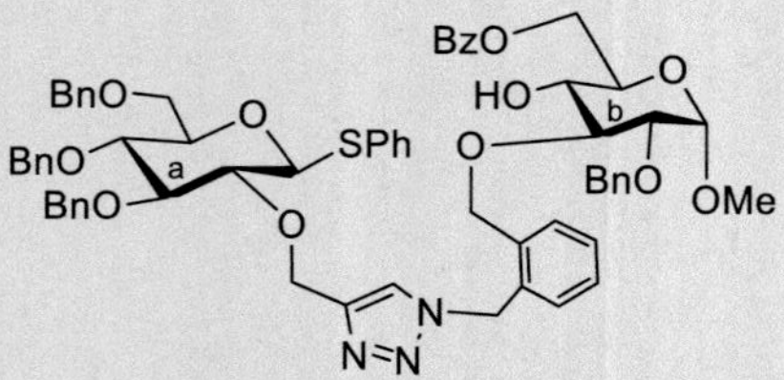 **71e**	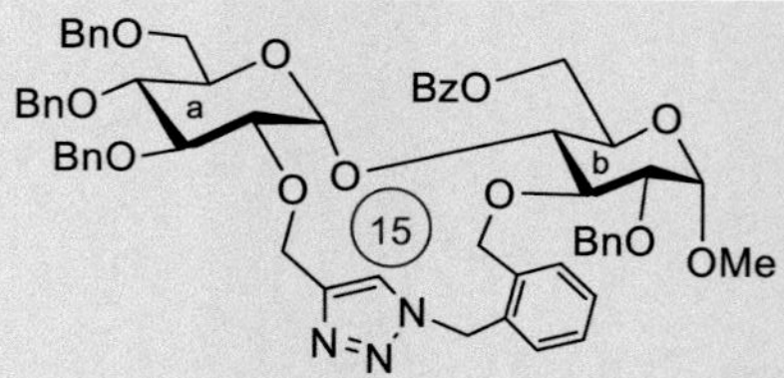 **72e**	Yield: 55%, only α-selectivity
6	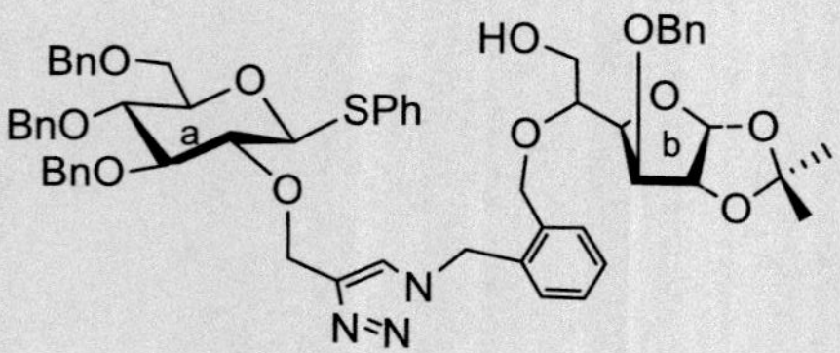 **71f**	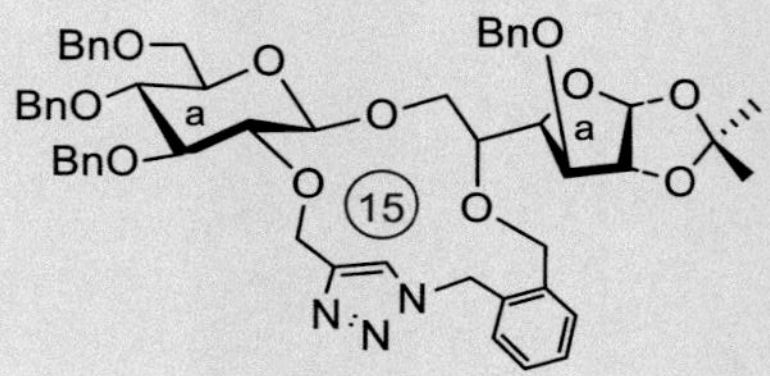 **72f**	Yield: 58%, selectivity: over 98% β-selectivity

7	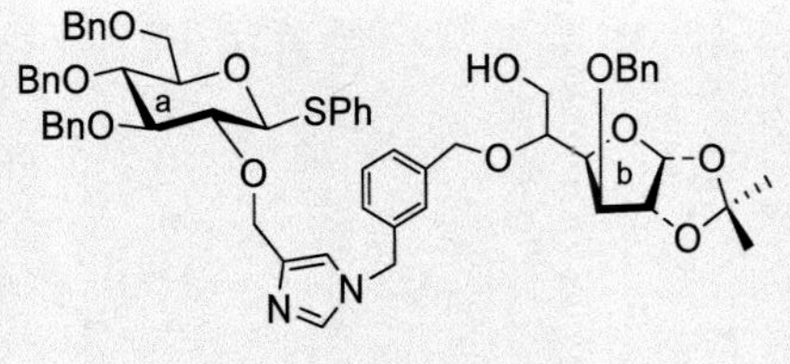 **71g**	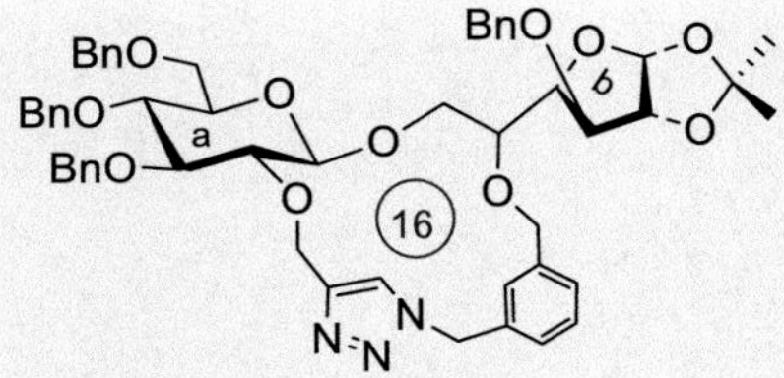 **72g**	Yield: 65%, selectivity: $\beta/\alpha = 3:1$
8	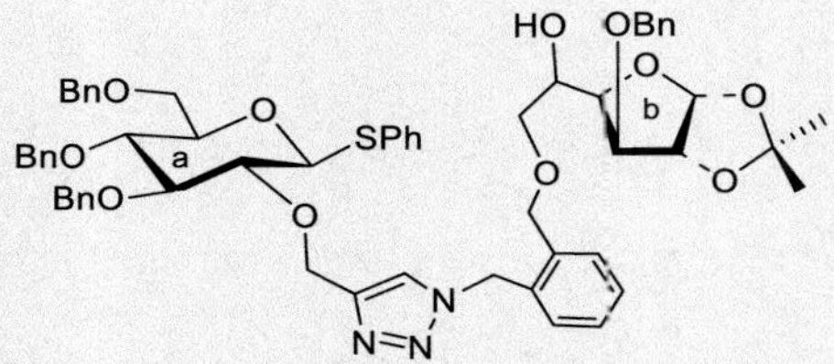 **71h**	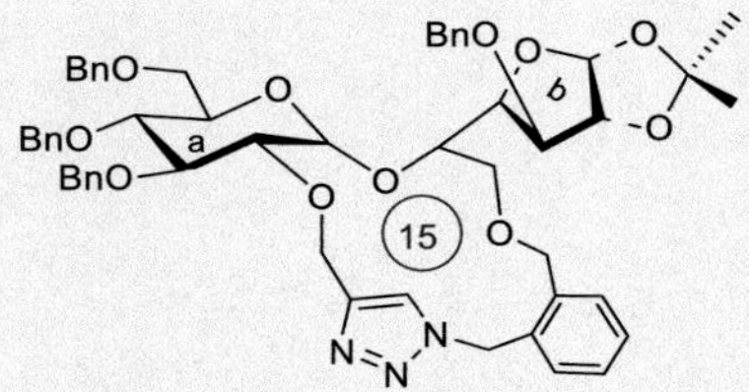 **72h**	Yield: 80%, selectivity: $\alpha/\beta = 3:2$
9	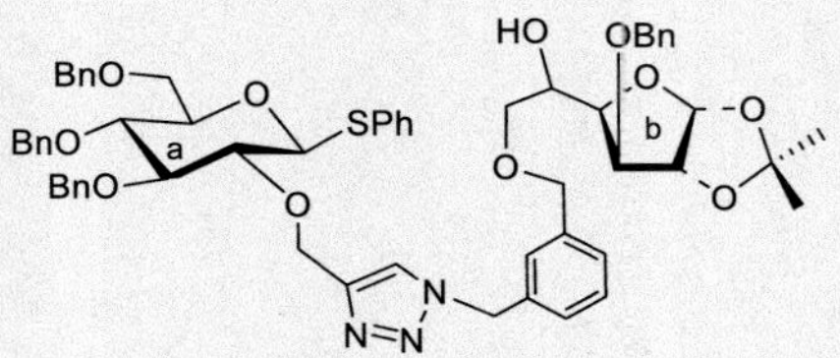 **71i**	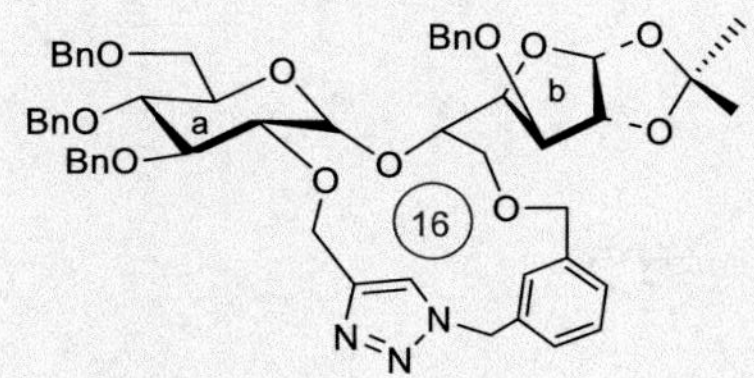 **72i**	Yield: 50%, selectivity: $\alpha/\beta = 2:1$

(*Continued*)

TABLE 2.2 (Continued)

Entry	Triazole-linked donor–acceptor disaccharides (71a–h)	IAD-mediated disaccharide-containing macrocycle (72a–h)	Yield and selectivity
10	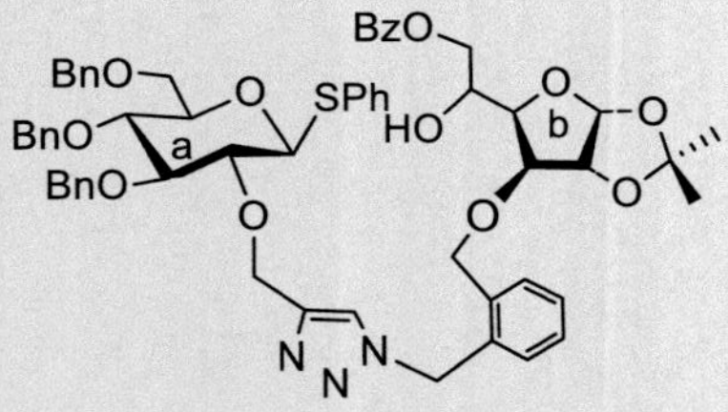 **71j**	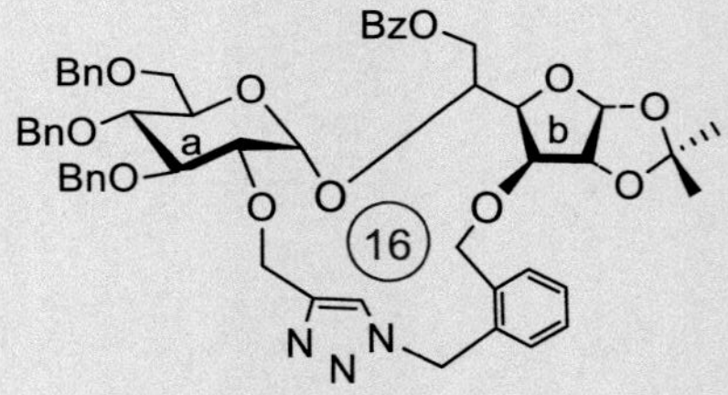 **72jα**	Yield: 20%, selectivity: only α-selectivity
11	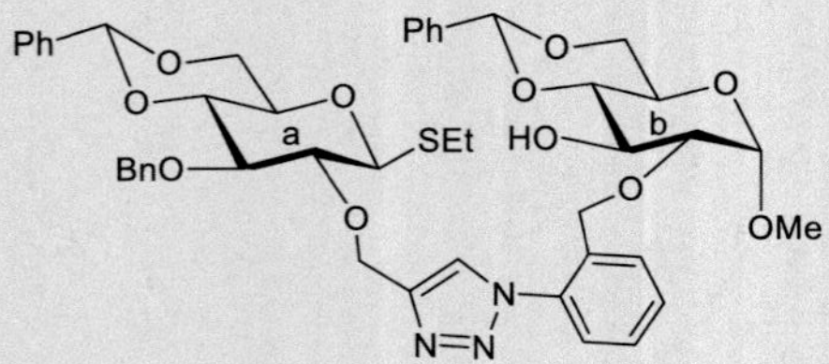 **71k**	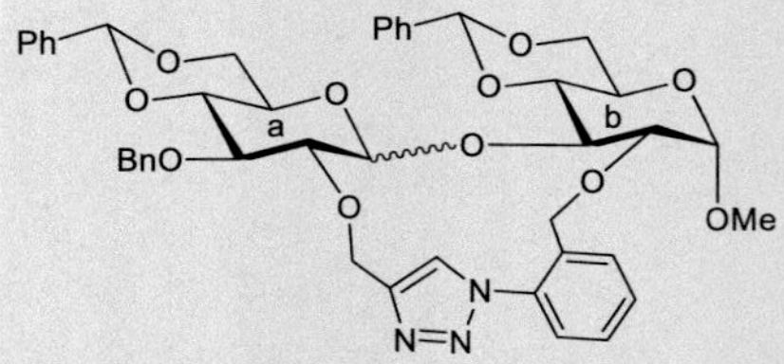 **72k**	Yield: 68%, selectivity: $\alpha/\beta = 1:25$
12	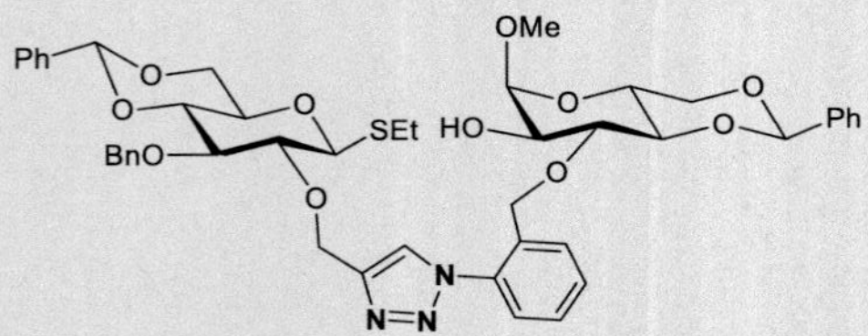 **71l**	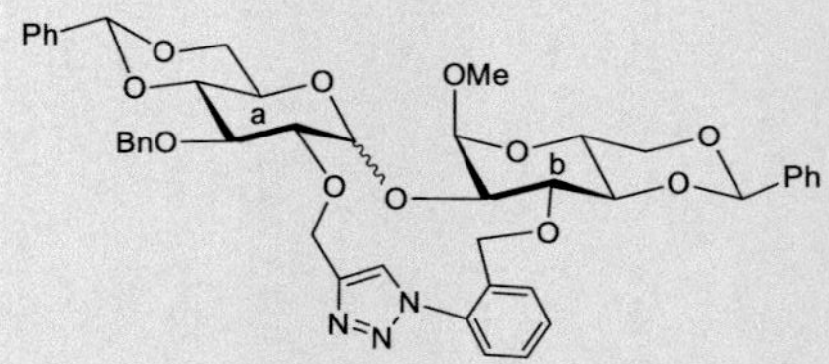 **72l**	Yield: 68%, selectivity: $\alpha/\beta = 8:1$

[a]Intramolecular glycosylation condition: NIS/TMSOTf, anhydrous DCM, inert condition, 0°C-r.t., 2–3 h; Anomeric selectivity was determined by ^{1}HNMR of crude IAD product, and yields referred to isolated after column chromatography (SiO_2).

anomeric) selectivity in this intramolecular glycosylation depends on various factors; however, the more deciding factor is presumably the size of the macrocyclic ring during macrocyclization as part of IAD. Fascinatingly, the diastereofacial factor directed in this IAD is independent of the configuration of the donor, for example, **68α** or **68β**.

Schmidt et al. further extend the protocol by utilizing the *o*-azidobenzyl residue required for the triazole formation to explore it as a rigid spacer in intramolecular glycosylations (Scheme 2.15).[67] Thus click conjugation of *O*-linked *o*-substituted aryl azido-based acceptor (**73a,b**) with glycosyl alkyne-based donor **74** could afford excellent yield of the required donor−triazole spacer−acceptor system **75a,b** (**Step 1**). The similar intramolecular glycosylation of donor−acceptor **75a** and **75b** help the desired IAD and afforded good yield of respective β-(1−3)- and α-(1−2)-linked disaccharide macrocycles **76a** and **76b** (as a part of 14-membered ring formation) with an anticipated selectivity (**Step 2**). In the end, the triazole as a rigid spacer was removed successfully from the macrocyclic **76a,b** under 'Birch Reduction' condition and finally complete the reaction sequence (**Step 3**) (Scheme 2.15A).[67] Due to the restricted conformational mobility and simple interaction between 1,2,3-triazole and phenyl ring, on the whole, affect the lower conformational mobility than 2-azidomethylbenzyl residue. Further, a rational approach for the possible anchimeric assistance in IAD is depicted in Scheme 2.15B. Even though the option about the stabilization of anomeric carbenium ion is part of the six-membered ring through *N*-3 of triazole, it may approach from both ways, that is, α- and/or the β-side to anomeric carbenium ion; however, the 2-*O*-linked (1-phenyl-triazol-4-yl)methyl part is practically nil anchimeric assistance in the IAD. Therefore the electron-donating nature of the triazole *N*-atom is reasonably less important in IAD (Scheme 2.15B).[67]

Despite the exactly same size macrocycle formation with donor−triazole spacer−acceptor system **75a** and **75b**, compound **71k** on IAD gave β(1−3)-linked disaccharide **72k** as major IAD product (α,β, in 1:25), while compound **71l** afforded the α-(1−2)-linked disaccharide **72l** (α/β, in 8:1 ratio) with unpredictably with an opposite selectivity (Scheme 2.15A).[67]

In the end, the rigid spacer in triazole-containing macrocycle disaccharide products **78** and **80** was successfully removed by reacting the macrocycle separately with sodium in the presence of liquid ammonia at −75°C (Birch conditions), and the resulting free disaccharides then subjected to acetylation with Ac_2O/pyridine finally afforded good yields of optically pure disaccharides (**79β** and **81α**) (Scheme 2.15C).[67] The successful removal of rigid spaces, for example, triazole positively supports further application in this direction of the investigation. Also, there is plenty of an opportunity to explore similar series of donor−acceptor systems simply by incorporating suitable monosaccharides including D-gluco-, D-galacto, and D-mannose (in furanose as well pyranose form) and their functional analogs in

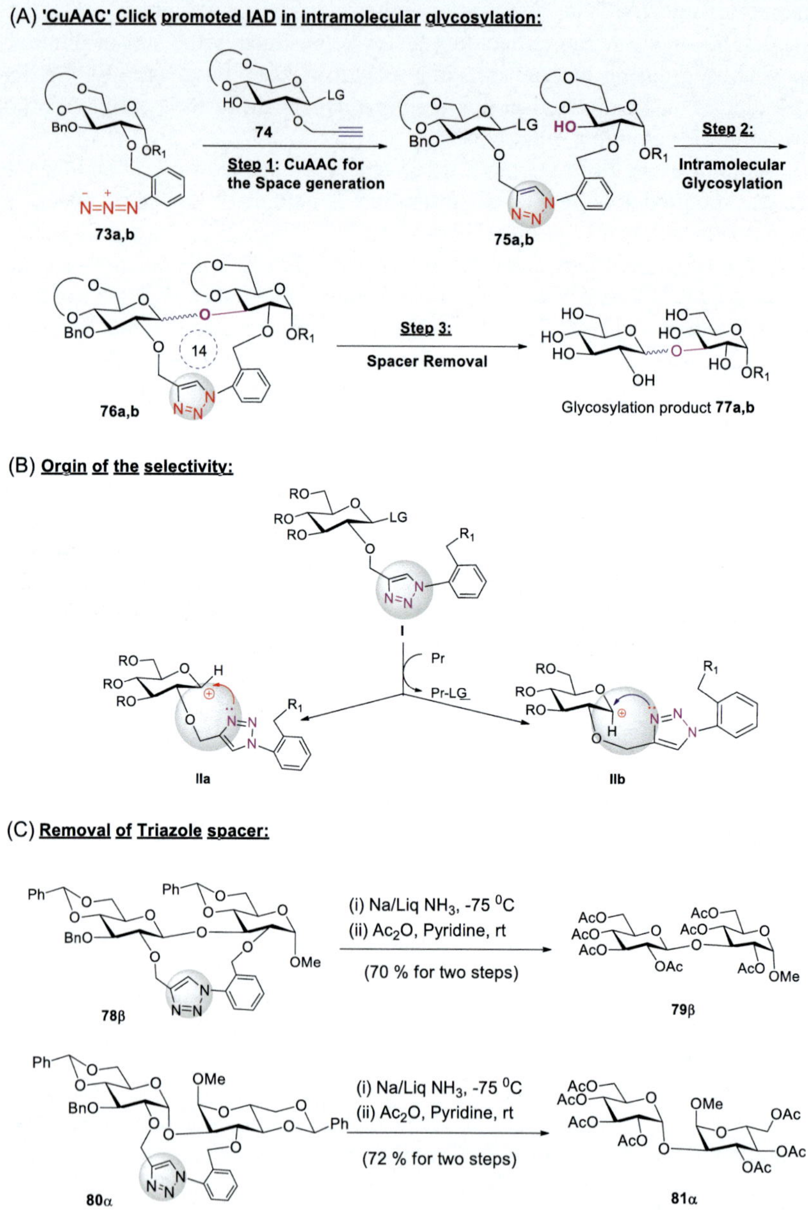

SCHEME 2.15 (A) Alternate route to develop desired rigid spacer required for the intramolecular glycosylation. (B) Triazole in possible anchimeric assistance in IAD. (C) Removal of 1,2,3-triazole spacer from macrocycles obtained through IAD with donor–acceptor system under 'Birch Reduction' condition. (LG = leaving group, Pr = promoter, IAD = intramolecular aglycan delivery).

intramolecular glycosylation for easy access of biologically relevant sugar-containing macrocyclic architectures and the oligosaccharides.

2.3 Intramolecular aglycon delivery concept

Intramolecular aglycon delivery (Scheme 2.1B), commonly known as IAD, was first reported by Barresi et al. for the stereospecific synthesis of β-mannopyranosides and further extended the methodology to the facile synthesis of additional 1,2-*cis*-glycosides.[68,69] The attachment of the glycosyl donor directly with the hydroxyl group of the glycosyl acceptor is the unique characteristic of intramolecular glycosylation through IAD from the other approaches where the acceptor linked away from the hydroxy group that has to be glycosylated. The adjacent position (C-2) next to the glycosyl donor or a distant position (C-6) both can serve as the tethering site with the glycosyl acceptor. Generally, this consists of a two-step process: in the very first step, the acceptor molecule tethered with glycosyl donor via ketal linkages followed by the activation of leaving group led to the formation of glycosylated product in a stereocontrolled manner. In this particular, the reaction of 2-isopropenylmannose donor **82** with glycosyl acceptor **83** and **87** in the presence of TsOH produces ketal **84** and **88** in 45% and 51% yield, respectively (Scheme 2.16). Further, NIS-promoted glycosidation resulted in good yields of respective disaccharides **85** and **89** with complete β-selectivity. Deprotection of compounds **85** and **89** using H_2, Pd/C led to the formation of disaccharides **86** and **90**, respectively. Although relatively moderate yields throughout the ketal production and glycosylation stages, this brilliant idea emerges in the development of the IAD approach as a highly effective

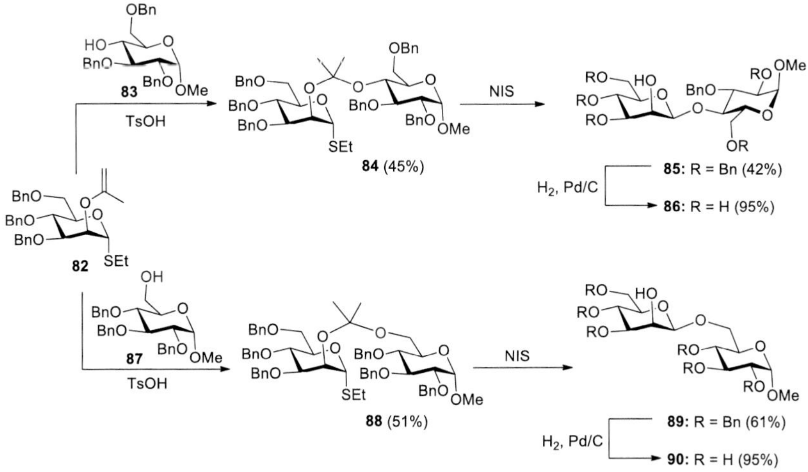

SCHEME 2.16 Intramolecular aglycon delivery through dimethylketal-tethered intermediates.

methodology. As the IAD has been discussed extensively,[16,17,23,70–72] we covered here only the fundamentals and the notable recent advances made in this growing area.

Ennis et al. demonstrated the stereoselective construction of both α-glucosides and β-mannosides in one or two steps by using *N*-iodosuccinimide to achieve both tethering and thioglycoside activation.[73,74] Particularly, NIS-mediated reaction of *manno* **91** and *gluco* donor **95** with galactosyl alcohol **92** at low temperature produces mixed ketal intermediates **93** and **96** in good to excellent yields, respectively (Scheme 2.17). These mixed ketals were next attempted to be intramolecularly glycosylated through NIS in THF, but only modest yields of products were obtained. However, the addition of 2,6-di-tert-butyl-4-methylpyridine (DTBMP) in DCM at room temperature significantly improves the yield of the products in the final glycosylation reaction. Following the above optimized clean intramolecular glycosylation with NIS and DTBMP, stereoselective disaccharides with the structures β-Man-(1→6)-Gal (**94**) and α-Glc-(1→6)-Gal (**97**) were produced in 63% and 86% yield, respectively. As a result, this strategy is very helpful for the stereoselective synthesis of *O*-linked glycopeptides.

Furthermore, an intermediate possessing less-hindered alkylidene acetal was investigated to enhance the effectiveness of the bridged intermediate formation. Fairbanks et al. reported allyl ether as the precursor to enol−ether, which was transformed into the acetal with help of iodonium reagents (Scheme 2.18).[75,76] Thus the vinyl ether **99** was obtained in 96% yield from the corresponding 2-*O*-allyl protected mannosyl donor **98** by the treatment of Wilkinson's catalyst and BuLi in accordance with the linking approach.

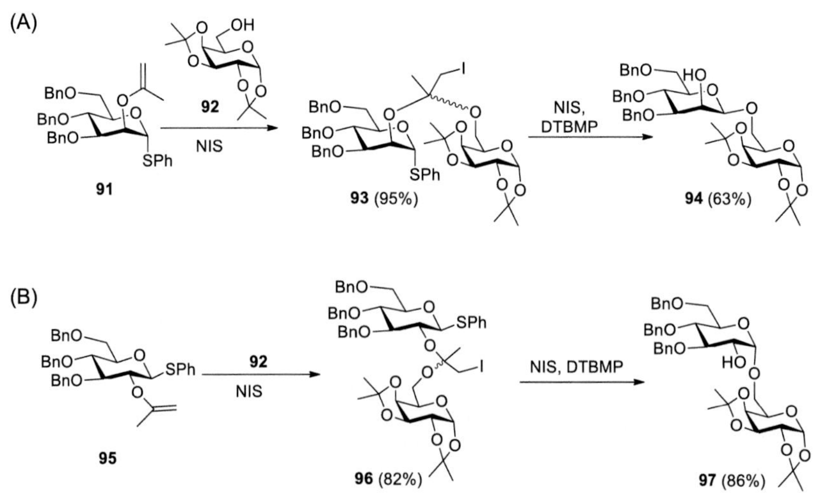

SCHEME 2.17 NIS and DTBMP-mediated IAD of *manno* donor **91** (part A) and *gluco* donor **95** (Part B) through 2-iodomethyl-1-methyl-ketal-tethered intermediates **93** and **96**.

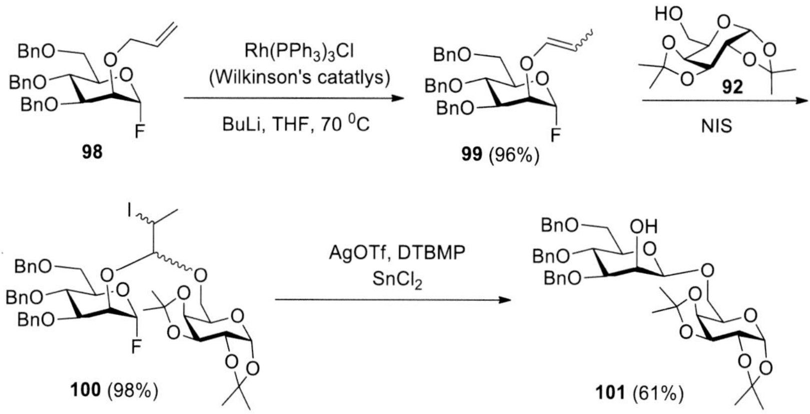

SCHEME 2.18 AgOTf and DTBMP-promoted IAD through 2-iodopropylidene-tethered intermediate **100**.

Afterward, NIS-mediated tethering of **99** with **92** produces mixed acetal 2-iodopropylidene-tethered intermediate **100** in 98% yield. Thereafter, activation of glycosyl fluoride **100** by silver triflate, tin(II) chloride, and 2,6-di-tert-butyl-4-methylpyridine (DTBMP) furnished the desired intramolecularly glycosylated product β-Man-(1→6)-Gal (**101**) in 61% yield. Additionally, some reports documented intramolecular glycosylation through allyl ether-mediated IAD to yield either α-glucosides or β-mannosides accordingly.

Fairbanks et al. also established that a combination of iodonium reagents I_2 and AgOTf in the presence of DTBMP was best for the synthesis of mixed acetal, although the subsequent glycosylation was best achieved by Me_2S_2/Tf_2O and DTBMP.[77–82] Ito et al. employed the PMB group at the C-2 position of the glycosyl donor to achieve tethering mixed acetal with a hydroxy group of the acceptor using DDQ to produce an alternative linker for the stereoselective synthesis of β-mannosides.[83] The synthesis of a number of complex oligosaccharides and glycoconjugates having difficult mannoside architectures was effectively accomplished in the early experiments using this PMB-based IAD approach.[84,85] Additionally, a highly impressive use of the IAD in polymer-supported reactions has been established for the construction of β-manno glycoside.[86] Bertozzi and coworker implemented PMB base linker such as 3,4-dimethoxybenzylidene tethering that worked best in application to the stereoselective synthesis of α,α-linked trehalose derivatives.[87,88]

Krog-Jensen et al. reported stereospecific synthesis of β-fructofuranoside (β-Fruf) **105**, which is present in a capsular polysaccharide from *Haemophilus influenzae*, to apply the PMB-mediated IAD to the 1,2-*cis*-furanosidic linkage.[89,90] The acetal intermediate **104** was obtained from thioglycoside β-Fruf donor **102** by the treatment of DDQ followed by free

3-hydroxymannosamine derivative **103** in dry DCM at ambient temperature. Subsequently, dimethyl(methylthio)sulfonium triflate (DMTST)-mediated activation directed the intramolecular transfer from acetal leading to the formation of β-fructofuranoside (β-Fruf) **105** in 76% yield (Scheme 2.19).

Gelin et al. demonstrated the first IAD onto a D-fucofuranosyl entity for the synthesis of α-D-fucofuranose-containing disaccharides found in the fucopyranose-containing oligosaccharides, produced by *Chaetoceros curvisetus* and *Eubacterium saburreum*.[91] The stereoselective construction of such distinct disaccharide molecules was efficiently achieved by PMB-mediated IAD. In general, the IAD approach involves the introduction of an orthogonal protective group at 2-*O* on the glycosyl donor. On this consideration, the designed fucofuranosyl donor **106** coupled with galactose acceptor **107** in the presence of DDQ to produce the desired intermediate **108** (Scheme 2.20). Finally, mild activation of *n*-pentenyl fucosyl donor by *N*-iodosuccinimide directly accesses the intramolecular transfer from acetal and resulted in the rare α-D-fucofuranose-containing disaccharide **109** in 34% yields.

A number of reports dealing with PMB-mediated IAD for the construction of challenging glycoside linkages have been widely adopted by the synthetic community.[92–94] Similarly in the extension of the PMB-based strategy, the implementation of 2-naphthylmethyl (NAP) group resulted in a significant enhancement in the IAD strategy to afford the desired 1,2-*cis*-glycosides.[95] This modification has made it possible to stereoselectively tether a wider variety of hindered glycosyl acceptors followed by intramolecular transfer producing high yields of product. This method's adaptability comes

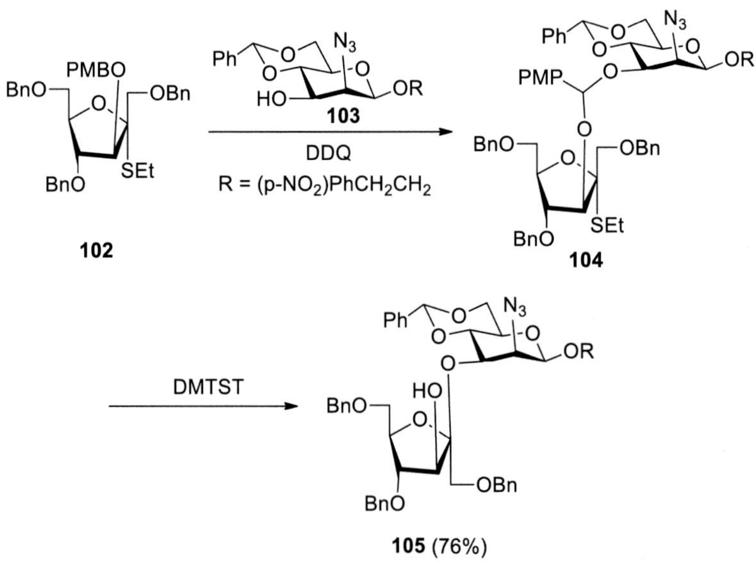

SCHEME 2.19 PMB-mediated IAD for β-fructofuranoside (β-Fruf) **105**.

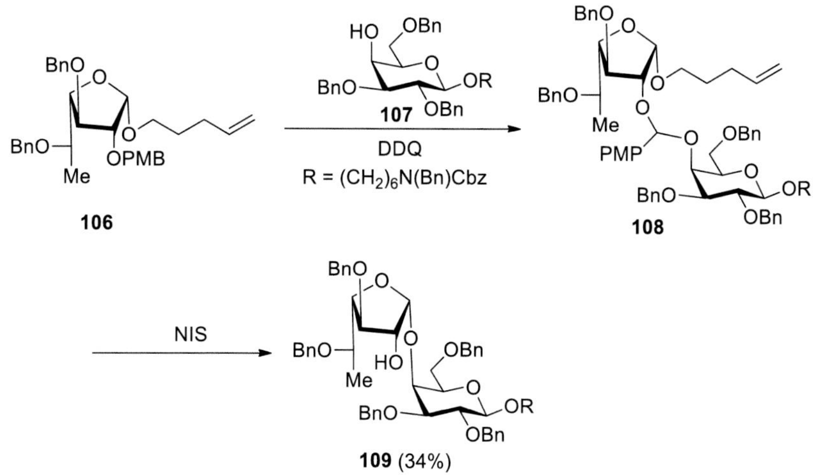

SCHEME 2.20 PMB-mediated IAD for α-D-fucofuranoside (Fucf) **109**.

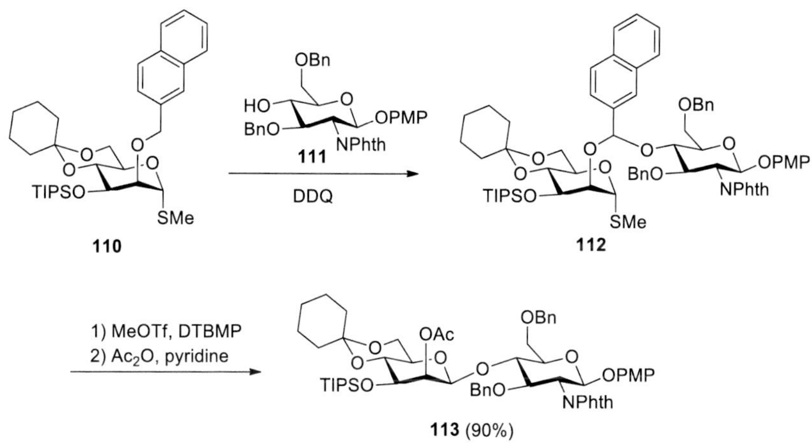

SCHEME 2.21 2-*O*-NAP-mediated IAD for the synthesis of disaccharide **113**.

from the fact that, in comparison to earlier IAD strategy that has already been established, it typically offers considerably higher yields. NAP ether has received great recognition as a hydroxy-protecting group in natural product synthesis.[96] In a model reaction for the synthesis of disaccharide **113** depicted in Scheme 2.21, which clearly proved that the reaction proceeded quantitatively. Thus the formation of mixed acetal **112** was achieved by the addition of DDQ to 2-*O*-NAP-protected thiomannoside **110** and acceptor **111** in 2 hours. Afterward, the mixed acetal **112** can be intramolecularly glycosylated in the presence of MeOTf and DTBMP followed by acetylation to produce disaccharide **113** in 90% yield with complete β-selectivity (Scheme 2.21).[97]

This methodology was initially reported for the stereoselective synthesis of 1,2-*cis*-glycosides such as β-mannosides, α-glucosides, and β-arabinofuranosides and later expanded for construction of β-rhamnosides and many other complexcoupling and targets.[98–107] An improvement in naphthyl methyl-ether-mediated IAD toward the stereoselective construction of β-L-rhamnosides was greatly achieved by Ishiwata and coworkers with various acceptors in good yield.[28,108] The authors highlighted that, in contrast to allyl-mediated IAD, their approach is more effective and rewarding in terms of high yield and excellent selectivity with both primary and secondary acceptors. A representative reaction for the construction of disaccharide **116** is depicted in Scheme 2.22, which clearly illustrated the quantitative nature of the reaction. The oxidative coupling of C2-*O*-naphthyl methyl rhamnosyl donor **114** and Glu-4-OH acceptor **83** in the presence of DDQ successfully produces mixed acetal derivative **115** in 89% yield (Scheme 2.22). The resultant acetal **115** further underwent acetic treatment with MeOTf and DTBMP followed by acetylation to produce the corresponding β-L-rhamnoside **116** in 72% yield as a sole product.

The efficacy of the 2-*O*-NAP methyl-ether tether IAD methodology with the 3-*O*-TMS protective group was later explored by the same group in both a step-by-step and one-pot fashion without the need to purify the mixed acetal intermediate. The yield of naphthylidene acetal **119** was found to be quite comparable in both conditions. However, the advantages of 3-*O*-TMS protecting group come from the in situ synthesis of **119** by capturing the benzylic cation produced in the reaction medium.[109] As a result, 3-*O*-TMS protected donor **117** was coupled with Glu-4-OH acceptor **83** under optimized conditions to afford the naphthylidene acetal **119**, which upon regioselective DIBAL-H reduction produces β-L-rhamnoside **120** in 87% yield

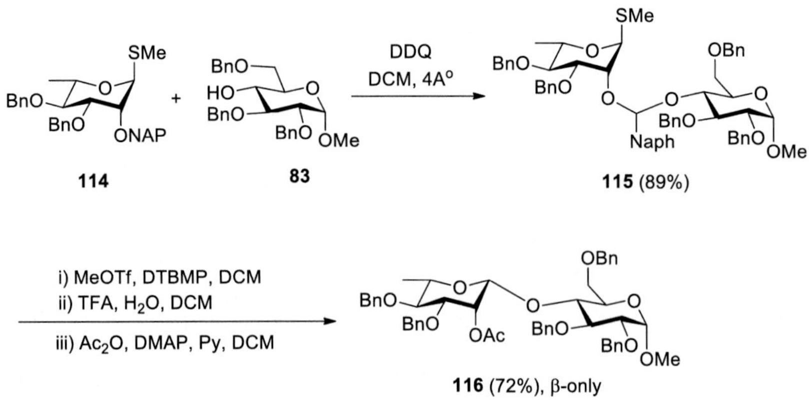

SCHEME 2.22 NAP-mediated IAD for synthesis of β-L-rhamnoside **116**.

(Scheme 2.23). Based on a ^{1}HNMR and nuclear overhauser effect experiment, the stereochemistry of the cyclic acetal carbon was determined to be *endo* as depicted in Fig. 2.1.

The regio- and stereoselective construction of dive0rse molecular architectures has illustrated the adaptability of silicon-tethered intramolecular reactions.[17,110] In this context, Stokes and Bols demonstrated the applicability of silyl acetal as a tether for construction of 1,2-*cis* β-mannosides.[111,112] Taking example, the reaction of phenyl-1-thio-3,4,6-tri-*O*-benzyl-mannoside **121** with chlorodimethylsiloxy derivative **122** produces the requisite silyl acetal **123** quantitatively. Further oxidation of phenyl thio group of **123** gave sulfoxide **124**, which upon activation with triflic anhydride and DTBMP (0.05 M in DCM) afforded β-mannopyranoside **125** in 73% yield (Scheme 2.24).

Similarly, Bols and coworkers expanded the utility of the silyl acetal-based IAD approach for the successful construction of 1,2-*cis*-α-glucoside.[113–115] Additional studies outlined its applicability in the production of stereocontrolled α-glucoside **129** in a two-step process. In this reaction, intramolecular glycosidation is accomplished by carrying out the reaction with aglycon tethered to the C-2 position of the glycosyl donor by silicon linkage. The silyl tethered mixed acetal **128** was obtained by coupling of donor **126** with acceptor **127** (Scheme 2.25). Further, activation of dimethylsilylene tethered disaccharide **128** with NIS afforded the desired α-galactoside **129** in good yield.[116]

The Bols group further explored the utility of this approach to long-range intramolecular glycosylation.[117] In this application, the tether attachment was

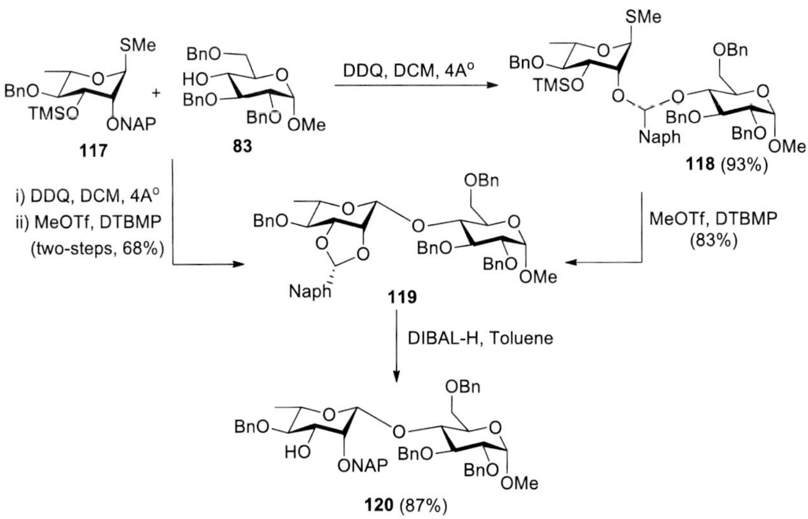

SCHEME 2.23 NAP-mediated IAD using crude mixed acetal for β-L-rhamnoside **120**.

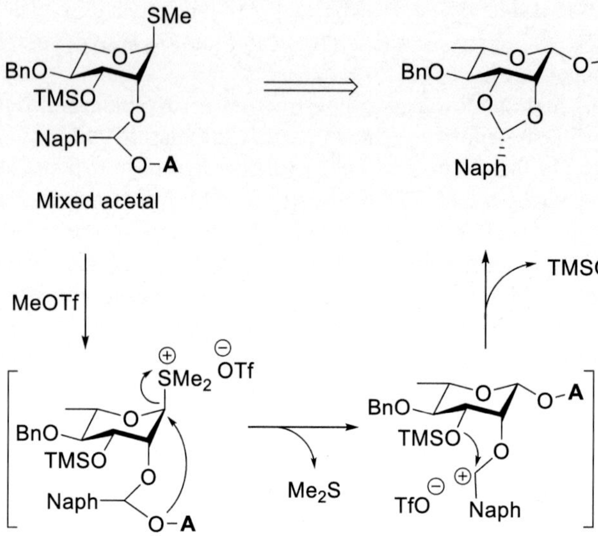

FIGURE 2.1 Possible mechanism of NAP-mediated IAD for β-L-rhamnoside **120**.

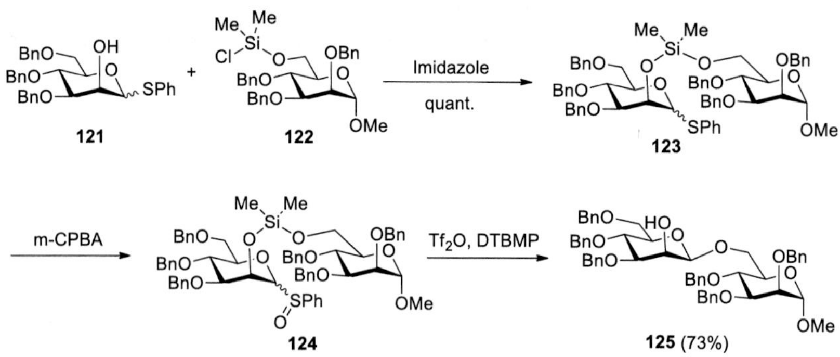

SCHEME 2.24 Silicon-tethered approach for the synthesis of β-mannoside **125**.

positioned away from the anomeric center, providing a more flexible method for producing either 1,2-*cis* or 1,2-*trans* linkages depending on the tether's position. Glycosylation of straightforward acceptor octanol was carried out as proof of the long-range concept study. In particular, the IAD of 3-*O*-tethered glucosyl donor **130** was not completely selective (α:β = 1:4) with a simple octanol acceptor resulting in the formation of a mixture of stereoisomers in the form of octyl glycoside **131**. However, IAD of 5-*O*-tethered ribofuranosyl donor **132** effectively produces the 1,4-*cis* adduct **133** in a completely stereoselective manner (Scheme 2.26).

Furthermore, Montgomery et al. extended the concept of the long-range IAD through silicon tethering based on early investigations by Stork and

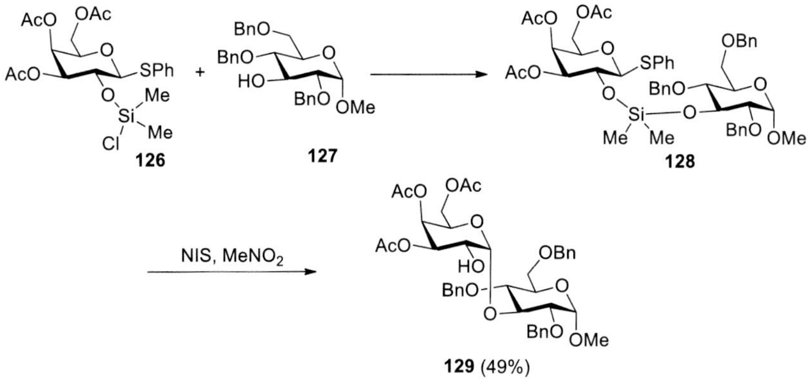

SCHEME 2.25 IAD through silicon-tethered approach for the synthesis of disaccharide **129**.

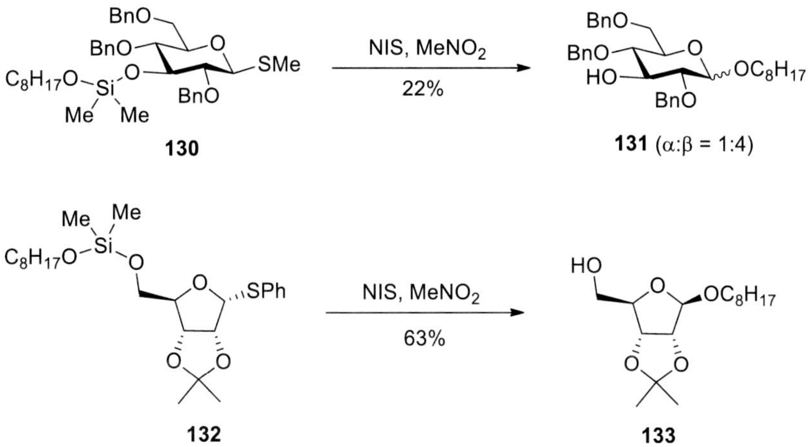

SCHEME 2.26 IAD through long-range silicon-tethered intermediates.

Bols.[118] In a recent report, authors demonstrated the silicon-tethered intramolecular aglycone delivery for stereoselective glycosylation by utilizing the catalytic dehydrogenative coupling of sugar silane with diverse alcohols.[119] The authors protected the 3,4-*trans*-diol with a cyclic *bis*-ketal to prevent the formation of cyclized products. IAD of donor **134** with primary alcohol as acceptors were found to be highly selective and resulted in the formation of diverse β-glycosides in excellent yields. Particularly, with primary glycosyl acceptor **92**, the reaction yield was marginally reduced (41% in two steps) due to the formation of homo-coupling products (Scheme 2.27). Secondary alcohol acceptors were significantly less effective, indicating the high substrate specificity of this methodology. An investigation with other thioglycoside donor series including 2-azido and 2-deoxy sugars produced a similar outcome. This approach exhibited promising stereoselectivity and tolerated a

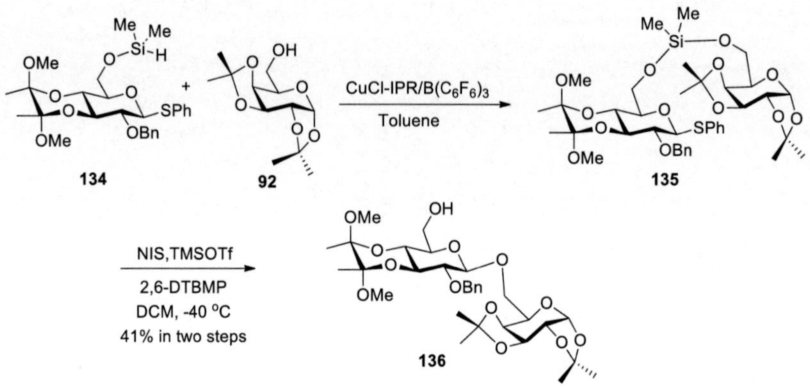

SCHEME 2.27 IAD through long rang silicon-tethered intermediate **135**.

significantly larger variety of acceptors in the construction of glycosidic bonds of various linkages such as β-manno, α-gluco, or β-gluco type.

Nakagawa et al. have successfully reported the glycosyl-acceptor-derived arylboronic ester-mediated aglycon delivery for the regio- and stereoselective synthesis of 1,2-*cis*-α-glycosides under mild conditions.[120] In this report, galacto-configured aryl-boronic ester derivative **137** was reacted with 1,2-anhydro donor **138** to afford the requisite disaccharide **139** in 70% yield (Scheme 2.28). The authors suggested that the reaction proceeds through transition state 1 (**TS-1**) where the oxacarbenium ion is attached to a tetra-coordinated boronate ester. Subsequently, glycosylation from less-hindered boron-oxygen face (C6-bond) resulted exclusively in the desired 1,6-linked disaccharide **139** with complete α-selectivity. However, the glycosylation with acceptors like 3,4-diol of galactose produces α-1,4-link product in 65% yield while 2,3-diol of mannoside afforded selectively 1,3-linked disaccharide in 70% yield. Further, gluco-configured boronic ester **140** underwent glycosylation under similar conditions and produced (1,4)-linked product **141** in 82% yield with high α-selectivity (Scheme 2.28). Again, the less-hindered boron-oxygen face (**TS-2**) which is C4-bond in the case of glucose drives the regioselectivity of the glycosylation reaction. Likewise, other glycosyl acceptors such as mannose, glucosamine, and glucal predominantly form 1,4-linked products in good yields with high α-selectivity.

The synthesis of β-mannoside from a 1,2-anhydromannose donor has been identified as a useful development in the boronic ester-mediated IAD technique.[121] This report describes the β-stereoselective mannosylation of 1,2-anhydromannose donor **144** with monoglycosyl acceptor **142** by utilizing the corresponding glycosyl-acceptor-derived borinic-ester catalyst and thereby its application to the total synthesis of biologically active natural product, Acremomannolipin A (Scheme 2.29). The glycosyl-acceptor-derived borinic ester is generated from corresponding glycosyl acceptors, which

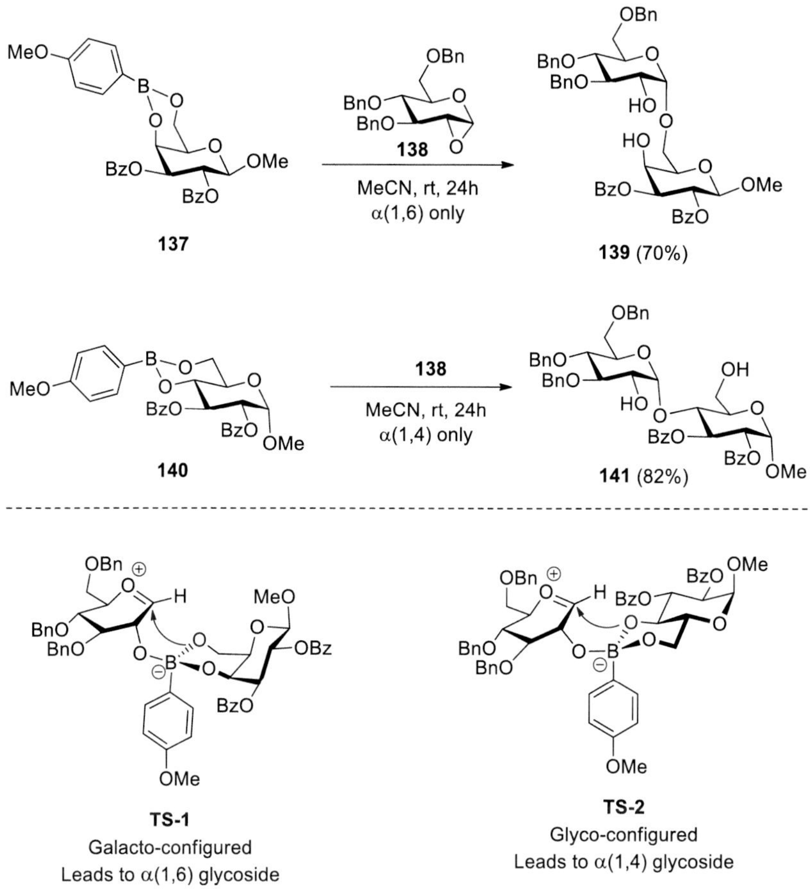

SCHEME 2.28 Boronic ester-mediated IAD with favored transition states deciding the regioselectivity of glycosylation.

activate the donor **144** without any further additive to afford the desired β-mannoside product **145a–c** in excellent yield. The reaction proceeds through tetra-coordinate borinate ester involving oxonium cation moiety which direct the intramolecular attack of a nucleophile to the β-face of the mannosyl donor. Additionally, the synthesis of a tetrasaccharide repeating unit of lipopolysaccharide (LPS) produced from *E. coli-O75* was greatly achieved using this glycosylation technique.[122]

2.4 Leaving group approach

In this approach (Scheme 2.1C), the leaving group of the glycosyl donor is attached to the glycosyl acceptor and as the leaving group is released, the

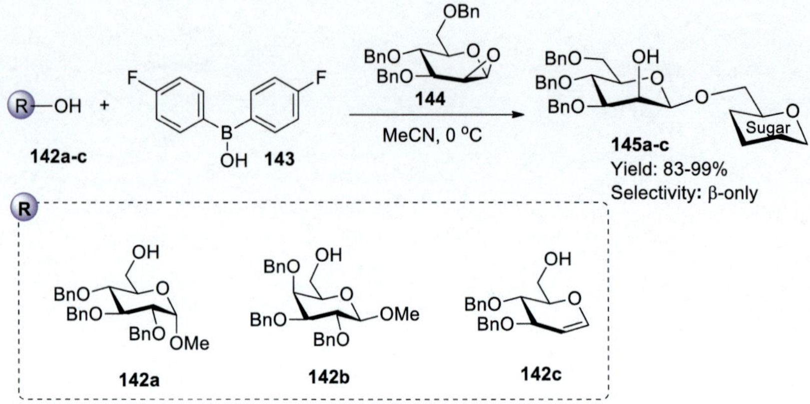

SCHEME 2.29 Aryl borinic-ester-mediated stereoselective synthesis of β-mannosides (**145a–c**) from 1,2-anhydromannose donor **144**.

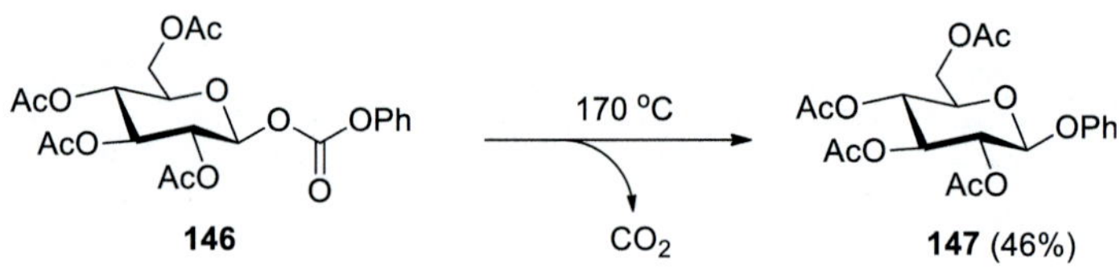

SCHEME 2.30 First example of decarboxylative glycosylation.

accepting atom is transferred to the anomeric carbon. The first example of leaving group-based intramolecular glycosylation was reported by Ishido using thermal decarboxylation of glycosyl carbonates[123] as well as 1,2-orthoester rearrangement by Lindberg[124] and Malysheva.[125] In this report, authors nicely utilize the thermal pyrolysis of aryl carbonates concept of ether formation to hemiacetal moiety such as glycosyloxy substituents. The decarboxylative pyrolysis of 2,3,4,6-tetra-*O*-acetyl-l-*O*-phenoxycarbonyl-β-D-glucopyranose **146** resulted in the formation of phenyl 2,3,4,6-tetra-*O*-acetyl-β-D-glucopyranoside **147** in 46% yield (Scheme 2.30). The process was also effective in the construction of purines and pyrimidine-derived *N*-glycoside. Despite the moderate yield of the reaction, the stereoselectivity was high and produced only β-anomers.

The glycoside bond formation cannot be avoided once the anomeric center and the nucleophile are nearby for bond formation as demonstrated by Schmidt et al. (Scheme 2.31).[126] Hence the most fascinating research on related work includes the design of good examples with a six-membered ring **149** or other convenient ring sizes that surprisingly were so far not investigated.

Laursen and coworkers documented a report of intramolecular glycosylation in which the donor and acceptor were linked via a 3,5-dinitrosalicylic

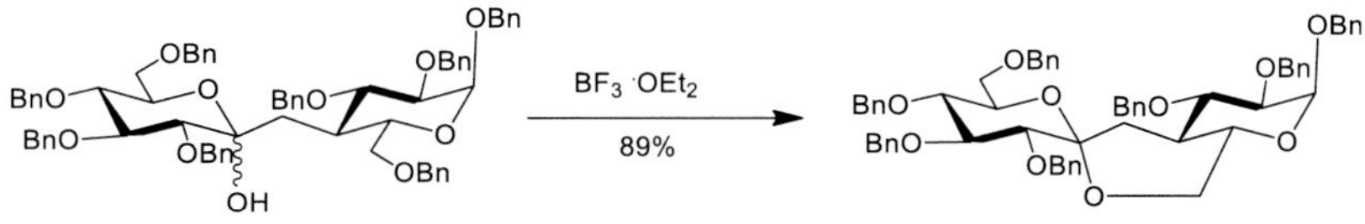

SCHEME 2.31 Schmidt's glycosylation when the anomeric center is close enough to the nucleophile for bond formation.[126]

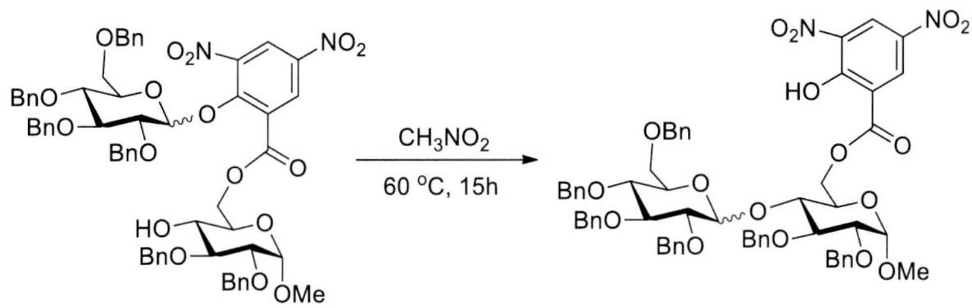

SCHEME 2.32 Methyl 3,5-dinitrosalicylate (DISAL) tethered glycosyl donor **150** favor intramolecular glycosylation.

acid derivative for the successful synthesis of 1,4-linked disaccharides under neutral conditions.[127] The intramolecular glycosylation of glycoside **150**, in which the glycosyl acceptor was attached through an ester bond at the *o*-position of the donor by methyl 3,5-dinitrosalicylate (DISAL) leaving group, showed the best result in nitromethane at 60°C. As a result, the corresponding 1,4-linked glycoside **151** was produced in moderate yield in addition to a low degree of stereoselectivity (Scheme 2.32). The competitive formation of the hemiacetal product due to hydrolysis reduces the yield of the product. Crossover investigation with competing secondary alcohols such as 1,2:5,6-di-*O*-isopropylidene-α-D-glucofuranose and cyclohexanol were tested, in which secondary alcohol of 1,2:5,6-di-*O*-isopropylidene-α-D-glucofuranose showed equivalent in nucleophilicity to the 4-OH of the tethered acceptor **150**. However, reactive cyclohexanol intermolecularly formed cyclohexyl glycoside as a major product compared to intramolecular glycosylation products.

Liu and coworkers established an intramolecular *O*-glycosylation protocol based on leaving group approach by utilizing the 1,2-*trans*-1-dihydroxyboryl benzyl *S*-glycosides donor for the synthesis of *O*-glycosides in moderate yields.[128] The authors suggested that the formation of boronic ester **154** would be a key intermediate if the glycosyl donor **152** is activated in the presence of glycosyl acceptor **153**. Further, an oxygen nucleophile attached to the boronate ester attacked the anomeric center on the opposite side of the sulfonium intermediate **154**, which resulted in the dissociation of the anomeric C–S bond and produced the stereoselective product **155** (Scheme 2.33). Boronic acid-tethered thio-glycoside donor **152** was subjected to activation in the presence of NBS with 3-methyl benzyl alcohol **153** and the result showed good selectivity (α:β = 4.3:1) of 1,2-*cis* glycoside **155**.

The applicability of the decarboxylative technique was further utilized by Ikegami and coworkers for the stereoselective synthesis of β-glycosides.[129] A glucose donor (mixed β-carbonates) that has been *O*-benzoyl protected produced high stereoselectivity of the product. The activation of mixed β-carbonate **156** in presence of catalytic TMSOTf exclusively produces β-disaccharide **157** in 81% yields (Scheme 2.34). To synthesize other β-disaccharides with excellent selectivity, galactose and phthalimido-protected glucosamine donors were efficiently coupled with 3-*O*- or 6-*O*-unprotected acceptors. Interestingly, all the pure β-carbonates were found to be stable at room temperature.

Furthermore, in another report of leaving group-based intramolecular glycoside bond formation, Schmidt and coworkers demonstrated that crowded β-hydroxy-substituted glucosyl carboxylates furnish diastereoselectively α- or β-glucosides with strong alkylating agents in the presence of a base.[130] This is one of the brilliant representative examples of glycoside bond formation for which intramolecularity was proposed. The β-hydroxy-substituted donor **158 was** easily obtained from the corresponding trichloracetimidate

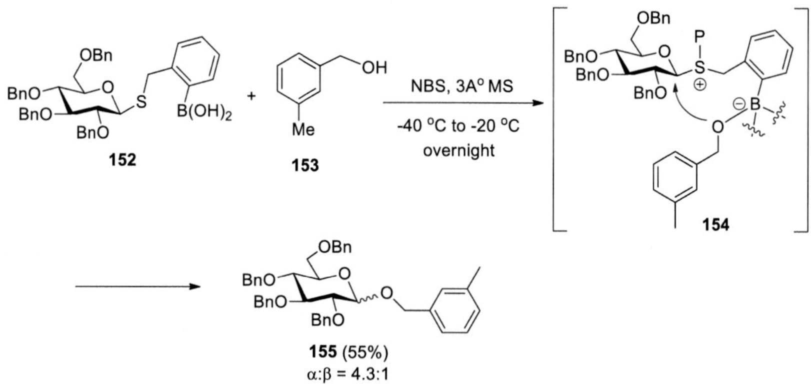

SCHEME 2.33 Boronic acid-derived intramolecular glycosylation.

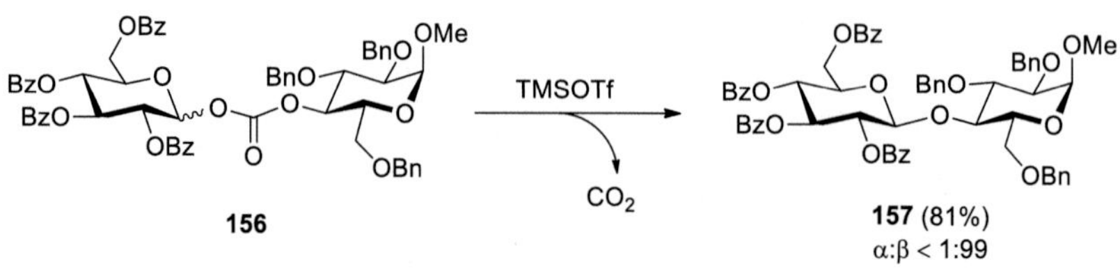

SCHEME 2.34 TMSOTf-promoted stereocontrolled decarboxylative glycosylation.

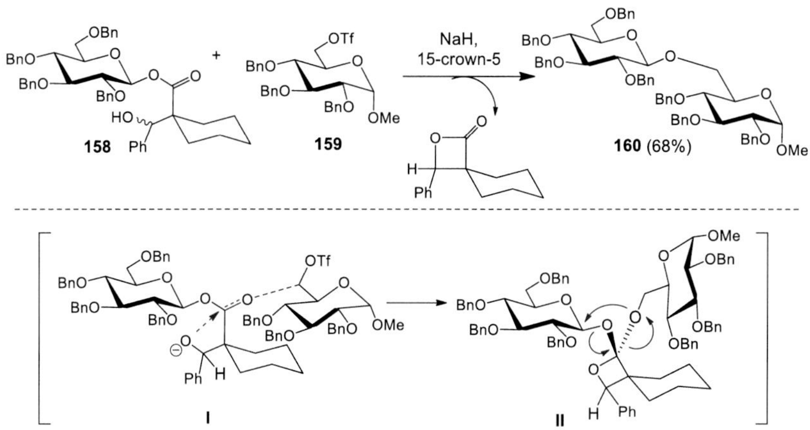

SCHEME 2.35 Stereoselective glycosylation through sterically crowded β-hydroxy-substituted glucosyl carboxylates **158** as donor.

and cyclohexane carboxylic acid followed by an aldol reaction with benzaldehyde. The coupling of **158** with triflate acceptor **159** gave the desired β-disaccharide **160** with high stereoselectivity (β:α = 10:1) (Scheme 2.35). The reaction preceded via possible intermediate **I** and **II**. Consequently, a donor with a neighboring nonparticipating group at 2-*O* achieved a remarkable outcome. However, based on glycosylations, the scope of the β-lactone-formation and its limitations are still unknown as it was only carried out for a few examples. So far, competition experiments have not been carried out to ascertain intramolecularity.

Schmidt et al. reported pentenyl-type activation to investigate an intramolecular glycosylation strategy in which acceptor moiety attached to the leaving group at the glycosyl donor.[131] Initially, 6-*O*-unprotected acceptor **87** was employed in the ester formation with the compound of 2-hydroxymethylbenzoic acid mannoside **161**. Subsequently, Tabbe reduction afforded the desired glycosyl donor **162** having enol–ether linkage in overall high yield (Scheme 2.36). Further, AgOTf/PhSeCl-promoted activation of donor **162** produces exclusively the α-mannoside **163** in 80% yield. In the glucopyranosyl donor set of experiments, regardless of the starting material's configuration, at the low temperature preferentially β-linked disaccharides were obtained.

Beyond early intramolecular glycosylations attained via the orthoester rearrangement by Lindberg[124] followed by Ishido's concept on decarboxylation of glycosyl carbonates,[123] Barresi-Hindsgaul[68] are often credited for the invention of the intramolecular glycosylation. However, it is a pioneering albeit less known research by Kusumoto et al. in 1986 that actually started the developments in this area.[36] The trisaccharides/tetrasaccharides synthesis through intramolecular glycosylation supported by rigid spacers has been

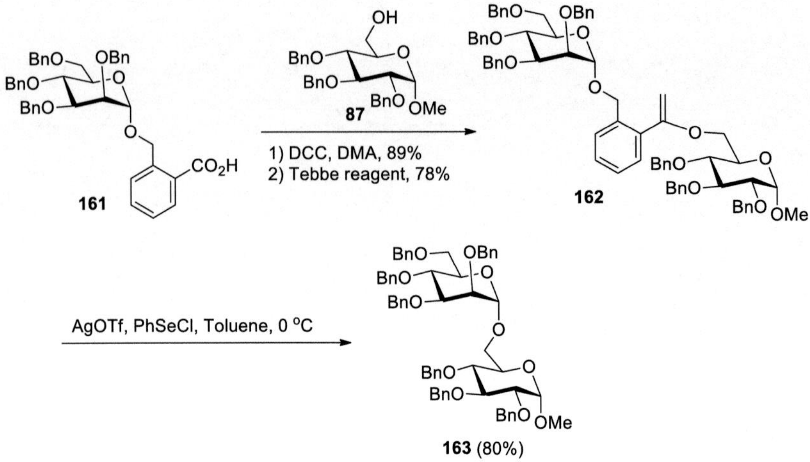

SCHEME 2.36 A pentenyl-type activation for the synthesis of α-mannoside **10**.

reported by Schmidt et al.[53] Treatment of α,α′-dibromo-*m*-xylylene with 6-*O*-unprotected thiomaltoside (as glycosyl donor), followed by 4-*O*-unprotected galactoside derivative (as acceptor) that finally after 3-*O*-MPM deprotection and subsequent intramolecular glycoside bond formation afforded β-linked macrocyclic trisaccharide in high yield. Intramolecular glycosylation of a glycosyl acceptor temporarily tethered to the 2-hydroxyl group of the glycosyl donor may provide a way to control the anomeric stereochemistry for the molecular tethering approaches to *cis*-1,2-glycosides.[132,133] Fairbanks accounts for the optimization, scope, and limitations of allyl IAD on the development of intramolecular glycosylation based on allyl-protecting groups.[71] Synthesis of cyclodextrins was also made successfully by using intramolecular glycosylation.[50]

2.5 Conclusions and future outlook

Over the past two decades, there is a significant improvement in intramolecular glycosylation methodologies to achieve notable stereoselectivity with high reaction yields. The synthesis of challenging macrocyclization products and required tedious glycosidic linkages has been greatly accomplished by utilizing all these well-conventional new tethers, templates, and spacers under optimized reaction conditions. The IAD methodology toward stereoselective construction of 1,2-*cis*-glycosides through various mixed acetal linkages has been broadly investigated. This chapter outlines the noteworthy development in intramolecular glycosylation techniques and their growing impact on stereoselective synthesis of biologically active and complex glycans with diverse functionalities. The resulting glycoconjugates in pure anomeric form has certainly a potential scope in future drug discovery programs.[134] There are more efforts to be required to understand the concepts governing the use of important tethers and templates to adapt them for the successful construction of complex glycoconjugates and also new targets.

Acknowledgments

The authors sincerely acknowledge Science and Engineering Research Board (Grant No.: CRG/2021/002776) and CSIR, New Delhi (Grant No.: 02(0345)/19/EMR-II) for the financial support and the Institute of Eminence (IoE)-BHU for the incentive grant.

References

1. Tiwari, V. K., Ed. *Carbohydrates in Drug Discovery and Development;* Elsevier Inc.: The Netherlands, 2020.
2. Cummings, R. D.; Pierce, J. M. *Chem. Biol.* **2014,** *21*, 1–15.
3. Varki, A. *Glycobiology* **1993,** *3*, 97–130.
4. Bertozzi, C. R.; Kiessling, L. L. *Science* **2001,** *291*, 2357–2364.
5. Dwek, R. A. *Chem. Rev.* **1996,** *96*, 683–720.
6. Chen, X.; Varki, A. *ACS Chem. Biol.* **2010,** *5*, 163–176.
7. Vankar, Y. D.; Schmidt, R. R. *Chem. Soc. Rev.* **2000,** *29*, 201–216.
8. Mishra, S.; Upadhaya, K.; Mishra, K. B.; Shukla, A. K.; Tripathi, R. P.; Tiwari, V. K. *Stud. Nat. Product. Chem.*, 49. Elsevier Inc.: Amsterdam (The Netherlands), 2016307–361.
9. Tiwari, V. K.; Mishra, R. C.; Sharma, A.; Tripathi, R. P. *Mini-Rev. Med. Chem.* **2012,** *12*, 1497–1519.
10. Zhu, X.; Schmidt, R. R. *Angew. Chem., Int. Ed.* **2009,** *48*, 1900–1934.
11. Schmidt, R. R.; Kinzy, W. *Adv. Carbohydr. Chem. Biochem.* **1994,** *50*, 21–123.
12. Aly, M. R. E.; Ibrahim, E. S. I.; Ashry, E. H. E.; Schmidt, R. R. *Carbohydr. Res.* **1999,** *316*, 121–132.
13. Schmidt, R. R.; Castro-Palomino, J. C.; Retz, O. *Pure Appl. Chem.* **1999,** *71*, 729–744.
14. Kulkarni, S. S.; Wang, C.-C.; Sabbavarapu, N. M.; Podilapu, A. R.; Liao, P.-H.; Hung, S.-C. *Chem. Rev.* **2018,** *118*, 8025–8104.
15. Cheng, J.; Yu, H.; Lau, K.; Huang, S.; Chokhawala, H. A.; Li, Y.; Tiwari, V. K.; Chen, X. *Glycobiology* **2008,** *18*, 686–697.
16. Yu, H.; Chen, X. *Org. Biomol. Chem.* **2016,** *14*, 2809–2818.
17. Oligosaccharides in Chemistry and Biology. In A Comprehensive Handbook *Vol. I;* Ernst, B., Hart, G., Sinay, P., Eds.; Wiley-VCH: Weinheim, 2000.
18. Khan, S. H.; O'Neill, R. A., Eds. *Modern Methods in Carbohydrate Chemistry;* Harwood Academic Publishers: Amsterdam, 1996.
19. Hettikankanamalage, A. A.; Lassfolk, R.; Ekholm, F. S.; Leino, R.; Crich, D. *Chem. Rev.* **2020,** *120*, 7104–7151.
20. Adero, P. O.; Amarasekara, H.; Wen, P.; Bohe, L.; Crich, D. *Chem. Rev.* **2018,** *118*, 8242–8284.
21. Bohé, L.; Crich, D. *Carbohydr. Res.* **2015,** *403*, 48–59.
22. Frihed, T. G.; Bols, M.; Pedersen, C. M. *Chem. Rev.* **2015,** *115*, 4963–5013.
23. Crich, D. *Acc. Chem. Res.* **2010,** *43*, 1144–1153.
24. Mydock, L. K.; Demchenko, A. V. *Org. Biomol. Chem.* **2010,** *8*, 497–510.
25. Singh, Y.; Geringer, S. A.; Demchenko, A. V. *Chem. Rev.* **2022,** *122*, 11701–11758.
26. Jung, K. H.; Muller, M.; Schmidt, R. R. *Chem. Rev.* **2000,** *100*, 4423–4442.

27. Ziegler, T. In *Handbook of Chemical Glycosylation;* Demchenko, A. V., Ed.; Wiley-VCH Verlag GmbH & Co: Weinheim, Germany, 2008; p 469.
28. Ishiwata, A.; Lee, Y. J.; Ito, Y. *Org. Biomol. Chem.* **2010,** *8*, 3596–3608.
29. Bols, M.; Skrydstrup, T. *Chem. Rev.* **1995,** *95*, 1253–1277.
30. Nawoj, M.; Grobelny, A.; Mlynarski, J. *Eur. J. Org. Chem.* **2000,** 47–51.
31. Jia, X. G.; Demchenko, A. V. *Beil. J. Org. Chem.* **2017,** *13*, 2028–2048.
32. Ito, Y.; Ohnishi, Y.; Ogawa, T.; Nakahara, Y. *Synlett* **1998,** 1102–1104.
33. Huchel, U.; Schmidt, R. R. *Tetrahedron Lett.* **1998,** *39*, 7693–7694.
34. Cumpstey, I. *Carbohydr. Res.* **2008,** *343*, 1553–1573.
35. Scheffler, G.; Behrendt, M. E.; Schmidt, R. R. *Eur. J. Org. Chem.* **2000,** 3527–3539.
36. Kusumoto, S.; Imoto, M.; Ogiku, T.; Shiba, T. *Bull. Chem. Soc. Jpn.* **1986,** *59*, 1419–1423.
37. Ziegler, T.; Lau, R. *Tetrahedron Lett.* **1995,** *36*, 1417–1420.
38. Mori, M.; Ito, Y.; Ogawa, T. *Tetrahedron Lett.* **1989,** *30*, 1273–1276.
39. Kuyama, H.; Nukada, T.; Ito, Y.; Nakahara, Y.; Ogawa, T. *Carbohydr. Res.* **1995,** *268*, C1–C6.
40. Zhang, Y.; He, H.; Chen, Z.; Huang, Y.; Xiang, G.; Li, P.; Yang, X.; Lu, G.; Xiao, G. *Angew. Chem., Int. Ed.* **2021,** *60*, 12597–12606.
41. Lemanski, G.; Ziegler, T. *Tetrahedron* **2000,** *56*, 563–579.
42. Lemanski, G.; Ziegler, T. *Helv. Chim. Acta* **2000,** *83*, 2655–2675.
43. Wakao, M.; Fukase, K.; Kusumoto, S. *Synlett* **1999,** 1911–1914.
44. Lau, R.; Schüle, G.; Schwanaberg, U.; Ziegler, T. *Liebigs Ann.* **1995,** 1745–1754.
45. Valverde, S.; Gómez, A. M.; Hernández, A.; Herradón, B.; López, J. C. *J. Chem. Soc., Chem. Commun.* **1995,** 2005–2006.
46. Valverde, S.; Gómez, A. M.; López, J. C.; Herradón, B. *Tetrahedron Lett.* **1996,** *37*, 1105–1108.
47. Valverde, S.; García, M.; Gómez, A. M.; López, J. C. *Synlett* **2000,** 22–26.
48. Yamada, H.; Imamura, K.; Takahashi, T. *Tetrahedron Lett.* **1997,** *38*, 391–394.
49. Ogawa, T.; Takahashi, Y. *Carbohydr. Res.* **1985,** *138*, C5–C9.
50. Wakao, M.; Fukase, K.; Kusumoto, S. *J. Org. Chem.* **2002,** *67*, 8182–8190.
51. Claude, P.; Lehmann, C.; Ziegler, T. *Beil. J. Org. Chem.* **2011,** *7*, 1609–1619.
52. Müller, M.; Huchel, U.; Geyer, A.; Schmidt, R. R. *J. Org. Chem.* **1999,** *64*, 6190–6201.
53. Müller, M.; Schmidt, R. R. *Eur. J. Org. Chem.* **2001,** 2055–2066.
54. Ziegler, T.; Lemanski, G.; Hürttlen, J. *Tetrahedron Lett.* **2001,** *42*, 569–572.
55. Paul, S.; Müller, M.; Schmidt, R. R. *Eur. J. Org. Chem.* **2003,** 128–137.
56. Tennant-Eyles, R. J.; Fairbanks, A. J.; Davis, B. G. *Chem. Commun.* **1999,** 1037–1038.
57. Tennant-Eyles, R. J.; Davis, B. G.; Fairbanks, A. J. *Tetrahedron: Asymmetry* **2000,** *11*, 231–243.
58. Tennant-Eyles, R. J.; Davis, B. G.; Fairbanks, A. J. *Tetrahedron: Asymmetry* **2003,** *14*, 1201–1210.
59. Greenwell, D. R.; Ibnouzaki, A. F.; Warriner, S. L. *Angew. Chem., Int. Ed.* **2002,** *41*, 1215–1218.
60. Pornsuriyasak, P.; Jia, X. G.; Kaeothip, S.; Demchenko, A. V. *Org. Lett.* **2016,** *18*, 2316–2319.
61. Jia, X. G.; Pornsuriyasak, P.; Demchenko, A. V. *J. Org. Chem.* **2016,** *81*, 12232–12246.
62. De Meo, C.; Farris, M.; Ginder, N.; Gulley, B.; Priyadarshani, U.; Woods, M. *Eur. J. Org. Chem.* **2008,** 3673–3677.
63. Lin, C.; Jiao, J.; Maisonneuve, S.; Mallétroit, J.; Xie, J. *Chem. Commun.* **2020,** *56*, 3261–3264.
64. Tiwari, V. K.; Kumar, A.; Schmidt, R. R. *Eur. J. Org. Chem.* **2012,** 2945–2956.

65. Tiwari, V. K.; Mishra, B. B.; Mishra, K. B.; Mishra, N.; Singh, A. S.; Chen, X. *Chem. Rev.* **2016,** *116*, 3086–3240.
66. Agrahari, A. K.; Bose, P.; Jaiswal, M. K.; Rajkhowa, S.; Singh, A. S.; Hotha, S.; Mishra, N.; Tiwari, V. K. *Chem. Rev.* **2021,** *121*, 7638–7956.
67. Kumar, A.; Geng, Y.; Schmidt, R. R. *Eur. J. Org. Chem.* **2012,** 6846–6851.
68. Barresi, F.; Hindsgaul, O. *J. Am. Chem. Soc.* **1991,** *113*, 9376–9377.
69. Barresi, F.; Hindsgaul, O. *Synlett* **1992,** 759–761.
70. Barresi, F.; Hindsga, O. *Can. J. Chem.* **1994,** *72*, 1447–1450.
71. Fairbanks, A. J. *Synlett* **2003,** 1945–1958.
72 Fairbanks, Ed. *Glycosylation Through Intramolecular Aglycon Delivery*, Vol., 2; Elsevier Science: Amsterdam, 2021; pp 413–434.
73. Ennis, S. C.; Fairbanks, A. J.; Slinn, C. A.; Tennant-Eyles, R. J.; Yeates, H. S. *Tetrahedron* **2001,** *57*, 4221–4230.
74. Ennis, S. C.; Fairbanks, A. J.; Tennant-Eyles, R. J.; Yeates, H. S. *Synlett* **1999,** 1387–1390.
75. Cumpstey, I.; Fairbanks, A. J.; Redgrave, A. J. *Org. Lett.* **2001,** *3*, 2371–2374.
76. Seward, C. M. P.; Cumpstey, I.; Aloui, M.; Ennis, S. C.; Redgrave, A. J.; Fairbanks, A. J. *Chem. Commun.* **2000,** 1409–1410.
77. Attolino, E.; Cumpstey, I.; Fairbanks, A. J. *Carbohydr. Res.* **2006,** *341*, 1609–1618.
78. Cumpstey, I.; Chayajarus, K.; Fairbanks, A. J.; Redgrave, A. J.; Seward, C. M. P. *Tetrahedron: Asymmetry* **2004,** *15*, 3207–3221.
79. Aloui, M.; Chambers, D. J.; Cumpstey, I.; Fairbanks, A. J.; Redgrave, A. J.; Seward, C. M. P. *Chem. Eur. J.* **2002,** *8*, 2608–2621.
80. Attolino, E.; Fairbanks, A. J. *Tetrahedron Lett.* **2007,** *48*, 3061–3064.
81. Cumpstey, I.; Fairbanks, A. J.; Redgrave, A. J. *Tetrahedron* **2004,** *60*, 9061–9074.
82. Chayajarus, K.; Chambers, D. J.; Chughtai, M. J.; Fairbanks, A. J. *Org. Lett.* **2004,** *6*, 3797–3800.
83. Ito, Y.; Ogawa, T. *Angew. Chem., Int. Ed. Engl.* **1994,** *33*, 1765–1767.
84. Dan, A.; Ito, Y.; Ogawa, T. *J. Org. Chem.* **1995,** *60*, 4680–4681.
85. Dan, A.; Lergenmüller, M.; Amano, M.; Nakahara, Y.; Ogawa, T.; Ito, Y. *Chem. Eur. J.* **1998,** *4*, 2182–2190.
86. Ito, Y.; Ogawa, T. *J. Am. Chem. Soc.* **1997,** *119*, 5562–5566.
87. Pratt, M. R.; Leigh, C. D.; Bertozzi, C. R. *Org. Lett.* **2003,** *5*, 3185–3188.
88. Leigh, C. D.; Bertozzi, C. R. *J. Org. Chem.* **2008,** *73*, 1008–1017.
89. Krog-Jensen, C.; Oscarson, S. *J. Org. Chem.* **1996,** *61*, 4512–4513.
90. Krog-Jensen, C.; Oscarson, S. *J. Org. Chem.* **1998,** *63*, 1780–1784.
91. Gelin, M.; Ferrieres, V.; Lefeuvre, M.; Plusquellec, D. *Eur. J. Org. Chem.* **2003,** 1285–1293.
92. Gannedi, V.; Ali, A.; Singh, P. P.; Vishwakarma, R. A. *Tetrahedron Lett.* **2014,** *55*, 2945–2947.
93. Bernlind, C.; Homans, S. W.; Field, R. A. *Tetrahedron Lett.* **2009,** *50*, 3397–3399.
94. Desire, J.; Prandi, J. *Carbohydr. Res.* **1999,** *317*, 110–118.
95. Kaeothip, S.; Ishiwata, A.; Ito, T.; Fushinobu, S.; Fujita, K.; Ito, Y. *Carbohydr. Res.* **2013,** *382*, 95–100.
96. Gaunt, M. J.; Yu, J.; Spencer, J. B. *J. Org. Chem.* **1998,** *63*, 4172–4173.
97. Ishiwata, A.; Munemura, Y.; Ito, Y. *Eur. J. Org. Chem.* **2008,** 4250–4263.
98. Kenfack, M. T.; Bleriot, Y.; Gauthier, C. *J. Org. Chem.* **2014,** *79*, 4615–4634.
99. Rai, D.; Kulkarni, S. S. *Org. Biomol. Chem.* **2020,** *18*, 3216–3228.
100. Ishiwata, A.; Kaeothip, S.; Takeda, Y.; Ito, Y. *Angew. Chem., Int. Ed.* **2014,** *53*, 9812–9816.

101. Ishiwata, A.; Ito, Y. *Trends Glycosci. Glycotechnol.* **2009,** *21*, 266–289.
102. Ishiwata, A.; Sakurai, A.; Nishimiya, Y.; Tsuda, S.; Ito, Y. *J. Am. Chem. Soc.* **2011,** *133*, 19524–19535.
103. Ishiwata, A.; Fujita, K.; Fushinobu, S.; Tanaka, K.; Ito, Y. *Bioorg. Med. Chem.* **2022,** *68*, 116849.
104. Kaeothip, S.; Ishiwata, A.; Ito, Y. *Org. Biomol. Chem.* **2013,** *11*, 5892–5907.
105. Ishiwata, A.; Ito, Y. *J. Am. Chem. Soc.* **2011,** *133*, 2275–2291.
106. Ishiwata, A.; Ito, Y. Intramolecular Aglycon Delivery toward 1,2-cis Selective Glycosylation. In *Selective Glycosylations: Synthetic Methods and Catalysts;* Bennett, C. S., Ed.; Wiley-VCH, 2017; pp 81–96.
107. Ishiwata, A.; Tanaka, K.; Ao, J.; Ding, F.; Ito, Y. *Front. Chem.* **2022,** *10*, 972429.
108. Lee, Y. J.; Ishiwata, A.; Ito, Y. *J. Am. Chem. Soc.* **2008,** *130*, 6330–6331.
109. Noyori, R.; Murata, S.; Suzuki, M. *Tetrahedron Lett.* **1981,** *37*, 3899–3910.
110. Stork, G.; Suh, H.; Kim, G. *J. Am. Chem. Soc.* **1991,** *113*, 7054–7055.
111. Stork, G.; Kim, G. *J. Am. Chem. Soc.* **1992,** *114*, 1087–1088.
112. Stork, G.; La Clair, J. L. *J. Am. Chem. Soc.* **1996,** *118*, 247–248.
113. Bols, M. *J. Chem. Soc., Chem. Commun.* **1992,** 913–914.
114. Bols, M. *J. Chem. Soc., Chem. Commun.* **1993,** 791–792.
115. Bols, M. *Acta Chem. Scand.* **1996,** *50*, 931–937.
116. Bols, M. *Tetrahedron* **1993,** *49*, 10049–10060.
117. Bols, M.; Hansen, H. C. *Chem. Lett.* **1994,** 1049–1052.
118. Buchan, Z. A.; Bader, S. J.; Montgomery, J. *Angew. Chem., Int. Ed.* **2009,** *48*, 4840–4844.
119. Walk, J. T.; Buchan, Z. A.; Montgomery, J. *Chem. Sci.* **2015,** *6*, 3448–3453.
120. Nakagawa, A.; Tanaka, M.; Hanamura, S.; Takahashi, D.; Toshima, K. *Angew. Chem., Int. Ed.* **2015,** *54*, 10935–10939.
121. Tanaka, M.; Nashida, J.; Takahashi, D.; Toshima, K. *Org. Lett.* **2016,** *18*, 2288–2291.
122. Nishi, N.; Nashida, J.; Kaji, E.; Takahashi, D.; Toshima, K. *Chem. Commun.* **2017,** *53*, 3018–3021.
123. Inaba, S.; Yamada, M.; Yoshino, T.; Ishido, Y. *J. Am. Chem. Soc.* **1973,** *95*, 2062–2063.
124. Höök, J. E.; Lindberg, B. *Acta Chem. Scand.* **1968,** *22*, 2157–2160.
125. Malysheva, N. N.; Torgov, V. I.; Klimov, E. M.; Kochetkov, N. K. *Izv. Akad. Nauk, SSSR, Ser. Khim.* **1974,** 2153–2155.
126. Preuss, R.; Jung, K. H.; Schmidt, R. R. *Liebigs Ann. Chem.* **1992,** 377–382.
127. Laursen, J. B.; Petersen, L.; Jensen, K. J. *Org. Lett.* **2001,** *3*, 687–690.
128. Liu, X.; Zhang, B.; Gu, X.; Chen, G.; Chen, L.; Wang, X.; Xiong, B.; You, Q.-D.; Chen, Y.-L.; Shen, J. *Carbohydr. Res.* **2014,** *398*, 45–49.
129. Azumaya, I.; Niwa, T.; Kotani, M.; Iimori, T.; Ikegami, S. *Tetrahedron Lett.* **1999,** *40*, 4683–4686.
130. Behrendt, M. E.; Schmidt, R. R. *Tetrahedron Lett.* **1993,** *34*, 6733–6736.
131. Scheffler, G.; Schmidt, R. R. *J. Org. Chem.* **1999,** *64*, 1319–1325.
132. Demchenko, A. V. *Curr. Org. Chem.* **2003,** *7*, 35–79.
133. Demchenko, A. V. *Synlett* **2003,** 1225–1240.
134. Mishra, N.; Tiwari, V. K.; Schmidt, R. R. Recent Trends and Challenges on Carbohydrate-based Molecular Scaffolding: General Consideration toward impact of carbohydrates in drug discovery and development. In *Carbohydrates in Drug Discovery and development;* Tiwari, V. K., Ed.; Elsevier Inc., 2020; pp 1–69.

Chapter 3

Thioglycoside-based glycosylations in oligosaccharide synthesis

Nitish Verma, Maude Cloutier and Charles Gauthier
Unité mixte de recherche (UMR) INRS-UQAC, Centre Armand-Frappier Santé Biotechnologie, Institut national de la recherche scientifique (INRS), Laval & Chicoutimi (Qc), Canada

3.1 Introduction

Besides traditional roles as energy sources and structural polymers, carbohydrates are involved in a plethora of crucial cellular events such as mediating cell–cell interactions, altering protein conformation, preventing protein degradation and aggregation, and modulating the pharmacokinetic properties of natural compounds.[1] Due to their diversity and importance as fundamental building blocks, it is mandatory to access them in useful quantities for furthering pharmacological and clinical studies.[2–4] However, carbohydrates and their conjugates are commonly found in microheterogeneous forms and their purification is often tedious from natural sources, which represent major hurdles for yielding them in sufficient purity and quantity. Over the past decades, significant research efforts have been undertaken by organic chemists around the world for implementing carbohydrate synthesis through both chemical[3] and enzymatic[5] approaches.

Regio- and stereoselective glycosylations stand as major challenges in the field of oligosaccharide synthesis.[6–8] In a regioselective glycosylation, a glycosyl donor is linked to a specific hydroxyl group of the glycosyl acceptor through the formation of a glycosidic bond. Because of the presence of multiple hydroxyl groups in the carbohydrate scaffold, masking the other hydroxyl groups with protecting groups has become a useful approach for targeting the glycosylation reaction at a specific position. To this end, a wide and diverse range of orthogonal protecting groups are now available for carbohydrate chemists, enabling the synthesis of suitably protected building blocks of different reactivities.[4,6,9,10] A second challenge is the construction of stereoselective glycosidic bonds.[11–14] Theoretically, the formation of two stereoisomers, that

Synthetic Strategies in Carbohydrate Chemistry. DOI: https://doi.org/10.1016/B978-0-323-91729-2.00005-7

is, α- and β-glycosides, is possible at the anomeric center (the so-called anomers). An ester-protecting group at the OH-2 position of a glycosyl donor is usually employed to achieve 1,2-*trans* stereoselective glycosylations through neighboring group participation (also called anchimeric assistance). On the other hand, solvent effects[15] in addition to temperature,[16] order of addition of reagents (preactivation),[17,18] additives,[19–21] and remote effect of protecting groups[8,22–24] are some of the most important reaction features to have been studied for enabling 1,2-*cis* stereoselective glycosylations.

Over the last two decades, different types of leaving groups have been developed at the anomeric center of glycosyl donors, among which thioaglycons,[2,25–27] trichloroacetimidates (TCAI),[28–30] *N*-phenyl-2,2,2-trifluoroacetimidates (PTFAI),[31,32] sulfoxides,[33–35] halides,[36–38] and *ortho*-alkynylbenzoates (OABz)[39–41] are the most commonly employed.[42] Among these, thioglycosides remain one of the most popular glycosyl donors in the literature. This is due to their ease of formation as well as their stability under both acidic and basic conditions together with their straightforward activation with soft electrophiles. Moreover, thioglycosides can exhibit dual functions acting as both protecting and leaving groups, which allows their use as glycosyl donors without the need of additional steps for activation.

In this chapter, we first aim to discuss some stepwise and one-pot strategies, which were reported for the preparation of thioglycoside building blocks. Additionally, we provide a discussion on the different promoters known to activate thioglycoside donors. Thioglycoside-based glycosylations have constantly attracted the attention of carbohydrate chemists. In consequence, several literature reviews, which are focused on thioglycosides, were previously published.[2,26,43] Herein, in addition to well-known traditional methods, we provide a comprehensive account for some recently reported examples of thioglycoside-based glycosylation strategies such as chemoselective, orthogonal, and preactivation-based approaches for the synthesis of various biologically relevant oligosaccharides. Mechanistic discussions on thioglycoside-based glycosylations, the use of thioglycosides for automated solid-phase oligosaccharides synthesis, and the introduction of thioglycosides as glycomimetics are excluded of our chapter content.

3.2 Preparation of thioglycosides

Many methods have been reported for the synthesis of thioglycoside building blocks. One of the most commonly used approaches relies on the use of peracetylated sugars as starting material. A list of conditions used for the synthesis of thioglycosides from peracetylated sugars is presented in Table 3.1. In the 1950s Lemieux et al. performed the mercaptolysis of glucose and galactose pentaacetates using $ZnCl_2$.[44] Their work highlighted that while α-anomers were resistant to the mercaptolysis, β-pentaacetates were rapidly converted to the corresponding ethyl tetraacetyl-1-thio-β-D-glycosides in

TABLE 3.1 Reported methods for the synthesis of thioglycosides from peracetylated sugars.

Conditions	References
$ZnCl_2$, RSH[a]	44,45
$BF_3 \cdot Et_2O$, RSH[b]	46,47
TMSOTf, $EtSSnBu_3$	48
$BF_3 \cdot Et_2O$, TMS-SMe	49
TMSOTf, TMS-SMe	49
$ZrCl_4$, RSH[c]	50,51
$FeCl_3$, RSH[d]	52
TMSOTf, dicyclohexylmethanethiol	53
ZnI_2, TMS-SMe	54
I_2, TMS-SMe	55
I_2, Me_2S_2-HMDS	55
I_2, EtSH	55
$TiCl_4$, $InCl_3$, RSH[e]	56
$SnCl_4$, PhSH	57
MoO_2Cl_2, RSH[f]	58
(i) I_2, Et_3SiH, (ii) thiourea, (iii) RI,[g] Et_3N	59
$Fe_{(s)}$, I_2, RSH[h]	60
$In(OTf)_3$, PhSH, ball mill	61
TfOH, RSH[i]	62
$InCl_3$, RSH[j]	63

[a] *R = Tol, Et.*
[b] *R = Ph, Bn, Et, Me.*
[c] *R = Tol, Et, Py.*
[d] *R = Et, Ph.*
[e] *R = Tol, Ph, Et.*
[f] *R = Tol, Ph, 4-PhCl, 4-PhOMe, 2-NAP, Et, Bn, cyclohexyl, t-Bu, $HS(CH_2)_3$, $(CH_2)_{10}CO_2Me$, $(CH_2)_2OTBS$.*
[g] *R = Et, $CH_2(CH_2)_8CH=CH_2$, Me, Ph.*
[h] *R = Tol, Et, Bn, cyclohexyl, t-Bu, $(CH_2)_{10}CO_2Me$, Ph, 4-PhOMe, Tol, 4-PhCl, 2-NAP.*
[i] *R = Tol, Ph, Et, Taz.*
[j] *Solvent-free conditions, R = Ph, Tol, 4-PhCl, cyclohexyl.*

good yields, which was explained through the formation of cyclic carbonium ion in case of β-pentaacetates by the involvement of neighboring C2-acetoxy. Nowadays, the most commonly used method entails the activation of acetate donors in the stoichiometric presence of $BF_3 \cdot Et_2O$ (used as a Lewis acid) and thiols, for example, ethanethiol, thiophenol, and *p*-thiocresol.[46,47] This method is quite reliable, however, with the drawback of harsh acidic conditions. It is thus commonly used in the initial steps of a synthetic scheme for the preparation of glycosyl building blocks, followed by the installation of protecting groups, which otherwise may not survive in an acidic medium. Alternatively, Ogawa used TMSOTf in combination with ethyl tributylstannyl sulfide ($EtSSnBu_3$) for the synthesis of thioglycosides from peracetylated sugars.[48] In addition to that, methyl 1,2-*trans*-1-thioglycopyranosides were synthesized by the reaction of per-*O*-acetylated glycopyranoses in the presence of methylthiotrimethylsilane and Lewis acids such as BF_3 and TMSOTf. Metal-based Lewis acids have also been exploited to synthesize thioglycosides. For instance, easy-to-handle $ZrCl_4$ can lead to alkyl 1,2-*trans*- and 1,2-*cis*-related thioglycosides by only modulating the reaction temperature.[50,51] On the other hand, $FeCl_3$ was used as a cheaper alternative to obtain thioglycosides.[52] The Boons group used stoichiometric amounts of TMSOTf (1.1 equiv) to synthesize thioglycosides, which are protected with sterically hindered dicyclohexylmethanethiol at the anomeric center and explored their ability to decrease aglycon transfer during chemoselective glycosylations.[53]

Because free thiols are usually pungent in odor, less smelly alternatives were therefore developed in which a silylated thiol was used in combination with I_2 as a cheap and easy-to-handle Lewis acid.[55] The decreased nucleophilicity of the sulfur functionality in silylated thiol as compared to free thiol minimizes the occurrence of side reactions such as acid-catalyzed epimerizations and thus usually improves yields and stereoselectivities. Furthermore, $InCl_3$ when used in combination with $TiCl_4$ led to short reaction times and better yields in addition to being readily removed after reaction during the aqueous work-up.[56] In contrast, an exclusive β-product was obtained with $SnCl_4$.[57] MoO_2Cl_2 was found to be the most reactive catalyst, among the various oxometallic species examined by the authors, for the thioglycosylation of peracetylated glycosides.[58] In addition, thiouronium intermediates, formed from in situ-generated glycosyl iodides, were also employed to provide 1,2-*trans*-thioglycosides predominantly.[59] Interestingly, iron(III) iodide was in situ-generated from elemental iodine and iron, enabling the catalysis of thioglycosylation through a simplified *modus operandi*.[60] Among sulfonic acids, TMSOTf,[49,53] triflic acid (TfOH),[62] and $In(OTf)_3$[61] have been employed, and excellent reaction yields were observed in the stoichiometric presence of TfOH, as studied by Demchenko and coworkers.[62] A novel method for the synthesis of thioglycosides containing various thioaglycons (Ph, Tol, 4-PhCl, and cyclohexyl) was recently reported,[63] in which a catalytic amount of readily available $InCl_3$ was used under solvent-free conditions while heating between 70°C and 140°C.

3.2.1 One-pot synthesis of thioglycosides from unprotected sugars

Alternative methods allowing for the formation of thioglycosides from unprotected sugars through one-pot protocols have been reported. In these processes, both the acetylation and thioglycosylation reactions were performed in the same flask without purification of the acetylated intermediates (Table 3.2). In Field's one-pot method,[64] acetylation of the free reducing sugar was first performed in the stoichiometric presence of Ac_2O and catalytic I_2 under solvent-free conditions. This was followed by the in situ formation of glycosidic *iodo* intermediates through the addition of I_2 and hexamethyldisilane (HMDS), which were reacted with disulfide to synthesize various thioglycosides. In another approach, the one-pot synthesis of alkyl/aryl thioglycosides from unprotected reducing sugars was performed. A stoichiometric quantity of Ac_2O was used in the presence of $BF_3 \cdot Et_2O$ to allow acetylation, followed by the addition of alkyl/aryl thiols.[66] Moreover, the thioglycosylation of unpurified peracetylated glycosides, synthesized by using readily available H_2SO_4 or TsOH, was also explored by using $SnCl_4$ and $BF_3 \cdot Et_2O$ with the addition of

TABLE 3.2 Reported methods for the synthesis of thioglycosides from unprotected sugars by one-pot reactions.

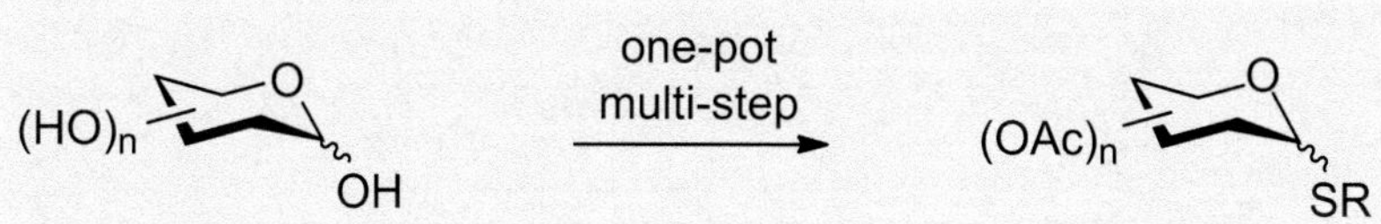

Conditions	References
(i) Ac_2O, I_2; (ii) I_2, HMDS; (iii) RSSR	64
(i) Ac_2O, $Cu(OTf)_2$; (ii) thiourea; (iii) RX, Et_3N	65
(i) Ac_2O, $BF_3 \cdot Et_2O$; (ii) RX	66
(i) Ac_2O, $La(OTf)_3$; (ii) HSR, $BF_3 \cdot Et_2O$	67
(i) Ac_2O, H_2SO_4; (ii) TolSH, $BF_3 \cdot Et_2O$ (or $SnCl_4$)	68
(i) Ac_2O, $SnCl_4$; (ii) TolSH	69
(i) Ac_2O, $In(OTf)_3$, ball mill; (ii) PhSH, $In(OTf)_3$, ball mill	61
(i) Ac_2O, $Cu(ClO_4)_2 \cdot 6H_2O$; (ii) TolSH, $BF_3 \cdot Et_2O$	70
(i) Ac_2O, $Mg(OTf)_2$; (ii) RSH, $BF_3 \cdot Et_2O$	71

p-thiocresol.[68] Notably, a sialyl thioglycoside donor was synthesized via this approach. Various other strategies, such as *S*-glycosyl isothiouronium salts,[65] $La(OTf)_3$,[67] $In(OTf)_3$,[61] and $Mg(OTf)_3$,[71] have also been disclosed. Other approaches involving diversely functionalized substrates as starting material such as propargyl *O*-glycosides,[72] glycals,[73] glycosyl bromides,[74] and hemiacetals have also been developed.[75]

3.3 Activation of thioglycoside donors with various promoters

Synthesized thioglycosides can be directly used as versatile donors to perform glycosylations of sugar- as well as nonsugar-derived acceptors. Thioglycosides can be activated by various reagent systems, such as thiophilic metal salts, organosulfur reagents, and halonium ions, as well as by single electron transfer reactions (Scheme 3.1 and Table 3.3).

Initial studies to activate thioglycosides were performed in the presence of thiophilic mercury salts [$HgSO_4$ and $Hg(OAc)_2$], due to their soft acid/soft base mode of interaction with thioglycosides.[25] In fact, a disaccharide was synthesized by glycosylation between a perbenzylated thioglucoside donor and a galactose-based acceptor in the presence of $HgSO_4$ as the promoter. Harsh reaction conditions are usually needed to activate thioglycoside donors with mercuric salts. Therefore these promoters were used for the glycosylation of very reactive (armed) thioglycoside donors. In these seminal works, the mercury-based promoters were used in stoichiometric amounts, which

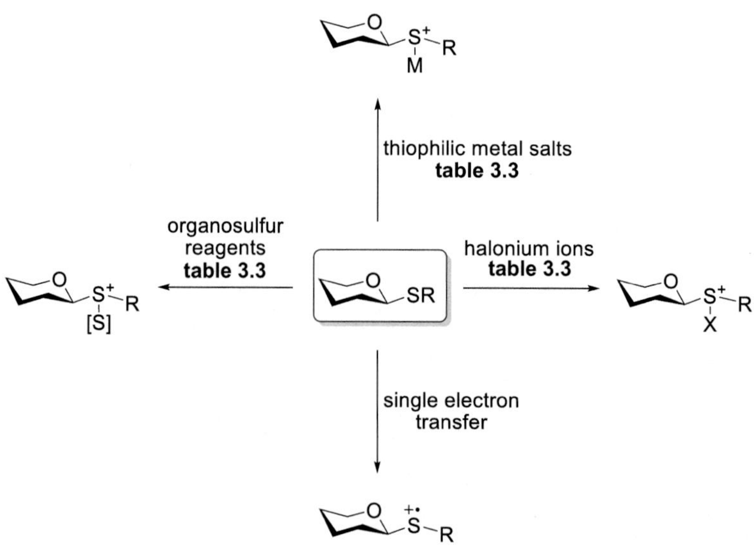

SCHEME 3.1 Different modes of activation of thioglycosides.

TABLE 3.3 List of various promoters for the activation of thioglycoside donors.

Conditions	References
$HgSO_4$ or $Hg(OAc)_2$	25
$Hg(OBz)_2$	76
PhHgOTf	77
N-Bromosuccinimide (NBS)	78
$HgCl_2$	79
MeOTf	80
$CuBr_2$, Bu_4NBr	81
Dimethyl(methylthio)sulfonium trifluoromethanesulfonate (DMTST)	82
$NOBF_4$	83
SO_2Cl_2, TfOH	84
PhSeOTf	85
MeSOTf or PhSOTf	86
Br_2, AgOTf or $Hg(CN)_2$	87
TBPA	88
NIS or NBS, TfOH	89–92
IDCP	93
IDCT	94
PhIO, Tf_2O	95
Tris(4-bromophenyl)ammoniumyl hexachloroantimonate (BAHA)	96
PhSeNPhth, TMSOTf	97
[Bis(trifluoroacetoxy)iodo]benzene	98
NBS or NIS, Ph_2IOTf or Bu_4NOTf or Bu_4NClO_4 or $LiClO_4$ or $LiNO_3$	99
I_2	100
PhIO, $Sn(OTf)_2$ or $SnCl_2$ or TMSOTf	101
$TrB(C_6F_5)_4$, $NaIO_4$	102
Selectfluor, $BF_3 \cdot Et_2O$	103
PhSeNPhth or PhIO, $Mg(ClO_4)_2$	104
I_2, $TrB(C_6F_5)_4$, DDQ	105
NIS or NBS, $TrB(C_6F_5)_4$	106

(Continued)

TABLE 3.3 (Continued)

Conditions	References
EtSNPhth/TrB(C_6F_5)$_4$	107
MPBT, Tf_2O	108
1-Benzenesulfinyl piperidine (BSP), Tf_2O	109
NISAc	110
NIS/NBS, TMSOTf	111
NIS, TrOTf or $HClO_4$	112
1-Fluoropyridinium triflates	113
Ph_2SO, Tf_2O	114
ICl or IBr, AgOTf	115,116
N-(*p*-phenylthio)-ε-caprolactam, Tf_2O	117
NIS, $HClO_4$-silica	118
BSM, Tf_2O	119
Me_2S_2, Tf_2O	120
Ipy_2BF_4, TfOH	121
NIS, TSEOTf	122
p-NO_2PhSCl, AgOTf	123
Bromodimethylsulfonium bromide (BDMS)	124
NIS, [bmim][OTf]	125
$R_4N^+AOT^-$, NIS	126
DMTPSB, AgOTf	127
Trichloroisocyanuric acid (TCCA), TMSOTf	128
N-(*p*-methylphenylthio)-ε-caprolactam, TMSOTf	129
Br_2, AgOTf	130
Phenyl(trifluoroethyl)iodonium triflimide	131
Thioperoxide, TMSOTf	132
In(OTf)$_3$, ball mill	61
$AgClO_4$, MeSBr	133
(*p*-Tol)$_2$SO/Tf_2O	134
PIFA, TfOH	135
NIS or NBS, M(OTf)x[a]	59,91,136–139

(Continued)

TABLE 3.3 (Continued)

Conditions	References
$AuCl_3$	140,141
$Ph_3Bi(OTf)_2$	142,143
1,3-Dibromo-5,5-dimethylhydantoin (DBDMH), TMSOTf or TfOH	144
$PdBr_2$, propargyl bromide	145
TolSCl, AgOTf	146,147
$CuBr_2$	148

[a]*CuOTf, $Cu(OTf)_2$, $Bi(OTf)_3$, $Yb(OTf)_3$, $Ga(OTf)_3$, $Sc(OTf)_3$, $Fe(OTf)_3$, $In(OTf)_3$, $Sn(OTf)_2$, $La(OTf)_3$.*

limits their application due to environmental toxicity concerns. Interestingly, the rapid metal-promoted activation of heterocyclic aryl thioglycosides was observed when a heteroatom was present in the aryl moiety.[149,150]

An ideal promoter to activate thioglycosides should possess specific features. It should be thiophilic enough, readily accessible, and benchtop stable with good shelf life. Importantly, it should be amenable to be used in catalytic amounts at a broad range of temperatures, and one should be able to use it to perform diverse glycosylation reactions in a one-pot manner. Halonium ion-based activators were thus developed in early works by van Boom[90] and Fraser-Reid[151] with the hope of solving some of these issues. Halonium activators are more thiophilic than mercuric-based salts, and, notably, they are effective over a broad range of temperatures.

Nowadays, the most employed promoters for activating thioglycosides are *N*-iodosuccinimide (NIS) (in combination with TfOH, AgOTf, or TMSOTf), MeOTf, IDCP, dimethyl(methylthio)sulfonium trifluoromethanesulfonate (DMTST), Ph_2SO, and *p*-TolSCl/AgOTf. However, there are some limitations to the use of each of these promoters. For instance, MeOTf can cause the methylation of the hydroxyl groups of less reactive acceptors.[152] On the other hand, the formation of *N*-succinimide glycosides as donor-side products can occur when a succinimide-based promoter is used to perform the glycosylation of less reactive acceptors.[153] Interestingly, methods have been developed in which the thioglycoside donors were first converted to bromo or iodo intermediates, which were then directly used to perform glycosylation with the addition of promoters such as I_2,[105] ICl or IBr/AgOTf.[115,116,154]

Me_2S_2[120] and Ph_2SO[114] are the most common organosulfur reagents used for glycosylation with thioglycoside donors.[82] They have been used in combination with MeOTf and Tf_2O, respectively, to generate in situ electrophilic promoter species such as DMTST and diphenyl sulfonium. A sterically hindered nonnucleophilic base is usually added to the reaction to quench the in situ-formed acid. Additionally, in some cases, triethyl phosphite[114] was used to quench sulfonium species to prevent the tandem activation of formed thioglycosylated products, especially during chemoselective glycosylations when both the donor and acceptor are thioglycosides. On a different note, care must be taken for the glycosylation of donor/acceptor pairs bearing alkene-containing groups, such as allyl and pentenyl, because the electron-rich sp^2 centers could compete with thioglycosides for the nucleophilic attack of electrophilic species.[155]

In the specific case of single electron activation, a radical cation is generated by the transfer of one electron from the sulfur atom to the electron acceptor, which is followed by a heterolytic cleavage leading to the formation of an oxocarbenium ion and disulfide species. Then, the hydroxyl group of the glycosyl acceptor can attack the activated donor, and a glycosylation product is obtained. This process of single electron transfer can occur photochemically,[156] electrochemically,[45,157] or chemically.[88] For more information regarding the different types of reagents enabling activation of thioglycosides, the reader is oriented toward the review published by Yu and coworkers.[43]

3.4 Thioglycosides for chemoselective glycosylations

A chemoselective glycosylation is a reaction in which both the glycosyl donor and acceptor are thioglycosides but where only the donor thioaglycon is chemoselectively activated due to difference in anomeric reactivities. Several parameters can be invoked for explaining the reactivity differences such as steric[158] and electronic[159] factors, anomeric configurations,[160] types of promotion system,[161] temperature,[162] and the nature of protecting groups.[163] However, if the reactivities of the glycosyl donor/acceptor pair are not well controlled, competitive reactions involving the transfer of the acceptor thioaglycon to the activated donor can occur, which could result in acceptor hydrolysis or polymerization and regeneration of the donor thioglycoside. This phenomenon is known as aglycon transfer.[164–167] Over the past years, efforts have been made to provide solutions to this challenge, which resulted in the classification of chemoselective glycosylations into three types (Scheme 3.2). In the type A

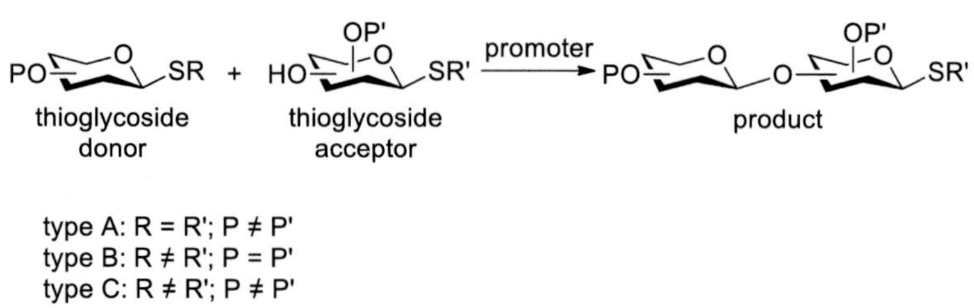

SCHEME 3.2 Types of chemoselective glycosylations.

glycosylation, anomeric thioaglycons are the same on both the donor and acceptor building blocks; however, different sets of protecting groups are present. On the other hand, in a type B glycosylation, the donor and acceptor building blocks share the same hydroxyl protecting group pattern, except for the anomeric position, which is protected with different thioaglycons. A type C glycosylation is defined as the chemoselective coupling between thioglycosides donor and acceptor, which differ in both their protecting group pattern and anomeric thioaglycon.

3.4.1 Type A chemoselective glycosylations

Some examples of type A chemoselective glycosylations are shown in Table 3.4. For instance, the synthesis of a rhamnose-containing disaccharide by chemoselective glycosylation was reported by Ley and coworkers (entry 1,

TABLE 3.4 Selected examples of type A chemoselective glycosylations.

Entry	Donor	Acceptor	Promoter	Conditions	Product (% yield)	References
1	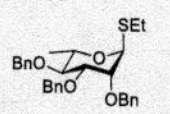 1	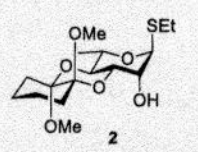 2	IDCP	DCE:Et_2O (1:4, v/v) 0 °C to rt	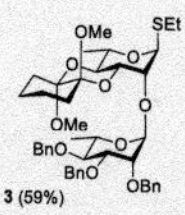 3 (59%)	158
2	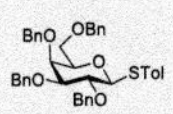 4	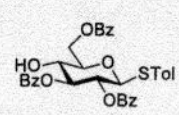 5	NIS TfOH	DCM −40 to −20 °C to rt	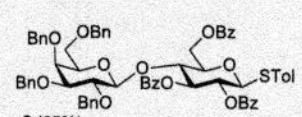 6 (65%)	159
3	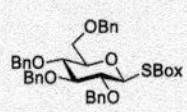 7	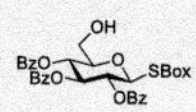 8	$Cu(OTf)_2$	DCE	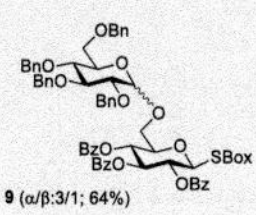 9 (α/β:3/1; 64%)	27
4	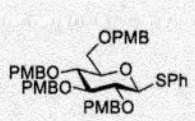 10	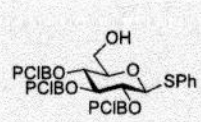 11	NIS TfOH	DCM −78 to 0 °C	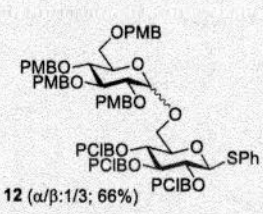 12 (α/β:1/3; 66%)	168
5	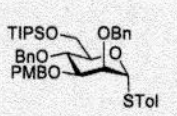 13	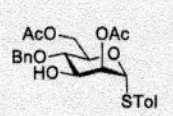 14	NIS TMSOTf	DCM, −20 °C	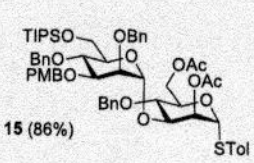 15 (86%)	169
6	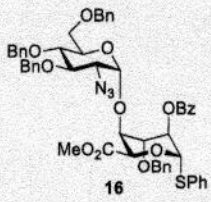 16	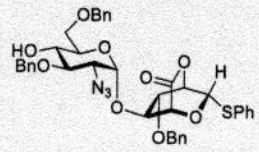 17	NIS AgOTf	DCM −25 °C to rt	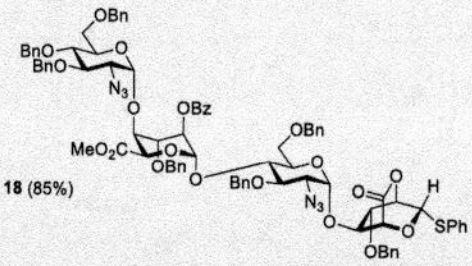 18 (85%)	170
7	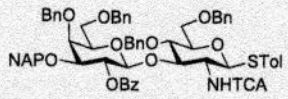 19	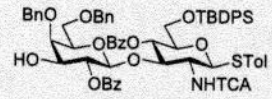 20	NIS TMSOTf	DCM, −50 °C	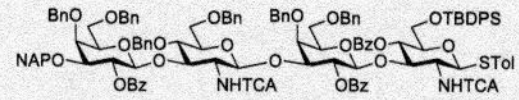 21 (86%)	163

Table 3.4).[158] A cyclohexane-1,2-diacetal protection was used to decrease the reactivity of rhamnosyl acceptor **2** for coupling with rhamnosyl donor **1**, which featured armed protecting groups. Glycosylation was promoted by iodonium dicollidine perchlorate (IDCP), which afforded disaccharide **3** in 59% yield with exclusive α-stereoselectivity. In an example depicted in entry 2, the relative reactivity values (RRVs) were measured and used as a rationale for chemoselective glycosylation. To this aim, two thioglycosides were allowed to competitively react with methanol in the presence of a limiting amount of promotor, and the RRVs were determined by calculating the amount of unreacted donors using high performance liquid chromatography (HPLC). The chemoselective coupling was performed between an armed thiogalactoside donor with high RRV (**4**, RRV = 1776) and a disarmed glucosyl acceptor with low RRV (**5**, RRV = 8.4) under the promotion of NIS/TfOH, which resulted in the formation of thiolactose disaccharide **6** in 65% yield.[159]

Alternatively, a 1-*S*-benzoxazolyl (SBox)-based chemoselective glycosylation was developed by Demchenko and coworkers.[27] In entry 3 of Table 3.4, perbenzylated glucosyl donor **7** bearing an SBox leaving group was reacted with acceptor **8**, which was disarmed by the presence of electron-withdrawing benzoyl groups, providing disaccharide **9** in 64% yield with a 3:1 α/β stereoselectivity.

Recently, chemoselective glycosylations between *para*-substituted benzyl-protected glycosyl donor and acceptor thioglycosides were performed.[168] As shown in entry 4, a reactivity-based glycosylation between armed *p*-methoxybenzylated (PMB) donor **10** and remotely disarmed *p*-chloro benzylated (PClB) acceptor thioglucoside **11** afforded disaccharide **12** in 66% yield, yet in moderate stereoselectivity (α/β = 1/3), which could be rationalized by the absence of a participating group at the C2 position of the glycosyl donor. Similarly, in entry 5, mannosyl disaccharide **15** was synthesized by glycosylation between reactive donor **13** and electronically disarmed acceptor **14**.[169]

An interesting example of a [2 + 2] chemoselective glycosylation is shown in entry 6, in which lactone-based conformationally disarmed thioglycoside disaccharide acceptor **17** was glycosylated with disaccharide donor **16** in the presence of NIS/AgOTf, providing heparan sulfate tetrasaccharide derivative **18** in 85% yield.[170] It is worth to mention that Lin and coworkers performed a RRV-guided reactivity-based glycosylation between reactive benzylated disaccharide donor **19** (RRV = 8351) and stereoelectronically disarmed acceptor **20** (RRV = 2060), yielding type I LacNAc tetrasaccharide **21** bearing an anomeric STol in 86% yield.[163]

3.4.2 Type B chemoselective glycosylations

For type B chemoselective glycosylations, Huang et al. conducted coupling between donor/acceptor pairs differing by their anomeric thioaglycon part.[171] For instance, in entry 1 (Table 3.5), galactosyl donor **22** with a *p*-OMe

TABLE 3.5 Selected examples of type B chemoselective glycosylations.

Entry	Donor	Acceptor	Promoter	Conditions	Product (% yield)	References
1	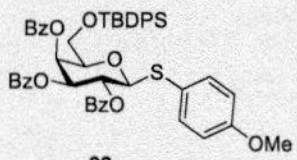 **22**	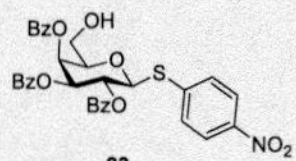 **23**	NIS, TfOH	DCM −60 to −40 °C	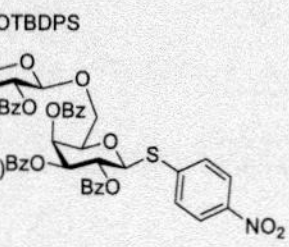 **24** (78%)	171
2	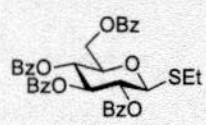 **25**	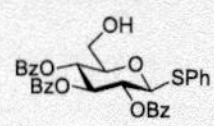 **26**	*N*-trifluoromethylthio-saccharin, TMSOTf	DCM 0 °C to rt	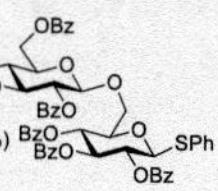 **27** (78%)	161
3	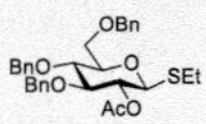 **28**	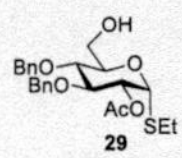 **29**	NIS TMSOTf	DCM −78 to −20 °C	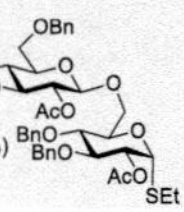 **30** (67%)	160

thiophenyl as thioaglycon was found to be more reactive than its *p*-nitro-substituted counterpart, as revealed from the downfield shifted anomeric proton of acceptor **23** ($\delta_{H1} = 5.15$ ppm). The [1 + 1] glycosylation between donor **22** and acceptor **23** (with *p*-NO_2-SPh aglycon) afforded disaccharide **24** in a good 78% yield. Additionally, it was found that SEt donors are more reactive than their SPh counterparts.[161] In entry 2, glucose donor **25** bearing an SEt thioaglycon was readily activated following the addition of the *N*-trifluoromethylthiosaccharin/TMSOTf promoter system. Glycosylation with acceptor **26** bearing SPh as a leaving group furnished disaccharide **27** in 78% yield with exclusive β-stereoselectivity.

As previously mentioned, the anomeric configuration also plays an important role for modulating the reactivity of coupling partners. For instance, β-donor **28** was shown to be activated more easily than its α counterpart **29** upon the addition of NIS/TMSOTf during chemoselective glycosylation, owing to the removal of stabilizing and arming capability of the 2-*O*-acetyl group of an α-configured donor (entry 3).[160]

3.4.3 Type C chemoselective glycosylations

Several interesting examples have been reported for the synthesis of di- and trisaccharides by using the combined effect of different anomeric thioaglycons in addition to the presence of protecting groups at different positions of the glycosyl residues. For instance, the stereoselective formation of disaccharide **33** occurred when the glycosylation of disarmed acceptor **32** was performed with seven-carbon glucosyl donor **31** in the presence of NIS/AgOTf at low temperature (Table 3.6, entry 1).[172] Likewise, the chemoselective

TABLE 3.6 Selected examples of type C chemoselective glycosylations.

Entry	Donor	Acceptor	Promoter	Conditions	Product (% yield)	References
1	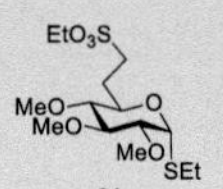 **31**	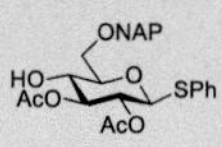 **32**	NIS AgOTf	DCM −75 to −65 °C	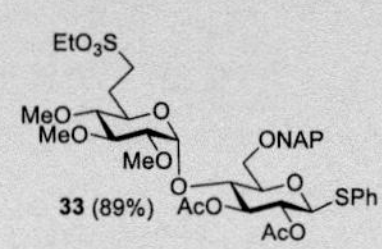 **33** (89%)	172
2	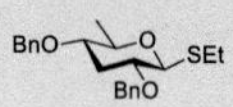 **34**	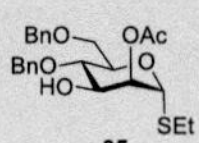 **35**	NIS, TfOH	DCM:Et_2O (1:3, v/v) −20 °C	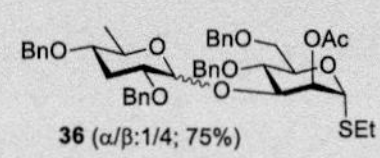 **36** (α/β:1/4; 75%)	173
3	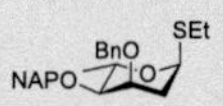 **37**	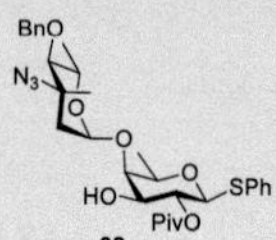 **38**	$AgPF_6$ TTBP	DCM −10 °C	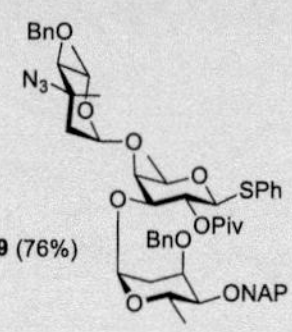 **39** (76%)	19

coupling of dideoxyparatose donor **34** with acceptor **35** (entry 2) led to the formation of a paratose-containing disaccharide anomeric mixture (**36**, α/β:1/4, 75% yield).[173]

As depicted in entry 3, a reagent-controlled α-glycosylation was developed by Bennett et al. for the synthesis of the branched trisaccharide of saccharomicin B, an heptadecaglycoside antibiotic containing 6-deoxysugar residues, which was isolated from the uncommon actinomycete *Saccharothrix espanaensis*.[19] Disaccharide acceptor **38** (with anomeric SPh) was glycosylated with SEt donor **37** in the presence of $AgPF_6$ and 2,4,6-tri-*tert*-butylpyrimidine (TTBP) at −10°C, which gave trisaccharide **39** in 76% yield exclusively in its α-anomeric form. The preferential activation of thioglycoside donor **37** was attributed to the difference of reactivities toward $AgPF_6$ between the 2-oxy and the 2-deoxy-thioglycosides.

3.4.4 One-pot chemoselective glycosylations

To avoid the tedious column chromatographic purification of intermediates following each glycosylation step, thioglycoside-based one-pot chemoselective glycosylations have been developed. In theory, the most reactive thioglycoside donor is first activated for coupling with a less reactive thioglycoside acceptor, resulting in the synthesis of an intermediate thioglycoside disaccharide donor, which is further activated to react with the least reactive acceptor (Scheme 3.3). This sequence can be performed in a tandem manner. The final glycosylation with the reducing-end acceptor usually leads

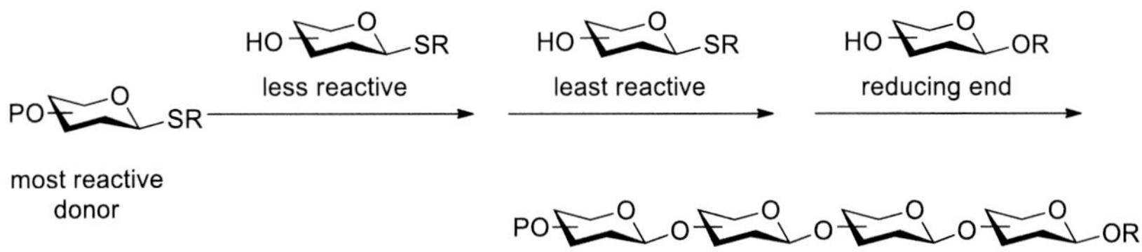

SCHEME 3.3 Principles of one-pot chemoselective glycosylations.

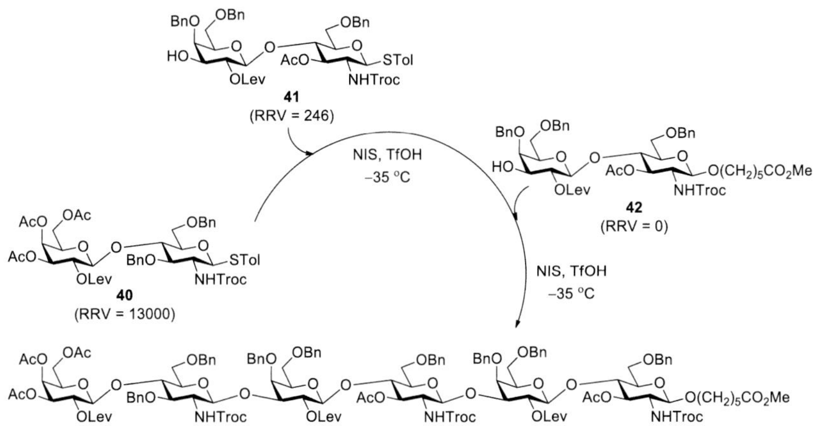

SCHEME 3.4 Synthesis of a type II LacNAc-repeating hexasaccharide by one-pot glycosylation.

to the desired oligosaccharide through one-pot reactions. Although it looks straightforward, the main challenge is the modulation of the reactivities of the thioglycoside building blocks so that significant reactivity differences can be reached to achieve successful chemoselective glycosylation between an armed donor and a disarmed acceptor. It is, therefore, crucial to keep sufficient reactivity for the acceptor thioaglycon in order that the resulting intermediate glycosylated product could still act as a reactive donor for the next glycosylation step.

For instance, Wong and coworkers reported the RRV-based one-pot synthesis of type II LacNAc-repeating oligosaccharides (Scheme 3.4).[174] An armed type II LacNAc donor (**40**, RRV = 13,000), bearing Troc and di-*O*-benzyl functionalities as amine and ether protecting groups, respectively, was first coupled with disarmed disaccharide acceptor **41**, having an *O*-acetyl group at O3 of the GlcNAc residue, resulting in the formation of an intermediate thioglycoside tetrasaccharide. This was followed by the addition of NIS/TfOH to activate the in situ-formed tetrasaccharide donor, which was reacted with reducing-end disaccharide acceptor **42**, producing hexasaccharide **43** in a 55% one-pot yield.

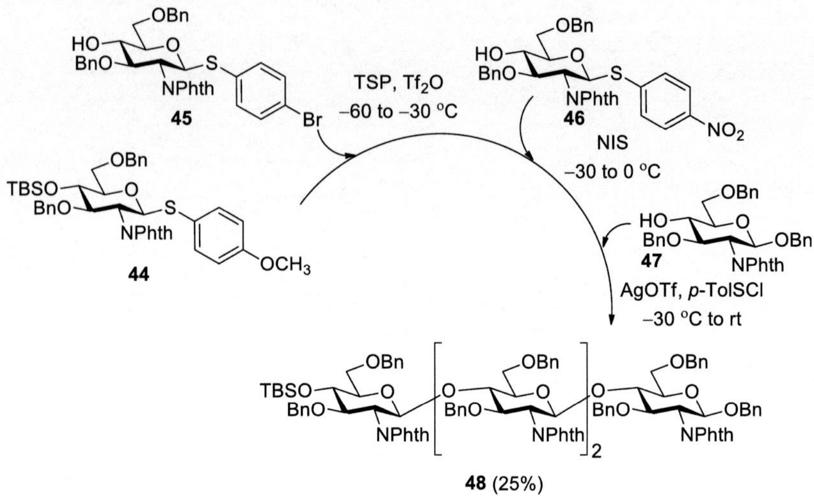

SCHEME 3.5 Substituted thioaglycon-based one-pot chemoselective glycosylation for the synthesis of a chitin tetrasaccharide.

In another interesting report, the Huang group performed a one-pot glycosylation to synthesize chitin-based oligosaccharides.[171] In this specific example, the reactivities of both the glycosyl donor and acceptor were solely modulated by the substitution pattern of the *para*-thiotolyl substituent. As shown in Scheme 3.5, silyl-protected armed donor **44** (with an electron-donating methoxy substituent on the thioaglycon) was first coupled with bromo-disarmed acceptor **45**, then subsequently with GlcNAc acceptor **46** bearing a nitro-substituted thioaglycon, and finally with reducing-end acceptor **47**, yielding GlcNAc-based chitin tetrasaccharide **48** in 25% overall yield.

In 2008, the Bols group developed the one-pot addition synthesis of a trisaccharide through a sequential activation of donors modulated by temperature (Scheme 3.6).[162] At low temperature, silyl-protected "superarmed" donor **49** was activated and glycosylated with armed acceptor **50** bearing a reactive 6-OH. The in situ-formed disaccharide was activated at higher temperature and reacted with acetyl-disarmed acceptor **51** (with free 4-OH), resulting in the formation of trisaccharide **52** in 64% yield.

With the development of versatile SBox-based thioglycosides for chemoselective glycosylations, Demchenko and coworkers demonstrated their utility for one-pot glycosylations as shown in Scheme 3.7.[175] Benzylated armed glucosyl donor **53** was selectively activated and coupled with benzoylated-disarmed acceptor **8** at 0°C, followed by reaction with acceptor **54** (bearing an anomeric α-methoxy group), which resulted in the synthesis of trisaccharide **55** in good yield (83%) through a one-pot glycosylation.

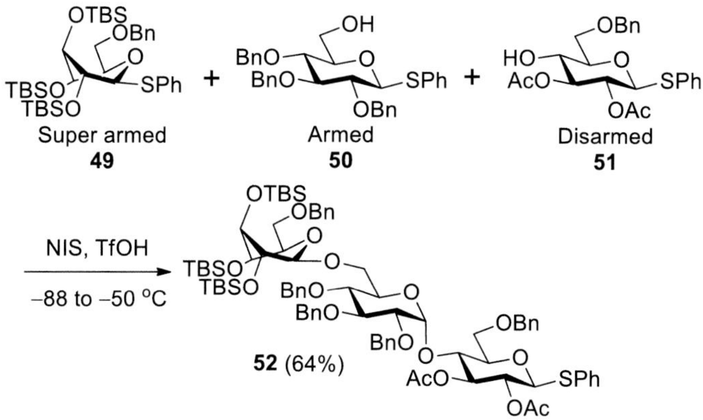

SCHEME 3.6 Temperature-controlled one-pot one-addition synthesis of trisaccharide.

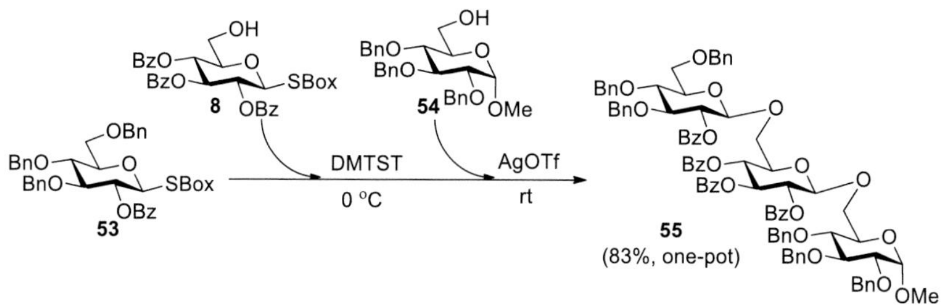

SCHEME 3.7 SBox-based one-pot chemoselective glycosylation for the synthesis of a trisaccharide.

3.5 Orthogonal glycosylations

An orthogonal glycosylation is a reaction in which the glycosyl donor is activated using conditions that are inert to the leaving group of the glycosyl acceptor. As depicted in Scheme 3.8, in the first glycosylation, a donor (with leaving group LG_1) can be activated with promoter 1 without any activating/influencing acceptor of the leaving group LG_2 and a disaccharide product can be obtained. This is followed by the addition of promoter 2 to activate the disaccharide (with leaving group LG_2) to react with a second acceptor. As described in the next section, there can be different combinations of orthogonal glycosylations.

3.5.1 Orthogonal glycosylations of thioglycoside acceptors with halide glycosyl donors

A widely explored type of orthogonal glycosylations involves the coupling of glycosyl-halide donors (fluoride, chloride, bromide, and iodide) and

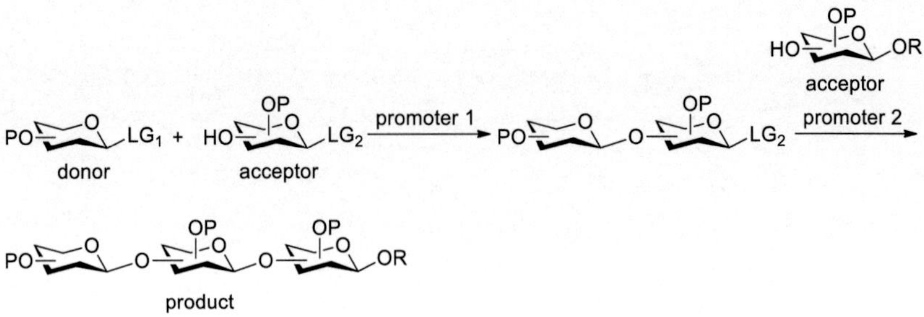

SCHEME 3.8 Principle of orthogonal glycosylation.

thioglycoside acceptors.[176] Mukaiyama and coworkers performed the one-pot synthesis of tumor-associated mucin-related F1α antigen. A highly convergent synthesis of a branched heptaglucoside (with phytoalexin-elicitor activity) was accomplished using the selective activation of a glycosyl-fluoride donor to couple with a thioglycoside acceptor.[112] In another report, the one-pot synthesis of core 2 class glycosyl amino acid was performed via orthogonal glycosylation between a peracetylated glucosamine-fluoride donor and a silylated thiogalactosamine acceptor building block.[177] The silylated acceptor was used to increase the nucleophilicity of the acceptor hydroxyl group to afford regioselective glycosylation at O6 in the presence of a free O3-OH.

An interesting example of orthogonal glycosylation between a glycosyl-fluoride donor and a thioglycoside acceptor was recently demonstrated by the Imamura group for the synthesis of type II LacNAc-repeating oligosaccharides.[178] As shown in Scheme 3.9, a [8 + 4] orthogonal glycosylation was performed between octasaccharide-fluoride donor **56** and thioglycoside tetrasaccharide acceptor **57** to synthesize dodecasaccharide **58**, protected with orthogonal protecting groups, that is, Troc and SPh, at the nonreducing and reducing ends, respectively, for further chain elongation. In another example, the orthogonal coupling of a thioglycoside acceptor with a glycosyl-chloride donor was done in the presence of AgOTf during the synthesis of a dodecasaccharide.[176,179]

In Table 3.7, some examples of orthogonal glycosylations between bromide glycosyl donors and thioglycoside acceptors are depicted. Bromide donors were rapidly activated at low temperatures (−30 and −74°C) with AgOTf (entries 1[180] and 2[181]) to react with GlcNAc acceptors bearing SEt as anomeric leaving groups, which resulted in the formation of Gal-β-(1→3)-GlcNAc disaccharides **61** and **64** in good yields. On the other hand, in entry 3, an electronically deactivated acceptor (**66**, with a *p*-nitro-substituted thioaglycon) was used for the orthogonal glycosylation with peracetylated galactosyl bromide **65**.[182] Additionally, sterically hindered thioaglycons [*S*-dimethyl phenyl (SDMP)] were employed to minimize the acceptor activation during

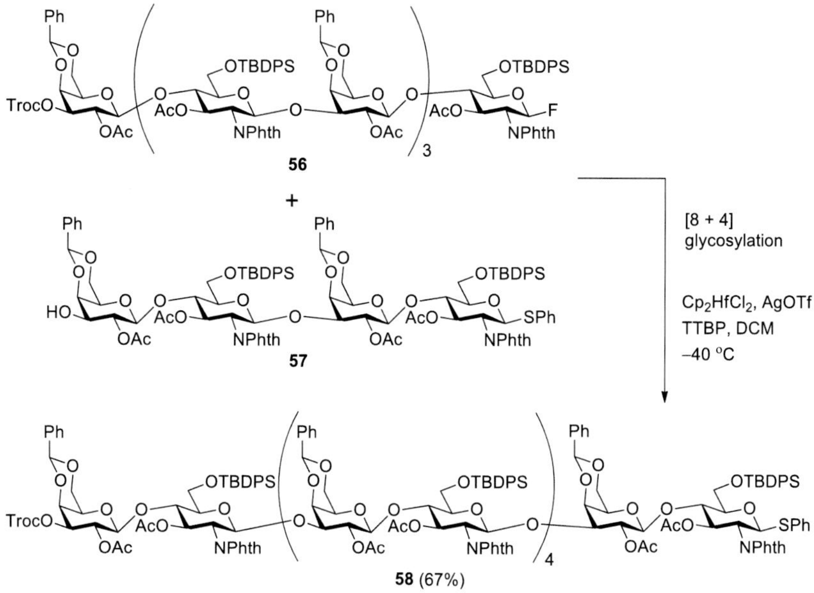

SCHEME 3.9 Orthogonal glycosylation between fluoride donor and thioglycoside acceptor to synthesize type II LacNAc-repeating oligosaccharide.

orthogonal reactions with bromide donors in the presence of AgOTf as promoter (entries 4[164] and 5[183]). As a relevant example of orthogonal glycosylation between a thioglycoside acceptor and a glycosyl-iodide donor, the Oscarson group employed triphenylphosphine oxide to selectively activate the glycosyl-iodide donor at room temperature.[184]

3.5.2 Orthogonal glycosylations between trichloroacetimidate donors and thioglycoside acceptors

Similarly, imidate donors were used to perform orthogonal glycosylations of thioglycoside acceptors (Table 3.8). For instance, TCAI donor **74** (with N_3 at the C2 position) was used to glycosylate acceptor **75** (bearing SDMP as an anomeric protecting group), providing disaccharide **76** in 82% yield without aglycon transfer (entry 1).[164] The reaction was promoted by TMSOTf, while α-stereoselectivity was achieved by using diethyl ether as the solvent. On the other hand, TMSOTf-catalyzed orthogonal coupling between TCAI galactosyl donor **77** and phthalimide-protected GlcNAc acceptor **78** led to the formation of orthoester **79** (entry 2), which was rearranged into the desired Gal-β-(1→3)-GlcNAc disaccharide upon treatment with catalytic amounts of TMSOTf.[185] Recently, Oscarson and colleagues reported the synthesis of the serotype-D fragment of glucuronoxylomannan capsular polysaccharide of *Cryptococcus neoformans*.[186] The authors performed an

TABLE 3.7 Examples of orthogonal glycosylations between bromide glycosyl donors and thioglycoside acceptors.

Entry	Donor	Acceptor	Promoter	Conditions	Product (% yield)	References
1	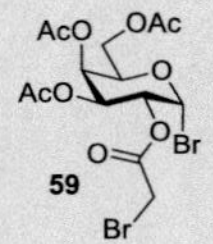 **59**	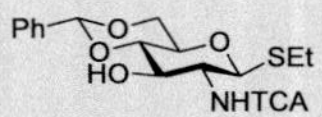 **60**	AgOTf 1,1,3,3-tetramethylurea	DCM −30 °C	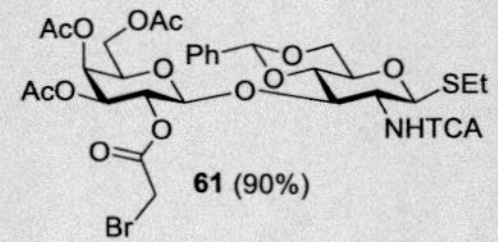 **61** (90%)	180
2	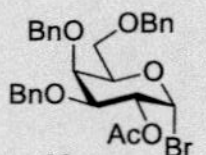 **62**	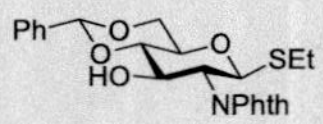 **63**	AgOTf 2,6-di-*tert*-butyl -pyridine	DCM:toluene (2:1, v/v) −74 °C	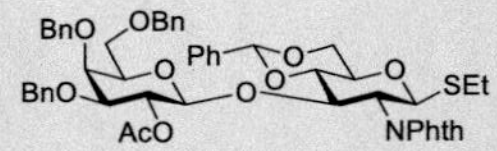 **64** (84%)	181
3	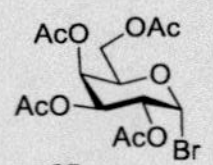 **65**	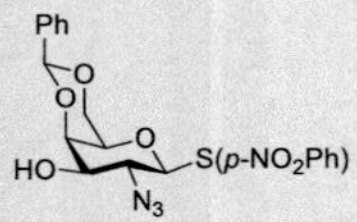 **66**	AgOTf	DCM −30 to −20 °C	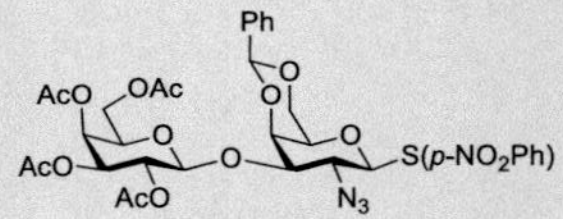 **67** (57%)	182
4	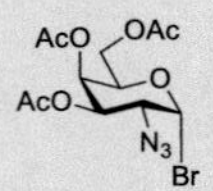 **68**	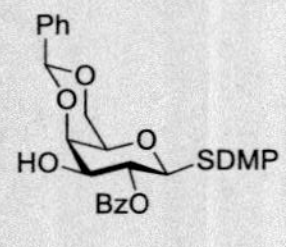 **69**	AgOTf di-*tert*-butylmethylpyridine	DCM −60 to 0 °C	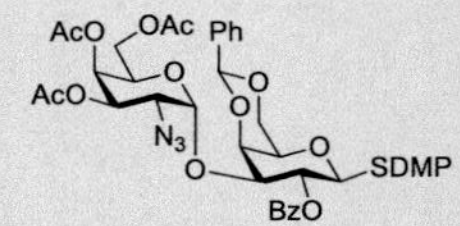 **70** (78%)	164
5	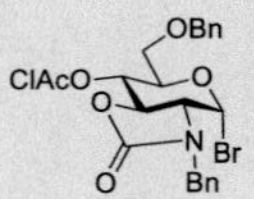 **71**	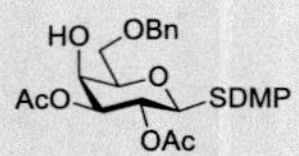 **72**	AgOTf di-*tert*-butylmethylpyridine	dioxane:toluene (3:1, v/v) 0 °C to rt	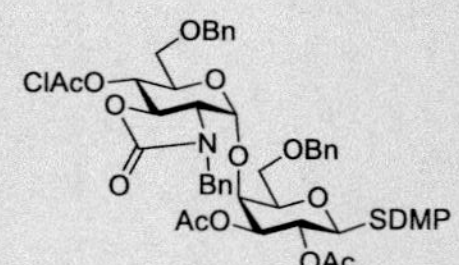 **73** (92%)	183

TABLE 3.8 Examples of orthogonal glycosylations between TCAI glycosyl donors and thioglycoside acceptors.

Entry	Donor	Acceptor	Promoter	Conditions	Product (% yield)	References
1	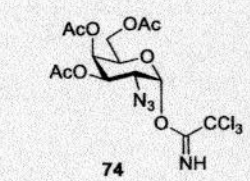 **74**	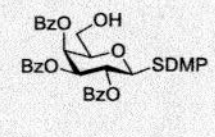 **75**	TMSOTf	Et_2O −20 to 0 °C	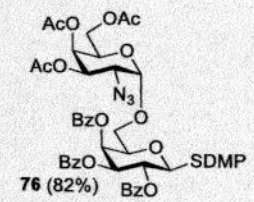 **76** (82%)	164
2	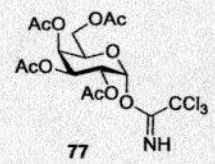 **77**	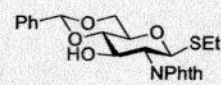 **78**	TMSOTf	DCM, 0 °C	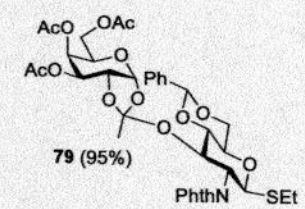 **79** (95%)	185
3	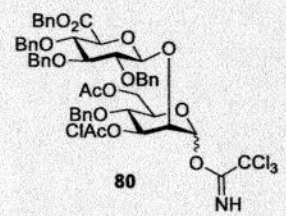 **80**	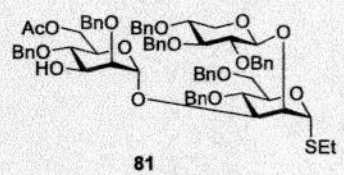 **81**	TMSOTf	Et_2O, −40 °C	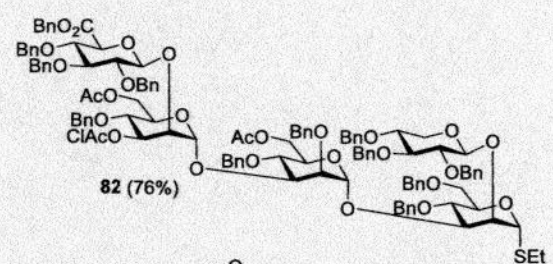 **82** (76%)	188
4	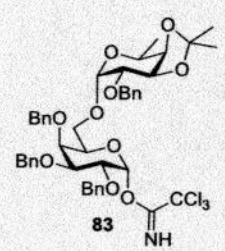 **83**	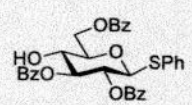 **84**	TMSOTf	DCM, −15 °C inverse procedure	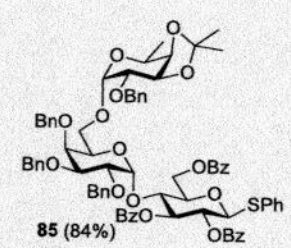 **85** (84%)	187
5	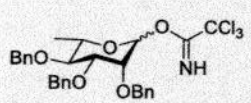 **86**	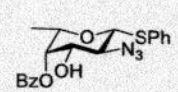 **87**	TMSOTf	DCM, 0 °C	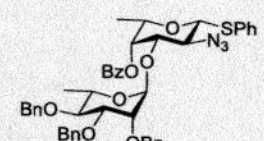 **88** (73%)	188
6	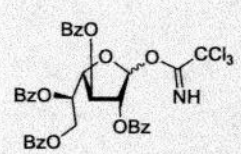 **89**	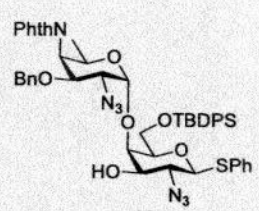 **90**	TMSOTf	DCM −30 °C	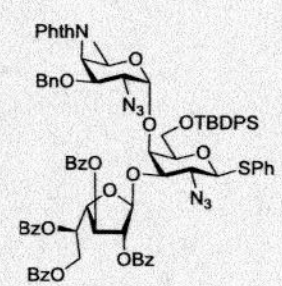 **91** (82%)	189

orthogonal coupling between disaccharide imidate donor **80** and trisaccharide SEt acceptor **81** through a reverse glycosylation procedure catalyzed by TMSOTf, producing exclusively desired α-pentasaccharide **82** in 76% yield (entry 3). Similarly, several other glycosides (di- to trisaccharides) were synthesized via TCAI (imidate)/SPh orthogonal couplings (entries 4–6).[187–189]

3.5.3 Orthogonal coupling of thioglycoside acceptors with *N*-phenyltrifluoroacetimidate donors

In this section, selected examples of orthogonal glycosylations between the more stable PTFAI donors and thioglycoside acceptors are presented. In entry 1 of Table 3.9, Codée et al. reported the synthesis of *Streptococcus pneumoniae* disaccharide by orthogonal glycosylation between imidate donor **92** and glucuronic acid thioglycoside **93** using TfOH as the promoter.[190] The resulting thioglycoside disaccharide **94** was used as a donor in a subsequent [2 + 1] glycosylation to access a bacterial trisaccharide. In another report, the same group synthesized disaccharide building block **97** *via* an orthogonal condensation between glucosamine donor **95** with glucuronic acceptor **96** in the presence of TfOH at 0°C (entry 2).[191] The synthesized thioglycoside disaccharide was transformed into an imidate donor, which was used in the automated solid-phase synthesis of hyaluronic acid oligosaccharides of defined lengths. Recently, for the synthesis of alginate fragments containing β-D-mannuronic and α-L-guluronic acids, the authors performed a stereoselective orthogonal glycosylation between silylidene-protected gulose imidate donor **98** and α-*S*-tolyl-mannuronic acid acceptor **99** to synthesize disaccharide **100** in 91% yield (entry 3).[192] Nguyen and colleagues reported the nickel catalyst system for the synthesis of 1,2-*cis*-2-aminosugars by using C(2)-*N*-substituted benzylidene trichloroacetimidates.[193] In entry 4, the issue of aglycon transfer was decreased by using thioglycoside acceptor **102**, with 2-trifluoromethylphenyl aglycon, to block the thioaglycon transfer from acceptor to *N*-phenyltrifluoroacetimidate donor **101** while producing desired α-disaccharide **103** in 61% yield.

3.5.4 Orthogonal glycosylations of thioglycoside acceptors with other glycosyl donors

Other examples of orthogonal glycosylations involving thioglycoside acceptors with donors bearing various leaving groups are listed in Table 3.10. A high yielding α-glycosylation was reported by the Wong group, in which the coupling between sialyl phosphate donor **104** (with dibutyl phosphate group) and thiogalactoside acceptor **105** under the promotion of TMSOTf at −78°C provided α-disaccharide **106** in excellent yield (87%) (entry 1).[194] A new class of donors, that are, *N*-(glycosyloxy)acetamides, was developed by the Ye group.[195] The successful orthogonal glycosylation of disarmed

TABLE 3.9 Examples of orthogonal glycosylations between PTFAI glycosyl donors and thioglycoside acceptors.

Entry	Donor	Acceptor	Promoter	Conditions	Product (% yield)	References
1	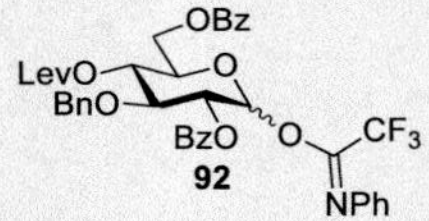 **92**	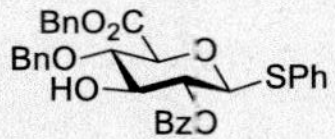 **93**	TfOH	DCM, 0 °C	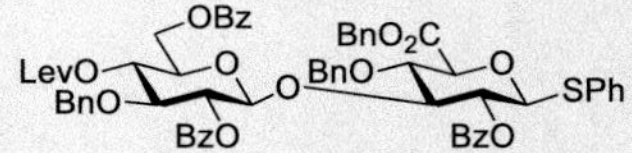 **94** (64%)	190
2	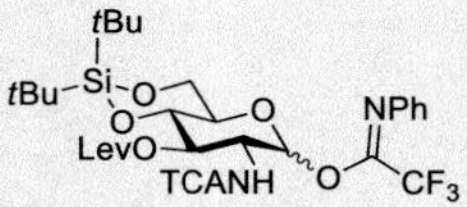 **95**	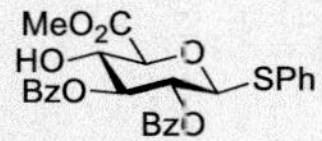 **96**	TfOH	DCM, 0 °C	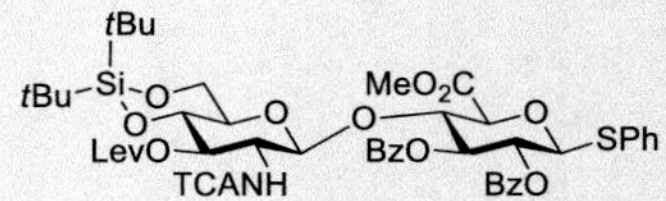 **97** (78%)	191
3	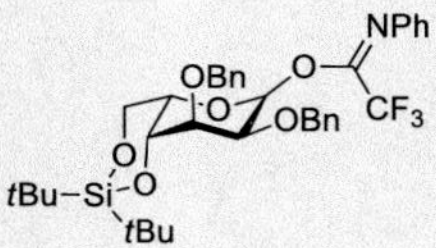 **98**	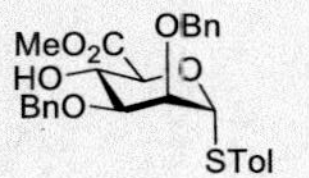 **99**	TMSOTf	DCM −78 to −20 °C	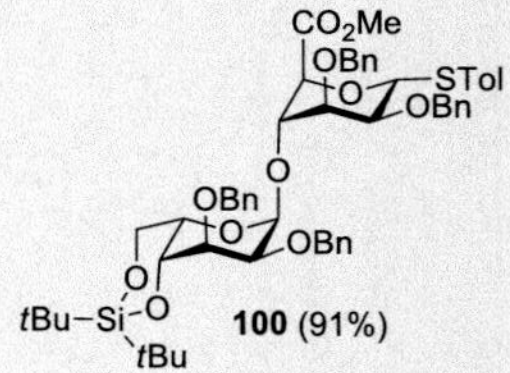 **100** (91%)	192
4	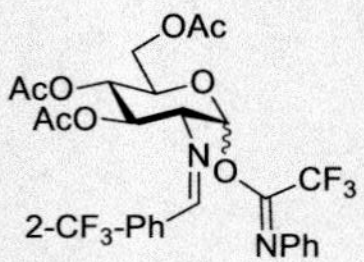 **101**	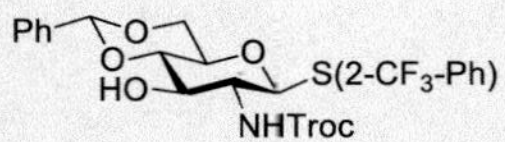 **102**	$Ni(4\text{-}F\text{-}PhCN)_4OTf_2$	DCM, 35 °C	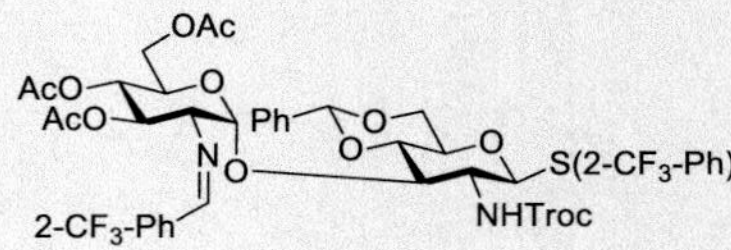 **103** (61%)	193

TABLE 3.10 Selected examples of orthogonal glycosylations between various glycosyl donors and thioglycoside acceptors.

Entry	Donor	Acceptor	Promoter	Conditions	Product (% yield)	References
1	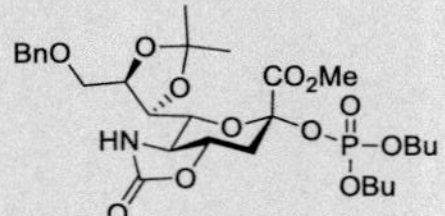 **104**	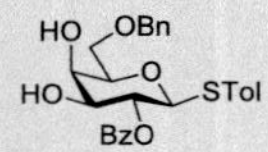 **105**	TMSOTf	DCM, −78 °C	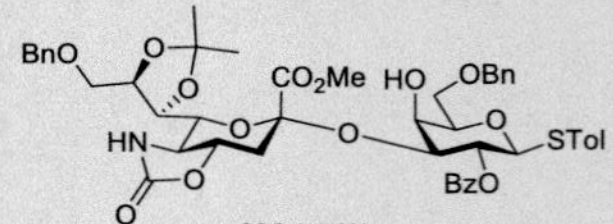 **106** (87%)	194
2	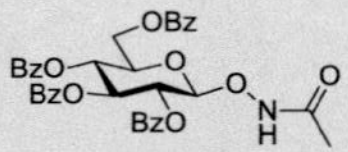 **107**	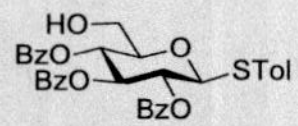 **108**	$SnCl_4$	DCE microwave (100 W, 40 °C)	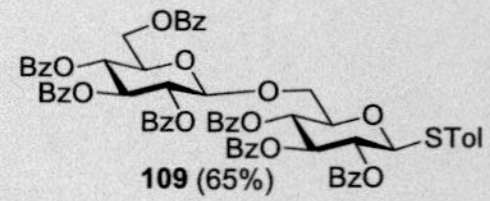 **109** (65%)	195
3	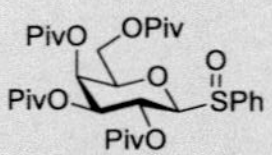 **110**	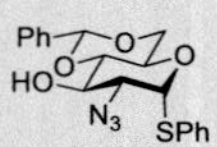 **111**	Tf_2O, di-*tert*-butylmethylpyridine	DCM −78 to −30 °C	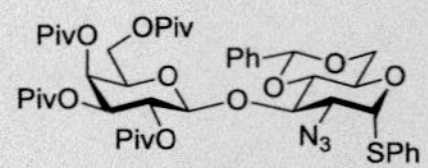 **112** (86%)	34
4	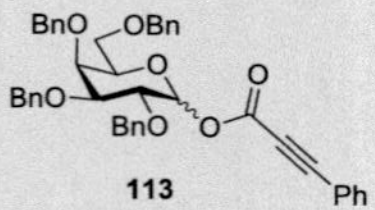 **113**	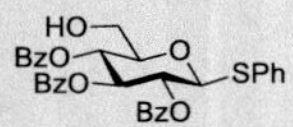 **26**	$AuCl_3$	DCM, 45 °C	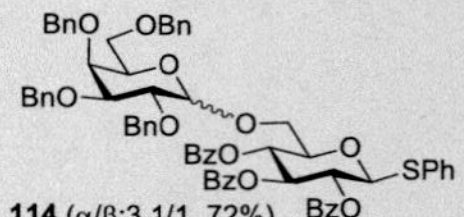 **114** (α/β:3.1/1, 72%)	196
5	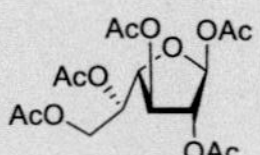 **115**	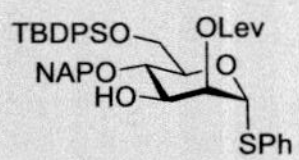 **116**	$SnCl_4$	DCM, −20 °C	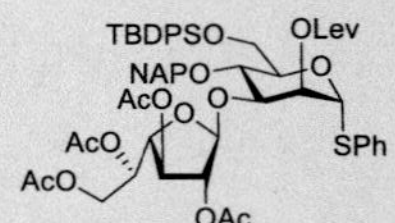 **117** (91%)	197

thioglucoside acceptor **108** with benzoyl-protected glucosyl donor **107** in the presence of $SnCl_4$ as the promoter under microwave irradiations produced desired disaccharide **109** in 65% yield with exclusive β-stereoselectivity (entry 2). Further orthogonal coupling between sulfoxide donor and thioglycoside acceptor was performed by Kahn and coworkers.[34] Hence, pivaloyl-protected galactose sulfoxide donor **110** reacted with azido thioglucoside **111** in the presence of Tf_2O at −78°C to produce type I Gal-β-(1→3)-GlcNAc disaccharide **112** with complete β-stereoselectivity (86%) due to the neighboring group participation of a pivaloyl ester at the C2 position (entry 3).

Recently, a new phenylpropiolate glycosyl donor (**113**, PPG) was shown to perform gold-catalyzed ($AuCl_3$) orthogonal coupling with perbenzoylated thioglucoside acceptor **26** to yield disaccharide **114** in 72% yield (entry 4).[196] The reported features of PPGs include rapidly synthesizable and benchtop-stable donors, which can be activated under mild conditions by using gold catalysts without any additives. Furthermore, the Boons group recently coupled furanose acetate donor **115** with thiomannoside acceptor **116** by using $SnCl_4$ as the promoter in dichloromethane (DCM) at −20°C to yield disaccharide product **117** (91%, entry 5).[197]

3.5.5 Thioglycoside-based one-pot orthogonal glycosylations

In a similar way to the one-pot chemoselective glycosylation strategy, researchers have used thioglycoside-based one-pot orthogonal approaches to generate rapidly some biologically important oligosaccharides. In Scheme 3.10 is depicted the orthogonal one-pot glycosylation of idraparinux, a next-generation anticoagulant drug.[198] First, glycosyl phosphate donor **118** was coupled with Glc−Glc disaccharide acceptor **119** using TMSOTf as a promoter to generate in situ a thioglycoside trisaccharide donor. This was followed by a second glycosylation step of the formed donor with reducing-end Ido-Glc disaccharide acceptor **120** in the exclusive presence of NIS. Because stoichiometric amounts of TMSOTf were used in the first glycosylation step, it was not necessary to add more TMSOTf during the second coupling, and only NIS was added to pursue the [3 + 2] glycosylation. This sequence resulted in the formation of pentasaccharide product **121** in 70% yield over two steps with complete α-stereoselectivity.

It is worth to mention that three orthogonal glycosylations were performed in a sequential manner to synthesize tetrasaccharide **125** through one-pot glycosylation (Scheme 3.11).[199] First, Demchenko's perbenzoylated SBox donor **122** was glycosylated with acceptor **123** (bearing an *O*ABz leaving group) in the presence of AgOTf, followed by the gold-mediated Yu's glycosylation of *n*-pentenyl acceptor **124** with the in situ-formed disaccharide donor coming from the first glycosylation, providing a trisaccharide donor in situ. The final coupling was performed with reducing-end acceptor **54** in the presence of NIS/TMSOTf at 0°C to generate linear tetrasaccharide **125** through a multistep one-pot orthogonal glycosylation strategy.

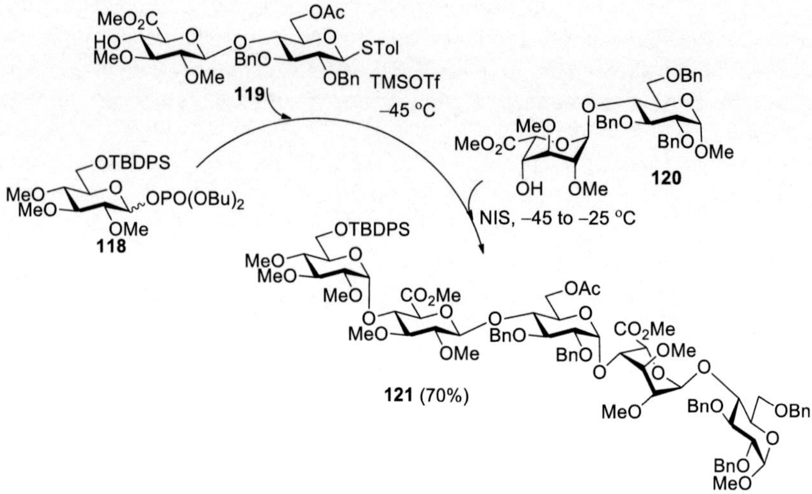

SCHEME 3.10 One-pot orthogonal glycosylation based on glycosyl phosphate donor and thioglycoside acceptor.

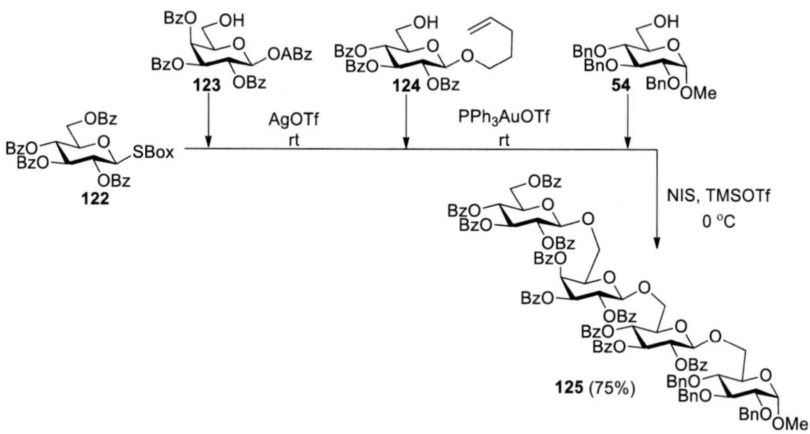

SCHEME 3.11 SBox-based one-pot orthogonal glycosylation.

Similarly, a one-pot relay glycosylation was reported by the Wan group.[200] The synthetic utility of this newly developed method was demonstrated by the one-pot synthesis of kankanoside F, which possesses vasorelaxant properties and was isolated from the parasitic plant *Cistanche tubulosa*. First, *S*-2-[(propan-2-yl)sulfinyl]benzyl (SPSB) donor glycoside **126** was coupled with acceptor **127** [bearing an *o*-ethylphenylsulfanyl (S*o*EP) as the leaving group] in the presence of Tf_2O (Scheme 3.12). This relay glycosylation was based on the in situ formation of cyclic thiosulfonium ion for the first glycosylation step,

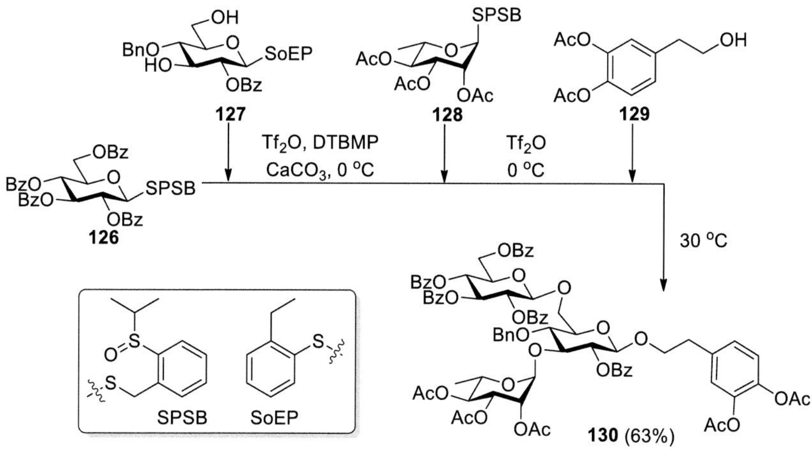

SCHEME 3.12 Wan's one-pot relay glycosylation.

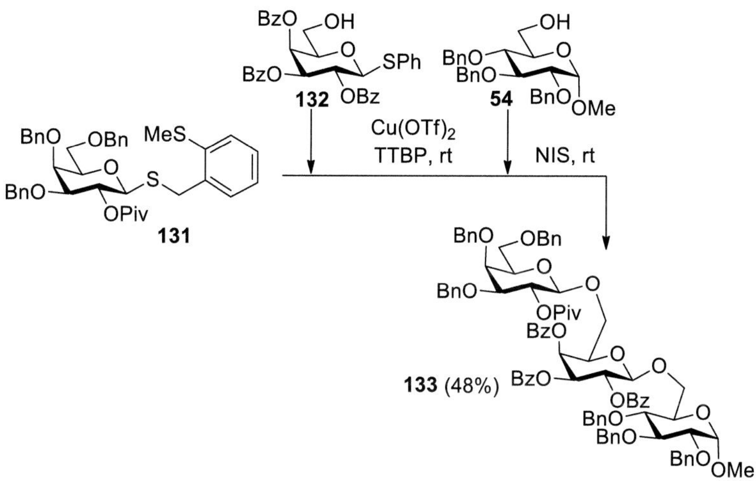

SCHEME 3.13 Orthogonal one-pot glycosylation based on *ortho*-(methylthio)benzyl thioglycoside donor.

which would act as the relay promoter to activate the in situ-formed disaccharide S*o*EP donor for subsequent glycosylation with the second acceptor **128** (bearing an SPSB leaving group). Finally, the third glycosylation was conducted in the same pot with the addition of phenylethanoid aglycon **129** and Tf_2O, which afforded the protected derivative of kankanoside F in a 63% one-pot yield.

At the same time, a protocol was developed for latent-active glycosylations[201] in which *ortho*-(methylthio)benzyl thioglycosides were used as donors (Scheme 3.13). First, donor **131** was orthogonally coupled with thiogalactoside acceptor **132** (carrying SPh as an aglycon) using $Cu(OTf)_2$ as the promoter, and a disaccharide donor was formed in situ. The selective activation of donor **131** [bearing *ortho*-(methylthio)benzyl thioaglycon] was justified based on the chelation of two sulfur atoms to the Cu metal. Following this, a second acceptor (**54**, with OMe at anomeric position) was added together with NIS at room temperature. As $Cu(OTf)_2$ was already present as the Lewis acid during the first glycosylation step, no additional acid was needed for the next step. Desired trisaccharide **133** was finally formed in a 48% one-pot yield over two steps.

Recently, glycals were used for orthogonal one-pot synthesis (Scheme 3.14).[202] First, a regio- and α-stereoselective 1,2-direct addition was performed with thioglucoside acceptor **108** in the presence of $Bi(OTf)_3$ as a dual catalyst, following which the second acceptor (**135**) and NIS were added at 0°C to pursue the next glycosylation through the activation of the in situ-formed thioglycoside disaccharide. Desired trisaccharide **136** was isolated in 65% one-pot yield.

In a recent report by the Lin group, a one-pot synthesis of type I LacNAc-repeating octasaccharide was performed (Scheme 3.15).[203] First, anomeric-OTBS-protected tetrasaccharide acceptor **141** was synthesized by thioglycoside-based one-pot chemoselective glycosylation.[163] In parallel, an orthogonal [2 + 2] glycosylation was performed between reactive benzylated trichloroacetyl oxazoline disaccharide donor **142** and thioglycoside acceptor **143** at −60°C in the catalytic presence of TMSOTf, without causing thioaglycon transfer. The in situ-formed thioglycoside tetrasaccharide donor was then activated with NIS and TfOH, which was coupled with tetrasaccharide acceptor **141** to access a tumor-associated type I LacNAc-octasaccharide (**144**) in a 45% one-pot yield over two steps.

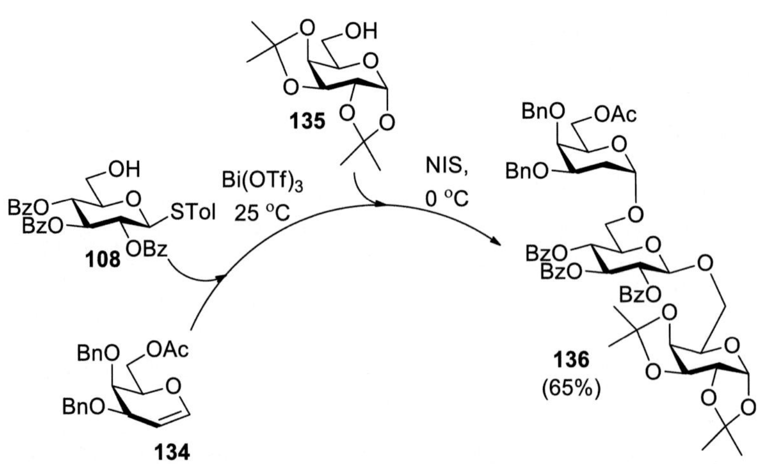

SCHEME 3.14 Thioglycoside-based one-pot orthogonal glycosylation with glycal donor.

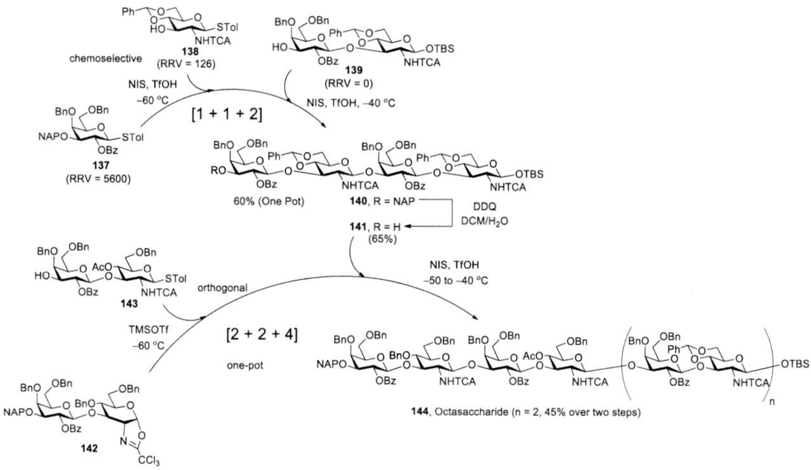

SCHEME 3.15 One-pot synthesis of type I LacNAc-repeating octasaccharide via orthogonal glycosylation between reactive trichloroacetyl oxazaoline donor and thioglycoside acceptor.

3.6 Preactivation-based glycosylation strategies

An ideal way to assemble oligosaccharides is to use a minimal set of donor and acceptor building blocks, without involving too many protection/deprotection steps.[2] Moreover, it is highly desirable to employ simple reagent-based glycosylation conditions.[3] These described features could be achieved by the development of preactivation-based glycosylation strategies.[17,147] First, upon addition of the promoter, thioglycoside donor **145** can be preactivated to form transient reactive intermediates (**146** and **147**), which could be stabilized by the formation of contact ion pair/solvent-separated ion pair or neighboring group participation (Scheme 3.16).[204] This can be followed by the addition of an acceptor thioglycoside (**148**) resulting in the formation of the glycosidic bond. Because the resulting glycoside product is equipped with an anomeric thioaglycon, the preactivation and glycosylation sequences could be repeated to achieve the synthesis of longer oligosaccharides. Because donor activation was performed in the absence of an acceptor, it is therefore not requested to modify the protecting group pattern in the acceptor building block to modulate its reactivity. Several unique oligosaccharides have been prepared by preactivation-based glycosylation approaches as described in the following section.

3.6.1 Preactivation glycosylations based on the AgOTf/p-ArSCl promoter system

p-ArSOTf, in situ-generated from the reaction of AgOTf with *p*-ArSCl, is a powerful reagent to activate highly disarmed donors. Some selected

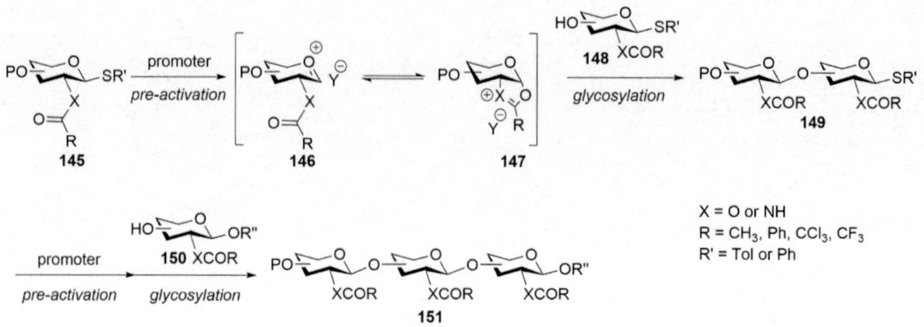

SCHEME 3.16 Scheme showing the principle of preactivation-based glycosylation.

examples of AgOTf/*p*-ArSCl-promoted preactivation glycosylations are shown in Table 3.11. In entry 1, the synthesis of a hyaluronic acid decasaccharide was accomplished by the Huang group by first preactivating donor **152** with the addition of AgOTf and *p*-TolSCl at −78°C.[205] TMSOTf was used as an additive to drive the equilibrium toward the formation of reactive oxazolinium cation intermediates and prevent the accumulation of stable and unreactive benzylidene-protected oxazolines. Decasaccharide **154** was obtained in 77% yield through a [4 + 6] glycosylation coupling. Similarly, AgOTf/*p*-TolSCl-based preactivation was accomplished between donor **155** and acceptor **156** to synthesize the challenging CDE fragment trisaccharide (**157**, 77% yield, entry 2) of an octasaccharide-containing glycopeptide, that is, syndecan-1 chondroitin sulfate glycopeptide, which plays important roles in mediating cell binding and signaling.[206] Furthermore, a stereoselective glycosylation can also be achieved through preactivation of the donor. For example, an α-stereoselective disaccharide formation was achieved by the Wang group for the synthesis of 2-deoxyglycosides containing disaccharides.[18] In entry 3, donor **158** was preactivated by stirring for 15 minutes with AgOTf and *p*-TolSCl at −78°C, followed by the addition of acceptor **159**, and continued stirring for three hours led to the formation of disaccharide product **160** in 70% yield (α only). Notably, the stereoselectivity of this reaction was attributed to the anomeric effect. Recently, Madsen et al. used commercially available *p*-NO_2PhSCl as a convenient solid reagent (easy to add) to preactivate donor **161**, followed by coupling with acceptor **162** at an optimized temperature (−65 to −55°C) to synthesize arabinoxylan oligosaccharides without intermolecular aglycon transfer (entry 4).[207] These oligosaccharides could be used as substrates for characterizing hemicellulose-degrading enzymes. Huang and coworkers performed the preactivation of donor **164** with AgOTf/*p*-TolSCl, followed by the addition of acceptor **165** to synthesize a key disaccharide intermediate (entry 5, **166**, 88% yield), which was used for the synthesis of longer bikunin-related chondroitin sulfate oligosaccharides.[208] The stereoselectivity of this reaction was governed by neighboring group participation from the C2-*O*-benzoyl group at the glycosyl

TABLE 3.11 Selected examples of AgOTf/*p*-ArSCl-promoted preactivation-based glycosylations.

Entry	Donor	Acceptor	Promoter	Conditions	Product (% Yield)	References
1	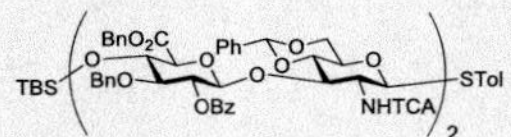 **152**	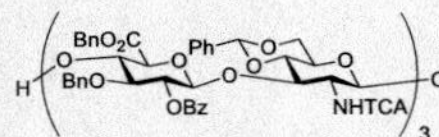 **153**	AgOTf, *p*-TolSCl TMSOTf	DCM:MeCN (19:1, v/v) −78 to −10 °C	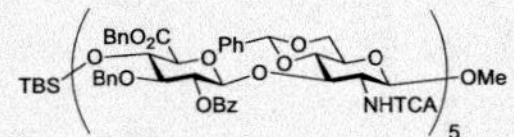 **154** (77%)	205
2	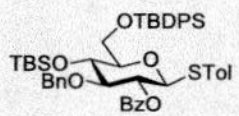 **155**	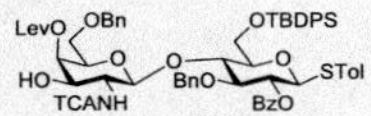 **156**	AgOTf, *p*-TolSCl TTBP	DCM −78 to 0 °C	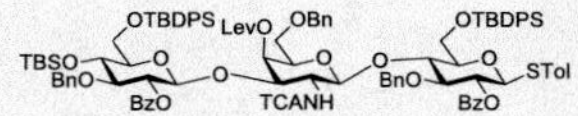 **157** (77%)	206
3	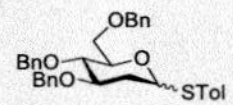 **158** (α/β:2/1)	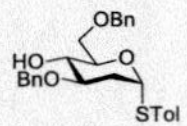 **159**	AgOTf *p*-TolSCl	DCM −78 °C	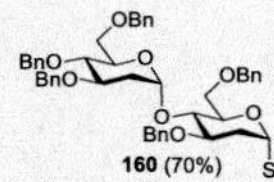 **160** (70%)	18
4	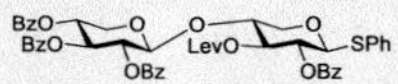 **161**	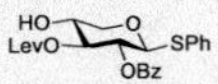 **162**	AgOTf, *p*-NO_2PhSCl	DCM −65 to −55 °C	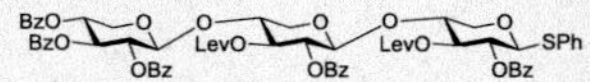 **163** (88%)	207
5	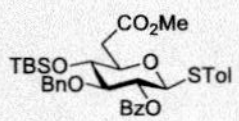 **164**	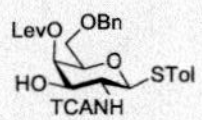 **165**	AgOTf, *p*-TolSCl TTBP	Et_2O:DCM (2:1, v/v) −78 °C	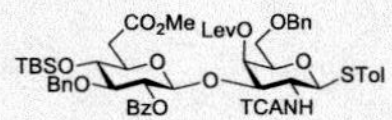 **166** (88%)	208
6	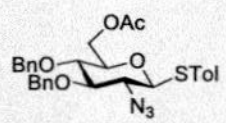 **167**	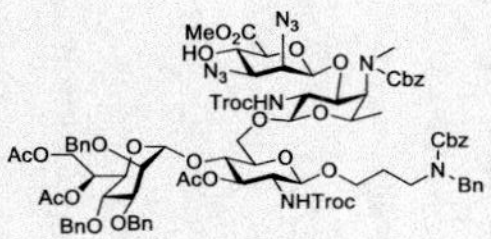 **168**	AgOTf *p*-TolSCl	Et_2O:DCM (1:1, v/v) −78 °C to rt	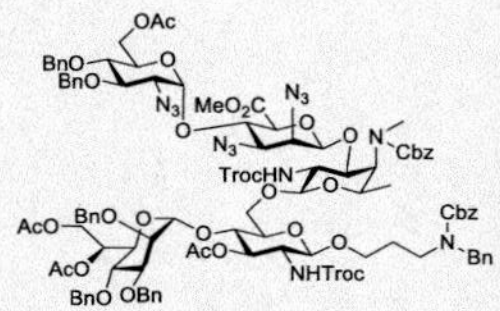 **169** (63%)	209

donor. Recently, a pertussis-LPS-like pentasaccharide (**169**) was synthesized for the first time through the preactivation of donor **167** (with 2-N_3 and 6-OAc) by using AgOTf and *p*-TolSCl, followed by subsequent coupling with acceptor **168**.[209] The α-stereoselective product was formed in 63% yield, following which deprotection steps were carried out to study its potential as an antipertussis vaccine.

Some examples of glycosylations based on the preactivation of thioglycosides by using reagents such as 1-benzenesulfinyl piperidine (BSP)[109] and Ph_2SO[114] in combination with Tf_2O and TTBP are listed in Table 3.12. These promoters are shelf stable, nonhygroscopic, and can rapidly activate both armed and disarmed donors for stereoselective glycosylations. The preactivation of disarmed donor **170** with Ph_2SO/Tf_2O followed by coupling with thioglycoside acceptor **111** and subsequent quenching with $(EtO)_3P$ was performed by van der Marel and colleagues to synthesize a Gal-β-(1→3)-GlcNAc-based disaccharide (**171**) in 64% yield (entry 1).[114] Importantly, the addition of $(EtO)_3P$ to quench the in situ-generated *S*-thiophenyl species was necessary to afford convenient yields by avoiding sequential activation of the formed thioglycoside product. Similarly, Yamago et al. preactivated a GlcNAc-based trisaccharide thioglycoside donor (**172**) with BSP/Tf_2O mixture at −60°C for coupling with monosaccharide acceptor **173** through a [3 + 1] glycosylation approach to synthesize tetra-GlcNAc product **174** in 83% yield (entry 2).[210] Additionally, a mixture of BSP and Tf_2O in the presence of TTBP was employed for the β-stereoselective glycosylation between benzylidene-protected mannose donor **175** and 2-*O*-propargyl protected acceptor thioglycoside **176** (entry 3).[24] Disaccharide product **177** (bearing anomeric SPh) was obtained in 88% yield (β/α = 16/1) and was used as a donor for the next glycosylation step. A 2-*O*-propargyl was incorporated to minimize the steric effect caused by the bulky *O*-3 substituent at the glycosyl donor to afford an increased stereoselectivity for the synthesis of a trisaccharide. Hyaluronic acid oligomers were also synthesized by a [2 + 3] preactivation strategy using BSP/Tf_2O.[211] In entry 4, disaccharide donor **178** was preactivated for 15 minutes at −60°C, followed by the addition of trisaccharide acceptor **179** (bearing an azidopropyl linker at the reducing end) and continued stirring at −15°C furnished desired pentasaccharide **180** in 48% yield. The modest yield of this reaction could be attributed to the formation of a less reactive benzylidene-protected trichloroacetyl oxazoline as a side product.

A *N*-*N*-dimethylformamide (DMF)-modulated α-stereoselective preactivation glycosylation was developed by Mong and coworkers.[212] An armed benzyl-protected fucose donor (**181**) was preactivated in the presence of stoichiometric amounts of DMF, NIS, and TMSOTf, followed by glycosylation with thiogalactoside acceptor **182**, resulting in the formation of disaccharide product **183** in 62% yield together with an excellent stereoselectivity (entry 5, α/β = 49/1). On the basis of nuclear magnetic resonance (NMR) studies, the proposed explanation for the α-stereoselectivity was the in situ formation of a glycosyl imidate intermediate upon reaction with DMF during the preactivation step.

TABLE 3.12 Other important examples of preactivation-based glycosylations.

Entry	Donor	Acceptor	Promoter	Conditions	Product (% Yield)	References
1	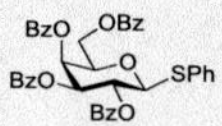 **170**	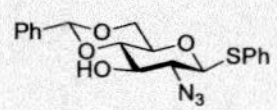 **111**	BSP, Tf_2O TTBP	DCM	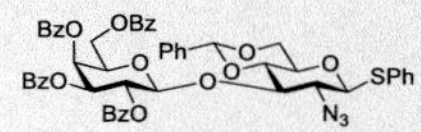 **171** (64%)	114
2	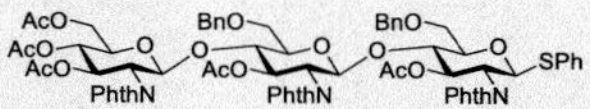 **172**	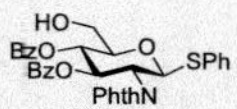 **173**	BSP, Tf_2O	DCM −60 °C	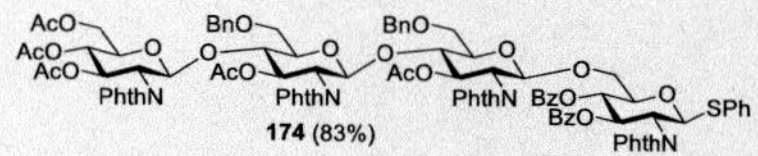 **174** (83%)	210
3	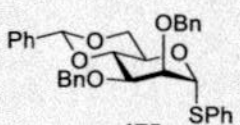 **175**	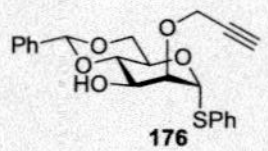 **176**	BSP, Tf_2O TTBP	−60 °C	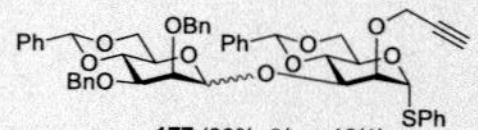 **177** (88%, β/α = 16/1)	24
4	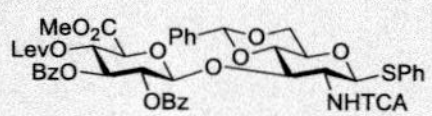 **178**	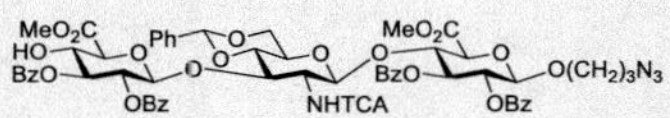 **179**	Ph_2SO, TTBP Tf_2O	DCM −60 to −15 °C	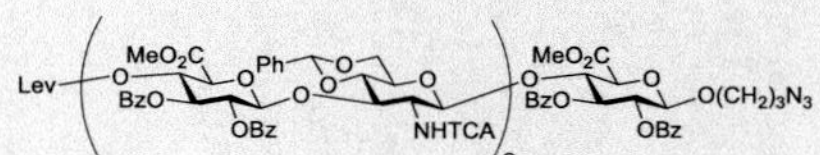 **180** (48%)	211
5	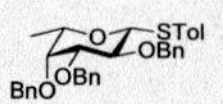 **181**	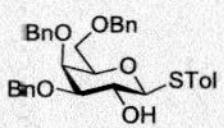 **182**	DMF, NIS TMSOTf	DCM −10 °C	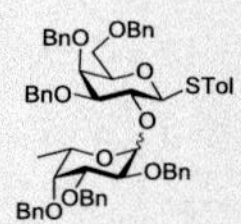 **183** (α/β:49/1; 62%)	212

3.6.2 Preactivation-based one-pot glycosylations

To simplify the process of oligosaccharide synthesis and reduce the number of purification steps, multiple glycosylations were performed in a single flask through preactivation of thiotolyl glycoside donors and acceptors. For instance, a novel preactivation-based iterative one-pot synthesis was reported by Huang and Ye (Scheme 3.17).[205] First, donor thiomannoside **184** was preactivated by using *p*-TolSCl and AgOTf at −60°C followed by the addition of acceptor thiomannoside **185** (with 2-OH) and continued stirring at room temperature. An in situ-formed thioglycoside disaccharide donor was then preactivated with the addition of *p*-TolSCl, and subsequent coupling with benzoyl-protected thioglucoside acceptor **108** generated the trisaccharide donor. This was followed by donor preactivation and glycosylation with reducing-end glucose acceptor **186** (with 6-OH) to allow coupling between a longer thioglycoside donor and a shorter disarmed acceptor. The tetrasaccharide product (**187**), containing both α and β-glycosidic linkages, was obtained in a 55% overall one-pot yield.

Iterative one-pot disarmed-armed glycosylations were also performed (Scheme 3.18).[213] Disarmed thiomannoside donor **188** (RRV = 1912) was first preactivated with DMF and NIS with the addition of TMSOTf, followed by coupling with armed thiomannoside acceptor **189** (with 2-OH, RRV = 12500). Finally, a [2 + 1] coupling was performed with the addition of reducing-end acceptor **190** along with the promoter NIS/TMSOTf to afford α-(1→2)-mannan trisaccharide **191** in a 35% one-pot yield.

The preactivation glycosylation and convergent synthesis of cancer-associated KH-1 antigen were reported by Guo and colleagues (Scheme 3.19).[214] A phthalimide-protected type 2 [Gal-β-(1→4)-GlcNAc] disaccharide donor was used. First, disaccharide donor **192** was preactivated at −78°C by the addition of *p*-TolSCl and AgOTf. After 15 minutes, disaccharide acceptor **193** was added along with TTBP as a nonnucleophilic base to prevent any acidic hydrolysis. A tetrasaccharide thioglycoside donor was formed, which was further allowed to react in situ with lactose-based acceptor **194** (bearing an azidoethyl linker at the anomeric center) through donor preactivation and glycosylation. The core linear

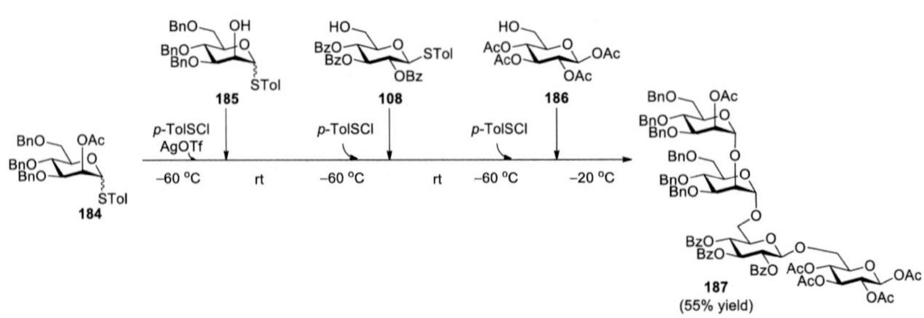

SCHEME 3.17 Iterative one-pot glycosylations based on thioglycosides.

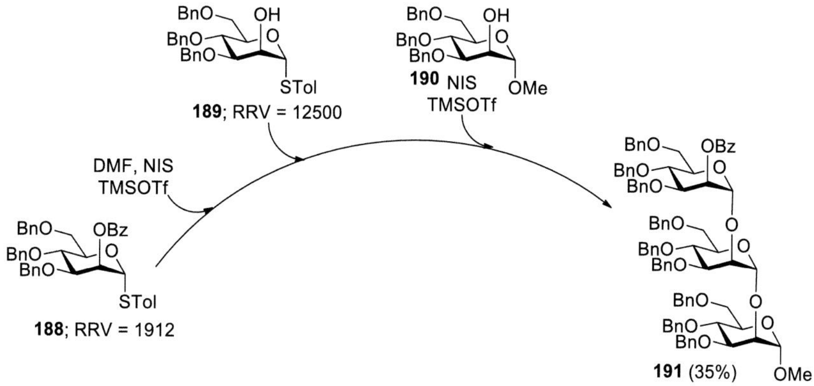

SCHEME 3.18 DMF-mediated preactivation-based one-pot glycosylation.

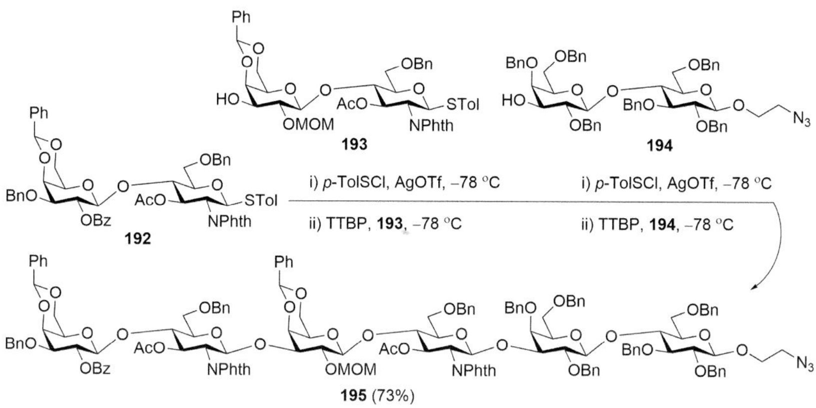

SCHEME 3.19 Synthesis of core type II LacNAc-repeating hexasaccharide of KH-1 antigen by AgOTf/*p*-TolSCl promoted preactivation-based one-pot glycosylation.

hexasaccharide (**195**) was formed through a [2 + 2 + 2] preactivation one-pot glycosylation in 73% yield. The resulting hexasaccharide was used to synthesize a KH-1 antigen derivative upon orthogonal removal of benzoyl and acetyl protecting groups and subsequent installation of three fucose residues in one step.

Recently, de-*N*-acetylated poly-β-(1→6)-*N*-acetyl-D-glucosamine (dPNAG) trisaccharide analogs were synthesized by a preactivation-based glycosylation.[215] These analogs were used to evaluate the role of substrate de-*N*-acetylation on substrate recognition by DspB, an enzyme that plays an important role in bacterial biofilm dispersal. As a representative example (Scheme 3.20), a Troc-protected GlcNAc donor thioglycoside (**196**) was preactivated with a mixture of *p*-TolSCl/AgOTf at −78°C followed by the addition of acceptor **197** (with 6-OH) at 0°C for 20 minutes. The temperature was decreased

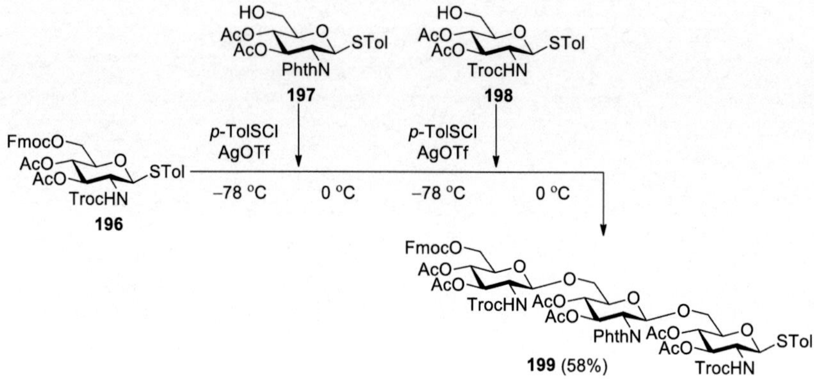

SCHEME 3.20 Synthesis of GlcNAc-containing trisaccharide by preactivation-based one-pot glycosylation.

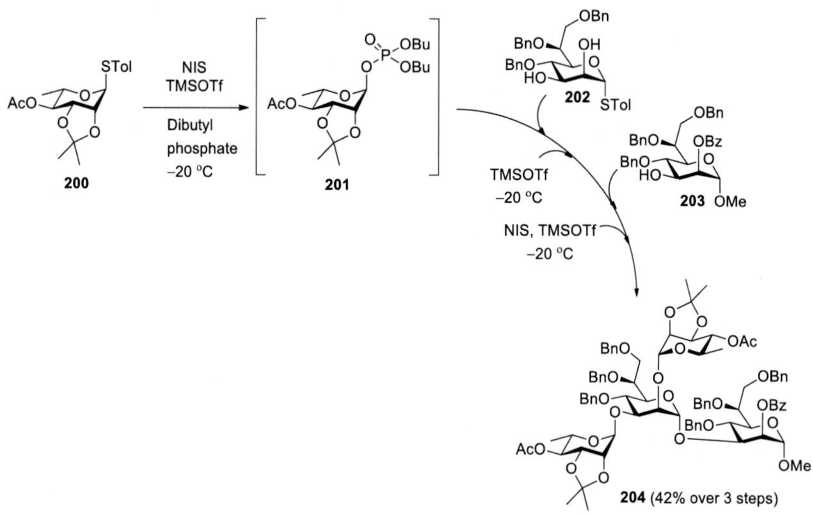

SCHEME 3.21 Preactivation of thioglycoside donor to glycosyl phosphate intermediate for the synthesis of a branched trisaccharide by one-pot glycosylation.

to −78°C after complete consumption of acceptor **197**, and a second aliquot of activation reagent was added to preactivate the in situ-formed disaccharide donor thioglycoside. The second acceptor (**198**) was then added, and the reaction was stirred at 0°C to synthesize trisaccharide **199** through a [2 + 1] glycosylation in a 58% one-pot yield.

An in situ-formed phosphate donor was recently used for the synthesis of the core oligosaccharide of the lipopolysaccharide of *Ralstonia solanacearum* through a preactivation one-pot glycosylation (Scheme 3.21).[216] Thiorhamnoside donor **200** was activated and transformed into phosphate donor intermediate **201**, which was allowed to react with thioglycoside acceptor **202** (with diol) in the presence of TMSOTf at −20°C in a stereoselective manner to form a trisaccharide with two α-glycosidic linkages. The stereoselectivity was ascribed to the presence of a conformationally strained isopropylidene protecting group at the rhamnose residue. The formed trisaccharide donor was then reacted with reducing-end Hep acceptor **203** in the presence of promoter (NIS and TMSOTf) at −20°C to yield branched tetrasaccharide **204** through one-pot glycosylation in 42% overall yield with the desired α-stereoselectivity. The stereoselective outcome was explained by the presence of two rhamnose residues, which were shielding the β-face of the trisaccharide anomeric center and therefore hindering the attack of Hep acceptor **203**.

3.7 Summary and future outlook

The main emphasis of this chapter was to provide selected examples of unique thioglycoside-based glycosylations for oligosaccharides synthesis. Multiple attractive features are associated with thioglycoside building blocks. They can be directly used as glycosyl donors and, concurrently, can be divergently converted to other glycosyl donors in one or two steps. Thioaglycons act as anomeric protecting groups, whereas a plethora of promotion systems have been implemented and studied for the activation of thioglycosides. Moreover, the glycosyl donor reactivities can be modulated by substituted thioaglycons or by using bulkier thioaglycons to avoid reactivity mismatches between the glycosyl donor and acceptor.

Like other glycosyl donors, the use of thioglycosides is, however, hampered by challenges such as laborious task of preparation of building blocks of different reactivities, aglycon transfer from thioglycoside acceptor to activated donor especially while performing glycosylations between a disarmed donor and an armed acceptor, and also issues to achieve exclusive stereoselective glycosylations. Notably, promoters to activate thioglycosides are highly electrophilic, and could lead to side reactions from the product containing nucleophilic centers. To solve those challenges, researchers have come up with ingenious solutions such as using artificial intelligence to predict the reactivity of synthesized[159,163,217] and virtual[218] building blocks (up to 50,000) by machine learning. Recently, Seeberger et al. thoroughly investigated the activation temperature of thioglycosides through novel and semiautomated assays.[219] Glycosylations conducted below this activation temperature circumvent deleterious side reactions. This represents significant efforts to use thioglycosides for oligosaccharides synthesis by overcoming some challenges. Besides, glycan assembly by using an automated glycan synthesizer is currently a hot topic in this area and, recently, challenging oligo- and polysaccharides (by the Seeberger[220] and Ye[221] groups) were synthesized by using carefully-tuned thioglycoside building blocks. For instance, the remarkable synthesis of the largest ever-synthesized polysaccharide (arabinans up to a 1080-mer, reported by the Ye group) empowers the usage of thioglycosides for challenging oligo- and polysaccharides.[221]

References

1. Essentials of glycobiology. In ; Varki, A.; Cummings, R. D.; Esko, J. D., et al., Eds. 4th edition Cold Spring Harbor Laboratory Press: Cold Spring Harbor (NY), 2022. Available from. Available from: https://www.ncbi.nlm.nih.gov/books/NBK579918/, http://doi.org/10.1101/9781621824213.
2. Codée, J. D. C.; Litjens, R. E. J. N.; van den Bos, L. J.; Overkleeft, H. S.; van der Marel, G. A. *Chem. Soc. Rev.* **2005,** *34*, 769–782.
3. Hsu, C.-H.; Hung, S.-C.; Wu, C.-Y.; Wong, C.-H. *Angew. Chem., Int. Ed. Engl.* **2011,** *50*, 11872–11923.
4. Kulkarni, S. S.; Wang, C.-C.; Sabbavarapu, N. M.; Podilapu, A. R.; Liao, P.-H.; Hung, S.-C. *Chem. Rev.* **2018,** *118*, 8025–8104.
5. Wen, L.; Edmunds, G.; Gibbons, C.; Zhang, J.; Gadi, M. R.; Zhu, H.; Fang, J.; Liu, X.; Kong, Y.; Wang, P. G. *Chem. Rev.* **2018,** *118*, 8151–8187.
6. Wang, C.-C.; Lee, J.-C.; Luo, S.-Y.; Kulkarni, S. S.; Huang, Y.-W.; Lee, C.-C.; Chang, K.-L.; Hung, S.-C. *Nature* **2007,** *446*, 896–899.
7. Boltje, T. J.; Buskas, T.; Boons, G.-J. *Nat. Chem.* **2009,** *1*, 611–622.
8. Komura, N.; Kato, K.; Udagawa, T.; Asano, S.; Tanaka, H.-N.; Imamura, A.; Ishida, H.; Kiso, M.; Ando, H. *Science* **2019,** *364*, 677–680.
9. Codée, J. D. C.; Ali, A.; Overkleeft, H. S.; van der Marel, G. A. *C. R. Chimie* **2011,** *14*, 178–193.
10. Ghosh, B.; Kulkarni, S. S. *Chem. Asian J.* **2020,** *15*, 450–462.
11. Ling, J.; Bennett, C. S. *Asian J. Org. Chem.* **2019,** *8*, 802–813.
12. Mukherjee, M. M.; Ghosh, R.; Hanover, J. A. *Front. Mol. Biosci.* **2022,** 9.
13. Chang, C.-W.; Wu, C.-H.; Lin, M.-H.; Liao, P.-H.; Chang, C.-C.; Chuang, H.-H.; Lin, S.-C.; Lam, S.; Verma, V. P.; Hsu, C.-P. *Angew. Chem., Int. Ed. Engl.* **2019,** *58*, 16775–16779.
14. Ishiwata, A.; Tanaka, K.; Ao, J.; Ding, F.; Ito, Y. *Front. Chem.* **2022,** 10. Available from: https://doi.org/10.3389/fchem.2022.972429.
15. Kafle, A.; Liu, J.; Cui, L. *Can. J. Chem.* **2016,** *94*, 894–901.
16. Nigudkar, S. S.; Demchenko, A. V. *Chem. Sci.* **2015,** *6*, 2687–2704.
17. Yang, B.; Yang, W.; Ramadan, S.; Huang, Xn *Eur. J. Org. Chem.* **2018,** *2018*, 1075–1096.
18. Verma, V. P.; Wang, C.-C. *Chem. Eur. J.* **2013,** *19*, 846–851.
19. Soliman, S. E.; Bennett, C. S. *Org. Lett.* **2018,** *20*, 3413–3417.
20. D'Angelo, K. A.; Taylor, M. S. *J. Am. Chem. Soc.* **2016,** *138*, 11058–11066.
21. Wang, L.; Overkleeft, H. S.; van der Marel, G. A.; Codée, J. D. C. *J. Am. Chem. Soc.* **2018,** *140*, 4632–4638.
22. Imamura, A.; Ando, H.; Korogi, S.; Tanabe, G.; Muraoka, O.; Ishida, H.; Kiso, M. *Tetrahedron Lett.* **2003,** *44*, 6725–6728.
23. Laroussarie, A.; Barycza, B.; Andriamboavonjy, H.; Tamigney Kenfack, M.; Blériot, Y.; Gauthier, C. *J. Org. Chem.* **2015,** *80*, 10386–10396.
24. Crich, D.; Jayalath, P. *Org. Lett.* **2005,** *7*, 2277–2280.
25. Ferrier, R.; Hay, R.; Vethaviyasar, N. *Carbohydr. Res.* **1973,** *27*, 55–61.
26. Oscarson, S. Thioglycosides. *Carbohydr. Chem. Biol.* **2000,** 93–116.

27. Kamat, M. N.; De Meo, C.; Demchenko, A. V. *J. Org. Chem.* **2007,** *72*, 6947–6955.
28. Schmidt, R. R.; Michel, J. *Angew. Chem., Int. Ed. Engl.* **1980,** *19*, 731–732.
29. Schmidt, R. R.; Zhu, X. Glycosyl trichloroacetimidates. In *Glycoscience: Chemistry and Chemical Biology;* Fraser-Reid, B. O., Tatsuta, K., Thiem, J., Eds.; Springer Berlin Heidelberg: Berlin, Heidelberg, 2008; pp 451–524.
30. Peng, P.; Schmidt, R. R. *Acc. Chem. Res.* **2017,** *50*, 1171–1183.
31. Yu, B.; Tao, H. *Tetrahedron Lett.* **2001,** *42*, 2405–2407.
32. Yu, B.; Sun, J. *Chem. Commun.* **2010,** *46*, 4668–4679.
33. Kahne, D.; Walker, S.; Cheng, Y.; Van Engen, D. *J. Am. Chem. Soc.* **1989,** *111*, 6881–6882.
34. Yan, L.; Kahne, D. *J. Am. Chem. Soc.* **1996,** *118*, 9239–9248.
35. Zeng, J.; Liu, Y.; Chen, W.; Zhao, X.; Meng, L.; Wan, Q. *Top. Curr. Chem.* **2018,** *376*, 27.
36. Koenigs, W.; Knorr, E. *Ber. Dtsch. Chem. Ges.* **1901,** *34*, 957–981.
37. Toshima, K. Glycosyl Halides. In *Glycoscience: Chemistry and Chemical Biology;* Fraser-Reid, B. O., Tatsuta, K., Thiem, J., Eds.; Springer Berlin Heidelberg: Berlin, Heidelberg, 2008; pp 429–449.
38. Singh, Y.; Geringer, S. A.; Demchenko, A. V. *Chem. Rev.* **2022,** *122*, 11701–11758.
39. Li, Y.; Yang, Y.; Yu, B. *Tetrahedron Lett.* **2008,** *49*, 3604–3608.
40. Li, Y.; Yang, X.; Liu, Y.; Zhu, C.; Yang, Y.; Yu, B. *Chem.—Eur. J.* **2010,** *16*, 1871–1882.
41. Chen, Y.; Wu, Z.; Li, X.; Hu, Y.; Yang, Y. *Org. Lett.* **2021,** *23*, 3216–3220.
42. Muru, K.; Gauthier, C. *Chem. Rec.* **2021,** *21*, 2990–3004.
43. Lian, G.; Zhang, X.; Yu, B. *Carbohydr. Res.* **2015,** *403*, 13–22.
44. Lemieux, R. *Can. J. Chem.* **1951,** *29*, 1079–1091.
45. Balavoine, G.; Gref, A.; Fischer, J.-C.; Lubineau, A. *Tetrahedron Lett.* **1990,** *31*, 5761–5764.
46. Ferrier, R. J.; Furneaux, R. H. *Carbohydr. Res.* **1976,** *52*, 63–68.
47. Ferrier, R.; Furneaux, R. 2-trans-*1-thioglycosides. General Methods*, 1. Elsevier, 1980251–253.
48. Ogawa, T.; Beppu, K.; Nakabayashi, S. *Carbohydr. Res.* **1981,** *93*, C6–C9.
49. Pozsgay, V.; Jennings, H. J. *Tetrahedron Lett.* **1987,** *28*, 1375–1376.
50. Contour, M.-O.; Defaye, J.; Little, M.; Wong, E. *Carbohydr. Res.* **1989,** *193*, 283–287.
51. Ding, Y. *Synth. Commun.* **1999,** *29*, 3541–3546.
52. Dasgupta, F.; Garegg, P. J. *Acta Chem. Scand.* **1989,** *43*, 471–475.
53. Geurtsen, R.; Holmes, D. S.; Boons, G.-J. *J. Org. Chem.* **1997,** *62*, 8145–8154.
54. Crout, D. G.; áMichael Lord, J. *J. Chem. Soc., Perkin Trans. 1* **1998,** 2287–2294.
55. Ravindranathan Kartha, K.; Field, R. A. *J. Carbohydr. Chem.* **1998,** *17*, 693–702.
56. Das, S. K.; Roy, J.; Reddy, K. A.; Abbineni, C. *Carbohydr. Res.* **2003,** *338*, 2237–2240.
57. Blom, P.; Ruttens, B.; Van Hoof, S.; Hubrecht, I.; Van der Eycken, J.; Sas, B.; Van hemel, J.; Vandenkerckhove, J. *J. Org. Chem.* **2005,** *70*, 10109–10112.
58. Weng, S.-S.; Lin, Y.-D.; Chen, C.-T. *Org. Lett.* **2006,** *8*, 5633–5636.
59. Valerio, S.; Iadonisi, A.; Adinolfi, M.; Ravida, A. *J. Org. Chem.* **2007,** *72*, 6097–6106.
60. Weng, S.-S. *Tetrahedron Lett.* **2009,** *50*, 6414–6417.
61. Kumar, V.; Taxak, N.; Jangir, R.; Bharatam, P. V.; Kartha, K. P. R. *J. Org. Chem.* **2014,** *79*, 3427–3439.
62. Escopy, S.; Singh, Y.; Demchenko, A. V. *Org. Biomol. Chem.* **2019,** *17*, 8379–8383.
63. Jensen, D. L.; Trinderup, H. H.; Skovbo, F.; Jensen, H. H. *Org. Biomol. Chem.* **2022,** *20*, 4915–4925.
64. Mukhopadhyay, B.; Kartha, K. R.; Russell, D. A.; Field, R. A. *J. Org. Chem.* **2004,** *69*, 7758–7760.

65. Tiwari, P.; Agnihotri, G.; Misra, A. K. *J. Carbohydr. Chem.* **2005,** *24*, 723–732.
66. Agnihotri, G.; Tiwari, P.; Misra, A. K. *Carbohydr. Res.* **2005,** *340*, 1393–1396.
67. Dasgupta, S.; Rajput, V. K.; Roy, B.; Mukhopadhyay, B. *J. Carbohydr. Chem.* **2007,** *26*, 91–106.
68. Chao, C.-S.; Chen, M.-C.; Lin, S.-C.; Mong, K.-K. T. *Carbohydr. Res.* **2008,** *343*, 957–964.
69. Yan, S.; Ding, N.; Zhang, W.; Wang, P.; Li, Y.; Li, M. *J. Carbohydr. Chem.* **2012,** *31*, 571–583.
70. Chatterjee, D.; Paul, A.; Yadav, S. *RSC Adv.* **2015,** *5*, 29669–29674.
71. Mukherjee, M. M.; Basu, N.; Chaudhury, A.; Ghosh, R. *RSC Adv.* **2016,** *6*, 109301–109314.
72. Vidadala, S. R.; Thadke, S. A.; Hotha, S.; Kashyap, S. *J. Carbohydr. Chem.* **2012,** *31*, 241–251.
73. Li, J.; Wang, M.; Jiang, X. *Org. Lett.* **2021,** *23*, 9053–9057.
74. Mukherjee, C.; Tiwari, P.; Misra, A. K. *Tetrahedron Lett.* **2006,** *47*, 441–445.
75. Tanaka, T.; Matsumoto, T.; Noguchi, M.; Kobayashi, A.; Shoda, S.-I. *Chem. Lett.* **2009,** *38*, 458–459.
76. Van Cleve, J. W.. *Carbohydr. Res.* **1979,** *70*, 161–164.
77. Garegg, P. J.; Henrichson, C.; Norberg, T. *Carbohydr. Res.* **1983,** *116*, 162–165.
78. Nicolaou, K.; Seitz, S.; Papahatjis, D. *J. Am. Chem. Soc.* **1983,** *105*, 2430–2434.
79. Tsai, T. Y.; Jin, H.; Wiesner, K. *Can. J. Chem.* **1984,** *62*, 1403–1405.
80. Lönn, H. *Carbohydr. Res.* **1985,** *139*, 105–113.
81. Sato, S.; Mori, M.; Ito, Y. *Carbohydr. Res.* **1986,** *155*, C6–C10.
82. Fügedi, P.; Garegg, P. J. *Carbohydr. Res.* **1986,** *149*, C9–C12.
83. Pozsgay, V.; Jennings, H. J. *J. Org. Chem.* **1987,** *52*, 4635–4637.
84. Lönn, H. *Glycoconj. J.* **1987,** *4*, 117–118.
85. Ito, Y.; Ogawa, T. *Tetrahedron Lett.* **1988,** *29*, 1061–1064.
86. Dasgupta, F.; Garegg, P. J. *Carbohydr. Res.* **1988,** *177*, C13–C17.
87. Kihlberg, J. O.; Leigh, D. A.; Bundle, D. R. *J. Org. Chem.* **1990,** *55*, 2860–2863.
88. Marra, A.; Mallet, J.-M.; Amatore, C.; Sinaÿ, P. *Synlett* **1990,** *1990*, 572–574.
89. Konradsson, P.; Mootoo, D. R.; McDevitt, R. E.; Fraser-Reid, B. *J. Chem. Soc., Chem. Commun.* **1990,** 270–272.
90. Veeneman, G.; Van Leeuwen, S.; Van Boom, J. *Tetrahedron Lett.* **1990,** *31*, 1331–1334.
91. Konradsson, P.; Udodong, U. E.; Fraser-Reid, B. *Tetrahedron Lett.* **1990,** *31*, 4313–4316.
92. Sasaki, M.; Tachibana, K.; Nakanishi, H. *Tetrahedron Lett.* **1991,** *32*, 6873–6876.
93. Veeneman, G.; Van Boom, J. *Tetrahedron Lett.* **1990,** *31*, 275–278.
94. Veeneman, G.; Van Leeuwen, S.; Zuurmond, H.; Van Boom, J. *J. Carbohydr. Chem.* **2008,** 783–796.
95. Fukase, K.; Hasuoka, A.; Kinoshita, I.; Kusumoto, S. *Tetrahedron Lett.* **1992,** *33*, 7165–7168.
96. Zhang, Y.-M.; Mallet, J.-M.; Sinaÿ, P. *Carbohydr. Res.* **1992,** *236*, 73–88.
97. Shimizu, H.; Ito, Y.; Ogawa, T. *Synlett* **1994,** 535–536.
98. Sun, L.; Li, P.; Zhao, K. *Tetrahedron Lett.* **1994,** *35*, 7147–7150.
99. Fukase, K.; Hasuoka, A.; Kinoshita, I.; Aoki, Y.; Kusumoto, S. *Tetrahedron* **1995,** *51*, 4923–4932.
100. Kartha, K. R.; Aloui, M.; Field, R. A. *Tetrahedron Lett.* **1996,** *37*, 5175–5178.
101. Fukase, K.; Kinoshita, I.; Kanoh, T.; Nakai, Y.; Hasuoka, A.; Kusumoto, S. *Tetrahedron* **1996,** *52*, 3897–3904.

102. Hiromi, U.; Teruaki, M. *Chem. Lett.* **1997,** *26*, 121–122.
103. Burkart, M. D.; Zhang, Z.; Hung, S.-C.; Wong, C.-H. *J. Am. Chem. Soc.* **1997,** *119*, 11743–11746.
104. Fukase, K.; Nakai, Y.; Kanoh, T.; Kusumoto, S. *Synlett* **1998,** *1998*, 84–86.
105. Kazuya, T.; Takayuki, T.; Hideki, J.; Teruaki, M. *Chem. Lett.* **2000,** *29*, 692–693.
106. Kazuya, T.; Takayuki, T.; Teruaki, M. *Chem. Lett.* **2000,** *29*, 124–125.
107. Jona, H.; Takeuchi, K.; Saitoh, T.; Mukaiyama, T. *Chem. Lett.* **2000,** *29*, 1178–1179.
108. Crich, D.; Smith, M. *Org. Lett.* **2000,** *2*, 4067–4069.
109. Crich, D.; Smith, M. *J. Am. Chem. Soc.* **2001,** *123*, 9015–9020.
110. Aloui, M.; Fairbanks, A. J. *Synlett* **2001,** *2001*, 0797–0799.
111. Qin, Z.-H.; Li, H.; Cai, M.-S.; Li, Z.-J. *Carbohydr. Res.* **2002,** *337*, 31–36.
112. Hashihayata, T.; Ikegai, K.; Takeuchi, K.; Jona, H.; Mukaiyama, T. *Bull. Chem. Soc. Jpn.* **2003,** *76*, 1829–1848.
113. Tsukamoto, H.; Kondo, Y. *Tetrahedron Lett.* **2003,** *44*, 5247–5249.
114. Codée, J. D.; Litjens, R. E.; den Heeten, R.; Overkleeft, H. S.; van Boom, J. H.; van der Marel, G. A. *Org. Lett.* **2003,** *5*, 1519–1522.
115. Meijer, A.; Ellervik, U. *J. Org. Chem.* **2004,** *69*, 6249–6256.
116. Ercegovic, T.; Meijer, A.; Magnusson, G.; Ellervik, U. *Org. Lett.* **2001,** *3*, 913–915.
117. Durón, S. G.; Polat, T.; Wong, C.-H. *Org. Lett.* **2004,** *6*, 839–841.
118. Mukhopadhyay, B.; Collet, B.; Field, R. A. *Tetrahedron Lett.* **2005,** *46*, 5923–5925.
119. Wang, C.; Wang, H.; Huang, X.; Zhang, L.-H.; Ye, X.-S. *Synlett* **2006,** *2006*, 2846–2850.
120. Tatai, J.; Fügedi, P. *Org. Lett.* **2007,** *9*, 4647–4650.
121. Huang, K. T.; Winssinger, N. *Eur. J. Org. Chem.* **2007,** *2007*, 1887–1890.
122. Schmidt, T. H.; Madsen, R. *Eur. J. Org. Chem.* **2007,** *2007*, 3935–3941.
123. Crich, D.; Cai, F.; Yang, F. *Carbohydr. Res.* **2008,** *343*, 1858–1862.
124. Xiong, D. C.; Zhang, L. H.; Ye, X. S. *Adv. Synth. Catal.* **2008,** *350*, 1696–1700.
125. Galan, M. C.; Brunet, C.; Fuensanta, M. *Tetrahedron Lett.* **2009,** *50*, 442–445.
126. Galan, M. C.; Tran, A. T.; Boisson, J.; Benito, D.; Butts, C.; Eastoe, J.; Brown, P. *J. Carbohydr. Chem.* **2011,** *30*, 486–497.
127. Peng, P.; Ye, X.-S. *Org. Biomol. Chem.* **2011,** *9*, 616–622.
128. Basu, N.; Maity, S. K.; Ghosh, R. *RSC Adv.* **2012,** *2*, 12661–12664.
129. Maity, S. K.; Basu, N.; Ghosh, R. *Carbohydr. Res.* **2012,** *354*, 40–48.
130. Daly, R.; Scanlan, E. M. *Org. Biomol. Chem.* **2013,** *11*, 8452–8457.
131. Chu, A.-H. A.; Minciunescu, A.; Montanari, V.; Kumar, K.; Bennett, C. S. *Org. Lett.* **2014,** *16*, 1780–1782.
132. He, H.; Zhu, X. *Org. Lett.* **2014,** *16*, 3102–3105.
133. Hasty, S. J.; Ranade, S. C.; Demchenko, A. V. *Rep. Org. Chem.* **2014,** *4*, 1–10.
134. Liu, G.-j; Zhang, X.-t; Xing, G.-W. *Chem. Commun.* **2015,** *51*, 12803–12806.
135. Kajimoto, T.; Morimoto, K.; Ogawa, R.; Dohi, T.; Kita, Y. *Eur. J. Org. Chem.* **2015,** *2015*, 2138–2142.
136. Zhang, Y.; Knapp, S. *J. Org. Chem.* **2016,** *81*, 2228–2242.
137. Nukada, T.; Berces, A.; Whitfield, D. M. *J. Org. Chem.* **1999,** *64*, 9030–9045.
138. Périon, R.; Lemée, L.; Ferrières, V.; Duval, R.; Plusquellec, D. *Carbohydr. Res.* **2003,** *338*, 2779–2792.
139. Mukherjee, S.; Mukhopadhyay, B. *Synlett* **2010,** *2010*, 2853–2856.
140. Vibhute, A. M.; Dhaka, A.; Athiyarath, V.; Sureshan, K. M. *Chem. Sci.* **2016,** *7*, 4259–4263.

141. Vibhute, A. M.; Dhaka, A.; Athiyarath, V.; Sureshan, K. M. *Chem. Sci.* **2016,** *7*, 6282 6282.
142. Goswami, M.; Ellern, A.; Pohl, N. L. *Angew. Chem., Int. Ed. Engl.* **2013,** *52*, 8441–8445.
143. Kabotso, D. E.; Pohl, N. L. *Org. Lett.* **2017,** *19*, 4516–4519.
144. Xu, F.-F.; Pereira, C. L.; Seeberger, P. H. *Beil. J. Org. Chem.* **2017,** *13*, 1994–1998.
145. Escopy, S.; Singh, Y.; Demchenko, A. V. *Org. Biomol. Chem.* **2021,** *19*, 2044–2054.
146. Crich, D.; Sun, S. *Tetrahedron* **1998,** *54*, 8321–8348.
147. Huang, X.; Huang, L.; Wang, H.; Ye, X.-S. *Angew. Chem. Int. Ed. Engl.* **2004,** *43*, 5221–5224.
148. Pooladian, F.; Escopy, S.; Demchenko, A. V. *Molecules* **2022,** *27* (21), 7354.
149. Hanessian, S.; Bacquet, C.; Lehong, N. *Carbohydr. Res.* **1980,** *80*, C17–C22.
150. Chen, Q.; Kong, F. *Carbohydr. Res.* **1995,** *272*, 149–157.
151. Konradsson, P.; Mootoo, D. R.; McDevitt, R. E.; Fraser-Reid, B. *J. Chem. Soc., Chem. Commun.* **1990,** 270–272.
152. Garegg, P. J.; Oscarson, S.; Szönyi, M. *Carbohydr. Res.* **1990,** *205*, 125–132.
153. Krog-Jensen, C.; Oscarson, S. *J. Org. Chem.* **1996,** *61*, 1234–1238.
154. Meijer, A.; Ellervik, U. *J. Org. Chem.* **2002,** *67*, 7407–7412.
155. Zuurmond, H. M.; van der Laan, S. C.; van der Marel, G. A.; van Boom, J. H. *Carbohydr. Res.* **1991,** *215*, C1–C3.
156. Griffin, G. W.; Bandara, N.; Clarke, M. A.; Tsang, W. S.; Garegg, P. J.; Oscarson, S.; Silwanis, B. A. *Heterocycles* **1990,** *30*, 939–947.
157. Amatore, C.; Jutand, A.; Mallet, J.-M.; Meyer, G.; Sinay, P. *J. Chem. Soc., Chem. Commun.* **1990,** 718–719.
158. Ley, S. V.; Priepke, H. W. M. *Angew. Chem., Int. Ed. Engl.* **1994,** *33*, 2292–2294.
159. Zhang, Z.; Ollmann, I. R.; Ye, X.-S.; Wischnat, R.; Baasov, T.; Wong, C.-H. *J. Am. Chem. Soc.* **1999,** *121*, 734–753.
160. Smith, R.; Müller-Bunz, H.; Zhu, X. *Org. Lett.* **2016,** *18*, 3578–3581.
161. Mc Carthy, C.; Zhu, X. *Org. Biomol. Chem.* **2020,** *18*, 9029–9034.
162. Pedersen, C. M.; Marinescu, L. G.; Bols, M. *Chem. Commun.* **2008,** 2465–2467.
163. Verma, N.; Tu, Z.; Lu, M.-S.; Liu, S.-H.; Renata, S.; Phang, R.; Liu, P.-K.; Ghosh, B.; Lin, C.-H. *J. Org. Chem.* **2021,** *86*, 892–916.
164. Li, Z.; Gildersleeve, J. C. *J. Am. Chem. Soc.* **2006,** *128*, 11612–11619.
165. Tu, Z.; Hsieh, H.-W.; Tsai, C.-M.; Hsu, C.-W.; Wang, S.-G.; Wu, K.-J.; Lin, K.-I.; Lin, C.-H. *Chem. Asian J.* **2013,** *8*, 1536–1550.
166. Tu, Z.; Liu, P.-K.; Wu, M.-C.; Lin, C.-H. *Isr. J. Chem.* **2015,** *55*, 325–335.
167. Arboe Jennum, C.; Hauch Fenger, T.; Bruun, L. M.; Madsen, R. *Eur. J. Org. Chem.* **2014,** *2014*, 3232–3241.
168. Heuckendorff, M.; Poulsen, L. T.; Jensen, H. H. *J. Org. Chem.* **2016,** *81*, 4988–5006.
169. Zhou, J.; Lv, S.; Zhang, D.; Xia, F.; Hu, W. *J. Org. Chem.* **2017,** *82*, 2599–2621.
170. Jeanneret, R. A.; Dalton, C. E.; Gardiner, J. M. *J. Org. Chem.* **2019,** *84*, 15063–15078.
171. Huang, L.; Wang, Z.; Huang, X. *Chem. Commun.* **2004,** 1960–1961.
172. Mező, E.; Herczeg, M.; Eszenyi, D.; Borbás, A. *Carbohydr. Res.* **2014,** *388*, 19–29.
173. Dhara, D.; Baliban, S. M.; Huo, C.-X.; Rashidijahanabad, Z.; Sears, K. T.; Nick, S. T.; Misra, A. K.; Tennant, S. M.; Huang, X. *Chem.–Eur. J.* **2020,** *26*, 15953–15968.
174. Mong, T. K.; Huang, C. Y.; Wong, C. H. *J. Org. Chem.* **2003,** *68*, 2135–2142.
175. Mydock, L. K.; Demchenko, A. V. *Org. Lett.* **2008,** *10*, 2107–2110.
176. Kaeothip, S.; Demchenko, A. V. *Carbohydr. Res.* **2011,** *346*, 1371–1388.
177. Tanaka, H.; Adachi, M.; Takahashi, T. *Tetrahedron Lett.* **2004,** *45*, 1433–1436.

178. Ozaki, H.; Asano, T.; Tanaka, H.-N.; Komura, N.; Ando, H.; Ishida, H.; Imamura, A. *Carbohydr. Res.* **2022,** *512*, 108502.
179. Matsuo, I.; Totani, K.; Tatami, A.; Ito, Y. *Tetrahedron* **2006,** *62*, 8262–8277.
180. Soliman, S. E.; Kováč, P. *Angew. Chem., Int. Ed. Engl.* **2016,** *55*, 12850–12853.
181. Chernyak, A.; Oscarson, S.; Turek, D. *Carbohydr. Res.* **2000,** *329*, 309–316.
182. Cheshev, P. E.; Kononov, L. O.; Tsvetkov, Y. E.; Shashkov, A. S.; Nifantiev, N. E. *Russ. J. Bioorg. Chem.* **2002,** *28*, 419–429.
183. Manabe, S.; Ishii, K.; Ito, Y. *J. Org. Chem.* **2007,** *72*, 6107–6115.
184. Gemma, E.; Lahmann, M.; Oscarson, S. *Carbohydr. Res.* **2005,** *340*, 2558–2562.
185. Zhu, T.; Boons, G.-J. *Carbohydr. Res.* **2000,** *329*, 709–715.
186. Guazzelli, L.; Crawford, C. J.; Ulc, R.; Bowen, A.; McCabe, O.; Jedlicka, A. J.; Wear, M. P.; Casadevall, A.; Oscarson, S. *Chem. Sci.* **2020,** *11*, 9209–9217.
187. Zhang, L.; Zhu, X. *Carbohydr. Res.* **2014,** *391*, 43–47.
188. Sanapala, S. R.; Kulkarni, S. S. *J. Am. Chem. Soc.* **2016,** *138*, 4938–4947.
189. Pathan, E. K.; Ghosh, B.; Podilapu, A. R.; Kulkarni, S. S. *J. Org. Chem.* **2021,** *86*, 6090–6099.
190. de Jong, A.-R.; Hagen, B.; van der Ark, V.; Overkleeft, H. S.; Codée, J. D. C.; Van der Marel, G. A. *J. Org. Chem.* **2012,** *77*, 108–125.
191. Walvoort, M. T. C.; Volbeda, A. G.; Reintjens, N. R. M.; van den Elst, H.; Plante, O. J.; Overkleeft, H. S.; van der Marel, G. A.; Codée, J. D. C. *Org. Lett.* **2012,** *14*, 3776–3779.
192. Zhang, Q.; vanRijssel, E. R.; Walvoort, M. T. C.; Overkleeft, H. S.; van der Marel, G. A.; Codée, J. D. C. *Angew. Chem., Int. Ed. Engl.* **2015,** *54*, 7670–7673.
193. Yu, F.; Nguyen, H. M. *J. Org. Chem.* **2012,** *77*, 7330–7343.
194. Chuang, H.-Y.; Ren, C.-T.; Chao, C.-A.; Wu, C.-Y.; Shivatare, S. S.; Cheng, T.-J. R.; Wu, C.-Y.; Wong, C.-H. *J. Am. Chem. Soc.* **2013,** *135*, 11140–11150.
195. Liu, M.; Li, B.-H.; Xiong, D.-C.; Ye, X.-S. *J. Org. Chem.* **2018,** *83*, 8292–8303.
196. Shaw, M.; Thakur, R.; Kumar, A. *J. Org. Chem.* **2019,** *84*, 589–605.
197. Ding, N.; Li, X.; Chinoy, Z. S.; Boons, G.-J. *Org. Lett.* **2017,** *19*, 3827–3830.
198. Dey, S.; Lo, H.-J.; Wong, C.-H. *J. Am. Chem. Soc.* **2019,** *141*, 10309–10314.
199. Zhang, Y.; Xiang, G.; He, S.; Hu, Y.; Liu, Y.; Xu, L.; Xiao, G. *Org. Lett.* **2019,** *21*, 2335–2339.
200. Xiao, X.; Zeng, J.; Fang, J.; Sun, J.; Li, T.; Song, Z.; Cai, L.; Wan, Q. *J. Am. Chem. Soc.* **2020,** *142*, 5498–5503.
201. Hedberg, C.; Jessen, K. S.; Hansson, R. F.; Heuckendorff, M.; Jensen, H. H. *Org. Lett.* **2020,** *22*, 7068–7072.
202. Kumar, M.; Gurawa, A.; Kumar, N.; Kashyap, S. *Org. Lett.* **2022,** *24*, 575–580.
203. Verma, N.; Renata, S.; Lin, C.-H. Reactive Oxazoline-based Synthesis of Type I LacNAc-repeating Oligosaccharides by Efficient One-pot Glycosylations. Unpublished work, 2022.
204. Yang, L.; Qin, Q.; Ye, X.-S. *Asian J. Org. Chem.* **2013,** *2*, 30–49.
205. Lu, X.; Kamat, M. N.; Huang, L.; Huang, X. *J. Org. Chem.* **2009,** *74*, 7608–7617.
206. Ramadan, S.; Yang, W.; Zhang, Z.; Huang, X. *Org. Lett.* **2017,** *19*, 4838–4841.
207. Underlin, E. N.; Böhm, M.; Madsen, R. *J. Org. Chem.* **2019,** *84*, 16036–16054.
208. Ramadan, S.; Li, T.; Yang, W.; Zhang, J.; Rashidijahanabad, Z.; Tan, Z.; Parameswaran, N.; Huang, X. *ACS Cent. Sci.* **2020,** *6*, 913–920.
209. Wang, P.; Huo, C.-x; Lang, S.; Caution, K.; Nick, S. T.; Dubey, P.; Deora, R.; Huang, X. *Angew. Chem., Int. Ed. Engl.* **2020,** *59*, 6451–6458.
210. Yamago, S.; Yamada, T.; Maruyama, T.; Yoshida, J.-I. *Angew. Chem., Int. Ed. Engl.* **2004,** *43*, 2145–2148.

211. Dinkelaar, J.; Codee, J. D.; van den Bos, L. J.; Overkleeft, H. S.; van der Marel, G. A. *J. Org. Chem.* **2007,** *72*, 5737–5742.
212. Lu, S.-R.; Lai, Y.-H.; Chen, J.-H.; Liu, C.-Y.; Mong, K.-K. T. *Angew. Chem., Int. Ed. Engl.* **2011,** *50*, 7315–7320.
213. Lin, Y. H.; Ghosh, B.; Tony Mong, K.-K. *Chem. Commun.* **2012,** *48*, 10910–10912.
214. Li, Q.; Guo, Z. *Org. Lett.* **2017,** *19*, 6558–6561.
215. Wang, S.; Breslawec, A. P.; Alvarez, E.; Tyrlik, M.; Li, C.; Poulin, M. B. *ACS Chem. Biol.* **2019,** *14*, 1998–2005.
216. Mong, K.-K. T.; Cheng, K.-C.; Lu, I. C.; Pan, C.-W.; Wang, Y.-F.; Shen, L.-C. *J. Org. Chem.* **2020,** *85*, 16060–16071.
217. Li, X.; Huang, L.; Hu, X.; Huang, X. *Org. Biomol. Chem.* **2009,** *7*, 117–127.
218. Cheng, C.-W.; Zhou, Y.; Pan, W.-H.; Dey, S.; Wu, C.-Y.; Hsu, W.-L.; Wong, C.-H. *Nat. Commun.* **2018,** *9*, 5202.
219. Tuck, O. T.; Sletten, E. T.; Danglad-Flores, J.; Seeberger, P. H. *Angew. Chem., Int. Ed. Engl.* **2022,** *61*, e202115433.
220. Joseph, A. A.; Pardo-Vargas, A.; Seeberger, P. H. *J. Am. Chem. Soc.* **2020,** *142*, 8561–8564.
221. Yao, W.; Xiong, D.-C.; Yang, Y.; Geng, C.; Cong, Z.; Li, F.; Li, B.-H.; Qin, X.; Wang, L.-N.; Xue, W.-Y. *Nat. Synth.* **2022,** *1*, 854–863.

Chapter 4

General strategy for the synthesis of *N*-glycosides

Ariza Khanam[1] and Pintu Kumar Mandal[1,2]

[1]*Medicinal and Process Chemistry Division, CSIR-Central Drug Research Institute, Lucknow, Uttar Pradesh, India,* [2]*Academy of Scientific and Innovative Research (AcSIR), Ghaziabad, Uttar Pradesh, India*

4.1 Introduction

Carbohydrates involving glycoconjugates are the most functionally and structurally diverse molecules in nature,[1–4] playing immense importance in several life processes like cell−cell communication and adhesion, embryogenesis, protein folding, blood group specificity, bacterial and viral infection, consequential immune responses, and cancer metastasis.[5–11] This structural diversity of carbohydrates is generally translated into a vast array of biological functions, allowing carbohydrates to be considered as "privileged" molecules. Carbohydrates can be linked to an aromatic aglycone through *O*-, *N*-, and *C*-glycosidic bonds to give *O*-, *N*-, and *C*-glycosides, respectively.[12–14]

N-Glycosides or aminosugars are widespread in nature,[15] possessing a variety of biological roles. *N*-Glycosides (1-aminosugars and derivatives) are a subclass that may serve as synthetic intermediates, besides playing a key role in *N*-linked glycoproteins involved in many recognition events or nucleosides and antibiotics (e.g., streptomycin, found in 1944).[16,17] After that, a series of other aminoglycosides or other antibiotics from natural sources are discovered[18] (Fig. 4.1).

In the midst of this, *N*-glycosides secured a privileged position due to their significant role in DNA/RNA synthesis, wherein a nucleobase (purine or pyrimidine) is attached to a ribofuranose sugar via an *N*-glycosidic bond. The most well-known *N*-glycosides are nucleosides, which can serve not only as essential components of DNA/RNA[19] but also as clinical drugs for tumor treatment, antiviral therapy, and immune regulation.[20–24] Also, they were found as the first organic molecules on our planet.[25] Another common category of *N*-glycosides are glycopeptides and glycoproteins, which are found attached to proteins through the nitrogen atom of asparagine.[26,27]

Synthetic Strategies in Carbohydrate Chemistry. DOI: https://doi.org/10.1016/B978-0-323-91729-2.00008-2

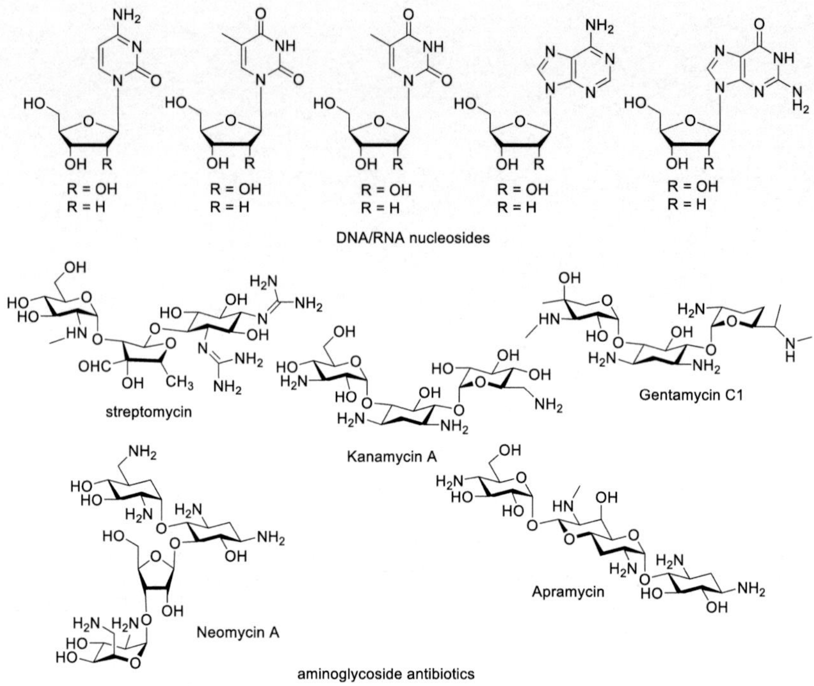

FIGURE 4.1 Structure of nucleosides and aminoglycoside antibiotics.

4.2 Nucleosides and their derivatives found in nature

As the building units of nucleic acids, nucleotides are formed by the combination of a nucleobase, either ribose sugar or deoxyribose sugar, as well as phosphate. Nucleoside, on the other hand, lacks phosphate at the C5′ position. In naturally existing nucleosides, the sugar (ribose or deoxyribose) moiety is attached to nucleobases via β-glycosidic bonds. The five common types of nucleosides are adenosine, guanosine, cytidine, thymidine, and uridine (Fig. 4.2).[28] Naturally modified nucleosides include a significant and diverse number of compounds (Fig. 4.3), some of them with slight changes mostly at the base or major structural modifications done by enzymes. Specifically, nucleosides having alkyne moiety like 5-ethynyluridine have been widely used for the labeling of nucleic acids, incorporating the inspection of RNA synthesis as well as envision of cellular localization.[29–33] In the medicinal field as well, many drugs contain synthetic nucleosides or nucleoside analogs for the treatment of disease, especially cancer and viral infection.[20,21,25,34,35] As a result, designing expeditious routes to obtain nucleosides is of paramount importance to organic and medicinal chemists.

N-Glycosides, excluding nucleosides, exist in natural products as well as pharmaceutical molecules and display several biological roles, such as

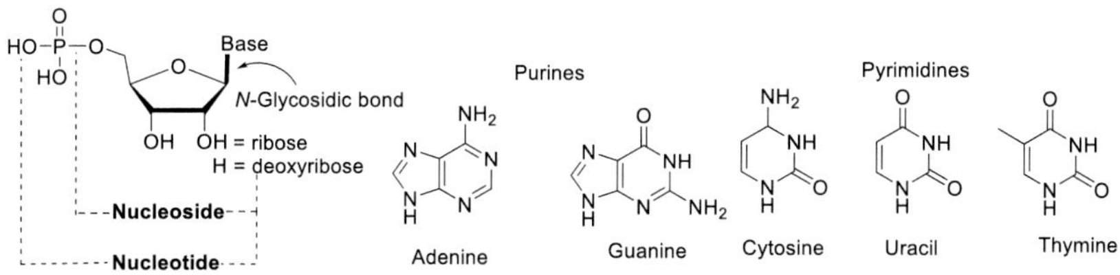

FIGURE 4.2 The structural compositions of nucleosides and nucleotides.

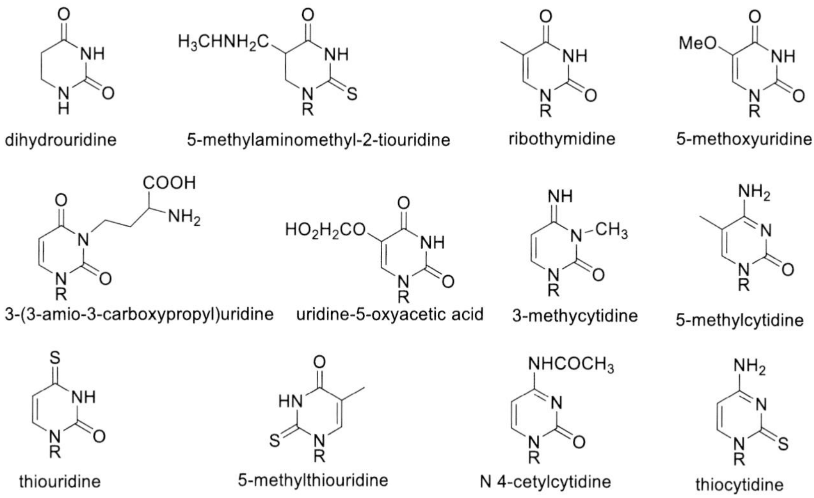

FIGURE 4.3 Some examples of naturally modified nucleosides.

glycogen phosphorylase inhibitors and galactosidase inhibitors.[36–40] Consequently, the insistence on such molecules in every area of life science has entailed the discovery of chemical glycosylation strategies for their production.[41–46] Considering their significance in carbohydrate chemistry, this chapter presents the summary of various methods used to synthesize *N*-glycosides, including nucleosides, and also points out new challenges and opportunities. We hope that this comprehensive collection of methods might serve as inspiration for glycochemists to discover new opportunities, both in terms of the application of existing techniques, as well as in the development of new chemical methodologies.

4.3 General strategy for the chemical synthesis of nucleosides and other *N*-glycosides

Nature flawlessly executes the *N*-glycosylation reaction enzymatically, but chemical glycosylation remains incommodious. Therefore the evaluation of a new route for the synthesis of *N*-glycosides is of paramount significance to organic as well as medicinal chemists. Based on the literature, three general glycosylation strategies dominate in *N-glycoside* synthesis. (1) replacing the various leaving group on sugar moiety involving the nucleophilic attack of heterocyclic base (such as purine, pyrimidine, quinolone, and asparagine) via *N*-glycosylation; (2) replacing the anomeric − OH group of sugar by *N*-containing groups; and (3) by direct installing amine group into sugar moiety or the formation of reactive glycosyl cation with charge delocalization across the ring oxygen to promote direct attack of *N*-containing nucleophile into sugar moiety (Scheme 4.1). All these strategies lead to the formation of biologically significant nucleosides or *N*-glycosides that prevail ubiquitously in DNA/RNA, glycoproteins, and other biological molecules.

4.3.1 Chemical synthesis of nucleosides via *N*-glycosylation

Conventionally, the synthesis of nucleoside has occurred via *N*-glycosylation of a nucleobase involving the nucleophilic attack of the heterocyclic base on sugar moiety (Scheme 4.2). Albeit the appearance of synthetic pathway of nucleosides may rather simple, it is complicated due to the low basicity of nucleobases and requirement of elaborate protecting group manipulations present on sugar moiety that make this reaction more challenging in terms of regio- and diastereoselectivity[47] and may also cause other side reaction like nonselective nucleophile attack (α-nucleoside formation) as well as nucleophilic attack of other functional groups, leading to product mixtures that difficult to separate. To overcome such hurdles, a wide range of strategies have been developed to produce nucleosides with good yield and better selectivity.

All the methods reported to date for the synthesis of nucleosides generally progress via three key intermediates (I, II, III) that further undergo attack by nucleobase. These key intermediates are 1,3-dioxolane cation formation via

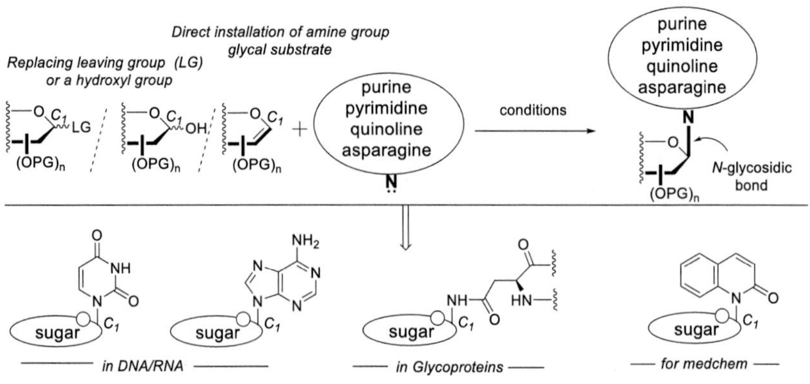

SCHEME 4.1 General strategy for the chemical synthesis of various *N*-glycosides (see text for explanations).

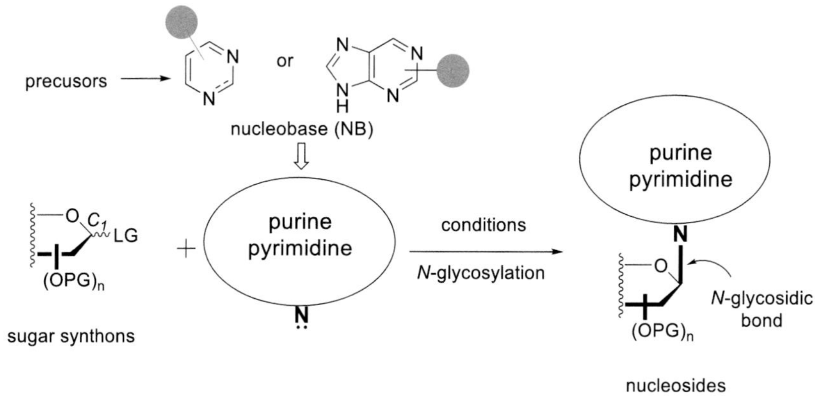

SCHEME 4.2 Synthesis of nucleosides attacked by nucleobase.

neighboring group participation, generation of a reactive intermediate, that is, glycosyl cation having delocalization of charge across the ring oxygen, and the employment of a good leaving group that sets the right configuration upon S_N2-type substitution at the anomeric position (Scheme 4.3). All these intermediates can be obtained by using diverse glycosyl donors containing one or more protecting groups obtained from different types of sugar such as ribose, deoxyribose, and pyranose in a number of synthetic steps (Scheme 4.3, bottom). Noteworthy, the following overview mentions only those strategies that conventionally synthesize the nucleoside in reasonable yield and selectivity.

4.3.1.1 Glycosyl acetates

The Vorbrüggen[48] extended silyl form of the Hilbert–Johnson nucleoside synthesis from glycosyl acetate donors is the most classical strategy for the synthesis of pyrimidine nucleosides specifically (Scheme 4.4). In this, a fully acetate-protected glycosyl donor is catalyzed by Lewis acid, generating an intermediate glycosyl cation. This reactive intermediate was attacked by silyl-protected nucleobase to furnish a desired product after global deprotection. Different leverages, such as exceptional substrate scope, easy adaptability, and reliability, have made this process more popular in the chemical industry. Howbeit, this process has a few disadvantages like the requirement of silylated nucleobases as well as harsh reaction conditions, and also the use of stoichiometric amounts of Lewis acid catalyst[49–52] often limits the choice of protecting groups. This method is not always applicable to the synthesis of purine and asparagine glycosides.

Recently, Styskala and colleague[53] reported a novel methodology for the synthesis of peracetylated N^7-nucleosides with a broad application range.

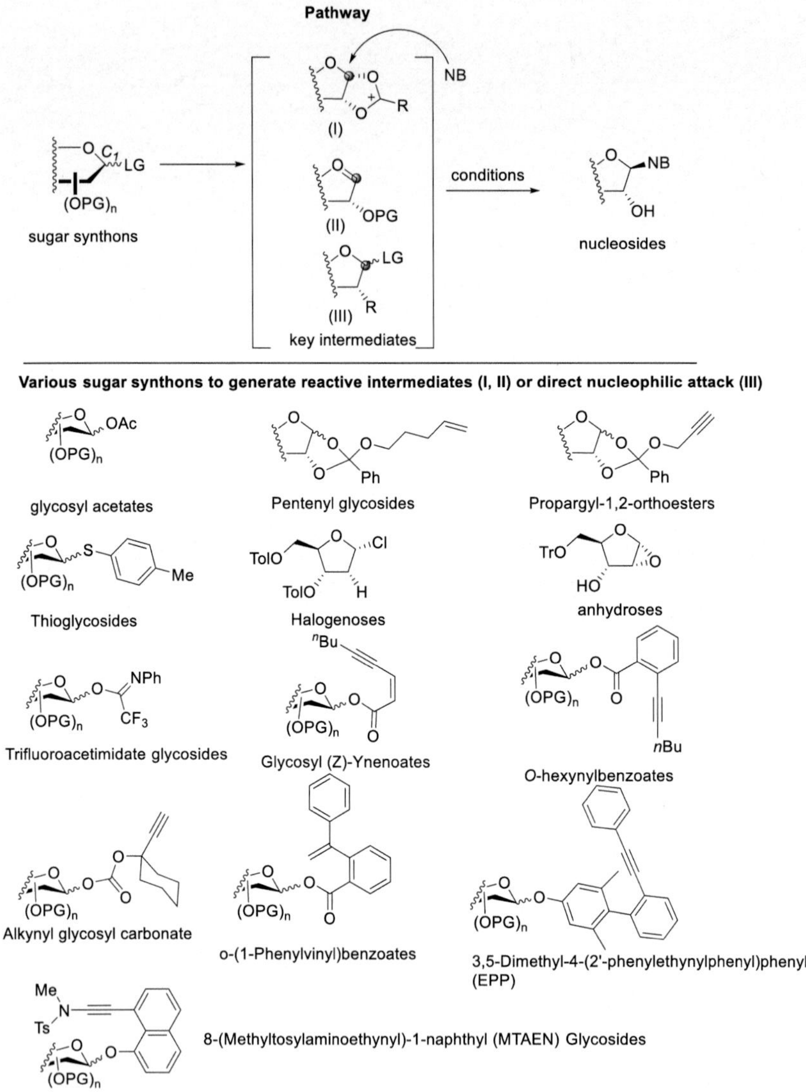

SCHEME 4.3 Synthesis of nucleosides via key intermediates and various sugar synthons to generate reactive intermediates.

The interaction of trimethylsilyl-protected purine bases (6-chloropurine or 2,6-dichloropurine) with anomeric 1-*O*-acyl glycosyl derivatives under the activation of $SnCl_4$ or $TiCl_4$ catalyst via Vorbrüggen reaction condition in dichloroethane (DCE) solvent leading to the formation of predominant N^7-nucleosides and in few cases as a single compound (Scheme 4.5).

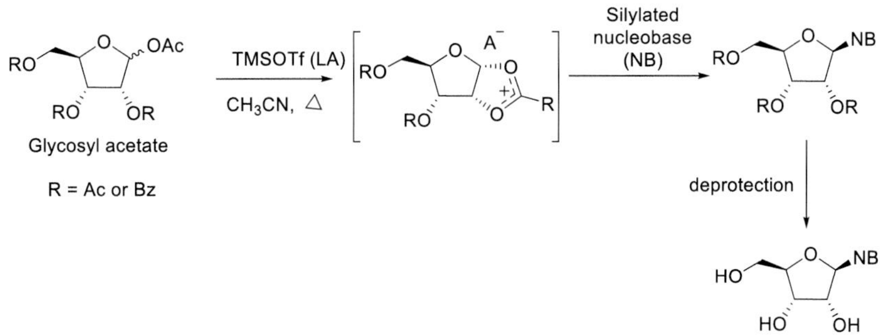

SCHEME 4.4 Vorbrüggen classic method from anomeric 1-*O*-acyl glycosyl donors.

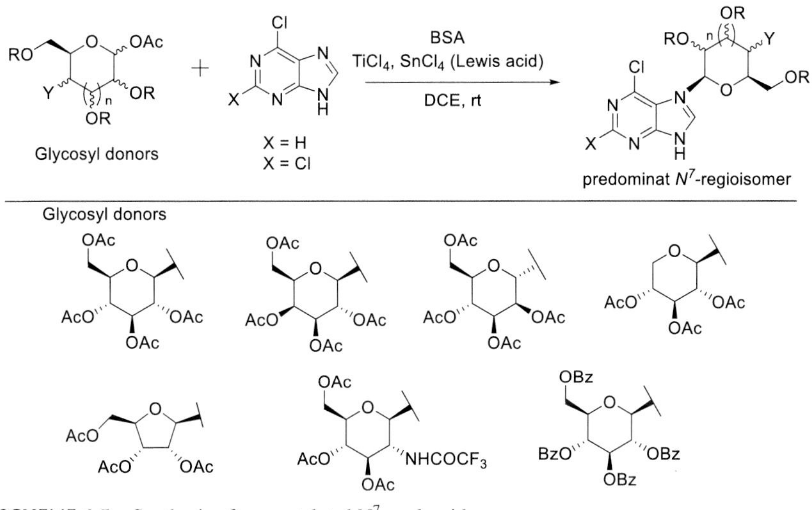

SCHEME 4.5 Synthesis of peracetylated N^7-nucleosides.

4.3.1.2 Glycosyl halide

Since the first method was prominent for the formation of 2-deoxy nucleosides involving methoxyriboside transformation to glycosyl chloride and its direct nucleophilic substitution with nucleobase (Scheme 4.6), innumerable methodology has been performed for nucleobase promotion, involving silylation as well as deprotonation by strong bases.[54–57] Nevertheless, the functional group resistance of this process is obstructed by the harsh conditions required for this conversion. Also, the yield and regioselectivity of synthesized purine nucleoside are usually limited. However, this halogenose-based glycosylation allows a convenient atom economy in contrast to other methods.

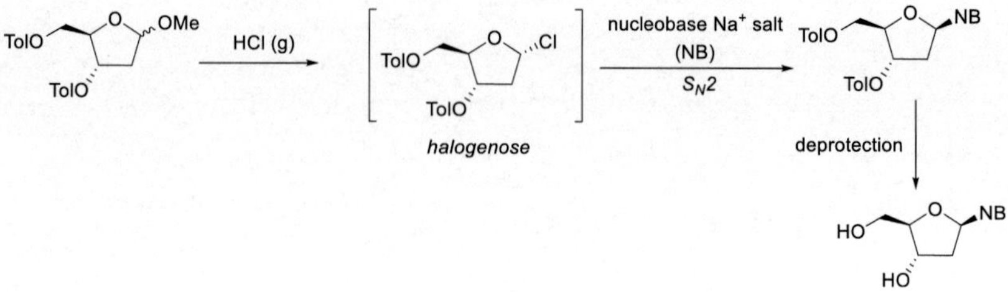

SCHEME 4.6 Nucleobase glycosylation with halogenoses.

4.3.1.3 *Thioglycosides*

The most versatile donor in carbohydrate synthesis is thioglycosides, which have been employed for several glycosylation processes. However, its recruitment in nucleoside synthesis is rare. One such process was reported by Lieu and colleague[58] who produced thioglycoside as a donor and treated it with di-*p*-tolyl sulfoxide and triflic anhydride. Similar to others, this process also involves in situ generation of a charged ring. This reactive intermediate then undergoes nucleophilic attack by activated trimethylsilyl (TMS) nucleobases to afford β-anomer of both furanosyl nucleosides and pyranosyl nucleosides (Scheme 4.7). Although the glycosylation step involves good-to-excellent yields, this process suffers from similar disadvantages such as the requirement of a lengthy sequence for sugar synthons synthesis, excessive use of protecting groups, and silylating agents. Notably, this approach is even applicable for 2′-deoxynucleosides, but in comparatively low yield as well as anomeric selectivity.

4.3.1.4 *Anhydroses*

A strategy based on Mitsunobu glycosylation conditions was proposed by Hocek and colleagues.[59,60] They prepared a mono-protected anhydrase from starting material riboside using tributylphosphine and 1,1'-(azodicarbonyl) dipiperidine (ADDP), which on subsequent treatment with deprotonated nucleobase provides the β-nucleoside (Scheme 4.8). This method has several leverages, such as the application of one type protecting group only, a broad substrate scope as well as good yield and complete stereoselectivity, but it has one disadvantage as it is applicable only at 2-hydroxy sugars.

4.3.1.5 *Pentenyl glycosides*

A more streamlined glycosyl donor carrying an *n*-pentenyloxy group, prepared from a perbenzoylated methyl glycoside, was brought by Fraser-Reid and coworkers.[61] This method pathway first formed *n*-pentenyl orthoester through the employment of a 2-benzoyl group of sugar. The treatment of this ester with an *N*-iodosuccinimide (NIS) and $Yb(OTf)_3$ followed by a silylated nucleobase furnished the corresponding β-nucleoside after global

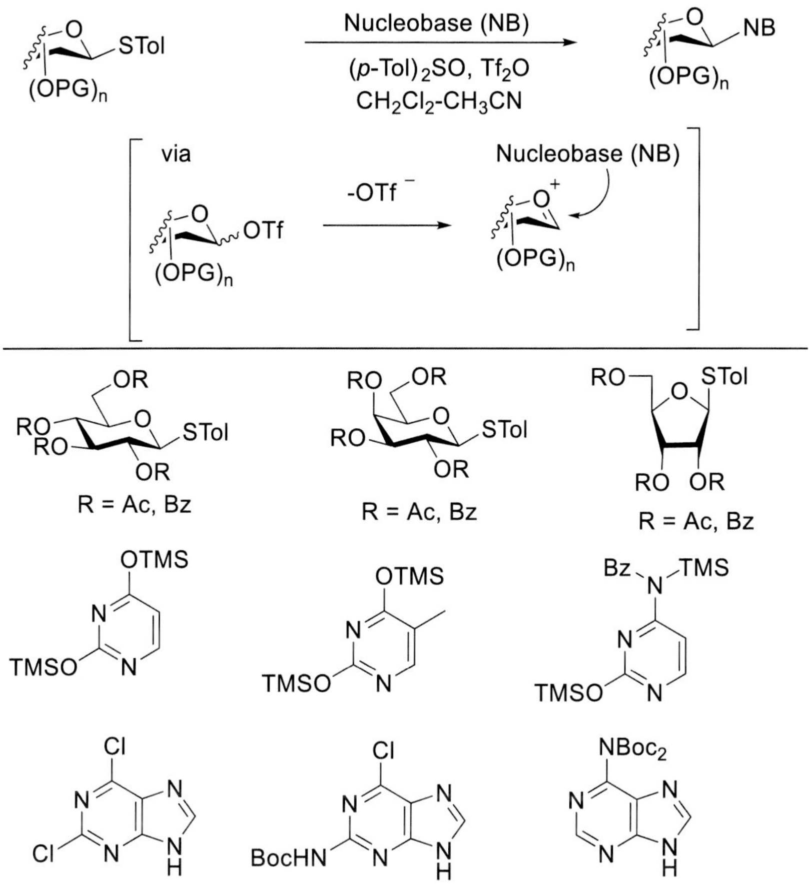

SCHEME 4.7 Nucleobase glycosylation with thioglycosides.

deprotection (Scheme 4.9) and the *n*-pentenyloxy group in the form of iodomethyl tetrahydrofuran get separated as a by-product. This method gives excellent stereoselectivity as well as regioselectivity. However, it consists of a large number of synthetic steps (involving nearly seven steps) and a rigid restriction to 2′-hydroxy nucleosides. Also, this process has limited substrate scope, and the uses of extensive protecting groups enhance the concern related to the efficiency perspective.

4.3.1.6 Propargyl-1,2-orthoesters

Influenced by the Fraser-Reid work, Rao and coworkers[62] further developed a methodology to enhance the yield of the desired nucleosides. They synthesized a propargyl-1,2-orthoester group containing glycosyl donor from

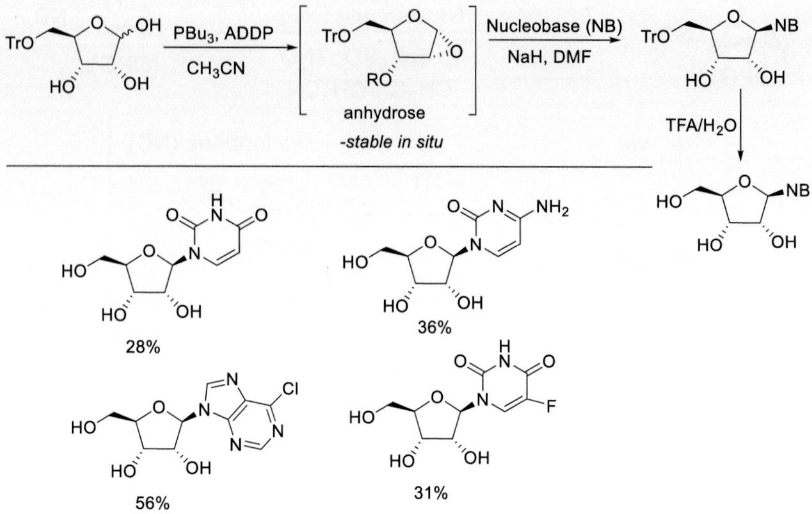

SCHEME 4.8 Nucleobase glycosylation with anhydroses.

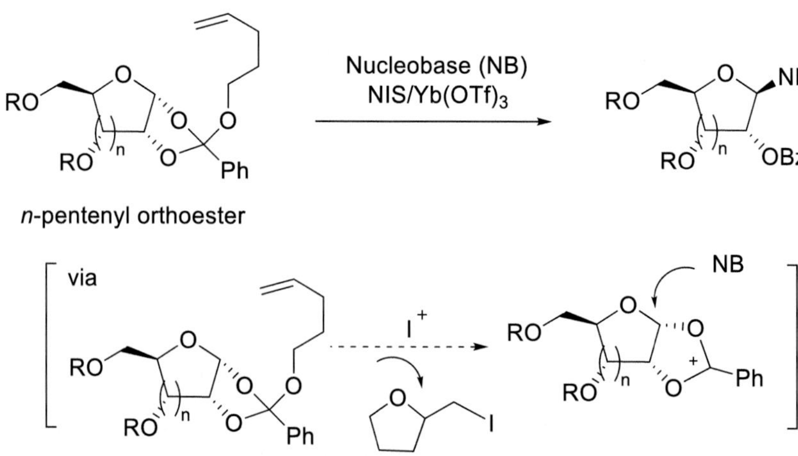

SCHEME 4.9 Nucleobase glycosylation with pentenyl glycosides.

perbenzoylated glycoside via base-mediated attack of the nucleophiles under moderate reaction conditions to promote subsequent glycosylation. Further, the incorporation of silylated pyrimidine nucleobase along with catalyst AuBr/AgOTf affords β-nucleoside exclusively (Scheme 4.10). However, such a method is applicable to only two nucleobases, apparently since the formation of the glycosyl donor is rather long and effortful.

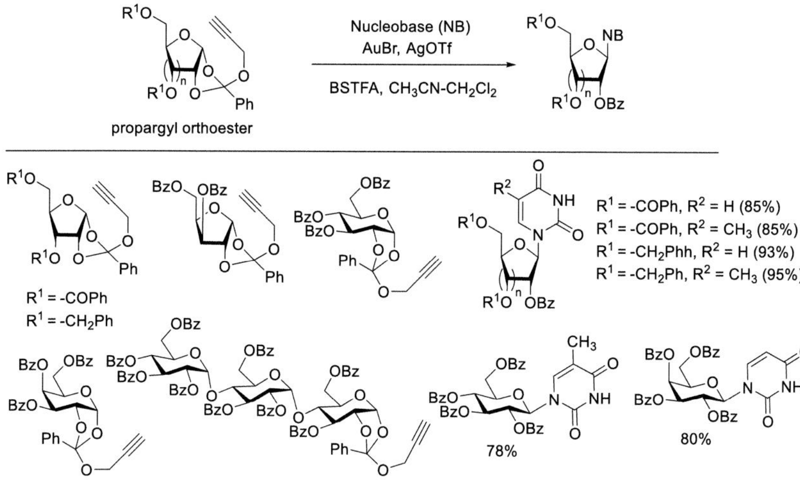

SCHEME 4.10 Nucleobase glycosylation with propargyl-1,2-orthoesters.

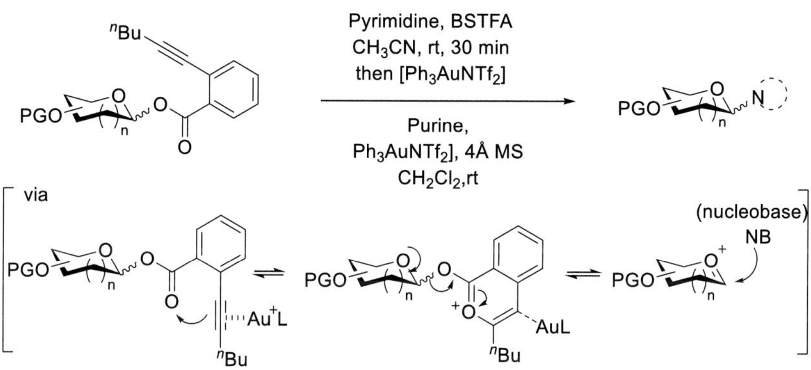

SCHEME 4.11 Nucleobase glycosylation with *O*-hexynylbenzoate donor.

4.3.1.7 *Glycosyl* ortho-*hexynylbenzoates*

As already mentioned, the formation of nucleoside using a purine base via the Vorbrüggen-type pathway is not always applicable. To solve this difficulty, Yu et al.[46] proposed an *ortho*-hexynylbenzoic esters group containing ribosyl donors catalyzed via gold catalyst (Scheme 4.11). In the presence of gold catalysis, the benzoic ester present at an anomeric position reversibly rearranges to form a glycosyl cation and an isocoumarin scaffold. The formation of glycosyl cation efficiently minimizes the struggle of the departing group with the approaching nucleophile toward the anomeric center and also enhances the reactivity of less nucleophilic nucleobases like purine. The merits of this method include moderate reaction conditions and avoiding the use of stoichiometric activating reagents. However, the requirement of the

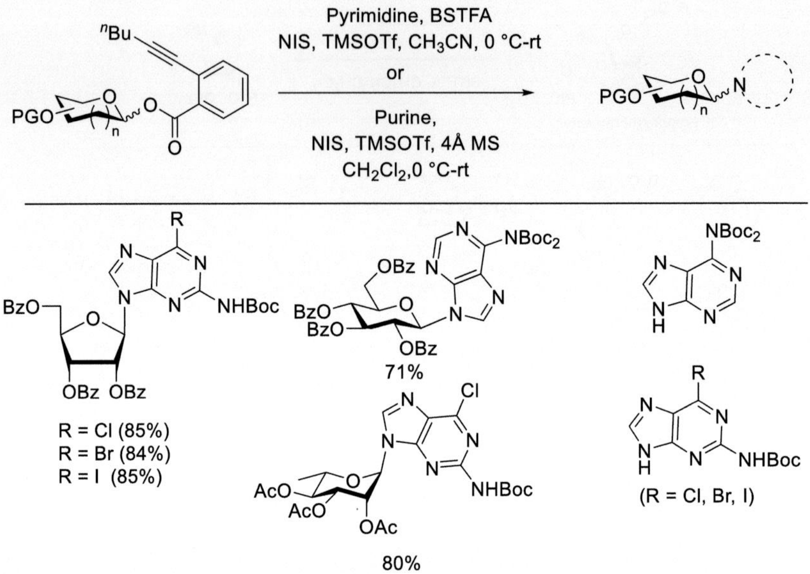

SCHEME 4.12 NIS/TMSOTf-promoted nucleobase glycosylation with *o*-hexynylbenzoates.

number of reaction steps, as well as excessive protecting groups for both sugar and the nucleobase moiety, makes this method rather complicated.

Further, Yang's group[63] explored a versatile *N*-iodosuccinimide/trimethylsilyl triflate (NIS/TMSOTf)-promoted glycosidation approach with the same glycosyl donor, that is, *ortho*-hexynylbenzoates, for the efficient synthesis of nucleosides (Scheme 4.12). The iodonium ion obtained from the cooperation of NIS and TMSOTf formed a complex with a triple bond (C≡C) of glycosyl *ortho*-hexynylbenzoate and induced the formation of sugar oxocarbenium ion as intermediate and 4-iodoisocoumarin derivative as departing group. Subsequently, the nucleophilic attack of silylated nucleobase on the intermediate occurred, which promotes the formation of respective nucleosides. Here, the bis(trimethylsilyl)trifluoroacetamide (BSTFA) reagent is mainly used for the silylation of nucleobase (purine and pyrimidine derivatives) to transform the insoluble nucleobase into soluble silylated nucleobase derivative. This protocol highlights several merits, such as mild reaction conditions, cheap promoters, extremely wide substrate scope, and good-to-excellent yields.

4.3.1.8 *Glycosyl (Z)-ynenoates*

A highly efficient gold(I)-catalyzed *N*-glycosylation method for the synthesis of a collection of nucleosides and deoxynucleosides with the shelf-stable glycosyl (*Z*)-ynenoates as donors in the presence of TfOH and BSTFA in acetonitrile were given by same group (Scheme 4.13).[64] Notably, the addition of

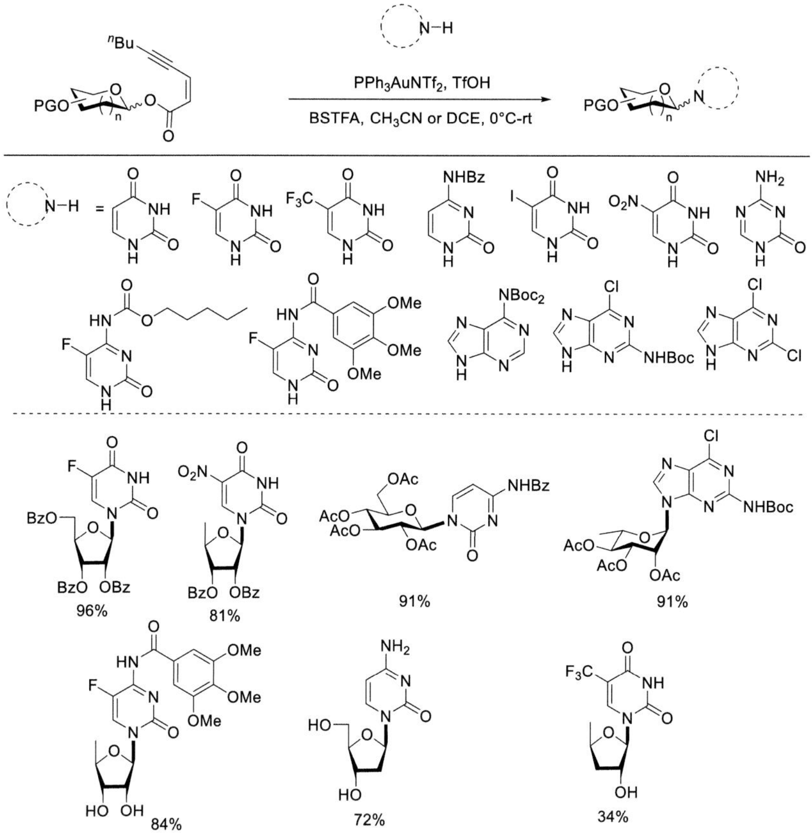

SCHEME 4.13 Gold(I)-catalyzed *N*-glycosylation with glycosyl (*Z*)-ynenoates.

BSTFA produces the soluble silylated pyrimidine derivatives that, in turn, coupled efficiently with perbenzoylated glycosyl (*Z*)-ynenoate donor and the addition of TfOH could significantly accelerate the *N*-glycosylation of glycosyl (*Z*)-ynenoates with nucleobases, which collectively promote the formation of corresponding nucleosides regioselectively in good-to-excellent yields.

4.3.1.9 Glycosyl o-(1-phenylvinyl)benzoate donor

Based on a similar method, Xiao and group put forward a highly efficient and versatile glycosyl *o*-(1-phenylvinyl)benzoate donor that avoids interaction of the departing group with the nucleobase via irreversible separation of a vinylbenzoic ester (Scheme 4.14).[65] In this procedure, primarily benzoylated glycosyl hemiacetal undergoes esterification with *o*-(1-phenylvinyl)benzoic acid to access a stable glycosyl *o*-(1-phenylvinyl)benzoate donor. Further activation of NIS with TMSOTf generates iodonium species that

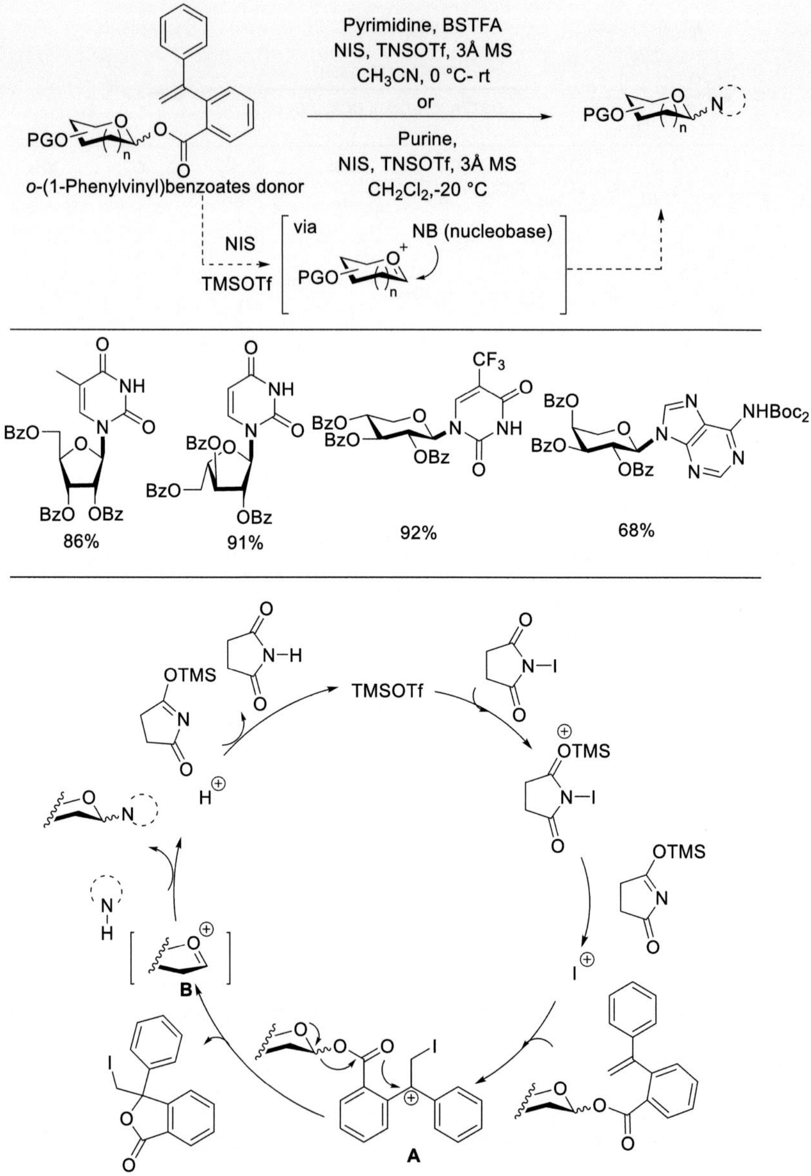

SCHEME 4.14 Glycosylation with *ortho*-(1-phenylvinyl)benzoate (PVB) glycosides.

undergo interaction with the aglycone part of glycosyl *o*-(1-phenylvinyl)benzoate (PVB) to generate the glycosyl oxocarbenium cation B and the isobenzofuran-1-one derivative as departing species. The oxocarbenium cation is finally attacked by a silylated nucleobase to afford the desired nucleoside. The excellent glycosylation yields and regioselectivity with broad substrate scopes under mild conditions make this method more fruitful. However, the requirement of a number of synthetic steps, along with protecting the group, makes this method rather challenging.

4.3.1.10 Alkynyl glycosyl carbonate

Another unified, direct, and catalytic strategy for the synthesis of *N*-glycosides that involves the silver-assisted gold-catalyzed activation of alkynyl glycosyl carbonate donors and their further treatment with glycosyl acceptors such as purines, pyrimidines, 2-hydroxyquinolines, and protected asparagine in the presence of BSTFA at ambient temperatures were set forth by Hotha and group (Scheme 4.15).[66] Notably, this protocol was also found suitable for the formation of Araf- nucleosides from Ribf- nucleosides via following oxidation–reduction sequence that are otherwise difficult to prepare. Moreover, the direct bonding of an asparagine-containing dipeptide to the glucopyranose and subsequent extrapolation to furnish the dipeptide disaccharide unit present in the cell membrane of chloroviruses is another feature of this protocol.

4.3.1.11 3,5-Dimethyl-4-(2′-phenylethynylphenyl)phenyl glycoside donors

Later Yu and his colleague[67] performed a similar strategy under a similar reaction on a new glycosyl donor, that is, 3,5-dimethyl-4-(2′-phenylethynylphenyl)phenyl (EPP) glycosides as alternative leaving groups. These EPP glycosides can be activated very easily in the presence of NIS/TMSOTf via an unprecedented dearomative activation mechanism to afford the desired β-nucleoside exclusively (Scheme 4.16).

4.3.1.12 N-phenyl trifluoroacetimidate glycosides

Thereafter, a very efficient and potent glycosyl donor, that is, trifluoroacetimidate obtained by the treatment of protected glycoside with *N*-phenyl trifluoroacetimidate (PTFAI), was explored by Liao and group.[68,69] The treatment of this donor with a silylated nucleobase in the presence of TMSOTf affords thermodynamically favored β-nucleoside with quantitative yields (Scheme 4.17). However, this process involves some applicability limitations in terms of the pyrimidine base. Despite its impressive yields, the excessive application of protecting groups, stoichiometric use of promoter, and unfavorable atom economy make this process debatable.

Filippov and coworker[70] reported a glycosylation reaction of ribofuranosyl *N*-phenyltrifluoroacetimidates and carboxamide side chains of asparagine and glutamine using promoter TMSOTf or $HClO_4$–SiO_2 to obtain the corresponding *N*-glycosides with α-selectivity (Scheme 4.17, bottom).[66] Moreover, ribosylation of asparagine using ribofuranosyl trifluoroacetimidate

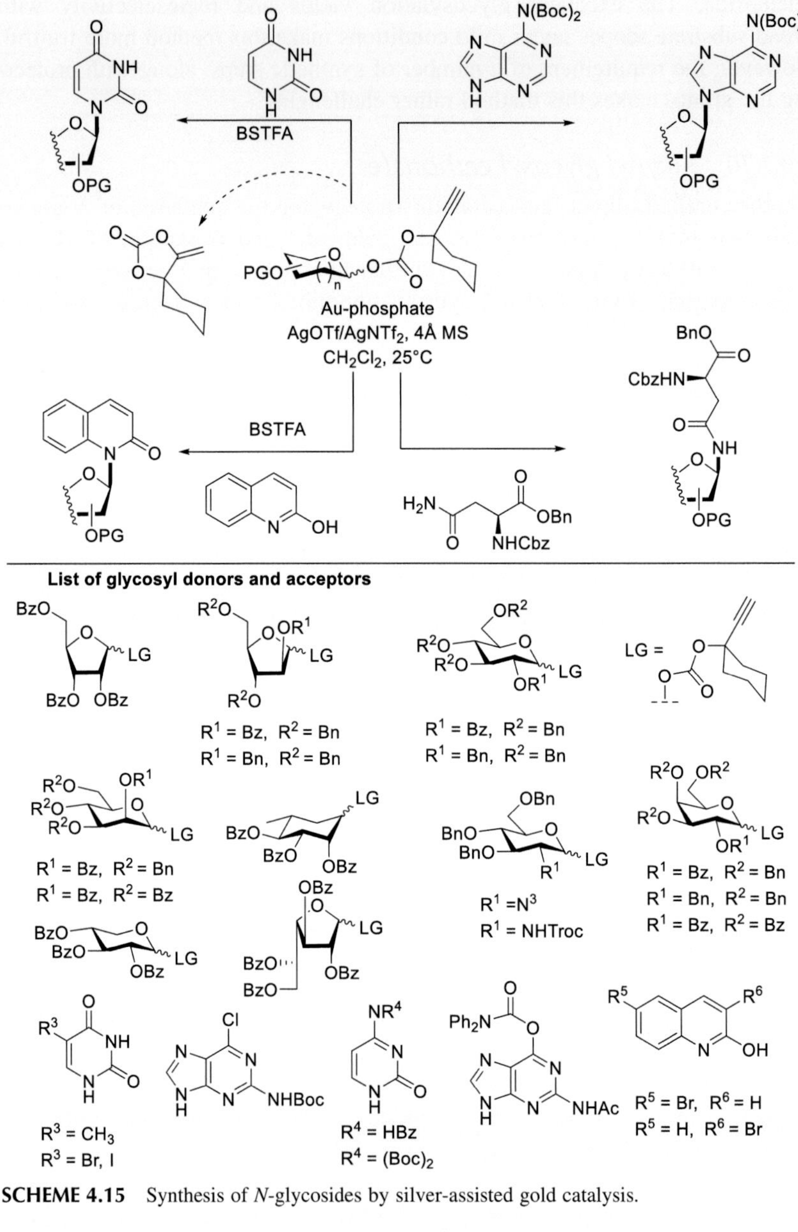

SCHEME 4.15 Synthesis of *N*-glycosides by silver-assisted gold catalysis.

donor and TMSOTf as an activator afforded the product an excellent α-selectivity. But in the case of glutamine, under similar reaction conditions, it resulted in the product with lower selectivity. Further, the glycosylation of glutamine using $HClO_4-SiO_2$ instead of TMSOTf as an activator proved to offer the desired product with great α-selectivity. This strategy has several

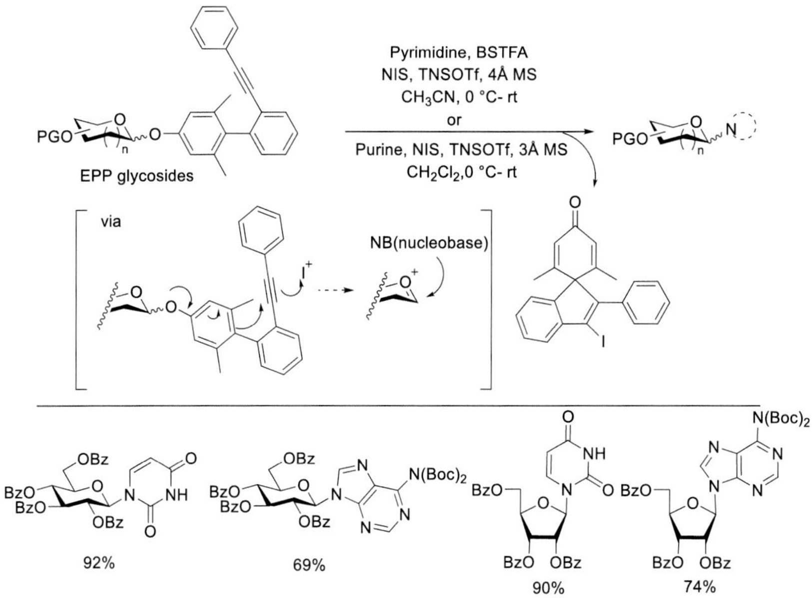

SCHEME 4.16 Nucleobase glycosylation with EPP donors.

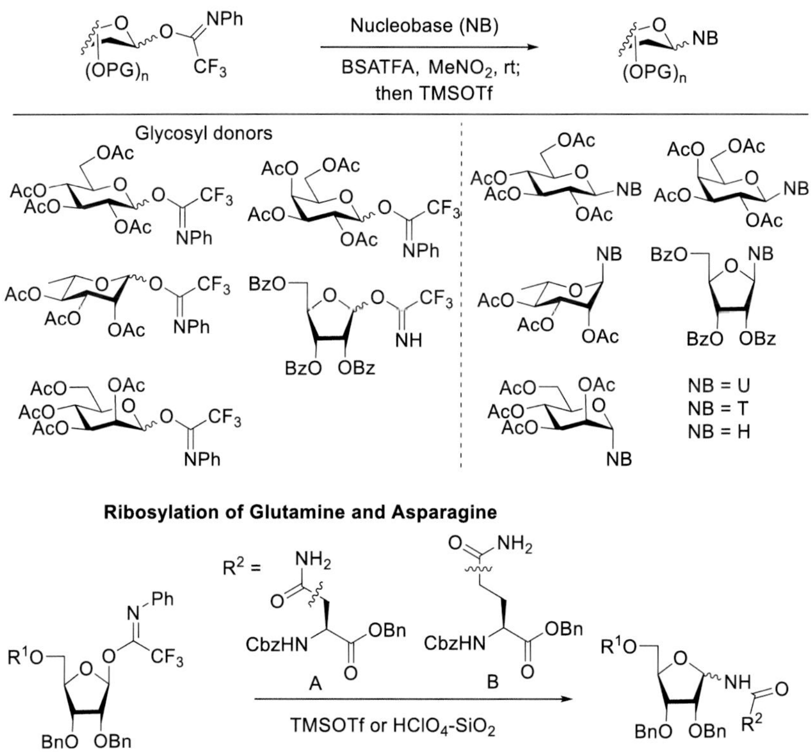

SCHEME 4.17 Nucleobase glycosylation with trifluoroacetimidates.

benefits such as a broad functional group as well as protecting group tolerance, good yield, and high selectivity and can conceivably be extended to the ribosylation of other amino acids. However, the requirement of a large number of synthetic steps with multiple protecting group manipulations makes this approach rather laborious.

Recently, Zhang et al. also developed a new glycosylation method using the same glycosyl *N*-phenyl trifluoroacetimidate (PTFAI) donors for stereoselective synthesis of nucleoside via a hydrogen bond-mediated aglycon delivery system using glycosyl PTFAI donors as well as charged thiourea (hydrogen-bond-donor) catalyst (Scheme 4.18).[71] In this protocol, the H-bond acceptor glycosyl imidate donors can be activated with the H-bond donor organo-(thio)urea catalysts under mild reaction conditions leading to the formation of oxocarbenium ion-type species and departure of secondary amide as a leaving group. The nature of leaving group is neither acidic nor nucleophilic enough to battle with the acceptor toward the activated donor.

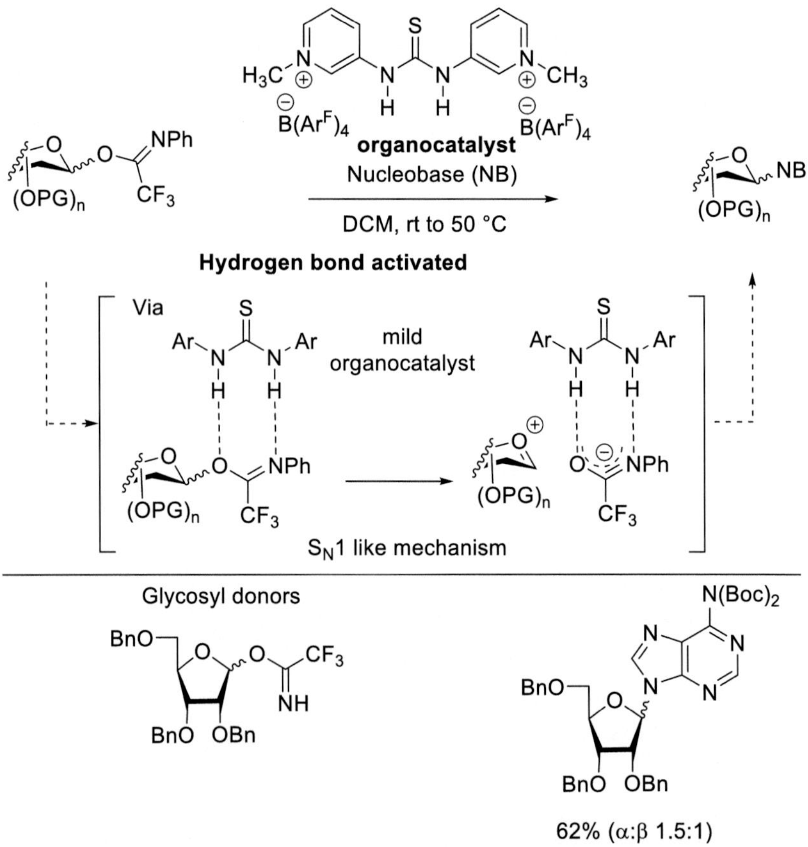

SCHEME 4.18 *N*-Glycosylation with glycosyl *N*-phenyl trifluoroacetimidate (PTFAI) donors.

The oxocarbenium ion formed can be harnessed to achieve stereoselective β-nucleoside. This protocol features broad substrate scope, good-to-excellent yields as well as regioselectivity and environmentally favorable mild catalysis conditions.

4.3.1.13 8-(Methyltosylaminoethynyl)-1-naphthyl glycosides

Zhou et al. developed another novel glycosyl donor, that is, 8-(methyltosylaminoethynyl)-1-naphthyl (MTAEN) glycosides.[72] Under the effect of TMSOTf, the glycosylation of MTAEN glycosyl donor with nucleobase took place regioselectively to furnish β-nucleosides exclusively (Scheme 4.19). This protocol exhibits the benefits of shelf-stable donors, mild catalytic conditions, wide substrate scope, and good yields.

4.3.2 Chemical synthesis of *N*-glycosides excluding nucleosides

4.3.2.1 Direct replacement of leaving group

This portion includes the strategies that promote the formation of *N*-glycosides by direct S_N2-type substitution of the leaving group present at the anomeric position of glycosyl donors. All such methods for *N*-glycosylation via nucleophilic displacement are commonly involved in a reactive glycosyl intermediate that undergoes a reaction with *N*-nucleophiles.

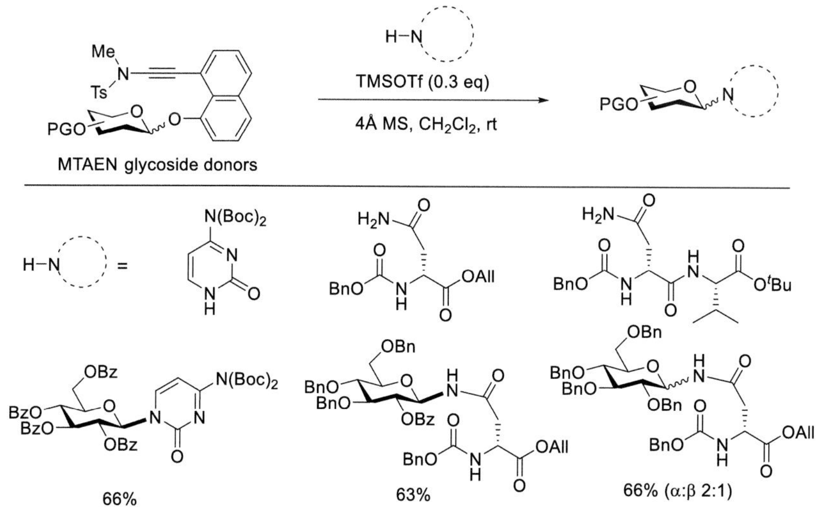

SCHEME 4.19 *N*-Glycosylation with 8-(methyltosylaminoethynyl)-1-naphthyl (MTAEN) glycosyl donor.

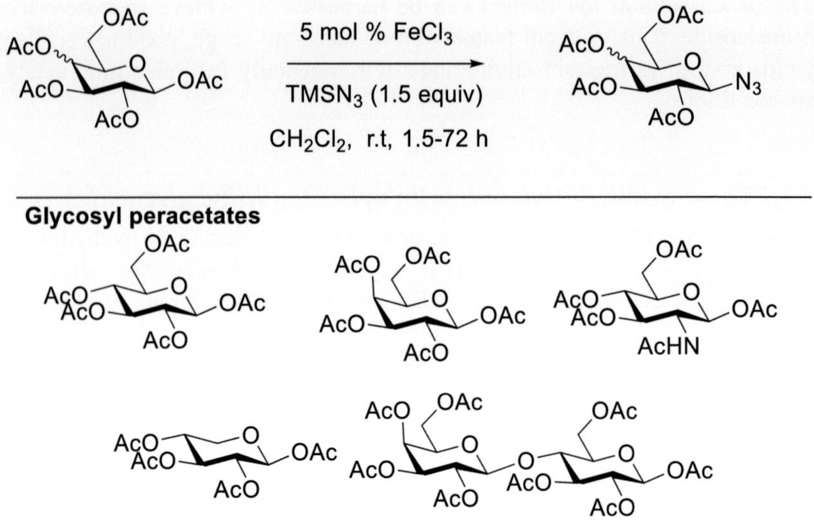

SCHEME 4.20 β-Glycosyl azidation by $FeCl_3$.

4.3.2.2 Glycosyl acetates

The first attempt to synthesize *N*-glycosides by substituting the acetate group of glycosyl β-peracetates with $TMSN_3$ was reported by Chen and coworkers[73] in 2011 using $FeCl_3$ as the catalyst to synthesize 1,2-*trans* glycosyl azides (Scheme 4.20).

Since then, several other groups have put their effort into this strategy that generally requires a 5–10 mol% of Lewis acid catalysts like AuCl, $ZnCl_2$, $InBr_3$ and $In(OTf)_3$, $SnCl_4$, and $FeCl_3$. The conditions that are commonly employed are listed in Scheme 4.21. However, the environmentally friendly $FeCl_3$ catalyzed stereoselective β-glycosyl azidation in 6 hours in 96% yield under the mentioned reaction conditions explored as a more prominent one. Additionally, $FeCl_3$ with copper powder can catalyze a reaction between azido glycosides and terminal alkynes through a 1,3-dipolar cycloaddition reaction to furnish the 1,2,3-triazoles, respectively, in good-to-excellent yields with exclusive regioselectivity.

4.3.2.3 Glycosyl halide

Direct glycosylation of glycosyl halide can be an extremely used strategy for the generation of glycosyl azide. Bertho's group[74] gave their initial contribution to the formation of glycosyl azides from acetylated halogenoses that involved the treatment of glycosyl bromide with sodium azide via homogeneous one-phase reactions in *N,N*-dimethylformamide (DMF) at elevated temperature (100°C) while phase-transfer catalysis permits milder conditions. Further, the modification in this reaction condition was made by Roy et al.[75] with the motive to simplify and permit milder conditions using phase-

SCHEME 4.21 β-Glycosyl azidation by Lewis acid catalysts.

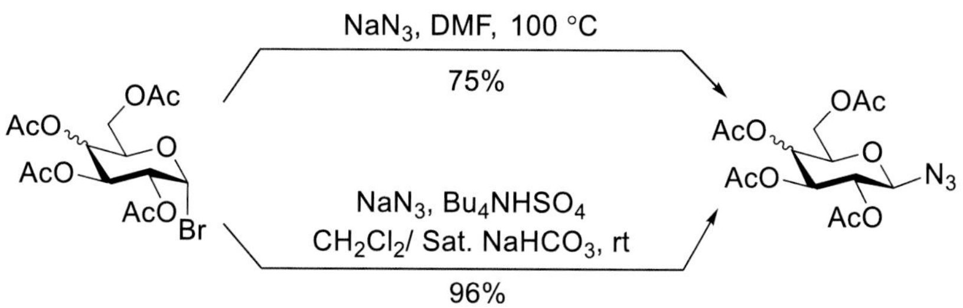

SCHEME 4.22 Preparation of glycosyl azides from peracetylated glycosyl halides.

transfer catalysis. They treated glycosyl bromide with sodium azide in the presence of tetrabutylammonium hydrogen sulfate using a two-phase system (CH_2Cl_2/sat. $NaHCO_3$) for 1–2 hours at room temperature to afford glycosyl azide with higher yield (Scheme 4.22).

4.3.2.4 Methyl glycoside

Colinas and Bravo[76] demonstrated methyl glycoside–promoted *N*-glycosides, involving glycosylation of benzylated methyl glycoside and sulfonamide in the presence of catalyst boron trifluoride etherate in dichloromethane (DCM). Unsuitably, due to the toxic behavior of boron trifluoride etherate and the less impact of stereoelectronic effects in the five-membered moiety responsible for poor stereoselectivity, they further extended their study to overcome this drawback, and in 2008, they reported a similar reaction in short reaction time as well as minimum workup using an inexpensive, nontoxic, and recyclable catalyst $HClO_4$-SiO_2 to afford the corresponding sulfonamidoglycosides with good-to-high yields (Scheme 4.23).[77] Based on obtained data, it was concluded that the hindered sulfonamides were effectively coupled to the methyl ribofuranosides, whereas opposite to this steric sulfonamide underwent glycosylation to form glycosylsulfonamide with poor yield. Additionally, this method was also adapted to methyl glycopyranosides to give the sulfonamidoglycosides, respectively, in good yields. Although there is no change in the stereochemistry of the synthesized products obtained, this catalyst shows no effect on the anomeric selectivity.

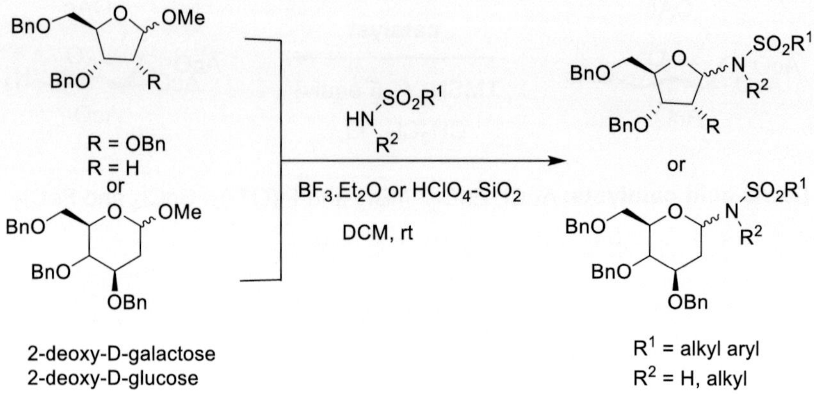

SCHEME 4.23 Sulfonamidoglycosylation of methyl glycosides.

4.3.2.5 *Glycosyl trichloroacetimidate*

Amidst diverse synthesized glycosyl donors, Schmidt's trichloroacetimidates are classical donors frequently applied in glycosylation due to their great reactivity and simple synthesis procedure.[78–81] Trichloroacetimidate donor is generally activated by Lewis acids, having pK_a value (≤5) such as TMSOTf and boron trifluoride etherate.[82,83] With this concept, Pedersen's group[45] firstly brought a potent two-component strategy for the effective synthesis of *N*-glycosides in the absence of promoter, that is, a self-promoted *N*-glycosylation just by treating trichloroacetimidates imidate donors with carbamate acceptors (having pK_a value ≤ 5) together in DCM solvent at room temperature (Scheme 4.24). Nevertheless, this protocol also manifests the limit as the selection of acceptors is absolutely reliant upon its pK_a value.

The stereochemistry of this process is quite complicated as the α-glycosyl imidate donors undergo glycosylation to form a product with β-stereoselectivity exclusively with high yields, whereas opposite to that the glycosylation with β-glycosyl imidate donors produced the corresponding product as a mixture of anomers. This stereochemical outcome is elaborated by the plausible mechanism involving intermediates formation via S_Ni-like glycosylation. Both pathways (A-F pathway for α-TCA and G-L pathway for β-TCA) include the formation of two conformers (shown in Scheme 4.25A, B, G, and J), howbeit out of four, only conformer **A** of α-TCA is found to be energetically favorable one leading to the formation of major product with β-selectivity.

A similar strategy in extended form was reported by Mandal's groups[84] who synthesized structurally modified and more biologically stable *N*-glycofuranosyl sulfonamides under the mentioned reaction conditions using α-glycofuranosyl trichloroacetimidates with similar tosyl- and nosyl- carbamates acceptors in good yield with β-stereoselectivity (Scheme 4.26).

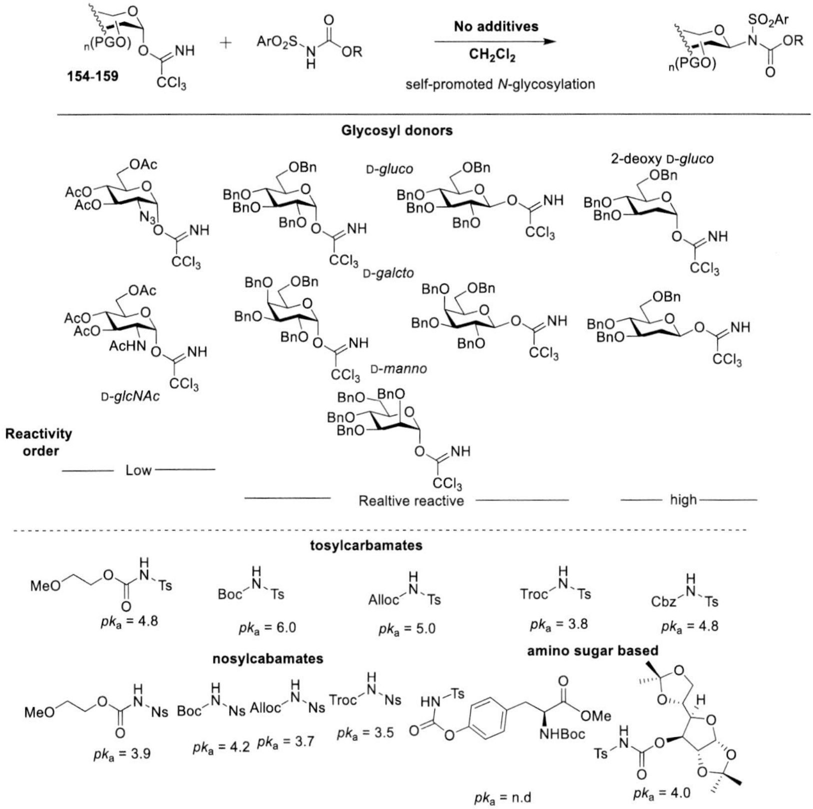

SCHEME 4.24 Self-promoted stereospecific *N*-glycosylations.

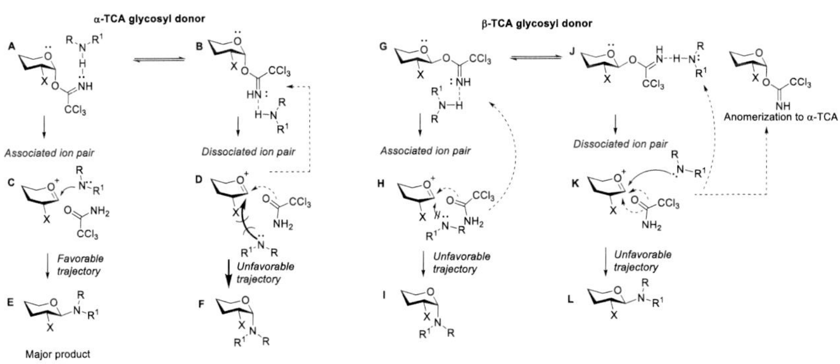

SCHEME 4.25 Mechanistic pathway during an S_Ni-like glycosylation.

The structure of synthesized glycosides was confirmed by assigning both homo- and heteronuclear 2D experiments, while its stereochemistry was approved by interpreting nOe's experiment. A strong interaction

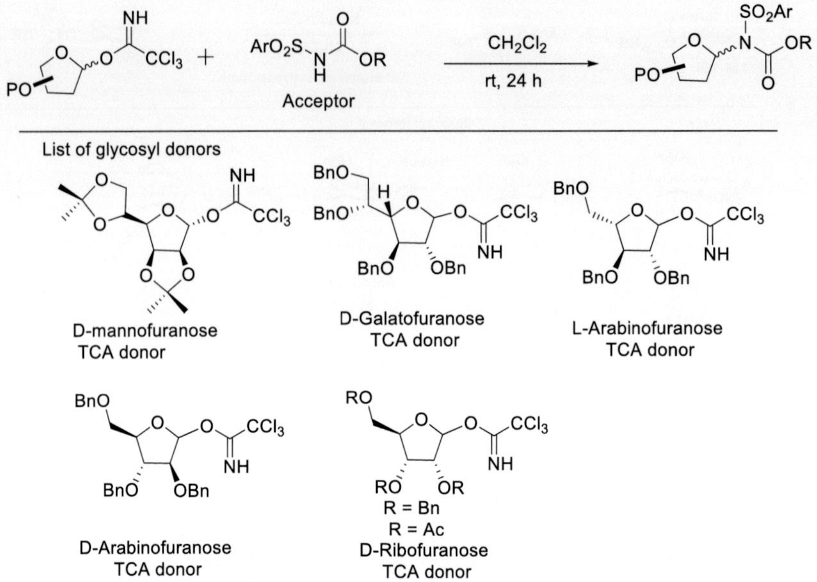

SCHEME 4.26 Self-promoted stereospecific *N*-glycofuranosyl sulfonamides.

between H-1 (δ_H 6.29) with H-4 (δ_H 4.49) and H-2 (δ_H 5.28) confirmed β stereochemistry (Scheme 4.27).

Based on the data study, it is observed that the α-imidates proceed with a high β-selectivity, whereas lower selectivity is obtained with β-imidate donors. On focusing toward the mechanistic pathway, the formation of a transition state in both α-imidate donors and β-imidate donors may cover different reaction pathways. The α-imidate glycosyl donors undergo attack by nucleophiles via the associated ion-pair mechanism to form *β-N*-glycosides exclusively. In contrast, the β-imidate glycosyl donors undergo a more dissociative reaction pathway that causes steric hindrance for the upcoming nucleophile leading to the generation of the products with a mixture of anomers.

A novel Brønsted-acid-salt catalyzed reaction, comprising Brønsted acid and 2-halogenated azole *N*-glycosylation method using glycosyl trichloroacetimidate, was demonstrated by Takemoto and coworkers[85] to form a unique *N*-acylorthoamide. To this end, this process involves the direct coupling of imidate donors with amides, including the asparagine residues of several peptides in the presence of halogen bond (XB) donor as well as Schreiner's thiourea as cooperative catalysts in good yields. A cocatalyst 2-iodoazolium salt would interact with thiourea moiety and increase the hydrogen bond donating capability of thiourea that promotes the leaving group resistance to the distinct functional group to generate intermediate *N*-acylorthoamides that thermodynamically support the generation of β-*N*-glycosides (Scheme 4.28).

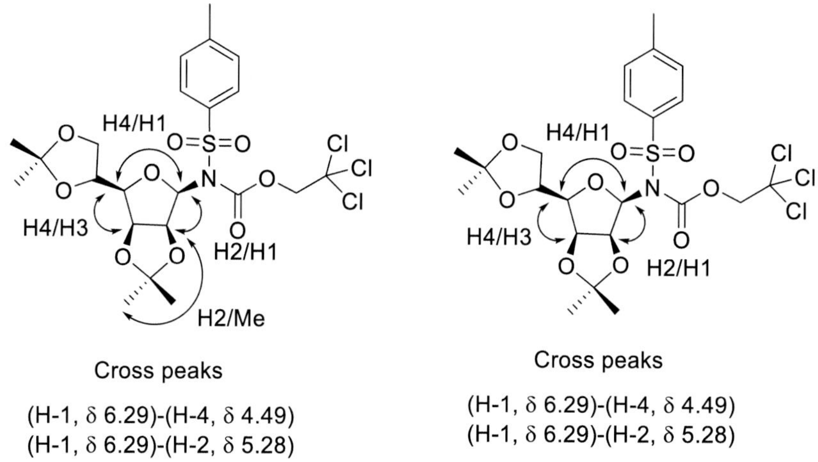

SCHEME 4.27 Structure identification.

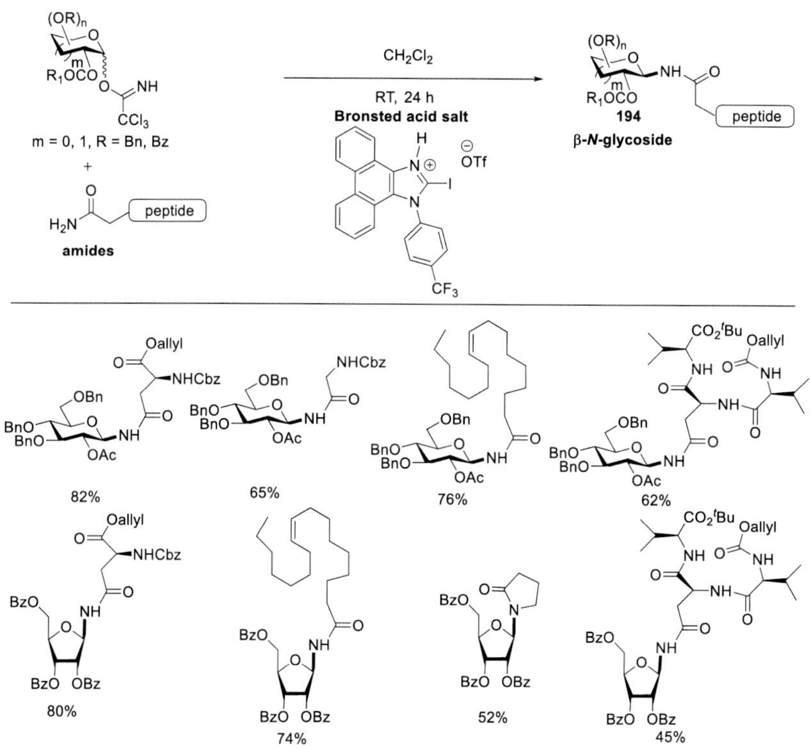

SCHEME 4.28 Brønsted-acid–salt catalyzed *N*-glycosylation.

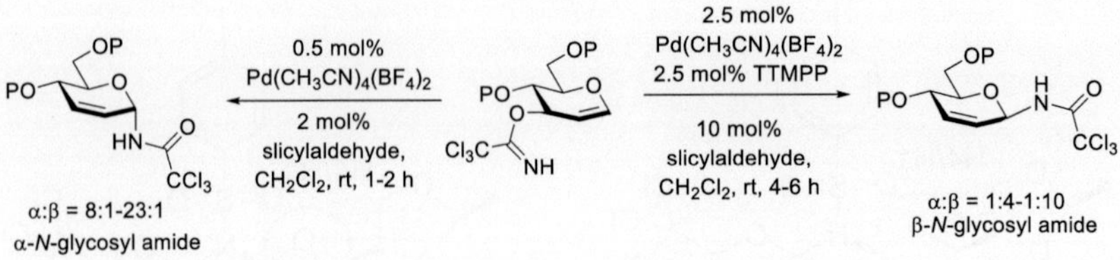

SCHEME 4.29 Formation of α- and β-*N*-glycosyl trichloroacetamides through glycal C-3 imidates rearrangement.

N-Glycosylation beneath moderate reaction conditions can also be accomplished via the activation of glycosyl trichloroacetimidate donors, which comes out as an admirable leaving group. In 2007 Nguyen group[41] was the first who disinterred the synthesis of stereoselective α- and β-*N*-glycosyl trichloroacetamides by using glycal C3-imidates rearrangement activated via palladium catalyst in the absence of acceptors and displayed the involvement of palladium–ligand complex in maintaining the anomeric selectivity (Scheme 4.29).

This protocol involved the reaction of glucal imidate with a catalyst [Pd (PhCN)2C12], bulky ligand tris(2,4,6-trimethoxyphenyl) phosphine (TTMPP), and salicylaldehyde that undergo rearrangement in CH_2Cl_2 at room temperature to produce the corresponding β-*N*-glycosyl trichloroacetamides exclusively in very good yields. On the other side, a similar reaction with cationic palladium $[Pd(CH_3CN)_4](BF_4)_2$ catalyst without using any ligand, produced α-*N*-glycosyl trichloroacetamide exclusively and supported that cationic palladium(II) heads toward the α-selectivity, while the neutral palladium(II) leads to the β-selectivity.

The proposed pathway of this method began with the interaction of cationic palladium $Pd(CH_3CN)_4(BF4)_2$, salicylaldehyde, and imidate nitrogen that ultimately leads to the generation of allylic cation that is further attacked by trichloroamide from the α-face affording the formation of α-*N*-glycosylamide. Howbeit, the coordination of formed Pd(PhCN)2C12-TTMPP-salicylaldehyde complex with the imidate double bond promotes the nucleophilic attack of nitrogen of imidate, leading to the π-complex that lastly underwent Grob-like fragmentation along with dissociation to form β-anomer (Scheme 4.30).

Thereafter in 2009, they further demonstrated the formation of stereoselective α-glycosyl ureas via 1,3-rearrangement of glycosyl trichloroacetimidates in the presence of nickel catalyst.[86] In this, the cationic nickel(II) displays a significant character to rearrange glycosyl trichloroacetimidate that leads to the generation of α-trichloroacetamide followed by its direct conversion into α-glycosyl ureas in the presence of amines and base (Scheme 4.31).

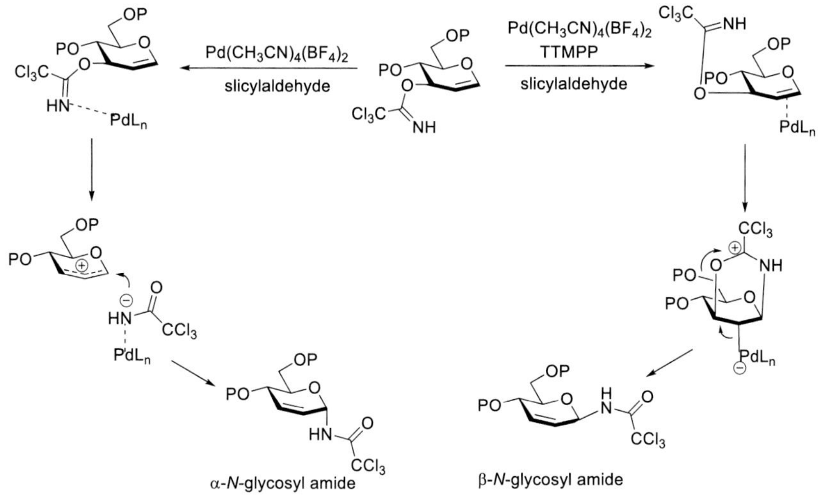

SCHEME 4.30 Proposed pathway.

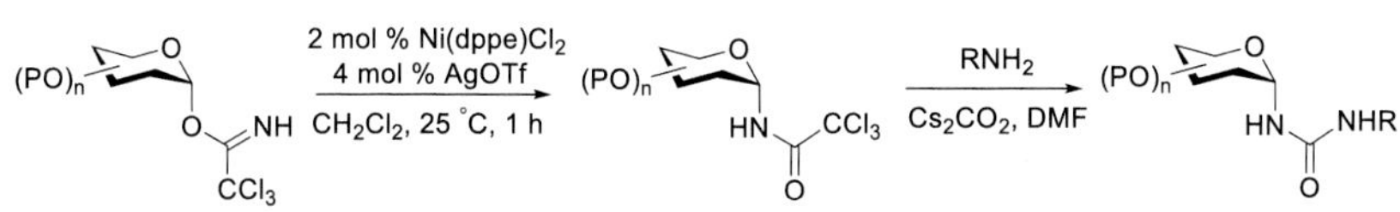

SCHEME 4.31 Nickel-catalyzed stereoselective α-trichloroacetamide.

Based on the study, a reaction pathway initiated by the coordination of nickel catalyst with the equatorial C2-ether functionality of the α-glycosyl trichloroacetimidate and the nitrogen atom of the substrate to produce complex, which further on ionization form oxocarbenium as an intermediate. A trichloroacetamide group by interaction with the anomeric center generates the complex (Pathway A). Consecutive isolation of the cationic nickel(II) catalyst will produce the aspected α-glycosyl trichloroacetamide. Always, the segregation can provide tight ion pair (Pathway B) that promotes the formation of α-trichloroacetamide majorly (Scheme 4.32).

4.3.3 Replacing a hydroxyl group of sugar by *N*-containing groups

Amidst a gamut of strategies for the stereoselective preparation of saccharides, replacing a hydroxyl group of unprotected sugar with *N*-containing groups come forth as a unique strategy involving the direct use of glycosyl hemiacetal that can be turned into a leaving group in situ and utilized directly for glycosylation. Despite the method having various merits such as the ease of preparation and the mild activation protocols, these glycosylation strategies have been less examined during the past few decades. As in many cases, the glycosyl donors, which is effectively applicable for the preparation

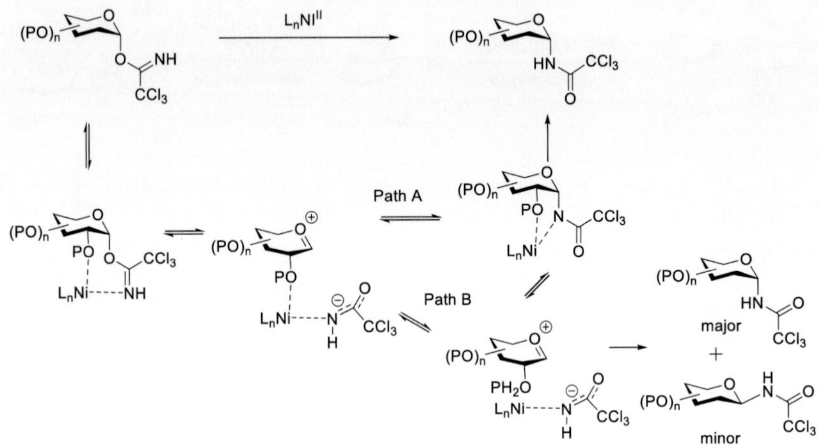

SCHEME 4.32 Mechanism pathway.

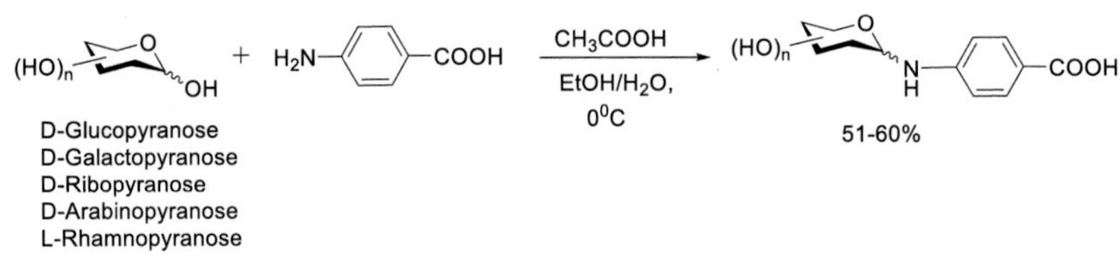

SCHEME 4.33 Synthesis of *N*-arylglycosylamines from glycosyl hydroxides and *p*-aminobenzoic acid.

of *O*-glycosides, are not that much efficient for the formation of *N*-glycosides due to the low basicity of *N*-nucleophiles.

In this process, the glycosyl hemiacetal is reacted with *N*-nucleophile such as hydroxyamine, aniline derivatives, phenyl hydrazines, and hydrazides. Earlier, it was not clear which structure was actually validated in the solid state, resulting in less attention being granted to this particular reaction. Later on, glycosylamine-type derivatives were found to be the correct ones.

Firstly, Sorokin[87] thought about this and brought a method to synthesize *N*-aryl glycosylamines by treating unprotected sugars with various derivatives of aniline as nucleophile. Thereafter, inspired by his work, many other groups also worked on this process that basically required the treatment of the sugar donor with a nucleophile in a protic solvent in the presence of a catalytic amount of acid.

Wang et al.[88] proposed a method that involves the treatment of both penta- and hexapyranoses with *p*-aminobenzoic acid as nucleophile catalyzed by using acetic acid in ethanol/water (5/1) solvent, at 0°C to furnish a number of *N*-glycopyranosylamines with α- or β-anomer (Scheme 4.33).

Further, Kublashvili group[89] performed a similar reaction condition using *o*- and *m*-aminobenzoic acids along with *p*-aminobenzoic acid as nucleophiles. As a result, they prepared the respective *N*-glycopyranosylamines with α- and β-mixture in every case. Further, Qian and group reported a strategy with the aim of enhancing the stereoselectivity of the synthesized *N*-arylglycosylamines. They coupled sugars with fluorinated anilines using a small amount of hydrochloric acid as a catalyst in water, and as a result, *N*-arylglycosylamines were obtained with β-stereoselectivity (Scheme 4.34).[90]

Thereafter, Winum's group was disclosed the method for the synthesis of *N*-arylglycosides with complete α-selectivity (Scheme 4.35).[91] In this method, they used sulfanilamides as a nucleophile and coupled them with different types of sugars in the presence of ammonium chloride in ethanol solvent to afford *N*-(*p*-sulfamoylphenyl)-α-D-glycopyranosylamines with α-anomer only.

Later, Bridiau and coworkers[92] further explored this field and developed an eco-friendly glycosylation protocol for the generation of *N*-glycosides with β-selectivity (Scheme 4.36). This process involved the coupling of unprotected glycosyl donor with different derivatives of aniline in an aqueous medium (having pH 6.5) at a temperature $< 50°C$ to obtain glycosylamines only; no formation of Amadori product occurred as it was produced in aqueous alkaline medium. The important highlight of this process involves the formation of products with complete β-selectivity that serve as synthetic precursors of tumor-associated carbohydrate antigen building blocks.

Further, an impressive and direct method for the synthesis of β-glycosyl azides was established by Shoda and coworkers.[93] This process involves the

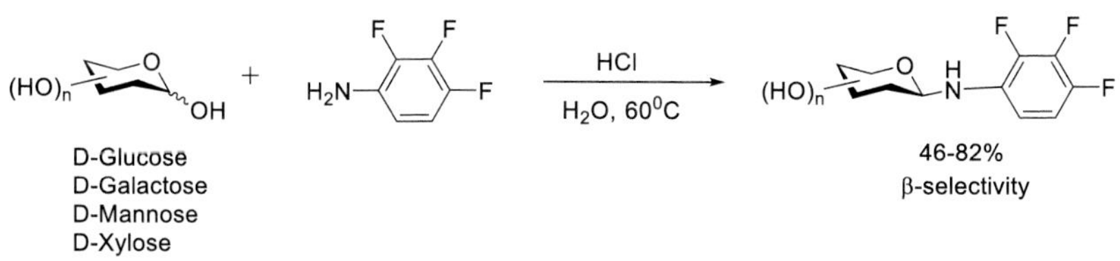

SCHEME 4.34 Synthesis of *N*-arylglycosylamines from glycosyl hydroxides with fluorinated anilines.

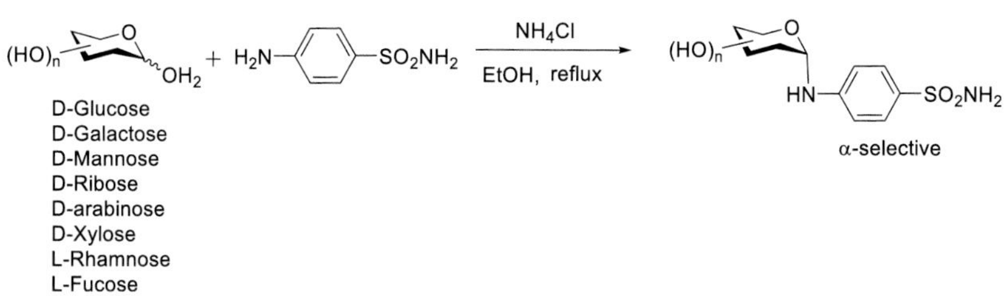

SCHEME 4.35 Synthesis of *N*-arylglycosylamines from glycosyl hydroxides.

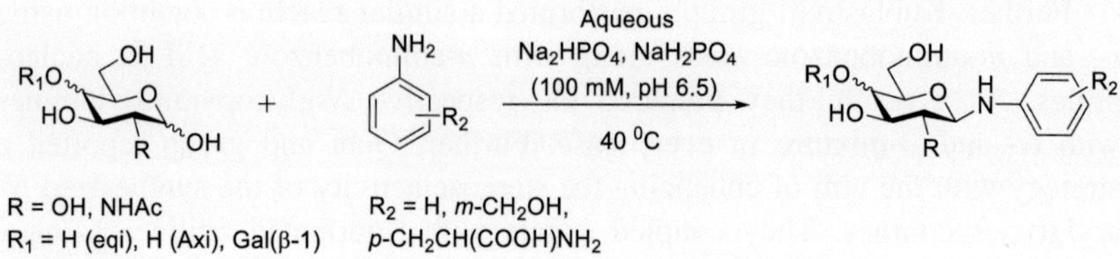

SCHEME 4.36 Synthesis of environmental-friendly β-*N*-aryl-glycosides.

formation of a reactive intermediate due to the preferential attack of the anomeric hydroxy group of unprotected sugars on 2-chloro-1,3-dimethylimidazolinium chloride (DMC selectively activates anomeric hydroxy group, as the pK_a of anomeric hydroxy group is comparatively less than that of other hydroxy groups and water). After screening various organic bases, *N,N*-diisopropylethylamine or 2,6-lutidine and NaN_3 in water at 0 °C afforded the corresponding β-azidosaccharides (Scheme 4.37). Moreover, the production of disialo complex-type decasaccharide was also possible by this process.

A direct glycosylation method using cyclic phosphonium anhydrides as an efficient promoter to catalyze less nucleophilic acceptors was reported by Walczak et al.[94] This reaction protocol involves the treatment of glycosyl hemiacetal with promoter cyclic phosphonium anhydrides (which in turn obtained from bis-phosphine oxides and trifluoromethanesulfonic anhydride) and *N*-nucleophiles in the presence of base 2,4,6-tri-tert-butylpyridine (TTBP) and solvent DCM at 0°C to furnish the desired *N*-glycosides with good yield and excellent selectivity (Scheme 4.38). This protocol involves several leverages, such as broad functional group tolerance, which can tolerate deactivated glycosyl donors and the undesired dimerization of glycosyl substrates.

To further simplify this protocol, Mandal's group[95] reported a direct glycosylation method involving a readily accessible, inexpensive, and nontoxic trichloroisocyanuric acid (TCCA) reagent in combination with PPh_3 as a combined promoter to activate glycosyl hemiacetal to form in situ a reactive intermediate, that is, alkoxy phosphonium ions that further may promote the synthesis of respective *N*-glycosides by the nucleophilic attack of the *N*-nucleophile (Scheme 4.39). The benefits of this method include the use of cheap and eco-friendly TCCA reagent, high yield, and complete stereoselectivity.

4.3.4 Direct installation of amine group into sugar moiety

Another common way to synthesize *N*-glycoside is the induction of an amine-based nucleophile into glycal-like sugar moiety. Glycals are highly reactive species readily undergoing variant reactions, including substitution, addition, and allylic rearrangement to generate a wide range of *C*-, *N*-, or *O*-

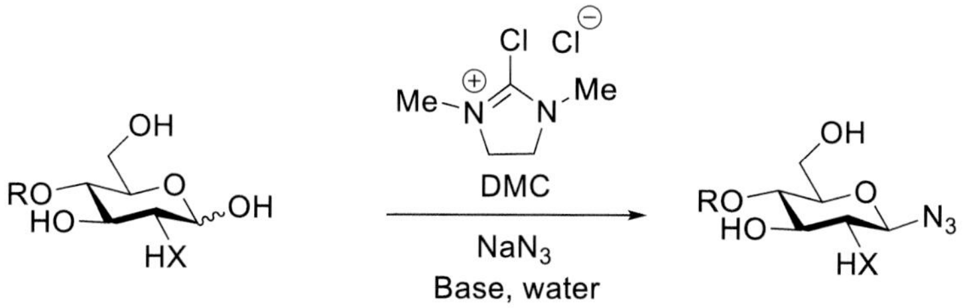

Substrate (mM)	DMC (equiv.)	Base (equiv.)	Yields (%)
Glc (250)	3	Et_3N (10)	83
Glcα1-4Glc (250)	3	DIPEA (9)	62
(Glcα1-4)$_5$Glc (25)	10	DIPEA (30)	65
Galβ1-4Glc (100)	5	DIPEA (15)	86
GlcNAcβ1-4GlcNAc (250)	3	2,6-lutidine (6)	77
(GlcNAcβ1-4)$_3$GlcNAc (20)	15	2,6-lutidine (30)	85
Disialo complex-type decasaccharide (36)	20	2,6-lutidine (40)	87

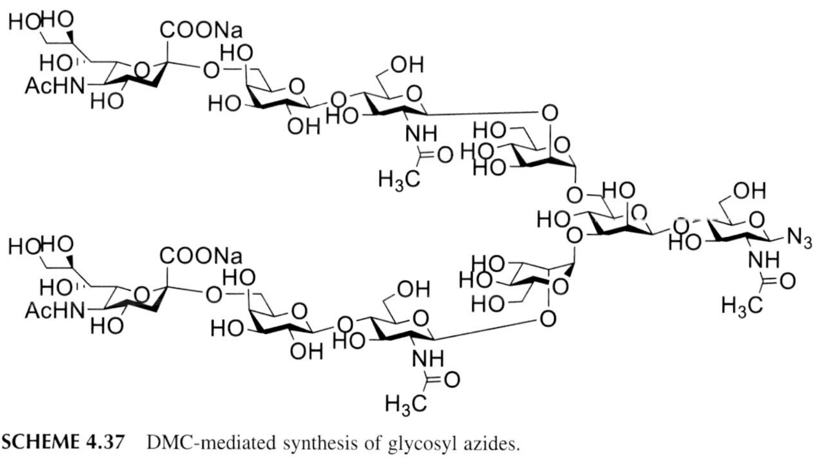

SCHEME 4.37 DMC-mediated synthesis of glycosyl azides.

glycosides. The most common reaction followed in glycal-like moiety is the Ferrier rearrangement that involves nucleophilic substitution (using various nucleophiles such as *O*-, *S*-, *N*-, and *C*-nucleophiles) with allylic rearrangement using promoters such as Lewis acids, oxidants, or various metal catalysts.

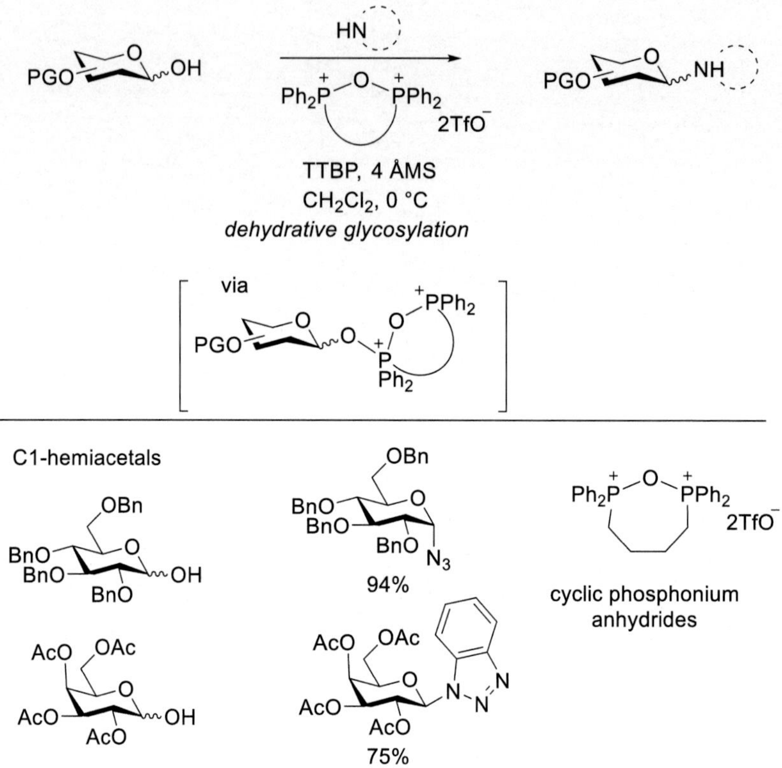

SCHEME 4.38 Direct glycosylation using cyclic phosphonium anhydrides with *N*-based nucleophiles.

4.3.4.1 Installation of azide group

Azide, among all, is the most commonly applied nucleophile for the synthesis of *N*-pseudoglycals. Related to this context, the Hayashi group initially described the use of $TMSN_3$ as a nucleophile in Ferrier rearrangement reactions.[96] They treated 3,4,6-tri-*O*-acetyl-D-glucal and $TMSN_3$ in the presence of Lewis acids catalyst like TMSOTf and $Yb(OTf)_3$ to furnish 2,3-unsaturated glycosyl azide but in lower yield with predominantly α-isomer (Entries 1, 2, Table 4.1).[97] To further improve the yield and stereoselectivity, Yadav et al.[98] discovered iodine as a catalyst to carry out the similar reaction of glycals with $TMSN_3$ in DCM solvent to furnish the desired products in good yield with better α-selectivity (Table 4.1, Entries 3,4). Besides, $InBr_3$[98] is another efficient catalyst used for the synthesis of glycosyl azides (Table 4.1, Entries 5).

In 2004 Chandrashekar group[99] first disclosed the utilization of tris(pentafluorophenyl)borane $B(C_6F_5)_3$ as a catalyst with several nitrogen nucleophiles via Ferrier azaglycosylation to form a variety of azapseudoglycals (Table 4.1, Entry 6). Another catalyst that can promote Ferrier glycosylation

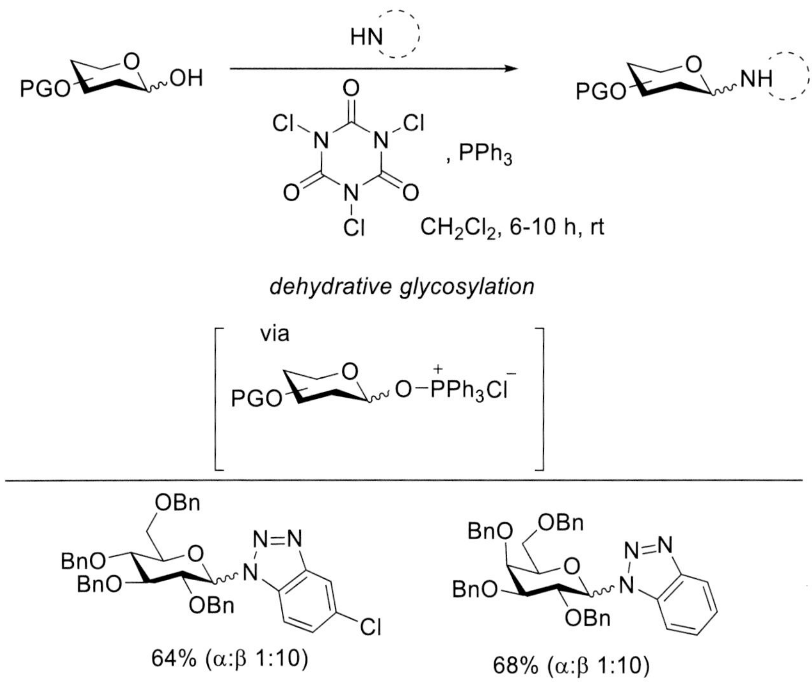

SCHEME 4.39 Direct glycosylations of C1-hemiacetals with *N*-based nucleophiles.

for the formation of pseudoglycals includes $ZrCl_4$[100] (Table 4.1, Entries 7). Later in 2006 Kamble and coworkers[106] turned the previous method into a more economically favorable one by replacing tris(pentafluorophenyl)borane with magnesium perchlorate as a catalyst and, under similar reaction conditions, produced the corresponding glycosyl azide with quite similar yield but in higher α-selectivity (Table 4.1, Entry 8). In contrast, Liu's group[101] explained heterogeneous catalysts such as $ZnCl_2/Al_2O_3$ as an effective catalyst for producing glycosyl azide with good yield as well as higher α-selectivity (92%, α:β = 86:14, Table 4.1, Entry 9). However, Bert Sels and group,[102] in 2010, set forth a more greener strategy that involves the application of more suitable microporous and mesoporous substances, such as ultra stable H-USY zeolite CBV-720, as promising catalysts that smoothly undergo Ferrier rearrangement to give the desired product. This catalyst is beneficial in terms of its low cost and effortless separation from the reaction mixture (Table 4.1, Entry 10).

Further, Vankar et al.[105] reported the carbon-Ferrier rearrangement on 3,4,6-tri-*O*-acetyl-D-glucal by using $TMSN_3$ as a nucleophile and CAN as a catalyst in dry acetonitrile solvent and, as a result, synthesized the mixture of substituted glycosyl azides (1:1) at C-1 and C-3 position in average yield (69%) (Table 4.1, Entry 13).

TABLE 4.1 Azidation of tri-*O*-acetyl-D-glucal with different catalysts.

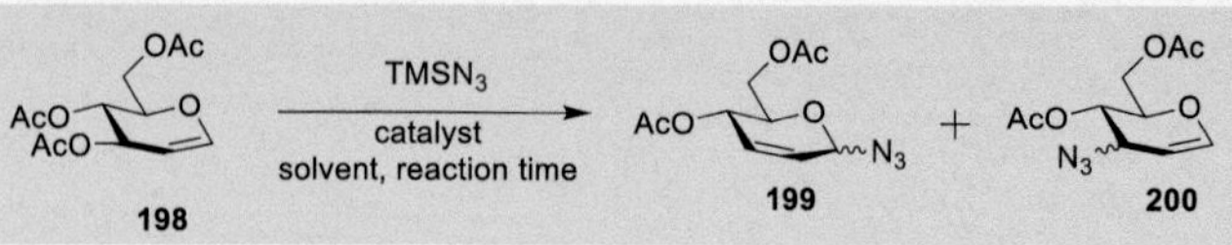

Entry	Reagents	Solvent/ time	Yield of 199 (α/β)	Yield of 200 (α/β)	References
1	TMSOTf (10 mol%)	CH_3CN, 1–21 h	28	62	[96]
2	$Yb(OTf)_3$ (10 mol%)	CH_3CN, 1 h	33	59	[96]
3	I_2 (8 mol%)	CH_2Cl_2, 8 h	87 (2.3:1)	–	[97]
4	I_2 (10 mol%)	CH_2Cl_2, 7 h	67 (3.5:1)	22 (2:1)	[97]
5	$InBr_3$ (5 mol%)	CH_2Cl_2, 7 h	85 (8:1)	–	[98]
6	$B(C_6F_5)_3$ (0.5 mol%)	CH_3CN, 2 h	94 (5:1)	–	[99]
7	$ZrCl_4$ (10 m0l %)	CH_3CN, 8 min	86 (6:1)	–	[100]
8	$Mg(ClO_4)_2$ (3 mol%)	CH_3CN, 7 h	92 (9:1)	–	[81]
9	$ZnCl_2/Al_2O_3$ (10 mol%)	CH_2Cl_2, 20 min	92 (6:1)	–	[101]
10	H-USY zeolite	THF, 3 h	40 (3:1)	40 (7:2)	[102]
11	$RuCl_3$ (5 mol %)	CH_3CN, 0.5 h	35 (8:1)	47 (3:1)	[103]
12	$Zn(OTf)_2$ (10 mol%)	$C_2H_4Cl_2$, 6 h	88 (5:1)	–	[104]
13	CAN (2 equiv.)	CH_3CN, 2 h	55 (5:1)	32 (7:1)	[105]

In 2015 Kashyap et al.[103] explained the use of ruthenium(III) chloride as a catalyst. Under mild reaction conditions, they treated glycal with trimethylsilylazide using catalyst ruthenium(III) chloride to obtain the respective 2,3-unsaturated glycosyl azides with α-selectivity but in lower yield (35%, α:β 89:11). They further work on it with the aim to improve the outcome and this time they used a nontoxic and mild reagent $Zn(OTf)_2$[104] as a catalyst to perform the similar reaction. As a result, the corresponding C-1 glycosyl azide was obtained with an 88% yield (α:β 83:17, Table 4.1, Entry 12).

4.3.4.2 Installation of sulfonamide

The synthesis of 2,3-unsaturated-*N*-glycosides or "*N*-pseudoglycals" using sulfonamides as a nucleophile is one of the fascinating areas of research for carbohydrate chemists. As the *N*-glycosyl derivatives, such as sulfonamide glycoconjugates, serve as a key element in many therapeutic agents such as antibiotic, antineoplastic, and antiviral compounds. The most familiar and direct method for the formation of 2,3-unsaturated glycosides is a type I Ferrier rearrangement. Moreover, in terms of *N*-nucleophiles, the synthesis of *N*-glycosyl sulfonamides has gathered more attention (Table 4.2).

TABLE 4.2 Sulfonamidoglycosylation of tri-*O*-acetyl-D-glucal.

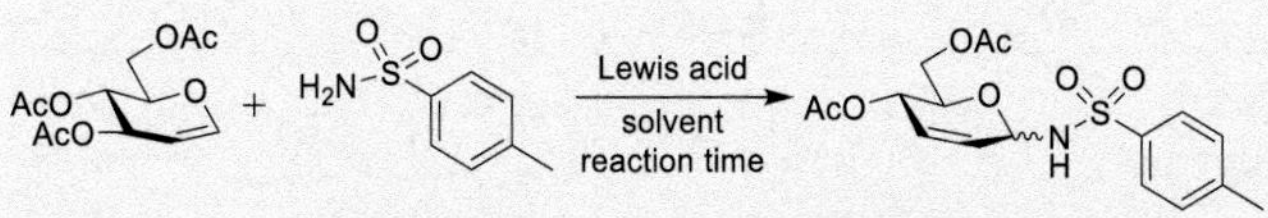

Entry	Reagents	Solvent/time	References
1	$B(C_6F_5)_3$	CH_3CN, 3 h	[99]
2	$BF_3.Et_2O$	CH_2Cl_2, 0.5 h	[107]
3	$HClO4.SiO_2$	CH_2Cl_2, 3 h	[107]
4	$HBF_4.SiO_2$	CH_2Cl_2, 0.5 h	[107]
5	$HBF_4.SiO_2$	CH_3CN	[108]
6	Amberlyst-15	CH_3CN, 1 h	[109]
7	$Hf(OTf)_4$	CH_3CN, 0.5 h	[110]
8	$Tm(OTf)_3$	CH_3CN, 0.6 h	[111]
9	$Cu(OTf)_2$	CH_2Cl_2, 2 h	[101]
10	$Yb(OTf)_2$	CH_2Cl_2, 2 h	[101]
11	$NOBF_4$	CH_3CN, 0.6 h	[112]
12	$RuCl_3$	CH_2Cl_2, 2 h	[101]
13	$RuCl_3$	CH_3CN, 0.6 h	[113]
14	$ZnCl_2$	CH_2Cl_2, 2 h	[101]
15	$ZnCl_2/Al_2O_3$	CH_2Cl_2, 0.3 h	[101]
16	$Zn(OTf)_2$	DCE, 6 h	[114]

Chandrashekar's group[99] was the first time to mention the Ferrier-type glycosylation of sulfonamides with tri-*O*-acetyl-D-glucal in the presence of a catalyst (pentafluorophenyl)borane [$B(C_6F_5)_3$] and solvent acetonitrile in 3 hours (Table 4.2, Entry 1). Later, Colinas et al. explored the Ferrier-type glycosylation of sulfonamides using Lewis acid catalyst $BF_3 \cdot OEt_2$ (Table 4.2, Entry 2) in DCM solvent with relatively less time of 30 minutes. Besides $BF_3 \cdot OEt_2$, other acid catalysts were also used for the activation of tri-*O*-acetyl-D-glucal and tri-*O*-acetyl-D-galactal and found suitable for the sulfonamidoglycosylation (Table 4.2, Entries 3, 4, 5, 6).[107–109] However, the use of acid as a catalyst is a relatively toxic and unstable reagent.

Thereafter, Liu and group reported the efficient strategy for the synthesis of *N*-glycosyl sulfonamides by treating acetylated glucal with the sulfonamides derivatives under the catalysis of Lewis acids such as $Cu(OTf)_2$, Yb $(OTf)_2$, $RuCl_2$, and $ZnCl_2$. Later, they also explored the use of $ZnCl_2/Al_2O_3$ (Entry 15) as an effective promoter for the Ferrier azaglycosylation of glycals and sulfonamides. However, the substrate scope in terms of *N*-aryl sulfonamides, catalytic loading of Lewis acids, and solvent optimization was not studied therein. In 2013 Misra's group developed the Ferrier sulfonamidoglycosylation of glycals using nitrosyl tetrafluoroborate ($NOBF_4$),[112] a nonmetallic organocatalyst.

In 2015 Peiran Chena and Bin Bia[111] were reported the use of lanthanide salts as catalysts in Ferrier rearrangement. They treated the glycals with sulfonamide in the presence of $Tm(OTf)_3$ in acetonitrile at 40°C to obtain *N*-glycosyl sulfonamides in moderate-to-good yields and α-stereoselectivity (Table 4.2, Entry 8). A similar procedure using $Hf(OTf)_4$ as a catalyst was reported by Chena and Zhanga[110] and produced the desired product with 90% yields and complete α-selectivity (Table 4.2, Entry 7). Thereafter, new azaglycosylation protocols using convenient and eco-friendly catalysts ruthenium chloride[113] and Zn(II) triflate[114] were described by Kashyap's group for the expeditious synthesis of *N*-glycosylsulfonamides from glycals (Entries 13, 16).

Although Colinas group did great work to ease this conversion, the mechanistic pathway of such methods leads to moderate stereoselectivities except for the protocol from the similar group, where they fascinatingly present the way to produce 2-deoxy-β-glycosyl sulfonamides stereoselectively in the presence of triphenylphosphine hydrobromide from tri-*O*-benzyl-D-galactal (Scheme 4.40).[115]

Analogously, Toste group[116] also proposed a strategy using [$ReOCl_3(SMe_2)$ (Ph_3PO)] as a catalyst to couple tri-*O*-benzyl-D-galactal with *p*-toluenesulfonamide in PhMe solvent afforded the corresponding *N*-glycosylsulfonamides in 81% yield and major β-selectivity ($\alpha/\beta = 1/3.3$) (Scheme 4.40).

To enhance the stereoselectivity and regioselectivity *N*-glycosides, Qiang Zhang group[44] put forth a new strategy with alternate reaction substrates such as 3,4-*O*-carbonate glycal and *N*-tosyl functionalized aliphatic and

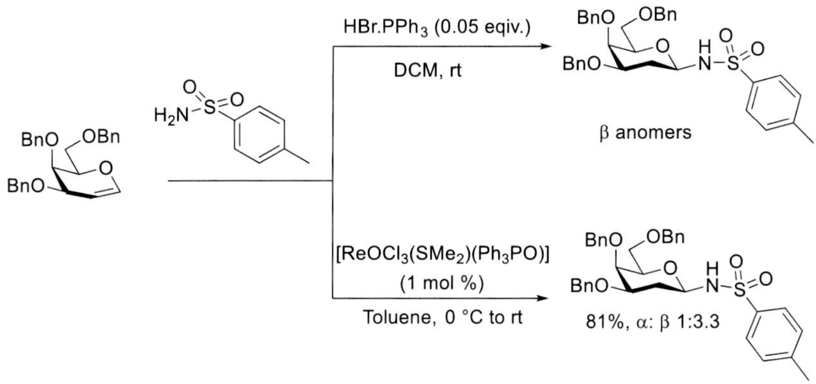

SCHEME 4.40 Synthesis of 2-deoxy-*N*-glycosylsulfonamides.

aromatic amines under $Pd(OAc)_2$ (5 mol%) catalysis in CH_2Cl_2 at 40°C to furnish the respective *N*-glycoside with good yield and excellent stereoselectivity (Scheme 4.41). This strategy is initiated with the reduction of the Pd (II) catalyst to Pd(0) that forms a complex with a double bond of glycal, which then undergoes oxidative addition followed by the attack of sulfonamides to form *N*-glycoside in excellent yields up to 98% with complete retention of stereochemistry ($\beta/\alpha > 30:1$, confirmed by X-ray diffraction (XRD) crystallographic analysis).

After studying the success in the synthesis of β-*N*-glycosides, the attention then turned toward the synthesis of α-selective *N*-glycosides. Liu's group[42] discovered a strategy for the synthesis of α-*N*-glycosides from enone sugars that undergo glycal rearrangement via $BF_3 \cdot Et_2O$-promoter (Scheme 4.42). They also tried this process with other Lewis acids such as TMSOTf, TESOTf, $SnCl_4$, $TiCl_4$, and $Sc(OTf)_3$ as catalysts, but in such cases they obtained incomplete conversion and poor selectivity. This process pathway basically involves the interaction of the acetoxy group with Lewis acid $BF_3 \cdot OEt_2$ leading to the generation of allyloxocarbenium ion I followed by double bond shifting and further the coordination of $BF_3 \cdot OEt_2$ with alkoxy group accelerates the attack of sulfonamide from the opposite side to form α-selective 2-enopyranoside II that at last underwent β-elimination to produce dihydropyranones (Scheme 4.42).

4.3.4.3 Installation of other groups

Although the major portion of the reported work for the synthesis of 2,3-unsaturated-*N*-glycosides or "*N*-pseudoglycals" is performed using $TMSN_3$ and sulfonamide as a nucleophile, there are many other reports also present that explored the Ferrier azaglycosylation using nucleophiles such as benzyl or *t*-butyl carbamate ($CbzNH_2$ or $BocNH_2$), benzyl or acetyl amide, aniline, and cyclic amide.

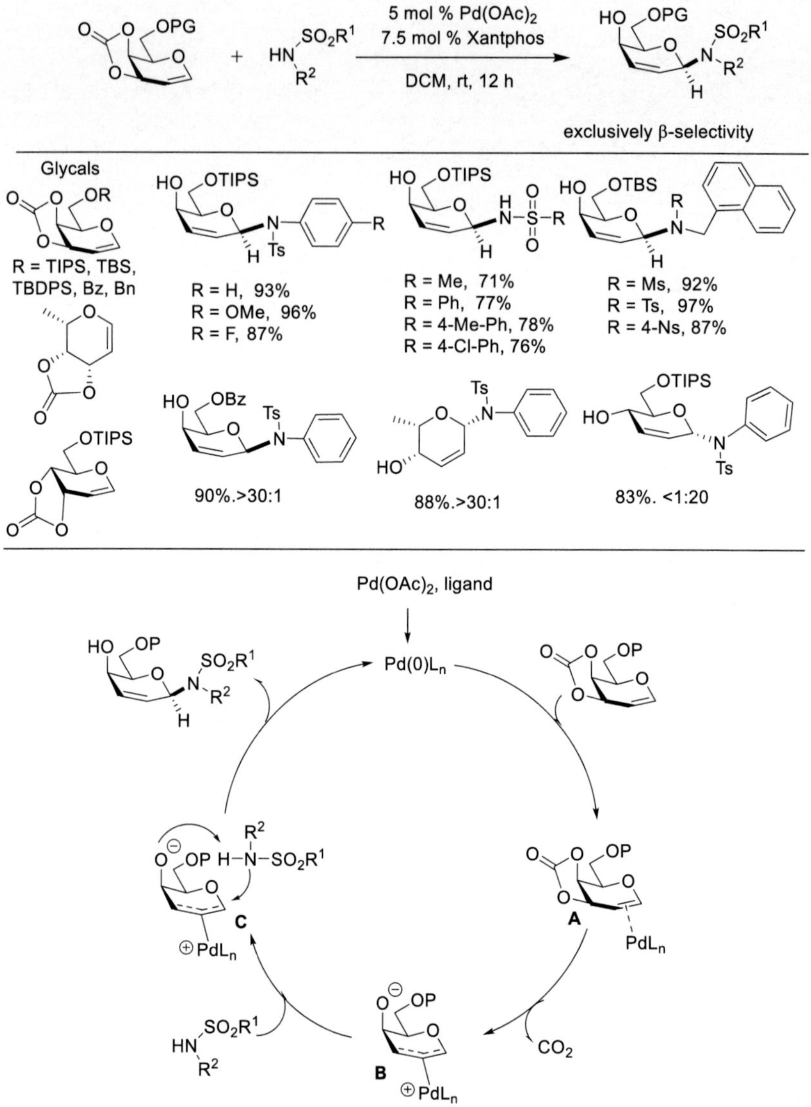

SCHEME 4.41 Pd-catalyzed 2,3-unsaturated β-*N*-glycosides.

Several works reported using benzyl carbamate ($CbzNH_2$) and *tert*-butyl carbamate ($BocNH_2$) as nucleophiles, in which the efficient coupling of tri-*O*-acetyl-D-glucal with such nucleophiles was performed using potent catalysts such as $[B(C_6F_5)_3]$ and $ZnCl_2/Al_2O_3$ having great catalytic activity in azaglycosidation reactions (Table 4.3). As an outcome of such experiments, it was found that $[B(C_6F_5)_3]$ catalyst promotes better α/β selectivity, whereas $ZnCl_2/Al_2O_3$ combination leads to higher yield comparatively. When $ZnCl_2$/

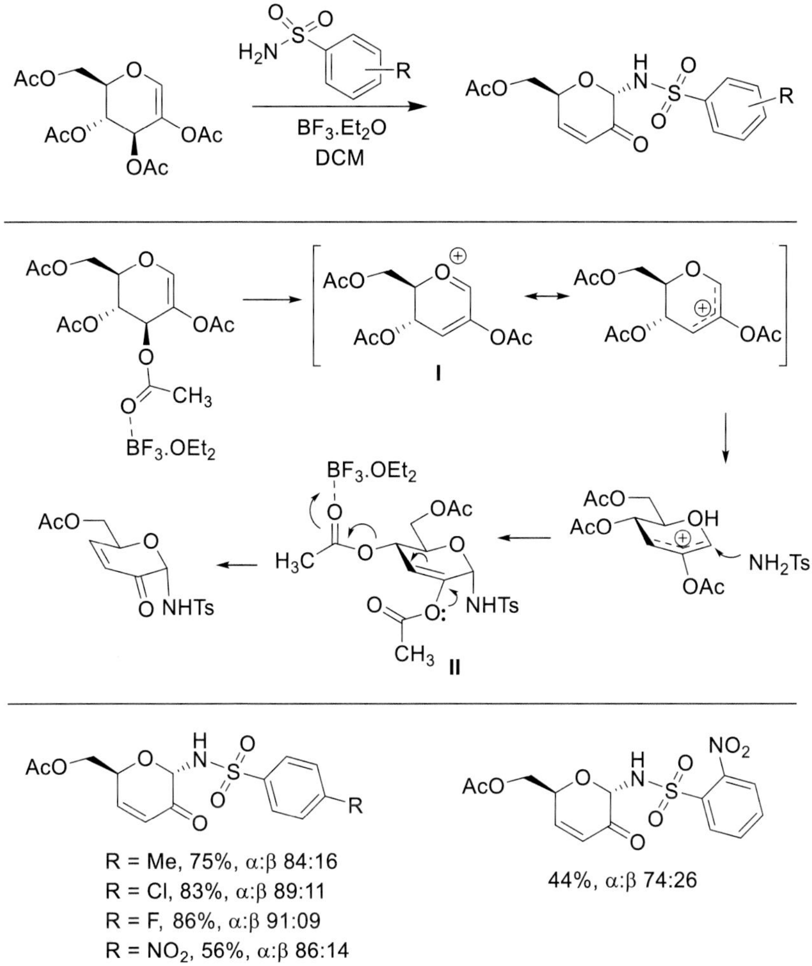

SCHEME 4.42 $BF_3 \cdot Et_2O$-promoted *N*-glycosylation.

Al_2O_3 combination was used to promote the coupling of tri-*O*-acetyl-D-glucal with benzamide nucleophiles, the yield of the obtained product got lower. Moreover, $Zn(OTf)_2$ was also used to catalyze the coupling of benzyl carbamate ($CbzNH_2$) with tri-*O*-acetyl-D-glucal but afforded the product with both lower yields and poor selectivity, while the coupling of tri-*O*-acetyl-D-glucal with benzamide nucleophile using the same $Zn(OTf)_2$ catalyst furnished the respective product with a good yield (80%).

Another nucleophile, like acetyl amide, was catalyzed by iodine to perform azaglycosylation with glucal and afforded the corresponding product with moderate yield, while anilines containing electron-withdrawing groups were also used as a nucleophile in Ferrier azaglycosylation reactions with

TABLE 4.3 Ferrier azaglycosylation of several *N*-nucleophiles.

Substrate	Nucleophile	Catalyst	Products	Yield (%) α/β ratio	Ref
	NH_2Cbz	$ZnCl_2/Al_2O_3$	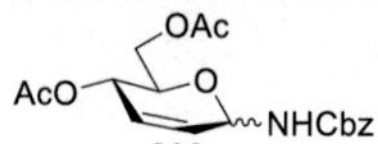 **208**	80 11.5/1	108
	NH_2Cbz	Zn(OTf)2	**208**	67 3/1	119
	NH_2Cbz	$B(C_6F_5)_3$	**208**	71 12/1	105
	NH_2Boc	$ZnCl_2/Al_2O_3$	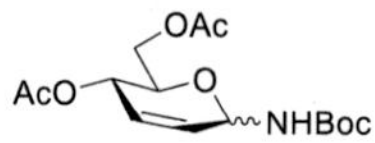 **209**	70 5.2/1	108
	NH_2Boc	$B(C_6F_5)_3$	**209**	56 15/1	105
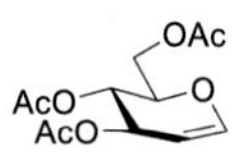 **198**	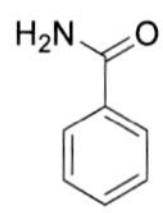	Zn(OTf)2	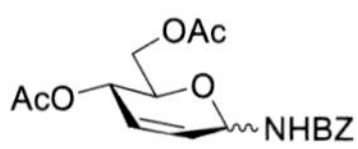 **210**	80 2.5/1	119
	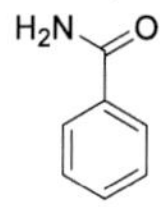	$ZnCl_2/Al_2O_3$	**210**	54 3/1	108
	NH_2AC	I_2	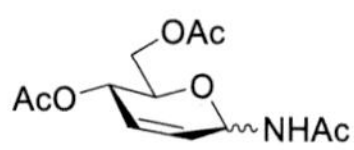 **211**	65 3/1	103
	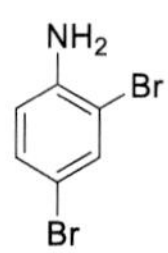	I_2	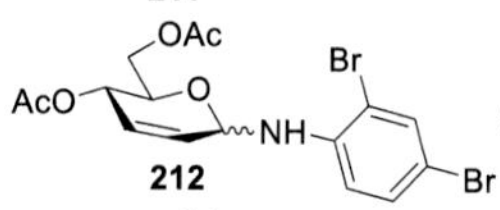 **212**	53 3/1	103
	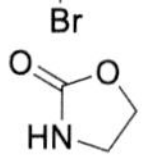	Hf(OTf)4	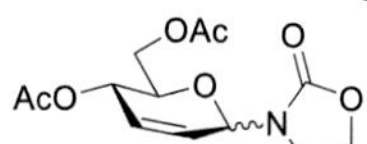 **213**	92 α-only	118

iodine as a catalyst in tetrahydrofuran (THF) at reflux but afforded the product with a lower yield. However, the Ferrier azaglycosylation reactions of cyclic amide nucleophile furnished the desired product with both excellent yield and complete α-selectivity (Table 4.3).

Later, Liu and coworkers[117] brought a very concise and effective methodology using 1,1′-carbonyldiimidazole (CDI) (1.5 equiv.) as a nucleophile and $Pd(PPh_3)_4$ as a catalyst in DCM solvent for the formation of *N*-heterocyclic glycosides through decarboxylative allylation reaction involving three steps, namely carbamation, decarboxylation, and allylation (Scheme 4.43). In this process, initially, glycal coordinates with CDI to form carbamate intermediate I. This intermediate I undergoes interaction with palladium catalyst via a

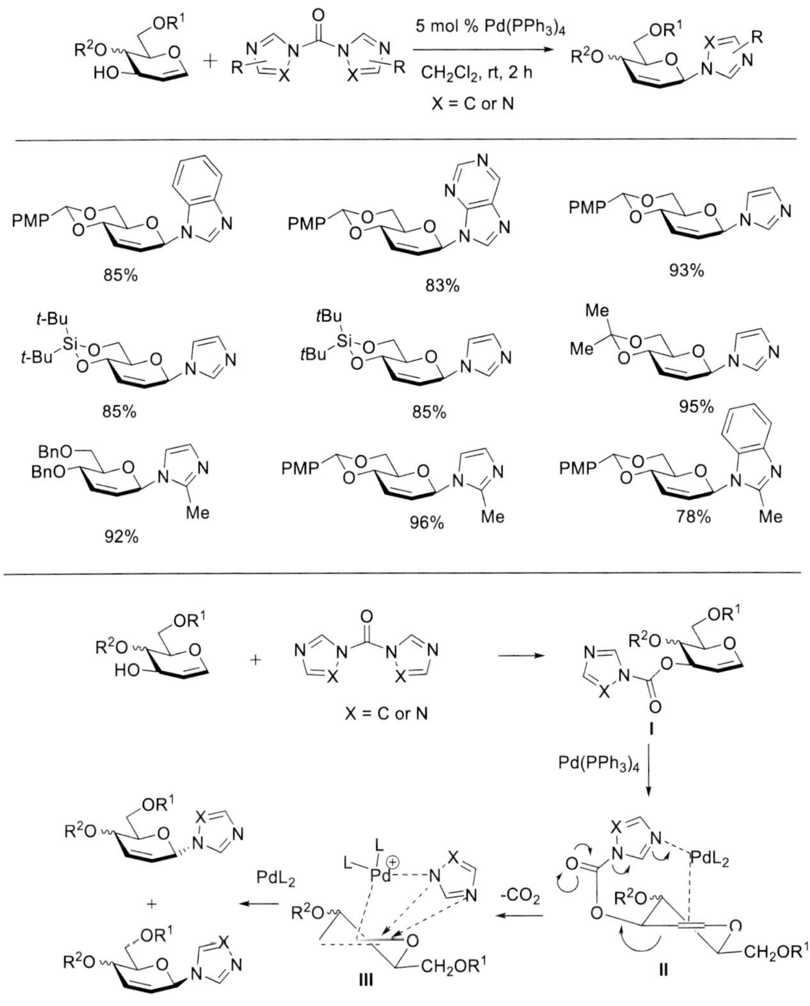

SCHEME 4.43 Pd-catalyzed 2,3-unsaturated β-*N*-glycosyl imidazole.

double bond of glycal to form key intermediate II, which further on decarboxylation, generates intermediate III, and finally nucleophilic addition on intermediate III afforded the desired product *N*-glycosyl imidazole in good yield and excellent β-selectivity (Scheme 4.43).

To further explore the field of *N*-heterocyclic glycosylation, a similar research group[118] came up with a new strategy based on glycal type donors and various *N*-nucleophile such as isatins, imides, and imidazole derivatives for the synthesis of *N*-heterocyclic glycosides via palladium-catalyzed allylation with both β- or α-selectivity separately by using directing group like picoloyl group at C3 and C1 positions of glucal donors (Scheme 4.44).

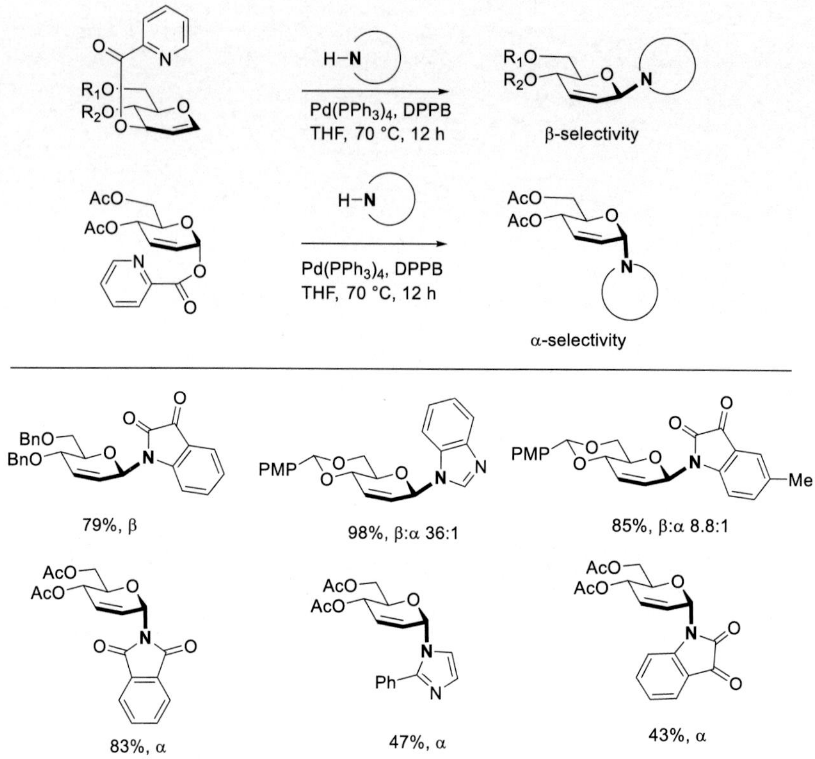

SCHEME 4.44 Palladium-catalyzed *N*-heterocyclic glycosides.

The palladium catalyst interacts with the β-face of glucal as well as picoloyl group simultaneously and generates intermediate II, further leading to the removal of the picoloyl group via classic Ferrier rearrangement to obtain intermediate III. Subsequently, the interaction of the *N*-nucleophile to the palladium formed intermediate IV that ultimately underwent nucleophilic addition to afford the desired *N*-glycoside with *β*-selectivity. Similarly, for α-type 2,3-unsaturated hexopyranoside donor, *N*-glycoside with major *α*-selectivity was produced (Scheme 4.45).

Zhang et al.[119] discovered a convergent peptide arginine glycosylation method for the synthesis of *N*-glycosides using palladium catalyst and xantphos (Scheme 4.46). This method involves arginine modification, which has been challenging because of the inertness of the guanidino side chain. In such process, a classic Tsuji – Trost reaction mechanism was followed; the glycal donor primarily interacted with palladium catalyst from a less hindered face and formed π-allyl Pd(0) as intermediate that further underwent reaction with arginine (pbf) in the presence of xantphos to furnish the desired *N*-glycoside in good yield and exclusive stereoselectivity as well as regioselectivity. This method highlights a mild reaction condition, high amino acid residue tolerance, and dual nature of pbf both as a

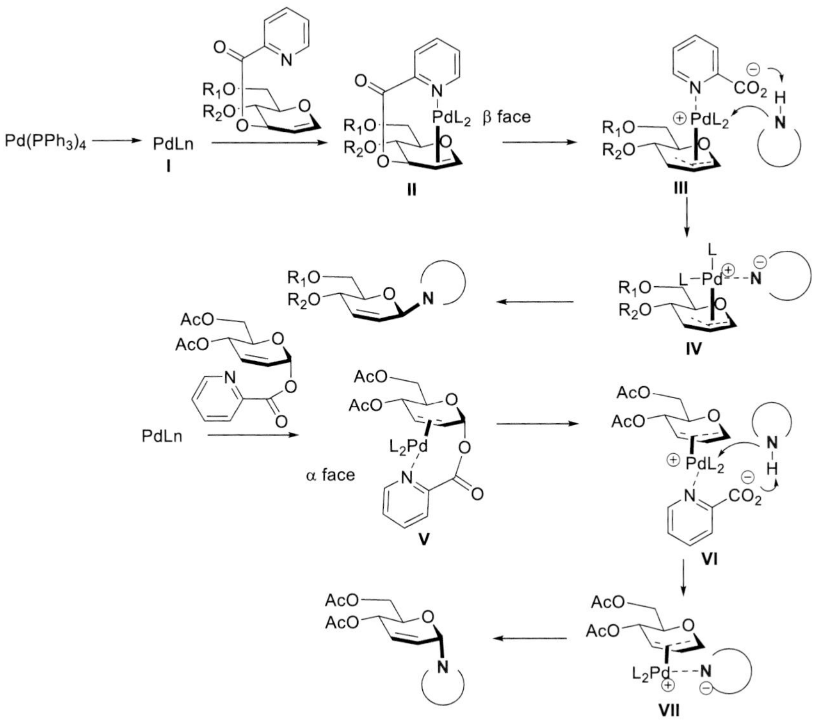

SCHEME 4.45 Plausible mechanism.

protecting and a specific activating group that further get easily removed. Compared with extant strategies, this method gives a facile alternative strategy for highly efficient glycosylation, which potentially could assist the biological studies toward arginine glycosides.

4.4 Conclusions and future outlook

After researching long years to produce an efficient and sustainable synthetic strategy to form *N*-glycosides and their analogs, diverse effective and powerful methodologies have been developed for the synthesis of biologically significant *N*-glycosides (including nucleosides). To date, researchers still continue their studies in this field due to the considerable therapeutic potential of *N*-glycoside analogs, their involvement in the synthesis of DNA/RNA, and other biologically active molecules. Moreover, the other common categories of *N*-glycosides are glycopeptides and glycoproteins, which are found attached to proteins through the nitrogen atom of asparagine. The *N*-linked glycoproteins serve as significant pharmacological agents, such as antibiotic, antineoplastic, or antiviral compounds, and antiproliferative agents. Moreover, nucleosides and their derivatives are crucial biomolecules that can act not only as essential

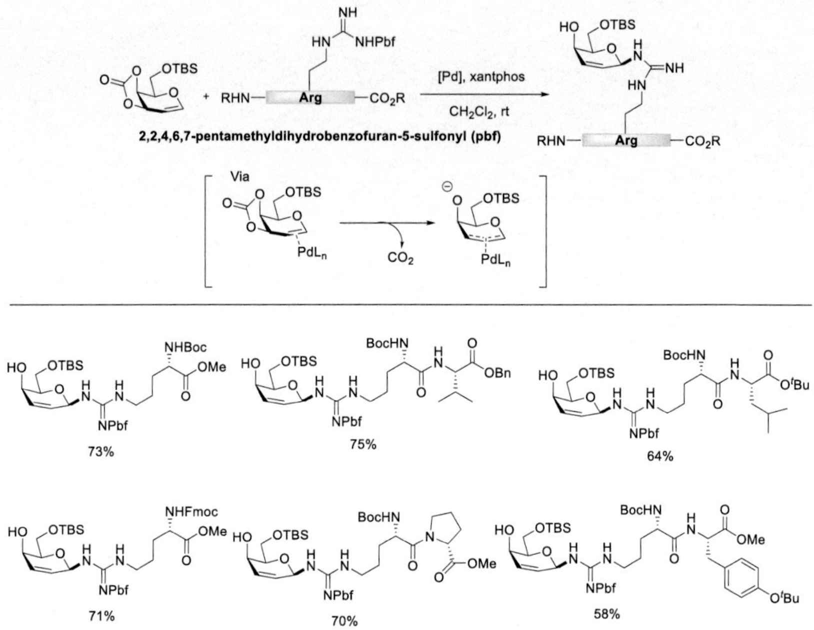

SCHEME 4.46 Palladium-catalyzed stereospecific arginine glycosylation.

components of DNA/RNA but also as clinical drugs for tumor treatment, antiviral therapy, and immune regulation.

However, the reported strategies for the synthesis of these compounds via *N*-glycosylation suffer from serious drawbacks. Several reported protocols do not provide access to aminosugars with complete stereoselectivity as well as regioselectivity, which can be problematic, often leading to the formation of mixtures that are difficult to separate and, consequently, the overall yields of final products are often low. The glycosylation step between nucleobase and carbohydrate is often difficult. Taking into account the protection, glycosylation, and deprotection sequences, the procedures are laborious and inefficient. Therefore it is not surprising that there is a constant need for convenient, inexpensive, but efficient methods of amino sugar synthesis.

To keep up with the expanding areas of glycoscience, it is essential to make *N*-glycosides more approachable to the general chemical and biomedical spectators. The advancement in the methodology of *N*-glycosylation and their wider validation is important to meet this need. The goal of optimizing *N*-glycosylation has been pursued in many ways. This significantly enhanced our understanding and ability to simplify the reaction conditions, diminish side reactions, control the stereochemical outcome of the starting material, and predict the output of many glycosylations. We hope that this comprehensive collection of methods might influence the glycochemists to discover new opportunities, both in terms of the application of existing techniques and in the development of new chemical methodologies.

Acknowledgments

A. K. thanks CSIR, New Delhi for providing Fellowship. The authors gratefully acknowledge financial support by DST-SERB (CRG/2022/003936), New Delhi, India. CDRI communication no. 10652.

References

1. Bashkin, J. K. *Chem. Rev.* **2000,** *100*, 4265–4711.
2. Varki, A.; Cummings, R.; Esko, J., Eds. *Essentials of Glycobiology;* CSHL Press: New York, 2002; p 17.
3. Wang, P.; Bertozzi, C. E. Glycochemistry: Principles, Synthesis, and Applications; Marcel Dekker: New York, 2001.
4. Hölemann, A.; Seeberger, P. H. *Curr. Opin. Biotechnol.* **2004,** *15*, 615–622.
5. Varki, A. *Glycobiology* **1993,** *3*, 97–130.
6. Boltje, T. J.; Buskas, T.; Boons, G.-J. *Nat. Chem.* **2009,** *1*, 611–622.
7. Varki, A. *Glycobiology* **2017,** *27*, 3–49.
8. Bertozzi, C. R.; Kiessling, L. L. *Science* **2001,** *291*, 2357–2364.
9. Okajima, T.; Irvine, K. D. *Cell* **2002,** *111*, 893–904.
10. Seeberger, P. H. *Nature* **2005,** *437*, 1239.
11. Phillips, M. L.; Nudelman, E.; Gaeta, F. C.; Perez, M.; Singhal, A. K.; Hakomori, S.; Paulson, J. C. *Science* **1990,** *250*, 1130–1132.
12. Yang, Y.; Yu, B. *Chem. Rev.* **2017,** *117*, 12281–12356.
13. Panza, M.; Pistorio, S. G.; Stine, K. J.; Demchenko, A. V. *Chem.Rev.* **2018,** *118*, 8105–8150.
14. Sangwan, R.; Khanam, A.; Mandal, P. K. *Eur. J. Org. Chem.* **2020,** 5949–5977.
15. Tanaka, K.; Fukase, K. In *Glycoscience: Chemistry and Chemical Biology;* 2nd ed. Fraser-Reid, B. O., Tatsuta, K., Thiem, J., Eds.; Vol. I – III; Springer: New York, 2009.
16. Dwek, R. A. *Chem. Rev.* **1996,** *96*, 683–720.
17. de Clerq, E. *Antivir. Res.* **2005,** *67*, 56–75.
18. Mathe, C.; Gosselin, G. *Antivir. Res.* **2006,** *71*, 276–281.
19. Roy, B.; Depaix, A.; Périgaud, C.; Peyrottes, S. *Chem. Rev.* **2016,** *116*, 7854–7897.
20. Jordheim, L. P.; Durantel, D.; Zoulim, F.; Dumontet, C. *Nat. Rev. Drug Discov.* **2013,** *12*, 447–464.
21. Xu, W.; Chan, K. M.; Kool, E. T. *Nat. Chem.* **2017,** *9*, 1043–1055.
22. Jao, C. Y.; Salic, A. *Proc. Natl. Acad. Sci. U. S. A.* **2008,** *105*, 15779–15784.
23. Allen-Mersh, T. G.; Earlam, S.; Fordy, C.; Abrams, K.; Houghton, J. *Lancet* **1994,** *344*, 1255–1260.
24. Schürmann, D.; Rudd, D. J.; Zhang, S.; De Lepeleire, I.; Robberechts, M.; Friedman, E.; Keicher, C.; Hüser, A.; Hofmann, J.; Grobler, J. A., et al. *Lancet HIV.* **2020,** *7*, 164–172.
25. Ruiz-Mirazo, K.; Briones, C.; de la Escosura, A. *Chem. Rev.* **2014,** *114*, 285–366.
26. Rayon, C.; Lerouge, P.; Faye, L. *J. Expt. Bot.* **1998,** *49*, 1463–1472.
27. Norris, P. *Curr. Top. Med.Chem.* **2008,** *8*, 101–113.
28. Ni, G.; Du, Y.; Tang, F.; Liu, J.; Zhao, H.; Chen, Q. *RSC Adv.* **2019,** *9*, 14302–14320.

29. Sticks, R. V.; Williams, S. J. *Carbohydrates: The Essential Molecules of Life*, 2nd ed.; Elsevier: Amsterdam, 2009.
30. Benoff, S. *Mol. Hum. Reprod.* **1997,** *3*, 599–637.
31. Cipolla, L.; Araújo, A. C.; Bini, D.; Gabrielli, L.; Russo, L.; Shaikh, N. *Expert. Opin. Drug Discov.* **2010,** *5*, 721–737.
32. Dube, D. H.; Bertozzi, C. R. *Nat. Rev. Drug Discov.* **2005,** *4*, 477–488.
33. Murrey, H. E.; Hsieh-Wilson, L. C. *Chem. Rev.* **2008,** *108*, 1708–1731.
34. De Clercq, E. *Annu. Rev. Pharmacol. Toxicol.* **2011,** *51*, 1–24.
35. De Clercq, E.; Holý, A. *Nat. Rev. Drug Discov.* **2005,** *4*, 928–940.
36. Trost, B. M.; Krische, M. J.; Berl, V.; Grenzer, E. M. *Org. Lett.* **2002,** *4*, 2005–2008.
37. Lahiri, R.; Ansari, A. A.; Vankar, Y. D. *Chem. Soc. Rev.* **2013,** *42*, 5102–5118.
38. Li, W.; Yu, B. *Chem. Soc. Rev.* **2018,** *47*, 7954–7984.
39. Köllmann, C.; Wiechert, S. M.; Jones, P. G.; Pietschmann, T.; Werz, D. B. *Org. Lett.* **2019,** *21*, 6966–6971.
40. Carroux, C. J.; Rankin, G. M.; Moeker, J.; Bornaghi, L. F.; Katneni, K.; Morizzi, J.; Charman, S. A.; Vullo, D.; Supuran, C. T.; Poulsen, S.-A. *J. Med. Chem.* **2013,** *56*, 9623–9634.
41. Yang, J.; Mercer, G. J.; Nguyen, H. M. *Org. Lett.* **2007,** *9*, 4231–4234.
42. Ding, F.; William, R.; Liu, X.-W. *J. Org. Chem.* **2013,** *78*, 1293–1299.
43. Ji, L.; Xiang, S.-H.; Leng, W.-L.; Le Mai Hoang, K.; Liu, X.-W. *Org. Lett.* **2015,** *17*, 1357–1360.
44. Dai, Y.; Zheng, J.; Zhang, Q. *Org. Lett.* **2018,** *20*, 3923–3927.
45. Nielsen, M. M.; Mała, P.; Baldursson, E. P.; Pedersen, C. M. *Chem. Sci.* **2019,** *10*, 5299–5307.
46. Zhang, Q.; Sun, J.; Zhu, Y.; Zhang, F.; Yu, B. *Angew. Chem. Int. Ed.* **2011,** *50*, 4933–4936 *Angew. Chem.* **2011**, *123*, 5035–5038.
47. Vorbrüggen, H.; Ruh-Pohlenz, C. *Org. React.* **2004,** 1–630.
48. Niedballa, U.; Vorbrüggen, H. *Angew. Chem. Int. Ed. Engl.* **1970,** *9*, 461–462.
49. Kim, H. O. *Arch. Pharm. Res.* **2001,** *24*, 508–513.
50. Parmenopoulou, V.; Chatzileontiadou, D. S. M.; Manta, S.; Bougiatioti, S.; Maragozidis, P.; Gkaragkouni, D.-N.; Kaffesaki, E.; Kantsadi, A. L.; Skamnaki, V. T.; Zographos, S. E., et al. *Bioorg. Med. Chem.* **2012,** *20*, 7184–7193.
51. Shirouzu, H.; Morita, H.; Tsukamoto, M. *Tetrahedron* **2014,** *70*, 3635–3639.
52. Moreau, C.; Kirchberger, T.; Swarbrick, J. M.; Bartlett, S. J.; Fliegert, R.; Yorgan, T.; Bauche, A.; Harneit, A.; Guse, A. H.; Potter, B. V. L. *J. Med. Chem.* **2013,** *56*, 10079–10102.
53. Tranova, L.; Styskala, J. *J. Org. Chem.* **2021,** *86*, 13265–13275.
54. Hocková, D.; Hocek, M.; Dvořáková, H.; Votruba, I. *Tetrahedron* **1999,** *55*, 11109–11118.
55. Ackermann, D.; Famulok, M. *Nucleic Acids Res.* **2013,** *41*, 4729–4739.
56. Freskos, J. N. *Nucleos. Nucleot.* **1989,** *8*, 549–555.
57. Kazimierczuk, Z.; Cottam, H. B.; Revankar, G. R.; Robins, R. K. *J. Am. Chem. Soc.* **1984,** *106*, 6379–6382.
58. Liu, G.; Zhang, X.; Xing, G. *Chem. Commun.* **2015,** *51*, 12803–12806.
59. Downey, A. M.; Richter, C.; Pohl, R.; Mahrwald, R.; Hocek, M. *Org. Lett.* **2015,** *17*, 4604–4607.
60. Downey, A. M.; Pohl, R.; Roithová, J.; Hocek, M. *Chem. Eur. J.* **2017,** *23*, 3910–3917.
61. Fraser-Reid, B.; Ganney, P.; Ramamurty, C. V. S.; Gómez, A. M.; López, J. C. *Chem. Commun.* **2013,** *49*, 3251–3253.

62. Rao, B. V.; Manmode, S.; Hotha, S. *J. Org. Chem.* **2015,** *80*, 1499–1505.
63. Liu, R.; Hua, Q.; Lou, Q.; Wang, J.; Li, X.; Ma, Z.; Yang, Y. *J. Org. Chem.* **2021,** *86*, 4763–4778.
64. Liu, R.; Chen, Y.; Zheng, J.; Zhang, L.; Xu, T.; Xu, P.; Yang, Y. *Org. Lett.* **2022,** *24*, 9479–9484.
65. Li, P.; He, H.; Zhang, Y.; Yang, R.; Xu, L.; Chen, Z.; Huang, Y.; Bao, L.; Xiao, G. *Nat. Commun.* **2020,** *11*, 405–414.
66. Chakraborty, S.; Mishra, B.; Das, P. K.; Pasari, S.; Hotha, S. *Angew. Chem. Int. Ed.* **2023,** *135*, e202214167.
67. Hu, Z.; Tang, Y.; Yu, B. *J. Am. Chem. Soc.* **2019,** *141*, 4806–4810.
68. Liao, J.; Sun, J.; Yu, B. *Tetrahedron Lett.* **2008,** *49*, 5036–5038.
69. Liao, J.; Sun, J.; Yu, B. *Carbohydr. Res.* **2009,** *344*, 1034–1038.
70. Kistemaker, H. A. V.; van der Heden.; van Noort, G. J.; Overkleeft, H. S.; van der Marel, G. A.; Filippov, D. V. *Org. Lett.* **2013,** *15* (9), 2306–2309.
71. Xiao, K.; Hu, Y.; Wan, Y.; Li, X. X.; Nie, Q.; Yan, H.; Wang, L.; Liao, J.; Liu, D.; Tu, Y., et al. *Chem. Sci.* **2022,** *13*, 1600–1607.
72. Zhou, S.-Y.; Hu, X.-P.; Liu, H.-J.; Zhang, Q.-J.; Liao, J.-X.; Tu, Y.-H.; Sun, J.-S. *Org. Lett.* **2022,** *24*, 653–657.
73. Salunke, S. B.; Babu, N. S.; Chen, C. T. *Chem. Commun.* **2011,** *47*, 10440–10442.
74. Bertho, A. *Ber. Dlsch. Chelll. Ges.* **1930,** *63*, 836–843.
75. Tropper, F. D.; Andersson, F. O.; Braun, S.; Roy, R. *Synthesis* **1992,** 618–620.
76. Colinas, P. A.; Bravo, R. D. *Tetrahedron Lett* **2005,** *46*, 1687–1689.
77. Colinas, P. A.; Nunez, N. A.; Bravo, R. D. *J. Carbohydr. Chem.* **2008,** *27*, 141–147.
78. Bennett, C. S. *Selective Glycosylations: Synthetic Methods and Catalysts;* Wiley-VCH Verlag GmbH & Co. KGaA, 2017.
79. Jung, K.-H.; Müller, M.; Schmidt, R. R. *Chem. Rev.* **2000,** *100*, 4423–4442.
80. Demchenko, A. V. *Handbook of Chemical Glycosylation: Advances in Stereoselectivity and Therapeutic Relevance;* Wiley-VCH Verlag GmbH & Co. KGaA, 2008.
81. Peng, P.; Schmidt, R. R. *Acc. Chem. Res.* **2017,** *50*, 1171.
82. Schmidt, R. R. *Angew. Chem.* **1986,** *98*, 213–236.
83. Schmidt, R. R.; Kinzy, W. *Adv. Carbohydr. Chem. Biochem.* **1994,** *50*, 21–123.
84. Khanam, A.; Mandal, P. K. *N. J. Chem.* **2021,** *45*, 15386–15391.
85. Kobayashi, Y.; Nakatsuji, Y.; Li, S.; Tsuzuki, S.; Takemoto, Y. *Angew. Chem. Int. Ed.* **2018,** *130*, 3708–3712.
86. Park, N. H.; Nguyen, H. M. *Org. Lett.* **2009,** *11*, 2433–2436.
87. Sorokin, B.; Anilide, D.; Glycosen, T. D. *J. Prakt. Chem.* **1888,** *37*, 291–317.
88. Wang, L.; Maniglia, C. A.; Mella, S. L.; Sartorelli, A. C. *J. Med. Chem.* **1983,** *26*, 1323–1326.
89. Kublashvili, R. *Chem. Nat. Compd.* **2003,** *39*, 586–588.
90. Qian, X.; Li, Z.; Liu, Z.; Song, G.; Li, Z. *Carbohydr. Res.* **2001,** *336*, 79–82.
91. Winum, J.-Y.; Casini, A.; Mincione, F.; Starnotti, M.; Montero, J.-L.; Scozzafava, A.; Supuran, C. T. *Bioorg. Med. Chem. Lett.* **2004,** *14*, 225–229.
92. Bridiau, N.; Benmansour, M.; Legoy, M. D.; Maugard, T. *Tetrahedron* **2007,** *63*, 4178–4183.
93. Tanaka, T.; Nagai, H.; Noguchi, M.; Kobayashi, A.; Shoda, S.-I. *Chem. Commun.* **2009,** 3378–3379.
94. Dyapa, R.; Dockery, L. T.; Walczak, M. A. *Org. Biomol. Chem.* **2017,** *15*, 51–55.
95. Khanam, A.; Javed.; Lal, M.; Mandal, P. K. *Asian J. Org.Chem.* **2023,** e202300044.

96. Kawabata, H.; Kubo, S.; Hayashi, M. *Carbohydr. Res.* **2001,** *333*, 153–158.
97. Yadav, J. S.; Reddy, B. V. S.; Rao, C. V.; Chand, P. K.; Prasad, A. R. *Synlett* **2001,** 1638–1640.
98. Yadav, J. S.; Reddy, B. V. S. *Synthesis* **2002,** 511–514.
99. Chandrasekhar, S.; Reddy, C. R.; Chandrashekar, G. *Tetrahedron Lett.* **2004,** *45*, 6481–6484.
100. Smitha, G.; Reddy, C. S. *Synthesis* **2004,** 834–836.
101. Ding, F.; William, R.; Gorityala, B. K.; Ma, J.; Wang, S.; Liu, X.-W. *Tetrahedron Lett.* **2010,** *51*, 3146–3148.
102. Levecque, P.; Gammon, D. W.; Jacobs, P.; Vos, D. D.; Sels, B. *Green. Chem.* **2010,** *12*, 828–835.
103. Srinivas, B.; Reddy, T. R.; Kashyap, S. *Carbohydr. Res.* **2015,** *406*, 86–92.
104. Reddy, T. R.; Rao, D. S.; Kashyap, S. *RSC Adv.* **2015,** *5*, 28338–28343.
105. Ansari, A. A.; Reddy, Y. S.; Vankar, Y. D. *Beilstein J. Org. Chem.* **2014,** *10*, 300–306.
106. Vinod, T.; Bandgar, P. K. B.; Khobragade, N. C.; Gacche, N. R.; Kamble, A. V. *Lett. Org. Chem.* **2006,** *3*, 658–663.
107. Colinas, P. A.; Bravo, R. D. *Carbohydr. Res.* **2007,** *342*, 2297–2302.
108. Colinas, P. A.; Bravo, R. D.; Vullo, D.; Scozzafava, A.; Supuran, C. T. *Bioorg. Med. Chem. Lett.* **2007,** *17*, 5086–5090.
109. Témpera, C. A.; Colinas, P. A.; Bravo, R. D. *Tetrahedron Lett.* **2010,** *51*, 5372–5374.
110. Chen, P.; Zhang, X. *Tetrahedron Lett.* **2017,** *58*, 309–312.
111. Chena, P.; Bi, B. *Tetrahedron Lett.* **2015,** *56*, 4895–4899.
112. Santra, A.; Guchhait, G.; Misra, A. K. *Synlett* **2013,** 581–586.
113. Reddy, T. R.; Chittela, S.; Kashyap, S. *Tetrahedron* **2014,** *70*, 9224–9229.
114. Reddy, T. R.; Battina, S. K.; Kashyap, S. J. *Carbohydr. Chem.* **2015,** *34*, 133–144.
115. Colinas, P. A.; Bravo, R. D. *Org. Lett.* **2003,** *23*, 4509–4511.
116. Sherry, B. D.; Loy, R. N.; Toste, F. D. *J. Am. Chem. Soc.* **2004,** *126*, 4510–4511.
117. Xiang, S.; He, J.; Ma, J.; Liu, X.-W. *Chem. Commun.* **2014,** *50*, 4222–4224.
118. Ji, L.; Xiang, S.-H.; Leng, W.-L.; Hoang, K. L. M.; Liu, X.-W. *Org. Lett.* **2015,** *17*, 1357–1360.
119. Yang, J.; Dai, Y.; Bartlett, R.; Zhang, Q. *Org. Lett.* **2021,** *23*, 4008–4012.

Chapter 5

The emerging polyhydroxyazepane-based *C*-glycosides for glycosidase inhibition

Sumit and Indrapal Singh Aidhen

Department of Chemistry, Indian Institute of Technology Madras, Chennai, Tamil Nadu, India

5.1 Introduction

Glycosidases and glycosyl transferases, in nature, are enzymes responsible for the hydrolysis and formation of glycosidic bonds. They play a key role in controlling the metabolism of carbohydrates (breakdown and reassembly) of *O*- and *N*-linked glycoproteins and glycolipids. Their involvement in several metabolic disorders and other carbohydrate-mediated diseases such as diabetes, lysosomal storage disorders, viral and bacterium infection, and cancer have made them surface as valuable drug targets for varied therapeutics.[1] With the discovery of nojirimycin **1**,[2] polyhydroxylated piperidine and pyrrolidine-based natural or synthetic compounds have surfaced as a new class of sugar mimics, commonly referred to as iminosugars. In these substrates, the pyrano or furano endocyclic oxygen has been replaced by a basic nitrogen atom (Fig. 5.1).

Although the iminosugars class also includes isoiminosugars (i.e., 1-azacarbasugars or 1-*N*-iminosugars) having a methylene unit in place of endocyclic oxygen and the anomeric carbon replaced with nitrogen, The International Union of Pure and Applied Chemistry (IUPAC) recommendations[3] distinguishes the two carefully. According to their recommendations, the name iminosugar should be used for compounds where ring oxygen has been replaced with nitrogen, while azasugar should be used for compounds wherein carbon has been replaced with nitrogen. The spectacular development of this class of compounds during the last two decades has resulted from their initial promise embedded in glycosidase inhibition activities.[4–6]

Synthetic Strategies in Carbohydrate Chemistry. DOI: https://doi.org/10.1016/B978-0-323-91729-2.00011-2

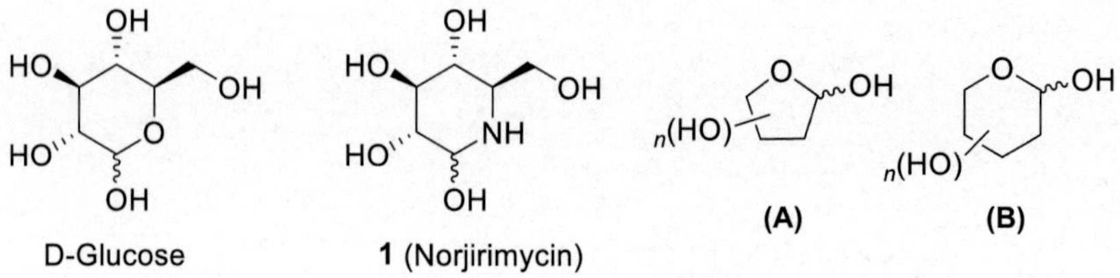

(A) polyhydroxylated furanose form of monosaccharide carbohydrate
(B) polyhydroxylated pyranose form of monosaccharide carbohydrate

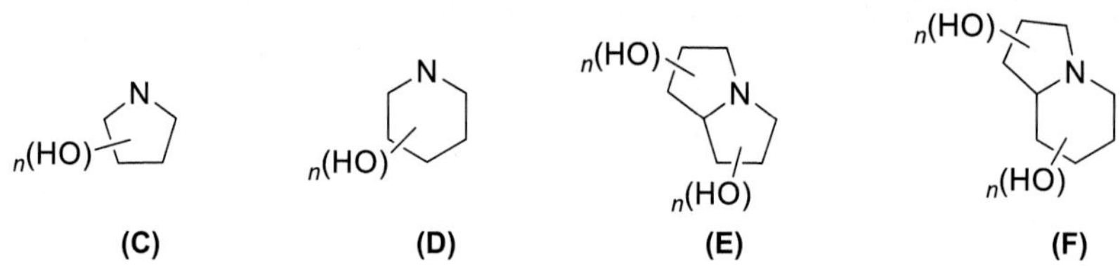

Imino/Aza-sugars analogues
(C) Pyrolidine based (D) Piperidine based (E) Pyrrolizidine based
(F) Indolizidine based

FIGURE 5.1 General structures for imino/aza-sugars.

The acidic hydrolysis is initiated with protonation of the exocyclic oxygen, formation of species **3**, and subsequent expulsion of aglycone leading to the formation of species **4a/4b**. The oxocarbenium ion-like transition state depicted by structure **4b** expects planar arrangement of C-5, O-5, C-1, and C-2 atoms, and this is being accommodated within a half chair or envelope conformation (Fig. 5.2).

The cleavage of exocyclic C–O bond during hydrolysis by glycosidases and subsequent release of the glycone and the aglycone part results either with retention of configuration or inversion of configuration at the anomeric site of glycose component was first proposed by Daniel E. Koshland, Jr.[7] in 1953 (Fig. 5.3). The designing of glycosidase inhibitors based on transition-state mimicry focuses on the geometry and charge. The geometry component is based on commonly accepted half-chair conformation for the transition state wherein anomeric carbon acquires sp^2 character to various degrees. This has been substantiated by isotope effects studies. The kinetic analysis data[8] indicates that iminosugars bearing a formally sp^2-hybridized "anomeric center" are likely to mimic the oxocarbenium ion-like transition state better as compared to those containing sp^3-hybridized carbon atom.

The progress in the design, development, and applications of mechanism-based glycosidase inhibitors have been thoroughly reviewed at different intervals of time.[9,10] Excellent review by Mikael Bols and coworkers in

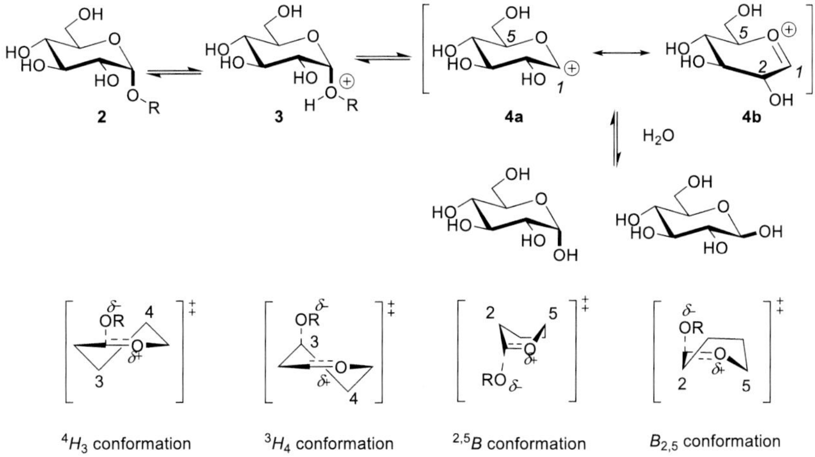

FIGURE 5.2 The initiation of the acidic hydrolysis pathway.

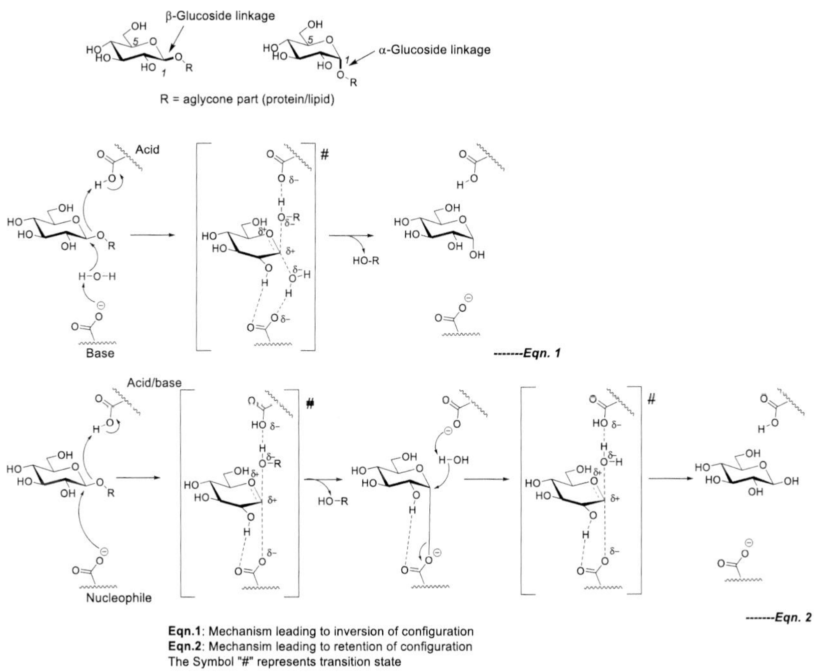

FIGURE 5.3 Classic Koshland-type catalytic mechanisms of glycosidases.

2002[11] elegantly delineates the species that were being mimicked by the compounds represented by iminosugars or azasugar, whether of natural or synthetic origin. The data collected, so far, indicate that potent glycosidase inhibitors are compounds that have nitrogen in place of the exo- or endocyclic oxygen or anomeric carbon, which mimics a charged transition state or intermediate. Mimicking the charge development seems to be more important than the shape. The protonation of the basic nitrogen in the iminosugars under physiological pH seems to be the key component for the mimicry.

5.2 Piperidine-based iminosugars as glycosidase inhibitors: background

The aza-analog **1** of D-glucose, wherein the ring oxygen is replaced by a nitrogen atom was isolated as a natural product, R-468, from the broths of *Streptomyces roseochromogenes* R-468 and *Streptomyces lavendulae* SF-425 during search for an antibiotic.[12] On exploring its other potential, it was found to be a potent inhibitor of α- and β-glucosidases.[13] Later, it was found in the fermentation broth of *Streptomyces nojiriensis*, hence the name nojirimycin. 1-deoxynojirimycin **6** (DNJ), the reduction product of nojirimycin and more stable than nojirimycin was also isolated as a natural product from the roots of mulberry trees as molanoline.[14] The in vivo α-glucosidase inhibition activity,[15] although moderate, the stability of 1-DNJ over nojirimycin inspired further modification in search of a molecule for therapeutic applications and the search culminated with the discovery of Miglitol (BAY m 1099)[16] and *N*-alkylated DNJ derivative Zavesca **8** (Miglustat).[17] Miglitol **7** (Glyset) was granted market clearance by the US Food and Drug Administration (FDA) in 1996 for the treatment of type II diabetes, while Zavesca **8** (Miglustat) was for the treatment of type I Gaucher's disease and Niemann-Pick type-C (NPC) disease (Fig. 5.4).

Gaucher disease[18] is a relatively rare hereditary disorder, most commonly affecting Jews.[19] The most common and the mildest form of the disease is characterized by bone pain, skeletal lesions, anemia, thrombocytopenia, and liver or spleen damage. At the molecular level, this results from defective catalytic activity of an enzyme β-glucosidase involved in the catabolism of glycosphingolipids in lysosomes.[20] Although the administration of Cerezyme, a recombinant form β-glucosidase as a supplement for the defective enzyme, surfaced as the first line of treatment, its prohibitively expensive cost is a serious issue ($1,50,000 per year). In this context discovery of

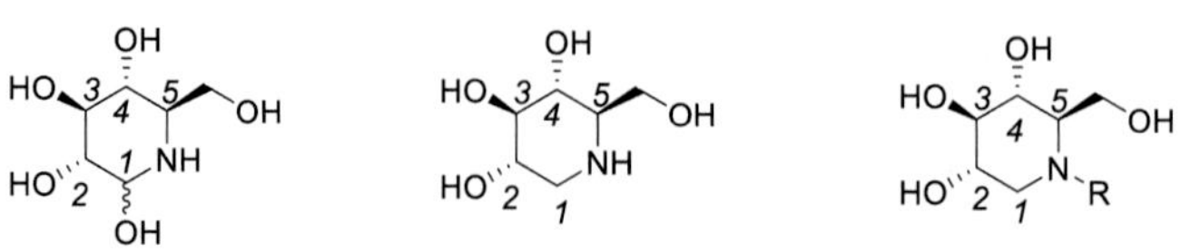

1 Nojirimycin | **6** D-1-deoxynojirimycin (DNJ) | **7**: R = CH_2CH_2OH Glyset (Miglitol) | **8**: R = $CH_2CH_2CH_2CH_3$ Zavesca (Miglustat)

FIGURE 5.4 Nojirimycin-inspired drugs Glyset **7** and Zavesca **8**.

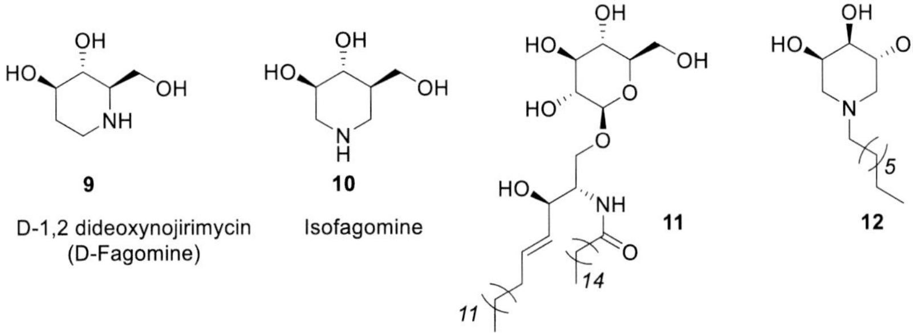

FIGURE 5.5 D-Fagomine **9** and Isofagomine **10**.

Zavesca, as an orally administered drug, came as a promising alternative, which works through a strategy called substrate reduction therapy (SRT). It blocks the biosynthesis of glycosphingolipids through inhibition of glucosylceramide synthase, thereby reducing the substrate, whose accumulation, because of faulty degradation is the very cause of Gaucher's disease. Another analog of 1-DNJ **6** was 1,2-dideoxynojirimycin **9** (Fig. 5.5). This was first isolated as a natural product D-fagomine **9** from the Japanese buckwheat seeds of *Fagopyrum esculentum* Moench in 1974.[21] It was found to display inhibitory activity against mammalian α-glucosidase and β-galactosidase[22,23] This initial activity associated with D-fagomine has inspired search and asymmetric syntheses of all other stereoisomers of D-fagomine **9** to harness its potential as an additional lead.[24] Synthetic efforts around D-fagomine resulted in the discovery of a regio-isomeric analog **10** first reported in 1994 as isofagomine,[25] which was found to be a potent glucosidase inhibitor.[26] Isofagomine **10** (AT2101, Plicera) was well tolerated under another strategy, referred to as "active-site-specific pharmacological chaperone" (ASSPC) for the treatment of Gaucher's disease. Pharmacological chaperones (PCs) in general are small molecules that bind to the active site of the enzyme at subinhibitory concentration restoring its correct folding and helping translocation to the lysosomes, while at higher concentrations, they behave as reversible inhibitors of the same enzyme. On translocation of the PC-bound enzyme to the lysosomes, the PCs get displaced by the natural substrate present in high concentrations and the intended biological function is achieved. In the case of Gaucher's disease, the biological function is the degradation of glucosylceramide (GlcCer) **11** through glucocerebrosidase (GCase). Inadequate clinical results with isofagomine as PC marred its further progress. The unmodified isofagomine (IFG) was a competitive inhibitor of GCase and failed presumably due to high hydrophilicity, which hampered efficient transport to the lysosomes. However, the aforementioned initial results with unmodified iminosugars did inspire intense activity from the synthetic chemistry community[27] with the hope to further understand and possibly arrive at lead molecules toward drug discovery through appropriate structural modification. In fact, Richard Storer at Summit Corporation based in Abingdon UK aptly coined an acronym Seglins (second-generation leads from iminosugars) for the efforts beyond the first-generation molecules.[28] Indeed simple *N*-alkylation afforded compounds,[29] with better metabolic properties as potential PCs. The *N*-alkylated iminosugar **12** showed measurable chaperone activity with human fibroblasts derived from Gaucher patients.[30]

5.2.1 Piperidine-based iminosugar *C*-glycosides inspiring entry to azepane-sugars

Principally, in parallel to *O/C*-glycosides, of pyranose-based sugars, iminosugars represented by 1-DNJ **6**, D-fagomine **9**, and isofagomine **10** also provide opportunities for the entry to corresponding glycosides, besides the basic nitrogen site. The intrinsic instability of aminoacetal functionality with the *O*-glycoside of iminosugars **13** renders them less attractive in comparison to *C*-glycosides of iminosugars **14** (Fig. 5.6). The incorporation of aglycone through the C–C bond at the pseudoanomeric position not only provided vast opportunities for structural modification in iminosugars but also lead to valuable insight into the structural features necessary for improving the biological activity.

Stereo-controlled synthesis of various α- and β-1-*C*-butyl-1-deoxynojirimycins **19a,b** involved stereoselective addition of organometallic reagent, *n*BuLi, onto advanced imine intermediate **16** derived through a series of

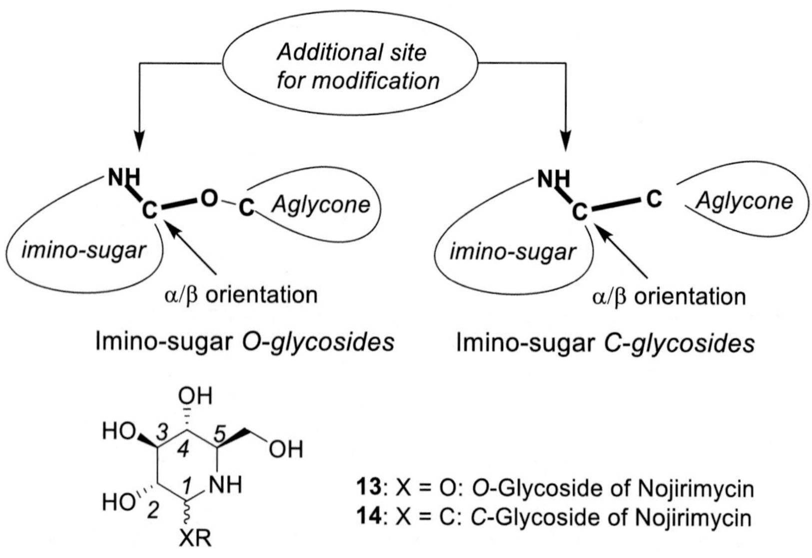

13: X = O: *O*-Glycoside of Nojirimycin
14: X = C: *C*-Glycoside of Nojirimycin

FIGURE 5.6 General structure for *O/C*-glycosides of iminosugars.

functional group interconversion starting from inexpensive L-sorbose based starting material **15**. The product obtained **17a** and **17b** as diastereomeric was predominantly in favor of the **17a** (98:2). Intramolecular reductive amination on the major unmasked aminosorbose hemiketal **17a** furnished the diastereomerically pure diol **18** with α-orientation of the butyl group at C-1. Debenzylation produced the requisite α-1-*C*-butyl-1-deoxynojirimycin derivatives **19a** (Fig. 5.7).[31]

The biological evaluation of 1-*C*-alkyl-1-deoxynojirimycin derivatives **20a–e**, representatives of *C*-glycosides of iminosugars **14** indicated that a simple shift of the alkyl chain residue from the endocyclic nitrogen to the "anomeric" carbon in an α-configuration could lead to more potent glucosidase inhibitors.[32] Further removal of the hydroxymethyl moiety in this structure led to the most interesting and prominent improvement of the activity by a factor of 100![33] This was attributed to the change from D-*gluco* **21a** to the D-*xylo* **21b** series and the conformational change from $^4C_1$ to $^1C_4$. The presence of α-1-*C*-nonyl-1-deoxyimino-D-xylitol (α-1-*C*-9-DIX) **21b** in $^1C_4$ conformation was confirmed through nuclear magnetic resonance (NMR) data, which confirmed all hydroxyl groups in axial orientation and the alkyl chain in equatorial orientation. The isomeric *N*-nonyl analog **23**, however, continues to have $^4C_1$ conformation and diminished activity as compared to **21a** (Fig. 5.8). Inspired by the significance of $^1C_4$ conformation and to further decipher the relevance of hydroxymethyl unit, Martin's group synthesized 1-*C*-alkyl-1,5-dideoxy-1,5-imino-L-iditol series **24a–c**, wherein the L-*ido* configuration ensured the $^1C_4$ conformation because of its strong preference and had the hydroxymethyl unit. The biological evaluation with the synthesized compounds **24a**, **24b**, and **24c** on human β-glucocerebrosidase (GCase), however, were not very promising.[34] The study indicated that the L-*ido* epimers were not as good inhibitors as the iminoxylitols and were even weaker than the parent D-*gluco* series.

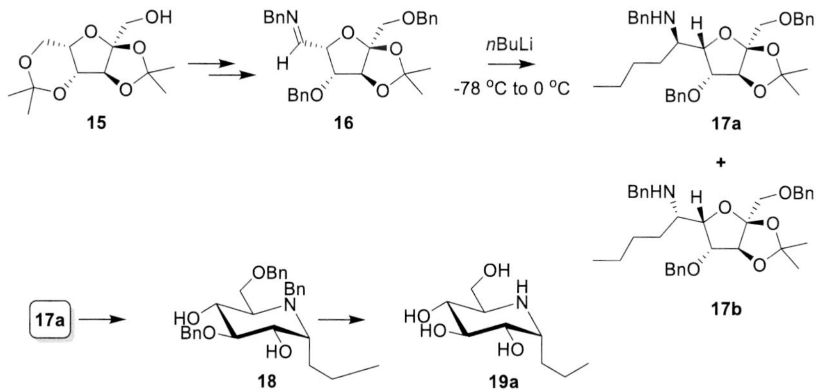

FIGURE 5.7 General approach for α- and β-1-*C*-butyl-1-deoxynojirimycin **19a**.

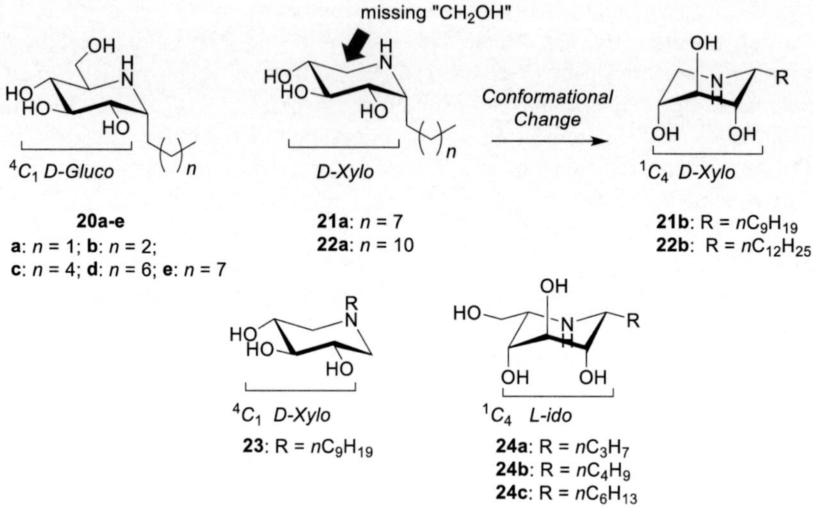

FIGURE 5.8 Structural variations in for α-1-*C*-alkyl-1-deoxynojirimycin.

In this context, the presumed significance of $^1C_4$ conformation was found to be incorrect and the relevance of hydroxymethyl unit seems to be essential for the recognition of α-glucosidases and not for β-glucocerebrosidase.

5.3 The emergence of polyhydroxyazepanes

The success of polyhydroxylated pyrrolidines, pyrrolizidines, and piperidine as glycosidase inhibitors[11,35,36] stimulated search and entry into azepane-based glycosidase inhibitors, because of their unexplored factor of higher conformational flexibility. Paulsen and Todt's work[37] converting 6-amino-6-deoxy-D-glucose **25** and 6-amino-6-deoxy-D-galactose **29** to the corresponding seven-membered ring with four secondary hydroxyls, compound **28** and **32**, initiated the entry and interest in azepane-based iminosugars. Depezay[38a] arrived at the corresponding manno-configured **35** using D-mannitol as starting material (Fig. 5.9).

5.3.1 Nomenclature and Numbering

Based on carbohydrate nomenclature, the tetrahydroxyazepane compounds **28**, **32**, and **35** originating from D-glucose **25**, D-galactose **29**, and D-mannitol **33** would be aptly referred as 1,6-dideoxy-1,6-imino-D-glucitol, 1,6-dideoxy-1,6-imino-D-galactitol, and 1,6-dideoxy-1,6-imino-D-mannitol respectively, because of their starting monosaccharide. The general name would be 1,6-dideoxy-1,6-imino-D/L-hexitol, wherein the configurational prefix will define the stem, hexitol. These compounds viewed as substituted azepane

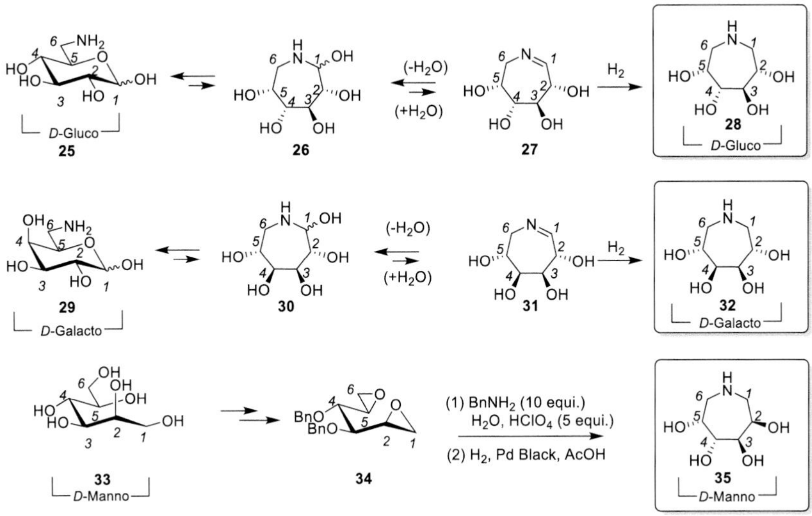

FIGURE 5.9 Tetrahydroxyazepanes (examples).

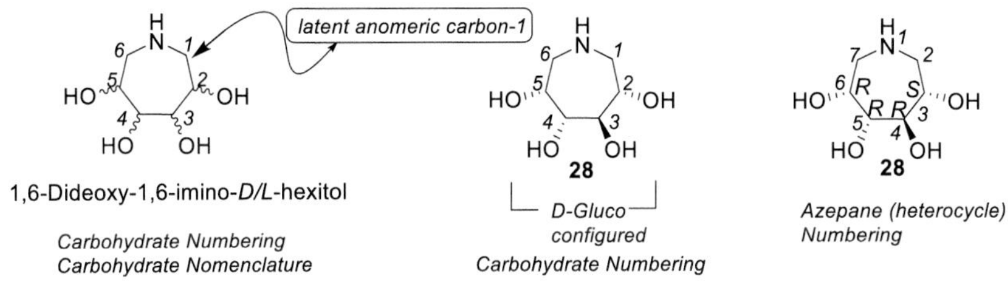

FIGURE 5.10 Tetrahydroxyazepanes nomenclature and numbering.

(a heterocycle) would necessitate numbering to start from the nitrogen atom. In this context, the stereo-orientation of the four secondary hydroxyls is represented by *R-S* descriptors (Fig. 5.10). For example, **28** would be (*3S,4R,5R,6R*)-3,4,5,6-tetrahydroxyazepane.

5.3.2 Various 1,6-dideoxy-1,6-imino-D/L-heptitols: synthesis and biological activities

Wong and coworkers used 3,4-di-*O*-allyl-D-mannitol **38** [39] in contrast to 3,4-di-*O*-benzyl-D-mannitol **34** used by Depezay,[38a] for the synthesis of 1,6-dideoxy-1,6-imino-D-mannitol (D-manno azepane) **44** and L-ido-azepane **45**. Two C_2-symmetrical bis-epoxides **39a** and **39b** were prepared from commercially available D-mannitol derivative **36**. Route-1 from **38** enabled inversion of configuration at C-2 and C-5 position for entry into L-ido configuration, while epoxide formation under Mitsunobu conditions retained the configuration at C-2 and C-5 and the configurational prefix, D-manno. Both C_2-symmetrical bis-epoxides **39a** and **39b** when treated with excess allyl amine (12 equiv.) and perchloric acid (6 equiv.) furnished the azepanes **40** and **42**, respectively, as major products. Intermolecular regioselective epoxide opening by allylamine followed by intramolecular 7-*endo-tetra* ring closure explains the formation of the major product, according to Baldwin's rules. The minor product in each case is a result of competing 6-*exo-tetra* closure. Final deprotection in the azepanes **40** and **42** afforded the tetrahydroxyazepanes **44** and **45**.[38b] The azepane **44** was found to inhibit seven different glycosidases with K_i in the micromolar range (Scheme 5.1).

The promise of nojirimycin **1** and 1-DNJ **6** as 1,5-dideoxy-1,5-iminohexitols and the absence of hydroxymethyl unit ($-CH_2OH$) in the aforementioned tetrahydroxyazepanes, justifiably prompted design, synthesis, and biological studies of the seven-membered higher homologs, 1,6-dideoxy-1,6-iminoheptitols **46**, featuring hydroxymethyl unit. If $-CH_2OH$ is visualized as substituent and tetrahydroxy-azepane as a heterocycle, either of the two methylene (A or B) can accommodate $-CH_2OH$ unit. The nomenclature necessitates the numbering of the heterocycle in a way, that makes the

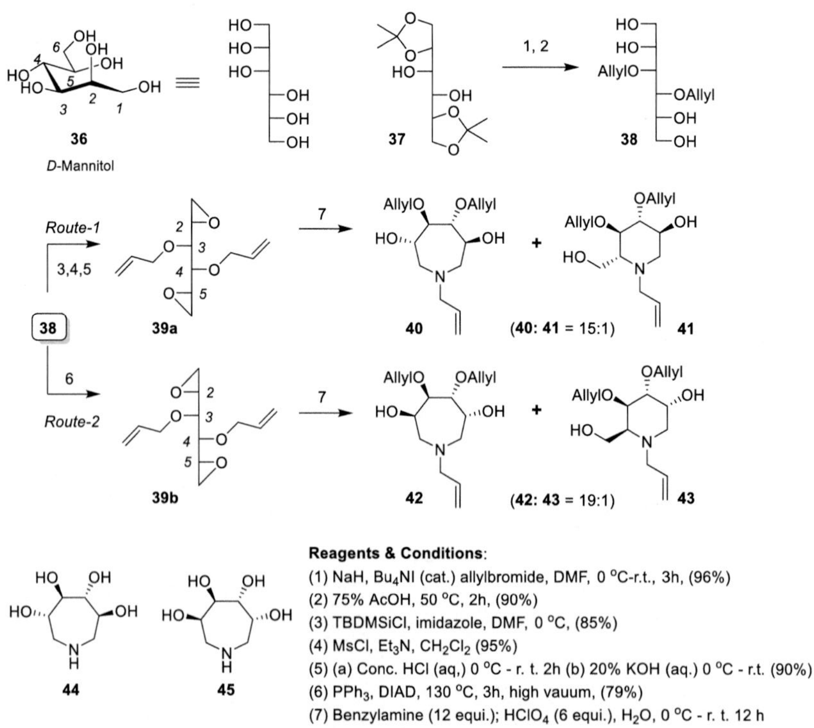

SCHEME 5.1 Depezay synthesis of tetrahydroxyazepane **44** and **45**.

carbon accommodating $-CH_2OH$ unit, to acquire the least number. Hence **47**, a specific representative example of 1,6-dideoxy-1,6-imino-D/L-heptitol **46** would be named as (2*R*,3*R*,4*R*,5*S*,6*S*)-2-hydroxymethyl-3,4,5,6-tetrahydroxyazepane and not as (3*S*,4*S*,5*R*,6*R*,7*R*)-7-hydroxymethyl-3,4,5,6-tetrahydroxyazepane.

D-Arabinose-derived alkene **48**[40] served as the starting material for several azepanes featuring hydroxymethyl units. Reductive amination of the ketone **49** with allylamine and acetic acid in the presence of $NaBH_3CN$ afforded the D-*arabino* and L-*xylo* *N*-allylaminohexenitols **50a** and **50b** in 58% yield, in the ratio 3:2. (Scheme 5.2). For applying ring-closing alkene metathesis (RCM) methodology on these dienes **50a** and **50b**, toward the construction of azepane ring, the amine nitrogen was protected as carbamate. The protected dienes **51** and **53** when subjected to RCM furnished the expected didehydroazepane **52** (91%) and **54** (84%), respectively. Both these didehydroazepanes served as key intermediates for entry into a variety of 1,6-dideoxy-1,6-iminoheptitols featuring hydroxymethyl unit[41] (see general structure **46** in Fig. 5.11).

Although with modest facial diastereoselectivity, the dihydroxylation of the double bond in didehydroazepane **52**, afforded the diols **55** and **56** in 96% yield (**55**: **56** = 3:7).[41] Debenzylation afforded the targeted 1,6-dideoxy-1,6-iminoheptitols **57** and **58** in quantitative yield. The same sequence, when applied to didehydroazepane **54** furnished another set of 1,6-dideoxy-1,6-

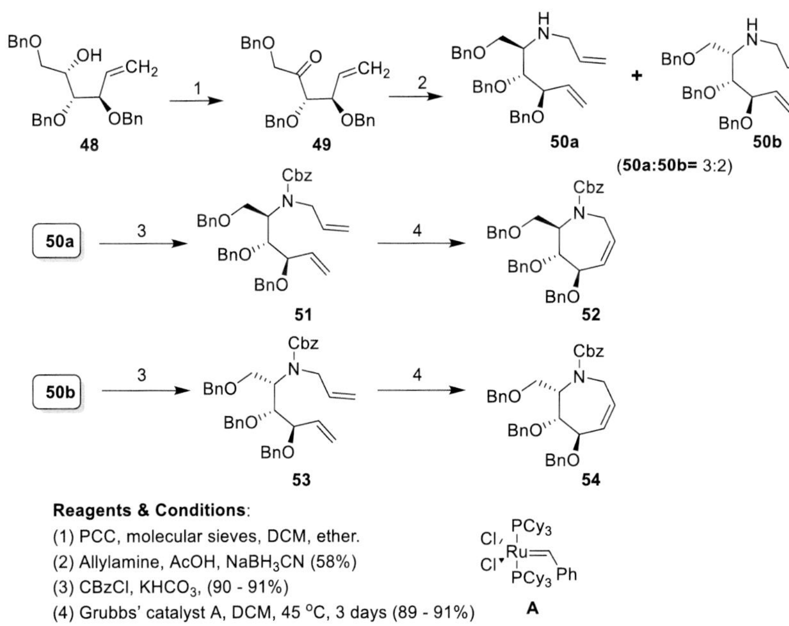

SCHEME 5.2 Synthesis of didehydroazepane **52** and **54**.

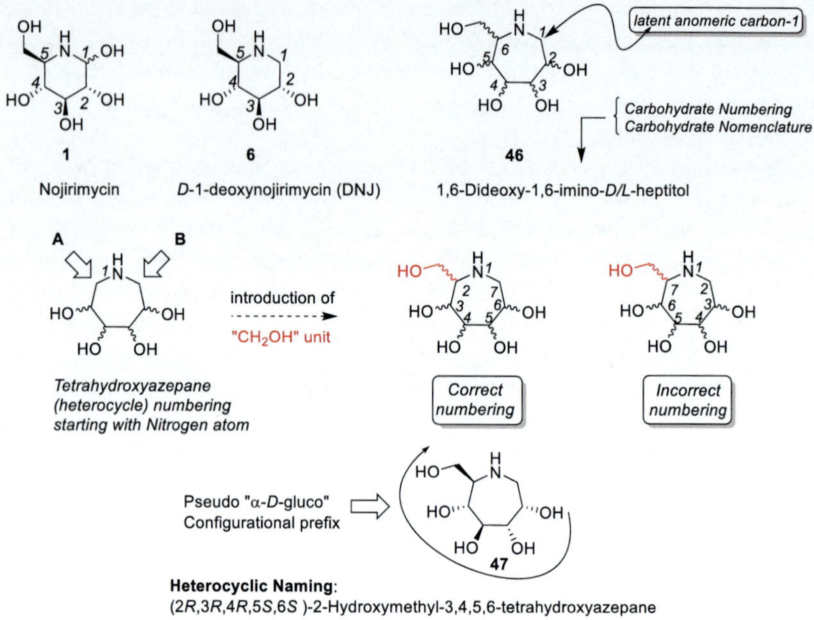

FIGURE 5.11 1,6-Dideoxy-1,6-imino-D/L-heptitol naming.

iminoheptitols **59** and **60** (Scheme 5.3). For the description of relative configurations of the stereocenters in these polyhydroxyazepanes (**57–60**), the use of configurational prefixes derived from hexoses has been made.

The synthesized polyhydroxyazepanes **57–60** were assayed for their inhibitory activity toward various glycosidases. Azepane **58** exhibited the best inhibitory activity toward α-galactosidases from coffee beans ($Ki = 2.2\ \mu M$, competitive). Azepane **57** also inhibits α-galactosidases from coffee beans ($Ki = 16\ \mu M$, competitive) along with good competitive inhibition ($Ki = 5.7\ \mu M$) of β-galactosidase from bovine liver.[41]

The didehydroazepanes **52** and **54** have played a pivotal as a building block enabling several transformations for generating a library of polyhydroxyazepanes. The epoxidation of these didehydroazepanes with *m*-chloroperbenzoic acid provided all the expected epoxides, **61–64**, as separable compounds (Scheme 5.4).[42a]

Nitrite ion being an ambident nucleophile is presumably attacking through the oxygen center followed by in situ hydrolysis of the alkyl nitrite ester intermediate. (For discussion only hydrolysis of alkyl nitrite esters in general.)[42b] Although this can explain the *trans*-orientation of the hydroxy groups, in the products, **65** and **66** at C-5 and C-6 carbons (as claimed), the regio-selective attack on the C-6 carbon in the epoxide is not easily discernible. Both these procedures gave a poor yield of the desired products. The epoxides **61** and **62** enabled access to 5,6-*trans*-oriented dihydroxylated

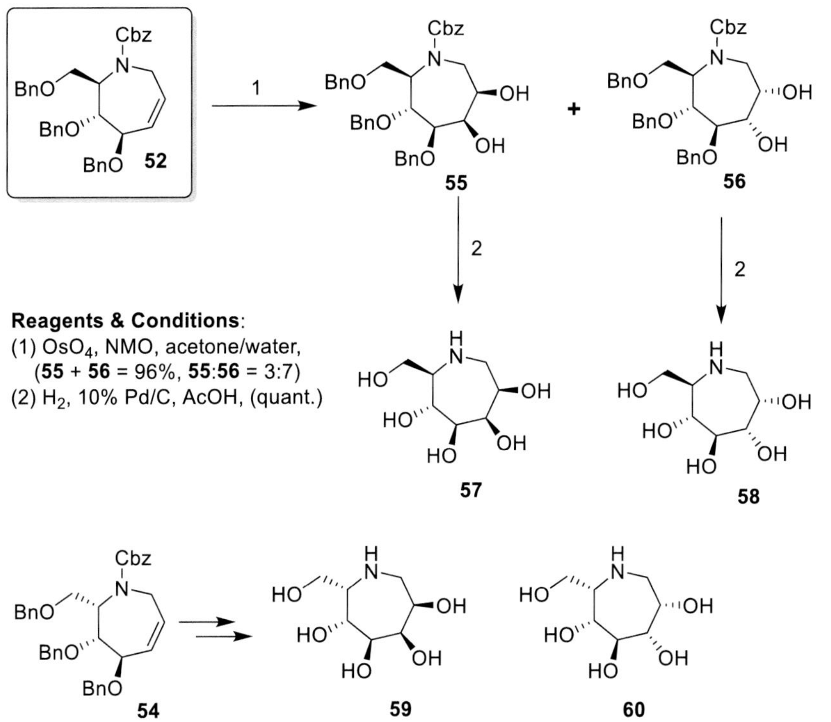

SCHEME 5.3 Synthesis of 1,6-dideoxy-1,6-imino-D/L-heptitols.

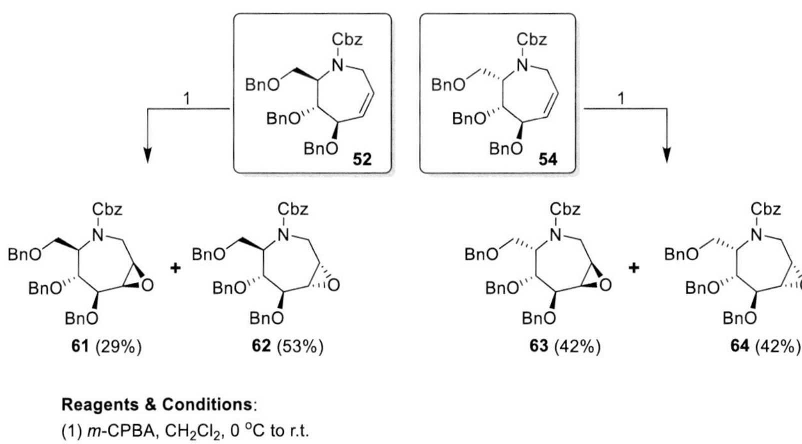

SCHEME 5.4 Epoxidation of didehydroazepanes **52** and **54**.

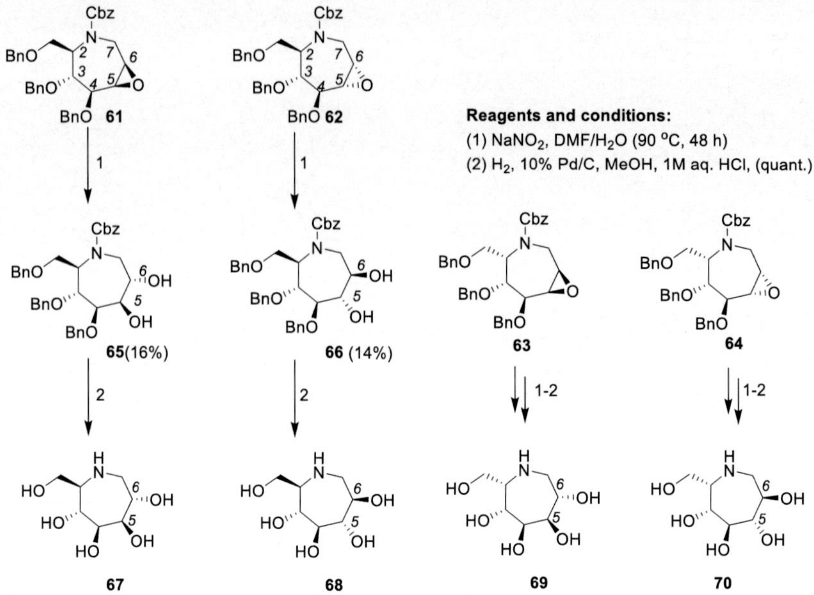

SCHEME 5.5 Synthesis of 5,6-*trans*-oriented azepanes **67**–**70**.

azepanes **67** and **68**, while the same chemistry applied on **63** and **64** has furnished the other diastereoisomeric azepanes **69** and **70** with *trans*-orientation of the hydroxyl groups at C-5 and C-6 carbons (Scheme 5.5).

Epoxides **61** and **62** also provided an opportunity for opening the epoxide functionality with lithium triethylborohydride (Super Hydride) (Scheme 5.6). No regioselectivity was observed during the delivery of hydride, resulting in the formation of separable tetrahydroxyazepanes **75**–**76** and **77**–**78**, respectively. The same synthetic operation on epoxides **63** and **64** furnished the other diastereoisomeric tetrahydroxyazepanes **79**–**80** and **81**–**82**, respectively.

Iminoheptitols **67**–**70** and **75**–**82** were assayed toward a range of commercially available glycosidases. They did not inhibit β-galactosidases from *Escherichia coli* and *Aspergillus oryzae* at 1 mM. The weak glycosidase inhibition observed with azepanes **67**–**70** has been attributed to the conformational change brought in by the *trans* diol unit at C-5 and C-6, making interaction of the substrate in the active site of the enzymes, unfavorable. The good and selective inhibition of α-L-fucosidase by azepanes **76** and **78** parallels the observation made with deoxymannojirimycin **83** (DMJ) and deoxynorijimycin (DNJ) **6**, respectively (Fig. 5.12).[42a]

Synthesis of a series of 1,6-dideoxy-1,6 iminoheptitols displaying an amino group at C-2 or C-3 position was also accessible through the opening of the epoxides **61** and **62** with the azide anion (Scheme 5.7).[43]

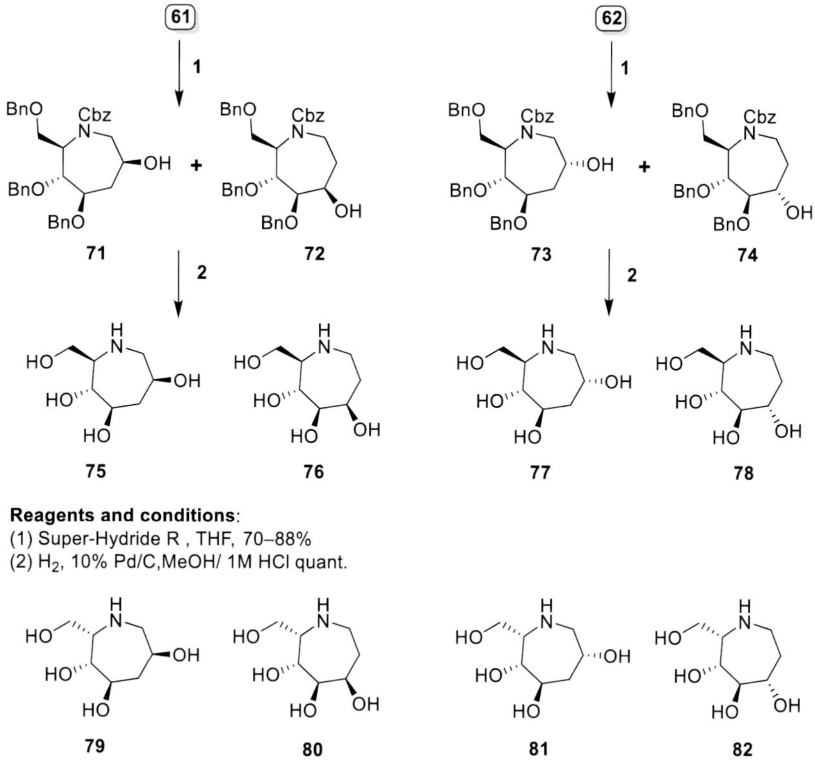

SCHEME 5.6 Synthesis of tetrahydroxy 1,6-dideoxy-1,6-imino-D/L-heptitols.

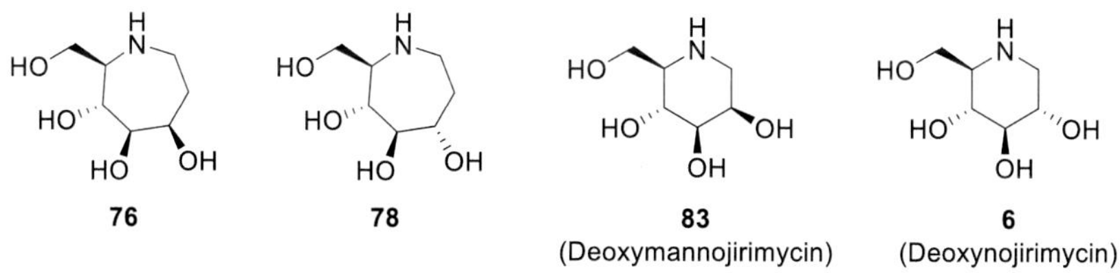

FIGURE 5.12 Analogy between DNJ, DMJ, and azepane-based homologs.

These epoxides on heating with sodium azide in a DMF–water mixture at 90°C gave a mixture of **84/85** and **86/87**, respectively. The regio-isomers on separation and hydrogenation, followed by deprotection afforded the targeted azepane derivatives **88–91** featuring a hydroxymethyl unit along with an amino group at the C-2/C-3 position. The same chemistry was applied on epoxides **63** and **64** for obtaining the other diastereoisomeric azepanes **92–95** bearing amino functionality. The eight iminoheptitols **88–95** were assayed for their inhibitory activity toward 24 commercially available

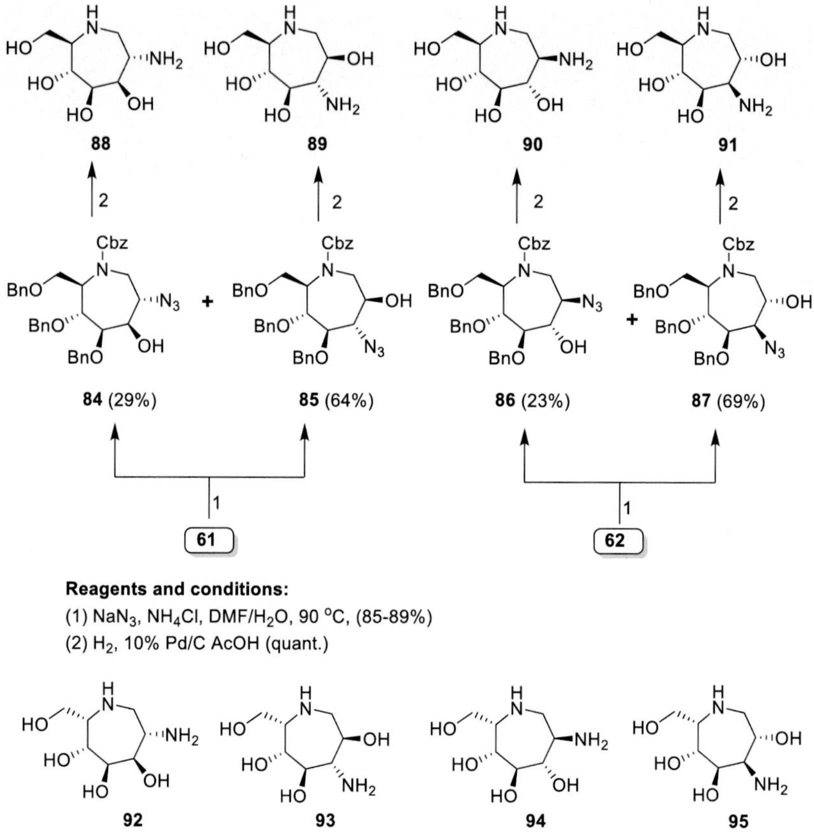

SCHEME 5.7 Route for 1,6-dideoxy-1,6-imino-D/L-heptitols with amino group.

glycosidases. No inhibition was observed with the following enzymes at 1 mM concentration: α-galactosidase from *E. coli*, β-galactosidases from *E. coli*, *Aspergillus niger*, and *A. oryzae*, α-glucosidases from rice, β-glucosidases from almonds and *Saccharomyces cerevisiae*, β-mannosidase from *Helix pomatia*, β-xylosidase from *A. niger*, α-*N*-acetylglucosaminidase from chicken liver, and β-*N*-acetylglucosaminidases from jack bean and bovine epididymis A and B.[43] The *N*-acetyl derivative of polyhydroxyazepanes **97–101** also became useful for evaluation against another important enzyme, *O*-GlcNAcase. The enzyme, *O*-GlcNAcase, catalyzes the cleavage of β-*O*-linked GlcNAc (*O*-GlcNAc) unit attached to glycoproteins through serine or threonine residues. The glycoproteins bearing the monosaccharide 2-acetamido-2-deoxy-β-D-gluco-pyranoside unit have been implicated in a wide range of cellular functions,[44] such as transcription, proteasomal degradation, and cellular signaling and have also found to play roles in the etiology of disease states,[45] including type II diabetes, Alzheimer, and cancer.

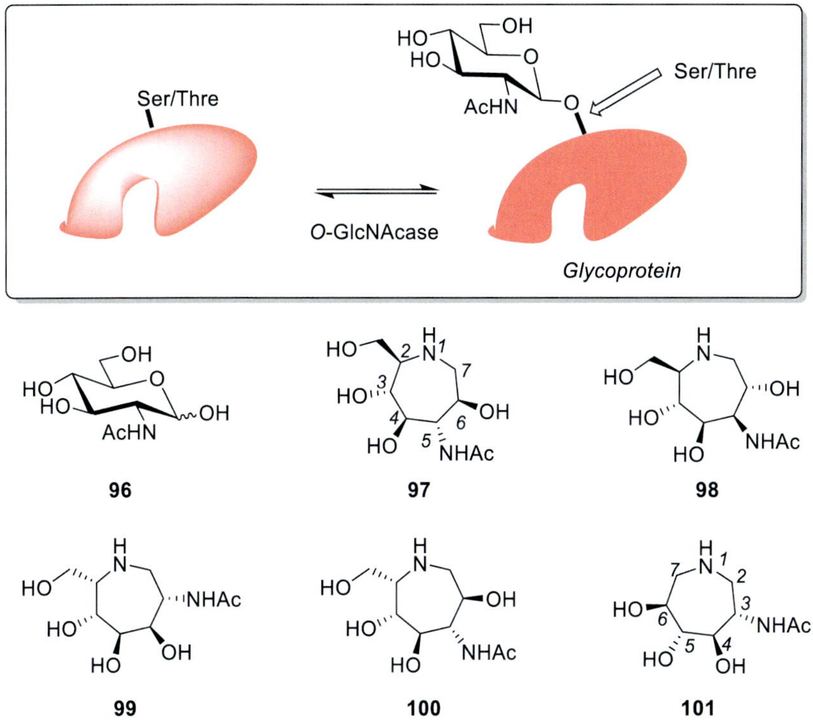

FIGURE 5.13 Azepane analogs **97–101** for *O*-GlcNAcase inhibition.

The cleavage mechanism catalyzed by *O*-GlcNAcase, reveals anchimeric assistance from the acetamido moiety of the glycone. Inspired by the functional importance of *O*-GlcNAcase, Bleriot's group undertook studies[46] for exploring polyhydroxyazepanes **97–101** synthesized in their group for inhibition of *O*-GlcNAcase (Fig. 5.13).

The studies revealed that the conformational flexibility intrinsic to polyhydroxylated azepanes seems to be operational, thereby enabling alternate conformations within the active site of the enzyme. The azepanes **97** and **101** were found to bind some 30- and 140-fold more tightly than (GlcNAc, **96**, $Ki = 1500$ μM).

The initial synthesis of polyhydroxyazepanes featuring hydroxymethyl unit, from Sinay's group, inspired the synthesis of two new azepanes **107** and **108**, by Dhavale et al.[47] (Scheme 5.8). Few other groups have also synthesized polyhydroxyazepanes featuring hydroxymethyl units at the C-2 position. Using L-serine **109** as starting material, Lin et al. synthesized oxazolidinyl azacycles **110a,b** and **111a,b** using RCM as a key step. Dihydroxylation of these oxazolidinyl azacycles using a catalytic amount of osmium tetraoxide (1 mol%) and NMO in acetone/water paved the way for

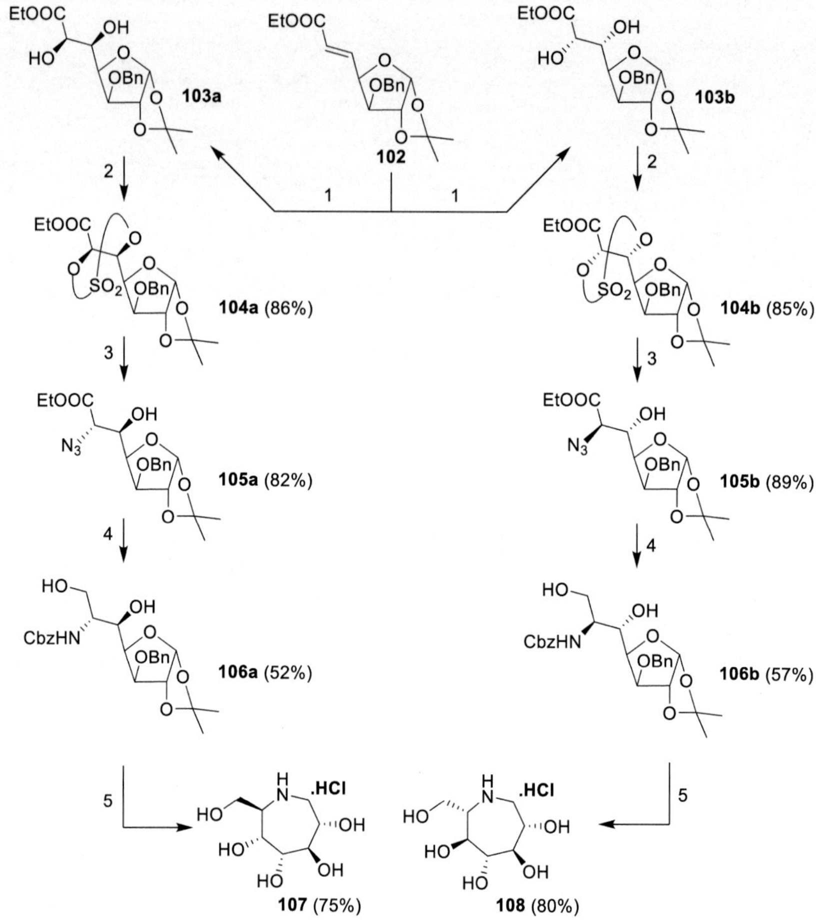

Reagents & Conditions:

(1) $K_2[OsO_2(OH)_4]$ (Cat.) $K_3Fe(CN)_6$, K_2CO_3, $CH_3SO_2NH_2$, *tert*-butanol/water (1:1) (80-85%)

(2) (a) $SOCl_2$, Py. CH_2Cl_2, 0 °C, 30 min (b) $NaIO_4$, $RuCl_3$. $3H_2O$, $CH_3CN:H_2O$ (3:1) 0 °C, 10 min.

(3) (a) NaN_3, acetone/water (4:1), 0 to 25 °C, 2h (b) 20% H_2SO_4, ether/water (6:1), 25 °C, 6 h

(4) (a) LAH, THF, 0 to 25 °C, 3 h (b) $HCOONH_4$, Pd/C, MeOH, 80 °C, 1 h
(c) CbzCl, MeOH/H_2O (9:1), 0 to 25 °C, 3.5 h

(5) (a) TFA/H_2O (2:1), 0 to 25 °C, 2.5 h (b) H_2. Pd/C, MeOH/HCl, 80 psi, 24 h

SCHEME 5.8 Synthesis of azepanes **107** and **108**.

the access to several polyhydroxyazepanes **112–123** featuring the hydroxymethyl unit at C-2 position (Fig. 5.14). Among these azepanes, compound **115** alone showed the most potent inhibition of β-glucosidase ($Ki = 26.3$ mM) and competitively binds to the enzyme's active site. Compounds **120–123** bearing *trans*-2,3 configuration (azepane numbering) displayed greater than 50% competitive inhibition for the β-glucosidase.[48]

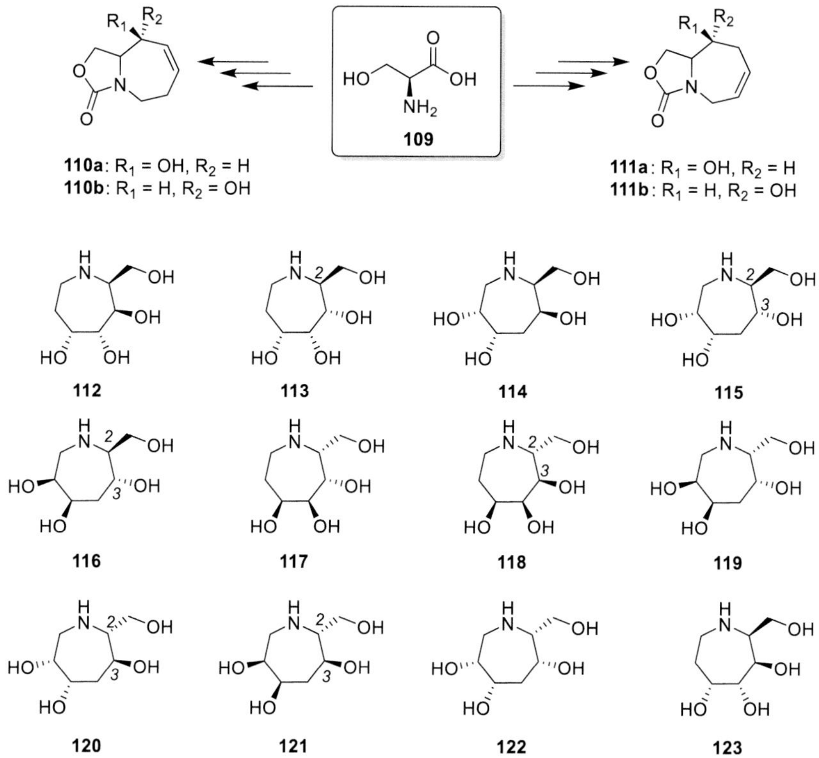

FIGURE 5.14 L-Serine-derived polyhydroxyazepanes **112–123**.

Estévez et al.[49a] used Henry's reaction for two carbon homologations on the protected furanose derivative **124** in an attempt to synthesize azepane derivatives **132** and **133** (Scheme 5.9). Reduction of the nitro group in the Henry reaction products **125a** and **125b** and protection of the resultant amine functionality enabled the formation of the advanced intermediates **126** and **128**, respectively. Removal of the silyl ether protection, followed by amine deprotection paved the way for an intramolecular reaction at the hemiacetal carbon and formation of the protected polyhydroxyazepane derivatives **130** and **131**. Final deprotection furnished the desired azepane products **132** and **133**.

Azepane **133** evinced a weak but selective inhibition (IC_{50} = 1.7 mM) against α-L-fucosidase. Gonzalez et al. also reported the synthesis of compounds **132** and **133** along with azepanes **134** and **135** (Fig. 5.15).[49b] Although azepane **134** was found to be biologically inactive, the polyhydroxyazepane derivative featuring tertiary amide functionality displayed modest inhibitory activity against β-*N*-acetylglucosaminidase from jack bean and bovine kidneys.

Rao et al. banked on didehydroazepane derivative **137** for the synthesis of polyhydroxyazepane derivatives **58** and **140–141** (Scheme 5.10).

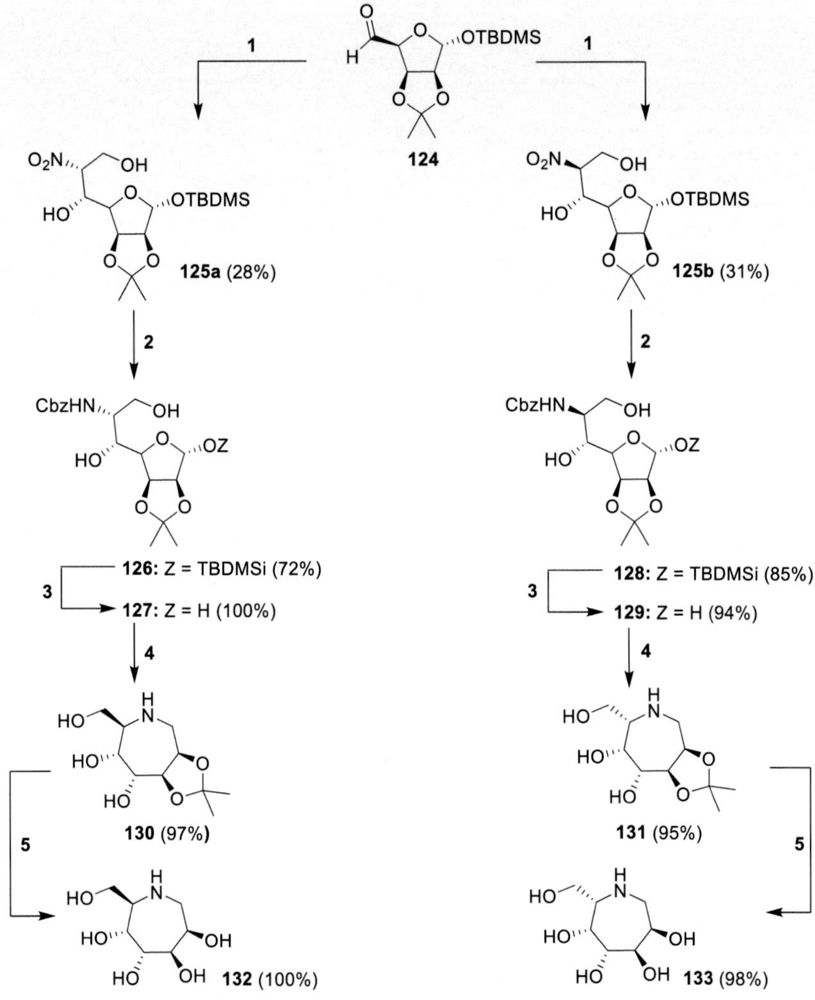

Reagents and conditions:
(1) $NO_2CH_2CH_2OH$, NaOMe, MeOH, rt, 4h
(2) (a) H_2, Raney nickel, MeOH, rt, 2 h;
(b) ClCbz, $NaHCO_3$, EtOAc, rt,4h
(3) TBAF, THF, rt, 15 h
(4) NH_4HCOO, Pd black, MeOH, AcOH, rt, 12 h then 50 °C,4h
(5) TFA/H_2O 1:1, rt, 15h

SCHEME 5.9 Polyhydroxyazepanes **132–133** using Henry reaction for C–C bond formation.

The diene **136** derived in eight steps from commercially available D-gluconolactone was subjected to ring-closing metathesis using either the first-generation or second-generation Grubb's catalyst for efficient synthesis of didehydroazepane **137** in high yields. Simple hydrogenation on **137** and

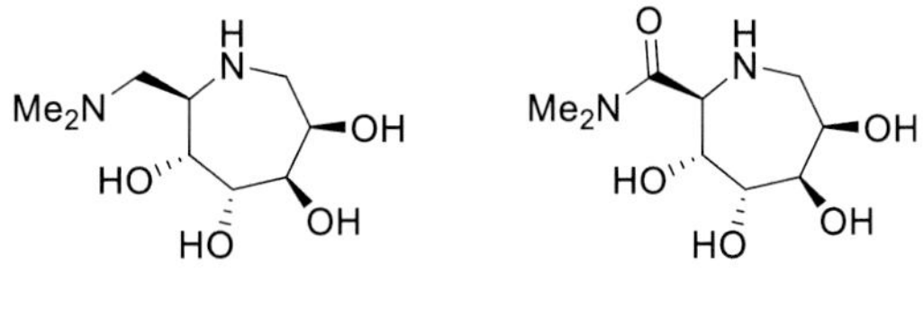

FIGURE 5.15 Polyhydroxyazepane featuring *tert*-amide functional group.

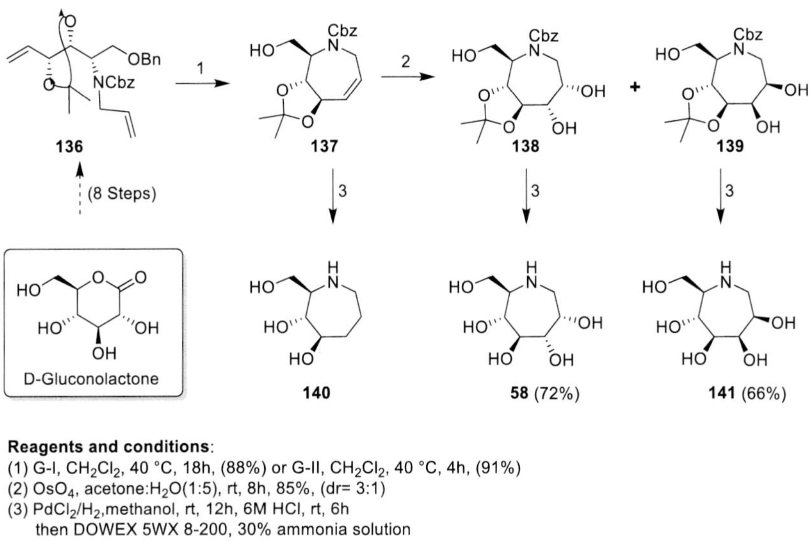

SCHEME 5.10 D-Gluconolactone-derived polyhydroxyazepanes.

isopropylidene deprotection with 6 M HCl in EtOH furnished the trihydroxyazepane **140** while dihydroxylation on **137** with OsO_4 followed by complete deprotection provided **58** and **141**.[50]

Three building blocks **142**, **143**, and **144** (Fig. 5.16) derived from D-quinic acid,[51a-c] through multistep synthetic schemes, have been used for the synthesis of several polyhydroxyazepanes (Scheme 5.11).[51d] The protected 1,4,5-cyclohex-2-ene-triols **145a** and **145b** obtained from building block **142** when subjected to dihydroxylation under $RuCl_3/NaIO_4$/phosphate buffer (pH 7) conditions afforded diols **146a** and **146b** respectively. The diols **146a** and **146b**, were oxidatively cleaved by $NaIO_4$ to furnish the corresponding dialdehydes which on reductive amino cyclization using benzylamine (6.1 equiv.) and $NaBH(OAc)_3$ (3.3 equiv.) allowed the construction of the azepanes **147a** and **147b** respectively. Hydrogenation with 10% Pd/C in MeOH-2N HCl afforded the targeted polyhydroxyazepanes **148a** and **148b**, respectively. The same chemistry when applied on the building block **143** provided access to polyhydroxyazepanes **149a** and **149b**.

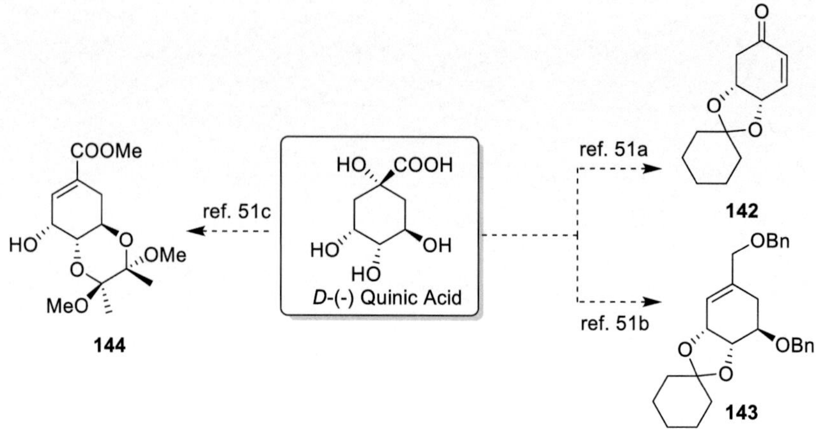

FIGURE 5.16 D-(−)-Quinic acid–derived building blocks for polyhydroxyazepanes.

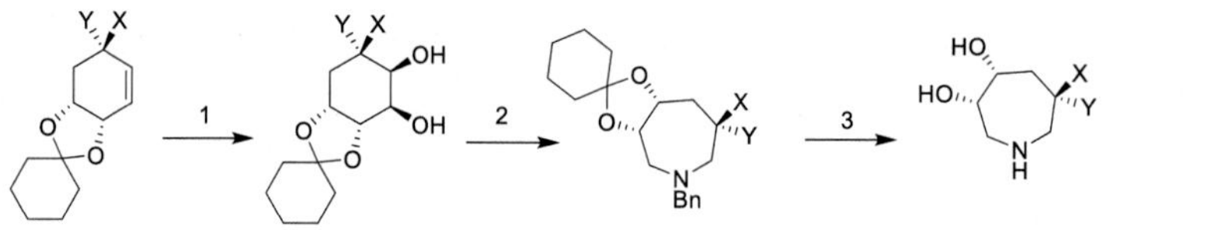

145a: X = H, Y = OBn
145b: Y = H, X = OBn

146a: X = H, Y = OBn (90%)
146b: Y = H, X = OBn (73%)

147a: X = H, Y = OBn (63%)
147b: Y = H, X = OBn (76%)

148a: X = H,Y = OH (67%)
148b: Y = H, X = OH (99%)

Reagents and Conditions:
(1) $RuCl_3$, $NaIO_4$, EtOAc: AcCN: H_2O: Na_2HPO_4 (3:3:1:1)
(2) (a) $NaIO_4$, MeOH (b) $BnNH_2$, $NaBH(OAc)_3$
(3) H_2, Pd/C, 2N HCl, MeOH

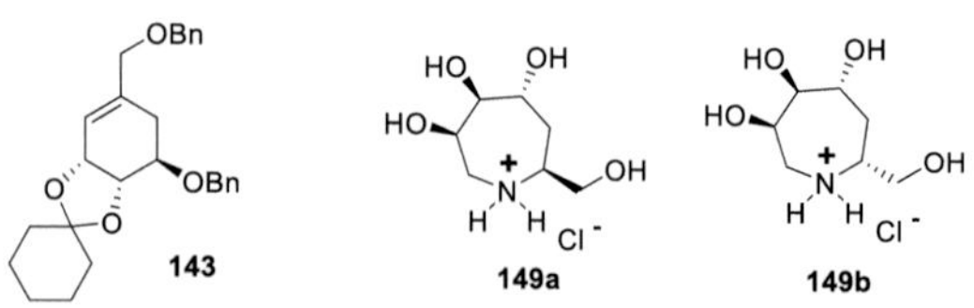

SCHEME 5.11 D-(−)-Quinic-derived polyhydroxyazepanes.

(−)-Shikimic acid–derived derivative **150a** on ozonolysis, although provided the desired dicarbonyl compound **151a**, an intramolecular closure to furanose derivative **152** was inevitable. The ozonolysis product in high yield (93%) was found to be a mixture of **151a** and **152** (α/β). Visualizing double reductive amination on the dicarbonyl component **151a**, the mixture was treated with sodium cyanoborohydride in the presence of ammonium bicarbonate. This attempt indeed afforded an inseparable mixture of azepanes **153** and **154** in 63% yield. Removal of isopropylidene deprotection enabled the formation of polyhydroxyazepanes **155** and **156** as an inseparable mixture.

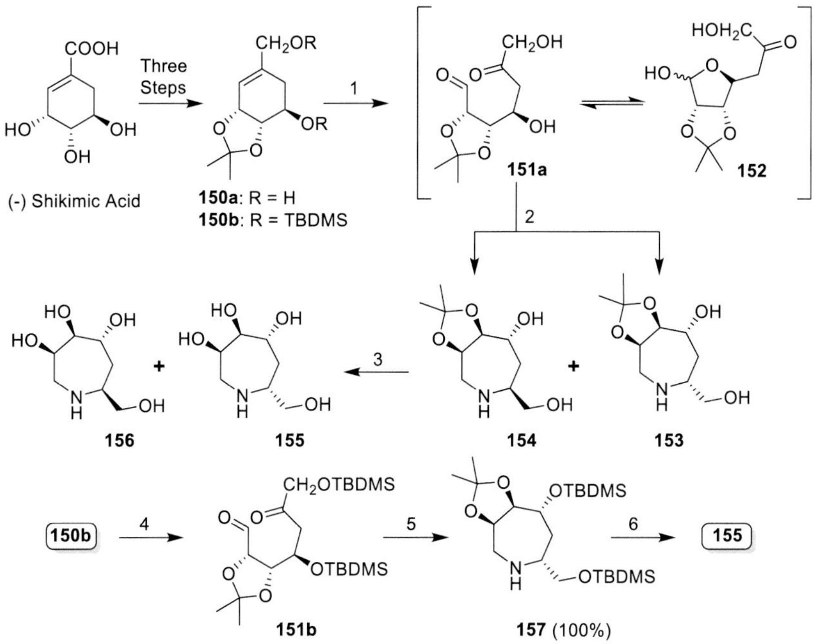

Reagents and Conditions:
(1) O_3, Ph_3P, Cl_2CH_2/MeOH, 78 °C, 23 min (93%);
(2) NH_4HCO_3, $NaCNBH_3$, MeOH, rt, 2 h, (63%)
(3) aq. acidic condition
(4) O_3, Ph_3P, Cl_2CH_2/MeOH, 78 °C, 15 min (75%);
(5) NH_4HCO_3, $NaCNBH_3$, MeOH, rt, 2 h, (78%)
(6) 6 N aq HCl, MeOH, 24 h, quantitative.

SCHEME 5.12 (−) Shikimic acid−derived polyhydroxyazepanes.

However, the same synthetic operation, when performed on *tert*-butyldimethylsilyl ether-protected derivative **150b**, not only avoided the intramolecular furanose formation, but also provided higher diastereoselectivity during double reductive amination step, yielding azepane **157** alone. On effecting complete deprotection under acidic conditions, the synthesis of **155** was obtained as the hydrochloride salt (Scheme 5.12).[52a]

D-Ribose-derived derivative **158** served as a key starting point for Ramapanicker's synthetic scheme toward polyhydroxyazepanes.[52b] For incorporation of the amino functionality, the scheme banks on asymmetric α-hydrazination in aldehyde, catalyzed by D-proline or L-proline. Two carbon homologations of **158** and suitable protection and functional group interconversions (Scheme 5.13) delivered the aldehyde **160** for effecting asymmetric α-hydrazination. Hydrogenolysis of the hydrazino groups in **161** and **164** using freshly prepared Raney-Ni and H_2 yielded the corresponding amino functionality. Cyclization through an intramolecular displacement of

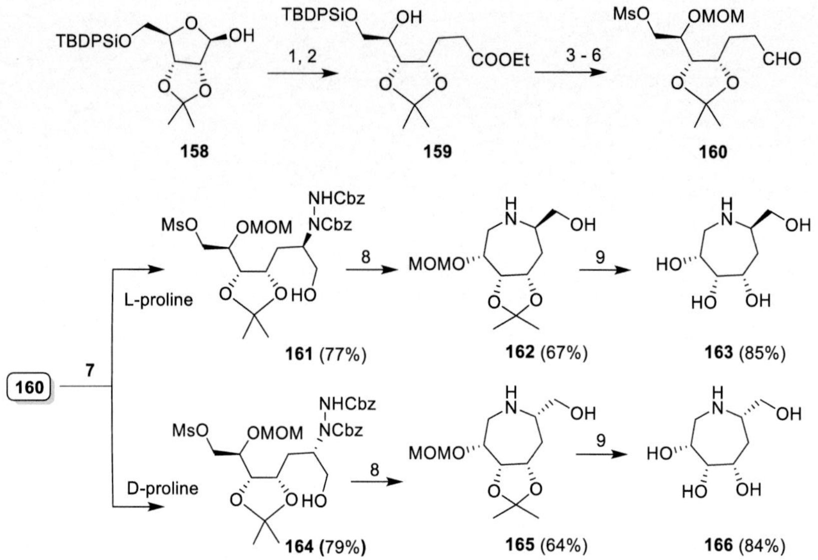

Reagents and Conditions:

(1) Ph_3P=CHCOOEt, CH_2Cl_2, r. t. 6h
(2) MeOH, Pd/C, r. t. 30 min (85%, over two steps)
(3) MOM-Cl, *i*-Pr_2NEt, 0 °C, r. t. 12 h (85%)
(4) TBAF, THF, r. t. 1h (92%)
(5) MsCl, Et_3N, CH_2Cl_2, 0 °C to r. t.(95%)
(6) DIBAL-H, -78 °C, 4 h (76%)
(7) CbzN=NCbz, MeCN, 3h, 0 °C then $NaBH_4$/EtOH
(8) H_2/Raney-Ni, MeOH,r.t., 16h
(9) EtOAc.HCl, r.t. 24h

SCHEME 5.13 D-Ribose-derived polyhydroxyazepanes.

mesylate afforded the formation of azepanes **162** and **165**. Acidic hydrolysis furnishes the polyhydroxyazepanes **163** and **166** as the hydrochloride salt.

Bol's isofagomine **10**, a potent β-glucosidase inhibitor mimicking the carbocationic form of the oxycarbenium-like transition state, at physiological pH, inspired structural modification for increasing the potency. Indeed, the potency of this compound was further improved by the insertion of a hydroxyl group at the C-2 position to afford noeuromycin **167**, a nanomolar β-glucosidase inhibitor.[53] The hydroxymethyl unit featuring in these compounds inspired the incorporation of the same in tetrahydroxyazepane-based glycosidase inhibitors. The free hydroxyl in **168** allowed incorporations of CH_2OH through oxidation, olefination, hydroboration, and oxidation sequence, and thereby access to azepanes **172** and **173** featuring hydroxymethyl unit (Scheme 5.14). Unfortunately, these azepane-based derivatives

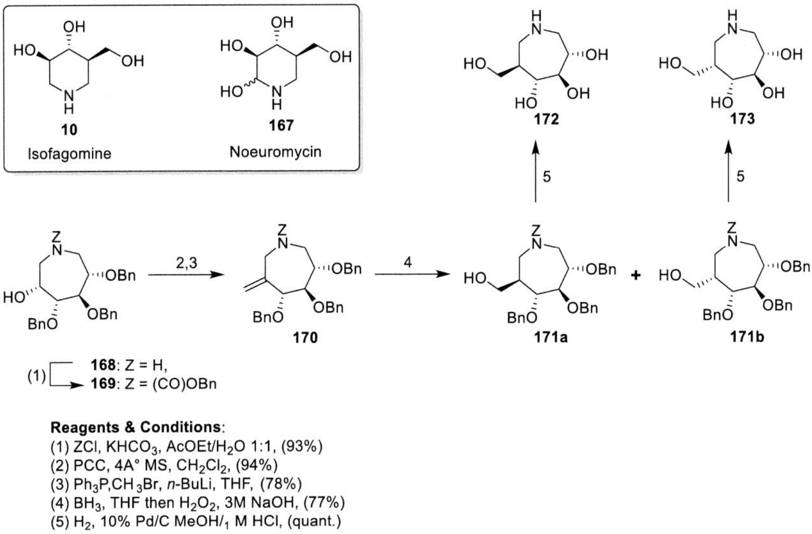

SCHEME 5.14 Azepane-based analogs featuring hydroxyl methyl unit.

were weaker than piperidine-based isofagomine **10** and noeuromycin **167** in their glycosidase inhibition activity. **172** performed selective and potent inhibition on almonds β-glucosidase ($Ki = 4\ \mu M$, uncompetitive) while **173** inhibited jack bean α-mannosidase ($Ki = 42\ \mu M$).[54] The intermediate **170** also enabled access to other analogs **175–176** featuring hydroxymethyl unit at C-3 position (Scheme 5.15). The intermediates **174a** and **174b** upon tosylation and then reduction using $LiBHEt_3$ solution in tetrahydrofuran (THF) provided **177** and **178**, which upon complete Pd-catalyzed debenzylation afforded branched derivatives of azepane **179** and **180**. The introduction of a methyl and tertiary hydroxy group was assumed to hold the sugar hydroxyl groups in the correct topological orientation in the hope to get more selective and potent glycosidase inhibitors. Both these series of azepanes **175–176** and **179–180** were assayed against 22 commercially available glycosidases, but the results were not promising with any of the glycosidases.[55]

5.4 Emerging *C*-glycosides of polyhydroxyazepanes

The introduction of alkyl groups either on C-2 or *N* position and assessing their biological activities, in comparison with Miglitol **7** became interesting and important, because the lipophilicity of the alkyl residues promoted better absorption, delivery, and uptake by cells.[27a,56a-c] The enone **142**[51a] obtained in four steps from D-(−)-quinic acid served as a requisite building block for incorporation of the alkyl group and synthesis of polyhydroxyazepanes with

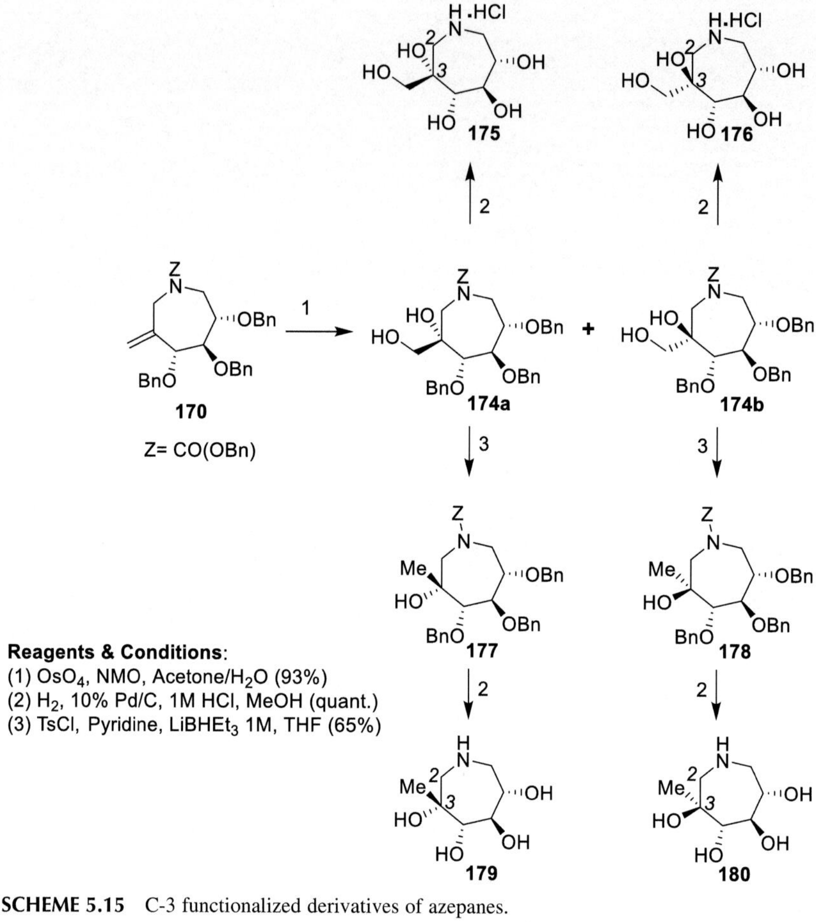

SCHEME 5.15 C-3 functionalized derivatives of azepanes.

alkyl unit on C-2 carbon (Scheme 5.16). The synthetic strategy is illustrated by the successful addition of MeLi and *n*-BuLi onto enone **142** in the presence of $CeCl_3$ at $-78°C$. The 1,2 additions of these alkyl units afforded separable epimeric mixtures **181–182** and **183**, respectively. These products on treatment with pyridinium chlorochromate (PCC) and silica gel enabled rearrangement and allylic oxidation furnishing the enones **184** and **185**. Luche reduction, followed by protection, afforded protected triols **188** and **189**. The double bond in these substrates, **188** and **189**, was subjected to dihydroxylation and oxidative cleavage. The reductive amino cyclization, with benzylamine and $NaCNBH_3$, leads to the formation of separable epimeric products **190a,b** and **191a,b**, respectively, in three steps. The final *N*-debenzylation on these compounds under hydrogenation conditions provided the desired azepanes **192–195** with methyl and *n*-butyl groups at the C-2 position.[57]

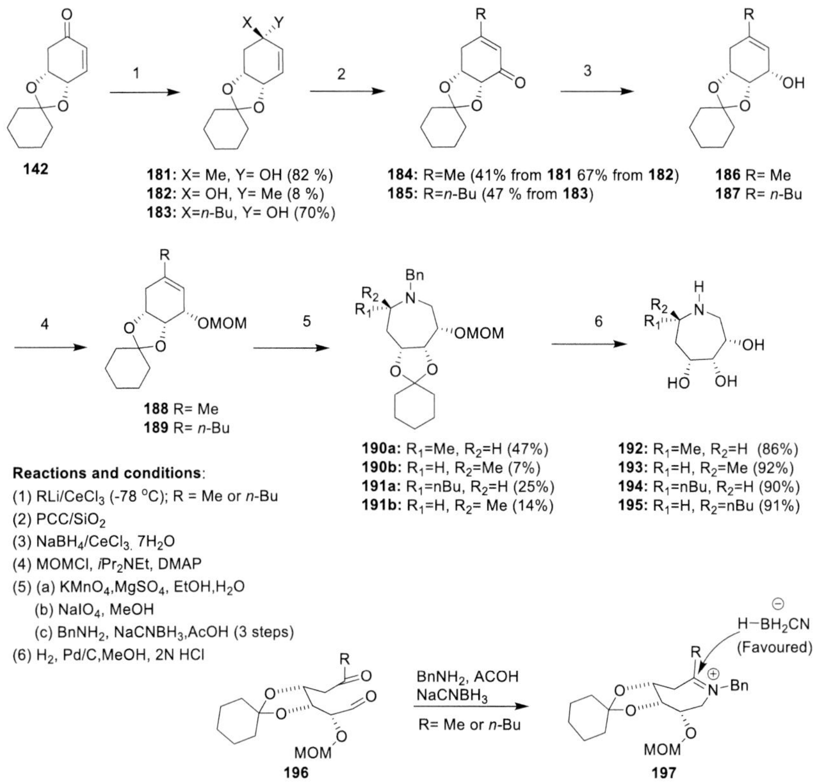

SCHEME 5.16 Synthesis C-2-alkyl-polyhydroxyazepanes **192–195**.

Another set of diastereomeric azepanes **199–202** (Fig. 5.17) were obtained through the same strategy using another building block **198**, also derived from D-(−)-quinic acid.[57]

The corresponding *N*-methylated and *N*-butylated analogs of **192–195** and **199–202** were also meaningful for studying the influence of the position of the alkyl residues.

These analogs, **211–214** and **218–219**, were prepared by a similar synthetic scheme banking on oxidative cleavage of the double followed by reduction amino cyclization, using methyl amine or butylamine as the amine component during reductive amino cyclization (Scheme 5.17). The synthesized library of *C*- or *N*-alkylated polyhydroxyazepanes, **192–195**, **199–202**, **211–214**, and **218–219** were evaluated for their biological activities as glycosidase inhibitors. The azepane **194** with the butyl substituted in 2*R*-configuration has better inhibition potency against β-galactosidase ($IC_{50} =$ 3 μM) while **195** showed no activity. Similar behavior was shown by methyl homologs **192–193**, wherein azepane **192** displayed moderate inhibition ($IC_{50=}$100 μM), while **193** was found to be inactive against β-galactosidase.[57]

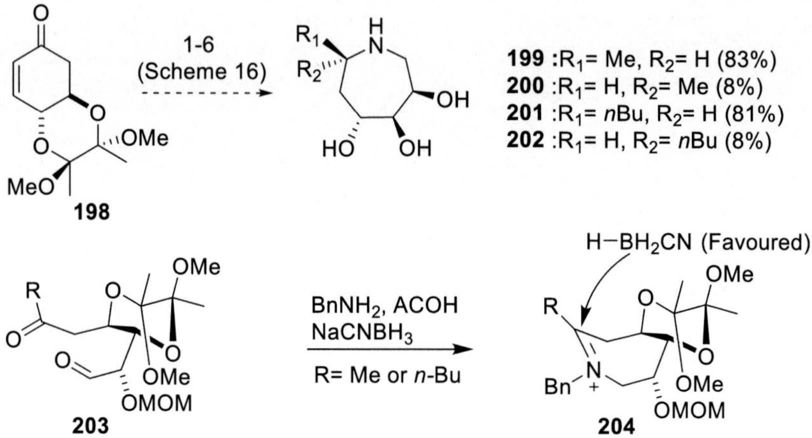

FIGURE 5.17 Synthesis *C*-2-alkyl-polyhydroxyazepanes **199–202**.

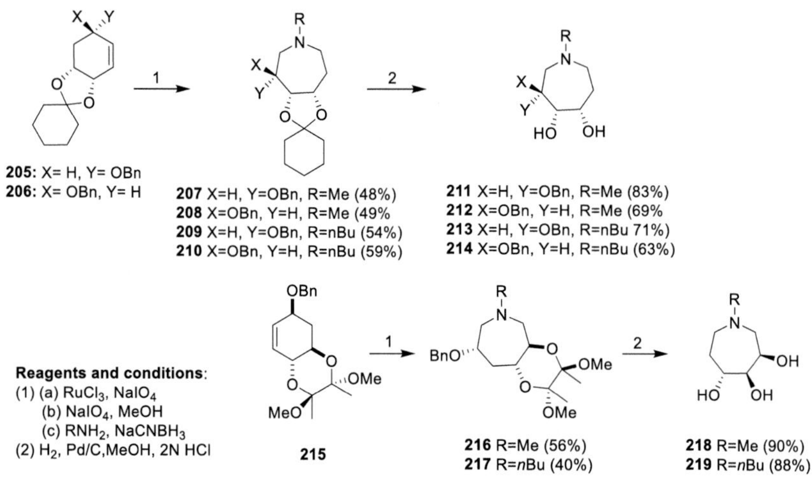

SCHEME 5.17 Synthesis of *N*-alkylated polyhydroxyazepanes **211–214** and **218–219**.

Sugar-derived seven-membered cyclic nitrones were unreported until efforts of Yu when he visualized the potential of these nitrones as building blocks for the synthesis of *C*-2 alkylated polyhydroxylated azepanes through diastereoselective nucleophilic addition of various organometallics reagents RMgX/RLi onto the nitrone functionality. As an illustration, Scheme 5.18 delineates the synthesis of D-glucose-derived nitrone **230**. It banked on intramolecular condensation of a hydroxylamine unit onto an aldehyde. Either end of open-chain D-glucose derivative **220** provided an opportunity for incorporating the hydroxylamine unit. Path A enabled the incorporation of the hydroxylamine group on the C-6 carbon whereas Path B allowed

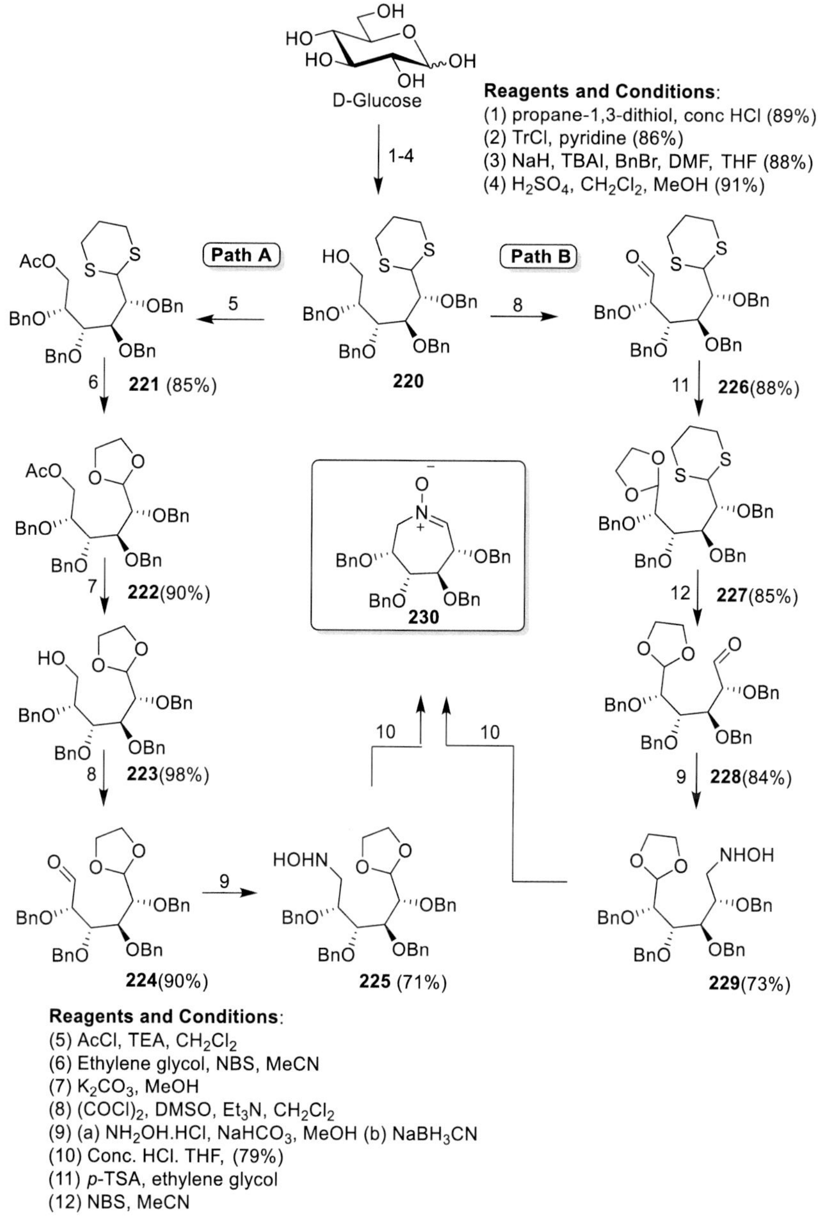

SCHEME 5.18 Synthesis of D-glucose-derived seven-membered ring nitrone.

incorporation on C-1 carbon. The final cyclization either on substrate **225** or on **229** provided the same nitrone **230**. Both routes are efficient and high yielding. The successful application of this approach with other monosaccharides such as D-gulose, D-altrose, and L-talose reflects the versatility of

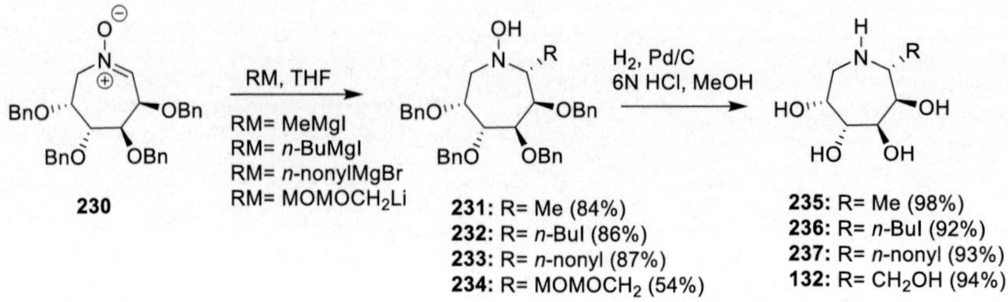

SCHEME 5.19 Synthesis of *C*-2 alkylated azepanes using sugar-derived nitrone **230**.

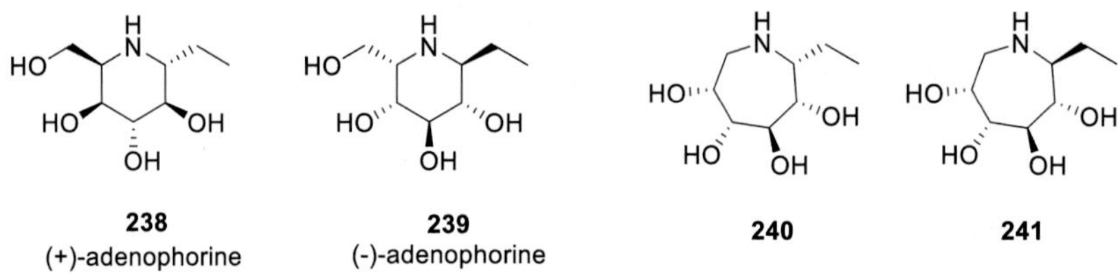

FIGURE 5.18 Structure of adenophorines and seven-membered analogs **240–241**.

the developed route for the synthesis of sugar-derived seven-membered nitrones. The D-mannitol-derived nitrone **230** has paved the way for the synthesis of *C*-2-alkylated polyhydroxyazepanes **235–237** and **132**.[58]

These compounds were assayed for their glycosidase inhibition activity against a range of different enzymes. None of the azepanes displayed satisfactory inhibition. Fairly good inhibitory activity against β-galactosidase was shown by azepane **236** (bovine liver, $IC_{50} = 505\ \mu M$) and **237** (bovine liver, $IC_{50} = 4.5\ \mu M$). Azepane **132** also inhibited β-glucosidase ($IC_{50} = 12\ \mu M$) (Scheme 5.19).[58]

The natural product (+)-adenophorine **238** isolated in 2000 along with its glucoside from *Adenophorae radix*, a traditional Chinese crude drug named "Sya-zin"[59] is a rare example of iminosugar *C*-glycoside, containing hydrophobic alkyl substituents at pseudoanomeric position. The structure was firmly established in 2003 by Davis[60] through the synthesis of its enantiomer (−)-adenophorine **239**, a L-iminosugar. Compared to (+)-adenophorine **238**, the (−)-adenophorine **239** exhibits a very distinctive inhibitory profile and was found to be a moderate α-fucosidase inhibitor (Fig. 5.18).

Although synthesis and biological studies of six-membered piperidine-based iminosugars *C*-glycosides have been extensively covered,[27a] Bleriot synthesized **239** through skeletal rearrangement in an easily accessible *C*-2-alkyl seven-membered iminosugar building block **242** (Scheme 5.20). The *N*-benzyl-protected azepane derivative **242** underwent skeletal rearrangement[61] during Mitsunobu conditions affording the corresponding piperidine product **245** (34%) and the fully protected azepane product **247** (30%). Contraction to piperidine ring results from a nucleophilic attack at the

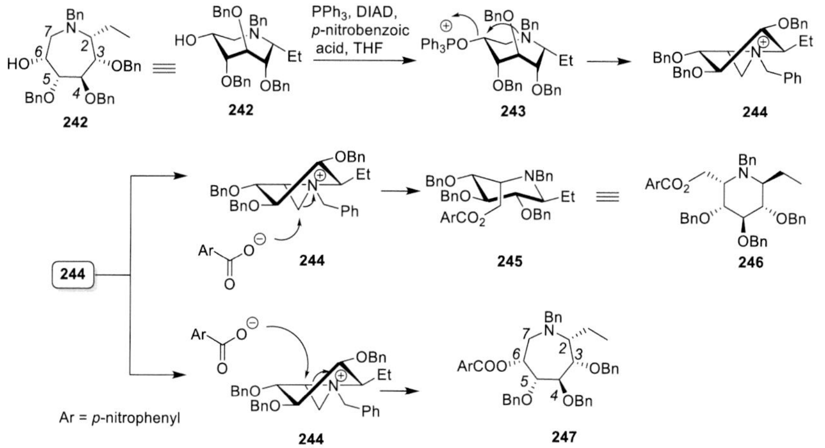

SCHEME 5.20 Skeletal rearrangement for synthesis of (−)-adenophorine.

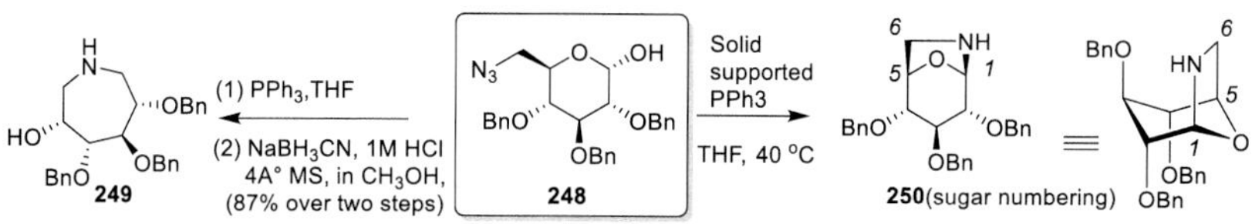

SCHEME 5.21 Synthesis of azidolactol precursor **250**.

aziridinium methylene carbon in the fused piperidine-aziridinium intermediate **244** while the azepane is obtained from a nucleophilic attack at the aziridinium methine carbon. Ester hydrolysis and debenzylation furnished the targeted (−)-adenophorine **239**. The *C*-2 ethyl azepane-based analogs **240** and **241** were weak glycosidase inhibitors[62] compared to (−)-adenophorine.

The reactivity of N, O acetals in general with organometallics is well documented,[63] the addition of organomagnesium reagents to the N, O acetal functionality present in [5, 6] bicyclic system were essentially unexplored, until the elegant efforts and results disclosed by Bleriot with common sugar-based azidolactol precursor **250** (Scheme 5.21).[62]

The stereochemical aspect associated with the addition of RMgX needs special attention. With the R unit incapable of π-stacking, such alkyl residues (butyl, hexyl, nonyl, and dodecyl) the addition occurs from the Re-face, giving the 2,6 *cis*-*C*-alkyl azepane derivatives **255a–d** (Scheme 5.22). However, the addition of allylmagnesiumbromide, wherein the allyl unit is capable of π-stacking with the phenyl moiety of benzyl ether protection, dictates the entry of the allyl group from the Si face of the imine functionality **252**,[62] thereby leading to the formation of 2,6 *trans*-*C*-allyl azepane derivative **259** after work up. This subtle difference in the addition trajectory of *n*-alkyl versus allyl has been elegantly utilized for the obtainment of the other *C*-2-alkyl azepanes

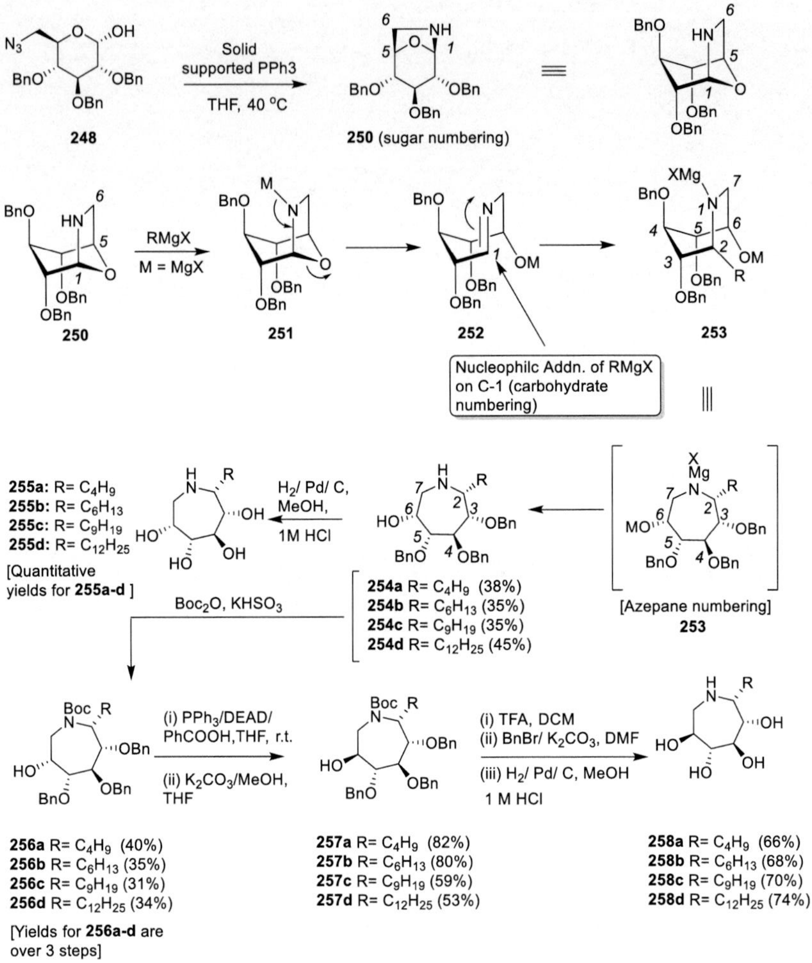

SCHEME 5.22 Synthesis of *C*-2 alkylated tetrahydroxyazepane **255a–d** and **258a–d**.

262a–c with inverted stereochemistry at C-2. The allyl-substituted building block enabled chain extension through a cross-metathesis elongation strategy. **259** on treatment with pent-1-ene, oct-1-ene, and undec-1-ene in the presence of Hoveyda–Grubbs catalyst led to the formation of **262a–c**.[64] Following the inversion of stereochemistry at the C-6 position in the *N*-Boc protected building block, the same chemistry allowed the formation of other diastereoisomeric *C*-2-alkyl azepanes **265a–c** (Scheme 5.23).

The azepane **249** was further alkylated at the nitrogen center to afford *N*-alkylated azepanes series **266a–d** and **269a–d** after several steps (Scheme 5.24).[64] In comparison to *C*- and *N*-alkylated six-membered iminosugars, the study of seven-membered *C*/*N*-alkylated azepanes for

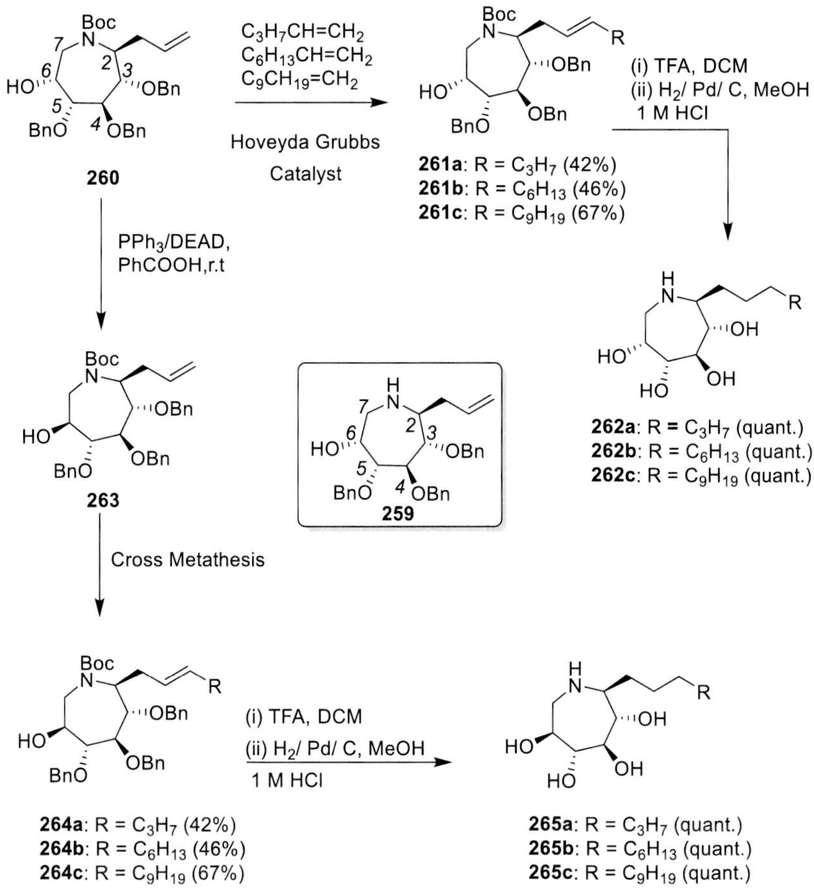

SCHEME 5.23 Synthesis of *C*-2 alkylated tetrahydroxyazepane **262a–c** and **265a–c**.

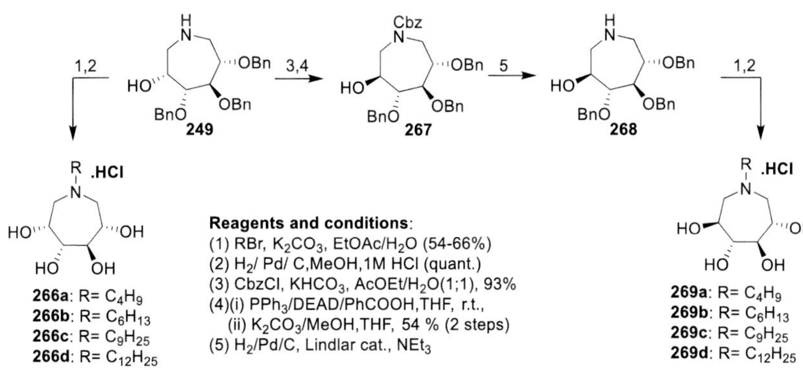

SCHEME 5.24 *C*- and *N*-alkylated seven-membered iminosugars.

glycosidase inhibition is only a recent advent through the major contribution of Bleriot's group. The library of *C*-2 and *N*-alkylated azepanes **255a–d**, **258a–d**, **262a–c**, **265a–c**, **266a–d**, and **269a–d** obtained from glucopyranose-derived azidolactol **250** has been evaluated for glycosidase inhibition. The alkyl chain at the C-2 position increased the inhibition activities of azepanes against β-glucosidase and β-galactosidase. The C-2-alkyl azepanes **255a–d** and **258a–d** with alkyl chain *cis* to the C_3–OH group exhibit nearly similar potency as compared to the *C*-2-alkyl azepanes **262a–c** and **265a–c** with alkyl chain being *trans* to C-3-OH. This observation suggested that the stereochemistry of the alkyl chain is not crucial for glucosidase inhibition. The *N*-alkyl substituted compounds, **266a–d**, and **269a–d** displayed reduced inhibition selectivity and showed broad inhibition against β-glucosidases, β-galactosidases, α-mannosidase, α-L-rhamnosidase, etc.

5.4.1 *C*-2 functionalized polyhydroxyazepane

Plantier-Royon-envisaged polyhydroxyazepanes with functionalized alkyl units such as (ethoxycarbonyl)methyl or carboxymethyl at the pseudoanomeric position, to invoke stronger binding in the active site of the glycosidases comprising Glu and Asp amino acids as acid/base catalysis pair. The partially functionalized azacycloheptenes **271**, featuring (ethoxycarbonyl)methyl at the pseudoanomeric were visualized as the requisite advanced intermediate for further hydroxylation and entry into targeted polyhydroxyazepanes **272**, featuring carboxymethyl unit at the pseudoanomeric position (Fig. 5.19).

The azacycloheptenes represented by general structure **271** were prepared through ring-closing metathesis on diene **270**, which was itself prepared in nine steps, using commercially available D-xylose as the starting material.[65] The *syn* dihydroxylation on azacycloheptenes of **271a** and **271b** under the standard Upjohn conditions (OsO_4, NMO) afforded the corresponding diols **273** and **275**, respectively. In each case, an inseparable mixture of diastereomeric *cis*-diols (6:4) was obtained. Subsequent protection of the diols as isopropylidene derivatives, in each case, enabled the separation of all the compounds, **274a,b** and **276a,b** (Scheme 5.25).

Finally, the removal of different protections in a one-step procedure by treatment with a mixture of 6 M HCl/THF 4:1 at 8°C enabled access to

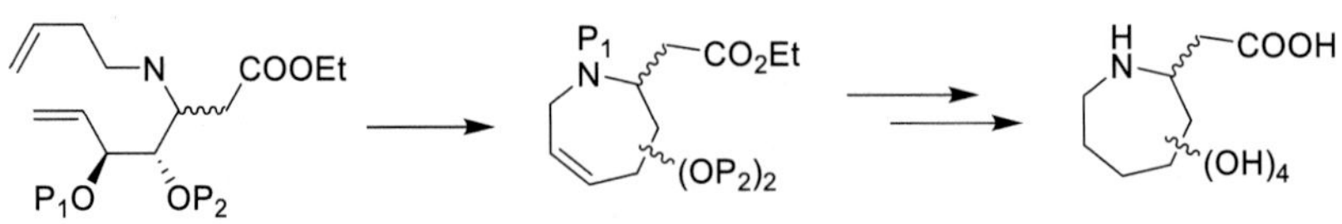

270 271 272

P1 and P2 = Suitable protections

FIGURE 5.19 Strategy for polyhydroxyazepanes featuring carboxymethyl unit.

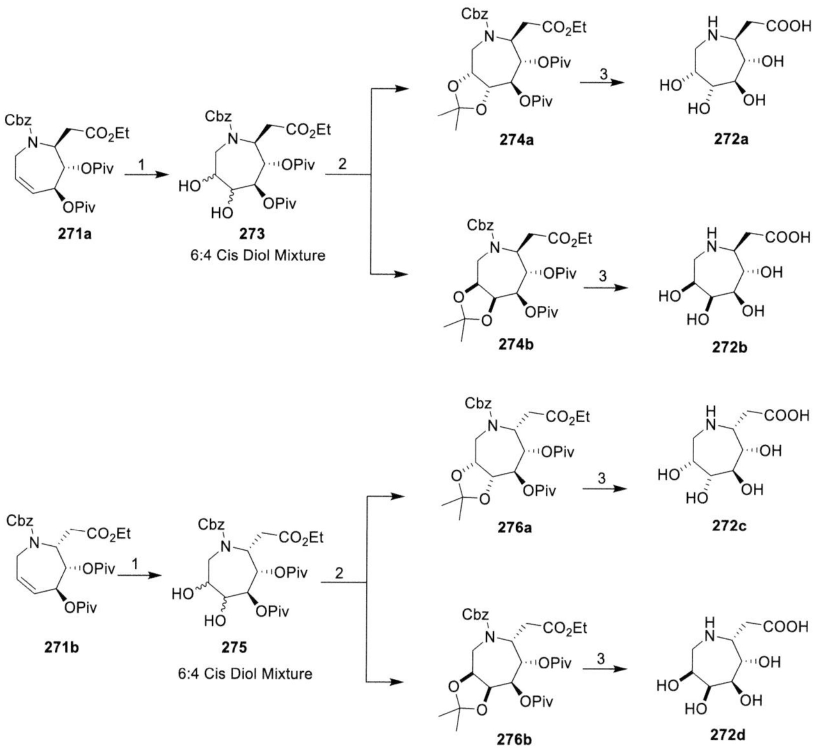

(1) OsO_4 2.5% in *tert*-BuOH, NMO, EtOAc/H_2O 4:1 (78% for **273**, 84% for **275**);
(2) Acetone/Dimethoxypropane (5:1), *p*-TSA cat., 96% for **274a**,b and 93% for **276a,b**.
(3) HCl 6 M/THF 4:1, 80 °C, 12 h (75% for **272a,** 65% for **272b**, 90% for **272c**, 83% for **272d**)

SCHEME 5.25 Synthesis of polyhydroxyazepanes featuring carboxymethyl unit.

polyhydroxyazepanes **272a,d** in good yields (65%–90%). Another set of polyhydroxyazepanes was made through the epoxidation of azacycloheptenes of **271a** and **271b**. Each azacycloheptene afforded a diastereoisomeric mixture of epoxides **277** and **279**, respectively. Ring opening of the epoxides with acetic anhydride in the presence of the catalytic amount of $Er(OTf)_3$ resulted in the formation of expected, protected polyhydroxyazepanes, **278a,b** and **280a,b**. The ratio of **280a–280b** being 95:5 suggested exclusive formation of **280a**. One-step deprotection under acidic conditions furnished the three polyhydroxyazepanes, **272e–g** (Scheme 5.26).

Seven of these polyhydroxylated azepanes (**272a–g**) were assessed as inhibitors of various commercially available glycosidases.[66] Azepanes **272a** and **272b** were moderate inhibitors ($IC_{50} = 279\ \mu M$ and $IC_{50} = 215\ \mu M$, respectively) while **272e** exhibited the most potent inhibition ($IC_{50} = 21\ \mu M$).

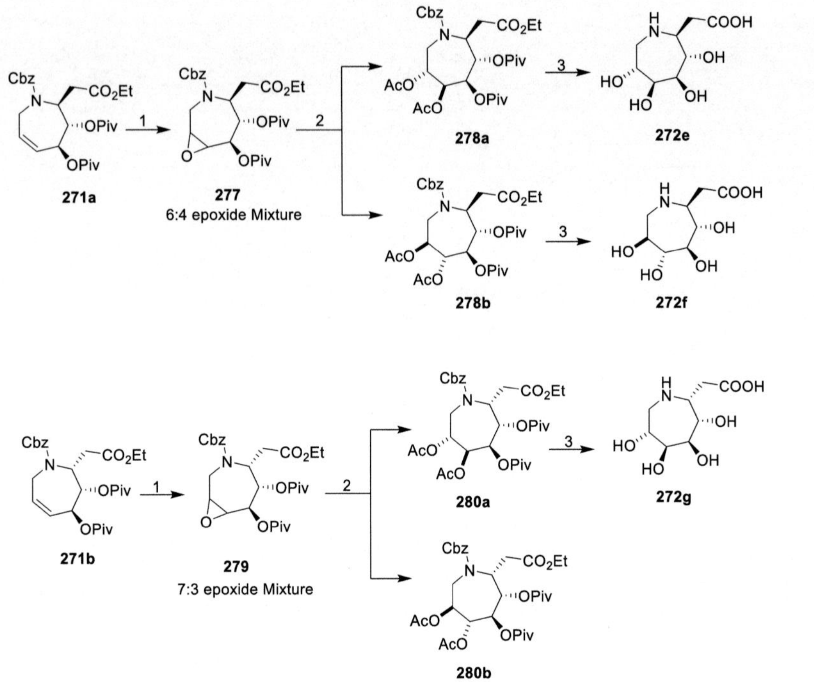

Reagent and conditions:

(1) Oxone (excess, $NaHCO_3$ excess, Bu_4NHSO_4 cat., EtOAc/H_2O 4:1 (78% for **277**, 84% for **279**)

(2) Ac_2O excess, $Er(OTf)_3$ cat. (87% for **278a, 278b**, (**278a: 278b** 70:30); 77% for **280a, 280b**, (**280a: 280b**, 95:5)

(3) HCl 6 M/THF 4:1, 80 °C, 12 h (50% for **272e**, 8% for **272f**, 83% for **272g**)

SCHEME 5.26 Synthesis of polyhydroxyazepanes featuring carboxymethyl unit.

5.5 Conclusions and future outlook

The glycoprocessing enzymes, glycoside hydrolases (glycosidases), and glycosyl synthases (glycosyl transferases) involved in the biosynthesis or processing of glycans and glycoconjugates have emerged as valuable targets for drug discovery, because of them being implicated in major global health problems, such as diabetes and cystic fibrosis. The lysosomal storage disorders represented by Gaucher, Pompe, GM1-gangliosidosis, and Fabry diseases, although rare, are also resulting from malfunctioning of these enzymes and merit equal attention. Disrupting the activity of these enzymes in pathogens has additionally paved the way for combating infections. Biologically active *C*-glycosides inspired by natural products are a broad and highly attractive area of research,[67] holding promise toward therapeutics. Iminosugars/Azasugars inspired by the natural product nojirimycin have emerged as a valuable class of compounds, as inhibitors of glycoprocessing enzymes. This class of compounds has inspired several synthetic pursuits and synthesis of a large number of analogs in search of potent and selective inhibitors[68] for therapeutic applications. The success of six-membered piperidine-based iminosugars has inspired entry into seven-membered polyhydroxyazepanes for the synthesis and evaluation as inhibitors. It was the conformational flexibility of azepane, possibly allowing favorable binding in the active site of the enzyme that inspired the synthesis of seven-membered analogs of deoxynojirimycin. The polyhydroxyazepanes with unusual distribution of hydroxy groups and conformational flexibility have generated new inhibitory profiles and have attracted several synthetic pursuits. The *C*-2 alkylated azepanes have displayed promise against β-glucosidase and β-galactosidase. This review broadly covers the available synthetic routes for the synthesis of polyhydroxyazepanes and subsequent entry into *C*-glycosides of seven-membered imino/aza-sugars, during the recent years. In the context that iminosugars in general are attracting possible applications in other domains such as innate immune response signaling routes, metastasis, and age-related neurodegenerative disorders, for example, Alzheimer's and Parkinson's diseases, the possible evaluation of polyhydroxyazepanes is being awaited. Given the fact that the combination of N-substitution and C-branching in the iminosugar scaffold is a valuable strategy for fine tuning of biological activity,[27a] much more work is anticipated, within this class of polyhydroxyazepanes, in search for lead molecules toward therapeutic applications.

References

1. (a) Bols, M.; Lopez, O.; Ortega-Caballero, F. *Comprehensive Glycoscience* **2007,** *1*, 815–884.
   (b) Ernst, B.; Magnani, J. L. *Nat. Rev. Drug Discovery* **2009,** *8*, 661–677.
2. (a) Inouye, S. E.; Tsuruoka, T.; Ito, T.; Niida, T. *Tetrahedron* **1968,** *24*, 2125.
   (b) Hughes, A. B.; Rudge, A. J. *Nat. Prod. Rep.* **1994,** 135.
3. McNaught, A. D. *Pure Appl. Chem.* **1996,** *68*, 1919–2008.
4. Stutz, A. E. *Iminosugars as Glycosidase Inhibitors: Nojirimycin and Beyond;* Wiley-VCH: Weinheim, 1999.
5. Compain, P.; Martin, O. R., Eds. *Iminosugars: from Synthesis to Therapeutic Applications;* John Wiley & Sons Ltd, 2007.
6. Asano, N. *Glycobiology* **2003,** *13*, 93–104.
7. Koshland, D. E.; Rev, Biol *Camb. Philos. Soc.* **1953,** *28*, 416–436.
8. Wicki, J.; Williams, S. J.; Withers, S. G. *J. Am. Chem. Soc.* **2007,** *129*, 4530–4531.
9. Vasella, A.; Davies, G. J.; Bohm, M. *Curr. Opin. Chem. Bio.* **2002,** *6*, 619–629.
10. Kallemeijn, W. W.; Witte, M. D.; Wennekes, T.; Aerts, J. M. F. G. *Adv. Carbo. Chem. Biochem.* **2014,** *71*, 297.
11. Lillelund, V. H.; Jensen, H. H.; Liang, X.; Bols, M. *Chem. Rev.* **2002,** *102*, 515–553.
12. (a) Inoue, S.; Tsuruoka, T.; Niida, T. *J. Antibiot.* **1966,** *19*, 288–292.
   (b) Inoue, S.; Tsuruoka, T.; Ito, T.; Niida, T. *Tetrahedron* **1967,** *23*, 2125–2144.
13. Niwa, T.; Inoue, S.; Tsuruoka, T.; Koaze, Y.; Niida, T. *Agric. Biol. Chem.* **1970,** *34*, 966–968.
14. Yagi, M.; Kouno, T.; Aoyagi, Y.; Murai, H. *Nippon. Nogei Kagaku Kaishi* **1976,** *50*, 571–572.

15. Junge, B.; Matzke, M.; Stltefuss, J. Chemistry and structure activity relationships of glucosidase inhibitors. In *Handbook of Experimental Pharmacology;* Kuhlmann, J., Puls, W., Eds.; Vol. 119; Springer-Verlag: New York, 1996; pp 411–482.
16. Sels, J. P. J. E.; huijberts, M. S. P.; Wolffenbuttel, B. H. R. *Exp. Opin.pharmacother.* **1999,** *1*, 149–156.
17. Desnick, R. J.; Astrin, K. H.; Schuchman, E. H. Therapies for lysosomal storage diseases. In *Emery and Rimoin's Principles and Practice of Medical Genetics and Genomics;* Pyeritz, R. E., Korf, B. R., Grody, W. W., Eds.; Academic Press, 2019; pp 205–227.
18. Zimran, A. *Gaucher's Disease;* Balliere Tindall, 1997.
19. Phetthong, T.; Tim-Aroon, T.; Khongkrapam, A.; Noojarem, S.; Kuptanon, C.; Winchajam, K.; Sathienkijkanchai, A.; Suphapeetiporn, K.; Charoenkwan, P.; Tantiworawit, A., et al. *Orphanet J. Rare Dis.* **2021,** *16*, 1–10.
20. Horowitz, M.; Zimran, A. *Hum. Mutat.* **1994,** *3*, 1.
21. Koyama, M.; Sakamura, S. *Agric. Biol. Chem.* **1974,** *38*, 1111.
22. Schofield, A. M.; Fellows, L. E.; Nash, R. J.; Fleet, G. W. J. *Life Sci.* **1986,** *39*, 645.
23. Kato, A.; Asano, N.; Kizu, H.; Matsui, K.; Watson, A. A.; Nash, R. J. *J. Nat. Prod.* **1997,** *60*, 312.
24. Davies, S. G.; Fletcher, A. M.; Roberts, P. M.; Thomson, J. E. *Tetrahedron* **2019,** *75*, 130727.
25. Jespersen, T. M.; Dong, W.; Skrydstrup, T.; Sierks, M. R.; Lundt, I.; Bols, M. *Angew. Chem., Int. Ed. Engl.* **1994,** *33*, 1778–1779.
26. Dong, W.; Jespersen, T. M.; Skrydstrup, T.; Bols, M.; Sierks, M. R. *Biochemistry* **1996,** *35*, 2788–2795.
27. (a) Compain, P.; Chagnault, V.; Martin, O. R. *Tetrahedron Asymmetry* **2009,** *20*, 672.
    (b) Horne, G.; Wilson, F. X. *Progress in Medicinal Chemistry* **2011,** *50*, 135–176.
28. Horne, G.; Wilson, F. X.; Tinsley, J.; Williams, D. H.; Storer, R. *Drug. Discov. Today* **2011,** *16*, 107–118.
29. (a) Sánchez-Fernández, E. M.; García-Fernández, J. M.; Ortiz Mellet, C. *Chem. Commun.* **2016,** *52*, 5497–5515.
    (b) Wrodnigg, T. M.; Stutz, A. E. *Curr. Enzyme Inhib.* **2012,** *8*, 47–99.
30. Parmeggiani, C.; Catarzi, S.; Matassini, C.; D'Adamio, G.; Morrone, A.; Goti, A.; Paoli, P.; Cardona, F. *ChemBioChem.* **2015,** *16*, 2054–2064.
31. Godin, G.; Compain, P.; Mason, G.; Martin, O. R. *J. Org. Chem.* **2002,** *67*, 6960–6970.
32. Yu, L.; Ikeda, K.; Kato, A.; Adachi, I.; Godin, G.; Compain, P.; Martin, O. R.; Asano, N. *Bioorg. Med. Chem.* **2006,** *14*, 7736.
33. Compain, P.; Martin, O. R.; Boucheron, C.; Godin, G.; Yu, L.; Ikeda, K.; Asano, N. *ChemBioChem* **2006,** *7*, 1356.
34. Schonemann, W.; Gallienne, E.; Compain, P.; Ikeda, K.; Asano, N.; Martin, O. R. *Bioorg. Med. Chem.* **2010,** *18*, 2645.
35. Harit, V. K.; Ramesh, N. G. *RSC Adv.* **2016,** *6*, 109528–109607.
36. Ramesh, N. G. In *Carbohydrates in Drug Discovery and Development;* Tiwari, V. K., Ed.; Elsevier, 2020; pp 331–381.
37. Paulsen, H.; Todt, K. *Chem. Ber.* **1967,** *100*, 512–520 For another synthesis of **GG** using D-(+)-glucurono-g-lactone as cheap starting material, see: Kalamkar, N. B.; Kasture, V. M.; Chavan, S. T.; Sabharwal, S. Dhavae, D. D. *Tetrahedron*, **2010**, *66*, 8522–8526.
38. (a) Poitout, L.; Merrer, Y. L.; Depezay, J.-C. *Tetrahedron Lett* **1994,** *35*, 3293–3296.
    (b) Merrer, Y. L.; Poitout, L.; Depezay, J.-C.; Dosbaa, I.; Geoffroy, S.; Foglietti, M.-J. *Bioorg. Med. Chem.* **1997,** *5*, 519–533.

39. Qian, X.; Moris-Varas, F.; Wong, C.-H. *Bioorg. Med, Chem. Lett.* **1996,** *6*, 1117–1122.
40. Ovaa, H.; Leeuwenburgh, M. A.; Overkleeft, H. S.; Van der Marel, G. A.; Van Boom, J. H. *Tetrahedron Lett.* **1998,** *39*, 3025–3028.
41. Li, H.; Bleriot, Y.; Chantereau, C.; Mallet, J. M.; Sollogoub, M.; Zhang, Y.; Garcia, E. R.; Vogel, P.; Barbero, J. J.; Sinay, P. *Org. Biomol. Chem.* **2004,** *2*, 1492–1499.
42. (a) Li, H.; Schutz, C.; Favre, S.; Zhang, Y.; Vogel, P.; Sinay, P.; Bleriot, Y. *Org. Biomol. Chem.* **2006,** *4*, 1653–1662.
    (b) Lglesias, E.; Garcia-Rio, L.; Leis, J. R.; Pena, M. E.; Williams, D. L. H. *J. Chem. Soc. Perkin Trans.* **1992,** *2*, 1673–1679.
43. Li, H.; Bleriot, Y.; Mallet, J.-M.; Rodriguez-Garcia, E.; Vogel, P.; Zhang, Y.; Sinay, P. *Tetrahedron Asymmetry* **2005,** *16*, 313–319.
44. (a) Cole, R. N.; Hart, G. W. *J. Neurochem.* **1999,** *73*, 418–428.
    (b) Vosseller, K.; Wells, L.; Lane, M. D.; Hart, G. W. *Proc. Natl. Acad. Sci. U.S.A.* **2002,** *99*, 5313–5318.
    (c) Zhang, F.; Su, K.; Yang, X.; Bowe, D. B.; Paterson, A. J.; Kudlow, J. E. *Cell* **2003,** *115*, 715–725.
45. (a) Yao, P. J.; Coleman, P. D. *J. Neurosci.* **1998,** *18*, 2399–2411.
    (b) Gao, Y.; Wells, L.; Comer, F. I.; Parker, G. J.; Hart, G. W. *J. Biol. Chem.* **2001,** *276*, 9838–9845.
    (c) Hart, G. W.; Housley, M. P.; Slawson, C. *Nature* **2007,** *446*, 1017–1022.
46. (a) Marcelo, F.; He, Y.; Yuzwa, S. A.; Niet, L.; Jimenez-Babero, J.; Sollogoub, M.; Vocadio, D. J.; Davies, G. D.; Bleriot, Y. *J. Amer. Chem. Soc.* **2009,** *131*, 5390–5392.
    (b) Li, H.; Marcelo, F.; Bello, C.; Vogel, P.; Butters, T. D.; Rauter, A. P.; Zhang, Y.; Sollogoub, M.; Bleriot, Y. *Bioorg. Med. Chem* **2009,** *17*, 5598–5604.
47. Dhavale, D. D.; Markad, S. D.; Karanjule, N. S.; Reddy, J. P. *J. Org. Chem.* **2004,** *69*, 4760–4766.
48. Lin, C. C.; Pan, Y.; Patkar, L. N.; Lin, H. M.; Tzou, D. L. M.; Subramanian, T.; Lin, C. C. *Bioorg. Med. Chem.* **2004,** *12*, 3259–3267.
49. (a) Estévez, A. M.; Soengas, R. G.; Otero, J. M.; Estévez, J. C.; Nash, R. J.; Estévez, R. J. *Tetrahedron: Asymmetry* **2010,** *21*, 21–26.
    (b) Ona, N.; Romero, A.; Carmen, A.; Bello, C.; Vogel, P.; Gonzalez, M. S. P. *Tetrahedron: Asymmetry* **2010,** *21*, 2092–2099.
50. Jagadeesh, Y.; Ramakrishna, K.; Rao, B. V. *Tetrahedron: Asymmetry* **2012,** *23*, 697–702.
51. (a) Colas, C.; Quiclet-Sire, B.; Cleophax, J.; Delaumeny, J. M.; Sepulchre, A.-M.; Gero, S. D. *J. Am. Chem. Soc.* **1980,** *102*, 857–858.
    (b) Shih, T.-L.; Kuo, W.-S.; Lin, Y.-L. *Tetrahedron Lett* **2004,** *45*, 5751–5754.
    (c) Alves, C.; Barros, M. T.; Maycock, C. D.; Ventura, M. R. *Tetrahedron* **1999,** *55*, 8443–8456.
    (d) Shih, T.-L.; Yang, R.-Y.; Li, S.-T.; Chiang, C.-F.; Lin, C.-H. *J. Org. Chem.* **2007,** *72*, 4258–4261.
52. (a) Gonzalez-Castro, M. A.; Poole, D. A.; Estevez, J. C.; Fleet, G. J. W.; Estevez, R. J. *Tetrahedron: Asymmetry* **2015,** *26*, 320–323.
    (b) Petakamsetty, R.; Jain, V. K.; Majhi, P. K.; Ramapanicker, R. *Org. Biomol. Chem.* **2015,** *13*, 8512–8523.
53. Liu, H.; Liang, X.; Sohoel, H.; Bulow, A.; Bols, M. *J. Am. Chem. Soc.* **2001,** *123*, 5116–5117.
54. Li, H.; Zhang, Y.; Vogel, P.; Sinay, P.; Bleriot, Y. *Chem. Commun.* **2007,** 183–185.
55. Li, H.; Zhang, Y.; Favre, S.; Vogel, P.; Sollogoub, M.; Bleriot, Y. *Carbohydr. Res.* **2012,** *356*, 110–114.

56. (a) Winchester, B. G. *Tetrahedron: Asymmetry* **2009,** *20*, 645–651.
    (b) Pearson, M. S. M.; Evain, M.; Mathé-Allainmat, M.; Lebreton, J. *Eur. J. Org. Chem.* **2007,** 4888–4894.
    (c) Pearson, M. S. M.; Plantier-Royon, R.; Szymoniak, J.; Bertus, P.; Behr, J.-B. *Synthesis* **2007,** 3589–3594.
57. Shih, T. L.; Liang, M. T.; Wu, K. D.; Lin, C. H. *Carbohydr. Res.* **2011,** *346*, 183–190.
58. Zhao, W.-B.; Nakagawa, S.; Kato, A.; Adachi, I.; Jia, Y.-M.; Hu, X.-G.; Fleet, G. W. J.; Wilson, F. X.; Horne, G.; Yoshihara, A., et al. *J. Org. Chem.* **2013,** *78*, 3208–3221.
59. Ikeda, K.; Takahashi, M.; Nishida, M.; Miyauchi, M.; Kizu, H.; Kameda, Y.; Arisawa, M.; Watson, A. A.; Nash, R. J.; Fleet, G. W. J., et al. *Carbohydr. Res.* **2000,** *323*, 73.
60. Maughan, M. A. T.; Davies, I. G.; Claridge, T. D. W.; Courtney, S.; Hay, P.; Davis, B. G. *Angew. Chem., Int. Ed.* **2003,** *42*, 3788.
61. Mondon, M.; Fontelle, N.; Désiré, J.; Lecornué, F.; Guillard, J.; Marrot, J.; Blériot, Y. *Org. Lett.* **2012,** *14*, 870.
62. Mondon, M.; Lecornue, F.; Gulllard, J.; Nakagawa, S.; Kato, A.; Bleriot, Y. *Bioorg. Med. Chem.* **2013,** *21*, 4803–4812.
63. (a) Bloch, R. *Chem. Rev.* **1998,** *98*, 1407–1438.
    (b) Enders, D.; Reinhold, U. *Tetrahedron: Asymmetry* **1997,** *8*, 1895–1946.
    (c) Steinig, A. G.; Spero, D. M. *Org. Prep. Proc. Int.* **2000,** *32*, 205–234.
64. Desire, J.; Mondon, M.; Fontelle, N.; Nakagawa, S.; Hirokami, Y.; Adachi, I.; Iwaki, R.; Fleet, G. W. J.; Alonzi, D. S.; Twigg, G., et al. *Org. Biomol. Chem.* **2014,** *12*, 8977–8996.
65. Goumain, S.; Taghzouti, H.; Portella, C.; Behr, J.-B.; Plantier-Royon, R. *Tetrahedron Lett.* **2012,** *53*, 4440–4443.
66. Taghzouti, H.; Goumain, S.; Harakat, D.; Portella, C.; Behr, J. B.; Royon, R. P. *Bioorg. Chem.* **2015,** *58*, 11–17.
67. Mannem, R. R.; Thoti, N.; Aidhen, I. S. In *Carbohydrates in Drug Discovery and Development;* Tiwari, V. K., Ed.; Elsevier, 2020; pp 97–153.
68. Conforti, I.; Marta, A. *Org. Bimol. Chem.* **2021,** *19*, 5439–5475.

Chapter 6

Sialic acid donors: stereoselective chemical and enzymatic *O*-glycosylations

Manoj K. Jaiswal, Vinay K. Pandey and Vinod K. Tiwari ☆

Department of Chemistry, Institute of Science, Banaras Hindu University, Varanasi, Uttar Pradesh, India

6.1 Introduction

Carbohydrates, being one of the most abundant natural products, are an enormously important chiral molecule with a wide range of applications in chemistry, biology, and medicine.[1−5] By analyzing the structure of the cell and its several multifunctional organelles, carbohydrates were found to have substantial effects on selectivity and chirality in a number of cellular chemical, metabolic, and recognition processes. These significant roles of carbohydrate moiety for the survival of the living system have been well documented for a long time.[6−10] Carbohydrate-containing molecules offer several opportunities for therapeutic intervention due to their vital role in a number of crucial biological activities, including inflammation, immunological response, metastasis, fertilization, and microbial infections.[11−16] The development of mechanism-based novel drug candidates with low toxicity, improved bioavailability, biocompatibility, hydrophilicity, and enhanced absorption, distribution, metabolism, and excretion (ADME) properties are prominent applications in which carbohydrate motifs are identified as noteworthy scaffolds.[1] In addition to the five- and six-carbon saccharides as one of the important constituents of the vertebrates, sialic acids (commonly known as *N*-acetylneuraminic acids, Neu5Ac) are a unique class of nine-carbon carboxylated monosaccharides and produced by higher animals such as Echinoderms, Hemichordata, Cephalochordata, and Vertebrata with profound applications in the chemical biology.[2−5] However, in the structural composition of glycoproteins and glycolipids of vertebrates, the sialic

☆ This chapter is dedicated to Prof. Xi Chen for her extraordinary contribution on "Chemoenzymatic Synthesis of Sialic Acid containing Glycans of Chemotherapeuic Potential."

Synthetic Strategies in Carbohydrate Chemistry. DOI: https://doi.org/10.1016/B978-0-323-91729-2.00016-1

acid residues are present at the terminal of *N*-linked and *O*-linked carbohydrate chains.[3] The negative charge and hydrophilic properties of sialic acid enable its role in a variety of normal and pathological processes, acting as binding sites for various pathogens and toxins. Additionally, as a terminal sugar motif in the capsular polysaccharides or lipooligosaccharides of harmful bacteria, these residues function as crucial recognition elements.[4] *N*-Acetylneuraminic acid (**1**, Neu5Ac), *N*-glycolylneuraminic acid (**2**, Neu5Gc), and 2-keto-3-deoxynonulosonic acid (**3**, KDN) (Fig. 6.1A) are three fundamental forms of sialic acids. Moreover, sialic acid-containing glycans regulate a variety of cellular functions, such as adhesion, migration, recognition, invasion, bacterial and viral invasion, tumor metastasis, and much more.[4] There are more than 50 structurally unique sialic acid analogs (free or sialosides) found in nature, and more than 15 of them have been detected in humans (Fig. 6.1B).[5]

The sialic acids and analogs are naturally occurring important facilitators and modulators of several cellular interactions, which are incorporated in a complex sequence of critical cell surfaces. In nature, sialic acid-containing oligosaccharides and glycoconjugates are primarily formed through sialyltransferase-catalyzed processes that transfer sialic acid from its activated sugar nucleotide, cytidine 5′-monophosphate-sialic acid (CMP-Sia), to suitable acceptors although *trans*-sialidases have also been used by parasites and bacteria. In sialoglycans or α-sialosides, sialic acids are linked to other sugars such as galactose (Gal), *N*-acetylgalactosamine (GalNAc), through coupling an oxocarbenium ion generated from an active sialyl donor with suitable gycosyl acceptors. A number of significant research effort has been performed to address the challenges associated with the structural complexity of sialic acid to establish simple and efficient sialylation techniques.[4,7] Based on several important parameters including leaving group, electronic, conformational, and steric considerations, a number of sialic acid donors are well explored to improve the required stereo- and/or regioselectivity of sialylation reactions. This chapter presents a concise and recent progress in the stereoselective synthesis of diverse α-sialosides with the aid of various sialic acid-containing donors in chemical and enzymatic glycosylation reactions. Also, a more focused study is considered for the synthesis of protected sialic acid donors, their opportunities in sialylations, and the challenges associated with their growing application in stereoselective glycosylation.

6.2 Some notable characteristics of sialic acids

Despite identifying and isolating sialic acid separately, Klenk and Blix were both widely acknowledged as the sialic acid discoverers. Ernst Klenk extracted the neuraminic acid derivative from brain glycolipids, however, Gunnar Blix extracted sialic acid from submaxillary mucin (sialos: saliva in Greek).[17] Sialic acids are mostly found in both human and avian tissues as lipooligosaccharides, lipopolysaccharides, and as the glycan component of

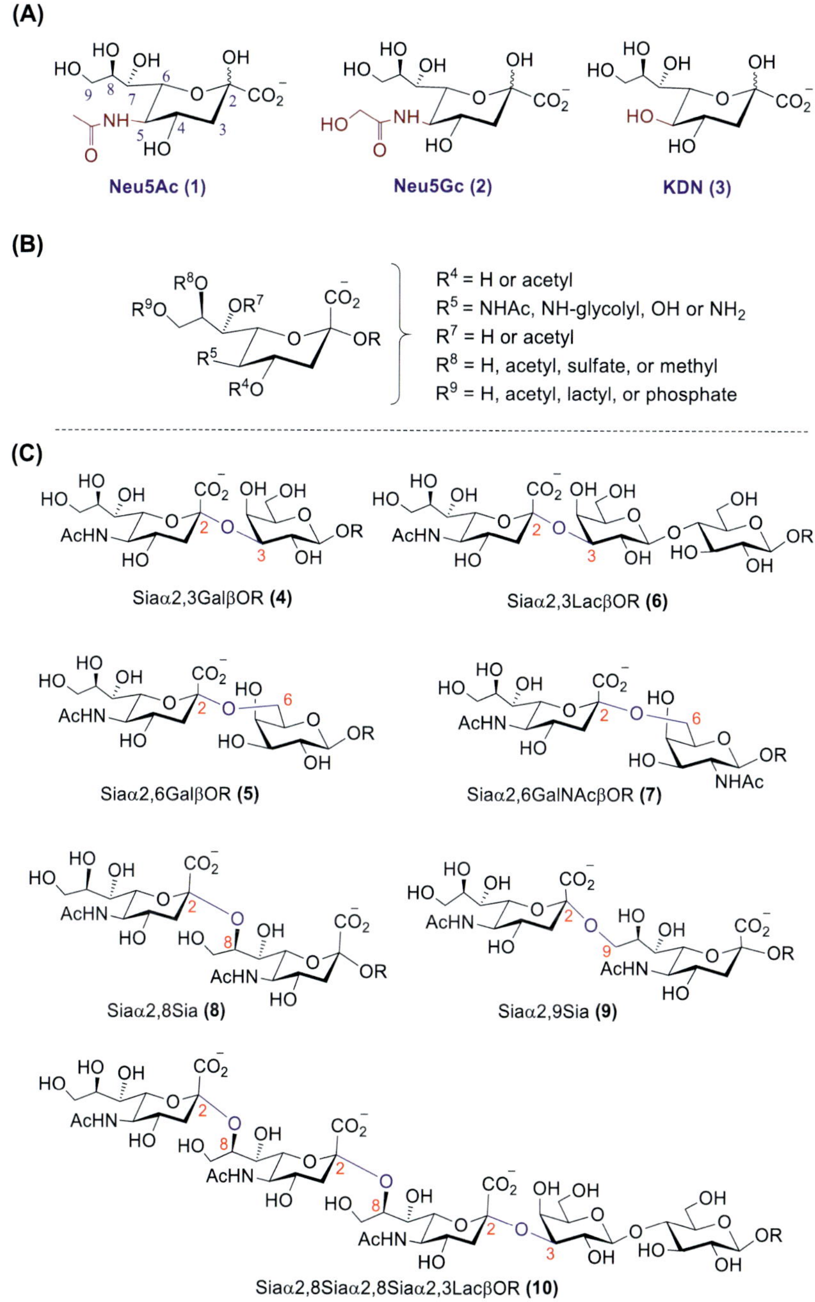

FIGURE 6.1 (A) Most common analogs of sialic acids; (B) modification at various positions leading to >50 different sialic acid-containing glycans in nature; (C) common glycosidic linkages in sialosides (**4–10**) (e.g., Sia-α2,3- or Sia-α2,6- or Sia-α2,8 or Sia-α2,9) in nature.

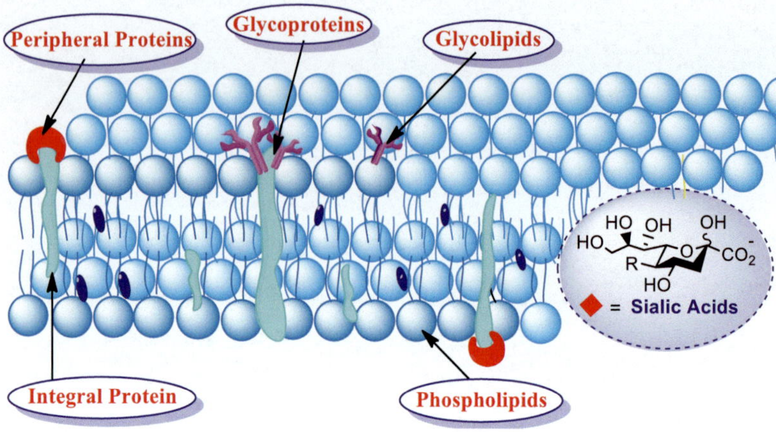

FIGURE 6.2 Schematic presentation of sialic acids as terminal sugars in the lipid bilayer.

glycoproteins. They majorly exist in wide glycosidic linkages, such as α-2,3 and α,2,6 to galactose (or lactose residue), as well as α-2,8- and α-2,9-linkages with other homosia molecules of Neu5Ac.

The most fascinating feature of sialic acid is its position at the terminal end in the lipid bilayer. Sialic acids and their derivatives are referred to as "Terminal Sugar" due to their tendency to attach at the distal end of glycan chains. In turn, being at the "outermost" position allows the molecule to participate in various cellular recognition and interaction with the other cells (Fig. 6.2).[2,18,19]

Another important characteristic of sialic acids has their notable structural complexity. Surprisingly, a wide range of molecular complexity with their functional influence is another notable factor of sialic acid and its analogs.[19] The most remarkable and distinct analogs of sialic acids include *N*-acetylneuraminic acid (**1**, Neu5Ac, *N*-acetyl-5-amino-3,5-dideoxy-D-glycero-D-galacto-2-nonilosonic acid), *N*-glycolylneuraminic acid (**2**, Neu5Gc, having a hydroxyl group in place of the hydrogen at the *N*-acetyl group of Neu5Ac), and deaminoneuraminc acid (**3**, KDN, 3-deoxy-D-glycero-D-galacto-2-nonilosonic acid, having a hydroxyl group in place of the *N*-acetamido group at the C-5 position of Neu5Ac) as the basic parent scaffold found abundantly in nature (Fig. 6.1A). The aforesaid analogs bearing hydroxyl groups positioned at C-4, C-7, C-8, and C-9 are able to further undergo easy additional substitution reactions including *O*-acetylation and *O*-methylation, although less frequently can also proceed for *O*-lactylation, *O*-sulfation, and *O*-phosphorylation.[2,19]

In the case of the structural functionality, some of the analogs perhaps contain other functional groups and diverse substituents instead of the alcoholic –OH groups that are most commonly present in the parent motif. However, the sia molecules majorly attach to the nonreducing end of carbohydrate chains of glycolipids and/or glycoproteins as acetal functionality.

These motifs are majorly seen in the terminal carbohydrate units of capsular polysaccharides, or lipooligosaccharides of pathogenic bacteria and glycoproteins, glycolipids of vertebrates.[2,19] Besides, sialic acids found in vertebrates foremost appear as (Sia-α2,8Sia)n homo polymers or, sometimes as linked "sia" molecules in the outermost units of glycan, for example, Sia-α2,3-Galβ OR (**4**), Sia-α2,3-LacβOR (**5**), Sia-α2,6-GalβOR (**6**), Sia-α2,6-GalNAcβOR (**7**), which are in nature quite abundant. Therefore a wide range of sialic acid molecular libraries has been generated due to the enormous possibilities of permutation in their functionality appearance with diverse substitutions (at C-4, C-5, C-7, C-8, and/or C-9 positions) leading to the development of versatile sets of sia analogs (Fig. 6.1B). Particularly, the sialic acid analogs occupy as an internal structural unit with Siaα-2,8-Sia (**8**), Siaα-2,9Sia (**9**), and Siaα2,8Siaα2,8Siaα2,3LacβOR (**10**) or altering α-2,8-/α-2,9-linkages (Fig. 6.1C) in the extracellular capsular polysaccharides and lipopolysaccharides (LPSs) of bacteria.[20] Besides, in the case of the gram-negative bacteria sialic unit comprises the terminal chain in the glycans of bacterial lipooligosaccharides.[21,22] However, several investigations have found that, in human blood cell surfaces, gastrointestinal mucins and saliva protein contain more than 15 analogs of the parent motif. Among those, Neu5Ac and their derivatives, KDN, and *O*-acetylated KDN have been found predominantly in the nature.[23–25]

The discussion has been focused on those factors and their relative functionality, which have influential effects on the biochemical attribution and medicinal impact to find out pharmacologically active entities as the lead candidate for further development. However, with the substituent variation of the parent core molecule, the final molecular glycan architectures attribute and their proposed functionality may vary; besides, the synthetic method or the enzymatic synthesis within the different organisms also affects the biochemical outcomes.

Another substantial characteristic of sialic acids is about their occurrence. Sialic acids are found in some bacteria, viruses, protozoa, and fungal glycoconjugates as well as in animals of the deuterostome lineage that make glycoproteins, glycolipid-like gangliosides, and glycosamino glycan in mammals and lower vertebrates and serve as ligands or receptors for cell–cell communication or host–parasite interaction.[26] In addition to certain bacteria, such as some strains of *Escherichia coli*,[27] *Neisseria meningitides*[28] imparts the virulence factors.[29] Consequently, several viruses, such as the Sindbis virus,[30] the Rous sarcoma,[31] and the vesicular stomatitis,[32] include sialic residue that contains membrane components.

Furthermore, *sialic acids have great biological significance*. In vertebrates, a diverse number and functionality-equipped sialic acids are discovered exhibiting ascendancy over versatile pathophysiological outcomes. Significant functions of sialic acid residues reported till date cover the cell–cell interaction,[33] differentiation,[34] inflammation,[35] infection,[36] fertilization,[37] malignancy,[38] and cell signaling.[39] Besides, the presence of different sialic analogs in human blood and serum, however, Neu5Ac predominates in the portion upto 80%.[40] Although efforts are still on to gain complete knowledge of the biochemical importance but the whole understating is missing due to their incoherent and unusual structural features from the other carbohydrate motifs along with back draws in the synthetic method. To avail this aim a more focused investigation is further required to complete understanding of the chemical and biological significance of sialic acids and their analogs in living system.

In addition to the above-mentioned characteristics of sialic acid and their analogs, the next one is *the biosynthesis of diverse sialosides in eukaryotic and prokaryotic systems*. Naturally, the biosynthesis of sialic acids in different organisms is distinct. However, sialyltransferase (ST)-mediated transformation is a key step in the sialoconjugate biosynthetic pathways in which a sialic acid is conveyed from CMP-sialic acid as an activated donor to other targeted acceptors like galactoside or derivative of Sia-OR. Besides, the deep, persuasive, and long investigation has well documented the biosynthesis of CMP-Neu5Ac (**14**) or their derivatives,[41] which in turn confirmed that genera and subclasses from bacteria to vertebrates have a direct influence over the synthesis procedure of sialic acids.[42]

In the case of eukaryotes cells, de novo synthesis of Neu5Ac with three consecutive steps (**Path A**, Scheme 6.1) dominates and occurs in the cytosol.[43] Whereas, in bacteria, the pathway is very distinct. In the presence of the aldolase enzyme, the Neu5Ac **1** is directly synthesized via condensation of ManNAc**11** with either phosphoenolpyruvate (PEP) or pyruvate (**Path B**, Scheme 6.1).[44,45] However, in bacteria there is no alteration of synthetic steps of the CMP-KDN biosynthetic pathway.[3,46] Besides, the CMP-Neu5Ac **14** in bacteria is most possibly produced by the conversion of the *N*-acetyl group to the *N*-glycolyl group in the presence of CMP-Neu5Ac hydroxylase-mediated oxidative reaction.[47]

Due to advancements in mammalian cell structure, more sophisticated synthetic paths get a chance to develop like *O*-acetylation, *O*-methylation, *O*-sulfation, and *O*-lactylation, which are evitable in the synthesis of sialoconjugates.[48] Metabolism of the synthesized sialic analogs also continues in cells under the activity of specific catalysis enzymes like sialic acid aldolases or *N*-acetylneuraminate lyases (NanAs). They bring aldol cleavage of Neu5Ac (**1**) to form pyruvate and ManNAc (**11**) as the final digested metabolites. Interestingly, sialic acid and analogs were synthesized purposely by nurturing the naturally occurring aldolase-catalyzed asymmetric aldol-type reactions. Indeed, these enzyme-mediated aldol reactions are comparatively promising and deeply studied for years by several research groups. In this typical method, the six-carbon monosaccharides (e.g., D-mannose, ManNAc, and their derivatives) are invested with pyruvic acid in specific pathways depicted in **Path B**, Scheme 6.1.

Chen Glyco Group at UC-Davis widely explored CMP-Neu5Ac (**14**) as a donor for the ST-induced stereoselective and regioselective enzymatic

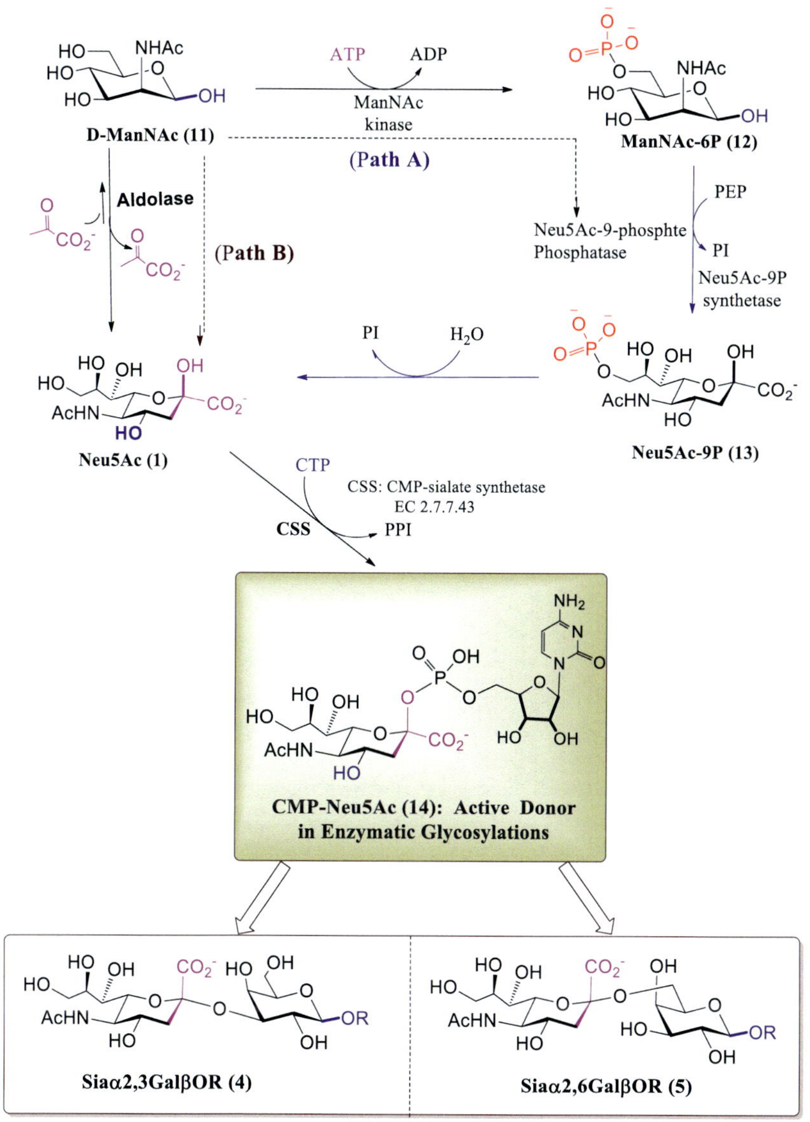

SCHEME 6.1 Biosynthesis of CMP-Neu5Ac (**14**) in eukaryotic and prokaryotic systems and their possible application in enzyme-induced stereoselective sialylations.

sialylation reactions.[49] This enzymatic glycosylation part is presented in detail in a separate section.

Fascinatingly, the enzymatic synthesis of sialylated glycans furnished sialosides in high-to-excellent yields and moreover with almost complete selectivity. Among the different synthesis methods of sialic acids, condensation of a neutral six-carbon D-ManNAc scaffold with the three-carbon pyruvate via enzymatic influence is a notable one in terms of ease of synthesis, reaction yield, wide substrate scope to access molecular diversity in products, selectivity (regioselectivity, chemoselectivity and, stereoselectivity), and more over the well-acceptable environmental benign green conditions. To achieve the targeted sialic acid analogs, a deep knowledge of enzymatic influence and molecular stability is required. One of attention drawing biochemical features of sialic acid is it activates as cytidine mononucleotides, that is, CMP-Sia in vertebrates; however, other monosaccharides are activated in the form of uridine or guanine dinucleotides (e.g., GDP-Man and Fuc, UDP-Glc, Gal, GlcNAc, GalNAc, GlcUA, and Xyl, etc.). Therefore the unusual feature due to their high-energy nucleotide sugar donor form rectifies this "sia" as extraordinary nine-carbon sugars with distinct biochemical functions.[19] Due to its interesting nature, a number of efforts are made around the globe to investigate more about the donor ability of sialic acids utilizing different glycosylation routes.

6.3 Common approaches in chemical sialylation

Sialic acids and their analogs as free sia, such as *O*-glycosides, *S*-glycosides, *C*-glycosides, heterosaccharides, glycoproteins, membrane glycosaccharide chains, and other related molecular architectures play a diverse and pivotal role in the biological systems.[50] One of the hardest parts is the investigation of the specific role played by sialic acid and its analogs in the biological systems.[51] The gap between the synthesis of sialylated glycans and their possible utilities in chemical biology is now narrowing in the course of the emergence of convenient synthetic methods. More such analogs with diverse functional groups can be developed in high yield by utilizing stereoselective chemical and enzymatic glycosylation methodologies.

6.3.1 Chemical synthesis and glycosylation

A number of chemical sialylation reports have been demonstrated to develop desired sialic acid-linked glycans.[52,53] However, the chemical synthesis of α-sialosides is still challenging for the researcher. Various synthons were attempted to nullify the side reactions (e.g., hydrolysis or elimination) in chemical glycosylation so that the α-anomer can be obtained as the sole glycosylation product (Scheme 6.2).[20,52–58] The sialylation of glycosyl donor **15** is believed to proceed via oxacarbenium ion **I**, which then underwent nucleophilic addition of acceptor **16** afforded desired sialylation product **17**.

The anomeric selectivity can be controlled in a variety of ways like the nature of leaving group, the promoter used, and the group present in donor/accepters. Due to the electron-withdrawing nature of the carboxylate group at *C*-1 and acetamido group at *C*-5 (for Neu5Ac, **1**), the positive charge of the anomeric carbon may get destabilization and thus also affect the

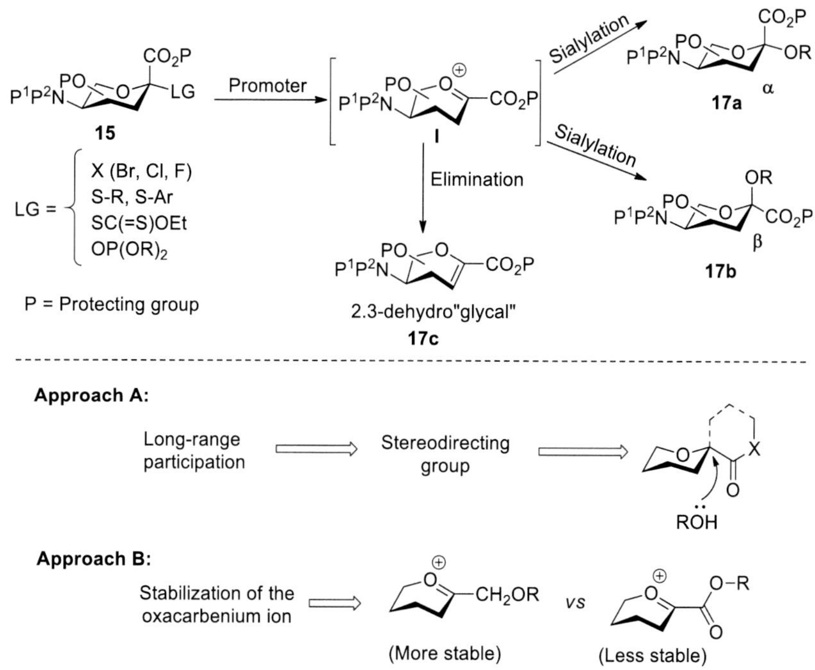

SCHEME 6.2 Mechanistic aspects of stereoselective chemical glycosylation reaction and common stereo-directing groups at C-1.

reactivity of the donor and as a consequence the anomeric selectivity in sialylation reaction.

Regarding modifications at the *C*-1 position of sailic acid, a number of groups have focused their investigation and reported versatile approaches for the nucleophilic addition to the α-face of oxacarbenium ion **I** to the successful installation of desired stereo-directing groups at the anomeric position. The methods may follow either through introducing a participating group at anomeric position C-1 (**Approach A**, Scheme 6.2) or through stabilizing the resulting oxacarbenium ion simply via reduction of the carboxylate group (**Approach B**, Scheme 6.2).

However, such sialylation reactions need further improvement to be utilized for the exclusive synthesis of desired α-sialosides. The chemical structure of some selected common sialyl donors (**18**) explored in chemical glycosylation (e.g., sialylation) with various acceptors is summarized in Scheme 6.3.[4]

Takahashi and coworker have utilized 2-methylthioethyl ester group at C-1 of sialic acid donor **21**, which on chemical glycosylation with acceptor, for example, 2-[2-(2-azidoethoxy)ethoxy]ethanol **22** using *N*-iodosuccinimide (NIS)/triflic acid (TfOH) promoter in acetonitrile afforded the respective

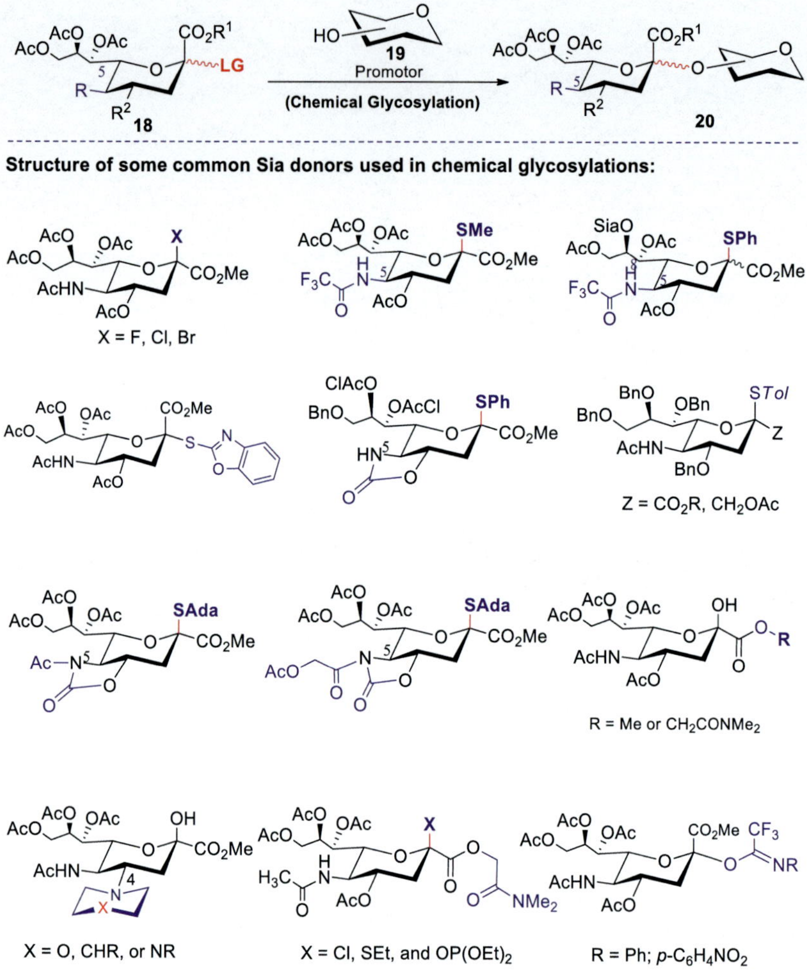

SCHEME 6.3 Some selected common sialic acid donors used for chemical glycosylation.

glycosylation product **23** in 50% yield (selectivity: α/β:4/1) (Scheme 6.4). A similar glycosylation reaction when performed in 1,2-dimethoxyethane (DME) yielded the desired glycosylation product **23** with improved α-stereoselectivity (α/β:20/1). Further, the authors examined the α- and β-selectivity of glycosidation reaction with primary and secondary alcohols of protected galactopyranoside. Glycosidation of donor **21** with free 3-OH and 6-OH of galactopyranoside under optimized conditions afforded corresponding lower yield of product **24** (21%–35%, α/β:5/1) and **25** (39%–51%, α/β:5/1).[59] Thus the method could not be fully generalized in terms of selectivity and yield.

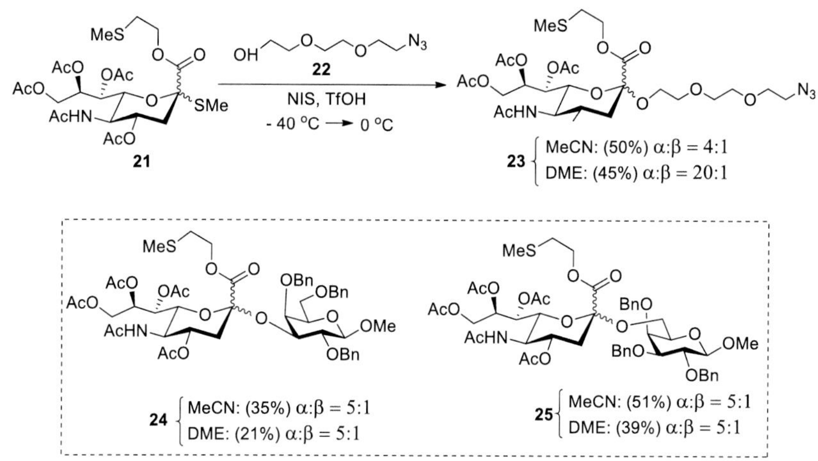

SCHEME 6.4 2-Methylthioethyl ester donor **21** for α-stereoselective glycosylation.

Wong et al. converted the sialyl donors to functionalized hydroxymethyl groups and explored the stereo-directing glycosylation under optimized conditions to give a high yield of β-product.[60] The authors suggested that the sialylation reactions performed in solvents such as ether, toluene, and dichloromethane (DCM) produced even more β-anomer than those carried out in acetonitrile. Sialylation with different promoters was also investigated. Promoters such as MeOTf and PhSOTf show reluctant to activate sialyl donors, while the use of dimethyl(methylthio)sulfonium trifluoro methanesulfonate (DMTST) or NIS and TfOH improve the α-selectivity of sialylated product.

Another more useful consideration to attain desired selectivity and reactivity in anomeric sialylation reaction is greatly influenced through positional variation or modifications at the C-2 position, that is, leaving groups concept. This approach is in general widely explored using halides, thiosialosides, trifluoroacetimidates, etc. as leaving groups for the stereoselective *O*-sialylation reactions. Sialyl chloride was first utilized by Kuhn et al. in 1966 as a suitable donor[61] and soon after became a popular donor system for access to *O*-sialosides.[52,62–66] Other halide systems like 2-bromo-[67] and 2-fluoro analogs[68–70] were also known but they are less effective than 2-chloro sia donor. The classical Ag(I)-induced activation (Koenigs–Knorr approach) of sialyl halide in glycosylation reaction proceeded with a faster rate and afforded product in high selectivities. Thus sialyl chloride (**26**, donor) on glycosylation with diol **27** (acceptor) using silver carbonate (as a promoter) in DCM afforded respective α(2→6)-glycoside **28** with high selectivity (Scheme 6.5).[71] Further investigation of C2-modified (chloro) sialyl donors were well documented for stereoselective construction of S-glycosyl xanthates, sialyl xanthate, thioglycosides, and amide-linked oligomers as

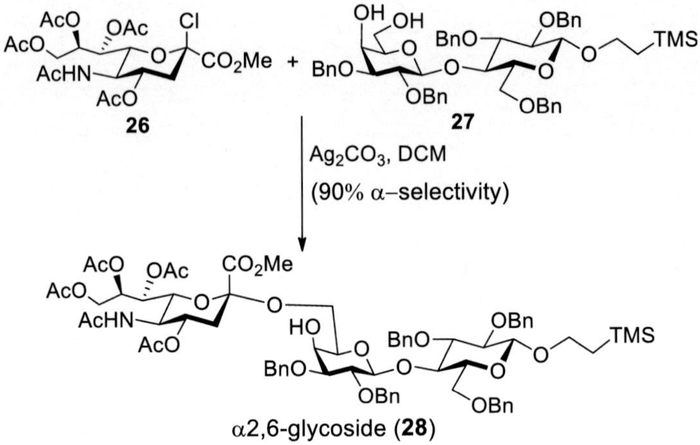

SCHEME 6.5 Glycosyl chloride **26** as a useful donor in stereoselective sialylation reactions.

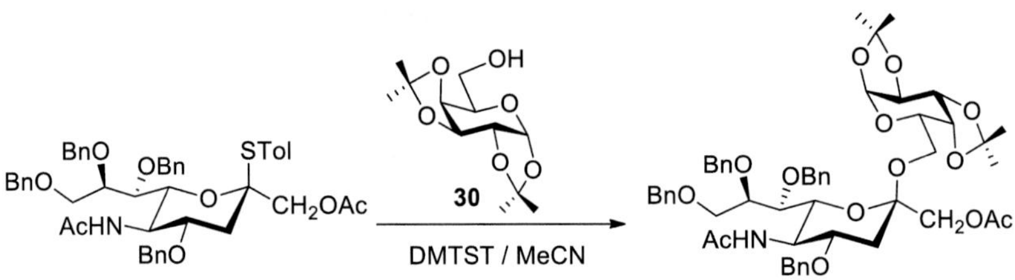

29 **31** (>90%) β:α > 10:1

SCHEME 6.6 A representative example of sialylations using thiosialoside **29**.

well as selective inhibitors of both influenza virus protein Hemagglutinin and Nuraminidase.[72–77] In addition to the donor, sometimes the choice of solvent used also controls the stereoselectivity, for example, similar glycosylation in DCM or 1,2-dichloroethane (DCE) resulted in α-stereoselectivity than acetonitrile.[78] Hg(II) salts (Helferich method) can also be used for a high yield of glycosylation.[55,71]

Thiosialosides, mainly due to their high stability, long shelf life, and also ease of its preparation, were well explored as suitable donors in the glycosylation reactions. A number of alkyl- and aryl- thioglycosides such as methyl-,[79–82] ethyl-,[74,83] and phenyl-thioglycosides[66,73] had displayed notable performance as the sialyldonors[52–57] in the desired sialylation reactions. DMTST or NIS/TfOH are the standard promoters used for the activation of these thioglycosides (Scheme 6.6).[84,85]

The selectivity of sialylation reaction in nitrile solvents was significantly increased by reacting C5-azido sialyl donors with the glycosyl acceptors comprising the primary hydroxyl groups.[86] In nitrile solvent, a more

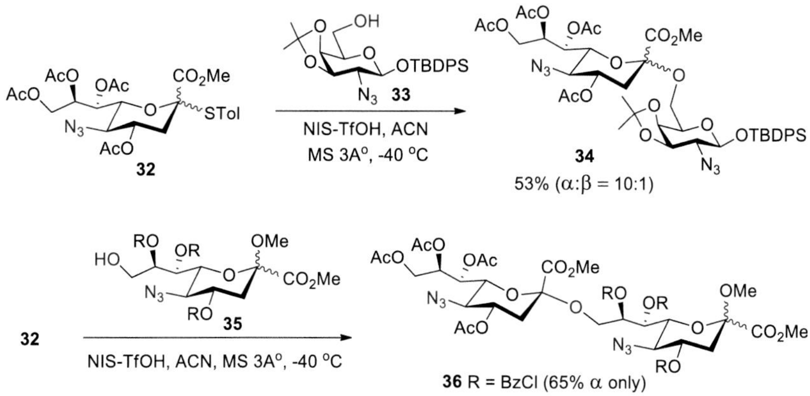

SCHEME 6.7 NIS-TfOH promoted sialylation using a C5-modified (azido) sialyl donor **32**.

stable acetonitrile intermediate was formed and reduced steric hindrance produced in the α-face by linear azido group likely to be responsible for higher α-selectivity. However, a considerable amount of acyl migration (from acceptor) and elimination (from donor) products were observed by performing the sialylation reaction in DCM solvent. In this particular example, Wong and coworker demonstrated that the α-selective sialylation reaction of C5-azido donor **32** with C6 hydroxyl of GalN **33** and C9 hydroxyl of NeuAc **35** produced $\alpha(2\rightarrow6)$-linked sialoside **34** and NeuAc$\alpha(2\rightarrow9)$NeuAcsioloside **36** in good yield (Scheme 6.7).

Likewise, Boons et al. demonstrated the efficacy of C5-modified *N*-trifluoroacetyl (*N*-TFA) methylthio glycoside sialyl donor **37** in stereoselective sialylation.[87] For example, the NIS/TfOH-mediated sialylation of thioglycoside **37** with lactosyl acceptor **38** under optimized conditions gave trisaccharide **39** in 84% yield as single isomer, that is, α-sialoside (Scheme 6.8). The conventional NMR spectroscopy for the assignment of anomeric configurations of sialoconjugates may not be applicable to TFA-protected derivatives. Therefore after sialylation, the TFA group can be converted to an acetyl group by hydrolyzing the sialosides in 1 M aqueous NaOH followed by acetylation.

Ando et al. greatly explored the potential of an *N*-Troc-protected sialic acid donor **40** on α-selectivity.[88,89] Usually, *N*-Troc-donors produce α-sialosides in high yields upon coupling with different glycosyl acceptors. The coupling of *N*-Troc donor **40** with Gal acceptor **41** in EtCN produces α-selective disaccharide **42** in 52% yields (Scheme 6.9). Compound **42** could be separated by simple crystallization using (*n*-hexane/EtOAc). Considering this advantage, **42** has been extensively employed for the construction of oligosaccharides containing the Neu-α(2,3) Gal structure.[90]

To control the stereoselectivity of glycosylation reaction, Demchenko et al. developed hydrogen bond-mediated aglycone delivery (HAD) using a picoloyl group.[91] *N*-Iodosuccinimide and catalytic triflic acid-mediated activation of

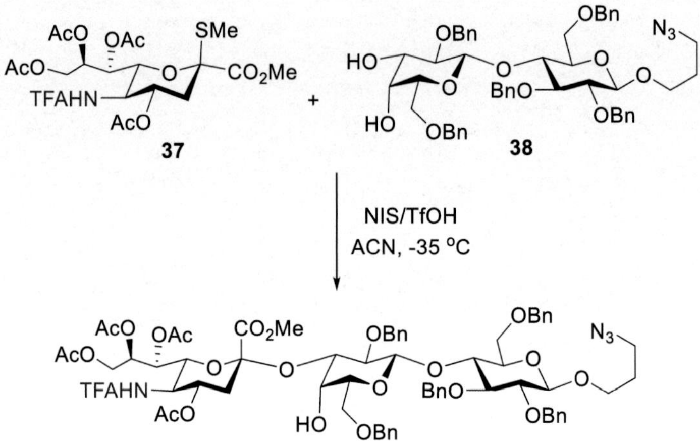

39 (84 %, α-only)

SCHEME 6.8 NIS promoted sialylation using an *N*-TFA sialyl donor **37**.

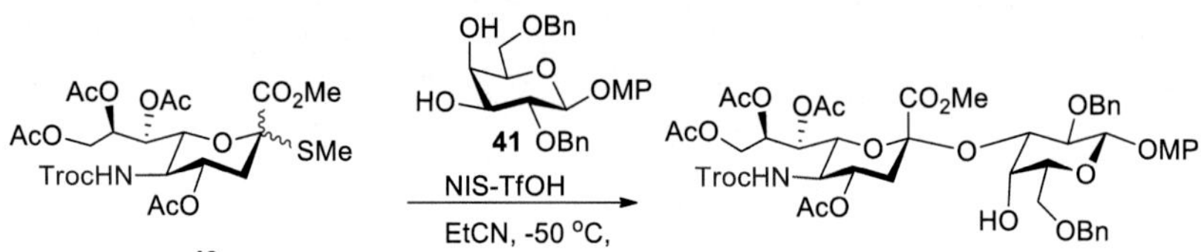

40

42 (52 %, α only)
(Isolated by crystallization)

SCHEME 6.9 NIS-mediated sialylation using *N*-trocsialyl donor **40**.

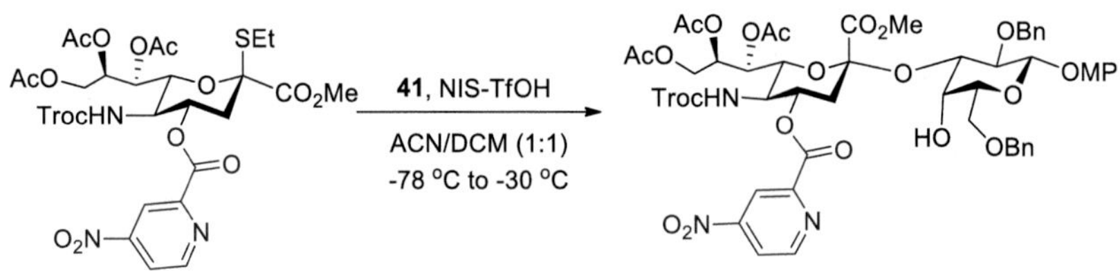

43

44 (56%, α:β=15:1)

SCHEME 6.10 Sialylation using 4-nitropicoloyl-protected donor **43**.

ethyl thioglycoside sialyl donor **43** bearing 4-nitropicoloyl moiety as the stereo-directing group was explored by Liu and coworkers.[92] In this reaction, the coupling of **43** with galactosyl acceptor **41** afforded disaccharide **44** (yield: 56%, α:β = 15:1); however, a lower yield was observed for the corresponding C4-*O*-picoloyl donor (Scheme 6.10). A protected trisaccharide present in the repeating unit of serotype Ia Group B *Streptococcus* capsular polysaccharide was effectively synthesized using the approach.

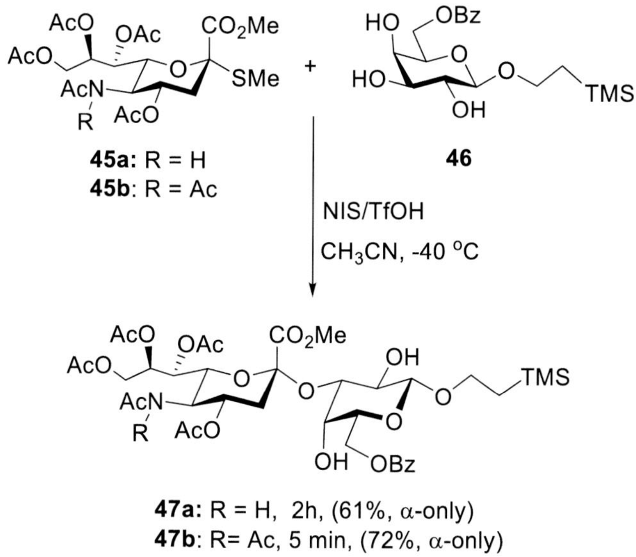

SCHEME 6.11 Stereoselective sialylation using methylthio sialyl donor **45a,b**.

Furthermore, Hasegawa and coworker fruitfully explored 2-thiomethyl glycosides of Neu5Ac as suitable donors in chemical sialylation.[93,94] Per-*O*-acetylated 2-methylthio-Neu5Ac **45a** on glycosylation with 2(trimethylsilyl)ethyl 6-*O*-benzoyl-*β*-D-galactopyranosyl acceptor **46** using NIS/TfOH promoter for activation in acetonitrile at −40°C for 2 hours and afforded α(2→3)-linked disaccharide **47a** in 61% yield with α-selectivity (Scheme 6.11).[95] DMTST can also be used as a promoter in this reaction, although the reaction yield was poor (**47a**, 52%, α-only). Interestingly, a 2-thioglycosyl donor of Neu5Ac having an additional *N*-acetylacetamido group present at *C*-5 (donor **47b**) was found to be considerably even more reactive than mono-*N*-acetylated thioglycosyl donor of Neu5Ac (donor **45a**).[96]

The stereoselective sialylation of oxazolidinone protected β-phenylthiosialyl donor **48** with free 3-OH galactopyranoside **49** in the presence of NIS/TfOH in anhydrous CH_3CN at −40°C gets complicated. Crich and coworkers explored adamantylthio donors in sialylation reaction.[97] The method has some merit as it is more reactive than phenylthio and functional residues can be incorporated with no nasty odor. For example, *β*-adamantyl thiosialyl donor **50** in glycosylation with acceptor **49** afforded the desired 2,3-linked disaccharide **51** in 87% yield with high α-selectivity (α/β = 10/1) (Scheme 6.12).[97]

Yamaguchi et al. demonstrated the stereoselective α-sialylation reactions with the use of silylene/oxazolidinone double-locked sialic acid building block in DCM.[98] Phenylthiosialoside protected by 4,5-oxazolidinone was used to produce such building blocks. The double-locked building blocks **52** and **54** were protected with *N*5,*O*7-di-tert-butylsilylene (DTBS) of 4*O*,5*N*-oxazolidinones, in

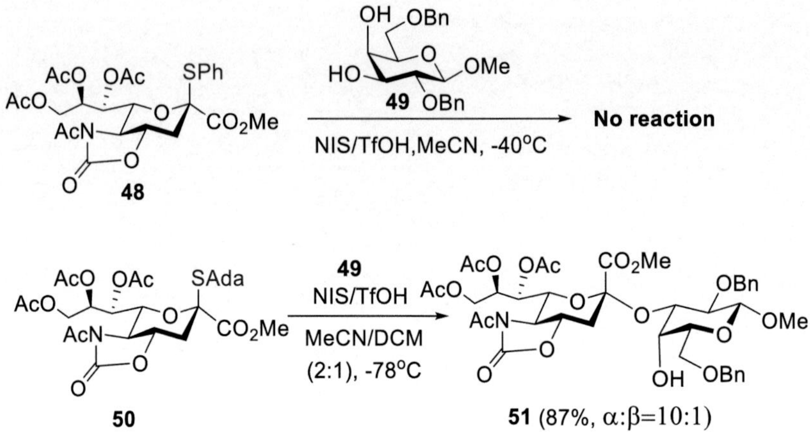

SCHEME 6.12 Sialylation reaction using *S*-adamantyl group in donor **50**.

which *O*8/*O*9 positions were protected with DTBS and tetraisopropyl disiloxanylidene (TIPDS), respectively. These "double-locked" donors furnished the desire coupling products **53** and **55** in good yields with complete α-selectivity in sialylation reactions against galactosyl acceptor **41** (Scheme 6.13). These sialyl donors **52** and **54** displayed higher α-selectivity than the "single-locked" donor **56**. Later, this methodology was greatly utilized for the synthesis of starfish ganglioside AG2 pentasaccharides.[99]

In addition to thio groups (e.g., SMe, SPh, *O*-Tol, etc.) as routine donors, Marra and coworkers explored anomeric xanthates in glycosylation reaction.[73] These xanthate donors were obtained in good yields from sialyl chloride by treating it with potassium ethoxydithiocarbonate in EtOH.[73] Like the thioglycosides, the common promoter can also be used for the activation of the xanthate group in stereoselective glycosylation reactions.[100] In addition to other factors like leaving groups, the choice of promoters (e.g., DMTST[101] or NIS/TfOH,[79,102,103] or phenylsulfenyl trifluoromethanesulfonate (PhSOTf) or 2,6-di (tert butyl)pyridine (DTBP) at low temperature) also affect the yield and selectivity in sialylation. De Meo and coworkers exemplified thioimidates (e.g., S-benzoxazolyl known as SBox group) as leaving the group in sialylation reactions.[104] Thioimidate acceptor has advantageous selective activation over thioglycosyl, for example, ethylthio leaving group (Scheme 6.14). Thus AgOTf-mediated glycosylation of galactopyranoside **59** with per-*O*-acetylated 2-*S*-benzoxazolyl Neu5Ac donor **58** afforded high yield (89%) of respective 2,6-linked disaccharide **60**. Subsequently, disaccharide **60** was subjected to further glycosylation with galactopyranoside **59** in the presence of NIS/TfOH to produce a good yield of desired β-2,4-linked GM3 analogs.[104]

Ando et al. reported a fully stereoselective sialylation approach by employing constrained macrobicyclic sialyl donors as the synthetic equivalent of bridgehead oxonium cations.[105] NIS-mediated glycosylation of constrained macrobicyclic sialyl donor **61** under optimized conditions afforded corresponding sialosides **62**

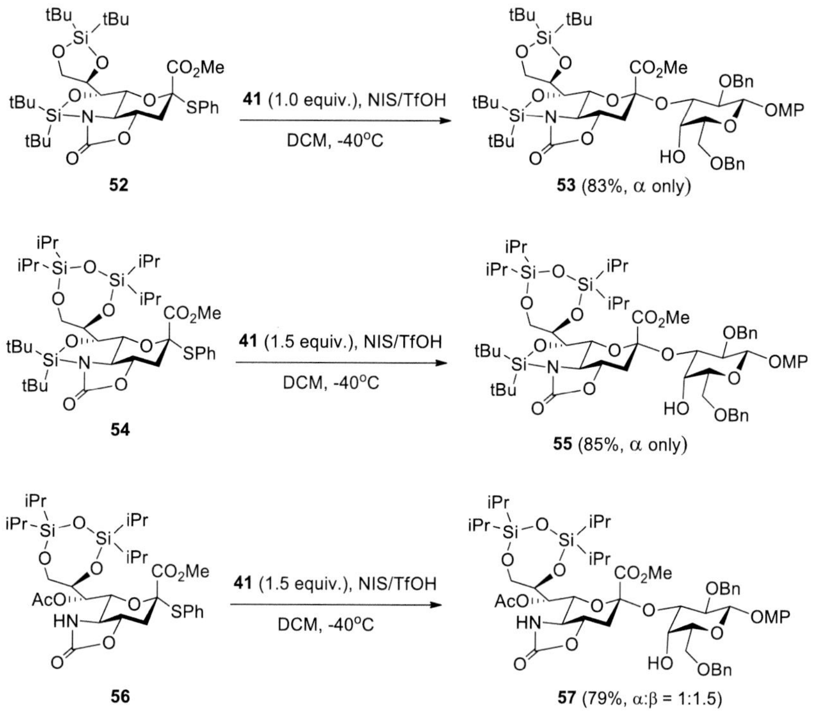

SCHEME 6.13 NIS-mediated sialylations using double-locked sialyl donors **52**, **54**, and **56**.

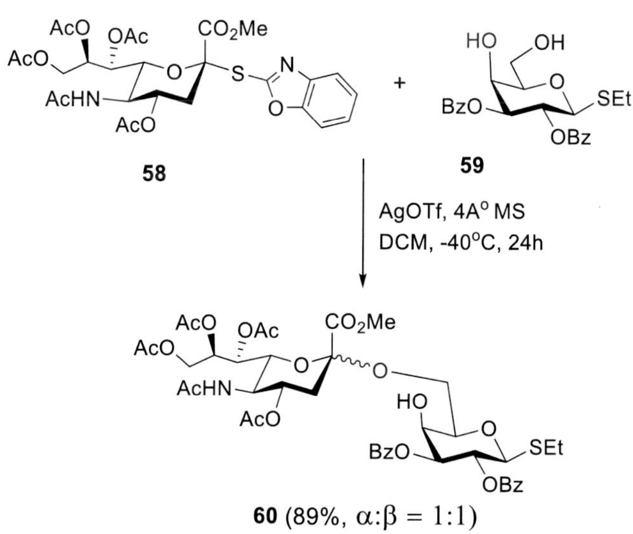

SCHEME 6.14 Selective sialylation using thioimidoyl as a suitable sialyl donor **58**.

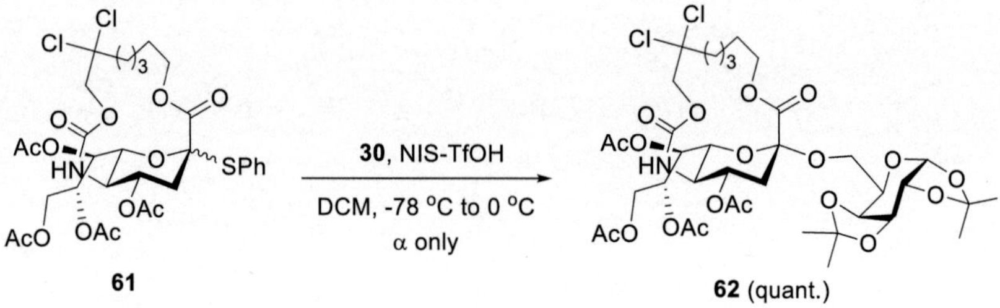

SCHEME 6.15 Sialylation using constrained macrobicyclic sialyl donor **61** to enable α-glycosides **62**.

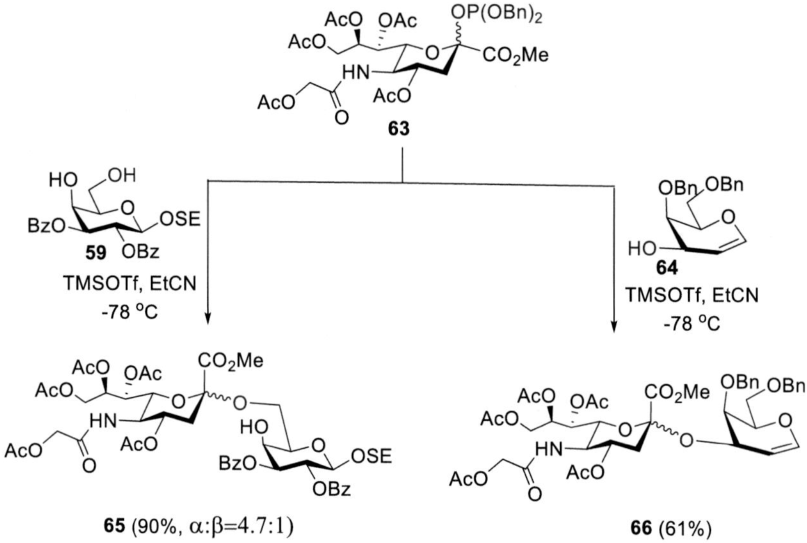

SCHEME 6.16 Sialylation reaction using phosphite donor **63**.

in almost quantitative yield with complete α-selectivity (Scheme 6.15). Additionally, this sialylation technique is effectively utilized for the synthesis of gangliosides and α-2,8-linked sialic acid pentamers.[106]

Through these representative examples, this is clear that the glycosylation reaction with thioglycoside donor of sialic acids has been quite useful to furnish the desired *O*-glycosides in good yields with notable anomeric selectivity. However, when we compare the reactivity with phosphite donors of sialic acids, in general, the latter is considered to be even more reactive than corresponding thioglycoside donors in chemical glycosylation. Thus the phosphite donor of sialic acids can be preferentially used for the development of their respective Sia-*O*-glycoside. For example, donor **63** on glycosylation separately with acceptor **59** or **64** in propionitrile at −78°C afforded high yields of Neu5Gc-glycosides **65** and **66**, respectively (Scheme 6.16).[107]

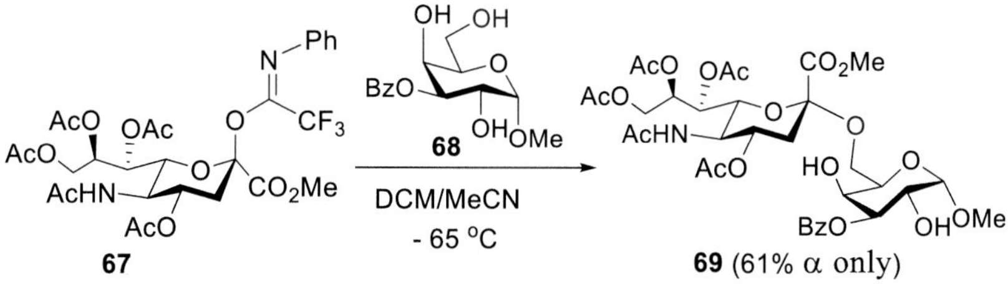

SCHEME 6.17 *N*-Phenyltrifluoroacetimidate **67** in stereoselective sialylation.

Cai and coworkers in early 2000 used *N*-phenyltrifluoroacetimidates as an oxygen-based leaving group, that is, the *O*-sialosidation approach for the anomeric sialation reaction.[108] Thus sia donor **52** is obtained from per-*O*-acetylated 2-hydroxy-Neu5Ac by reacting with 2,2,2-trifluoro-*N*-phenylacetimidoyl chloride in the presence of K_2CO_3.[108] Interestingly, imidate **67** on glycosylation with 1,2,3,4-di-*O*-isopropylidene-α-D-galactopyranosyl via TMSOTf prompted activation in anhydrous dichloromethane resulted to the respective disaccharide with β-selectivity (α/β: 1/2.4) and that selectivity can be switched to α-selectivity (α/β: 1.6/1) once the reaction is performed in CH_3CN. Thus the choice of solvent affects the leaving group ability of the donor molecule. However, glycosylation of donor **67** with methyl 3-*O*-benzoyl-α-D-galactopyranosyl acceptor **68** afforded product **69** in 61% yields with complete α-selectivity (Scheme 6.17).

Further, Ito et al. introduced an auxiliary group particularly at the C3 position featuring C3-SePh and C3-SPh substitution, respectively, to develop sialyl donors for anticipating its participation from the β-face of the anomeric center as an α-stereo-directing group.[109,110] The sialylation of **70** with galactose acceptor **30** in the presence of AgOTf-$SnCl_2$ was α-selective and produced Neu-α(2,6)Gal **72** in 46% yield (Scheme 6.18). C3-SPh donor **71** was efficiently employed to address the formation of 2,3-ene selenides as the major side product. However, the C2-bromo sialyl donor **71** having SPh group at C3 coupled with **30** in the presence of $Hg(CN)_2$-$HgBr_2$ to delivered selectively α-sialoside **73** in 72% yield. After sialylation, the C3-SPh moiety can be selectively removed by reduction with Ph_3SnH/AIBN to achieve the natural C3-deoxy structure of sialosides.

Similarly, Schmidt and coworkers investigated the anchimeric assistance of C3-modified sialyl donors with bromo and phenoxythiocarbonyloxy group in stereoselective sialylation reactions.[111] Silver triflate promoted coupling of donor **75** having C3-phenoxythiocarbonyloxy moiety as the assisting group with lactosyl acceptor **74** afforded trisaccharide **77** in 89% yield with α-selectivity (Scheme 6.19). C3-Phenoxythiocarbonyloxy donors usually form a stable five-membered oxathiolanium intermediate, which leads to direct the nucleophilic attack of acceptor from α-face. In addition, C3-Br phosphite donor **76** was also found to be α-selective and proceeds through bromonium ion intermediates.

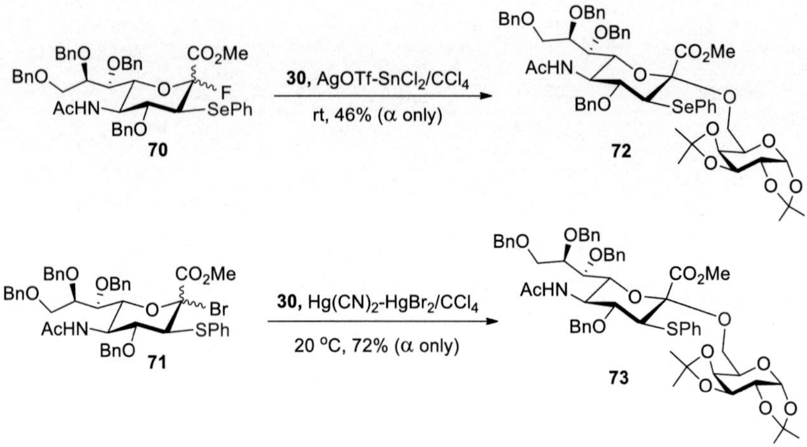

SCHEME 6.18 Sialylation reaction leading to sialosides **72** and **73** using C3-modified (C3-SePh and C3-SPh) sialyl donors **70** and **71**.

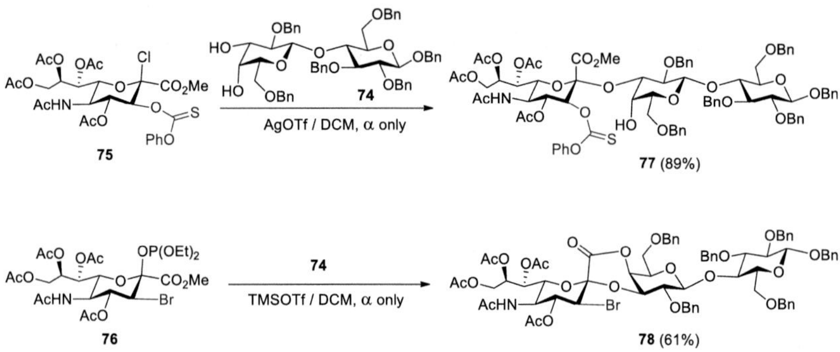

SCHEME 6.19 Stereoselective sialylation reactions using C3-modified thiobenzoyloxy **75** and C3-modified bromo **76** donors.

In other aspects, the biological role of sialomimetic conjugates has also been considered for the effective utilization of this scaffold in drug discovery and development including *Vibrio cholera* sialidase.[112] The authors utilized glycosylation to deliver 6-*C*-methyl-Neu2en5Ac **82a**, 6-*C*-(hydroxymethyl)-Neu2en5Ac **82b**, 6-*C*-methyl-Neu5Ac **84a**, 4-methyl umbellyferyl glycosides **83a,b** starting from peracetates **79a,b** (Scheme 6.20).[113] 4-Methylumbelliferyl glycosides **80a,b** and olefines **81a,b** were obtained from respective peracetates **79a,b** via silver carbonate promoted activation of glycosyl chloride followed by rapid sialylation in the presence of tetrabutylammonium salt of methylumbelliferone (Scheme 6.20).[113] Further deacetylation of resulting peracetates using NaOMe/MeOH followed by saponification of methyl ester afforded respective methylumbelliferyl glycosides **83a,b**.

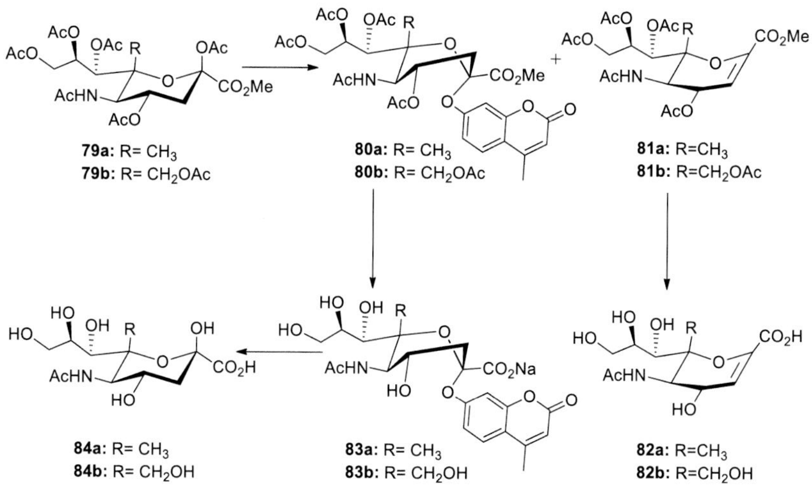

SCHEME 6.20 Ag_2CO_3 promoted sialylation for the 6-*C*-methyl (**84a**) and 6-*C*-(hydroxymethyl) analog (**84b**) of sialic acid and branched-chain Neu2enSAc analogs (**82a,b**).

The enzyme inhibition data suggested weak inhibitors of *V. cholera* sialidase for the branched-chain **82a,b**, whereas the sialosides **83a,b** were identified as very poor substrates.[114]

The stereoselective synthesis of different sialosides and their applications in the preparation of gangliosides and other sialoglycans is ever-increasing.[115] In addition to the above-mentioned representative examples, thioalkyl or thioarylsialyl donors system with appropriate substitution such as *N,N*-diacetyl *N*-phthalimide, *N*-trifluoroacetyl, *N*-TCA, *N*-Fmoc, *N*-Troc, or azido group at C-5 position of sialic acids have also been well explored in glycosylation with suitable acceptors. Such a group once introduced at the *C*-5 position in sialyl donors could increase the reactivity of donors and often furnished the desired sialosides with high anomeric selectivity and notable ease. In chemical glycosylation, sialic acid having *C*-9 free hydroxyl group when used as an acceptor with the sialyl chloride as a donor then the reaction is comparatively easy, fast, and straightforward and afforded the respective α-2,9-linked dimeric structure in reasonably good yields.[116]

Also, among these common *N*-protecting groups at the *C*-5 position of sialic acid, some of them can be well utilized (via deprotection followed by standard derivatization of *C*-5 NH_2 group) to furnish a high yield of respective Neu5Gc analogs.[117] Due to the complex structures of sialic acids and also the involvement of a number of synthetic steps including standard protection–deprotection, in general, the chemical method of synthesis is time consuming, expensive, and thus inherently difficult to exemplify in the laboratory and industry. Only a few general methods are reported for easy access to diverse sialic acid-based monosaccharides. For example, Cu(I)-catalyzed stereo-divergent propargylation of unprotected aldoses can afford sialic acids.[118] Even chemical synthesis of simple monosaccharides, for example, KDN involves 10 steps (with overall ~4%) starting from (−)-3-dehydroshikimic acid or (−)-quinic acid,[119] while Neu5Gc was obtained in three steps in 22% overall yield from D-arabinose.[120]

6.3.2 Glycodiversification through enzymatic glycosylation

Natural sugars available on the mammalian cell surface or their structurally modified sialic acid analogs at a specific position of sialic acids including acetylation, methylation, phosphorylation, sulfation, and lactylation have displayed flexible biological significance.[2–5,18] The sulfated analogs found in proteoglycans, glycoproteins, and glycolipids are thought to participate a vital role in explicit molecular recognitions phenomenon. To get a good amount of these structurally modified sialoside analogs, a chemical method including sialylations has some technical issues.

One of the most important reactions in organic chemistry is certainly the carbon–carbon bond-forming methodology. The enzyme-induced stereoselective version of the C–C bond-forming reaction is greatly explored for various purposes.[121,122] A number of aldolases were widely explored during the past many years.[123–129] Toward this end, the sialic acid aldolase-mediated aldol addition of six-carbon monosaccharides including D-mannose, ManNAc, ManNGc, or their derivatives **85** with three-carbon pyruvic acid can easily result in the respective nine-carbon monosaccharides **86** as natural and nonnatural analogs of sialic acid. For example, by utilizing the C2- and/or C6-substituted mannose derivatives as suitable precursors under above-mentioned enzymatic condition, C5 and/or, C9-modifications in sialic acid structures could be easily achieved in high-to-excellent yields (Scheme 6.21A).[130–132]

PmNanA has wide broad substrate scope and higher expression level and is thus known to be a superior biocatalyst led to sialic acid analogs over EcNanA.[134] Chen group reported the structural basis for substrate specificity and mechanism of *P. multocida N*-acetyl neuraminate lyase. The authors determined wild-type PmNanA and its K164A mutant, in the absence of substrates, and furthermore trapped in a Schiff base between Lys164 (**87**) and pyruvate, respectively (Scheme 6.21B).[133] The sialic acid is bind to the active site and each hydroxyl group of the monosaccharides (in open-chain ketone form) may undergo noncovalent interactions (hydrogen bonding) with the enzyme PmNanA and its K164A mutant as well the residues that eventually determine the required specificity.[133]

Naturally occurring 5-methyl sialic acid monosaccharide analogs like 8-*O*-methyl Neu5Ac (**89a**) and 8-*O*-methyl Neu5Gc (**89b**) can be directly obtained in good yields from the respective six-carbon monosaccharide 5-*O*-methyl ManNAc (**88a**) and 5-*O*-methyl ManNGc (**88b**).[135,136] In 2008 a capillary electrophoresis assay was reported by Chen and coworkers to

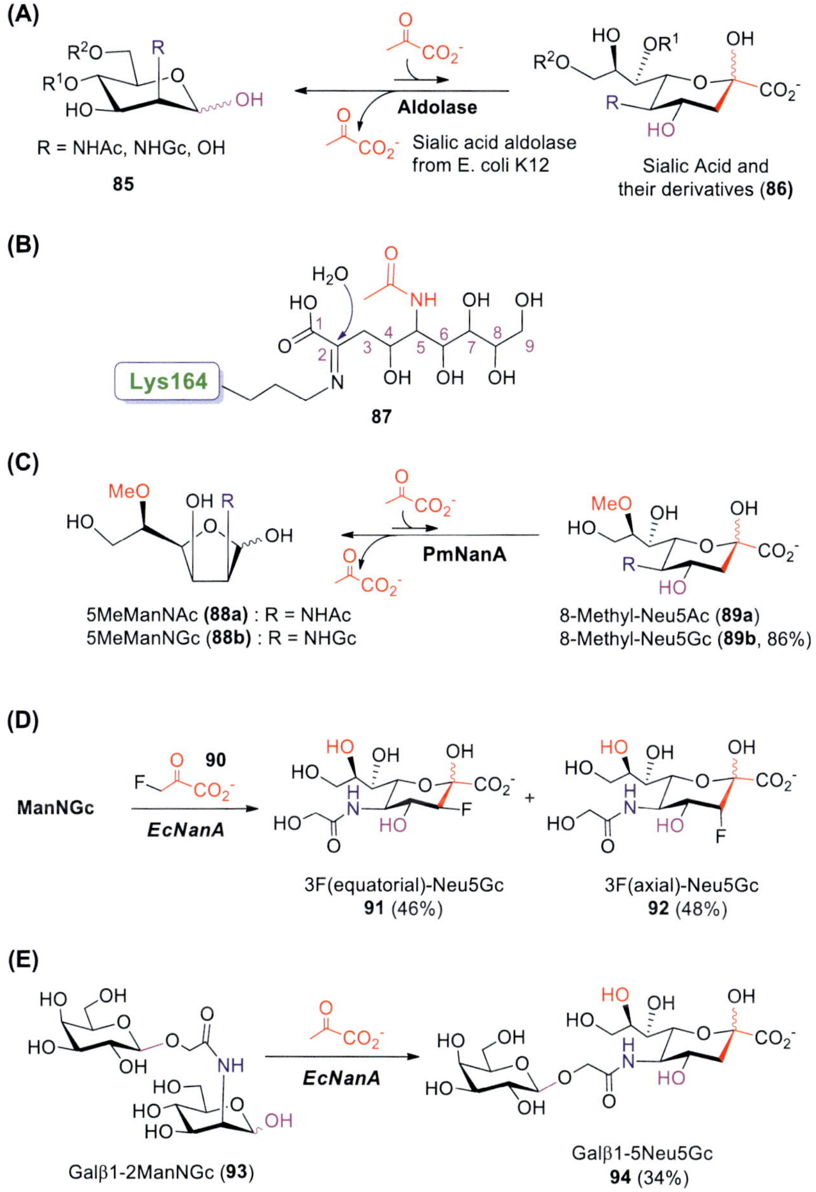

SCHEME 6.21 (A) Aldolase-catalyzed enzymatic synthesis of sialic acid monosaccharides. (B) The structural basis for substrate specificity with the mechanism of PmNanA.[133]

straightforward characterization of the activities of NanAs in both the directions of Neu5Ac cleavage and Neu5Ac synthesis. Thus 5-*O*-methyl-ManNAc **88a** is used as a well-suitable substrate for PmNanA, which on enzymatic reaction with pyruvic acid furnishes the respective Neu5Ac8Me **89a** in high yield (Scheme 6.21C).[135] Likewise, PmNanA-catalyzed reaction of ManNGc5Me (**88b**) using sodium pyruvate (5 equiv.) in Tris–HCl buffer (100 mM, pH 7.5) at 37°C for 24 hours led to the formation of Neu5Gc8Me (**89b**) in 86% yield.[136] The resulting sialic acid analogs can be easily purified in high purity by the combined anion exchange chromatography and gel filtration column.

Although Neu5Gc-containing glycans have been found in human cancer cells and in a range of human tissues, human cells are unable to synthesize Neu5Gc due to its dietary incorporation.[137] Over 20 Neu5Gc analogs have been found in nonhuman vertebrates, and to evaluate their biological roles, a number of reports are available to focus their synthesis.[138] Furthermore, disaccharide with ManNGc at the reducing end, for example, Galβ-1,2-ManGc (**93**), was subjected to EcNanA-catalyzed aldol reaction with pyruvic acid to deliver the respective disaccharide Galβ1–5Neu5Gc (**94**) in 34% yield (Scheme 6.21E).[139]

The incorporation of fluorine atoms in carbohydrate-based molecules to address their impact in medicinal chemistry is well established,[140] and thus the synthesis of fluorine-containing sialic acid monosaccharide is of utmost importance. Interestingly, 3F analogs of Neu5GC could be obtained through the EcNanA-catalyzed aldol reaction of ManNGc and 3-fluoropyruvate **90**. The enzymatic reaction afforded the respective 3F-Neu5GC analogs as a mixture of 3F(axial)Neu5Gc (**91**) and 3F(equatorial) Neu5Gc (**92**) (in ~1:1. ratio), which after flash chromatography easily separated and isolated in the pure form (Scheme 6.21D).[141]

Once, the aldolase-mediated required nine-carbon monosaccharides were obtained from the six-carbon monosaccharides, then issue for their effective utilization in enzymatic sialylation reaction to get the desired α-2,3-, α-2,6-, or α-2,8-linked sialosides. In all-purpose, the substrate or ligand specificity of sialic acid-recognizing proteins may be examined in the following three ways:

(a) The routine modification in negatively charged nine-carbon sialic acid analogs could be considered via aldolase-mediated reaction of six-carbon monosaccharides namely Man, ManNAc, and ManNGc or their functional analogs including *O*-methyl, acetate, azido, deoxy, etc. group at 2-, 4-, 5-, 6-, 2,4-, 2,5-, 2,6-, 4,5-, 4,6-, 2,4,5-, 2,4,6-, 4,5,6-, or 2,4,5,6-position as the donors.

(b) The next important aspect to be considered is the type of glycosidic linkage, that is, α-2,3- or, α-2,6-linked sialic acid with penultimate sugar.

(c) In the end, the structure of the penultimate sugars such as Gal or GalNAc or lactose or their derivatives would be another major criterion needed to evaluate the substrate specificity of the sialic acid-recognizing proteins.[49]

Furthermore, neuraminic acids on reaction in the presence of a second enzyme, that is, NmCSS gave a high yield of respective cytidine 5-monophospho-nonulosonate analogs, for example, cytidine 5′-monophospho-3,7-dideoxy-7-azido-D-glycero-D-galacto-nonulosonic acid, which was recently claimed for the effective treatment of a Neisseria gonorrhoeae infection.[142] At first, the NanAs-mediated aldol addition of ManNAc or their derivatives **95** and pyruvate to form neuraminic acid analogs **96** (**Step 1**). This was then proceeded for the required activation using the second enzyme CMP-sialic acid synthetase to give the desired donor **97** (**Step 2**). At the end of the sequence, this donor in the presence of PmST1 or Nm2,3ST sialyltransferase, a third enzyme is subjected to transfer the donor **97** to the respective acceptors **98** (e.g., galactose, GalNAc, Lac, and their functionalized derivatives) to furnish the high yield of desired α-2,3-linked sialosides **99** (Scheme 6.22A).[49,135–138,143]

Interestingly, these three-step enzymatic reactions were performed in a one-pot fashion without the isolation of the intermediate involved and moreover the desired α-2,3-linked sialoside **99** was obtained in complete regio- and stereoselective manner, that is, α-2,3-selectivity.[49,135–138] Similarly, this method is greatly utilized for another very important class of naturally linked sialosides, that is, α-2,6-linked sialosides (**100**). The method involves the similar three-step aldolase-mediated reaction of ManNAc derivative **95** with pyruvate to form Neu5Ac or their derivatives **96** (**Step-1**) followed by CMP-sialic acid synthetase-mediated activation (**Step-2**) and finally, enzymatic glycosylation of the resulted donor **97** with suitable acceptors **98** (e.g., gal, GalNAc, Lac, and their derivatives) using Pdα-2,6-ST (**Step-3**) (Scheme 6.22B). The devised one-pot multienzyme (OPME) protocol is effective for a diverse range of donors, for example, ManNAc and their modified analogs and the acceptors, for example, gal or lac analogs for the successful delivery of corresponding natural as well as unnatural α-2,3- or α-2,6-linked sialosides.[49,135–138,144]

In the context of virology, Neu5Ac of the host cell has immense significance in cell infection and also in virus survival. Neu5Ac residue can get recognized without difficulty by influenza virus hemagglutinin and binds for host cell captivation with high affinity. Thus for the development of an effective antiinfluenza drug candidate, Neu5Ac can be considered as the most dependable notable scaffold. This way, a pendant α-sialoside **101** was developed and used for the study of protein–carbohydrate binding interaction as a promising viral inhibitor (Fig. 6.3).[139,145] Further, a potent inhibitor of divalent influenza virus hemagglutinin, for example, ethyl β-D-galactoside cored-α-2,3-linked nona saccharide sialoside **102** (Fig. 6.3) was developed through the chemoenzymatic route. The sialoside **102** solely governs the protein–ligand interaction during virus-cell recognition and adhesion phenomena.[146,147]

Similarly, the OPME-mediated enzymatic sialylation reaction of 5-azido ManNAc or their modified derivatives (as donors) and Gal/Lac derivative (as acceptor) furnish excellent yield of corresponding α-2,3- or α-2,6-linked 8-

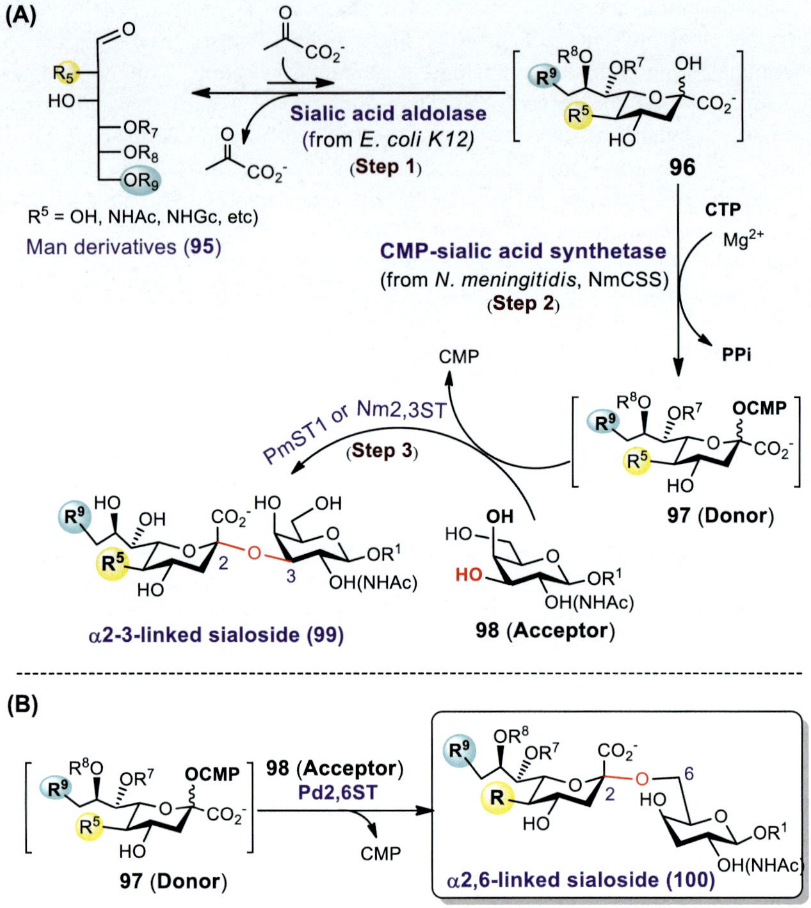

SCHEME 6.22 Chen's one-pot synthesis of α-2,3-linked sialoside **99** and α-2,6-linked sialoside **100**.

azido sialic acid functional analogs. Likewise, 5-*O*-modified Man **103a** or ManNAc **103b** on OPME induce sialylation with acceptor β-LacProN$_3$ **104** gave high yields of desired 8-azido or 8-deoxy or 8-*O*-methyl α-2,3-linked sialosides (**105a,b**) or α-2,6-linked sialosides (**106a,b**) (Scheme 6.23).[148] This OPME strategy provides an excellent way to generate a diverse range of *C*8-modified sialic acids-containing glycans of notable application in chemistry, biology, and medicine.

Considering the biological relevance of 3-fluoro-sialoside, the same group further exemplified the similar optimized OPME protocol for easy access to C3-fluorinated α-2,3- and 2,6-linked sialoconjugates.[141] Thus chemotherapeutic potent fluorine group can be easily incorporated at the *C*-3 position of sialic acid via established aldolase-mediated synthesis, and both

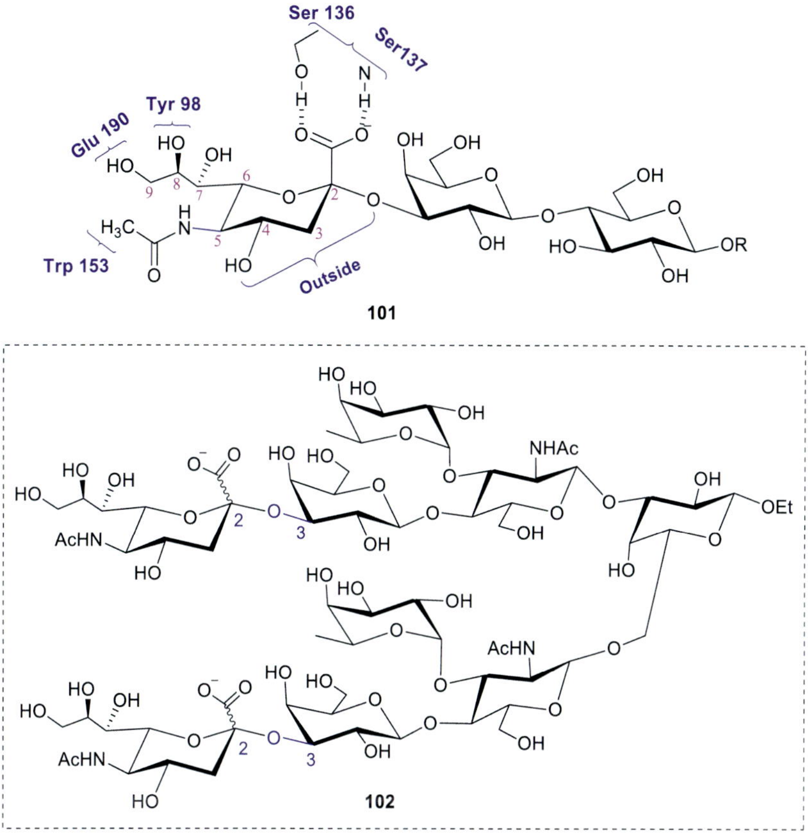

FIGURE 6.3 Schematic diagram for the plausible binding interaction of first (α-2,3) sialyl lactose **101** with Influenza A virus X-31 hemagglutinin and structure of α-2,3-linked sialoconjugate **102** as a potent inhibitor of Influenza virus hemagglutinin.[139,145]

analogs **107** and **108** were obtained in good-to-excellent yields. Further, CMP-3FNeu5Ac and CMP-3FNeu5Gc were individually used as donors in OPME glycosylation in the presence of NmCSS and *Pasteurella multocida*PmST1 (Scheme 6.24).[141] For example, fluorinated 3F(*equatorial*)Neu5Ac (**107a**) and 3F(*equatorial*)Neu5Gc (**107b**) obtained from ManNAc **11** on enzymatic glycosylation using NmCSS/α-2,3ST afforded high yield of R-α2,3-linked sialoside-3F(*equatorial*)Neu5Ac (**109a**) and R-α2,3-linked sialoside-3F(*equatorial*)Neu5Gc (**109b**), respectively. Similar enzymatic glycosylation of **107a** or **107b** using NmCSS/Pd-α2,6-ST furnished respective fluorinated R-α2,6-linked sialosides **110a** or **110b**, respectively. Likewise, CMP-3F(*axial*)Neu5Ac and CMP-3F(*axial*)-Neu5Gc were also explored as donors under similar optimized enzymatic glycosylation using PmST1 or NmCSS/Pd-α2,6-ST and successfully afforded the title fluorinated

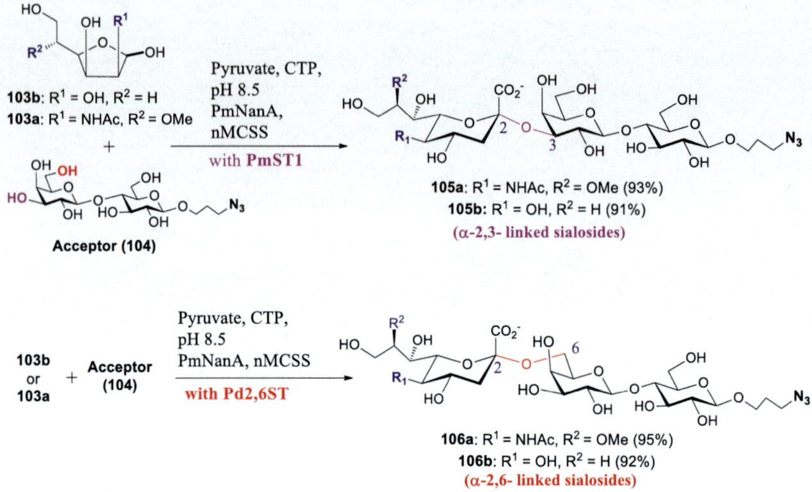

SCHEME 6.23 OPME route for *C*8-modified saialic acid-containing glycans (**105,106**).[148]

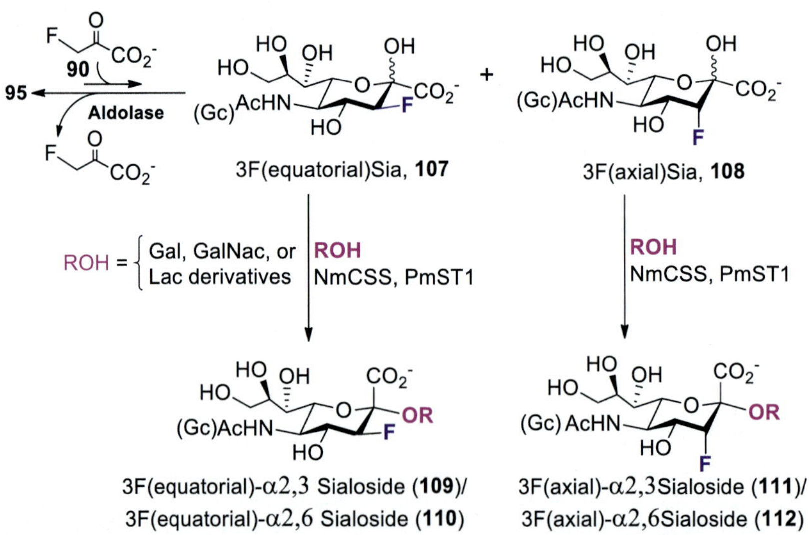

SCHEME 6.24 Chemoenzymatic synthesis of fluorinated α-2,3- and 2,6-linked sialosides (**109–112**).[141]

R-α2,3-linked 3F(*axial*)Neu5Ac (**111a**) and 3F(*axial*)Neu5Gc analogs (**111b**), or R-α2,6-linked 3F(*axial*)Neu5Ac (**112a**) and 3F(*axial*)Neu5Gc (**112b**), respectively (Scheme 6.24).[141]

Despite enzymatic synthesis for α-2,6-linked 3F-sialosides and α-2,3-linked 3F-sialosides, the chemical synthesis of 3-fluoro-sialic acid-containing sialosides is still difficult. Wong et al. reported an alternate route for the high-yielding synthesis of $3F^{ax}$-Neu5Ac-based fluorinated glycoform, for example, $3F_{ax}$-Neu5Ac-α-2,6-Gal.[149] The direct introduction of the fluorine atom in sioloconjugates for the synthesis of α-2,6-linked $3F^{ax}$-Neu5Ac, using $3F^{ax}$-Neu5Ac-based sialyl donor is challenging under different conditions due to the strong negative effect of $3F^{ax}$ group on donor reactivity. In this order, Wong describes the indirect route for $3F^{ax}$-Neu5Ac-based sialyl donor, in which the C3-equatorial hydroxyl (3-OH^{eq}) group was replaced with fluoride after the sialylation reaction. The sialoconjugate 3-OH^{eq}-Neu5Ac-α2,6-Gal-STol **115** achieved in 35% yield with excellent α-selectivity (α/β = 13:1) from 3-OH^{eq}-Neu5Ac bromide donor **113** with galactose acceptor **114** under AgOTf-Na_2HPO_4 promoter system (Scheme 6.25). Afterward, fluorination of **115** resulted in the desired $3F^{ax}$-Neu5Ac-α2,6-Gal-STol disaccharide donor **116**, which underwent subsequent glycosidation with different acceptors in the presence of NIS/TMSOTf.

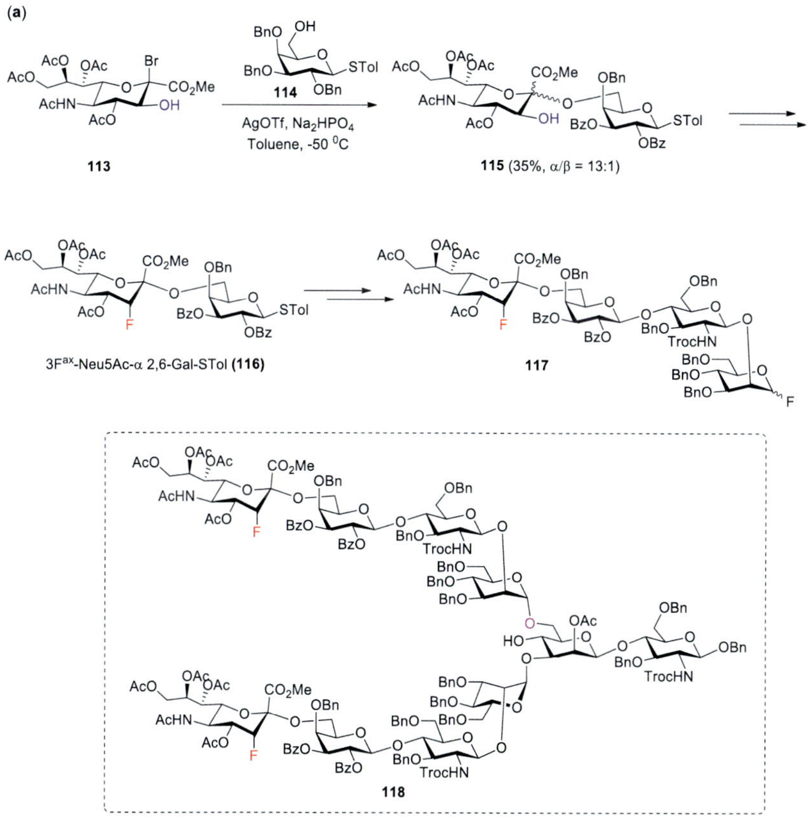

SCHEME 6.25 Chemical approach for sialidase-resistant $3F^{ax}$-Neu5Ac $3F^{ax}$ Neu5Ac-α2,6-Gal glycoside **116** and the possible glycan assembly for complex sialoconjugate decasaccharide **118**.[149]

Further, a fully protected sialidase-resistant complex molecular architecture namely 2,6-F-linked sialoconjugate decasaccharide (**118**) could be obtained via chemical glycosylation of tetrasaccharide donor **117** with a suitable hexasaccharide acceptor in presence of AgOTf-Cp_2HfCl_2 promoter (Scheme 6.25).[149] The resulting $3F^{ax}$-Neu5Ac-based glycoform could be further explored to find out more potent sialo-linked glycoproteins.

All the above-described examples clearly confirm that Chen's chemoenzymatic OPME protocol is one of the most convenient ways to get a homogeneous library of diverse α-2,3- or α-2,6-linked sialic acid-containing molecular architectures of great chemotherapeutic potential and utilities in chemical biology.[114,136,149–153]

Furthermore, a number of disialylated oligosaccharides such as GD3 ganglioside and GT3 displayed potential roles in biological systems particularly for the carbohydrate–protein interactions studies. The two-step multienzyme protocol is then fascinatingly explored for the expeditious synthesis of diverse Siaα-2,8Sia-linked disialyl glycans of great impact in chemical biology.[114] In **Step-1**, desired α-2,3- or α-2,6-linked monosialylated oligosaccharides (**99** or **100**) were obtained in good-to-excellent yields by utilizing the OPME route as described in Scheme 6.22. The monosialylated glycans GM3 **119** thus obtained were then explored as a suitable acceptor in **Step-2** for the α-2,8-sialyltransferase activity CstII mutant and afforded a high yield of title GD3 disialyl ganglioside **120** and GT3 oligosaccharides **121** (Scheme 6.26).[114]

CstII mutant (CstIIΔ32I53S) cloned from a synthetic gene has the specific α-2,8-sialidase activity. Interestingly, this has also the ability to transfer a sialic acid from a GD3 ganglioside to a different GM3 oligosaccharide, that is, possess α-2,8-*trans*-sialidase activity.[135] Both GD3 disialyl ganglioside **120** and trisialyl oligosaccharides **121** could be a precious probe required to examine the complete understanding of their biological importance.

Sialic acids are widely present at the terminals of oligosaccharides on gangliosides and glycoproteins. They are usually attached to the galactose residue by 2,3- or 2,6-linkage or polymerized by 2,8- or 2,9-linkage.[154,155]

The resulting sialoconjugates can be examined at various levels of diversity, including the modification of the sialic acid core structure, the linkage between Sia and the underlying sugar residue, the nature and arrangement of the underlying sugar residues, and the structural characteristics of the underlying sugars.[156] For analysis at each level of diversity, distinct experimental approaches were applied. To assign structures of sialic acid, the first level of complexity arises from the different α-linkages between the C-2 carbon of sialic acid and underlying sugar residues by specific sialyltransferases particularly using CMP-sialic acid donars.[157] α-2,3 or α-2,6-Linkages to other sugar units or α-2,8 to another sialic acid are the most common linkages found in sialoconjugates (Scheme 6.27). One-dimensional ^{13}C NMR spectroscopy is considered as a gold standard method for the analysis of sialic acid linkages. Thus Chen et al. reported one-pot chemoenzymatic synthesis

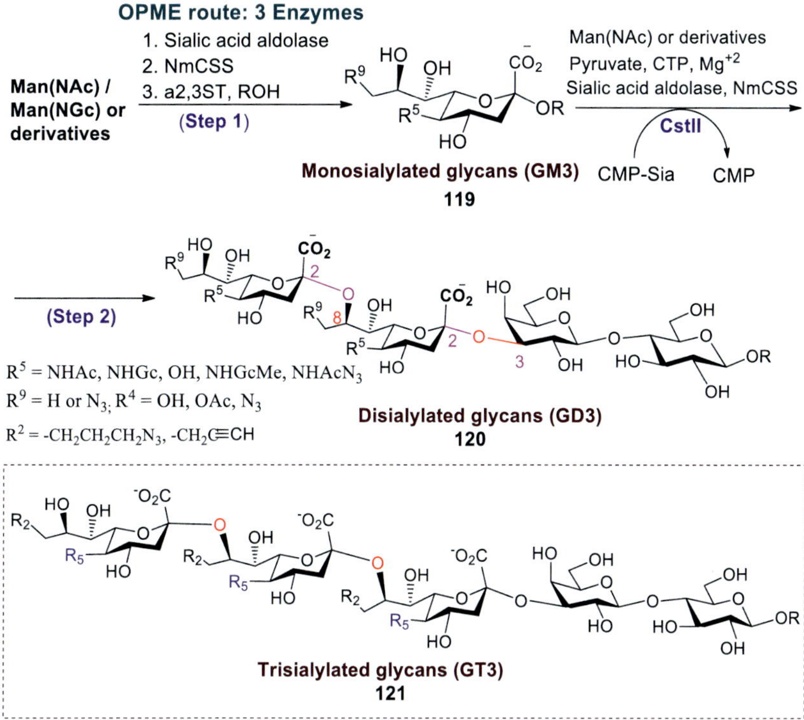

SCHEME 6.26 Chen's chemoenzymatic protocol for easy access of disialylated GD3 **120** and GT3 analogs **121**.[114]

of SialylLewis^x for cloning and characterization of α-2,3-sialyltransferase (ST3Gal-I).[158] ^{13}C NMR spectrum confirmed that the attachment of Neu5Ac in the sialylated product **123** was to the C-3 of the terminal Gal residue in Le^{x}βProN$_3$ **122**. Sialylation resulted in a significant downfield shift of the substituted carbon from 72.58 to 75.82 ppm (a downfield shift of 3.24 ppm of C-3 signal) in the starting trisaccharide **122** to the sialylated product **123** However, a little upfield shift from 71.16 ppm to 69.43 ppm was detected nearby C-4 signal in the Gal residue of sialylated product **123**. Similarly, in another experiment, Chen group again demonstrated the attachment of Nue5Ac in the sialylated product **125** was to the C-6 of the terminal Gal residue in LNnTβProN$_3$ (**124**) using simple ^{13}C NMR spectroscopy.[159] The substituted carbon significantly shifted downfield from the 60.91 ppm of initial tetrasaccharide **124** to 63.22 ppm of the sialylated product **125** (a downfield shift of 2.31 ppm) as a result of sialylation. In addition, Neu5Ac is a common structural component of GD1c and b-series ganglioside GD3 that have an α-2,8-linked disialyl moiety. Further, ^{13}C NMR data suggested that the second Neu5Ac in the sialylated product GD3MU (**127**) was attached to the C-8 of the terminal Neu5Ac residue in GM3MU (**126**). Sialylation at C-8

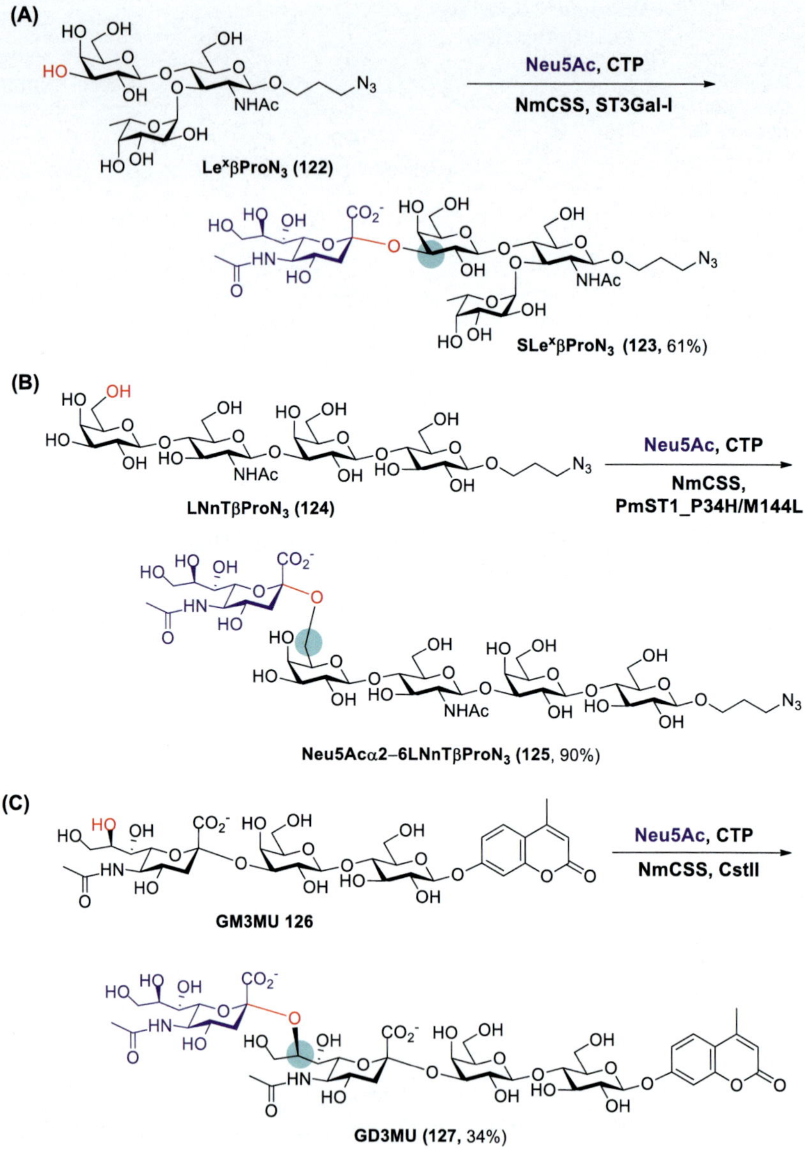

SCHEME 6.27 Characterization of the common linkages in sialosides including (A) α-2,3-linked sialosides (C-3 of Gal with 2-Sia), (B) α-2,6 linked sialoside (C-6 of Gal or GalNAc), and (C) α-2,8-linked sialoside (C-8 of Sia) using ^{13}C NMR to underlying sugars epitopes in sialic acid-containing glycans.

of Nue5Ac of starting tetrasaccharide **126** caused a downfield shift in the C-8 signal of sialylated product GD3MU (**127**) by 6.20 ppm from 71.96 to 78.16 ppm.[135]

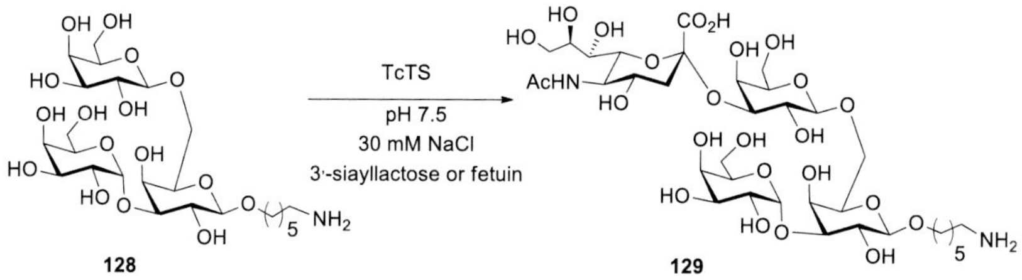

SCHEME 6.28 Selective sialylation of branched trisaccharide **128** in the presence of TcTS.

Giorgi and coworkers reported the selective monosialylation of branched trisaccharide, 6-aminohexyl α-D-Galp(1→3) [β-D-Galp(1→6)]β-D-Galp, as acceptor substrate for TcTS. This trisaccharide **128** is the smallest structure characterized by the presence of an antigenic unit D-Gal*p*(α1→3) and a β-D-Gal*p* branch that act as an acceptor of sialic acid during TcTS reaction and afforded the monosialylated analog **129** in regioselective (2,3-linkage) and stereoselective (α-selectivity) manner (Scheme 6.28).[160]

Despite the dedicated efforts, not much is known about the structure-activity relationship (SAR) of such modified sialic acid-containing glycans and therefore the field is ever-challenging and demanding. Although this enzymatic method has a number of notable features such as convenience to perform, high-to-excellent reaction yields, avoidance of protecting group, one-pot condition, environment benign conditions (enzymes are biodegradable and reaction can be performed in aqueous media), and their stereospecific nature (α-selectivity in sialosides) and regiospecificity (regioselectivity, i.e., α-2,3-, α-2,6-, or α-2,8-linkage between donor–acceptor system), the method has few limitations. For example, finding the right enzyme in high yield is not an easy task. The prime aim to find out such capable enzymes includes an easy expression of the active and soluble enzyme (Obtainable) and next it can tolerate the substrate modification (Flexible). And, also the aldolase-mediated synthesis is not well suitable to furnish the sialic acid monosaccharides having a suitable group placed at the C-4 position of sialic acid analogs.

6.3.3 Some glycodiversification through modular triazole forming coupling reactions

Cu(I)-mediated click conjugation of azide with terminal alkynes to furnish regioselective 1,4-disubstituted triazoles has been well explored in carbohydrate chemistry with various emerging applications, particularly drug discovery and development.[12,13,161] The chemoenzymatic construction of sialic acid-containing structurally diverse macrocyclic oligosaccharides of different sizes under modular click conditions has been devised by Chen et al.[162] This method involves the chemoenzymatic synthesis of azido-alkyne derivative **130** with propargyl group at reducing end and azido group containing sialic

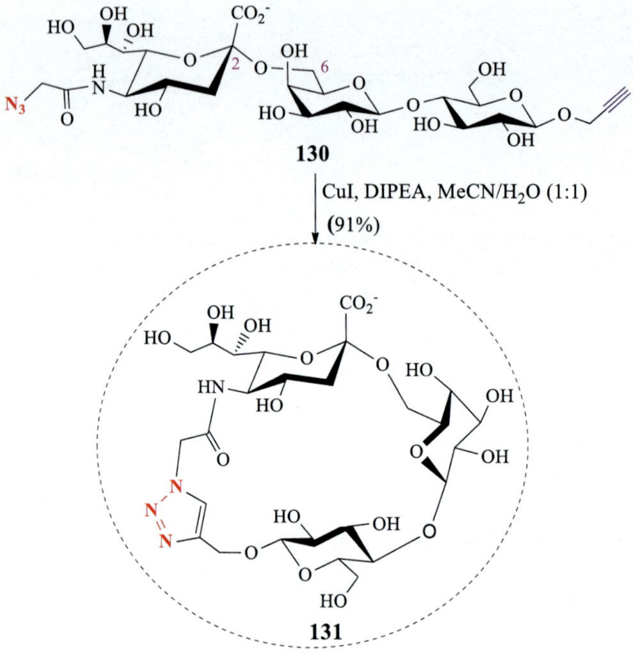

SCHEME 6.29 Intramolecular click-mediated macrocyclization for α-2,6-linked sialic acid-containing macrocycle **131**.[162]

acid to the nonreducing end. Intramolecular Cu(I)-catalyzed azide alkyne cycloaddition (CuAAC) of **130** in the presence of CuI/*N*,*N*-diisopropylethylamine (DIPEA) directed to the formation of α-2,6-linked sialic acid-containing macrocycle **131** in excellent yield (Scheme 6.29).[162] The chemoenzymatic method describes that the stereospecific macrocyclization of carbohydrates of different sizes is achieved in a controlled manner from simple-to-complex sugar-based molecular architectures of diverse applications in chemical biology.

Considering the impact of multivalent glycoconjugates in a diverse array of biological recognition, Field and coworkers devised an efficient route for the development of cyclo-oligomeric neoglycoconjugates of chemotherapeutic value. To accomplish cyclooligomerization, the azido-alkyne derivative **132** was preferentially used having azido group at the *C*-6 position and propargyl ether at *C*-1 under modular CuAAC irradiation conditions. The isomeric cyclic oligosaccharides (dimer **133a**, and trimer **133b**) obtained from CuAAC-based cyclooligomerization could serve as templates for enzymatic sialylation with TcTS in the presence of 2′-(4-methylumbelliferyl)-α-*D*-*N*-acetylneuraminic acid (MUNANA) (Scheme 6.30).[163]

Another very interesting yet, challenging approach could be considering "*the opportunities of sialic acid in glycan microarrays*" in chemical

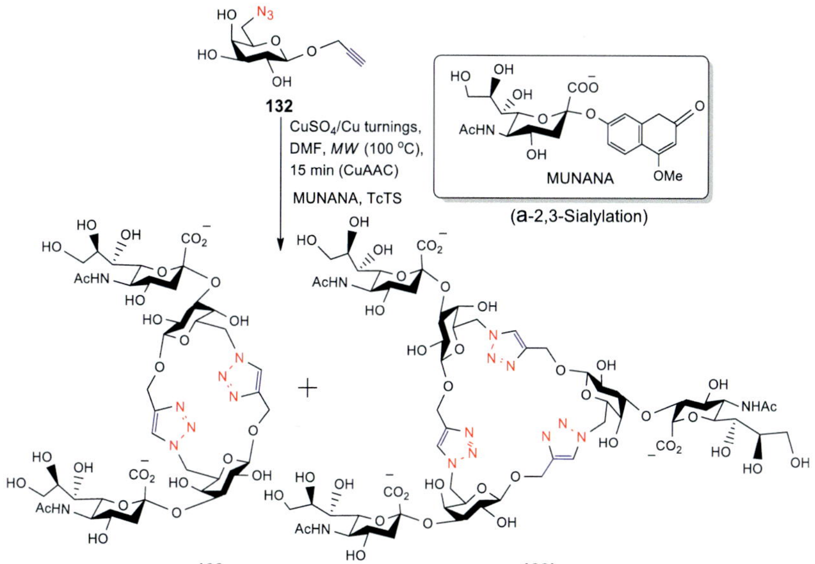

SCHEME 6.30 Cyclo-oligomerization of azido-alkyne **132** followed by sialylation leading to 1,6-linked cyclic pseudooligosaccharides (**133a**,**b**).[163]

biology.[164–168] A number of notable efforts were made for easy access to the wide range of biologically relevant simple-to-complex glycans, however because of their structural diversity, the appropriate protocol required to facilitate the high-throughput screening of glycan-binding proteins is still limited. Subsequently, this resulted in a rather challenging task and can be considered as a barrier toward the investigation of the possible role of glycoconjugates in disease progression. Fascinatingly, glycan microarrays overcome such issues to some extent and are successfully utilized as powerful tools simply by immobilization (either through noncovalent immobilization or covalent immobilization) of glycans into a high-density solid surface (using their minimal quantity) and consequently allowed for the rapid assessment of required binding interactions.[169–174] In the noncovalent immobilization method, the hydrophobic absorption or other interactions (e.g., electrostatic or fluorophilic) were considered, while in covalent immobilization in glycan microarrays, a reactive functional group is needed to functionalize the surface using some expeditious variant of click reactions.[175–177] Further, an interesting protocol for labeling cell-surface glycans, for example, alkynated proteins in crude cell lysates is exemplified through triazole forming ligation with the picolyl azides using CuAAC catalyst in the presence of BTTPS or BTTP ligand. In live mammalian cells, a mannose derivative **134** may originate, metabolize, and consequently be easily incorporated in cell-surface glycoconjugates. This is effectively recognized as early as the two-cell stage and could be considered

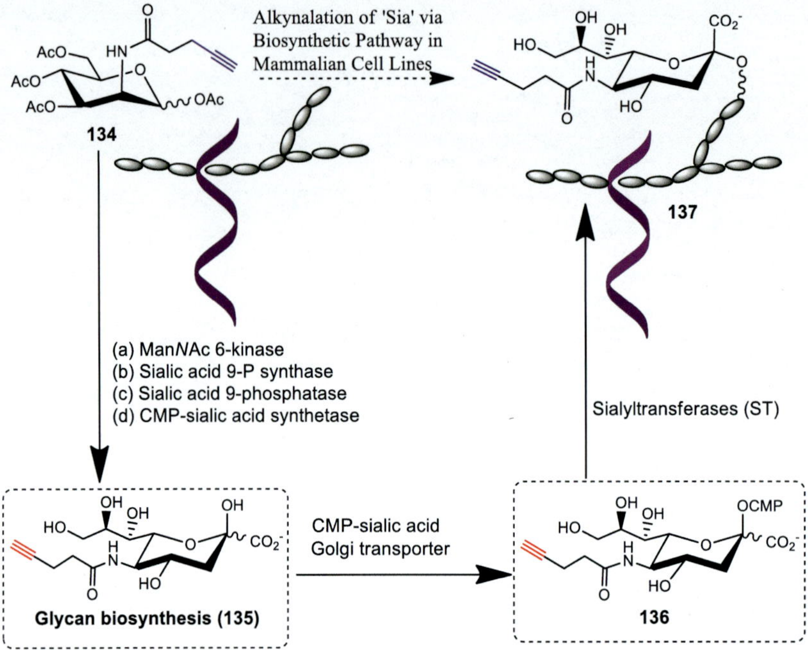

SCHEME 6.31 Alkynalation of sia via a biosynthetic pathway in mammalian cell lines.[178]

as a promising tool for the detection via monitoring the kinetics of sia biosynthesis in Jurkat cells (Scheme 6.31).[178]

With the increase in interesting and dependable development of synthetic methods for constructing desired complex sialoconjugates useful in chemical biology, the gap between goals and achieving the target has narrowed down. Interestingly, the substrate specificity of sialidase has a pivotal role in the conjugation process of Neu5Ac. Further, the sialidase activity with a wide range of sialic acid structures has gained prime attention for their biological importance. In the search for novel drug molecules, one of the major biomolecules sia has been investigated for more than three decades for its wide pharmacological significance.

Although sialic acid and its analogs can be recognized by multiple glycan-binding proteins, the complete understanding and explanation of their molecular recognition are still in dark. Only a few reports are available on the sialic acid recognition process by using novel sialylated glycan microarray embedded with modified sialic acids on different glycan backbones, for example, OPME-mediated microarrays to evaluate the interactions of modified sialic acids with proteins and viruses.[179] To achieve α-2,3- and α-2,6-linked sialyl glycans **139**, alkylamine-devised fluorophore at the reducing end and β-linked galactose-terminated glycans at the nonreducing end are sialylated and, further, is turned to develop 77 sialyl glycans **140** (Scheme 6.32).[179] Biological investigation

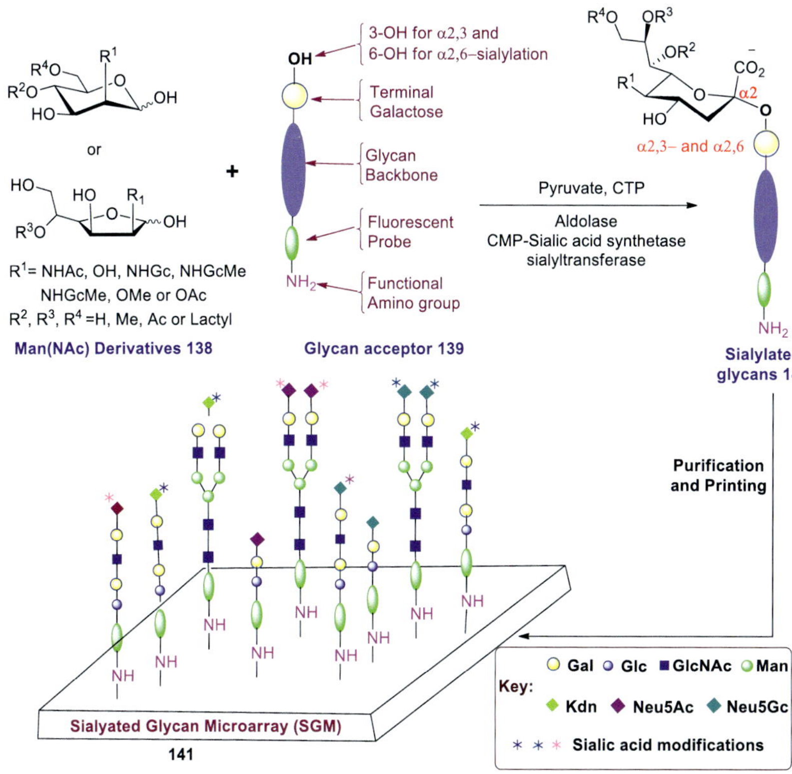

SCHEME 6.32 The design and preparation of a sialylated glycan microarray (SGM).[179]

concluded that the glycans terminating with Neu5Ac or Neu5Gc interacted with three human para-influenza viruses and with the Influenza A virus.[179] Further, Cumming and coworkers nicely exemplified a variety of sialic acid-binding proteins to cross-comparison of the two sialoglycan microarrays. This microarray makes stronger support for the superiority of information.[180]

The OPME sialylation reactions was further exemplified with diazido and triazido-mannose analogous which were found as an effective chemoenzymatic synthons to afford high-to-excellent yield of the desired azido-appended sialosides. Further, the *N*-acetyl sialosides were achieved from the respective azido groups via *N*-acetylation, which was proved to be sialic acid-binding stable probes including plant lectin MAL II.[181]

Most probably, the macromolecular receptor of sialic acid having diverse functionality encases immense influence over a number of important cellular processes. Hence, it can be assumed that the study of the carbohydrate–protein interaction and their role in physiological and pathological variance beholds a deep understanding of many unknown biochemical processes. Thus the sialic acids and their significance in biochemistry in turn, in physiology and medicine, have potential value; therefore, these molecules are one of the focused pharmacophores over three consecutive decades. Several potent high-throughput polar immobilization tools were well established for the fabrication of surface-based glycan microarrays. Similar monolayer-assisted carbohydrate–lectin interactions can be used for the successful recognition of any specific proteins of notable pathological significance. In addition, these tools can also be used for the interaction study of carbohydrates with other glycan-binding candidates such as antibodies, cells, bacteria, and viruses.[182] The protocol can emerge as a diagnostic tool for cancer diagnosis and also therapeutics. Overall, the scope of sialic acid-containing molecules in future drug discovery and development is highly appealing.[183] Very recently, Capicciotti and coworkers nicely used a photo-cross-linking tool and demonstrated a selective and exo-enzymatic cell-surface glycan labeling for the successful capturing of glycan–protein interactions. The devised exo-enzymatic tool can be very useful for the swift identification of required binding partners involved in the carbohydrate–protein interaction.[184]

6.4 Conclusions and future outlook

The development of chemical and enzymatic methods for the synthesis of sialoglycan is driven by the understanding of the important aspects of sialoglycans in chemical biology and cell physiology. Sialic acid-based high-yielding, regioselective, and stereoselective glycosylation technique is constantly growing. A number of sialyl donors have been developed to achieve the challenging task of α-sialoside synthesis. Sialic acid-containing molecules/glycans have been extensively studied in chemical biology to address their important role in biological systems. The aforementioned techniques were effectively used to construct various challenging sialoglycans, gangliosides, and sialoglycoproteins. The enzymatic OPME methodology could be regarded as a straightforward and practical one for easy access to desired sialosides. It is certainly considered as a valuable and promising motif in the assessment of carbohydrate–protein interactions in cell membranes, which in turn influence the future of drug discovery and development. The production of oligosaccharides containing sialic acid will require automated methods, combinatorial libraries, and solid-phase chemistry in the future.

Acknowledgments

The authors gratefully acknowledge to Science and Engineering Research Board (Grant No. CRG/2021/002776) and CSIR, New Delhi (Grant No. 02(0345)/19/EMR-II) for the financial support and the Institute of Eminence (IoE)—Banaras Hindu University for the incentive grant. We authors would like to express our sincere thanks to Prof. Xi Chen, UC-Davis and Prof. Richard R. Schmidt, Universitat Konstanz, Germany for their valuable suggestions during the preparation of the chapter.

References

1. Tiwari, V. K., Ed. *Carbohydrates in Drug Discovery and Development;* Elsevier Inc.: The Netherland, 2020.
2. Dwek, R. A. *Chem. Rev.* **1996,** *96*, 683–720.
3. Angata, T.; Varki, A. *Chem. Rev.* **2002,** *102*, 439–469.
4. Chen, X.; Varki, A. *ACS Chem. Biol.* **2010,** *5*, 163–176.
5. Bose, P.; Agrahari, A. K.; Singh, A. S.; Jaiswal, M. K.; Tiwari, V. K. Sialic Acid-Containing Molecules in Drug Discovery and Development. In *Carbohydrates in Drug Discovery and Development;* Tiwari, V. K., Ed.; *Elsevier*, 2020; pp 213–266.
6. Bertozzi, C. R.; Kiessling, L. L. *Science* **2001,** *291*, 2357–2364.
7. Varki, A. *Glycobiology* **1993,** *3*, 97–130.
8. Varki, A.; Schnaar, R. L.; Schauer, R. Sialic Acids and Other Nonulosonic Acids. In *Essentials of Glycobiology;* Varki, A., Cummings, R. D., Esko, J. D., Stanley, P., Hart, G. W., Aebi, M., et al., Eds.; . 3rd ed. Cold Spring Harbor; Laboratory Press: New York, NY, 2015; pp 179–195.
9. Mishra, S.; Upadhayay, K.; Mishra, K. B.; Tripathi, R. P.; Tiwari, V. K. *Stud. Nat. Prod. Chem.* **2016,** *49*, 307–361.
10. Tiwari, V. K.; Mishra, R. C.; Sharma, A.; Tripathi, R. P. *Mini-Rev. Med. Chem.* **2012,** *12*, 1497–1519.
11. Tiwari, V. K. *Chem. Rec.* **2021,** *21*, 3029–3048.
12. Agrahari, A. K.; Bose, P.; Jaiswal, M. K.; Rajkhowa, S.; Singh, A. S.; Hotha, S.; Mishra, N.; Tiwari, V. K. *Chem. Rev.* **2021,** *121*, 7638–7956.
13. Tiwari, V. K.; Mishra, B. B.; Mishra, K. B.; Mishra, N.; Singh, A. S.; Chen, X. *Chem. Rev.* **2016,** *116*, 3086–3240.
14. Zhu, X.; Schmidt, R. R. *Angew. Chem. Int. Ed.* **2009,** *48*, 1900–1934.
15. Mishra, A.; Mishra, N.; Tiwari, V. K. *Curr. Org. Synth.* **2016,** *13*, 176–219.
16. Singh, S. K.; Mishra, N.; Kumar, S.; Jaiswal, M. K.; Tiwari, V. K. *Chem. Sel.* **2022,** e202201314.
17. Blix, G.; Gottschalk, A.; Klenk, E. *Nature* **1957,** *179*, 1088.
18. Honke, K.; Taniguchi, N. *Med. Res. Rev.* **2002,** *22*, 637–654.
19. Schauer, R. *Glycoconjigate J.* **2000,** *17*, 485–499.
20. Kiefel, M. J.; Itzstein, M. V. *Chem. Rev.* **2002,** *102*, 471–490.
21. Hood, D. W.; Cox, A. D.; Gilbert, M.; Makepeace, K.; Walsh, S.; Deadman, M. E.; Cody, A.; Martin, A.; Mansson, M.; Schweda, E. K., et al. *Mol. Microbial.* **2001,** *39*, 341–350.
22. Wakarchuk, W. W.; Watson, D.; Michael, F. S.; Li, J.; Wu, Y.; Brisson, J.-R.; Young, N. M.; Gilbert, M. *J. Biol. Chem.* **2001,** *276*, 12785–12790.
23. Schauer, R. *Curr. Opin. Struct. Biol.* **2009,** *19*, 507–514.
24. Robbe, C.; Capon, C.; Maes, E.; Rousset, M.; Zweibaum, A.; Zanetta, J.-P.; Michalski, J.-C. *J. Biol. Chem.* **2003,** *278*, 46337–46348.
25. Zanetta, J.-P.; Pons, A.; Iwersen, M.; Mariller, C.; Leroy, Y.; Timmerman, P.; Schaue, R. *Glycobiology* **2001,** *11*, 663–676.
26. Sullivan, C. W.; Volcani, B. E. *Arch. Biochem. Biophys.* **1974,** *163*, 29–45.
27. DeWitt, C. W.; Rowe, J. A. *Nature* **1959,** *184*, 381–382.
28. Watson, R. G.; Marinetti, G. V.; Scherp, H. W. *J. Immunol.* **1958,** *81*, 337–344.
29. Barry, G. T.; Abbot, V.; Tsai, T. *J. Gen. Microbiol.* **1962,** *29*, 335–352.
30. Burge, B. W.; Strass, J. H. *J. Mol. Biol.* **1970,** *47*, 449–477.
31. Krantz, M. J.; Lee, Y. C.; Hung, P. P. *Nature* **1974,** *248*, 684–686.

32. Schloemer, R. H.; Wagner, R. R. *J. Virol.* **1974,** *14*, 270–281.
33. Crocker, P. R. *Curr. Opin. Struct. Biol.* **2002,** *12*, 609–615.
34. Eckhardt, M.; Bukalo, O.; Chazal, G.; Wang, L.; Goridis, C.; Schachner, M.; Gerardy-Schahn, R.; Cremer, H.; Dityatev, A. *J. Neurosci.* **2000,** *20*, 5234–5244.
35. Simanek, E. E.; McGarvey, G. J.; Jablonowski, J. A.; Wong, C.-H. *Chem. Rev.* **1998,** *98*, 833–862.
36. Kim, C. U.; Lew, W.; Williams, M. A.; Liu, H.; Zhang, L.; Swaminathan, S.; Bischofberger, N.; Chen, M. S.; Mendel, D. B.; Tai, C. Y., et al. *J. Am. Chem. Soc.* **1997,** *119*, 681–690.
37. Hughes, R. C. *Biochim. Biophys. Acta* **1999,** *1473*, 172–185.
38. Malykh, Y. N.; Schauer, R.; Shaw, L. G. *Biochimie* **2001,** *83*, 623–634.
39. Munday, J.; Floyd, H.; Crocker, P. R. *J. Leukoc Boil* **1999,** *66*, 705–711.
40. Corfield, A. P.; Wember, M.; Schauer, R.; Rott, R. *Eur. J. Biochem.* **1982,** *124*, 521–525.
41. Hinderlich, S.; Stäsche, R.; Zeitler, R.; Reutter, W. *J. Biol. Chem.* **1997,** *272*, 24313–24318.
42. Angata, T.; Nakata, D.; Matsuda, T.; Kitajima, K.; Troy, F. A. *J. Biol. Chem.* **1999,** *274*, 22949–22956.
43. Kean, E. L. *Glycobiology* **1991,** *1*, 441–447.
44. Comb, D. G.; Roseman, J. *J. Biol. Chem.* **1960,** *235*, 2529–2537.
45. Vann, W. F.; Tavarez, J. J.; Crowley, J.; Vimr, E.; Silver, R. P. *Glycobiology* **1997,** *7*, 697–701.
46. Angata, T.; Kitazume, S.; Terada, T.; Kitajima, K.; Inoue, S.; Troy, F. A., II; Inoue, Y. *Glycoconj. J.* **1994,** *11*, 493–499.
47. Kawano, T.; Koyama, S.; Takematsu, H.; Kozutsumi, Y.; Kawasaki, H.; Kawasimal, S.; Kawasaki, T.; Suzuki, A. *J. Biol. Chem.* **1995,** *270*, 16458–16463.
48. Lewis, A. L.; Nizet, V.; Varki, A. *Proc. Nat. Acad. Sci.* **2004,** *101*, 11123–11128.
49. Yu, H.; Chokhawala, H. A.; Huang, S.; Chen, X. *Nature Protocol.* **2006,** *1*, 2485–2492.
50. Ng, S.-S.; Dain, J. A. *Eds. The Natural Occurrence of Sialic Acids: Biological Roles of Sialic Acid;* Springer, 197659–102.
51. Boons, G. J.; Demchenko, A. V. *The Chemistry of Sialic Acid. Carbohydrate-based Drug Discovery;* Wiley-VCH, 200355–102.
52. Boons, G. J.; Demchenko, A. V. *Chem. Rev.* **2000,** *100*, 4539–4565.
53. Halcomb, R. L.; Chappell, M. D. *J. Carbohydr. Chem.* **2002,** *21*, 723–768.
54. DeMeo, C.; Boons, G. J.; Demchenko, A. V. Synthesis of Glycosides of Sialic Acid: Chemical Aspects (Scope and Limitations). In *Comprehensive Glycoscience-From Chemistry to Systems Biology;* Kamerling, J. P., Ed.; *Elsevier: Amsterdam*, 2007The Netherlands.
55. Ress, D. K.; Linhardt, R. J. *Curr. Org. Synth.* **2004,** *1*, 31–46.
56. Magnusson, G.; Nilsson, U. J.; Fraser-Reid, B.; Kuniaki, T.; Thiem, J. Eds *Glycoscience: Chemistry and Chemical Biology*, 2. *Springer: Berlin*, 20011543–1587.
57. Navaluri, C.; Crich, D. Stereocontrolled Synthesis of Sialosides. In *Glycochemical Synthesis Strategies and Applications;* Hung, S.-C., Zulueta, M. M. L., Eds.; *John Wiley & Son, Inc*: Hoboken, 2016New Jersey.
58. Ando, T.; Ando, H.; Kiso, M. *Trends Glycosci. Glycotechnol.* **2001,** *13*, 573–586.
59. Takahashi, T.; Tsukamoto, H.; Yamada, H. *Tetrahedron Lett.* **1997,** *38*, 8223–8226.
60. Ye, X.-S.; Huang, X.; Wong, C. H. *Chem. Commun.* **2001,** 974–975.
61. Kuhn, R.; Lutz, P.; MacDonald, D. L. *Chem. Ber.* **1966,** *99*, 611–617.
62. Okamoto, K.; Goto, T. *Tetrahedron* **1990,** *46*, 5835–5857.

63. Deninno, M. P. *Synthesis (Mass)* **1991,** 583–593.
64. Paulsen, H.; Tietz, H. *Angew. Chem. Int. Ed Engl.* **1982,** *21*, 927–928.
65. Marra, A.; Sinay, P. *Carbohydr. Res.* **1989,** *190*, 317–322.
66. Kirchner, E.; Thiem, F.; Dernick, R.; Heukeshoven, J.; Thiem. *J. Carbohydr. Chem.* **1988,** *7*, 453–486.
67. Byramova, N. E.; Tuzikov, A. B.; Bovin, N. V. *Carbohydr. Res.* **1992,** *237*, 161–175.
68. Sharma, M. N.; Eby, R. *Carbohydr. Res.* **1984,** *127*, 201–210.
69. Kunz, H.; Waldmann, H. *J. Chem. Soc. Chem. Commun.* **1985,** 638–640.
70. Mugunthan, G.; Kartha, K. P. R. *Tetrahedron Lett.* **2012,** *53*, 5631–5634.
71. Pazynina, G.; Tuzikov, A.; Chinarev, A.; Obukhova, P.; Bovin, N. *Tetrahedron Lett.* **2002,** *43*, 8011–8013.
72. Tropper, F. D.; Andersson, F. O.; Cao, S.; Roy, R. *J. Carbohydr. Chem.* **1992,** *11*, 741–750.
73. Marra, A.; Sinay, P. *Carbohydr. Res.* **1989,** *187*, 35–42.
74. Martichonok, V.; Whitesides, G. M. *J. Org. Chem.* **1996,** *61*, 1702–1706.
75. Gregar, T. Q.; Gervay-Hague, J. *J. Org. Chem.* **2004,** *69*, 1001–1009.
76. Sun, X.-L.; Kanie, Y.; Guo, C.-T.; Kanie, O.; Suzuki, Y.; Wong, C.-H. *Eur. J. Org. Chem.* **2000,** 2643–2653.
77. Shpirt, A. M.; Kononov, L. O.; Torgov, V. I.; Shibaev, V. N. *Russ. Chem. Bull.* **2004,** *53*, 717–719.
78. Orlova, A. V.; Shpirt, A. M.; Kulikova, N. Y.; Kononov, L. O. *Carbohydr. Res.* **2010,** *345*, 721–730.
79. Nifantiev, N. E.; Tsvetkov, Y. E.; Shashkov, A. S.; Kononov, L. O.; Menshov, V. M.; Tuzikov, A. B.; Bovin, N. V. *J. Carbohydr. Chem.* **1996,** *15*, 939–953.
80. Tsvetkov, Y. E.; Schmidt, R. R. *Carbohydr. Lett.* **1996,** *2*, 149–156.
81. Simeoni, L. A.; Byramova, N. E.; Bovin, N. V. *Bioorg. Khim.* **1999,** *25*, 62–69.
82. Cheshev, P. E.; Khatuntsvesa, E. A.; Gerbst, A. G.; Tsvetkov, Y. E.; Shashkov, A. S.; Nifant'ev, N. E. *J. Bioorg. Chem.* **2003,** *29*, 372–381.
83. Cao, S.; Meunier, S. J.; Andersson, F. O.; Letellier, M.; Roy, R. *Tetrahedron: Asymmetry* **1994,** *5*, 2303–2312.
84. Oscarson, S. Thioglycosides. In *Carbohydrates in Chemistry and Biology;* Ernst, B., Hart, G. W., Sinay, P., Eds.; 1; *Wiley-VCH*: *Weinheim, New York*, 2000; pp 93–116.
85. Oscarson, S. General Synthetic Methods: Reactions at the Anomeric Center: S-Glycosylation. In *Glycoscience: Chemistry and Chemical Biology;* Fraser-Reid, B., Tatsuta, K., Thiem, J., Eds.; 1; *Springer*: *New York*, 2001; pp 643–671.
86. Yu, C.-S.; Niikura, K.; Lin, C.-C.; Wong, C.-H. *Angew. Chem., Int. Ed.* **2001,** *40*, 2900–2903.
87. Meo, C. D.; Demchenko, A. V.; Boons, G. J. *J. Org. Chem.* **2001,** *66*, 5490–5497.
88. Fuse, T.; Ando, H.; Imamura, A.; Sawada, N.; Ishida, H.; Kiso, M.; Ando, T.; Li, S.-C.; Li, Y.-T. *Glycoconjugate J.* **2006,** *23*, 329–343.
89. Ando, H.; Koike, Y.; Ishida, H.; Kiso, M. *Tetrahedron Lett.* **2003,** *44*, 6883–6886.
90. Fujikawa, K.; Nakashima, S.; Konishi, M.; Fuse, T.; Komura, N.; Ando, T.; Ando, H.; Yuki, N.; Ishida, H.; Kiso, M. *Chem. Eur. J.* **2011,** *17*, 5641–5651.
91. Yasomanee, J. P.; Demchenko, A. V. *J. Am. Chem. Soc.* **2012,** *134*, 20097–20102.
92. Liu, D.-M.; Wang, H.-L.; Lei, J.-C.; Zhou, X.-Y.; Yang, J.-S. *Eur. J. Org. Chem.* **2020,** 575–585.
93. Kanie, O.; Kiso, M.; Hasegawa, A. *J. Carbohydr. Chem.* **1988,** *7*, 501–506.
94. Hasegawa, A.; Kiso, M. In Carbohydrates-Synthetic Methods and Applications in Medicinal Chemistry*;* Ogura, H., Hasegawa, A., Suami, T., Eds.; Nodansha, VCH: New York, 1992; pp 243–266.

95. Hasegawa, A.; Nagahama, T.; Ohki, H.; Hotta, K.; Ishida, H.; Kiso, M. *J. Carbohydr. Chem.* **1991,** *10*, 493–498.
96. Demchenko, A. V.; Boons, G. J. *Tetrahedron Lett.* **1998,** *39*, 3065–3068.
97. Crich, D.; Li, W. *J. Org. Chem.* **2007,** *72*, 7794–7797.
98. Hanashima, S.; Sato, K.; Ito, Y.; Yamaguchi, Y. *Eur. J. Org. Chem.* **2009,** 4215–4220.
99. Hanashima, S.; Yamaguchi, Y.; Ito, Y.; Sato, K.-I. *Tetrahedron Lett.* **2009,** *50*, 6150–6153.
100. Dziadek, S.; Brocke, C.; Kunz, H. *Chem. Eur. J.* **2004,** *10*, 4150–4162.
101. Marra, A.; Sinay, P. *Carbohydr. Res.* **1990,** *195*, 303–308.
102. Schmidt, R. R.; Kovac, P., Eds. *ACS Symposium Series*, 560; *American Chemical Society: Washington, DC*, 1994; pp 276–296.
103. Greilich, U.; Brescello, R.; Jung, K. H.; Schmidt, R. R. *Liebigs Ann.* **1996,** 663–672.
104. De Meo, C.; Parker, O. *Tetrahedron: Asymmetry* **2005,** *16*, 303–307.
105. Komura, N.; Kato, K.; Udagawa, T.; Asano, S.; Tanaka, H.-N.; Imamura, A.; Ishida, H.; Kiso, M.; Ando, H. *Science* **2019,** *364*, 677–680.
106. Takahashi, M.; Shirasaki, J.; Komura, N.; Sasaki, K.; Tanaka, H.-N.; Imamura, A.; Ishida, H.; Hanashima, S.; Murata, M.; Ando, H. *Org. Biomol. Chem.* **2020,** *18*, 2902–2913.
107. Hanashima, S.; Tomiya, T.; Ishikawa, D.; Akai, S.; Sato, K. *Carbohydr. Res.* **2009,** *344*, 959–965.
108. Cai, S.; Yu, B. *Org. Lett.* **2003,** *5*, 3827–3830.
109. Ito, Y.; Ogawa, T. *Tetrahedron Lett.* **1987,** *28*, 6221–6224.
110. Ito, Y.; Ogawa, T. *Tetrahedron Lett.* **1988,** *29*, 3987–3990.
111. Castro-Palomino, J. C.; Tsvetkov, Y. E.; Schmidt, R. R. *J. Am. Chem. Soc.* **1998,** *120*, 5434–5440.
112. Storz, T.; Vasella, A. *Helv. Chim. Acta* **1998,** *81*, 1896–1907.
113. Vasella, A.; Wyler, R. *Helv. Chim. Acta* **1990,** *73*, 1742–1763.
114. Yu, H.; Cheng, J.; Ding, L.; Khedri, Z.; Chen, Y.; Lau, K.; Tiwari, V. K.; Chen, X. *J. Am. Chem. Soc.* **2009,** *131*, 18467–18477.
115. Vibhute, A. M.; Komura, N.; Tanaka, H.-N.; Imamura, A.; Ando, H. *Chem. Rec.* **2021,** *21*, 3194–3223.
116. Ogawa, T.; Sugimoto, M. *Carbohydr. Res.* **1984,** *128*, c1–c4.
117. Sherman, A. A.; Yudina, O. N.; Shashkov, A. S.; Menshov, V. M.; Nifantev, N. E. *Carbohydr. Res.* **2001,** *330*, 445–458.
118. Xiao-Feng, W.; Shimizu, Y.; Kanai, M. *ACS Cent. Sci.* **2016,** *2*, 21–26.
119. Banwell, M. G.; Hungerford, N. L.; Jolliffe, K. A. *Org. Lett.* **2004,** *6*, 2737–2740.
120. Hong, Z.; Liu, L.; Hsu, C. C.; Wong, C. H. *Angew. Chem. Int. Ed Engl.* **2006,** *45*, 7417–7421.
121. Brovetto, M.; Gamenara, D.; Méndez, P. S.; Seoane, G. *Chem. Rev.* **2011,** *111*, 4346–4403.
122. Sugiyama, M.; Greenberg, W.; Wong, C.-H. *J. Syn. Org. Chem., Japan.* **2008,** *66*, 605–615.
123. Fong, S.; Machajewski, T. D.; Mak, C. C.; Wong, C.-H. *Chem. Biol.* **2000,** *7*, 873–883.
124. Royer, S. F.; Haslett, L.; Crennell, S. J.; Hough, D. W.; Danson, M. J.; Bull, S. D. *J. Am. Chem. Soc.* **2010,** *132*, 11753–11758.
125. Gonzalez-Garcia, E.; Helaine, V.; Klein, G.; Schuermann, M.; Sprenger, G. A.; Fessner, W.-D.; Reymond, J.-L. *Chem. Eur. J.* **2003,** *9*, 893–899.
126. Wada, M.; Hsu, C.-C.; Franke, D.; Mitchell, M.; Heine, A.; Wilson, I.; Wong, C.-H. *Bioorg. Med. Chem.* **2003,** *11*, 2091–2098.

127. Woodhall, T.; Williams, G. J.; Berry, A.; Nelson, A. *Angew. Chem. Int. Ed.* **2005,** *44*, 2109–2112.
128. Hsu, C.-C.; Hong, Z.; Wada, M.; Franke, D.; Wong, C.-H. *Proc. Natl. Acad. Sci. USA* **2005,** *102*, 9122–9126.
129. Helaine, V.; Gastaldi, C.; Lemaire, M.; Claps, P.; Guerard-Helaine, C. *ACS Catal* **2022,** *12*, 733–761.
130. Yu, H.; Yu, H.; Karpel, R.; Chen, X. *Bioorg. Med. Chem.* **2004,** *12*, 6427–6435.
131. Yu, H.; Chokhawala, H.; Karpel, R.; Yu, H.; Wu, B.; Zhang, J.; Zhang, Y.; Jia, Q.; Chen, X. *J. Am. Chem. Soc.* **2005,** *127*, 17618–17619.
132. Ding, L.; Yu, H.; Lau, K.; Li, Y.; Muthana, S.; Wang, J.; Chen, X. *Chem Commun.* **2011,** *47*, 8691–8693.
133. Huynh, N.; Aye, A.; Li, Y.; Yu, H.; Cao, H.; Tiwari, V. K.; Shin, D.-W.; Chen, X.; Fisher, A. J. *Biochemistry* **2013,** *52*, 8570–8579.
134. Li, Y.; Yu, H.; Cao, H.; Lau, K.; Muthana, S.; Tiwari, V. K.; Son, B.; Chen, X. *Appl. Microb. Biotech.* **2008,** *79*, 963–970.
135. Cheng, J.; Yu, H.; Lau, K.; Huang, S.; Chokhawala, H. A.; Li, Y.; Tiwari, V. K.; Chen, X. *Glycobiology* **2008,** *18*, 686–697.
136. Yu, H.; Chen, X. *Org. Biomol. Chem.* **2016,** *14*, 2809–2818.
137. Xiao, A.; Li, Y.; Li, X.; Santra, A.; Yu, H.; Li, W.; Chen, X. *ACS Catal.* **2018,** *8*, 43–47.
138. Bai, Y.; Yu, H.; Chen, X. *Curr. Protoc.* **2021,** *1*, e91.
139. Sauter, N. K.; Hanson, J. E.; Glick, G. D.; Brown, J. H.; Crowther, R. L.; Park, S. J.; Skehel, J. J.; Wiley, D. C. *Biochemistry* **1992,** *31*, 9609–9621.
140. Linclau, B.; Arda, A.; Reichardt, N.-C.; Sollogoub, M.; Unione, L.; Vincet, S. P.; Jimenez-Barbro, J. *Chem. Soc. Rev.* **2020,** *49*, 3863–3888.
141. Chokhawala, H. A.; Cao, H.; Yu, H.; Chen, X. *J. Am. Chem. Soc.* **2007,** *129*, 10630–10631.
142. I.C. Schoenhoffen. US Patent No. 2021/0401863 A1, 2021, December 30, 2021.
143. Yu, H.; Huang, S.; Chokhawala, H.; Sun, M.; Zheng, H.; Chen, X. *Angew. Chem. Int. Ed.* **2006,** *45*, 3938–3944.
144. Huang, S.; Yu, H.; Chen, X. *Angew. Chem. Int. Ed.* **2007,** *46*, 2249–2253.
145. Sabesan, S.; Duus, J. O.; Fukunaga, T.; Bock, K.; Ludvigsen, S. *J. Am. Chem. Soc.* **1991,** *113*, 3236–3246.
146. Swarte, V. V.; Joziasse, D. H.; Mebius, R. E.; Van den Eijnden, D. H.; Kraal, G. *Cell Adhes. Commun.* **1998,** *6*, 311–322.
147. Shawn, A.; Frees, D.; Kosch, W.; Way, W.; Paulson, J. C.; Sabesan, S.; Halcomb, R. L.; Huang, D.-H.; Ichikawa, Y.; Wong, C.-H. *J. Am. Chem. Soc.* **1995,** *117*, 66–79.
148. Yu, H.; Cao, H.; Tiwari, V. K.; Li, Y.; Chen, X. *Bio. Org. Med. Chem. Lett.* **2011,** *21*, 5037–5040.
149. Lo, H. J.; Krasnova, L.; Dey, S.; Cheng, T.; Liu, H.; Tsai, T. I.; Wu, K. B.; Wu, C. Y.; Wong, C. H. *J. Am. Chem. Soc.* **2019,** *141*, 6484–6488.
150. Li, R.; Kooner, A. S.; Muthana, S. M.; Yuan, Y.; Yu, H.; Chen, X. *J. Org. Chem.* **2020,** *85*, 16157–16165.
151. Yu, H.; Gadi, M. R.; Bai, Y.; Zhang, L.; Li, L.; Yin, J.; Wang, P. G.; Chen, X. *J. Org. Chem.* **2021,** *86*, 8672–8682.
152. Santra, A.; Li, Y.; Ghosh, T.; Li, R.; Yu, H.; Chen, X. *ACS Catal.* **2019,** *9*, 211–215.
153. Li, W.; Santra, A.; Yu, H.; Slack, T. J.; Muthana, M. M.; Shi, D.; Liu, Y.; Chen, H. *J. Org. Chem.* **2019,** *84*, 6697–6708.
154. Kirsch, S.; Muthing, J.; Peter-Katalinic, J.; Bindila, L. *Biol. Chem.* **2009,** *390*, 657–672.

155. Kammeijer, G. S. M.; Jansen, B. C.; Kohler, I.; Heemskerk, A. A. M.; Mayboroda, O. A.; Hensbergen, P. J.; Schappler, J.; Wuhrer, M. *Scientific Rep.* **2017,** *7*, 3733.
156. Lewis, A. L.; Varki, A. *Evolutionary Considerations in Studying the Sialome: Sialic Acids and the Host-Pathogen Interface; John Wiley & Sons*: *New York*, 2010.
157. Cohen, M.; Varki, A. *OMICS: A J. Integrative Biol.* **2010,** *14*, 455–464.
158. Sugiarto, G.; Lau, K.; Yu, H.; Vuong, S.; Thon, V.; Li, Y.; Huang, S.; Chen, X. *Glycobiology* **2011,** *21*, 387–396.
159. McArthur, J. B.; Yu, H.; Zeng, J.; Chen, X. *Org. Biomol. Chem.* **2017,** *15*, 1700–1709.
160. Giorgi, M. E.; Lopez, R.; Agusti, R.; Marino, C.; de Lederkremer, R. M. *Carbohydr. Res.* **2017,** *450*, 30–37.
161. Tripathi, R. P.; Dwivedi, P.; Sharma, A.; Kushwaha, D.; Tiwari, V. K. Triazolyl Glycoconjugates in Medicinal Chemistry. In *Click Chemistry in Glycoscience: New Developments and Strategies;* Witczak, Z. J., Bielski, R., Eds.; Wiley-VCH, 2013; pp 293–323. 2013.
162. Muthana, S.; Yu, H.; Cao, H.; Cheng, J.; Chen, X. *J. Org. Chem.* **2009,** *74*, 2928–2936.
163. Campo, V. L.; Carvalho, I.; DaSilva, C. H. T. P.; Schenkman, S.; Hill, L.; Nepogodiev, S.; Field, R. A. *Chem. Sci.* **2010,** *1*, 507–514.
164. Muthana, S. M. Glycan Microarray: Towards Drug Discovery and Development. In Carbohydrates in Drug Discovery and Development*, Chapter 6;* Tiwari, V. K., Ed.; Elsevier, 2020; pp 213–266.
165. Song, X.; Xia, B.; Lasanajak, Y.; Smith, D. F.; Cummings, R. D. *Glycoconjugate J.* **2008,** *25*, 15–25.
166. Song, X.; Ju, H.; Lasanajak, Y.; Kudelka, M. R.; Smith, D. F.; Cummings, R. D. *Nat. Methods* **2016,** *13*, 528–534.
167. Song, X.; Xia, B.; Stowell, S. R.; Lasanajak, Y.; Smith, D. F.; Cummings, R. D. *Chem. Biol.* **2009,** *16*, 36–47.
168. Yan, M.; Lasanajak, Y.; Smith, D. F.; Song, X. *Glycobiology* **2018,** *28*, 1067–1068.
169. Song, E. H.; Pohl, N. L. *Curr. Opin. Chem. Biol.* **2009,** *13*, 626–632.
170. Oyelaran, O.; Gildersleeve, J. C. *Curr. Opin. Chem. Biol.* **2009,** *13*, 406–413.
171. De Paz, J. L.; Seeberger, P. H. *Methods Mol. Biol.* **2012,** *808*, 1–12.
172. Rillahan, C. D.; Paulson, J. C. Glycan Microarrays for Decoding the Glycome. *Annual Rev. Biochem.* **2011,** *80*, 797–823.
173. Lonardi, E.; Balog, C. I.; Deelder, A. M.; Wuhrer, M. Natural Glycan Microarrays. *Exp. Rev. Proteom.* **2010,** *7*, 761–774.
174. Park, S.; Gildersleeve, J. C.; Blixt, O.; Shin, I. Carbohydrate Microarrays. *Chem. Soc. Rev.* **2013,** *42*, 4310–4326.
175. Calarese, D. A.; Lee, H. K.; Huang, C. Y.; Best, M. D.; Astronomo, R. D.; Stanfield, R. L.; Katinger, H.; Burton, D. R.; Wong, C. H.; Wilson, I. A. *Proc. Natl. Acad. Sci. USA* **2005,** *102*, 13372–13377.
176. Manimala, J. C.; Li, Z.; Jain, A.; VedBrat, S.; Gildersleeve, J. C. *Chem. Biochem.* **2005,** *6*, 2229–2241.
177. Lee, M. R.; Shin, I. *Org. Lett.* **2005,** *7*, 4269–4272.
178. Jiang, H.; Zheng, T.; Lopez-Aguilar, A.; Feng, L.; Kopp, F.; Marlow, F. L.; Wu, P. *Bioconjugate Chem.* **2014,** *25*, 698–706.
179. Song, X.; Yu, H.; Chen, X.; Lasanajak, Y.; Tappert, M. M.; Air, G. M.; Tiwari, V. K.; Cao, H.; Chokhawala, H. A.; Zheng, H., et al. *J. Biol. Chem.* **2011,** *286*, 31610–31622.

180. Padler-Karavani, V.; Song, X.; Yu, H.; Hurtado-Ziola, N.; Huang, S.; Muthana, S.; Chokhawala, H. A.; Cheng, J.; Verhagen, A.; Langereis, M. A., et al. *J. Biol. Chem.* **2012,** *287*, 22593–22608.
181. Kooner, A. S.; Diaz, S.; Yu, H.; Santra, A.; Varki, A.; Chen, X. *J. Org. Chem.* **2021,** *86*, 14381–14397.
182. Vacchini, M.; Edwards, R.; Guizzardi, R.; Palmioli, A.; Ciaramelli, C.; Paiotta, A.; Airoldi, C.; Ferla, B. L.; Cipolla, L. *Curr. Med. Chem.* **2019,** *26*, 6349–6398.
183. Bose, P.; Jaiswal, M. K.; Singh, S. K.; Singh, R. K.; Tiwari, V. K. *Carbohydrate Res* **2023,** *527*, 108804.
184. Babulic, J. L.; Capicciotti, C. J. *Bioconjugate Chem* **2022,** *33*, 773–780.

Chapter 7

Common protection–deprotection strategies in carbohydrate synthesis

Bhaswati Ghosh and Suvarn S. Kulkarni
Department of Chemistry, Indian Institute of Technology Bombay, Powai, Mumbai, Maharashtra, India

7.1 Introduction

Carbohydrates form a highly functionalized class of natural biopolymers. Every monosaccharide contains a number of stereocenters and bears multiple hydroxyl and sometimes amine functionalities. These are, in turn, can be decorated with sulfate or phosphate groups, acyl or lactic acid esters, or ether moieties. Other than these, carboxylate functions can also be found. Most frequently, the amine groups are found acetylated, but various amide functional groups are also found, as well as *N*-sulfates and alkylated amines. The diversity of the functional groups on a carbohydrate ring continues to be one of the biggest challenges in synthetic carbohydrate chemistry.[1–3]

This chapter describes the nuances in the reactivity of the numerous functional groups on a carbohydrate moiety and ways to apply these in the design of effective protecting group strategies. The protecting group on a carbohydrate plays the pivotal role in modulating the reactivity of the (mono)saccharide, and this section will describe how protecting group can be used to control stereoselective transformations (most importantly, glycosylation reactions) and reactivity-driven one-pot synthetic strategies. Applications and strategies used in automated synthesis are also included in this chapter.

7.2 Common protection–deprotection strategies

7.2.1 Possible protection at the primary position of saccharides

There is much effort in functionalizing every hydroxyl group of saccharides through systematic protecting group strategies.[4,5] Primary hydroxyl group

Synthetic Strategies in Carbohydrate Chemistry. DOI: https://doi.org/10.1016/B978-0-323-91729-2.00006-9

being sterically less hindered allows selective functionalization at this position. Synthetic strategies are developed in such a way that the sterically hindered reagents mainly prefer the primary position over secondary positions. The common protecting groups to mask the primary position of the sugar derivatives are trityl,[6] tosyl,[7] and silyl[8] groups using bulky reagents. For example, the treatment of trityl chloride using pyridine as a solvent on unprotected mono or disaccharides favors regioselectively primary hydroxyl protection. The classical deprotection of trityl ether requires mild temperature and acidic condition (acetic acid or trifluoroacetic acid). If any other acid-sensitive group is already present in the sugar moiety, for example, acetals, it limits the use of the trityl-protecting group.

The application of silyl protecting groups is widely relevant when orthogonal protecting group strategies are desired. Triisopropylsilyl (TIPS), *t*-butyldimethylsilyl (TBDMS), and *tert*-butyldiphenylsilyl (TBDPS) protecting groups are usually used for the protection of primary hydroxyl groups on carbohydrates. However, trimethylsilyl (TMS) and triethylsilyl (TES) are typically nonselective for the primary position. Silyl ethers offer good stability over a range of reaction conditions. The silyl protection is usually achieved by using equimolar amounts of chlorosilanes in the presence of a base, mostly imidazole.[9] Silyl ether deprotection is also quite straightforward, which makes it even more attractive when it comes to the primary protecting group. They are labile to fluoride ion[10] and acidic conditions.[11]

Regioselective 6-*O*-sulfonylation represents another class of protection in the primary hydroxyl group. These are importantly employed to access halides[12] and azides[13] at that position via nucleophilic displacements or to achieve 6-deoxy sugars (e.g., D-fucose or D-rhamnose derivatives) through reduction. Halogenated and azide precursors are very useful intermediates in glycochemistry. The tosyl group displacements are mainly reported with the iodide group and sometimes with the bromide group. Tosylation can be achieved using tosyl chloride on unprotected sugar under basic conditions (usually pyridine) to obtain the sugar tosylate, which acts as a better-leaving group. In the subsequent step, it could be reacted with some strong nucleophilic group to achieve the desired functionalization of sugar.

The two main strategies for acylation of primary alcohol are chemical methods and chemoenzymatic methods. Controlled use of acylating agents is recommended to avoid multiacylations. Pivaloate group was initially used due to its bulkiness, but the deprotection of this group requires strong basic conditions that limit the use in primary hydroxyl protection on sugar.[14] Acetate group can be introduced at the 6-position of the sugar moiety by using a slight excess of equimolar acyl chloride at a very low temperature.[15] Other selective acylations are also reported including tin-based esterifications to provide the desired 6-*O*-benzoates in good yields and regioselectivities.[16] Use of ethyl acetate as a solvent in the presence of concentrated sulfuric acid lead to transesterification with complete regioselectivity at the primary position.[17]

Conversion of C6-OH to the corresponding halides is also common in carbohydrate chemistry with the employment of CCl_4 and CBr_4 in the presence of triphenylphosphine. Diethylaminosulfur trifluoride (DAST) is useful for fluorination with a short reaction time to avoid potential decomposition.[18]

Several reports demonstrate selective oxidation of the primary alcohol of sugars to uronic acid derivatives using a Pt metal catalyst. The most used method is 2,2,6,6-tetramethylpiperidinyl-1-oxyl (TEMPO) oxidation that involves the use of bis(acetoxy)iodobenzene (BAIB) with or without $NaHCO_3$ and sometimes in conjunction with NaOCl.[19] Electrochemical oxidations are also reported, which happen to be more selective using Pt anode and also in the presence of TEMPO.[20] To stop the oxidation at the aldehyde stage, often nonaqueous solvents are employed. However, in the case of unprotected carbohydrates, solubility becomes an issue. In such cases, alternative solvents have been examined and *N,N*-dimethylformamide (DMF) was found to be compatible.[21]

The 1,6-anhydro sugars indirectly provide the means of protection of C6-OH without having to use any other protecting group. Ring opening of an 1,6-anhydro sugar via acetolysis makes the primary hydroxyl group accessible for further coupling. Although the selective alkylation of the primary alcohol is often considered a difficult task, the deprotection of 6-*O*-benzyl group is well documented in the literature (Fig. 7.1). Usually, benzyl groups are

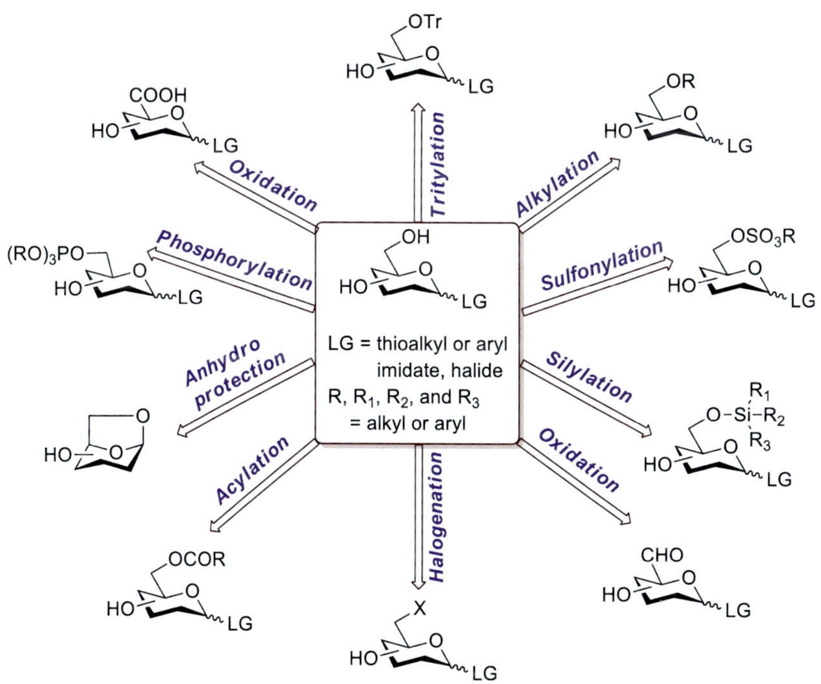

FIGURE 7.1 Regioselective functionalization of primary position of the saccharides.

removed by hydrogenolysis, but the primary benzyl ethers are labile in nature, which makes them suitable for selective deprotection keeping the secondary benzyl groups intact. $CrCl_2$/LiI was employed for the regioselective removal of the 6-*O*-benzyl group; however, later it was found to be highly substrate dependent.[22] An alternative strategy that has become more attractive is the acetolysis of the C6-*O*-benzyl group.[23] In this reaction, acetic anhydride is used under protic acid or Lewis acid conditions. Later, the primary acetate group is removed by using NaOMe in methanol (Zemplén conditions).

7.2.2 Protecting groups for the secondary hydroxyl groups of carbohydrates

Regioselective secondary alcohol protection is a daunting task. However, steric interactions, electronic effects, metal complexation, and kinetic versus thermodynamic regioselectivity concepts are prevalent. For example, the C-2 OH group is close to the anomeric center, and it shows more acidic character and less nucleophilicity. The anomeric position is differentiated by acetal chemistry. Many efforts are being made to protect the rest of the hydroxyl groups. It is easy to differentiate the primary OH group from the rest, as discussed earlier. However, the secondary hydroxyls present at C-2, C-3, and C-4 positions have similar reactivity and therefore are difficult to differentiate.

Secondary hydroxyl groups are often protected by temporary protecting groups for selective removal at a later stage and by permanent protecting groups, which are removed in the final stage of the synthesis. Anomeric stereochemistry in the oligosaccharide synthesis often becomes a constraint in choosing the protecting group. Ester-protecting groups are widely used for providing anchimeric assistance during glycosylation and an ether protecting group is considered as nonparticipating group for 1,2-*cis*-glycosidic linkage formation. The permanent protecting groups used for this purpose are benzyl ethers whereas the derivatives of benzyl ethers (arylmethyl ethers) act as temporary protecting groups. The protection and deprotection strategy are summarized in Table 7.1.

The silyl groups are employed frequently where the secondary hydroxyl groups are concerned. Less bulkier silyl ethers such as TMS, TES, and TBDMS ethers are mostly employed to protect the secondary hydroxyl group. Although the TMS group is labile in many reaction conditions, it is widely used in one-pot tandem reactions.[29]

Acetates and benzoates are abundantly used in carbohydrate chemistry because they are considered as simple ester-protecting groups. Pivaloyl (Piv), chloroacetyl (ClAc), levulinoyl (Lev), picoloyl (Pico), and so on are notable examples of other ester-type protecting groups. Pivaloyl group limits its wide application in carbohydrate chemistry as it requires harsh basic conditions for its removal. Other groups such as chloroacetyl (ClAc),[30] levulinoyl (Lev),[30] picoloyl (Pico)[31] groups are seen for routine use in carbohydrate chemistry not

TABLE 7.1 Protection and deprotection of various ether functional groups.

Protecting groups	Protection conditions	Deprotection conditions
Benzyl (Bn)	1. NaH, BnBr, DMF/THF 2. BnCl, KOH, dioxane	1. H_2, Pd/C, MeOH 2. Na, NH_3 (liq.), THF
p-Methoxy benzyl (PMB)	NaH/DIPEA, PMBCl, DMF/THF	DDQ or CAN, CH_2Cl_2/H_2O
p-Bromo benzyl (PBB)	$ArCH_2Br$, NaH, DMF	$ZnCl_2$, CH_2Cl_2[24]
p-Chloro benzyl (PCB)	$ArCH_2Cl$, NaH, DMF	$SnCl_4$, CH_2Cl_2[24]
p-Nitro benzyl (PMB)	*p*-NO_2ArCH_2Br, 2,4,6-collidine, AgOTf	In, NH_4Cl, MeOH/*i*-PrOH, 85°C[25]
o-Nitro benzyl (OMB)	*o*-NO_2ArCH_2Br, NaH, DMF	1. hν, MeOH, or MeCN[26] 2. H_2, Pd/C, MeOH
2-Naphthyl methyl benzyl	2-NapCH_2Br, NaH, DMF[27]	DDQ, CH_2Cl_2/H_2O[27]
2-Pyridyl methyl	PyrCH_2Br, NaH, DMF	H_2, Pd/C, MeOH[28]

only because they require mild conditions for their protection and removal but also, due to their orthogonal nature.

Regioselective protections are challenging due to the discrimination of two or more secondary hydroxyl groups that are generally present in a sugar unit. Often axial hydroxy groups are more sterically hindered than the equatorial ones. Tin ketals offer an effective approach to differentiate between the neighboring axial and equatorial hydroxyl groups although it uses the toxic tin metal in stoichiometric or catalytic amounts.[32] There are several examples reported over last four decades, where tin reagents are used stoichiometrically or catalytically to form tin ketals for the regioselective acyl/alkyl protection of secondary hydroxyl groups (Scheme 7.1A and B). Taylor and coworkers have introduced boronic acid catalysis to replace toxic tin with benign boron nonmetal.[33] α-*O*-Methyl-fucopyranoside **6** can be regioselectively alkylated or acylated using a catalytic amount of diphenylborinic ethylamine ester and benzyl bromide or benzoyl chloride (Scheme 7.1C). The reaction proceeds via borinate intermediate that reacts in a highly regioselective manner to protect the equatorial alcohol at C3. Metal triflates are also useful as catalysts for regioselective protection.[34]

7.2.3 Protecting group at the anomeric position of carbohydrates

Anomeric center has the utmost significance in carbohydrate chemistry. The protection at the anomeric center is temporary. It is very much dependent

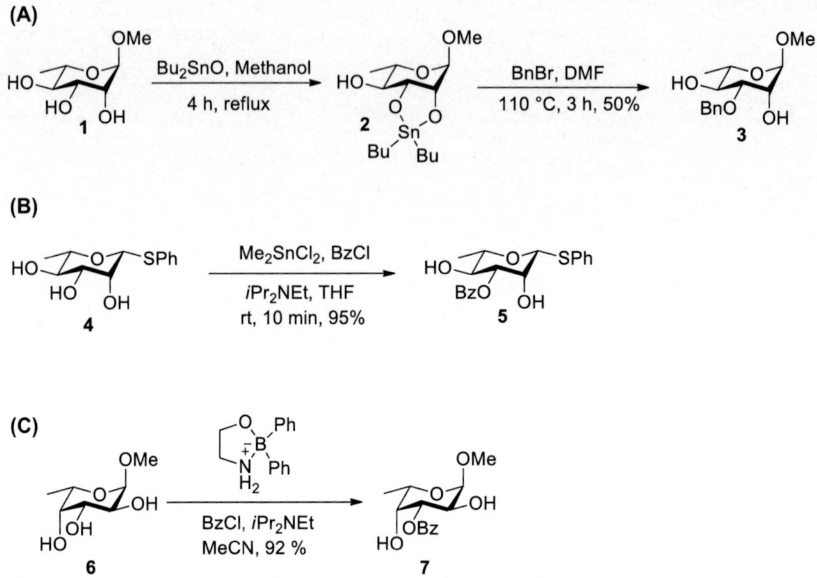

SCHEME 7.1 Regioselective protection of secondary hydroxyl groups of the saccharides.

upon the nature and the location of the glycosidic bond therefore the protecting group term could be best replaced by the leaving group.

A small methyl group is the simplest protection at the anomeric position, which is achieved by Fischer glycosylation in methanol using strong acid as a catalyst. There are other examples of modified Fischer glycosylation involving ethanol, pentanol, allyl alcohol, or propargyl alcohol as solvents. The alkyl groups once introduced at the anomeric position of oligosaccharides are difficult to remove as the harsh conditions would also cleave the glycosidic linkages. In their place, the aryl *O*-glycosides have become a popular choice of anomeric protection. The protection is carried out by treating the fully acetylated sugar with the appropriate phenol in the presence of Lewis acids. Anomeric aryloxy group can be swiftly removed by hydrolysis with trifluoroacetyl (TFA)[35] or by oxidative cleavage with CAN.[36]

The most widely used glycosyl esters are anomeric acetates, employing mostly acetic anhydride in the presence of Lewis acid. Usually, the first step of oligosaccharide synthesis is the per-*O*-acetylation of carbohydrates. Sometimes, the regioselective esterification at the anomeric position is also taken into consideration. Other than acetates, benzoates,[37] and levulinyl[38] groups are also used for anomeric protection. Hydrazine acetate is commonly used to hydrolyze the anomeric acetate, ammonia for levulinyl group, and methyl amine for benzoates.

The selective protection of anomeric centers with cyclic acetal, ketals, and orthoesters is worth mentioning. Anomeric 1,2-cyclic group can withstand a variety of reaction conditions as it requires more forcing conditions for removal. Acetonide protection is commonly done in anhydrous acetone using Lewis acid as a catalyst. Anomeric 1,2 orthoesters are also useful when there is a requirement of differentiating other hydroxyl groups present in the sugar moiety. They can be easily prepared by activation of acetylated glycosyl bromides in the presence of alcohol and the hindered base 2,6-lutidine. Orthoesters are resistant to basic conditions, therefore other hydroxyl groups can be protected with groups that require basic conditions without affecting the orthoester.

Bulky silyl ethers frequently serve as temporary anomeric protecting groups. In this context, thexyl dimethyl silyl (TDS)[39] and tertbutyl dimethylsilyl (TBS or TBDMS) groups are used to serve the purpose. The anomeric TDS ether is found to be tolerant under a wide range of reaction conditions. Even TBS groups are relatively resistant to acidic hydrolysis. Other than these, bulkier TBDPS and TIPS groups are used occasionally to protect the anomeric position of sugar.

Thioglycosides serve a dual purpose of anomeric protection functions and as a leaving group as they are found to be stable under a variety of reaction conditions. Another advantage of using thioglycosides is that they are perfectly orthogonal with the other protecting group present in the moiety. *N*-Bromosuccinimide in aqueous acetone causes hydrolysis of thioglycosides. *N*-Iodo succinimide can be used in the presence of silver triflate and aqueous dichloromethane (DCM). Although *N*-glycosyl derivatives are less common in literature, the growing interest of this class of compounds in chemical biology warrants their mention in this section. *N*,*O*-Dialkoxyamine for *N*-glycosides are widely used as anomeric protecting group.[40] Hydrolysis of the *N*-glycoside protecting group is done by using *N*-chlorosuccinimide in aqueous THF or trimethylsilyl chloride in THF/methanol. The possible anomeric protection in carbohydrates is depicted in Fig. 7.2.

7.2.4 Cyclic protecting group of carbohydrates

The use of a cyclic protecting group often helps to distinguish at least two hydroxyls from the others. The main discussion in this section is the protection of cyclic acetals in hexoses. A pictorial representation of cyclic protecting groups is given in Table 7.2. The cyclic protecting groups are often seen in trehalose kind of substrates. The advantage associated with this cyclic kind of protecting group is not only protecting two hydroxyl groups at the same time but also it offers rigid ring conformations that play an important role in stereoselectivity in ensuing glycosylation reactions.

In this context, the benzylidene type of cyclic group has immense importance in carbohydrate chemistry. This is also a great way to achieve diverse functionality. Regioselective, reductive cleavage of benzylidene acetals gives the access to O6 hydroxyl group of that particular sugar moiety where the

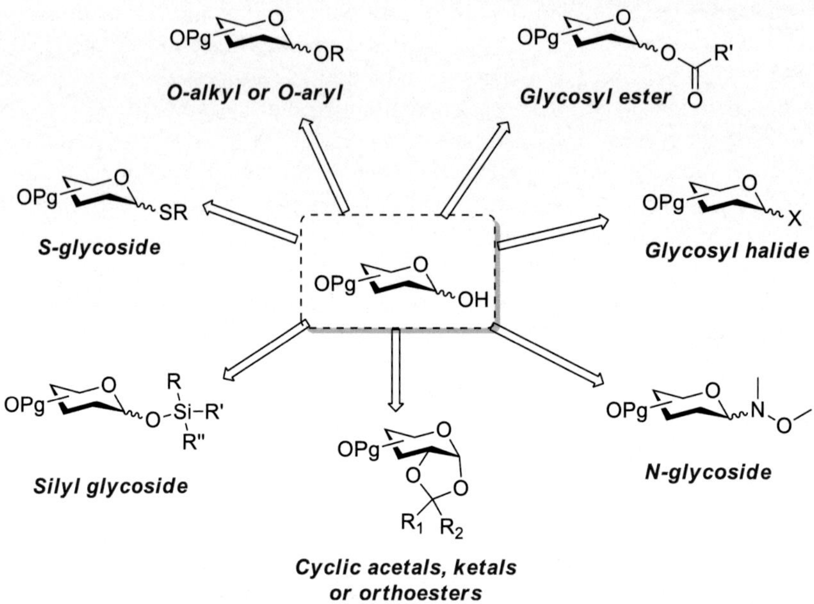

FIGURE 7.2 Possible anomeric protection in carbohydrates.

TABLE 7.2 Different cyclic protecting groups.

Methylene acetal[41]	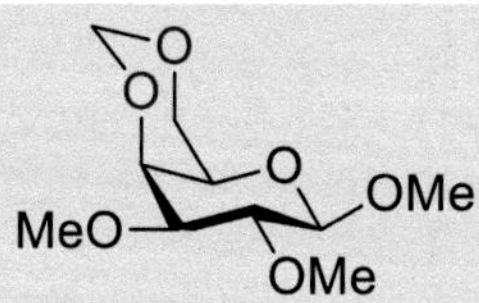
Phenylsulfonylethylidene[42]	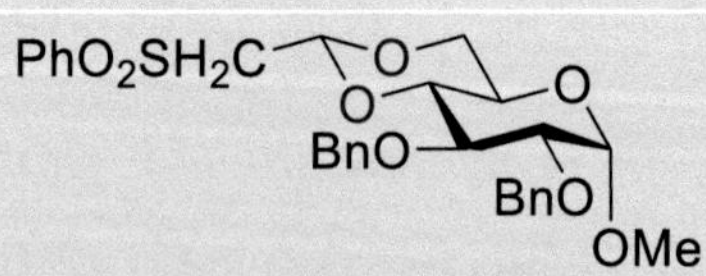
Cyclohexylidene[43]	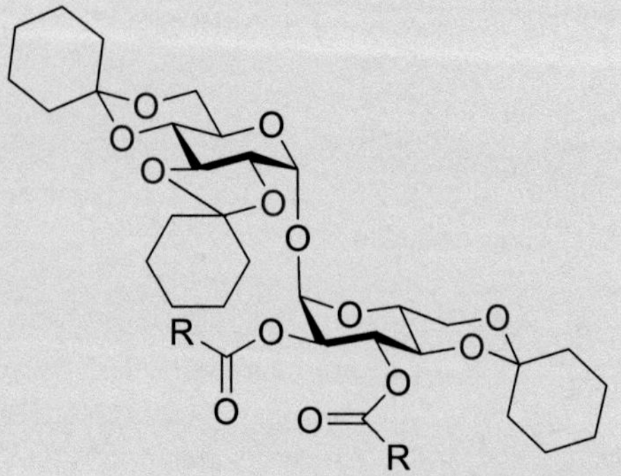

(Continued)

TABLE 7.2 (Continued)

Isopropylidene (acetonide)	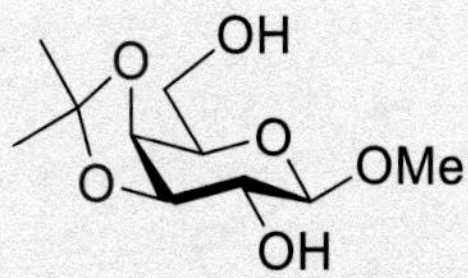
Cyclohexane-1,2-diacetal (CDA)[44]	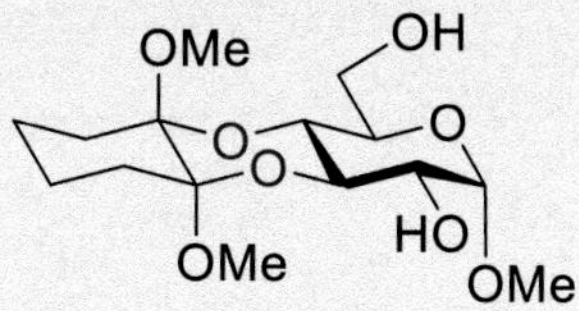
Butane-1,2-diacetal (BDA)[45]	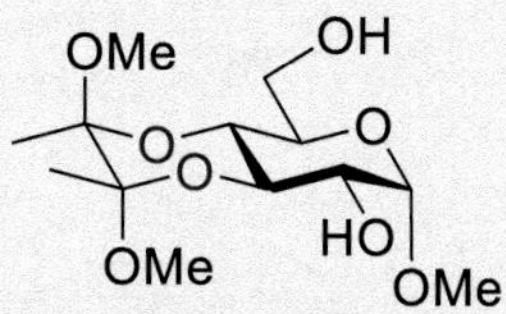
Benzylidene acetals	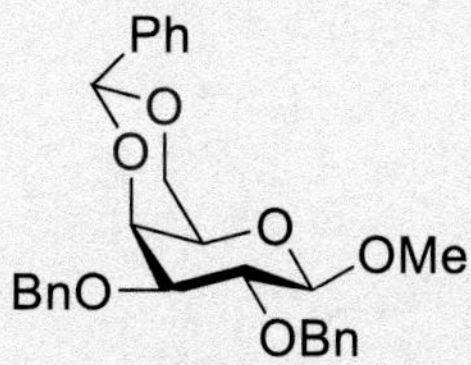
p-methoxybenzylidene acetals	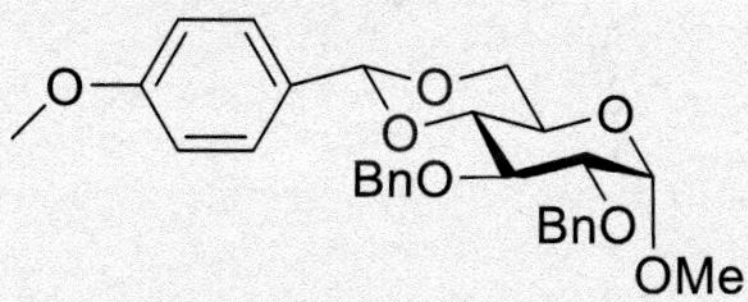
Halobenzylidene acetals	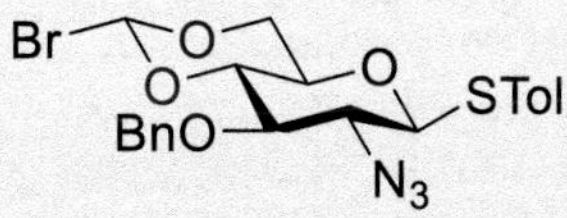
Nitrobenzylidene acetals[46]	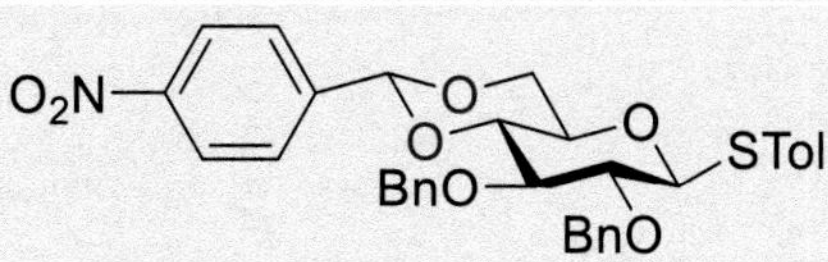
Polyaromatic methylene acetals	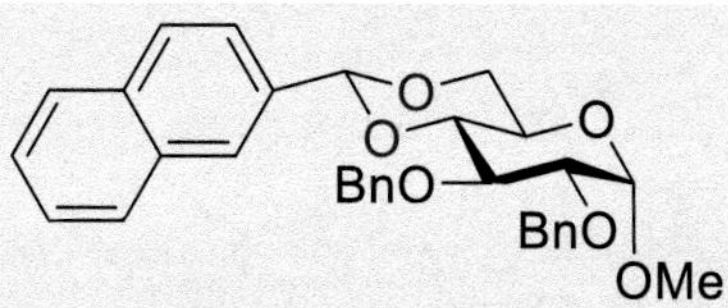

(*Continued*)

TABLE 7.2 (Continued)

Silylene acetals[47]	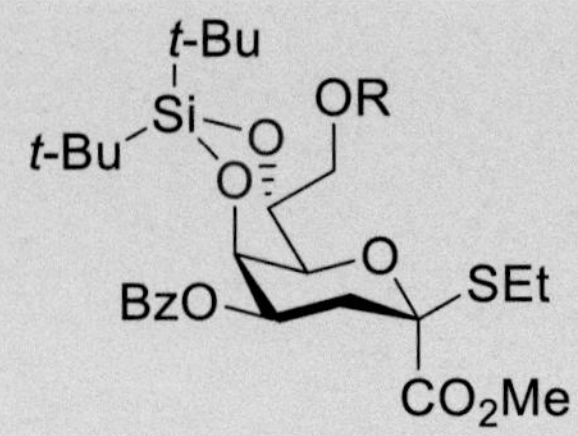
Cyclic carbonate	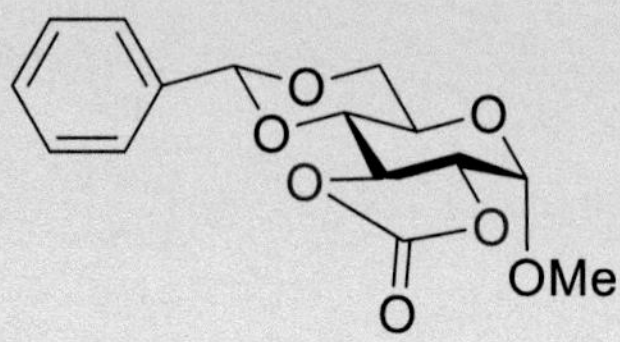

primary hydroxy can be converted to acid, azide, halide, or deoxy functionality by suitable reaction conditions.[48] On the other hand, often primary group is protected with a suitable protecting group (e.g., TBDPS) or regioselectively glycosylated at O6 in the presence of secondary O4 hydroxy due to huge reactivity difference between O4 and O6 hydroxy groups. Not only this, benzylidene can be regioselectively opened from O4 or O6 position using suitable reaction conditions.[31b,49]

7.2.5 *N*-protecting groups for 2-amino-2-deoxy-glycosides

2-Amino-2-deoxy glycoside represents an important class of natural oligosaccharides. Some representative examples are chitin, a linear homopolymer of 2-acetamido-D-glucosamine that is the component of the cell wall in fungi; heparin, a linear sulfated polysaccharide possessing alternating 2-acetamido-D-glucosamine and L-iduronic acid units, and the lipid-A region of lipopolysaccharides that contain two D-glucosamine residues with β-1,6-linkage, known to influence several inflammatory processes. Most common glycosidic linkages found here are β- but α-linkages too can occur. Whenever a participating group is present at C-2 in the form of ester or carbamate, the formation of 1,2-*trans* glycosidic linkages is constructed through anchimeric assistance but the nonparticipating group at C-2 such as azide (N_3), alone does not guarantee complete 1,2-*cis* stereoselectivity.

Many *N*-protecting groups are developed for directing the stereochemical outcome of the glycosylation reaction. There are myriads of *N*-protecting groups developed over the years but we will discuss only the most common *N*-protecting groups used in carbohydrate chemistry on a regular basis.[50] Most common groups for the construction of 1,2-*trans* glycosidic linkages

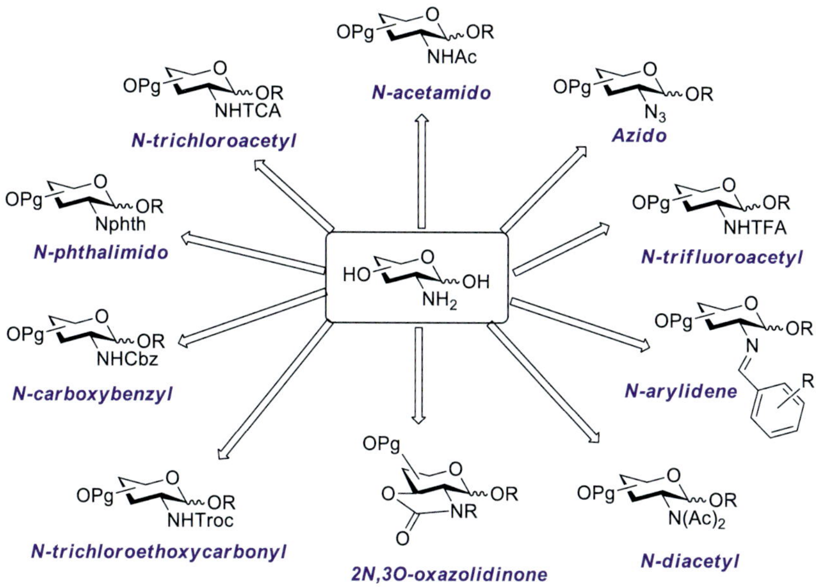

FIGURE 7.3 Numerous amines protecting function in carbohydrates.

are *N*-trichloroethoxycarbonyl (*N*-Troc), *N*-trichloroacetyl (*N*-TCA), *N*-phthaloyl (*N*-Phth), *N*-carboxybenzyl (*N*-Cbz) and *N*-trifluoroacetyl (*N*-TFA). 1,2-*Cis* stereoselective glycosylation of 2-amino-2-deoxy sugars remains one of the principal challenges in complex oligosaccharide synthesis. In 2001 Kerns and coworkers developed 2,3-oxazolidinone-protected thioglycoside donors that possess high α selectivity.[51] Nguyen and coworkers used *p*-methoxybenzylidene group as an *N*-protecting group for 2-deoxy-glucosamine to make C (2)-*N*-substituted benzylidene D-glucosamine trichloroacetimidate donors for their 1,2-*cis*-glycosidic linkage construction study.[52]

Although tremendous efforts were made to succeed in achieving 1,2-*cis* stereoselectivity, the use of 2-azido-2-deoxy glycoside was reported long back in this regard.[53] The use of azide group gained popularity not only because it is nonparticipating in nature but also it is easily reducible to get the desired functionality. Below is the schematic representation of *N*-protected 2-amino-2-deoxy sugars. Some common route for the protection of amine functionality in carbohydrates is depicted in Fig. 7.3.

7.2.6 One-pot multistep regioselective protection of carbohydrates catalyzed by acids

One-pot protection of monosaccharides relies on the reactivity difference of the various hydroxyl groups present in the sugar moiety. The very first example of this kind is when β-methyl-D-glucopyranoside is successively treated with trityl chloride and acetic anhydride in pyridine, the selectively 6-*O*-tritylated and fully protected glucoside is formed in a 65% yield as a crystalline material.[54]

Silyl glycosides are the popular choice for one-pot multistep regioselective protection on carbohydrates due to the increased solubility in organic media as well as their ready preparation from their unprotected precursor in almost quantitative yield. The most common catalyst used with per-*O*-trimethylsilylated α-methyl glucopyranoside is TMSOTf.[55] Later, $Cu(OTf)_2$[56] has also been used. The regioselective one-pot protection on per-*O*-trimethylsilylated α-methyl glucopyranoside/β-thioglycoside designed by Hung and coworkers paved access to monosaccharide building blocks with a variety of protecting group patterns.[29b] These transformations are efficiently performed by TMSOTf, and the scope of various protecting groups has been shown. The overall transformation in this method saves the traditional step-by-step protection and deprotection reactions, associated workups, and purification. This one-pot method has been extended to a sequential step, opening the way for divergent pathways. Benzylidene-protected glucopyranoside is achieved through one-pot three steps method, which was further regioselectively opened at the O4 or O6 positions. Additionally, greater diversity was attained when one-pot acylation/alkylation is performed at the O2 position.

In the glucopyranose series, the easy formation of 4,6-*O*-benzylidene alone in the one-pot multistep series is attributed to the 2,3 *trans* configuration. Whereas, in mannopyranose series due to 2,3 *cis* configuration, the preferred 4,6-*O*-benzylidene formation often occurs with the formation of the endo and exo 2,3-*O*-benzylidene cyclic acetals. This makes a further one-pot modification more difficult to control. It is nevertheless possible to cleanly produce the 4,6-*O*-benzylidene only. Sequential TMSOTf-catalyzed monobenzylidenation and deacetylation followed by regioselective benzylidene ring opening with dimethylethylsilane furnish the 6-OH derivative. Mannose dibenzylidene is also found to be a useful intermediate for selective ring opening. The 2-OH derivative is rather ingeniously derived from the intermediary 2,3-exo-4,6-*O*-dibenzylidene by a regioselective opening of only the exo-5-membered acetal by diisobutylaluminum hydride.[55]

In this regard, the galactopyranose series is quite straightforward. 4,6-*O*-Benzylidene galactose is easily prepared by one-pot multistep synthesis. Later treating the benzylidene-protected galactopyranoside with *p*-anisaldehyde in the presence of triethylsilane furnishes regioselective O3-PMB-protected galactopyranoside.

An important consequence is improved efficiency of the overall transformation, thus circumventing the traditional step-by-step protection–deprotection approaches associated with successive workups, as well as time and solvent-consuming intermediate purifications. These catalyzed protocols also reduce wastes, notably by facilitating the use of stoichiometric reagents or amounts close to stoichiometry. An unavoidable step for success is, however, the fine-tuning of the operating conditions for each particular carbohydrate substrate. The silylated substrates provide the most flexible routes for differentiating the hydroxyl groups because such routes have been developed using highly regioselective reductive benzylation of pyranosides from either benzylidenes or isolated trimethylsilyl ethers. A common way to access several hydroxyl groups from respective TMS glucosides is depicted in Scheme 7.2.

Acyl migration under the influence of acidic, basic, neutral, or enzymatic conditions or even in refluxing conditions is well known in carbohydrate chemistry. This property of the acyl group is often taken advantageously in case of synthetic manipulations of partially acylated polyols and site-selective acylation of multihydroxy-containing compounds. Acyl group migration takes place to the adjacent hydroxyl group intramolecularly via 5- or 6-membered orthoester-type intermediate. The examples discussed next are O2→O3 silyl

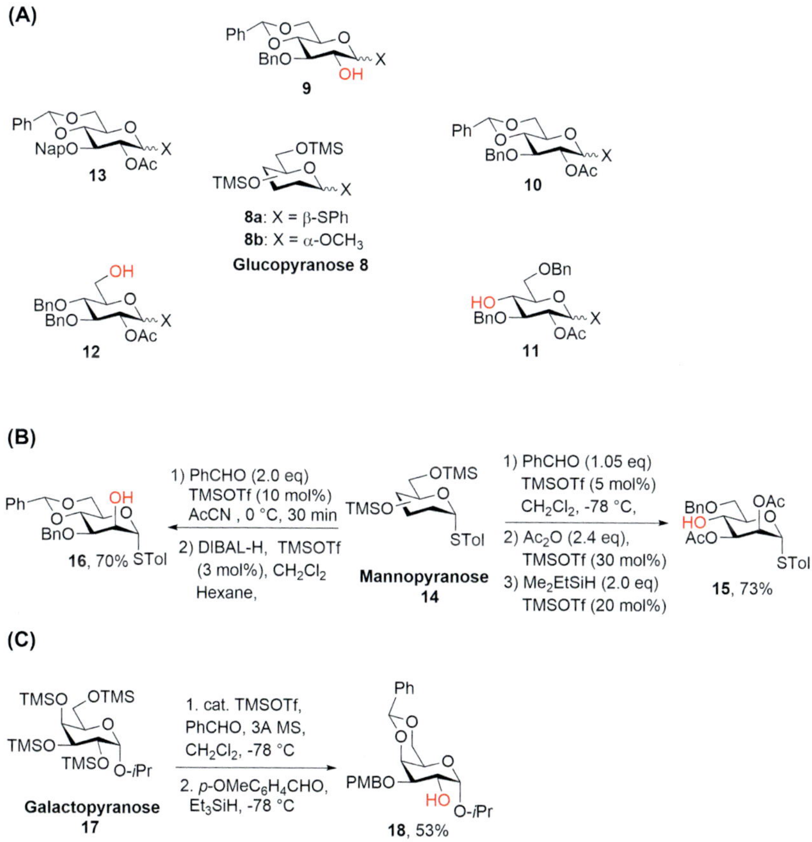

SCHEME 7.2 Regioselective one-pot protection strategy to access various alcohols using TMS glycosides.

group migration reactions in protected glucopyranosides and O3→O4 acyl group migration reactions in protected galactopyranosides.[57]

Wide applications of acyl migration in protecting group manipulation have been seen when treatment of tetra-butyl ammonium fluoride (TBAF) is done on *O*-silyl protected primary hydroxyl group having a close ester-protecting group.[58] This manipulation is used by many groups as the yield is excellent.

7.2.7 Acyl migrations in carbohydrate chemistry

A common route for the TBDMS deprotection and in situ acetate migration is depicted in Scheme 7.3:

7.2.8 *De novo* asymmetric synthesis of oligosaccharides using atom-less protecting groups

In this section, we will talk about the *de novo* asymmetric synthesis that tries to minimize the use of protecting groups. Enone functionality is the key to the atom-less protecting groups approach. This approach relies on mainly three concepts, the use of *de novo* asymmetric synthesis of simple glycosyl

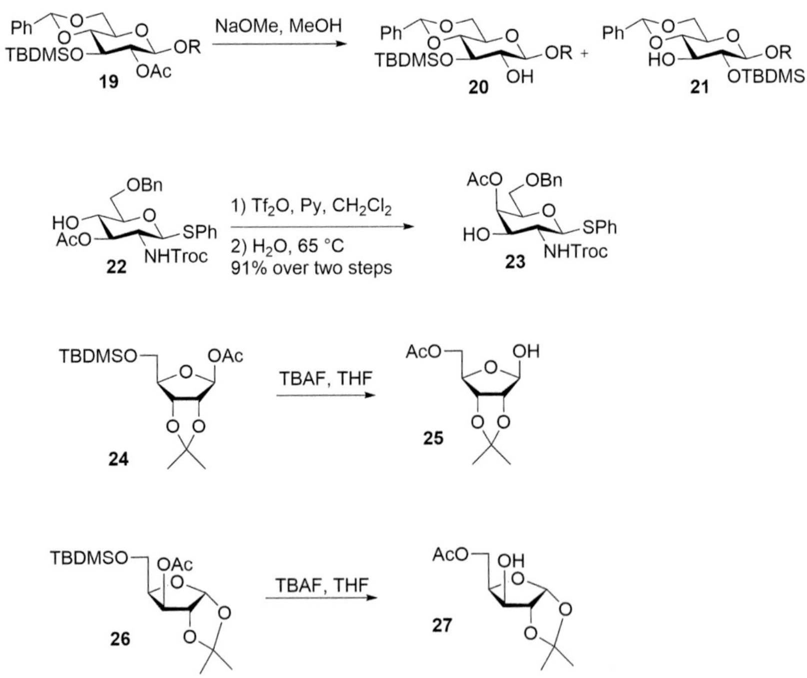

SCHEME 7.3 TBS deprotection and in situ acetate migration.

donors, the use of a Pd-π-allyl-catalyzed glycosylation reaction for installing the glycosidic bonds, and the use of a C-2 to C-4 enone functional group in the moiety as a precursor to the desired C-2 to C-4 hydroxyls of sugars. *De novo* synthetic approaches on carbohydrates were first utilized by Masamune and Sharpless to prepare possible hexoses using iterative asymmetric epoxidation of allylic alcohols.[59] Danishefsky demonstrated asymmetric hetero-Diels–Alder reaction for the synthesis of various sugars and extended to oligosaccharide synthesis.[60] MacMillan reported proline-catalyzed iterative aldol reaction to obtain several hexoses.

O'Doherty has made notable contributions to the development of *de novo* asymmetric syntheses of hexoses.[61] His approach involves an iterative asymmetric dihydroxylation of dienoates to produce sugar lactones and Achmatowicz's approach to oligosaccharides that led to atom-less protecting groups. In this context, the enone functionality in the pyranones was converted into simple C-4 alcohol and C-3,4 diols. The C4 alcohol **29** when subjected to diimide reduction of allylic alcohol affords amecitose **29a**. Mitsunobu inversion of the C-4 alcohol **29** gave inverted alcohol **30**, which upon diimide reduction furnished rhodinose **30a**. Upon Myers' 1,3-allylic reduction of the allylic alcohol **33**, pyrans with a C-3,4 double bond can be produced which upon a highly diastereoselective dihydroxylation affords digitoxose **34** (Scheme 7.4).

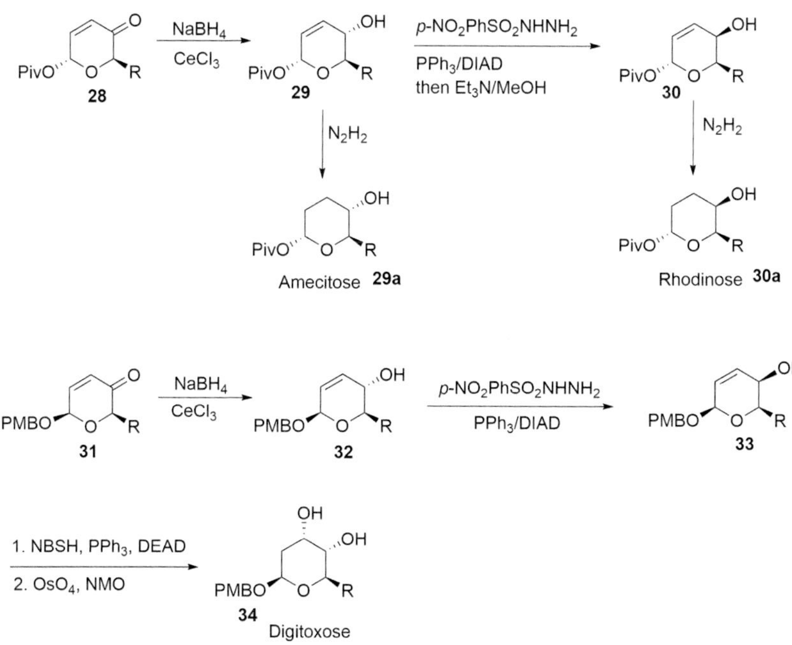

SCHEME 7.4 *De novo* asymmetric approach to deoxy-pyranones.

7.2.9 Protecting group strategies for sialic acid derivatives

Sialic acid is composed of different functional groups including multiple hydroxy, amide and an anomeric carboxylic acid. The very first protection on sialic acid is done on the carboxylic acid group and the most common group is ester protection mainly methyl ester, which is installed using sulfonic acid-type resin on methanol and subsequent per-*O*-acetylation. Or after per-*O*-acetylation on sialic acid, the carboxylic acid group is protected by diazomethane.[62] Other protections such as allyl ester, thioester, amide formation, and lactonization are also reported.

*N*5-amino group plays a dominant role in sialic acid chemistry. Available sialic acid comes with acetamido group, which in practice is first cleaved to generate the amine and then the amine protection is done. The common method of rupturing the acetamido group is refluxing in methanol in the presence of methanesulfonic acid or cleavage using barium hydroxide in water at 90°C.[63] For the free *N*5 amine, several protections are reported such as trifluoroacetamide function, the *N*-(9-fluorenylmethyl), *N*-allyl, and *N*-trichloroethyl carbamates, the *N*-phthalimido group, the *N*-*tert*-butoxycarbonylacetamide, and the azido function. Dual protection of *O*4 and *N*5 as oxazolidinone and dual protection of *O*7 and *N*5 as an oxazinone are also reported, which offer selectivity in the construction of α-glycosidic linkage.[64] The hydroxy groups are acetylated, alkylated, or silylated. Sometimes acetal protection is also seen to render rigidity to it.[65] Representative protecting group manipulation in most common sialic acid derivatives, Neu5Ac is depicted in Fig. 7.4.

7.2.10 Protecting group strategies toward glycofuranoses

Carbohydrates can cyclize in two different forms viz. six-membered rings (pyranose forms, pyranosides) or five-membered derivatives (furanose forms, furanosides). The main development of furanose forms of carbohydrates is clearly the structural elucidation of nucleic acids. Pentofuranoses and

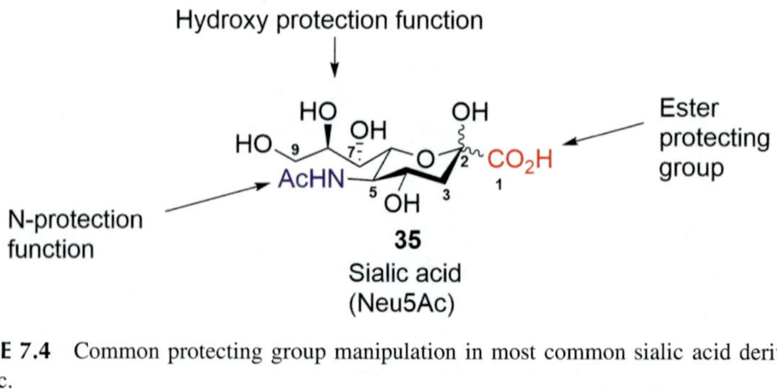

35

FIGURE 7.4 Common protecting group manipulation in most common sialic acid derivatives, Neu5Ac.

hexofuranoses exist in microorganisms such as archaea, fungi, and bacteria. To obtain a furanose ring, the major difficulty to address is to block and then ensure the control of the ring size and also to find specific and differentiating analytical data for these highly functionalized cycles.

Blocking the anomeric position of the carbohydrate is one of the most common approaches to achieve glycofuranoses. A notable example is Fischer glycosylation that proceeds through an oxacarbenium intermediate and results in furanoses and pyranoses both depending upon the nature of reactants, catalysts used, and the reaction conditions provided such as temperature, concentration, and reaction time. Some methodology uses direct protection of hexoses by ketalization, a common example involves protection of D-glucose and D-mannose by ketalization provides the thermodynamic product 1,2:5,6-diketalglucofuranose and 2,3:5,6-diketalmannofuranose leaving the 3-OH position free for glucose and anomeric position for mannose respectively. Ample examples of functionalization of 3-hydroxy of 1,2:5,6-diketalglucofuranose can be seen in carbohydrate chemistry. Sometimes 3-OH of compound **37** is protected with different ether-protecting groups, such as allyl, benzyl, and naphthylmethyl, to access different important carbohydrate building blocks, or sometimes the 3-OH functionalization can be seen in medicinal use.[66]

Unlike D-glucose and D-mannose, D-galactose is not straightforward in terms of ketalization, the orientation of the galactose structure does not favor the furanose form. The most common approaches used to prepare galactofuranoses are benzoylation and silylation (Scheme 7.5).[67]

7.2.11 Protecting group strategies toward sulfated glycosaminoglycans

Glycosaminoglycans (GAGs) are a family of complex polyanionic macromolecules linked to a core protein through a tetrasaccharidic linkage region to form what are called proteoglycans (PGs). GAGs are categorized into two types depending on the nature of the hexosamine (gluco or galactosamine): heparin/heparin sulfate (Hep/HS) (glucosaminoglycans) and chondroitin sulfate/dermatan sulfate (CS/DS) (galactosaminoglycans). The general approach is being taken to carry out O-sulfation of the free primary hydroxyl group or secondary hydroxyl group with SO_3.py or SO_3.Et_3N complex, while the rest of the hydroxyl groups are protected with some permanent protecting group (Fig. 7.5).[68]

7.2.12 Orthogonally protected building blocks for automated glycan assembly

Solid-phase oligosaccharide synthesis (SPOS) has an immense potential to assemble oligosaccharide in a short period of time. Oligosaccharide synthesizers can carry out temperature-controlled reactions and washes in a fully automated

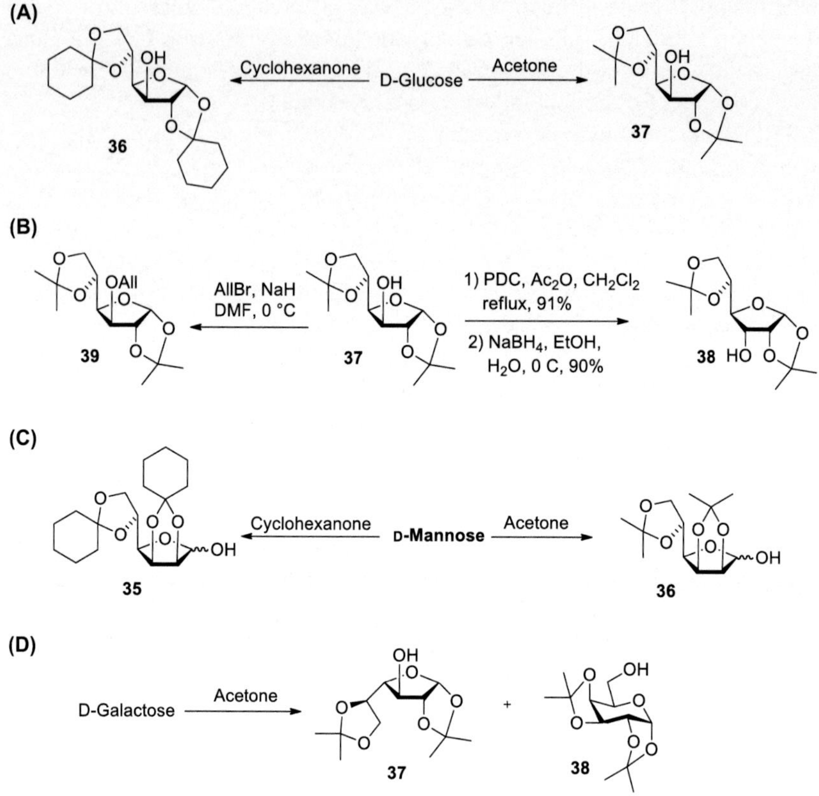

SCHEME 7.5 Ketalization of sugars leading to furanose forms.

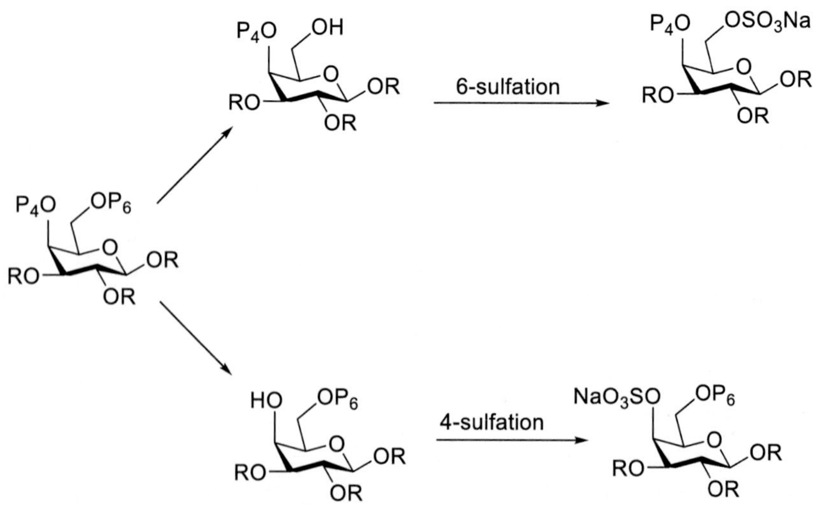

FIGURE 7.5 Sulfation at O4 and O6 of D-galactopyranoside.

TABLE 7.3 Building blocks for automated glycan assembly.

Permanent protecting groups	Deprotection	Nature of protecting group
Bn	H_2, Pd/C	Nonparticipating
N_3	H_2, Pd/C	Nonparticipating
Bz	NaOMe in MeOH	Participating
NHTCA	H_2, Pd/C	Participating
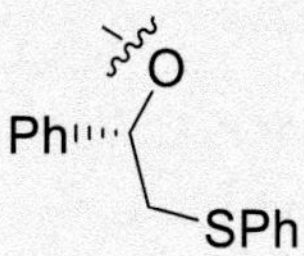	$BF_3.OEt_2/Ac_2O$ in DCM	Participating (1,2-*cis* selective)
Temporary protecting groups	**Deprotection**	**Nature of protecting group**
Fmoc	NEt_3 in DMF	Participating
Lev	$N_2H_4 \cdot AcOH$ in THF/H_2O	Participating
Nap	DDQ in $DCM/MeOH/H_2O$	Nonparticipating
TBDMS	TBAF in THF	Nonparticipating
Alloc	$Pd(PPh_3)_4$ in THF/AcOH	Participating
TOM	TBAF in THF	Nonparticipating

manner, saving the operator numerous manual operations. Many large and complex glycans have been produced by automated glycan assembly.[69] Again, protecting group is the key in SPOS for choosing building blocks. The most common protecting groups in SPOS are discussed in Table 7.3.

Compared to the development of peptide and oligonucleotide area, solid-phase oligosaccharide is much more complex due to the diversity of linkages in the glycans as well as the stereoselectivity of the glycosidic linkages.

7.3 Conclusions and future outlook

Protecting group chemistry can make or break any (oligo)saccharide synthesis effort. Much progress has been made over the years to understand and exploit reactivity differences between the functional groups on a carbohydrate and many efficient protecting group strategies, and schemes are now available. Even though these schemes may present multistep synthesis routes, they often involve optimized chemistry, assuring reliable synthetic outcomes. Nonetheless, there is a demand for ever-shorter synthetic routes, and the development of one-pot operations to introduce multiple protecting groups is therefore of high importance. The demand for more efficiency can also be met by the development of better and more effective protective groups.

References

1. Codée, J. D. C.; Ali, A.; Overkleeft, H. S.; van der Marel, G. A. *C.R. Chim.* **2011,** *14*, 178–193.
2. Guo, J.; Ye, X. S. *Molecules* **2010,** *15*, 7235–7265.
3. Wittmann, V. *Angew. Chem. Int. Ed.* **2006,** *45*, 3399–3400.
4. (a) Kocienski, P. J. *Protecting Groups*, 3rd ed.; Thieme: New York, 2005.
   (b) Wuts, P. G. M.; Greene, T. W. *Greene's Protective Groups in Organic Synthesis;* Wiley: Hoboken, NJ, 2007.
   (c) Vidal, S., Ed. *Protecting Groups: Strategies and Applications in Carbohydrate Chemistry;* Wiley-VCH, 2019.
   (d) Sahloul, K.; Lowary, T. L. *J. Org. Chem.* **2015,** *80*, 11417–11434.
5. (a) Kulkarni, S. S.; Wang, C. C.; Sabbavarapu, N. M.; Podilapu, A. R.; Liao, P. H.; Hung, S. C. *ACS Chem. Rev.* **2018,** *118*, 8025–8104.
   (b) Kulkarni, S. S. In *Selective Glycosylations: Synthetic Methods and Catalysts;* Bennett, C., Ed.; Wiley-VCH, 2017; pp 255–273. Chapter 12.
   (c) Ghosh, B.; Kulkarni, S. S. *Chem. Asian J.* **2020,** *15*, 450–462.
   (d) Protective Group Strategies; In *The Organic Chemistry of Sugars;* Oscarson, S., Ed.; *CRC Press*: Boca Raton, 2006; pp 53–85.
6. (a) Ogawa, T.; Yamamoto, H. *Carbohydr. Res.* **1985,** *137*, 79–88.
   (b) Liu, L. D.; Liu, H. W. *Tetrahedron Lett* **1989,** *30*, 35–38.
   (c) Patil, P. R.; Ravindranathan Kartha, K. P. *J. Carbohydr. Chem.* **2008,** *27*, 279–293.
7. (a) Almant, M.; Moreau, V.; Kovensky, J. *Chem. Eur. J* **2011,** *17*, 10029–10038.
   (b) Gabrielli, L.; Calloni, I.; Donvito, G. *Eur. J. Org. Chem* **2014,** 5962–5967.
   (c) Branchaud, B. P.; Meier, M. S. *J. Org. Chem* **1989,** *54*, 1320–1326.
8. (a) Bartoszewicz, A.; Kalek, M.; Stawinski, J. *Tetrahedron* **2008,** *64*, 8843–8850.
   (b) Barros, M. T.; Petrova, K. T.; Correia-da-Silva, P.; Potewar, T. M. *Green Chem* **2011,** *13*, 1897–1906.
   (c) Lellouche, J. P.; Koeller, S. *J. Org. Chem* **2001,** *66*, 693–696.
9. Craig, D.; Munasinghe, V. R. N.; Tierney, J. P. *Tetrahedron* **1999,** *55*, 15025–15044.
10. (a) Falconer, R. A. *Tetrahedron Lett* **2002,** *43*, 8503–8505.
    (b) Berkowitz, D. B.; Bose, M.; Pfannenstiel, T. J.; Doukov, T. *J. Org. Chem* **2000,** *65*, 4498–4508.
11. (a) Ellis, D.; Norman, S. E.; Osborn, H. M. I. *Tetrahedron* **2008,** *64*, 2832–2854.
    (b) Fokt, I.; Skora, S.; Conrad, C. *Carbohydr. Res* **2013,** *368*, 111–119.
12. (a) Branchaud, B. P.; Meier, M. S. *J. Org. Chem* **1989,** *54*, 1320–1326.
    (b) Haskins, W. T.; Hann, R. M.; Hudson, C. S. *J. Am. Chem. Soc* **1946,** *68*, 628–632.
13. (a) Almant, M.; Moreau, V.; Kovensky, J. *Chem. Eur. J* **2011,** *17*, 10029–10038.
    (b) Gabrielli, L.; Calloni, I.; Donvito, G. *Eur. J. Org. Chem* **2014,** 5962–5967.
14. (a) Nicolaou, K. C.; Caulfield, T. J.; Kataoka, H.; Stylianides, N. A. *J. Am. Chem. Soc* **1990,** *112*, 3693–3695.
    (b) Okada, Y.; Mukae, T.; Okajima, K. *Org. Lett* **2007,** *9*, 1573–1576.
15. (a) Ishihara, K.; Kurihara, H.; Yamamoto, H. *J. Org. Chem.* **1993,** *58*, 3791–3793.
    (b) Thompson, A. J.; Williams, R. J.; Hakki, Z. *Proc. Natl. Acad. Sci. USA* **2012,** *109*, 781–786.

16. (a) Morcuende, A.; Valverde, S.; Herradón, B. *Synlett* **1994,** 89–91.
(b) Demizu, Y.; Kubo, Y.; Miyoshi, H. *Org. Lett* **2008,** *10*, 5075–5077.
17. Liang, P. H.; Lu, Y. J.; Tang, T. H. *Tetrahedron Lett.* **2010,** *51*, 6928–6931.
18. (a) Somawardhana, C. W.; Brunngraber, eg, *Carbohydr. Res* **1983,** *121*, 51–60.
(b) Miethchen, R.; Zur, C. *J. Prakt. Chem* **1995,** *337*, 189–195.
19. Huang, L.; Teumelsan, N.; Huang, X. *Chem. Eur. J.* **2006,** *12*, 5246–5252.
20. (a) Nguyen, B. H.; Perkins, R. J.; Smith, J. A.; Moeller, K. D. *Beil. J. Org. Chem* **2015,** *11*, 280–287.
(b) Schnatbaum, K.; Schäfer, H. J. *Synthesis* **1999,** 864–872.
21. Angelin, M.; Hermansson, M.; Dong, H.; Ramström, O. *Eur. J. Org. Chem* **2006,** 4323–4326.
22. Falck, J. R.; Barma, D. K.; Venkataraman, S. K. *Tetrahedron Lett.* **2002,** *43*, 963–966.
23. (a) Lu, W.; Navidpour, L.; Taylor, S. D. *Carbohydr. Res* **2005,** *340*, 1213–1217.
(b) Wang, M.; Xu, Z.; Tu, P. *Bioorg. Med. Chem. Lett* **2004,** *14*, 2585–2588.
24. Plante, O. J.; Buchwald, S. L.; Seeberger, P. H. *J. Am. Chem. Soc.* **2000,** *122*, 7148–7149.
25. Nakatsuka, T.; Tomimori, Y.; Fukuda, Y.; Nukaya, H. *Bioorg. Med. Chem. Lett.* **2004,** *14*, 3201–3203.
26. Zhu, C. J.; Yi, H.; Chen, G. R.; Xie, J. *Tetrahedron* **2008,** *64*, 10687–10693.
27. (a) Lloyd, D.; Bylsma, M.; Bright, D. K.; Chen, X.; Bennett, C. S. *J. Org. Chem* **2017,** *82* (16), 3926–3934.
(b) Volbeda, A. G.; Kistemaker, H. A. V.; Overkleeft, H. S.; van der Marel, G. A.; Filippov, D. V.; Codee, J. D. C. *J. Org. Chem* **2015,** *80*, 8796–8806.
28. Smoot, J. T.; Pornsuriyasak, P.; Demchenko, A. V. *Angew. Chem. Int. Ed* **2005,** *44*, 7123–7126.
29. (a) Kulkarni, S. S.; Gervay-Hague, J. *Org. Lett* **2008,** *10*, 4739–4742.
(b) Wang, C. C.; Lee, J. C.; Luo, S. Y. *Nature* **2007,** *446*, 896–899.
(c) Wang, C. C.; Kulkarni, S. S.; Lee, J. C. *Nat. Protoc* **2008,** *3*, 97–113.
(d) Sarpe, V. A.; Kulkarni, S. S. *J. Org. Chem* **2011,** *76* (16), 6866–6870.
30. Ghotekar, B. K.; Podilapu, A. R.; Kulkarni, S. S. *Org. Lett.* **2020,** *22*, 537–541.
31. (a) Yasomanee, J. P.; Demchenko, A. V. *J. Am. Chem. Soc.* **2012,** *134*, 20097–20102.
(b) Behera, A.; Rai, D.; Kushwaha, D.; Kulkarni, S. S. *Org. Lett.* **2018,** *20*, 5956–5959.
32. (a) David, S.; Hanessian, S. *Tetrahedron* **1985,** *41*, 643–663.
(b) Demizu, Y.; Kubo, Y.; Miyoshi, H.; Maki, T.; Matsumura, Y.; Moriyama, N.; Onomura, O. *Org. Lett* **2008,** *10*, 5075–5077.
(c) Sanapala, S. R.; Kulkarni, S. S. *J. Am. Chem. Soc* **2016,** *138*, 4938–4947.
33. Dimakos, V.; Taylor, M. S. *Chem. Rev.* **2018,** *118*, 11457–11517.
34. Sanapala, S. R.; Kulkarni, S. S. *Org. Lett.* **2016,** *18*, 3790–3793.
35. Kim, Y. J.; Wang, P.; Navarro-Villalobos, M. *J. Am. Chem. Soc.* **2006,** *128*, 11906–11915.
36. (a) Hou, D.; Skogman, F.; Lowary, T. L. *Carbohydr. Res* **2008,** *343*, 1778–1789.
(b) Bodine, K. D.; Gin, D. Y.; Gin, M. S. *J. Am. Chem. Soc* **2004,** *126*, 1638–1639.
(c) Biswas, S.; Ghotekar, B. K.; Kulkarni, S. S. *Org. Lett.* **2021,** *23*, 6137–6142.
37. Al-Mughaid, H.; Grindley, T. B. *J. Org. Chem.* **2006,** *71*, 1390–1398.
38. Yang, Y.; Oishi, S.; Martin, C. E.; Seeberger, P. H. *J. Am. Chem. Soc.* **2013,** *135*, 6262–6271.
39. (a) Roussel, F.; Knerr, L.; Schmidt, R. R. *Eur. J. Org. Chem* **2001,** 2067–2073.
(b) Ronchi, P.; Scarponi, C.; Salvi, M. *J. Org. Chem* **2013,** *78*, 5172–5183.
40. (a) Dasgupta, S.; Nitz, M. *J. Org. Chem* **2011,** *76*, 1918–1921.
(b) Huang, S.; Liao, J.; Zhao, Q. *J. Carbohydr. Chem* **2013,** *32*, 158–168.

41. Maryanoff, B. E.; Costanzo, M. J.; Nortey, S. O. *J. Med. Chem.* **1998,** *95*, 1315–1343.
42. (a) Chéry, F.; Rollin, P.; De Lucchi, O.; Cossu, S. *Tetrahedron Lett* **2000,** *41*, 2357–2360.
    (b) Chéry, F.; Rollin, P.; De Lucchi, O.; Cossu, S. *Synthesis* **2001,** 2, 286–292.
43. Sarpe, V. A.; Kulkarni, S. S. *Org. Lett.* **2014,** *16*, 5732–5735.
44. (a) Ley, S. V.; Osborn, H. M. I.; Priepke, H. W. M.; Warriner, S. L. *Org. Synth* **1998,** *75*, 170–173.
    (b) Shing, T. K. M.; Luk, T. *Tetrahedron: Asymmetry* **2009,** *20*, 883–886.
45. (a) Douglas, N. L.; Ley, S. V.; Osborn, H. M. I. *Synlett* **1996,** 793–795.
    (b) Wang, C. C.; Tang, Y.; Yang, K. *Tetrahedron* **2013,** *69*, 8284–8290.
    (c) Kongkathip, B.; Kongkathip, N.; Rujirawanich, J. *Synth. Commun* **2014,** *44*, 2248–2255.
46. (a) Ohlin, M.; Johnsson, R.; Ellervik, U. *Carbohydr. Res* **2011,** *346*, 1358–1370.
    (b) Panchadhayee, R.; Misra, A. K. *Synlett* **2010,** 1193–1196.
    (c) Herczeg, M.; Demeter, F.; Mezo, E. *Eur. J. Org. Chem.* **2015,** 5730–5741.
47. (a) Barett, A. G. M.; Pena, M.; Willardsen, J. A. *J. Org. Chem* **1996,** *61*, 1082–1100.
    (b) Hoberg, J. O. *Carbohydr. Res* **1997,** *300*, 365–367.
    (c) Mori, Y.; Hayashi, H. *J. Org. Chem* **2001,** *66*, 8666–8668.
48. Blanco, J. J.; Fernandez, J. G.; Gadelle, A.; Defaye, J. *Carbohydr. Res.* **1997,** *303*, 367–372.
49. Thota, N.; Bjorkling, F.; Kankala, S. *ChemistrySelect* **2019,** *4*, 7976–7980.
50. Enugala, R.; Carvalho, L. C. R.; Pires, M. J. D.; Marques, M. M. B. *Chem. Asian J.* **2012,** *7*, 2482–2501.
51. Benakli, K.; Zha, C.; Kerns, R. J. *J. Am. Chem. Soc.* **2001,** *123*, 9461–9462.
52. Mensah, E. A.; Yu, F.; Nguyen, H. M. *J. Am. Chem. Soc.* **2010,** *132*, 14288–14302.
53. (a) Rai, D.; Kulkarni, S. S. *Org. Lett.* **2023,** *25*, 1509–1513.
    (b) Ghosh, B.; Bhattacharjee, N.; Podilapu, A. R.; Puri, K.; Kulkarni, S. S. *Org. Lett* **2022,** *24*, 3696–3701.
54. Helferich, B.; Becker, J. *Ann* **1924,** *440*, 1–18.
55. Procopiou, P. A.; Baugh, S. P. D.; Flack, S. S.; Inglis, G. G. A. *Chem. Commun.* **1996,** 2625–2626.
56. Shie, C. R.; Tzeng, Z. H.; Kulkarni, S. S. *Angew. Chem. Int. Ed* **2005,** *44*, 1665–1668.
57. (a) Adak, S.; Emmadi, M.; Kulkarni, S. S. *RSC Adv* **2014,** *4*, 7611–7616.
    (b) Emmadi, M.; Kulkarni, S. S. *J. Org. Chem* **2011,** *76*, 4703–4709.
58. Chevallier, O. P.; Migaud, M. E. *Beilstein J. Org. Chem.* **2006,** *2* (14). No.
59. (a) Ko, S. Y.; Lee, A. W. M.; Masamune, S. *Science* **1983,** *220*, 949–951.
    (b) Ko, S. Y.; Lee, A. W. M.; Masamune, S. *Tetrahedron* **1990,** *46*, 245–264.
60. Gordon, D. M.; Danishefsky, S. J. *Carbohydr. Res.* **1990,** *206*, 361.
61. Harris, J. M.; Li, M.; Scott, J. G.; O'Doherty, G. A. In *Strategies and Tactics in Organic Synthesis;* Harmata, M., Ed.; Vol. 5; Elsevier, 2004; pp 221–253.
62. (a) Paulsen, H.; Tietz, H. *Carbohydr. Res* **1984,** *125*, 47–64.
    (b) Myers, R. W.; Lee, R. T.; Lee, Y. C. *Anal. Biochem* **1980,** *101*, 166–174.
63. Fujita, S.; Numata, M.; Sugimoto, M. *Carbohydr. Res.* **1992,** *228*, 347–370.
64. (a) Tanaka, H.; Nishiura, Y.; Takahashi, T. *J. Am. Chem. Soc* **2006,** *128*, 7124–7125.
    (b) Kancharla, P. K.; Crich, D. *J. Am. Chem. Soc* **2013,** *135*, 18999–19007.
65. (a) Wu, Y. F.; Tsai, Y. F.; Guo, J. R. *Org. Biomol. Chem* **2014,** *12*, 9345–9349.
    (b) Yuan, X.; Ress, D. K.; Linhardt, R. J. *J. Org. Chem* **2007,** *72*, 3085–3088.
66. (a) Einhorn, C.; Luche, J. L. *Carbohydr. Res* **1986,** *155*, 258–261.
    (b) Mulani, S. K.; Guh, J. H.; Mong, K. K. T. *Org. Biomol. Chem* **2014,** *12*, 2926–2937.

(c) Guo, J.; Frost, J. W. *J. Am. Chem. Soc.* **2002,** *124*, 10642–10643.

(d) Kienle, M.; Eisenring, P.; Stoessel, B.; Horlacher, O. P.; Hasler, S.; van Colen, G.; Hartkoorn, R. C.; Vocat, A.; Cole, S. T.; Altmann, K. H. *J. Med. Chem.* **2020,** *63*, 1105–1131.

67. (a) Cattiaux, L.; Sendid, B.; Collot, M. *Bioorg. Med. Chem* **2011,** *19*, 547–555.

(b) Dureau, R.; Legentil, L.; Daniellou, R.; Ferrieres, V. *J. Org. Chem* **2012,** *77*, 1301–1307.

68. (a) Tamura, J. I.; Nishihara, J. *Bioorg. Med. Chem. Lett* **1999,** *9*, 1911–1914.

(b) Tamura, J. I.; Nishihara, J. *J. Org. Chem* **2001,** *66*, 3074–3083.

(c) Jacquinet, J. C. *Carbohydr. Res* **2006,** *341*, 1630–1644.

(d) Goto, F.; Ogawa, T. *Tetrahedron Lett* **1992,** *33*, 5099–5102.

(e) Rio, S.; Beau, J. M.; Jacquinet, J. C. *Carbohydr. Res* **1994,** *255*, 103–124.

69. (a) Seeberger, P. H. *Chem. Soc. Rev* **2008,** *37*, 19–28.

(b) Seeberger, P. H. *Acc. Chem. Res* **2015,** *48*, 1450–1463.

Chapter 8

Synthetic chemistry of D-fructose

Tanmaya Pathak
Department of Chemistry, Indian Institute of Technology Kharagpur, Kharagpur, West Bengal, India

8.1 Introduction

Among the constituent monosaccharides, the most common, inexpensive, and easily accessible sugars, D-glucose, D-ribose, D-xylose, D-galactose, and so on, are widely used in chemical modifications.[1,2] Chemists have long been fascinated by the preparation of enantiopure compounds, which is very essential in natural product synthesis as well as in medicinal chemistry.[1,2] Intense effort has gone into the development of chiral auxiliaries, chiral reagents, and chiral catalysts that have in many cases been modified to a high level of efficiency. A lot of these synthetic tools are based on amino acids, terpenes, and alkaloids. Carbohydrates, the most abundant and one of the important biomass precursors[3] in the world used as "chiral pool," have received considerably less attention. Carbohydrate molecules exist in different architectures with several polar functional groups. By modifying the functional groups, reactive and coordinative sites can be introduced into the carbohydrate framework. Exploiting this concept, several valuable synthetic tools have been developed.[1–3]

However, the utilization of sugar molecules as "chiral pools" in synthetic chemistry is crucially dependent on the availability of partially protected carbohydrate molecules. The nonfood utilization of comparatively inexpensive, ton-scale accessible ketose, namely D-fructose, is considerable, and the market size is estimated to reach $6.4 billion by 2027.[4] Attempts were also made in the past[5] and research continued to improve the utility of ketoses as useful starting materials.[6,7] For the high industrial demand of D-fructose for food and beverage applications, the reversible isomerization of glucose to fructose is carried out on a large industrial scale in an aqueous phase with the enzyme D-glucose/xylose isomerase (GI, EC 5.3.1.5). For several reasons, enzyme activity is still low from an economic point of view. One alternative

Synthetic Strategies in Carbohydrate Chemistry. DOI: https://doi.org/10.1016/B978-0-323-91729-2.00017-3

route, for example, makes use of commercially available zeolite catalysts containing only silicon and aluminum for getting unprecedentedly high yields of fructose from glucose in alcohol and aqueous media using as depicted in Scheme 8.1.[8]

Unfortunately, despite its easy availability, the chemistry of D-fructose is considerably complicated and less well understood because, in solution, D-fructose exists (Scheme 8.2) in two pyranoid (**1Pα + 1Pβ**) and two furanoid (**1Fα + 1Fβ**) tautomers with very little acyclic tautomer **1A** (Scheme 8.2).[5] However, a later study on the stabilization of the acyclic tautomers of reducing carbohydrates indicated that the stabilization of the acyclic tautomers of reducing carbohydrates such as D-fructose could be achieved by creating a hydrophobic microenvironment around the carbonyl group.[9]

8.2 D-Fructose in synthetic chemistry

8.2.1 Di-D-fructose dianhydrides

A unique family of stereoisomeric spiro-tricyclic disaccharides, di-D-fructose dianhydrides (DFAs) is formed from unprotected sucrose and/or D-fructose-rich materials under thermal and/or acidic activation conditions. These are mono- or dispirocyclic acetals, obtained by condensation of two fructose molecules, with the formation of two reciprocal glycosidic linkages resembling motifs found in many natural products, including steroidal saponins, polyether ionophores, macrolide antibiotics, insect pheromones, and toxic metabolites from algae and fungi. Spiroketalization is a reversible process under normal caramelization conditions. Consequently, the isomeric distribution of DFAs in the mixture can be rationalized in terms of kinetic and thermodynamic control. The DFA formation involves a fructosyl oxacarbenium cation **2b** from D-fructose **2a** that undergoes in situ glycosylation to the corresponding keto-disaccharide **3** by reacting with a second fructose molecule.

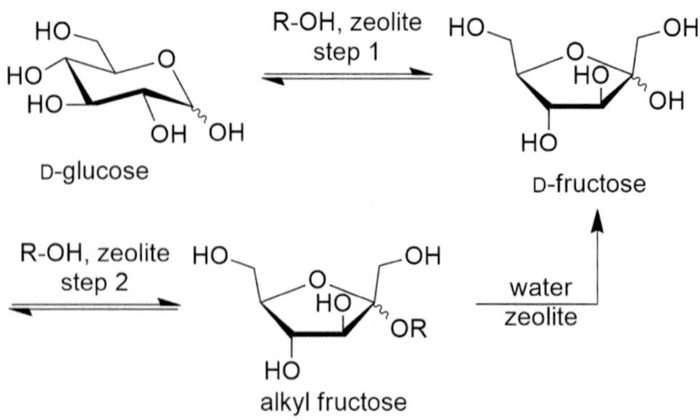

SCHEME 8.1 Conversion of glucose to fructose using zeolites.

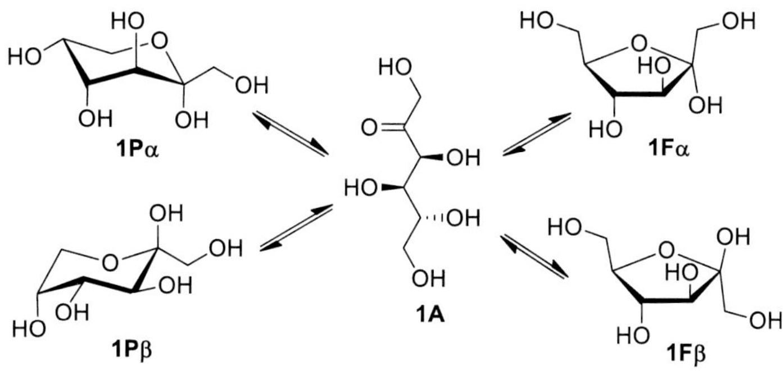

SCHEME 8.2 Tautomeric forms of D-fructose in solution.

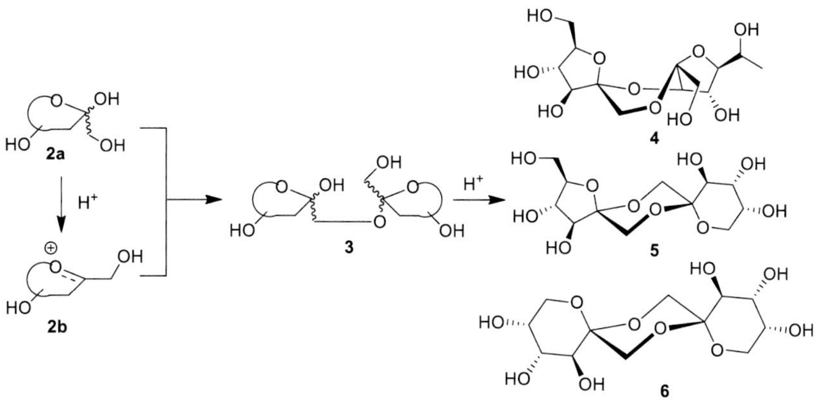

SCHEME 8.3 Acid-catalyzed formation of di-D-fructose dianhydrides (selected structures).

Further intramolecular spiroketalization to close the central 1,4-dioxane ring to afford DFAs **4–6** is a reversible process (Scheme 8.3).[10]

DFA derivatives bearing diverse functional groups on their primary positions (i.e., halogens, azide, thioether, amine, or amide) have shown remarkable properties as liquid crystals. Some members of this library are excellent precursors of hydrophilic polymers and noncationic surfactants. Recently, the utility of DFAs in food products and their remarkable nutritional features have attracted considerable interest from the food industry. DFAs behave as low-caloric sweeteners and show bifidogenic, anticarciogenic, and antitooth decaying effects.[10] Several DFA isomers are involved in intestinal metal cation absorption.[10,11]

8.2.2 Partial protection of D-fructose

Methods were developed over the years to generate a plethora of partially protected D-fructoses, both in pyranoside and furanoside forms (Scheme 8.4).

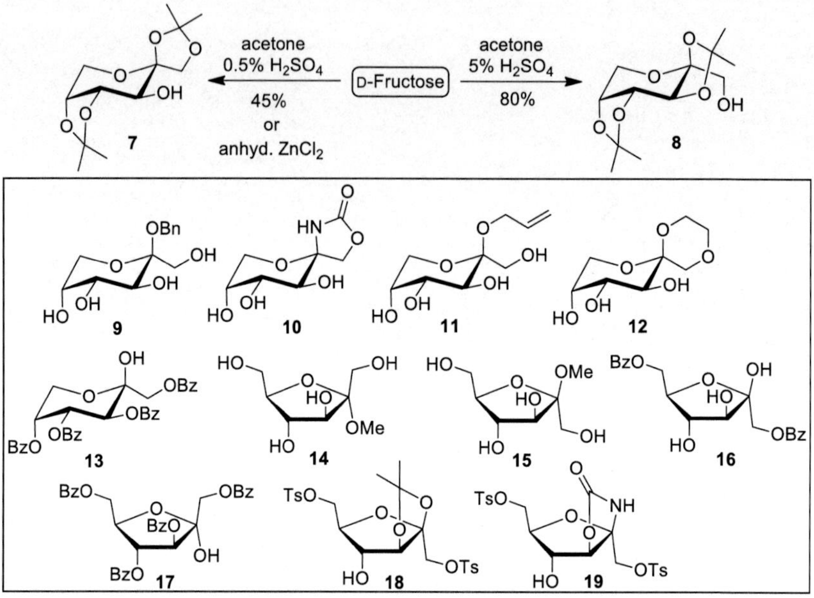

SCHEME 8.4 Partially protected D-fructose pyranosides and furanosides.

However, a slight variation in the reaction conditions led to a mixture. For example, it was observed that **7** (with free secondary alcohol) is the product of the kinetically controlled condensation of D-fructose with acetone; it is observed first, probably because of the higher reactivity of the primary hydroxyl group on C-l. Diacetal **8** (with free CH_2OH) was the major product at equilibrium and thus was thermodynamically more stable than **7**. In acetone containing 5% of sulfuric acid, both compounds were converted into an equilibrium mixture (3:47) of 1,2:4,5- and 2,3:4,5-diacetal within 5 minutes at room temperature, but, the anhydrous zinc chloride catalyzed reaction, the 1,2:4,5-diacetal **7** did not rearrange to the 2,3:4,5-diacetal.[12] Compound **7** was also obtained in 45% yield by treating D-fructose in 0.5% H_2SO_4 in acetone for 2 hours at room temperature (Scheme 8.4).[13]

However, the search for a better method for selective protection continued and several partially protected derivatives **9–19** were synthesized.[5] Thus sucrose, directly on reactions with 2,2-dimethoxypropane under acidic conditions, afforded the pentosyl derivative in varying yields in three different solvents.[14] Interestingly, D-sucrose **20** underwent acetonation in 1,4-dioxane to afford **21** in 68% yield (after acetylation), which was generated along with six different isomers minor products (Scheme 8.5).[15]

An efficient procedure for the synthesis of α-methyl 1,3-O-isopropylidene-fructofuranoside has been developed by treating D-fructose with acidic MeOH overnight to afford a 1:1 anomeric mixture that could be separated

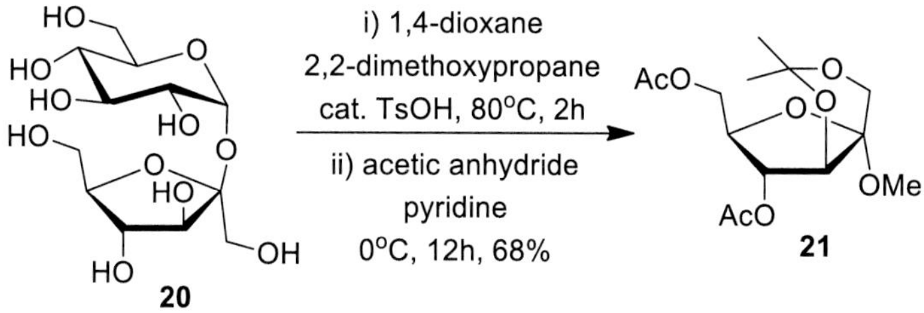

SCHEME 8.5 Synthesis of protected D-fructose from sucrose.

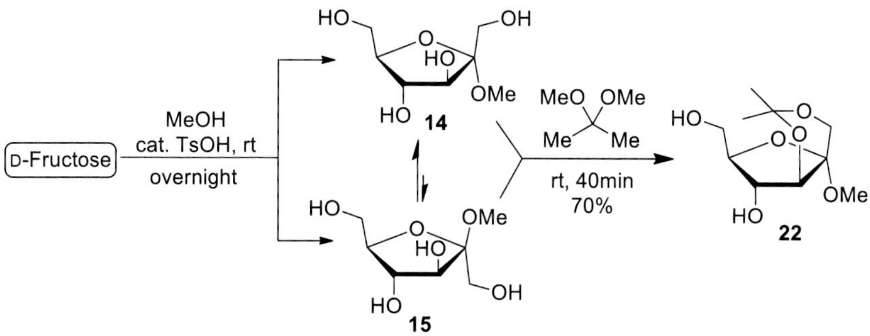

SCHEME 8.6 Interconversion of fructofuranosides during acetonation.

with difficulties. Surprisingly, each anomer on treatment with 2,2-dimethoxypropane at room temperature afforded a single compound **22** (Scheme 8.6) probably because the *trans-oriented* 1,3-dihydroxyl group in β-methyl fructofuranoside **15** equilibrated under mildly acidic conditions, to the α-anomer **14**. Thus the final product **22** was obtained by avoiding the tedious separation of anomers.[16]

8.2.3 Epoxides derived from D-fructose

In general, carbohydrate-derived epoxides are extensively used as synthetic intermediates because of their easy accessibility. Epoxides react with a variety of nucleophiles under neutral, acidic, or basic conditions due to their ring strain affording regioselective products.[17a,b,c] Due to their ring strain, epoxides react with a variety of nucleophiles under neutral, acidic, or basic conditions. The reactions are almost always regioselective and the regioselectivity is dependent on the structural aspects including steric hindrance around the epoxide carbons. However, interest in this class of compounds has been renewed with the recognition of epoxides as a class of "spring-loaded" groups suitable for "click" reactions.[18] Over the years, several furanose and pyranose epoxides **24–30** shown in Scheme 8.7 were synthesized from D-fructose, and those molecules were utilized as important synthetic

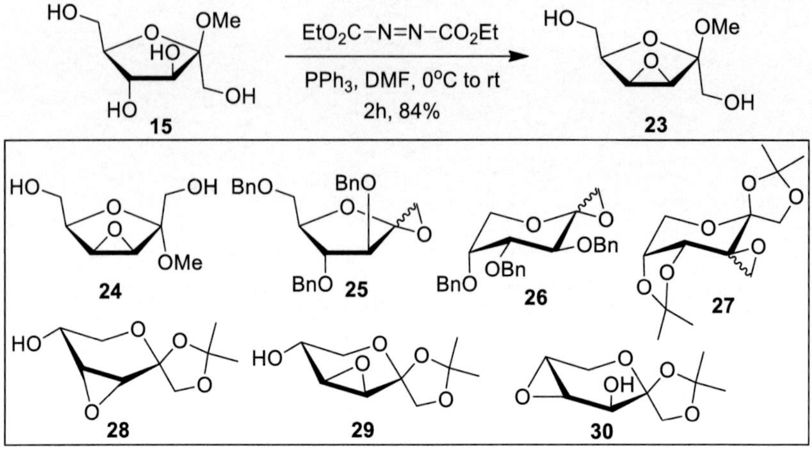

SCHEME 8.7 Epoxide synthesis under Mitsunobu reaction.

intermediates.[17a,b,c] Although the traditional leaving group-based strategy may be used for epoxide synthesis, the unprotected methylfructofuranoside was directly converted to the β-epoxide, where the methyl glycoside **15** was converted to β-epoxide **23** (Scheme 8.7) under Mitsunobu reaction conditions.[19a] However, the search for more efficient methods using phosphorous-based reagents continued.[19b]

Although amino- and thio-sugars in general were easily accessible from epoxides of D-glucose and other aldoses, these reactions usually required high temperatures and long reaction hours.[20a] The epoxide of D-fructose failed to react with neat primary amines or primary amines in acetonitrile even at an elevated temperature. In the past, metal perchlorates, such as Zn $(ClO_4)_2 0.6H_2O$, $Mg(ClO_4)_2 0.6H_2O$, and $NaClO_4 0.6H_2O$, have been used as catalysts for the opening of epoxide rings. However, $LiClO_4 0.6H_2O$ was shown to possess better regioselectivity in the epoxide opening reactions; the strong electron-withdrawing counteranion (ClO_4^-) makes the central metal ion (Li^+) more oxophilic, which helps in the better coordination process. It was expected that Li^+ in acetonitrile would strongly coordinate with the oxygen of the epoxide and increase the electrophilicity of the carbon atoms of the epoxide ring; the process would facilitate the opening of the ring by amines. The reactions of amines with epoxides of D-fructose and L-sorbose (see later) would lead to the preparation of analogs of fructosamine, which was considered as "a truly unique carbohydrate structure."[20b] Thus appropriately protected D-fructofuranoside and fructopyranoside derived epoxides were used as substrates for epoxide ring opening reactions.

To synthesize epoxides **33** and **34**, the easily accessible epoxide **24** was selectively tritylated to afford **32**, which was benzylated to produce fully protected epoxide **33**. To decrease the steric crowding for nucleophilic attack on the epoxide ring, the C-6 trityl group was deprotected under acidic conditions to produce the 6-hydroxylated epoxide **34** (Scheme 8.8).[20a]

Dibenzylated hexofuranoside **31** on treatment with benzylamine and cyclohexylamine in the presence of lithium perchlorate ($LiClO_4$, $6H_2O$) produced single diastereomeric aminoalcohol derivatives **35a** and **35b**, respectively (Scheme 8.9). Comparatively bulky 2,2-dimethylethanolamine did not react with epoxide 31 under similar reaction conditions.

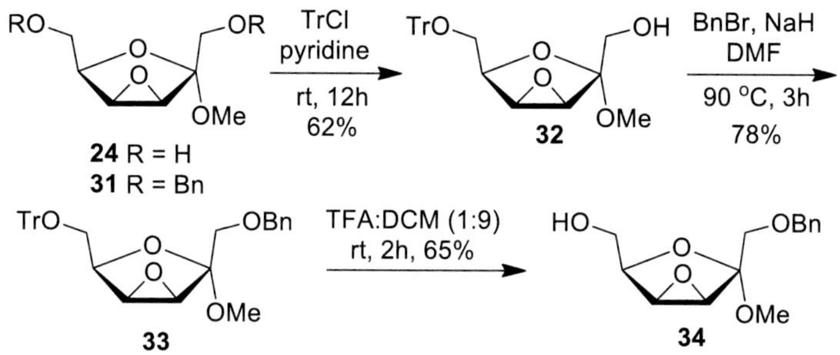

SCHEME 8.8 Synthesis of 3,4-anhydro-D-fructofuranoside **33** and **34**.

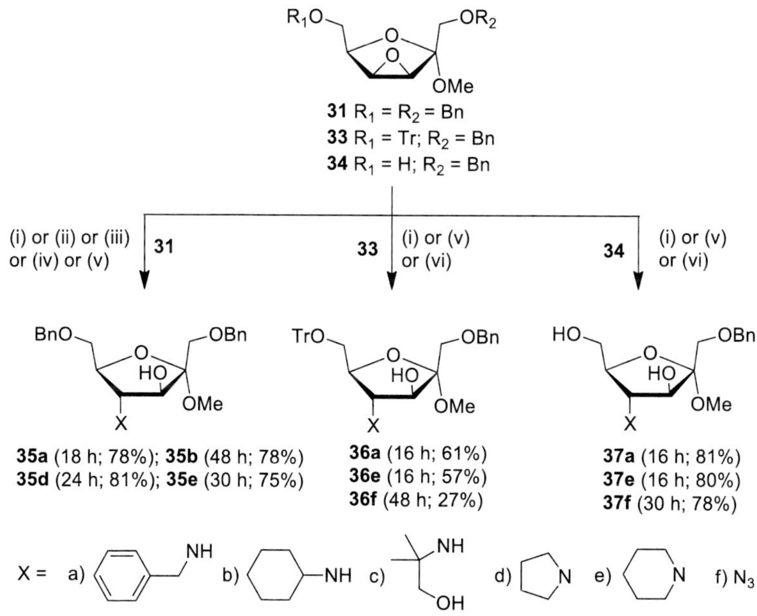

Reagents and conditions: (i) benzylamine, $LiClO_4$, MeCN, 80 °C; (ii) cyclohexylamine, $LiClO_4$, MeCN, 80 °C; (iii) 2-amino-2-methyl-1-propanol, $LiClO_4$, MeCN, 80 °C; (iv) pyrrolidine, 70 °C; (v) piperidine, 70 °C; (vi) LiN_3, EtOH, 80 °C.

SCHEME 8.9 Reactions of 3,4-anhydro-D-fructofuranoside with amines and azide.

Secondary amines such as pyrrolidine and piperidine, which are more efficient nucleophiles, reacted with epoxide **31** in the absence of solvent and $LiClO_40.6H_2O$ to produce single diastereomeric aminoalcohols **35d** and **35e**, respectively. LiN_3 did not react with **31** even in refluxing ethanol. The epoxide ring of fructofuranoside **33** and **34** was regioselectively opened by benzylamine and piperidine as well as by LiN_3 to generate **36a**, **36e–f**, and **37a**, **37e–f**, respectively. In the case of epoxide **34**, the epoxide ring was opened more easily to afford **37a,e,f** perhaps because of the less crowded C-4 (Scheme 8.9).

The corresponding pyranosyl epoxide derivative of D-fructose, on treatment with benzylamine, cyclohexylamine, and 2,2-dimethylethanolamine in the presence of $LiClO_40.6H_2O$ afforded single diastereomeric aminoalcohols **39a–c**, respectively, in good yields. Neat pyrrolidine and piperidine reacted with hexopyranoside **38** to afford aminoalcohols **39d** and **39e**, respectively; LiN_3 yielded the azidoalcohol **39f** (Scheme 8.10).[20a]

To expand the scope of this study, 1,2-*O*-(1-methylethyllidene)-3,4-anhydro-α-L-piscofuranose **40** derived from L-sorbose (see later), was benzylated to afford the benzyl protected epoxide.[41] Benzylamine, cyclohexylamine, 2,2-dimethylethanolamine, pyrrolidine, piperidine, and LiN_3 regioselectively afforded single diastereomers **42a–f** in moderate-to-good yields (Scheme 8.11). Since secondary amines easily reacted with epoxides, a bisnucleophilic secondary amine piperazine (0.5 equiv.) reacted with oxiranes **31**, **38**, and **41** in the presence of $LiClO_40.6H_2O$ in acetonitrile at 80°C. The expected dimers **43**, **44**, and **45** were obtained in good yields (Scheme 8.12).[20a]

8.2.4 Chiral ketones from D-fructose

1,2;4,5-Diisopropylidene-D-erythro-2,3-hexodiuro-2,6-pyranoside **46**, a chiral ketone was synthesized from the protected D-fructose **7**. Ketone **46** was

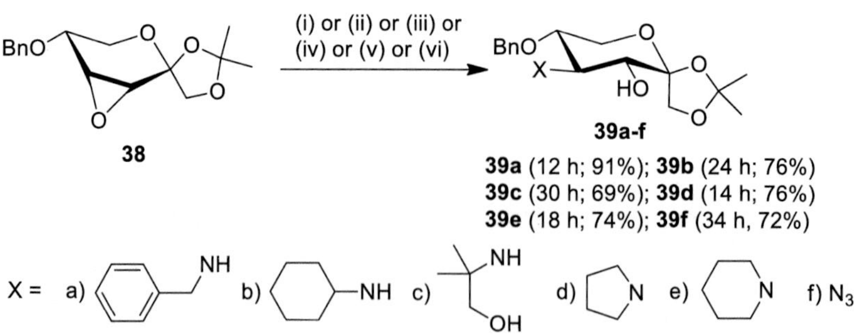

Reagents and conditions: (i) benzylamine, $LiClO_4$, MeCN, 80 °C; (ii) cyclohexylamine, $LiClO_4$, MeCN, 80 °C; (iii) 2-amino-2-methyl-1-propanol, $LiClO_4$, MeCN, 80 °C; (iv) pyrrolidine, 70 °C; (v) piperidine, 70 °C; (vi) LiN_3, EtOH, 80 °C.

SCHEME 8.10 Reactions of 3,4-anhydro-D-fructopyranoside with amines and azide.

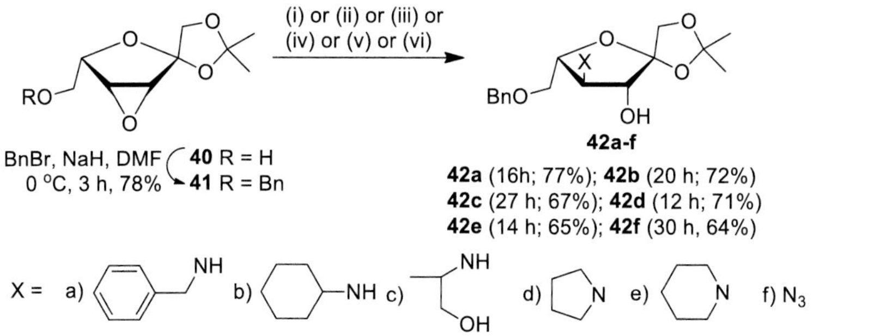

Reagents and conditions: (i) benzylamine, $LiClO_4$, MeCN, 80 °C; (ii) cyclohexylamine, $LiClO_4$, MeCN, 80 °C; (iii) 2-amino-2-methyl-1-propanol, $LiClO_4$, MeCN, 80 °C; (iv) pyrrolidine, 70 °C; (v) piperidine, 70 °C; (vi) LiN_3, EtOH, 80 °C.

SCHEME 8.11 Reactions of 3,4-anhydro-L-tagatofuranoside with amines and azide.

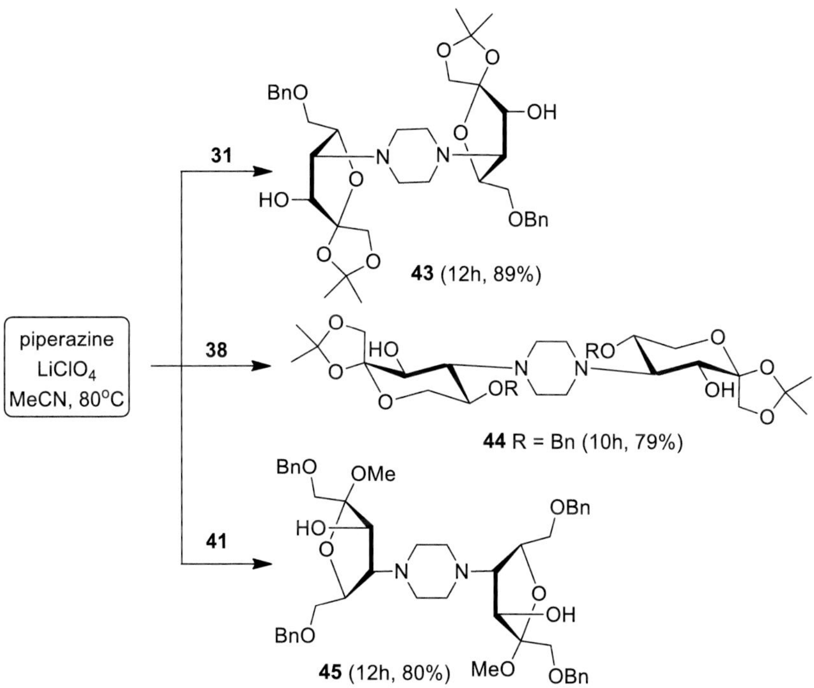

SCHEME 8.12 Piperazine-linked pseudodisaccharides.

reduced to **47** and the acid treatment re-equilibrated it to the partially protected D-psicose derivative **48**; the latter was converted to the target (+)-hydantocidin **49** in 19% overall yield from D-fructose (Scheme 8.13).[21]

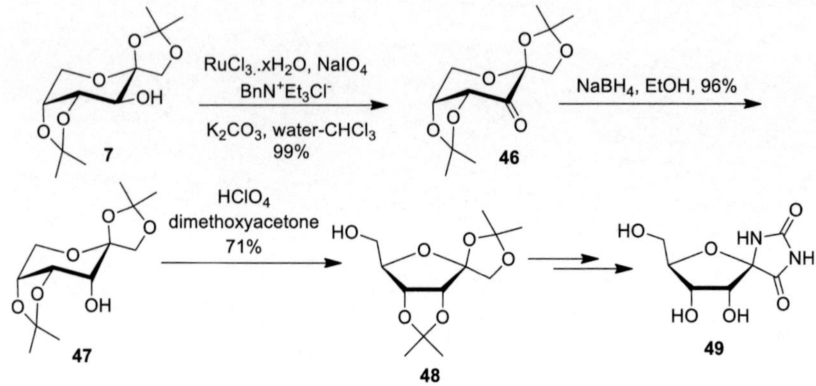

SCHEME 8.13 Synthesis of (+)-hydantocidin from D-fructose.

However, ketohexose was rapidly developed into a new class of asymmetric epoxidation catalysts for the epoxidation of a wide range of olefins. Originally, the bisprotected ketone **46** was used as a catalyst in combination with oxone as the oxidant for the effective asymmetric epoxidation of *trans*- and trisubstituted olefins under mild conditions. Thus the olefin **50** was converted to epoxide **51** using ketone **46** as the catalyst (Scheme 8.14). The enantioselectivity was high and a variety of functional groups was tolerated.[22a] Extensive studies led to excellent results for other unsaturated systems.[22b–e] Chiral 1,3-oxazolidine-2-thiones (OZTs) **55** and **56** were used as chiral auxiliaries in asymmetric reactions. These compounds were synthesized from D-fructose via epoxide intermediates.[23] Several carbocyclic analogs **57a–d** (Scheme 8.14) of D-fructose were also synthesized and studied for their catalytic activities.[24]

The catalytic efficiency of one such ketone, **52** was utilized in the asymmetric preparation of an intermediate **59** from the olefinic compound **58** in the "first asymmetric" total synthesis of (+)-coriandrone A **60** (Scheme 8.15).[25]

8.2.5 1,5-Anhydro-D-fructose

1,5-Anhydro-D-fructose (AF) **62** was first identified in fungi and red algae. It was produced by the degradation of glycogen, starch, and maltosaccharides with α-1,4-glucan lyase (EC 4.2.2.13). The access to the monoketosaccharide **62** on a large scale has now been achieved either by enzymatic or by chemical processes. The simplest method for the preparation of **62** was the basic deacetylation (MeOH/NaOMe) of tetra-*O*-acetyl-2-hydroxy-D-glucal **61**, the acetylated enol ether at low temperature for the deacetylation (Scheme 8.16).[26a,b]

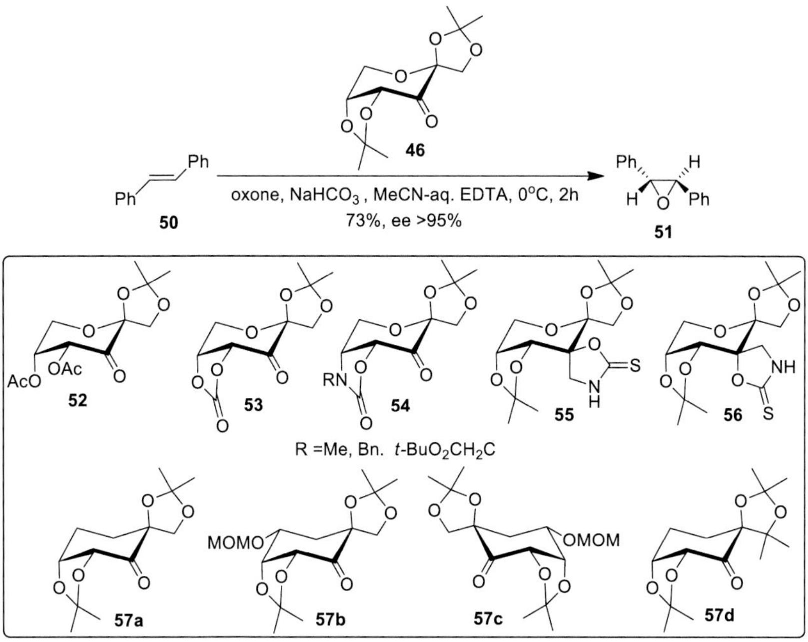

SCHEME 8.14 Chiral ketone **46** catalyzed epoxidation of *trans*-stilbene.

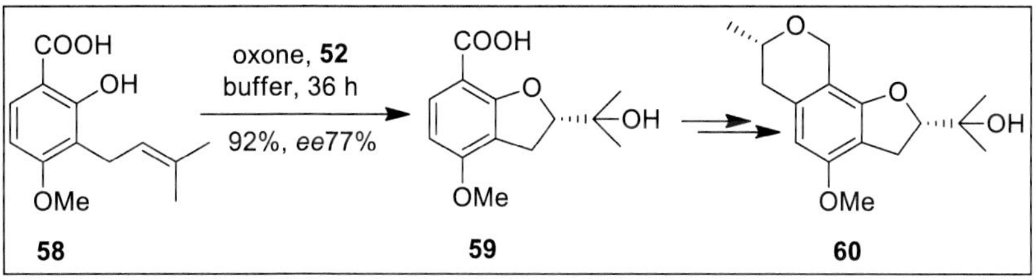

SCHEME 8.15 Synthesis of (+)-coriandrone A via ketone **52** mediated asymmetric induction.

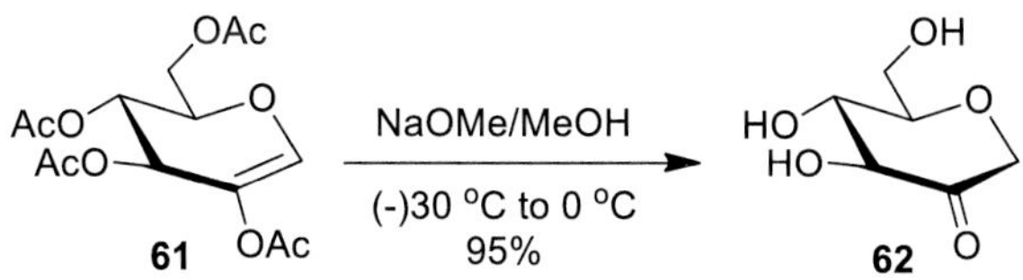

SCHEME 8.16 Conversion of D-glucal to 1,5-anhydro-D-fructose.

A five-step synthetic method started from the suitably protected D-Fructose **8**, which was converted to tosylate **63a** or mesylate **63b**. The regioselective removal of the 4,5-*O*-isopropyidene moiety keeping the 2,3-acetal intact was achieved by using oxalic acid in 85% aqueous acetonitrile at reflux temperature furnishing desired products **64a** in poorer yield whereas under optimized conditions, the corresponding mesylate **64b** was obtained in better yield. Intramolecular anhydro ring closure was performed with sodium hydride in dimethylformamide (DMF) for the tosylate or by employing aqueous sodium hydroxide at gentle reflux for 15 minutes to afford the acetal **65**. Deprotection of **65** with 80% aqueous acetic acid afforded the desired **62** in excellent yield. This sequence of reactions from D-fructose can be used at a scale of 10–100 g in each step (Scheme 8.17).[27a]

Enzymes like dehydratases, isomerases, and reductases can convert **62** to a group of secondary products, such as ascopyrones **66a–b**, **67**,[27a] microthecin **68**, and 5-epipentenomycin-1 **69**, with antibacterial and antioxidant activities.[26a,b] Chemical modifications can also give similar compounds as well as natural products like 1-deoxymannonojirimycin **70**[27b] as well as an important mannosidase inhibitor clavulazine **71** (Scheme 8.17).[26a,b,28] Based on its potential benefits and minimal toxicity, it was anticipated that AF **62** would be used as a dietary supplement to support general health. However, the effects of 1,5-AF on the gut microbiota are yet to be established beyond doubt.[29]

8.2.6 Aza-heterocycles from D-fructose

The reaction of D-fructose with ammonia under pressure (40 atm) in a pressure vessel and at 75°C afforded 4(5)-(D-arabinotetritol-1-yl) imidazole **72** in

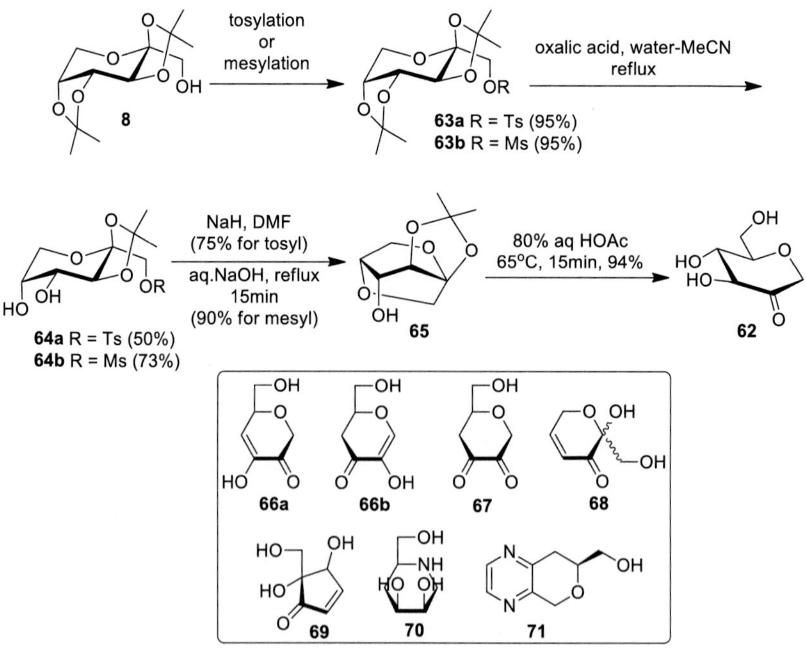

SCHEME 8.17 Conversion of D-fructose to 1,5-anhydro-D-fructose.

38% yield (Scheme 8.18).[30] Later D-fructose reacted with formamidine acetate in the presence of ammonium carbonate to afford **72** in 50% yield.[31] This type of reaction was utilized to convert D-fructose into room temperature ionic liquids (RTILs).[32a] Thus D-fructose, on treatment with ammonia and formaldehyde in the presence of $CuCO_3$ was converted to hydroxymethyleneimidazole **73**. These unique reactions obviously proceeded through the retroaldolization of D-fructose to glyceraldehyde and dihydroxyacetone. A large-scale (10–20 g) benzylation of **72** afforded an inseparable mixture of **75ab**. Methylation of **74a–b** followed by anion metathesis afforded a wide variety of RTILs **75ab**. The Heck reaction of methyl acrylate with iodobenzene using $Pd(OAc)_2$ catalyst in **75ab** ($X = NTf_2$) was quite rapid and afforded methyl cinnamate in >95% yield (Scheme 8.18). These fructose-derived RTILs may in some cases act as substitutes for conventional [bmim] series of ILs.[32b]

Retrosynthetic analysis of 1,4-dideoxy-1,4-imino-D-arabinitol **79** (DAB) and australine **81** led to (3 S,4 R)-1,3,4-tris(benzyloxy)-hex-5-en-2-one **78** as the starting material (Scheme 8.19). Building block **78** was synthesized from the methyl glycosides **14/15** of D-fructose via iodo derivatives **76αβ**, which were protected as tribenzyl compound **77αβ**. Sonication of **77αβ** with zinc in a tetrahydrofuran (THF)–water mixture led to a fragmentation reaction to afford the unsaturated ketone **78**. Thus D-fructose afforded DAB **79** in 7 steps in 11% overall yield. Synthetic manipulation of **78** afforded the intermediate **80**, which could be converted to australine **81** in three steps (Scheme 8.19).[33]

A multicomponent reaction of D-fructose, aniline, and 1,3-dicarbonyl compounds in deep eutectic solvent made of choline chloride/urea at 80°C

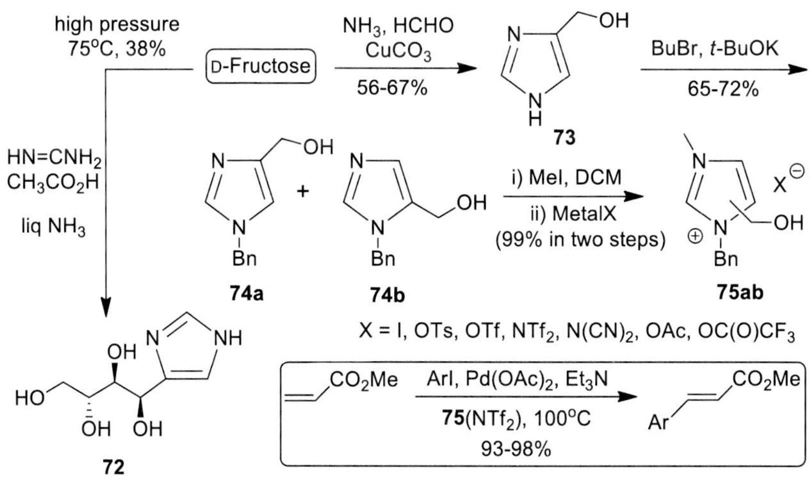

SCHEME 8.18 Conversion of D-fructose to imidazoles and RTILs.

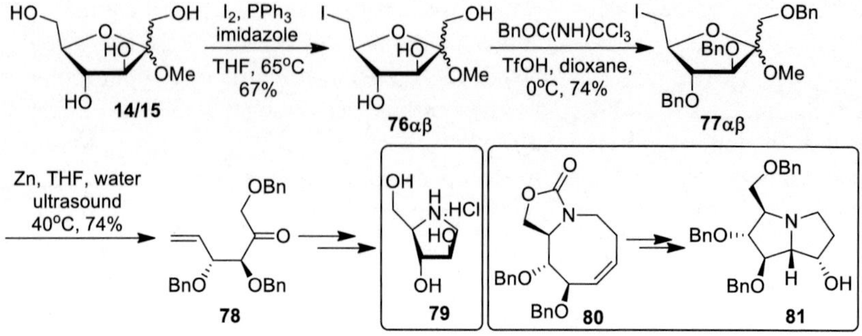

SCHEME 8.19 Conversion of D-fructose to different aza-heterocycles.

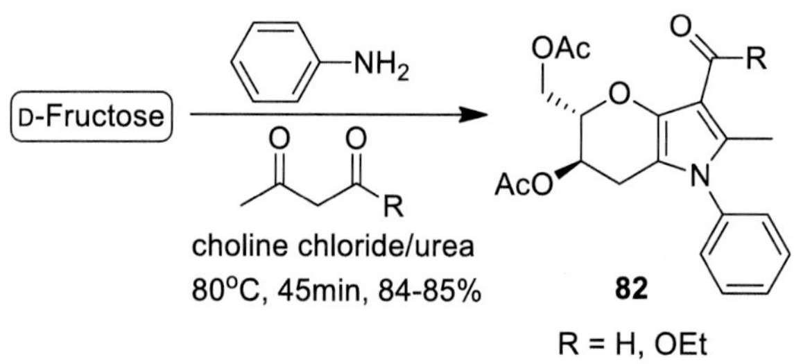

SCHEME 8.20 Conversion of D-fructose to different aza-heterocycles.

afforded the corresponding annulated pyrroles in good yield. The reaction was believed to be initiated by the formation of enaminoketone formed by the reactions of aniline and 1,3-diketone; the reaction between the enaminoketone and the sugar was triggered by the hydrogen bonding of urea with the anomeric hydroxyl group followed by subsequent cyclodehydration (Scheme 8.20).[34]

D-Fructose was identified as an environmentally benign C_2 synthon.[35,36] Thus in a gram scale synthesis, *ortho*-phenylenediamine was reacted with D-fructose using $BF_3.OEt_2$ as a catalyst and *tert*-butyl hydroperoxide as the oxidant at 90°C for 1 hour in acetonitrile to produce **83** a in 86% along with a trace of 1*H*-benzo[*d*]imidazole **84**. Several other derivatives were also synthesized (Scheme 8.21). As mentioned earlier, the reaction proceeded through a retroaldol strategy.[36] D-Fructose, on the other hand, reacted with 2-amino-*N*-phenylbenza-mide under copper-catalyzed oxidative C–C cleavage conditions to afford a quinazolinone scaffold **85** (Scheme 8.21).[37]

2-(1-Hydroxyethyl)-imidazoline **92**, a high-value and versatile building block for synthesizing a wide range of bioactive targets. D-fructose on reactions with ethylenediamine in the presence of $InCl_3$ at 130°C formed a hemiaminal, which underwent retroaldol reaction to afford a mixture of two C_3

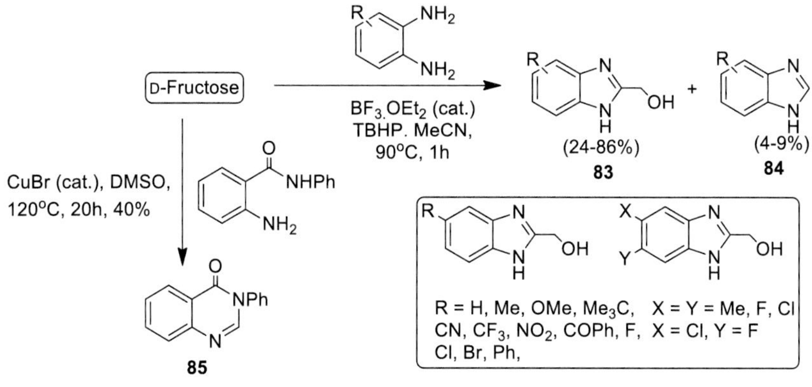

SCHEME 8.21 Conversion of D-fructose to different aza-heterocycles.

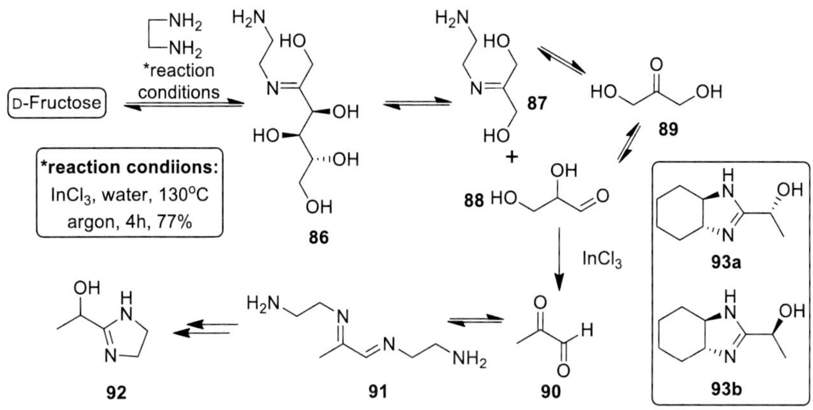

SCHEME 8.22 Conversion of D-fructose to imidazoline derivatives.

fragments **87** and **88**; compound **87** degraded further to dihydroxyacetone **89**, which also equilibrated with glyceraldehyde **88**. In the presence of $InCl_3$, **88** and **89** dehydrated to pivalic acid **90**. Two equivalents of ethylenediamine reacted with **90** and the intermediated underwent cyclization followed by 1,2-hydride shift to afford **92** in high yield even in gram scale reactions. $InCl_3$ catalyzed reactions between D-fructose and *trans*-1,2-diaminocyclohexane in 17% water afforded 2-(1-hydroxyethyl)-hexahydro-benzimidazole **93a** and **93b** as a diastereomeric mixture in 66% yield (Scheme 8.22).[38]

8.2.7 Vinyl sulfone-modified fructose derivatives

Although, amino nucleophiles required catalytic amount of $LiClO_4$, a more efficient nucleophile *p*-tolylthiol in the presence of tetramethyl guanidine

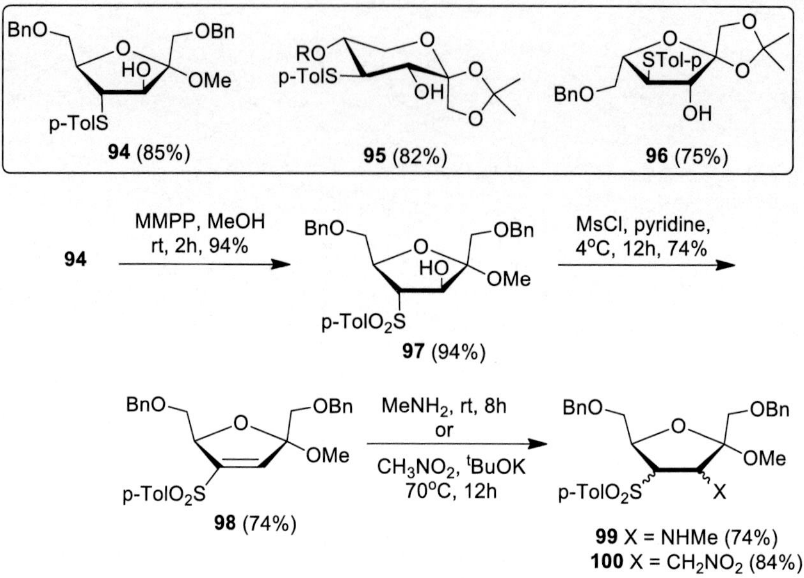

SCHEME 8.23 Synthesis of vinyl sulfone-modified fructofuranoside.

opened the epoxide rings of **31**, **38**, and **40** to afford thiol derivatives **94**, **95**, and **96**, respectively, at 90°C (Scheme 8.23).

Since vinyl sulfone-modified carbohydrates were used extensively for the synthesis of a variety of modified sugar derivatives,[39] converted to vinyl sulfone-modified carbohydrate **98** was prepared and its reactivity was studied. Thus **94** was oxidized to sulfone **97** using magnesium bis(monoperoxyphthalate) hexahydrate (MMPP). The sulfone on reactions with methanesulfonyl chloride (MsCl) underwent mesylation and concomitant elimination to generate the vinyl bond in **98**. The D-fructose-derived Michael acceptor **98** reacted with methylamine and the potassium salt of nitromethane to yield Michael adducts **99** and **100**, respectively. However, the configurational identities of these single diastereomers could not be established (Scheme 8.23).

The formation of single diastereomers **99** and **100** led to the designing of another vinyl sulfone-modified carbohydrate **103**, expected to generate cyclopropanated derivatives of D-fructose via Michael Initiated Ring Closure.[40a,b] Thus the epoxide ring of **34** was opened by *p*-tolylthiol to generate **101**; oxidation of **101** to the sulfonylated diol **102** followed by the mesylation of the latter led to the formation of the desired mesylated vinyl sulfone **103** (Scheme 8.24). Benzylamin/^{t}BuOK, potassium salt of dimethylmalonate/^{t}BuOK, 2-mercaptoethanol/^{t}BuOK, and NaOMe/MeOH reacted with the vinyl sulfone **103** in THF to furnish the cyclopropanated

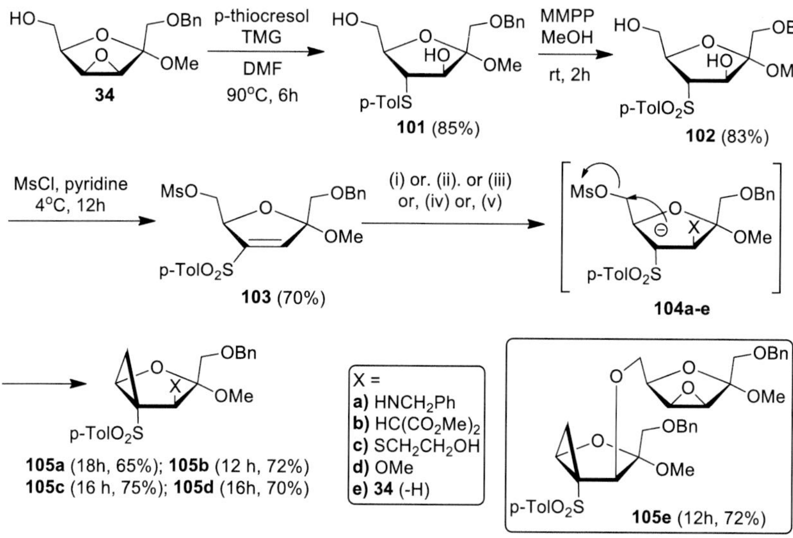

Reaction conditions: (i) benzylamine, ${}^tBuOK$, THF, rt; (ii) dimethylmalonate, ${}^tBuOK$, THF rt; (iii) 2-mercaptoethanol, ${}^tBuOK$, THF, rt; (iv) NaOMe, MeOH, rt; (v) **34**, NaH, DMF, rt

SCHEME 8.24 Cyclopropanation of vinyl sulfone-modified fructofuranoside.

sugar derivatives **105a–d**, respectively, via intermediates **104a–d**; ${}^tBuOK$ was required in first three cases, the methoxide ion was sufficiently basic to abstract the C-4 proton required for the concomitant cyclopropane ring formation for **105d**. A bulky D-fructose-derived epoxide **34**, easily reacted with **103** in the presence of NaH to produce the disaccharide **105e**. The anomeric configuration of **103** dictated the diastereoselectivity of the addition of all nucleophiles at C-3 (Scheme 8.24).

8.2.8 Vinyl selenone-modified fructose derivatives

Vinyl selenone-modified carbohydrates have rarely been utilized partly because of instability.[41a,b] However, stable vinyl selenone-modified carbohydrates were synthesized and it was established that the aryl selenoyl functionality of vinyl selenone played a dual role as an efficient electron-withdrawing group as well as a leaving group allowing Michael addition followed by elimination in tandem fashion.[42]

The construction of vinyl selenone moiety on D-fructose was therefore attempted. Thus sodium phenyl selenide generated from diphenyl diselenide and sodium borohydride, regioselectively opened the epoxide ring of dibenzyl derivative **31** to afford the β-hydroxy selenide derivative **106**. The free hydroxyl group of compound **106** was mesylated to generate compound **107**. Crude **107**, because of its instability was immediately treated with ${}^tBuOK$ in

DMF to produce vinyl selenide **108**. The latter compound on treatment with MMPP in methanol afforded vinyl selenone **109** in 47% overall yield starting from **106** (Scheme 8.25).[43]

Vinyl selenone **109** reacted with aqueous ammonia in DMF and ethanolamine to afford single diastereomeric sugar-derived aziridines **110a** and **110b**, respectively. However, aqueous methylamine in DMF, cyclohexylamine, isopropylamine, 2-amino-2-methyl-1-propanol, and *tert*-butylamine reacted with **109** at room temperature to afford separable mixtures of aziridines **110c–g/111c–g**, respectively. As expected, all nucleophiles were added in Michael fashion followed by the substitution/elimination of selenone moiety to form diverse scaffolds (Scheme 8.26). In addition to usual analytical data, structures of aziridines **111** were established via alternative synthesis. Thus epoxide **31** was treated with cyclohexylamine and the aminol **35b** was cyclized under Mitsunobu reaction conditions to afford the

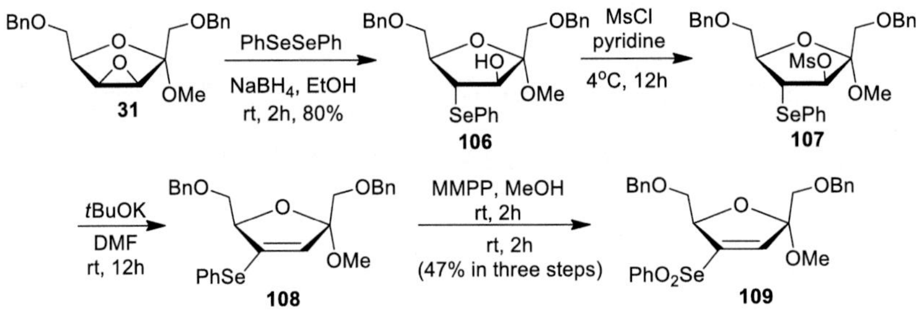

SCHEME 8.25 Synthesis of vinyl selenone-mofide furanosyl fructose.

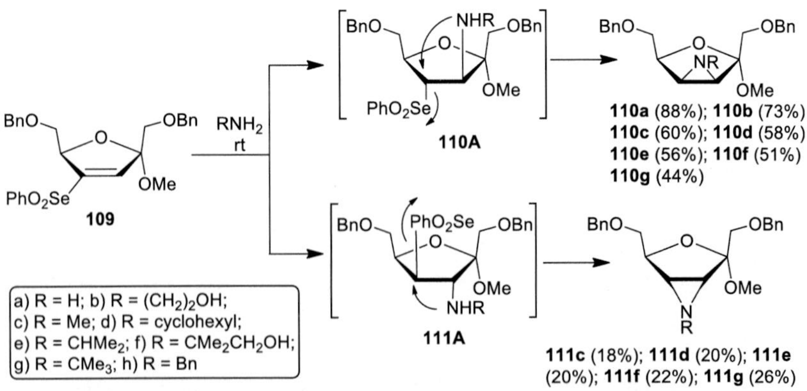

Reagents and conditions: 110a: 25% ammonia, DMF, 1-2h; **110b:** neat ethanolamine, 3-4h; **110c/111c:** 40% methylamine, DMF, 2-4h; **110d/111d:** neat cyclohexylamine; **110e/111e:** neat isopropylamine, 3-6h; **110f/111f:**neat 2-amino-2-methyl-1-propanol, 6h; **110g/111g:** neat *tert*butylamine, 8h;15h: neat benzylamine, 4h

SCHEME 8.26 Synthesis of aziridinated carbohydrates from vinyl selenone.

corresponding *psico* aziridine derivative, which was identical to compound **111d** (1:1 mix ^{1}H NMR) synthesized from vinyl selenone **109** (Scheme 8.27). From this alternative synthesis and NMR data analysis, **111c–g** were confirmed as *psico* aziridines and **110a-g** were *tagato* aziridines.[43]

Vinyl selenone-modified hexopyranoside **115** was synthesized from pyranoside epoxide **38** via intermediates **112**, **113**, and **114** in good overall yield (Scheme 8.28). Both the vinyl selenones **109** and **115** were found to be stable and could be stored for a long period below 30°C. Hexopyranoside

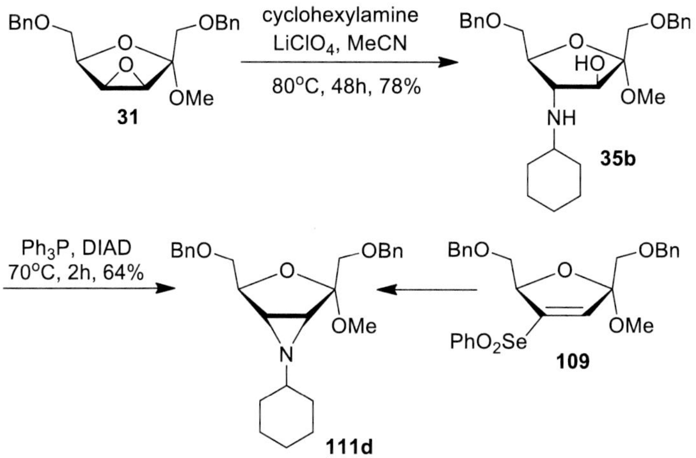

SCHEME 8.27 Alternative Synthesis of aziridinated carbohydrate.

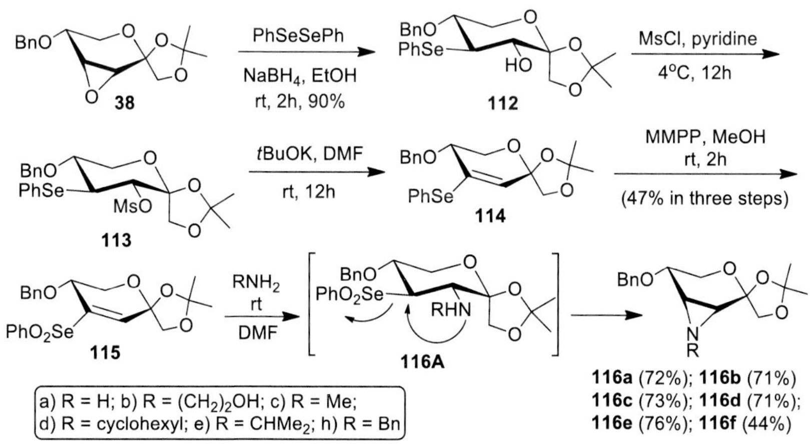

Reagents and conditions: 116a: 25% ammonia, DMF, 1-2h; **116b:** neat ethanolamine 3-4h; **116c:** 40% methylamine, DMF, 2-4h; **116d:** neat cyclohexylamine; **116e:** neat isopropylamine, 3-6h; **116f:** neat benzylamine, 4h

SCHEME 8.28 Synthesis and reactions of vinyl selenone-modified fructopyanoside.

115 on separate treatment with aqueous ammonia in DMF, ethanolamine, methylamine in DMF, cyclohexylamine, isopropylamine, and benzylamine at room temperature afforded single diastereomeric aziridines **116a–f**, respectively. The formation of aziridine presumably proceeded through the addition of primary amines to C-3 followed by an intramolecular attack at C-4 (**116 A**; Scheme 8.28). The structure of compound **116f** was unambiguously established by X-ray analysis of its single crystal.[43]

Malononitrile and ethyl cyanoacetate efficiently reacted with **109** in the presence of ^{t}BuOK in THF to afford single diastereomeric cyclopropanated fructose derivatives **118a–c**, respectively (Scheme 8.29). Probable intermediates **117A** and **117B** dictated the configuration of the products. The pyranoside selenone **115** also reacted with malononitrile, dimethylmalonate, and ethyl cyanoacetate under similar conditions to furnish cyclopropane derivatives **119a–c**, respectively, in good yields. The potassium salt of nitromethane, however, did not produce the expected cyclopropane derivative. In this case, the *epi*-nitro intermediate **120A** generated from the conventional addition product **120** led to the formation of the unusual olefinic product **121** (Scheme 8.29).[43]

Ambident nucleophiles such as methyl acetoacetate, ethyl acetoacetate, dimethylmalonate, and dimidone on reactions with **109** in the presence of ^{t}BuOK yielded single diastereomeric dihydrofuran derivatives **122a–d**, respectively. Ethyl nitroacetate also on treatment with **109** under similar reaction conditions produced single diastereomeric dihydroisoxazole derivative **123** (Scheme 8.30). However, pyranoside **115** did not react with methyl

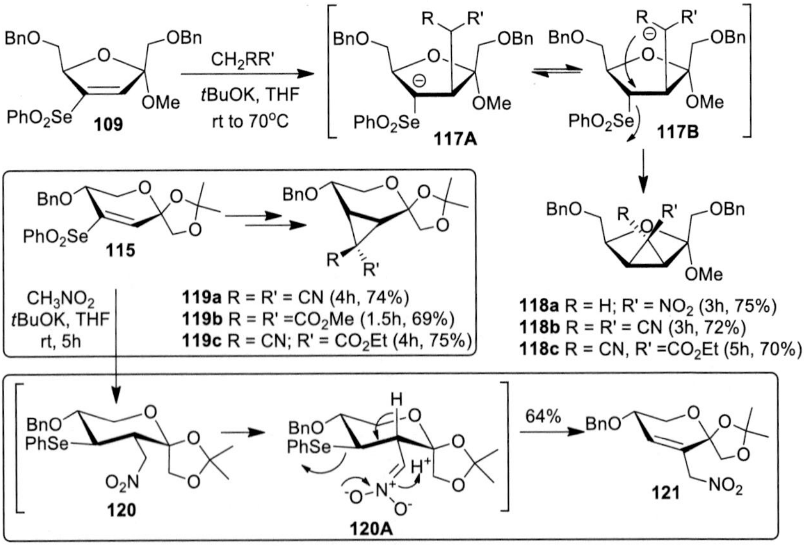

SCHEME 8.29 Cyclopropanated carbohydrates from vinyl selenone-modified fructoses.

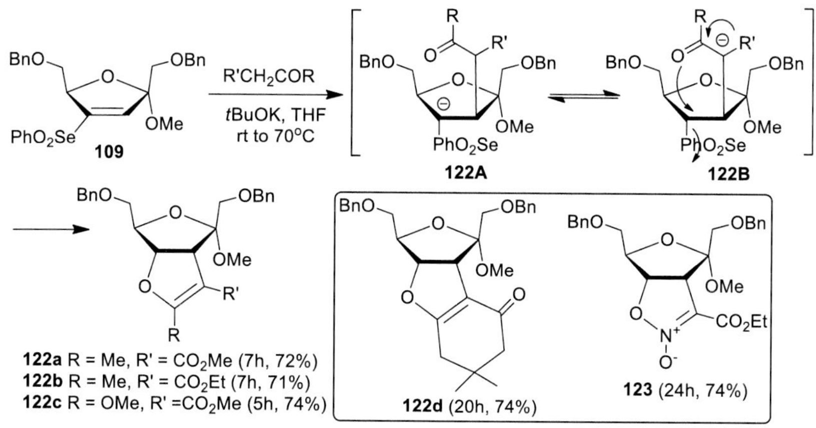

SCHEME 8.30 Dihydrofurans and dihydroisoxazole from vinyl selenone-modified fructose.

acetoacetate, ethyl acetoacetate, dimidone, and ethyl nitroacetate under similar reaction conditions.[43] Acetylacetone reacted with **109** to afford the unusual cyclopropane derivative **124**, which was probably formed via a four-membered ring intermediate **124B** from **124A** and the former collapsed to **124C**; the expected dihydrofuran derivative **125** was also formed in approximately 1:1 ratio; pyranoside **115** on similar treatment, produced only the unusual cyclopropane derivative **126** (Scheme 8.31).[43]

In short, in the addition reactions represented earlier, the incoming nucleophile was delivered to the C2 position of vinyl slenone-modified carbohydrates from a direction opposite to that of the anomerinc oxygen. A background experiment was therefore conducted to establish that the stereoelectronic repulsion imposed by the C-2 oxygen was more dominating than CH_2OBn of **109** or CH_2-O-C$(Me)_2$ residue of **115**. Thus the vinyl selenone **109** on reactions with potassium hydroxide or sodium sulfide in water at room temperature afforded single diastereomeric "up" oxirane **127a** and thiirane **128a**, respectively. Vinyl selenone **115** under similar reaction conditions generated only "down" oxirane **128a** and thiirane**128b** respectively. These experiments established beyond doubt that the C2-methoxy group of **109** and the C2-O residue of **115** preferentially directed the diastereoselectivity of addition of nucleophiles to the Michael acceptors (Scheme 8.32).[43]

8.2.9 Isonucleosides from D-fructose

A significant number of pyrrole, imidazole, triazole, and tetrazole nucleosides have been shown to exhibit a broad spectrum of biological activity.[44a,b] However, their isonucleoside derivatives are less known,[45] and examples of glyco isonucleosides are also rare. Vinyl sulfone-modified carbohydrates were used in the past to synthesize isonucleosides but the desulfonylation of

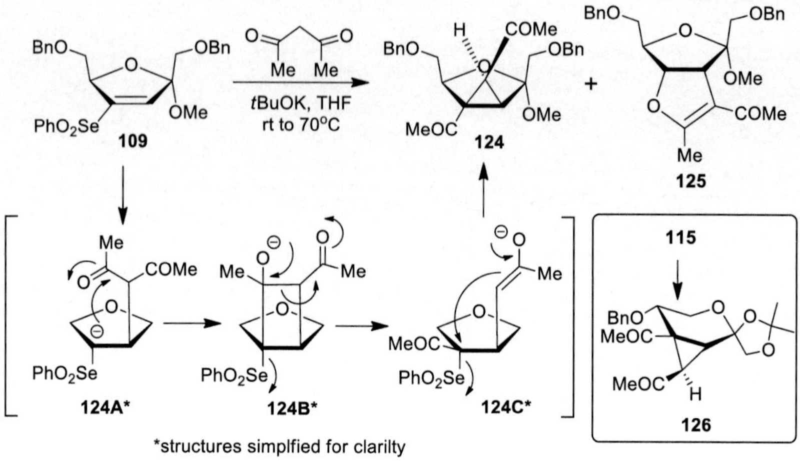

SCHEME 8.31 Unusual cyclopropanyl derivatives vinyl selenone-modified fructose.

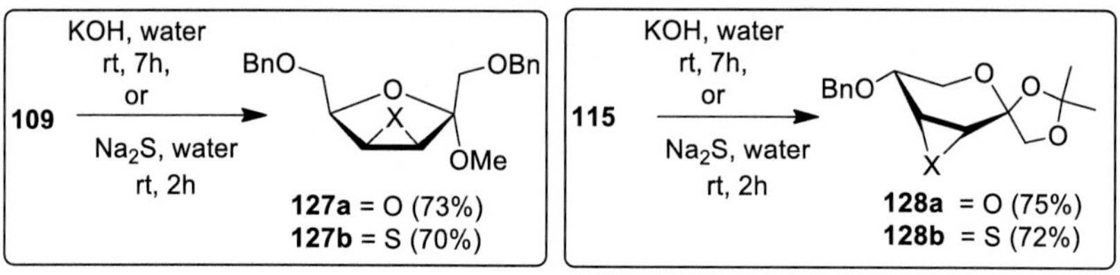

SCHEME 8.32 Oxirane and thiirane derivatives from vinyl selenone-modified fructose.

the isonucleoside remained a serious problem.[45] Therefore pyrrole, imidazole, 1,2,4-triazole, and 1-*H*-tetrazole were reacted with the Michael acceptor **109**, and the products were isolated as D-fructose-derived isonucleosides **129a–d**, respectively (Scheme 8.33). The pyranoside vinyl selenone **115** on similar treatment with imidazole, 1,2,4-triazole, 1-*H*-tetrazole, and the nucleobase thymine afforded isonucleosides **130b–e**, respectively. Both Michael acceptors reacted through the expected intermediates **129A** and **130A**, which underwent elimination by the inorganic base used for the generation of nucleophiles (Scheme 8.33). The hexopyranoside **115** efficiently reacted with, imidazole, 1,2,4-triazole, and 1-*H*-tetrazole in the presence of 1,1,3,3-tetramethylguanidine (TMG) to yield isonucleosides **130b–d** in comparable yields (Scheme 8.33).

To avoid the double bond formation, compound **109** was reacted with thymine/TMG, which did not produce any additional product; instead, a furan derivative **131** was obtained. Compound **131**, a new derivative of 2,5-

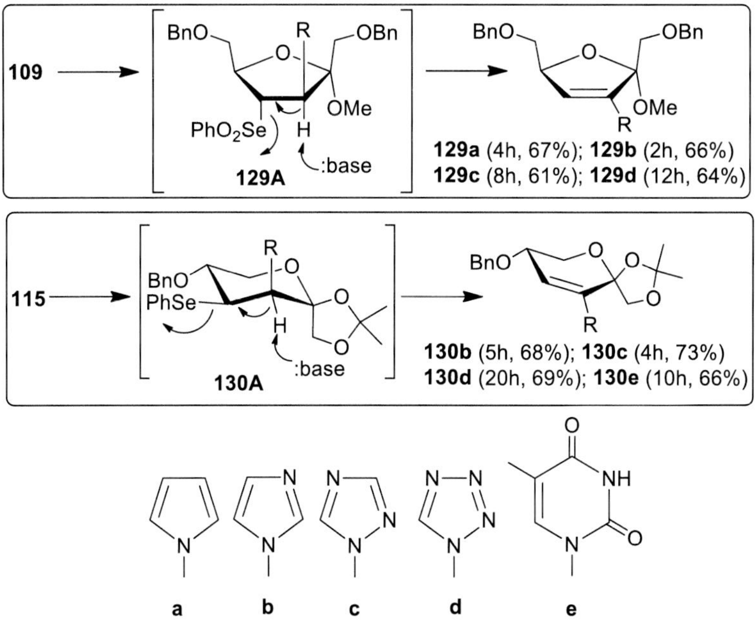

Reagents and conditions: 129a: pyrrole, *t*-BuOK, THF, 70°C; **129b/130b:** imidazole, water, 80°C; **129c/130c:** 1,2,4-triazole, K_2CO_3, THF, 70°C; **129d/129d:** 1-*H*-tetrazole, K_2CO_3, THF, 70°C; **130e:** thymine, *t*-BuOK, THF, rt

SCHEME 8.33 Isonucleosides from vinyl selenone-modified fructose.

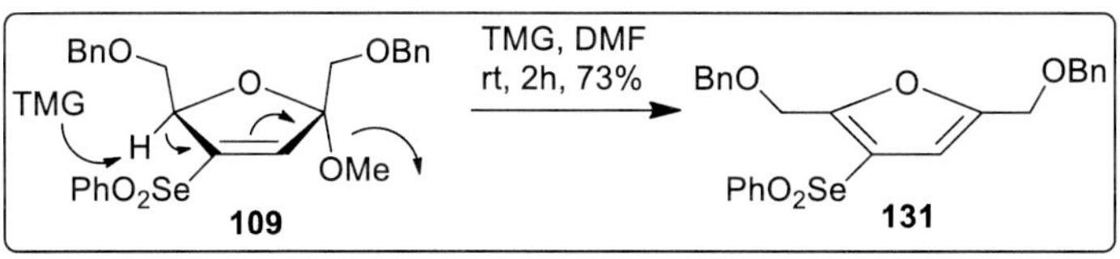

SCHEME 8.34 A furan analog from the vinyl selenone-modified D-fructose.

dimethylfuran (see later) was also obtained in the absence of thymine (Scheme 8.34).

Interestingly, thymine and uracil in the presence of TMG took a different route. The reaction was initiated by the attack of nucleobases from the α-face of the vinyl selenone with instantaneous elimination of the benzyloxy group to form intermediate **132 A** followed by shifting of the double bond and elimination of phenyl selenone group to produce **132B**. Addition of a molecule of water from a direction opposite to that of the nucleobases afforded a new class of olefinic isonucleosides **133a** and **133b** in good overall yield (Scheme 8.35).[43]

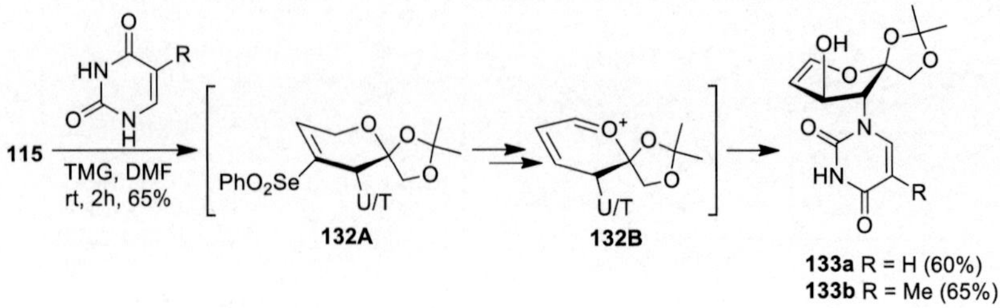

SCHEME 8.35 Reactions of nucleobases with vinyl selenone-modified D-fructopyranoside.

8.2.10 Biomass and fructose

As fossil resources deplete, the presently prevailing economic advantage of petrochemicals is going to change within a few decades.[5] In the near future, utilization of renewable raw materials will gain importance in the synthesis of chemical substances. It is expected that in 2025, up to 30% of raw materials for the chemical industry will be produced from renewable sources like biomass. Among all other biomolecules, carbohydrates are by far the most abundant organic compounds on earth and represent the major portion of the annually renewable biomass.[46a,b] As a result of continued efforts, 5-hydroxymethylfurfural (HMF) **134** has been identified as a future platform chemical.[47] Although HMF was first reported in 1875 as an intermediate in the formation of levulinic acid from sugar and sulfuric acid,[47a] different reagents and reaction conditions including ionic liquids are being probed to find high-yielding methodologies.[47b,c,48a,b,c] HMF is now considered as a petrochemical derived from D-fructose by acid-catalyzed intramoleculer dehydration.[48a,b,c] Mechanisms involving cyclic intermediates have also been suggested (Scheme 8.36).[48a,b,c]

However, this remains the classical route, with 6-carbon sugars (hexoses) such as fructose undergoing acid-catalyzed poly-dehydration. It was reported that the reactivity of glucose is lower than fructose (ketose) because of the lower abundance of acyclic glucose compared to acyclic fructose. Since enolization was shown to be the rate-determining step for HMF **134** formation and therefore fructose is expected to react faster than glucose.[48a,b,c]

Although D-fructose is a sugar found naturally in fruits, fruit juices, some vegetables, and honey, it can be directly associated with biomass sources. For example, inulins are a group of naturally occurring polysaccharides produced by many types of plants, industrially most often extracted from chicory. Inulin is a heterogeneous collection of fructose polymers (Scheme 8.37).[48a,b,c,49] Inulin is a polydisperse β(2−1) fructan and the fructose units in this mixture of linear fructose polymers and oligomers are each linked by β(2−1) bonds. Because of β(2−1) bonds inulin is not digested like a typical carbohydrate.[50a,b] The direct conversion of inulin to 5-HMF **134** was carried out in two steps using ionic liquids. Thus [BMIM][HSO_4] was used for the

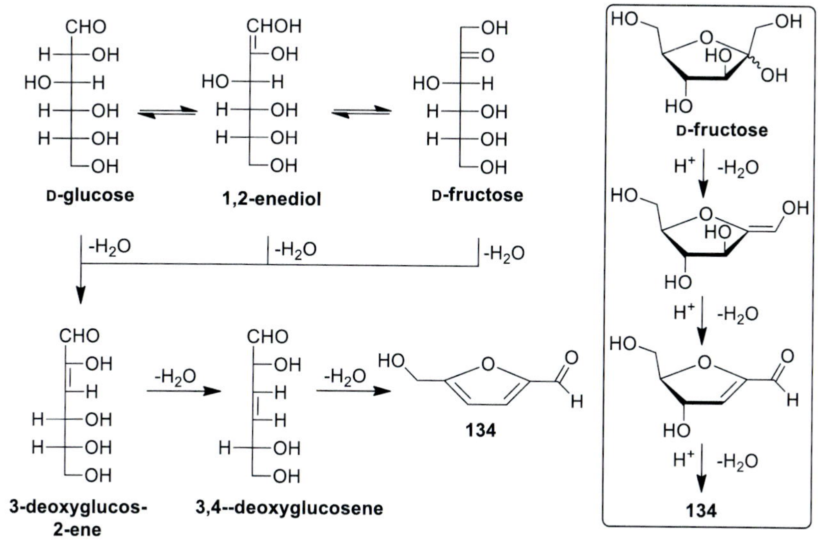

SCHEME 8.36 Mechanisms of D-fructose to **134** under acidic conditions.

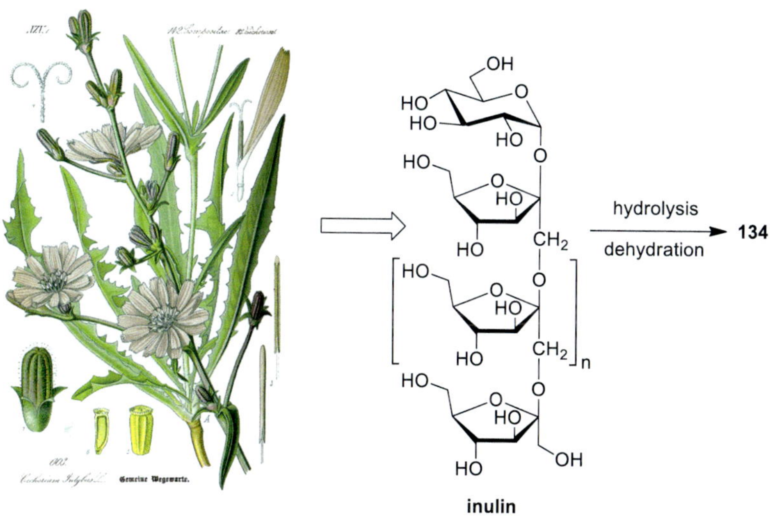

SCHEME 8.37 Inulin, a fructose polymer from chicory plant.[49]

conversion of inulin to D-fructose, and [BMIM][Cl] and ion exchange resin was used to transform fructose to 5-HMF in high yield of **134**.[51] HMF **134** stands out among the platform chemicals and research on this furan derivatives has sky-rocketed in recent years. Since HMF **134** retains all six carbon atoms of the hexoses and it can be prepared in high selectivity especially

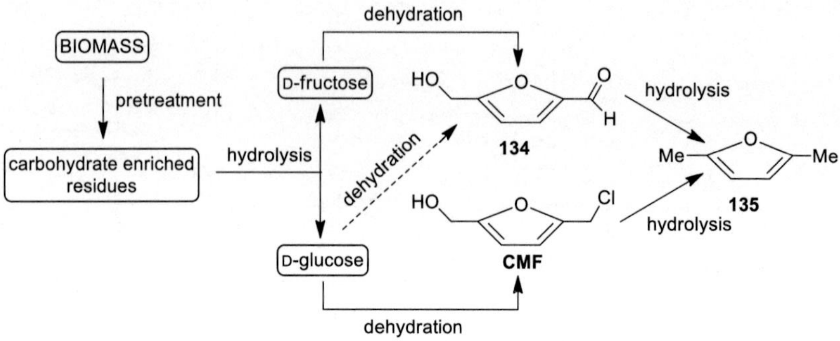

SCHEME 8.38 Process for the generation of a biofuel 2,5-dimethylfuran (DMF) **135**.

from fructose, a wide range of petrochemical can be generated from it.[48b,c] Another related compound, 2,5-dimethylfuran **135** obtained so far from petrochemicals, has also received significant attention as an alternative fuel because of its low oxygen content, high energy content, and high octane value.[52a,b] **135** is now generated from D-fructose, a major constituent of carbohydrate biomass, via 5-hydroxymethylfurfural **134**; the other intermediate for this transformation is 5-chloromethylfurfural (CMF) (Scheme 8.38).[53]

However, acid-catalyzed rehydration of HMF produced levulinic acid and formic acid. HMF has been used in the synthesis of readily accessible intermediate chemicals like 2,5-bis(hydroxymethyl)furan, 5-hydroxymethyl-2-furoic acid, and 2,5-dicarboxylic acid. Moreover, HMF is primarily considered for the synthesis of various compounds having industrial importance. Selected compounds of future commercial interests are depicted in Scheme 8.39.[19,20] A new class furan analog **131** has been synthesized from vinyl selenone-modified D-fructose (Scheme 8.34).[43] Several bio-based furans were converted to functionalized aromatic compounds (F2A process) using Diels–Alder reactions. Different types of furanic derivatives were evaluated for these transformations, and it has been noted that the substituents attached to the furan ring with more pronounced electron-donating properties may decrease the aromaticity of the furan ring and facilitate the cycloaddition reaction.[54] These transformations have been extensively reviewed.[5,52a,b,54,55]

8.3 L-Sorbose in synthetic chemistry

Any discussion on D-fructose is incomplete without the mention of L-sorbose, the 5-epimer of D-fructose; this ketose is also a readily available natural sugar and one of the rare ones in the L-series.[7] L-Sorbose is readily available in relatively large scale due to its technical production by regiocontrol dehydrogenation of D-sorbitol by using *Gluconobacter oxydans* or *Acetobacter* species. The chemistry of L-sorbose (like D-fructose) is considerably complicated and less well studied. Therefore the utilization of L-sorbose as organic

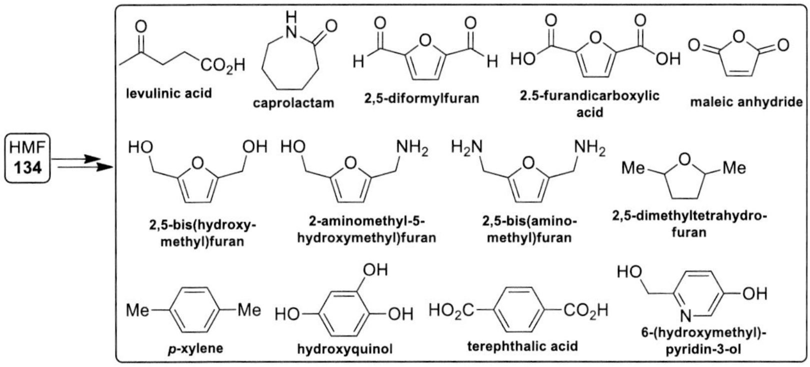

SCHEME 8.39 Furanics and aromatics generated from HMF **135**.

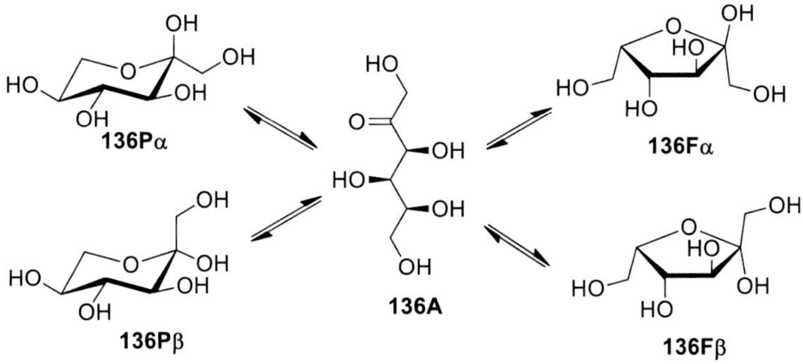

SCHEME 8.40 Tautomeric forms of L-sorbose in solution.

raw materials is relatively less popular in synthetic chemistry and thus, L-sorbose (annual production 6 × 10^4 metric ton/year in the world)[6] is less well understood than those of the more popular aldohexoses.[7] In solution, L-sorbose exists in two pyranoid (**136Pα** + **136Pβ**) and two furanoid (**136Fα** + **136Fβ**) tautomers with very little acyclic tautomer **136 A** (Scheme 8.40). Therefore, glycosylations, acylations, alkylations, or any other simple derivatization for the preparation of partially protected L-sorbose, usually generate a mixture of products.[7]

L-Sorbose has been largely used for the synthesis of vitamin C. Reichstein–Grüssner process was the first industrial process that generated L-sorbose from D-glucose in two steps using hydrogenation followed by microbiological oxidation. Acetonation of L-sorbose afforded 2,3,4,6-di-*O*-isopropylidene-α-L-sorbofuranose **137**, which was oxidized to di-*O*-isopropylidene-2-ketogulonic acid **138**. Acidic deprotection resulted into the rearrangement of unprotected 2-ketogulonic acid to L-ascorbic acid (vitamin C) (Scheme 8.41).[56] There are several biotechnological processes available for the production of vitamin C via L-sorbose.[56,57] One such fermentation

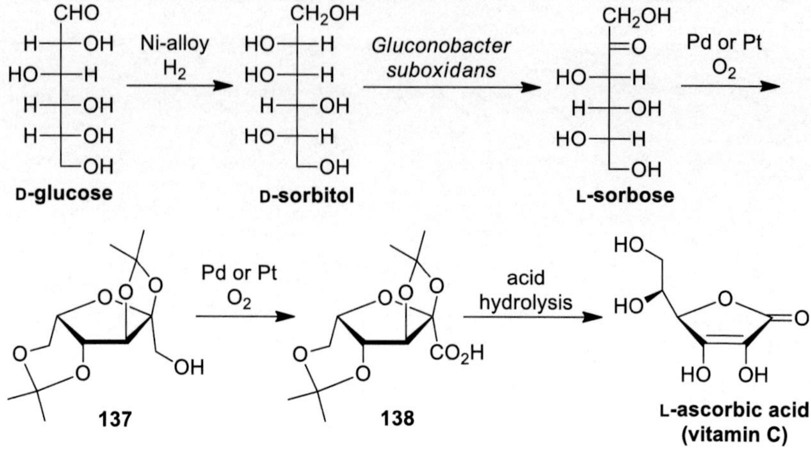

137 138

SCHEME 8.41 L-Sorbose mediated industrial preparation of L-ascorbic acid (vitamin C).

process involved three oxidation reactions from D-sorbitol to 2-ketogulonic acid in a single process step that was facilitated by a mixed culture of *Gluconobacter suboxydans* IFO3255 and *Ketogulonicigenium vulgare*. *Gluconobacter* mediates the oxidation of sorbitol to sorbose. All of the industrial bioprocesses resulted in the formation of 2-ketogulonic acid, which rearranged to vitamin C.[56]

An interesting study to prepare L-fructose, a synthetic sugar from L-sorbose revealed that slight variation in reaction conditions afforded the desired protections. Thus the bisisopropylidene derivative **139** was obtained in 51% yield only when the reaction was $SnCl_2$ catalyzed. Mesylation afforded **140**, which was partially deprotected to **141**. Crude **141** was treated under alkalined conditions to afford 1,2-*O*-isopropylidene-α-L-fructofuranose **143**. Deprotection of **143** afforded the desired L-fructose **144 F** (represented in the furanoside form) in 62% overall yield in four steps from **140**) (Scheme 8.42).[58]

In another approach, L-sorbose on treatment with acetone in the presence of copper(II) sulfate afforded 1,2-*O*-isopropylidine pyranoside **145**.[59] JICS Selective tosylation of most acidic 2-hydroxyl group under phase transfer condition generated the monotosylate **146**, which under epoxidation (**147**), ring opening (**148**), and deprotection afforded L-fructose **144 P**, represented in the pyranoside form (Scheme 8.43).[60]

A comparative study on the pattern of *endo*- and *exo*-hydroxyglycal esters was carried out. Thus D-fructose on treatment with benzoyl chloride followed by HBr treatment afforded the bromo derivative **149**. Base-induced elimination of HBr from **149** afforded **149*exo*** olefin whereas the iodo derivative generated in situ afforded **149*endo*** on heating. The corresponding L-sorbose derivative **150** compound under iodination and base treatment afforded an

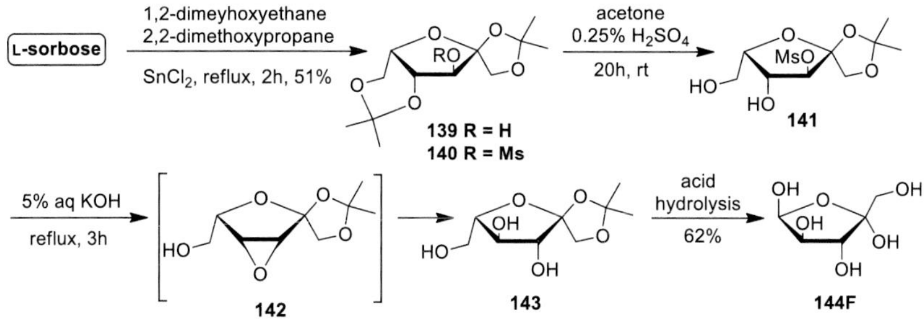

SCHEME 8.42 Synthesis of L-fructose from L-sorbose.

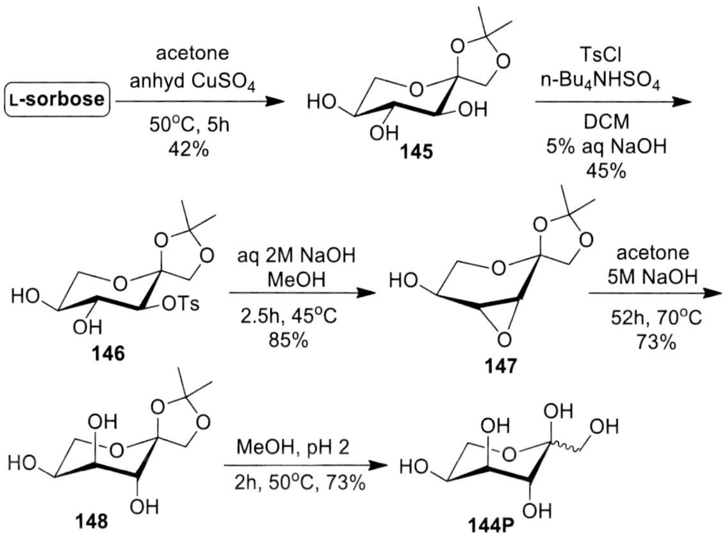

SCHEME 8.43 Synthesis of L-fructose from L-sorbose.

inseparable mixture of **150*exo*** and **150*endo*** compounds from which the **150*exo*** compound was removed as hydroxylamine derivative **150a** (Scheme 8.44). The olefinic compounds were functionalized further to afford a wide range of compounds. Compounds **149a–c** were obtained from **149*exo*** whereas **149*endo*** afforded **149c–e**. The *endo* analog **150*endo*** afforded olefins **150b–c** (Scheme 8.44).[61]

L-Sorbose was converted to 2-ethyl-l,6-dioxaspiro[4.4]nonane **153** via partially protected alcohol **137**, which was oxidized the aldehyde **151**; the crude aldehyde was subjected to Wittig reaction affording the olefin **152**. Mutistep synthetic manipulation afforded the target compound **153** (Scheme 8.45).[62]

Crude aldehyde **151** generated from **137** was reacted with (diethylamino)-sulfur trifluoride (DAST) to afford 1-deoxy-1,1-difluoro-2,3:4,6-di-O-

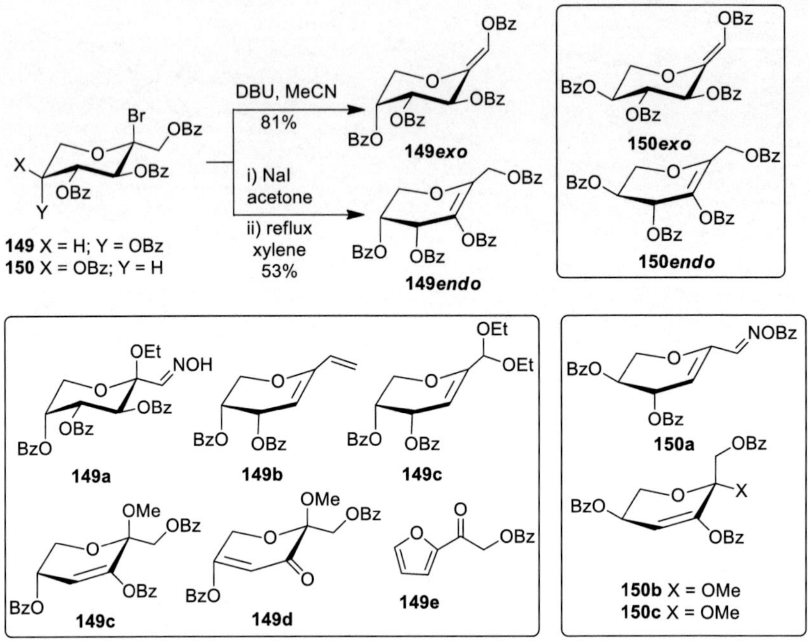

SCHEME 8.44 Synthesis of L-fructose from L-sorbose.

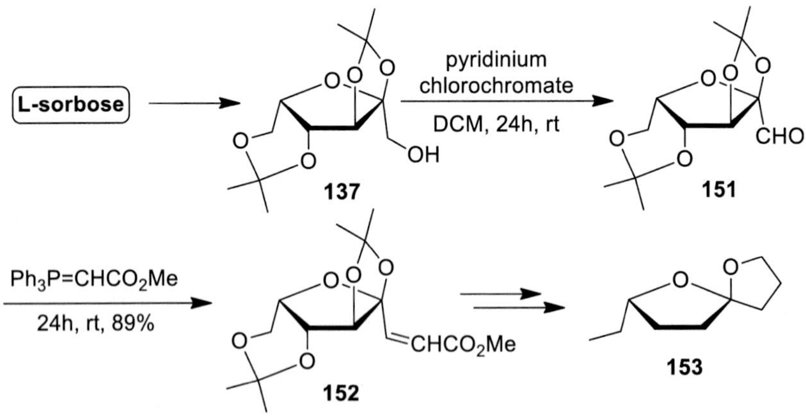

SCHEME 8.45 Synthesis of L-fructose from L-sorbose.

isopropylidene-α-L-sorbofuranose **154** in 36% yield (Scheme 8.46). Selective removal of the 4,6-O-isopropylidene group of **154** followed by selective tosylation of the primary alcohol produced **155**. Azidolysis of **155** afforded **156** and the latter in three steps afforded the hydrochloride salt **157** (Scheme 8.46). The free amino analog of **157**, 1,5,6-trideoxy-6,6-difluoro-1,5-imino-

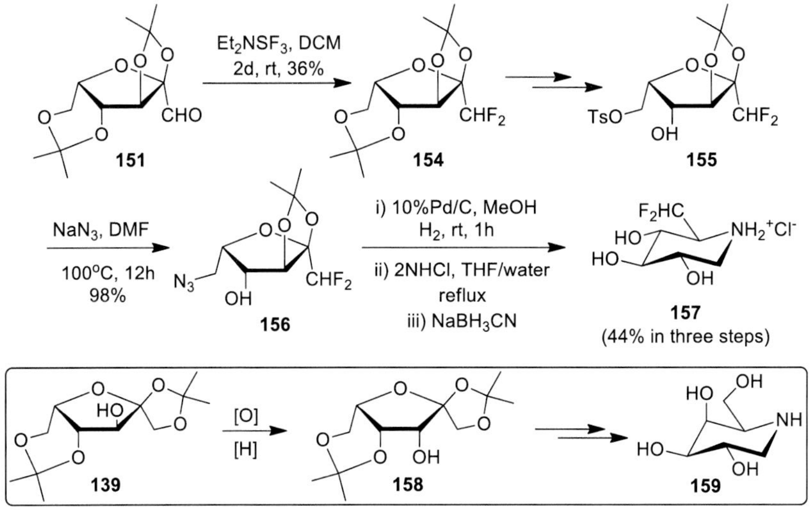

SCHEME 8.46 Synthesis of L-fructose from L-sorbose.

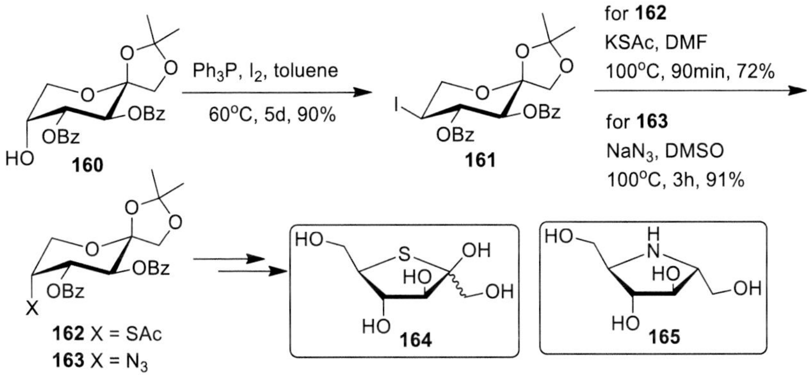

SCHEME 8.47 Synthesis of L-fructose from L-sorbose.

D-glucitol (1,6-dideoxy-6,6-difluoronojirimycin) was shown to be a micromolar level competitive inhibitor of yeast α-glucosidase and a noncompetitive inhibitor of almond β-glucosidase.[63] In a related approach, partially protected sorbofuranoside **139** was converted to 1,2:3,4-di-*O*-isopropyliden-α-L-tagatofuranose **158** via oxidation/reduction strategy. Partial deprotection and of azidation of **158**, followed by well-established synthetic protocols afforded 1,5-dideoxy-1,5-imino-D-galactitol (1-deoxygalactonojiycin) **159** (Scheme 8.46).[64]

The easy interconversion of D-fructose to L-sorbose has provided opportunity for various synthetic applications. Partially protected β-D-fructopyranoside **160** was iodinated using triphenylphosphine/I_2 to afford the 3,4-di-*O*-

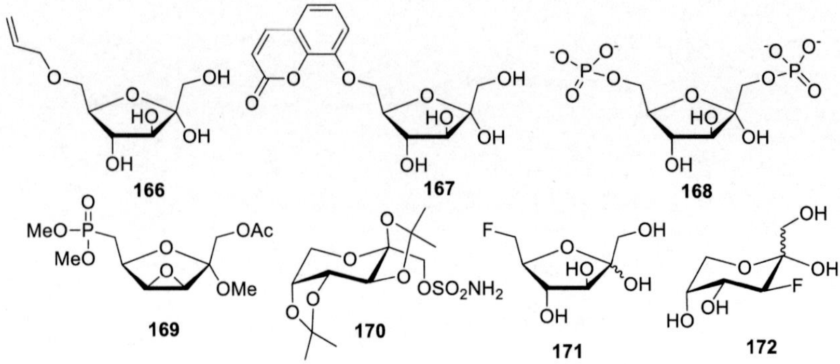

FIGURE 8.1 Structure of some selected bioactive fructose containing molecules.

benzoyl-5-deoxy-5-iodo-1,2-O-isopropylidene-β-L-sorbopyranose **161**. Reactions of **161** with two different sets of reagent systems afforded the D-fructopyranoside thio derivative **162** and azido derivative **163**. Both these compounds were converted to 5-thio-D-fructose **164** as well as the precursor of 2,5-deoxy-2,5-imino-D-mannitol **165**. The interaction of **164** with the D-fructose transporter GLUT5, was found to be weaker than that of D-fructose (Scheme 8.47).[65]

8.4 Conclusion and future prospective

Even after the publication of the review article entitled "Toward improving the utility of ketoses as organic raw materials"[5] about 25 years back, the potential of D-fructose, an easily accessible and inexpensive six-carbon source is yet to be fully realized. It should be noted that in the absence of appropriate key words, it is a difficult task to search publications related to the synthetic applications of D-fructose (and L-sorbose). Therefore with the help of several scattered examples on the usefulness of D-fructose (and L-sorbose) derivatives compiled from literature, an attempt has been made to draw the attention of synthetic chemists to the fact that these carbohydrates have the potential to act as an unlimited source of functionalized organic molecules required for the synthesis. Despite the utility of D-fructose (and L-sorbose) in industry, their chemistry is still remaining unexplored. Fortunately, a considerable number of reactions dealing with selective protection of D-fructose (and L-sorbose) have been developed. Some of these reactions are scalable and the required starting materials can be produced on a gram scale Moreover, highly reactive epoxides, one of the most important functionalized derivatives of carbohydrates in general, have also been prepared from D-fructose (and L-sorbose) in both pyranoside and furanoside forms. These two ketose sugars have found applications in synthetic chemistry as an important multifunctional precursor useful for the generation of asymmetric catalysts as well as densyl functionalized small organic molecules having synthetic and biological value. The intrinsic chirality of the sugar backbone can produce a complex compound that can act as a highly effective chiral controller in asymmetric synthesis. Depending on the reaction conditions, these ketoses can also act as "two carbon" precursors via a retroaldol route. Derivatives of D-fructose have found applications in biology and medical science as well. For example, the *O*-allyl analog **166** (Fig. 8.1) was shown as an inhibitor of the transporter enzyme GLUT5. 6-Coumarine-D-fructose **167** was used as fluorogenic markers.

Fructose-1,6-diphosphate **168** is used as a clinical drug in the prevention and treatment of acute hypophosphatemia and phosphate depletion and also has substantial cytoprotective effects in a variety of ischemia-reperfusion injury scenarios. 6-Deoxy-6-dimethoxy phosphonate **169** was synthesized from D-fructose as the intermediate for the preparation of various D-tagato- and fructofuranoside derivatives as inhibitors of sialyl transferase or sialidase. 2,3:4,5-Bis-*O*-(1-methylethylidene)-β-D-fructopyranosesulfamate (Topiramate, brand name Topa-max) **170** is an anticonvulsant (antiepilepsy) drug approved by FDA.[43] Fructose transport via the specific hexose transporter GLUT5 and its metabolism play important role in breast cancer growth and proliferation. 6-Deoxy-6-fluoro-D-fructose (6-FDF) **171** is a substrate for GLUT5 and 6-deoxy-6-[^{18}F]fluoro-D-fructose (6-[^{18}F]FDF) was evaluated as a PET tracer for imaging GLUT5 expression.[66a] Deoxy-3-fluoro-D-fructose (3-FDF) **172** (Fig. 8.1) has the potential for the imaging of GLUT5 expressing breast cancer tissue using PET.[66b] Fluorescent fructose derivatives have also been utilized for imaging breast cancer cells.[67] Encouragingly, with the advent of applications of biomass as an alternative source of "petroleum-based" chemicals, research has proliferated in fructochemistry as an inexpensive source of furanics such as **134** and **135**. L-Sorbose has also generated **135** in choline chloride melts in moderate yields.[68] Fundamental studies for overcoming the challenges for the selective transformation of ketoses to these intermediate platform chemicals like furans still continue.[69] Moreover, new results generated by research groups with varied interests are continuously opening up unchartered areas, such as visible-light-catalyzed synthesis of 2,5-dimethylfuran,[70] new catalysts for glucose to fructose isomerization[71a] or Zn(II)/Cd(II)-based metal-organic frameworks as catalysis of fructose/glucose to HMF conversion,[71b] synthesis of L-deoxyhexoses,[72] new biosurfactants,[73] and synthetic monosaccharide channels for the transport of fructose.[74] Although these recent publications raise hope and expectation, the vast potential of D-fructose and related ketoses as chirally pure starting materials in organic synthesis and their downstream products are yet to be realized.

Acknowledgments

The author thanks all the past and present MSc project students, research scholars, and for their contributions toward the development of research on carbohydrate chemistry in his laboratory. Financial support from Science and Engineering Research Board (SERB), Department of Science and Technology, Government of India, Department of Biotechnology, Ministry of Science and Technology, New Delhi, The Council of Scientific and Industrial Research, India, and the Indo-French Centre for the Promotion of Advanced Research is gratefully acknowledged.

References

1. Boysen, M. M. K. *Chem. Eur. J.* **2007,** *13*, 8648–8659.
2. Collins, P. M.; Ferrier, R. J. *Monosaccharides: Their Chemistry and Their Roles in Natural Products;* Wiley: Chichester, 1996.
3. Kamm, B. *Angew. Chem. Int. Ed.* **2007,** *46*, 5056–5058.
4. Available from: https://www.industryarc.com/Research/Global-Fructose-Market-Research-513787.
5. Lichtenthaler, F. W. *Carbohydr. Res.* **1998,** *313*, 69–89.
6. Wolfgang, W. *Fructose. Ullmann's Encyclopedia of Industrial Chemistry;* Wiley-VCH: Weinheim, 2004. Available from: http://doi.org/10.1002/14356007.
7. Zebiri, I.; Balieu, S.; Guilleret, A.; Reynaud, R.; Haudrechy, A. *Eur. J. Org. Chem.* **2011,** 2905–2910.
8. Saravanamurugan, S.; Paniagua, M.; Melero, J. A.; Riisager, A. *J. Am. Chem. Soc.* **2013,** *135* (14), 5246–5249.
9. Mossine, V. V.; Barnes, C. L.; Chance, D. L.; Mawhinney, T. P. *Angew. Chem. Int. Ed.* **2009,** *48*, 5517–5520.
10. García-Moreno, M. I.; Benito, J. M.; Carmen Ortiz Mellet, C. O.; Garcia Fernández, J. M. *Molecules* **2008,** *13*, 1640–1670.
11. Mellet, C. O.; Garcia Fernandez, J. M. *Top. Curr. Chem.* **2010,** *294*, 49–77.
12. Brady, R. F., Jr. *Carbohydr. Res.* **1970,** *15*, 35–40.
13. Kang, J.; Lim, G. J.; Yoon, S. K.; Kim, M. Y. *J. Org. Chem.* **1995,** *60*, 564–577.
14. Cortes-Garcia, R.; Hough, L.; Richardson, A. C. *J. Chem. Soc., Perkin Trans.* **1981,** *1*, 3176–3181.
15. Hanaya, T.; Sato, N.; Yamamoto, H. *Carbohydr. Res.* **2005,** *340*, 2494–2501.
16. Yu, K.; Xu Zhao, X.; Wenjie Wu, W.; Zhangyong Hong, Z. *Tetrahedron Lett.* **2013,** *54*, 2788–2790.
17. (a) Williams, N. R. *Adv. Carbohydr. Chem. Biochem.* **1970,** *25*, 109–179.
    (b) Fraser-Reid, B. O.; Tatsuta, K.; Thiem, J., Eds. *Glycoscience: chemistry and chemical biology*, vol. 1; Springer: Berlin – Heidelberg, New York, 2001; pp 291–304.
    (c) Pino-Gonzalez, M. S.; Romero-Carrasco, A.; Ona, N. *Carbohydr. Chem* **2013,** *39*, 150–180.
18. Kolb, H. C.; Finn, M. G.; Sharpless, K. B. *Angew. Chem. Int. Ed.* **2001,** *40*, 2001–2021.
19. (a) Guthrie, R. D.; Jenkins, I. D.; Yamasaki, R.; Skelton, B. W.; White, A. H. *J. Chem. Soc. Perkin Trans.* **1981,** *1*, 2328–2334.
    (b) Peter Sunde-Brown, P.; Jenkins, I. D.; Blanchard, H.; Houston, T. A. *Tetrahedron* **2022,** *120*, 132877.
20. (a) Das, A.; Bhaumik, A.; Pathak, T. *Carbohydr. Res.* **2020,** *487*, 107870.
    (b) For a review on fructosamine, see Mossine, V. V.; Mawhinney, T. P. *Adv. Carbohydr. Chem. Biochem.* **2010,** *64*, 291–402.
21. Mio, S.; Kumagawa, Y.; Sugai, S. *Tetrahedron* **1991,** *47*, 2133–2144.
22. (a) Tu, Y.; Wang, Z.-X.; Shi, Y. *J. Am. Chem. Soc.* **1996,** *118*, 9806–9807.

(b) Wang, Z.-X.; Tu, Y.; Frohn, M.; Zhang, J.-R.; Shi, Y. *J. Am. Chem. Soc* **1997,** *119*, 11224–11235.

(c) Shu, L.; Ship, Y. *Tetrahedron* **2001,** *57*, 5213–5218.

(d) Zhang, Z.; Tang, J.; Wang, X.; Shi, H. *J. Mol. Catal. A: Chem* **2008,** *285*, 68–71.

(e) Burke, C. P.; Shi, Y. *Angew. Chem. Int. Ed* **2006,** *45*, 4475–4478.

23. Tardy, S.; Vicente, J. L.; Tatibouet, A.; Dujardin, G.; Rollin, P. *Synthesis* **2008,** 3108–3120.
24. Wang, Z.-X.; Miller, S. M.; Anderson, O. P.; Shi, Y. *J. Org. Chem.* **2001,** *66*, 521–530.
25. Wang, W.; Xue, J.; Tian, T.; Jiao, Y.; Li, Y. *Org. Biomol. Chem.* **2013,** *11*, 6686–6690.
26. (a) Andersen, S. M.; Lundt, I.; Marcussen, J.; Yu, S. *Carbohydr. Res.* **2002,** *337*, 873–890.
    (b) Lundt, I.; Shukun, Y. *Carbohydr. Res* **2010,** *345*, 181–190.
27. (a) Andreassen, S. M.; Lundt, I. *Carbohydr. Res.* **2006,** *341*, 1692–1696.
    (b) Maier, P.; Andersen, S. M.; Lundt, I. *Synthesis* **2006,** 827–830.
28. Fiskesund, R.; Abeyama, K.; Yoshinaga, K.; Abe, J.-i.; Yuan, Y.; Yu, S. *Planta Med.* **2010,** *76*, 1635–1641.
29. Ito, T.; Totoki, T.; Takada, S.; Otsuka, S.; Maruyama, I. *Sci. Rep.* **2021,** *11*, 19648.
30. Streith, J.; Boiron, A.; Frankowski, A.; Le Nouen, D.; Rudyk, H.; Tschamber, T. *Synthesis* **1995,** 944–946.
31. Brust, A.; Cuny, E. *Green. Chem.* **2013,** *15*, 2993–2998.
32. (a) Handy, S. T.; Okello, M.; Dickenson, G. *Org. Lett.* **2003,** *5*, 2513–2515.
    (b) For a review on ionic liquids from natural sugars, see Chiappe, C.; Marra, A.; Mele, A. *Top. Curr. Chem.* **2010,** *295*, 177–195.
33. Lauritsen, A.; Madsen, R. *Org. Biomol. Chem.* **2006,** *4*, 2898–2905.
34. Rokade, S. M.; Garande, A. M.; Ahmad, N. A. A.; Bhate, P. M. *RSC Adv.* **2015,** *5*, 2281–2284.
35. Raja, D.; Philips, A.; Palani, P.; Lin, W.-Y.; Devikala, S.; Senadi, G. S. *J. Org. Chem.* **2020,** *85*, 11531–11540.
36. Raja, D.; Philips, A.; Sundaramurthy, D.; Senadi, G. C. *Chem. Asian J.* **2021,** *16*, 3754–3759.
37. Philips, A.; Raja, D.; Arumugam, A.; Lin, W.-Y.; Senadi, G. C. *Asian J. Org. Chem.* **2021,** *10*, 1795–1800.
38. Yu, F.; Darcel, C.; Fischmeister, C. *ChemSusChem* **2022,** *15*, e202102361.
39. Bose, A.; Pathak, T. *Adv. Carbohydr. Chem. Biochem.* **2020,** *78*, 1–134.
40. (a) Das, I.; Pal, T. K.; Pathak, T. *J. Org. Chem.* **2007,** *72*, 9181–9189.
    (b) Atta, A. K.; Pathak, T. *J. Org. Chem* **2009,** *74*, 2710–2717.
41. (a) Martin, O. R.; Rafka, R. J.; Szarek, W. A. *Carbohydr. Res.* **1989,** *185*, 77–89.
    (b) Heath, C. E.; Gillam, M. C.; Callis, C. S.; Patto, R. R.; Abelt, C. J. *Carbohydr. Res* **1998,** *307*, 371–373.
42. Bhaumik, A.; Pathak, T. *J. Org. Chem.* **2015,** *80*, 11057–11064.
43. Bhaumik, A.; Das, A.; Pathak, T. *Asian J. Org. Chem.* **2016,** *5*, 1048–1062.
44. (a) Yu, K.; Zhao, X.; Wu, W.; Hong, Z. *Tetrahedron Lett* **2013,** *54*, 2788–2790.
    (b) Kwon, H. B.; Itzstein, M. V.; Jung, K.-Y. *Tetrahedron Lett* **2009,** *50*, 2543–2544.
45. Sanki, A. K.; Bhattacharya, R.; Atta, A. K.; Suresh, C. G.; Pathak, T. *Tetrahedron* **2008,** *64*, 10406–10416.
46. (a) Kuster, B. F. M. *Starch* **1990,** *42*, 314–321.
    (b) Chheda, J. N.; Huber, G. W.; Dumesic, J. A. *Angew. Chem. Int. Ed.* **2007,** *46*, 7164–7183.

47. (a) Grote, A.; Freiherrn, V.; Tollens, B. *Liebigs Ann* **1875,** *175*, 181–204.
(b) Roman-Leshkov, Y.; Barrett, C. J.; Liu, Z. Y.; Dumesic, J. A. *Nature* **2007,** *447*, 982–985.
(c) Stahlberg, T.; Fu, W.; Woodley, J. M.; Riisager, A. *ChemSusChem* **2011,** *4*, 451–458.
48. (a) Rosatell, A. A.; Simeonov, S. P.; Fradea, R. F. M.; Carlos, A. M.; Afonso, C. A. M. *Green Chem* **2011,** *13*, 754–793.
(b) van Putt, R.-J.; van der Waal, J. C.; de Jong, E.; Rasrendra, C. B.; Heeres, H. J.; de Vries, J. G. *Chem. Rev.* **2013,** *113*, 1499–1597.
(c) Menegazzo, F.; Ghedini, E.; Signoretto, M. *Molecules* **2018,** *23*, 2201.
49. Available from: https://commons.wikimedia.org/wiki/File:Illustration_Cichorium_intybus0_clean.jpg?uselang = en#Licensing.
50. (a) Niness, K. R. *J. Nutr.* **1999,** *129*, S1402–S1406.
(b) Roberfroid, M. B. *J. Nutr* **2007,** *137*, 2493S–2502S.
51. Qi, X.; Watanabe, M.; Aida, T. M.; Smith, R. L., Jr *Green. Chem.* **2010,** *12*, 1855–1860.
52. (a) Zhu, C.; Wang, H.; Cai, C.; Bi, K.; Bijuan Cai, B.; Song, X.; Qiying Liu, Q.; Ma, L. *ACS Sustainable Chem. Eng* **2019,** *7*, 16026–16040.
(b) Dutta, S.; Bhat, N. S. *ACS Omega* **2021,** *6*, 35145–35172.
53. Anh Tuan Hoang, A. T.; Aykut, I. O.; Nizetic, S. *J. Energy Inst.* **2021,** *94*, 360–386.
54. For a review on furanics to aromatics, see Averochkin, G. M.; Gordeev, E. G.; Skorobogatko, M. K.; Kucherov, F. A.; Ananikov, V. P. *ChemSusChem* **2021,** *14*, 3110–3123.
55. Xia, H.; Xu, S.; Hua, H.; An, J.; Li, C. *RSC Adv.* **2018,** *8*, 30875–30886.
56. Eggersdorfer, M.; Laudert, D.; Letinois, U.; McClymont, T.; Medlock, J.; Netscher, T.; Bonrath, W. *Angew. Chem. Int. Ed.* **2012,** *51*, 12960–12990.
57. Sugisawa, T.; Miyazaki, T.; Hoshino, T. *Biosci. Biotechnol. Biochem.* **2005,** *69*, 659–662.
58. Chen, C.-C.; Whistler, R. L. *Carbohydr. Res.* **1988,** *175*, 265–271.
59. Patil, J. R.; Bose, J. L. *J. Ind. Chem. Soc.* **1966,** *43*, 161–168.
60. Gizaw, Y.; BeMiller, J. N. *Carbohydr. Res.* **1995,** *266*, 81–85.
61. Boettcher, A.; Lichtenthaler, F. W. *Tetrahedron: Asymmetry* **2004,** *15*, 2693–2701.
62. Cubero, I.; Lopez-Espinosa, M. T. P.; Kari, N. *Carbohydr. Res.* **1994,** *261*, 231–242.
63. Szarek, M. A.; Wu, X.; Szarek, W. A. *Carbohydr. Res.* **1997,** *299*, 165–170.
64. Fumeaux, R. H.; Tyler, P. C.; Whiteuse, L. A. *Tetrahedron Lett.* **1993,** *34*, 3609–3612.
65. Tatibouet, A.; Lefoix, M.; Nadolny, J.; Martin, O. R.; Rollin, P.; Yang, J.; Holman, G. D. *Carbohydr. Res.* **2001,** *333*, 327–334.
66. (a) Wuest, M.; Trayner, B. J.; Grant, T. N.; Jans, H.-S.; Mercer, J. R.; Murray, D.; West, F. G.; McEwan, A. J. B.; Wuest, F.; Cheeseman, C. I. *Nucl. Med. Biol.* **2011,** *38*, 461–475.
(b) Soueidan, O.-M.; Trayner, B. J.; Grant, T. N.; Henderson, J. R.; Wuest, F.; West, F. G.; Cheeseman, C. I. *Org. Biomol. Chem* **2015,** *13*, 6511–6521.
67. Levi, J.; Cheng, Z.; Gheysens, O.; Patel, M.; Chan, C. T.; Wang, Y.; Namavari, M.; Gambhir, S. S. *Bioconjugate Chem.* **2007,** *18*, 628–634.
68. Russ, C.; Begli, A. H.; Koenig, B. *Synth. Commun.* **2013,** *43*, 2452–2456.
69. Caratzoulas, S.; Davis, M. E.; Gorte, R. J.; Gounder, R.; Lobo, R. F.; Nikolakis, V.; Sandler, S. I.; Snyder, M. A.; Tsapatsis, M.; Vlachos, D. G. *J. Phys. Chem. C.* **2014,** *118*, 22815–22833.
70. Hu, W.; She, J.; Fu, Z.; Yang, B.; Zhang, H.; Jiang, D. *RSC Adv.* **2021,** *11*, 23365–23373.
71. (a) Hou, Q.; Laiq Ur Rehman, M.; Bai, X.; Qian, H.; Lai, R.; Xia, T.; Yu, G.; Tang, Y.; Xie, H.; Ju, M. *Fuel* **2023,** *338*, 127308.

(b) Patel, U.; Parmar, B.; Dadhania, A.; Suresh, E. *Inorg. Chem.* **2021,** *60*, 9181–129191.

72. Hsu, Y.-H.; Chang, C.-C. *J. Org. Chem.* **2022,** *87*, 13308–13314.
73. Cabezas, J. T.; Waglay, A.; Karboune, S. *Food Biosci.* **2022,** *50A*, 102026.
74. Lee, H.-G.; Dhamija, A.; Das, C. K.; Park, K. M.; Chang, Y.-T.; Schafer, L. V.; Kim, K. *Angew. Chem. Int. Ed.* **2023,** *62*, e202214326.

Chapter 9

Recent synthetic strategies to access diverse iminosugars

Camilla Matassini and Francesca Cardona
Department of Chemistry "Ugo Schiff" (DICUS), Sesto Fiorentino, Italy

9.1 Introduction

Iminosugars are nitrogenated glycomimetics bearing a nitrogen atom replacing the endocyclic oxygen and/or the anomeric carbon of carbohydrates. In the case of anomeric carbon replacement, they are often referred to with the term "azasugars," although the most correct IUPAC name, as pointed out recently, should be aza-carbasugars.[1] Anyway, from the chemical point of view, both structures are polyhydroxylated piperidines. Nevertheless, within the term iminosugars, the researchers aim to include a wide class of polyhydroxylated mono- and bicyclic alkaloids and not only polyhydroxylated piperidines, which resemble more closely the natural hexose structure.

As structurally altered analogs of sugars, glycomimetics in general, and iminosugars in particular, offer the opportunity to mimic the biological action of carbohydrates while circumventing their drawbacks, and for this reason, they are more and more promising as new drugs.[2]

Natural iminosugars are known since 1967, with the first isolation of nojirimycin from *Streptomyces roseochromogenes* and *Streptomyces lavendulae*.[3] The more stable 1-deoxynojirimycin was isolated in 1976 from white mulberry root barks.[4] Since then, many other iminosugars with different structures, namely polyhydroxylated pyrrolidines, piperidines, pyrrolidizines, indolizidines, and nortropanes have been isolated and hundreds of new structures have been synthesized.

Synthetic organic chemistry constitutes a powerful tool to access natural iminosugars, thus offering not only the possibility to obtain quantities of material as pure as required for biological assays (since the sample from natural sources often contain many impurities), but also the opportunity to confirm or propose a new structure for a natural compound.

Moreover, and even more importantly, synthetic organic chemistry offers the possibility to synthesize structural analogs of natural compounds, as well

Synthetic Strategies in Carbohydrate Chemistry. DOI: https://doi.org/10.1016/B978-0-323-91729-2.00003-3

as new heterocyclic skeletons. This is definitely the case with iminosugars, as demonstrated by the diverse classes of compounds, in addition to natural compounds, which are reported in this chapter, including monocyclic and bicyclic skeletons different from those found in nature.

Structural modifications of iminosugars aim not only to enhance affinity and selectivity toward the biological target but also to improve their drug-like properties, which are important for their in vivo applications. Beyond their biological interest, iminosugars are an amazing playground for organic chemists, since the syntheses are often highly challenging due to the presence of several chirality centers that give the researchers the opportunity to investigate new stereoselective reactions. In this regard, the chiral pool approach from carbohydrates is certainly the most straightforward and has the advantage of following at least 1 of the 12 principles of green chemistry (e.g., the use of renewable sources).

Excellent reviews on the total syntheses of iminosugars have appeared in the literature. However, each review is often focused on a particular class of iminosugars[5] and/or on a different reaction employed to access their skeleton,[6] alternatively iminosugar biological properties are explored.[1,7]

Differently, this chapter, far to pretend to be exhaustive, aims to give an overview of a selection of recent papers (since 2015) aimed at the synthesis of diverse classes of iminosugars, thus giving a glance at the most modern organic chemistry involved in the total syntheses of these extremely appealing and biological promising compounds. We select in this chapter those manuscripts in which a new synthetic procedure for the construction of the heterocyclic skeleton is described, highlighting the key step in which the nitrogen atom is introduced and the key carbohydrate employed as the enantiopure starting material for their synthesis.

9.2 Monocyclic iminosugars

9.2.1 Four-membered iminosugars (azetidines)

With respect to five-, six-, and seven-membered analogs, the synthesis of four-membered azetidine iminosugars is less exploited. However, in the last 7 years, some interesting synthetic strategies for their preparation appeared in the literature.

The group of Dhavale and coworkers developed a procedure for the obtainment of a series of azetidine iminosugars bearing an additional hydroxymethyl substituent at the α-position of the ring oxygen, in view of their potential in the inhibition of glycosidase enzymes. In particular, the authors described the preparation of α-geminal hydroxyalkyl, 2-carboxylic acid, and *N*-carboxylic-α,γ-trihydroxymethyl azetidine iminosugars,[8] and an example of this strategy is shown in Scheme 9.1 for the synthesis of iminosugars **7**, **8**, and **9**. Their approach started from 1,2:5,6-di-*O*-isopropylidene-oxo-α-D-

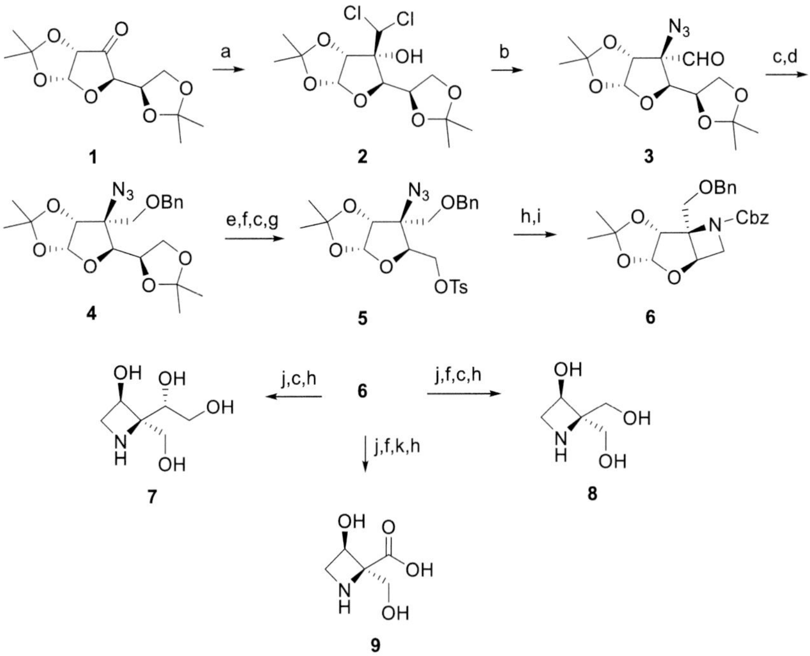

SCHEME 9.1 Reagents and conditions: (a) LDA, CH_2Cl_2, THF, − 78°C, 60%; (b) NaN_3, TBAI, DMF, 110°C, 82%; (c) $NaBH_4$, MeOH:water (9:1), 0°C, 93%; (d) BnBr, NaH, TBAI, 0°C to rt, 92%; (e) AcOH:water (3:2), 55°C, 91%; (f) $NaIO_4$, acetone:water (9:1) 0°C; (g) TsCl, Py, 60°C, 88%; (h) H_2, 10% Pd/C, MeOH; (i) CbzCl, $NaHCO_3$, MeOH:water (9:1), 89%; (j) TFA:H_2O; (k) $NaClO_2$, 30% H_2O_2, CH_3CN:H_2O (1:1), 0°C, 69%.

glucofuranose (**1**), which was treated with in situ generated dichloromethyl lithium. Attack of the nucleophile from the ß-face (opposite to the bulky 1,2-acetonide group) afforded the 3-dichloromethyl-D-allose **2** as a single diastereoisomer. Insertion of the nitrogen atom was achieved through treatment with NaN_3 and tetra-*n*-butylammonium iodide (TBAI) via in situ formation of a chlorooxirane intermediate followed by S_N2 attack of the azide ion from the ß-face, which yielded compound **3** featured by the presence of nitrogen atom. Reduction with $NaBH_4$ followed by benzylation of the primary OH afforded compound **4**. The subsequent steps involved the removal of the 5,6-acetonide group in acidic conditions, followed by periodate cleavage, reduction of the formed aldehyde and tosylation of the primary OH group to form **5**. Catalytic hydrogenation (100 psi) afforded the furanose fused azetidine ring skeleton **6** via azide reduction followed by an intramolecular S_N2 displacement of the tosylate at C-5 and protection of nitrogen as Cbz carbamate. Then, hydrolysis of the 1,2-acetonide group with TFA/H_2O gave an anomeric mixture of hemiacetals that afforded azetidine **8** through oxidation with $NaIO_4$ followed by $NaBH_4$ reduction and final hydrogenation.

Avoiding the oxidation step, the same sequence afforded azetidine iminosugar **7**. The azetidine **9** bearing the carboxylic moiety was finally obtained through hydrolysis of the 1,2-acetonide group in **6** and treatment with $NaIO_4$ followed by Pinnick oxidation.

Incorporation of fluorine atom(s) into bioactive compounds enhances their pharmacological properties, since it produces compounds with increased metabolic stability and bioavailability. For this reason, fluorination has become a standard practice in pharmaceutical chemistry.[9]

A strategy for the synthesis of fluoroazetidine iminosugars was reported by Fleet and coworkers in 2015.[10] The authors chose as the starting material to get all the 4-fluoroaxetidines the diacetone of 3-fluoroglucose **11**, which was formed in 97% yield from diacetone allose **10**, following a procedure previously reported in the literature.[11] An example of their strategy applied to the synthesis of the meso fluoroazetidine **17** is reported in Scheme 9.2. Hydrolysis of the acetonide in **11** followed by peracetylation afforded the tetraacetate **12** as a 1:1 anomeric mixture. Then, standard carbohydrate manipulations (4 steps) allowed the synthesis of the pyranose ditriflate **13**. This latter underwent high-yield nucleophilic displacement with benzylamine that introduced the nitrogen atom and concomitantly formed the bicyclic azetidine **14**. Acetolysis of **14** by acetic anhydride in the presence of boron trifluoride etherate gave the mixed acetal **15**

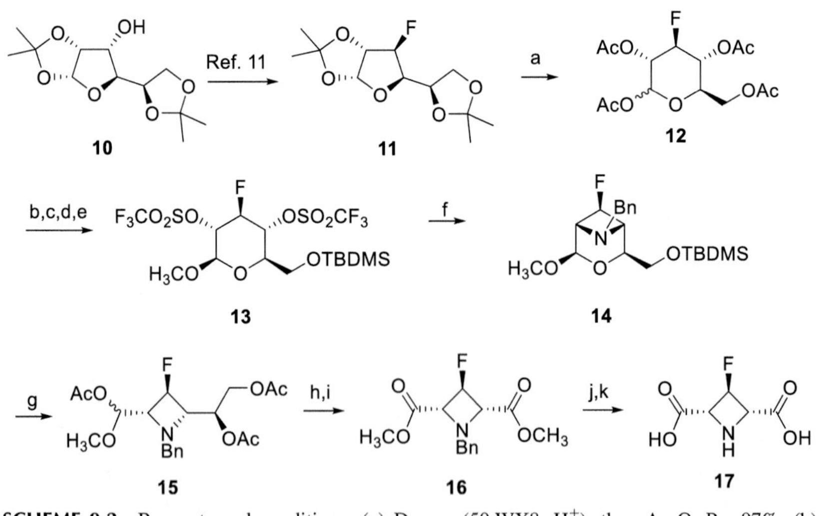

SCHEME 9.2 Reagents and conditions: (a) Dowex (50 WX8, H^+), then Ac_2O, Py, 97%; (b) HBr, AcOH, then $AgCO_3$, MeOH, 87%; (c) NaOMe, MeOH, 96%; (d) TBDMSCl, imidazole, 98%; $(CF_3SO_2)_2O$, Py; (e) $BnNH_2$, DIPEA, CH_3CN, 84%; (f) $Et_2O.BF_3$, Ac_2O, 100%; (g) DIBALH, toluene, then $NaBH_4$, MeOH, 97%; (h) H_2, Pd/C, H_2O, dioxane, 88% (i) $NaIO_4$, H_2O, dioxane, then I_2, K_2CO_3, MeOH, 69%; (j) K_2CO_3, H_2O, dioxane, 57%; (k) H_2, Pd/C, H_2O, dioxane, 45%.

quantitatively. A sequential reduction of the **15** with diisobutylaluminium hydride (DIBALH) followed by sodium borohydride in MeOH, oxidative cleavage of the 1,2-diol moiety with sodium periodate, oxidation of the crude aldehyde by iodine in methanol and esterification gave the diester **16**. Final ester hydrolysis and nitrogen deprotection through catalytic hydrogenation afforded the target compound **17**.

9.2.2 Five-membered iminosugars (pyrrolidines)

Transformation of the anomeric carbon of a furanose-type carbohydrate into a nitrogen-containing moiety that can undergo nucleophile addition with an organometallic reagent is a common strategy for the synthesis of pyrrolidine iminosugars substituted at the carbon atom adjacent to nitrogen. In this general framework, we can mention the work of Martin and coworkers, who reported the preparation of *N-tert*-butanesulfinyl glycosylamines that reacted as latent imine equivalents with a variety of carbon nucleophiles leading to precursors for the diastereoselective synthesis of iminosugar-*C*-glycosides.[12]

This strategy is depicted in Scheme 9.3 for the synthesis of compound **23**. The *N-tert*-butanesulfinyl glycosylamine **20** was obtained by heating tri-*O*-benzyl-arabinofuranose (**18**) and (*S*)-sulfinamide (**19**) in the presence of Ti $(OEt)_4$. The organometal addition of BnMgCl on **20** gave excellent diastereoselectivity (*dr* 7:3) the compound **21** in which the newly formed stereocenter had the *S* configuration. Cyclization was then achieved through mesylation of the free OH group followed by basic treatment to yield **22**, which was then subject to *N*- and *O*-deprotection to finally give the target iminosugar-*C*-glycoside **23**.

The same researchers described that the treatment of *N-tert*-butanesulfinyl glycosylamine of type **20** with $NaBH_4$ as the nucleophile allowed to achieve

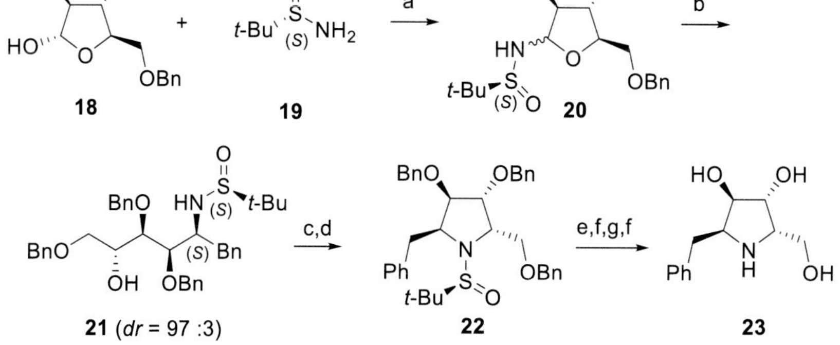

SCHEME 9.3 Reagents and conditions: (a) $Ti(OEt)_4$, PhMe, 70°C; (b) BnMgCl, 83 %, *dr* 97:3; (c) NEt_3, MsCl; (d) *t*BuOK, 67%; (e) HCl, MeOH; (f) Amberlite IRA-400 (OH^- form), 98%; (g) H_2, $Pd(OH)_2$, AcOH, 20°C, 3d.

the synthesis of 5-membered iminosugars not substituted at the carbon adjacent to nitrogen,[13] while the addition of Bu_3SnLi as the nucleophile gave 1-*C*-stannylated iminosugars amenable for subsequent cross-coupling reactions.[14]

A straightforward synthesis of pyrroline-type iminosugars **29** was described by Ramesh and coworkers in 2020,[15] exploiting the three-step conversion of tri-*O*-benzyl-D-glucal **24** into the triol **25** following a modified procedure of Danishefsky previously reported by the authors,[16] which allowed introduction of nitrogen at C-2. The triol **25** was then subjected to oxidative cleavage with $NaIO_4$, smoothly going to completion to form the aminoaldehyde **26** in very good yield, which existed only in the hemiaminal form **27** as a mixture of diastereoisomers in a 2:1 ratio. Deoxygenation of **27** was carried out with triethylsilane in the presence of $BF_3.Et_2O$ to afford **28** in 87% yield. Final reprotection under Birch conditions afforded the known 1,4-dideoxy-1,4-imino-L-xylitol **29** in 79% yield (Scheme 9.4).

The synthesis of two polyhydroxylated pyrrolidines as 1,4-dideoxy-imino-D-arabinitol (DAB) analogs bearing a hydrazide moiety was described by Sydnes and coworkers.[17] The synthesis started with 2,3,4-tri-*O*-benzyl-L-xylofuranose (**30**), which was condensed with hydroxylamine to provide an aldoxime intermediated followed by dehydration under Appel conditions, which afforded nitrile **31** in 52% yield (Scheme 9.5). Triflation of the free hydroxy group at C-4 followed by treatment with Boc-hydrazide promoted the stereoselective hydrazidation with inversion of the stereochemistry at C-4. Removal of the Boc protecting group of **32** with TFA promoted cyclization of the 5-membered ring and formation of the hydrazide imide **33** after

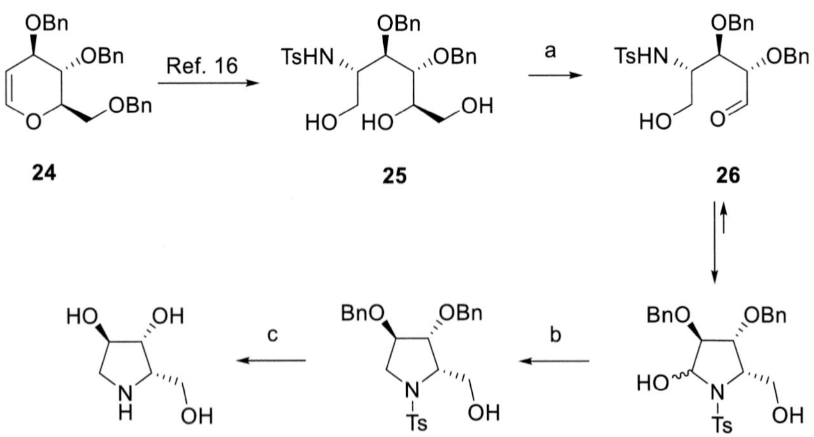

24 **25** **26** **29** **28** **27**

SCHEME 9.4 Reagents and conditions: (a) $NaIO_4$, aq. $NaHCO_3$, 30°C, 8 h, CH_2Cl_2, 80%; (b) Et_3SiH, $BF_3.Et_2O$, 0°C–33°C, CH_2Cl_2, 7 h, 87%; (c) Na, Liq. NH_3, − 78°C, 3 h, 79%.

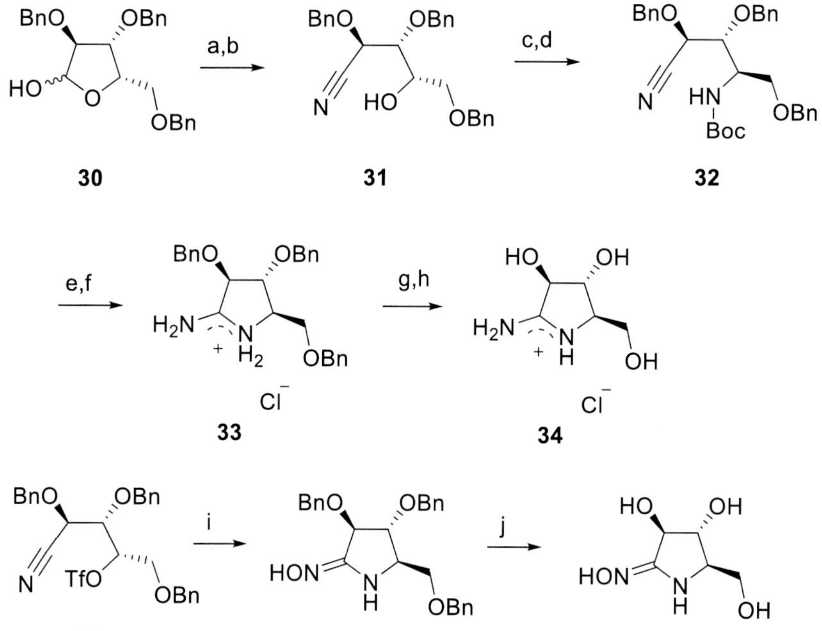

30 31 32 33 34 35 36 37

SCHEME 9.5 Reagents and conditions: (a) NH_2OH, EtOH; (b) CBr_4, PPh_3, CH_3CN, rt, 52%; (c) Tf_2O, Py, CH_2Cl_2, 0°C; (d) NH_2NHBoc, THF, rt, 54%; (e) TFA, CH_2Cl_2, rt; (f) HCl, 88%; (g) H_2, Pd/C, EtOH, TFA, rt, (h) HCl, 19 h, 90%; (i) NH_2OH, EtOH, rt, 17 h, 36%; (j) BCl_3, CH_2Cl_2, −78°C, 3 h, 0°C, 15 h, 67%.

treatment with HCl. Final hydrogenation for 19 hours formed the target compound **34** (Scheme 9.5). Interestingly, when the hydrogenation step was prolonged to 48 hours, the corresponding *N*-ethyl hydrazide imide was formed as the major product. If hydroxylamine was employed instead of Boc-hydrazide, an amide oxime moiety could be introduced in place of the amidine moiety.[18] The reaction of triflate **35** with hydroxylamine in ethanol resulted in the formation of **36** in 36% yield (a 2% of the D-ribo epimer was formed, not shown). The final removal of the *O*-benzyl groups with boron trichloride prevented the reduction of the oxime moiety and afforded the target compound **37** in good yield.

In 2017 Timmer and coworkers combined the Vasella-reductive amination and the carbamate annulation reaction and applied this methodology to a ketose starting material, with the aim of accomplishing the total and expeditious synthesis of five-membered tetrahydroxylated iminosugars.[19] The Fischer glycosidation of D-fructose (**38**) gave compound **39**, which underwent regioselective iodination at the sterically more accessible C-6 to afford compound **40** in 65% over 2 steps (Scheme 9.6). A careful optimization of the Vasella-reductive amination conditions (excess Zn, NH_4OAc, AcOH,

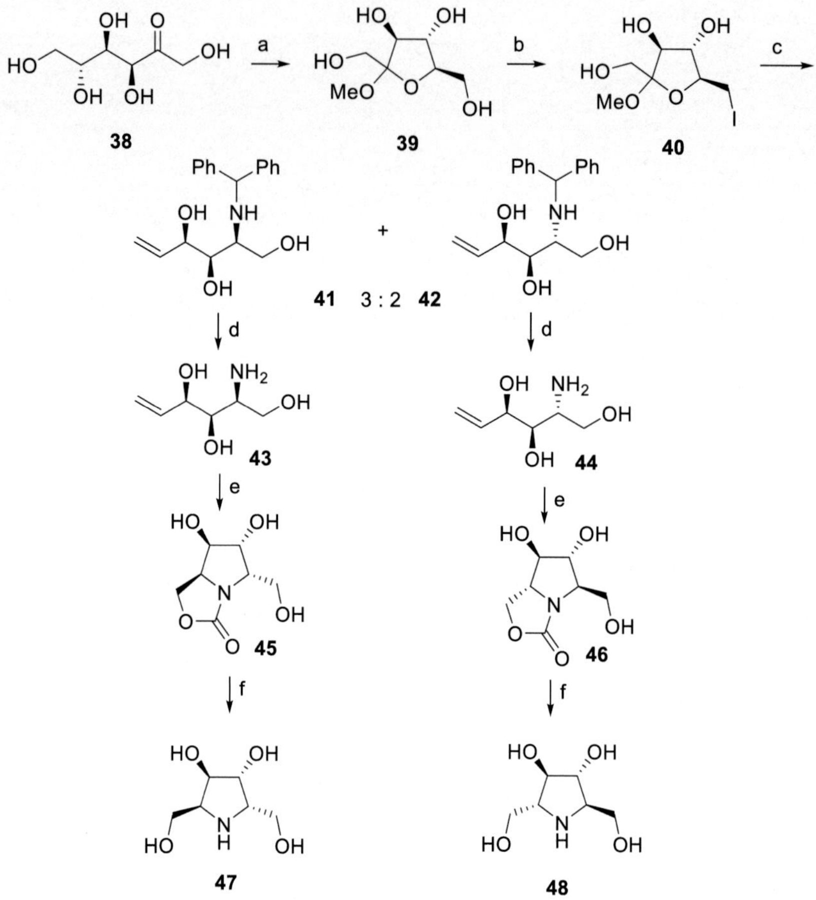

SCHEME 9.6 Reagents and conditions: (a) MeOH, H_2SO_4, 20 min, rt; (b) PPh_3, I_2, imidazole, THF, 65% (2 steps); (c) Zn, Ph_2CHNH_2, $NaCNBH_3$, AcOH, EtOH, reflux, 18 h, 85%; (d) TFA; Et_3SiH, 54%–55%; (e) $NaHCO_3$, I_2, 18 h, rt; (f) NaOH, EtOH, reflux, 4 h, 98% of **47**, 16% of **48**.

EtOH, reflux, 168 hours) on **40** led to the introduction of the nitrogen atom by means of reaction with diphenylamine as the amine nucleophile. This resulted in the formation of a 3:2 mixture of the alkenyl amines **41** and **42**, which could be separated by column chromatography. After amine deprotection to afford **43** and **44**, carbamate annulation with $NaHCO_3$ and I_2 according to a protocol developed previously by the authors,[20] the base-mediated hydrolysis of the carbamates **45** and **46**, finally led to the target tetrahydroxypyrrolidines **47** and **48**, respectively. In particular, with this strategy, the 2,5-dideoxy-2,5-imino-L-iditol (**47**) was obtained in six steps and 18% overall yield with minimal use of protecting groups (Scheme 9.6). With a similar strategy, the authors described the synthesis of 2,5-dideoxy-2,5-imino-D-altritol in seven steps and 22% overall yield from readily available D-tagatose.[21]

Finally, the metal-free Overman rearrangement of trichloroimidates was described in 2021 for the synthesis of the nitrogenated iminosugar derivative **55** and its six-membered analog **56** (Scheme 9.7).[22] Starting from L-arabinose (**49**), witting olefination and reaction with CCl_3CN in 1,8-diazabicyclo[5.4.0]undec-7-ene (DBU) afforded the tetrakisimidate **50**, which, upon reaction with dibutylhydroxytoluene (BHT) and *t*-BuPh at 200°C in a sealed tube, underwent four sequential Overman rearrangements with the formation of four new C−N bonds and three contiguous stereocenters, two of which were stereospecific through chirality transfer of the secondary alcohols, while the third was installed with a 2.2:1 diastereoselectivity. The poor diastereoselectivity derived from the fourth rearrangement was not a problem since both epimers gave the same final product through epimerization at a later stage. The terminal olefin of **51** was converted into aldehyde **52** through standard dihydroxylation followed by diol cleavage with $Pb(OAc)_4$, and the final palladium-catalyzed hydrogenation yielded an inseparable mixture of

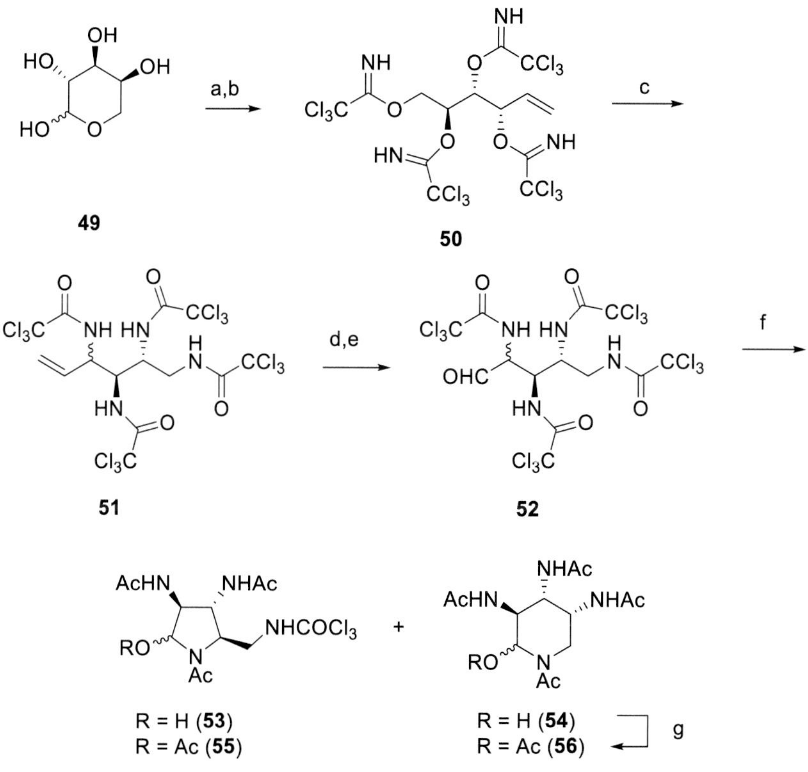

SCHEME 9.7 Reagents and conditions: (a) $MePPh_3Br$, *t*-BuOK, THF, 60°C; (b) CCl_3CN, DBU, CH_2Cl_2, − 20°C, 25% (over 2 steps); (c) BHT, *t*-BuPh, 200°C, 77%, *dr* 2.2:1; (d) OsO_4 (5 mol%), NMO, CH_2Cl_2/H_2O, rt; (e) $Pb(OAc)_4$, CH_2Cl_2, rt; (f) H_2 (1 atm), Pd/C (200 wt.%), NEt_3, MeOH, rt; (g) Ac_2O, Py, rt, 20% for **55**, 6% for **56** over 4 steps.

hemiaminal derivatives **53** and **54** via epimerization at the α-position of the aldehyde in the acyclic form. Acetylation of the free OH allowed separation of D-arabinofuranose-type peracetylated **55** and the D-arabinose-type peracetylated **56** in 20% and 6% yield, respectively, over four steps.

9.2.3 Six-membered iminosugars (piperidines)

Great efforts have been recently devoted to the building of piperidine iminosugars functionalized on the carbon adjacent to nitrogen, due to the emerging biological properties of these iminosugars.

In 2017 the group of Martin and coworkers developed an innovative procedure for the synthesis of two epimers of 4-*epi*-isofagomine carrying a pentyl group at C-5a, namely (5a*R*)- and (5a*S*)-5a-*C*-pentyl-4-*epi*-isofagomine.[23] The key step of the strategy relied on the addition of nitrohexane to a keto-pentopyranoside intermediate derived from D-lyxose. Inspired by the previous concise synthesis of isofagomine (IFG) developed by Andersch and Bols in 2001,[24] they employed a longer chain nitroalkane to simultaneously introduce the nitrogen atom and the C-pentyl fragment on the final molecules, which were designed to interact with lysosomal β-galactosidase enzyme and then tested as pharmacological chaperone on cell lines bearing mutations associated with GM1-gangliosidosis and type B Morquio disease. In this regard, a five-carbon atom chain was introduced as a good compromise to provide some lipophilic character to the molecule without rising problems of toxicity that are often encountered with longer alkyl chains.[25]

The synthesis started with D-lyxose, which was converted into ketone **57** by adaptation of a known procedure. The reaction in neat nitrohexane in the presence of NEt_3 formed two new stereocenters. Among the four new compounds that could be formed, in principle, the addition led to a 7:3 mixture of the two epimers **58** at the nitromethane groups and a single configuration at the ring carbon atom, which was favored by both kinetic (addition to the less hindered face of the **57**) and thermodynamic factors (the nitrohexyl groups in the more stable equatorial position). Two consecutive and selective hydrogenations for the reduction of the nitro moiety, the cleavage of the benzyl groups, and an intramolecular reductive amination afforded an inseparable mixture of two iminosugars **59** epimeric at C-5a (Scheme 9.8).

The two epimers **59** were then separated by the protection of the nitrogen atom as a benzyl carbamate, and removal of the hydroxymethyl group was achieved by means of a radical deoxygenation with Bu_3SnH of a cyclic thiocarbonate to yield the final compounds **60** and **61** (see Ref. 23 for further specific details).

A one-pot iminium-ion-based strategy was developed by Baskaran and coworkers in 2018 for the highly selective synthesis of novel C-aryl glycoside iminosugars.[26] The in situ generated iminium ion **63** was formed from the D-ribose-derived tosylate **62** and reacted in a smooth one-pot alkylation

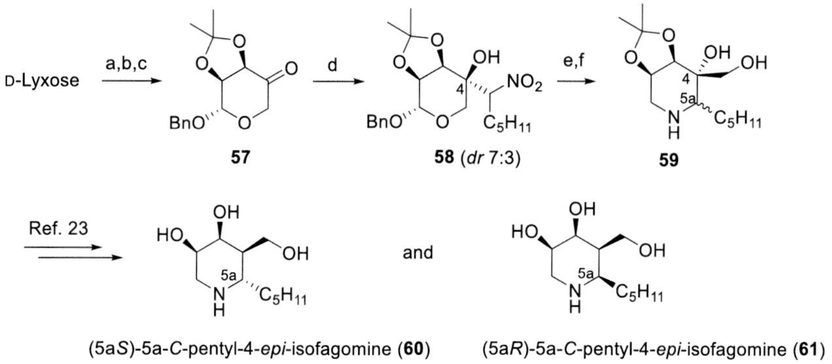

SCHEME 9.8 Reagents and conditions: (a) BnOH, *p*-TSA, 60°C, 78 h, 78%; (b) 2,2-dimethoxypropane, acetone, *p*-TSA, rt, 18 h, 65 %; (c) Dess-Martin periodinane, CH_2Cl_2, rt, 16 h, 99%; (d) 1-nitrohexane, NEt_3, rt, 4 d, 52%; H_2, Raney Ni, AcOH, iPrOH, rt, 5 d, 85%; (f) H_2, Pd $(OH)_2$/C, AcOH, iPrOH, rt, 3d, quant.

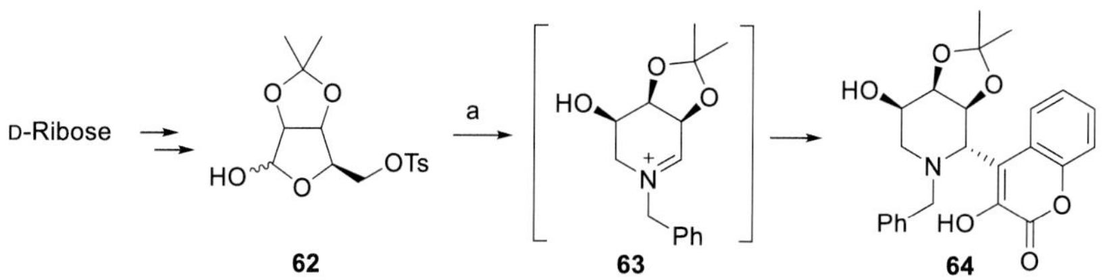

SCHEME 9.9 Reagents and conditions: (a) $BnNH_2$, CH_2Cl_2, 2 h, then 4-hydroxycoumarin, NEt_3, rt, 9 h, 82%.

with electron-rich aromatic systems such as 4-hydroxycoumarin, yielding the corresponding ß-*C*-coumarinyl glycoside iminosugar **64** as a single diastereoisomer in very good yield (Scheme 9.9).[26] The stereoselectivity was rationalized by the authors invoking a preferential addition of the nucleophile to the iminium ion to the *Si* face (*pseudo* equatorial approach) rather than a more sterically hindered *Re* face attack (*pseudo*axial approach). The scope of this one-pot transformation was further tested with various heteroaromatic systems and amines (six examples).

A nucleophilic addition onto an iminium ion was also used as the strategy for the introduction of a propyl group at the carbon adjacent to nitrogen in the deoxynojirimycin derivative **69** (Scheme 9.10). In this case, Mellet and coworkers exploited the chemistry of their sp^2-derived iminosugars, from which the iminium ion was generated.[27] Compound **67** was prepared from 1,2-*O*-isopropylidene-α-D-glucoronolactone (**65**) through compound **66** employing an optimized double inversion sequence using triflate as the leaving group and trifluoroacetate and azide as the nucleophile. After peracetylation, the reaction with the nucleophile AllTMS in acetonitrile at 80°C using

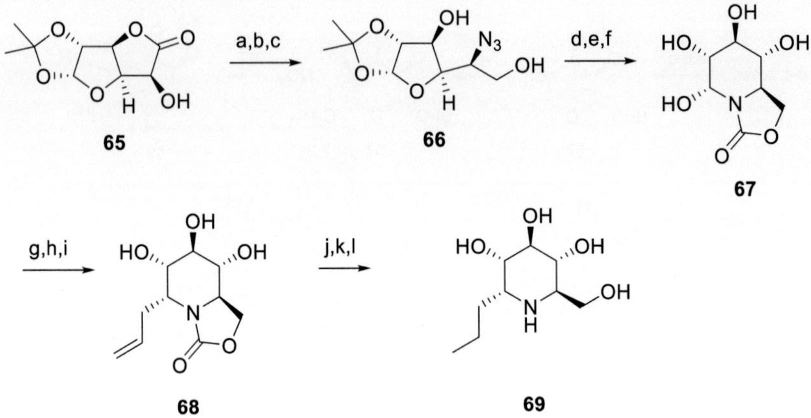

SCHEME 9.10 Reagents and conditions: (a) Tf_2O, CH_2Cl_2, Py, then $NaOCOCF_3$, 80%; (b) Tf_2O, CH_2Cl_2, Py, then NaN_3, 85%; (c) DIBAL-H, $NaBH_4$, 80%; (d) Py, quant.; (e) $(CCl_3)_2CO$, 90%; (f) TFA-H_2O, quant.; (g) Ac_2O/Py, 80%; (h) AllTMS, BF_3.Et_2O, 91%; (i) NaOMe/MeOH, quant.; (j) BnBr, 97%; (k) $Ba(OH)_2$, 93%; (l) H_2/Pd/C, quant.

boron trifluoro etherate as a promoter provided the corresponding allyl α-*C*-glycoside **68** in excellent yield after acetyl deprotection. After perbenzylation of the hydroxy groups, hydrolysis of the cyclic carbamate with barium hydroxide followed by Pd(0) catalyzed hydrogenation delivered the known DNJ iminosugar derivative **69** (Scheme 9.10).

Cardona, Matassini, and coworkers reported an efficient and straightforward strategy for the synthesis of trihydroxypiperidines alkylated at C-2, during their ongoing search for new pharmacological chaperones for glucerobrosidase (also known as GCase), the defective enzyme in Gaucher disease.[28] The authors developed a stereodivergent access to both epimeric *N*-hydroxylamines **72**, which exploited the stereoselective addition of Grignard reagents onto nitrone **71**, in turn obtained from the D-mannose derived aldehyde **70**,[29] in the presence or in the absence of BF_3 as the Lewis acid.[30] The stereoselectivity of the addition (*dr* ranging from 4:1 to 9:1) was always in favor of the *S*-configured stereocenter when no Lewis acid was used and of the *R*-configured one when BF_3 was employed. The observed stereoselectivity was rationalized on the basis of a preferred nucleophilic attack to the *Re* face of nitrone **71** in the presence of the coordinating Lewis acid, while with no Lewis acid the preferred nucleophilic attack to the *Si* face of nitrone **71** was postulated invoking a Cram-chelate transition state. Final catalytic hydrogenation followed by acidic treatment afforded the target trihydroxypiperidines **73** alkylated at C-2 (Scheme 9.11). Further exploitation of this strategy employing new nitrones-bearing lipophilic chains allowed the synthesis of bis-alkylated trihydroxypiperidines.[31]

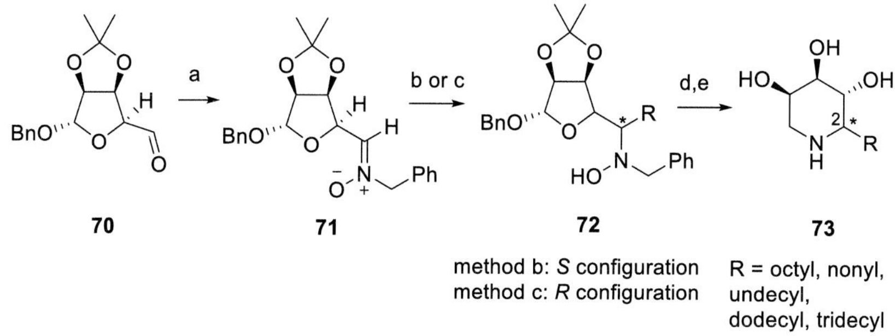

SCHEME 9.11 Reagents and conditions: (a) *N*-benzyl hydroxylamine, CH_2Cl_2, rt, 85%; (b) RMgBr, THF, −78°C, 52%–70%; (c) RMgBr, THF, $BF_3.Et_2O$ −78°C, 70%–87%; (d) Pd/C, H_2, CH_3COOH, MeOH, rt, 2d, then Ambersep 900-OH, 78%–93%; (e) HCl, MeOH, then Ambersep 900-OH, 76%–100%.

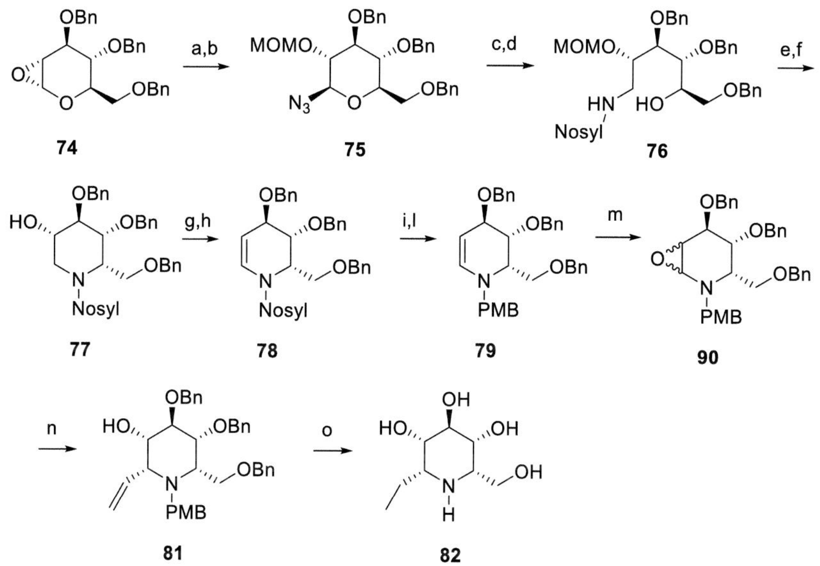

SCHEME 9.12 Reagents and conditions: (a) $Ce(NH_4)_2(NO_3)_6$, NaN_3, CH_3CN, H_2O, rt, 10 min, 80%; (b) MOM-Cl, DIPEA, CH_2Cl_2, 0°C to rt, 24 h, 85%; (c) $LiAlH_4$, THF, 0°C to reflux; (d) Nosyl-Cl, NEt_3, CH_2Cl_2, rt, 2 h, 75%; (e) PPh_3; DIAD, toluene, rt, 5 h, 78%; (f) TFA, CH_2Cl_2, rt, 48 h, 83%; (g) Tf_2O, Py, CH_2Cl_2, 0°C, 15 min; (h) DBU, THF, reflux, 1 h, 71%; (i) K_2CO_3, PhSH, CH_3CN, rt, 10 h; (j) PMB-Cl, K_2CO_3, NaI, THF, rt, 2 h, 71%; (k) Oxone, aq $NaHCO_3$, acetone, CH_2Cl_2, 0°C, 3 h; (l) vinylMgBr, THF, 0°C to rt, 6 h, 65%; (m) H_2, Pd/C, MeOH, rt, 24 h 68%.

In 2018 the group of Vankar and coworkers exploited the use of iminoglycals as precursors for the preparation of piperidine iminosugars, as demonstrated with the synthesis of (-)-1-*epi*-adenophorine (**82**) (Scheme 9.12).[32] Introduction of the nitrogen atom was achieved through sodium azide nucleophilic opening of the D-glucal-derived epoxide **74** in the presence of ceric ammonium nitrate (CAN), followed by protection of the hydroxy group at C-2 to yield compound **75**. Reduction of the azide moiety with $LiAlH_4$ along with amine protection with nosyl chloride afforded compound **76**. An intramolecular Mitsunobu reaction followed by methoxymethyl (MOM) deprotection with TFA led to compound **77**, whose hydroxy group was reacted with triflic anhydride and pyridine to form the corresponding triflate that afforded the iminoglycal **78** by means of elimination of the triflate group with DBU. The protecting group on the ring nitrogen was changed from nosyl to *p*-methoxy benzyl (PMB) to obtain **79**. Smooth oxidation of **79** yielded the unstable epoxide **80**, which was directly treated with vinylMgBr to introduce the vinyl group at C-2. Finally, catalytic hydrogenation of compound **81** afforded the target iminosugar (-)-1-*epi*-adenophorine (**82**) (Scheme 9.12).

In 2015 Saxena and coworkers presented an interesting and new reductive cyclization/Ugi multicomponent reaction sequence that was applied to an azido-hemiacetal derived from commercially available D-ribose (**83**) en route to the synthesis of L-1-deoxyallonojirimycin (**90**) (Scheme 9.13).[33] The synthesis of the key azido-hemiacetal **85** was achieved from **83** through the protection of the *cis*-hydroxy group with cyclohexanone, selective tosylation of the primary hydroxy groups, benzoyl protection of the anomeric oxygen, introduction of the nitrogen atom at C-5 with sodium azide on tosylate **84** and anomeric oxygen deprotection. With the key intermediate **85** in hand, the authors developed a new methodology for azide reduction followed by cyclization, which employed $BF_3.Et_2O/NaI$ in CH_3CN instead of the previously reported $FeSO_4 \cdot 7H_2O/NH_3$ in dichloromethane.[34] The reaction formed in situ a cyclic imine that underwent a Ugi reaction with pent-4-enoic acid to yield the bis-amide **86**. Cyclohexyl protection of the *cis*-hydroxy groups was then replaced with benzyl group protection to afford compound **87**. Selective hydrolysis of one of the amide groups with I_2 in THF/H_2O afforded amide **88**, which was then reduced to aldehyde **89** with triflic anhydride and 2-fluoro-pyridine and triethylsilylhydride in CH_2Cl_2. Final hydrogenation with catalytic Pd/C in MeOH afforded the target L-1-deoxyallonojirimycin (**90**) (Scheme 9.13). The isomer L-1-deoxyaltronojirimycin was also obtained, exploiting Mitsunobu inversion of configuration of the hydroxy group of the intermediate **86**, then subjected to a similar synthetic strategy.

An interesting and effective method for the synthesis of *N*-substituted trihydroxypiperidine derivatives was also described by Shao and coworkers. The reaction of the D-ribose-derived tosylate with several amines formed a cyclic iminium ion intermediate that reacted in situ with the α-carbon of enolized ketones through a preferred *Si* facial attack with a Mannich-type reaction to afford trihydroxypiperidines substituted at C-2. However, the work did not describe the synthesis of the target deprotected iminosugars.[35]

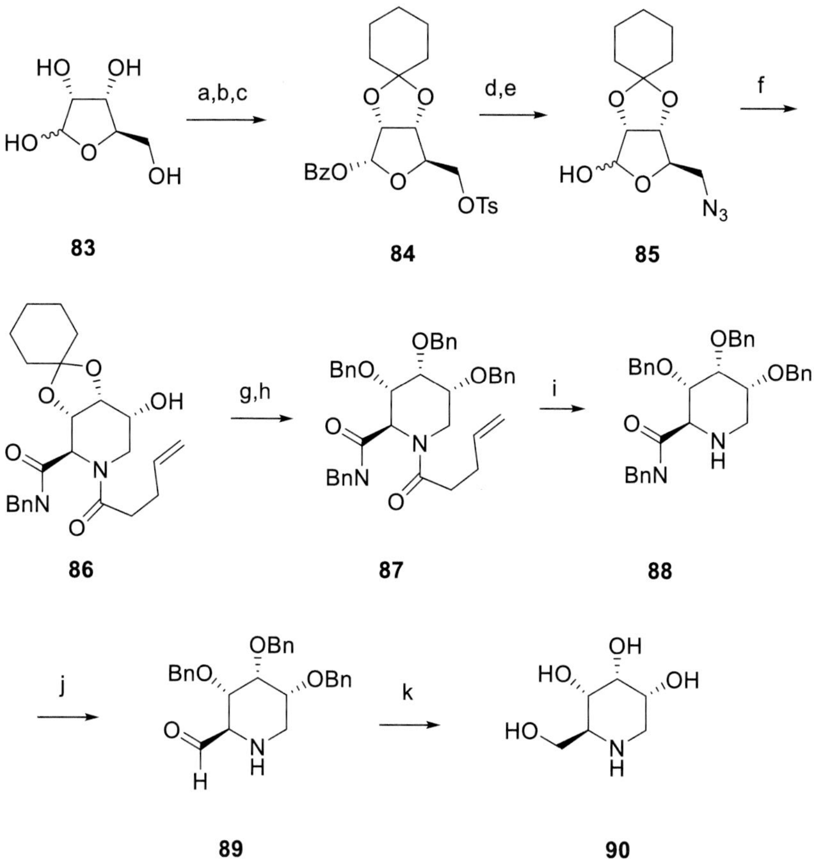

SCHEME 9.13 Reagents and conditions: (a) cyclohexanone, *p*-TSA; (b)TsCl, Py; (c) BzCl, Py; (d) NaN_3, DMF; (e) NaOMe (cat.), MeOH; (f) $BF_3.Et_2O$, pent-4-enoic acid, NaI, CH_3CN; (g) HF(Py), $CHCl_3$; (h) BnBr, NaI, DMF; (i) I_2, THF/H_2O; (j)Tf_2O, 2-FPyr, Et_3SiH, CH_2Cl_2; (k) H_2, Pd/C, MeOH.

9.2.4 Seven-membered iminosugars (azepanes) and eight-membered iminosugars (azocanes)

Azepane iminosugars were synthesized for the first time by Paulsen and coworkers in 1967.[36] In 2004 Martin and coworkers first reported the synthesis and biological evaluation of eight-membered azocane iminoalditols.[37] However, azepane and azocane iminosugars, have definitely received very little attention probably due to the extremely challenging obtainment of the final larger ring cyclization.

In 2016 Shaw and coworkers published the synthesis of seven- and eight-member higher-ring analogs of D-fagomine from L-xylose.[38] The synthesis began with commercially available L-xylose, which was selectively methylated at the anomeric oxygen followed by benzylation of the remaining hydroxy groups and anomeric deprotection to yield **30**, which was subjected to Wittig olefination, and the resulting **91** was converted into the phthalimido compound **92** in Mitsunobu conditions that inverted the stereochemistry at C-4 contemporarily allowing nitrogen insertion (Scheme 9.14). The phtalimido moiety was hydrolyzed with $MeNH_2$ and the resulting amine was *N*-alkyated with benzyl bromide or 4-bromobut-1-ene to afford, respectively, dienes **93** and **96**, suitable for the following cyclization step that was conducted through ring-closing metathesis with Grubbs II catalyst. Final hydrogenation of **94** and **97** afforded the target azepane **95** and azocane **98**, respectively (Scheme 9.14).

9.3 Bicyclic iminosugars

9.3.1 Pyrrolizidines

Among the synthetic strategies reported in these last years for the synthesis of bicyclic pyrrolizidine iminosugars, the great majority relies on the

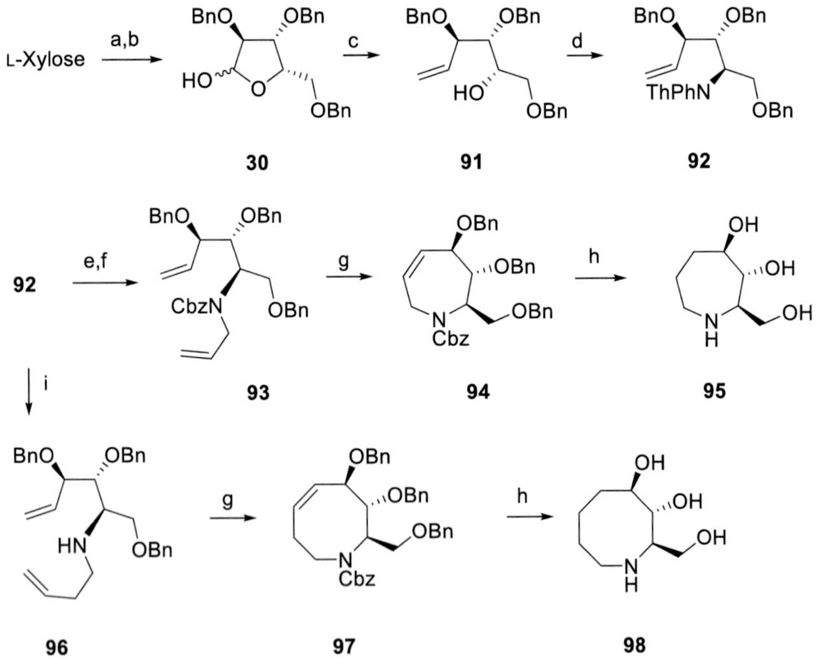

SCHEME 9.14 Reagents and conditions: (a) MeOH, AcCl, 0°C to rt, then NaH, BnBr, DMF, 0°C to rt, 12 h; (b) $CH_3CN–H_2O$-TFA (5:4:2), 80°C, 4 h, 75%; (c) Ph_3PCH_3Br, *n*-BuLi, THF, − 20°C, 3 h, 72%; (d) PPh_3, DIAD, phtalimide, THF, − 20°C to rt, 4 h, 68%; (e) 40% $MeNH_2$, 60°C, 3 h then CbzCl, NEt_3, CH_2Cl_2, 0°C, 2 h, 65%; (f) allyl bromide, NaH, THF, 0°C 3 h, 88%; (g) Grubbs II (5 mol%), CH_2Cl_2 (0.01 M), reflux, 6–10 h, 89%–92%; (h) H_2, Pd/C, MeOH -1 N HCl, 48–72 h, 90%; (i) 40% aq. $MeNH_2$, 60°C, 3 h then 4-bromobut-1-ene, K_2CO_3, DMF, 18 h, 48% overall two steps; (j) CbzCl, NEt_3, CH_2Cl_2, 0°C, 2 h, 85%.

exploitation of the chemistry of pentaatomic cyclic nitrones. For instance, the D-arabinose-derived nitrone **100** already possesses the correct stereochemistry of the hydroxy groups at ring B of the target pyrrolizidine iminosugars **101** (Scheme 9.15), rendering the reported strategies highly straightforward. Nitrone **100** was synthesized from commercially available tri-*O*-benzyl-D-arabinose (**18**): nitrogen insertion was achieved by conversion of the sugar aldehyde into the corresponding silylated oxime, and the correct configuration was installed by a double inversion of configuration at C-4 via the key iodide intermediate **99** (Scheme 9.15).[39]

Goti, Reissig, and coworkers in 2017 employed nitrone **100** for the stereodivergent synthesis of the natural compound (-)-hyacinthacine B_4 (**107**), as well as of two epimers of (+)-hyacinthacine C_5, which also allowed suggestion of the likely structure of the isolated natural product.[40] Addition of lithiated benzyloxyallene **102** to nitrone **100** was controlled by steric and stereoelectronic effects yielding the adduct **103** with exclusive *trans* stereoselectivity with respect to the vicinal benzyloxy group at C-3 of the nitrone, thus installing the correct stereochemistry at the bridgehead carbon of the natural alkaloid. The intermediate hydroxylamine **103** was not isolated but it directly afforded, upon cyclization, a 1:1 mixture of the two 1,2-oxazine epimers **104** (apparently, the methyl group was too far from the newly formed C−C bond and therefore its orientation had no effect on rate and diastereoselectivity) (Scheme 9.16). The synthesis continued on the two separate epimers **105**, which finally afforded the same stereoisomer **106**, after N−O cleavage by samarium iodide, mesylation of the free hydroxy group and mesylate displacement by nitrogen. Final hydrogenation and ketone reduction yielded the target compound **107**.

In 2022 Petricci and coworkers employed nitrone **100** to develop an innovative strategy for the synthesis of pyrrolizidine alkaloids such as

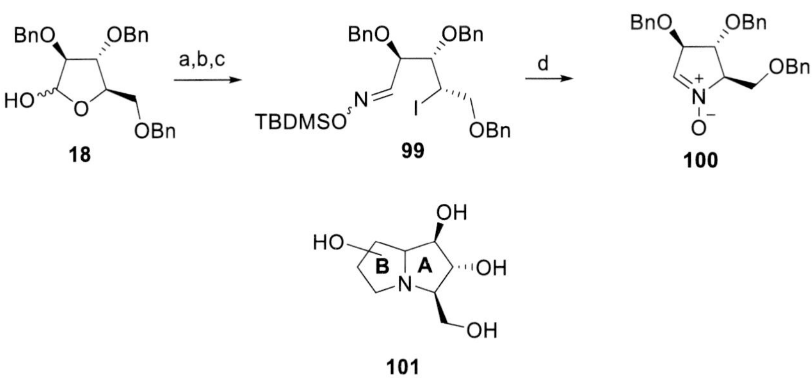

SCHEME 9.15 Reagents and conditions: (a) $NH_2OH.HCl$, Py, rt, 24 h; (b) TBDPSCl, Py, rt, 18 h; (c) I_2, PPh_3, imidazole, toluene, 60°C, 2 h; TBAF, THF, rt, 18 h.

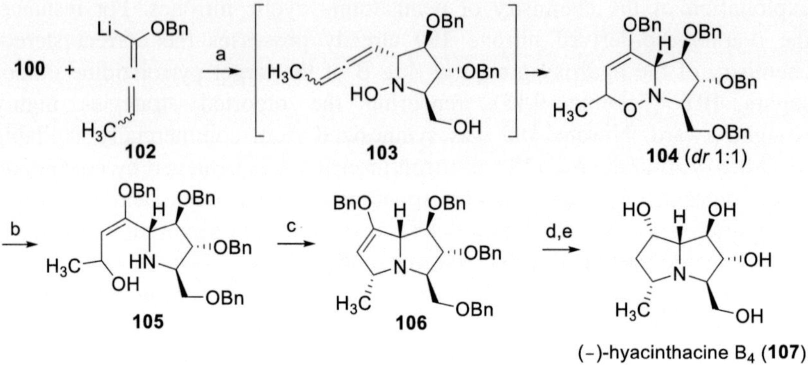

SCHEME 9.16 Reagents and conditions: (a) THF, −78°C, 5 h then CH_2Cl_2, rt, 24 h; (b) SmI_2, THF, rt, 16 h; (c) MsCl, NEt_3, CH_2Cl_2, rt, 16 h; (d) H_2, Pd/C, aq HCl, MeOH, rt, 48 h; (e) $NaBH_4$.

hyacinthacine A_2 using an intramolecular hydryoaminomethylation protocol (HAM) under microwave dielectric (*MW*) heating condition.[41]

In the same year, also Yu and coworkers explored the chemistry of nitrone **100** to prepare a series of C-7 fluorinated casuarine and australine derivatives, another important subclass of highly hydroxylated pyrrolizidine with six contiguous stereocenters. In particular, as the key step of their strategy, they employed an organoselective α-fluorination of the nitrone-derived aldehyde **110** in the presence of proline-derived chiral catalysts.[42] This work follows a previous one from the same research group, where the authors encountered many difficulties in the synthesis of all the desired fluorinated epimers.[43]

Nitrone **100** was efficiently converted on a large scale to the protected alkene **108** (Scheme 9.17) through organometal addition of an allyl zinc derivative followed by nitrogen protection.[44] This intermediate was employed by the same authors for the synthesis of four diastereoisomers belonging to the family of casuarines, including casuarine itself.[44]

Dihydroxylation of **108** gave a mixture of diols **109** (*dr* 1.8:1), which, on periodate oxidation, afforded the key aldehyde intermediate **110**. The availability of both proline catalysts (*S*)- and (*R*)-**114** allowed the synthesis, after Wittig reaction, of both alkenes **111** and its epimer in three steps and 77% overall yield (only **111** is shown as a representative example in Scheme 9.17). Hydroboration followed by hydrogen peroxide oxidation afforded the primary alcohol **112**, which, upon mesylation and removal of the Cbz protecting group, cyclized to form the pyrrolizidine skeleton. Final hydrogenation in highly acidic conditions gave the target 7-*epi*-fluoro-australine (**113**) in 97% yield. The same manuscript also describes the synthesis of 7-fluoroaustraline as well as four different isomers of C7-fluorinated

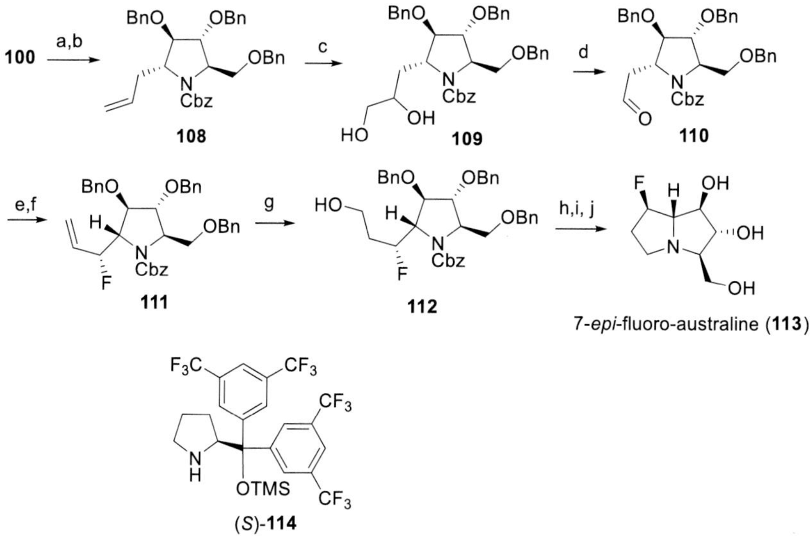

SCHEME 9.17 Reagents and conditions: (a) Zn, allyl bromide, THF, sat. NH_4Cl; (b) CbzCl, $NaHCO_3$, THF/H_2O, 75% over two steps; (c) OsO_4, NMO, acetone, 94%, *dr* 1.8:1; (d) $NaIO_4$, $MeOH/H_2O$; (e) NFSI (1 equiv), (*S*)-**114** (10%), MTBE; (f) $P^+Ph_3CH_3Br^-$, KO*t*Bu, THF, 0–5°C, 77% for the three steps; (g) BH_3.THF, THF, −20°C, then aq. NaOH, H_2O_2, 74%; (h) 10% Pd/C, H_2, AcOH, MeOH; (i) MsCl, NEt_3, CH_2Cl_2, 0°C, 80% over two steps; (j) 10% Pd/C, H_2, 1 N HCl, MeOH.

casuarines and C7-difluorocasuarines, together with a preliminary biological investigation as inhibitors of commercially available glycosidases.[42]

The synthesis of unprecedented branched pyrrolizidines was accomplished *via* organometal addition reaction on the ketonitrone **117** (Scheme 9.18), which was previously synthesized from the corresponding hydroxylamine through oxidation procedures by Goti, Cardona, and coworkers, together with its aldonitrone regioisomer.[45]

Behr and coworkers developed a new regioselective synthesis of nitrone **117** starting from 2,3,5-tri-*O*-benzyl-D-arabinose (**18**), involving direct oximation with NH_2OH and reduction with $NaBH_3CN$ (95% for the two steps) to form **115**. The hydroxyl group of the oxime moiety was selectively protected over the secondary hydroxy group to yield compound **116**, which was finally oxidized with the strongly acidic Jones reagent to afford the ketonitrone **117** (Scheme 9.18). The nucleophilic addition of vinyl magnesium bromide to **117** yielded, after in situ reduction of the formed hydroxylamine intermediate, the pyrrolidine **118** as the major diastereoisomer, resulting from an attack of the organometallic reagent to the *Re* face of **117**. The synthesis was completed *via* *N*-allylation, which formed the intermediate **119**, as the suitable substrate of the subsequent ring-closing metathesis in the presence of Grubb's II catalyst. Final catalytic hydrogenation afforded the final pyrrolizidine **120** (Scheme 9.18).[46]

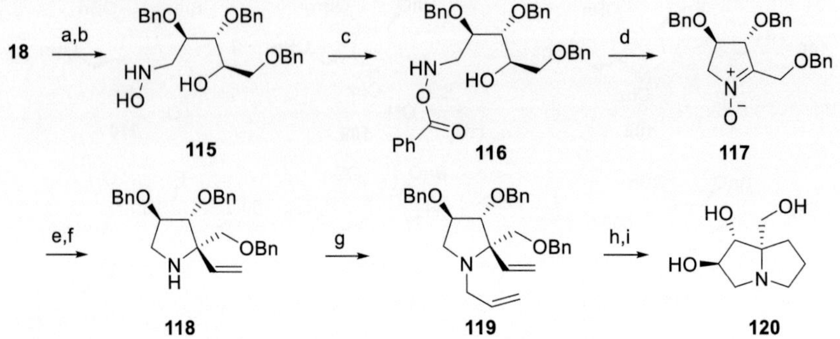

SCHEME 9.18 Reagents and conditions: (a) $NH_2OH.HCl$, MeONa, MeOH; (b) $NaBH_3CN$, MeOH/HCl; (c) PhCOOH, carbonyldiimidazole, CH_2Cl_2, 83%; (d) CrO_3-H_2SO_4 (Jones reagent), acetone, − 20°C, 54%; (e) vinyl magnesium bromide, − 78°C, *dr* 91:9; (f) Zn powder, AcOH; (g) allyl-Br, K_2CO_3, cat. KI, CH_3CN; (h) Grubbs II cat. CH_2Cl_2, reflux; (i) H_2, Pd/C, MeOH, aq HCl (6 M).

9.3.2 Indolizidines

Sugar-derived pentaatomic cyclic nitrones have been also used for the synthesis of indolizidine iminosugars in the last few years. In this regard, a recent study has been conducted in our laboratory with the aim of investigating the regio- and diastereoselectivity in the 1,3-dipolar cycloaddition reaction of D-arabinose, D-xylose, and D-ribose-derived nitrones with several dipolarophiles.[47]

In particular, the cycloaddition of D-ribose-derived nitrone **121** with buten-1-ol (**122**) gave a 1:1.5 mixture of the two diastereoisomeric isoxazolidines **123a** and **123b** in 70% yield, derived from an *exo-anti* and an *exo-syn* approach of the dipole and the dipolarophile (where *exo* is referred to the direction of the approach of the partners and *anti* or *syn* is referred to the relative position to the vicinal OBn group on the nitrone), which were not separated at this stage. Mesylation of the primary hydroxyl groups of the mixture **123a,b** formed in situ the tricylic salt **124a,b**, which was subjected to catalytic hydrogenation to simultaneously achieve the N–O bond cleavage and debenzylation, and separated via flash column chromatography after acetylation of the hydroxy groups. Finally, each peracetylated indolizidine, **125a** and **125b**, was passed through a basic ion-exchange resin to yield the target indolizidine iminosugars **126a** and **126b** (Scheme 9.19).

A complementary strategy was pursued by Py and coworkers in 2015.[48] They built the six-membered ring of the indolizidine skeleton with the aid of a newly synthesized carbohydrate-derived piperidine-type nitrone and exploited the organometal addition to this latter, followed by the ring-closing metathesis reaction, to get the second ring. Their strategy is briefly described in Scheme 9.20. To target the castanospermine analogs **132a** and **132b**, the nitrone **129**, possessing a partial D-glucose configuration, was prepared by

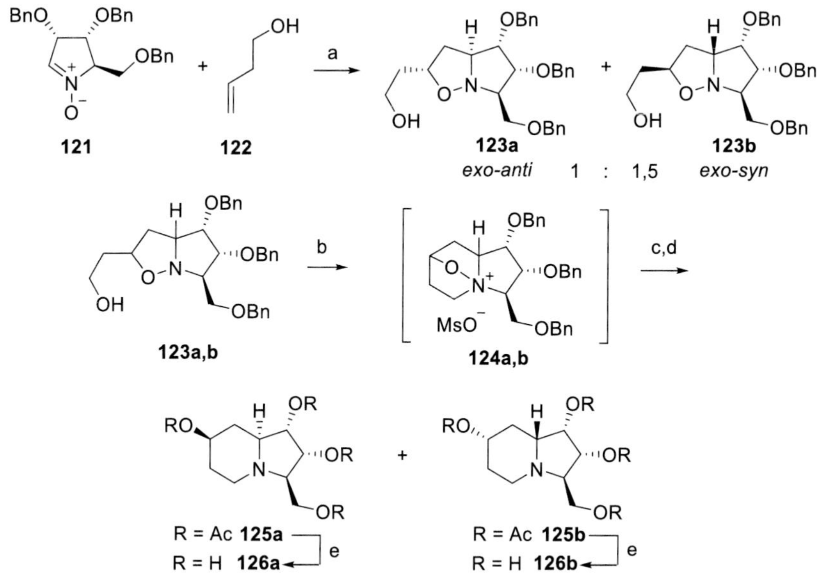

SCHEME 9.19 Reagents and conditions: (a) toluene, 60°C, 72 h, 70%; (b) MsCl, NEt_3, dry CH_2Cl_2, rt., 45 min; (c) H_2, Pd/C, MeOH, HCl, rt, 72 h; (d) Ac_2O, Py, rt, 18 h, 17% for **125a**, 24% for **125b**; (e) Ambersep 900 OH, MeOH, rt, 18 h, 100%.

treating the readily available 1,3,4,5-tetra-*O*-benzyl-L-sorbopyranose (**127**) with $PPh_3.Br_2$ and pyridine to afford intermediate **128**. This latter was then converted into nitrone **129** in a single step by reaction with an excess of hydroxylamine hydrochloride and triethylamine in THF. With nitrone **129** in hand, the authors undertook the stereodivergent synthesis of the two alkylated piperidines **130a** and **130b** exploiting the stereoselective addition of Grignard reagent in the presence or absence of a Lewis acid, followed by N–O cleavage with Zn in acidic media. While without Lewis acid the preferred attack occurred *anti* to the vicinal OBn group for steric effects, in the presence $ZnCl_2$ as the Lewis acid exclusive formation of **130b** (with the vinyl group *syn* with respect to the vicinal OBn group) was observed. The stereochemical outcome was explained by the authors invoking stereoelectronic effects (see the original manuscript for further details).[48] For the completion of the synthesis, allylation of nitrogen on both isomers afforded good yields of compound **131a** and **131b**, respectively, which were subjected to ring-closing metathesis and catalytic hydrogenation to reduce the double bond and deprotect the hydroxyl groups, thus finally yielding the two castanospermine analogs **132a** and **132b**. Upon biological testing, compound **132b** showed to be an excellent and selective α-glucosidase inhibitor (IC_{50} = 52 nM), with an interesting and peculiar noncompetitive mode of inhibition.

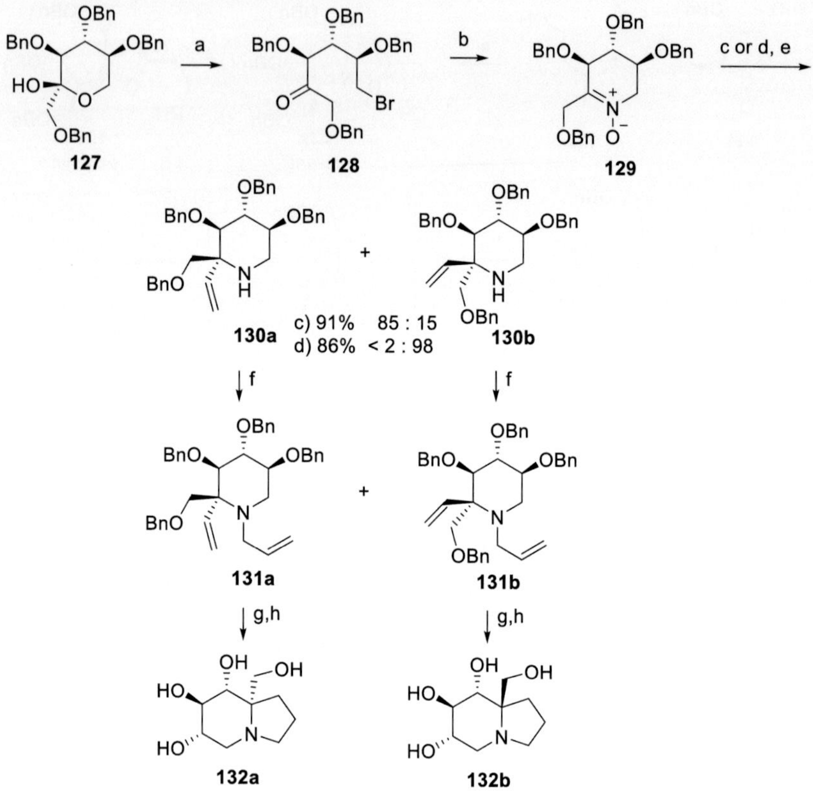

SCHEME 9.20 Reagents and conditions: (a) $PPh_3.Br_2$, Py, THF, 65°C, 3 h, 87%; (b) $NH_2OH \cdot HCl$, NEt_3, THF, rt, 3d, 81%; (c) vinyl magnesium bromide, CH_2Cl_2, −78°C; (d) $ZnCl_2$, vinylMgBr, CH_2Cl_2, −78°C; (e) Zn, EtOH/AcOH 4:1 80°C, sonication; (f) allyl bromide, K_2CO_3, cat. KI, CH_3CN, 80°C; (g) Grubbs II cat., CH_2Cl_2, reflux; (h) H_2, $Pd(OH)_2/C$, HCl, MeOH.

Jarosz and coworkers reported an original strategy for the synthesis of indolizidine alkaloids in 2016, and the application of their strategy to the synthesis of natural (+)-lentiginosine (**141**) is shown as a title example in Scheme 9.21.[49] Their synthesis started from the D-mannitol derived **133**, which was subjected to acetonide deprotection to afford **134**. This latter, without purification, was reacted with $NaIO_4$, and the resulting hemiacetal reduced with $NaBH_4$ to yield triol **135**. Another oxidative cleavage provided hemiacetal **136**, which was reacted with hydroxylamine to introduce nitrogen. The crude oxime was subjected to the Appel conditions (CBr_4/PPh_3), which afforded the desired ω-bromo nitrile **137** in good overall yield (60% over 3 steps). With compound **137** in hand, the authors studied the diastereoselective addition of allyl magnesium bromide followed by the in situ reduction of the formed imine with zinc borohydride, a mild reducing agent

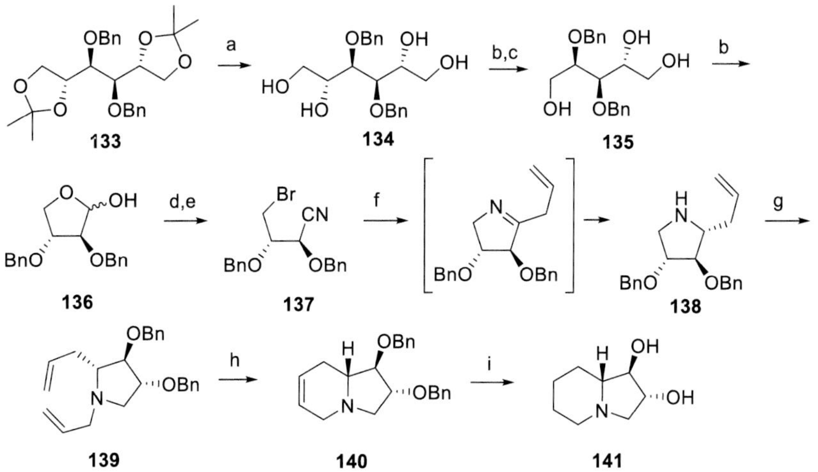

SCHEME 9.21 Reagents and conditions: (a) conc. HCl, MeOH, 65°C, 1 d; (b) $NaIO_4$, sat. $NaHCO_3$, CH_2Cl_2, rt, 2 h; (c) $NaBH_4$, MeOH, rt, 1 h, 59% over 3 steps; (d) $NH_2OH \cdot HCl$, Py, rt, 24 h; (e) CBr_4, Ph_3P, CH_3CN, rt, 24 h, 60% over 3 steps; (f) allyl-MgBr, $Zn(BH_4)_2$, toluene, 0°C, 1 h, 67%, *dr* 4:1; (g) allyl bromide, CH_3CN, K_2CO_3, rt, 24 h, 55%; (h) TFA, Grubbs-Hoveyda II cat. (5 mol%), toluene, 60°C, 12 h, 94%; (i) H_2, $Pd(OH)_2/C$, MeOH, rt, 12 h, 91%.

successfully applied to reduction of imines, and resulted in the formation of the pyrrolidine **138** as the major diastereoisomer (*dr* 4:1). Subsequent *N*-allylation furnished **139**, the suitable substrate for the ring-closing metathesis step, which yielded the indolizidine **140** in excellent 94% yield. Final hydrogenation of **140** afforded (+)-lentiginosine (**141**) (Scheme 9.21).

9.3.3 Quinolizidines

Quinolizidines are constituted by two fused six-membered rings with a bridgehead nitrogen atom. The synthesis of quinolizidine iminosugars, as well as their occurrence in nature, is far less common than the previously mentioned pyrrolizidines and indolizidines. However, in these last 5 years, some recent methodologies were proposed for their synthesis and are briefly described in this section.

Py and coworkers employed the above-described L-sorbose-derived nitrone **129** for the synthesis of polyhydroxylated quinolizidines bearing a hydroxymethyl group at the ring junction, with a similar strategy to that proposed for the synthesis of indolizidines **132a** and **132b** described in Scheme 9.20. In this case, the addition of allylmagnesium Grignard instead of vinylmagnesium Grignard to nitrone **129** formed always the diastereoisomeric *N*-hydroxypyrrolidine **142a** as the major diastereoisomer over **142b**. The diastereomeric ratio could be increased up to 70:30 in favor of **142a** upon the addition of $ZnCl_2$ as the Lewis acid.[50] The preferred formation of the stereoisomer **142a** bearing the ally group in *cis* with respect to the vicinal OBn group was interpreted by the authors invoking a *chair-like* transition state with antiperiplanar relationship between the new developing C–C bond and the nonbonding electronic pair on the nitrogen atom.[48] Hydroxylamines **142a** and **142b** were separated and transformed into the corresponding piperidines **143a** or **143b** by zinc-mediated reduction of the N–O bond. The following *N*-allylation furnished **144a** or **144b**, which were subjected to the ring-closing metathesis step to yield **145a** or **145b**. The major isomer **145a** was then dihydroxylated at the double bond affording two diastereoisomer **146a** and **146b** in a 60:40 ratio, which finally yielded the target quinolizidines **147a** and **147b** by benzyl deprotection, while direct hydrogenation of **145a** or **145b** afforded the less hydroxylated **148a** or **148b** (Scheme 9.22).

Rowicki and coworkers described the concise synthesis of two polyhydroxy quinolizidines **155** and **156** from 2,3-*O*-isopropylidene-D-ribofuranose (**149**).[51] The authors exploited the intramolecular 1,3-dipolar cycloaddition (1,3-DC) of the nitrone formed in situ in the reaction of **149** with hydroxylamine **150** (Scheme 9.23), which occurred in toluene at 105°C for 24 hours.

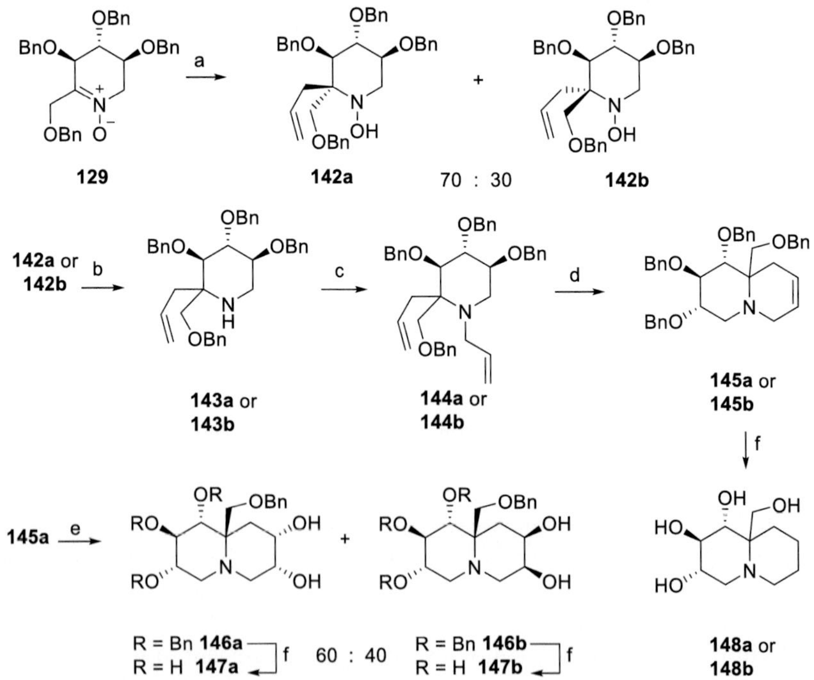

SCHEME 9.22 Reagents and conditions: (a) allylMgBr, $ZnCl_2$, CH_2Cl_2, $-78°C$, *dr* 70:30 77%; (b) Zn, AcOH/AcOH, ultrasound irradiation, 80°C; (c) allylBr, K_2CO_3, cat. KI, CH_3CN, 90°C; (d) Grubbs II, CH_2Cl_2; (e) OsO_4, NMO, THF/H_2O, *dr* 60:40; (f) H_2 (5 bar), $Pd(OH)_2/C$, MeOH/HCl.

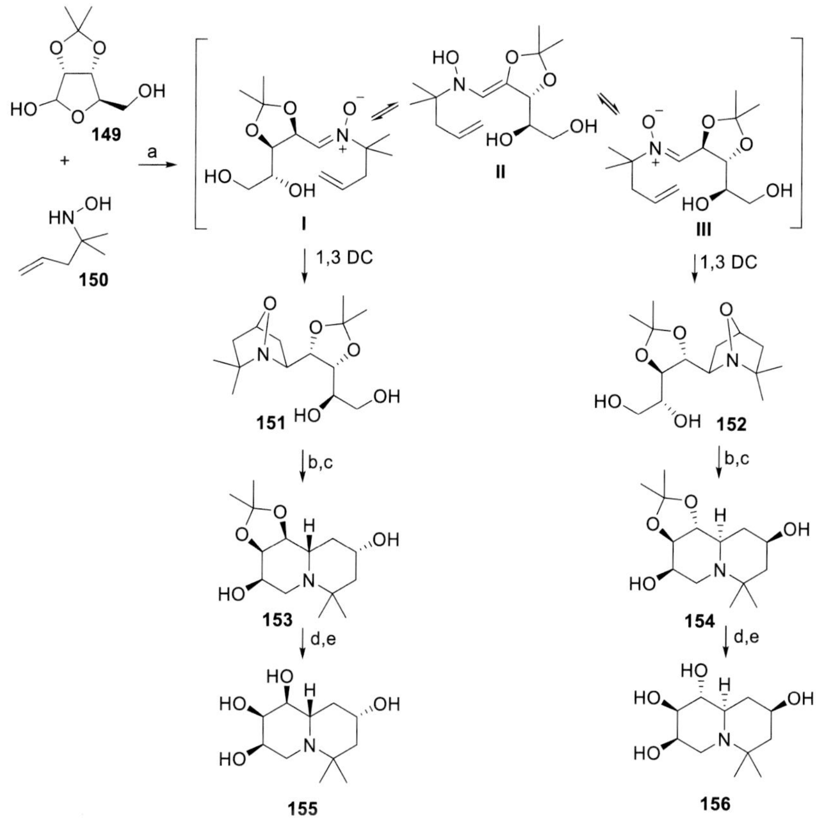

SCHEME 9.23 Reagents and conditions: (a) toluene, 105°C, 24 h, 64%, **151**:**152** = 1.1:1; (b) TsCl, Bu_2SnO, NEt_3, CH_2Cl_2, rt, 16 h; (c) H_2, Pd/C, NEt_3, *i*PrOH, 50°C, 2 d; (d) HCl, H_2O/MeOH, rt, 24 h; (e) Amberlite IRA-67, EtOH, 2°C, 3d.

While the formation of the bicyclic bridged isoxazolidine **151** was the predicted main product of the cycloaddition, the formation of compound **152** could only be explained by invoking an epimerization of the intermediate nitrone occurring through nitrone-enehydroxylamine tautomerism (intermediates I–III, Scheme 9.23). The synthesis of the target compounds proceeded with tosylation of the primary alcohols in **151** and **152** with tosyl chloride catalyzed by dibutytin(IV) oxide followed by catalytic reduction of the crude tosylates to yield **153** and **154**, respectively. Finally, acetal hydrolysis followed by neutralization of the resulting hydrochloride salts with an ion-exchange resin afforded the two diastereomeric tetrahydroxy quinolizidines **155** and **156** bearing a quaternary carbon adjacent to nitrogen, respectively (Scheme 9.23). The same authors published later the synthesis of pentahydroxy quinolizidines analogs bearing two hydroxymethyl groups at the carbon adjacent to nitrogen, by employing the same straightforward synthetic strategy but different starting materials (namely a 2-deoxy-D-ribose derivative and a hydroxylamine bearing two hydroxymethyl groups protected as pivaloyl esters).[52]

9.3.4 Nortropanes

Nortropanes are bicyclic heterocycles with a bridgehead nitrogen atom. In the last few years, the synthesis of iminosugars with a nortropane scaffold was the object of interest, among others, of Kato, Hirono, and coworkers, who synthesized and studied this class of compounds aiming at finding a new class of nortropane-type iminosugars for improving postprandial hyperglycemia.[53] They employed the pentaatomic cyclic nitrone-derived from D-xylose ***ent*-100** as the substrate of the intramolecular [3 + 2] cycloaddition developed by Goti and coworkers.[54] The reaction of ***ent*-100** with the allyl Grignard reagent formed in situ from allyl magnesium bromide (**157**)

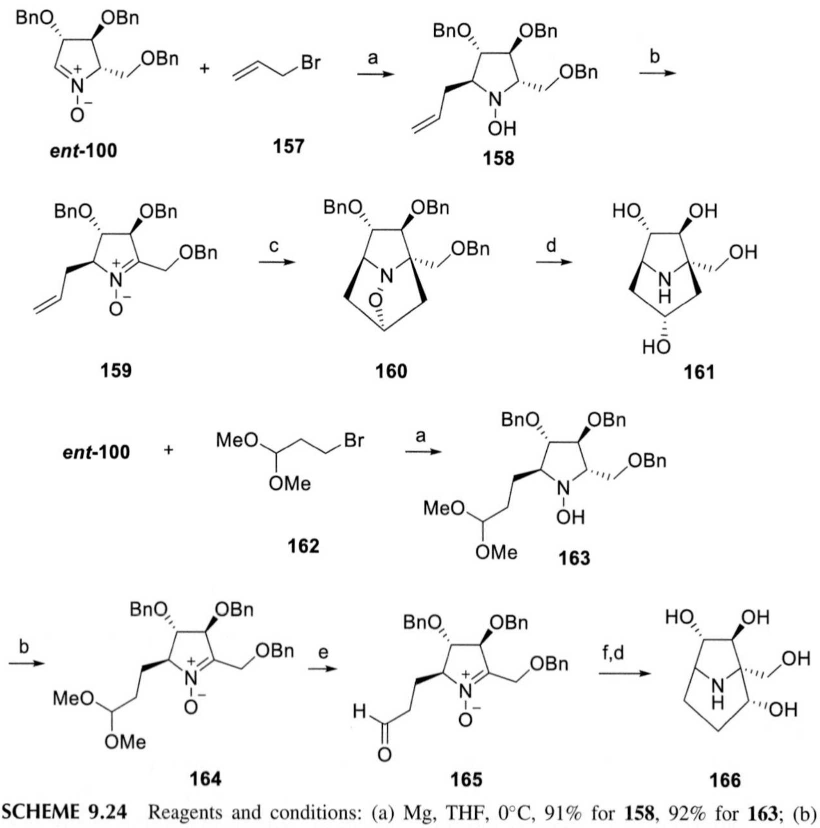

SCHEME 9.24 Reagents and conditions: (a) Mg, THF, 0°C, 91% for **158**, 92% for **163**; (b) MnO_2, 60% for **159**, 43% for **164**; (c) toluene, reflux, 87%; (d) Pd/C, H_2, HCl, MeOH; (e) cat. *p*-TsOH, acetone:H_2O 4:1; (f) SmI_2 (2 equiv), H_2O (80 equiv), THF, − 78°C, 65%.

afforded the hydroxylamine **158** with high *trans*-selectivity in agreement with previous reports. Oxidation with activated MnO_2 afforded nitrone **159** as the major isomer. Upon heating at reflux in toluene, the intramolecular cycloaddition took place affording the bicyclic compound **160**, whose N–O bond was cleaved by catalytic hydrogenolysis with concomitant debenzylation, which led to the nortropane alkaloid **161** (Scheme 9.24). The synthesis of the analog **166** was accomplished by employing the Grignard reagent obtained from **162**, which afforded **163** as the sole diastereoisomer. After oxidation with MnO_2 to yield nitrone **164** and conversion of the acetal into aldehyde **165**, the intramolecular reductive coupling reaction promoted by SmI_2[55] followed by catalytic hydrogenolysis yielded the nortropane alkaloid **166** (Scheme 9.24). Compound **161** strongly suppressed postprandial hyperglycemia in vivo (studies on mice), thus paving the way for the future development of this promising compound.[53]

9.4 Conclusions and future outlook

In conclusion, the synthesis of iminosugar-based glycomimetics from carbohydrates still offers great opportunities to synthetic organic chemists to exploit carbohydrate manipulation, functional organic group transformation, and modern organic synthesis in general, with the aim of obtaining biologically active compounds. Indeed, the introduction of a nitrogen atom in the iminosugar skeleton poses additional challenges connected to the purification of the final compounds, which are often not only highly hydrophilic, but also alkaline compounds, and therefore cannot be successfully purified through standard flash column chromatography, but preferably using ion-exchange resins.

The key introduction of the nitrogen atom in these syntheses is achieved by means of a nucleophilic displacement of a suitable leaving group (halogen, epoxide, or a sulfonate) with a nucleophilic nitrogen-containing group such as an amine, an azide, or a nitro group, or via Mitsunobu conditions with nitrogenated nucleophiles. Interestingly, more than one nitrogen atom can be also installed employing a metal-free Overman rearrangement of trichloroimidates.

Focusing on the cyclization strategies, the reductive amination or the double reductive amination reaction involving the anomeric carbon of the starting carbohydrates still represents one of the most employed reactions for the synthesis not only of piperidine iminosugars, as recently described,[6b] but also all kind of mono- and bicyclic structures.

Regarding the synthesis of bicyclic iminosugars, cyclic nitrones (five- and six-membered) from carbohydrates are often employed for building the first ring, since they can be further reacted with organometallic nucleophiles. These additions are often highly selective due to the presence of the protected hydroxy groups on the nitrone, and allow the introduction of a proper additional moiety, which is then further elaborated in the subsequent steps. Alternatively, cyclic nitrones can be reacted, as 1,3-dipoles, in highly stereoselective 1,3-dipolar cycloaddition reactions with suitable dipolarophiles, a powerful reaction that allows the formation of up to three new stereocenters and further synthetic elaboration of the formed isoxazolidines. Finally, the ring-closing metathesis reaction is still also a powerful method for the construction of the second ring.

Many of the new strategies follow some fundamental principles of green chemistry, such as the use of high atom economy reactions (e.g., rearrangement and cycloadditions), the employ of starting materials that can be in principle derived from renewable sources (such as carbohydrates), the development of stereoselective strategies and the use of heterogeneous catalysis. However, the replacement of toxic solvents and reagents, the use of mild conditions, and the avoidance of protection/deprotection steps are some aspects that could be improved for developing more sustainable syntheses of iminosugars. We deem that the synthesis of carbohydrate-based iminosugars will still offer some surprises in store for us and will play a pivotal role in the important total syntheses of these extremely fascinating organic compounds with a huge potential as drugs.

References

1. Conforti, I.; Marra, A. *Org. Biomol. Chem.* **2021,** *19*, 5439–5475.
2. (a) Compain, P.; Martin, O. R. *Iminosugars: From Synthesis to Therapeutic Applications;* Wiley-VCH: Weinheim, 2007187–298.
   (b) Horne, G.; Wilson, F. X. *Prog. Med. Chem.* **2011,** *50*, 135–176.
   (c) Nash, R. J.; Kato, A.; Yu, C. Y.; Fleet, G. W. J. *Future Med. Chem.* **2011,** *3*, 1513–1521.
3. Ishida, N.; Kumagai, K.; Niida, T.; Tsuruaka, T.; Yumato, H. *J. Antibiot.* **1967,** *20*, 66–71.
4. Yagi, M.; Kuono, T.; Aoyagi, Y.; Murai, H. *J. Agric. Chem. Soc. Jpn.* **1976,** *50*, 571–572.
5. (a) Stocker, B. L.; Dangerfield, A. D.; Win-Mason, A. L.; Haslett, G. W.; Timmer, M. S. M. *Eur. J. Org. Chem.* **2010,** 1615–1637.
   (b) Uhrig, M. L.; Mora Flores, W. E.; Postigo, A. *Chem. Eur. J.* **2021,** *27*, 7813–7825.
   (c) Dehoux-Baudoin, C.; Génissom, Y. *Eur. J. Org. Chem.* **2019,** 4765–4777.
   (d) Carroll, A. W.; Pyne, S. G. *Curr. Org. Synth.* **2019,** *16*, 498–522.
   (e) Wood, A.; Prichar, K. L.; Clarke, Z.; Houston, T. A.; Fleet, G. W. J.; Simone, M. I. *Eur. J. Org, Chem.* **2018,** 6812–6829.
6. (a) Dhara, D.; Dhara, A.; Bennet, J.; Murphy, P. V. *Chem. Rec.* **2021,** *21*, 2958–2979.
   (b) Clemente, F.; Matassini, C.; Cardona, F. *Eur. J. Org. Chem.* **2020,** 4447–4462.
   (c) Davies, S. G.; Fletcher, A. M.; Roberts, P. M.; Thomson, J. E. *Tetrahedron* **2019,** *75*, 130727.
   (d) Chadda, R.; McArdle, P.; Murphy, P. V. *Synthesis* **2017,** *49*, 2138–2152.
   (e) Ferjancic, Z.; Saicic, R. N. *Eur. J. Org. Chem.* **2021,** 3241–3250.
   (f) Prichard, K. L.; O'Brien, N.; Ghorbani, M.; Wood, A.; Barnes, E.; Kato, A.; Houston, T. A.; Simone, M. I. *Eur. J. Org. Chem.* **2018,** 6830–6842.
   (g) Szcześniak, P.; Furman, B. *Org. Biomol. Chem.* **2021,** *19*, 6842–6846.

7. (a) Esposito, A.; D'Alonzo, D.; De Fenza, M.; De Gregorio, E.; Tamanini, A.; Lippi, G.; Dechecchi, M. C.; Guaragna, A. *Int. J. Mol. Sci.* **2020,** *21*, 3353.
   (b) Prichard, K.; Campkin, D.; O'Brine, N.; Kato, A.; Fleet, G. W. J.; Simone, M. I. *Chem. Biol. Drug Des.* **2018,** *92*, 1171–1197.
8. Gavale, K. S.; Chavan, S. R.; Khan, A.; Joshi, R.; Dhavale, D. D. *Org. Biomol. Chem.* **2015,** *13*, 6634–6646.
9. (a) Purser, S.; Moore, P. R.; Swallow, S.; Gouverneur, V. *Chem. Soc. Rev.* **2008,** *37*, 320–330.
   (b) Qjima, L. *Fluorine in Medicinal Chemistry and Chemical Biology;* Wiley – Blackwell: Chichester, 2009.
   (c) Kirsch, P. *Modern Fluoroorganic Chemistry;* Wiley – VCH: Weinheim, 2013.
10. Liu, Z.; Jenkinson, S. F.; Vermaas, T.; Adachi, I.; Wormald, M. R.; Hata, Y.; Kurashima, Y.; Kaji, A.; Yu, C.-Y.; Kato, A., et al. *J. Org. Chem.* **2015,** *80*, 4244–4258.
11. Backer, D. H.; Horton, D.; Tindal, C. G. *Carbohydr. Res.* **1972,** *24*, 192–197.
12. Cocaud, C.; Nicolas, C.; Bayle, A.; Poisson, T.; Pannecoucke, X.; Martin, O. R. *Eur. J. Org. Chem.* **2015,** 4330–4334.
13. Jaszczyk, J.; Li, S.; Cocaud, C.; Nicolas, C.; Martin, O. R. *Carb. Res.* **2019,** *486*, 107855.
14. Li, S.; Jaszczyk, J.; Pannecoucke, X.; Poisson, T.; Martin, O. R.; Nicolas, C. *Adv. Synth. Catal.* **2021,** *363*, 470–483.
15. Harit, V. K.; Ramesh, N. G. *Carbohydr. Res.* **2020,** *492*, 107988.
16. Harit, V. K.; Ramesh, N. G. *J. Org. Chem.* **2016,** *81*, 11574–11586.
17. Haarr, M. B.; Lopéz, Ó.; Pejov, L.; Fernández-Bolaños, J. G.; Lindbäck, E.; Sydnes, M. O. *ACS Omega* **2020,** *5*, 18507–18514.
18. Haarr, M. B.; Lopéz, Ó.; Fernández-Bolaños, J. G.; Lindbäck, E.; Sydnes, M. O. *Synthesis* **2022,** *54*, 2916–2926.
19. Hunt-Painter, A. A.; Moggré, G.-J.; Tyler, P. C.; Stocker, B. L.; Timmer, M. S. M. *ChemistrySelect* **2017,** *2*, 8028–8032.
20. Dangerfield, E. M.; Timmer, M. S. M.; Stocker, B. L. *Org. Lett.* **2009,** *11*, 535–538.
21. Hunt-Painter, A. A.; Stocker, B. L.; Timmer, M. S. M. *Tetrahedron* **2018,** *74*, 1307–1312.
22. Okuyama, Y.; Kidena, M.; Sato, E.; Kawano, S.; Ishii, K.; Maie, K.; Miura, K.; Simizu, S.; Sato, T.; Chida, N. *Angew. Chem. Int. Ed.* **2021,** *60*, 5193–5198.
23. Front, S.; Biela-Banaś, A.; Burda, P.; Ballhausen, D.; Higaki, K.; Caciotti, A.; Morrone, A.; Charollais-Thoenig, J.; Gallienne, E.; Demotz, S., et al. *Eur. J. Med. Chem.* **2017,** *126*, 160–170.
24. Andersch, J.; Bols, M. *Chem. Eur. J.* **2001,** *7*, 3744–3747.
25. Mellor, H. R.; Nolan, J.; Pickering, L.; Wormald, M. R.; Platt, F. M.; Dwek, R. A.; Fleet, G. W. J.; Butters, T. D. *Biochem. J.* **2002,** *366*, 225–233.
26. Prasad, S. S.; Reddy, N. R.; Baskaran, S. *J. Org. Chem.* **2018,** *83*, 9604–9618.
27. Herrera-González, I.; González-Cuesta, M.; García-Moreno, M. I.; García-Fernández, J. M.; Ortiz Mellet, C. *ACS Omega* **2022,** *7*, 22394–22405.
28. Martínez-Bailén, M.; Clemente, F.; Matassini, C.; Cardona, F. *Pharmaceuticals* **2022,** *15*, 823.
29. (a) Mirabella, S.; Fibbi, G.; Matassini, C.; Faggi, C.; Goti, A.; Cardona, F. *Org. Biomol. Chem.* **2017,** *15*, 9121–9126.
   (b) Parmeggiani, C.; Matassini, C.; Cardona, F.; Goti, A. *Synthesis* **2017,** 2890–2900.
30. (a) Clemente, F.; Matassini, C.; Goti, A.; Morrone, A.; Paoli, P.; Cardona, F. *ACS Med. Chem. Lett.* **2019,** *10*, 621–626.
   (b) Clemente, F.; Matassini, C.; Faggi, C.; Giachetti, S.; Cresti, C.; Morrone, A.; Paoli, P.; Goti, A.; Martínez-Bailén, M.; Cardona, F. *Bioorg. Chem.* **2020,** *98*, 103740.
31. Clemente, F.; Matassini, C.; Giachetti, S.; Goti, A.; Morrone, A.; Martínez-Bailén, M.; Orta, S.; Merino, P.; Cardona, F. *J. Org. Chem.* **2021,** *86*, 12745–12761.

32. Chennaiah, A.; Dahiya, A.; Dubbu, S.; Vankar, Y. D. *Eur. J. Org. Chem.* **2018,** 6574–6581.
33. Azad, C. S.; Saxena, A. K. *Org. Chem. Front.* **2015,** *2*, 665–669.
34. Kamal, A.; Laxman, E.; Arifuddin, M. *Tetrahedron Lett.* **2000,** *41*, 7743–7746.
35. Yuan, W.; Xia, J.; Zhang, X.; Liang, P.; Zhang, J.; Jiao, W.; Shao, H. *Tetrahedron* **2016,** *72*, 3994–4000.
36. Paulsen, H.; Todt, K. *Chem. Ber.* **1967,** *100*, 512–520.
37. Godin, G.; Garnier, E.; Compain, P.; Martin, O. R.; Ikeda, K.; Asano, N. *Tetrahedron Lett.* **2004,** *45*, 579–581.
38. Das, P.; Ajay, S.; Shaw, A. K. *Synthesis* **2016,** *48*, 3753–3762.
39. (a) Martella, D.; D'Adamio, G.; Parmeggiani, C.; Cardona, F.; Moreno-Clavijo, E.; Robina, I.; Goti, A. *Eur. J. Org. Chem.* **2016,** 1588–1598.
(b) Wang, W.-B.; Huang, M.-H.; Li, Y.-X.; Rui, P.-X.; Hu, X.-G.; Zhang, W.; Su, J.-K.; Zhang, Z.-L.; Zhu, J.-S.; Xu, W.-H., et al. *Synlett* **2010,** *3*, 488–492.
(c) Cardona, F.; Faggi, E.; Liguori, M.; Cacciarini, M.; Goti, A. *Tetrahedron Lett* **2003,** *44*, 2315–2318.
40. Pecchioli, T.; Cardona, F.; Reissig, H.-U.; Zimmer, R.; Goti, A. *J. Org. Chem.* **2017,** *82*, 5835–5844.
41. Petricci, E.; Zurzolo, S.; Matassini, C.; Muramai, S.; Cardona, F.; Goti, A.; Taddei, M. *Molecules* **2022,** *27*, 4762.
42. Li, Y.-X.; Wang, J.-Z.; Shimadate, Y.; Kise, M.; Kato, A.; Jia, Y.-M.; Fleet, G. W. J.; Yu, C.-Y. *J. Org. Chem.* **2022,** *87*, 7291–7307.
43. Li, Y.-X.; Shimada, Y.; Sato, K.; Kato, A.; Zhang, W.; Jia, Y.-M.; Fleet, G. W. J.; Xiao, M.; Yu, C.-Y. *Org. Lett.* **2015,** *17*, 716–719.
44. Li, Y.-X.; Wang, J.-Z.; Kato, A.; Shimadate, Y.; Kise, M.; Jia, Y.-M.; Fleet, G. W. J.; Yu, C.-Y. *Org. Biomol. Chem.* **2021,** *19*, 9410–9420.
45. (a) Lupi, F.; Marzo, T.; D'Adamio, G.; Cretella, S.; Cardona, F.; Messori, L.; Goti, A. *ChemCatChem* **2017,** *9*, 4225–4230.
(b) Matassini, C.; Parmeggiani, C.; Cardona, F.; Goti, A. *Org. Lett.* **2015,** *17*, 4082–4085.
(c) D'Adamio, G.; Parmeggiani, C.; Goti, A.; Cardona, F. *Eur. J. Org. Chem.* **2015,** 6541–6546.
46. Massicot, F.; Messire, G.; Vallée, A.; Vasse, J.-L.; Py, S.; Behr, J.-B. *Org. Biomol. Chem.* **2019,** *17*, 7066–7077.
47. Martella, D.; D'Adamio, G.; Parmeggiani, C.; Cardona, F.; Moreno-Clavijo, E.; Robina, I.; Goti, A. *Eur. J. Org. Chem.* **2016,** 1588–1598.
48. Boisson, J.; Thomasset, A.; Racine, E.; Cividino, P.; Banchelin Sainte-Luce, T.; Poisson, J.-F.; Behr, J.-B.; Py, S. *Org. Lett.* **2015,** *17*, 3662–3665.
49. Malik, M.; Jarosz, S. *Org. Biomol. Chem.* **2016,** *14*, 1764–1776.
50. Vieria Da Cruz, A.; Kanazawa, A.; Poisson, J.-F.; Behr, J.-B.; Py, S. *J. Org. Chem.* **2017,** *82*, 9866–9872.
51. Rowicki, T.; Malinowski, M.; Łapczyński, S.; Gryszel, M.; Czerwińska, K.; Madura, I.; Sas, W. *Asian J. Org. Chem.* **2015,** *4*, 733–736.
52. Malinowski, M.; Rowicki, T.; Guzik, P.; Wielechowska, M.; Sobiepanek, A.; Sas, W. *Eur. J. Org. Chem.* **2016,** 3642–3649.
53. Kato, A.; Zhang, Z.-L.; Wang, H.-Y.; Jia, Y.-M.; Yu, C.-Y.; Kinami, K.; Hirokami, Y.; Tsuji, Y.; Adachi, I.; Nash, R. J., et al. *J. Org. Chem.* **2015,** *80*, 4501–4515.
54. Delso, I.; Tejero, T.; Goti, A.; Merino, P. *J. Org. Chem.* **2011,** *76*, 4139–4143.
55. Masson, G.; Py, S.; Vallée, Y. *Angew. Chem. Int. Ed.* **2002,** *41*, 1772–1775.

Chapter 10

Vinyl sugar enol ethers in organic synthesis

Junaid Shafi Banday[1], Irshad Ahmad Zargar[1], Nandagopal Hudait[1,2] and Debaraj Mukherjee[1,2]

[1]*Natural Products & Medicinal Chemistry Division, CSIR-Indian Institute of Integrative Medicine (IIIM), Jammu, Jammu and Kashmir, India,* [2]*Chemical Sciences, Bose Institute, Salt Lake City, Kolkata, West Bengal, India*

10.1 Introduction

Vinyl ethers are enol ethers in which the double bond is adjacent to an oxygen atom, and there is delocalization of the lone pair of electrons of the oxygen atom over the double bond as shown in Fig. 10.1A. When these vinyl ethers are a part of the monosaccharide ring, the resulting sugars are called as vinyl sugar ethers. Vinyl sugar ethers can be either endocyclic or exocyclic depending on whether the double bond is present within or outside the sugar ring (Fig. 10.1B–F). Further endocyclic vinyl sugar ethers can be either enopyranose or enofuranose depending on whether the endocyclic double bond is part of pyranose or furanose ring, respectively (Fig. 10.1C–F). Moreover, the double bond can exist either at the C1–C2 position of the pyranose/furanose ring, the C2–C3 position of the pyranose ring, and the C3–C4 position of the furanose ring. Accordingly, these four types of endocyclic vinyl sugar ethers do exist in the literature (Fig. 10.1C–F). Glycal is the special name given to the endocyclic vinyl sugar ether when the double bond is positioned between C1 and C2 of the pyran/furan ring, these glycals have been utilized extensively for diversity-oriented synthesis. This chapter

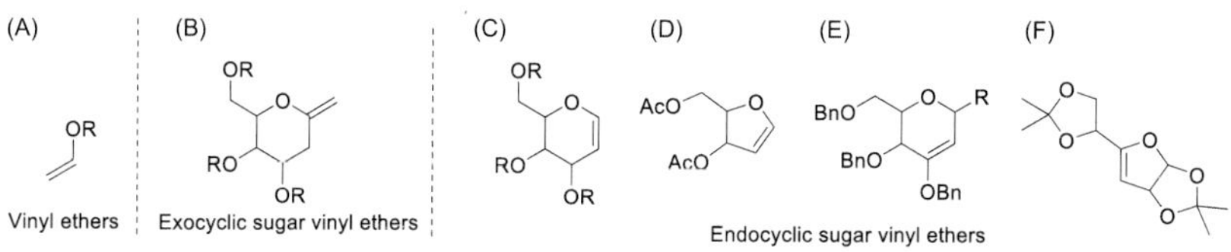

FIGURE 10.1 Various vinyl sugar ethers reported in the literature. (A) Simple vinyl ethers; (B) exocyclic sugar vinyl ethers; and (C–F) endocyclic sugar vinyl ethers.

Synthetic Strategies in Carbohydrate Chemistry. DOI: https://doi.org/10.1016/B978-0-323-91729-2.00002-1

will cover in detail the synthesis and reactivity of the endocyclic vinyl sugar ethers, particularly pyranose glycals.

Glycals are important chiral synthons used extensively in carbohydrate chemistry for synthesizing oligosaccharides and complex natural products of biological importance.[1] Over the past few decades numerous methodologies have been developed on glycals resulting in diversity-oriented synthesis.[2] This is attributed to the fact that these glycals are stable enough, easy to handle, commercially available at cheaper rates, and easy to synthesize. Moreover, these glycals show diverse reactivity due to the presence of endocyclic oxygen, enol ether-type unsaturation, and various conformations that it adopts based on different protecting groups that can be installed and tailored selectively.[1] Considering their importance in carbohydrate chemistry as a chiral synthon their synthesis has attracted much attention over the past few decades.

10.2 Synthesis and reactivity of pyranoid glycals

10.2.1 Synthesis of pyranoid glycals

Glycals are stable donors that can be synthesized from various glycosyl compounds via reductive elimination (Fig. 10.2). The first ever glycal synthesis was reported by Hermann Emil Fischer and Karl Zach in 1913.[3] They reported the synthesis of glucal from peracetylated glycosyl bromide via zinc-mediated reductive elimination under acidic conditions (Fig. 10.2A).[4] Ever since there have been improvements in the Fischer–Zach methodology like the replacement of bromine atom at the anomeric center with other leaving groups and the use of different reducing agents like sodium, lithium

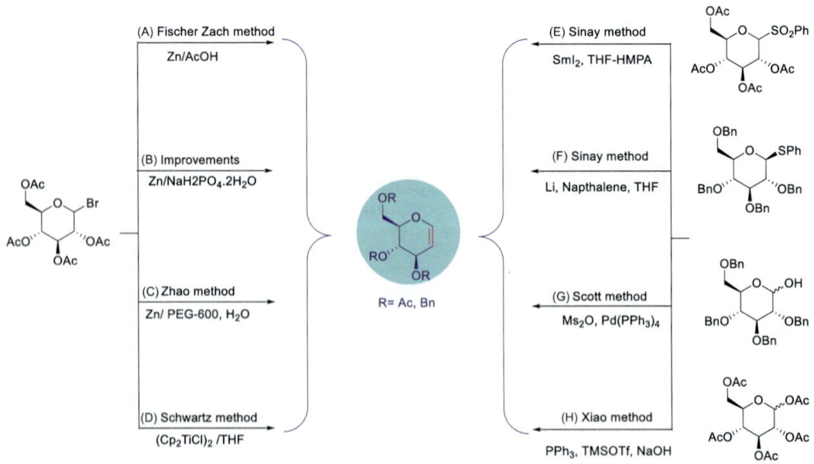

FIGURE 10.2 Different synthetic routes to access pyran glycals.

naphthalene, and Zn–Ag solution (Fig. 10.2B).[3] Toward this endeavor, Zhao et al. reported the use of zinc in aqueous PEG-600 as a modified variant of the Fischer–Zach method (Fig. 10.2C).[3] Further Schwartz et al. reported titanium-based reagent bis(cyclopentadienyl)titanium (IV) dichloride for the synthesis of glycals from acetobromo sugar via a free radical mechanism (Fig. 10.2D).[5] Sinay et al. utilized lithium naphthalenide under cryogenic condition for synthesizing glycals by using 3,4,6 benzylated-1-thio sugars (Fig. 10.2E).[6] The same group reported glycal synthesis from glycosyl phenyl sulfones via reductive elimination using samarium(II) iodide and hexamethylphosphoramide (Fig. 10.2F).[7] All the methods discussed earlier involve three-step synthesis of glycals viz acetylation of free sugar, replacement of anomeric acetate with bromine, chlorine, sulfone, or thio group, and reductive elimination to generate endocyclic double bond. In this context, Scott et al. reported the palladium-catalyzed synthesis of glycals from hydroxyl-based sugars using methane sulfonic anhydride (Fig. 10.2G).[8] Although these are the practical and reliable methods reported for the synthesis of glycals but are limited to pyran-based glycals only. Toward this context, very recently Xiao et al. reported a more robust and reliable method for the synthesis of glycals via E1cB elimination mediated by P(V) intermediate using triphenylphosphine under acidic conditions that can be accessed to both pyranoid and furanoid glycals (Fig. 10.2H).[9]

10.2.2 Synthesis of aryl *C*-glycosides

Glycosides are compounds in which aglycon component is attached to sugar moiety through anomeric center representing an important class of carbohydrates quite often found in nature. This linkage can be either through carbon, nitrogen, oxygen, or sulfur. One such linkage that is found in nature[10] includes aryl glycosides where an anomeric center is linked to an aglycon partner through an aryl group. Many naturally occurring compounds like vitexin,[11] orientin,[12] bergenin,[13] and papulacandin D[14] belong to this class of compounds (Fig. 10.3). Their importance can be further realized from the fact that certain FDA-approved drug molecules like dapagliflozin, canagliflozin, and empagliflozin contain aryl moiety (Fig. 10.4).[15] The past two decades have witnessed significant growth in the development of synthetic

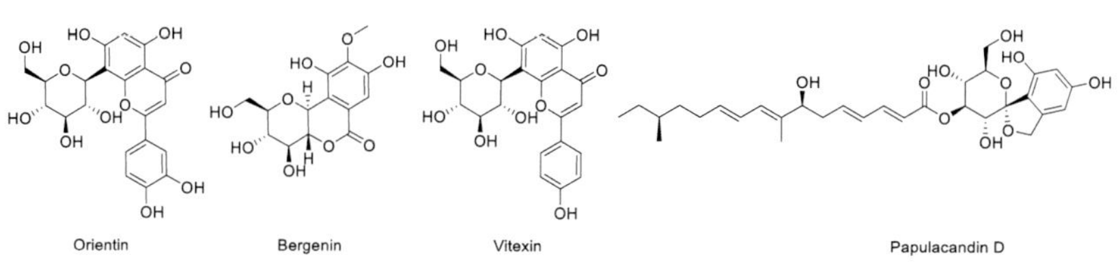

FIGURE 10.3 Naturally occurring aryl *C*-glycosides.

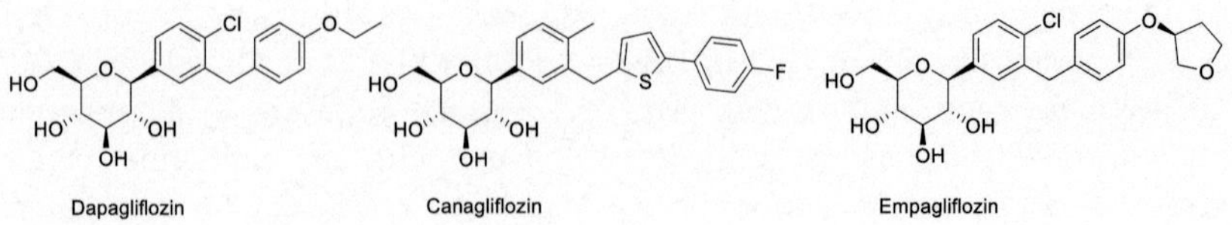

FIGURE 10.4 FDA-approved *C*-aryl drugs.

methodologies to access aryl *C*-glycosides stereoselectively. Broadly, these can be classified into the following four categories.

- Electrophilic substitution reactions between glycals and arenes.[16]
- Glycosylation reactions catalyzed by transition metals.[16]
- Synthesis of carbohydrate units by de novo method.[17]
- Migratory reactions involving *O*—*C*-aryl migration.[18]

Among these methods, metal-catalyzed reactions particularly Heck reactions have received distinct attention because of the use of stable and moisture-friendly reagents.[19,20] The product formation can be either β-hydride/hetero-atom elimination whose fate is exclusively controlled by the nature of protection present on glycal. For instance, when glycals are protected with a good leaving group like acetate, 2,3-deoxy-*C*-glycosides are formed, while 2-deoxy-*C*-glycosides are obtained exclusively when protected with a poor leaving group like benzyl or silyl; moreover, nature of ligand and electron density on aryl ring plays a key role in allowing such transformation to occur. Both these deoxy aryl *C*-glycosides are of considerable importance in particular the presence of double bonds at the C2 and C3 position as it can be further transformed through postmodification to generate diverse and useful scaffold.

Toward this endeavor, Maddaford et al. reported palladium-catalyzed aryl *C*-glycoside synthesis using aryl-boronic acids as carbon-based nucleophiles from glucal triacetate via carbon Ferrier-type reaction.[17] These aryl-boronic acids are used as a coupling partner due to some advantages like stability toward moisture, low toxicity, and being commercially available at cheaper rates. Initial attempts to synthesize aryl *C*-glycoside via Lewis acid-promoted carbon Ferrier reaction failed completely. The authors next focused on the generation of π-aryl-Pd complexes through *trans*-metalation of phenylboronic acids using a palladium catalyst. Treating glucal triacetate with phenylboronic acid and palladium acetate in acetonitrile (ACN) led to the synthesis of aryl *C*-glycoside in excellent yield with complete stereocontrol (Scheme 10.1). Screening of different catalysts revealed palladium acetate in catalytic amount was sufficient enough for the transformation to occur while the use of other catalysts resulted in poor yield or was unsuccessful. Similarly, examination of the nonpolar fraction revealed the formation of biphenyl as a minor product that was formed via the second *trans*-metalation of PhPdOAc. With optimized conditions in hand, the series of aryl *C*-glycosides were synthesized using both electron-withdrawing, electron-donating,

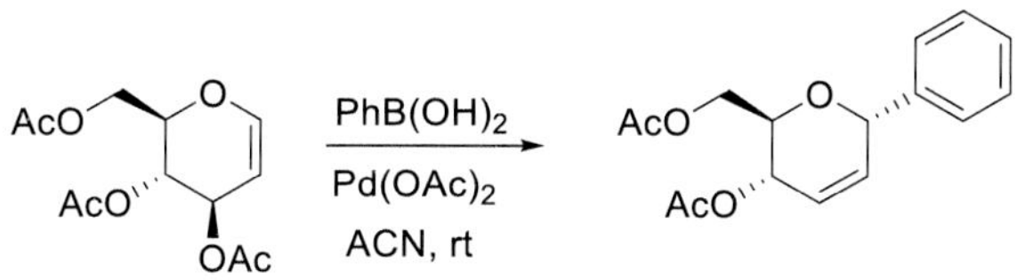

SCHEME 10.1 Pd-catalyzed *C*-arylation using aryl-boronic acids and peracetylated glycals.

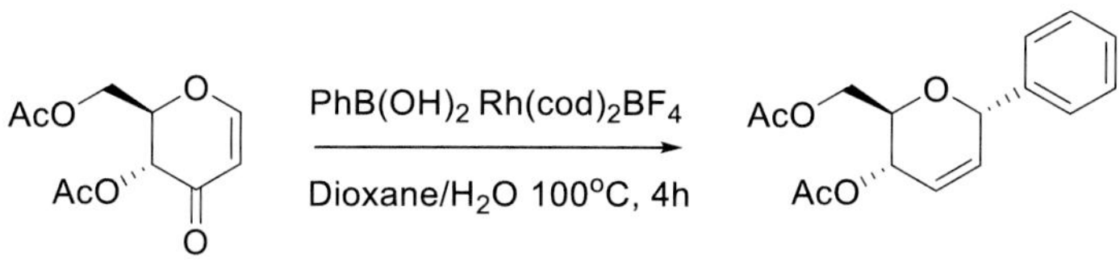

SCHEME 10.2 Rh-catalyzed *C*-arylation using aryl-boronic acids and enones derived from glycals.

and sterically congested boronic acids with lower yields obtained in case boronic acid possesses electron-withdrawing groups. Likewise, D-galactal and L-rhamnal yielded α- and β-aryl *C*-glycosides, respectively, with good-to-moderate yields obtained.

The same group in the subsequent work reported the synthesis of 2-deoxy aryl *C*-glycosides via rhodium(I)-catalyzed 1,4-addition of aryl-boronic acids to enones derived from glycals.[21] Initial attempts to perform 1,4-addition of phenylboronic acids using various rhodium-based catalysts in the presence of phosphine-based ligands and Lewis acids failed completely, however, upon using cationic rhodium complex $Rh(cod)_2BF_4$ in a mixture of dioxane/water at 100°C, resulted in the synthesis of 1,4-addition product with good yields obtained (Scheme 10.2). The products were characterized through spectroscopic analysis as 2-deoxy α-aryl *C*-glycoside. A variety of 2-deoxy α-aryl *C*-glycoside with modest to good yields was synthesized using electron-donating, vinyl, electron-withdrawing, and sterically congested boronic acids and enone obtained from galactal. It is pertinent to mention that complete stereocontrol is maintained with both glucal and galactal enones with $^4C_1$(D) and $^1C_4$(D) conformation adopted by them, respectively. This 1,4-addition reaction is completely stereospecific as evident that α-isomer is observed on the addition of organopalladium reagent to carbohydrate enone. Further, under anhydrous conditions, only 5% product formation is observed while the addition of a catalytic amount of potassium hydroxide completes the reaction in 15 minutes.

Apart from the use of aryl-boronic acid as aglycon partner for the synthesis of aryl glycosides, Liu et al. reported the synthesis of 2-deoxy aryl *C*-glycosides via palladium-catalyzed decarboxylative Heck coupling using benzoic acids and glycals.[22] Screening of reaction condition revealed the desired products were obtained in excellent yield with complete stereocontrol upon using a catalytic amount of palladium acetate, silver carbonate, and triphenylphosphine in (1:20) mixture of dimethyl sulfoxide/dimethylformamide (DMSO/DMF) at 80°C for 4 hours (Scheme 10.3). With optimized reaction conditions in hand the substrate scope was varied by using different glycals and moderate-to-good yields were obtained with complete stereocontrol regarding less of the nature of the protecting group on glycals. Noteworthy aryl *C*-glycoside derived from disaccharide also proceeded smoothly in moderate yield.

Although aryl-boronic acid as a coupling partner has certain advantages, however, while performing palladium-catalyzed cross-coupling, ring opening of sugar moiety remains a major concern.[23] Various alternatives to aryl-boronic acids have been explored in this context. Liu et al. reported aryl *C*-glycoside synthesis via cleavage of preactivated C–N bond using glycals and activated anilines.[16] Initial attempts to perform coupling between glycals and phenyl hydrazine using palladium acetate failed however after proper screening of catalyst, ligands, temperature, and solvent resulted in the formation of α-isomer with excellent yields obtained (Scheme 10.4). With optimized conditions in hand, the authors expanded the substrate scope of glycals. Noteworthy acetate and ethoxycarbonyloxyl protection at the C3 position yielded the desired product while other protecting groups like benzoyl, pivaloyl, or *tert*-butoxycarbonyloxy failed probably due to restriction by solvent toward reactivity. Other glycals prepared from galactose, L-6-deoxyglucose, and ribose reacted well with phenyl hydrazine to provide the corresponding aryl *C*-glycosides in moderate-to-high yields. Among various phenyl hydrazine tested ortho-substituted phenyl hydrazine showed poor reactivity except *o*-chlorophenylhydrazine possibly due to steric hindrance that restricts the aryl-palladium complex to approach the glycals thereby inhibiting migratory insertion. With meta- and para-substituted phenyl hydrazine moderate-to-good yields of the desired products were obtained whereas

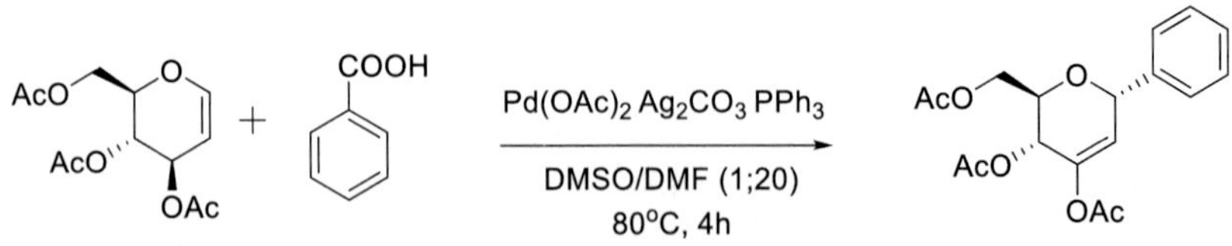

SCHEME 10.3 Pd-catalyzed *C*-arylation using aryl benzoic acids and glycals.

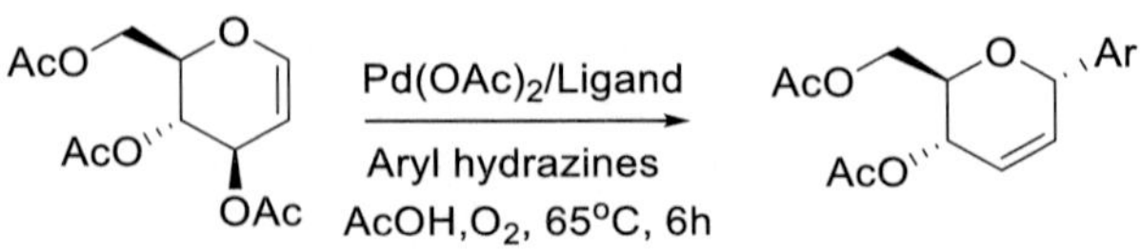

SCHEME 10.4 Pd-catalyzed *C*-arylation using aryl hydrazine.

strong electron-donating groups, like alkoxyl, did not yield any desired product possibly due to the decomposition of hydrazine substrates under standard conditions. Although most of the transition metal-catalyzed Heck-type *C*-glycosylation result in α-selectivity with few reports specifying the reason, Ye's group in 2009 reported the selectivity depends on the stereochemistry of the C3 group on glycal showing steric hindrance between the bulky palladium species and the protection present on C3 facilitates the palladium species to approach glycals from the side opposite to the C3 group.[20] This hypothesis was however ruled out as β-aryl *C*-glycosides were synthesized with anomeric mixture α:β ranging in between 1:1 and 1:0.6 thus showing that selectivity depends on other factors as well other than the nature of protecting group at the C3 position.

Transition metal-catalyzed reaction between protected glycals and boronic acids result in the synthesis of aryl *C*-glycosides via 3 β-H elimination or 3 β-O elimination; however, performing such reaction ring opening via 5 β-H elimination remains a major sort of concern resulting in the loss of valuable starting material and decrease in the product formation. Ye group in 2014 showed that these 5 β-H elimination products can be formed as a major product using appropriate reaction conditions and subsequent cyclization leading to the synthesis of aryl *C*-glycosides.[24] Initially, the reaction was carried out between benzylated glycals and phenylboronic acid using palladium acetate however no product formation was observed, subsequently using an oxidant resulted in the formation of ring-opened alcohol product. Screening of different oxidants showed bis(acetoxy)iodobenzene (BIAB) as an oxidant afforded ring-opened product albeit in lower yield. Upon investigating the nickel catalyst, it was observed that using $Ni(acac)_2$ in the presence of potassium phosphate in toluene at 110°C under air gave the required product in better yield than earlier. Use of other nickel catalysts did not yield the desired product, pertinent to mention that the use of phosphorus or nitrogen ligands destroyed the reaction. Next bases were screened, and it was found that comparable results were obtained with potassium and sodium carbonate, while lower yields were obtained with triethyl amine and no product formation was observed upon using stronger bases like cesium carbonate, sodium hydroxide, and potassium tertiary butoxide. Microwave irradiation and the use of molecular sieves along with higher temperatures yielded the desired product in excellent yield (Scheme 10.5). With optimized conditions in hand, the substrate scope was varied using different boronic acids, and good yields of the desired product were obtained regardless of the nature of the

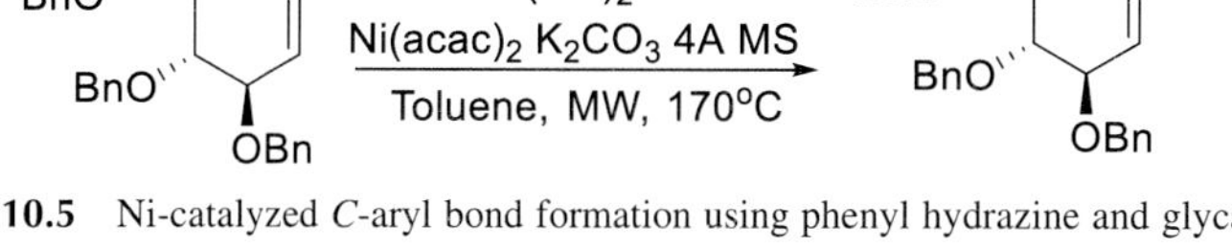

SCHEME 10.5 Ni-catalyzed *C*-aryl bond formation using phenyl hydrazine and glycals.

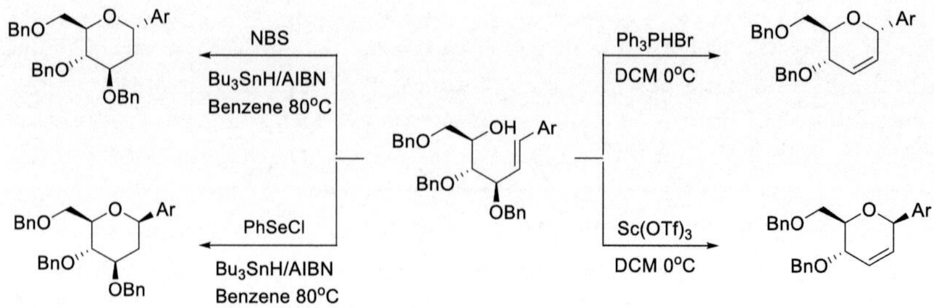

SCHEME 10.6 Ring cyclization to access α/β-aryl *C*-glycosides.

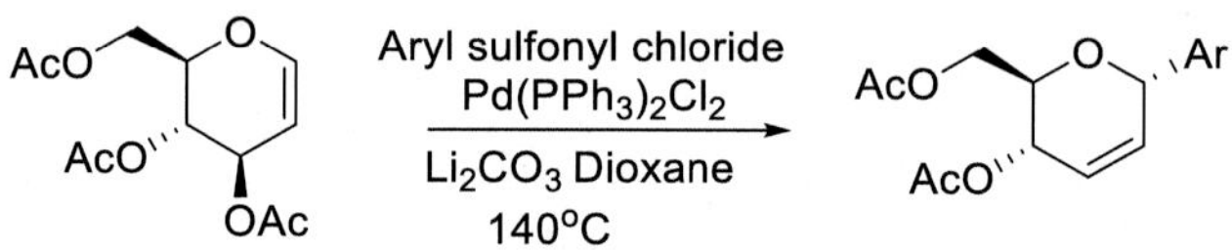

SCHEME 10.7 Pd-catalyzed *C*-arylation using aryl sulfonyl chlorides.

boronic acids. Similar results were observed upon using per benzylated glycals derived from galactose and xylose.

Next reaction condition was screened to access the stereoselective formation of *C*-aryl glycosides from ring-opened products. Screening of Lewis acids and protonic acids revealed α- and β-aryl glycosides were formed upon using Ph_3PHBr and $Sc(OTf)_3$-mediated cyclization, respectively. Similarly, α-aryl glycosides were obtained upon using phenyl selenium chloride-mediated cyclization under cryogenic conditions, whereas *N*-bromo succinimide (NBS)-mediated cyclization yielded β-aryl glycosides. Subsequent treatment with Bu_3SnH/azobisisobutyronitrile (AIBN) resulted in the synthesis of 2-deoxy aryl *C*-glycosides (Scheme 10.6).

More recently Mukherjee et al. reported palladium-catalyzed desulfitative *C*-arylation of glycals using aryl sulfonyl chlorides under basic conditions.[25] Screening of reaction condition revealed the desired aryl *C*-glycosides were obtained in good yields upon coupling glycal with aryl sulfonyl chloride in the presence of palladium catalyst lithium carbonate and 1,4 dioxane as solvent (Scheme 10.7). The reaction worked well with both electron-donating and electron-withdrawing sulfonyl chlorides yielding aryl *C*-glycosides in moderate yields with 99% selectivity observed. The substrate scope was extended by using glycals other than glucal with D-glucal and D-galactal α-selectivity, whereas with L-Rhamnal and L-Fucal β selectivity was observed. Next, the role of protecting groups on glycals was studied, and it was observed that the benzoyl protecting group had similar reactivity as that of acetate, whereas in the case of ether functionality ketone-type *C*-glycoside instead of the normal *C*-Ferrier product was obtained as reported in the literature.[26]

Glycals whose endocyclic double bonds have been functionalized to generate important synthon for constructing important scaffolds. One such

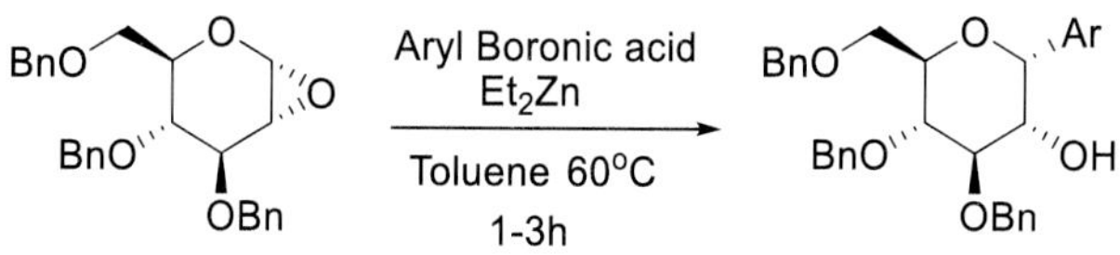

SCHEME 10.8 Organo zinc promoted arylation of 1,2-anhydro sugars using aryl-boronic acids.

transformation involves the conversion of double bonds into epoxides thus resulting in the synthesis of 1,2-anhydro sugars.[27] The same group utilized 1,2-anhydro sugar for synthesizing aryl *C*-glycosides via organo zinc promoted arylation formed through facile boron zinc exchange using aryl-boronic acid and diethyl zinc.[28] The exchange occurred at 60°C using toluene as solvent (Scheme 10.8). Subsequent addition of sugar epoxide resulted in the synthesis of aryl *C*-glycosides whose coupling constant value ranged between 4.2 and 5.9 Hz thus confirming the formation of α-isomer. The use of other aprotic solvents like dichloromethane (DCM) and tetrahydrofuran (THF) did not yield the desired product. Although the product formation was accelerated upon using different Lewis acids, decreased yields and selectivity were observed. Surprisingly, the use of zinc bromide resulted in the synthesis of β-aryl glycoside with a lower yield obtained. The methodology worked well with other boronic acids containing both electron-withdrawing and -donating groups with better yields reported in the latter case. Similarly, using 1,2-anhydro sugars derived from sugars other than glucal too worked well giving good yields of the desired product. It is pertinent to mention that α-aryl *C*-mannoside cannot be synthesized directly from this method however the stereochemistry of the free hydroxyl group at the C2 position could be inverted via oxidation reduction leading to the synthesis of α-aryl *C*-mannoside.

10.2.3 Synthesis of *C*-glycosides

Crotti et al. reported the synthesis of β-*C*-glycosides from vinyl oxirane derived from glycal.[26] Addition of organo-lithium reagent to vinyl oxirane at 0°C in diethyl ether resulted in the synthesis of corresponding *C*-glycoside via regioselective 1,4-addition with complete β selectivity observed (Scheme 10.9). Use of other organo-lithium reagents gave the desired products in good yield. The configuration was assigned on the basis of Nuclear Overhauser Effect (NOE) correlation between H1 and H5 protons and higher chemical shift value for C5 carbon observed in ^{13}C nmr. The use of other carbon-based nucleophiles showed complex behavior with vinyl oxirane giving different results with respect to regio- and stereoselectivity. The regio- and stereoselectivity obtained with organo-lithium reagents may be attributed to the bidentate chelation of metal with both exo- and endocyclic oxygen atoms, thereby facilitating organo-lithium to attack at the anomeric center from β-side selectively.

Germain et al. reported the synthesis of 2,3-unsaturated α-*C*-glycosides via Ferrier-type alkynylation reaction between glycals and iodo-alkynes

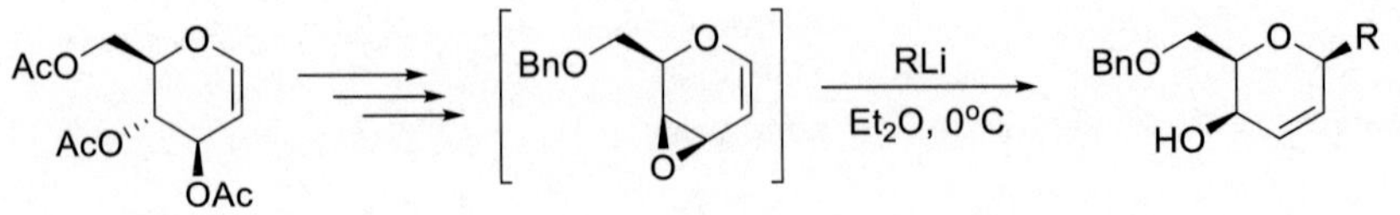

SCHEME 10.9 Synthesis of *C*-glycosides via addition of organo-lithium reagent to vinyl oxirane.

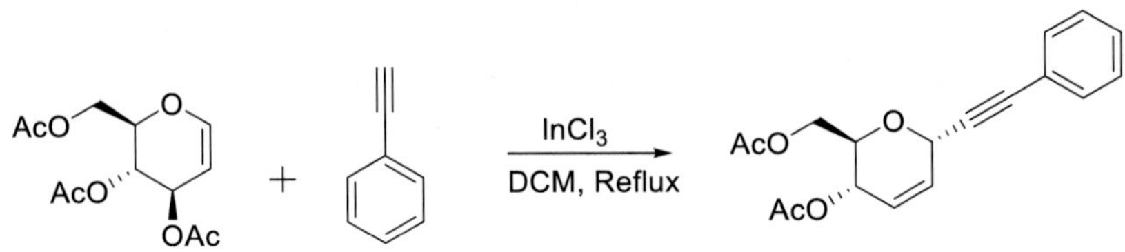

SCHEME 10.10 Synthesis of *C*-glycosides via Ferrier-type alkynylation reaction.

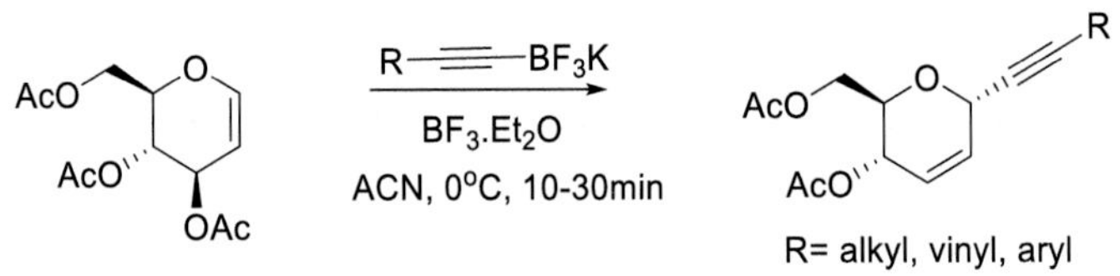

SCHEME 10.11 Synthesis of *C*-glycoside potassium alkynyltrifluoroborates.

under Barbier condition.[29] Refluxing iodophenyl acetylene with glycal triacetate in the presence of metallic indium resulted in the synthesis of *C*-alkynyl glycoside with 9:1 α/β-selectivity obtained (Scheme 10.10). Noteworthy, the presence of iodine atoms in alkyne is important as reactivity lacked when terminal alkyne was used under basic conditions. Further, the reaction took place efficiently in chlorinated solvent contrary to the Barbier process occurring in an aqueous medium. Various iodo-alkynes were treated with glycal triacetate resulting in the synthesis of *C*-glycosides under refluxing conditions with α-isomers formed predominantly. Similar results were obtained upon using glycosidic electrophiles and corresponding *C*-glycosides were obtained in good yields. The methodology worked well for the synthesis of disaccharides by coupling glycal with glycosyliodoalkyne.

Stefani et al. reported diastereoselective synthesis of *C*-glycosides from glycals via a Ferrier-type rearrangement mediated by Lewis acid using potassium alkynyltrifloroborate.[30] Screening of reaction condition revealed the desired product was formed in excellent yield upon using BF_3Et_2O as a source of Lewis acid in ACN at 0°C under the anhydrous condition with α-isomer formed predominantly (Scheme 10.11). The use of other Lewis acids either decreased the yield or did not result in the formation of the desired product. Further, the utility of the method developed was verified by synthesizing a series of glycosides using different potassium-based alkynyl trifluoroborates. The configuration was assigned based on the spectral data and was further correlated with the literature reports.

Liu et al. reported the synthesis of *C*-glycoside from 2-nitro glucals using *N*-heterocyclic carbene catalyst.[31] Screening of reaction condition revealed the Stetter type β-*C*-glycoside was formed exclusively when diisopropylethylamine was used as a base, whereas 2-nitro-eliminated *C*-glycoside was obtained exclusively upon using cesium carbonate (Scheme 10.12). Among the different catalysts employed, precatalyst E gave the best results. Next, the substrate scope was expanded by using different *N*-containing heterocycles, and good-to-moderate yields of the desired products were obtained. The presence of a hydroxyl group on *N*-heterocycles did not interfere with *C*-glycosylation thus indicating *C*-glycosylation is more favorable than *O*-glycosylation. Similarly, good yields of desired products were obtained upon using glycals other than glucal. Next *C*-glycosides were synthesized from quinolines with formyl groups at C2, C3, and C8 positions giving good yields of the desired products. The protocol was further extended to aliphatic and aromatic aldehydes, and corresponding glycosides were obtained in moderate yields.

Mukherjee et al. reported copper-mediated *C*-glycoside synthesis using unactivated alkynes.[32] The synthesis of *C*-glycoside is well documented in the literature, however, occurs in multiple steps. Moreover, a basic condition is needed for the activation of alkynes as well as an acidic condition needed for the Ferrier glycosylation to occur. The use of ascorbic acid in combination with transition metals and its ability to reduce Cu(II) to Cu(I) is reported in the literature.[33] Inspired by this transformation, the treatment of glycal triacetate with alkyne in the presence of copper triflate and ascorbic acid resulted in the formation of *C*-alkyl glycoside as a sole product (Scheme 10.13). The structure and stereochemistry of the isolated product were determined through spectral data and further correlated with the literature reports. The use of copper triflate under basic conditions did not yield the desired product while the

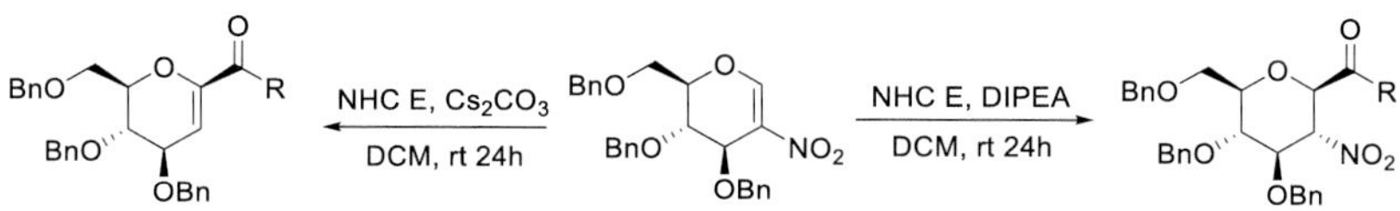

SCHEME 10.12 Synthesis of *C*-glycosides using *N*-heterocyclic carbine.

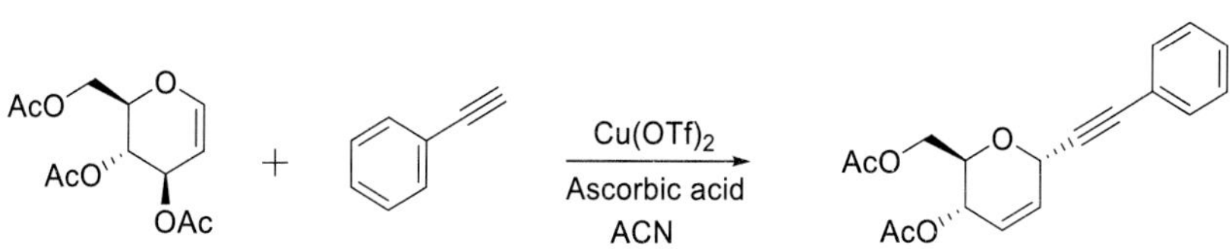

SCHEME 10.13 Copper-mediated synthesis of *C*-glycosides using unactivated acetylenes.

best results were obtained upon performing the reaction in ACN. Various phenyl acetylenes bearing electrons withdrawing and donating groups reacted smoothly to give the desired α-*C*-glycosides in good yields. Similar results were obtained with glycals other than glucal. The exact mechanism for the entire catalytic process is not clear however based on the literature reports of copper triflate chemistry, it can be assumed the generation of triflic acid during the course of the reaction is the active catalyst responsible for the eliminative formation of oxocarbonium ion, thereby suggesting Cu(I)OTf plays a key role in the formation of copper acetylene the attacking *C*-nucleophile.

In continuation of the work on *C*-glycoside synthesis, the same authors reported regio- and diastereoselective synthesis of halogenated *C*-vinyl glycosides using aryl acetylenes promoted by Lewis acid via tandem glycosylation halogenation reaction.[34] Initial treatment of glycal with phenylacetylene at room temperature in the presence of ferric bromide resulted in the formation of an inseparable mixture of four diastereomeric products. Lowering the temperature gratifyingly resulted in the formation of a single product with excellent yield and selectivity (Scheme 10.14). The structure of the major isomer formed was assigned on the basis of spectral analysis. Further, the substrate scope was varied using different phenyl acetylenes and glycals with moderate-to-good yields of the desired product obtained.

Liu et al. reported stereoselective *C*-glycosylation by performing gold-catalyzed reaction carried out between glycals and propargylic carboxylate to generate α-*C*-vinyl glycosides via tandem intermolecular 1,3-acyloxy migration/Ferrier rearrangement.[35] Screening of different gold catalysts revealed the desired products were obtained in good yields upon using a catalytic amount of PPh_3AuOTf, which was generated in situ from PPh_3AuCl and AgOTf (Scheme 10.15). Noteworthily, their individual use was found to be inactive; further, complete Z selectivity of the olefin was observed with all

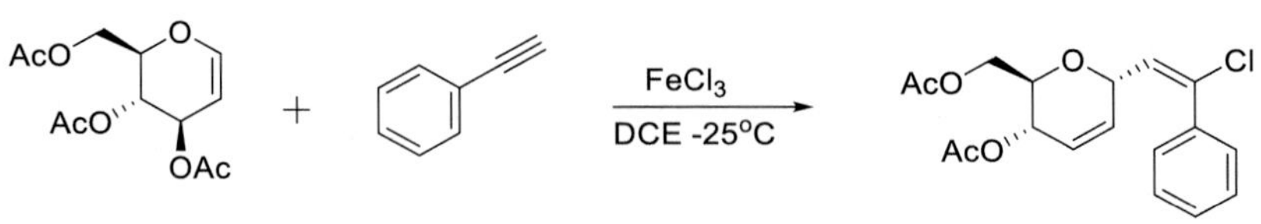

SCHEME 10.14 Lewis acid-mediated tandem glycosylation halogenation of glycals using unactivated acetylenes.

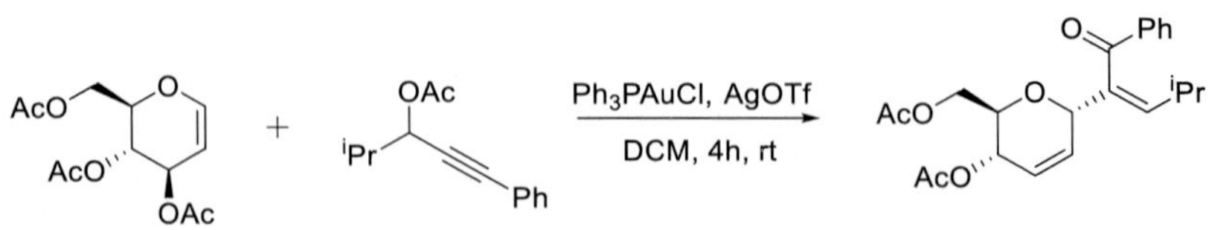

SCHEME 10.15 Gold-catalyzed *C*-glycoside synthesis.

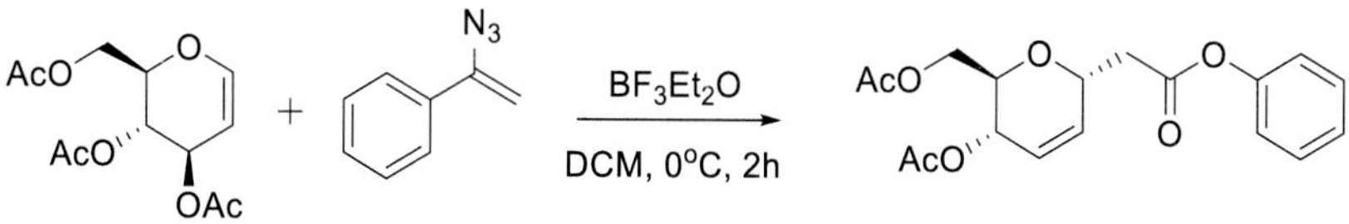

SCHEME 10.16 Lewis acid-catalyzed synthesis of *C*-glycosyl amides.

the desired products obtained whose stereochemistry was confirmed through spectral data and X-ray analysis. With optimized reaction conditions, the substrate scope was explored by treating different glycals and propargylic carboxylate affording the desired products in moderate-to-good yields.

Recently Mukherjee et al. reported the synthesis of α-*C*-glycosyl amides from glycals using aromatic vinyl azide under Lewis acidic condition through the formation of iminodiazonium intermediate that undergoes Schmidt-type denitrogenative 1,2-migration, thereby forming nitrilium ion, which upon subsequent hydrolysis, yields the desired products.[36] Screening of reaction conditions revealed the formation of desired products was achieved by using BF_3Et_2O as Lewis acid and DCM as solvent (Scheme 10.16). Further, the reaction was compatible with acetate and benzoate-protected groups while the use of benzyl protecting group hydrolyzed the vinyl azide itself. A series of *C*-glycosyl amides with complete stereocontrol were synthesized in good yields by taking different glycals and substituted aromatic vinyl azide.

10.2.4 Synthesis of nucleosides

Nucleosides are glycosylamines a central class of carbohydrate family often found in nature as phosphorylated derivatives with diverse biological functions like secondary messengers, enzyme cofactors, genetic information, storage, transcription, and translation.[37–39] Apart from this, in therapeutic areas, nucleosides and their analogs represent important lead roles.[40,41] Certain nucleosides and their derived compounds are FDA-approved drugs used for the treatment of cancer and viral infections.[42–44] In nucleosides, the nucleobase is attached to the anomeric center either through the nitrogen atom or the carbon atom thereby forming *N*/*C*-nucleosides, respectively, the latter being stable to enzymatic and acid hydrolysis. Various *C*-nucleosides are naturally occurring antibiotics (Fig. 10.5). Combined together the low abundance of naturally occurring *C*-nucleosides and their importance as biological probes, there has been much attention toward their synthesis.[42]

In this context, Mukherjee et al. reported direct palladium-catalyzed *C*-nucleoside synthesis via direct coupling of glycals and *N*-protected nucleobases without preactivation[45] (Scheme 10.17). Screening of reaction condition revealed Ferrier-type *C*-glycosylation product with complete α-selectivity observed upon using catalytic palladium acetate and equimolar ratio of silver oxide and pivalic acid in THF at 80°C (Scheme 10.17A). Glycals protected

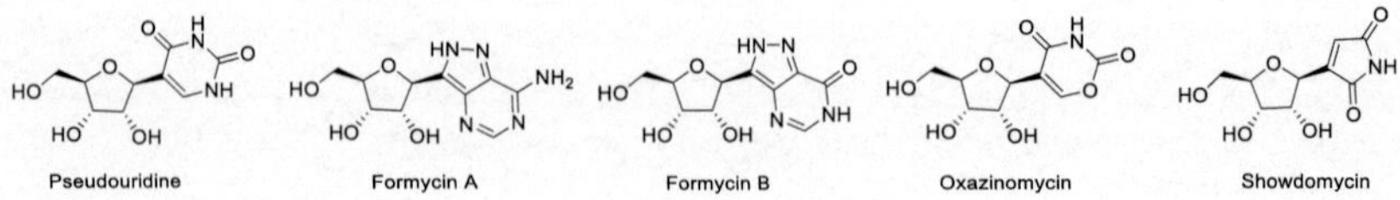

FIGURE 10.5 Naturally occurring *C*-nucleoside antibiotics.

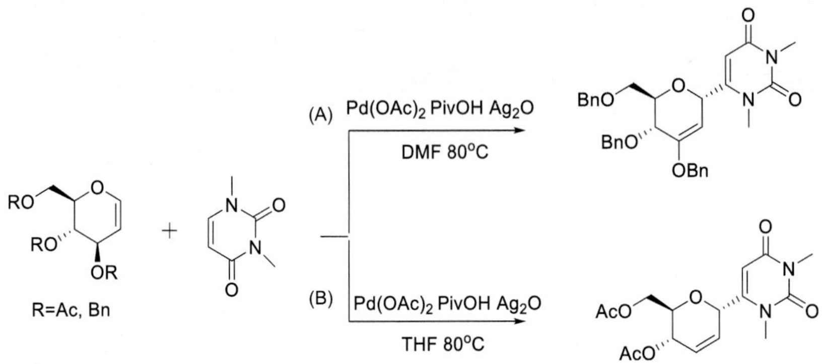

SCHEME 10.17 Palladium-catalyzed *C*-nucleoside synthesis from glycals and unactivated nucleobase.

with acetate and benzoate gave exclusively Ferrier-type product. The better yield was obtained upon using long chain alkyl as *N*-protection on nucleobase while the reaction failed upon using unprotected nucleobase possibly due to solubility issues. Similarly, reaction conditions were screened to obtain *C*-nucleoside via β-hydride elimination that included coupling of suitably protected glycal with unactivated uracil using catalytic palladium acetate and the equimolar ratio of silver oxide and pivalic acid in DMF at 80°C (Scheme 10.17B). Nuclear magnetic resonance (NMR) studies and control experiments were performed to know the plausible mechanism involved wherein it was found that initial electrophilic palladation occurs regioselectively at the C5 position of nucleobase rather than the C1 position of glycal thereby generating active organopalladium species, which then attacks the glycal regio and stereoselectively to afford *C*-nucleosides.

More recently the same group reported the synthesis of orthogonally protected nucleosides from epoxides derived from glycals using nonsilylated nucleobases under solvent-free conditions catalyzed by Lewis acid.[46] Refluxing the nucleobase in hexamethyldisilane in the presence of trimethylsilyl trifluoro methane sulfonate followed by addition of glycal epoxide led to the synthesis of orthogonally protected nucleosides (Scheme 10.18). Both α/β nucleosides were synthesized in a stereoselective fashion via *trans* opening of the glycal epoxide. A series of nucleosides was synthesized using different nucleobases and glycal epoxides obtained from different sugars. Screening for biofilm inhibition activity against four bacterial pathogens

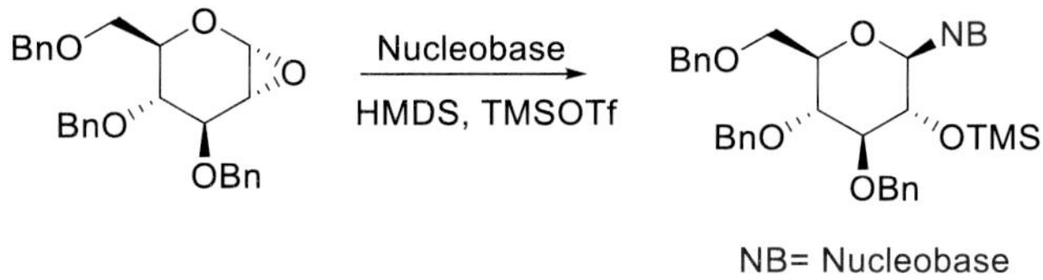

SCHEME 10.18 Synthesis of *N*-nucleosides under solvent-free condition.

was carried out wherein one analog showed promising activity against *Salmonella typhimurium* having 1.22 μM biofilm inhibition.

10.2.5 Synthesis of *N*-glycosides from pyranoid glycals

Other than *N*-nucleosides *N*-linked glycosides and glycoconjugates do exist in many natural products with promising biological activity.[47–49] Many *N*-linked heterocyclics like Akashine A and Staurosporine find potential use in pharmaceuticals (Fig. 10.6).[50,51] Moreover, the biological activity of certain natural products increases upon coupling with sugar moiety via *N*-glycosylation.[52–55]

Bravo et al. reported Lewis acid-catalyzed sulfonamido glycosylation of glycals using sulfonamide resulting in the synthesis of β-sulfonamido glycosides with complete stereocontrol.[56] Screening of different Lewis acids revealed the reaction proceeded smoothly upon using a catalytic amount of triphenylphosphine hydrobromide in dichloromethane (Scheme 10.19). With optimized conditions in hand, a series of compounds were synthesized using different sulfonamides with complete selectivity in excellent yield. The synthesized sulfonamides were tested against human hepatocellular carcinoma cell lines Hep-G2 showing potent inhibition.

Liu et al. reported palladium-catalyzed Ferrier-type *N*-glycoside synthesis from glycals using various *N*-heterocyclic nucleophiles like isatins, imides, and imidazole analogs.[57] Various palladium catalysts, ligands, and solvents were screened by taking isatin and glucal with picolyl protection at the C3 position as a model substrate to get the desired product in excellent yield with complete stereocontrol (Scheme 10.20). Next, the substrate scope was explored by using different glycal donors and *N*-heterocyclic nucleophiles resulting in the synthesis of desired products in excellent yield. Further, glycosylated indirubins were synthesized thus showcasing the applicability of this method developed.

More recently Goti et al. reported metal-free *N*-glycoside synthesis via (3,3)-sigmatropic allyl cyanate/isocyanate rearrangement of glycals in the presence of nucleophiles afforded β-*N*-glycosides in good yields.[58] The glucal-derived carbamate synthesized from glucal in three steps via one-pot upon hydrolysis yielded corresponding cyanate that underwent spontaneous [3,3]-sigmatropic rearrangement leading to the synthesis of isocyanate followed by nucleophilic attack resulting in the synthesis of corresponding β-*N*-glycosides

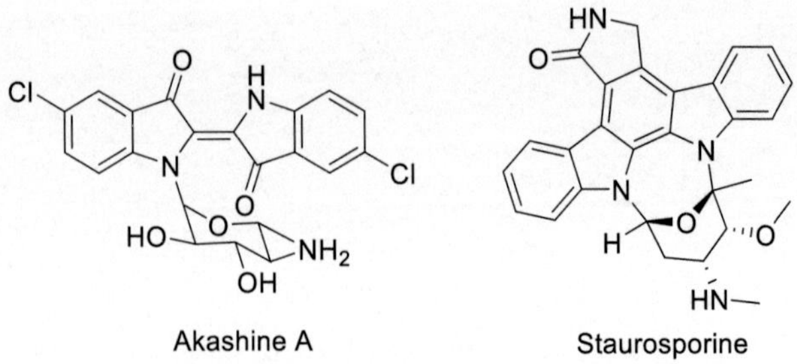

FIGURE 10.6 Naturally occurring *N*-glycosylated bis-indole compounds.

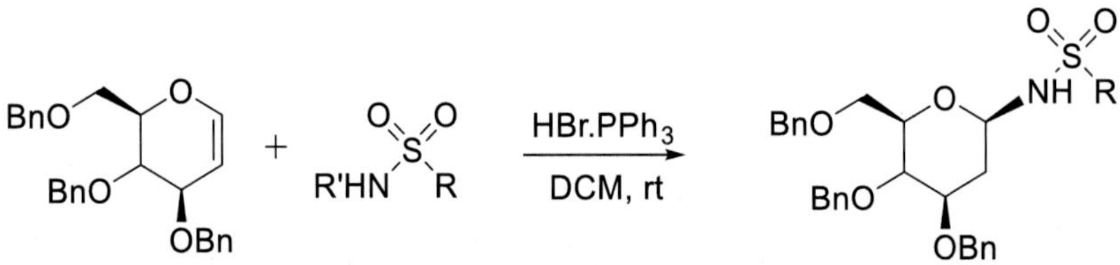

SCHEME 10.19 Lewis acid-catalyzed synthesis of sulfonamides from glycals.

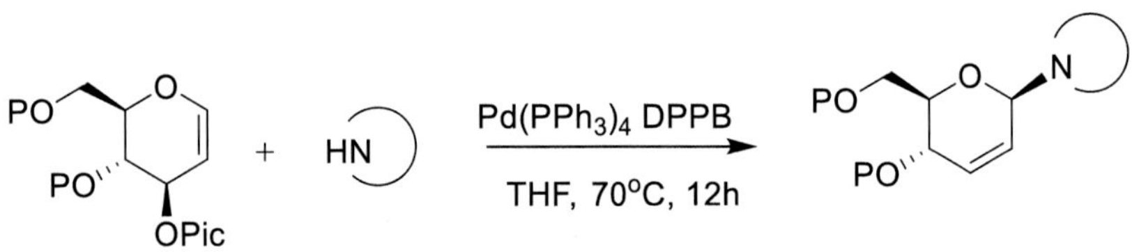

SCHEME 10.20 Palladium-catalyzed *N*-heterocyclic synthesis.

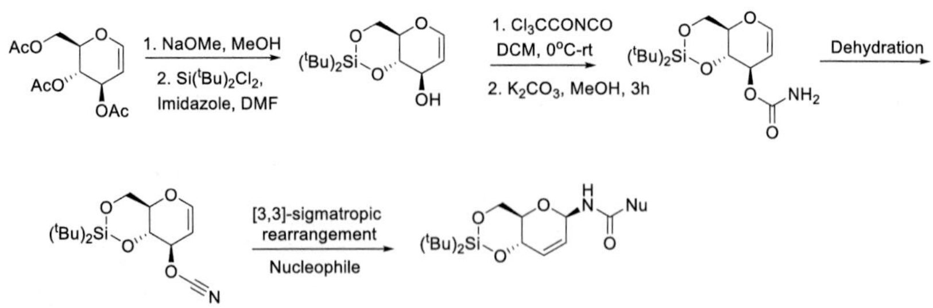

SCHEME 10.21 *N*-Glycoside synthesis via [3,3]-sigmatropic rearrangement.

(Scheme 10.21). It is pertinent to mention that the dehydration of the carbamate failed completely upon using standard appel's conditions however succeeded upon using trifluoroacetic anhydride in the presence of benzyl alcohol. With optimized conditions in hand, the substrate scope for *N*-glycoside synthesis was extended by using D-glucal and D-galactal in the presence of various *O*, *N*, and *C*-nucleophiles.

10.2.6 Synthesis of *O*-glycosides from pyranoid glycals

Over the past two decades, various glycosylation methods have been developed for synthesizing complex glycoconjugates and oligosaccharides. Owing to the versatile reactivity of π system involved in glycal, these glycal-based carbohydrate donors have been widely explored for synthesizing C2-branched or C2-derivatized oligosaccharides. The glycal assembly has been used as a convenient method for synthesizing 2-hydroxy gluco-, galacto-, and manno-pyranosides from corresponding *R*-1,2-anhydropyanosides.

Having reported earlier, the synthesis of C2-hydroxyglucopyranosides from glycals via selective transfer of oxygen atom from the sulfoxide reagent[59] and owing to the presence of α-mannopyranoside-based oligosaccharides widely in nature as host of *N*-linked protein glycoforms, the Gin group pursed to address one-pot synthesis of mannopyranoside from glycals via sulfonium-mediated oxidative glycosylation.[60] Prior method for synthesizing involves multiple steps with inversion of the C2 hydroxyl group following glycal oxidation.[61] It was hypothesized while performing such transformations stereochemistry involved in the oxidative glycosylation could be affected by the structure of sulfoxide reagent. To test their hypothesis, dibenzothiophene-5-oxide (DBTO) was used as a sulfoxide reagent in combination with triflic anhydride. This resulted in the stereoselective synthesis of 2-hydroxymannopyranosides with complete stereocontrol (Scheme 10.22). The synthetic utility of this method was verified by using various glucal donors incorporating different protecting groups that were found to be incompatible with the method. Apart from this, hydroxyl-based glycosyl acceptor too worked well. In conclusion, direct selective synthesis of 2-hydroxymannopyranosides from glycals has been reported using DBTO and Tf_2O as glycal activating agents, in addition to the appropriate selection of sulfoxide reagent for C2 stereocontrolled oxidative glycosylation.

Glycals have characteristic endo double bonds, which can be functionalized to generate different biologically important scaffolds. Apart from this, the presence of multiple hydroxyl groups on glycals possessing different reactivities can be altered to generate a new synthon that can be tailored to access diverse scaffolds. In this context, Crotti et al. reported the synthesis of vinyl oxirane derived from D-glucal via five-step synthesis.[62] This epoxide is of particular interest because its nucleophilicity can be tested by various

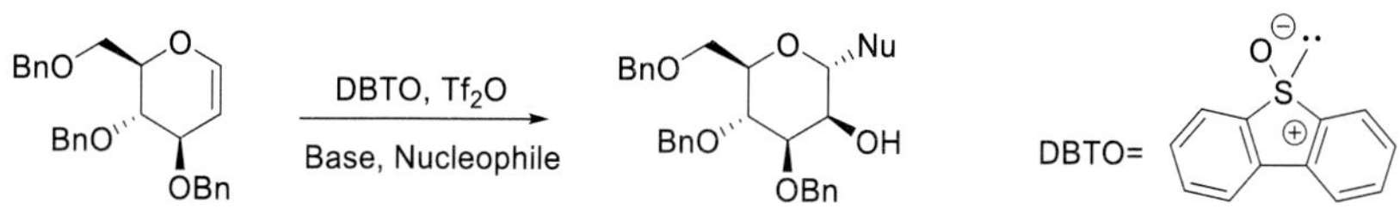

SCHEME 10.22 Synthesis of 2-hydroxy-α-mannopyranoside from glycals.

nucleophiles. It is important to mention like 1,2-anhydro sugars, this vinyl oxirane too is unstable, however, differs significantly in terms of reactivity as it can undergo 1,2/1,4-additions depending upon the reaction condition employed. Carrying out the transformation under basic conditions at room temperature afforded an equimolar mixture of anomeric 2-unsaturated methyl α/β glycoside indicating the nucleophilic attack taking place regioselectively in 1,4-addition pathways with inversion of configuration at C4 position but devoid of stereocontrol at anomeric position. This selectivity at the anomeric center was increased by the use of bulky solvents/nucleophiles. Inspired by such results the authors got further interested to make the process, more practical and reliable by making the entire process free from glycosylation. In the modified procedure, the epoxide was synthesized under inert and dry conditions followed by the addition of methanol as glycosyl acceptor thus affording 2-unsaturated methyl β-glycoside exclusively (Scheme 10.23). The use of other glycosyl acceptors worked well to afford corresponding β-glycosides with complete stereocontrol. The methodology also worked well for the synthesis of disaccharides using diacetone D-glucose as a glycosyl acceptor yielding β-linked disaccharide. Notably, acid-liable functional groups tolerated well with the reaction conditions. Although this glycosylation method seems to be similar to that of Ferrier-type rearrangement, the two glycosylation methods are different in the sense that α-glycosides are formed with Ferrier-type rearrangement that occurs in the presence of Lewis acid while the above protocol results in β-selectivity and takes place under basic or neutral condition. Moreover, the glycosidic product obtained possesses a galacto configuration.

Apart from having multiple hydroxyl groups present at various carbon center, certain modified carbohydrate units containing amino functionalities at different positions are found in numerous oligosaccharides and glycoconjugates.[63] In this context, the same group reported the stereoselective introduction of nitrogen functionality at the C4 position of glycal with combatant *O*-glycosylation from α-*N*-(mesyl)-aziridine derived from glycals resulting in the synthesis of 2,3,4-trideoxy-4-*N*-(substituted-amino)-hex-2-enopyranosides.[64] Glycal-based *N*-mesyl aziridine is unstable and could not be isolated and thus synthesized in situ via base-catalyzed cyclization of *N*,*O*-dimesylate intermediate under anhydrous condition. Upon treatment with methanol resulted in the formation of corresponding 4-*N*-(mesylamino)-2,3-unsaturated-α-*O*-methyl glycoside with complete stereocontrol via regioselective

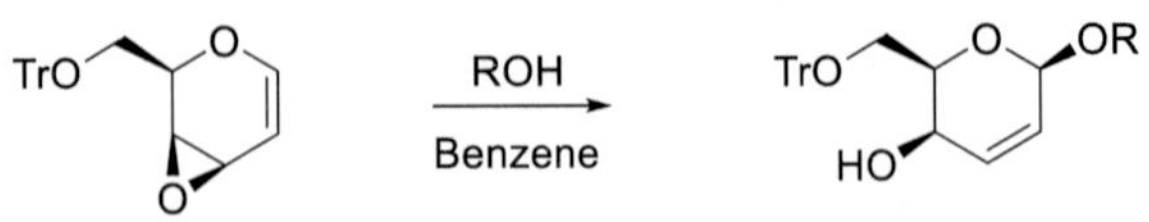

SCHEME 10.23 Synthesis of 2-unsaturated α-*O*-glycosides from vinyl oxirane derived from D-glucal.

S_N2 reaction (Scheme 10.24). Various *O*-glycosylated products were obtained in good yields upon using different glycosyl acceptors. Further, the reaction was carried out using alcohol itself as a glycosyl acceptor although the glycosylation reaction was completely 1,4-regioselective, however, decreased stereoselectivity was obtained. NOE correlation between H1 and H5 proton was carried out to determine the absolute configuration and structure of the glycosylated product.

Over the past few decades, the palladium-catalyzed reaction has gained interest in constructing C–C, C–O, and C–N bonds. In this context, glycal-based glycosyl donors have been utilized in π-allylpalladium strategies for the stereoselective synthesis of *O*-glycosides[65]; however, the low reactivity of both glycal donors and alcohol nucleophiles remains a challenge. Lee and coworkers reported Zn(II) to activate both alcohols for nucleophilic addition as well as glycal donor for ionization.[66] In this context, Nguyen et al. reported palladium-catalyzed Ferrier-type *O*-glycosylation using glycal donors with a trichloroacetimidate leaving group at the C3 position and phosphine-based biaryl ligand resulting in the synthesis of α-*O*-glycosides independent of protecting group manipulation and Lewis acid[67] (Scheme 10.25). The Pd(II)/L catalyst is believed to activate the glycal π-system for stereoselective attack by the external oxygen nucleophile, and the C3 trichloroacetimidate group serves as the leaving group as well as directs Pd(II) to the double bond of the glycal. The anomeric selectivity relies on the complexation of the palladium-ligand catalyst donor. Various glycal donors with different protecting groups at the C3 position were assessed to test the feasibility of this method. The desired *O*-glycosides with α-selectivity were obtained in all the cases suggesting that the nature of protecting group had little effect on anomeric selectivity rather bulky biaryl phosphine ligand was solely responsible for anomeric selectivity.

Liu et al. reported palladium-catalyzed intermolecular glycosylation using glycal-derived allylic carbonates as the glycosyl donors.[68] Initial treatment of

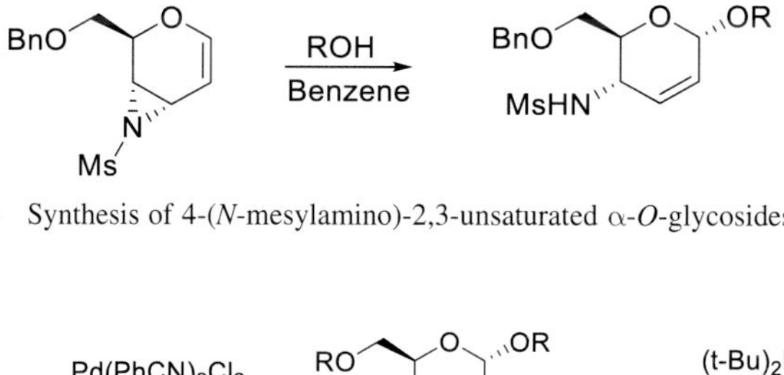

SCHEME 10.24 Synthesis of 4-(*N*-mesylamino)-2,3-unsaturated α-*O*-glycosides.

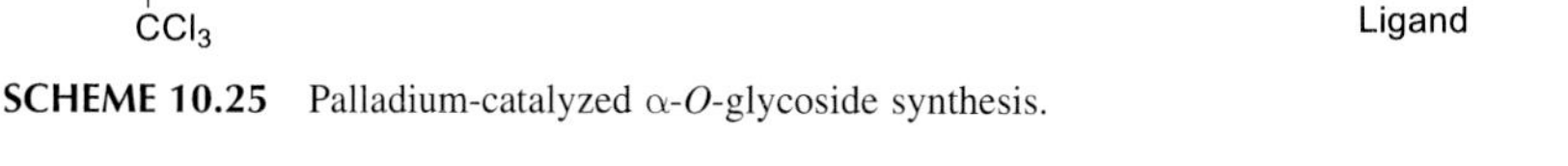

SCHEME 10.25 Palladium-catalyzed α-*O*-glycoside synthesis.

glycal carbonate with 3,4,5-trimethoxy phenol in the presence of palladium catalyst and ligand yields the desired product albeit in lower yield along with the formation of by-product as a major product (Scheme 10.26). Considering the low nucleophilicity of protonated phenol base was added to increase the yield of the desired product. Screening of different bases showed cesium carbonate was most effective in terms of both increasing the chemical yield as well as anomeric selectivity. Interestingly, the concentration of base plays a major role in determining the selectivity at the anomeric center. α-Isomer was observed in case less than 0.5 equivalent of the base was used, whereas β-isomer was formed exclusively when the concentration of base used was greater than 0.5 equivalents. Similarly, the presence of the acetate group at the C3 position resulted in the formation of a racemic mixture with a 45% yield obtained. With optimized conditions in hand, the substrate scope was extended, and better yields were obtained upon using strong electron-donating phenol compared to weak electron-donating phenols. A trace amount of desired product was observed in the case of phenols having electron-deficient groups, whereas no product formation was observed in the case of phenols bearing electrons withdrawing groups. Further, substrate scope was extended by using aliphatic alcohols with good yields obtained. Similarly, sugar-based alcohols were employed resulting in the iterative synthesis of di- and trisaccharides in good yields with complete stereocontrol.

Galan et al. reported the palladium-catalyzed synthesis of α-*O*-glycosides from disarmed glycals via a mechanism believed to proceed through alkoxy-palladation type.[69] Screening the reaction condition revealed the *O*-glycosides were formed in excellent yields with complete stereocontrol upon using 10 mol% of $Pd(MeCN)_2Cl_2$ in DCM at room temperature for 5 hours (Scheme 10.27). Other Pd catalysts were either detrimental to the reaction rates or eroded the α-selectivity, while no reaction was observed upon using palladium acetate. Having set the reaction conditions, the authors further

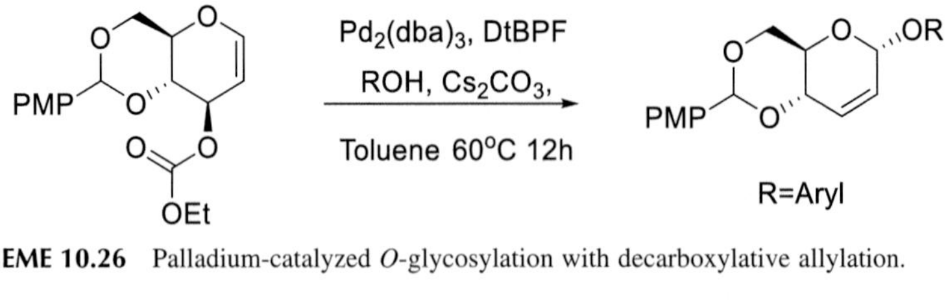

SCHEME 10.26 Palladium-catalyzed *O*-glycosylation with decarboxylative allylation.

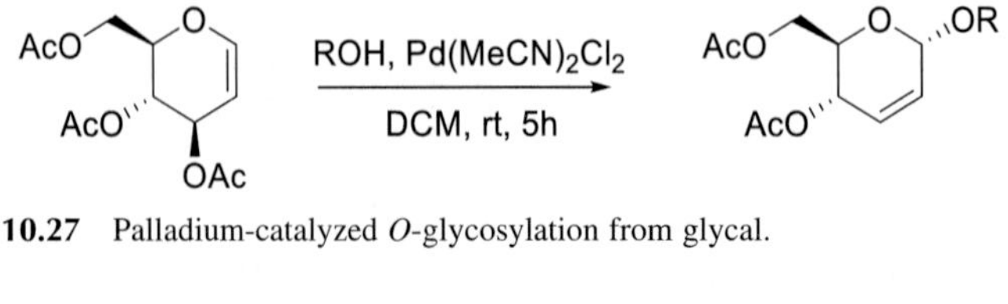

SCHEME 10.27 Palladium-catalyzed *O*-glycosylation from glycal.

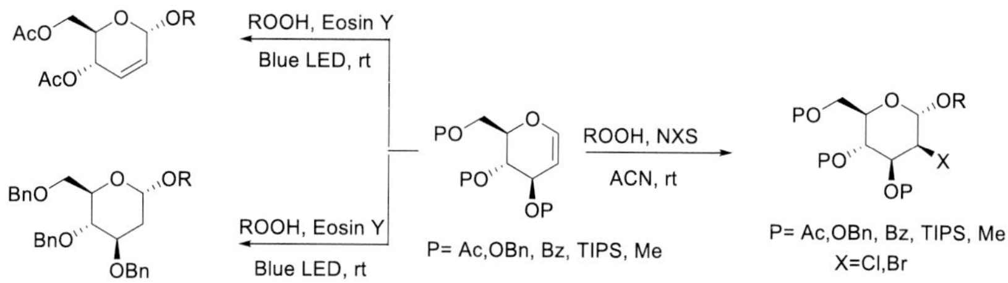

SCHEME 10.28 Synthesis of 2-iodo, 2-deoxy, and 2,3-unsaturated glycosides.

explored the substrate scope by varying different hydroxyl-based nucleophiles. Series of *O*-glycosides were synthesized in excellent yields with preferred α-selectivity. Similarly, moderate-to-good yields with high α-selectivity were obtained upon using glycals other than glucal.

Mukherjee et al. reported the synthesis of 2-iodo-glycosides, 2-deoxyglycosides, and 2,3-unsaturated glycosides using organic peroxides as glycosyl acceptors under optimized reaction conditions.[70,71] Screening the reaction condition, it was observed that upon using glycal, tertiary butyl hydrogen peroxide, and *N*-iodo succinimide at an ambient temperature in ACN afforded the *O*-glycosylated product in excellent yield with *trans* diaxial selectivity (Scheme 10.28). Series of 2-iodo-glycosides with α-selectivity were synthesized under the optimized reaction condition by using different protected glycals, except in the case of D-Xylal, which afforded an equimolar mixture of α- and β-anomers. Similarly, *N*-bromo succinimide gave a good yield of the glycosylated product, whereas no product formation was observed in the case of *N*-chloro succinimide. Further, substrate scope was explored using Cumin hydroperoxide as a glycosyl acceptor, and a good yield of the desired product was obtained. Having set the condition for the synthesis of 2-halo *O*-glycosides, the authors got further interested in using organic peroxides under photolytic conditions. Upon using superarmed glycals, tertiary butyl hydrogen peroxide and commercially available eosin Y in ACN yielded 2-deoxy-α-alkoxy glycosides exclusively, whereas the use of disarmed sugars under similar conditions resulted in the synthesis of a mixture of 2–3-unsaturated α-tertiary butyl glycosides and 2-deoxy glycosides in good yields.

More recently, the Galan group reported Cu(I) catalyzed direct synthesis of 2-deoxyglycerides from glycals.[71] Treating superarmed glycals with an alcohol-based nucleophile in the presence of 5 mol% $Cu(OTf)_2.C_6H_6$ and toluene at 45°C for 1.5 hours resulted in the synthesis of the desired product (Scheme 10.29). The substrate scope was varied by using different sugar-based alcohols and other alcohol-based glycosyl acceptors under the optimized reaction conditions, the reaction proceeded smoothly, and good-to-excellent yields were obtained along with α-selectivity. Apart from this, various protecting groups are tolerated well under the catalytic system. Similarly, the scope of the reaction was further explored by preparing various armed and disarmed glycal

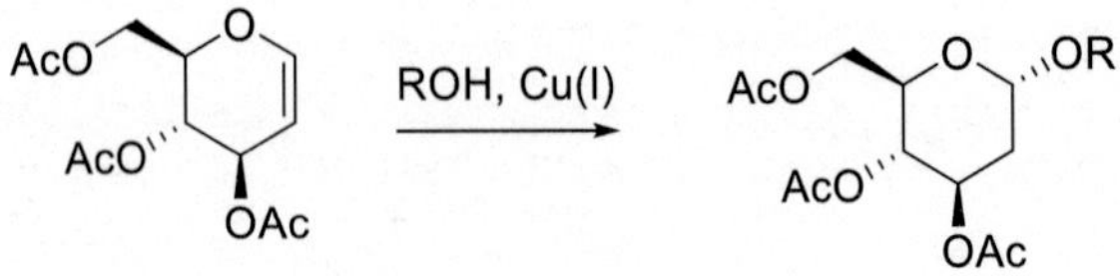

SCHEME 10.29 Synthesis of 2-deoxyglycosides from glycals.

donors, good yields of the glycosylated product were observed in all the cases except in the case of peracetylated glycal that afforded Ferrier-type glycoside as the major product. Detailed mechanistic study of the reaction along with NMR studies showed Cu(I) was an effective catalyst toward stereoselective glycosylation. The interaction between the Cu catalyst and both glycosyl donor and acceptor was further supported through cyclic voltammetry experiments.

10.2.7 C2 functionalization of pyranoid glycals

Other than well exploration of the anomeric position of glycals its C2 position too has been explored to generate diverse scaffolds. Liu et al. reported the synthesis of C2 alkenylated product via a direct cross-coupling reaction between glycals and activated alkenes.[72] Treating 3,4,6-tri-*O*-acetyl-D-glucal with ethyl acrylate in the presence of palladium acetate and copper triflate under aerobic condition at 70°C in an equimolar ratio of acetic acid and dimethylacetamide (DMA) for 24 hours resulted in the synthesis of corresponding C2 alkenylated product with complete E configuration in excellent yield (Scheme 10.30). Reaction condition was screened using different metal catalysts, cooxidants, and solvents. A series of C2 alkenylated compounds was synthesized using different glycals and alkenes.

Vankar et al. reported the synthesis of C2 alkenylated products via Heck coupling using 2-halo glycals derived from glycals and alkenes.[73] The 2-halo glycals were synthesized by reacting glycals with *N*-halo succinimide and silver nitrate under inert conditions in moderate-to-excellent yield. Further Heck coupling was carried out using different alkenes under the optimized condition thus generating a library of C2 alkenylated products in good yields (Scheme 10.31).

Steffani et al. reported the palladium-catalyzed synthesis of C2 alkynyl sugars using palladium acetate and cesium carbonate. Further, these conjugated enyne synthons were utilized for the construction of bicyclic glucal derivatives via an electrophile-promoted 5-endo-dig-cyclization pathway with moderate yield obtained[74] (Scheme 10.32). With optimized conditions in hand, a library of alkynyl derivatives were synthesized using different alkynes.

Lubin-Germain et al. reported palladium-catalyzed synthesis of 2-amidoglycals from 2-iodoglycals and amines via aminocarbonylation reaction.[75] Screening of reaction condition revealed the desired products were formed upon using a catalytic amount of palladium acetate. Pertinent to mention here that molybdenum hexacarbonate was used as a carbonyl source (Scheme 10.33).

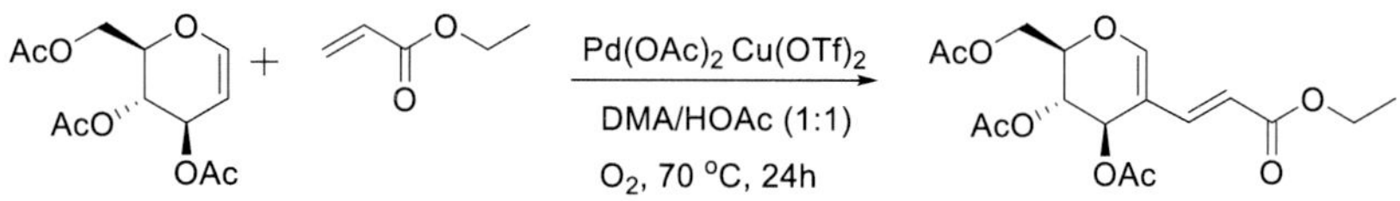

SCHEME 10.30 Pd-catalyzed C2 alkenylation of glycals.

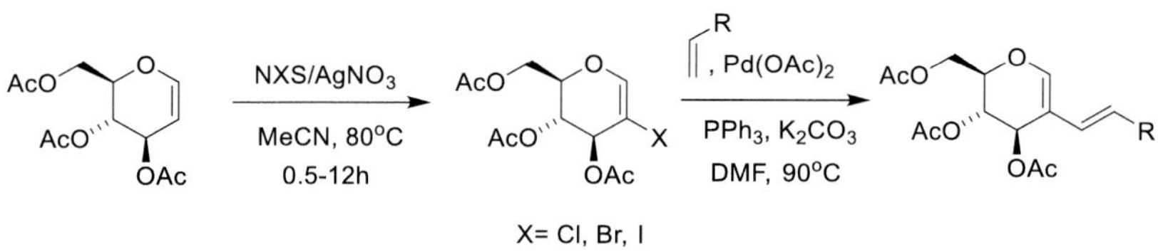

SCHEME 10.31 C2-alkenylated products from 2-iodoglycals via Heck coupling.

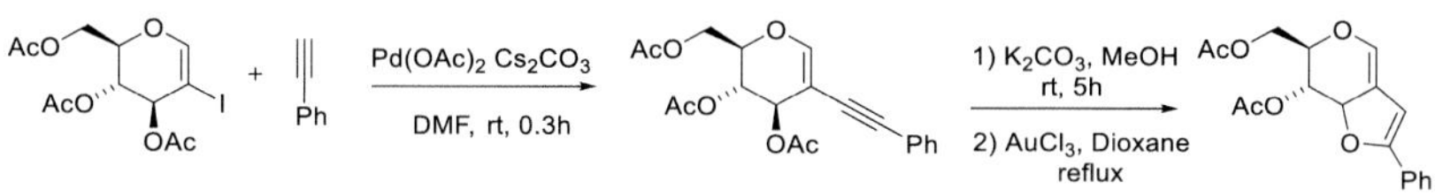

SCHEME 10.32 Synthesis of bicyclic glucal scaffolds.

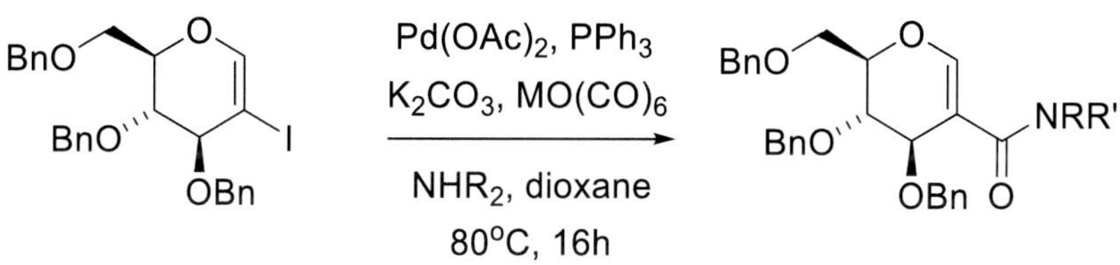

SCHEME 10.33 Synthesis of 2-amidoglycals.

With optimized conditions in hand, a series of 2-amidoglycals were synthesized using different substituted amines and protected glycals.

Recently, Mukherjee et al. reported a direct palladium-catalyzed sp^2-sp^3 cross-coupling reaction between glycals and cycloalkene resulting in the synthesis of C2-branched sugars.[76] Screening of reaction condition revealed the desired C2-allylic product was obtained upon reacting glycal with cyclohexene in a stoichiometric amount of palladium acetate and silver acetate in a mixture of DMF/DMSO (20:1) at 80°C for 24 hours with moderate-to-excellent yields obtained (Scheme 10.34). A library of compounds was synthesized under optimized conditions using different protected glycals and a variety of cycloalkenes.

Later in 2018, the same authors reported palladium-catalyzed sp^2-sp^2 cross-dehydrogenative coupling between C2 glycals and terminal alkene.[77] Screening of reaction condition was carried out resulting in a C2-alkenylated product with complete E stereoselectivity upon treating glycals with methyl

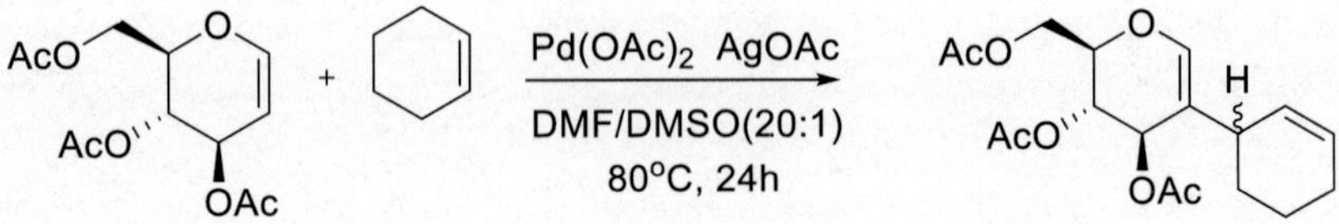

SCHEME 10.34 Palladium-catalyzed sp^2–sp^3 cross-coupling between glycals and cycloalkenes.

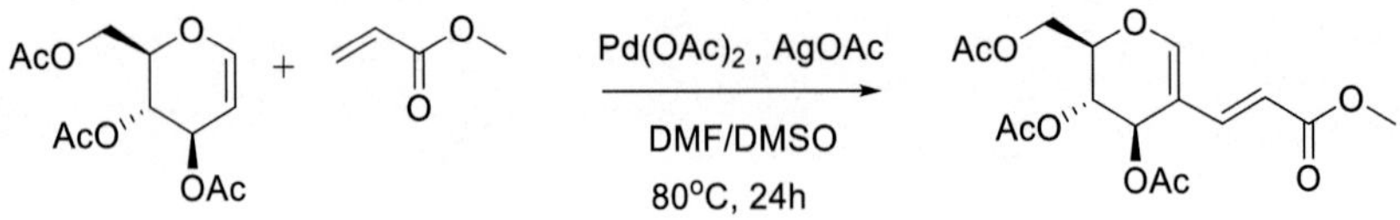

SCHEME 10.35 Palladium-catalyzed sp^2–sp^2 coupling between glycals and acrylates.

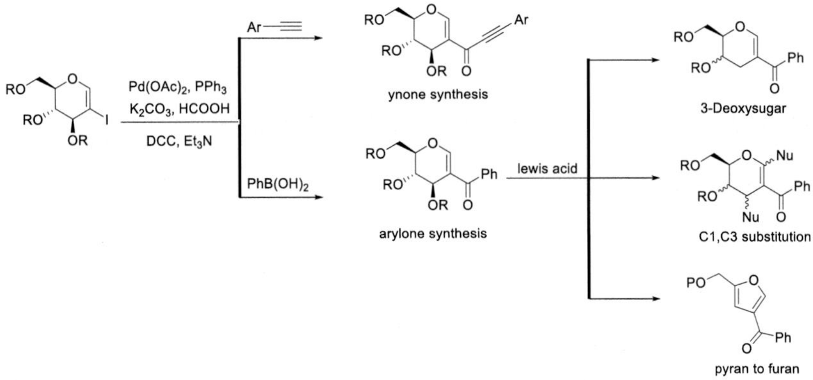

SCHEME 10.36 Synthesis of ynones and arylones from 2-iodoglycal.

acrylate in the presence of palladium acetate and silver acetate in a mixture of DMF/DMSO at 80°C for 24 hours (Scheme 10.35). After optimization of the reaction condition library of compounds was generated using different protected glycals and a variety of alkene partners resulting in the generation of desired C2-functionalized glycals. Postmodification was carried out on synthesized dienes via Diels–Alder cycloaddition leading to the synthesis of oxadecalin cores.

Recently, the same group reported the synthesis of sugar-based arylones and ynones via palladium-catalyzed carbonylative cross-coupling reaction from 2-iodoglycals using formic acid as a source of carbon monoxide.[78] Further, C2 branched glycals were reacted with different nucleophiles under the optimized condition for the synthesis of various C1 and C3 branched sugars. Further, under Lewis acid conditions, acetylated protected arylone sugar was transformed into substituted furan derivatives (Scheme 10.36). The synthetic utility was varied by synthesizing various C2-branched glycals under the optimized condition with moderate-to-good yields obtained.

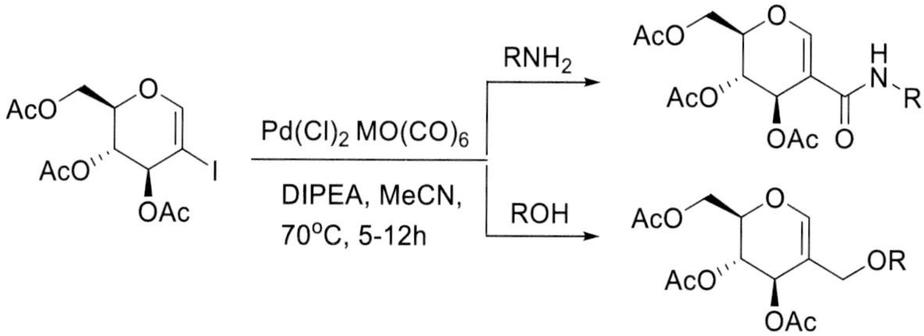

SCHEME 10.37 Synthesis of 2-amidoglucals and glucal esters.

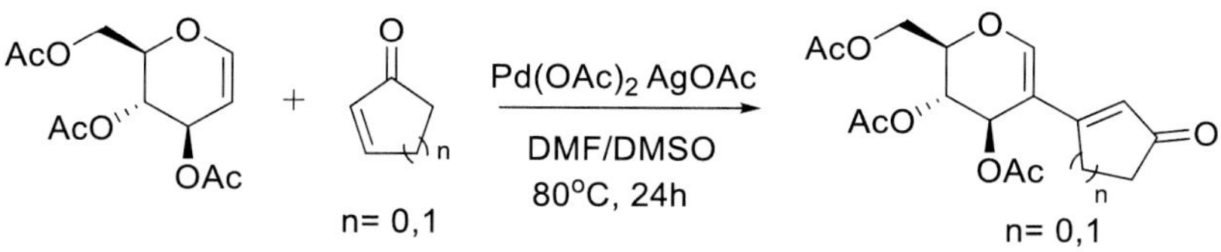

SCHEME 10.38 Synthesis of C2-functionalized glycals.

Stefani et al. reported a palladium-catalyzed carbonylative coupling reaction resulting in the synthesis of amido glucals and glucal esters from 2-iodoglucal using different substituted anilines and phenols, respectively[79] (Scheme 10.37). Reaction condition was screened by using different palladium catalysts and bases to obtain the desired products. With optimized conditions, the substrate scope was extended by using different glycals, anilines, and phenols. The reaction worked well with both electron-rich, electron-neutral, electron-poor, and heteroaromatic substituents.

Very recently, Hussain et al. reported the synthesis of C2-branched glycals from commercially available glycals and cyclohexenones via palladium-catalyzed C–H activation.[80] Treating glucal with cyclohexanones in the presence of palladium acetate and silver acetate in DMF:DMSO at 80°C for 24 hours resulted in the synthesis of corresponding C2-functionalized product with excellent yields obtained (Scheme 10.38). A library of compounds was synthesized under optimized conditions by using different protected glycals and a variety of cycloalkenones. Further, postmodification of the synthesized C2-branched glycals underwent cyclization upon treatment with benzyne via cyclo addition with concomitant ring opening of sugar moiety leading to the synthesis of biologically important chiral phenanthrenes.

10.2.8 One-pot C1 and C2 functionalization of pyranoid glycals

Apart from the exclusive functionalization of the C1 and C2 positions of glycals, various methods have been reported wherein both positions can be functionalized simultaneously despite being reversal to each other in terms

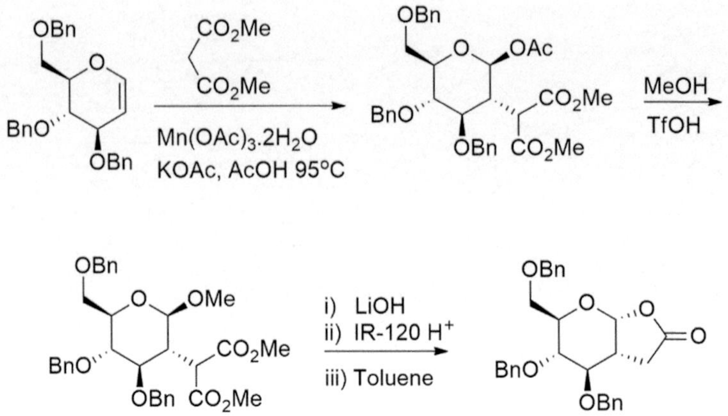

SCHEME 10.39 Linker method for the synthesis of a butyrolactone from glycals.

of reactivity. The preceding discussion will briefly elaborate on various transformations carried out on glycals to generate diverse carbohydrate motifs. Many natural products contain sugar-based γ-butyrolactone as an integral part with promising biological activity such as antimicrobial,[81] cytotoxic,[82,83] fungicidal,[84] and selective insecticidal.[85] Toward this endeavor, Linker and coworkers utilized glycals to access carbohydrate 1,2-lactones via multistep synthesis with lower yields obtained overall[86] (Scheme 10.39). Further, the reaction is not accessible to ester-protected glycals as basic hydrolysis of methyl ester is required. To address the shortcomings associated with this method, Mukherjee et al. reported a one-step synthesis of butyrolactone from glycals with complete regio- and stereocontrol via manganese(III) acetate-mediated radical lactonization[87] (Scheme 10.40A). The product formation was confirmed by spectral analysis with the disappearance of the olefinic bond and the appearance of the methylene group at δ2.54 and 2.40 in proton NMR and 40.6 ppm in carbon NMR. Various glycals with different protecting groups like silyl, benzyl, and ester proceeded smoothly to generate desired 1,2-lactones. Similar reaction conditions were applied to pseudo-glycals to generate 2,3-lactones (Scheme 10.40B).

Mukherjee et al. reported diastereoselective 1,2-dichlorination of glycals by using eco-friendly reagents with excellent selectivity obtained.[88] Treating D-glucal with *N*-chlorosuccinimide and triphenylphosphine in DCM at room temperature for the appropriate time yielded 1,2-dichlorinated products with excellent diastereoselectively in good yield (Scheme 10.41). Screening of reaction conditions like decrease in reaction time, change in temperature, and use of different reagent systems resulted in the formation of the desired product in excellent yield. With optimized reaction conditions in hand, the substrate scope was varied by using different glycals and good-to-excellent yields, regardless of the nature of the protecting group on glycals was obtained.

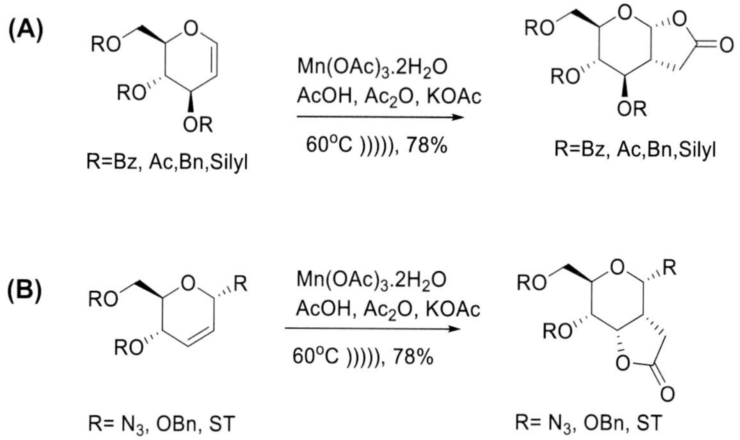

SCHEME 10.40 (A) $Mn(OAc)_3$-mediated synthesis of 1,2 lactones from glycals and (B) pseudoglycals from 2,3 lactones.

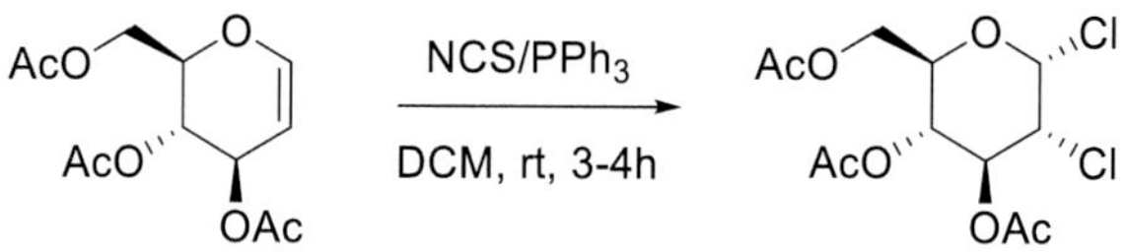

SCHEME 10.41 Synthesis of halogenated sugar.

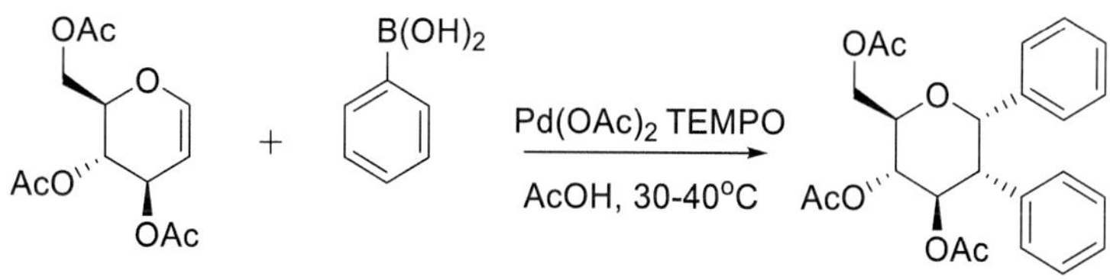

SCHEME 10.42 TEMPO-promoted diarylation of glycals using aryl-boronic acids.

Other than well exploration of aryl-boronic acids as a coupling partner for synthesizing mono-arylated aryl glycosides, the same group reported the synthesis of di-aryl-glycosides via Domino Heck Suzuki reaction using aryl-boronic acids and glycals.[89] Treating phenylboronic acid with glucal triacetate in the presence of the catalytic amount of palladium acetate and a stoichiometric amount of 2,2,6,6-tetramethylpiperidine 1-oxyl (TEMPO) in acetic acid furnished the diarylated-*C*-glycoside product in excellent yield (Scheme 10.42). The stereochemistry was found to be 1,2-α,α in the case of glycals and 2,3-β,β in the case of pseudoglycals. A series of boronic acids having different electronic and steric properties were reacted under the optimized condition, and good yields of the desired product with excellent diastereoselectivities were

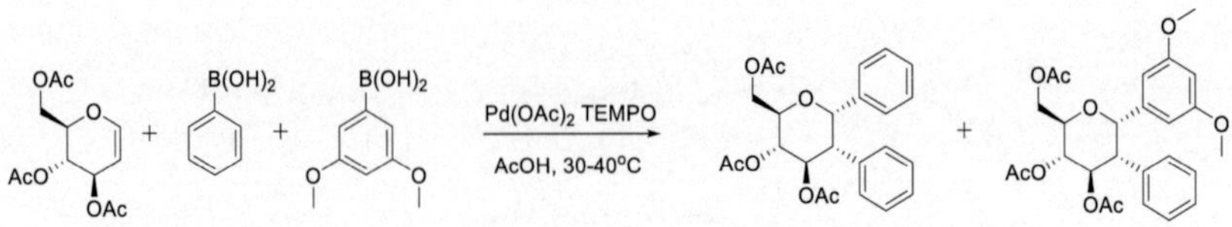

SCHEME 10.43 TEMPO-promoted diarylation of glycals using different aryl-boronic acids.

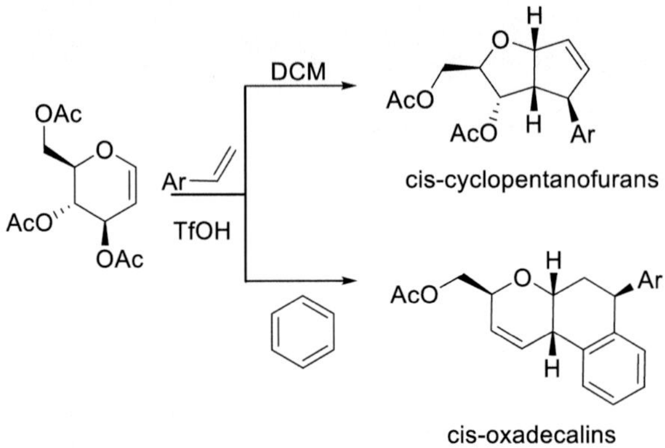

SCHEME 10.44 Synthesis of fused oxabicyclic scaffolds.

observed. Next, the authors focused on the regioselective synthesis of diarylated products by using two different boronic acids having different reactivities resulting in the synthesis of diarylated glycosides with complete stereocontrol albeit in lower yield (Scheme 10.43). The regio- and stereochemistry of the diarylated products was confirmed by 2D NMR studies.

The same group reported the synthesis of oxabicyclic scaffolds via an unprecedented domino reaction carried under acidic conditions between glycals and styrene.[90] It is pertinent to mention that the nature of solvent, temperature, and concentration of triflic acid plays a key role in product formation. Using 2 equivalents of triflic acid in nonparticipating solvent DCM at 0°C resulted in the synthesis of *cis*-cyclopentanofurans via unprecedented ring-opening and ring-closing sequence, whereas chiral benzo-fused oxadecalins were obtained exclusively upon using a participating solvent like benzene using 30 mol% triflic acid at 40°C via cascade Ferrier *C*-glycosylation and double Friedel–Crafts reaction (Scheme 10.44). Substrate scope was explored using different glycals, styrene, and aromatic solvents yielding the desired product formation in moderate yields with high stereoselectivity. Control and isotopic labeling experiments were performed to propose the plausible mechanism of the reaction further revealing the glycosylation step to be the slowest step involved in these cascade reactions.

Very recently, the same group reported nickel-catalyzed domino synthesis of *cis*-fused dihydro-benzofurans and indoles from glycals and *o*-iodophenols/

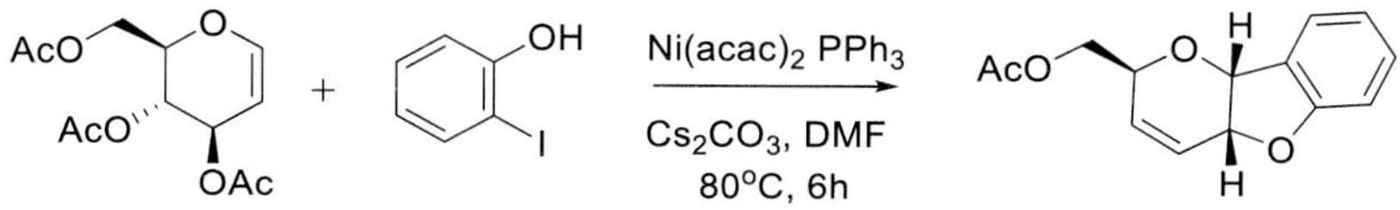

SCHEME 10.45 Ni-catalyzed synthesis of *cis*-fused dihydro-benzofurans/indoles.

aniline.[91] Screening the reaction condition showed the desired products were formed in good yields upon using $Ni(acac)_2$ as a catalyst, triphenylphosphine as a ligand, and cesium carbonate as a base under heating conditions (Scheme 10.45). A library of compounds was synthesized under the optimized reaction condition. Control experiments were performed to know the mechanistic studies involved in the reaction wherein it was found that initial Heck-type C–C formation takes place followed by cyclization under basic conditions via S_N^2 reaction.

10.3 Synthesis and reactivity of furanoid glycals

10.3.1 Synthesis of furanoid glycals

Ness and Fletcher reported the first synthesis of furanoid glycals in 1963 from commercially available D-Ribose (Scheme 10.46A).[92] However, the yield of the final furanoid glycal was very low (43%) as many aromatic side products were also formed during the course of the reaction. To overcome this, the same group reported the formation of aromatic by-products during the synthesis of furanoid glycals can be minimized by the presence of a better-protecting group with higher nucleophilicity. They reported the synthesis of furanoid glycals from D-Ribose having anisoyl protection at the C3 and C5 positions[93] (Scheme 10.46B).

In 1977 Jordaan et al. reported the synthesis of furanoid glycals from suitably protected furanosyl bromides via reduction using sodium or potassium in aprotic solvents[94] (Scheme 10.47A). The same group reported an alternative route for the synthesis of furanoid glycal by treating furanosyl halide with sodium naphthalide in THF to avoid the use of metals[95] (Scheme 10.47B).

In 1978 Ireland et al. reported the synthesis of 3-hydroxylated furanoid glycals from D-erythronolactone and D-ribonic acid-γ-lactone via reductive fragmentation of 2,3-*O*-isopropylidene-protected furanosyl chloride[96] (Scheme 10.48).

Schlosser et al. reported the synthesis of furan glycal from D-mannose in 10-step synthesis. The key step in the synthesis involved anomeric debenzylation using Li/liq NH_3 followed by chlorination using PPh_3 and CCl_4 and reduction with Li/liq NH_3 (Scheme 10.49).[97]

Synthesis of threo- and erythro-furanoid glycals from 2-deoxyribose via selenoxide-based elimination was reported by Kassou and Castillon in

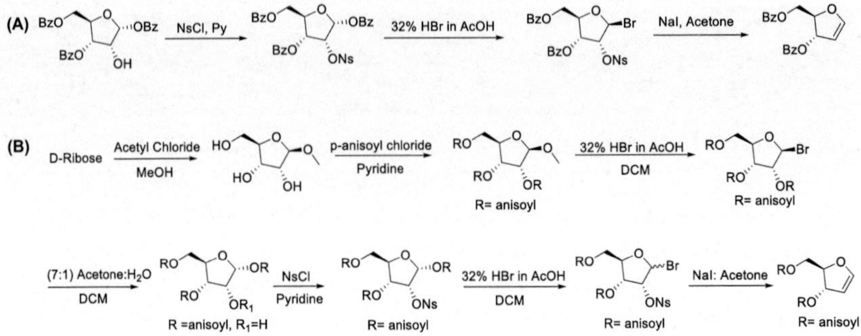

SCHEME 10.46 (A) Ness et al. synthesis of furanoid glycal (B) modified route.

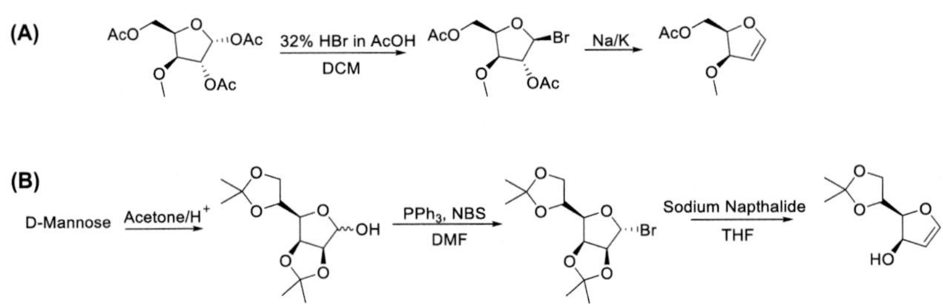

SCHEME 10.47 (A) Jordaan et al. synthesis of furanoid glycals; (B) alternative route.

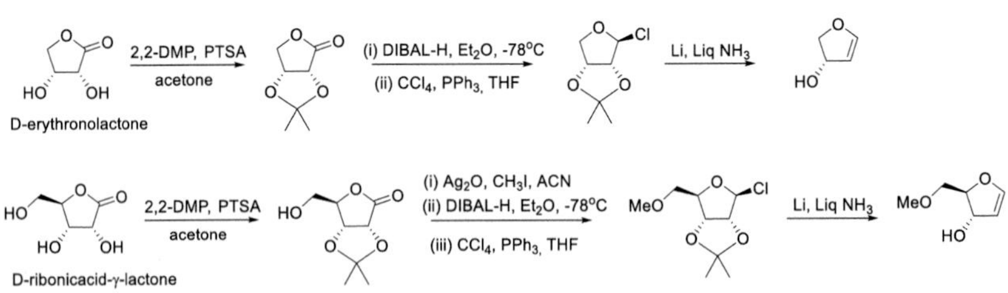

SCHEME 10.48 Ireland et al. synthesis of 3-hydroxylated furanoid glycal.

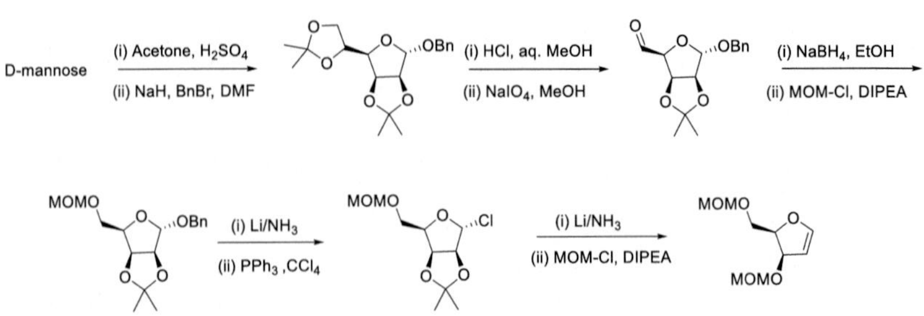

SCHEME 10.49 Schlosser route for synthesis of furanoid glycal.

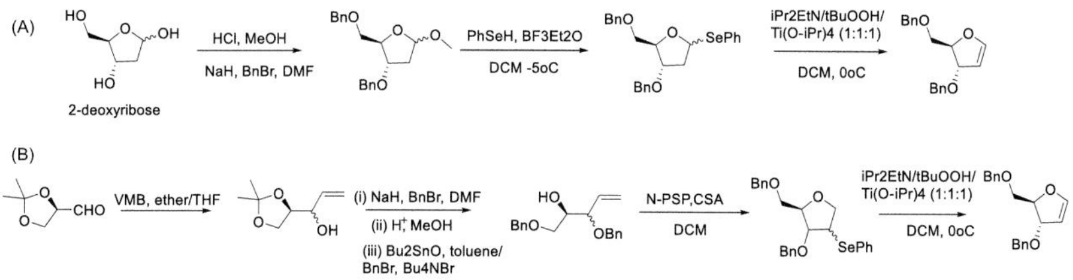

SCHEME 10.50 (A) Synthesis of furanoid glycals from 2-deoxyribose. (B) Synthesis of furanoid glycals from pentene 1,2,3-triol.

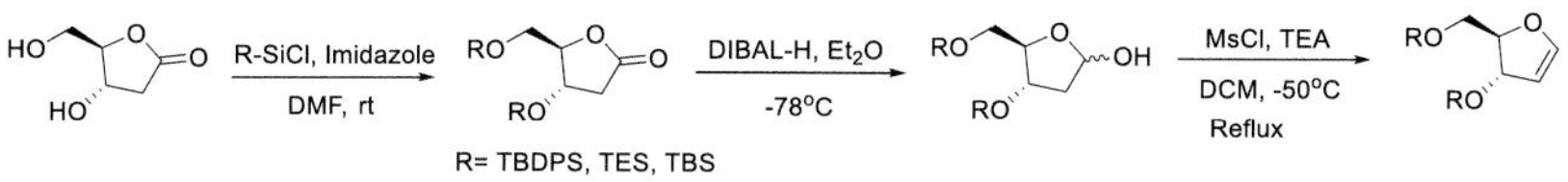

SCHEME 10.51 Townsend route for furanoid glycal.

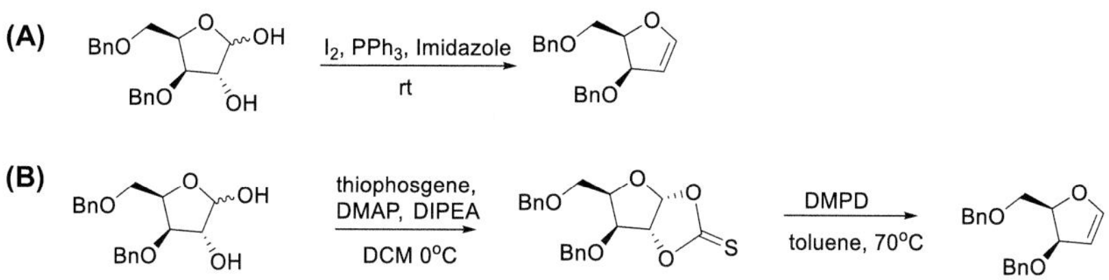

SCHEME 10.52 (A) Synthesis of furanoid glycals using Garegg–Sameuelsson reaction. (B) Synthesis of furanoid glycals from 1,2-thiocarbonates.

1994[98] (Scheme 10.50A). The same group applied this methodology for the synthesis of differentially protected erythro- and threo-furanoid glycals from 4-pentene 1,2,3-triol. The key step in the whole process involves selenium induced 5-endo trig cyclization followed by selenoxide elimination[94] (Scheme 10.50B).

In 1996 Townsend et al. reported gram-scale synthesis of ribofuranoid glycals from 2-deoxy-D-ribono-1,4-lactone in five steps via sequential protection, reduction, and in situ elimination of methane sulfonic acid[99] (Scheme 10.51).

Diaz et al. in 1997 reported the synthesis of furanoid glycals from vicinal diols. Vicinal diols on reaction with I_2-PPh_3-Im gave furanoid glycal via Garegg–Sameuelsson reaction[100] (Scheme 10.52A). The same group reported the conversion of vicinal diols into 1,2-thiocarbonates, which upon reaction with dimethyl-4-phenylenediamine (DMPD) undergoes elimination to generate furanoid glycals (Scheme 10.52B).

Recently, Lopez et al. reported the synthesis of furanoid glycals via controlled oxidation of thio-glycoside with per benzoic acid followed by butyl lithium-mediated elimination[101] (Scheme 10.53).

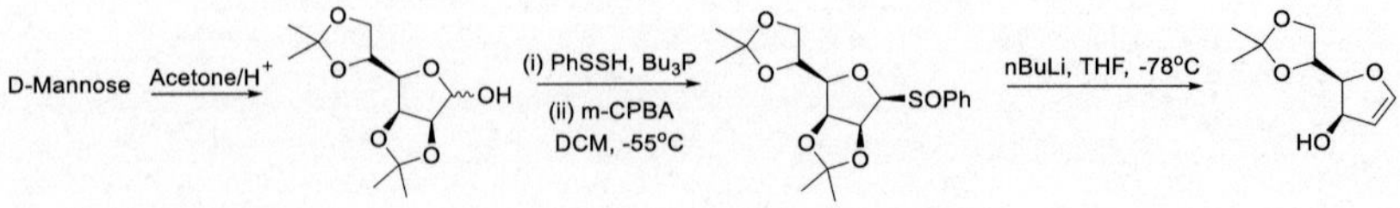

SCHEME 10.53 Lopez route for the synthesis of furanoid glycal.

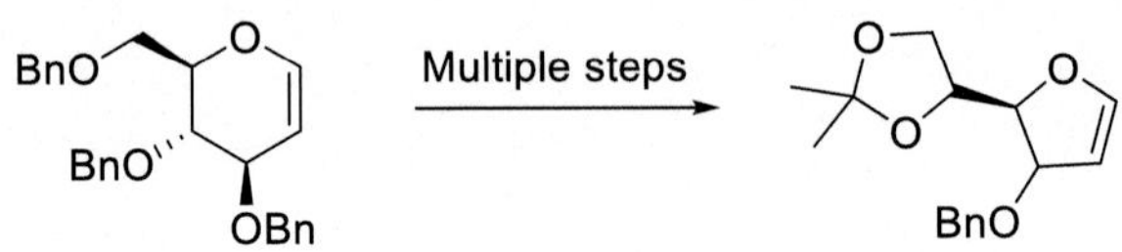

SCHEME 10.54 Synthesis of furanoid glycals from pyranoid glycals.

More recently, Shaw et al. developed a simple protocol for the synthesis of stereochemically pure furanoid glycals from commercially available pyranoid glycal[102] (Scheme 10.54).

10.3.2 Synthesis of natural products from furanoid glycals

Chiron approach synthesis has gained tremendous attention for the synthesis of complex natural products and target molecules for drug designing. Toward this endeavor, carbohydrates particularly furanoid glycals have been used as an important synthon in the total synthesis of complex natural products and nucleosides due to the presence of well-defined stereochemistry and inexpensive starting material. Ireland et al. reported the total synthesis of Lasalocid A from furan-based glycal as a key starting material (Fig. 10.7A).[103] Schlosser and coworkers reported the synthesis of erythro- and threo-sphingoshine from furanoid glycals (Fig. 10.7B and C).[97] Correia ad coworkers reported the synthesis of (−)-isoaltholactone from furanoid glycals derived from L-glutamic acid in seven steps with 25% over all yield. The key step in the synthesis involved Heck–Matsuda arylation (Fig. 10.7D).[104]

EJ Corey et al. reported the synthesis of 6-epi-leukotriene C and D from mannose-derived furanoid glycal (Fig. 10.8A).[105] The same glycal was utilized by Chida et al. for the synthesis of (+)-myriocine (Fig. 10.8B).[106] Chandrasekharan et al. documented the synthesis of fused perhydrofuro[2,3-β]pyrano/furano-γ-butyrolactone from glycals via NIS-mediated ring opening/closing of 1,2-cycloproponated sugars Fig. 10.8C.[107] Both isomers of exo-Brevicomin (Fig. 10.8D and E) and endo-Brevicomin (Fig. 10.8F and G) were synthesized from C3-protected furanoid glycal by Shaw and coworkers. The key intermediate olefin synthesized during the reaction sequence was utilized for the synthesis of both isomers of cardiobutanolide (Fig. 10.8H and I) and Goniofufurone (Fig. 10.8J).[108]

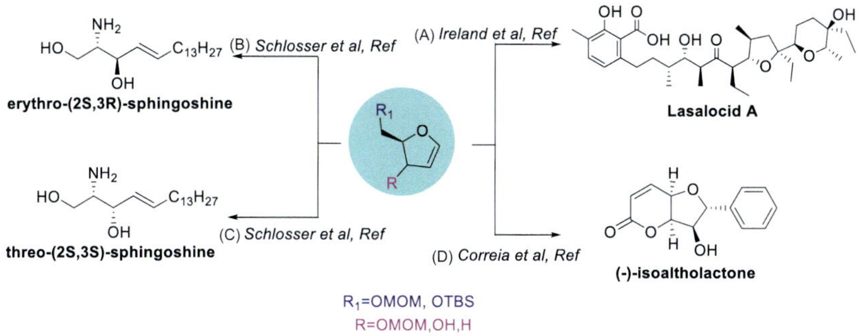

FIGURE 10.7 Synthesis of natural products from furanoid glycals.

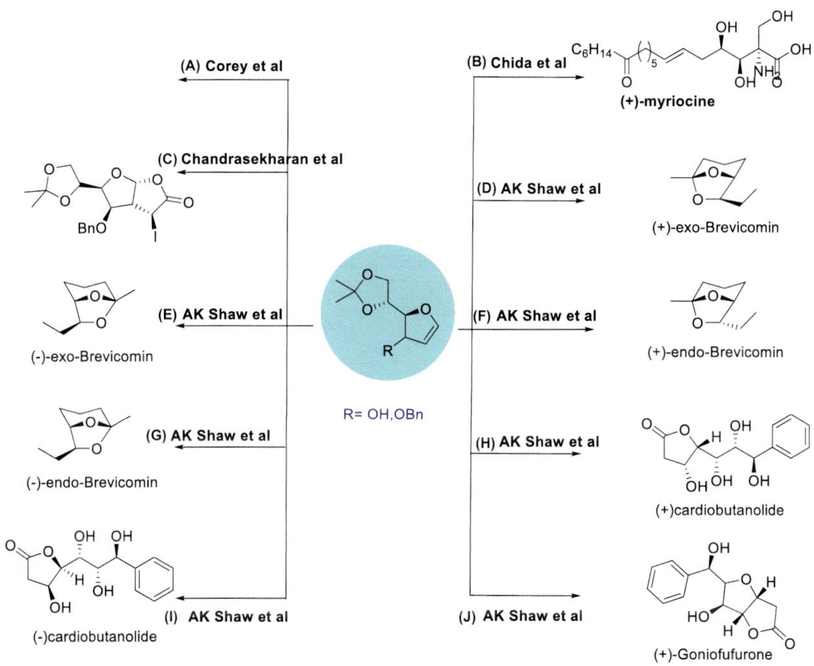

FIGURE 10.8 Synthesis of natural products from mannose-derived furanoid glycals.

10.3.3 Synthesis of nucleosides from furanoid glycals

Apart from their application in the total synthesis of natural products, furanoid glycals have been very well explored for the synthesis of nucleosides. The preceding discussion will first discuss the synthesis of *C*-nucleosides followed by *N*-nucleosides from various furanoid glycals. In 1983 Daves et al. reported *C*-nucleoside synthesis via a coupling reaction between furanoid glycal and preactivated nucleobase using a stoichiometric amount of palladium acetate (Scheme 10.55A).[109] In 1992 the same group reported the

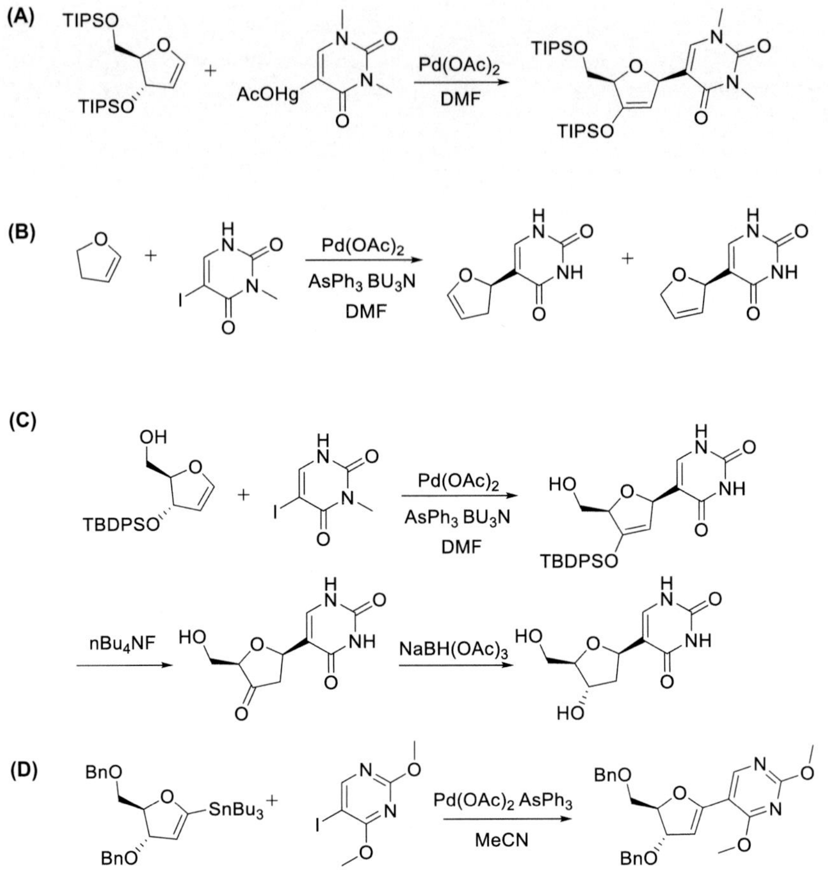

SCHEME 10.55 Synthesis of *C*-nucleosides.

synthesis of nucleosides via palladium-mediated coupling between furanoid glycals and iodo derivative of nucleobase using triphenyl arsine and tributyl amine (Scheme 10.55B).[110] Further, they were successful in synthesizing 2-deoxy pseudo-uridine in three steps with 63% over all yield obtained (Scheme 10.55C). In another report, they were successful in the synthesizing *C*-nucleoside via palladium-mediated coupling between iodoaglycon and stannylated furanoid glycal prepared by lithiation of corresponding furan glycal and n butyl tin chloride[111] (Scheme 10.55D).

Townsend et al. reported the synthesis of stereospecific pyrazine *C*-nucleoside via Pd(0) mediated cross-coupling reaction carried out between iodo aglycon and furanoid glycal[112] (Scheme 10.56).

Mclaughlin et al. reported the synthesis of pyrimidine-based *C*-nucleoside via regio- and stereospecific Heck reaction between furanoid glycals and pyridine heterocycles using $Pd(dba)_2$ and $(C_6F_5)_3P$ in ACN resulting in β-*C*-nucleosides[113] (Scheme 10.57).

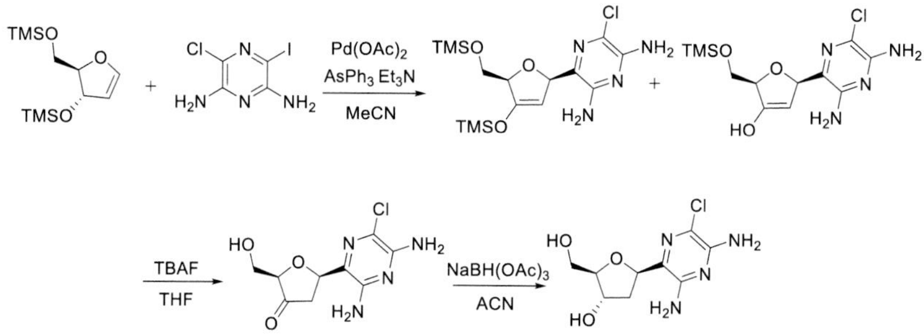

SCHEME 10.56 Palladium-mediated *C*-nucleoside synthesis.

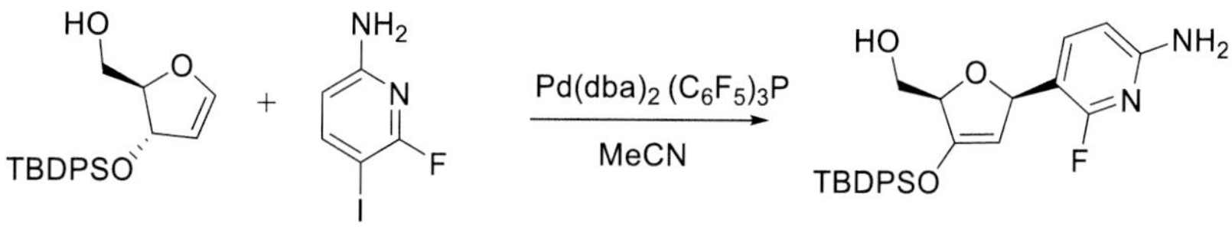

SCHEME 10.57 Synthesis of *C*-nucleoside via Heck coupling.

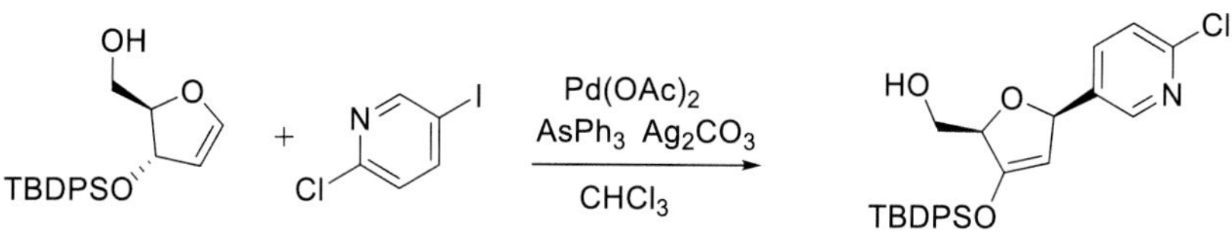

SCHEME 10.58 Palladium-catalyzed *C*-nucleoside synthesis.

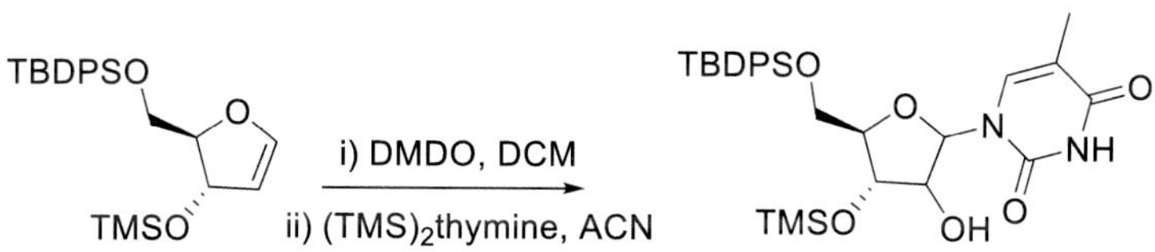

SCHEME 10.59 *N*-Nucleoside synthesis from 1,2-anhydro sugars.

Recently, Hocek et al. reported the synthesis of 6-substituted pyridine-3-yl *C*-nucleosides via a Heck reaction between furanoid glycal and 2-chloro-5-iodopyridine using palladium acetate, arsenic triphenylphosphine, and silver carbonate in chloroform[114] (Scheme 10.58).

In 1990 Danishefsky et al. reported the stereoselective synthesis of *N*-nucleosides via a Vorbrüggen-type reaction.[115] Furanoid epoxide was synthesized stereoselectively via in situ generation of dimethyl dioxirane using a biphasic system of sodium bicarbonate and oxone. Subsequent treatment of silylated nucleobase resulted in the synthesis of *N*-nucleoside. Noteworthy from α-epoxides, β-nucleosides were synthesized selectively and vice versa (Scheme 10.59).

Liotta et al. reported the synthesis of purine and pyrimidine-based nucleosides from furanoid glycals prepared in a straightforward fashion from the corresponding lactone via the appropriate addition of an electrophilic sulfur reagent. Subsequent addition of silylated nucleobase in the presence of tin chloride led to the synthesis of corresponding nucleoside in a regio- and stereoselective fashion with good yields obtained[116] (Scheme 10.60).

In 1993 Castillon et al. reported the synthesis of 2-deoxy nucleosides from furanoid glycals in sequential three-step synthesis involving the treatment of glycal with selenium reagents followed by glycosylation using silver triflate as an activator and hydrogenolysis using tributyltin hydride under refluxing condition[117] (Scheme 10.61).

In the same year, Robles et al. utilized furanoid glycals for the synthesis of nucleosides via *N*-iodo succinimide-mediated glycosylation.[118] The reaction proceeds through in situ formation of cyclic iodonium ion followed by stereoselective *trans* opening (Scheme 10.62).

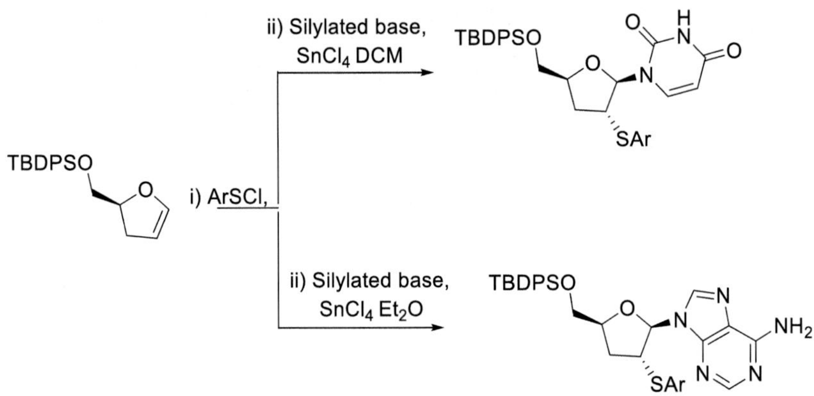

SCHEME 10.60 *N*-Nucleoside synthesis via the addition of sulfur electrophiles to furanoid glycals.

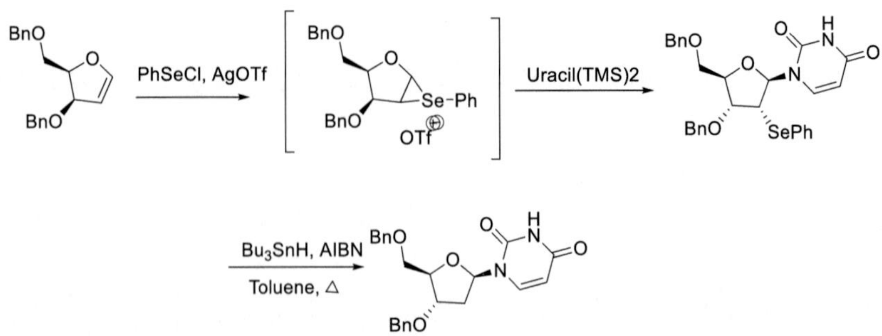

SCHEME 10.61 Synthesis of 2-deoxy nucleosides from furanoid glycals.

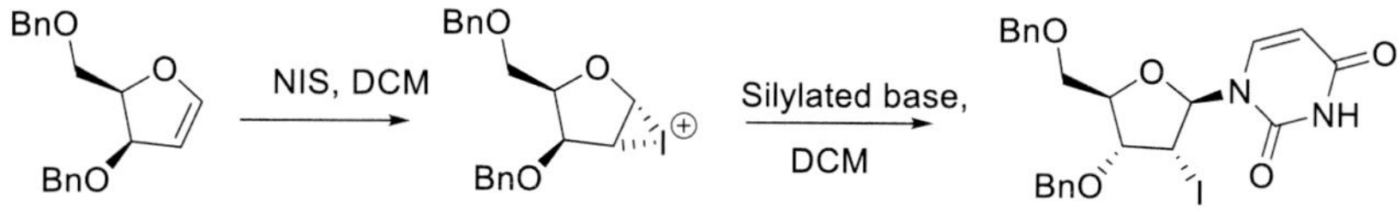

SCHEME 10.62 NIS-mediated *N*-nucleoside synthesis.

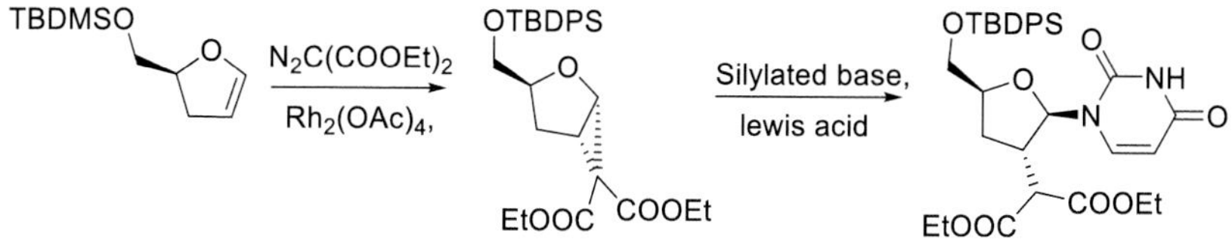

SCHEME 10.63 Rhodium-catalyzed *N*-nucleoside synthesis.

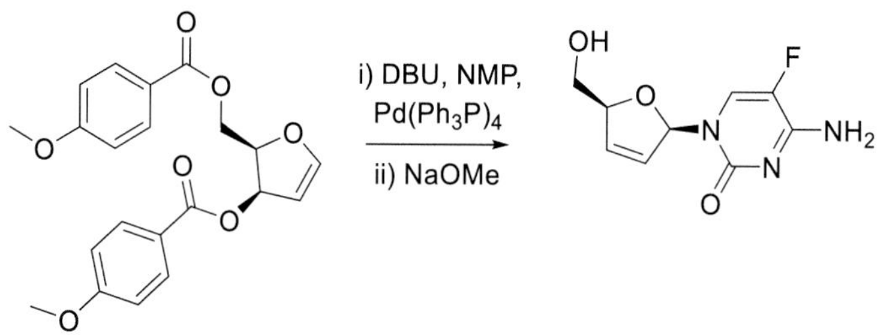

SCHEME 10.64 *N*-Nucleoside synthesis via Ferrier-type allylic coupling.

In 1999 Kim et al. reported rhodium-catalyzed cyclopropanation of 3-deoxy furanoid glycals using diethyl diazomalonate and utilized further for the synthesis of nucleosides via Lewis acid-mediated glycosylation using silylated nucleobases[119] (Scheme 10.63).

In 2005 Choudhury and Pierce et al. reported palladium-catalyzed nucleoside synthesis.[120] Screening the reaction condition revealed Ferrier-type allylic coupling took place between glycals and unprotected nucleobase upon using a catalytic amount of tetrakis-triphenyl phosphine palladium and *N*-methylpyrrolidone (NMP). Noteworthy the reaction occurred in the absence of base as well; however, low yields were obtained as aromatization of glycal occurred, which was circumscribed by performing the reaction under basic conditions (Scheme 10.64).

10.3.4 Synthesis of *C*-aryl glycosides from furanoid glycals

In addition to this, furanoid glycals have been utilized efficiently for synthesizing *C*-arylated glycosides. In this context, Daves et al. reported the regio- and stereoselective *C*-aryl synthesis via palladium-mediated coupling

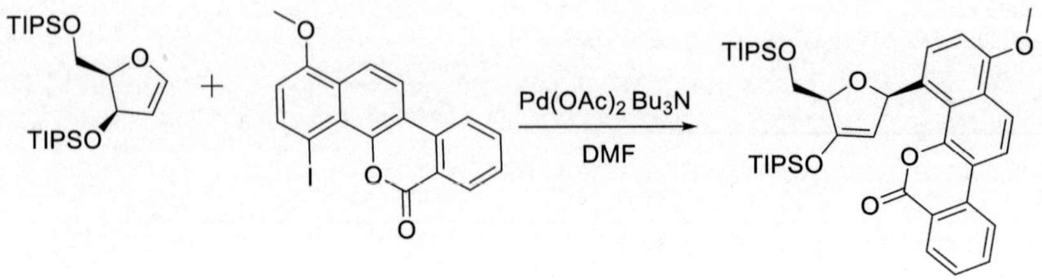

SCHEME 10.65 Palladium-catalyzed β-aryl glycoside synthesis.

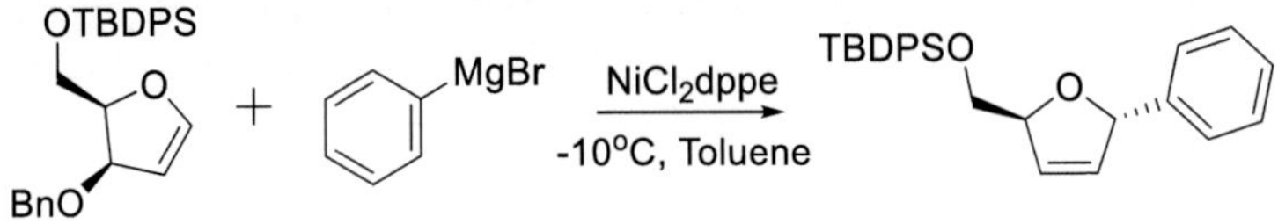

SCHEME 10.66 Nickel-catalyzed α-aryl glycoside synthesis.

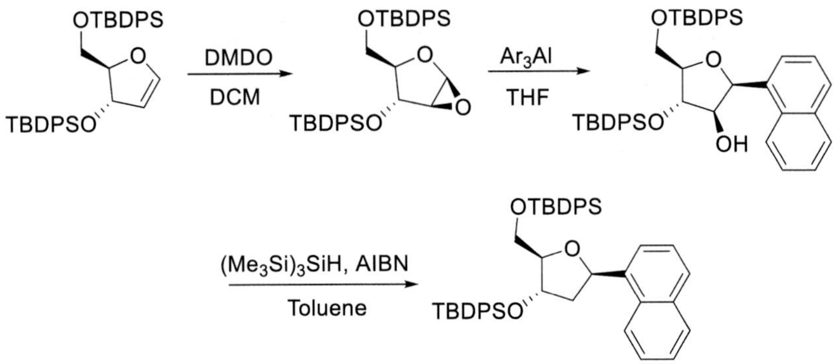

SCHEME 10.67 Synthesis of β-aryl 2-deoxy glycoside.

reaction of iodo anthracycline derivative and furanoid glycal using tertiary amine in DMF (Scheme 10.65).[121]

Tingoli et al. reported the synthesis of α-aryl glycosides via coupling of furanoid glycals with a slight excess of aryl Grignard reagent using low variant Ni(0) catalyst at lower temperature[122] (Scheme 10.66). Noteworthy temperature plays a key role in deciding the fate of the reaction as a slight increase in temperature leads to the formation of ring-opened products or degradation of stating material is observed. Moreover, the reduction of nickel catalyst is important for the successful formation of aryl *C*-glycoside.

Seitz et al. reported the synthesis of β-aryl 2-deoxy glycosides in three steps from suitably protected furanoid glycals in three steps viz epoxidation, glycosylation, and deoxygenation.[123] The key step in the synthesis involved the stereoselective addition of aryl-aluminum reagent to glycal epoxide followed by deoxygenation of β-hydroxy under standard Barton conditions (Scheme 10.67).

10.4 Conclusions and future outlook

In summary, this chapter discussed the synthesis of pyranoid and furanoid glycals and their reactivity toward the diverse-oriented synthesis. This chapter has discussed various synthetic routes for synthesizing glycal and their reactivity toward the generation of *C*-arylation, various *C*/*N*/*O*-glycosides, *C*/*N*-nucleosides, and natural products. Initially, pyranoid glycals are discussed in detail followed by furanoid glycals. The idea originated from the pioneering contribution by Emil Fischer for vinyl ethers that were afterward popularized by Ferrier.

References

1. Kinfe, H. H. *Org. Biomol. Chem.* **2019,** *17*, 4153–4182.
2. Danishefsky, S. J.; Bilodeau, M. T. *Angew. Chem. Int. Edn.* **1996,** *35*, 1380–1419.
3. Somsák, L. *Chem. Rev.* **2001,** *101*, 81–136.
4. Fischer, E. *Untersuchungen Über Kohlenhydrate und Fermente (1884–1908)* **1909,** 355–361.
5. Cavallaro, C. L.; Schwartz, J.; Scheme, A. *Glycoscience* **2008,** 699. Available from: https://doi.org/10.1007/978-3-540-30429-6_15.
6. Lancelin, J.-M.; Morin-Allory, L.; Sinay, P. *Chem. Communi* **1984,** *6*, 355–356.
7. De Pouilly, P.; Chbdi, A.; Mallet, J.-M.; Sirmy, P. *Tetrahedron Lett.* **1992,** *33*, 8065–8068.
8. Jones, G. S.; Scott, W. J. *J. Am. Chem. Soc.* **1992,** *114*, 1491–1492.
9. Liu, F.; Huang, H.; Sun, L.; Yan, Z.; Tan, X.; Li, J.; Luo, X.; Ding, H.; Xiao, Q. *Chem. Sci.* **2022,** *13*, 5588–5596.
10. Bililign, T.; Griffith, B. R.; Thorson, J. S. *Nat. Prod. Rep.* **2005,** *22*, 742–760.
11. Hwa Kim, J.; Chun LeC, B.; Hui Kim, J.; Sub Sim, G.; Hwan LeC, D.; Eun LeC, K.; Pyo Yun, Y.; Bae PyC, H. *Arch. Pharmacal. Res.* **2005,** *28*, 195–202.
12. Zhang, Y.; Jiao, J.; Liu, C.; Wu, X.; Zhang, Y. *Food Chem.* **2008,** *107*, 1326–1336.
13. Magaji, M. G.; Musa, A. M.; Abdullahi, M. I.; Ya'u, J.; Hussaini, I. M. *Avicenna J. Phytomed.* **2015,** *5*, 587.
14. Denmark, S. E.; Regens, C. S.; Kobayashi, T. *J. Am. Chem. Soc.* **2007,** *129*, 2774–2776.
15. Aguillon, A. R.; Mascarello, A.; Segretti, N. D.; de Azevedo, H. F. Z.; Guimaraes, C. R.; Miranda, L. S.; de Souza, R. O. *Org. Proc. Res. Dev.* **2018,** *22*, 467–488.
16. Bai, Y.; Kim, L. M. H.; Liao, H.; Liu, X. W. *J. Org. Chem.* **2013,** *78*, 8821–8825.
17. Ramnauth, J.; Poulin, O.; Rakhit, S.; Maddaford, S. P. *Org. Lett.* **2001,** *3*, 2013–2015.
18. Furuta, T.; Nakayama, M.; Suzuki, H.; Tajimi, H.; Inai, M.; Nukaya, H.; Wakimoto, T.; Kan, T. *Org. Lett.* **2009,** *11*, 2233–2236.
19. Lei, M.; Gao, L.; Yang, J. S. *Tetrahedron Lett.* **2009,** *50*, 5135–5138.
20. Li, H. H.; Ye, X. S. *Org. Biomol. Chem.* **2009,** *7*, 3855–3861.
21. Ramnauth, J.; Poulin, O.; Bratovanov, S. S.; Rakhit, S.; Maddaford, S. P. *Org. Lett.* **2001,** *3*, 2571–2573.
22. Xiang, S.; Cai, S.; Zeng, J.; Liu, X. W. *Org. Lett.* **2011,** *13*, 4608–4611.
23. De La Figuera, N.; Forns, P.; Fernàndez, J. C.; Fiol, S.; Fernández-Forner, D.; Albericio. *Tetrahedron Lett.* **2005,** *46*, 7271–7274.
24. Liu, C. F.; Xiong, D. C.; Ye, X. S. *J. Org. Chem.* **2014,** *79*, 4676–4686.
25. Kusunuru, A. K.; Yousuf, S. K.; Tatina, M.; Mukherjee, D. *Eur. J. Org. Chem.* **2015,** *3*, 459–462.

26. Xiong, D. C.; Zhang, L. H.; Ye, X. S. *Org. Lett.* **2009,** *11*, 1709–1712.
27. Halcomb, R. L.; Danishefsky, S. J. *J. Am. Chem. Soc.* **1989,** *111*, 6661–6666.
28. Plancq, B.; Lafantaisie, M.; Companys, S.; Maroun, C.; Ollevier, T. *Org. Biomol. Chem.* **2013,** *11*, 7463–7466.
29. Lubin-Germain, N.; Hallonet, A.; Huguenot, F.; Palmier, S.; Uziel, J.; Augé, J. *Org. Lett.* **2007,** *9*, 3679–3682.
30. Vieira, A. S.; Fiorante, P. F.; Hough, T. L.; Ferreira, F. P.; Ludtke, D. S.; Stefani, H. A. *Org. Lett.* **2008,** *10*, 5215–5218.
31. Vedachalam, S.; Tan, S. M.; Teo, H. P.; Cai, S.; Liu, X. W. *Org. Lett.* **2012,** *14*, 174–177.
32. Kusunuru, A. K.; Tatina, M.; Yousuf, S. K.; Mukherjee, D. *Chem. Commun.* **2013,** *49*, 10154–10156.
33. Mandoli, A. *Molecules* **2016,** *21*, 1174.
34. Tatina, M.; Kusunuru, A. K.; Yousuf, S. K.; Mukherjee, D. *Chem. Commun.* **2013,** *49*, 11409–11411.
35. Huang, N.; Liao, H.; Yao, H.; Xie, T.; Zhang, S.; Zou, K.; Liu, X. W. *Org. Lett.* **2018,** *20*, 16–19.
36. Rasool, F.; Ahmed, A.; Hussain, N.; Yousuf, S. K.; Mukherjee, D. *Org. Lett.* **2018,** *20*, 4036–4039.
37. Paoletta, S.; Sabbadin, D.; Von Kügelgen, I.; Hinz, S.; Katritch, V.; Hoffmann, K.; Abdelrahman, A.; Straßburger, J.; Baqi, Y.; Zhao, Q., et al. *Comput. Mol. Des.* **2015,** *29*, 737–756.
38. Kilfoil, P. J.; Chapalamadugu, K. C.; Hu, X.; Zhang, D.; Raucci, F. J.; Tur, J.; Brittian, K. R.; Jones, S. P.; Bhatnagar, A.; Tipparaju, S. M., et al. *Mol. Cell. Cardiol.* **2019,** *137*, 93–106.
39. Chen, J.; Zhao, Y.; Liu, Y. *Am. J. Respir. Crit. Care Med.* **2010,** *181*, 928–934.
40. McGuigan, C.; Hinsinger, K.; Farleigh, L.; Pathirana, R. N.; Bugert, J. *J. J. Med. Chem.* **2013,** *56*, 1311–1322.
41. Peng, L.-P.; Nagarajan, S.; Rasheed, S.; Zhou, C.-H. *RSC Adv.* **2015,** *5*, 51961–51965.
42. Ramesh, D.; Vijayakumar, B. G.; Kannan, T. *Eur. J. Med. Chem.* **2020,** *207*, 112801.
43. Jordheim, L. P.; Durantel, D.; Zoulim, F.; Dumontet, C. *Nat. Rev. Drug. Disc* **2013,** *12*, 447–464.
44. Ramesh, D.; Vijayakumar, B. G.; Kannan, T. *ChemMedChem* **2021,** *16*, 1403–1419.
45. Rasool, F.; Mukherjee, D. *Org. Lett.* **2017,** *19*, 4936–4939.
46. Ahmed, A.; Rasool, F.; Singh, G.; Katoch, M.; Mukherjee, D. *Eur. J. Org. Chem.* **2020,** *28*, 4408–4416.
47. Rossi, L. L.; Basu, A. *Bioorg. Med. Chem. Lett.* **2005,** *15*, 3596–3599.
48. Suthagar, K.; Fairbanks, A. J. *Org. Biomol. Chem.* **2016,** *14*, 1748–1754.
49. Norris, P. *Curr. Top. Med. Chem.* **2008,** *8*, 101–113.
50. Zheng, J.; Urkalan, K. B.; Herzon, S. B. *Angew. Chem. Int. Edn.* **2013,** *52*, 6068–6071.
51. Magkus, P. D.; Christopher, E.; Sear, N. L. *Tetrahedron* **1983,** *39* (22), 3725–3729.
52. Leclerc, S.; Garnier, M.; Hoessel, R.; Marko, D.; Bibb, J. A.; Snyder, G. L.; Greengard, P.; Biernat, J.; Wu, Y. Z.; Mandelkow, E. M., et al. *J. Biol. Chem.* **2001,** *276*, 251–260.
53. Sassatelli, M.; Bouchikhi, F.; Messaoudi, S.; Anizon, F.; Debiton, E.; Barthomeuf, C.; Prudhomme, M.; Moreau, P. *Eur. J. Med. Chem.* **2006,** *41*, 88–100.
54. Libnow, S.; Hein, M.; Michalik, D.; Langer, P. *Tetrahedron Lett.* **2006,** *47* (38), 6907–6909.
55. Pergola, C.; Gaboriaud-Kolar, N.; Jestädt, N.; König, S.; Kritsanida, M.; Schaible, A. M.; Li, H.; Garscha, U.; Weinigel, C.; Barz, D., et al. *J. Med. Chem.* **2014,** *57*, 3715–3723.
56. Colinas, P. A.; Bravo, R. D. *Org. Lett.* **2003,** *5*, 4509–4511.

57. Ji, L.; Xiang, S. H.; Leng, W. L.; Le Mai Hoang, K.; Liu, X. W. *Org. Lett.* **2015,** *17*, 1357–1360.
58. Goti, A.; Mirabella, S.; Petrucci, G.; Faggi, C.; Matassini, C.; Cardona, F. *Org. Lett.* **2020,** *22*, 9041–9046.
59. Di Bussolo, V.; Kim, Y. J.; Gin, D. Y. *J. Am. Chem. Soc.* **1998,** *120*, 13515–13516.
60. Kim, J. Y.; Di Bussolo, V.; Gin, D. Y. *Org. Lett.* **2001,** *3*, 303–306.
61. Seeberger, P. H.; Eckhardt, M.; Gutteridge, C. E.; Danishefsky, S. J. *J. Am. Chem. Soc.* **1997,** *119*, 10064–10072.
62. Bussolo, V. D.; Caselli, M.; Pineschi, M.; Crotti, P. *Org. Lett.* **2002,** *4*, 3695–3698.
63. Banoub', J.; Boullanger, P.; Lafont, D. *Org. Lett.* **2005,** *7*, 1299–1302.
64. Di Bussolo, V.; Romano, M. R.; Pineschi, M.; Crotti, P. *Org. Lett.* **2005,** *7*, 1299–1302.
65. Babu, R. S.; Zhou, M.; O'Doherty, G. A. *J. Am. Chem. Soc.* **2004,** *126*, 3428–3429.
66. Kim, H.; Men, H.; Lee, C. *J. Am. Chem. Soc.* **2004,** *126*, 1336–1337.
67. Schuff, B. P.; Mercer, G. J.; Nguyen, H. M. *Org. Lett.* **2007,** *9*, 3173–3176.
68. Xiang, S.; He, J.; Tan, Y. J.; Liu, X. W. *J. Org. Chem.* **2014,** *79*, 11473–11482.
69. Sau, A.; Galan, M. C. *Org. Lett.* **2017,** *19*, 2857–2860.
70. Rasool, F.; Bhat, A. H.; Hussain, N.; Mukherjee, D. *ChemistrySelect* **2016,** *1*, 6553–6557.
71. Palo-Nieto, C.; Sau, A.; Jeanneret, R.; Payard, P. A.; Salamé, A.; Martins-Teixeira, M. B.; Carvalho, I.; Grimaud, L.; Galan, M. C. *Org. Lett.* **2020,** *22*, 1991–1996.
72. Bai, Y.; Zeng, J.; Cai, S.; Liu, X. W. *Org. Lett.* **2011,** *13*, 4394–4397.
73. Suresh, D.; Vankar, Y. D. *Org. Lett.* **2014,** *16*, 1172–1175.
74. Shamim, A.; Vasconcelos, S. N. S.; Ali, B.; Madureira, L. S.; Zukerman-Schpector, J.; Stefani, H. A. *Tetrahedron Lett.* **2015,** *56*, 5836–5842.
75. Bordessa, A.; Ferry, A.; Lubin-Germain, N. *J. Org. Chem.* **2016,** *81*, 12459–12465.
76. Hussain, N.; Tatina, M. B.; Rasool, F.; Mukherjee, D. *Org. Biomol. Chem.* **2016,** *14*, 9989–9992.
77. Hussain, N.; Babu Tatina, M.; Mukherjee, D. *Org. Biomol. Chem.* **2018,** *16*, 2666–2677.
78. Hussain, N.; Bhardwaj, M.; Ahmed, A.; Mukherjee, D. *Org. Lett.* **2019,** *21*, 3034–3037.
79. Darbem, M. P.; Kanno, K. S.; De Oliveira, I. M.; Esteves, C. H. A.; Pimenta, D. C.; Stefani, H. A. *N. J. Chem.* **2019,** *43*, 696–699.
80. Maheshwari, M.; Pandey, R. P.; Hussain, N. *Chem. Commun.* **2023,** *59*, 627–630.
81. Husain, A.; Hasan, S. M.; Lal, S.; Alam, M. M. *Ind. J. Pharma. Sc.* **2006,** 68. Available from: https://doi.org/10.4103/0250-474X.27841.
82. Oh, S.; Jeong, I. H.; Shin, W. S.; Wang, Q.; Lee, S. *Bioorg. Med. Chem. Lett.* **2006,** *16*, 1656–1659.
83. El-Subbagh, H. I.; Abu-Zaid, S. M.; Mahran, M. A.; Badria, F. A.; Al-Obaid, A. M. *J. Med. Chem.* **2000,** *43*, 2915–2921.
84. Rauter, A. P.; Ferreira, M. J.; Font, J.; Virgili, A.; Figueredo, M.; Figueiredo, J. A.; Ismael, M. I.; Canda, T. L. *Carbohydr. Chem.* **1995,** *14*, 929–948.
85. Justino, J.; Rauter, A. P.; Canda, T.; Wilkins, R.; Matthews, E. *Pest. Manag. Sc.* **2005,** *61*, 985–990.
86. Yin, J.; Linker, T. *Angew. Chem.* **2009,** *121*, 1885–1887.
87. Yousuf, S. K.; Mukherjee, D.; Mallikharjunrao, L.; Taneja, S. C. *Org. Lett.* **2011,** *13*, 576–579.
88. Hussain, A.; Mukherjee, D. *Tetrahedron* **2014,** *70*, 1133–1139.
89. Kusunuru, A. K.; Jaladanki, C. K.; Tatina, M. B.; Bharatam, P. V.; Mukherjee, D. *Org. Lett.* **2015,** *17*, 3742–3745.
90. Bhardwaj, M.; Rasool, F.; Tatina, M. B.; Mukherjee, D. *Org. Lett.* **2019,** *21*, 3038–3042.
91. Bhardwaj, M.; Rasool, B.; Mukherjee, D. *Chem. Commun.* **2022,** *58*, 7038–7041.

92. Ness, R. K.; Hewitt, G.; Fletcher, Jr. J. *Org. Chem.* **1963,** *28*, 435–437.
93. Hagalb, M.; Ness, R. K. *J. Org. Chem.* **1965,** *30*, 158–162.
94. Stephen Eitelman, B. J.; Jordaan, A. *Chem. Commun.* **1977,** *16*, 552–553.
95. Stephen Eitelman, B. J.; Hall, R. H.; Jordaan, A.; Fischer, E.; Zach, K.; ber Kgl preuss, S.; Bischofberger, K.; Hall, R. H.; Ness, R. K.; Fletcher, H. G., et al. *J. Chem. Soc. Perkin Trans.* **1978,** *1*, 595–600.
96. Ireland, R. E.; Thaisrivongs, S.; Noel Vanier, l; Wilcox, C. S. *J. Org. Chem.* **1980,** *45*, 48–61.
97. Obayashi, M.; Schlosser, M. *Chem. Lett.* **1985,** *14*, 1715–1718.
98. Bravo, F.; Kassou, M.; Castillón, S. *Tetrahedron Lett.* **1999,** *40*, 1187–1190.
99. Walker, J. A.; Chen, J. J.; Wise, D. S.; Townsend, L. B. *J. Org. Chem.* **1996,** *61*, 2219–2221.
100. Robles Diaz, R.; Melgarejo, R.; Cubero, I.; Plaza Ldpez-Espinosa, M. T. *Carbohydr. Res.* **1997,** *300*, 375–380.
101. Gomez, A. M.; Casillas, M.; Barrio, A.; Gawel, A.; Lopez, J. C. *Eur. J. Org. Chem.* **2008,** *23*, 3933–3942.
102. Pal, P.; Kumar, B.; Shaw, A. K. *Eur. J. Org. Chem.* **2009,** *13*, 2179–2184.
103. Nader, B.; Franck, R. W.; Weinreb, S. M. *J. Am. Chem. Soc.* **1980,** *102*, 1153–1155.
104. Meira, P. R. R.; Moro, A. V.; Correia, C. R. D. *Synthesis* **2007,** *15*, 2279–2286.
105. Corey, E. J.; Goto, G. *Tetrahedron Lett.* **1980,** *21*, 3463–3466.
106. Oishi, T.; Ando, K.; Chida, N. *Chem. Commun.* **2001,** *19*, 1932–1933.
107. Haveli, S. D.; Sridhar, P. R.; Suguna, P.; Chandrasekaran, S. *Org. Lett.* **2007,** *9*, 1331–1334.
108. Pal, P.; Shaw, A. K. *Tetrahedron* **2011,** *67*, 4036–4047.
109. Hacksell, U.; Daves, G. D., Jr *J. Org. Chem.* **1983,** *48*, 2870–2876.
110. Zhang, H. C.; Daves, G. D., Jr *J. Org. Chem.* **1992,** *57*, 4690–4696.
111. Zhang, H.-C.; Brakta, M.; Davea, G. D. *Tetrahedron Lett.* **1993,** *34*, 1571–1574.
112. Chen, J. J.; Walker, J. A.; Liu, W.; Wise, D. S.; Townsend, L. B. *Tetrahedron Lett.* **1995,** *36*, 8363–8366.
113. Sun, Z.; Ahmed, S.; McLaughlin, L. W. *J. Org. Chem.* **2006,** *71*, 2922–2925.
114. Joubert, N.; Pohl, R.; Klepetářová, B.; Hocek, M. *J. Org. Chem.* **2007,** *72*, 6797–6805.
115. Chow, K.; Danishefsky, S. *J. Org. Chem.* **1990,** *55*, 4211–4214.
116. Wang, J.; Wurster, J. A.; Wilson, L. J.; Liotta, D. *Tetrahedron Lett.* **1993,** *34*, 4881–4884.
117. El, A.; Dlax, L. Y.; Casdlldn, S. *Tetrahedron Lett.* **1993,** *34*, 2821–2822.
118. Robles, R.; Rodriguez, C.; Izquierdo, I.; Plaza, M.-L. T.; Mota, A. *Tetrahedron: Asymmetry* **1997,** *8*, 2959–2965.
119. Lim, J.; Kim, Y. H. J. *Chem. Soc., Perkin Trans. 1* **1999,** *1*, 3239–3241.
120. Choudhury, A.; Pierce, M. E.; Nguyen, D.; Storace, L.; Confalone, P. N. *Tetrahedron Lett.* **2005,** *46*, 8099–8102.
121. Farr, R. N.; Outten, R. A.; Cheng, J. C. Y.; Daves, G. D., Jr *Organometallics* **1990,** *9*, 3151–3156.
122. Tingoli, M.; Panunzi, B.; Santacroce, F. *Tetrahedron Lett.* **1999,** *40*, 9329–9332.
123. Singh, I.; Seitz, O. *Org. Lett.* **2006,** *8*, 4319–4322.

Chapter 11

Synthesis of carbohydrate-containing macrocycles

Anji Chen and Guijun Wang
Department of Chemistry and Biochemistry, Old Dominion University, Norfolk, VA, United States

11.1 Introduction

Carbohydrate-based macrocyclic compounds have gained an intense interest in the scientific community due to their unique properties, drug likeness, ion complexation capacity, and as ligands for catalysis. Several excellent reviews covering different aspects of carbohydrate-based macrocycles have been reported; these include the synthesis and applications of glycomacrocycles by Xie et al.,[1] glycomacrocycles for medicinal chemistry applications[2] and biologically active sugar-containing macrocycles[3]; cyclic oligosaccharides as therapeutics[4] and the synthesis of macrocyclic drugs.[5] In this chapter, the most recent preparations and studies on carbohydrate-embedded macrocycles are reviewed, including publications from 2011 to 2021. The topics included in this chapter are divided into the following main categories including (1) natural product-based glycomacrocycles; (2) sugar-based crown ether, azacrown ether, and other crown ether-like macrocycles; (3) synthesis of macrocycles by C−C bond formatting reactions; (4) synthesis of sugar-embedded macrocycles by macrolactamization and macrolactonization; (5) synthesis of macrocycles via click or glycosylation macrocyclizations; (6) design and synthesis of macrocycles for biomedical applications, followed by a concluding remarks section for the chapter.

The general methods for the formation of macrocyclic compounds will be first reviewed, and then the detailed structure classes of macrocycles will be discussed. In the next section, we will discuss the main types of macrocycles and their preparation methods and general applications. These are divided into natural product-based and nonnatural product-based glycomacrocycles. For natural product-based glycomacrocycles, we will first review the structures of these compounds and representative synthesis will be reviewed next. For the synthetic nonnatural glycomacrocycles, the structures of the strategies for the macrocyclizations will be used to group them together.

Synthetic Strategies in Carbohydrate Chemistry. DOI: https://doi.org/10.1016/B978-0-323-91729-2.00012-4

11.2 Natural product-based glycomacrocycles

Carbohydrate-based macrocyclic natural products have a variety of biological properties. Many natural existing macrocycles containing sugar moieties as part of the cyclic structure have complex structures. In this section, the natural product-based macrocycles that contain sugar moiety in their main framework are discussed, and the compounds with sugars linked to the main skeleton of the macrocycles by glycosidic linkages or other linkages are omitted here.

11.2.1 Major classes of sugar-based naturally occurring macrocycles

Many of the naturally occurring sugar-embedded macrocycles are macrolactones or macrolides. Adding the macrocyclic structures increase the conformation rigidity of the compound. Recently, the naturally occurring constrained sugars or glycomacrocycles have been reviewed by Rodriguez et al.[6] The structures of several sugar-containing macrocycles as well as their biological activities and macrocyclization methods were discussed.[6] A majority of these natural glycomacrocycles are macrolactones, either monolactone or symmetrical dimeric lactones. Examples of several natural products are included in Fig. 11.1. These will be discussed in the next section.

The structures of five macrocyclic dilactones **1–5** related to Clemoarmanoside A were reported by Peng et al.[7] Mucorolactone **6**, a sugar glycoside-containing macrocycle, was isolated from fungus Mucor sp.[8] Cyclic steroid-containing sugar macrocycles were isolated from starfish *Echinaster luzonicus*, luzonicosides B–E (Fig. 11.2).[9] The total synthesis of luzonicosides A and D was reported by Yu et al.[10]

11.2.2 Synthesis of Clemoarmanoside A and Clemahexapetoside A and their analogs

A series of cyclic dimers of glycosylated benzoate-based macrodilactones have been isolated from folk medicinal plants; several total syntheses for these compounds have been reported and are included in the recent review by Xie et al.[1] Yu's group reported the total syntheses of the cyclic dimers of 4-(glycosyloxy)benzoates, Clemoarmanoside A and Clemahexapetoside A, as well as several cyclic trimeric, tetrameric, and pentameric macrocycles.[11] The synthesis was also included in the recent review; therefore, only one example of the synthesis of Clemoarmanoside A is given here as the highlight. Selective removal of the anomeric acetyl group from 1,2,3,4,6-penta-*O*-acetyl-D-allopyranose **7** by benzylamine followed by trifluoroacemidate formation afforded the intermediate **8** in 82% yield (Scheme 11.1). The glycosylation of **8** with methyl vanillate afforded the β-D-allopyranosides **9** or **10** in good yields.

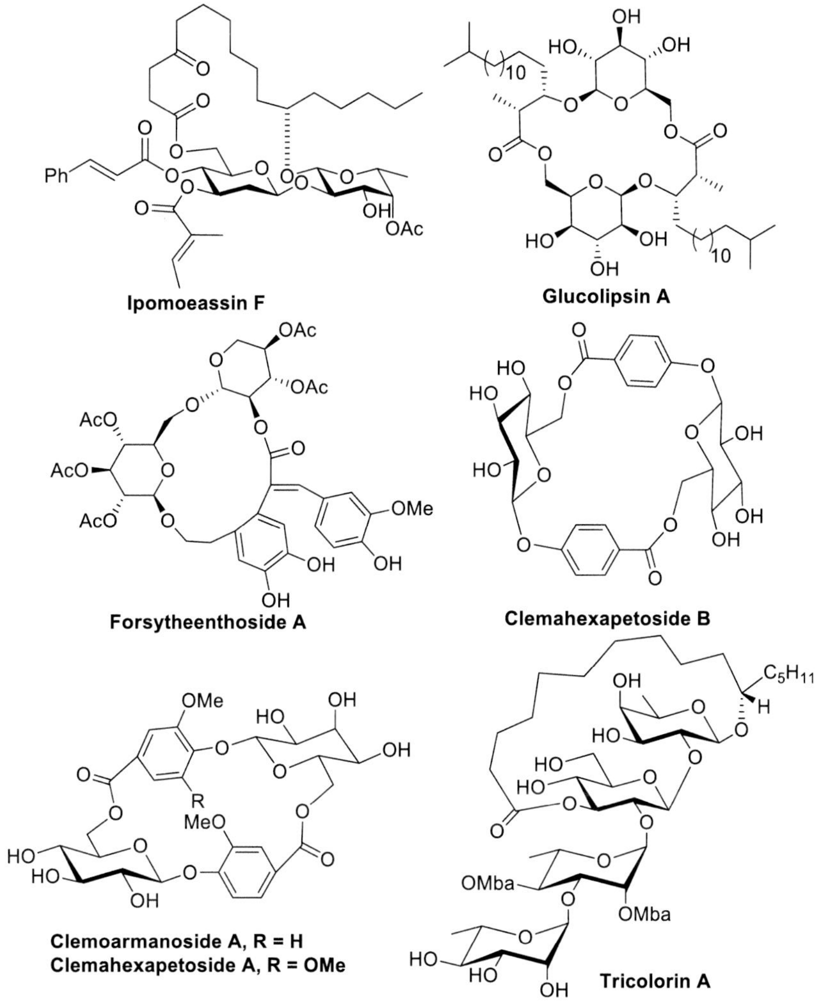

FIGURE 11.1 Structures of some sugar-embedded macrocyclic natural products.

Deacetylation using K_2CO_3 in THF/MeOH and followed by selective trityl protection at the C-6 position, benzylation at 2,3,4-positions, and hydrolysis of the methyl ester afforded the functionalized benzoic acid derivatives **11a** and **11b**, respectively, in good yields over four steps. Esterification of **11** with the β-glycoside intermediate **12** afforded the intermediates **13** or **14**, followed by removal of the trityl group using TFA, and cleaving the allyl group using $Pd(PPh_3)_4$ afforded the intermediate acids **15** or **16**. The macrolactonization of the carboxylic acid with the primary alcohol in **15** or **16** was achieved under Yamaguchi's conditions using 2,4,6-trichlorobenzoyl chloride,

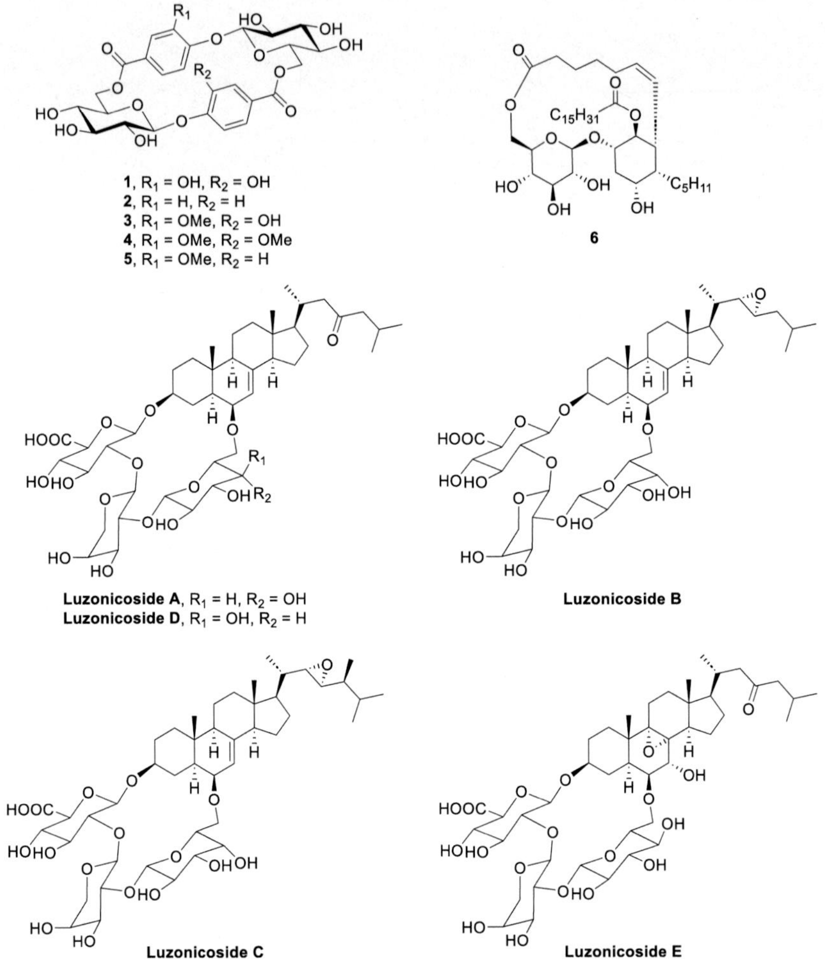

FIGURE 11.2 The chemical structure of macrocyclic dilactones **1–5** and mucorolactone **6** and luzonicosides A–E.

Et_3N, and DMAP in a diluted toluene solution at 2.5 mM. The macrolactonization afforded the cyclic dilactones **17–18** in good yields (78% and 88%, respectively); and final global catalytic hydrogenation afforded the corresponding target compounds Clemoarmanoside A and Clemahexapetoside A smoothly.[11]

Using the same strategy, the team also synthesized cyclic oligomers by repeating the esterification and linking three-, four-, and five-glucosylated benzoates together to produce the precursors **19–21** for macrocyclization and afforded the cyclic trimer **22**, tetramer **23**, and pentamer **24** in good-to-excellent yields (Scheme 11.2).

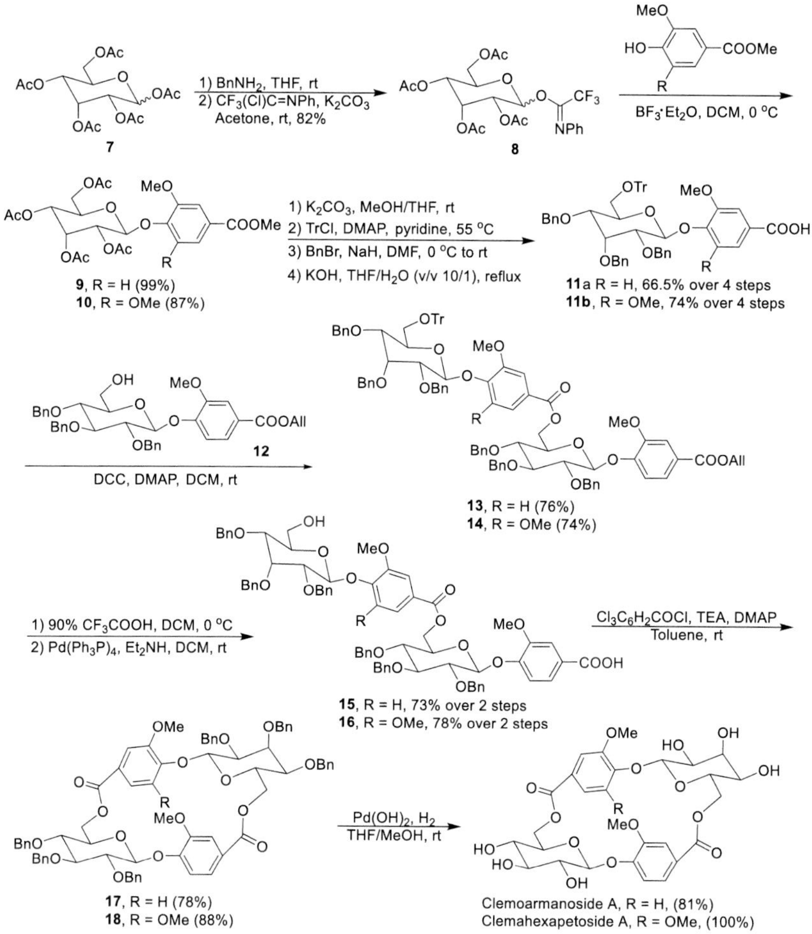

SCHEME 11.1 Total synthesis of Clemoarmanoside A and Clemahexapetoside A.[11]

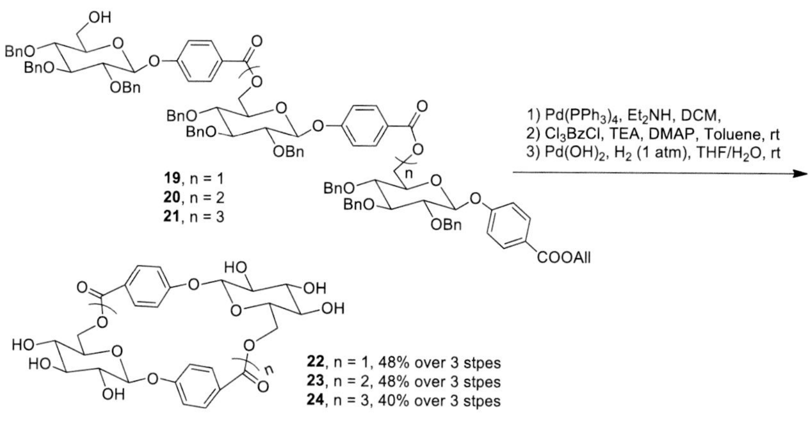

SCHEME 11.2 Synthesis of cyclic trimer **22**, tetramer **23**, and pentamer **24**.[11]

11.2.3 Synthesis of Forsythenethoside A

Forsythenethoside A contains a disaccharide unit embedded in a 15-membered ring macrocycle, this compound was isolated from the fruits of *Forsythia suspensa.*[12] The compound exhibited neuroprotective activities. The total synthesis of the Forsythenethoside A was accomplished by Yu et al.[13] The retrosynthetic analysis of Forsythenethoside A is shown in Fig. 11.3. The disaccharides can be prepared by glycosylation reactions and the macrocyclization carried out by oxidative coupling reaction. Yu et al. carried out the synthesis in the longest linear steps of 14 steps and 5% overall yield, and the key step of macrocyclization was carried out by intramolecular oxidative coupling reaction (Scheme 11.3).[13]

The synthesis started from intermediate **25**, which was synthesized from 3,4-dihydroxyphenylacetic acid in five steps with moderate yields.[13] The 4,6-protection with *p*-anisaldehyde dimethyl acetal followed by acetylation of the C2 and C3 positions afforded the intermediate **26**. Reductive ring opening at 4,6-positions using $NaBH_3CN$ afforded the intermediate **27a** with the free C-4 OH group. Esterification using caffeoyl chloride afforded intermediate **28**, and the C-6 *p*-methoxylbenzyl (PMB) protective group was then removed using 2,3-dichloro-5,6-dicyano-1,4-benzoquinone (DDQ) to afford intermediate **29**. The glycosylation reaction with intermediate **27b** using *N*-iodosuccinimide (NIS) afforded intermediate **30**, which was converted to compound **31** after the removal of allyl protective groups. This compound sets the stage for oxidative coupling macrocyclization and afforded the macrocycle **32** in moderate yield. Final global deacetylation was accomplished by treatment with acetyl chloride to afford the Forsythenethoside A.

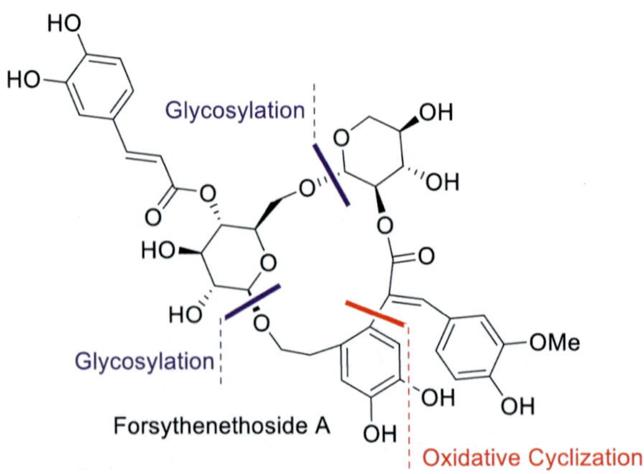

FIGURE 11.3 Retrosynthetic analysis for Forsythenethoside A.

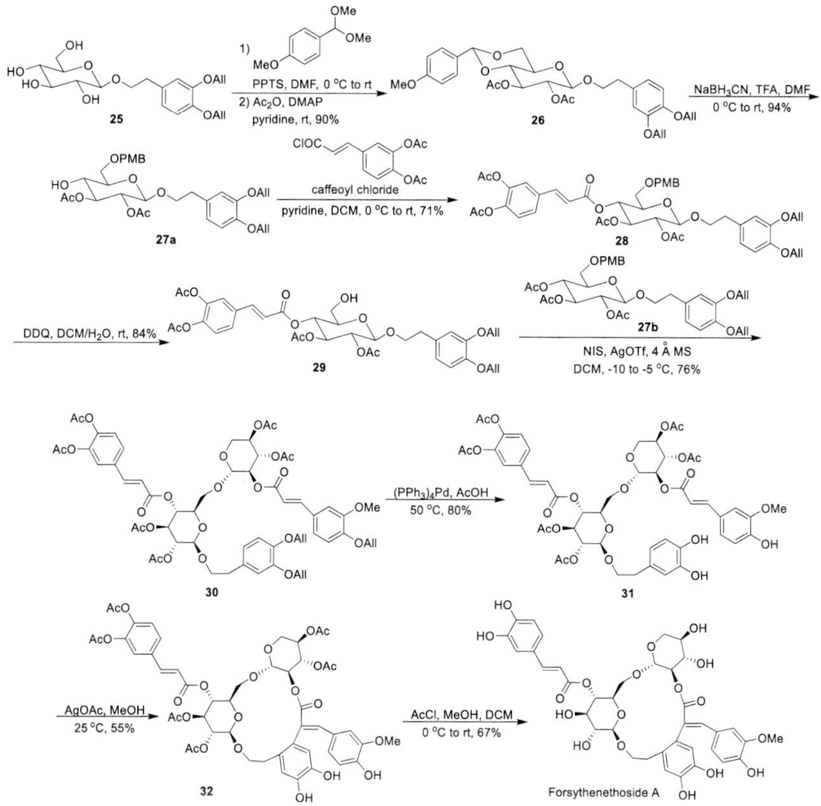

SCHEME 11.3 Total synthesis of the natural product Forsythenethoside A.[13]

11.2.4 Synthesis of Ananatoside A and analogs

Ananatoside is a 15-membered macrodilactone natural product isolated from microbial surfactants generated by the bacterium *Pantoeaananatis* (Fig. 11.4). The structure of this macrocyclic dilactone containing glucolipid was elucidated recently.[14] The total synthesis of Ananatoside A and related macrodilactone-containing rhamnolipids was completed by Cloutier and Gauthier et al.[15] Through the synthesis of several rhamnolipid analogs, especially the macrodilactones, they had demonstrated that macrolactonization of these lipids enhanced their cytotoxicity and hemolytic potential. The total synthesis of Ananatoside A is shown in Scheme 11.4 and Scheme 11.5. The compound has a relatively simple structure that can be prepared by esterification or glycosylation to complete the macrocyclization after all the suitable intermediates are installed. The protected glycosides were prepared from D-glucose by installing suitable protecting groups.

The C-2 position is introduced with the protective groups levulinoyl (Lev) in Scheme 11.4 and (2-azidomethyl)benzoyl (AZMB) in Scheme 11.5

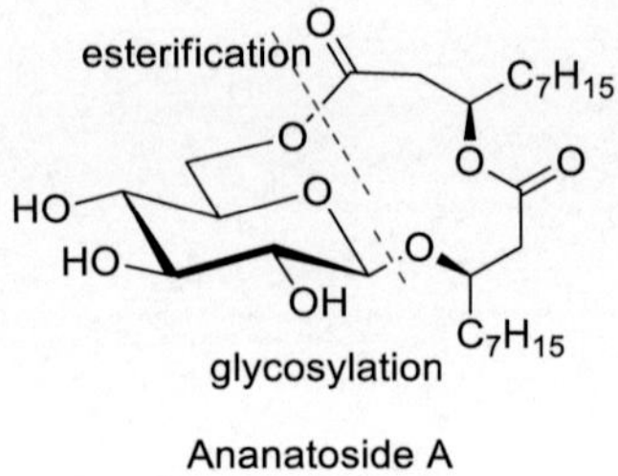

FIGURE 11.4 The chemical structure of Ananatoside A.

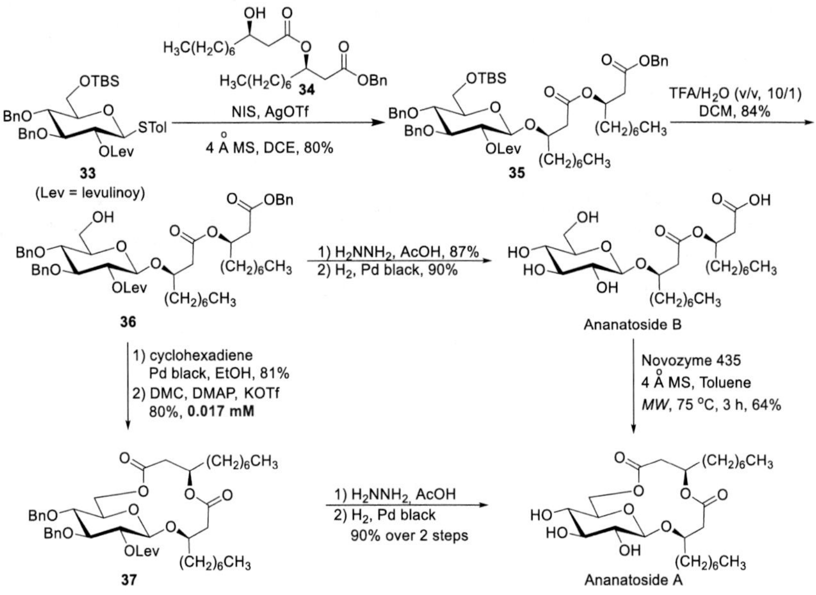

SCHEME 11.4 Synthesis of Ananatoside A by macrolactonization for ring closure.[15]

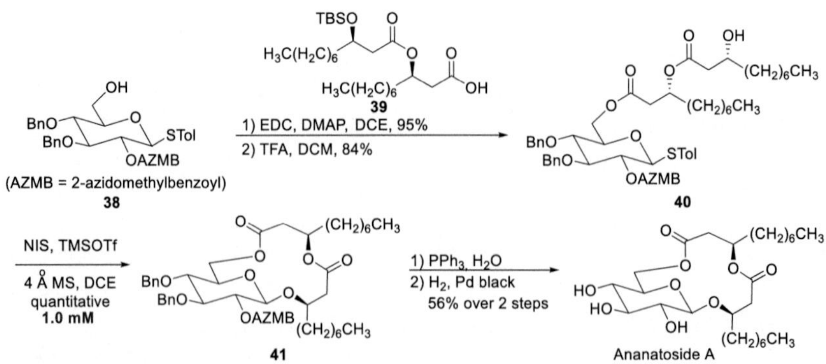

SCHEME 11.5 Synthesis of Ananatoside A by glycosylation for ring closure.[15]

as the neighboring participating groups to ensure the β-selectivity for the glycosylation or 1,2-*trans*-selectivity during glycosylation. Lev can be selectively cleaved in the presence of ester, which is compatible with the macrolactonization step, and the AZMB can be orthogonally cleaved without affecting macrolide functionality. As shown in Scheme 11.4, the protected thioglycoside **33** was glycosylated with the alcohol accepter **34** using NIS/AgOTf to afford the β-glycoside **35** in 80% yield. The *t*-butyldimethylsilyl (TBS) group was cleaved using TFA to afford the intermediate **36**, which was then converted to the unprotected glycoside, Ananatoside B, by deprotection of the Lev group using hydrazine and global catalytic debenzylation. The Ananatoside B underwent intramolecular enzymatic esterification (lactonization) of the carboxylic acid with the 6-OH group to afford the natural macrolide Ananatoside A directly in 64% yield. Alternatively, chemical lactonization was carried out in super diluting conditions from the intermediate **36**, after selective debenzylation of the benzyl ester to free carboxylic acid and intramolecular lactonization using DMC, DMAP, and KOTf afforded the macrolactone **37**. The further delevulinoylation and debenzylation produced the Ananatoside A in good yield.

The second route for the synthesis of Ananatoside A is shown in Scheme 11.5. Starting from the C2 AZMB (2-azidomethylbenzoyl) protected thioglycoside **38**, esterification of the C-6 primary hydroxyl group followed by selected TBS removal using TFA afforded the intermediate ester **40**. The intramolecular glycosylation of **40** with the hydroxyl group led to the formation of macrocycle **41**. Deprotection of AZMB by PPh_3 and debenzylation by catalytic hydrogenation afforded the natural product Ananatoside A.

Using the same glycosylation ring-closing methods, the authors also synthesized several rhamnose-containing macrolides, a few examples of the synthesis are shown in Scheme 11.6. Using the suitably protected thiorhamnosides **42**, **47**, and **50** by similar methods, the corresponding macrolactonized rhamnolipids were obtained. Esterification of **42** with **39** followed by TBS removal afforded intermediate **43**, intramolecular glycosylation using NIS/TMSOTf afforded the intermediate **44** as a mixture of the α/β isomers in 15:85 ratio, deprotection of Lev and Bn afforded the 1−2 macrolactones **45** and **46** in good yields, the beta isomer was the major product. Similarly, the 1−3 macrolactones **48** (α-anomer) and **49** (β-anomer) were prepared from the intermediate **47**, with the α-anomer as the major product. The 1−4 macrolactone **51** was obtained as the single isomer from intermediate **50** in good overall yield.

11.2.5 Synthesis of Ipomoeassin F and analogs

Several disaccharide bridged macrocycles, Ipomoeassin F and analogs, have been synthesized recently by Shi et al.[16−23] Ipomoeassins exhibit antitumor activities and other biological properties. Ipomoeassins A − F (Fig. 11.5) were isolated from the leaves of *Ipomoea squamosa*. They belong to a large family of glycolipids, resin glycosides, also known as glycoresins. Resin glycosides

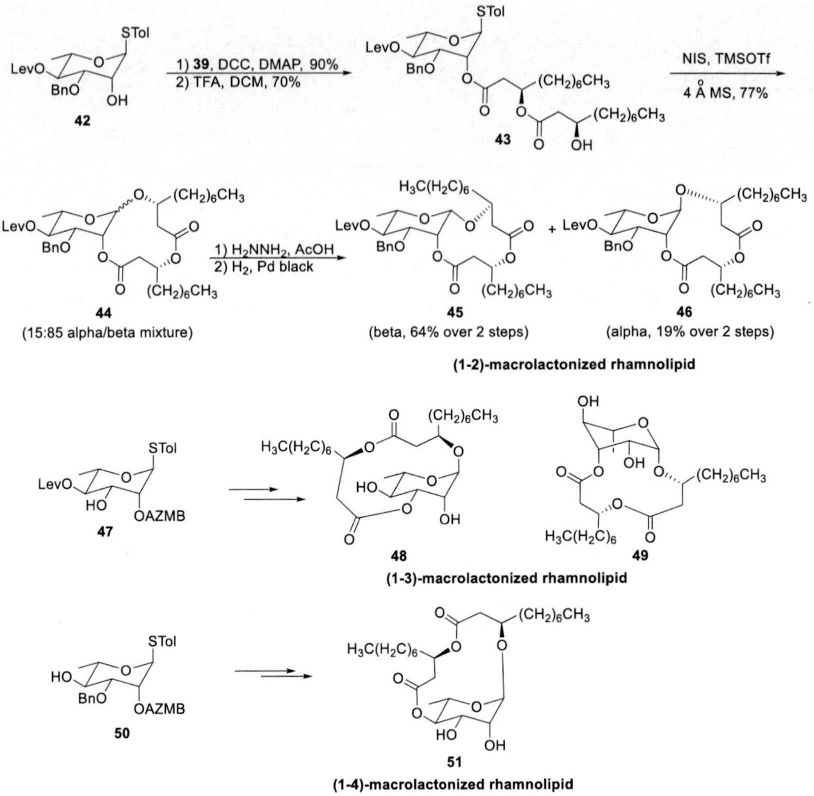

SCHEME 11.6 Synthesis of rhamnose-containing macrodilatones **45–46**, **48–49**, **51**.[15]

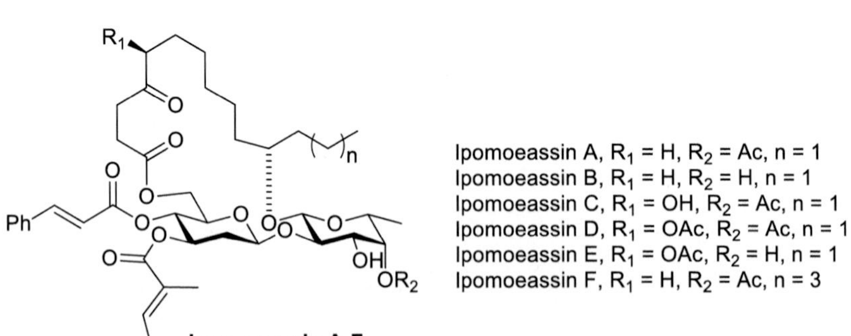

FIGURE 11.5 The chemical structures of Ipomoeassins A–F.

are natural products isolated from morning glory plants and have been explored extensively for medical applications. The resin glycosides containing macrocyclic architectures with embedded carbohydrates exhibited important biological activities. The Ipomoeassins D and F exhibited potent antitumor cytotoxicity, and this has inspired several total syntheses of the macrolides.

The medicinal properties of these compounds have drawn the attention of synthetic chemists for their total synthesis,[16] and the properties of these compounds have been covered in recent reviews.[2] Ipomeassin D and F exhibited potency against multiple cancer cell lines, especially Ipomoeassin F is the most potent in the family. Three different total syntheses of Ipomoeassin F have been achieved independently by Postema,[24] Fürstner,[25] and Zong.[16] Zong et al. reported an improved total synthesis of the Ipomoeassin F in 3.8% yield of linear 17 steps in 2015.[16] They were able to produce gram quantities of material, which enables further biological studies on the compound and their analogs and the elucidation of structure-activity relationship (SAR), which is important for drug development. Ring-closing metathesis (RCM) strategy was also employed for the macrocyclizations, but different intermediates and routes were used. A retrosynthesis analysis for Ipomoeassin F is shown in Scheme 11.7. The retrosynthesis shows that Ipomoeassin F can be

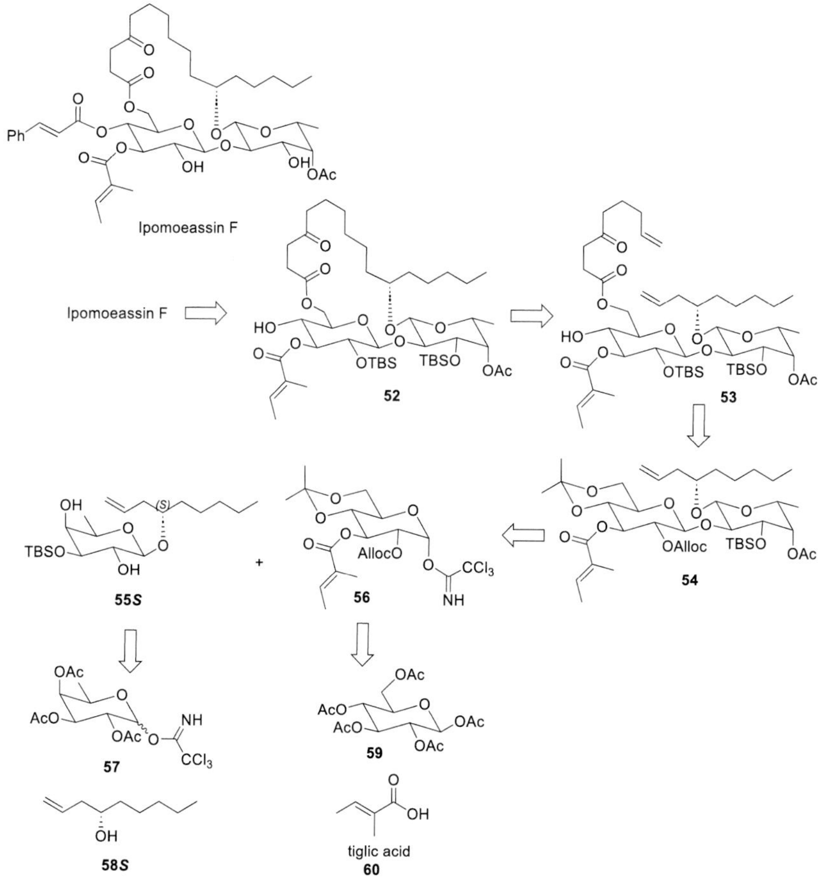

SCHEME 11.7 Retrosynthetic analysis of Ipomoeassin F.[16]

prepared by esterification and deprotection of **52**, and the key intermediate **53** can be the precursor of **52** by macrocyclization through RCM reaction. The intermediate **53** can be traced back to the fully protected intermediate **54**. This disaccharide intermediate can be prepared from the protected monosaccharides **55*S*** and **56**, which can be prepared using the simple building blocks **57–60**.

The synthesis of the intermediates is shown in Schemes 11.8 and 11.9, and the completed total synthesis is shown in Scheme 11.10. The glycosyl donor **56** was synthesized from the commercially available glucose pentaacetate compound **59**, glycosylation with *p*-methoxylphenol and $BF_3{\cdot}Et_2O$ followed by deacetylation using NaOMe and methanol afforded α-glycoside **61**. The 4,6-positions were then protected with isopropylidene acetal to form **62** and

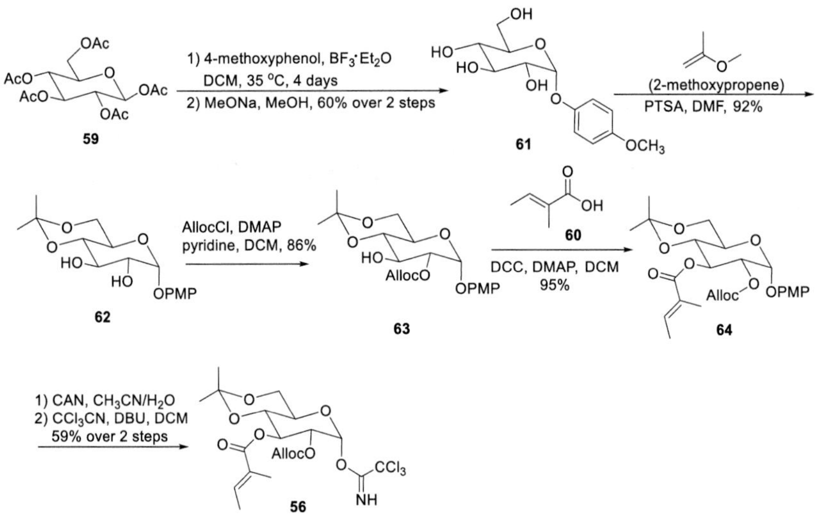

SCHEME 11.8 Synthesis of glycosyl donor **56**.[16]

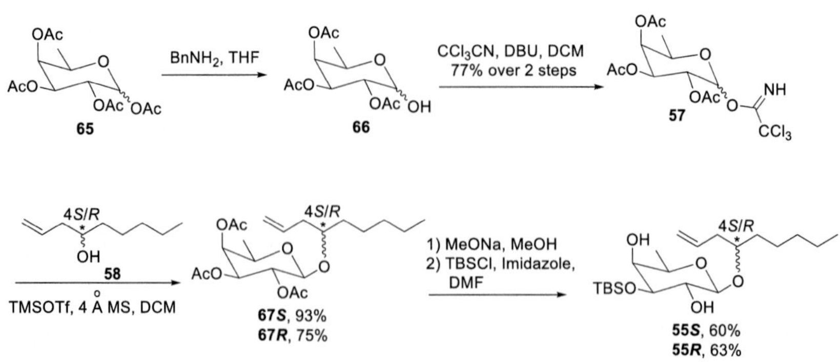

SCHEME 11.9 Synthesis of glycosyl acceptors **55*S*** and **55*R***.[16]

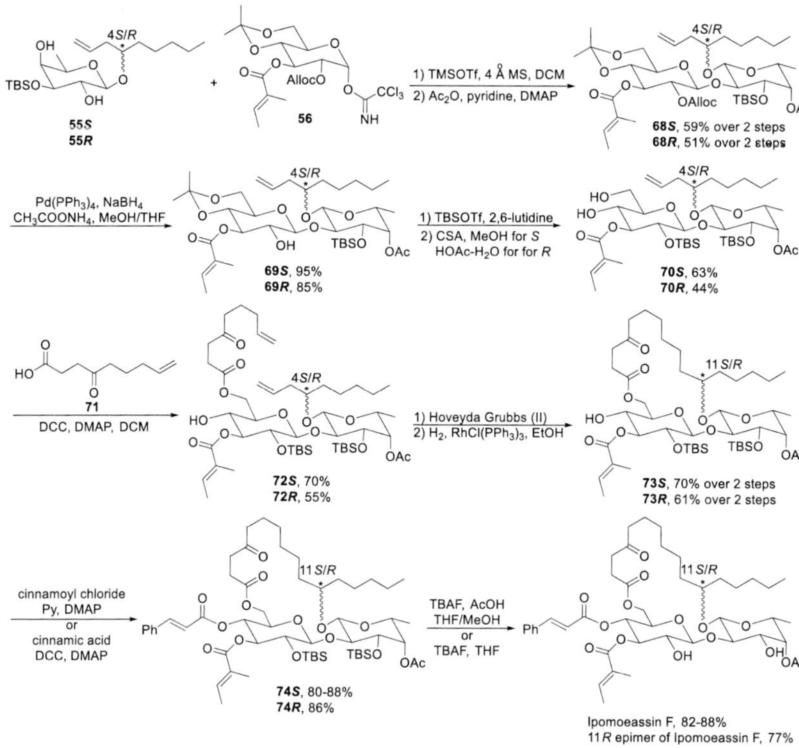

SCHEME 11.10 Completion of the total synthesis of Ipomoeassin F and its 11*R* epimer.[16]

selective protection of the 2-position with alloc group afforded the intermediate **63**. Esterification of the 3-hydroxyl group with tiglic acid **60** afforded the intermediate **64**. The anomeric *p*-methoxyphenyl group was then removed using CAN followed by a reaction with trichloroacetonitrile led to the intermediate **56**.

The synthesis of fucoside intermediates **55** starting from the fucosyl peracetate **65** is shown in Scheme 11.9. Selective deacetylation of the anomeric acetyl group using benzylamine to form interemdate **66** and followed by trichloriacetamidation afforded the fucosyl donor trichloro acetimidate **57**. Glycosylation of **57** with either the (*S*)- or (*R*)-4-hydroxy-1-nonene **58** afforded the β-glycosides **67**, which were then converted to the TBS-protected intermediates **55** in two steps.

Scheme 11.10 shows the completion of the total synthesis of Ipomoeassin F. The glycosyl donor **56** was reacted with the TBS-protected fucosides **55*S*** and **55*R*** using TMSOTf to afford the β-linked disaccharide regioselectively, and the free 4-OH was acetylated to form the intermediates **68**. Selective deprotection of the Alloc group was carried out under transfer hydrogenation reaction using ammonium acetate and $Pd(PPh_3)_4$ catalyst in good yield.

The free hydroxyl group in **69** was then protected using TBSOTf, followed by deprotection of the isopropylidene acetal under acetic conditions to afford the diol intermediates **70**. The primary hydroxyl group was then esterified with 4-oxo-8-nonenoic acid **71** to afford the intermediates **72**, which are the key precursors for the RCM reaction. RCM using the Hoveda Grubb (II) catalyst afforded the macrolide intermediate as a mixture of E/Z isomer, which was treated by catalytic hydrogenation using Wilkinson's catalyst to afford the macrolides **73**. Esterification using cinnamic acid or cinnamoyl chloride led to the intermediates **74**. Deprotection of the TBS groups in **74** using TBAF afforded the target, Ipomoeassin F, and its 11*R* epimer in good yields. The macrolides were evaluated for their anticancer activities, and the epimer showed a significant loss of potency in comparison to the natural product with 11*S* configuration. They have also demonstrated that Ipomoeassin F can induce apoptosis in human U937 cells.

Shi's group further mapped the SAR profile for Ipomoeassin F by investigating the importance of different parts of the molecule.[18] They synthesized various analogs and evaluated their activities; the structures important for anticancer activities are identified. Several structures are shown in Fig. 11.6. Among the analogs prepared and tested, the macrocycle **75j** showed similar potency but can be prepared more readily.

Further SAR studies on fucose-truncated monosaccharide analogs **76–78** were designed and synthesized by Zong et al.[19] In these compounds, the fucoside moiety was completely or partially removed from the disaccharide portion of the original natural product (Fig. 11.7). The macrolactones are glucose derivatives cyclized from C-1 to C-6 positions. However, the biological evaluation indicated that the 6-membered ring of the fucoside moiety is essential to the bioactivity, and it was rationalized that the presence of the fused six-membered ring of fucose moiety changed the overall conformation of the molecule, thereby affecting bioactivity.

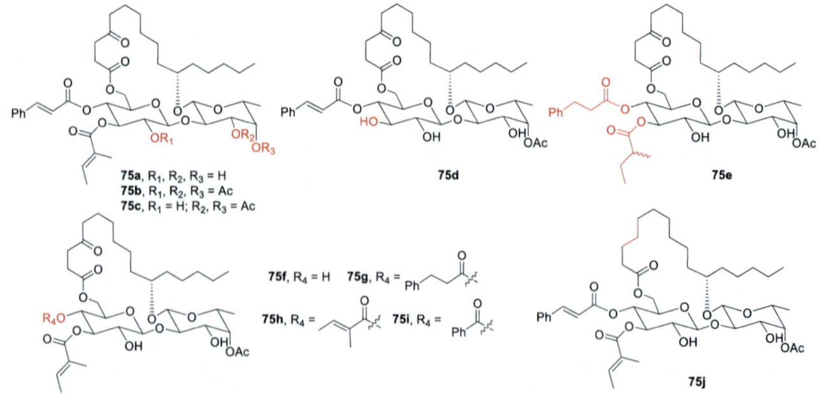

FIGURE 11.6 Structures of Ipomoeassin F analogs **75a–j** synthesized for SAR studies.[18]

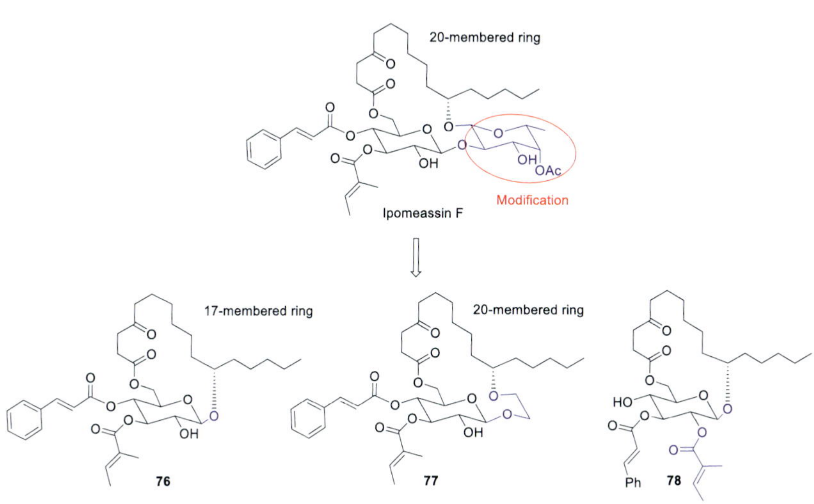

FIGURE 11.7 The chemical structure of fucose-truncated monosaccharide analogs **76–78**.[19]

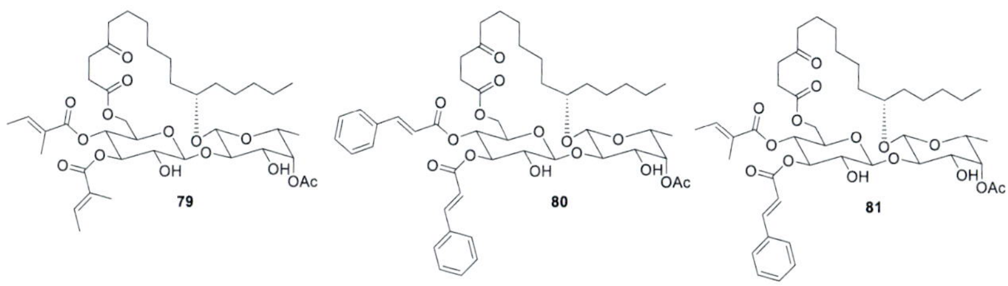

FIGURE 11.8 The side-chain analogs **79–81** of Ipomoeassin F.[21]

The influence of the side-chain esters on the bioactivity was further evaluated by the group. They designed an efficient route toward the synthesis of the analogs **79–83** (Fig. 11.8 and Scheme 11.11). An improved synthetic route was also developed for the total synthesis of Ipomoeassin F and the side-chain analogs. The retrosynthesis for **82–83** is shown in Scheme 11.11.[21] These compounds can be prepared from the intermediates **84–85**. These macrocycles can be prepared from compound **86** by RCM, hydrogenation, cinnamic/tiglic ester formation, removal of Lev, and hydrolysis of cyclic ketal. This intermediate can be traced back to acid **87** and disaccharide **88**. The disaccharide can be prepared by glycosylation reaction of **89** with **55*S***, similarly to the method employed in Scheme 11.10.

The syntheses of these macrocycles are shown in Scheme 11.12.[21] Glycosylation reaction of Lev-protected glucose derivative **89** with the 2-hydroxyl group in **55*S*** followed by acetylation of 4-OH afforded the disaccharide **90**. Removal of the 4,6-isopropylidene acetal protective group afforded intermediate **88**. Selective esterification at the C-6 position of **88**

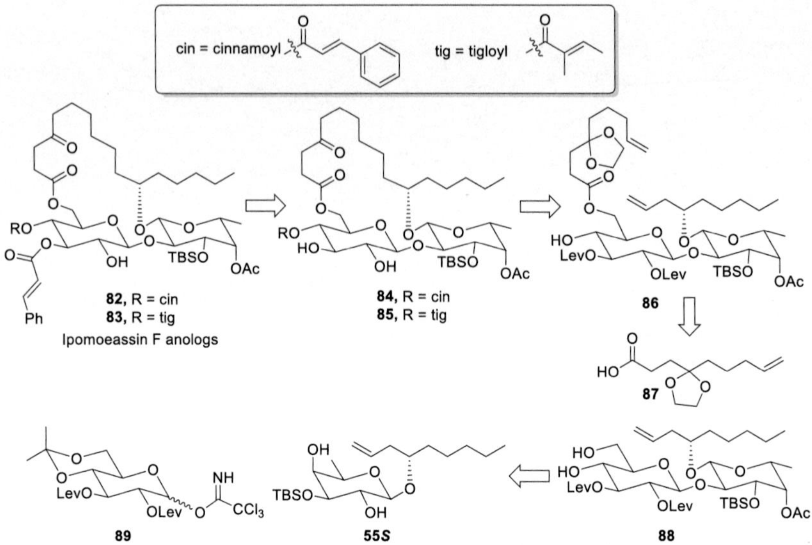

SCHEME 11.11 Retrosynthetic analysis of Ipomoeassin F and its two analogs **82** and **83**.[21]

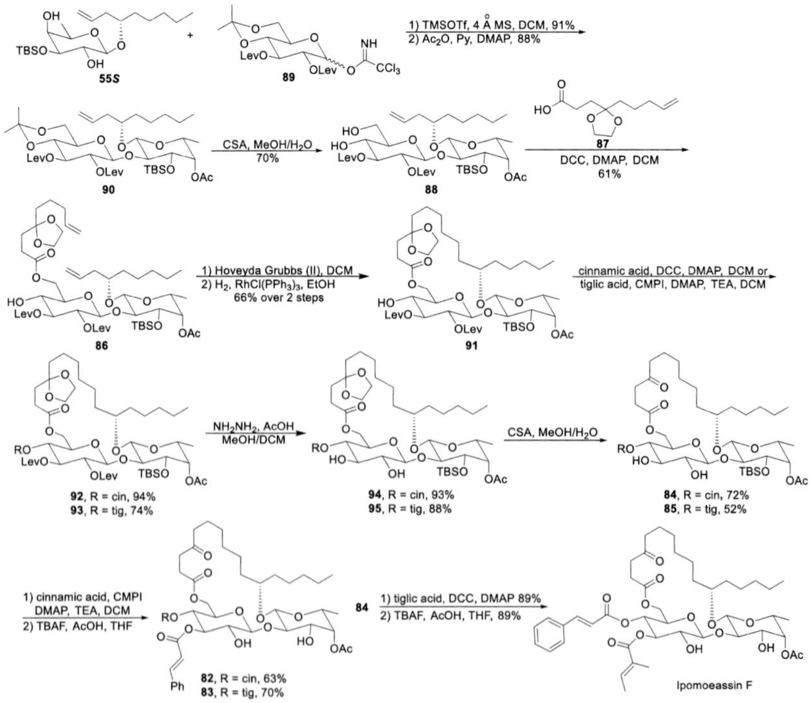

SCHEME 11.12 Total synthesis of Ipomoeassin F and its two analogs **82–83**.[21]

with the cyclic acetal-protected 4-oxo-8-nonenoic acid **87** led to the intermediate **86**, in which the dialkenes are then converted to the macrocycles **91** by RCM reaction and hydrogenation. Esterification reactions using different acid derivatives afforded the intermediates **92** or **93**. The Lev protective groups were then removed using hydrazine acetate to give **94** or **95**, respectively, and the ketone moiety needs to be protected to avoid reactions with hydrazine. Removal of the acetal using camphorsulfonic acid (CSA) afforded the corresponding products **84** or **85**. Selective esterification with cinnamic acid using 2-chloro-1-methylpyridinium iodide (CMPI), Mukaiyama's reagent, followed by treatment with TBAF to remove the TBS afforded the cinnamates **82** or **83**. Esterification of **84** with tiglic acid followed by TBS removal afforded the natural product Ipomoeassin F. The synthesis started from the commercially available D-glucose penta-acetate, and the Ipomoeassin F was obtained in 4.0% overall yield with 17 steps of the longest linear sequence. The bioassays indicated that the positions and identities of the two esters were crucial.

Due to the important anticancer activities and to further study its biological and pharmacological properties, Zong et al. designed and synthesized a series of Ipomoeassin F analogs or derivatives and studied their interactions with the binding target Sec61α (protein transport protein Sec61 subunit α isoform 1). These structures are shown in Fig. 11.9.[22] The derivatives **96** and **97** contain biotin and **98–100** contain fluorescent functional groups; these

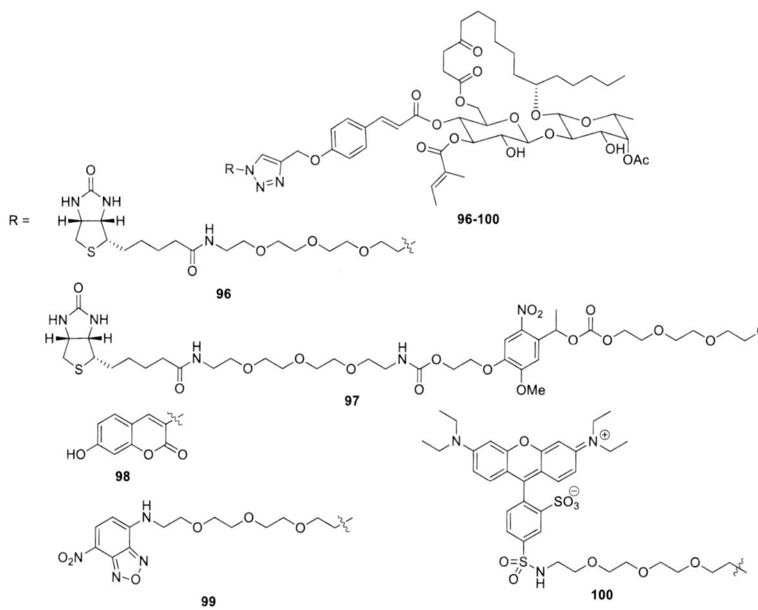

FIGURE 11.9 Structures of Ipomoeassin F analogs **96–97** and fluorescent derivatives **98–100**.[22]

were prepared for cell imaging studies. They found that the coumarin derivative and nitrobenzoxadiazole (NBD) derivatives retained cytotoxicity, however, were not suitable to use for cell imaging. The rhodamine derivative **100** showed some loss of potency but was suitable for cell imaging studies as a fluorescent probe.

Besides the studies on SAR of the macrocycle, sugar, side chains, and size of the rings, Zong et al. also synthesized and evaluated the cytotoxicity of a series of 5-oxa/aza analogs of Ipomoeassin F.[23] In this series of modifications, the 5-ketone moiety was modified by introducing an ester or amide function. As shown in Scheme 11.13, the synthesis of 5-oxa/aza analogs **111** and **112** of Ipomoeassin F became simplified, which is advantageous for drug development. They found that changing the 5-CH_2 with O or NH led to a small activity loss; however, *N*-methylation was able to restore the potency. The synthesis started from the intermediate **70*S*** prepared previously in Scheme 11.10,[16] esterification with different fatty acids **101–102**, which contain the C-5 modified functionality, afforded the intermediates **103** or **104**, respectively. Using similar RCM and hydrogenation reactions, the macrocycles **107–108** were prepared smoothly. The 4-OH was then esterified with cinnamic acid moiety, followed by deprotection of the silyl group using TBAF to afford the targeted macrocycles **111** (5-oxa analog) and **112** (5-aza-analog).

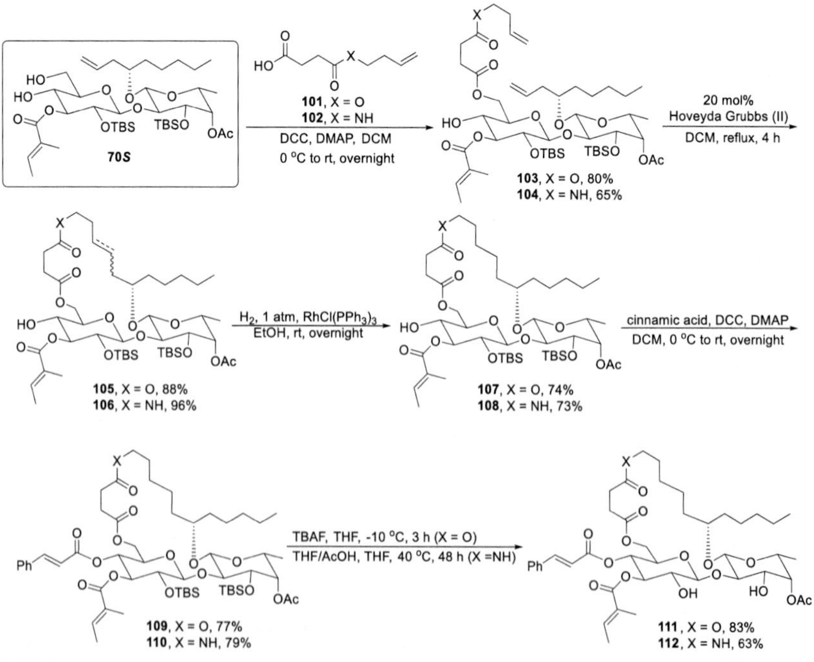

SCHEME 11.13 Total synthesis of 5-oxa/aza analogs **111** and **112** of Ipomoeassin F.[23]

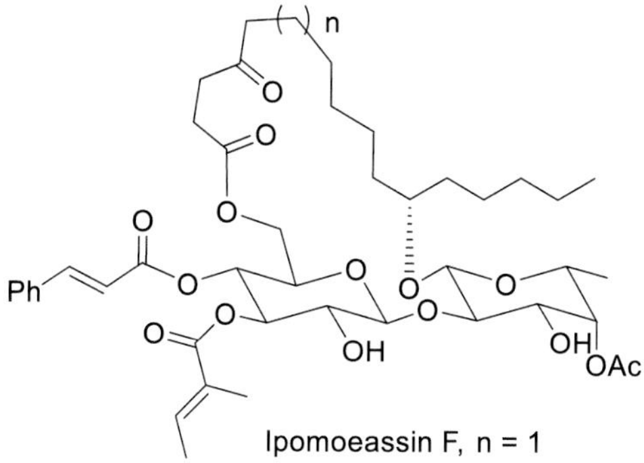

FIGURE 11.10 The chemical structure of ring size varying analogs **113** and **114**.[20]

Zong et al. also synthesized ring size varying analogs of Ipomoeassin F, as shown in Fig. 11.10, the 19-membered ring analog **113** ($n = 0$) and 22-membered analog **114** ($n = 3$) were synthesized and evaluated, and they found that the ring expanded analog **114** exhibited enhanced cytotoxicity.[20] The synthesis for Ipomoeassin F and compound **114** are shown in Scheme 11.14. The compound **113** was synthesized by a similar method using a different 4-oxo-alkenenic acid and 4-oxooct-7-enoic acid, for the step of RCM with the corresponding disaccharide derivative.

As shown in Scheme 11.14, the synthesis started from the same intermediates **55*S*** and **89** (Scheme 11.10), glycosylation followed by acetylation and deprotection of the Lev group afforded intermediate **115**, selective tiglation using CMPI of the 4′-OH followed by TES protection of the 3′-OH, and removal of the isoproylidene acetal using TFA afforded the intermediate **116** in good yield. Selective esterification of the primary hydroxyl group with the corresponding oxa-carboxylic acids using 1-ethyl-3-(3-dimethylaminopropyl) carbodiimide (EDC) and HOBt afforded the intermediates **117** and **118**. Ring-closing (RCM) followed by hydrogenation using Wilkinson's catalyst afforded the corresponding macrocycles **119** and **120**, respectively. The final macrocycles Ipomoeassin F or **114** were obtained by treating **119** or **120** with cinnamic acid and Mukaiyama reagent, followed by deprotection of the silyl protective groups using TBAF and AcOH.

11.2.6 Synthesis of resin glycoside Calysolin IX

Calysolin IX is one of the most complex resin glycosides isolated to date.[26,27] The compound contains tetra-saccharide core structure linked through a macrolactone to form the 27-membered ring macrocyclic glycoside (Fig. 11.11).

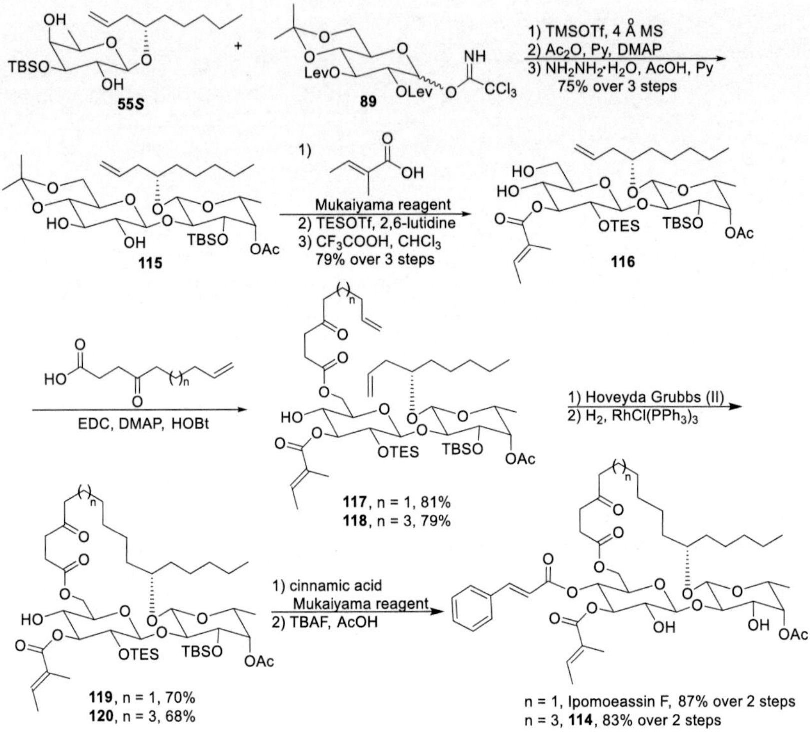

SCHEME 11.14 Synthesis of Ipomoeassin F and the 22-membered analog **114**.

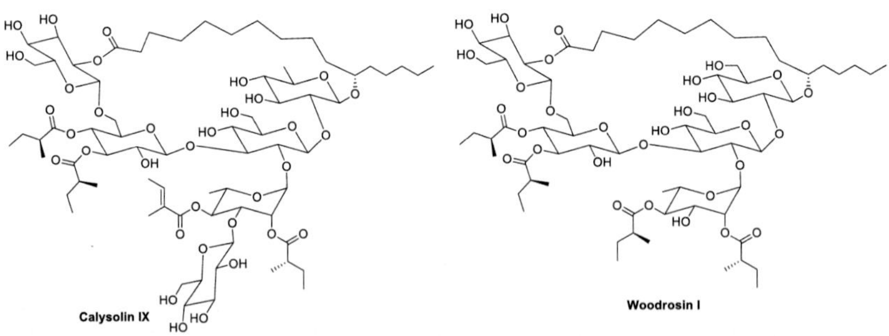

FIGURE 11.11 Chemical structure of Calysolin IX and Woodrosin I.

The synthesis of the macrocyclic core structure **131** of Calysolin IX by intramolecular glycosylation method is shown in Scheme 11.15.[28] The left side disaccharide **123** of the macrocycle was prepared by glycosylation using the PMB-protected 2-glucal **121** and the glucose derivative **122**. The deoxy glucal **121** was oxidized to the corresponding epoxide stereoselectively using Oxone followed by zinc chloride–mediated oxirane ring opening by the 6-OH

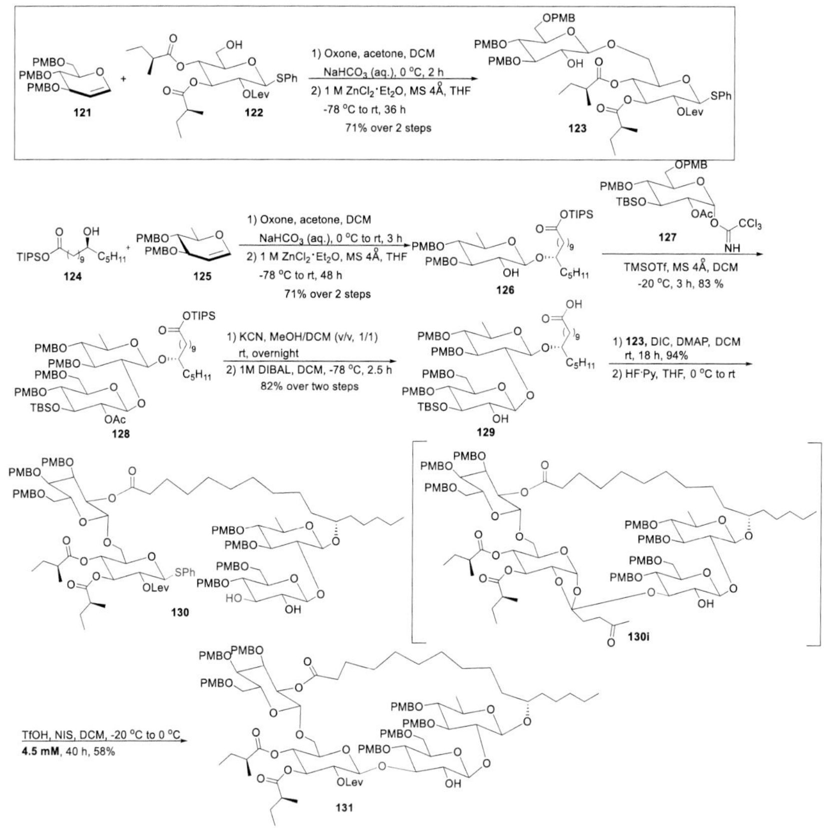

SCHEME 11.15 Synthesis of the macrocyclic core of the Calysolin IX.[28]

in **122** afforded the intermediate **123** in 71% yield. The right-side disaccharide moiety **129** was prepared starting from the chiral jalapinolic ester **124**, glycosylation with the protected 6-deoxylglucal **125** by a similar protocol as described earlier afforded the β-glycoside **126**, which was then coupled with the trichloroacetimidate **127** using TMSOTf as catalyst to form the disaccharide β-glycoside **128** in 83% yield. Triisopropylsilyl (TIPS) ester hydrolysis using KCN in methanol/dichloromethane (DCM) followed by deacetylation using DIBAL afforded **129** which contains both free carboxylic acid and hydroxyl group. The deprotection reaction using diisobutylaluminum hydride (DIBAL) was necessary since other conditions using basic reagents lead to the TBS migrated product and resulted in a mixture. The bulky 3′-OTBS protective group makes the free 2′-OH less prone to intramolecular lactonization; therefore, the COOH can be used to form ester intermolecularly with the disaccharide **123**. The esterification of the carboxylic acid in compound **129** with the free hydroxyl group in compound **123** was carried out successfully using diisopropylcarbodiimide (DIC) and DMAP (cat). The esterification conditions were screened using different amounts of compound **123**, using 5 equiv. of the alcohol **123** in 18 hours afforded the ester in 94% yield, the authors decided to use 2.5 equiv. of alcohol , which afforded the ester in 78% yield. The removal of the TBS protective group in the ester afforded the intermediate **130** with two free hydroxyl groups at the glucose ring, 2′-OH and 3′-OH. The intramolecular glycosylation of the 3′-OH should be the more accessible site for regioselective glycosylation. The 2"-levulinic protection was necessary for the thioglycoside activation by forming the orthoester intermediate **130i**. This orthoester formation step is crucial for the 3′-OH to react regioselectively due to the proximity. Using TfOH/NIS in DCM at −20 to 0°C, the orthoester intermediate **130i** was obtained. Then, macrocyclization of the intermediate **130i** by glycosylation reaction was explored under several different conditions, with the best results shown in Scheme 11.15 using TfOH (0.15 equiv.)/NIS, DCM, −20°C to 0°C, which led to the macrocycle **131** in 58% yield. The concentration of the substrate **130** was done at 4.5 mM, which required prolonged time (40 hours) for the orthoester conversion to the final glycoside **131**; the reaction can be done at 10 mM, which required a shorter time (16 hours) in 45% yield to the macrocycle **131**, however, further increase of concentrations to 45 mM of substrate **130** led to significant yield reduction to only 13% of macrocyclization product.

11.2.7 Synthesis of macrocycles by interrupted Pummerer reaction-mediated glycosylation

Wan's group reported the synthesis of several resin glycosides using interrupted Pummerer reaction-mediated (IPRm) glycosylation methods.[29] These include the total synthesis of several bioactive resin glycosides, murucoidins IV and V,[30] Tricolorin A,[31] and Merremoside D.[32] These are oligosaccharides connected through macrolactones. The formation of the oligosaccharides can be carried out utilizing the traditional glycosylation method, typically through activation of the trichloro acetoimidate glycosyl donor or thioglycoside donors. Wan et al. developed an interrupted Pummerer reaction-mediated (IPRm) glycosylation method, which allows the synthesis of oligosaccharides readily. The general mechanism is shown in Fig. 11.12.

11.2.7.1 Synthesis of murucoidins IV and V

The resin glycosides murucoidin IV and V were isolated from the flowers of *Ipomoea murucoides*. murucoidin V exhibits significant multidrug resistance (MDR) reversal activity. The two compounds share similar structures with the only difference in the macrolactone ring being either 18 or 19 membered, respectively. The retrosynthetic plan for murucoidins IV and V is included

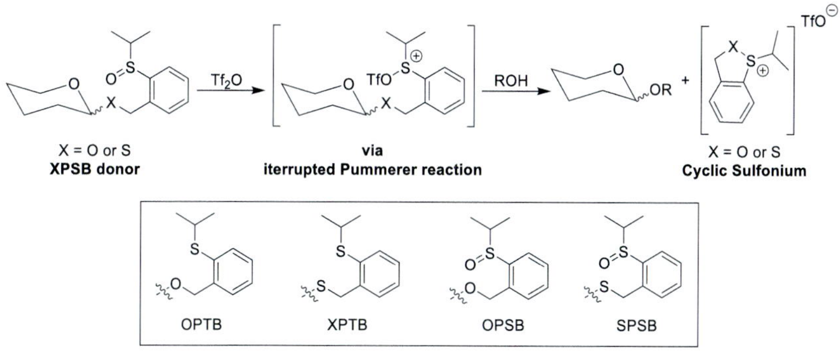

FIGURE 11.12 Interrupted Pummerer reaction-mediated (IPRm) glycosylation.

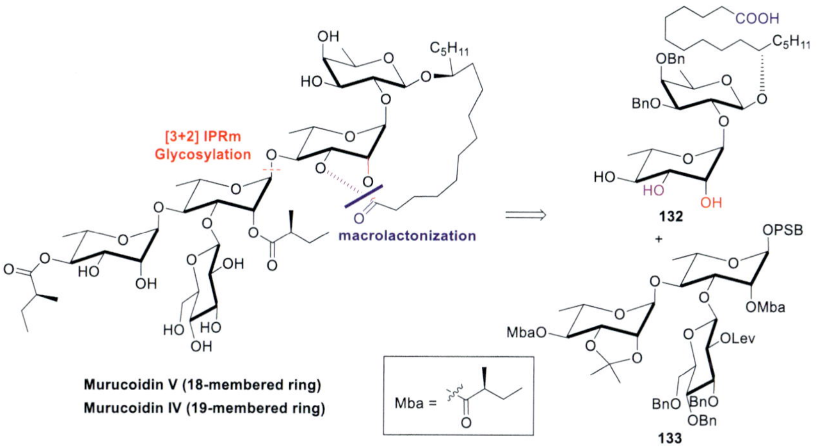

FIGURE 11.13 Retrosynthetic analysis of murucoidins IV and V.[30]

in Fig. 11.13. For the synthesis of murucoidin, they adopted the 3 + 2 IPRm glycosidic coupling and macrolactonization method. The trisaccharide **133** was synthesized first, and it is coupled with the disaccharide unit **132**.

The synthesis of trisaccharide **133** is shown in Scheme 11.16. The O-2-[(propan-2-yl)sulfinyl]benzyl (OPSB) glycosyl donor **134** was activated employing Tf_2O and 2,6-di-t-butyl-4-methyl pyridine (DTBMP) at −40°C, and this activated glycosyl donor reacting with the acceptor O-2-[(propan-2-yl)thio]benzyl (OPTB) glycoside **135** afforded the disaccharide **136i** in 82% yield; after removal of PMB protective group using DDQ led to the formation of disaccharide **136**. Glycosylation with the OPSB glucoside **137** under similar conditions furnished the trisaccharide **138**, and the OPTB group in **138** was then oxidized to the active OPSB glycoside donor **133**.

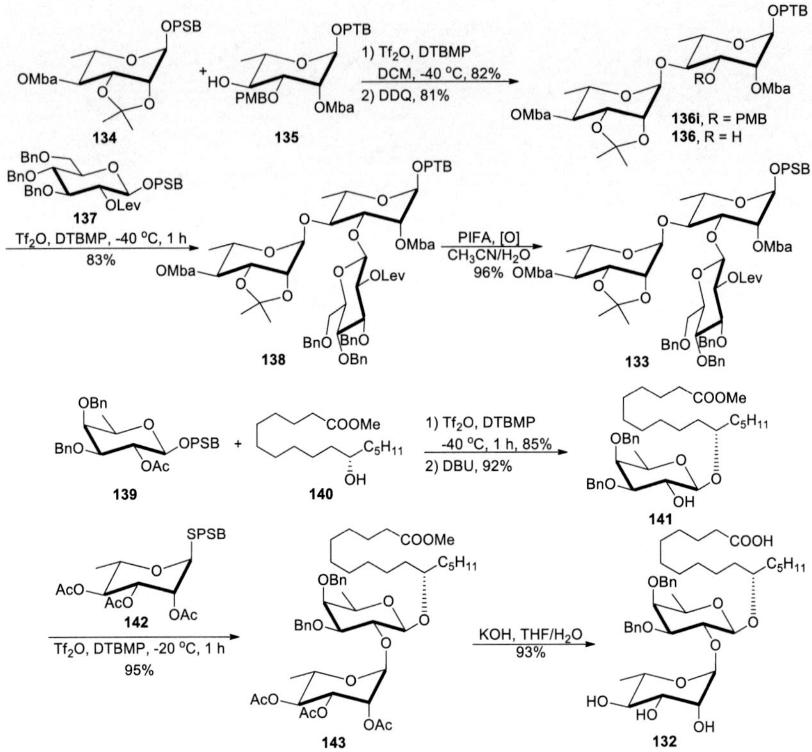

SCHEME 11.16 Synthesis of disaccharide and trisaccharide portions of murucoidins IV and V.[30]

The disaccharide **132** was synthesized starting from the OPSB glycoside **139**, glycosylation with the alcohol **140** and followed by removal of the acetyl group at C-2 position using 1,8-diazabicyclo [5.4.0]undec-7-ene (DBU) afforded the intermediate **141**. Coupling **141** with the *S*-2-(2-propylsulfinyl)-benzyl (SPSB) glycosyl donor **142** afforded the disaccharide **143**, which upon basic hydrolysis afforded the compound **132**.

With both parts of the key intermediates in hand, they tried different methods for macrocyclization. The successful strategy is shown in Scheme 11.17. The acid in **132** was protected with a thioester to obtain compound **144**; the *cis*-dihydroxyl group was then temporarily masked with phenylboronic acid to form the intermediate **144i**; the IPRm glycosylation reaction with the OPSB trisaccharide **133** afforded the pentasaccharide **145** in good yield. The transient cyclic boronic ester was stable during the glycosylation and was cleaved on silica gel during chromatography. The thioester was deprotected and the key intermediate **146** was obtained; macrolactonization under Corey–Nicolaou conditions using phenyl disulfide and triphenyl phosphine afforded 19- and 18-membered ring macrolactones **147** in 54% yield and **148** in 42% yield. Global deprotection of

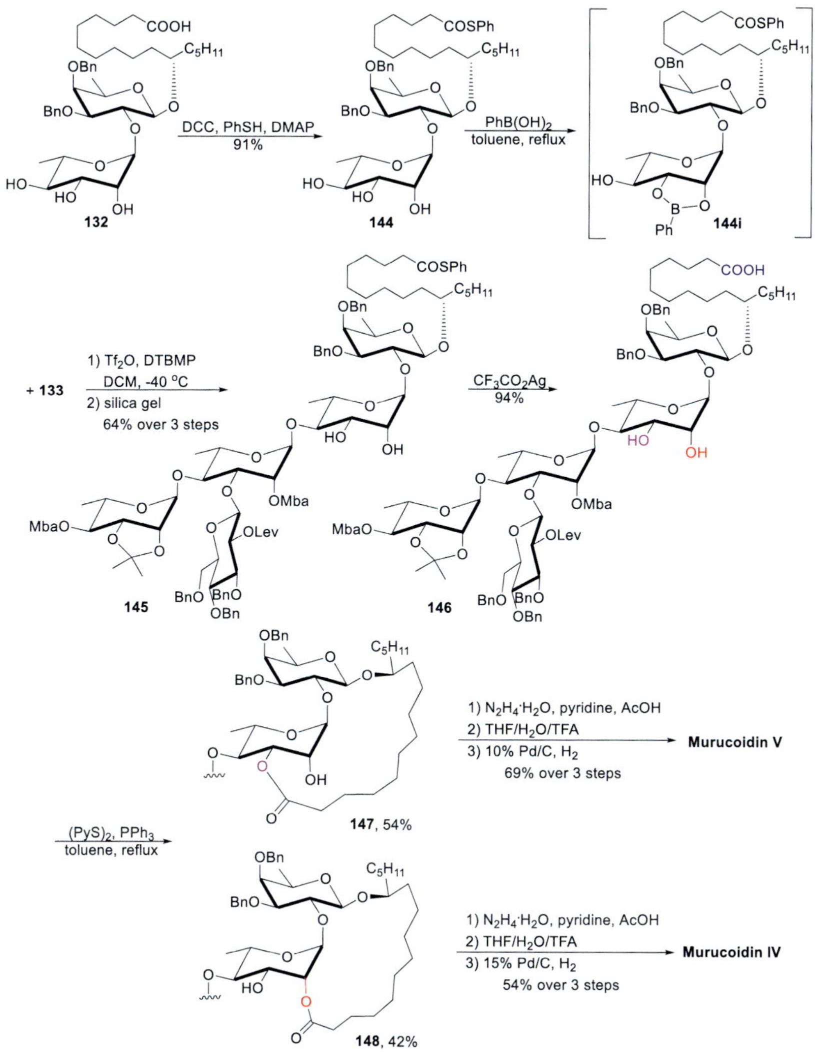

SCHEME 11.17 Synthesis of murucoidins IV and V by macrolactonization.[30]

Lev with hydrazine, Mba with TFA, and Bn by hydrogenation afforded the corresponding natural products murucoidins IV and V.

11.2.7.2 Synthesis of Tricolorin A

Wan's team also synthesized tricolorin using the one-pot IPRm glycosylation method (Scheme 11.18).[31] The IPRm glycosylation of The S-2-[(propan-2-yl)sulfinyl]benzyl (SPSB) glycosyl donor **149** coupling with alcohol acceptor **140** afforded the intermediate **150** similarly as described in

Scheme 11.16 for the murucoidins synthesis. The three acetyl groups were removed, and 3,4-diol was protected with isopropylidene acetal to give intermediate **151**, which has a free 2-hydroxyl group. This was then reacted with the SPSB glycoside **152** through the IPRm glycosylation condition to form the disaccharide **153**, under basic hydrolysis conditions, the benzoate and the methyl ester were cleaved to give the intermediate **154**. Under a similar macrolactonization reaction described in Scheme 11.17 under Corey–Nicolaou protocol, using 2,2′-pyridyl disulfide and triphenylphosphine refluxing in toluene for 3 days afforded the C3′- cyclized macrocycle **155** in 74% yield and the C2′-cyclized macrolacone **156** in 14% yield. The glycosylation coupling of **155** with the SPSB disaccharide **157** afforded the tetrasaccharide glycoside **159** in 76% α- and 21% β-isomer; when using the OPSB glycoside **158**, however, the only product obtained was the desired α-isomer in 92% yield. The benzylidene acetal and isopropylidene acetal in **159α** were removed by treatment with DL-1,4-dithiothreitol (DTT) and CSA to afford the intermediate **160** in 90% yield, global debenzylation by catalytic hydrogenation afforded Trichlorin A in 98% yield (Scheme 11.18).

11.2.7.3 Synthesis of Merremoside D

Using a one-pot relay IPRm glycosylation method, the team also completed the total synthesis of Merremoside D.[32] As shown in Scheme 11.19, the glycosylation of SPSB intermediate **161** with thioglycoside **162** and a long chain alcohol **140** generated the intermediate **163** in a one-pot fashion. Hydrolysis of **163α** produced the intermediate **164**, which contains three free hydroxyl groups and one free carboxylic acid group. This compound was then subjected to macrolactonization reactions. The authors found that using 1.5 equiv. of (*S*)-BTM (benzotetramisole) along with 1.0 equiv. of Piv_2O as esterification reagents, the regio-selective macrolactonization product **165** was obtained in good yield. The regioselective esterification using (*R*)-BTM and Lev_2O furnished the macrolactone **166**. Another one-pot relay glycosylation reaction using SPSB glycoside **167** and thioglycoside **168**, the resulting disaccharide intermediate was then reacted with **166** in a one-pot reaction to afford the tetrasaccharide **169**, which upon deprotection of Lev and isopropylidene acetal finally generated the desired natural product Merromoside D. This method is shortened by several steps using the chiral catalysts for site-selective esterification.

Sharif et al. reported the *de novo* synthesis of Merremoside D.[33] The key steps of the synthesis are shown in Scheme 11.20. The intermediate **170** was synthesized from the pyranone glycosyl donor **171** and the jalapinoic ester **140** in four steps. The Pd-catalyzed glycosylation reaction of **170** with **171** gave the disaccharide **172**, which was then converted to

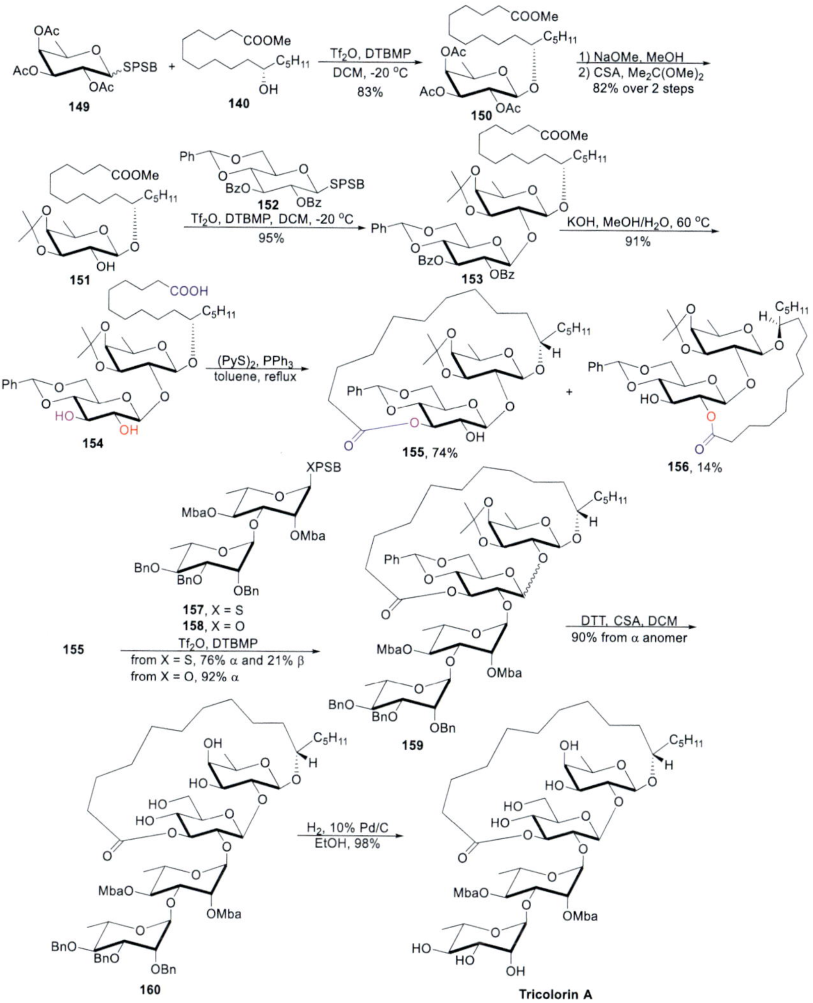

SCHEME 11.18 Synthesis of the disaccharide macrocycle part of Tricolorin A.[31]

intermediate **173** by protecting the free OH with benzyl and OsO_4-mediated dihydroxyation of the alkene and ester hydrolysis. The intermediate was subjected to the Cory-Nicolaou macrolactonization protocol and converted to the macrocycles **174** and **175** in a 2:1 ratio. The C2′-cyclized product **175** was the desired intermediate for the natural product, and this was then used for the construction of the tetrasaccharide intermediate **178** by glycosylation reaction. After deprotecting the acetonide with TFA, the removal of chloroacetyl groups using thiourea provided the natural product Merremoside D in good yield.

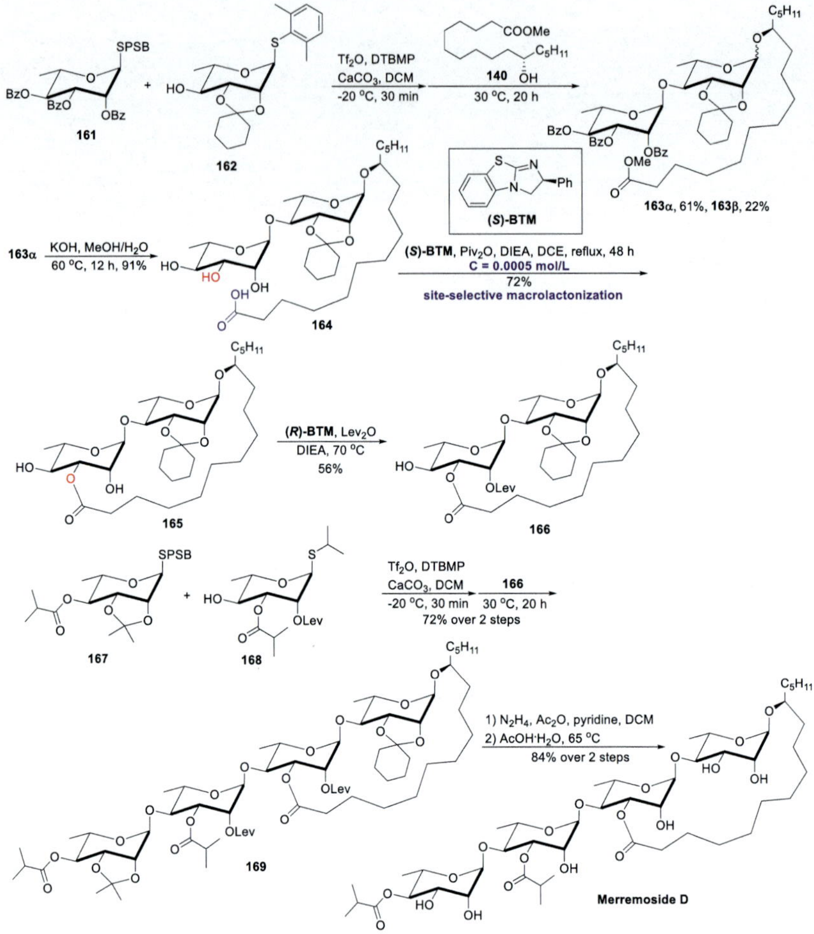

SCHEME 11.19 Synthesis of Merremoside D by one-pot relay glycosylation.[32]

11.2.8 Synthesis of Batatin VI

Batatin VI is a complex resin glycoside dimer, and each contains tetrasaccharide moieties. Zhu et al. reported the total synthesis of sugar macrolactone nature product Batatin VI.[34] As shown in Scheme 11.21, the known compound **152** was reacted with the fucoside derivative **179** to form the disaccharide **180**; upon hydrolysis, the free acid derivative **181** with a free 2′-OH was obtained. The authors explored different strategies for macrolactonization and used the Keck macrolactonization method, in which the activated ester was generated *in situ*, several conditions were explored and the best performing conditions were using *N,N′*-dicyclohexylcarbodiimide (DCC), pyridine, pyridinium *p*-toluenesulfonate (PPTS) in 1,2-dichloroethane (DCE) under refluxing

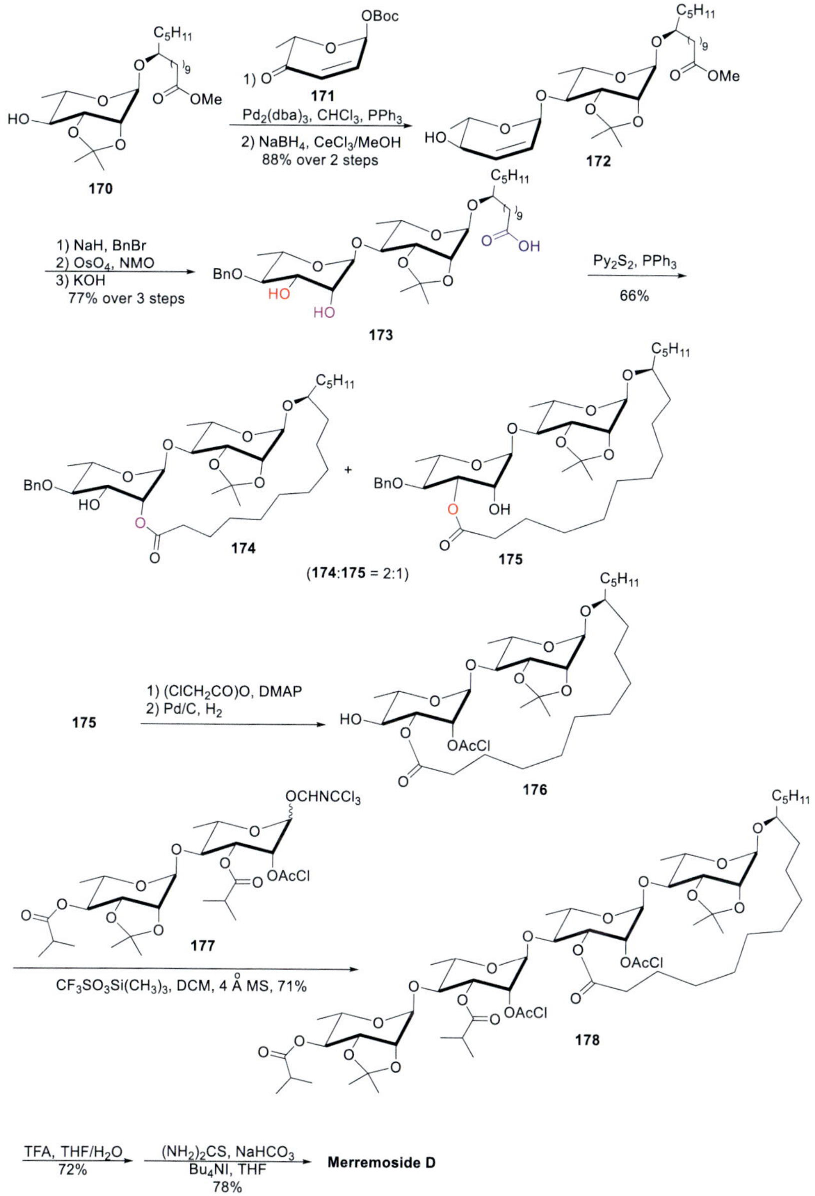

SCHEME 11.20 Synthesis of Merremoside D.[33]

for 3 hours, the macrolactone **181** was obtained in 95% yield. The desilylation was carried out using TBAF to give the glycosyl acceptor **182**, which then reacted with the glycosyl donor **183** to generate the tetrasaccharide derivative **184** in 85% yield. The glycosyl donor **183** was synthesized from the

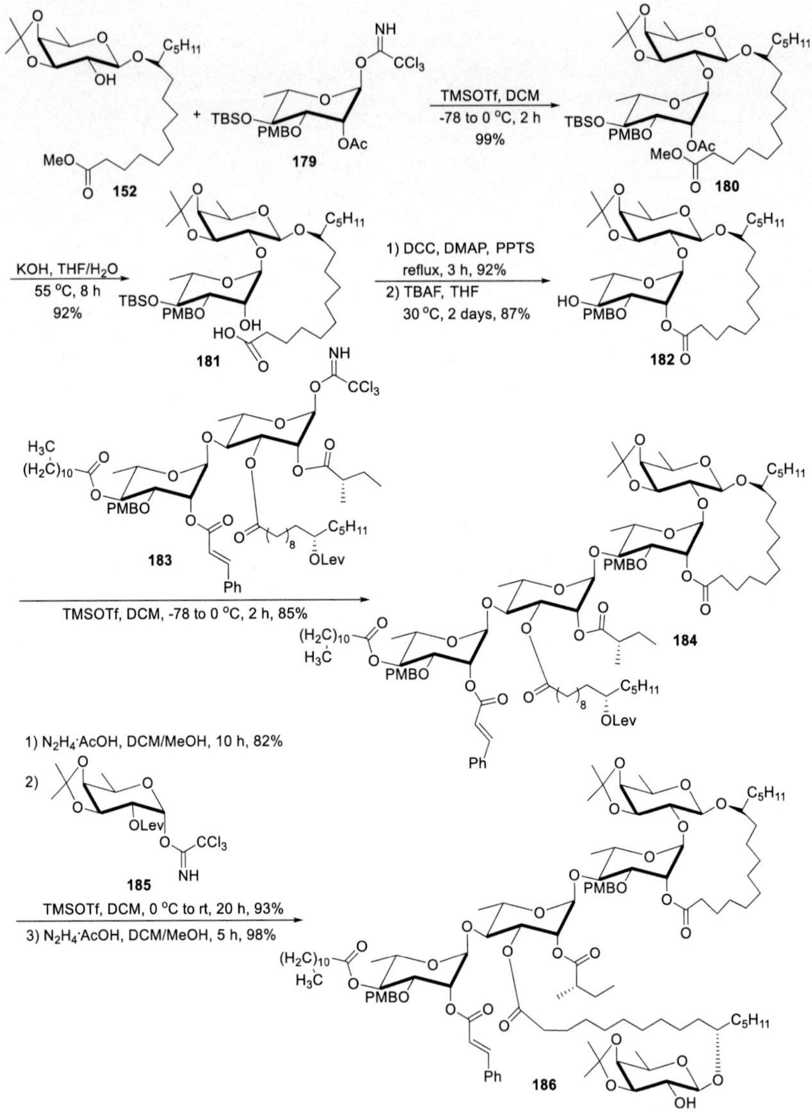

SCHEME 11.21 Synthesis of top half of Batatin VI.[34]

corresponding monosaccharides in five steps.[34] The side-chain Lev protecting group in **184** was removed by treating with hydrazine and then was coupled with the fucosyl donor **185** to form the corresponding glycoside and after the Lev deprotection led to the key intermediate **186** in overall good yield.

To complete the total synthesis of Batatin VI (Scheme 11.22), they synthesized the tetrasaccharide dimer **188** by glycosylation of compound **186** with the trisaccharide derivative **187**. The protecting groups were then

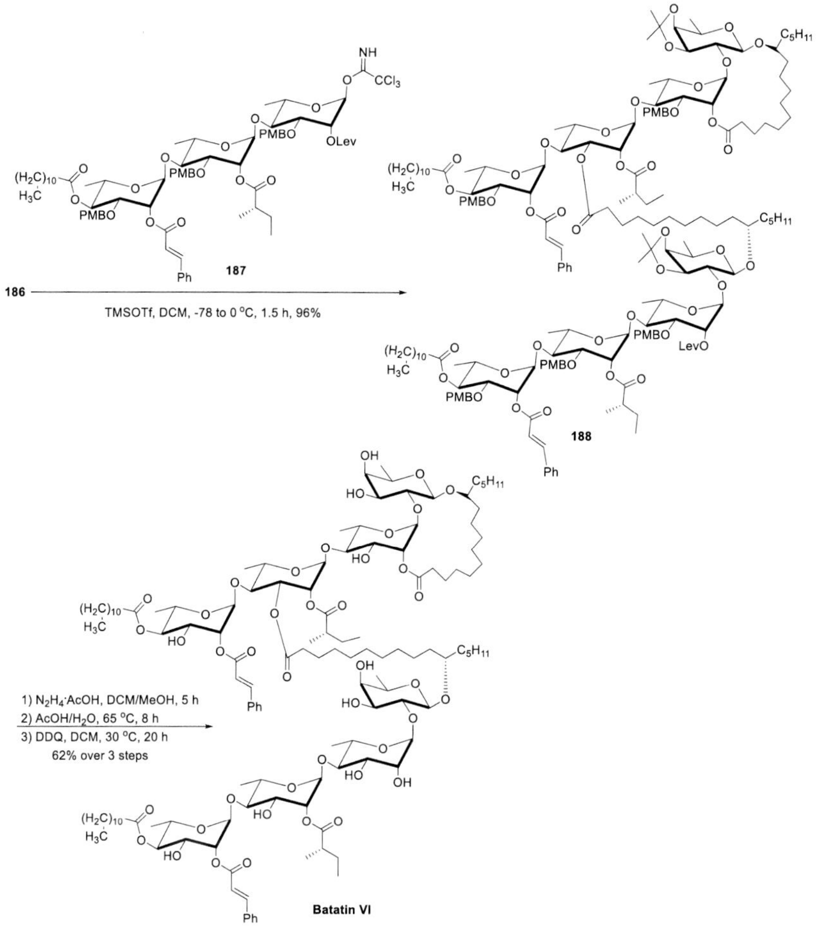

SCHEME 11.22 Completion of the total synthesis of Batatin VI.[34]

removed sequentially in three steps to afford the natural product Batatin VI in good yields.

11.2.9 Synthesis of ellagitannins and other related strained glycomacrocycles

Ellagitannins are a large family of natural products, which typically comprise D-glucose, galloyl group(s), and one or more axial chiral hexahydroxydiphenoyl (HHDP) groups. More than 1000 compounds in the family have been reported. The glucose is acylated with galloyl groups and hexahydroxydiphenoyl (HHDP) and dehydrohexahydroxydiphenoyl (DHHDP) groups. A few general examples of ellagitannins that will be reviewed are shown in Fig. 11.14.

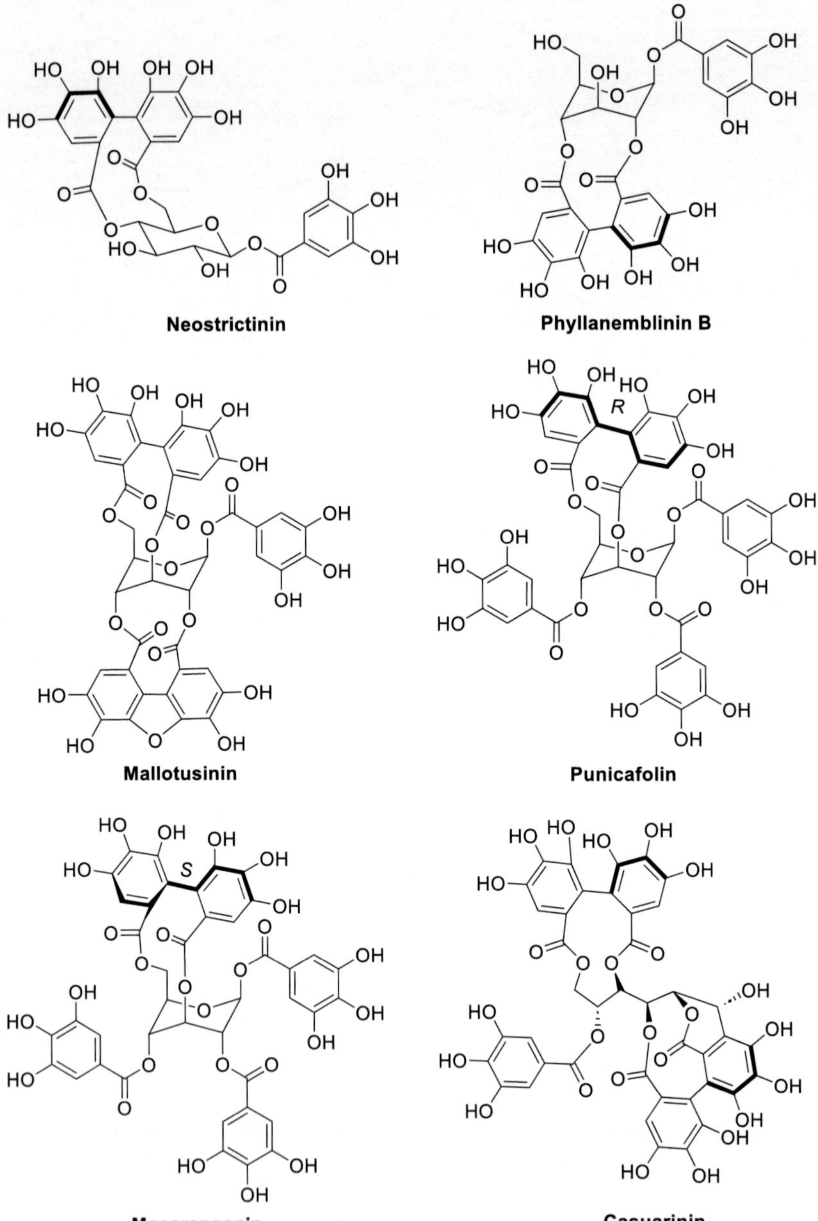

FIGURE 11.14 General structures of ellagitannins.

11.2.9.1 Synthesis of Neostrictinin by macrolactonization

A total synthesis of Neostrictinin was reported by Ikeuchi et al.[35] The synthesis is shown in Scheme 11.23, starting from the biphenyl anhydride **189** (a*R* major 87% *ee*) reacting with the glucose diol **190** selectively at the 6-hydroxyl group afforded the intermediate **191**. Several macrolactonization conditions were explored for compound **191**, a combination of EDCI and DMAP as well as two other conditions failed to produce the macrolacone, oxalyl chloride and DMAP afforded only 10% of the ester product, however Mukaiyama esterification condition produced satisfactory macrolactonization. Using CMPI and 2,6-lutidine in toluene at 100°C led to **192** in 29% isolated yield and a 3.5:1 diastereomeric ratio of a*R*/a*S*; further optimization using 5.0 equiv. of CMPI and 10.0 equiv. of 2,6-lutidine afforded the compound **192** in improved yield of 55% and a 7.5:1 diastereomeric ratio, with major a*R* isomer as the desired diastereomer. After they established the macrolactonization protocol, they started with 99% optically pure compound **189**, coupling reaction with **190** afforded the intermediate **191**, under the optimized condition the product **192** (*R*-isomer) was obtained as single isomer in 69% isolated yield. The ethylthio at the anomeric position was hydrolyzed using NIS and TfOH to the free glucoside derivative **193**. Esterification with tri-*O*-benzyl galloyl chloride **194** afforded a 1:1 mixture of the α- and β-anomers **195**. The **195**β was converted to the desired product

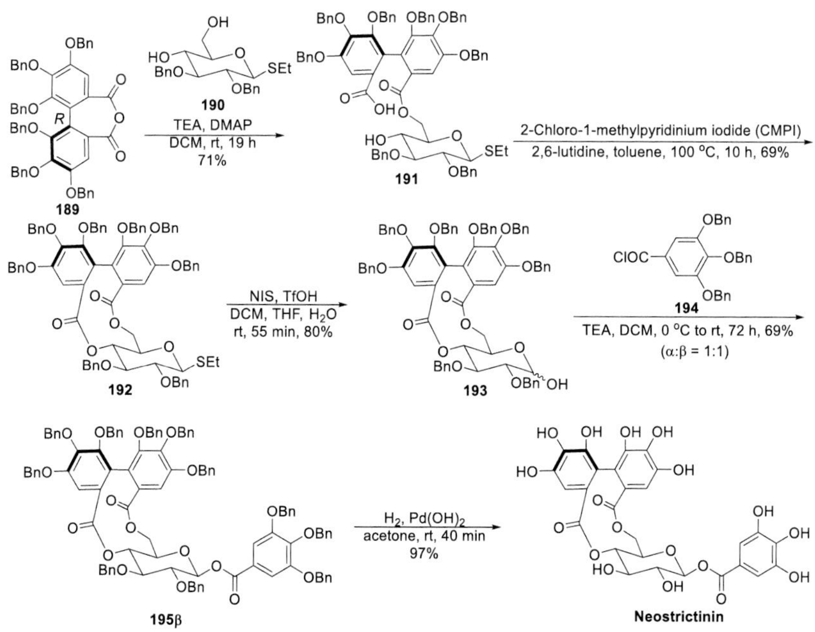

SCHEME 11.23 Total synthesis of Neostrictinin.[35]

Neostrictinin by catalytic hydrogenation debenzylation. The epi-Neostrictinin was also synthesized using the same protocol from the **195α** isomer.

11.2.9.2 Synthesis of Phyllanemblinin B

Matsumoto recently synthesized the natural product Phyllanemblinin B, which contains a β-galloyl at anomeric position and 2,4-*O*-hexahydroxydiphenoyl (HHDP) group on a glucose derivative.[36] This ellagitannin is a stereoisomer of Neostrictinin (4,6-linked macrocycle) with C2,4-linked macrocycle through the 2,4-*O*-HHDP bridge. In this synthesis, the HHDP group was synthesized by oxidative coupling of two galloyl groups, which is also the typical method for the preparation of the HHDP group for ellagitannins. The authors explored the formation of the HHDP group previously too and envisioned that using an open-chained glucose derivative would be an efficient method for the total synthesis. The total synthesis of Phyllanemblinin B is shown in Scheme 11.24. Starting from the known compound **196**, protection of the 2- and 4-hydroxyl groups using methoxylmethyl (MOM) chloride afforded the intermediate **197**. Hydrolysis of the phenyl thioacetal using *N*-bromosuccinimide (NBS) and water to release the free aldehyde at C-1 followed by Wittig olefination reaction with methyl ylide and subsequent protection of the 4-OH with PMB group afforded the open chain glucose derivative **198** in 50% overall yield.

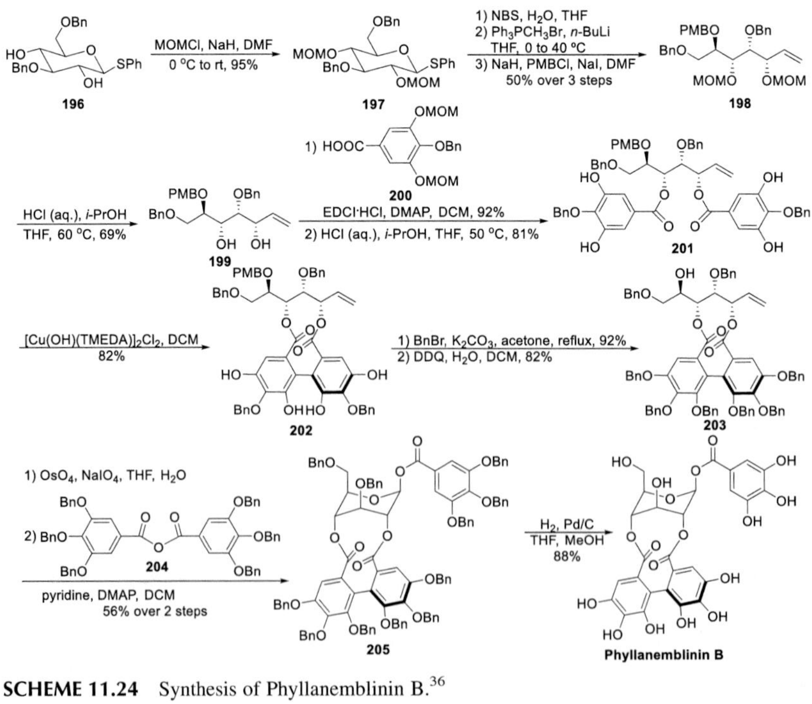

SCHEME 11.24 Synthesis of Phyllanemblinin B.[36]

The MOM ether was deprotected to give the free diol **199**, which was then acylated with MOM-protected galloyl derivative **200** and removal of the MOM groups afforded intermediate **201**. This compound was then subjected to oxidative C–C bond formation reactions, and they found that using $CuCl_2$ and butylamine in methanol only afforded 6% of the desired product **202**, however oxidation using $[Cu(OH)(TMEDA)]_2Cl_2$ in dichloromethane resulted in **202** in 82% yield. The stereochemistry of the compound was established as a single diastereomer by converting **202** to a known compound and comparing their optical rotations, the compound was deemed to be the *R*-isomer. With the correct stereoisomer in hand, they completed the total synthesis by perbenzylation and removal of the PMB group of **202** to afford the intermediate **203**. The terminal double bond was then oxidized using osmium tetroxide and cleaved with sodium periodate to regenerate the C-1 aldehyde moiety, further acylation at O-1 by *O*-benzylated gallic acid anhydride **204** afforded the pyranose ring-closed compound **205** in 56% over these two steps. Finally, global catalytic hydrogenation afforded the target compound Phyllanemblinin B in good yield.

11.2.9.3 Synthesis of Mallotusinin

In a related study, Yamashita reported the total synthesis of mallotusinin, which belongs to monomeric ellagitannins, typically comprising D-glucose and galloyl group(s).[37] The synthesis started from the known 1,2,4-*ortho*-acetylglucose **206**[38] for mallotusinin and corilagin. As shown in Scheme 11.25, diacylation with protected galloyl acid **207** afforded the intermediate **208**, oxidative intramolecular phenol coupling mediated by $CuCl_2/BuNH_2$ resulted in the 3,6-*O*-(*R*)-HHDP-bridged macrocycle as a single isomer, which after full benzylation afforded the intermediate **209**. The stereochemistry was confirmed by methanolysis of **209** and comparison with known enantiopure *R*-isomer. Next, they treated **209** with benzotrifluoride (BTF) containing 4 Å molecular sieves, thiophenol, and indium bromide, which afforded the anomeric phenylthiolation product with the orthoester cleaved to afford the intermediate **210a**, which was acetylated to form the compound **210**. Glycosyl esterification with the tri-*O*-Bn-gallic acid **211** similarly as described earlier provided the macrocycle **212** with β selectivity in 82% yield as a single isomer. The selective removal of the acetyl groups using hydrazine afforded the intermediate **213**, which upon global debenzylation led to the natural product corilagin. Alternatively, the intermediate **213** can be coupled with the bridged diacid **214** to afford the dilactone **215**, after global debenzylation will lead to the natural product mallotusinin. Several di-esterication conditions of the diol **213** with the diacid were explored; however, the best yield obtained for the dilactone was only 15% when using EDCI and 4-morpholinopyridine; all other conditions explored failed to generate the dilactone **215** directly. So they turned to a two-step process of forming the monoester intermediate **215i** using EDCI, 4-pyrrolidinopyridine (PPY), or DCC and PPY, the intermediate **215i** then

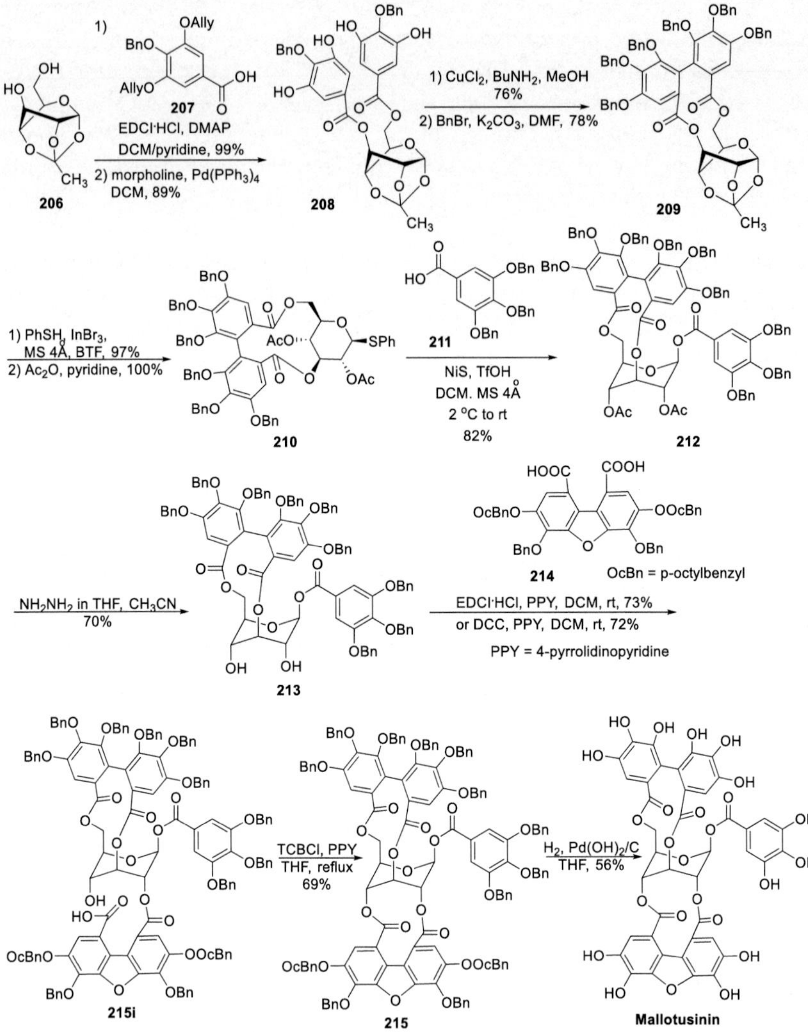

SCHEME 11.25 Synthesis of Mallotusinin.[37]

underwent a second-step macrolactonization reaction to afford the key dilactone **215**. The intramolecular lactonization condition was screened and the best result was using TCBCl, PPY in THF refluxing to afford the dilactone in 69% yield. Final global debenzylation produced Mallotusinin.

11.2.9.4 Synthesis of Macaranganin

Shibayama reported the total syntheses of punicafolin and macaranganin in seven steps by catalytic regioselective acylation.[39] The two ellagitannins contain 3,6-HHDP-bridged glucose and are esterified with gallic acid at

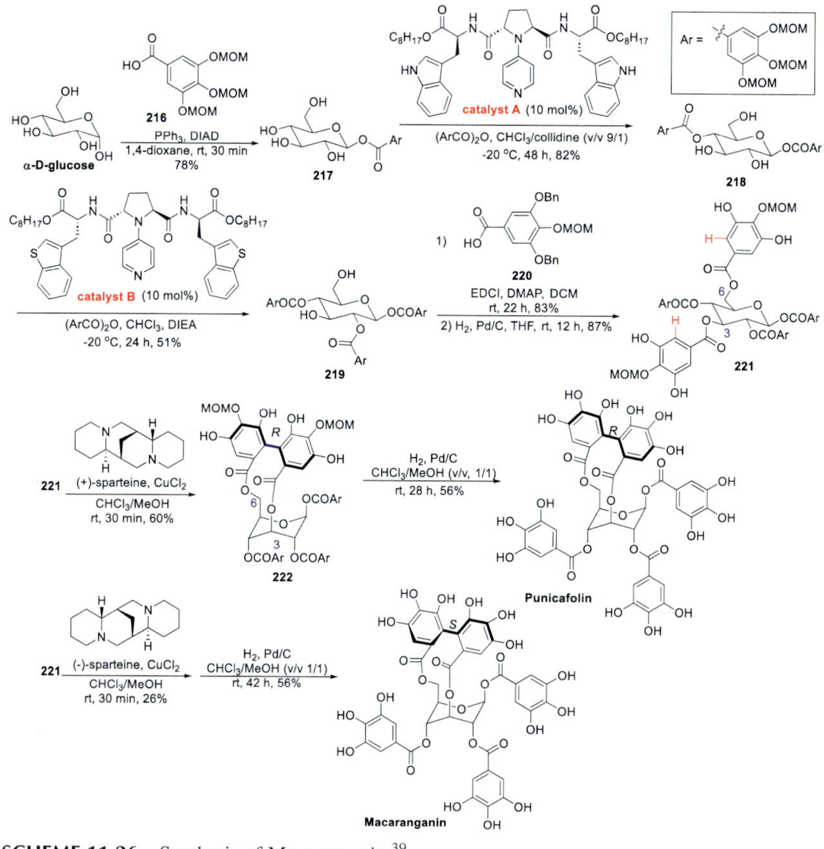

SCHEME 11.26 Synthesis of Macaranganin.[39]

1,2,4-positions. As shown in Scheme 11.26, the synthesis started from α-D-glucose, esterification with the MOM-protected gallic acid **216** under Mitsunobu conditions afforded the β-gallate **217** in 78% yield, and selective esterification using catalyst **A** afforded 1,4-digallate **218**. This was then further converted to 1,2,4-trigallate **219** using catalyst **B**. Esterification of the 3,6-OHs with the protected gallic acid **220** followed by catalytic hydrogenation afforded the key intermediate **221**. Oxidative phenol coupling using copper chloride and (+)-sparteine afforded the 3,6-HHDP-bridged macrocycle **222** as the single a*R* isomer; deprotection of the MOM and benzyl groups under catalytic hydrogenation using methanol as solvent afforded the natural product punicafolin. The oxidative coupling reaction using the (−)-sparteine led to the macrocycle as the pure a*S* isomer and after deprotection, the natural product Macaranganin was obtained. The synthesis is short and efficient since no protecting groups were used for glucose and both stereoisomers of 3,6-HHDP bridge were constructed from a common intermediate.

11.2.9.5 Total synthesis of Casuarinin

Wakamori et al. reported the total synthesis of another elligitannin, Casuarinin.[40] As shown in Scheme 11.27, the synthesis started from the ethyl thio glucoside **223**, esterification with the protected gallic acid **207** led to the intermediate **224**. The 4,6-benzylidene acetal in **224** was removed and subsequent esterification of the diol with the gallic acid derivative **200** afforded the tetraester **225**. The four MOM groups were removed under acetic conditions and the resulting intermediate underwent an oxidative phenyl coupling reaction to afford the HHDP-bridged compound **226**. The four free hydroxyl groups in the molecule was protected with Bn group, then the four allyl groups were deprotected to afford intermediate **227**. This was converted to the HHDP-bridged intermediate **228** in a few steps, oxidative coupling, and MOM protection followed by thiolethyl hydrolysis. The anomeric aldehyde then reacted with benzylamine and 5-OH was gallated using the protected gallic acid to

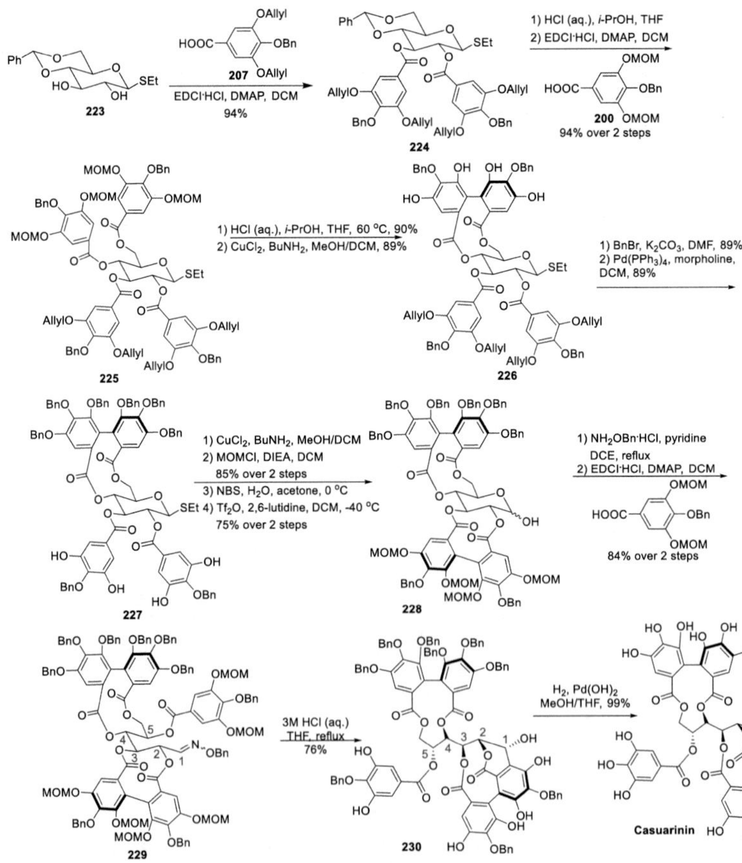

SCHEME 11.27 Total synthesis of Casuarinin.[40]

afford the intermediate **229**. The final step involves the formation of C–C bonding between the C-1 and the HHDP bridge at the 2-position. Compound **229** was treated with 3M HCl to form the C-glycoside **230** as a single diastereomer in 76% yield through a Friedel–Crafts-type *C*-glycosylation. Catalytic hydrogenation of **230** afforded the Casuarinin in good yield.

11.2.10 Synthesis of glycoterpenoid and steroid derived glycomacrocyles

A few macrocyclic natural product analogs containing steroids and sugar were synthesized. These are shown in Schemes 11.28 and 11.29.[41–43] Scheme 11.28 shows an example of the synthesis of a glycoterpenoid.[41] Starting from D-glucurono-3,6-lactone, treatment with NaOMe followed by acetylation afforded methyl-1,2,3,4-tetra-*O*-acetyl-β-D-glucopyranuronate **231**. This was then converted to the bromide **232** by treating with HBr/HOAc. Compound **232** reacting with diacid **233** in the presence of tetrabutylammonium bromide (TBAB) afforded the intermediate **234**. The methyl ester was converted to

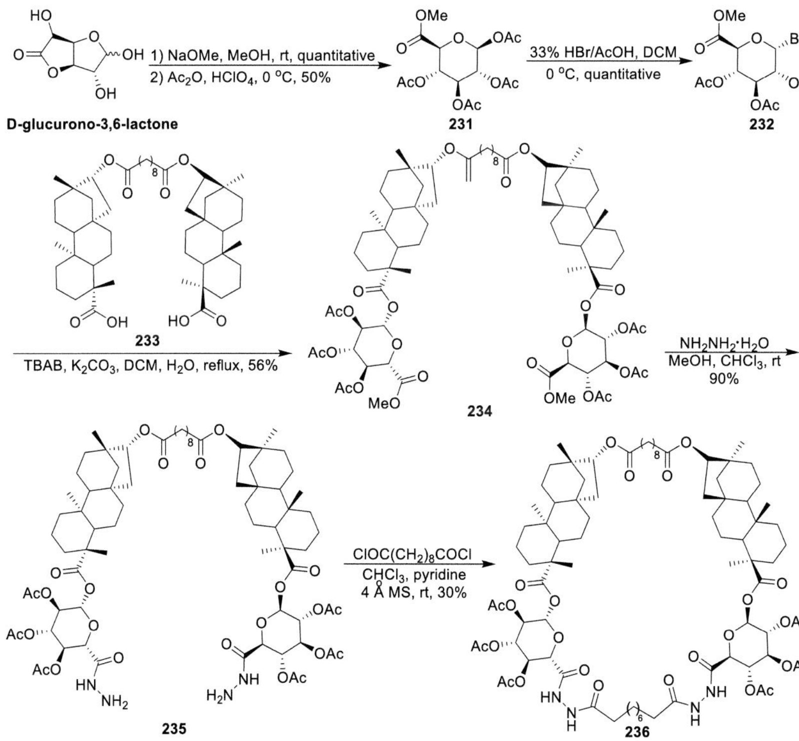

SCHEME 11.28 Synthesis of glycoterpenoid.[41]

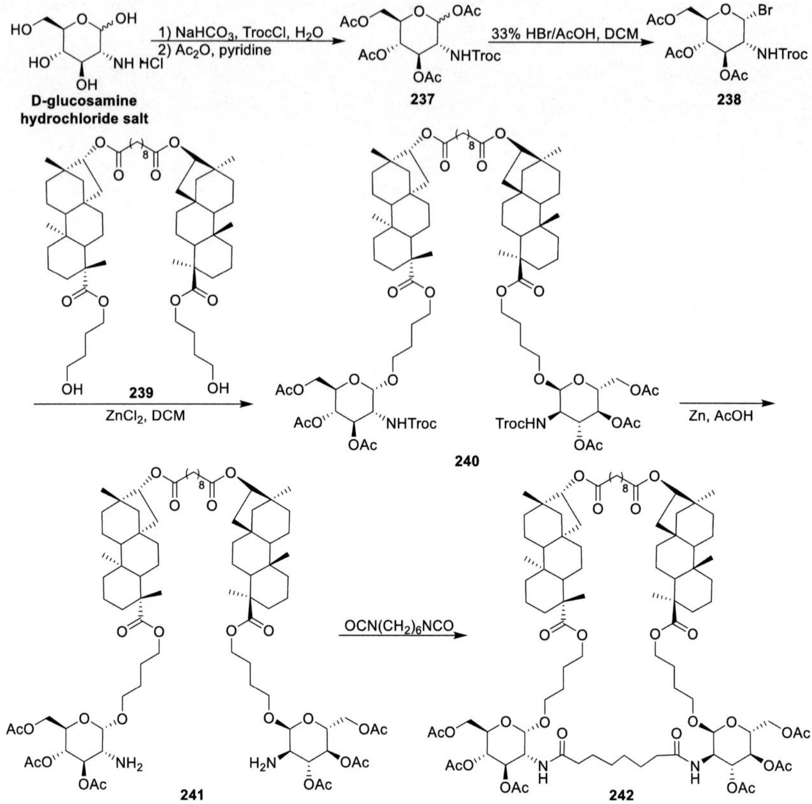

SCHEME 11.29 Synthesis of glycoterpenoid.[42]

dihydrazide **235**, which was double acylated with diacid chloride to form the glycoterpenoid macrocycle **236**.

The preparation of another glycoterpenoid is shown in Scheme 11.29.[42] D-Glucosamine was protected with Troc group and peracetylated to form intermediate **237**, the 1-OAc was converted to the α-bromo sugar **238**, glycosylation reaction with the diol **239** afforded the intermediate **240**. The Troc group was deprotected to form the amine derivative **241**, and the condensation with diisocycanate afforded the macrocycle **242**.

11.3 Sugar-based crown ether, azacrown ether, and other crown ether-like macrocycles

Carbohydrate-based crown ethers and azacrown ethers are special types of chiral macrocycles; their structures typically are composed of medium-sized cyclic molecules that are useful for chiral recognition, cation complexation, catalytic reactions, and so on. Many related research topics have been extensively reviewed already. Chiral crown ethers prepared from sugars have exhibited a capacity for chiral differentiation for enantiomers as summarized in the recent review by Bakó et al.[44] There are also many recent reviews on azacrown ether for asymmetric synthesis and applications.[45,46] Using sucrose as the chiral template, Jarosz et al. have synthesized several macrocyclic receptors and nitrogen-containing cyclic aggregates and these have been reviewed recently.[47,48] In this chapter, we will only highlight some recent developments of these systems.

11.3.1 Monosaccharide-based crownether derivatives formed by vicinal diol

Baker et al. reported the synthesis of several sugar- and sugar alcohol-based azacrown ethers.[49–53] Crown ether-based glycomacrocycles (no hetero-atoms),[44,54–56] several thia-containing and heterocycle-containing crown ether-based glycomacrocyles, were also reported recently.[56–61] The typical synthesis involves straightforward S_N2 reactions to close the ring, sometimes an amine is used to link two chains and form a ring. Glucose-based macrocyclic ether from C-1 to C-4 cyclization exhibited utilities for chiral (enantiomeric) recognition of amino-acid esters.[54]

Rapi et al. have reported the synthesis of various monosaccharide-derived azacrown ethers in recent years,[62–66] They discussed the applications of various azacrown ether-based glycomacrocycles for applications in various asymmetrical syntheses.[64,67–69] The typical preparation of the crown ether system is shown in Scheme 11.30, the 4,6-benzylidene acetal protected galactose derivative **243** was alkylated at the 2- and 3-OH to give the intermediate **244**, which was converted to the di-iodide **245**. Another double S_N2 reaction using an amine to displace the iodo groups in compound **245** led to the formation of the aza-crown ether derivatives **246**. Another type of mono-aza crown ether sugar derivative with nitrogen attached to the sugar ring at the C-3 position is shown in Scheme 11.31. Regioselectively opening of

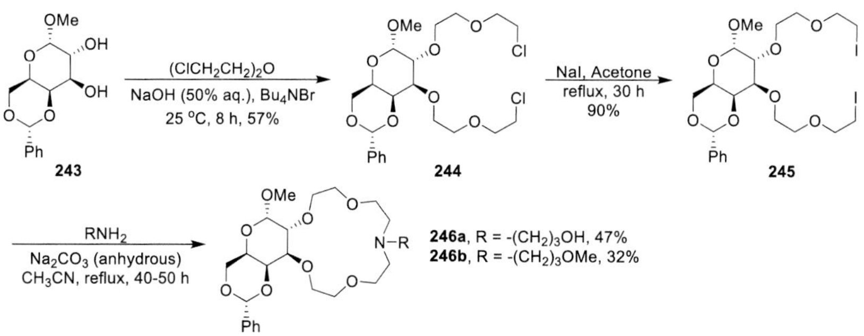

SCHEME 11.30 Synthesis of α-D-galactose-based azacrown ether **246a** and **246b**.[70]

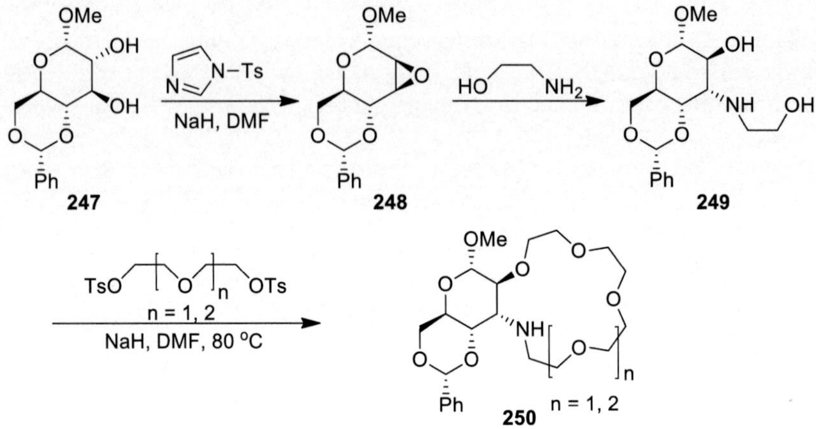

SCHEME 11.31 Synthesis of altrose-based azacrown ethers **250**.[63]

epoxy sugar **248** with ethanolamine afforded the amino sugar **249**, which served as the key intermediate for the sugar azacrown ethers **250** using tri or tetraethylene glycol bis-tosylates as the ring-closing reagents.

The two schemes discussed earlier summarized the general methods utilized for crown ether synthesis. Starting from different sugar-starting materials and using different nucleophiles, a large class of sugar-based crown ethers have been prepared. Here we include a few examples of the final structures of crown ethers in the next section, the synthesis will not be be discussed since they were synthesized by a similar strategy.

Several glucose-based crown ether derivatives **251–253** shown in Fig. 11.15 were studied as chiral phase transfer catalysts. Bakó reported the catalytic properties of the crown ether **251** for asymmetric Michael addition reactions.[69] The malonate and enone conjugate addition reactions proceeded with moderate yields and up to 60%–97% *ee*.[71] The crown ether analog **252** was shown to be able to catalyze Michael's addition of *N*-protected α-aminophosphonatesonto acrylic acid derivatives with *trans*-β-styrene in good enantiomeric selectivity (up to *ee* 96%).[72] Glucopyranoside-based lariat ether **253** containing a side arm of $-(CH_2)_3O(CH_2)_2OMe$ gave enantioselectivity of 99% for catalyzing Darzens condensation of phenacyl chloride and benzaldehyde.[62]

As shown in Fig. 11.16, six novel L-threitol-based crown ethers **254a–f** were prepared starting from 2,3-isopropylidene-protected L-threitol. Asymmetric Darzens condensation reactions catalyzed by monosaccharide-based crown ethers were studied. D-glucose-based chiral crown ether **251** was proved to be a much more efficient catalyst in terms of enantioselectivities (up to 96% *ee*) compared to D-mannose-based chiral crown ether **255**, which generally resulted in 20%–30% *ee*.

Crown ethers bearing D-glucopyranoside, D-mannopyranoside, and D-altropyranoside units were tested in a few asymmetric phase transfer

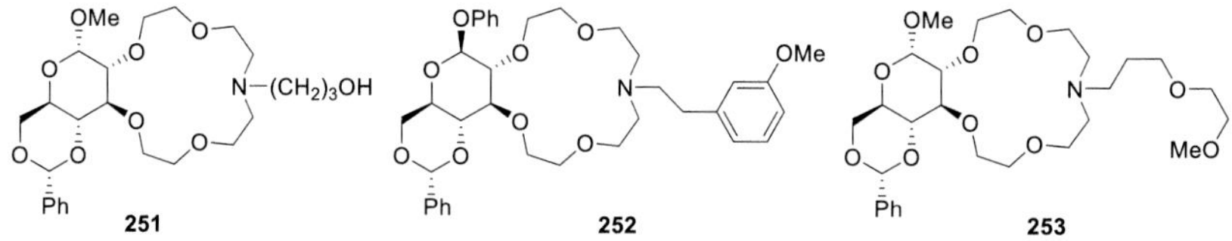

251 252 253

FIGURE 11.15 Structures of glucose-based crown ether derivatives **251–253**.

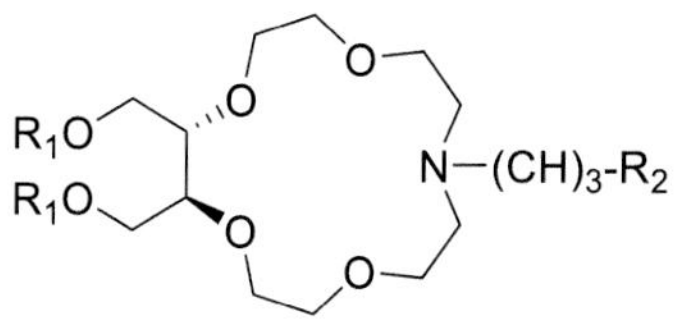

254a, R_1 = Me, R_2 = OH
254b, R_1 = Me, R_2 = OMe
254c, R_1 = *n*-Bu, R_2 = OH
254d, R_1 = *n*-Bu, R_2 = OMe
254e, R_1 = Bn, R_2 = OH
254f, R_1 = Bn, R_2 = OMe

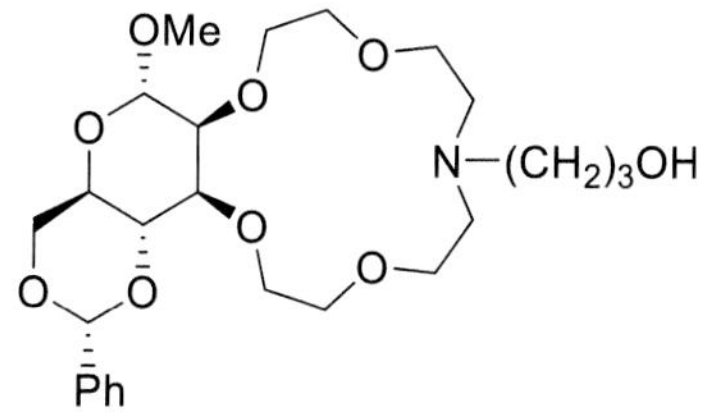

255

FIGURE 11.16 Structures of L-threitol-based crown ethers **254a–f**[66] and D-mannose-based chiral crown ether **255**.[67]

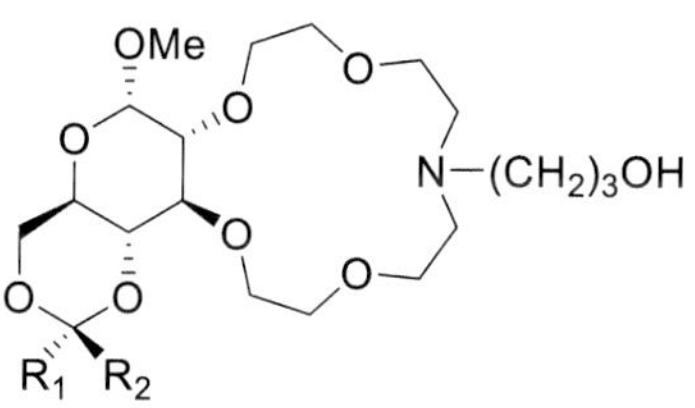

251, R_1 = Ph, R_2 = H
256a, R_1 = 1-naphthyl, R_2 = H
256b, R_1 = R_2= Me

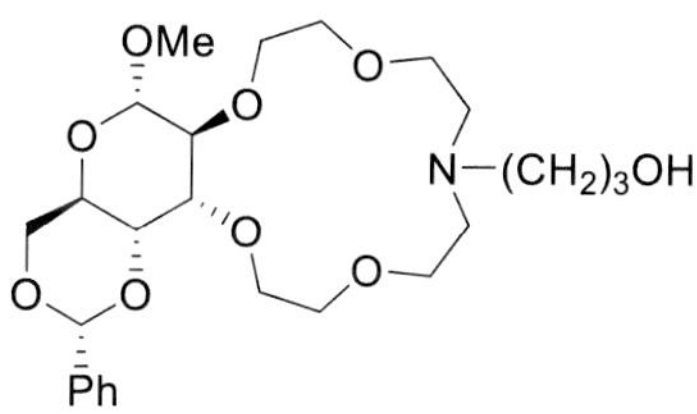

256c
Altrose

FIGURE 11.17 Sugar-based chiral glucose crown ethers **256a–b** and altrose-based chiral crown ether **256c**.[49,68]

cyclopropanation reactions (Fig. 11.17). It was found that, in all model reactions, utilizing D-mannose-based crown ether **255** and D-altrose-based crown ether **256c** gave the lowest *ee* values compared to D-glucose-based crown ethers **251** and **256a-b**.

Four ribose-based chiral crown ethers were prepared by Rapi et al. (Fig. 11.18).[63] It was observed that using ribo-hexopyranoside-based lariat ether **257c** under phase transfer catalytic condition afforded an enantioselectivity of 80% in the Michael addition of diethyl acetamidomalonate to *trans*-β-nitrostyrene.

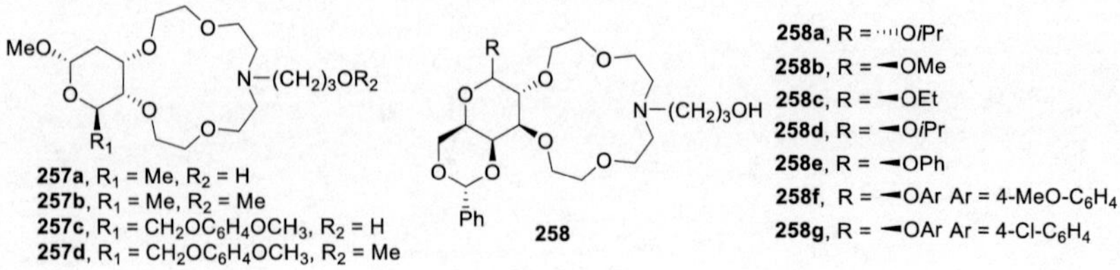

FIGURE 11.18 Chiral crown ethers derived from ribo-hexopyranoside **257a–d,**[63] D-galactose **258a–g.**[65]

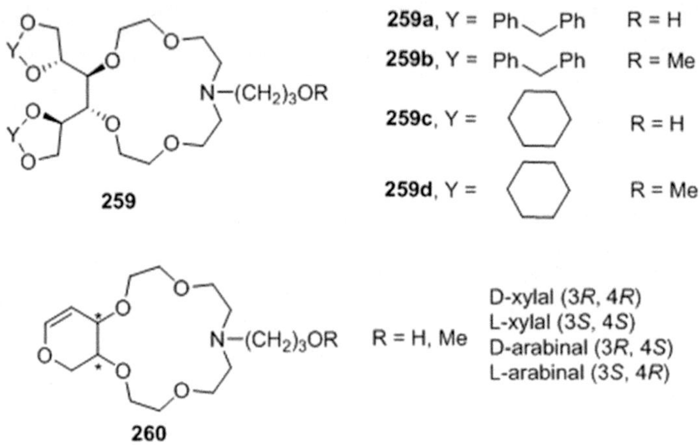

FIGURE 11.19 Structures of D-mannitol-based mono-aza-15-crown-5 type lariat ethers **259a–d,**[51] xylal- and arabinal-based mono-azacrown ethers **260.**[50]

In another study, several α- and β-D-galactopyranoside-based chiral crown ether **258a–g** were synthesized (Fig. 11.18).[65] In all screened asymmetric reactions (Darzens condensation, epoxidation of *trans*-chalcone, and asymmetric cyclopropanation formation), they found that the catalysts with β-configuration performed better than the α-anomers regarding the asymmetric induction.

As shown in Fig. 11.19, D-mannitol-based mono-aza 15-crown-5-type macrocyclic compounds **259a–d** were elaborated as phase transfer catalysts in asymmetric Michael reaction. Some synthesized xylal and arabinal-based chiral crown ether **260** also performed well as phase transfer catalysts in several asymmetric syntheses.

11.3.2 Synthesis of other crown ether-like macrocycles

Kim et al. synthesized and analyzed a series of macrocyclic C-aryl glycosides as sodium-dependent glucose transporter 2 (SGLT2) inhibitors for diabetes treatment.[73] The synthesis of macrocycle, which showed good in vitro activity, is shown in Scheme 11.32. Cyclization of glucose derivative **261** by S_N2

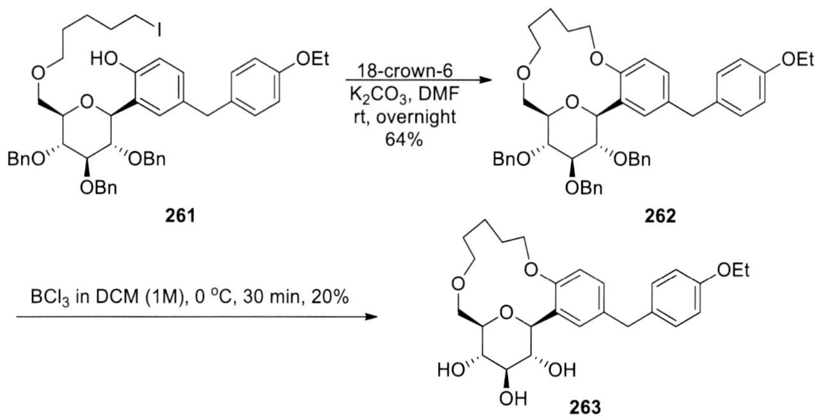

SCHEME 11.32 Preparation of ether-based macrocycle **263** by S_N2 displacement.[73]

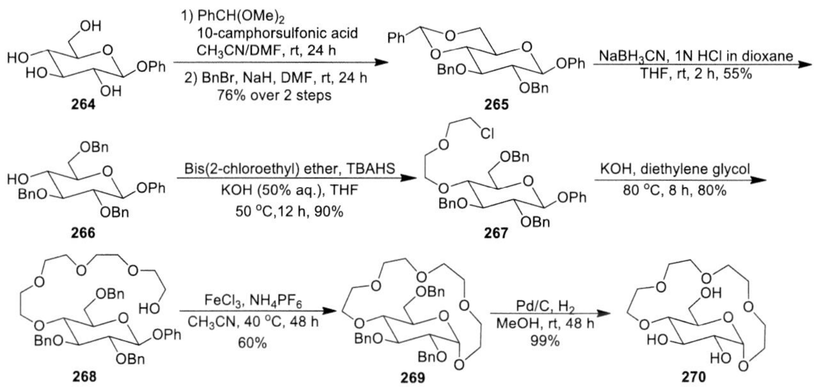

SCHEME 11.33 Synthesis of glucose-derived macrocyclic receptor **270**.[54]

displacement reaction afforded the macrocycle **262**, and debenzylation using BCl_3 afforded the SGLT2 inhibitor **263**, with $IC_{50} = 0.899$ nm against SGLT2.

A glucose-based crown ether **270** linking at C1 and C4 positions was studied by Lewandowski group for chiral recognition of amino-acid methyl esters in aqueous media.[54] As shown in Scheme 11.33, 4,6-benzylidene protection of *O*-phenyl-β-D-glucose **264** followed by benzylation of remaining free hydroxyl groups afforded compound **265**, which underwent reductive cleavage of the benzylidene acetal to give 4-OH compound **266**. The 4-hydroxyl group was derivatized by two consecutive S_N2 reactions to form the intermediate **268**. Intramolecular glycosylation of **268** was carried out using $FeCl_3$ and NH_4PF_6 under high dilution, giving exclusively α-anomer **269** as the product. Global deprotection of benzyl by catalytic hydrogenation provided the macrocyclic receptor **270**.

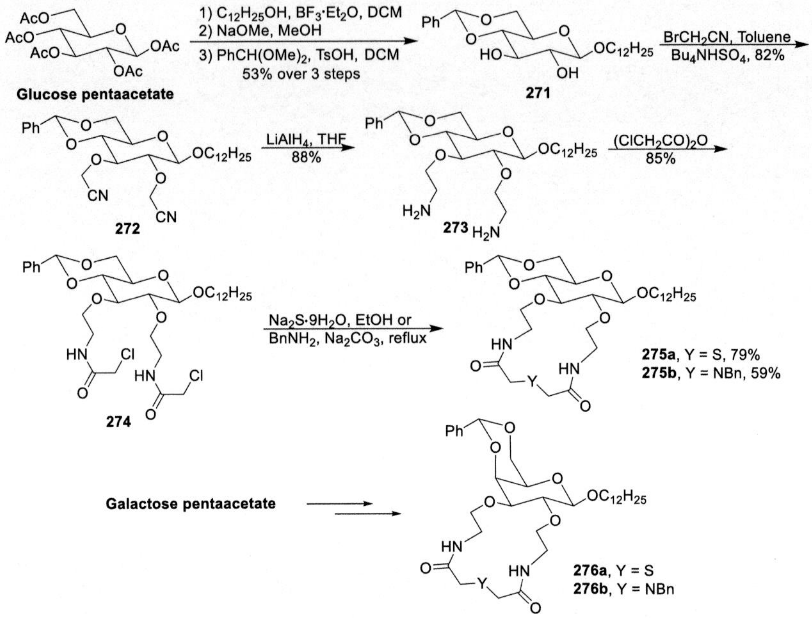

SCHEME 11.34 Synthesis of macrocycles **275–276** incorporating a glucose or galactose-based glycolipid at the 2,3-positions.[59]

Saha et al. reported the synthesis of glucose-based macrocycles containing nitrogen or thia linkages (Scheme 11.34).[59] The synthesis started from the commercially available glucose pentaacetate, glycosylation with dodecanol and BF_3 etherate, followed by deacetylation, and 4,6-protection afforded the intermediate **271**. The 2,3-diol were alkylated with bromacetonitrile to form intermediate **272**, and the reduction of the cyano group afforded the diamine **273**. Further derivatization of **273** into chloroacetamide **274** and subsequent cyclization reaction using either $Na_2S{\cdot}9H_2O$ or benzylamine completed the macrocycle synthesis. The galactose-based macrocycles **276a** and **276b** were prepared in the similar manner.

Fig. 11.20 depicts the structures of four glucose-derived aza and thia macrocycles **277–278** on 2,3 and 4,6-positions; these were synthesized by the same method as described in Scheme 11.34.[60] These mixed heteromacrocyclic glycolipids have been explored as phase transfer catalysts in asymmetric Michael addition reactions.

Glucose crown ether derivatives containing an acridine unit have been prepared as shown in Scheme 11.35.[58] The starting material di-tosylate **279** was reacted with 4,5-dihydroxyacridinone in dimethylformamide (DMF) and Na_2CO_3 through S_N2 reactions to afford the acridinone **280**. Reduction of the intermediate afforded the glucose-based acridine macrocycle **281**. Similarly, using the di-iodide glucose derivative **282**, the acridine analog **283**

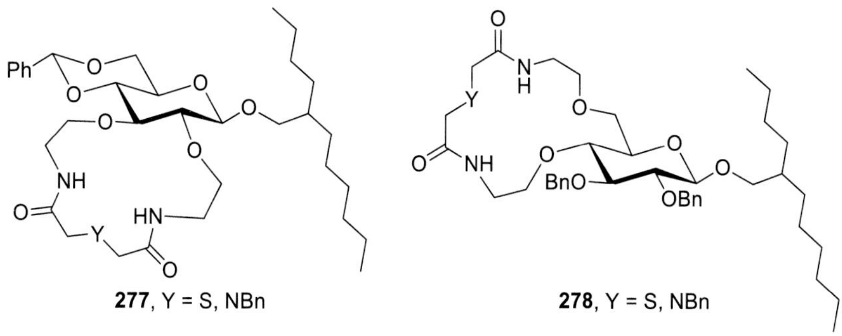

FIGURE 11.20 Glucose-derived aza and thia macrocycles **277–278** on 2,3 and 4,6-positions.[60]

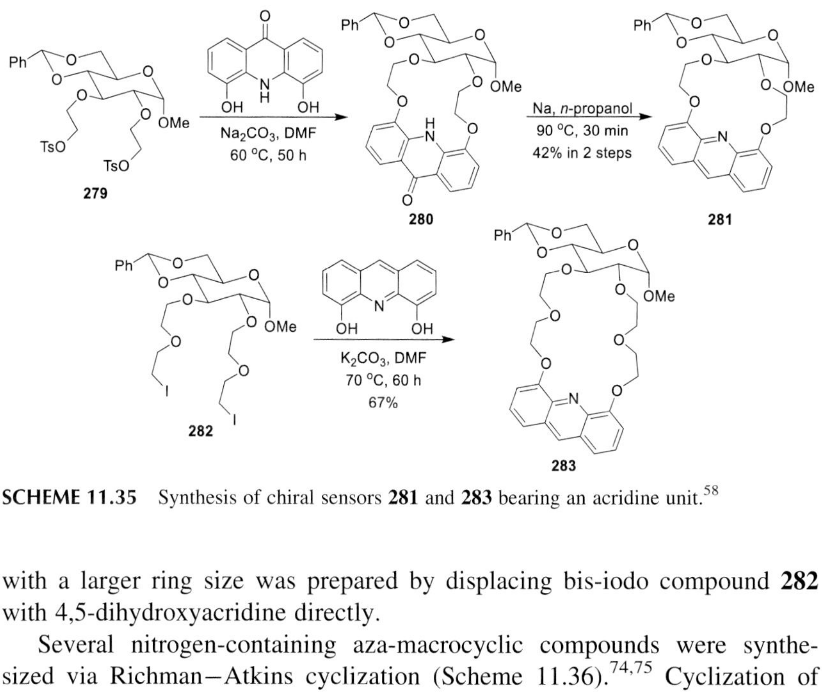

SCHEME 11.35 Synthesis of chiral sensors **281** and **283** bearing an acridine unit.[58]

with a larger ring size was prepared by displacing bis-iodo compound **282** with 4,5-dihydroxyacridine directly.

Several nitrogen-containing aza-macrocyclic compounds were synthesized via Richman–Atkins cyclization (Scheme 11.36).[74,75] Cyclization of bis-*p*-tol-sulfonamide **284** with diethylene glycol-derived bis-tosylate afforded sugar macrocycle **285**. Reaction of diethylene tri-*p*-tol-sulfonamide and sugar bis-tosylate **279** gave sugar azacrown ether derivative **286**. Transformation of ethylene bis-sulfonamide and sugar dichloro compound **287** produced sugar macrocycle **288**. Following the same approach, treatment of dichloride precursor **287** with various bis-toluene sulfonamides can produce three more sugar macrocycles **289–291**.

Ling et al. prepared the glucose-based thia-crown ether **295** in a few steps (Scheme 11.37).[57] The commercially available α-D-methyl glucopyranoside

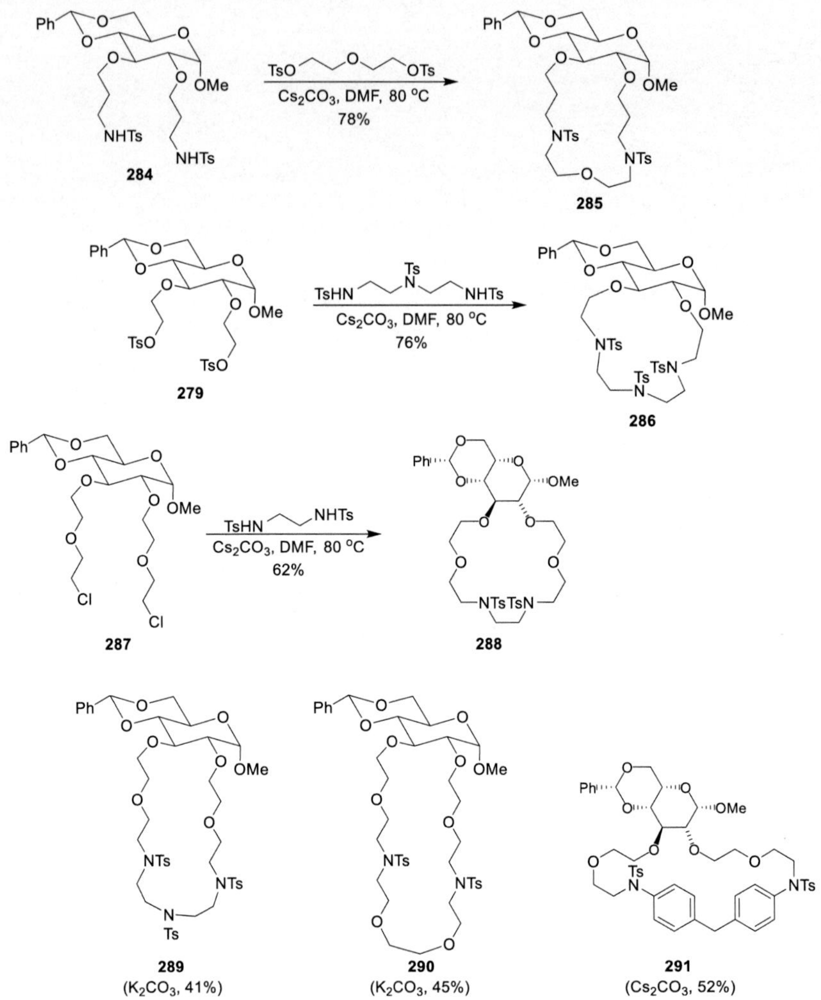

SCHEME 11.36 Synthesis of carbohydrate-based aza-macrocycles by Richman–Atkins cyclization.[75]

was converted to the thia-crown ether **292**, which was then transformed to the azido intermediate **293** through acetal deprotection, selective tosylation at primary alcohol, and azide formation. The pyrene unit was introduced at C-4 position and subsequent click reaction with calix[4]arene bis-alkyne afforded the glucose-based thia-crown ether-appended calix[4]arene **295**. Compound **295** was utilized as a chemosensor for Fe^{2+} and Fe^{3+} ions.

A polymeric thia-crown ether has also been prepared (Scheme 11.38).[56] Starting from the same thia-crown ether intermediate **292**, reductive cleavage of benzylidene protecting group led to intermediate **296**, propargylation at 6-OH afforded the alkyne **297**, which was conjugated to polyvinyl azide

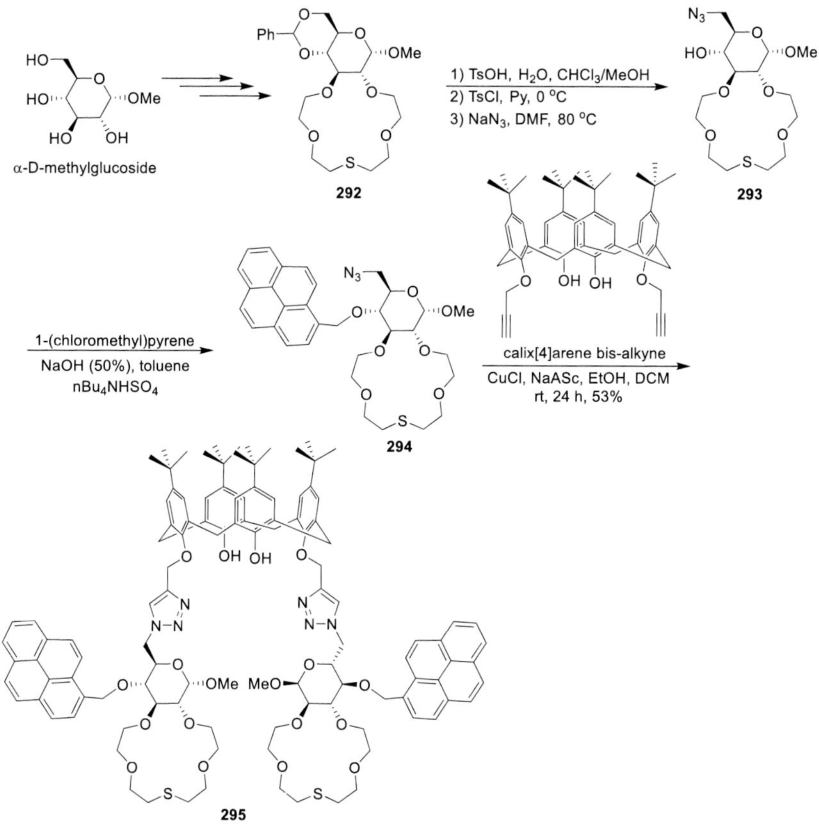

SCHEME 11.37 Synthesis of sugar thia-crown ether-appended calix[4]arene **295**.[57]

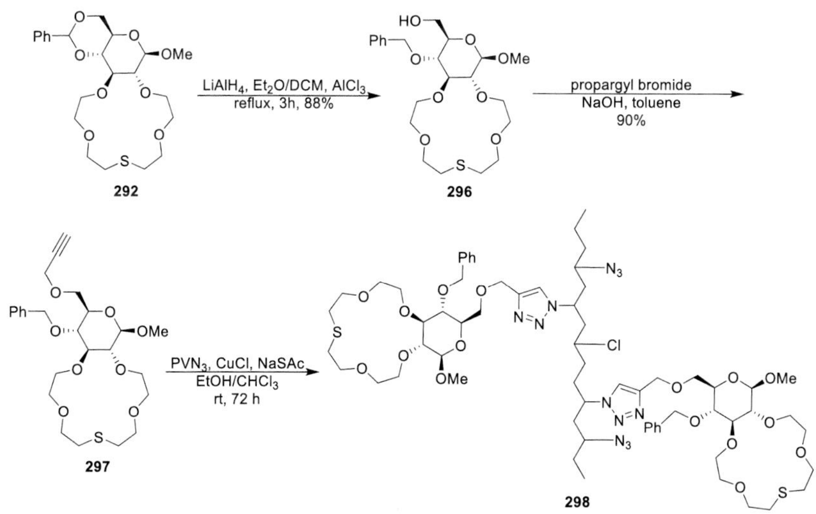

SCHEME 11.38 Synthesis of polymeric thia-crown ether **298**.[56]

(PVN_3) backbone by CuCl-catalyzed click reaction to form the polymeric thia-crown ether **298**. This polymer was found to be useful for metal (Co, Cd, Pb) ion removal.

Pintal et al. have synthesized several sugar-based cryptands that contain bis-azacrown ether and saccharide moieties as shown in Scheme 11.39.[76] The peracetyl-6-*O*-tosyl-1-azido-β-D-cellobiose **299** was reduced using triphenylphosphine in the presence of CO_2 and cryptand **300** to form the bis-urea bis-tosylate **301**. The di-tosylate was converted to bis-azide **302**. Similar reaction conditions as earlier led to the formation of tetra-urea-linked large macrocycle **303** in 17% yield. Using different sugar azides and similar methods, cellobiose cryptand derivative **305** and D-glucose cryptand derivative **307** were synthesized successfully from cryptand **304**. The macrocyclization of diamine with anomeric amine yields for the last step was 26% and 18% for **305** and **307**, respectively.

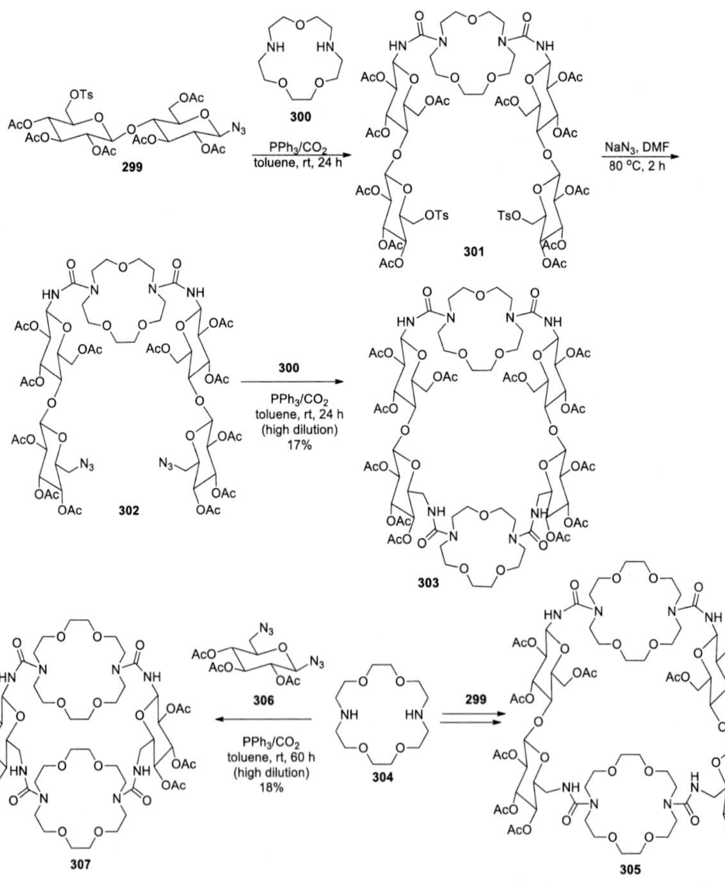

SCHEME 11.39 Synthesis of novel saccharide azacrown cryptands **303**, **305**, and **307**.[76]

11.3.3 Sucrose-based crown ether and other macrocycles

Jarosz et al. have carried out the synthesis of a variety of sucrose-based macrocycles, and many of these syntheses have been reviewed in their recent reviews.[47,77,78] Since the reviews have covered different methods for the synthesis of sucrose-containing macrocycles, here we mainly focus on the most recent reports in the area. Jarosz et al. synthesized a variety of different sucrose-derived macrocycles,[79–83] and molecular cages or capsules containing the sucrose moiety.[84–86] They have also studied their applications for host–guest binding and for chiral differentiations.[84,87–96] Typically, the C-6 and C-6′ positions of sucrose are selectively functionalized to generate a suitable leaving group, which can then be converted to other functional groups through S_N2 displacement. The following section will discuss a few examples of sucrose-derived crown ether type of macrocycles.

The primary hydroxyl groups on sucrose can be selected as the site for macrocyclization, and several different macrocycles were obtained by S_N2 reactions of an amine or alcohol nucleophile with a halogen (iodo and bromo) or sulfonate (tosylate and mesylate) leaving group. Mono iodo amino sucrose intermediate can spontaneously cyclize to afford the macrocyclic compounds (Schemes 11.40, 11.41). Sucrose was selectively silylated with the TBS group at both C6-OH positions for the glucose and fructose moieties. Global benzylation of the remaining hydroxyl groups afforded the intermediate **308**. Diol **309** was prepared by removing both silyl-protecting groups with TBAF in good yield. Selective silyation using 1.1 equiv. of TBDMSCl (TBSCl) afforded the intermediate **310** in 56% yield together with a small amount of disilylated compound **308** (13%) and unreacted diol **309** (25%). Derivatization of the hydroxyl group on the glucose moiety in **310** was conducted with tert-butyl bromoacetate to afford ester **311**. Deprotection of the silyl group on the fructose 6-OH group and subsequent Swern oxidation gave aldehyde **312** in good overall yield. Reductive amination of compound **312** with *N*-benzyl-2-(benzylamino)acetamide produced the amide ester **313**. Reduction of compound **313** with $LiAlH_4$ under THF reflux condition for 1.5 hour afforded the intermediate **314**. Treatment of the amino alcohol **314** with Ph_3P, imidazole, and I_2 afforded the iodo intermediate **315i**, spontaneous cyclization of which furnished the desired macrocycle **315** (Scheme 11.40).[82] Following a similar approach described earlier, compound **308** was converted to the sucrose-based macrocycle **317** with benzylamino group on the glucose end. The key intermediate mono alcohol **316** bearing 6-OH on fructose moiety was obtained from the disilylated compound **308** by applying 1.1 equiv. of TBAF.

Scheme 11.41 describes the synthesis of a series of azacrown ether type of sucrose-based macrocycles.[81] The primary hydroxyl group on the glucose unit of compound **309** was selectively alkylated with chloroacetonitrile to give the intermediate **318**, and the free hydroxyl group on the fructose was then alkylated with *t*-butyl bromoacetate to provide the cyano-ester **319**.

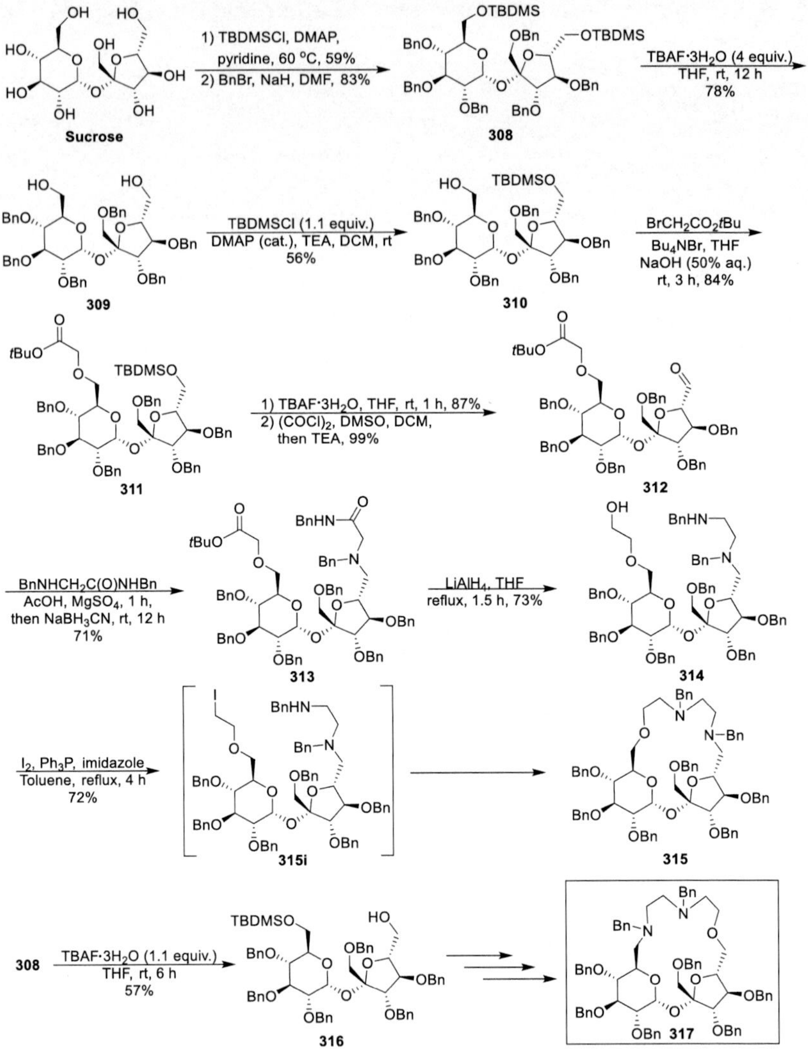

SCHEME 11.40 Synthesis of sucrose-based macrocyclic compound **315** and **317**.[82]

Double reduction of cyano and ester functional groups in **319** furnished the amino alcohol **320**. Iodination of the hydroxyl group in **320** followed by spontaneous cyclization of intermediate **321i** produced the sucrose-based azacrown ether **321**. The secondary amine nitrogen was then alkylated with different alkyl halides to generate a series of sucrose azacrown ether macrocyclic compounds **322a–f**.

Scheme 11.42 shows examples of sucrose-based macrocycles prepared by double S_N2 reactions of two nucleophiles with two leaving groups.

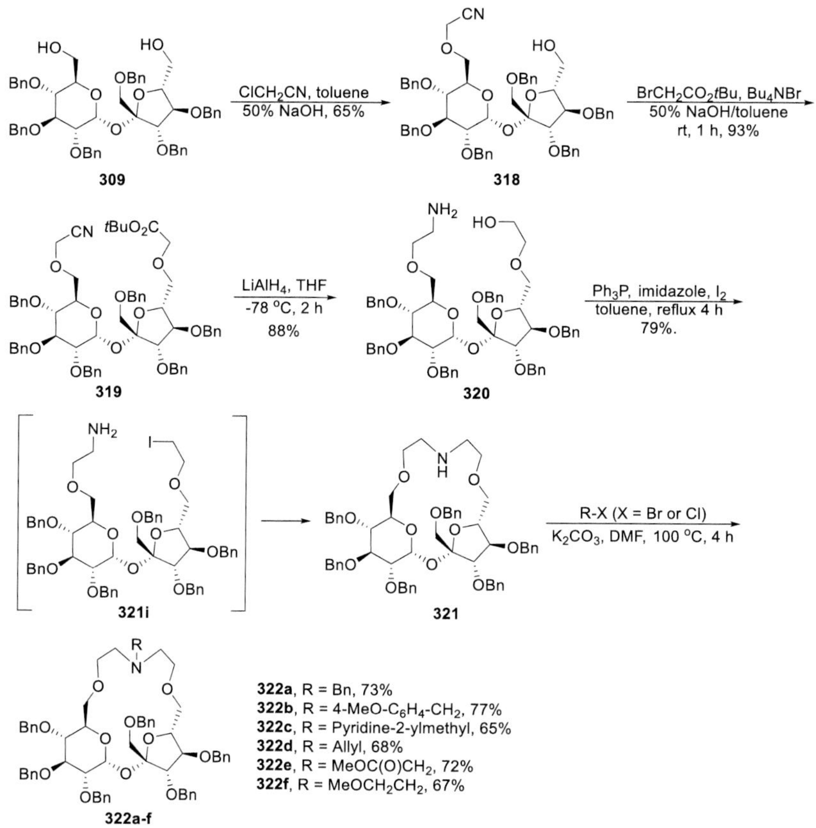

SCHEME 11.41 Synthesis of sucrose-based aza compounds **322a–f**.[81]

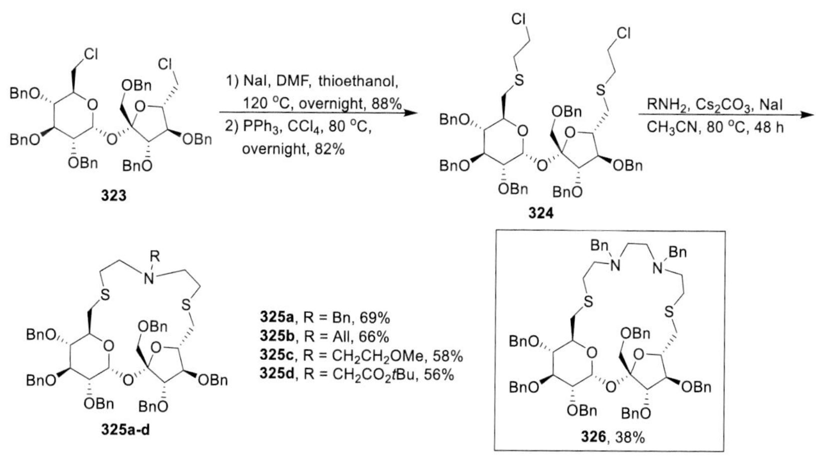

SCHEME 11.42 Synthesis of sulfur-containing macrocycles **325a–d** and **326** with sucrose scaffold.[94]

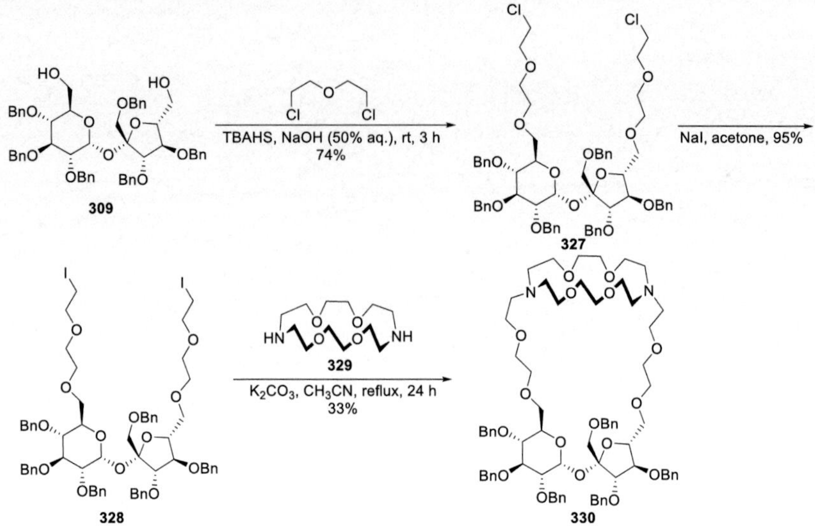

SCHEME 11.43 Synthesis of sucrose-containing cryptands **330**.[95]

In Scheme 11.42, two sulfur-containing sucrose-based macrocycles **325** and **326** were synthesized from the dichloro compound **323**.[94] Treating compound **323** with thioethanol in the presence of NaI followed by Appel reaction provided the dichloro compound **324**. Nucleophilic substitutions of **324** with mono-amines or di-amines led to the macrocycles **325** or **326**, respectively.

Another macrocyclization by double S_N2 reaction is shown in Scheme 11.43.[95] Functionalizing diol **309** with bis(2-chloroethyl) ether afforded dichloro compound **327**. It was converted to the di-iodide **328**, which was then reacted with commercially available di-aza-crown ether **329** to produce the sucrose-containing cryptand **330** in 33% yield.

An azobenzene-containing sucrose macrocycle *trans*-**333** was synthesized.[96] This compound exhibited photo-responsive properties and was found to function as a cation host. As shown in Scheme 11.44, selective trityl protection of 6-OH groups on both glucose and fructose ends in sucrose and subsequent benzyl protection of the remaining free hydroxyl groups afforded compound **331**. The trityl group in **331** was removed and then alkylated with bis(2-chloroethyl)ether to form the bis-chloro derivative **327**, which was further converted to bis-iodo compound **328**. 2,2′-Dihydroxyazobenzene **332** was used as a nucleophilic reagent and successfully displaced the iodo groups in compound **328** to yield the final macrocycle *trans*-**333**. The trans isomer can be converted to the *cis*-isomer **334** by green LED irradition.

Sucrose-derived macrocyclic compounds that contain another bridge were also synthesized. These compounds can be prepared from a precursor with three leaving groups with a tridentate nucleophile. Scheme 11.45 shows

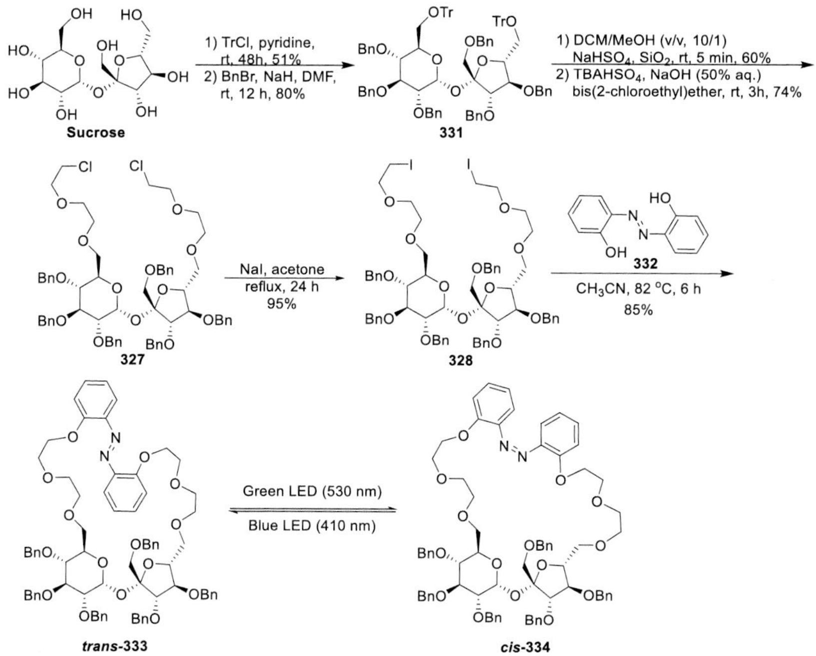

SCHEME 11.44 Synthesis of visible-light responsive sucrose-containing macrocycles **333** and **334**.[96]

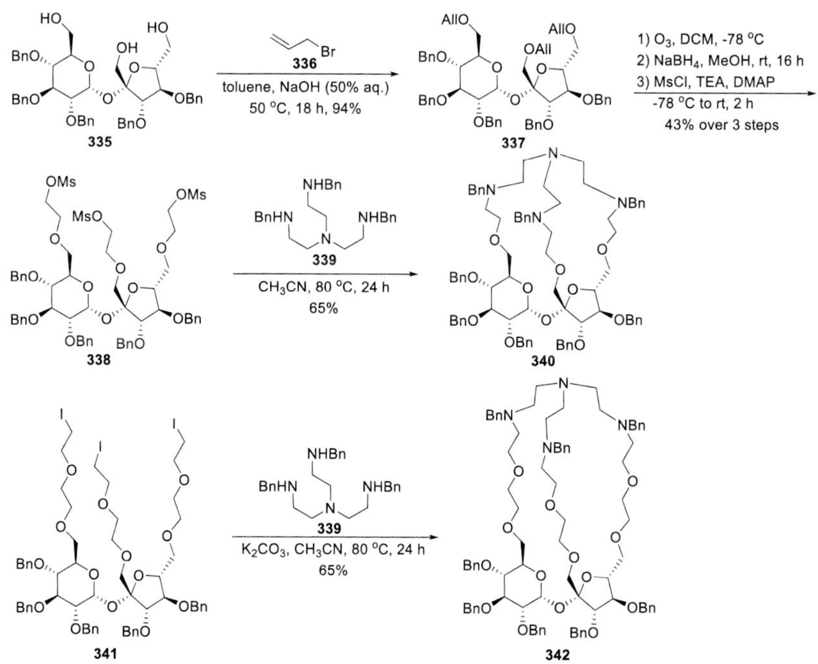

SCHEME 11.45 Synthesis of sucrose-containing cryptands **340** and **342**.[95]

examples of macrocyclization between three nucleophiles and three leaving groups through S_N2 reaction. In Scheme 11.45,[95] alkylation of tri-ol **335** with allyl bromide followed by ozonolysis, reduction, and mesylation produced the tri-mesylate compound **338**. Treatment of compound **338** with tripod amine **339** gave the sucrose-containing macrocycle cryptands **340**. A macrocyclic cryptand **342** with a longer linker was successfully prepared from the tri-iodo sucrose derivative **341**.

Scheme 11.46 shows the synthesis of capsule-like cryptands **346** and **347** by simultaneously displacing three leaving groups with three nucleophiles.[84] The synthesis was initiated from sucrose, silylation of all primary

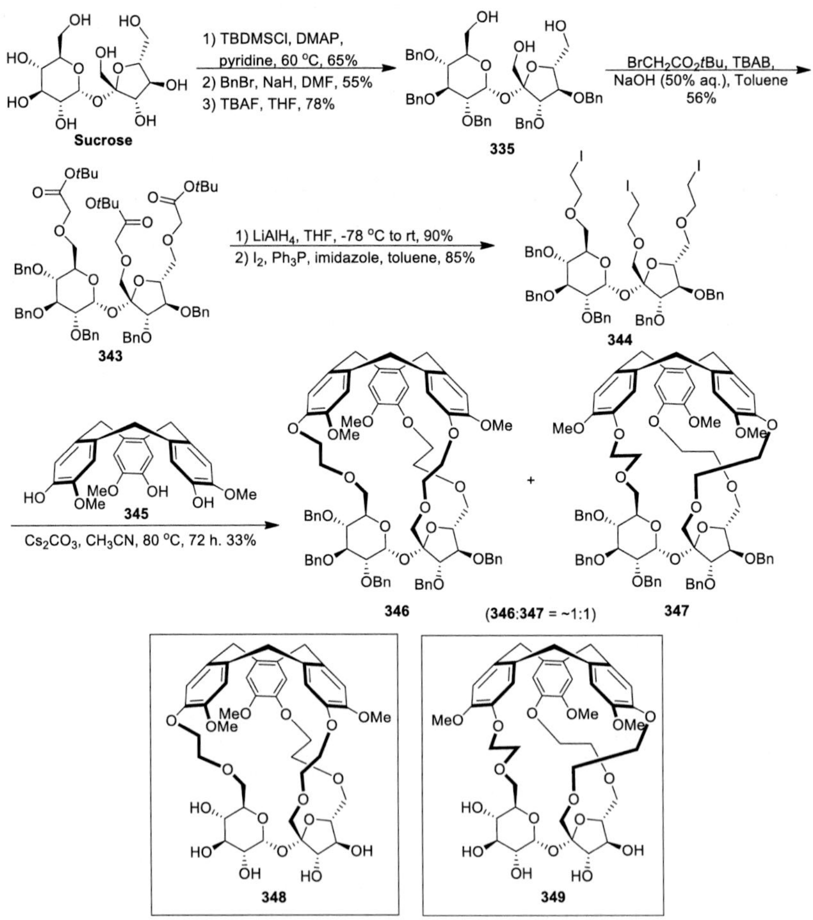

SCHEME 11.46 Synthesis of cyclotriveratrylene sucrose-based capsules **346–349**.[84]

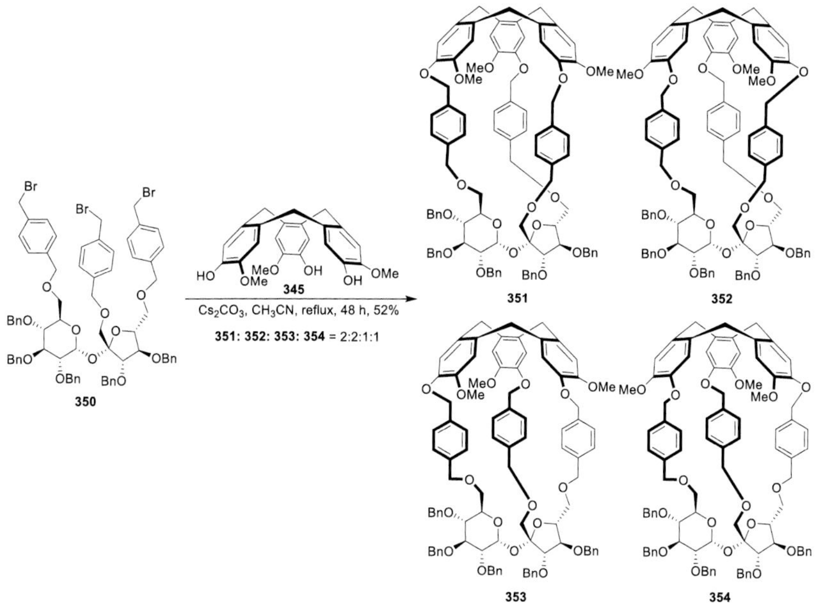

SCHEME 11.47 Synthesis of sucrose-based molecular cages **351–354** containing cyclotriveratrylene.[85]

hydroxyl groups, then benzylation of all secondary hydroxyl groups, and subsequent desilylation with TBAF furnished the trihydroxy compound **335**. Alkylation of the hydroxyl groups by *t*-butyl bromoacetate formed intermediate **343** in 56% yield. The *t*-Bu esters in **343** were reduced to primary alcohols, which were then converted to tri-iodide **344** by Mitsunobu reaction. The tri-iodide **344** served as a key precursor in triple S_N2 reactions with phenols in racemic cyclotriguaiacylene **345**. Two diastereomeric cryptands **346** and **347** were obtained in a 1:1 ratio. Pd-catalyzed hydrogenation of cryptands **346** and **347** resulted in water-soluble compounds **348** and **349**, respectively.

The trimeric benzyl bromide sucrose derivative **350** was also reacted with **345** through a triple nucleophilic substitution ring closure reaction (Scheme 11.47).[85] The Cs_2CO_3-mediated macrocyclization took place in refluxing acetonitrile for 2 days at highly dilute concentrations to generate four diastereomers **351–354**. These molecular cages were separated by preparative HPLC. Similarly, as shown in Scheme 11.48, macrocyclization between naphthalene spacer incorporated tribromide **355** and **345** was performed to afford four stereoisomers **356–359**.[86] These naphthalene-containing compounds **356–359** are fluorescent sensors for the detection of acetylcholine or choline.

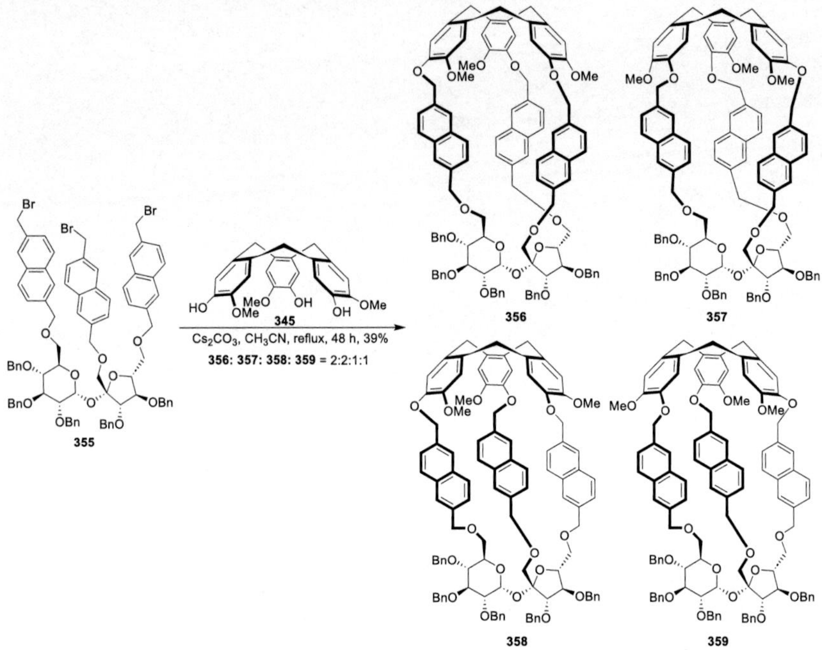

SCHEME 11.48 Synthesis of fluorescent sucrose molecular cages **356–359**.[86]

11.4 Synthesis of macrocycles by C–C bond forming reactions

Many sugar-based macrocycles have been synthesized through C–C bond formation reactions; in this section, we will only cover those glycomacrocycles using C–C bond formation for the macrocyclization step or as the last step for the ring-closing reactions. Several general methods will be discussed, the first method is to employ RCM reaction catalyzed by Grubbs catalysts to form a macrocycle with a double bond, typically this double bond is reduced to form saturated hydrocarbons by catalytic hydrogenation reactions. The second strategy utilizes the alkynyl bromide coupling reaction, Sonogashira and Heck reaction, and several medium ring-sized macrocycles were synthesized by this method. The third method employed the aldol condensation method to close the ring.

11.4.1 Synthesis of macrocycles by RCM strategy

Ring closing metathesis reaction (RCM) and catalytic hydrogenation strategy are frequently used in macrocyclization reactions. Jarikote et al. reviewed the synthesis of sugar-embedded macrocycles by the metathesis method.[97] Grubbs catalysts and Grubbs–Hoveyda catalysts are often utilized, and the common structures of the catalysts **360–362** utilized in the next section are shown in Fig. 11.21.

In the next section, a few recent examples of macrocycles synthesis by RCM strategy will be briefly discussed. Maurya et al. developed a general strategy for the synthesis of 13 macrocycles by click chemistry to link two sugar derivatives and RCM for the cyclization reaction.[98] They prepared allyl and alkynyl functionalized sugar building blocks from D-glucose, D-xylose, and L-arabinose.

Using RCM catalyzed by Grubbs catalysts followed by reduction of the resulting olefin (RCM-Re), a series of 12-, 13-, 17-, and 21-membered macrocycles **368, 369, 370, 371**, respectively, were synthesized.[99] As shown in Scheme 11.49, the starting material **363** with an olefin moiety was subjected

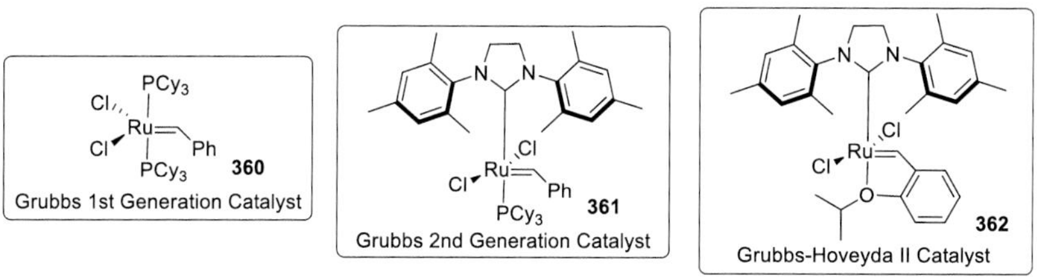

FIGURE 11.21 Structures of the Grubbs catalysts used in RCM macrocyclization reactions.

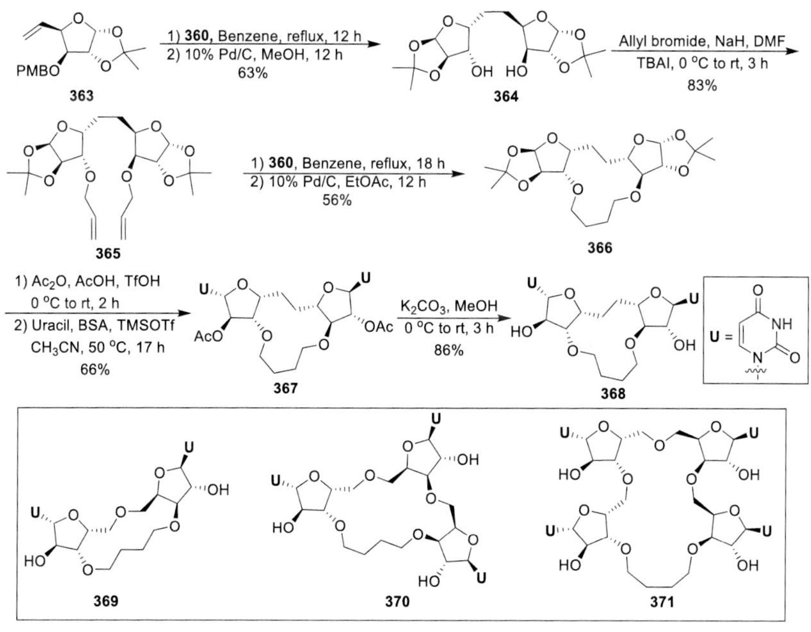

SCHEMES 11.49 Synthesis of sugar-based macrocycles **368–371** by RCM-Re.[99]

to RCM conditions using Grubbs first generation catalyst **360** and hydrogenation to afford the intermediate **364**, the free hydroxyl groups were alkylated with allyl bromide to afford the intermediate **365**, and another RCM/H_2 reaction led to the symmetrical 12-membered ring macrocycle **366**. The macrocycle **366** was converted to the nucleosides **368** in a few steps. First, the isopropylidene acetal was removed and the diols were acetylated in a one-pot reaction, the resulting sugar acetate was glycosylated with uracil to afford the macrocycle **367**. Deacetylation of the intermediate afforded the macrocyclic nucleoside **368**. The other macrocycles **369–371** were synthesized in a similar manner.

Using a similar macrocyclization strategy of RCM/hydrogenation, as shown in Scheme 11.50, several 14-membered ring macrocyclic compounds **377** and **378** were synthesized; these compounds exhibited antiangiogenesis activity.[100] Starting from the α-D-methylglucoside, perbenzylation followed by treatment of the intermediate with 2N HCl led to an open-ring aldehyde intermediate, which was subjected to reductive amination with allyl amine to afford the allyl amine derivative **372**. Acylation of intermediate **372** with

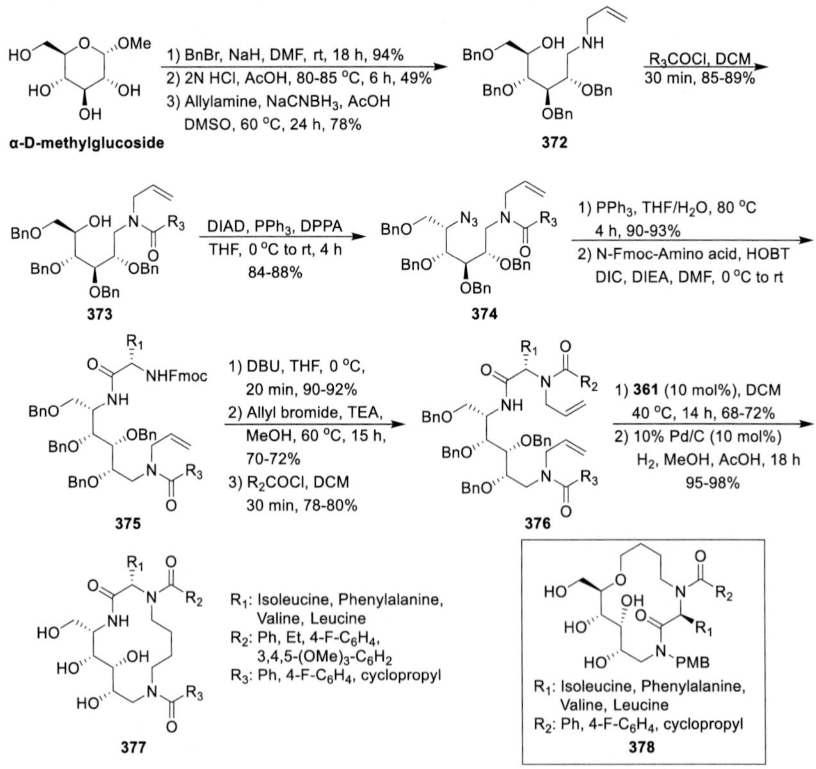

SCHEME 11.50 Synthesis of 14-membered ring macrolides **377** and **378**.[100]

different acid chlorides R_3COCl afforded the intermediate **373**. The hydroxyl group in this compound was converted to the corresponding azide with inverted stereochemistry under the Mitsunobu condition to give azide **374**. Reduction of the azido group and subsequent amide coupling with different *N*-Fmoc-protected amino acids furnished compound **375**. This was converted to bis-olefin **376** by deprotection of Fmoc to free amine and allylation of the amine, followed by acylation with different acid chlorides R_2COCl was prepared. The diallyl compound **376** was subjected to RCM reaction with Grubbs second generation catalyst **361**, and subsequent hydrogenation led to the formation of the 14-membered macrolides **377** in good yields. By varying different amino acids for R_1 and different acyl chlorides for R_2 and R_3, they have synthesized six different macrocycles with structure **377**, and using a similar approach, another series of eight macrocycles **378** was also prepared.

By the RCM method, a few other macrocycles **382**[101] and **383**[102] with a free hydroxyl group on the sugar ring were synthesized. As shown in Scheme 11.51, α-D-methylglucoside was perbenzylated first, then the ring-opened anomeric aldehyde underwent Wittig–Horner reaction with triethyl phosphonoacetate and NaH to afford the intermediate **379**. The intermediate was obtained as an inseparable anomeric mixture, which was formed by 5-OH conjugate addition to the substituted vinyl carboxylate. Selective debenzylation of the C-6 *O*-benzyl group, followed by allylation and methyl ester hydrolysis

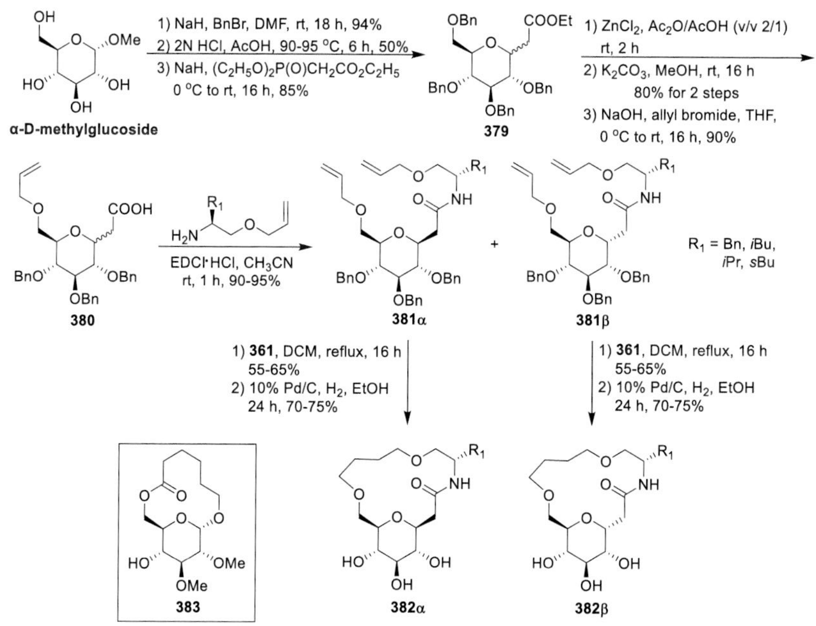

SCHEME 11.51 D-Glucose-derived macrocycles **382**[101] and **383**[102] synthesized by linking C1–C6 positions via RCM method.

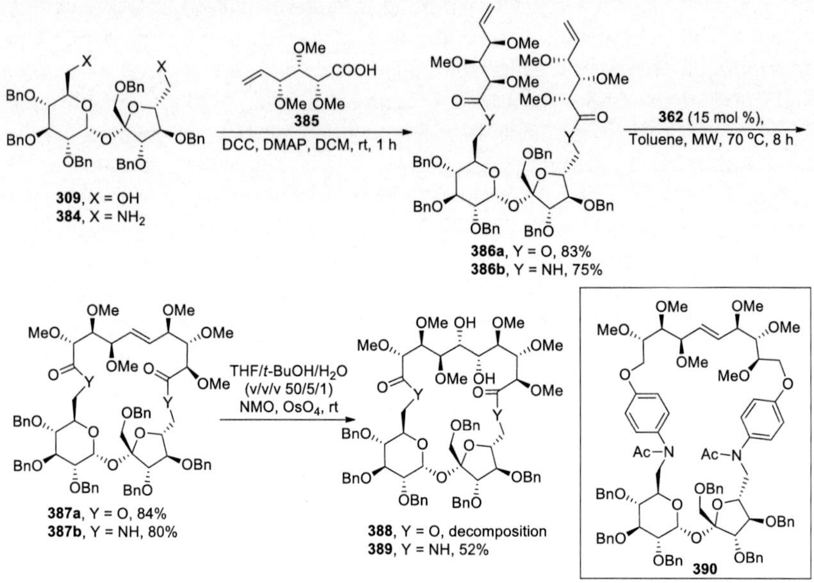

SCHEME 11.52 Synthesis of sucrose-based macrocyclic compound **389**[87] and structure of **390**.[93]

gave compound **380** bearing free carboxylic acid at C-1 position and alkene at C-6 position. Various *O*-allylated chiral amines reacting with acid **380** resulted in a separable mixture of the corresponding amides, **381α** and **381β**. Both compounds were then subjected to RCM/hydrogenation conditions using catalyst **361** to give macrocycles **382α** and **382β** in good yields. In another study using the RCM method, sugar-embedded macrolactone **383** was synthesized by RCM and the compounds showed weak antibiotic activity.[102]

Sucrose-based macrocyclic compounds **389**[87] and **390**[93] were also synthesized by RCM approach. As shown in Scheme 11.52, reacting sucrose diol **309** with hexanoic acid derivative **385** afforded diester compound **386a**. Amidation of sucrose diamine **384** with acid **385** gave the diamide **386b**. Both macrocycle precursors **386a** and **386b** were carried through RCM reaction in the presence of Grubbs–Hoveyda II catalyst **362** to furnish the sucrose-based macrocycles **387a** and **387b**, respectively. Dihydroxylation of the alkene group in diamide **387b** was achieved while the decomposition was observed using diester **387a** as the substrate under the same condition, only the hydrolyzed product **309** was isolated possibly due to the presence of base *N*-methylmorpholine (NMO). In another report by Tiara et al.,[93] the sucrose-based macrocycle **390** was prepared by derivatizing the C6 and C6′ positions with different monosaccharide linkages with an alkene functional group, and a similar RCM method was used for the macrocyclization.

Murphy's group has reported the synthesis of several series of macrocycles by RCM reactions.[103,104] Jarikote et al. synthesized four macrocycles that were designed as echinimycin mimetics. These glycophanes were

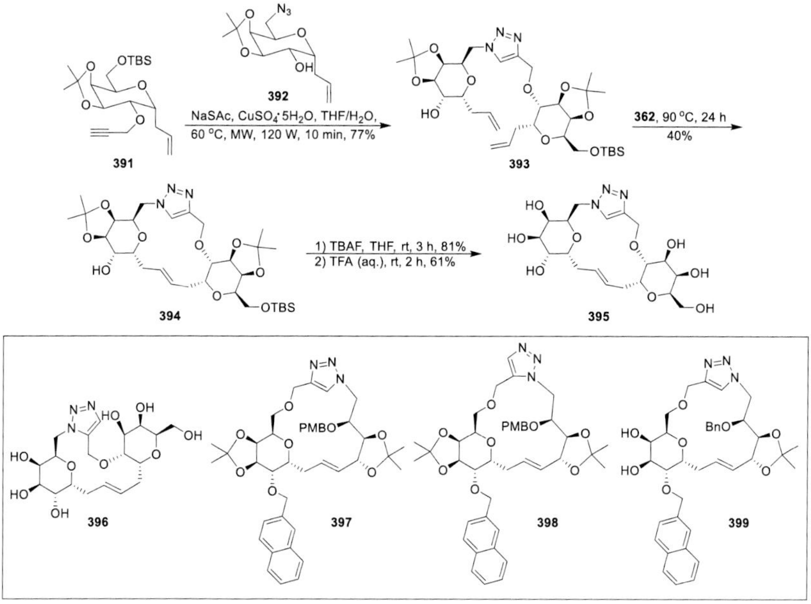

SCHEME 11.53 Synthesis of macrocycles **395–399** by RCM method.[104]

synthesized by grafting quinozaline to glucuronic acid derivatives and using the RCM to form the macrocycles.[103] As shown in Scheme 11.53, with a combination of click chemistry and RCM, several glycomacrocycles **395–399** were prepared.[104] Bis-olefin substituted disaccharide **393** was assembled from mono sugar alkyne **391** and azide **392** with 1,4-triazole as the linker and RCM using Grubbs–Hoveyda II catalyst **362** gave macrocycle **394**. Silyl group removal by TBAF and isopropylidene cleavage under acidic conditions afforded the polyhydroxylated macrocycle **395**. Several other analogs **396–399** bearing 1,5-triazole and naphthalene group installed at the C-2 position were synthesized in a similar fashion.

11.4.2 Synthesis of sugar-fused macrocyclic lactones via Sonogashira and Heck reaction

For medium and large ring macrocycles containing one sugar, various cyclization methods have been reviewed by Hussain et al.[105] Hussain et al. have reported the construction of sugar macrocycles by Sonogashira reaction and Heck coupling.[106–108] Synthesis of sugar-fused macrocyclic compounds **405** and **409** were selected as examples to illustrate the methodology (Schemes 11.54 and 11.55). Propargylation of di-isopropylidene-protected glucose **400**, which upon selective deprotection (C5 and C6 positions) and trityl primary alcohol protection gave compound **402** with a free 5-OH group. Then benzylation, trityl deprotection, and esterification with 2-bromobenzoic acid produced

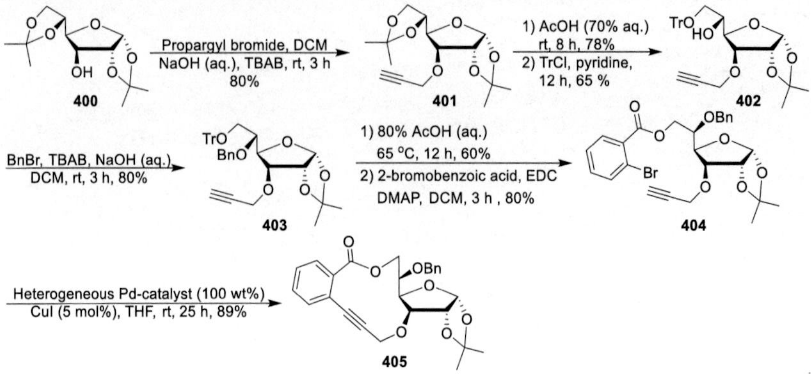

SCHEME 11.54 Synthesis of macrocycle **405** by Sonogashira reaction.[106]

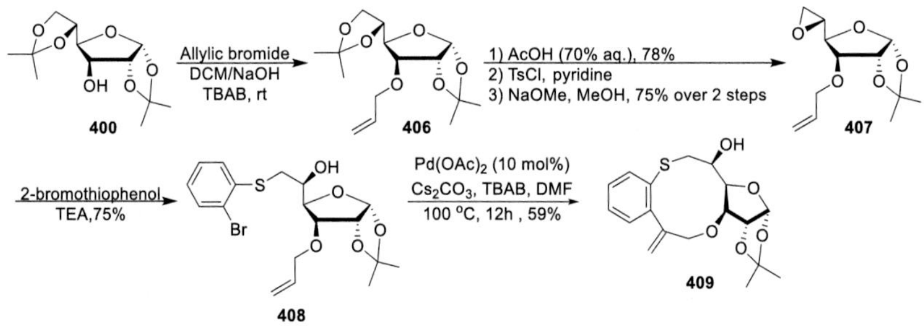

SCHEME 11.55 Synthesis of macrocycle **409** by Heck reaction.[108]

macrocycle precursor **404**, this intermediate contains an alkyne and aryl bromo functional groups. Under Sonogashira coupling condition, the intermediate **404** was converted to the cyclic compound **405**.

In Scheme 11.55, sulfur-containing macrocycles were synthesized starting from the glucose diacetonide **400**, an allyl group was introduced at the C-3 position to give the intermediate **406**. Selective deprotection of the 5,6-isopropyrlidene group and the formation of epoxide led to the intermediate **407**. Opening the sugar epoxy compound **407** with 2-bromothiophenol afforded sulfide **408** bearing the required reactive sites (aryl bromide and alkene), which was then cyclized under Heck condition to give the sulfur-containing sugar macrocycle **409**.

11.4.3 Synthesis of macrocycles via acetylene coupling reaction

Sugar-fused macrolactone **415** was synthesized by Peczuh et al. via intramolecular Stille coupling reaction for the macrocyclization (Scheme 11.56).[109] D-Glucose derivative **410** was esterified to the bromoalkyne **412**, which was then derivatized to tributyl tin intermediate **413**. The 4-hydroxyl group was esterified to introduce the vinyl iodide moiety in compound **414**, and

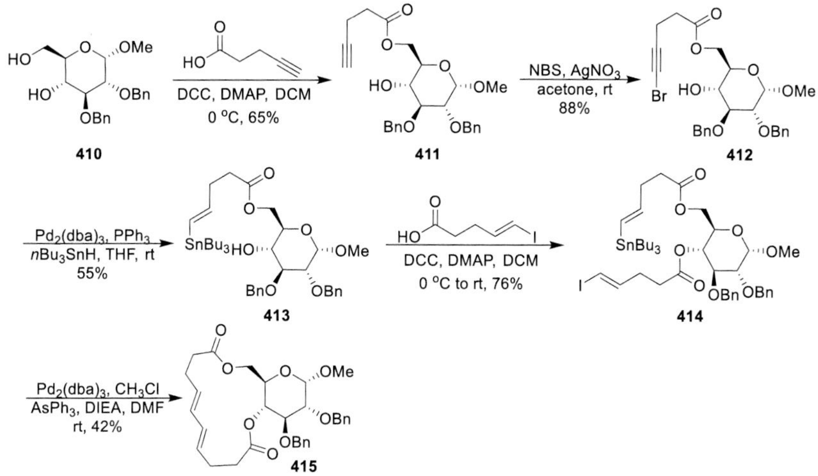

SCHEME 11.56 Synthesis of a carbohydrate-fused [15]-macrodilactone **415**.[109]

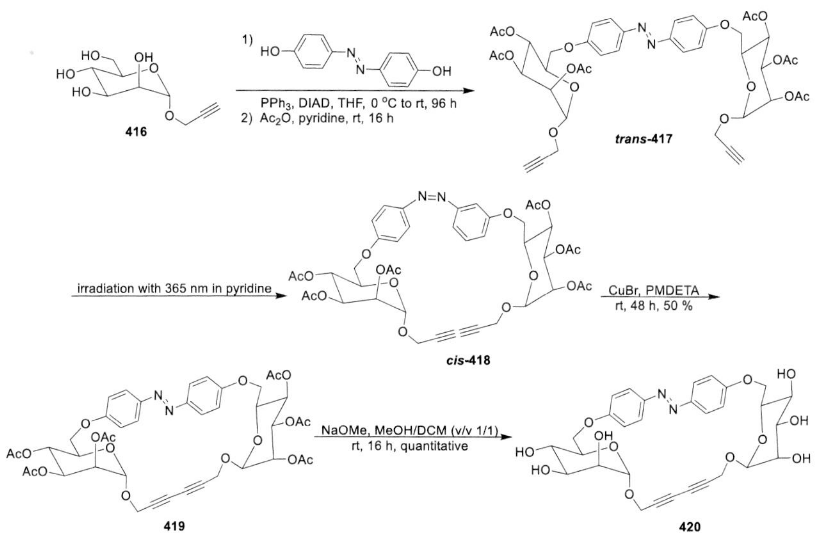

SCHEME 11.57 Synthesis of glycoazobenzene diacetylene-containing macrocycle **420.**[110]

then an intramolecular coupling reaction afforded the 4,6-glucose-fused macrolactone **415**.

Synthesis of diacetylene-containing macrocycle **420** was initiated from the D-mannose derivative **416** (Scheme 11.57).[110] Treatment of 4,4′-dihydroxyazobenzene with the primary hydroxyl groups in **416** under Mitsunobu condition followed by peracetylation afforded the di-alkyne compound ***trans*-417**. Taking advantage of the inherent photo-switching properties of the azobenzene moiety, ***trans*-417** was converted to *cis*-isomer **418**, which positioned the propargyl groups in closer proximity. This transformation facilitated the intramolecular CuBr-catalyzed Glazer coupling under high dilution to produce the diacetylene **419** in moderate yield. Removal of the acetyl group resulted in the final compound **420** in quantitative yield.

11.4.4 Macrocyclization via aldol condensation reactions

Intramolecular aldol condensation was utilized as the ring-closing method to achieve the synthesis of macrocyclic enone **426**, which was further derivatized to phenyl pyrazole **427**, aminopyrimidine **428**, and isoxazole **429**.[111] These three compounds showed promising anticancer activity. As shown in Scheme 11.58, azido compound **422** was prepared from 2-hydroxyacetophenone **421** in two

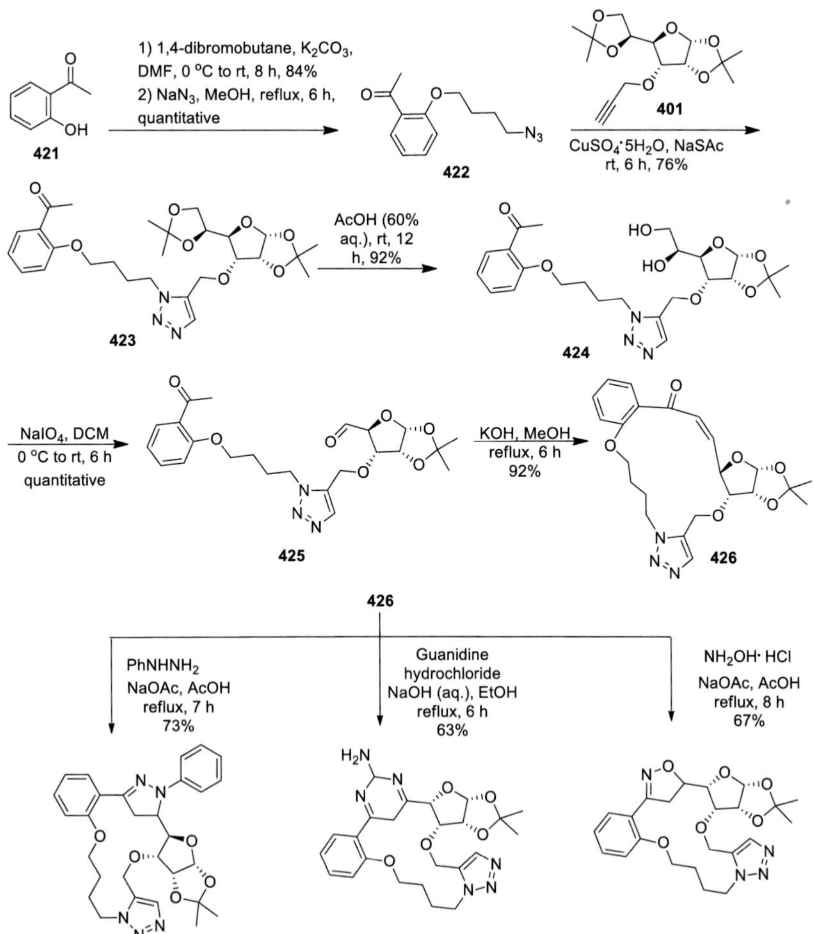

SCHEME 11.58 Triazole-containing sugar macrocycles **427–429** cyclized by intramolecular aldol condensation.[111]

steps. Click reaction of glucose propargyl derivative **401** and azide **422** gave the triazole compound **423**. Subsequent isopropylidene group removal under acidic conditions and oxidation with $NaIO_4$ furnished the aldehyde **425**, which was subjected to intramolecular aldol condensation to afford the key macrocycle intermediate **426** in good yield.

11.5 Synthesis of sugar-embedded macrocycles by reductive amination, macrolactamization, and macrolactonization

In this section, we will review several macrocycle syntheses by the formation of lactone or lactam or a tertiary amine as the macrocyclization step. These can be synthesized via different macrocyclization methods, and we will discuss each example based on the ring-closing reactions and methods.

11.5.1 Synthesis of aza-linked sugar-based macrocycles

11.5.1.1 Synthesis of macrocycles by intramolecular reductive amination

Zhou et al. synthesized and tested a library of sugar-derived macrocycles using a double reductive amination method for the macrocyclization step. These compounds were screened for applications as antileukemic agents.[112] The synthesis of one example is fully illustrated in Scheme 11.59. Galactose C-1-allyl derivative **430** was selectively protected with aTBS group at the

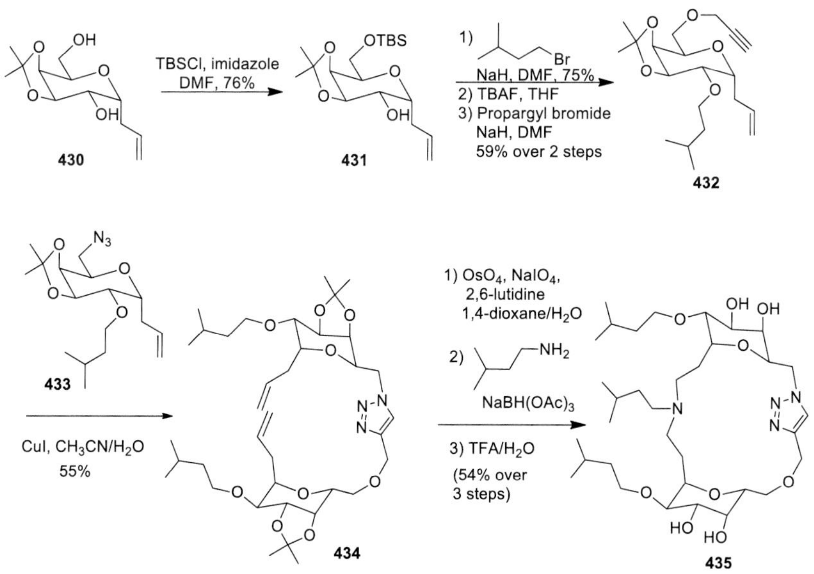

SCHEME 11.59 Synthesis of disaccharide-containing macrocycle **435** by reductive amination reactions.[112]

C-6 position to afford compound **431**. Alkylation of the free 2-OH in compound **431** followed by silyl deprotection and propargylation at the 6-OH generated the galactose alkyne **432**. Using copper(I) catalyzed azide-alkyne cycloaddition (CuAAC) reaction, the 1,4-triazole-linked intermediate **434** was prepared by galactose azide **433** and alkyne **432**. OsO_4-mediated oxidative cleavage of **434** afforded a bis-aldehyde intermediate, subsequent double reductive amination with 3-methylbutan-1-amine and finally acetal deprotection produced the disaccharide-containing macrocycle **435**. Applying the same strategy, a series of other macrocycles (**436-439**) were prepared; these are shown in Fig. 11.22. Macrocycle **438** (when R_4 = 2-Naphthylmethyl, R_5, R_6 = CMe_2) in Fig. 11.22 exhibited partial 5-HT2 antagonist properties.[112] Neigi et al. also reported the synthesis of a series of macrocycles by similar

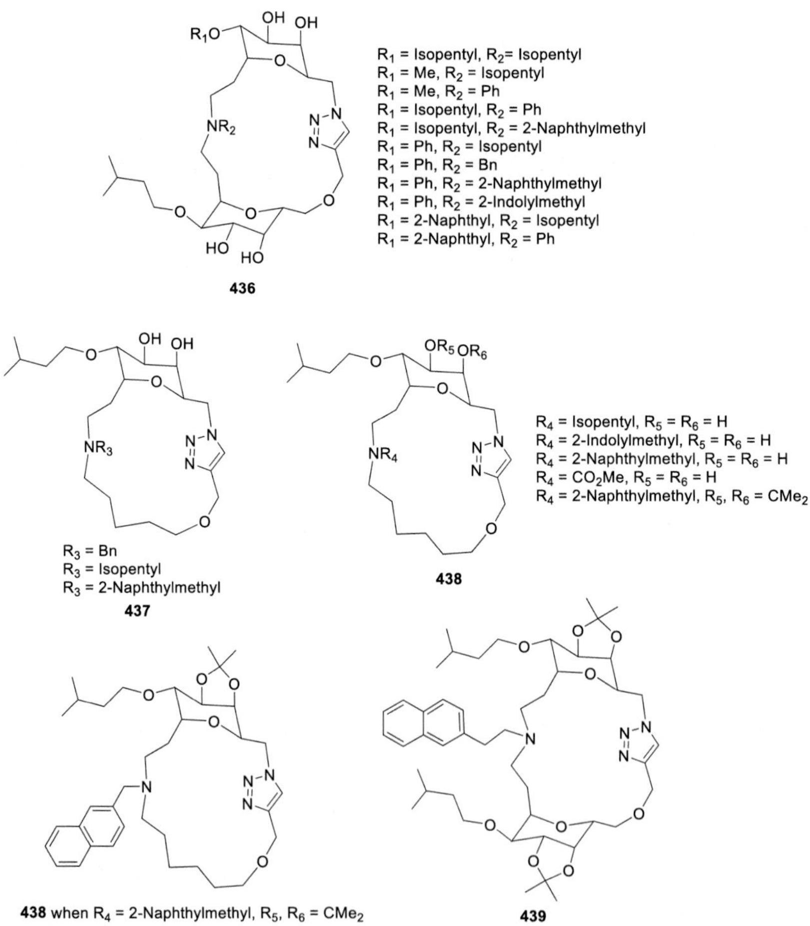

FIGURE 11.22 The structures of the macrocycles **436–439** synthesized and analyzed.[112]

strategies and studied their properties for protein–protein interaction.[113] The most potent compound structure **439** is shown in Fig. 11.22.

11.5.1.2 Synthesis of macrocycles by intermolecular reductive amination

Several sugar-containing cyclam analogs were reported by Bordes et al.[114,115] As shown in Scheme 11.60, the azide **440** was subjected to Staudinger-aza-Wittig condition to produce the hemiaminal **441**, which reacted with allyl magnesium bromide and the resulting intermediate was benzylated to give azepane **442**. Ring contraction of **442** using mesylate chloride afforded the chloromethylpiperidine **443** and subsequent azide displacement generated azido piperidine **444**. Oxidative cleavage of allyl moiety in compound **444** resulted in azidoaldehyde **445**. Staudinger reaction of the azido to the amino group and the reaction intermediate were subjected to

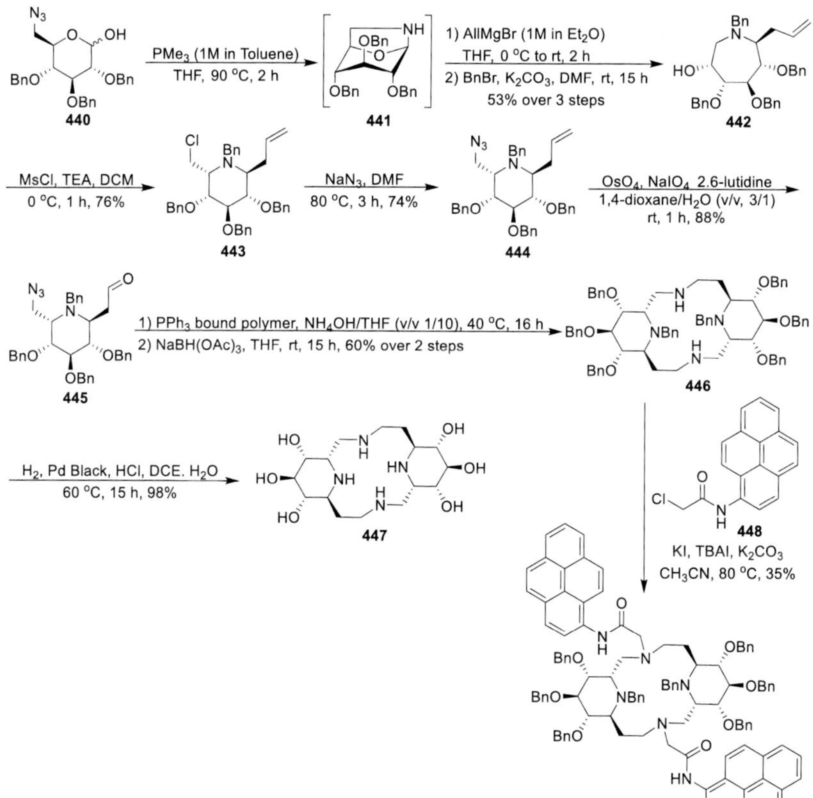

SCHEME 11.60 Synthesis of polyaza sugar-based macrocycle **446** and **447**[114] by double reductive amination and imino sugar-derived fluoroionophore **449**.[115]

double reductive amination reactions to produce the sugar-containing cyclam **446** successfully. Global catalytic hydrogenation of compound **446** afforded deprotected sugar cyclam **447** in almost quantitative yield. Alternatively, the intermediate **446** was alkylated with 2-chloro-*N*-(pyren-1-yl)acetamide **448** to compound **449**, which can selectively bind to Hg^{2+}.[115]

11.5.2 Synthesis of macrocycles by macrolactamization

Ghorai et al. described the synthesis of pentose sugar macrocycles **456–458** with triazole linkage.[116,117] The synthesis of macrocycle **456** is shown in Scheme 11.61.[116] The click reaction of sugar azido ester **450** and Cbz-protected

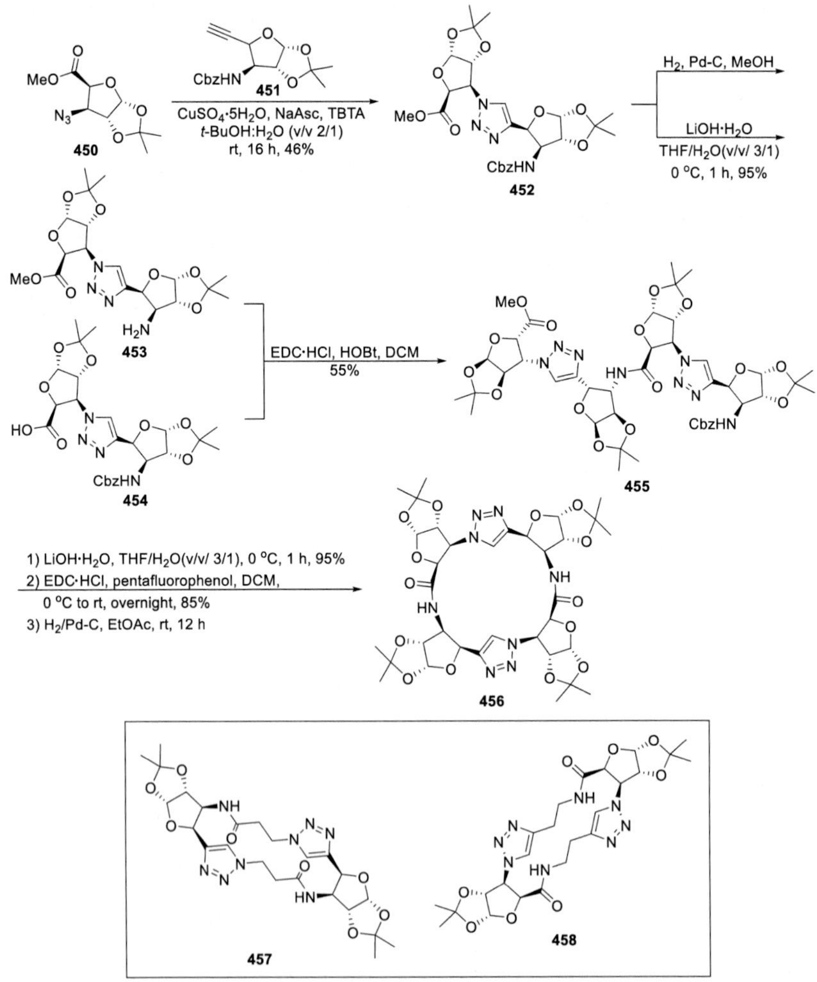

SCHEME 11.61 Synthesis of triazole-based peptidomimetic macrocycles **456–458**.[116,117]

alkyne **451** gave the 1,4-triazole-linked disaccharide derivative **452**. This intermediate was then hydrogenated to remove the Cbz group to form the free amine **453** or hydrolyzed to form the free acid **454**. Standard peptide coupling of these two compounds led to the dipeptide **455**, which is the key precursor for intramolecular macrocyclization. The macrocyclization was carried out by hydrolysis of the methyl ester to the free carboxylic acid, which was activated using EDC and pentafluorophenol (PFP) to pentafluorophenyl ester intermediate, upon hydrogenation removal of the Cbz group, the PFP ester spontaneously cyclized with the amino group to the form the macrocycle **456**. By similar protocols, two other triazole-linked cyclopeptides **457** and **458** were synthesized. These two macrocycles are regio-isomers prepared from the corresponding sugar precursors and β-alanine derivatives.[117]

A 16-membered ring macrocycle **464** containing an unusual carbonic carbamic anhydride moiety was synthesized via intramolecular macrocyclization (Scheme 11.62).[118] The macrocycle synthesis started from the D-gulose derivative **459**, Michael addition using benzylamine and subsequent

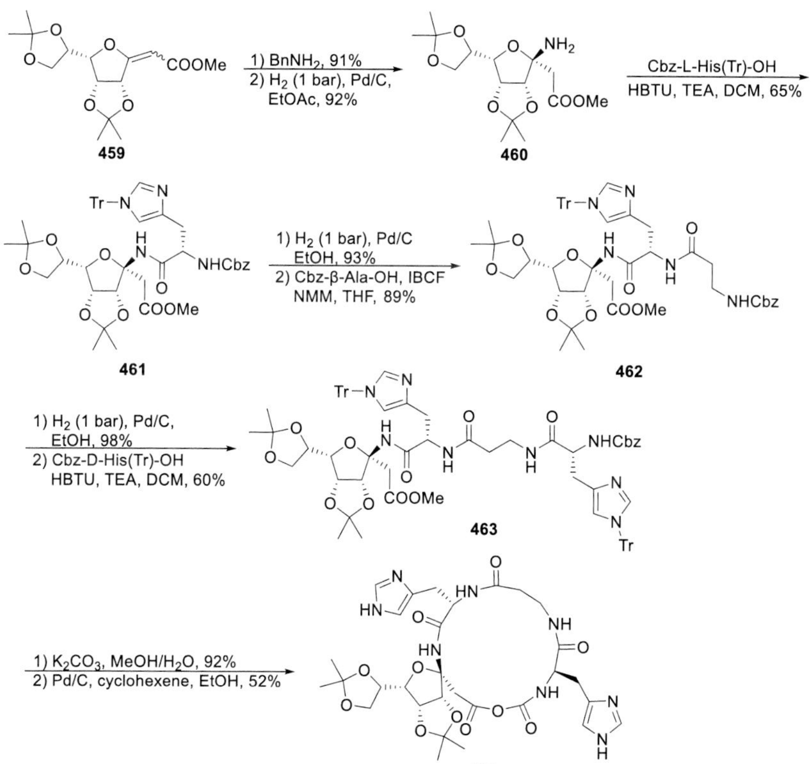

SCHEME 11.62 Synthesis of macrocycle **464** containing sugar amino acid.[118]

hydrogenation afforded the β-amino ester **460**. Coupling of **460** and Cbz-L-His(Tr)-OH using hexafluorophosphate benzotriazole tetramethyl uronium (HBTU) produced dipeptide **461**. The Cbz group in **461** was then removed, and the resulting amine was coupled with Cbz-β-Ala-OH to produce the intermediate **462**. These steps were repeated with a different amino acid, Cbz-D-His(Tr)-OH, to afford the macrocycle precursor **463**. The intramolecular cyclization occurred upon saponification of the linear tetrapeptide **463** to afford a carboxylate group and which reacted spontaneously with the carbobenzyloxy (Cbz) group, and subsequent trityl deprotection by hydrogenation afforded the macrocycle **464**. The macrocycle **464** formed complex with Cu (II) ions in aqueous solution.

Cyclic sugar macro dilactones **469** and **470** were synthesized and studied for their carboxylate ion binding properties.[119] As shown in Scheme 11.63, the glucose-derived diol **465** was esterified by Boc-Gly to form the intermediate **466**, and deprotection of Boc group using TFA afforded TFA salt of compound **467** in good yield. Cyclization of compound **467** with succinic or pyridine dicarboxylic resulted in macrocyclic compounds **468** and **470**, respectively. Macrocyclic compound **468** was hydrogenated to remove benzyl groups and afforded the trihydroxy macrocycle **469**.

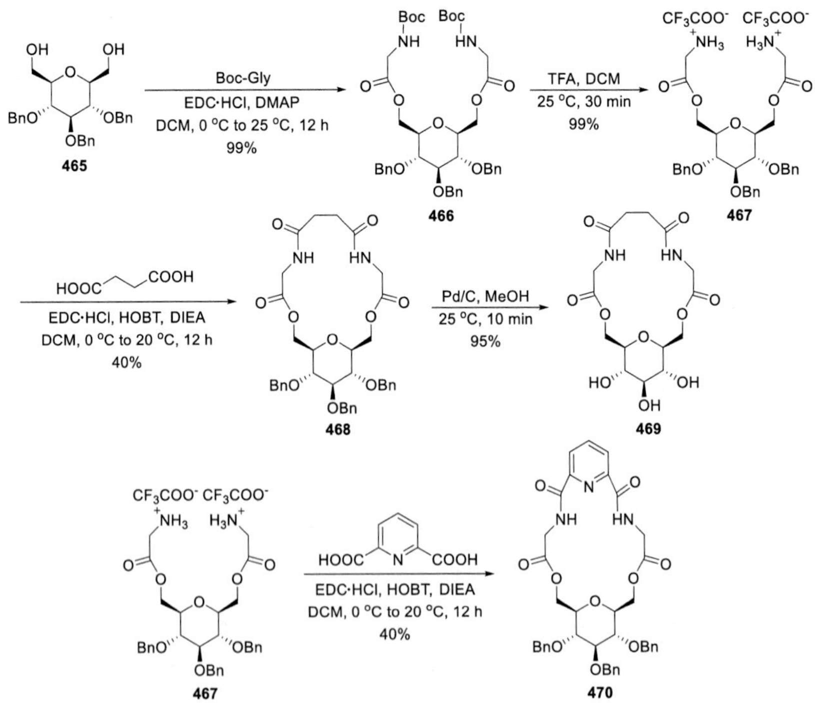

SCHEME 11.63 Synthesis of cyclic sugar amino acid containing compounds **469** and **470**.[119]

11.5.3 Synthesis of macrocycles by macrolactonization

By click chemistry and intramolecular lactonization, several novel macrolactones have been synthesized by the Wang group from cyclization of the C1–C6 positions of *N*-acetyl-D-glucosamine derivatives and C2–C6 cyclization of functionalized D-glucosamine derivatives.[120] As shown in Scheme 11.64, the reaction of 3,4-dibenzoylated tethered azides **471** at the anomeric position with alkynoic acids afforded 1,4-triazole-linked carboxylic acids **472**. Macrolactonization of compound **472** occurred between 6-OTs and the carboxylic acid group giving C1- and C6-linked mono-triazole marcolactones **473** and **474**. Applying the same method and using the triazole-containing alkynoic acids, bis-triazole-containing macrolactones **477** and **478** were synthesized successfully. The synthesis of C2- and C6-linked macrolactone was initiated from the sugar azide **479**, acetal deprotection, tosylation, and benzoylation transformed **479** to the key intermediate tosylate **482**, which was treated with undec-10-ynoic acid utilizing click reaction to generate the macrolactone precursor **483**. Under similar macrolactonization conditions, glycomacrolactone **484** was obtained in good yield. Macrolactonization condition was studied for the cyclization of bis-triazole compound **485**. Reaction in different bases (Na_2CO_3, K_2CO_3, Cs_2CO_3) in anhydrous DMF, different ratios of the monomer **486** versus the dimer **487** were obtained.

Komura et al. synthesized a series of sialic acid-derived macrolactones **494** and **496**, and these were utilized for the synthesis of α-glycosides (Scheme 11.65).[121] By the formation of macrocycles from the anomeric center of sialic acid, the macrocycle indues conformational constrain to the system, and therefore the macrocycles act as strained sialic acid donors in glycosylation reaction and can lead to stereoselective glycosylation from the less hindered side. The synthesis of a few examples of these macrocycles are shown in Scheme 11.65, the amine derivative of sialic acid **488** was coupled with the tethering units *O*-succinimide (OSu) carbonates **489** with different chain lengths resulting in the corresponding carbamates **490a–f**, respectively. Peracetylation of these compounds afforded the peracetates **491**, and selective hydrolysis of methyl ester using lithium iodide produced the free carboxylic acids **492**. The TBDMS (TBS) group was deprotected next to afford compound **493** with a free carboxylic acid and a free primary hydroxyl group. The macrolactonization of this intermediate was carried out using the Mitsunobu reaction to afford the strained bicyclic sialyl donors **494** in good yields. Applying the same synthetic strategy, dichloro-containing sialyl donor **496** was synthesized. The macrocycle was transformed to α-glycosides **497–500** in good yield by reacting with different glycosyl acceptors.

Another example of glycomacrolactones is shown in Scheme 11.66 for the preparation of sucrose-based macrolactones.[89] The 6-OH group at the glucose end of the sucrose derivative **501** was transformed to the phosphonate **502a**

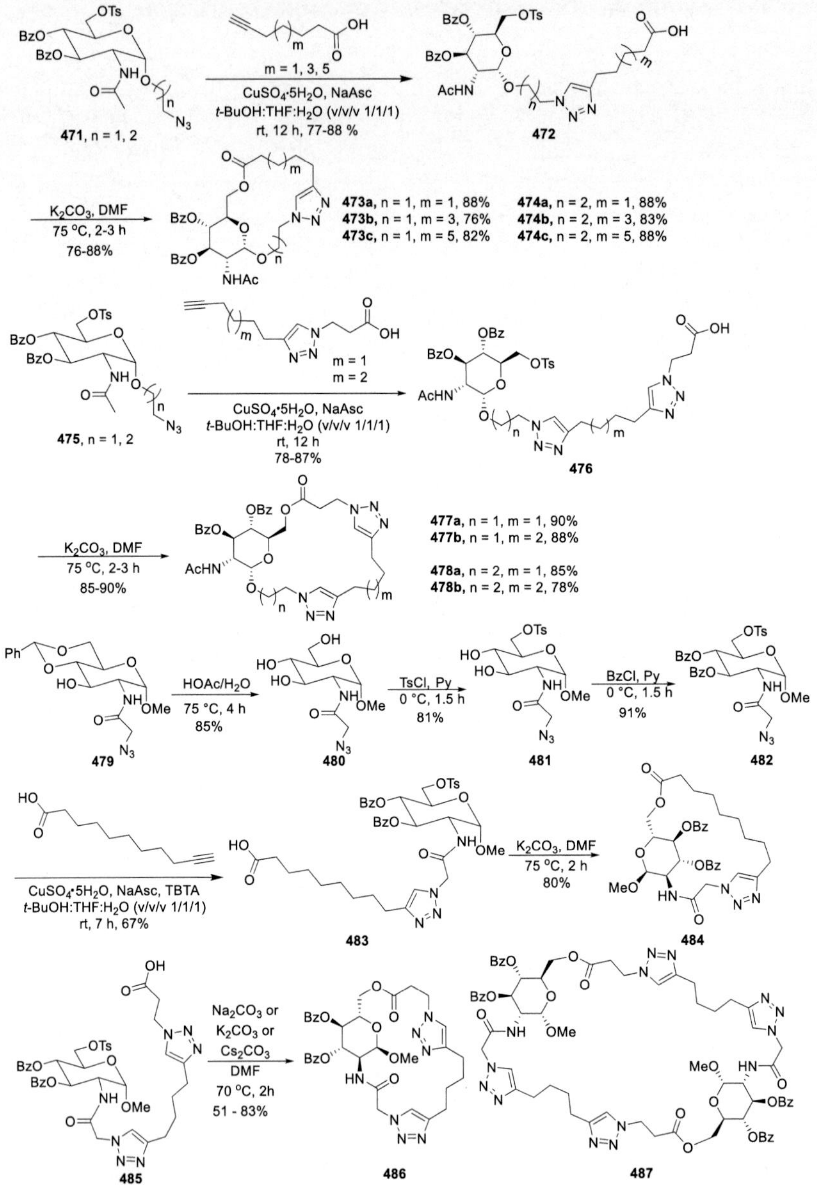

SCHEME 11.64 Synthesis of sugar-based macrolactones **473–474** and **477–478** by S_N2 reactions at C1–C6 linkage and **484**, **486–487** by S_N2 reactions at C2–C6 linkage.[120]

and separately to the aldehyde **502b** via Swern oxidation. Under phase transfer catalyst (PTC) condition using 18-crown-6, the phosphonate **502a** and aldehyde **502b** coupled through the Wittig reaction produced the enone **503** smoothly. Reduction of carbonyl group and methylation followed by silyl group deprotection afforded diol **504**, which was subjected to esterification conditions with diacyl chloride and triethylamine (TEA) to generate the permethylated macrocycle **505**. Perbenzylated analog **506** was prepared by the same strategy.

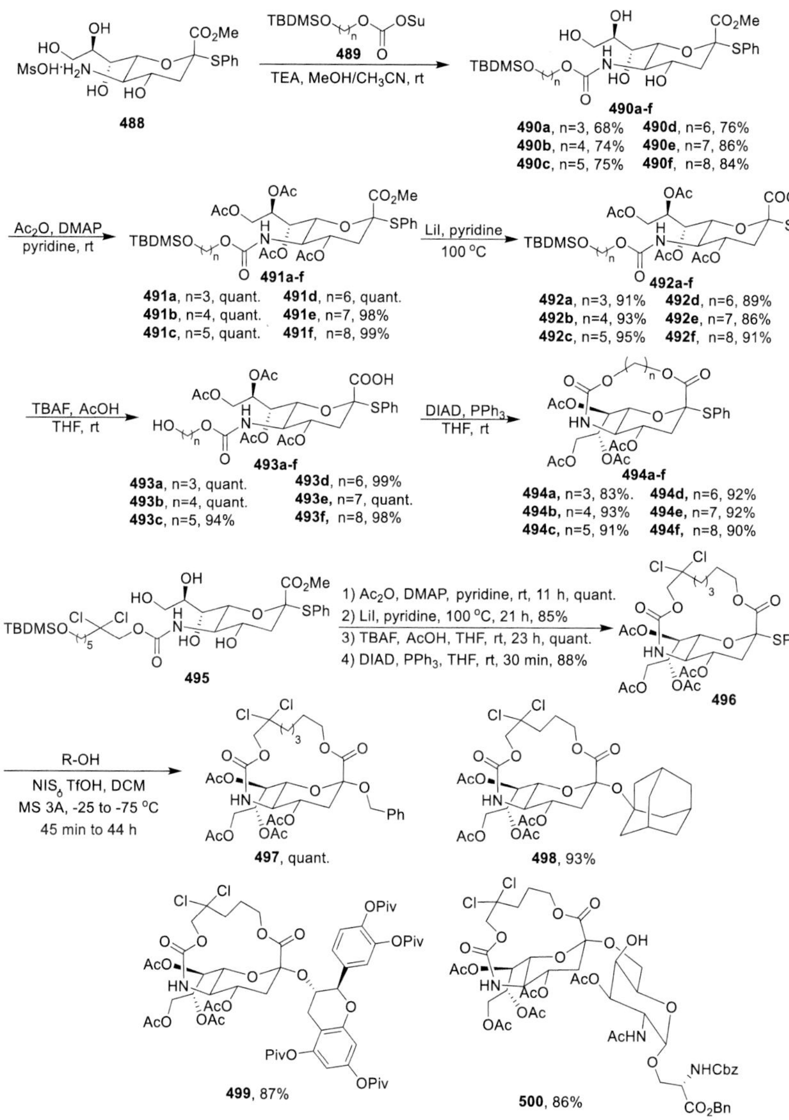

SCHEME 11.65 Synthesis of constrained sialic acid donors **494a–f**, **496** and glycosides **500a–d**.[121]

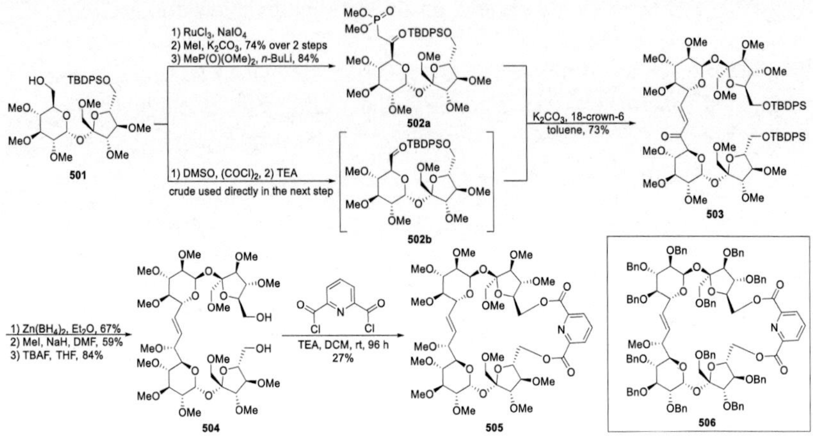

SCHEME 11.66 Synthesis of macrocyclic derivatives **505** and **506** with di-sucrose scaffold.[89]

11.6 Synthesis of macrocycles via click macrocyclizations

A large class of triazole-containing sugar-based macrocycles have been designed and synthesized over recent years; these macrocycles have also been the topic of many excellent reviews. Click reactions for the synthesis of macrocycles have been reviewed or covered as part of the topics in several recent reviews.[122–129] Triazole-containing macrocycles that include a sugar moiety were also discussed in another focused review that centered on the applications of 1,2,3-triazolophanes as molecular hosts, ion sensors, and so on.[123] The following papers have been reviewed in *Chem. Rev.* 2021[128] for using CuAAC reactions for macrocycles,[130–132] cyclic peptide mimicking sugar-containing macrocycles,[116,117,133,134] cyclodextrin mimics,[135,136] and macrocycles for ion binding and chiral recognition.[88,137] Herein, we mainly cover the recent development in these areas and focus on synthesis and applications of the macrocycles that have not been reviewed extensively. These are roughly categorized into pentose-based macrocycles and hexose-based macrocycles.

11.6.1 Pentose sugar-embedded macrocycles via click macrocyclizations

11.6.1.1 Synthesis of macrocycle by multiple component click reactions

Prasanna et al. reported the synthesis of triazole-linked glycomacrocycles by tandem double 1,3-dipolar cycloaddition reactions (Scheme 11.67).[138] Rection of *N*-alkyl azidoisatin **507a**, glucose-derived O-propargyl enone **507b**, and secondary amino acid **507c** under CuI, *N,N*-diisopropylethylamine (DIEA) in dry CH_3CN condition afforded the 1,4-triazole-incorporated sugar

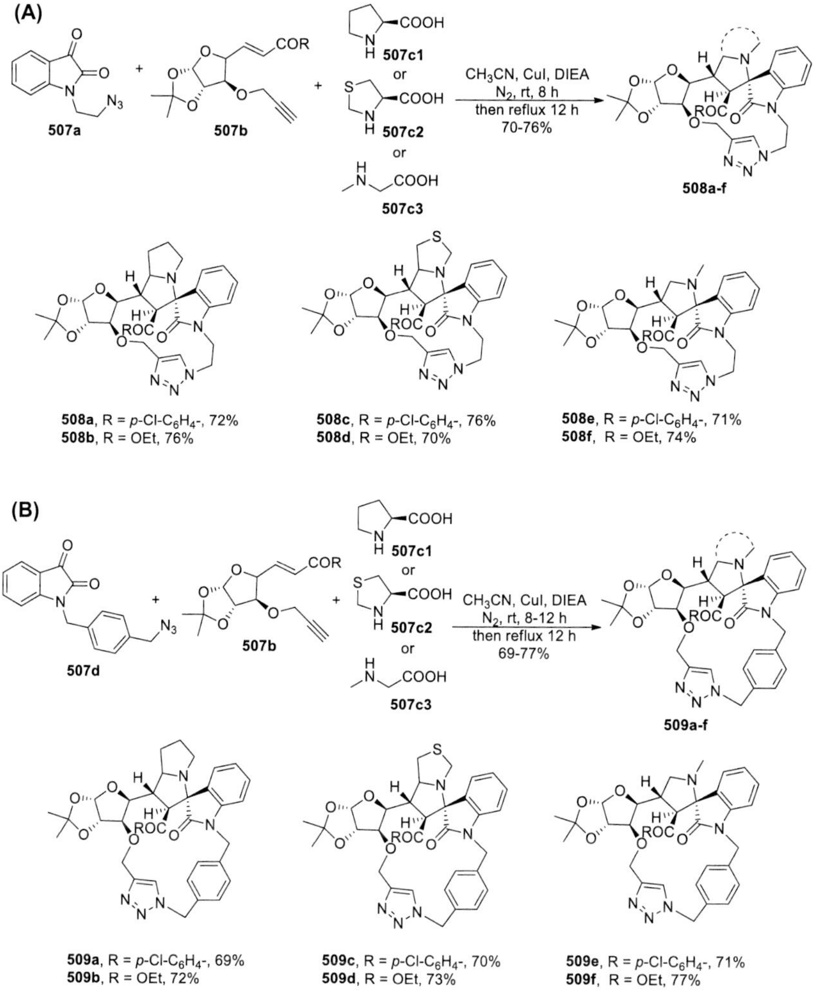

SCHEME 11.67 Synthesis of macrocycles **508a–f** and **509a–f** via tandem one-pot double 1,3-dipolar cycloaddition reaction.[138]

macrocycles **508a–f**. The developed method was applied to the synthesis of a series of macrocycles **509a–f** using the *N*-benzyl azidoisatin **507d**.

Another intramolecular macrocyclization was carried out using 1,3-dipolar cycloaddition of azomethine ylides to synthesize a series of triazole-linked glycomacrocycles (Scheme 11.68).[139] One-pot three-component reaction of *O*-propargyl nitro olefin **510**, salicylaldehyde-derived *O*-alkyl azides **511** or **512**, and secondary amino acid **507c** in the presence of CuI and DIEA in dry CH_3CN furnished 1.4-triazole-linked glycomacrocycles **513–514** in good yield.

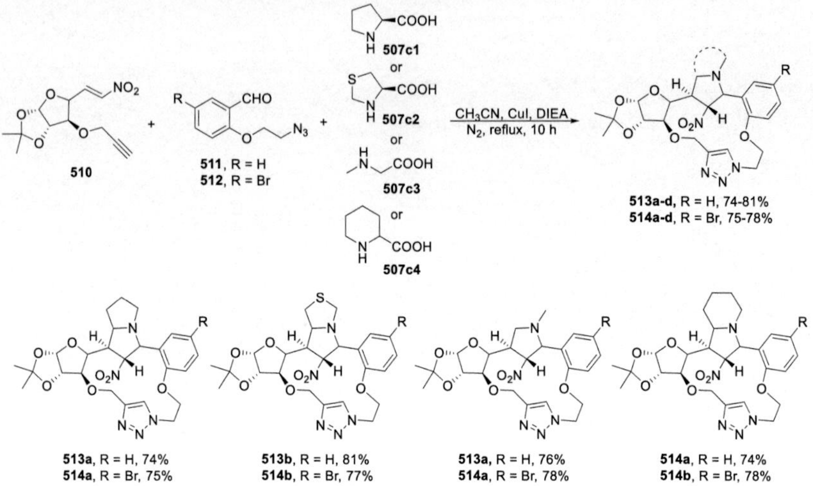

SCHEME 11.68 Synthesis of macrocycles **510d** and **511–514** via tandem one-pot double CuAAC.[139]

11.6.1.2 Synthesis of pentose sugar-based macrocycles by click reactions

Allam et al. synthesized a galacturonic acid-derived macrocycle via click macrocylization.[131] As shown in Scheme 11.69, D-galacturonic acid **515** reaction with methanol under acidic resin gave an anomeric mixture of the furanose ester **516** (α:β = 3:7) in quantitative yield. Acetylation with Ac_2O in the presence of HOAc and H_2SO_4 afforded tetraacetate **517**. $BF_3 \cdot Et_2O$-catalyzed glycosylation with propargyl alcohol produced exclusively β-furanose anomer **518** in good yield due to the neighboring group participation effect. Acetyl group removal and subsequent amide coupling reaction furnished macrocycle precursor **520** smoothly. Intramolecular click reaction of **520** was performed under copper sulfate and sodium ascorbate in water condition, leading to monotriazole furanose macrocycle **521** in 67% isolated yield.

Two *C*2-symmetric macrocycles **526** and **528** were synthesized by double click reaction and found applications in host–guest interaction and chiral recognition (Scheme 11.70).[88] Starting from D-glucosamine hydrochloride, the 2,5-anhydro-D-mannitol **522** was prepared with 80% yield over 3 steps. Through perbenzylation, partial acetolysis followed by removal of acetyl groups, bis-benzyl-protected sugar diol **523** was synthesized. The two hydroxyl groups in compound **523** were functionalized separately to afford bis-propargyl compound **524** and bis-azide **525**. Cycloaddition through double-click macrocyclization reaction afforded the macrocycle **526** in 30% yield in a one-pot reaction. *S*-phenyl ethyl ammonium acetate was used as a template for the click reactions. The same methodology was applied to the synthesis of a larger macrocycle **528**

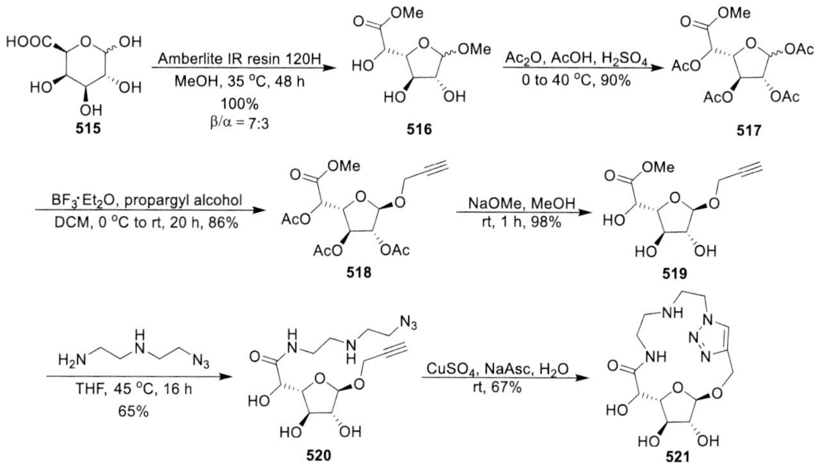

SCHEME 11.69 Synthesis of galacturonic acid-derived macrocycle **521** via click macrocylization.[131]

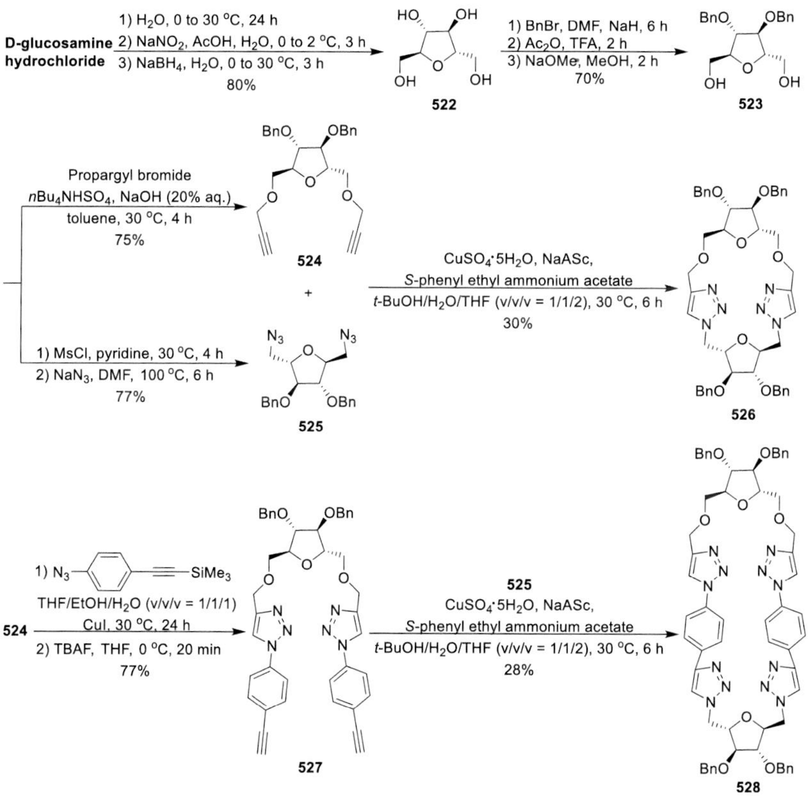

SCHEME 11.70 Synthesis of Sugar-based novel chiral macrocycles **526** and **528**.[88]

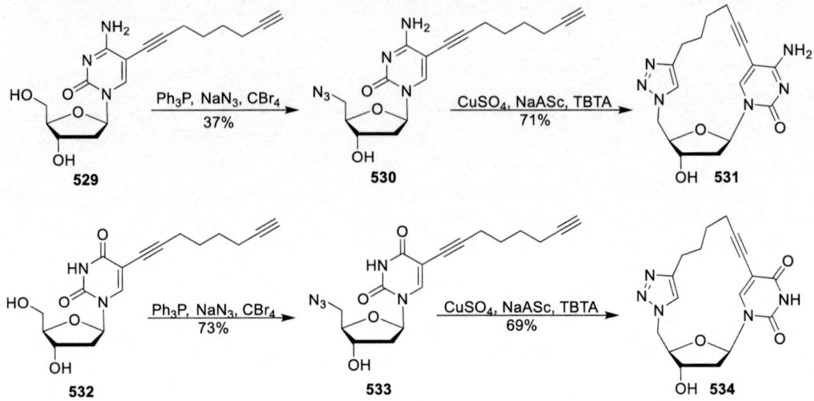

SCHEME 11.71 Synthesis of nucleoside macrocycles **531** and **534** using click chemistry.[137]

bearing a phenyl linker between bis-triazolyl sugar moieties. The bis-alkyne **524** was coupled with 4-trimethylsilylethynylphenyl azide followed by deprotection of trimethylsilyl group to furnish bis-triazolyl compound **527**. Using similar macrocyclization methods described earlier, the cyclo-addition reactions using bis-azide **525** with the bis-alkyne **527** afforded another sugar macrocycle **528** in 28% yield.

Liu et al. reported intramolecular click cyclization toward the synthesis of nucleoside macrocycles **531** and **534** in the presence of tris(benzyltria-zoylmethyl)amine (TBTA).[137] As shown in Scheme 11.71, the primary hydroxy group in the unprotected deoxycytidine derivative **529** was converted to the azido derivative **530** using triphenylphosphine, carbon tetrabromide, and sodium azide. The intramolecular click cyclization in the presence of TBTA afforded the nucleoside macrocycle **531** in good yield (71%). The same reaction methods were applied to deoxyuridine derivative **532**. It was converted to the macrocyclic nucleoside **534** smoothly.

In another study, triazole-fused medium-ringed pentose derivative macrocycles **538** were also synthesized via intramolecular azide and alkyne cyclo-addition reactions (Scheme 11.72).[140] The starting material 1,2,5,6-di-*O*-isopropylidene-α-D-allofuranose **535** was converted to azido sugar **536** using triflate anhydride and 2-azidophenol by nucleophilic substitution. Functional group transformations next afforded the sugar azido-alkyne compound **537**. Intramolecular cyclization of **537** under heating conditions in dry DMF gave monomeric product **538a** and dimeric product **538b**.

Dimerization reaction via Cu-catalyzed azide-alkyne cycloaddition reaction constructed triazole based macrocycles **544**, which can self-assemble to form a nanorod structure (Scheme 11.73).[134] D-Glucose-derived azido compound **539** reacted with DPPA/TEA to give acyl azide **540**, which was treated with *N*-hydroxy succinimide (NHS) to afford the desired carbamate derivative **541** via Curtius rearrangement. The linear azido-alkyne sugar precursors **543** were

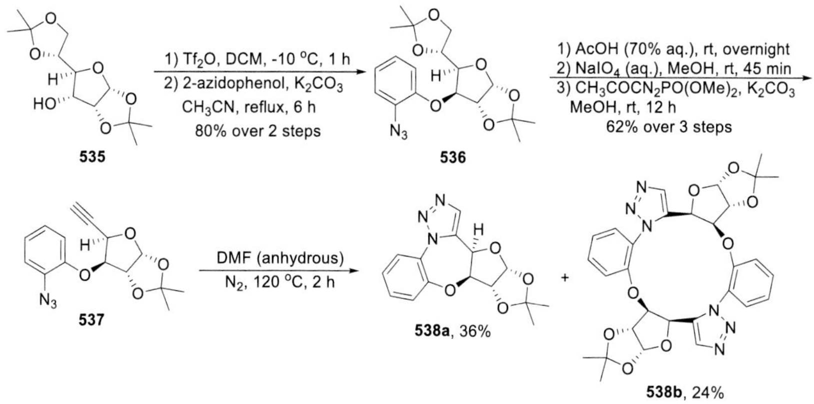

SCHEME 11.72 Synthesis of triazole-fused benzo-heterocycle **538b**.[140]

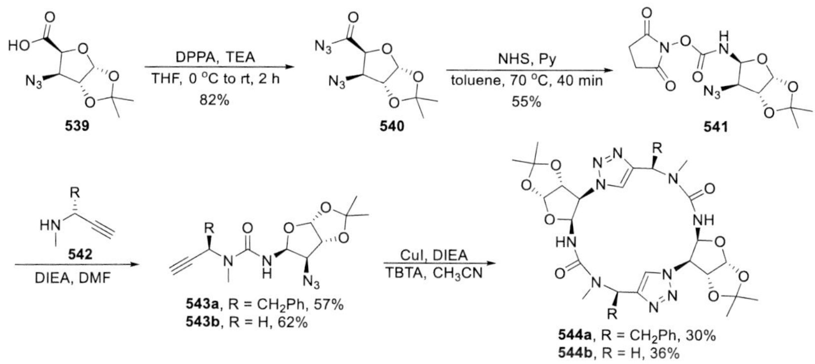

SCHEME 11.73 Synthesis of triazole/urea-based peptidomimetic macrocycles **544**.[134]

prepared by coupling *N*-methylated amino alkynes **542** with intermediate **541** under basic conditions. Intermolecular double-click macrocyclizations of the azido-alkyne derivatives **543** in the presence of a catalytic amount of TBTA, CuI, and DIEA provided the desired macrocycles **544**.

Vancomycin-like glycomacrocycle synthesis was reported by Singh et al. by Cu-catalyzed azide-alkyne cycloaddition reaction, and the prepared macrocycle showed promising antibacterial activity (Scheme 11.74).[141] Synthesis of 44-membered macrocycles **549** commenced with sugar alkyne **411** in a linear sequence fashion. Click reactions in *t*-BuOH and H_2O (v/v 1:1) with $CuSO_4 \cdot 5H_2O$ and sodium ascorbate gave **546a** and **546b**. Selectively removal of one set of isopropylidene groups of **546a** in acetic acid aqueous solution followed by oxidation of the resulting diol and then reduction of the carbonyl group afforded the sugar diol **547**, which was converted to bis-alkyne **548** smoothly. Intermolecular double-click macrocyclizations of the bis-azides **545**

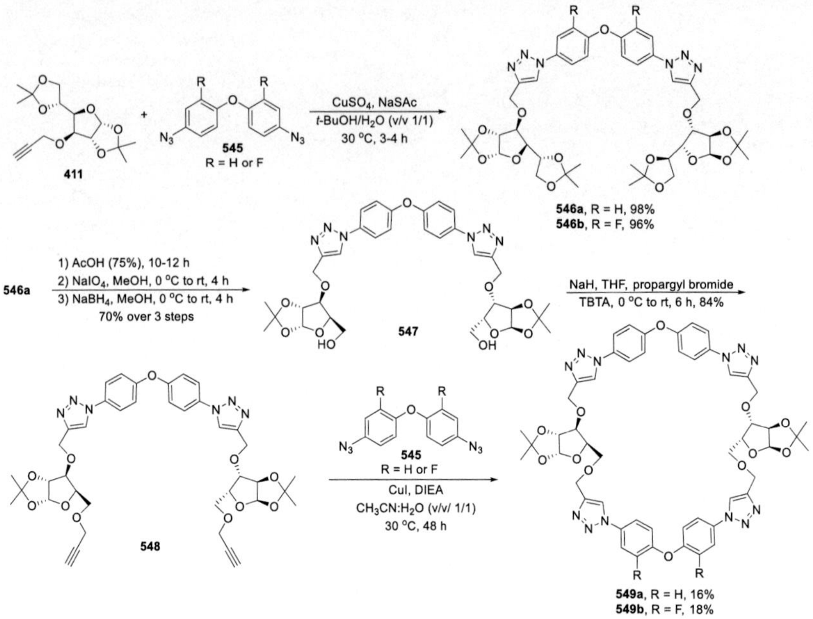

SCHEME 11.74 Synthesis of sugar-containing tetra-triazole derivatives **549**.[141]

with bis-alkyne **548** in water and acetonitrile (1:1, v/v), DIEA and CuI produced the Vancomycin-like glycomacrocycles **549**.

Besides the formation of 1,2,3-triazoles by cycloaddition reactions, isooxazoline-containing macrocycles were also synthesized by Hossain et al.[142] As outlined in Scheme 11.75, the *O*-alkenyl oximes **550** were made from glucose in several steps, which were then subjected to NaClO (4% aq.) in DCM and TEA condition to promote the dimerization by nitrile oxide-alkene cycloaddition to afford symmetrical isooxazoline incorporated glycomacrocycles **551**. Another set of asymmetric isooxazoline incorporated macrocycles **555–557** was prepared by intramolecular nitrile oxide-alkene cycloaddition of the disaccharide linear macrocycle precursors **554**, which were obtained from etherification of glucose-based alcohols **552** and glucose-derived mesylates **553** followed by standard functional group transformations.

Glycomacrocycles **564–569** containing isooxazoline moiety were synthesized via similar 3 + 2 cycloaddition reactions of nitrile oxide and alkene in an intramolecular fashion (Scheme 11.76).[143] The phthalate-linked alkenyl sugar derivatives **560** were prepared from isopropylidene-protected glucose or allose **558** in two steps. Subsequent selective removal of isopropylidene group, oxidative cleavage of resulting diol followed by oxime formation afforded the oxime alkenyl precursors **562**. Treatment of compounds **562** with NaClO (4% aq.), DCM, and TEA gave the isooxazoline-grafted glycomacrocycles **564–569** via intramolecular nitrile oxide-alkene cycloaddition.

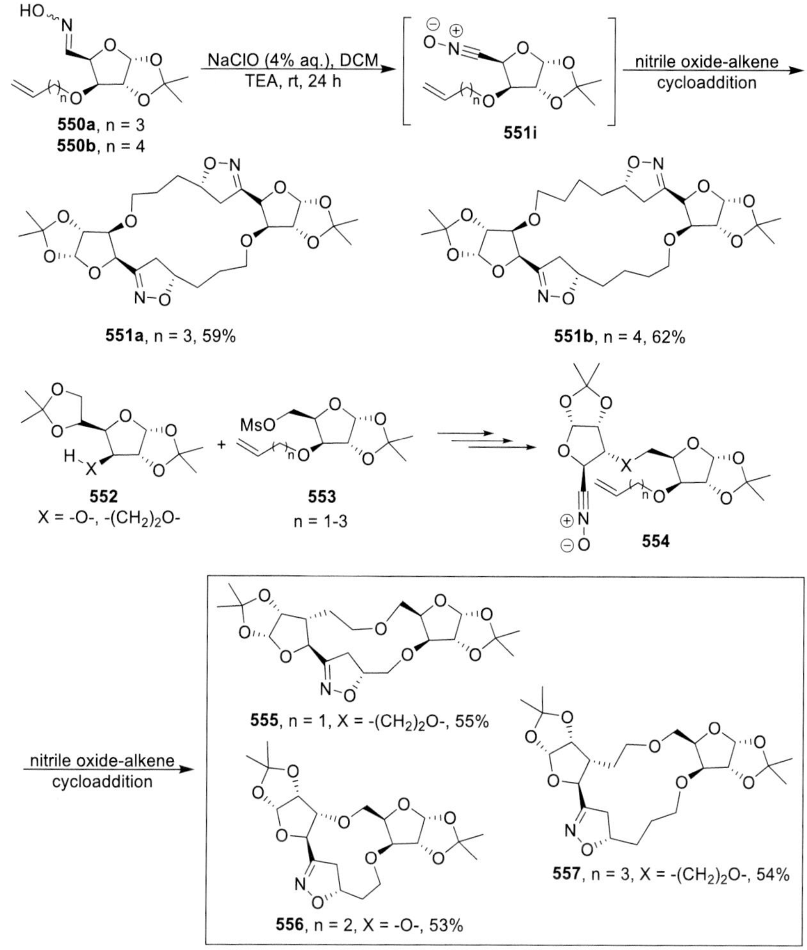

SCHEME 11.75 Synthesis of sugar-containing macrocycles **551** and **555–557** via nitrile oxide-alkene cycloaddition.[142]

11.6.2 Hexose sugar-embedded macrocycles via click macrocyclization

A large number of different triazole-embedded sugar macrocycles have been synthesized using various hexose as the starting materials. For example, D-glucose-based macrocycles linked through C-2 and C-3 positions to produce crown ether-like macrocycles by click reactions.[61] The glucose derivatives were converted to C1–C6 linked macrocycles by "click stitching" strategy.[101]

Ajay et al. synthesized a series of pyran-based 14-, 15-, and 16-memebered ring macrocycles **576** (D-glucose based) and **577** (D-mannose

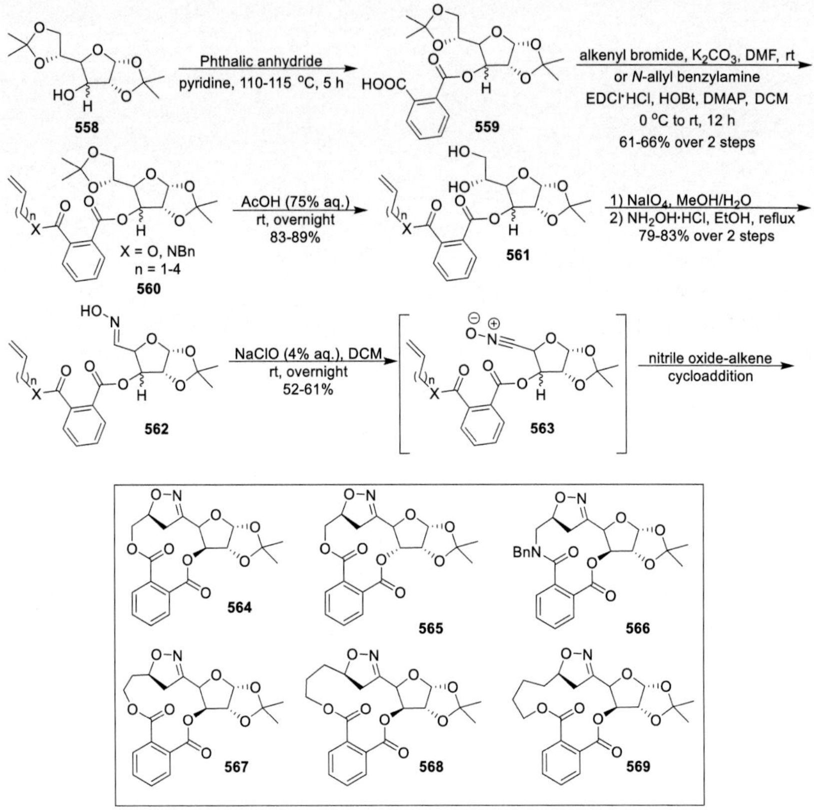

SCHEME 11.76 Synthesis of sugar-containing macrocycles **564–569** via nitrile oxide-alkene cycloaddition.[143]

based) via click chemistry.[130] As shown in Scheme 11.77, glucose butenones **570** was converted to the sugar tosylate **571** by tosylation and acetylation, and this was then converted to sugar azides **572**. Cu-catalyzed alkyne azide cycloaddition of azides **572** with various alkyne tosylates afforded the desired triazolyl butenones **573**. Sugar macrocyclic precursor **574** was obtained by reacting compound **573** with sodium azide in DMF. TBAHS-promoted intramolecular cycloaddition constructed the sugar macrocyclic 1,4,5-trisubstituted triazoles **575**. Removal of acetyl groups in **575** led to polyhydroxy compounds **576**. D-Mannose-based macrocycle **577** was prepared by the same methodology.

Xie et al. reported the synthesis of BODIPY-containing sugar-based macrocycles **584–587** by click chemistry (Scheme 11.78).[144] Starting from the 6-azido derivative of C-allyl glucoside **578**, azide reduction followed by Boc protection afforded intermediate **579**. The allyl group was oxidized to

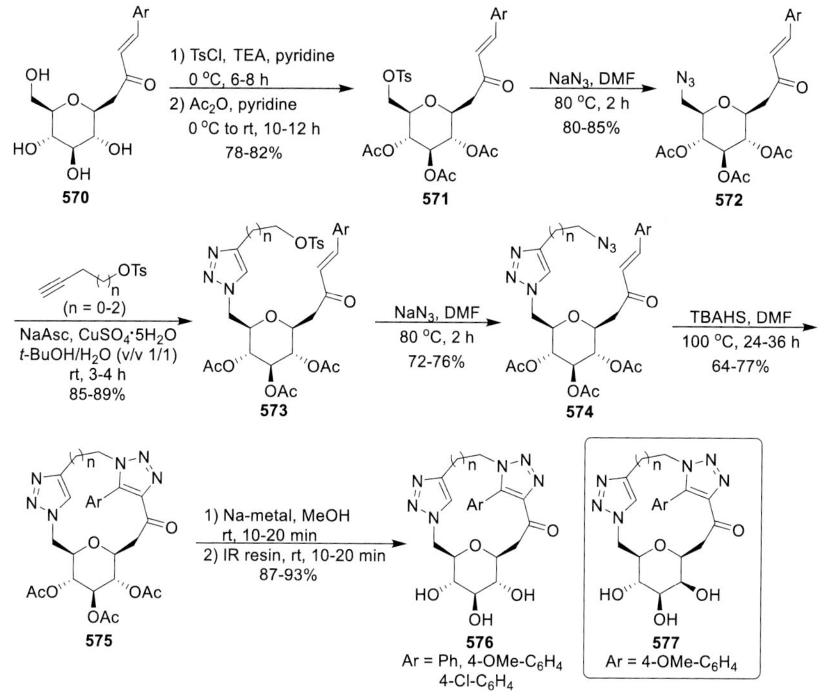

SCHEME 11.77 Synthesis of pyran-based 1,4-linked 14-, 15-, and 16-memebered ring macrocycles **576–577** via click chemistry.[130]

the corresponding diol and further transformation to form the intermediate **581**. Macrocycle precursor **583** was obtained by reacting azido sugar **581**, after removing the Boc group, with BODIPY-derived acid **582** under standard peptide coupling conditions. Copper-catalyzed intramolecular click reaction of **583** in the presence of $CuSO_4$ and sodium ascorbate in a mixture of DCM/H_2O gave BODIPY-containing sugar-based macrocycle **584**. By the same developed method, larger ring-sized macrocycles **585–587** bearing various amino-acid moieties were synthesized.

Wang group designed and synthesized a series of *N*-acetyl glucosamine-based bis-triazole macrocycles **592–595** and found their applications for accelerating $CuSO_4$-mediated click reaction (Scheme 11.79). $BF_3 \cdot Et_2O$-catalyzed glycosylation of *N*-acetyl-D-glucosamine with chloro ethanol gave the α-anomer as the major product, which was transformed to the sugar azide **588** and isolated as pure α-anomer by column purification. One-pot selective tosylation of **588** and subsequent acylation afforded sugar diacetate **589**. Mono click reaction of compound **589** with bis-terminal alkynes furnished the sugar triazolyl alkynes **590**. The 6-tosyl group was displaced with NaN_3 to produce the azido alkynes **591**. These key intermediates were

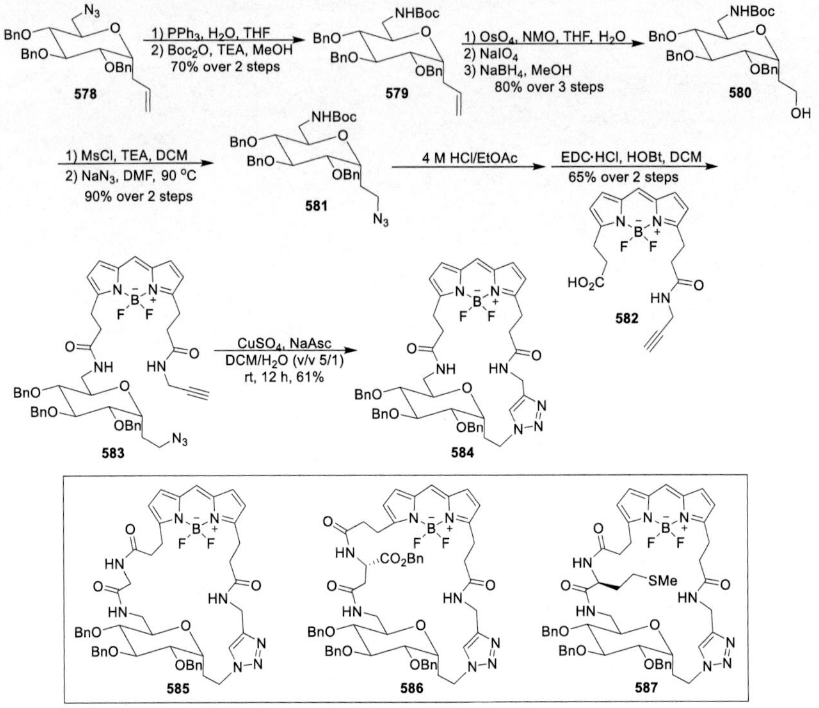

SCHEME 11.78 Synthesis of BODIPY-containing sugar-based macrocycles **584–587** by click chemistry.[144]

subjected to intramolecular click cyclization conditions using CuI (20 mol%) at 0°C in acetonitrile to generate 1,6-linked *N*-acetyl glucosamine-based macrocycles **592**. This condition was applied for the synthesis of several other macrocycles **593** with longer chain lengths at the anomeric position and benzoyl-protected analogs **594** and **595**.

Lepage et al. reported the synthesis of cyclodextrin analogs **597–599** starting from the same intermediate **596** using click chemistry (Scheme 11.80).[135] The intermolecular macrocyclization of the azido alkyne **596** was carried out using 10 equiv. of $CuSO_4$ and 20 equiv. of sodium ascorbate in a one-pot fashion to afford dimeric macrocycle **597** in 42% yield together with minor trimeric cyclized product **598** in 9% yield. The "clickable" terminal alkyne molecule **599** was prepared in 75% yield by removing the TIPS protecting group with 4.0 equiv. of AgF.

Pathigoolla synthesized reverse cyclodextrin mimetics **604** using click chemistry (Scheme 11.81).[136] The peracetylated galactose azide **600** was converted to the 4,6-benzylidene acetal-protected azide **601** in a few steps. Reductive ring opening and propargylation of the 4-hydroxyl group afforded

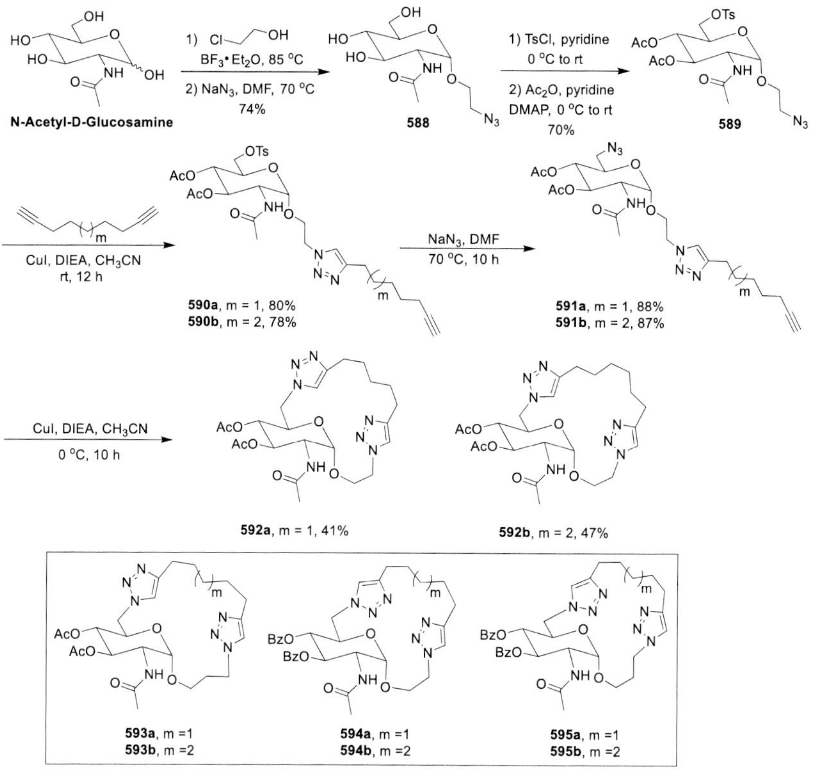

SCHEME 11.79 Synthesis of 1,6-linked *N*-acetyl glucosamine-based macrocycles **592–595**.[132]

the 1-azido and 5-acetylene D-galactose intermediate **602**. This azido alkyne was used as the building block for the click macrocyclization reactions and afforded product **603** as a mixture of four macrocycles. The mixture of cyclooligomers was isolated by chromatography to afford cyclic dimer **603a** (15%), trimer **603b** (28%), tetramer **603c** (15%), and cyclic pentamer **603d** (3%). Global debenzylation afforded the corresponding sugar triazole macrocycles **604**.

Lo Conteet al. synthesized triazole-linked sugar macrocycles **610–612** by CuAAC reactions and explored their binding properties with fluorescent probe 8-anilino-1-naphthalene-sulfonate (ANS).[145] As shown in Scheme 11.82, the alkynyl C-glycoside building blocks, 6-azide TBS-protected alkyne **605**, and 6-tosylate sugar alkyne **606** were used as the starting materials. They were synthesized from glucose derivatives in several steps. The azide **605** and alkyne **606** were coupled under MW-assisted CuAAC condition to give disaccharide azide **607** and subsequent azidation, and silyl group deprotection gave the disaccharide alkyne **608**. Applying the previously mentioned MW-assisted

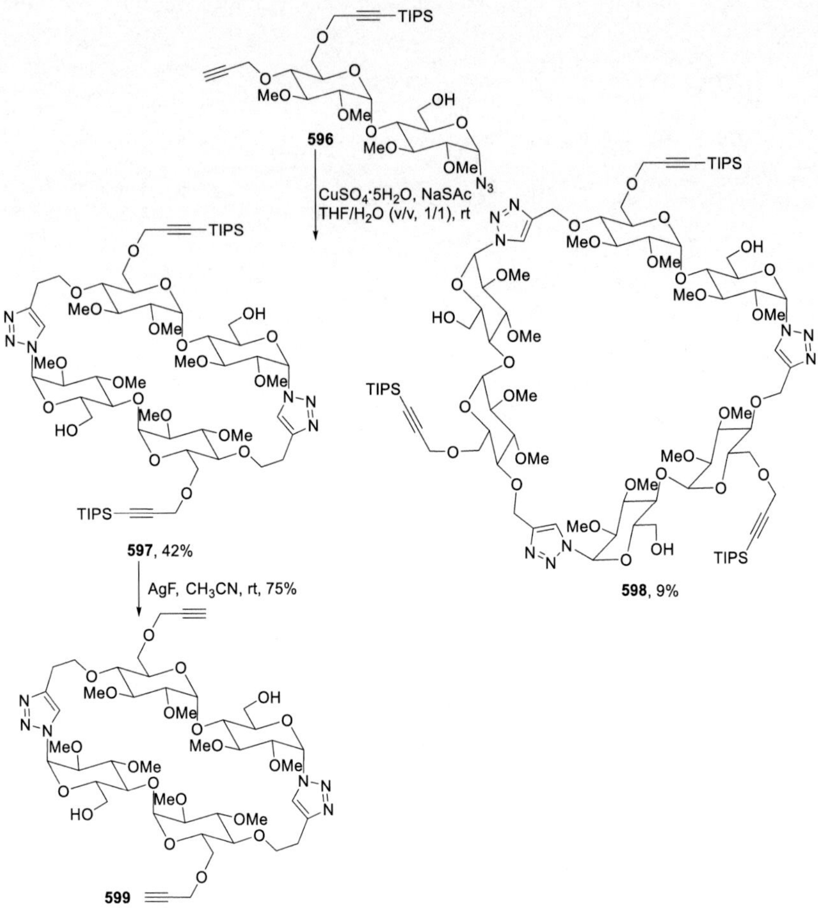

SCHEME 11.80 Synthesis of cyclodextrin analogs **597–599** by click chemistry.[135]

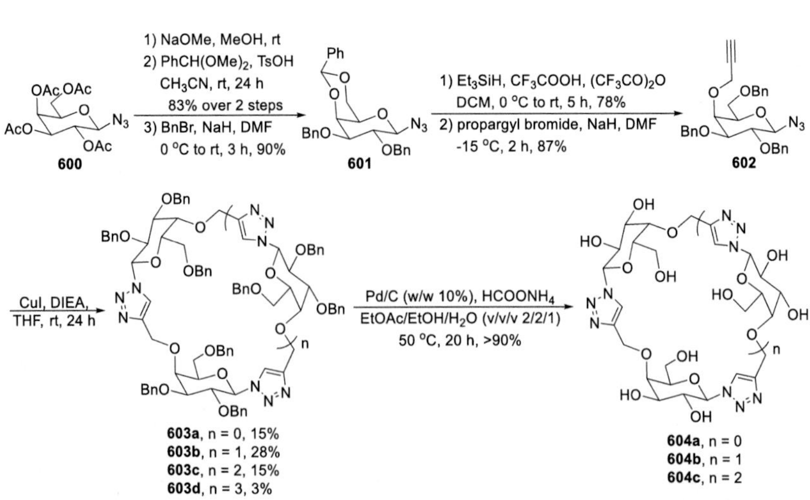

SCHEME 11.81 Synthesis of reverse cyclodextrin mimics **604** using click chemistry.[136]

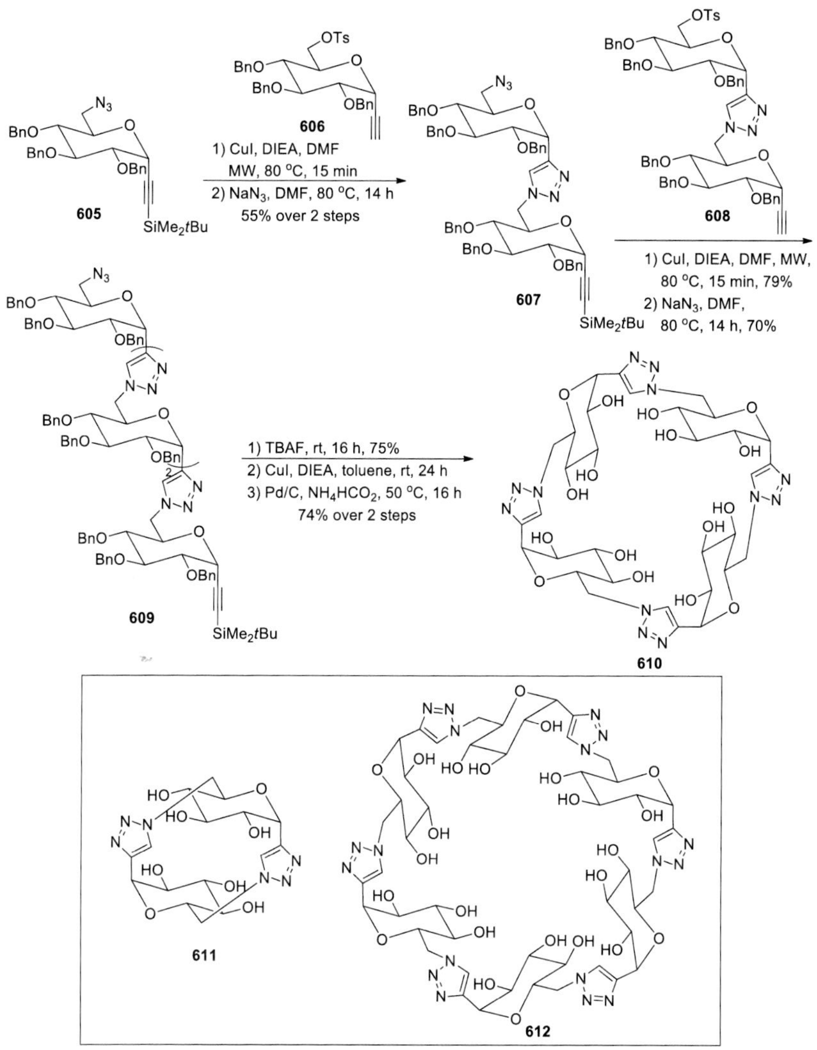

SCHEME 11.82 Synthesis of triazole-linked cyclic glucose-linked macrocycles **610–612**.[145]

click condition, reaction of azide **607** with alkyne **608** and subsequent azide formation afforded tetrasaccharide derivative **609**. The silyl group in compound **609** was removed to reveal the free alkyne; this intermediate then underwent intramolecular CuAAC reaction to form the benzyl-protected macrocycle, after catalytic hydrogenation furnished the triazole-linked sugar macrocycle **610**. Utilizing the same developed method, cyclic dimeric compound **611** and hexameric macrocycle **612** were prepared.

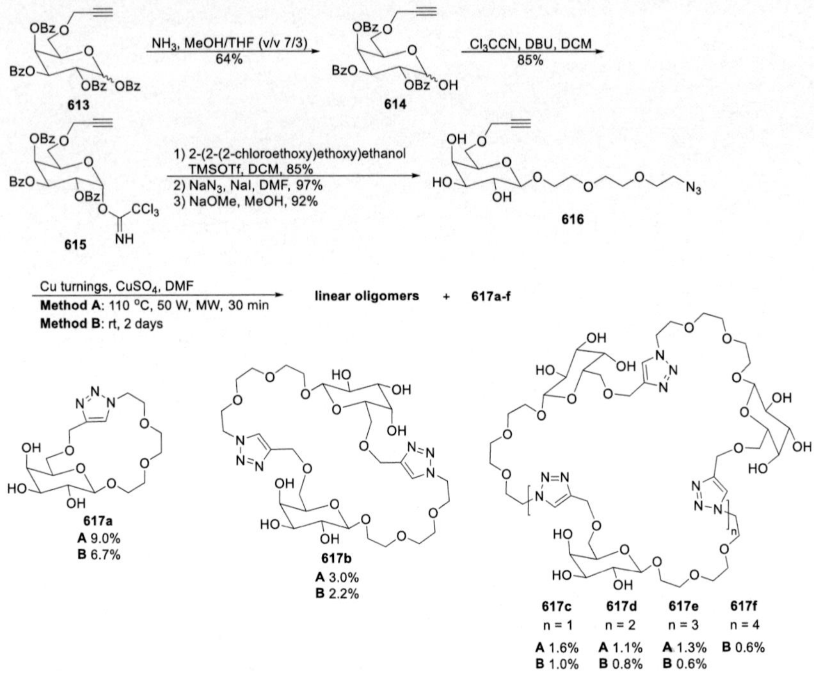

SCHEME 11.83 Synthesis of β-1−6-linked galactose-based triazole-containing macrocycles **617**.[146]

Campo et al. have synthesized a series of galactose derivatives using click reactions, they prepared linear oligomers and macrocyclic compounds **617** using the intermediate **616** (Scheme 11.83).[146] The azido-alkyne intermediate **616** was synthesized from D-galactose derivative **613** in a few steps by selective benzoate hydrolysis at anomeric position and subsequent conversion to the trichloroacetamide **615**. TMSOTf-catalyzed glycosylation with 2-(2-(2-chloroethoxy)ethoxy)ethanol of glycosyl donor **615** followed by azidation and Bz group deprotection provided key intermediate **616**. Intermediate **616** contains an azido group attached to the anomeric position with a triethylene spacer and an alkyne group attached to the C-6 position. This compound was subjected to CuAAC reactions to afford a series of macrocycles and linear oligomers. The mixture was isolated using reverse phase HPLC to obtain the β-1−6 linked cyclic monomer, dimer, trimer, tetramer, pentamer, and cyclic hexamer **617**. These compounds were evaluated for their antiparasitic properties against *Trypanosoma cruzi macrophage*. Intramolecular click reaction methods with $CuSO_4$/Cu turnings in DMF either at 110°C by microwave irradiation (Method A) or at room temperature (Method B) were tested out. The reaction progress was monitored by TLC, and the reaction using Method A finished after 30 minutes, while reaction completion was observed after 2 days for Method B.

11.6.3 Synthesis of sucrose-based macrocycle via click chemistry

Several sucrose-based macrocycles were synthesized using click chemistry as well.[79,91] A few examples are included in Scheme 11.84.[91] Sucrose-based bis-propargyl derivative **618** and bis-azide derivative **621** were prepared starting from sucrose in a few steps. The reaction between D-mannitol azide **619** and sucrose bis-alkyne **618** under the click condition produced macrocycle **620** in 26% yield, while the click reaction between sucrose bis-azide **621** and *O*-propargylated D-mannitol **622** furnished **623** in 34% yield.

11.6.3.1 Synthesis of macrocycle by glycosylation for macrocyclization

11.6.4 Macrocyclization by glycosylation

Several macrocycles were synthesized using glycosylation methods; sometimes, these macrocycles were further transformed into other products as reported recently.[147,148] Tiwari et al. synthesized several disaccharide-containing macrocycles using click chemistry and intramolecular glycosylation reactions.[149] Using a similar strategy of selective propargylation of the hydroxyl group, followed by click chemistry with a benzylic azide and intramolecular glycosylation reaction, they have synthesized several other macrocycles. As shown in Scheme 11.85, Schmidt et al. reported the synthesis of macrocycles by intramolecular glycosylation and click chemistry.[148] The macrocyclization was carried out by a glycosylation reaction between a hydroxyl group with an ethyl thio glycoside in adjacent sugar moieties intramolecularly. The 4,6-benzylidene-protected glucopyranoside **624** was treated with 2-azidobenzyl bromide under phase transfer catalysis conditions,

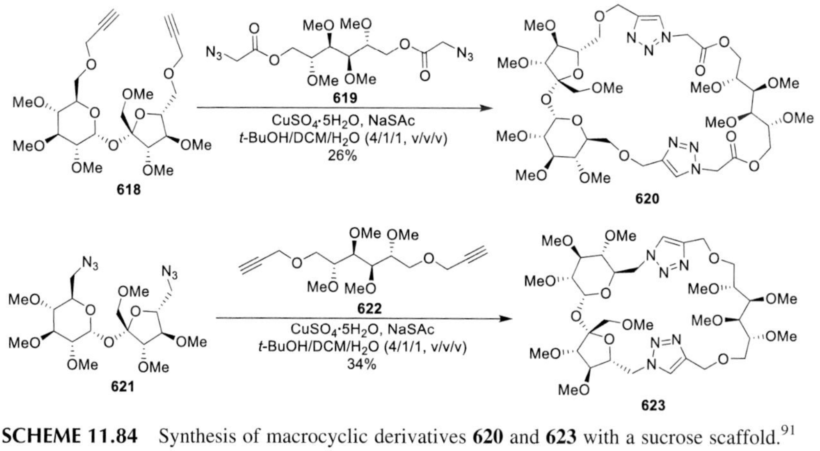

SCHEME 11.84 Synthesis of macrocyclic derivatives **620** and **623** with a sucrose scaffold.[91]

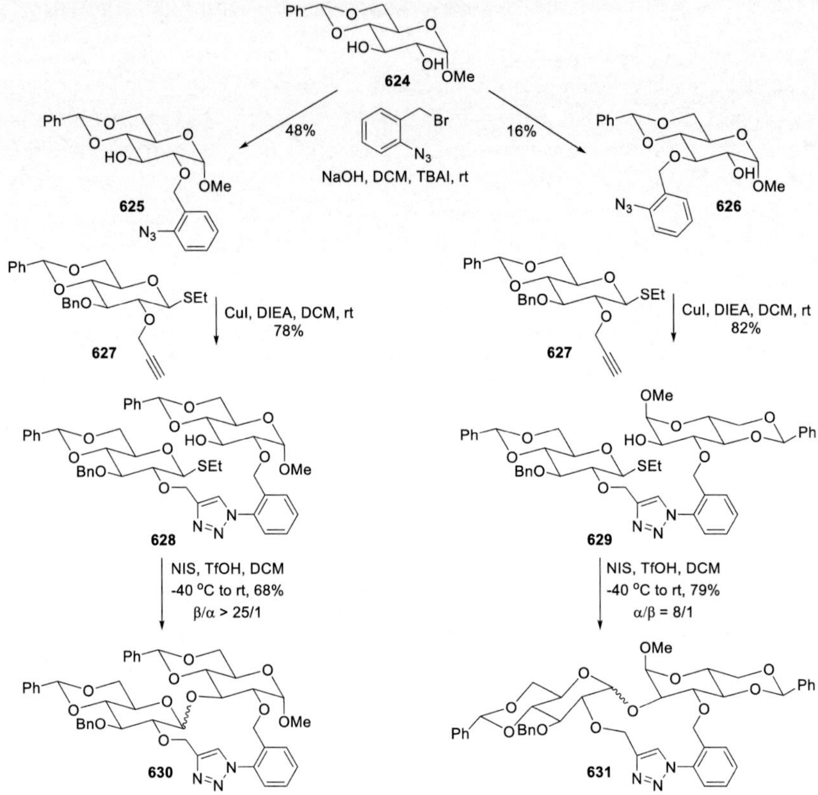

SCHEME 11.85 Synthesis of disaccharide macrocycles **630** and **631**.[148]

producing a mixture of desired isomeric sugar azides **625** and **626** with an isolated yield of 48% and 16%, respectively. Utilizing click reaction, sugar azides **625** and **626** reacted with glycosyl donor **627** giving macrocycle precursors **628** and **629** bearing donor and acceptor sites in one molecule, which underwent standard glycosylation conditions using NIS, TfOH giving disaccharide macrocycles **630** and **631** smoothly. Exclusive β-(1−3)-linked macrocycle **630** was obtained (β/α > 25/1) while mainly α-(1−2)-linkage product formation of macrocycle **631** was observed (α/β = 8/1).

11.6.5 Synthesis of cyclodextrin analogs by glycosylation

The synthesis of various cyclodextrin analogs composed of ring cyclic oligosaccharides including 3-monosugar, 4-monosugar, and 5-monosugar cyclic oligosaccharides has been reviewed by Jayaraman et al. recently.[150,151] The therapeutic applications of cyclodextrin (CD) and tetrameric cyclic glucose and other CD analogs are also reviewed recently.[4] In this section, we will

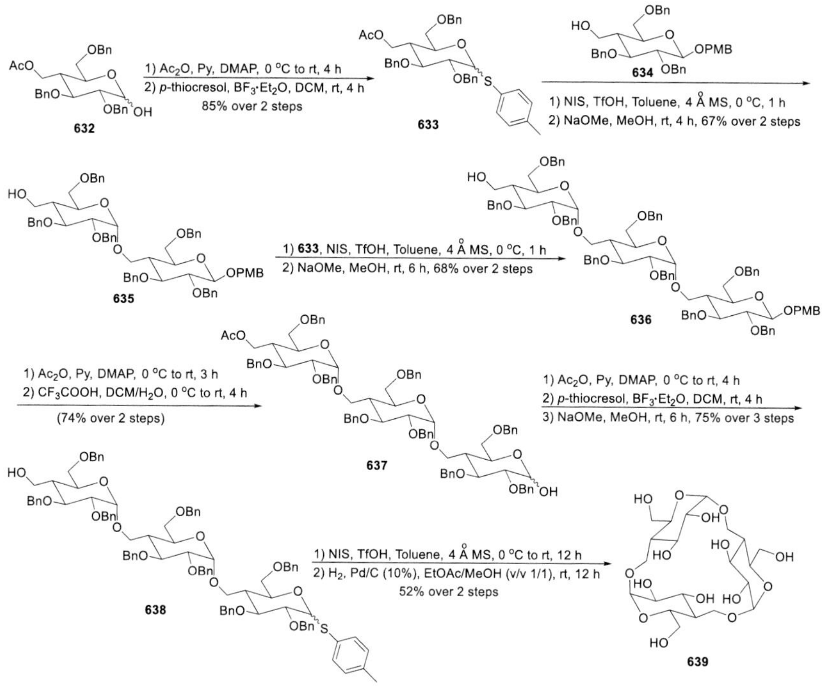

SCHEME 11.86 Synthesis of trimeric glucose-linked macrocycle **639**.[152]

highlight the general synthesis methods for different cyclo oligosaccharides using glycosylation for the macrocyclization step.

Maiti et al. synthesized a cyclic trisaccharide via glycosylation reaction of the C4-hydroxyl group with C1-tolylthio group intramolecularly.[152] As shown in Scheme 11.86, the C-4 extended glucose derivative **632** was converted to the thioglycoside **633** using *p*-thiocresol, glycosylation of the thioglycoside with a protected glycosyl acceptor **634** afforded the disaccharide **635** with a free hydroxyl group after deacetylation. Compound **635** was used as the glycosyl donor and, by a similar method, the trisaccharide **636** was obtained. Repeating the steps described earlier, the intermediate was converted to compound **638** with a free hydroxyl group and anomeric thio group, subsequent intramolecular glycosylation gave a mixture of the anomers. The major anomer was characterized to be α-anomer, which was isolated by column purification and further subjected to debenzylation to produce the cyclic trisaccharide **639** in moderate isolated yield.

Using a similar strategy described in Scheme 11.86, several other macrocycles have been synthesized; these include the synthesis of backbone-modified cyclic tetrasaccharide **644** (Scheme 11.87),[153] and cyclic pentasaccharide **648** (Scheme 11.88).[154] The glycosyl donor **640** was coupled with the alcohol **634** to give the disaccharide **641**, which was transformed to the

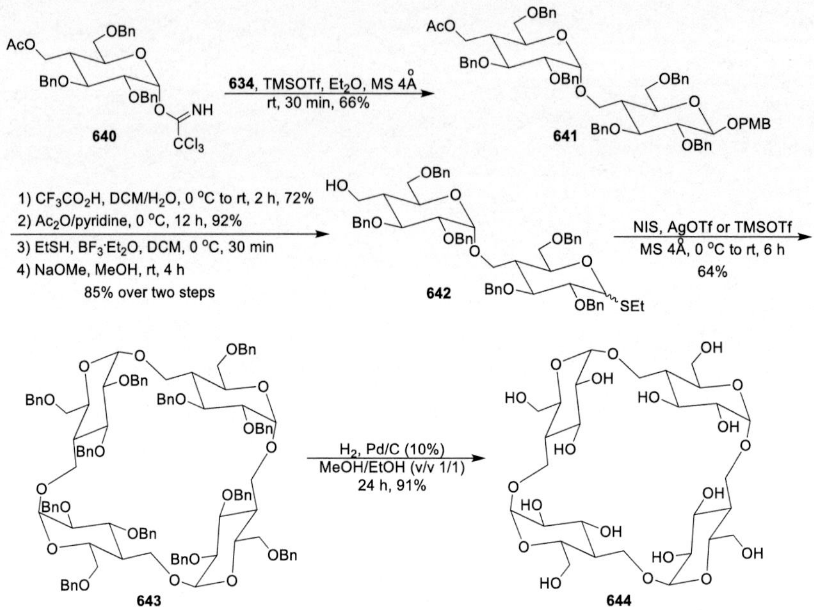

SCHEME 11.87 Synthesis of tetra-glucose-linked macrocycle **644**.[153]

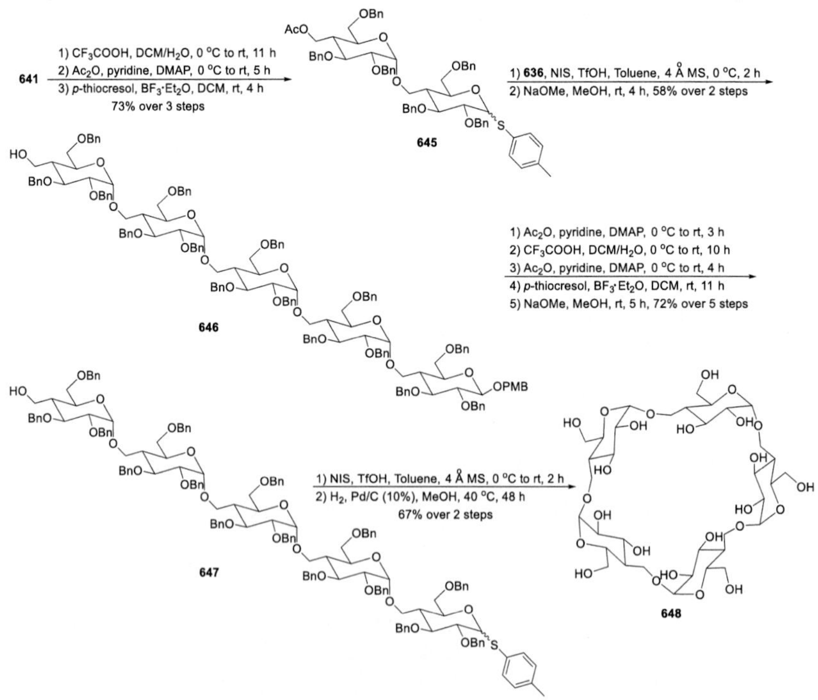

SCHEME 11.88 Synthesis of penta-glucose-linked macrocycle **648**.[154]

key intermediate **642** by converting anomeric ethylthio glycoside and a free hydroxyl group at the other end of the same molecule. This was subjected to glycosylation conditions using NIS and TMSOTf to afford the intermolecular dimerization product, the tetramer **643** in 64% yield, along with a small amount of dimer (11%) through intramolecular cyclization. After catalytic hydrogenation to remove all benzyl groups, the final isolated cyclic compound **644** was obtained and it was shown to be the α-anomer and the stereochemistry was confirmed by coupling constant based on^1 H NMR spectra.

Using a similar approach, this time through the formation of a linear penta-saccharide, the precursor **646** was synthesized from the same intermediate **641**. The pentasccharide was then transformed into an intermediate with a free hydroxyl group acting as glycosylation acceptor and tolylthio glycoside at the other end. Intramolecular glycosylation, followed by hydrogenation reaction afforded the cyclic penta saccharide **648** in good yield and as the single isolated α glycosylation product.

Liu et al. synthesized cyclogentiotriose via the final step glycosylation reaction of a macrolactone precursor as depicted in Fig. 11.23.[155] The mechanism of glycosylation will allow the formation of glycoside with beta selectivity through the reaction of a macrolactone. Subsequently, Rychnovsky's reductive acetylation gave the α-acetoxy ether. TfOH-mediated cyclization of the α-acetoxy ether resulted in a ring-closing glycosylated product with moderate-to-high β-selectivity. Applying this methodology (Fig. 11.23), cyclogentiotriose **660** was synthesized with preferred β-stereoselectivity.

As shown in Scheme 11.89, $BF_3 \cdot Et_2O$-mediated glycosylation of donor acetimidate **649** and acceptor **650** afforded the disaccharide thioglycoside **651** in good yield, and subsequent glycosylation in the presence of NIS with TfOH produced the trisaccharide **653**.[155] Four steps of transformation of trisaccharide **653** including acetyl and benzoyl hydrolysis, followed by benzylation of the free hydroxyls, regioselective benzylidene group opening and acetylation produced trisaccharide **654**. The desired amide **655** was prepared in 42% yield over three steps by removing the 4-methoxyphenyl (MP) group by ceric ammonium nitrate (CAN), deprotecting the acetyl group on the primary hydroxyl position followed by lactol ring opening under the oxidative amination condition using I_2 and $BnNH_2$. Protecting both hydroxyl groups in **655** by triethylsilyl (TES) group and activating amide group under neat conditions (Boc_2O, TEA, and DMAP) gave the Boc-protected imide **656** in good yield. Selective desilylation on the primary hydroxyl group by PPTS followed by LiOOH

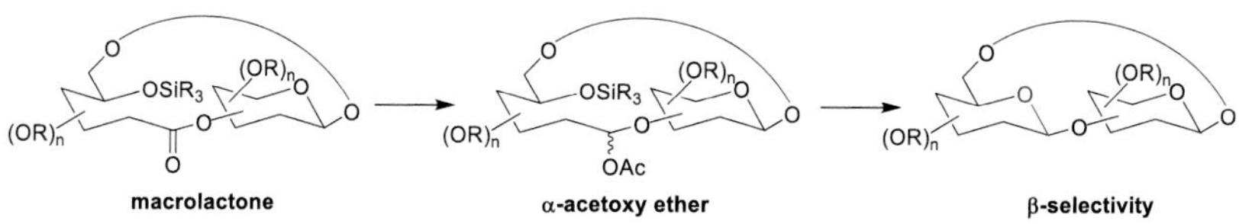

FIGURE 11.23 Two-step intramolecular glycosylation in linear and cyclic system.

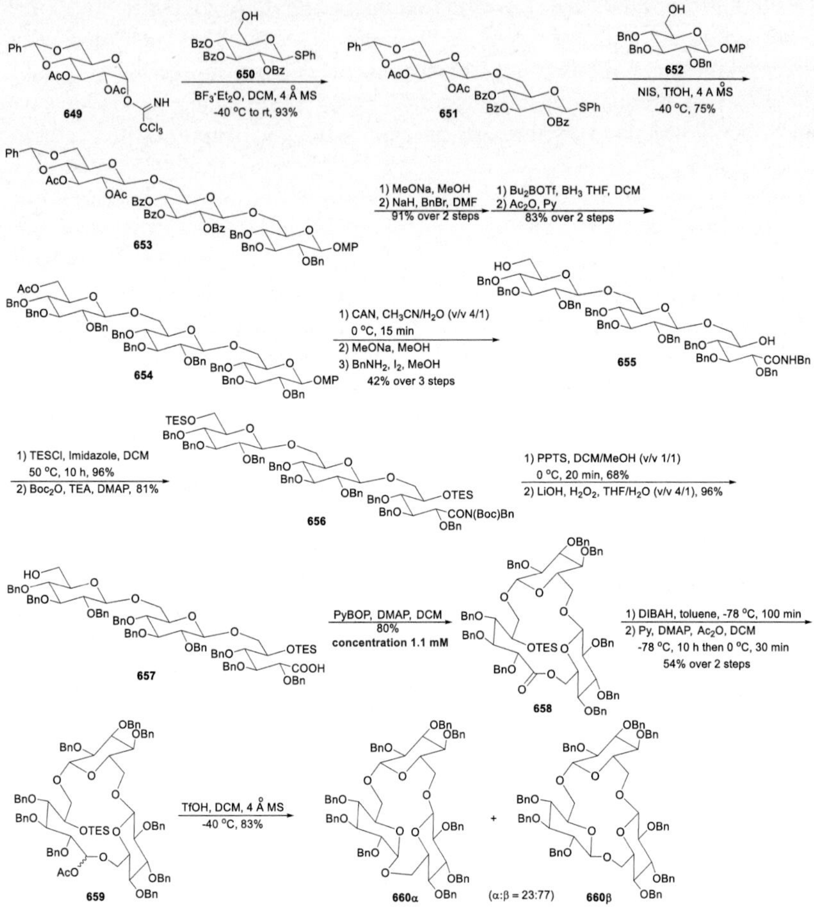

SCHEME 11.89 Synthesis of benzylated cyclogentiotriose **660**.[155]

(in situ generated from LiOH and H_2O_2) mediated cleavage of the imide led to the intermediate **657**. This intermediate contains both hydroxyl and carboxylic acid groups, and it was then subjected to macrolactonization condition in the presence of benzotriazol-1-yloxytripyrrolidinophosphonium hexafluorophosphate (PyBOP) and DMAP, providing the macrolactone **658** smoothly. The DIBAL-H mediated reductive acetylation of macrolactone **658** was carried out to afford the α-acetoxy ether **659** as a mixture of diastereomers. Final glycosylation using TfOH was performed to provide two cyclized products **660** with β-isomer preferred in 3:1 ratio.

Electrochemical synthesis of cyclic 4,5,6-meric glycomacrocycles via intramolecular electrochemical glycosylation reaction was discovered by the Itoh group.[156] As a comparison model (Scheme 11.90), the intramolecular cyclization of tetrasaccharide **661** under conventional chemical glycosylation conditions using NIS and TsOH was carried out, giving a mixture of α- and

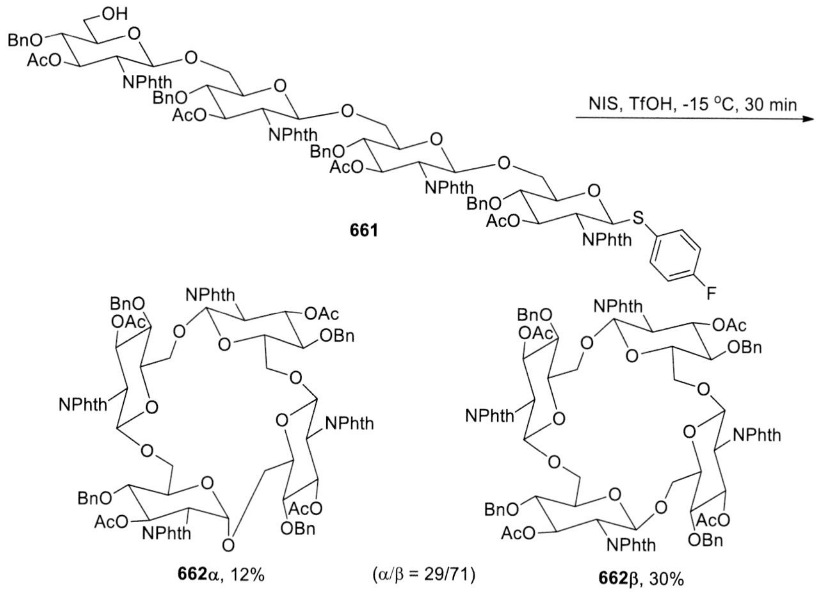

SCHEME 11.90 Synthesis of macrocycles **662** under the conventional glycosylation condition.[156]

β-isomeric cyclic compounds **662α** and **662β** (in ratio 29:71). The intramolecular cyclization featuring β-1−6 linkage formation was achieved under anodic oxidation conditions.

An electrochemical cell charged with 0.008 M linear tetrasaccharide **661** under electrochemical conditions gave exclusively β-selectivity with 81% yield of **662β** (Scheme 11.91). Expanding the substrate scope, the longer oligosaccharide precursors **663** and **664** were efficiently converted to β-selective cyclic pentasaccharide **665** and hexasaccharide **666** in 93% and 78% yield, respectively.

A series of cyclo-tetra-glucosamine was synthesized by Nifantiev and coworkers.[157–159] Some examples of the synthesis using tetrameric-glucose derivatives are included in Scheme 11.92. The tetrasaccharide contains a terminal-free C-6-OH and an anomeric thioethyl group, all other hydroxyls are fully protected with benzoate except the C-2 with an *N*-phthalimide to ensure stereoselectivity during glycosylation. Cyclization of the global benzoyl-protected tetrasaccharide **667a** furnished exclusively β-(1−6)-linked macrocycle **668a** in good yield. The tetrasaccharide **667b** consisting of alternate 2-OBz protected glucose and 2-*N*-Phth protected glucosamine units gave stereospecifically β anomer **668b** in 87% yield under the same glycosylation condition. Altering the sugar connecting sequence, intramolecular glycosylation of tetrasaccharide **667c** afforded the β-selective anomer **668c** in good yield. Similarly, the cyclization of tetrasaccharide **669**, whose sugar acceptor portion has α-(1−6)-linkage also produced β anomer **670** exclusively.

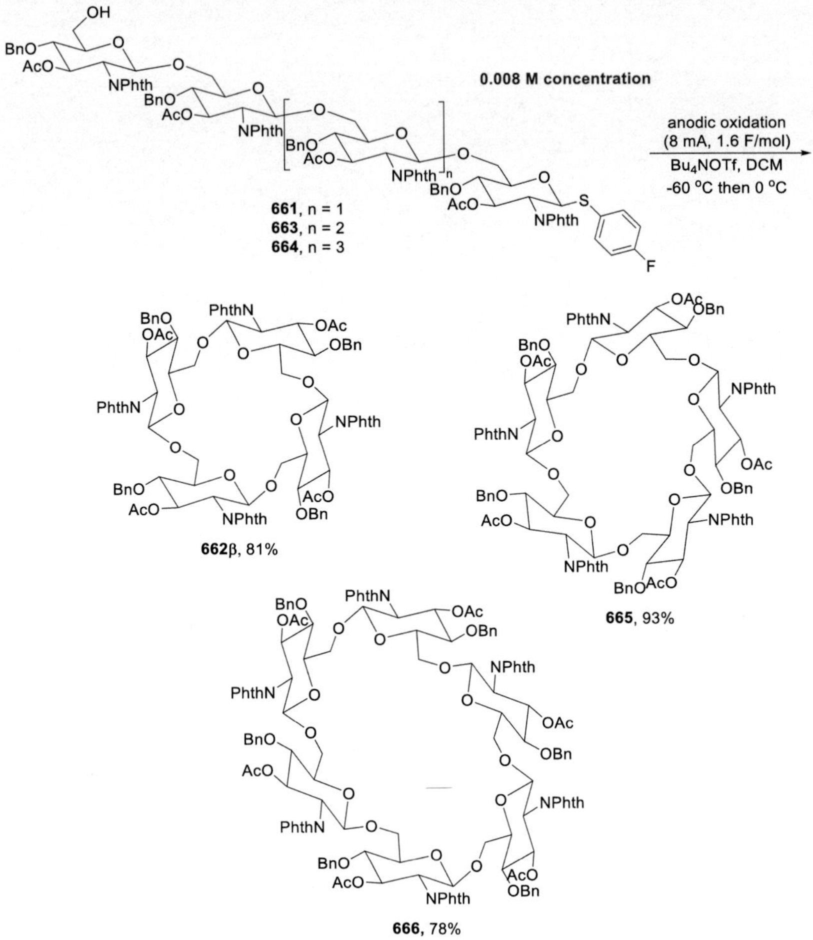

SCHEME 11.91 Stereoselective synthesis of cyclic oligosaccharides **662β**, **665**, and **666** via intramolecular electrochemical glycosylation.[156]

Ikuta et al. synthesized the smallest cyclodextrins (CDs), CD3 and CD4, which are strained cyclic trimeric and tetrameric glucose.[38] The structures of **671 (CD3)** and tetrameric compound **672 (CD4)** are shown in Fig. 11.24. These CDs were difficult to access, and their strategy relies on the introduction of a bridge between the 3 and 6 positions with ethylene dibenzylbromide **674** (EDB). The syntheses are shown in Schemes 11.93–11.95.

As shown in Scheme 11.93, esterification of 1,2,4-*ortho*-acetylglucose **206** with ethylene dibenzylbromide (EDB) **673** gave the EDB-bridged compound **674**, which kept O-3 and O-6 locked in the same direction to induce the α-selectivity in glycosylation step. $InBr_3$-mediated β-selective introduction of arylthio groups **675** followed by the opening of the orthoester **674**

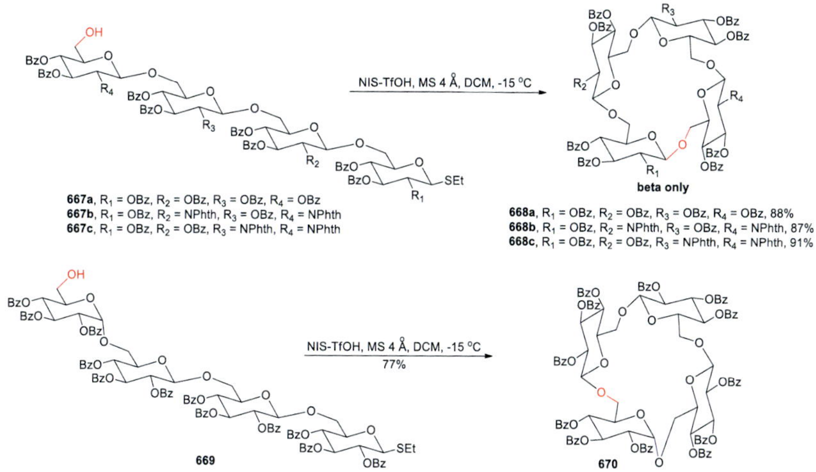

SCHEME 11.92 Synthesis cyclo-tetrasaccharide derivatives **668** and **670**.[157]

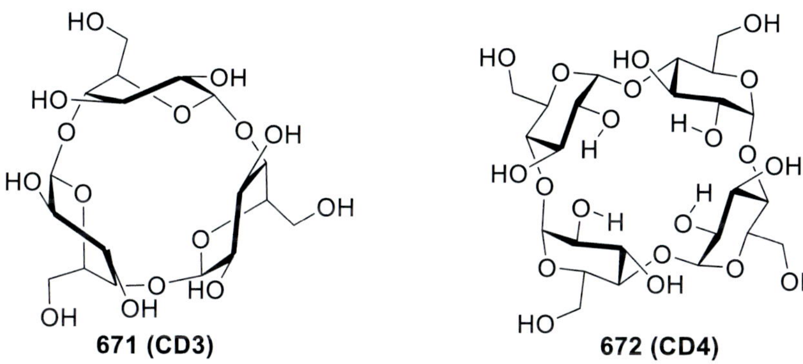

FIGURE 11.24 Structures of cyclodextrin mimics **671** (CD3) and **672** (CD4).

and deacetylation afforded compounds **676** and **677**, which were converted into glycosyl acceptor **678** and donor **679** with fluoride as the leaving group by standard functional group transformation, respectively. The glycosylation of compound **678** with compound **679** furnished the disaccharide **680** with exclusive α selectivity. Then glycosyl acceptor **681** and donor **682**, derived from **680**, underwent glycosylation condition to give linear tetrasaccharide **683** (Scheme 11.94). Repeating the glycosyl acceptor and donor preparation strategy from compound **683** followed by intramolecular glycosylation afforded the glycosylated product **685**. Deallylation of compound **685** and subsequent EDB group removal by hydrogenolysis completed the synthesis of the cyclic tetramer **CD4** (Scheme 11.94). The synthesis of cyclic trimeric **CD3** commenced with the intramolecular glycosylation of monomer **679** and

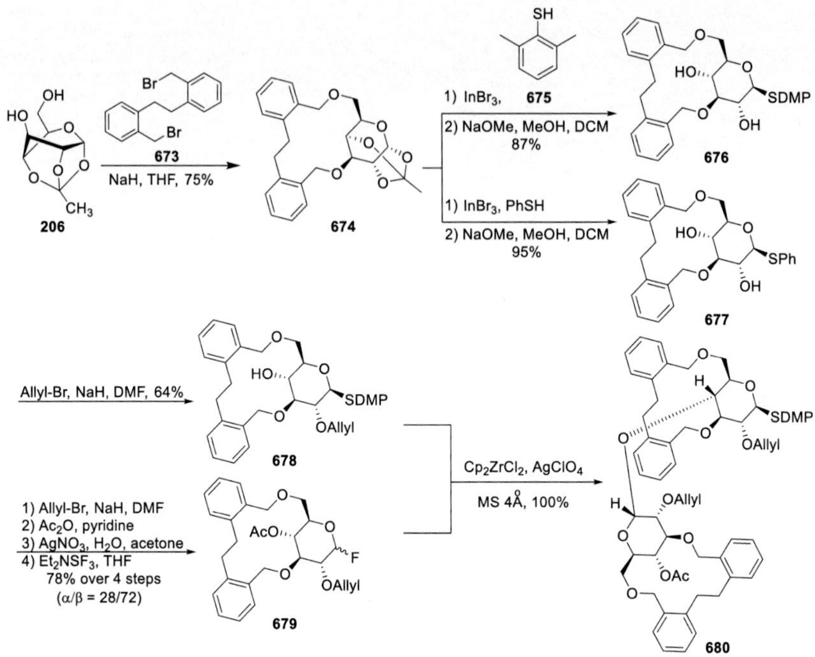

SCHEME 11.93 Synthesis of key di-saccharide intermediate **680**.[38]

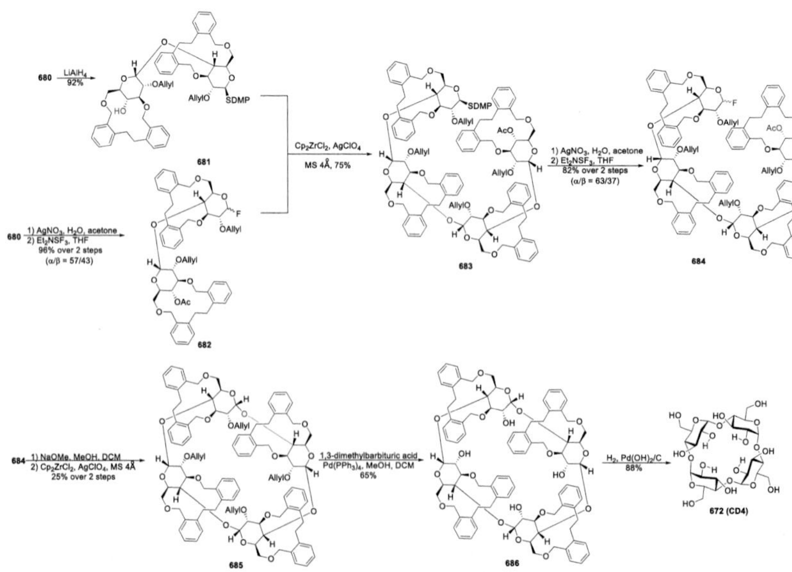

SCHEME 11.94 Synthesis of cyclodextrin mimics **672** (CD4).[38]

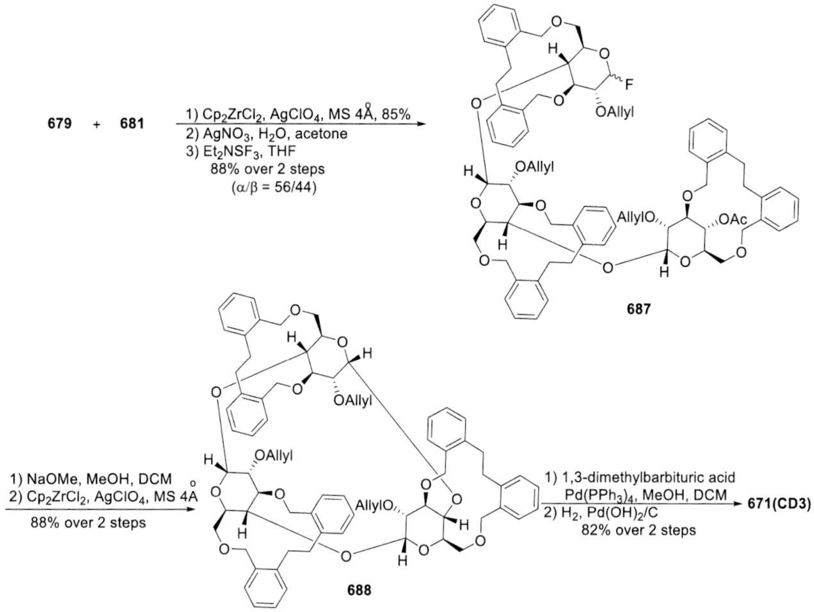

SCHEME 11.95 Synthesis of cyclodextrin mimics **671** (CD3).[38]

dimer **681**. The same methodology for **CD4** synthesis was applied, which included glycosyl acceptor and donor preparation, intramolecular glycosylation, deallylation, and EDB group removal, and produced the cyclic trimer **CD3** smoothly with perfect α-selectivity (Scheme 11.95).

Someya et al. reported a one-pot polyglycosylation of monosaccharides to form cyclic oligosaccharides (Scheme 11.96).[160] They explored various conditions using the thioglycoside **689** as the substrate for the formation of cyclic sugars of different numbers of repeating units from one to eight. The optimized conditions leading to the most cyclic oligosaccharide is using NIS/TMSOTf in DCM. The cyclic oligosaccharides consist of tri- to penta-galactosides containing repeating β-(1−6) glycosidic linkage as major products. The debenzoylation of the cyclic tetragalactoside **690c** was performed using $MeNH_2$ in MeOH to give the deprotected glycomacrocycle **691** in good yield.

11.6.6 Synthesis of azobenzene-containing glycomacrocycles by glycosylation

Lin et al. reported the synthesis of azobenzene-containing sugar-based macrocycles through intramolecular glycosylation.[161] The glycosylation step typically proceeds in favor of α-isomer. A few examples are shown in Fig. 11.25 and Fig. 11.26. Starting from the *E*-isomer, the glycosylation macrocyclization generally afforded higher yields than starting from the *Z*/*E* mixture with *Z* as the

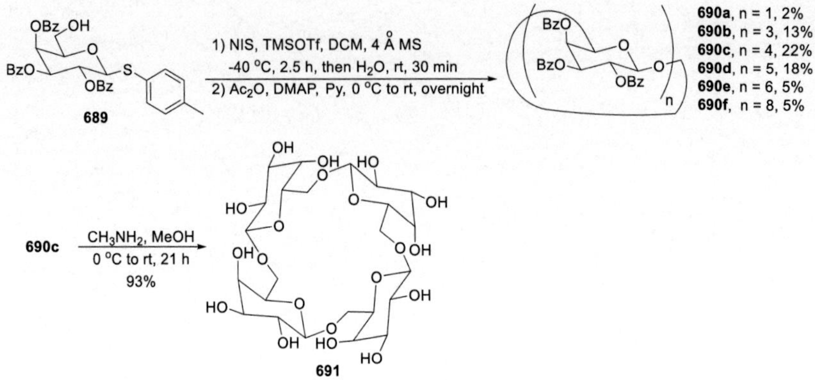

SCHEME 11.96 Synthesis of cyclodextrin analogs **690a–f** and **691**.[160]

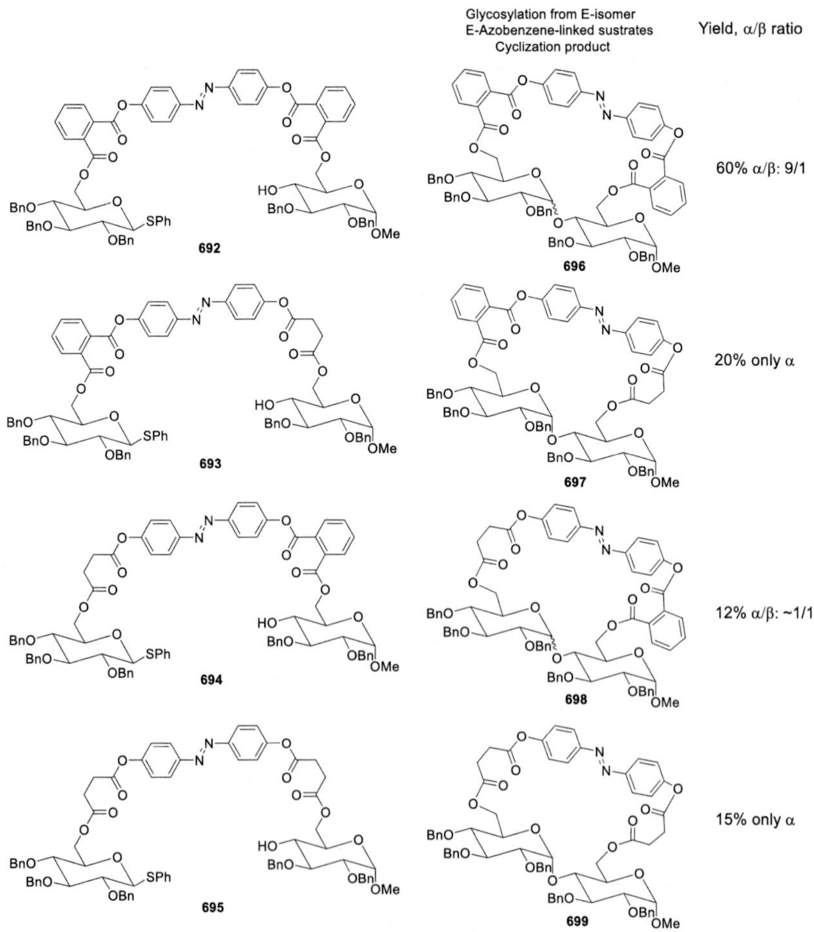

FIGURE 11.25 Synthesis of para-azobenzene-containing sugar macrocycles **696–699**.[161]

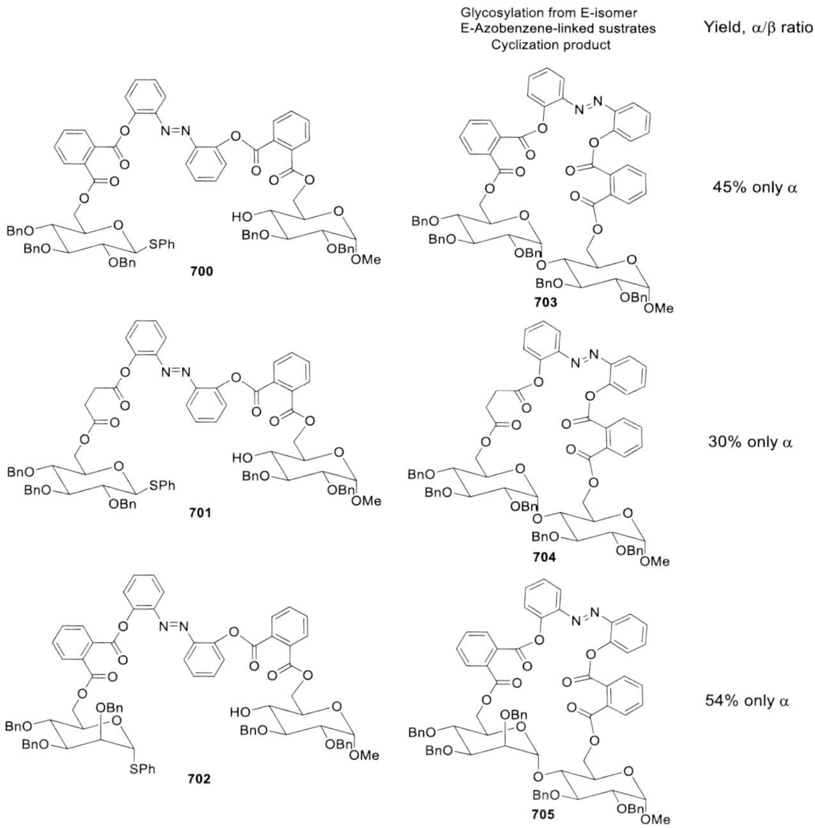

FIGURE 11.26 Synthesis of *ortho* azobenzene-containing sugar macrocycles **703–705**.[161]

major isomer. Both *E* and *Z*-azobenzene isomers led to the formation of α-glycoside as the major product. The authors concluded that the azobenzene moieties are good templates for the synthesis of α-glycoside by intramolecular glycosylation reactions. The 6,6′-linkages when incorporating a flexible linker resulted in poorer yields for the glycosylation step in comparison to those without the spacer, which may affect the optimal conformation of the precursor for the glycosylation reaction.

11.6.7 Azobenzene-containing sugar-based macrocycles

Juan Xie's group had reported the synthesis of several glyco-azobenzene macrocyles,[161–165] and these were also discussed in their recent review on azobenzene-based glycomycrocycles.[165] Lin et al. synthesized monosugar-based glyco-azobenzene macrocycle, starting from glucose or mannose phenyl thiol glycoside, the macrocycles **706–708** (Fig. 11.27) were synthesized

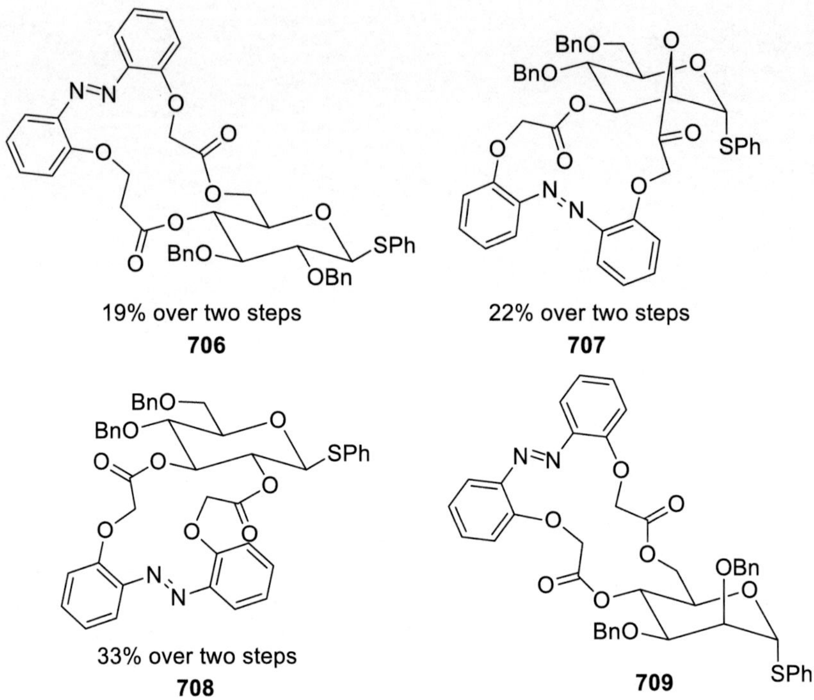

FIGURE 11.27 Structures of photoswitchable azobenzene-based glycomacrocycles **706–709**.

from the monosugar in two steps in 19%–33% yield for the macrocyclization step of alkylation reaction using 2,2′-dihydroxyazobenzene with the sugar derivatives.[163] The mannose-based macrocycle **709** was prepared by alkylation of the hydroxyl azobenzene by alkylation reaction using 4, 6-dibromoacetylated sugar precursor.[162] It was found to be able to form an organogel in cyclohexane and in ethanol, the gels were further studied for their interesting optical-electronic properties. Using similar methods, they have synthesized several other glycomacrocycles via double S_N2 macrocyclization using azophenol with glucose derivatives to form azobenzene-containing macrocycles.[163]

Despras et al. reported the synthesis of two glycomacrocycles **717** and **719** containing azobenzene moieties starting from D-glucose and D-maltose (Scheme 11.97). The C-6 and 6′ position hydroxyl groups were converted to an isothiocyanate group (for example, in **715**), the macrocyclization with piperazine can lead to the thiourea-linked macrocycles. The macrocyclization step can be controlled by the conformation of the azobenzene, the *trans-E* isomer afforded little or no cyclization products, but the *cis-Z* isomer afforded the macrocycles in good yields. Therefore light can be

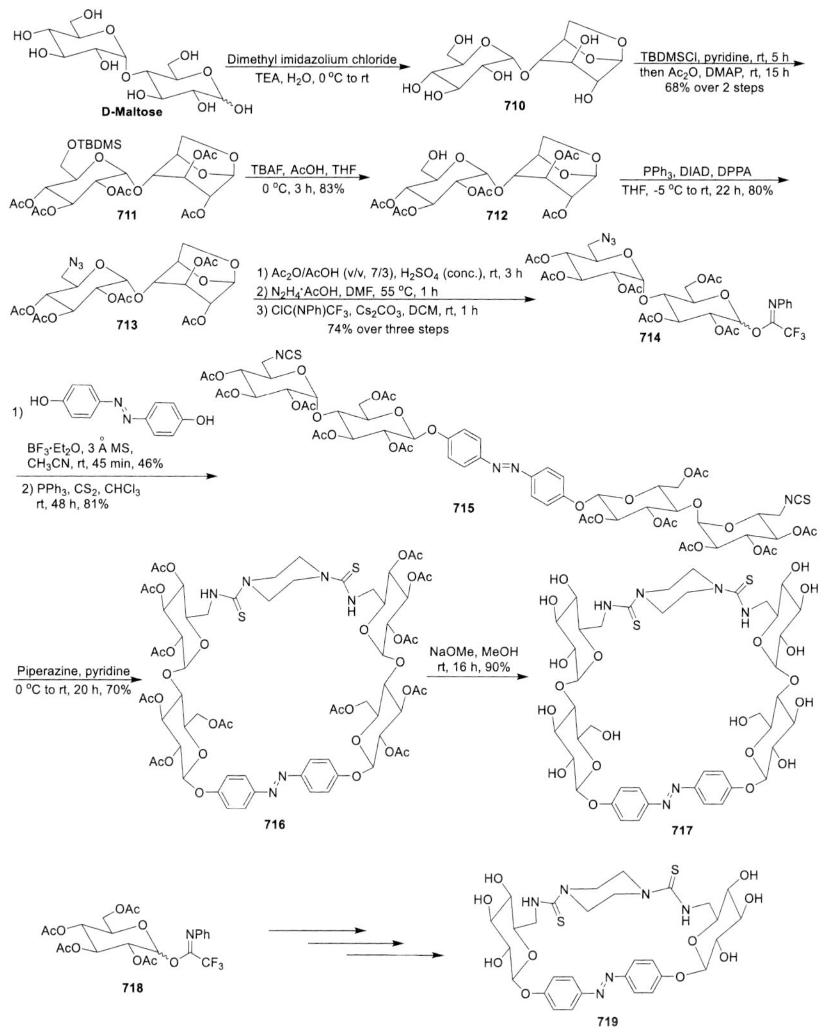

SCHEME 11.97 Synthesis of glycomacrocycles **717** and **719** containing azobenzene.[166]

used to control the cyclization reaction step to afford the desirable "Glycoazobenzene macrocycles."[166] The smaller glucose-linked glycoazobenzene macrocycle **719** was obtained in 48% yield after irradiation of the precursor to convert the azobenzene to *cis*-conformation; without irradiation, there was no product obtained. For the maltose derivatives, they were able to obtain the macrocycle **717** in 70% yield. The chiroptical properties of the unprotected sugar azobenzene macrocycles were further studied using NMR and CD spectrometry.

11.6.8 Synthesis of pyranose-based macrocycle libraries by intramolecular S_NAr reaction

Comer et al. reported a fragment-based domain shuffling method to mix and match different fragments for a macrocycle and reassemble them to obtain diversified macrocycle libraries. The feature of the macrocyclization was an intramolecular S_NAr reaction as shown in Scheme 11.98.[167] The glycal derivatives **720** and **724** were used as the building blocks of macrocycles **723** and **726**. Treatment of glycal-derived amines **720** and **724** with γ-amino acid **721** utilizing silica-bound dicyclohexylcarbodiimide (Si-DCC) and HOBt followed by removing both the Boc and *t*-butyldiphenylsilyl (TBDPS) groups in the presence of TFA afforded amino alcohols **722** and **725** smoothly. Coupling the amino alcohols **722** and **725** with fluoro and cyano-substituted benzoic acids separately produced the amides, which were then subjected to S_NAr macrocyclization conditions using cesium carbonate at 110°C in DMF to afford pyranose-based macrocycles **723** and **726**.

11.6.9 Synthesis of cyclooligosaccharide derivatives for gene delivery applications

Jimenez et al. reported the synthesis of trehalose-derived Janus cyclooglio-saccharides **728** and **730** through thiourea linkage (Scheme 11.99).[168] The heterodimeric thiourea-linked sugar-based macrocycle **728** was prepared in 81% yield by reaction of the diamine **727a** with the diisothiocyanate **727b**. Compound **728** contains six side chains with terminal amino groups, which were further converted to 12 cationic amino groups by reacting with the isothiocyanate in **729** through thiourea linkage. The cationic Janus cyclic trehalose derivatives **728** and **730** were explored for pH-responsive nonviral gene delivery.

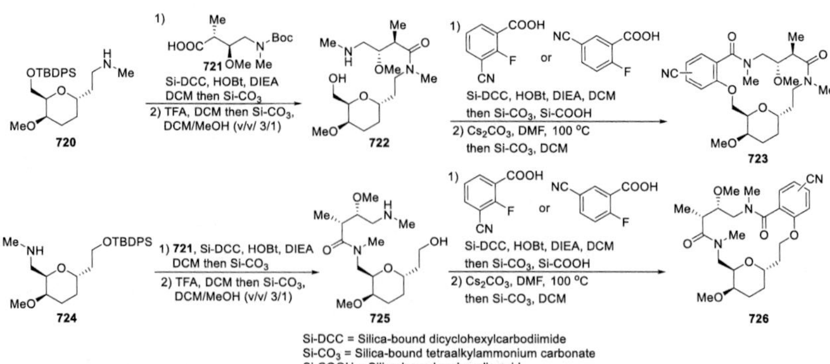

Si-DCC = Silica-bound dicyclohexylcarbodiimide
$Si\text{-}CO_3$ = Silica-bound tetraalkylammonium carbonate
Si-COOH = Silica-bound carboxylic acid

SCHEME 11.98 Synthesis of pyran-based macrocyclic lactams **723** and **726**.[167]

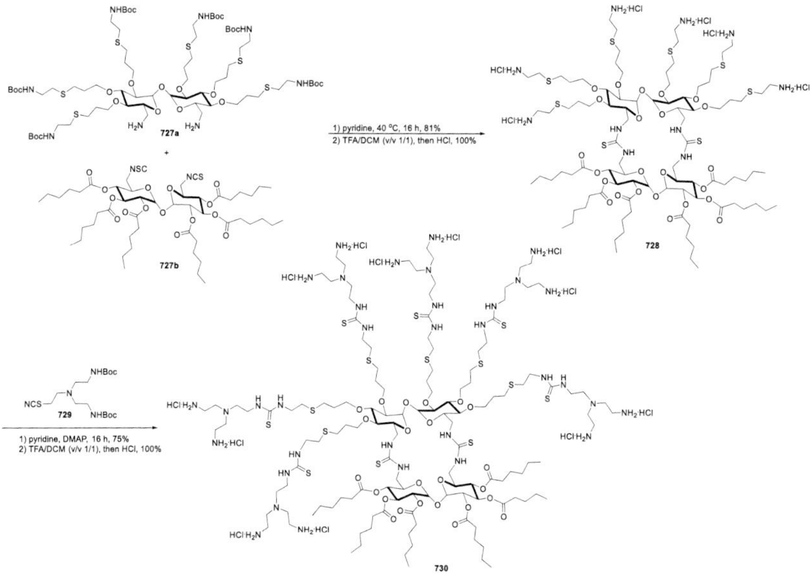

SCHEME 11.99 Synthesis of thiourea-linked Janus cyclotrehalans **728** and **730**.[168]

Using a similar macrocyclization method, Garcia Fernandez et al. reported the synthesis of several other dendritic compounds using trehalose dimer **736** or trimer **737** linked through thioureas as the core structures, these are used for complexing DNAs.[169] The synthesis of some of these compounds is included in Schemes 11.100 and 11.101. The key building blocks **733–735** for macrocyclic sugar thioureas formation were prepared starting from trehalose derivative **731** bearing six allyl and two azido groups (Scheme 11.100). The branched arm in hexa-alkene **731** was extended via thiol-ene reaction to afford the trehalose derivative **732**, with six terminal NHBoc groups. The azido group in compound **732** was further functionalized to give the diamine **733** and the diisocyanate **734** separately. The linear tetrasaccharide **735** was obtained by treating diisocyanate **734** in water/pyridine. As shown in Scheme 11.101, the thiourea-linked cyclo-dimeric compound **736** and cyclo-trimeric compound **737** were generated by macrocyclization between diamine **733** and either the disaccharide diisothiocyanate **734** or the linear tetrasaccharide **735**, respectively, followed by Boc deprotection using TFA. To prepare generations 1 and 2 (named as **G1** and **G2)** dendrimers, the corresponding **G0** derivatives were coupled with the isothiocyanate-armed branching compound **729** and subsequent Boc hydrolysis. These dendrimers contain polycationic amino groups at the periphery and can be used to complex DNA for gene delivery.

Enzymatic synthesis has also been reported for the construction of sugar-based macrocycles, for instance, cyclodextrins have been synthesized from

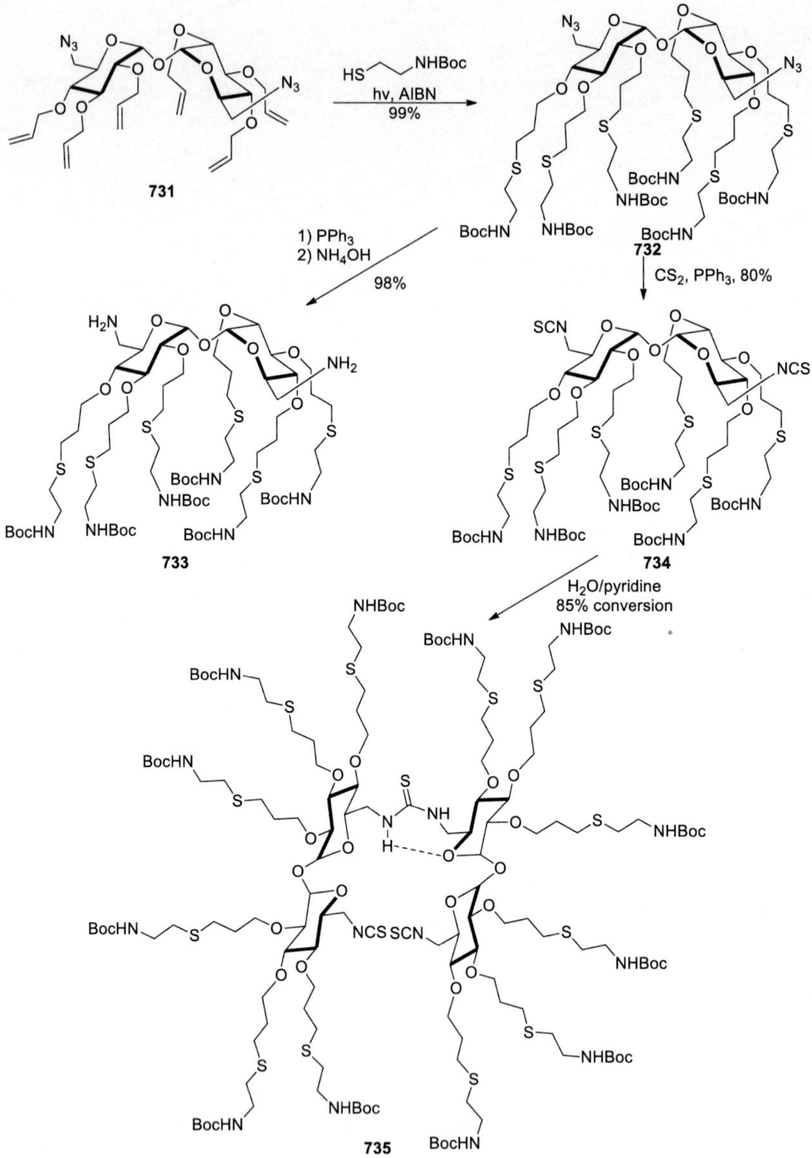

SCHEME 11.100 Synthesis of key precursors **733–735** for macrocyclic sugar thioureas formation.[169]

maltose using CGTase.[170] Using cyclic tetraglucosmine as the core structure, a series of multivalent glycoconjugates were synthesized and evaluated as inhibitors of Lectin LecA.[171] Cyclooligoglucosamine was also functionalized with hydrophobic tails and studied for ion-transport system.[172,173]

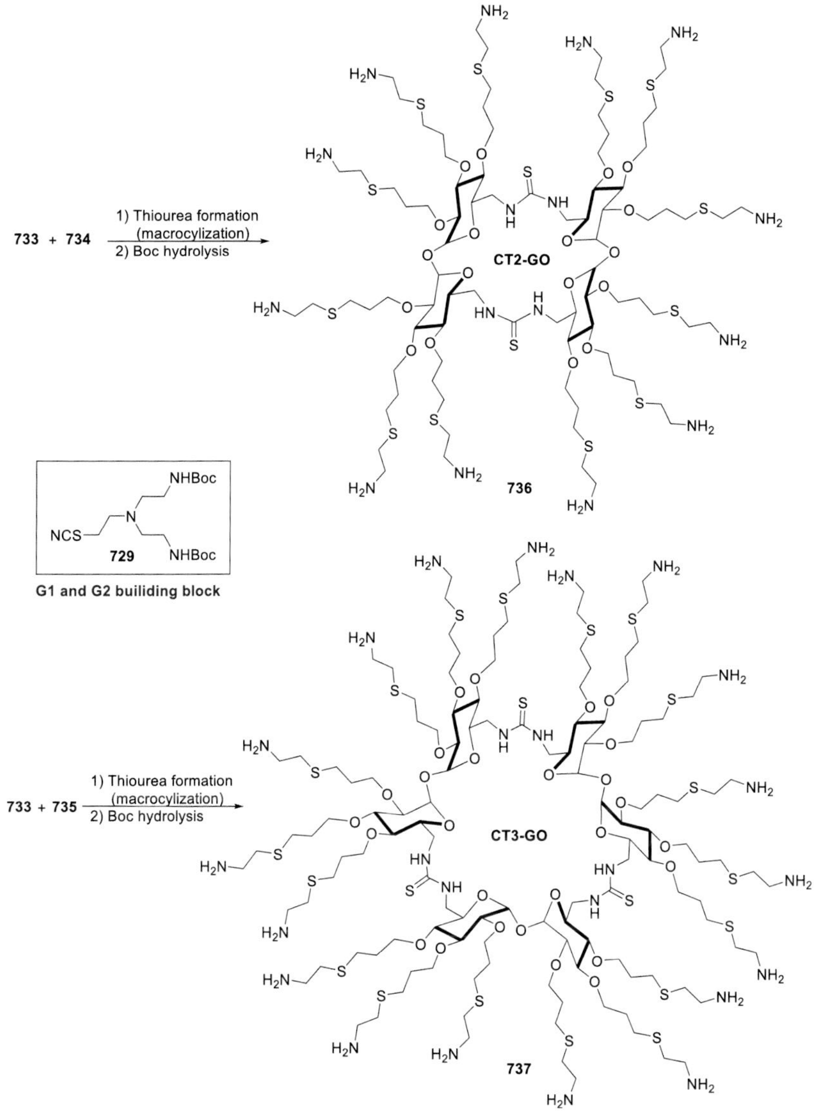

SCHEME 11.101 Synthesis of dendritic compounds **736** and **737** using macrocyclic sugar thioureas as the core structure.[169]

11.7 Conclusion and future perspective

In summary, we reviewed the most recent developments in the synthesis of carbohydrate-based macrocycles in the last 10 years, with an emphasis on the macrocyclization method. The chapter is aimed at providing state-of-the-art research summaries on the topic of carbohydrate-based macrocycles.

Besides the first part of the chapter on natural product-based glycomacrocycles, the remainder sections are divided based on the macrocyclization strategy mainly. In the first part of the chapter, we report the recent natural product-based glycomacrocycles and their total synthesis. The total synthesis of various classes of natural products are reviewed with an emphasis on unique methods. In the second section, we mainly reviewed the structures and synthesis of crown ether, azacrown ether, and other related macrocycles that have a cavity that can be used for catalysis and ion recognition. The third section covers the synthesis of macrocycles through carbon−nitrogen and carbon−oxygen bond formations that include macrolactamization and macrolactonizations. The fourth section discusses examples of the synthesis of macrocycles through C−C bond formation. The fifth section mainly covers macrocycles prepared by click reactions, this is followed by glycosylation reactions for macrocyclizations, and in the last section, we included a few other examples of different functional macrocycles. The research topic is an active research area, and many excellent research works are being carried out, with the aim of creating new glycomacrocycles with built-in functions, and ease of synthesis, as well as the ability to synthesize complex natural product like macrocycles for medicinal applications. We hope to summarize the recent work and provide a comprehensive platform and tools for researchers who are beginning to conduct research in the exciting field and stimulate new research ideas that contribute to the synthesis and development of glycomacrocycles.

Acknowledgement

We are grateful to the National Science Foundation Grant (USA) CHE 1808609.

References

1. Xie, J.; Bogliotti, N. *Chem. Rev.* **2014,** *114*, 7678−7739.
2. Singh, K.; Tripathi, R. P. *Curr. Med. Chem.* **2020,** *27*, 3386−3410.
3. Dash, A. K.; Hussain, N.; Mukherjee, D. Biologically active carbohydrate-containing macrocycles. In *Carbohydrates in Drug Discovery and Development;* Tiwari, V. K., Ed.; Elsevier Inc., 2020; pp 481−522.
4. Matencio, A.; Caldera, F.; Cecone, C.; Lopez-Nicolas, J. M.; Trotta, F. *Pharmaceuticals* **2020,** *13*, 281.
5. Yu, X.; Sun, D. *Molecules* **2013,** *18*, 6230−6268.
6. Rodriguez, J.; O'Neill, S.; Walczak, M. A. *Nat. Prod. Rep.* **2018,** *35*, 220−229.
7. Peng, D.; Lin, X.; Jiang, L.; Huang, J.; Zeng, G.; Deng, X.; Zhou, Y. *Nat. Prod. Res.* **2019,** *33*, 427−434.
8. Toure, S.; Barthelemy, M.; Sorres, J.; Genta-Jouve, G.; Dusfour, I.; Eparvier, V.; Stien, D. *Org. Lett.* **2018,** *20*, 3780−3783.
9. Kicha, A. A.; Kalinovsky, A. I.; Malyarenko, T. V.; Ivanchina, N. V.; Dmitrenok, P. S.; Menchinskaya, E. S.; Yurchenko, E. A.; Pislyagin, E. A.; Aminin, D. L.; Huong, T. T. T., et al. *J. Nat. Prod.* **2015,** *78*, 1397−1405.

10. Zhu, D.; Geng, M.; Yu, B. *Angew. Chem. Int. Ed.* **2022,** *61*, e202203239.
11. Li, Y.; Sun, J.; Gong, Y.; Yu, B. *J. Org. Chem.* **2011,** *76*, 3654–3663.
12. Shao, S.-Y.; Feng, Z.-M.; Yang, Y.-N.; Jiang, J.-S.; Zhang, P.-C. *Org. Biomol. Chem.* **2017,** *15*, 7034–7039.
13. Hu, Z.; Silipo, A.; Li, W.; Molinaro, A.; Yu, B. *J. Org. Chem.* **2019,** *84*, 13733–13743.
14. Gauthier, C.; Lavoie, S.; Piochon, M.; Martinez, S.; Milot, S.; Déziel, E. *Carbohydr. Res.* **2019,** *471*, 13–18.
15. Cloutier, M.; Prevost, M.-J.; Lavoie, S.; Feroldi, T.; Piochon, M.; Groleau, M.-C.; Legault, J.; Villaume, S.; Crouzet, J.; Dorey, S., et al. *Chem. Sci.* **2021,** *12*, 7533–7546.
16. Zong, G.; Barber, E.; Aljewari, H.; Zhou, J.; Hu, Z.; Du, Y.; Shi, W. Q. *J. Org. Chem.* **2015,** *80*, 9279–9291.
17. Zong, G.; Shi, W.Q. Total Synthesis of Ipomoeassin F and Its Analogs for Biomedical Research. In *Strategies and Tactics in Organic Synthesis*, Vol. 13; Strategies and Tactics in Organic Synthesis, 2017; pp 81–110.
18. Zong, G.; Aljewari, H.; Hu, Z.; Shi, W. Q. *Org. Lett.* **2016,** *18*, 1674–1677.
19. Zong, G.; Hirsch, M.; Mondrik, C.; Hu, Z.; Shi, W. Q. *Bioorg. Med. Chem. Lett.* **2017,** *27*, 2752–2756.
20. Zong, G.; Hu, Z.; Duah, K. B.; Andrews, L. E.; Zhou, J.; O'Keefe, S.; Whisenhunt, L.; Shim, J. S.; Du, Y.; High, S., et al. *J. Org. Chem.* **2020,** *85*, 16226–16235.
21. Zong, G.; Whisenhunt, L.; Hu, Z.; Shi, W. Q. *J. Org. Chem.* **2017,** *82*, 4977–4985.
22. Zong, G.; Hu, Z.; O'Keefe, S.; Tranter, D.; Iannotti, M. J.; Baron, L.; Hall, B.; Corfield, K.; Paatero, A. O.; Henderson, M. J., et al. *J. Am. Chem. Soc.* **2019,** *141*, 8450–8461.
23. Zong, G.; Sun, X.; Bhakta, R.; Whisenhunt, L.; Hu, Z.; Wang, F.; Shi, W. Q. *Eur. J. Med. Chem.* **2018,** *144*, 751–757.
24. Postema, M. H. D.; TenDyke, K.; Cutter, J.; Kuznetsov, G.; Xu, Q. *Org. Lett.* **2009,** *11*, 1417–1420.
25. Nagano, T.; Pospisil, J.; Chollet, G.; Schulthoff, S.; Hickmann, V.; Moulin, E.; Herrmann, J.; Mueller, R.; Fuerstner, A. *Chem. - Eur. J.* **2009,** *15*, 9697–9706.
26. Ono, M.; Kawakami, G.; Takigawa, A.; Kabata, K.; Okawa, M.; Kinjo, J.; Yokomizo, K.; Yoshimitsu, H.; Nohara, T. *Chem. Pharm. Bull.* **2014,** *62*, 839–844.
27. Ono, M.; Takigawa, A.; Kanemaru, Y.; Kawakami, G.; Kabata, K.; Okawa, M.; Kinjo, J.; Yokomizo, K.; Yoshimitsu, H.; Nohara, T. *Chem. Pharm. Bull.* **2014,** *62*, 97–105.
28. Nawoj, M.; Grobelny, A.; Mlynarski, J. *Eur. J. Org. Chem.* **2020,** *2020*, 47–51.
29. Zhao, X.; Zeng, J.; Meng, L.; Wan, Q. *Chem. Rec.* **2020,** *20*, 743–751.
30. Fang, J.; Zeng, J.; Sun, J.; Zhang, S.; Xiao, X.; Lu, Z.; Meng, L.; Wan, Q. *Org. Lett.* **2019,** *21*, 6213–6216.
31. Sun, J.; Fang, J.; Xiao, X.; Cai, L.; Zhao, X.; Zeng, J.; Wan, Q. *Org. Biomol. Chem.* **2020,** *18*, 3818–3822.
32. Xiao, X.; Zeng, J.; Fang, J.; Sun, J.; Li, T.; Song, Z.; Cai, L.; Wan, Q. *J. Am. Chem. Soc.* **2020,** *142*, 5498–5503.
33. Sharif, E. U.; Wang, H.-Y. L.; Akhmedov, N. G.; O'Doherty, G. A. *Org. Lett.* **2014,** *16*, 492–495.
34. Zhu, S.-Y.; Huang, J.-S.; Zheng, S.-S.; Zhu, K.; Yang, J.-S. *Org. Lett.* **2013,** *15*, 4154–4157.
35. Ikeuchi, K.; Ueji, T.; Matsumoto, S.; Wakamori, S.; Yamada, H. *Eur. J. Org. Chem.* **2020,** *2020*, 2077–2085.
36. Matsumoto, S.; Wakamori, S.; Nishii, K.; Tanaka, T.; Yamada, H. *Synlett* **2020,** *31*, 1389–1393.

37. Yamashita, K.; Kume, Y.; Ashibe, S.; Puspita, C. A. D.; Tanigawa, K.; Michihata, N.; Wakamori, S.; Ikeuchi, K.; Yamada, H. T. *Chem. - Eur. J.* **2020,** *26*, 16408–16421.
38. Ikuta, D.; Hirata, Y.; Wakamori, S.; Shimada, H.; Tomabechi, Y.; Kawasaki, Y.; Ikeuchi, K.; Hagimori, T.; Matsumoto, S.; Yamada, H. *Science* **2019,** *364*, 674–677.
39. Shibayama, H.; Ueda, Y.; Tanaka, T.; Kawabata, T. *J. Am. Chem. Soc.* **2021,** *143*, 1428–1434.
40. Wakamori, S.; Matsumoto, S.; Kusuki, R.; Ikeuchi, K.; Yamada, H. *Org. Lett.* **2020,** *22*, 3392–3396.
41. Andreeva, O. V.; Sharipova, R. R.; Garifullin, B. F.; Strobykina, I. Y.; Kataev, V. E. *Chem. Nat. Comp.* **2015,** *51*, 689–692.
42. Garifullin, B. F.; Strobykina, I. Y.; Sharipova, R. R.; Kravchenko, M. A.; Andreeva, O. V.; Bazanova, O. B.; Kataev, V. E. *Carbohydr. Res.* **2016,** *431*, 15–24.
43. Andreeva, O. V.; Sharipova, R. R.; Strobykina, I. Y.; Kravchenko, M. A.; Strobykina, A. S.; Voloshina, A. D.; Musin, R. Z.; Kataeva, V. E. *Russ. J. Org. Chem.* **2015,** *51*, 1324–1333.
44. Péter Bakó, G. K.; Zsolt Rapi.; László Toke. *Curr. Org. Chem.* **2012,** *16*, 297–304.
45. Orbán, I.; Bakó, P.; Rapi, Z. C. *Chemistry* **2021,** *3*, 550–577.
46. Schettini, R.; Sicignano, M.; De Riccardis, F.; Izzo, I.; Della Sala, G. *Synthesis* **2018,** *50*, 4777–4795.
47. Jarosz, S.; Potopnyk, M. A.; Kowalski, M. *Carbohydr. Chem.* **2014,** *40*, 236–256.
48. Potopnyk, M. A.; Jarosz, S. *Adv. Carbohydr. Chem. Biochem.* **2014,** *71*, 227–295.
49. Nemcsok, T.; Rapi, Z.; Bagi, P.; Guan, Y. H.; Orbán, I.; Keglevich, G.; Bakó, P. *Tetrahedron* **2020,** *76*, 130965.
50. Nemcsok, T.; Rapi, Z.; Bagi, P.; Keglevich, G.; Bakó, P. *Chirality* **2020,** *32*, 107–119.
51. Nemcsok, T.; Rapi, Z.; Keglevich, G.; Gruen, A.; Bakó, P. *Chirality* **2018,** *30*, 407–419.
52. Orbán, I.; Varga, B.; Bagi, P.; Hegedus, L.; Bakó, P.; Rapi, Z. *Molecules* **2021,** *26*..
53. Pálvölgyi, Á.; Rapi, Z.; Ozohanics, O.; Tóth, G.; Keglevich, G.; Bakó, P. *Res. Chem. Intermed.* **2017,** *44*, 1627–1645.
54. Dominique, P.; Schnurr, M.; Lewandowski, B. *Chem. Commun.* **2021,** *57*, 3476–3479.
55. Montesarchio, D.; Coppola, C.; Boccalon, M.; Tecilla, P. *Carbohydr. Res.* **2012,** *356*, 62–74.
56. Sabah, K. J.; Sead, F. F.; Mohammed, H. J.; Abedul-Hussien, I. H. *Carbohydr. Res.* **2020,** *495*, 108057.
57. Ling, I.; Hashim, R.; Sabah, K. J. *RSC Adv.* **2015,** *5*, 88038–88044.
58. Rapi, Z.; Bakó, P.; Keglevich, G.; Baranyai, P.; Kubinyi, M.; Varga, O. J. *Incl. Phenom. Macrocycl. Chem.* **2014,** *80*, 253–261.
59. Sabah, K. J.; Hashim, R. *Tetrahedron Lett.* **2013,** *54*, 1534–1537.
60. Sabah, K. J.; Zahid, N. I.; Hashim, R. *Res. Chem. Int.* **2021,** *47*, 2653–2667.
61. Salman, A. A.; Tabandeh, M.; Heidelberg, T.; Duali Hussen, R. S. *Carbohydr. Res.* **2015,** *406*, 41–45.
62. Rapi, Z.; Bakó, P.; Drahos, L.; Keglevich, G. *Heteroat. Chem.* **2015,** *26*, 63–71.
63. Rapi, Z.; Bakó, P.; Keglevich, G.; Szollosy, A.; Drahos, L.; Hegedus, L. *Carbohydr. Res.* **2013,** *365*, 61–68.
64. Rapi, Z.; Grun, A.; Nemcsok, T.; Hessz, D.; Kallay, M.; Kubinyi, M.; Keglevich, G.; Bakó, P. *Tetrahedron: Asymmetry* **2016,** *27*, 960–972.
65. Rapi, Z.; Nemcsok, T.; Bagi, P.; Keglevich, G.; Bakó, P. *Tetrahedron* **2019,** *75*, 3993–4004.
66. Rapi, Z.; Nemcsok, T.; Palvoelgyi, A.; Keglevich, G.; Gruen, A.; Bakó, P. *Chirality* **2017,** *29*, 257–272.

67. Rapi, Z.; Bakó, P.; Keglevich, G.; Szollosy, A.; Drahos, L.; Botyanszki, A.; Holczbauer, T. *Tetrahedron: Asymmetry* **2012,** *23*, 489−496.
68. Rapi, Z.; Nemcsok, T.; Grun, A.; Pálvölgyi, A.; Samu, G.; Hessz, D.; Kubinyi, M.; Kallay, M.; Keglevich, G.; Bakó, P. *Tetrahedron* **2018,** *74*, 3512−3526.
69. Rapi, Z.; Demuth, B.; Keglevich, G.; Grun, A.; Drahos, L.; Soti, P. L.; Bakó, P. *Tetrahedron: Asymmetry* **2014,** *25*, 141−147.
70. Rapi, Z.; Grün, A.; Keglevich, G.; Stirling, A.; Bakó, P. *N. J. Chem.* **2016,** *40*, 7856−7865.
71. Bakó, P.; Rapi, Z.; Grun, A.; Nemcsok, T.; Hegedus, L.; Keglevich, G. *Synlett* **2015,** *26*, 1847−1851.
72. Pham, S. T.; Rapi, Z.; Bakó, P.; Petnehazy, I.; Stirling, A.; Jaszay, Z. *N. J. Chem.* **2017,** *41*, 14945−14953.
73. Kim, M. J.; Lee, S. H.; Park, S. O.; Kang, H.; Lee, J. S.; Lee, K. N.; Jung, M. E.; Kim, J.; Lee, J. *Bioorg. Med. Chem.* **2011,** *19*, 5468−5479.
74. Rathjens, A.; Thiem, J. C. R. *Chim* **2011,** *14*, 286−300.
75. Rathjens, A.; Thiem, J. *Carbohydr. Res.* **2017,** *438*, 18−25.
76. Pintal, M.; Charbonniere-Dumarcay, F.; Marsura, A.; Porwanski, S. *Carbohydr. Res.* **2015,** *414*, 51−59.
77. Jarosz, S. *J. Carbohydr. Chem.* **2015,** *34*, 365−387.
78. Jarosz, S.; Sokołowska, P.; Szyszka, Ł. *Tetrahedron Lett.* **2020,** *61*..
79. Lewandowski, B.; Jarosz, S. *Synth. Commun.* **2011,** *41*, 2161−2168.
80. Potopnyk, M. A.; Cmoch, P.; Jarosz, S. *Org. Lett.* **2012,** *14*, 4258−4261.
81. Potopnyk, M. A.; Lewandowski, B.; Jarosz, S. *Tetrahedron: Asymmetry* **2012,** *23*, 1474−1479.
82. Potopnyk, M. A.; Jarosz, S. *Eur. J. Org. Chem.* **2013,** *2013*, 5117−5126.
83. Potopnyk, M. A.; Jarosz, S. *Monatsh Chem.* **2013,** *144*, 437−443.
84. Szyszka, L.; Cmoch, P.; Butkiewicz, A.; Potopnyk, M. A.; Jarosz, S. *Org. Lett.* **2019,** *21*, 6523−6528.
85. Szyszka, L.; Cmoch, P.; Gorecki, M.; Ceborska, M.; Potopnyk, M. A.; Jarosz, S. *Eur. J. Org. Chem.* **2021,** *2021*, 897−906.
86. Szyszka, L.; Gorecki, M.; Cmoch, P.; Jarosz, S. *J. Org. Chem.* **2021,** *86*, 5129−5141.
87. Leczycka, K.; Jarosz, S. *Tetrahedron* **2015,** *71*, 9216−9222.
88. Singh, A.; Khatri, V.; Malhotra, S.; Prasad, A. K. *Carbohydr. Res.* **2016,** *421*, 25−32.
89. Gajda, N.; Jarosz, S. *ARKIVOC* **2017,** 76−86.
90. Leczycka-Wilk, K.; Dabrowa, K.; Cmoch, P.; Jarosz, S. *Org. Lett.* **2017,** *19*, 4596−4599.
91. Chaciak, B.; Dabrowa, K.; Swider, P.; Jarosz, S. *N. J. Chem.* **2018,** *42*, 18578−18584.
92. Leczycka-Wilk, K.; Ulatowski, F.; Cmoch, P.; Jarosz, S. *Org. Biomol. Chem.* **2018,** *16*, 6063−6069.
93. Tiara, K.; Potopnyk, M. A.; Jarosz, S. *Beil. J. Org. Chem.* **2018,** *14*, 634−641.
94. Gajewska, A.; Osuch-Kwiatkowska, A.; Jarosz, S. *Carbohydr. Res.* **2019,** *486*, 107825.
95. Sokolowska, P.; Kowalski, M.; Jarosz, S. *Beil. J. Org. Chem.* **2019,** *15*, 210−217.
96. Sokolowska, P.; Dabrowa, K.; Jarosz, S. *Org. Lett.* **2021,** *23*, 2687−2692.
97. Jarikote, D. V.; Murphy, P. V. *Eur. J. Org. Chem.* **2010,** *2010*, 4959−4970.
98. Maurya, S. K.; Rana, R. *Beil. J. Org. Chem.* **2017,** *13*, 1106−1118.
99. Das, S. N.; Rana, R.; Chatterjee, S.; Suresh Kumar, G.; Mandal, S. B. *J. Org. Chem.* **2014,** *79*, 9958−9969.
100. Dasari, B.; Jogula, S.; Borhade, R.; Balasubramanian, S.; Chandrasekar, G.; Kitambi, S. S.; Arya, P. *Org. Lett.* **2013,** *15*, 432−435.

101. Jogula, S.; Dasari, B.; Khatravath, M.; Chandrasekar, G.; Kitambi, S. S.; Arya, P. *Eur. J. Org. Chem.* **2013,** *2013*, 5036–5040.
102. Desmond, R. T.; Magpusao, A. N.; Lorenc, C.; Alverson, J. B.; Priestley, N.; Peczuh, M. W. *Beil. J. Org. Chem.* **2014,** *10*, 2215–2221.
103. Jarikote, D. V.; Li, W.; Jiang, T.; Eriksson, L. A.; Murphy, P. V. *Bioorg. Med. Chem.* **2011,** *19*, 826–835.
104. Fox, K. A.; Chadda, R.; Cardona, F.; Barron, S.; McArdle, P.; Murphy, P. V. *Tetrahedron* **2020,** *76*, 131495.
105. Hussain, A.; Yousuf, S. K.; Mukherjee, D. *RSC Adv.* **2014,** *4*, 43241–43257.
106. Hussain, A.; Rao, L. M.; Sharma, D. K.; Tripathi, A. K.; Singh, B.; Mukherjee, D. *RSC Adv.* **2013,** *3*, 19899–19904.
107. Hussain, A.; Yousuf, S. K.; Kumar, D.; Lambu, M.; Singh, B.; Maity, S.; Mukherjee, D. *Adv. Synth. Catal.* **2012,** *354*, 1933–1940.
108. Hussain, A.; Yousuf, S. K.; Sharma, D. K.; Mallikharjuna Rao, L.; Singh, B.; Mukherjee, D. *Tetrahedron* **2013,** *69*, 5517–5524.
109. Si, D.; Peczuh, M. W. *Carbohydr. Res.* **2016,** *434*, 113–120.
110. Hain, J.; Despras, G. *Chem. Commun.* **2018,** *54*, 8563–8566.
111. Srinivas, A.; Rao, E. K. *Acta Chim. Slovenica* **2021,** *68*, 404–413.
112. Zhou, J.; Reidy, M.; O'Reilly, C.; Jarikote, D. V.; Negi, A.; Samali, A.; Szegezdi, E.; Murphy, P. V. *Org. Lett.* **2015,** *17*, 1672–1675.
113. Negi, A.; Reilly, C. O.; Jarikote, D. V.; Zhou, J.; Murphy, P. V. *Eur. J. Med. Chem.* **2019,** *176*, 292–309.
114. Bordes, A.; Poveda, A.; Troadec, T.; Franconetti, A.; Arda, A.; Perrin, F.; Menand, M.; Sollogoub, M.; Guillard, J.; Desire, J., et al. *Org. Lett.* **2020,** *22*, 2344–2349.
115. Bordes, A.; Poveda, A.; Fontelle, N.; Arda, A.; Guillard, J.; Ruan, Y. B.; Marrot, J.; Imaeda, S.; Kato, A.; Desire, J., et al. *Carbohydr. Res.* **2021,** *501*, 108258.
116. Ghorai, A.; Gayen, A.; Kulsi, G.; Padmanaban, E.; Laskar, A.; Achari, B.; Mukhopadhyay, C.; Chattopadhyay, P. *Org. Lett.* **2011,** *13*, 5512–5515.
117. Ghorai, A.; Padmanaban, E.; Mukhopadhyay, C.; Achari, B.; Chattopadhyay, P. *Chem. Commun.* **2012,** *48*, 11975–11977.
118. Felten, A.-S.; Pellegrini-Moise, N.; Selmeczi, K.; Henry, B.; Chapleur, Y. *Eur. J. Org. Chem.* **2013,** *2013*, 5645–5654.
119. Singh, A.; Maikhuri, V. K.; Verma, V.; Chhatwal, R. J.; Sharma, D.; Prasad, A. K. *Synth. Commun.* **2020,** *50*, 2787–2795.
120. Adhikari, S. B.; Chen, A.; Wang, G. *Molecules* **2021,** *26*.. Available from: https://doi.org/10.3390/molecules26113394.
121. Komura, N.; Kato, K.; Udagawa, T.; Asano, S.; Tanaka, H.-N.; Imamura, A.; Ishida, H.; Kiso, M.; Ando, H. *Science* **2019,** *364*, 677–680.
122. Aragão-Leoneti, V.; Campo, V. L.; Gomes, A. S.; Field, R. A.; Carvalho, I. *Tetrahedron* **2010,** *66*, 9475–9492.
123. Mohan, A.; Sankararaman, S. *Isr. J. Chem.* **2012,** *52*, 92–104.
124. Sokolova, N. V.; Nenajdenko, V. G. *RSC Adv.* **2013,** *3*, 16212–16242.
125. Xu, L.; Li, Y.; Li, Y. *Asian J. Org. Chem.* **2014,** *3*, 582–602.
126. Barlow, T. M. A.; Tourwe, D.; Ballet, S. *Eur. J. Org. Chem.* **2017,** *2017*, 4678–4694.
127. Tiwari, V. K.; Mishra, B. B.; Mishra, K. B.; Mishra, N.; Singh, A. S.; Chen, X. *Chem. Rev.* **2016,** *116*, 3086–3240.
128. Agrahari, A. K.; Bose, P.; Jaiswal, M. K.; Rajkhowa, S.; Singh, A. S.; Hotha, S.; Mishra, N.; Tiwari. *Chem. Rev.* **2021,** *121*, 7638–7956.

129. Singh, A. S.; Mishra, K. B.; Mishra, A.; Tiwari, V. K. *Click Chemistry in Organic Synthesis;* Wiley-VCH Verlag GmbH & Co. KGaA, 2016161–201.
130. Ajay, A.; Sharma, S.; Gupt, M. P.; Bajpai, V.; Hamidullah.; Kumar, B.; Kaushik, M. P.; Konwar, R.; Ampapathi, R. S.; Tripathi, R. P. *Org. Lett.* **2012,** *14*, 4306–4309.
131. Allam, A.; Dupont, L.; Behr, J.-B.; Plantier-Royon, R. *Eur. J. Org. Chem.* **2012,** *2012*, 817–823.
132. Chen, A.; Samankumara, L. P.; Dodlapati, S.; Wang, D.; Adhikari, S.; Wang, G. *Eur. J. Org. Chem.* **2019,** *2019*, 1189–1194.
133. Ghorai, A.; Achari, B.; Chattopadhyay, P. *Tetrahedron* **2016,** *72*, 3379–3387.
134. Ghorai, A.; Reddy K, S.; Achari, B.; Chattopadhyay, P. *Org. Lett.* **2014,** *16*, 3196–3199.
135. Lepage, M. L.; Schneider, J. P.; Bodlenner, A.; Compain, P. *J. Org. Chem.* **2015,** *80*, 10719–10733.
136. Pathigoolla, A.; Sureshan, K. M. *Chem. Commun.* **2014,** *50*, 317–319.
137. Liu, J.; Liu, J.; Leonard, P.; Muller, S. L.; Seela, F.; Liu, J.; Seela, F.; Daniliuc, C. *Beil. J. Org. Chem.* **2018,** *14*, 2404–2410.
138. Prasanna, R.; Purushothaman, S.; Raghunathan, R. *Org. Biomol. Chem.* **2014,** *12*, 9375–9383.
139. Rao, J. N. S.; Raghunathan, R. *Tetrahedron Lett.* **2015,** *56*, 2669–2673.
140. Das Adhikary, N.; Chattopadhyay, P. *J. Org. Chem.* **2012,** *77*, 5399–5405.
141. Singh, K.; Sharma, G.; Shukla, M.; Kant, R.; Chopra, S.; Shukla, S. K.; Tripathi, R. P. *J. Org. Chem.* **2018,** *83*, 14882–14893.
142. Hossain, J.; Natarajan, R.; Banerjee, A. K.; Majumdar, S. *Tetrahedron* **2017,** *73*, 3139–3146.
143. Majumdar, S.; Hossain, J.; Natarajan, R.; Banerjee, A. K.; Maiti, D. K. *RSC Adv.* **2015,** *5*, 106289–106293.
144. Yu, Y.; Bogliotti, N.; Tang, J.; Xie, J. *Eur. J. Org. Chem.* **2013,** *2013*, 7749–7760.
145. Lo Conte, M.; Grotto, D.; Chambery, A.; Dondoni, A.; Marra, A. *Chem. Commun.* **2011,** *47*, 1240–1242.
146. Campo, V. L.; Ivanova, I. M.; Carvalho, I.; Lopes, C. D.; Carneiro, Z. A.; Saalbach, G.; Schenkman, S.; Santana da Silva, J.; Nepogodiev, S. A.; Field, R. A. *Tetrahedron* **2015,** *71*, 7344–7353.
147. Jia, X. G.; Demchenko, A. V. *Beilstein J. Org. Chem.* **2017,** *13*, 2028–2048.
148. Kumar, A.; Geng, Y.; Schmidt, R. R. *Eur. J. Org. Chem.* **2012,** *2012*, 6846–6851.
149. Tiwari, V. K.; Kumar, A.; Schmidt, R. R. *Eur. J. Org. Chem.* **2012,** *2012*, 2945–2956.
150. Maiti, K.; Samanta, G. C.; Jayaraman, N. *ARKIVOC* **2021,** 113–136.
151. Jayaraman, N. *Chem. Rec.* **2021,** *21*, 3049–3062.
152. Maiti, K.; Jayaraman, N. *J. Org. Chem.* **2016,** *81*, 4616–4622.
153. Daskhan, G. C.; Jayaraman, N. B. *Chem. Commun.* **2014,** *50*, 8554–8557.
154. Samanta, G. C.; Maiti, K.; Jayaraman, N. *ACS Omega* **2018,** *3*, 7466–7473.
155. Liu, H.; Li, X. *Tetrahedron Lett.* **2014,** *55*, 5525–5528.
156. Manmode, S.; Tanabe, S.; Yamamoto, T.; Sasaki, N.; Nokami, T.; Itoh, T. *ChemistryOpen* **2019,** *8*, 869–872.
157. Titov, D. V.; Gening, M. L.; Gerbst, A. G.; Chizhov, A. O.; Tsvetkov, Y. E.; Nifantiev, N. E. *Carbohydr. Res.* **2013,** *381*, 161–178.
158. Gening, M. L.; Titov, D. V.; Grachev, A. A., et al. *Eur. J. Org. Chem.* **2010,** 2465–2475.
159. Gening, M. L.; Tsvetkov, Y. E.; Titov, D. V.; Gerbst, A. G.; Yudina, O. N.; Grachev, A. A.; Shashkov, A. S.; Vidal, S.; Imberty, A.; Saha, T., et al. *Pure Appl. Chem.* **2013,** *85*, 1879–1891.

160. Someya, H.; Seki, T.; Ishigami, G.; Itoh, T.; Saga, Y.; Yamada, Y.; Aoki, S. *Carbohydr. Res.* **2020,** *487*, 107888.
161. Lin, C.; Jiao, J.; Maisonneuve, S.; Malletroit, J.; Xie, J. *Chem. Commun.* **2020,** *56*, 3261–3264.
162. Lin, C.; Maisonneuve, S.; Metivier, R.; Xie, J. P. *Chem. Eur. J.* **2017,** *23*, 14996–15001.
163. Lin, C.; Maisonneuve, S.; Theulier, C.; Xie, J. *Eur. J. Org. Chem.* **2019,** 1770–1777.
164. Kim, Y.; Mafy, N. N.; Maisonneuve, S.; Lin, C.; Tamaoki, N.; Xie, J. *ACS Appl. Mater. Interfaces* **2020,** *12*, 52146–52155.
165. Maisonneuve, S.; Lin, C.; Xie, J. *Vietnam. J. Chem.* **2020,** *58*, 417–422.
166. Despras, G.; Hain, J.; Jaeschke, S. O. *Chem. Eur. J.* **2017,** *23*, 10838–10847.
167. Comer, E.; Liu, H.; Joliton, A.; Clabaut, A.; Johnson, C.; Akella, L. B.; Marcaurelle, L. A. *Proc. Natl. Acad. Sci. U. S. A.* **2011,** *108*, 6751.
168. Jimenez Blanco, J. L.; Ortega-Caballero, F.; Blanco-Fernandez, L.; Carmona, T.; Marcelo, G.; Martinez-Negro, M.; Aicart, E.; Junquera, E.; Mendicuti, F.; Tros de Ilarduya, C., et al. *Chem. Commun.* **2016,** *52*, 10117–10120.
169. Carbajo-Gordillo, A. I.; Jimenez Blanco, J. L.; Benito, J. M.; Lana, H.; Marcelo, G.; Di Giorgio, C.; Przybylski, C.; Hinou, H.; Cena, V.; Ortiz Mellet, C., et al. *Biomacromolecules* **2020,** *21*, 5173–5188.
170. Larsen, D.; Beeren, S. R. *Chem. Commun.* **2021,** *57*, 2503–2506.
171. Gening, M. L.; Titov, D. V.; Cecioni, S.; Audfray, A.; Gerbst, A. G.; Tsvetkov, Y. E.; Krylov, V. B.; Imberty, A.; Nifantiev, N. E.; Vidal, S. S. *Chem. - Eur. J.* **2013,** *19*, 9272–9285.
172. Roy, A.; Saha, T.; Gening, M. L.; Titov, D. V.; Gerbst, A. G.; Tsvetkov, Y. E.; Nifantiev, N. E.; Talukdar, P. *Chem. - Eur. J.* **2015,** *21*, 17445–17452.
173. Saha, T.; Roy, A.; Gening, M. L.; Titov, D. V.; Gerbst, A. G.; Tsvetkov, Y. E.; Nifantiev, N. E.; Talukdar, P. *Chem. Commun.* **2014,** *50*, 5514–5516.

Chapter 12

Recent advances in synthesis of diverse glycopeptides and glycohybrids

Rajdeep Tyagi[1], Kavita Singh[1], Vinay Kumar Mishra[2] and Ram Sagar[1,2]
[1]*Glycochemistry Laboratory, School of Physical Sciences, Jawaharlal Nehru University, New Delhi, India,* [2]*Department of Chemistry, Institute of Science, Banaras Hindu University (BHU), Varanasi, Uttar Pradesh, India*

12.1 Introduction

Glycopeptides are sugar moieties bonded to polypeptides that show structural variety and impact physical, chemical, and biological properties, as well as their functions. Naturally occurring glycoproteins and glycopeptides are not very stable, they also possess poor bioavailability and microheterogeneity. Therefore to overcome such problems, glycoproteins and glycopeptides were linked with oligosaccharides and used for medical applications.[1] They are distinctive due to their polyfunctionality nature.[2] The biological features such as cell adhesion, cell proliferation, protein folding and trafficking, secretion, cell targeting, signal transduction, and several intercellular processes are all attributed to the complexity of the carbohydrate moiety connected to the peptide.[3,4]

Glycan moiety incorporation into peptides is essential for controlling protein activities and demonstrates the variety of glycosidic linkages.[5] O-glycosidic bonds, C-glycosidic bonds, S-glycosidic bonds, and N-glycosidic bonds are the most common types of glycosidic bonds presents in glycopeptides. There are several well-known glycopeptide-based drugs in the market, including vancomycin, oritavancin, teicoplanin, dalbavancin, and telavancin. Because N- and O-glycosidic linkages are easily broken down by biological enzymes, such glycopeptides become less stable and lose their functional properties. In comparison to O-glycosides, thioglycosidic linkage is more stable to chemical degradation and enzymatic hydrolysis when the anomeric oxygen is replaced with a sulfur atom.[6] C-glycosidic bonds are more capable

Synthetic Strategies in Carbohydrate Chemistry. DOI: https://doi.org/10.1016/B978-0-323-91729-2.00010-0

to tolerate acids, bases, and enzymatic hydrolyses and are considered the most stable glycopeptides.[7]

Association of glycan parts with the anticancer peptides makes them more suitable as anticancer drugs because cancer cell requires more energy due to their high growth rate and possesses overexpressed glut transporter/receptors (GLUT1-GLUT4), which allowed the transport of sugar moiety or molecules containing sugar moiety across the cell membrane. Therefore developing new methods for the synthesis of glycopeptides and glycohybrids becomes a very important area of research. Vancomycin and teicoplanin are the first generation of glycopeptide antibiotics that are used for treating infections caused by gram-positive bacteria while telavancin, oritavancin, and dalbavancin are the second generation of antibiotics and possess better pharmacokinetics properties than their precursors.[8–10] New bacterial strains are known to be resistant to traditional known antibiotics, necessitating the development of novel glycopeptides-based antibiotics with potent antibacterial effects that combat the antibacterial resistance problem is the need of hour.[11–13] Thus researchers around the globe are developing new or modified chemistries to construct new glycopeptide and glycohybrids. This chapter discusses the recent developments for the synthesis of novel O-linked, S-linked, C-linked, and N-linked glycopeptides and new glycohybrids in the form of glycoconjugates, which may later be developed as potential drugs for treating various diseases brought by bacteria, viruses, or fungi as well as for creating anticancer medicines or vaccines.[14–17] The general methods used for the chemical synthesis of glycoprotein and glycopeptides are summarized in Fig. 12.1 and Fig. 12.2.

12.2 Synthesis of O-linked glycopeptides

O-linked glycopeptides are majorly synthesized via the Koenigs–Knorr method in which an activator compound ($Hg(CN)_2$, $AgNO_3$, Ag_2CO_3, AgOTf, $AgClO_4$, Ag_2O, $Cu(OTf)_2$) stimulates the anomeric center of glycosyl unit, which is coupled with an acceptor unit, followed by elimination of halide, thereby resulting in successful glycosylation.[18,19] The classical method employs the oxide/carbonates of silver to couple glycosyl halide and acceptor; however, the reaction takes many hours and days to accomplish. In this context, efforts have been made to accelerate the conventionally used silver oxide-catalyzed Koenigs–Knorr reactions using Lewis acids. Singh et al. employed TMSOTf to facilitate rapid glycosylation of halides with various acceptor units. The researchers reported the completion of the reaction in a few minutes with neutral pH conditions.[19] Some works have reported "minimally competent" Lewis acids, such as $InBr_3$, which gets regenerated during the reaction, and thereby requires less than 1 equiv. and also reduces the amount of side products.[20] Recently, researchers have utilized boronic acid with a low amount of catalytic imidazole that removes the tedious

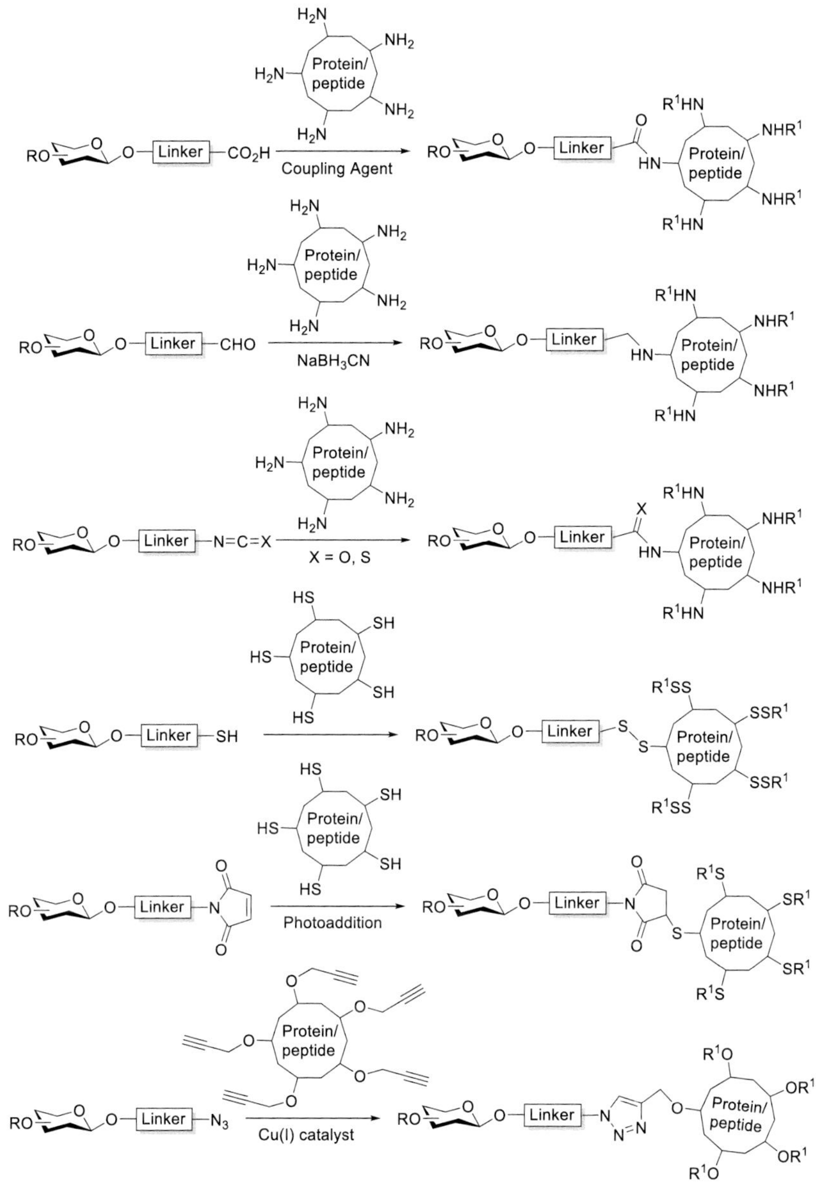

FIGURE 12.1 General methods used for the chemical synthesis of O-glycoprotein and glycopeptides.

process of protection/deprotection and furnishes regioselective 1,2-*trans* glycosidic linkage. However, all these synthetic routes require a metal catalyst; hence, a metal-free catalyzed approach has been demonstrated in which the

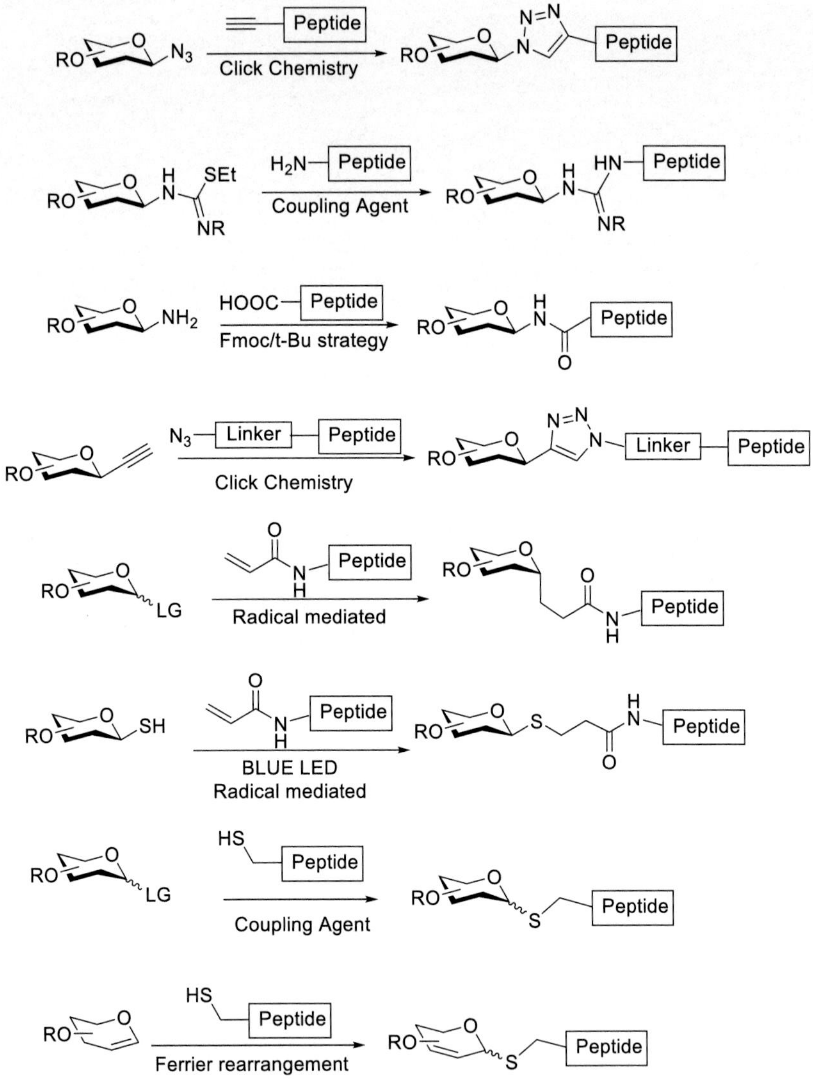

FIGURE 12.2 General methods used for the chemical synthesis of N-, C-, and S-glycoprotein and glycopeptides.

reaction has been carried out in the presence of urea, wherein glycosyl halide is activated through hydrogen bond interaction and offered high anomeric stereoselectivity in glycosylation process.[18]

Dziadek et al. reported the synthesis of α(2,6)-sialyl-T glycosylated MUC1 eicosapeptide possesses O-linkage between sugar and nonsugar moieties. N-Fmoc-protected α(2,6)-sialyl-T-threonine building block **8** was synthesized for solid-phase glycopeptide synthesis using N-Fmoc and *tert*-butyl ester-protected T_N-antigen threonine derivative **1** as a starting material for the assembly of (2,6)-sialyl-T-antigen by treating it with α,α-dimethoxytoluene using a catalytic amount of *p*-toluenesulfonic acid in acetonitrile gives 4,6-benzylidene acetal **2**. Coupling of 6-O-benzyl-protected galactosyl bromide **3** using $Hg(CN)_2$ with compound **2** to form disaccharide **4**, which on treating with aqueous acetic acid at 80°C, removal of benzylidene acetal group takes place results in sialyl acceptor **5** having a free hydroxyl group at C-6. Xanthate **6** of the *N*-acetyl neuraminic acid benzyl ester reacts with compound **5** using methylsulfenyl bromide and AgOTf in the mixture of CH_3CN/CH_2Cl_2 afforded **7** followed by acidolysis of *tert*-butyl ester with trifluoroacetic acid (TFA) and anisole (10:1) result in N-Fmoc protected (2,6)-sialyl-T-threonine building block **8** possessing free hydroxyl group at C-4. Desired glycopeptide was synthesized by preparing resin-bound nonapeptide **10** from proline-loaded resin **9** by the solid phase peptide synthesis (SPPS) method followed by coupling with glycosylated threonine derivative **8** using solid-phase glycopeptide synthesis (SPGS) method gives desired MUC1 glycopeptide **11** followed by deprotection of benzyl and acetyl group from glycan moieties afforded **12** (Scheme 12.1).[21]

Lindhorst et al. reported a method to synthesize a bivalent glycopeptide ligand with the capacity to bridge two putative carbohydrate binding sites on mannose-specific bacterial lectin, FimH. Desired O-linked glycopeptide was synthesized in two parts, that is, eastern and western parts, and lastly both the parts were coupled together. The eastern part was synthesized by taking azidoethyl mannoside **13** as starting material that on catalytic hydrogenation transformed into amine **14** followed by peptide coupling with *N*-Boc-protected pentaglycine (Gly_5Boc) using hydroxybenzotriazole (HOBt)/DCC in the presence of *N*,*N*-diisopropylethylamine (DIPEA) and dimethylformamide (DMF) as a solvent led to the formation of *N*-Boc-protected glycopeptide **15** followed by treatment with TFA gives eastern part amine **16**. The western part was synthesized by treating compound **13** with triethyl phosphite using tosylic acid and TFA in acetonitrile gives bis-orthoester **17**, under acidic conditions it results in the required 2,4-dibenzoate **18** and diol 2,6-dibenzoate **19** as a by-product. Treating later with trichloroacetimidate **20** using Lewis acid boron trifluoride etherate in dichloromethane (DCM) gives mannotrioside **21** followed by reduction of azide group and coupling with *N*-Boc-protected pentaglycine (Gly_5Boc) using HOBt/DCC in the presence of DIPEA and DMF gives **22**. Acid treatment of **22** with TFA gives desired mannotrioside peptide **23**. Coupling of compound **8** and compound **23** using squaric acid DES (diethyl ester) for ligation afforded desired bivalent glycopeptide **24** (Scheme 12.2).[22]

Karch et al. reported the synthesis of glycosylated Fmoc-β^3-homo-threonine conjugates for mucin-like glycopeptide. To form the desired glycopeptide, TentaGel S resin **25** preloaded with Fmoc-proline was used to get the desired glycopeptide. Initially, the first 13 amino acids of the MUC1 sequence were coupled, then glycosylated β^3-homo-threonine building block **26** was linked followed by coupling of five Fmoc-amino acids of TR-sequence using the standard coupling reaction conditions. Finally, triethylene glycol spacer **27** was coupled followed by the removal of resin using a mixture of TFA, triisopropyl silane (TIS), and H_2O resulting in glycan-protected

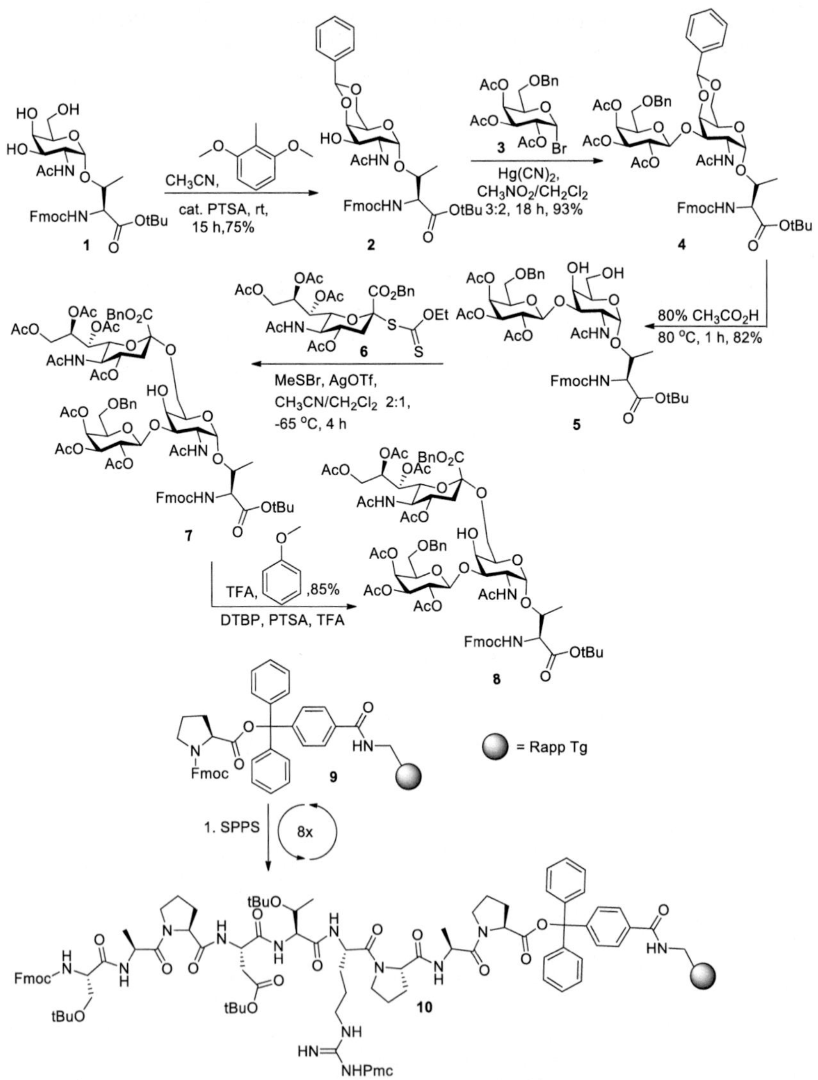

SCHEME 12.1 Route to synthesize α(2,6)-sialyl-T glycosylated MUC1 eicosapeptide.[21]

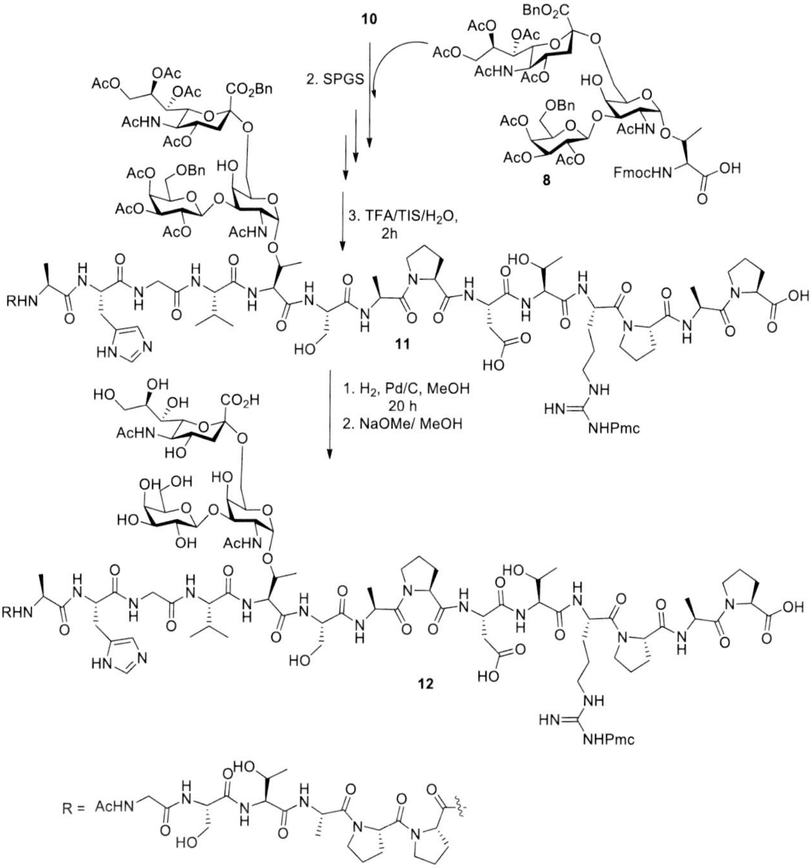

SCHEME 12.1 Continued

glycopeptide, which on deprotection using a catalytic amount of NaOMe in methanol afforded the desired glycopeptide **28** (Scheme 12.3).[23]

Mo et al. reported the use of α-Dystroglycan to synthesize phosphorglycopeptides having O-linkage. Hyper-acid-sensitive Sieber amide resin was used to synthesize the desired glycopeptide by coupling phosphormannosides with the Fmoc-protected threonine derivative. Initially, disaccharide **31** was prepared by coupling of thiogalactoside **29** with appropriate glycosyl acceptor **30** in the presence of NIS/TMSOTf. It was converted into trisaccharide **35**, by treating later with tetrabutylammonium fluoride (TBAF) buffered in acetic acid in tetrahydrofuran (THF) gives compound **32** followed by treating with trichloroacetonitrile in the presence of cesium carbonate gives corresponding trichloroacetimidate **33**, which in reacting with thiomannosyl acceptor **34** afforded trisaccharide **35**. Threonine derivative **36** was coupled with compound **35** in the presence of NIS/AgOTf to form

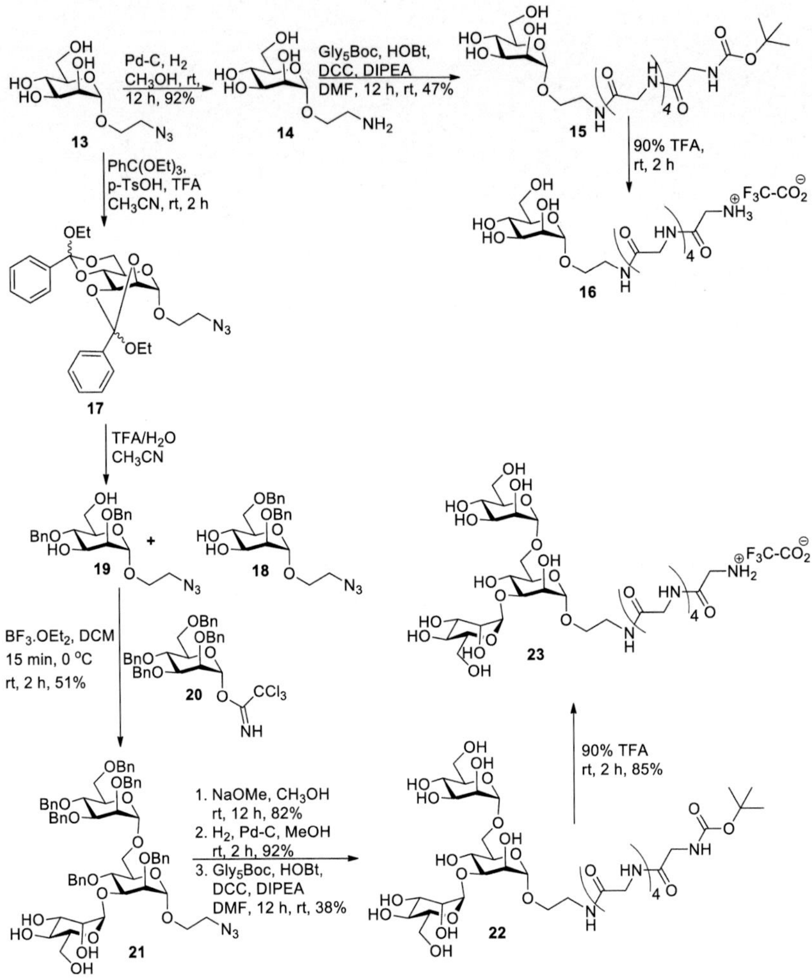

SCHEME 12.2 Synthesis of bivalent O-linked glycopeptide.[22]

glycosylated amino acid **37**. Phosphorylation and selective removal of Lev ester of compound **37** give alcohol derivative **38**, which on using the appropriate reaction conditions afforded phosphor-triester **39**. Nap ether and naphthalene acetal were oxidatively removed followed by reductive removal of Troc moieties, and acetylation of free amine and hydroxyl group afforded compound **40**, which on deallylation resulted in glycosylated threonine **41**. Threonine **41** was incorporated into hyper-acid-sensitive Sieber amide resin using the standard Fmoc method gives glycopeptide **42**, and acid treatment later with TFA in DCM removes the resin giving phosphoglycopeptide **43**. Removal of trityl protecting group followed by removal of phosphate and

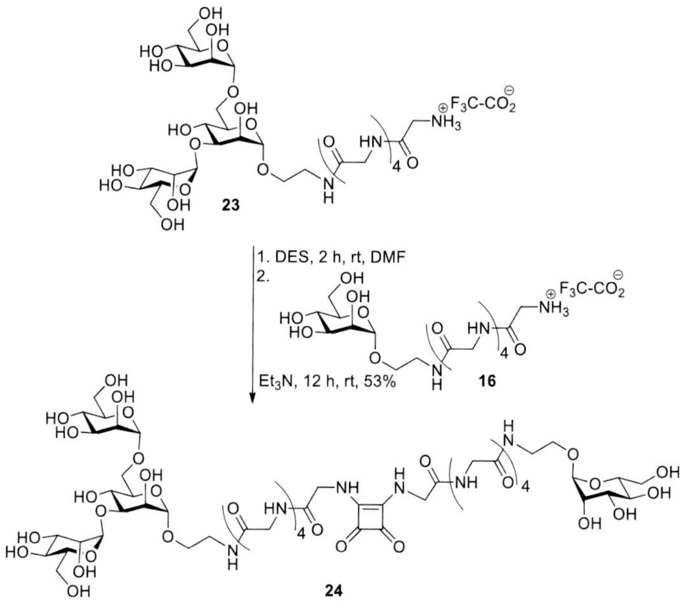

SCHEME 12.2 Continued

threonine side-chain protecting groups along with cleavage of acetyl ester gives the desired glycopeptide **44**.[24]

The use of promoters in the glycosylation process has been encouraged due to advantages such as nontoxicity, no requirement of silver salts, and ready and inexpensive nature.[25–27] NIS/TfOH promoter provides an effective and cheap source of iodonium ion, which endorses glycosylation process.[26,27] Lewis X pentasaccharide, which contains Fmoc-threonine amino acid and is a glycopeptide building block, was synthesized by Sardar et al. This leaves the door open for it to be used as a precursor in the solid-phase synthesis of a number of glycopeptides. The synthesis of pentasaccharide follows [3 + 2] convergent glycosylation pathway between Lewis X trisaccharide donor and core 1-diol acceptor and proceeds in the presence of 1-benzenesulfinylpiperidine and triflic anhydride (Tf_2O) promoter[28] at $-60°C$.[29] Iodine is also used as a promoter for the synthesis of glycosides from glycals **45** where a hydroxyl group containing amino acids couples with the double bond of glycals using the Ferrier reaction. Fmoc-protected derivatives of Ser, Try, and Thr couple with glycals in DCM using iodine as a promoter leading to the formation of O-glycosylated building blocks **46–48** (Scheme 12.5).[30]

Another method used to prepare O-linked glycopeptide is by using an allyl-linked derivative of glycan at the anomeric position. Reductive ozonolysis of this allylic double bond results in a species that reacts further with an amino group of peptide chain.[31,32] Interestingly, as compared to

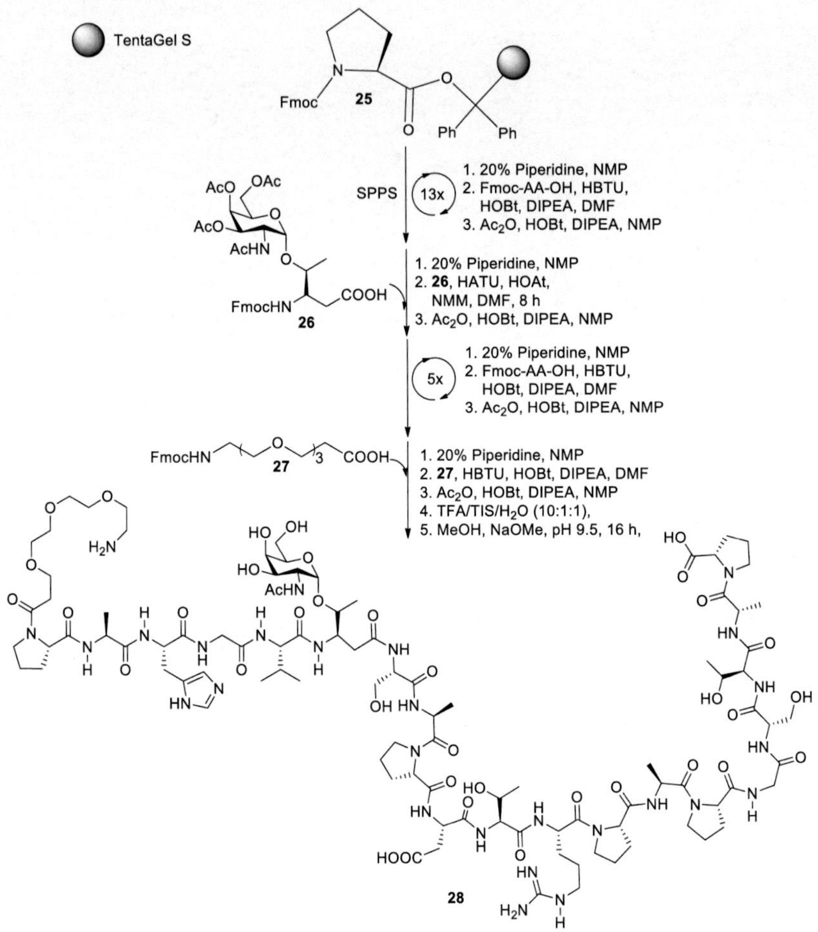

SCHEME 12.3 Synthesis of Fmoc-β^3-homo-threonine containing O-linked mucin-like glycopeptide.[23]

Fmoc-protected amino acid derivatives, the utilization of cheap Schiff bases with excellent glycosyl acceptors and strong nucleophilic properties makes them more efficient. The Schiff base approach is ideal for glycosylation of aromatic amino acids as well as Ser and Thr. In this approach, $BF_3 \cdot OEt_2$ is employed as a promotor to synthesize O-linked glycopeptides, resulting in a high yield compared to other promoters such as AgOTf.[33] By interacting with the appropriate glycosyl bromide, Mitchell et al. offered a technique for the selective synthesis of O-linked glycopeptides from Fmoc-protected amino groups of Ser and Thr.[34]

The ability of antifreeze glycoproteins (AFGPs) to prevent the formation of ice is crucial for aquatic animals to survive in the deep ocean. It is found in fish blood serum and protects cell membranes by preventing bodily fluids from crystallizing. AFGPs are mostly made up of O-linkages, which can be made by applying microwave irradiation to a SPGS technique. Microwave technology allows for the synthesis of massive building blocks while also reducing reaction time to a few minutes with high yield and purity.[35,36] Synthesis of O-linked glycopeptide was described by L. Corcilius et al. and possessed antifreezing properties (Scheme 12.4). Synthesis of hydroxyproline building block possessing α-linked GalNAc can be achieved using galactosamine hydrochloride salt **49**. Azide group was introduced on sugar by coupling **49** with imidazole-based diazotransfer reagent **50** and the alcohol moiety was acetylated using acetic anhydride in pyridine using catalytic amount of 4-dimethylaminopyridine (DMAP) results in 68% of **51**. The reaction of this acetylated molecule with ethylenediamine and acetic acid results in anomeric deacetylation, which provides hemiacetal **52** in assessable yield. This was

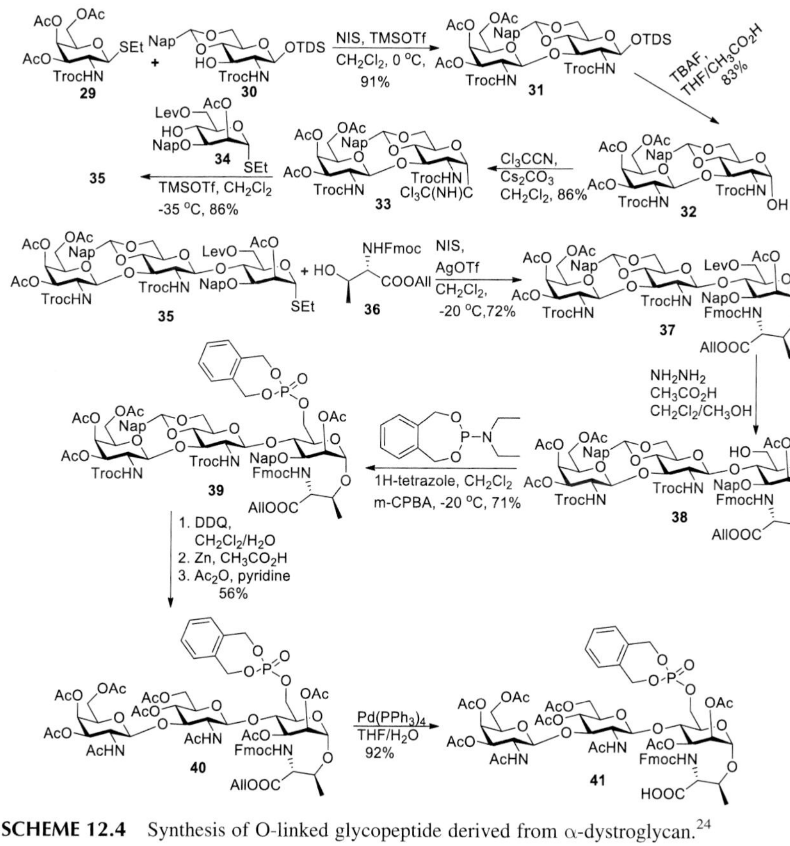

SCHEME 12.4 Synthesis of O-linked glycopeptide derived from α-dystroglycan.[24]

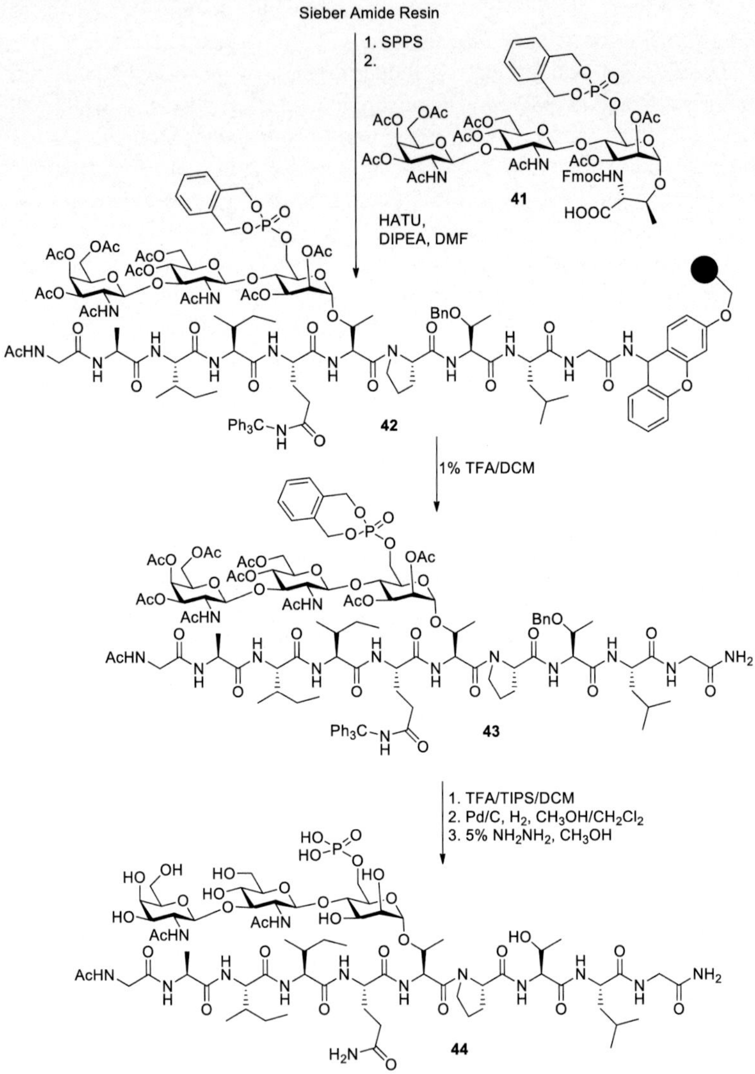

SCHEME 12.4 Continued

further reacted with trichloroacetonitrile and potassium carbonate to produce **53** in 71% yield. This trichloroacetimidate derivative reacts with Fmoc-Hyp-O*t*Bu at very low temperature in a mixture of diethyl ether and DCM using TMSOTf as Lewis acid resulting in the synthesis of Fmoc-HyPro-O*t*Bu building block **54**. Other Lewis acids like boron trifluoride etherate can be used with peracetylated glycosylated donors to synthesize glycosylated hydroxyproline building blocks (Scheme 12.6).[37,38] It is important to note that glycosylation in AFGPs takes place at the Thr side chain.[39,40] AgOTf-promoted O-

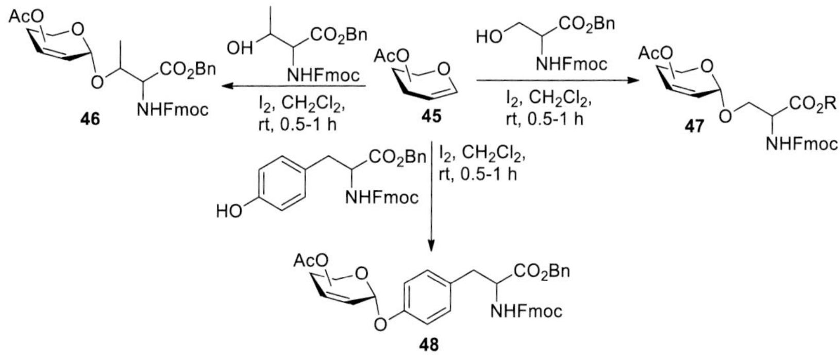

SCHEME 12.5 Synthesis of I_2-promoted O-glycosylated amino acid building blocks.[30]

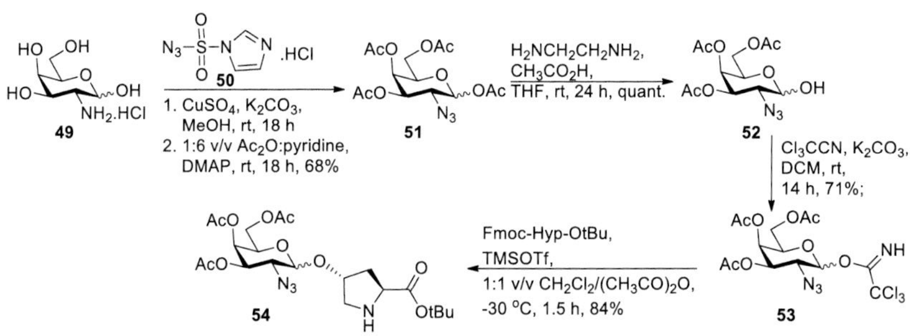

SCHEME 12.6 Synthesis of hydroxyproline-derived O-glycosylated building block.[37]

linked glycopeptide synthesis of tyrosine derivatives in DCM at low temperatures with glycosyl bromide derivatives results in stereoselective glycosides, while in acetonitrile, it produces a mixture of glycosides.[41]

Another method to synthesize O-linked glycopeptide includes the stereoselective total synthesis of β-1,2-linked triarabinosylated CLV3 peptide ([Ara_3]CLV3) which was reported by Shinohara et al. The nonreducing end of the arabinose chain of [Ara_3]CLV3 is located close to the peptide backbone, and it extends toward the C-terminal end of the peptide. Arabinosyl donor, ethyl thioarabinoside **55**, was used as a starting material; it is coupled with β-1,2-linked Hyp diarabinofuranoside **56** in the presence of 2,3-dichloro-5,6-dicyano-1,4-benzoquinone (DDQ) using CH_2Cl_2 as a solvent resulting in an intermediate acetal **57**, which on treatment with iodonium dicollidine perchlorate (IDCP) gives β-1,2-linked Hyp triarabinofuranoside **58** afforded by intramolecular aglycon delivery. Benzyl deprotection of compound **58** followed by acetyl protection leads to the formation of compound **59** that on incorporation into peptide resin at Hyp7 by using standard SPPS-Fmoc method and removal of resin by acid treatment along with deacetylation afforded desired [Ara_3]CLV3 glycopeptide **60** (Scheme 12.7).[42]

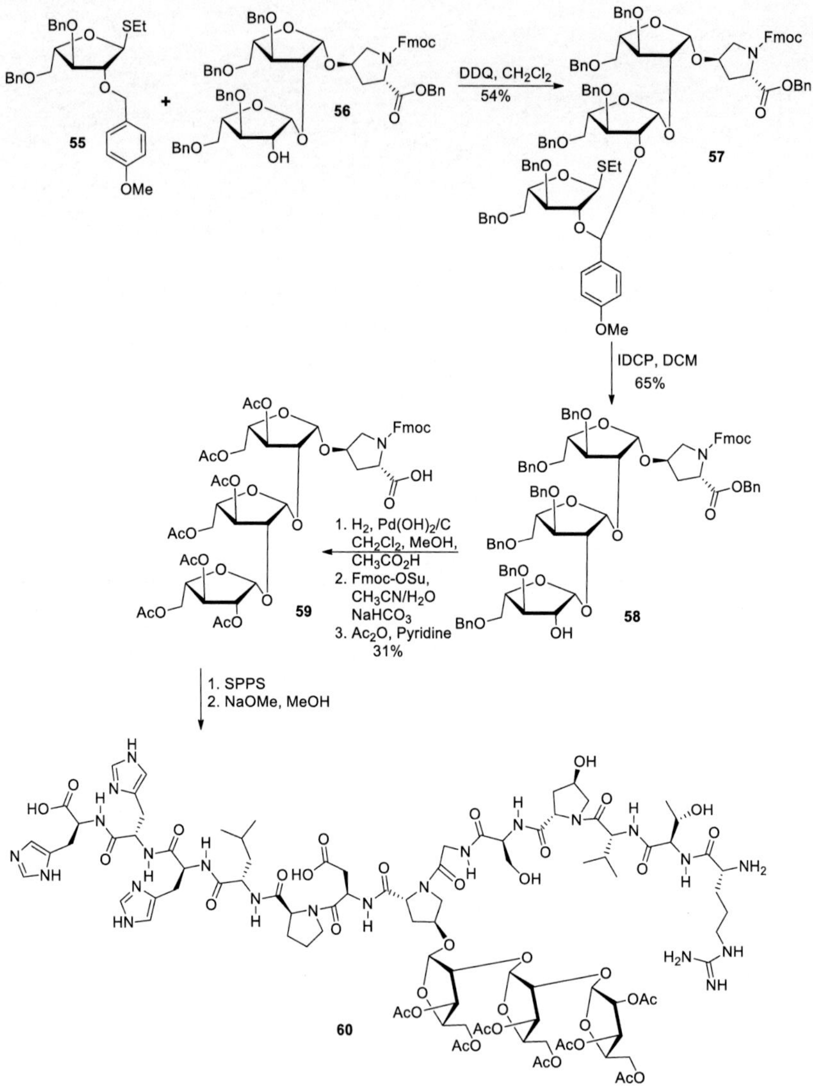

SCHEME 12.7 Synthesis of [Ara$_3$]CLV3 O-linked glycopeptide.[42]

Kumar et al. reported a cost-effective green method for the synthesis of complex glycopeptides using carbonyl diimidazole in a planetary ball mill by dry grinding under solvent-free conditions. This method avoids the use of expensive and hazardous coupling agents and solvents. Using the bromo derivative **61** as a starting point, *p*-aminophenyl glucoside **63** was synthesized by grinding it in a planetary ball mill, treating it with *p*-nitrophenol, and then reducing the nitrophenyl glucoside **62** with H-cube. Coupling of compound **10** with amino-protected lysine **64** using carbonyl diimidazole in the ball mill and ground at 550 rpm for 2 hours results in desired glycopeptide **65** and urea **66** (Scheme 12.8).[43]

Yoshida et al. reported a kinetically controlled α-selective method to synthesize O-linked glycopeptide using 2-nitroglycals. Bifunctional chiral thiourea was used as a catalyst, which is responsible for the high α-selectivity. 3,4,6-Tri-*O*-benzyl-2-nitro-D-galactal **67** was treated with peptides **68** and **71** using thiourea catalyst **69** in the presence of dichloromethane as a solvent at room temperature to obtain the desired O-linked glycopeptide **70** and **72**, respectively (Scheme 12.9).[44]

Mucin-1 (MUC-1), a potential antigen for the anticancer vaccine, contains Thomsen − Friedenreich (Tf) antigen and the Sialyl Thomsen-nouveau (STn) antigens that are O-linked glycans. Wu et al. synthesized MUC1 glycopeptides SAPDT*RPAP (T* is the glycosylation site) having α-linked Tf and the STn antigens, respectively. The Fmoc-protected STn antigen **77** and Fmoc-protected Tf antigen **80** were used as building blocks to synthesize MUC1-STn **73** and MUC1-Tf **74** glycopeptides, respectively. For MUC1-STn **73**, galactosamine threonine ester **1** was sialylated using sialyl donor **75** in the presence of p−TolSCl/AgOTf promoter, which resulted in the formation of **76** Fmoc-Neu5Ac-α-2,6-GalNAc-α-Thr along with β-isomer **76**β. The Fmoc-protected STn antigen **77** was obtained after acidification of **76**. This building block entered the SPPS cycle wherein mono-Fmoc-protected 1,4-diaminobutane **78** was loaded into *p*-nitrophenyl carbonate functionalized wang resins. Two coupling agents (2-(1H-benzotriazol-1-yl)-1,1,3,3-tetramethyluronium hexafluorophosphate (HBTU)/HOBt were used to couple Fmoc-protected amino acids and peptide chains. Next, glycosyl amino acid building block **77** was coupled using 1-[bis(dimethylamino)methylene]-1H-1,2,3-triazolo[4,5-b]pyridinium 3-oxide hexafluorophosphate (HATU)/1-

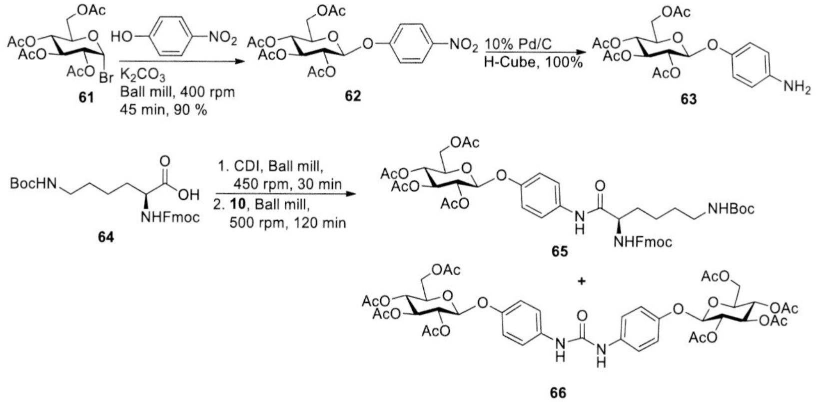

SCHEME 12.8 Synthesis of O-linked glycopeptide using ball-mill method.[43]

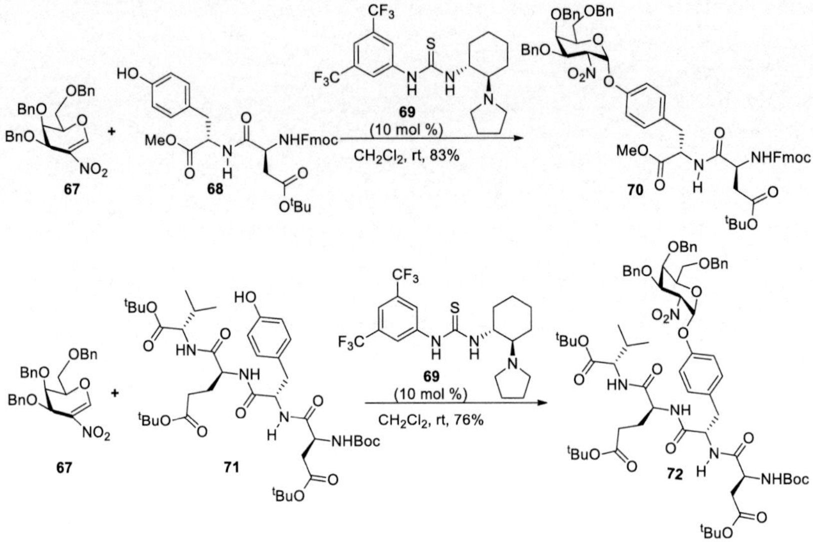

SCHEME 12.9 Synthetic route to synthesize O-linked glycopeptide via kinetically controlled α-selective method.[44]

hydroxy-7-azabenzotriazole (HOAt) agents, which followed capping of the N-terminus with acetyl group. Lastly, glycopeptide was cleaved from resins and incubated with adipate bis(4-nitrophenyl) ester **79**, which after HPLC purification yielded the desired MUC1-STn glycopeptides **73**. Next, for the synthesis of MUC1-Tf **74** glycopeptide, disaccharide donor **81** and acceptor **82** were glycosylated in the presence of p-TolSCl/AgOTf promoters to give Fmoc-protected Tf antigen **83** and its β-anomer **83β**. Further, the reduction of azide using Zn gave **84** followed by acid treatment to give the desired building block, Fmoc-protected Tf antigen **80**. The obtained **80** participated in SPPS methodology and resulted in MUC1-Tf **74** glycopeptide (Scheme 12.10).[45]

In their next work, the synthesis was modified to obtain β-linked isomer, which showed enhanced stability toward glycosidase than the α-glycosyl bond. For this purpose, firstly 3-hydroxyl group of galactosamine in Troc containing **85** was protected as *tert*-butyldimethylsilyl (TBS) ether **86**, which glycosylated Fmoc-protected threonine, subjected to further deprotection to give β-linked acceptor **87**. The glycan chain was extended using peracetylated Gal donor **88** to yield disaccharide threonine **89**. Further acetylation yielded **90**, which finally gave the desired glycosyl amino acid β-Tf **91** after protective group manipulation. The glycopeptide MUC1-β-Tf **92** along with α-isomer **92α** was assembled using solid phase peptide synthesis (SSPS) strategy involving Fmoc chemistry employing (2-(1*H*-benzotriazol-1-yl)-1,1,3,3-tetramethyluronium hexafluorophosphate (HBTU)/HOBt promoters and 1-bis(dimethylamino) methylene]-1H-1,2,3-triazolo4,5-b]pyridinium 3-oxide hexafluorophosphate (HATU)/1-hydroxy-7-azabenzotriazole (HOAt) coupling agents (Scheme 12.11).[46]

Zemlyakov et al. reported a method to synthesize O-linked glycopeptide using α-phenyl-D-glucosaminide peracetate **93**. Deacetylation of compound **93** followed by protection of hydroxyl group at C-4 and C-6 using isopropylidene gives acetal **94** having free hydroxyl group at C-3, which on treatment with L-2-bromopropanoic acid in the presence of sodium hydride as a base result into the formation of 4,6-*O*-isopropylidene-*N*-acetyl-D-muramic acid α-phenylglycoside **95**. It was coupled with L-alanyl-D-isoglutamine benzyl

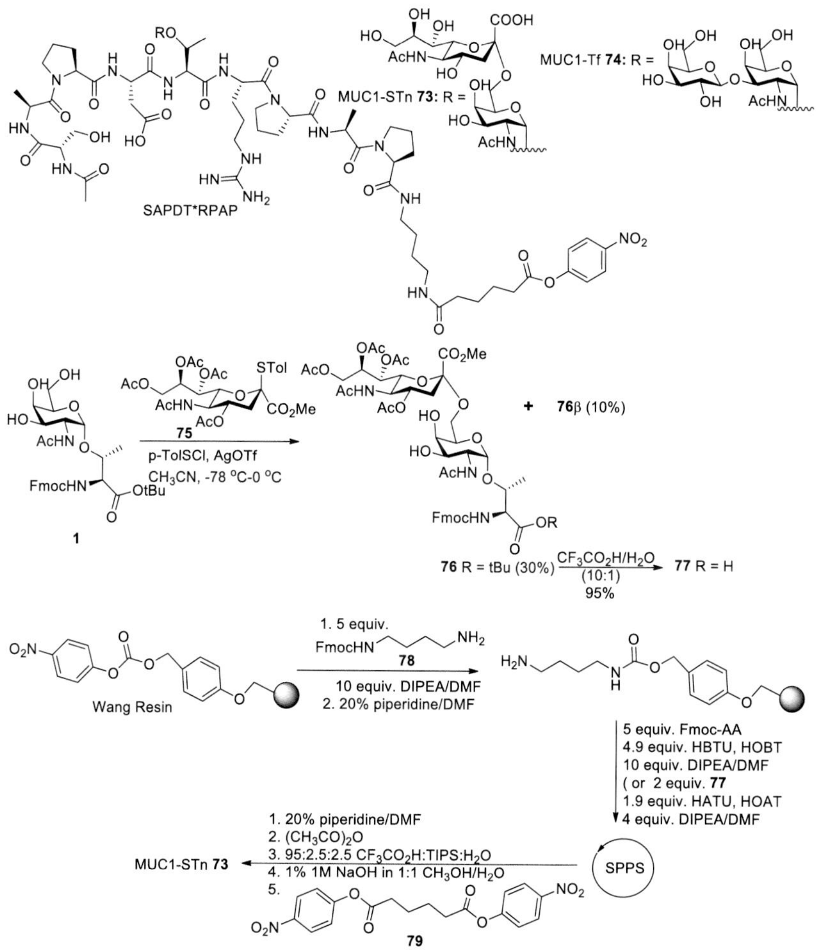

SCHEME 12.10 Synthetic route for the synthesis of MUC1-STn **73** and MUC1-Tf **74** O-linked glycopeptides.[45]

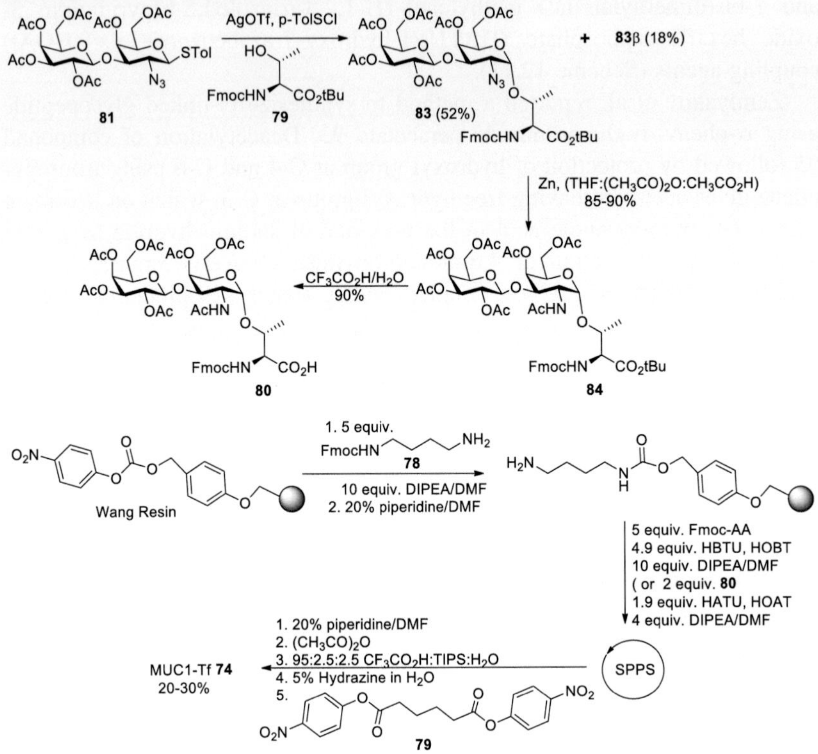

SCHEME 12.10 Continued

ester using *N*-hydroxysuccinimide (HONSu) and DCC gives isopropylidene-protected glycopeptide **96**, which on deprotection using acetic acid followed by hydrogenolysis of benzyl protection afforded desired O-linked glycopeptide *N*-acetylmuramyl-L-alanyl-D-isoglutamine α-phenylglycoside **97** (Scheme 12.12).[47,48]

A chemoenzymatic approach to synthesize O-linked glycopeptides containing core m1, core m2, and core m3 mannosyl glycans unit was reported by Li et al. 4,6-Benzylidene-protected mannopyranoside **98** was used as a starting material containing free 2-OH. Benzoylation of both free hydroxy groups takes place in two steps giving compound **99**. Removal of 2-*O*-Bz in compound **99** using sodium methoxide in methanol maintaining the pH of the solution 9–10 afforded **100** used for the synthesis of 1,2-linked core m1 glycans while regioselective borane-reduction of 4,6-*O*-benzylidene acetals in **100** results in 6-OH product **101**, which is used as a precursor for core m2 glycans. Interestingly, treatment of compound **99** with a mixture of Et_3SiH and TFA using dichloromethane as a solvent resulted in the reductive ring opening of benzylidene acetal afforded 4-OH product **102**, which is used as

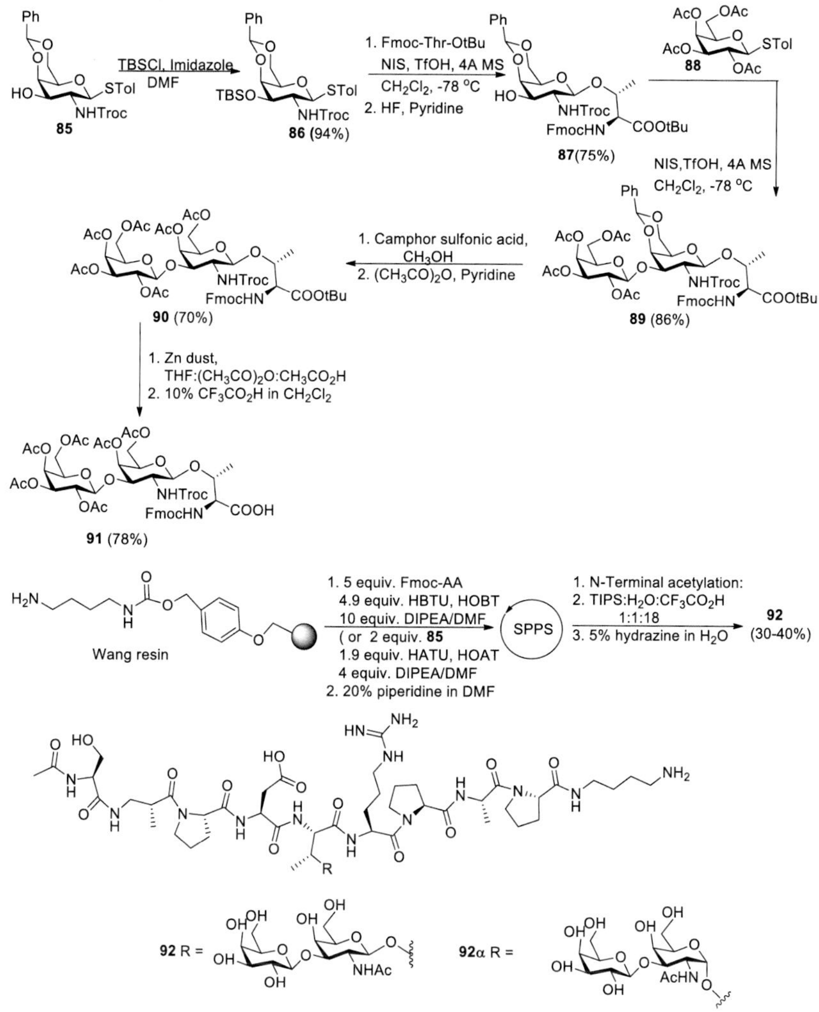

SCHEME 12.11 Synthesis of O-linked MUC1-β-Tf glycopeptide.[46]

a precursor for the formation of 1,4-linked core m3 glycan. To synthesize the β1,2-linked core m1 disaccharide product and β1,4-linked core m3 disaccharide product, glycosyl trichloroacetimidate **103** served as glycosyl donor while for core m2 trisaccharide product, glycosyl *N*-phenyltrifluoroacetimidate **107** was employed. In a mixture containing glycosyl acceptor **101** and Lewis acid TMSOTf as a catalyst, glycosyl donor **107** was added resulting in core m2 trisaccharide **108**. Treating **100** with glycosyl trichloroacetimidate **103** using a catalytic amount of triflic acid at low temperatures −40°C to

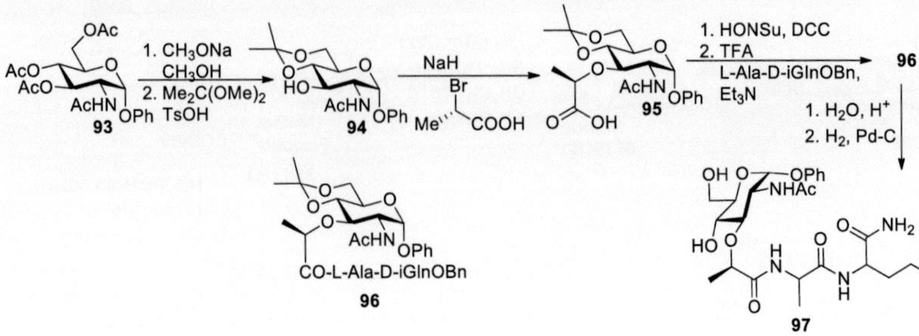

SCHEME 12.12 Synthetic route to synthesize O-linked *N*-acetylmuramyl-L-alanyl-D-isoglutamine α-phenylglycoside.[47]

−60°C in dichloromethane followed by treating it with acetic acid at 90°C gives core m1 disaccharide **104** while product **102** that reacts with glycosyl trichloroacetimidate **103** using a catalytic amount of TMSOTf at low temperature 0°C in dichloromethane gives core m3 disaccharide **112**. Protecting groups from obtained products **104**, **108**, and **112** were removed using the appropriate reaction conditions, and released hydroxyl groups were acetylated using acetic anhydride, and the methoxyphenyl group present at the anomeric position was removed by using CAN. Obtained products were further transformed using appropriate reaction conditions shown in Scheme 12.13 into trichloroacetimidate glycosyl donors **105**, **109**, and **113**, which on further treatment with orthogonally protected serine **110** followed by removal of PNB group using appropriate acid afforded glycosyl amino acid **106**, **111**, and **114**. 2-Chlorotrityl chloride resin was used to synthesize desired glycopeptides by SPPS under microwave conditions for 10 minutes at 50°C using DMF as a solvent. Fmoc group was removed by using mild conditions in a mixture of morpholine/DMF (1:1, v/v) to maintain the integrity of the glycan chain giving glycopeptides **115–117** (Scheme 12.13). For enzymatic alteration, bacterial β1,4-galactosyltransferase from *Neisseria meningitidis* (NmLg*t*B) for β1,4-galactosylation and an α2,3-sialyltransferase 1 B144D mutant from *Pasteurella multocida* (PmST1M144D) to catalyze β1,4-linked galactosides to form α2,3-sialylation linkage was used and purified from *Escherichia coli*. Enzymatic modifications of synthesized glycopeptides **115–117** are shown in Scheme 12.13 resulting in the formation of glycopeptides **118–123**.[49]

Another work demonstrates the synthesis of dynamic glycopeptide amphiphilic dendrimers composed of hydrophobic peptides as an arm with β-cyclodextrin (CD) **124** and hydrophilic saccharides as dendrons. O-linked glycopeptide synthesis reported by Bi et al. composed of 14 hydrophilic and bioactive saccharides as dendrons were linked to C-2 and C-3 positions of β-cyclodextrin whereas 7 oligopeptide arms at the C-6 position of β-cyclodextrin by an acylhydrazone dynamic covalent bond. β-Cyclodextrin treated with molecular iodine gives iodine-derived CD (CDI) **125**, which on treatment with *p*-hydroxybenzaldehyde in DMF at 80°C introduced a formyl group CD-CHO **126**. 6-Bromohexanoylchloride was used to form a bromo derivative of CD, which on treatment with sodium azide results in azide-derived β-cyclodextrin CDN_3 **127**. The CD-centered hydrophilic glycodendrons (CDSac) **128** with 14 saccharides on one side and 7 formyl groups on the other were produced by coupling alkyne derivatives of glycans with CDN_3 using copper as a catalyst. Finally, C-linked dynamic amphiphilic glycopeptide dendrimers that were desired were produced by coupling CDSac

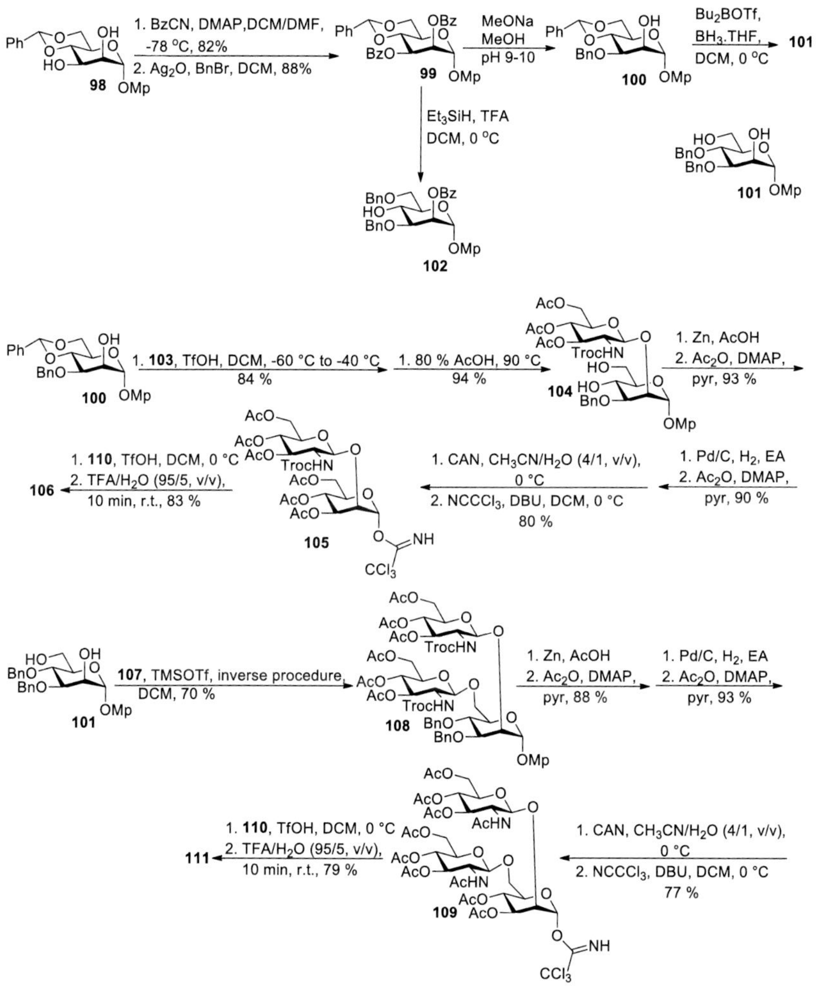

SCHEME 12.13 Route to synthesize core m1, core m2, and core m3 O-linked glycopeptides and alteration in glycopeptide using coenzymatic approach.[49]

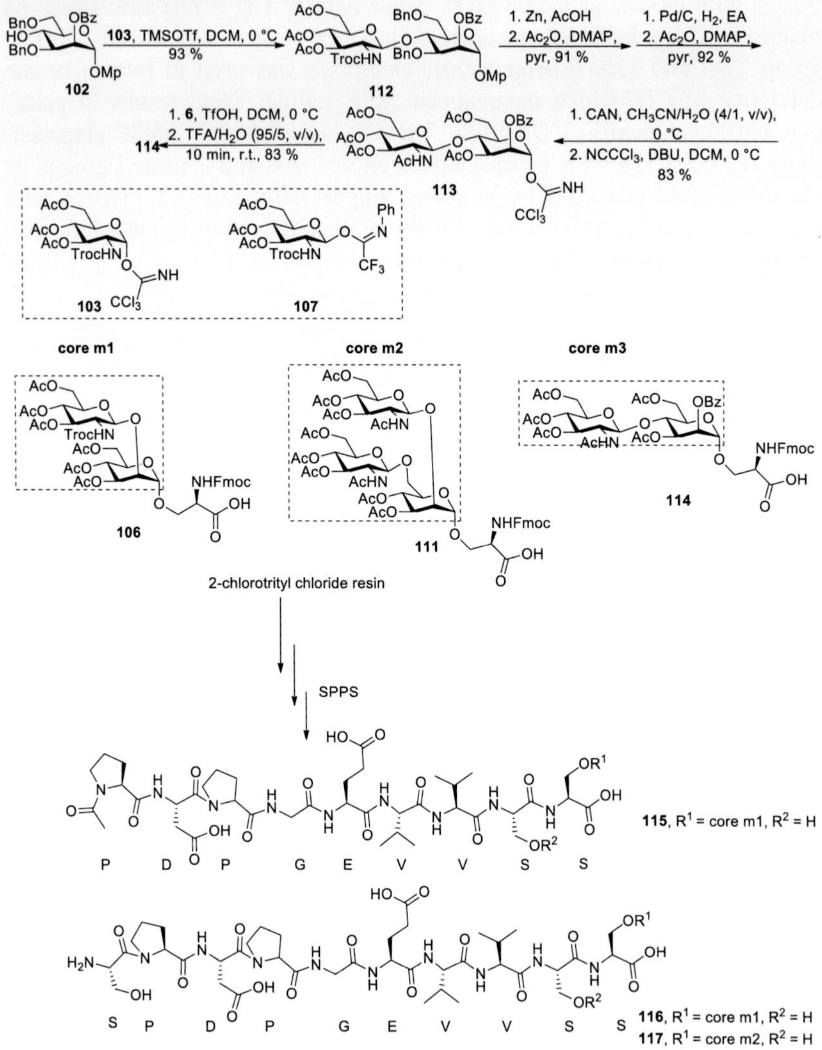

SCHEME 12.13 Continued

and acylhydrazine-ended hydrophobic phenylalanine dipeptides **129** and tetrapeptides **130** (Phe_2NHNH_2 and Phe_4NHNH_2) via dynamic covalent connections.

The several types of saccharides (mannose, glucose, galactose, maltose, lactose, and maltoheptaose) and peptides were changed separately to create a library of glycopeptide dendrimers (Scheme 12.14).[50]

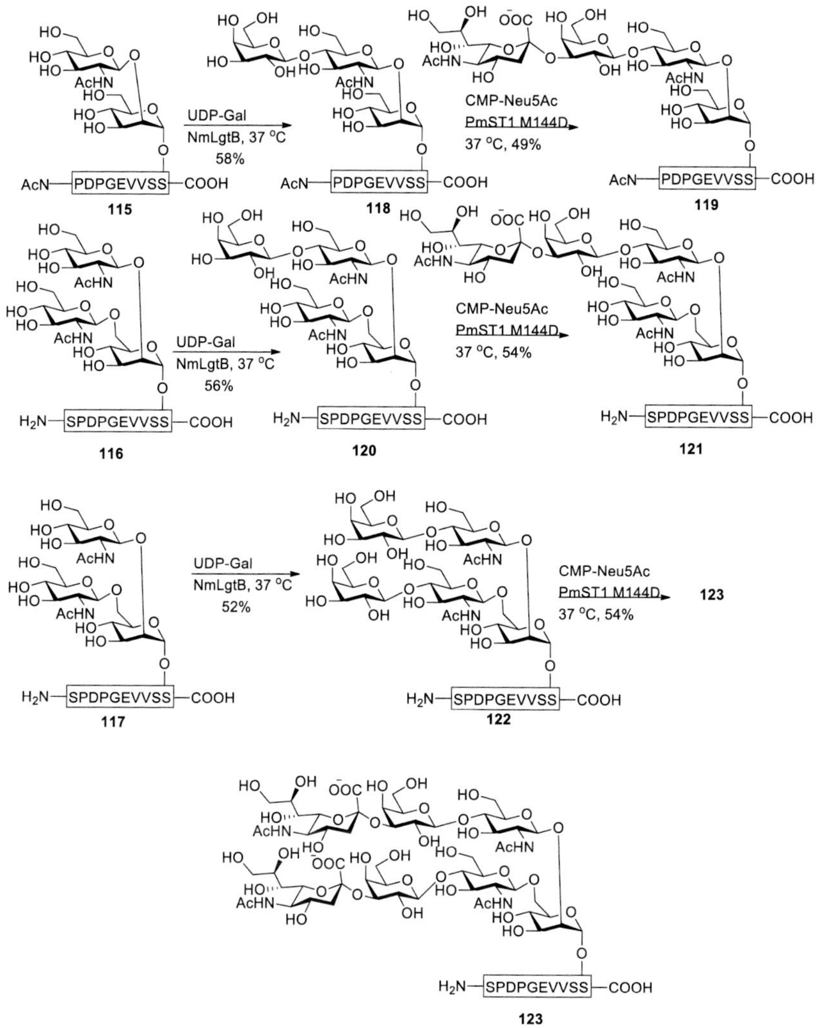

SCHEME 12.13 Continued

12.3 Synthesis of S-linked glycopeptides

Synthesis of S-linked glycopeptides is crucial for research purposes because of their biological investigations, as their chemical stability increases and they demonstrate enzymatic resistance. They can also be used as therapeutic agents. The only difference between S-linked and O-linked glycopeptides is that the anomeric oxygen in O-linked is replaced with sulfur in S-linked. The glycopeptide linkage was stabilized against chemical degradation by the

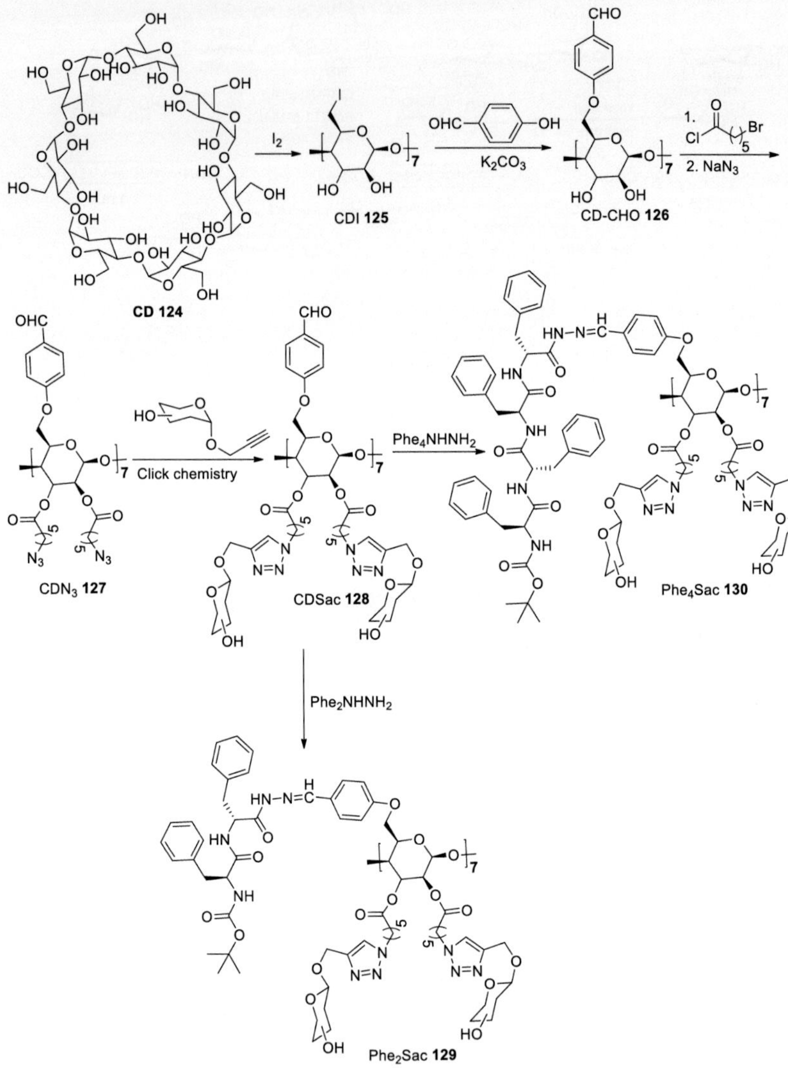

SCHEME 12.14 Synthesis of O-linked glycopeptide containing β-cyclodextrin (CD).[50]

presence of anomeric sulfur connected to the anomeric carbon of sugar. Co-translational or posttranslational strategies can be used to make S-linked glycopeptides; the latter is referred to as a convergent strategy.[51,52]

Zhu and Schmidt work described the synthesis of S-linked glycopeptide via the posttranslational method, which follows the coupling of 1-thiol sugar derivatives with a peptide consisting of β-bromoalanine or γ-bromohomoalanine. β-Bromoalanine **132** couples with α-GlcNAc **131** thiol in EtOAc/H_2O in a slight basic medium having pH around 8.5 using an aqueous solution of $NaHCO_3$ and tetrabutylammonium hydrogen sulfate (TBASH) results in 87% of α-thioglycoside **133**. The use of 10% Na_2CO_3 was avoided as it can result in β-elimination of bromoalanine derivative (Scheme 12.15).[51]

The Tamm-Horsfall glycoprotein (THP), an essential S-linked glycopeptide that is found in typical human urine, was synthesized as a facsimile. It has several important activities, including electrolyte balance in the nephron, blocking hemagglutination, and treating a variety of medical diseases (urinary tract infection, interstitial nephritis, stone formation, etc.). By employing activated molecular sieves as a base, the sugar moiety **134** was coupled to a peptide **135** containing a cysteine unit in which the thiol group acts as a nucleophilic by nature. This coupling to form S-β-linked glycopeptide **136** follows S_N2 mechanism. To achieve chemo- and stereoselective cysteine functionalization, there was no need to activate the thiol group or use a reactive nucleophilic substrate in this procedure (Scheme 12.16).[52,53]

Diphenylalanine scaffolds were utilized to synthesize S-linked glycopeptides possessing a higher solubility than diphenylalanine reported by Roytman et al. 2-Chlorotrityl resin was used to synthesize bromo-dipeptide derivative **137**, which coupled with β-GlcNAc thiol to give corresponding S-linked glycopeptides. Diphenylalanine containing bromo derivative **137** was synthesized using the standard Fmoc-SPPS method on 2-chlorotrityl resin. Fmoc-Phe-OH was treated with 2-chlorotrityl resin in the presence of DIPEA followed by treatment with 20% piperidine, the same procedure was followed to couple another unit of Fmoc-Phe-OH followed by treatment with bromoacetic acid in 1,3-diisopropylcarbodiimide (DIC) results in compound **137**. Nucleophilic attack of the free thiol group of β-GlcNAc thiols **138**, **134**,

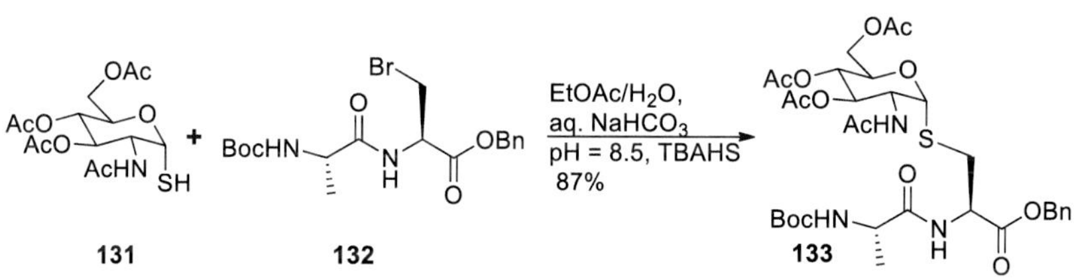

SCHEME 12.15 Synthesis of S-linked glycopeptide using thiosugars in basic conditions.[51]

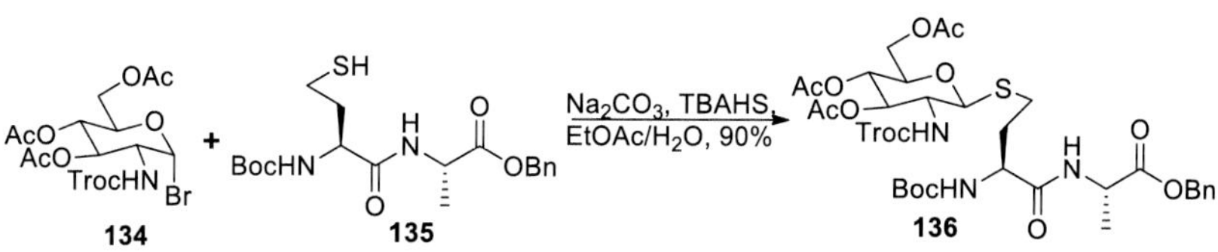

SCHEME 12.16 Synthesis of S-linkage glycopeptide using brominated sugars and peptide containing free thiol group.[52]

139, and **140** at the electrophilic position of compound **137** gives protected glycopeptide, which on treatment with hydrazine in DMF followed by resin cleavage by using TFA afforded desired glycopeptides **141–144** (Scheme 12.17).[54]

The Ferrier method is also used to synthesize S-linked glycopeptide in which direct coupling of cysteine **147** with tri-*O*-acetyl-D-glucal **145** takes place in anhydrous CH_3CN using $BF_3 \cdot OEt_2$ as a catalyst at 0°C results in an anomeric mixture of desired glycopeptide **148** in 86% yield in a few minutes. Other sugar moieties, such as acylated D-galactal **146** and acylated L-rhamnal **150**, were also employed, and the same procedures were used to create the necessary S-linked peptides **149** and **151**, respectively. Also, dimeric cysteine was found to be very useful to synthesize desired peptides; bis-carbobenzyloxylated cystine **152** was reacted with Val-OBn in HCl followed by reduction with tri-*n*-butylphosphane to break S–S bond and result in

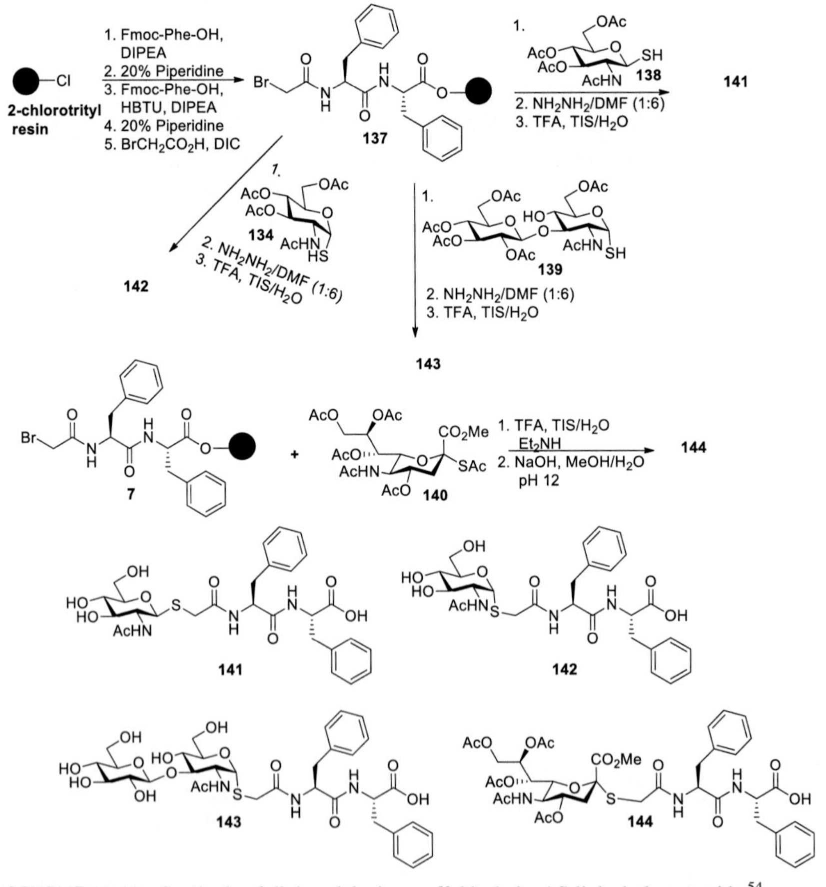

SCHEME 12.17 Synthesis of diphenylalanine scaffolds derived S-linked glycopeptide.[54]

dipeptide **153** with free −SH group. Coupling of **153** with tri-*O*-acetyl-D-glucal **145** in anhydrous CH_3CN using the catalytic amount of $AlCl_3$ at a very low temperature results in the desired S-glycopeptide **154** (Scheme 12.18).[55]

Due to their resistance to endogenous glycosidases and their H-bond acceptor ability, which aids in ligand binding, sulfur atoms in S-linked glycopeptides are responsible for improved glycosidic linkage durability. In a different study, S-linked STn antigens were employed to interact with the activated ester adipic acid p-nitrophenyl diester, which not only couples with keyhole limpet hemocyanin (KLH), but also with bovine serum albumin (BSR) to generate S-glycopeptide-protein conjugates.[56]

Hsieh et al. provide a convergent method for producing S-glycosylated Sublancin-168, a 37 amino acid peptide with bactericidal activity against gram-positive bacteria, utilizing a chemical ligation-based methodology. The four cysteine residues in sublancin-168 form disulfide linkages and the remaining one is linked to sugar. 1-Thiosugars and β-bromoalanine were combined in a phase-transfer-catalyzed reaction to make thioglycosyl amino acid building blocks, which were then put into a microwave-assisted SPPS chain elongation approach to create thioester ligation partners. The mercaptophenylacetic acid (MPAA) thiol catalyst and tris(2-carboxyethyl)-phosphine (TCEP) reductant were used to chemically ligate the glycopeptides and thioester (C-terminated), fold them using glutathione, and purify them using HPLC.[57]

Brimble et al. synthesized glycocin F using a native chemical ligation strategy, which is an S-linked glycopeptide. Its antimicrobial property makes it very useful for biological studies. Apart from S-linkage, it also possesses O-linkage where GlcNAc moieties are attached to Cys-43 and serine-18, respectively. Both glycopeptides were synthesized separately using *N*-acetylated sugar **155** chlorinated at an anomeric position and reacted with

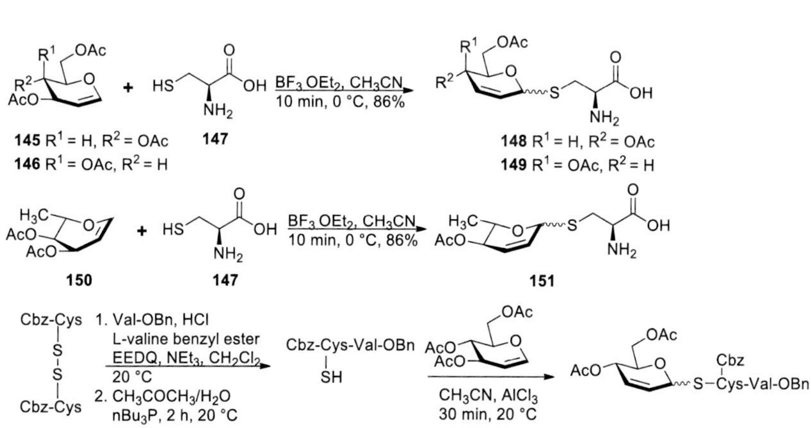

SCHEME 12.18 Synthesis of S-glycopeptides via Ferrier glycosylation.[55]

N-Fmoc amino acid. First, O-glycoside **157** was synthesized by reacting chlorinated sugar with protected L-Serine **156** using $HgBr_2$, and S-glycoside was synthesized by reacting chlorinated sugar with thiourea in acetone/reflux, which on further treatment with sodium metabisulfite in DCM/water mixture under refluxing conditions results in 1-thio-sugar **138**, which reacts with protected β-bromoalanine **158** using TBAHS in $EtOAc/H_2O$ at 50°C gives **159**. Alternatively, we can use direct coupling of 1-thio-sugar using the same procedure. Two different peptide fragments were synthesized and coupled using native chemical ligation, which, as determined by high-resolution mass spectrometry (HRMS), created a disulfide bond upon oxidation (Scheme 12.19).[58]

Using $Ca(OH)_2$ as a promotor, peptides with thiol groups as the acceptor, and fluoride sugars as the donor, Zhong et al. present a green S-glycosylation technique. Tripeptide **160** couples with glucopyranosyl fluoride **161** using the base $Ca(OH)_2$ in the presence of H_2O at room temperature results in desired S-linked glycopeptides **162** in 90% yield (Scheme 12.20).[59]

Kaur et al. demonstrated a metal-free visible-light-mediated thiol-ene reaction in an aqueous medium that uses a photoredox catalyst to initiate the radical process to generate thiyl radicals upon light irradiation. Thioglycoside **138** on coupling with N-terminus functionalized dipeptide **163** in the presence of a photocatalyst (9-mesityl-10-methylacridinum tetrafluoroborate) under blue light-emitting diode (LED) irradiation in CH_3CN/H_2O (1:1) mixture afforded S-linked glycopeptide **164** (Scheme 12.21).[60]

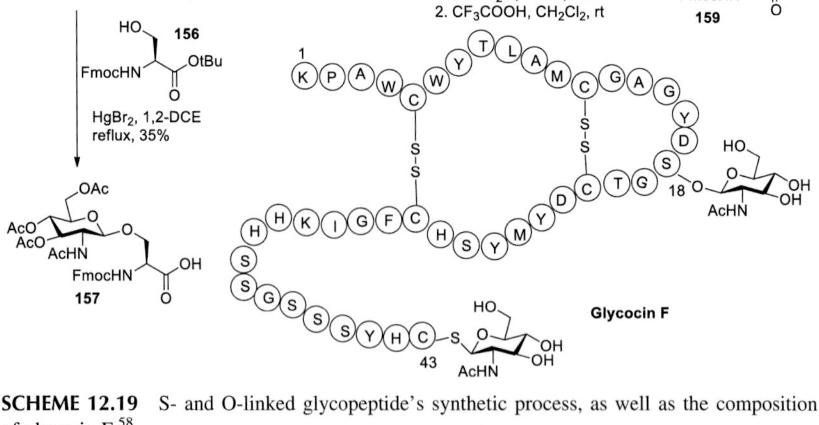

SCHEME 12.19 S- and O-linked glycopeptide's synthetic process, as well as the composition of glycocin F.[58]

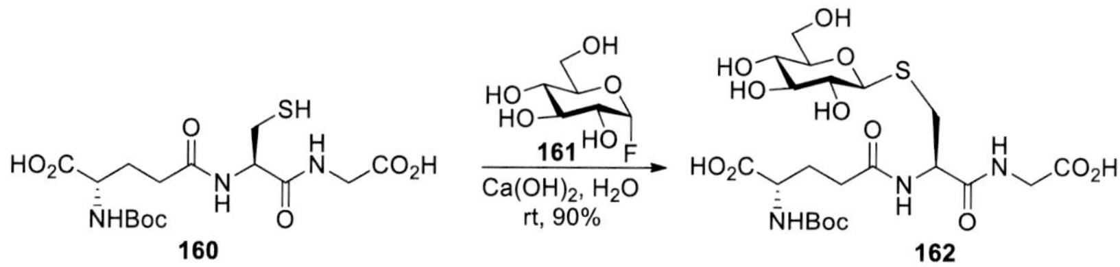

SCHEME 12.20 Synthesis of thioglycopeptide in H_2O.[59]

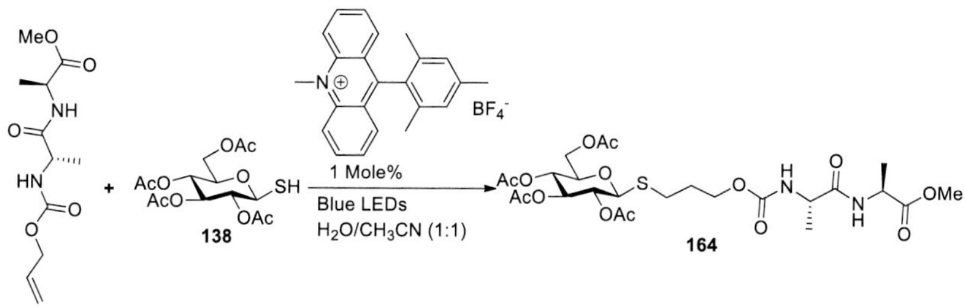

SCHEME 12.21 Synthesis of visible-light-mediated S-linked glycopeptide.[60]

Another study describes the chemoselective solution- and solid-phase synthesis of S-linked glycopeptides containing disulfide linkage shown by Banisalman et al. Diluted solution of per-*O*-acetylated thioglucose **138** (1 equiv.) added dropwise to a solution comprising DEAD (2.0 equiv.) in THF gives glucosulfenylhydrazine **165**. Coupling of tripeptide **166** (1 equiv.) through cysteine residue with **165** (5 equiv.) in THF at room temperature results in desired S-linked glycopeptide **167** (Scheme 12.22).[61]

12.4 Synthesis of C-linked glycopeptides

As they are not deglycosylated in vivo and have crucial qualities including stability in solution, solid-phase peptide synthesis, and biological activities, carbon-linked glycopeptide synthesis is vital. Replacement of anomeric oxygen in O-glycopeptide with carbon results in C-linked glycopeptide. A simple method was used to synthesize C-linked glycopeptide. The synthesis of olefin product **170**, which is a mixture of *cis* and *trans* product, is achieved through the Wittig reaction when benzoyl-protected glycosyl aldehyde **168** couples with β-alanyl anion **169** using BuLi as a base. Reduction of alkene −C = C− bond by diimide results in C-glycosyl oxazolidinone **171**. This compound **171** when treated with BOC$_2$O in Et$_3$N/DMAP resulted in BOC protection, which on further treatment with CsCO$_3$ in MeOH formed **172**. TFA was used followed by Fmoc-OSu to replace Boc with Fmoc to give the

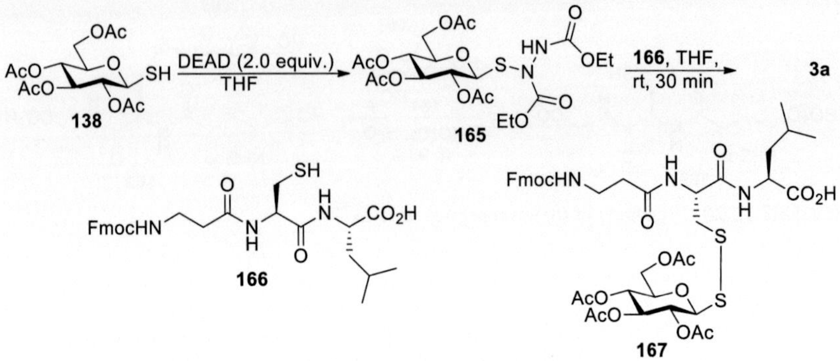

SCHEME 12.22 Route to synthesize the disulfide-linked glycopeptide.[61]

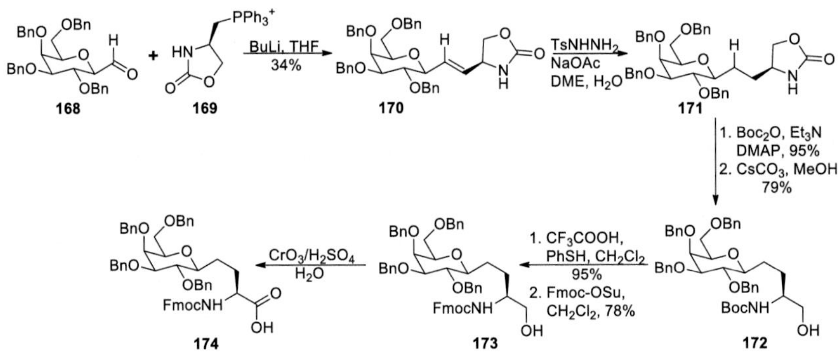

SCHEME 12.23 Synthetic route to synthesize C-glycopeptide building block.[62]

compound **173**. Oxidation of later with Jones reagent results in corresponding C-linked glycopeptide building block **174** (Scheme 12.23).[62]

Manabe et al. developed a C–C bond at the anomeric carbon of the sugar moiety, that is, with mannose and the C-2 position of the indole ring coupled to the tryptophan amino acid derivative, in a different way. Starting with an indole derivative **176** at the active C-2 site, building blocks that were formed by connecting 1,2-anhydro-mannose **175** via S_N2 mechanism using BuLi and $BF_3 \cdot OEt_2$ as a Lewis acid results in C-glycoside **177**. Several building blocks were prepared and incorporated into the peptide sequence to get the desired C-glycopeptide. To make C-glycopeptide, commercially available L-tryptophanol **178** was utilized. The amine group was first converted to azide by treating it with TfN_3, and the hydroxyl group was protected with TBS ether. Protected indole ring **179** was created by treating it with BuLi and benzenesulfonyl chloride. When this chemical is combined with 1,2-anhydro-mannose **175**, a compound is formed, which is then benzylated and acid is used to remove the TBS ether, yielding **180**. Using a mixture of iodobenzene diacetate and 2,2,6,6-tetramethyl-1-piperidinyloxy (TEMPO), the hydroxyl group in this group was oxidized to carboxylic acid, yielding **181**. The required C-glycopeptide **182** is obtained by deprotecting the indole ring and reducing the azide group into a Boc-protected peptide (Scheme 12.24).[63]

Vangala et al. developed a straightforward procedure to create C-linked glycopeptide by creating sugar furanoid *trans* vicinal diacid and inserting it into the tetrapeptide sequence (H-Phe-Trp-Lys-Thr-OH) as a γ-turn inducer. Vicinal diacids possess unique properties like selective protection of either of the carboxyl groups, act as a precursor for β-amino acid with a sugar framework, and act as a synthon for sugar α-amino acid. Diacetonide D-glucose **183** treated with pyridinium chlorochromate (PCC) to get the oxidation of free hydroxyl group gives ketone **184**, which on treatment with $CHCl_3$/ LHMDS in THF at low temperature afforded trichloromethyl alcohol **185** followed by converting later into α-azido acid in the presence of NaN_3 and NaOH using dioxane-water as a solvent, and condensation of α-azido acid **186** with benzylamine in the presence of EDC-HOBt gives amide **187**. Acid treatment of compound **187** gives diol **188** followed by oxidative cleavage using $NaIO_4$ in THF-H_2O and oxidation with oxone results in *trans*-vicinal diacid **189**. 2-Chlorotrityl chloride resin was used for the introduction of vicinal diacid **189** into a tetrapeptide sequence (H-Phe-Trp-Lys-Thr-OH). Initially, amino acid Fmoc-Thr(Bu^t) was attached using DIPEA in DMF followed by Fmoc deprotection using 20% piperidine in DMF. Coupling of other amino acids (Fmoc-Lys-(Boc), Fmoc-Trp, and Fmoc-Phe) was done using HBTU, HOBt, and base DIPEA followed by incorporation of

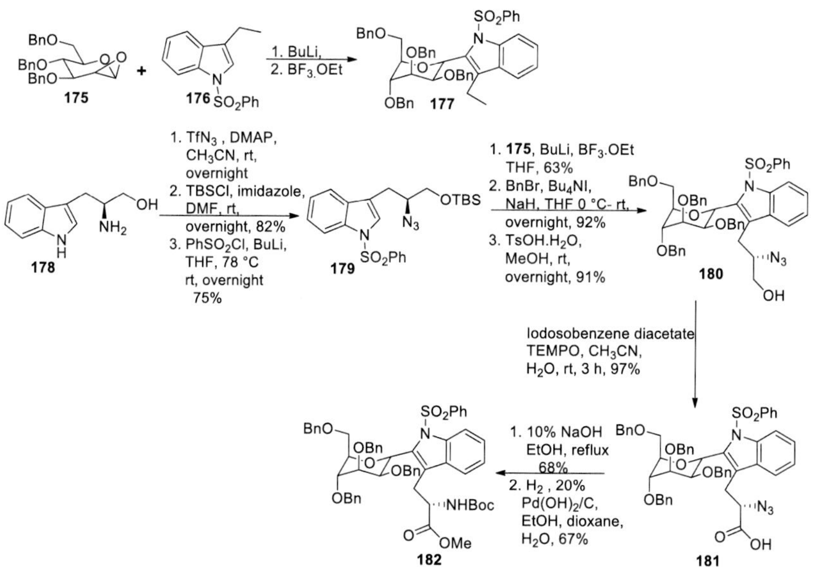

SCHEME 12.24 Synthetic route for the synthesis of C-mannosyl-tryptophan.[63]

compound **189** into N-terminus of tetrapeptide using the same reaction conditions that result in glycopeptide **190**, and acid treatment of later with TFA in DCM results in desired glycopeptide **191** (Scheme 12.25).[64]

Wu et al. demonstrated a C-H activation strategy to synthesize C-alkyl/aryl glycopeptide in a different way. Because of region-, chemo-, and diastereoselectivity, this method proved particularly valuable for research. Building blocks for C-alkyl/aryl glycosyl amino acids were created using palladium-catalyzed C(sp^3)-H/C(sp^2)-H activation made possible by peptide isosteric click-triazoles. C(sp^3)-H glycosylation of triazolyldimethylmethyl

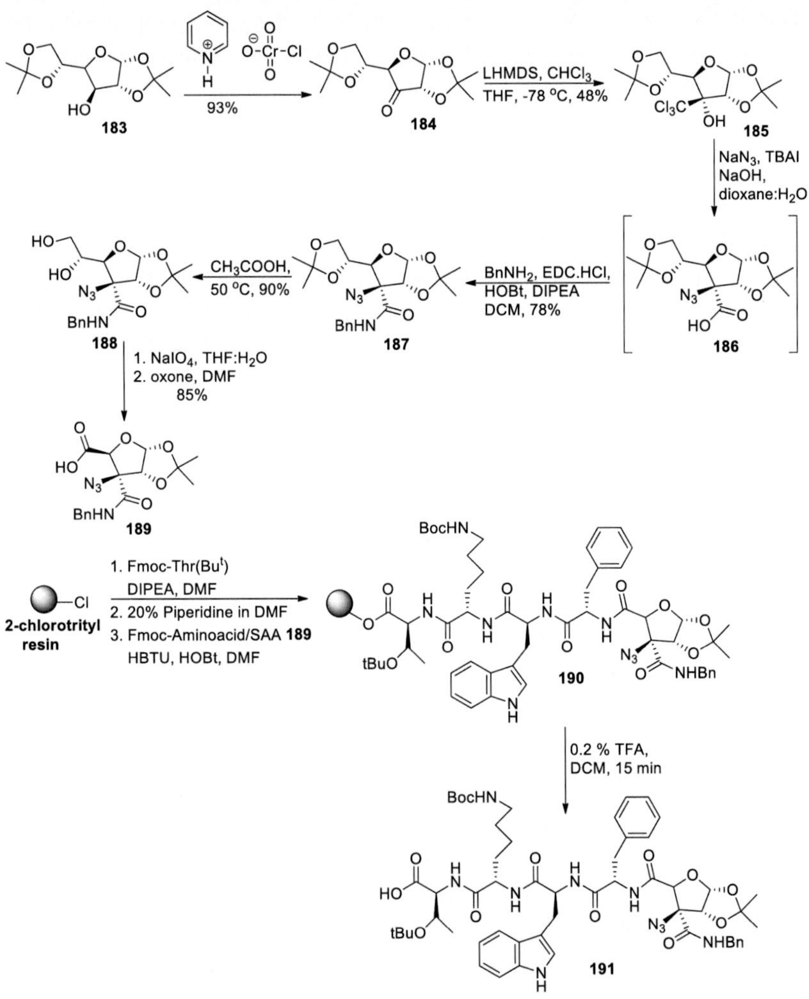

SCHEME 12.25 Synthetic route to synthesize C-linked glycopeptide using sugar furanoid *trans*-vicinal diacid.[64]

(TAM) amide **192** was coupled with sugar moiety **193** containing halide directly attached to olefinic carbon in the presence of catalyst $Pd(TFA)_2$ and additive AgOAc was added to it in polar solvent 1,4-dioxane to afford the desired glycopeptide **194** (Scheme 12.26).[65]

Another work reported by Jiang et al. revealed the synthesis of C-linked glycopeptides by Ti-catalyzed stereoselective glycosyl radical functionalization using a wide range of glycosyl chlorides and activated alkenes/alkynes under mild reductive conditions. Tetrabenzyl-protected α-mannosyl chloride **195** was treated with N-terminus of α-amino acid and oligopeptide derivatives **196** using 10 mol% of $CpTiCl_3$ (cyclopentadienyltitanium(IV) trichloride) and 2 equiv. of Mn as reductant and $Et_3N \cdot HCl$ as a proton donor using THF as a solvent at an ambient temperature to afford the desired glycopeptides **197** (Scheme 12.27).[66]

Using the earlier reaction conditions, the following compounds **198–203** were formed.

Interestingly, glycosyl radical generated from glycosyl sulfoxides in aqueous medium used as an intermediate to synthesize C-linked glycopeptides were reported by Shang et al. Unprotected glycosyl sulfoxides **204** (air and moisture stable) were used as starting material to synthesize desired

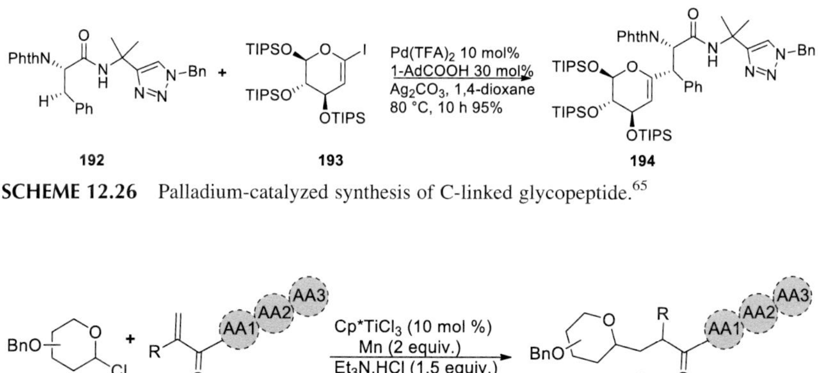

SCHEME 12.26 Palladium-catalyzed synthesis of C-linked glycopeptide.[65]

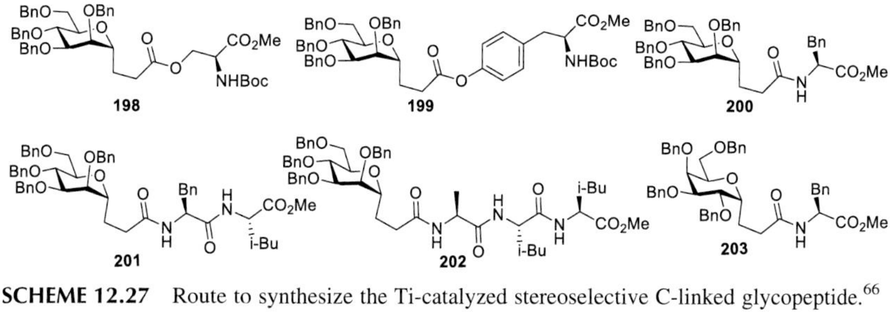

SCHEME 12.27 Route to synthesize the Ti-catalyzed stereoselective C-linked glycopeptide.[66]

C-linked glycopeptides. Aryl iodide moiety in glycosyl sulfoxides **204** got converted to an aryl radical using water-soluble borane complex **206** as a radical mediator, which reacted smoothly with peptide derivatives **205** in CH_3CN-H_2O mixture at room temperature to give the desired C-linked glycopeptide **207** (Scheme 12.28).[67] Using the earlier reaction conditions, a library of the following compounds **207a–e** was formed as shown in Scheme 12.28.

Wang et al. reported a new method to synthesize insect adipokinetic hormone Cam-HrTH-I **220** as a C-linked glycopeptide via palladium-catalyzed C-H glycosylation. Tryptophan methyl ester (H-Trp-Ome) **208** was used as a starting material to give compound **209** via amide coupling with isoquinoline-1-carboxylic acid (i_1QA) auxiliary group using DMF as a solvent at room temperature followed by treatment with tetrabenzyl-protected mannosyl chloride **210** (2 equiv.) using a catalytic amount of $Pd(OAc)_2$, 1.5 equiv. of KOAc, and 0.3 equiv. of Ac-lle-OH additive in toluene at 110°C for 12 hours, which afforded two compounds **211** and **212** in 85% and 5% yield, respectively. Further modifications on **211** were carried out by removing the auxiliary group and protecting the amine group with Boc_2O and triethyl amine in acetonitrile to give compound **213**, which on saponification with LiOH in a $MeOH/H_2O/THF$ mixture resulted in compound **215** with free acid. Hydrogenolysis followed by saponification of **213** afforded compound **214**. Amide coupling of **215** using HATU, HOAT in the presence of DIPEA as a base, and DMF as a solvent with side chain–blocked dipeptide H-Gly-Thr(Bn)-NH_2 **216** afforded **217**, which got converted into tetrapeptide **218** using standard amide coupling with an asparagine and Boc deprotection.

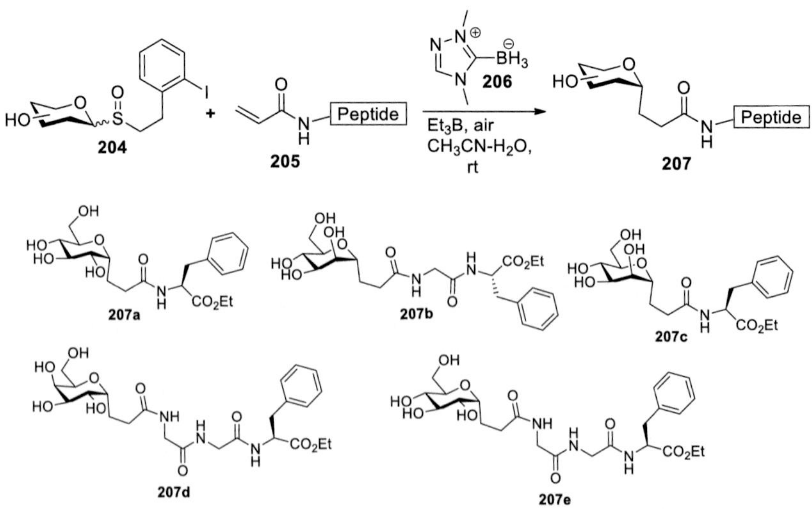

SCHEME 12.28 Synthesis of glycosyl radical-mediated C-linked glycopeptide.[67]

Hexamer peptide **219** was synthesized via the standard SPPS method using Fmoc-protected building blocks coupled with tetrapeptide **218** followed by benzyl deprotection to give the desired C-linked glycopeptide Cam-HrTH-I **220** (Scheme 12.29).[68]

In a different study, the mannosyl halide and haloarene underwent a reductive cross-coupling process facilitated by nickel in the presence of the Hantzsch ester (**HE**) as a photo reductant to synthesize 2-(α-D-mannopyranosyl)-L-tryptophan or Trp(Man) **224** contrary to other works using Ni/photocatalyst dual catalysis approach. Trp(Man), belonging to C-mannoside, exists in unusual $^{1}C_{4}$ conformation instead of $^{4}C_{1}$, followed majorly by other

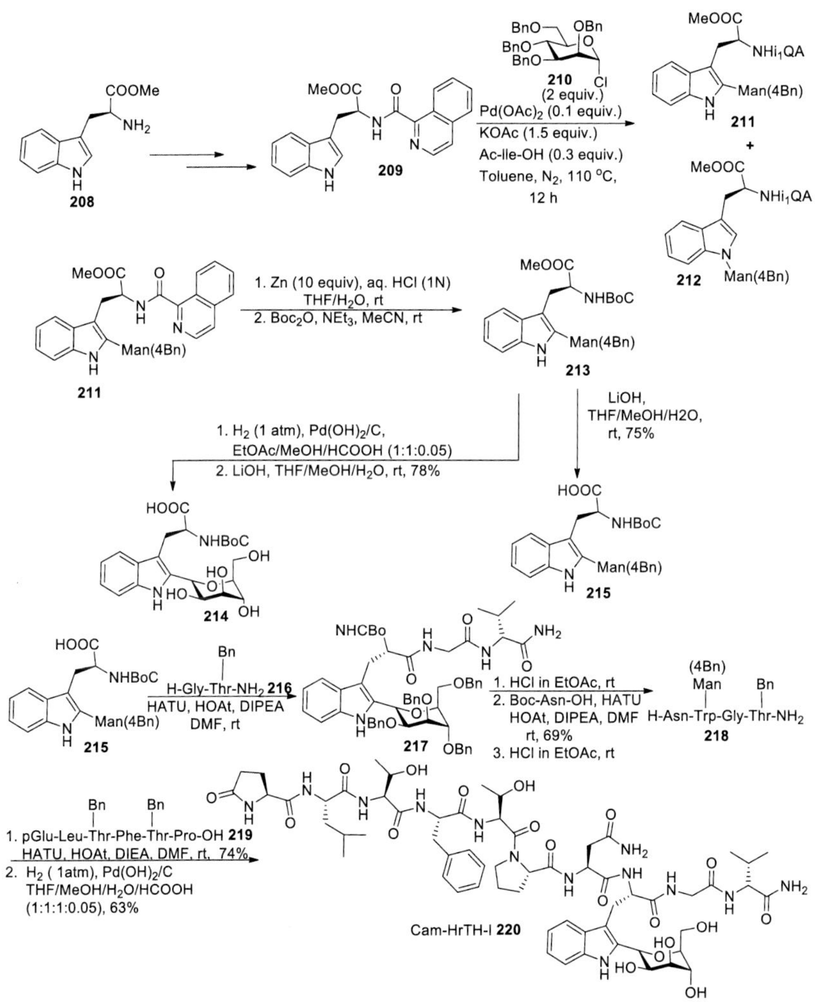

SCHEME 12.29 Synthesis of palladium-catalyzed C-linked glycopeptide Cam-HrTH-I.[68]

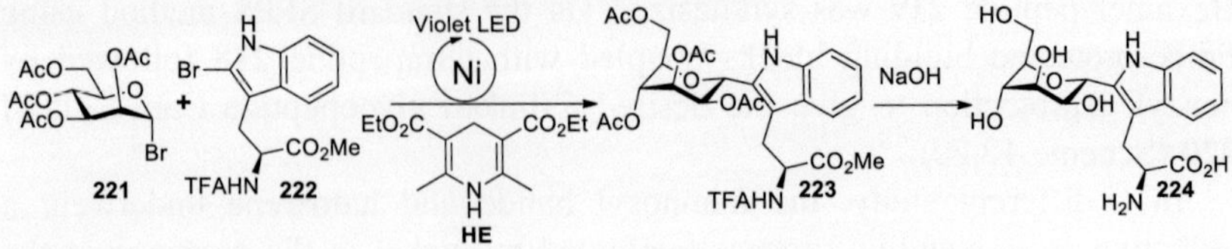

SCHEME 12.30 Synthesis of Ni-catalyzed C-mannosylated glycopeptides.[69]

mannosides. The reaction proceeded using mannosyl bromide **221** and 2-bromo-L-tryptophan **222** in the presence of photoexcited **HE** reductant to yield **223** in 78% yield, which is treated under basic conditions to yield Trp (Man) **224**. In this reaction, the photoexcited **HE** acts as a reductant and produces low-valence nickel species and glycosyl radical, which initiates the cross-coupling catalytic reaction (Scheme 12.30).[69]

A new approach that utilizes the concept of acyl transfer auxiliaries was reported by Nyandoro et al. C-linked glycopeptide was synthesized using this concept where glycosyl nitrile **225** was used as a starting material to get sugar-linked auxiliary **229**. Free CN group in **225** was reduced by catalytic hydrogen in the presence of di-*tert*-butyl dicarbonate (Boc_2O) to give **226**, which on treatment with compound **227** containing free CHO group using a catalytic amount of TFA in DCM followed by the reductive amination using $NaB(Oac)_3H$ gives **228**. Treating the later compound with triethylsilane and TFA in DCM results in **229**. Lastly, coupling of peptide **230** (H-STGWNETIG-OH) with **229** results in desired C-linked glycopeptide **231** (Scheme 12.31).[70]

12.5 Synthesis of N-linked glycopeptides

12.5.1 Glycopeptides containing asparagine

Glycosyl amine is the smallest glycoconjugate wherein amine groups in glucopyranose are linked to carboxyl groups in amino acid to form N-glycosidic linkage.[71]

N-glycosyl asparagine is widely used as a structural unit in the synthetic process of N-glycopeptide. In general, for synthesizing N-glycosylated asparagine, aspartic anhydride derivatives are employed due to their strong acylating ability along with the advantage of no prerequisites of the protection–deprotection step. Ibatullin and coworkers reported the synthesis of N-glycosylated asparagine derivative **233a,b** by employing N-Fmoc-protected aspartic anhydride **233** derived from aspartic acid **232** with per-O-acetylated glycosyl amine (Scheme 12.32).[72]

The glycosyl amine has been reported to react with carboxyl substituent on aspartic acid to form glycosylated asparagine through an amide linkage. N-Fmoc-aspartic anhydride, prepared from aspartic anhydride, reacts with

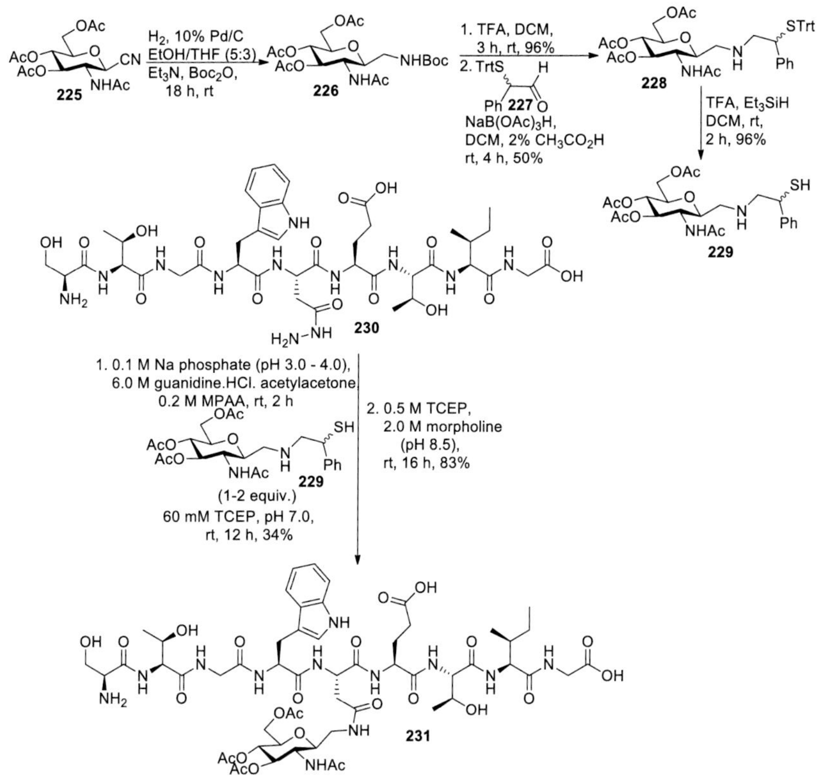

SCHEME 12.31 Route to synthesize C-linked glycopeptide using sugar-linked acyl transfer auxiliaries.[70]

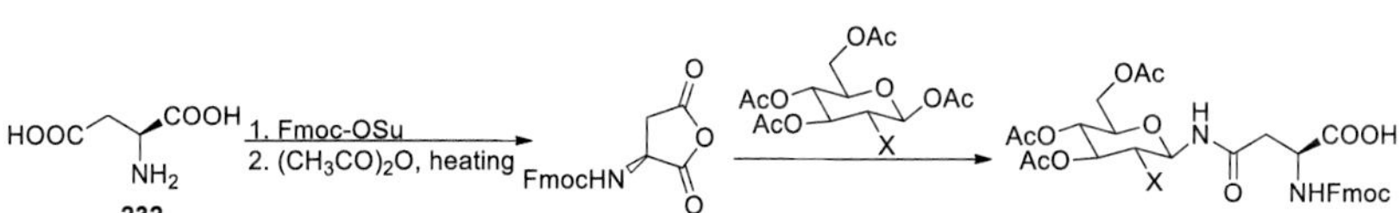

SCHEME 12.32 Synthesis of N-glycosyl asparagine.[72]

per-O-acetylated glycosyl amines and undergoes aminolysis to form N-glycosylated Asn derivates, which acts as a building block for solid as well as liquid phase synthesis of glycopeptides. The regioselectivity of the reaction has been reported to depend on the polarity of the solvent used. Isoasparagine derivative was found to be the main product in the case of less polar solvents; however, polar solvents yielded the desired glycosylated Asn as the major product, with the highest yield obtained using dimethyl sulfoxide (DMSO). The utility of the reaction was validated by synthesizing N-Fmoc-Asp derivatives of various sugars such as cellobiose, N-acetylglucosamine, maltose, and glucose.[73] The carbon at second positions in the first and second glucopyranose rings on the majority of N-glycosides are coupled with an acetamido group, which serves as a highly specific recognition site to serve a wide range of biological functions. To study their role in the recognition of molecules, Urge et al. synthesized N-linked monosaccharide/oligosaccharide exhibiting biological activity in the presence and absence of acetamido groups at second carbons of first and second glucopyranose rings. For this, a reducing sugar **234** was aminated using a saturated solution of ammonium bicarbonate to yield 1-glycosylamine **235**, which was further allowed to interact with Fmoc-Asp(OH)-O^{t}Bu in DMF−water mixture using diisopropylcarbodiimide chemistry to yield N-glycosides of Fmoc-Asn-OH structural unit **236** for SPGS. The treatment of chitobiose, 2-acetamido-2-deoxy-D-glucose, and peracetylated chitobiose with NH_4HCO_3 led to the formation of β-glycosylamines, and no epimerization was observed at C2 of pyranose. However, sugars with the absence of an acetamido group such as D-glucose, cellobiose, and maltose resulted in the formation of 1-glycosylamine (Scheme 12.33).[73]

Unlike the convergent approach used in the synthesis of β-glycosyl linkages, the unnatural α-glycosyl linkage is formed using a linear approach. The high anomeric ratio of 9:1 of α/β was achieved by Colombo et al. to synthesize N-Fmoc-N-(2,3,4,6-tetra-*O*-acetyl-α-D-gluco-pyranosyl)-1-asparagine **242** structural unit. Firstly, oxazoline **238** was prepared by treatment of 2,3,4,6-tetra-*O*-acetyl-β-D-glucosyl azide **237** with PPh_3 followed by acylation to yield protected glycosyl amino acid **239**. Further catalytic hydrogenation resulted in the formation of **240**, and treatment with pyridine and Fmoc-Osu **241** yielded the desired Fmoc derivative **242** (Scheme 12.34).[74] The existing strategies that are being used for amino-glycan bond formation are limited by several factors such as the poor yield and selectivity of reaction in convergent strategy, deprotection of carbohydrate at the final stage, and harsh conditions to detach glycopeptide from resins in solid-phase synthesis. Thus a new synthetic strategy has been reported for glycopeptide synthesis in the solution phase with subsequent reaction workup in the solid phase wherein glycosyl asparagine conjugates are used as structural units for the synthesis of glycopeptides. However, coupling between glycosyl amines and the carboxyl group of the aspartate side chain might lead to unwanted anomeric epimerization of the glycosyl amine at the glycan-reducing end. Therefore Xue et al. developed a stereoselective approach to avoid

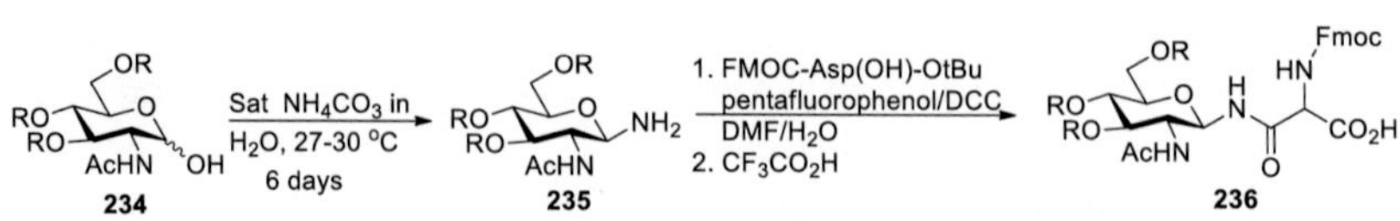

SCHEME 12.33 Synthesis of N-linked Fmoc-Asn-OH building block.[73]

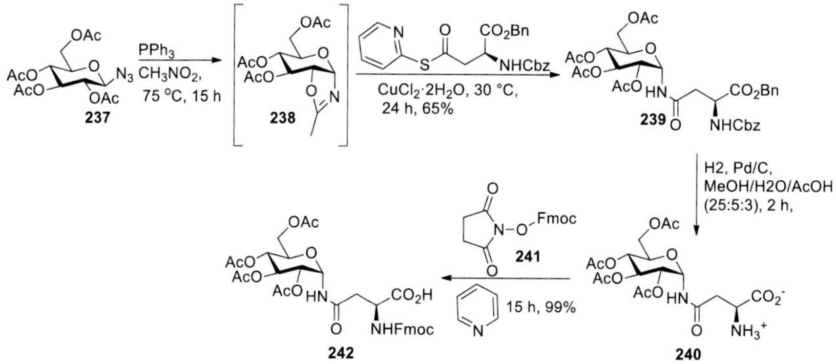

SCHEME 12.34 N-Fmoc-N-(2,3,4,6-tetra-O-acetyl-α-D-gluco-pyranosyl)-1-asparagine building block.[74]

epimerization and ease the process of purification of glycosyl amino acids. In this method, the glycosyl asparagine conjugate and short N-linked glycopeptides were synthesized using an intermediate β-configured glycosyl azide. The free oligosaccharide (e.g., Maltotriose **243**) was firstly aminated using ammonium hydrogen carbonate to yield glycosyl amine **244**, which participated in a stereospecific diazo transfer reaction to form free as well as protected glycosyl azide **245**. The poor solubility of glycosyl amine in DMF resulted in poor yield, thus DMSO was employed as a suitable solvent. This intermediate **245** underwent a Staudinger reaction with γ-diphenylphosphinothioester **246** of the aspartic acid side chain of aspartate in the presence of dry DMF to form N-glycosyl Asparagine conjugate **247**, which acts as a structural unit in the synthesis of N-linked glycopeptides (Scheme 12.35).[75]

In another work, the efficiency of three coupling agents was studied in building glycosidic bonds between amino acid and amino-glucopyranose. The researchers utilized three reagents, namely, N,N,N′,N′-tetramethyl-O-(benzotriazole-1-yl) uranium (TBTU), diisopropylcarbodiimide (DIC), and proplylphosphonic acid cyclic anhydride (T3P) for coupling between Fmoc-L-amino acids (Fmoc-Phe-OH, Fmoc-L-Asp(O*t*Bu)-OH, or Fmoc-L-Lys(Boc)-OH) and 2,3,4,6-tetra-*O*-acetyl-β-D-glucopyranosyl amine **248** to yield N-glycosylamino acids **249**, **250**, and **251**, respectively. To compare, Fmoc-Phe-OH was used as a model amino acid (Scheme 12.36).[71] For this, D-glucose and three different L-amino acids were used, that is, phenylalanine, aspartic acid, and lysine having nonpolar aromatic, acidic, and basic side chains to yield six new N-glycosyl amino acids. For step iv, DMF was used as a solvent to mix Fmoc-Phe-OH with TBTU, HOBt, and DIPEA followed by the addition of **248** to recover compound **249** after workup and purification. Next, Fmoc-Phe-OH was added to DIC employing CH_2Cl_2 as solvent under suitable reaction conditions to yield **249**. For method c, a mixture of Fmoc-Phe-OH, DIPEA, and EtOAc was taken in an old bath followed by the

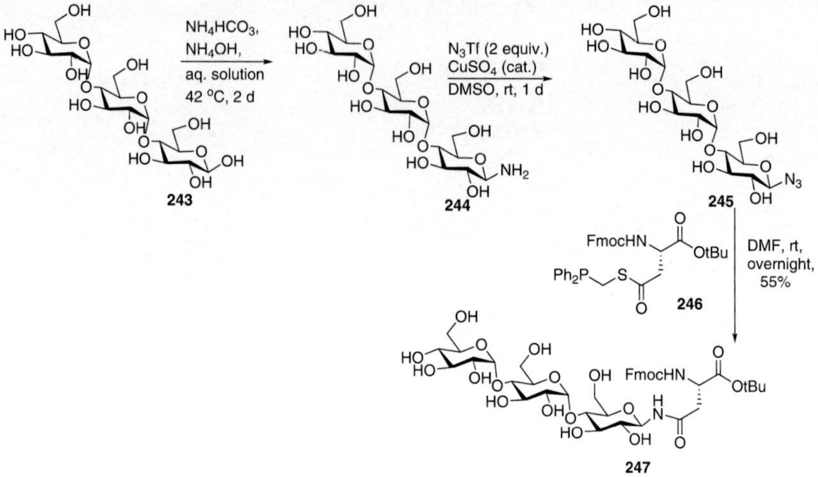

SCHEME 12.35 Synthesis of glycosyl asparagine conjugate and short N-linked building block.[75]

addition of T3P 50% EtOAc. After stirring and reaction workup, compound **249** was obtained. After examining all the adopted methodologies, T3P was selected for synthesizing other N-glycosyl amino acids **250** and **251**. Further, the deprotection of O*t*Bu and Boc was carried out using TFA/DCM to yield products **252** and **253**. The Fmoc group was removed by adding 4-methylpiperidine to **249**, **252**, and **253** to yield deprotected compounds **254**, **255**, and **256**. Lastly, deacetylation using hydrazine hydroxide resulted in desired compounds **257**, **258**, and **259**. The purification of both protected/unprotected N-glycosyl amino acids was performed using RP-SPE chromatography. Out of the three coupling agents used, T3P showed promising results for amide bond formation in terms of reduced side products, easy purification methods, high yields with good purity, low racemization, and excellent reaction selectivity. T3P serves the dual function of coupling agent as well as water scavenger; firstly, amino acid is activated and forms ester, which results in the formation of easily removable, nonhazardous, and water-soluble by-product, that is, propane phosphonic acid. Moreover, this coupling agent was classified a "green" reagent due to its nontoxic and nonallergic/sensitizing nature, the other two being toxic and explosive.[71,76,77]

Sureshbabu et al. show another method for the synthesis of N-linked glycopeptides by incorporation of urea moiety through the reaction between protected glycosyl amine and Fmoc-peptidyl isocyanates or carbamates. Urea moieties form strong H-bonding that makes glycopeptidylureas more water soluble than natural glycopeptides. Initially, protected glycosyl amines **248** and **262** were synthesized from acyl/benzyl-protected sugar moieties **260** and **261** at anomeric carbon C-1 by treating it with HBr in acetic acid followed

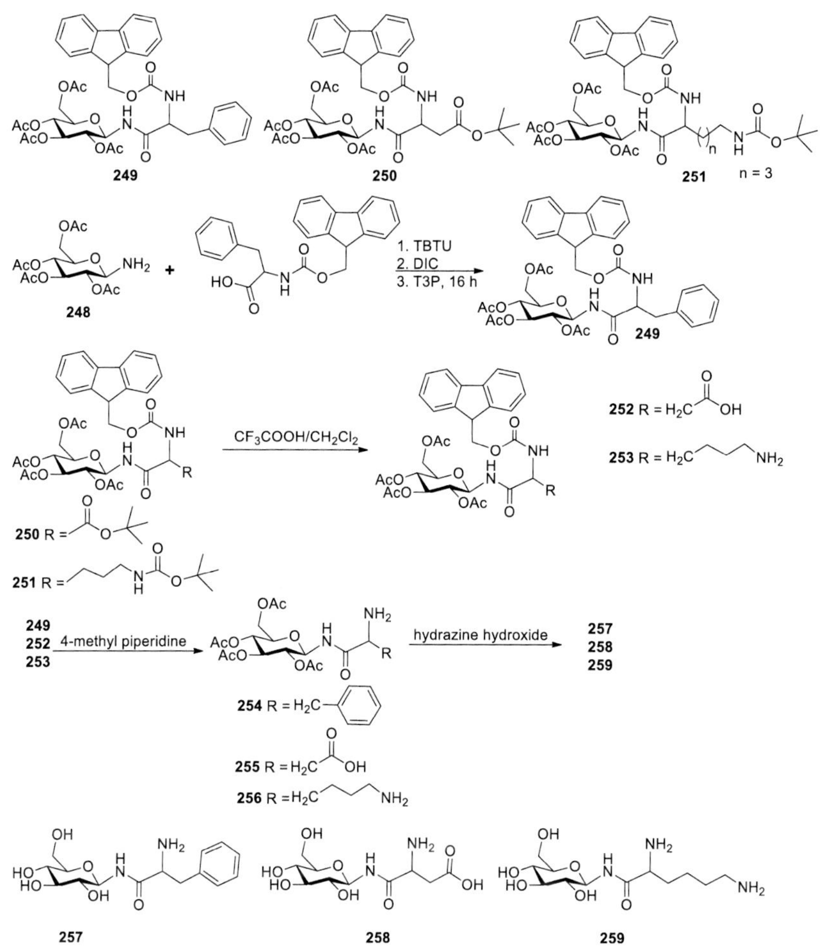

SCHEME 12.36 Synthesis of N-linked glycopeptide using proplylphosphonic acid cyclic anhydride.[71]

by nucleophilic substitution reaction using sodium azide forming azide derivative, which on reduction results in protected glycosyl amines **248** and **262**. In the next step, peptidyl isocyanates were synthesized from Fmoc-peptide acids **265a–d** by treating later with freshly prepared Bis-TMS-amino acid and then converted to corresponding acid azides via mixed anhydride followed by Curtius rearrangement gives corresponding Fmoc-peptidyl isocyanates **266a–d**. Coupling of glycosyl amines **248** and **262** with peptidyl isocyanates in the presence of N-methylmorpholine at 0°C in DCM gives desired neoglycopeptidylureas **263** and **264**. Fmoc-peptidyl carbamates were also used to synthesize neoglycopeptidylureas **263** and **264**; Fmoc-peptidyl isocyanates **266a–d** react with 2,4,5-trichloro phenol in the presence of N-methylmorpholine to give compound **267a–d**, which on further treatment with glycosyl amines **248** and **262** using *N*-methylmorpholine gives the desired compounds neoglycopeptidylureas **263** and **264** (Scheme 12.37).[78]

Using the earlier reaction conditions, the following compounds **263a–d** were formed.

Asahina et al. demonstrated a one-step deprotection strategy to design GlcNAc-Asn unit-bearing TFA-susceptible O-protecting groups **268** contrary to the previously used highly reactive benzyl-protected GlcNAc unit of Fmoc-Asn (GlcNAc)-OH, which necessitates the use of two-step deprotection in SPPS. In this work, a glycopeptide unit was synthesized wherein TFA-sensitive protecting groups were used for glycosyl amino acid protection, which was further subjected to a single deprotection step with TFA combined with the N-alkylcysteine thioesterification and used for SPPS of N-acetylglucosaminylated emmprin thioester. To synthesize GlcNAc-Asn unit **268**, MPM was used for the protection of the hydroxyl group of glycosyl

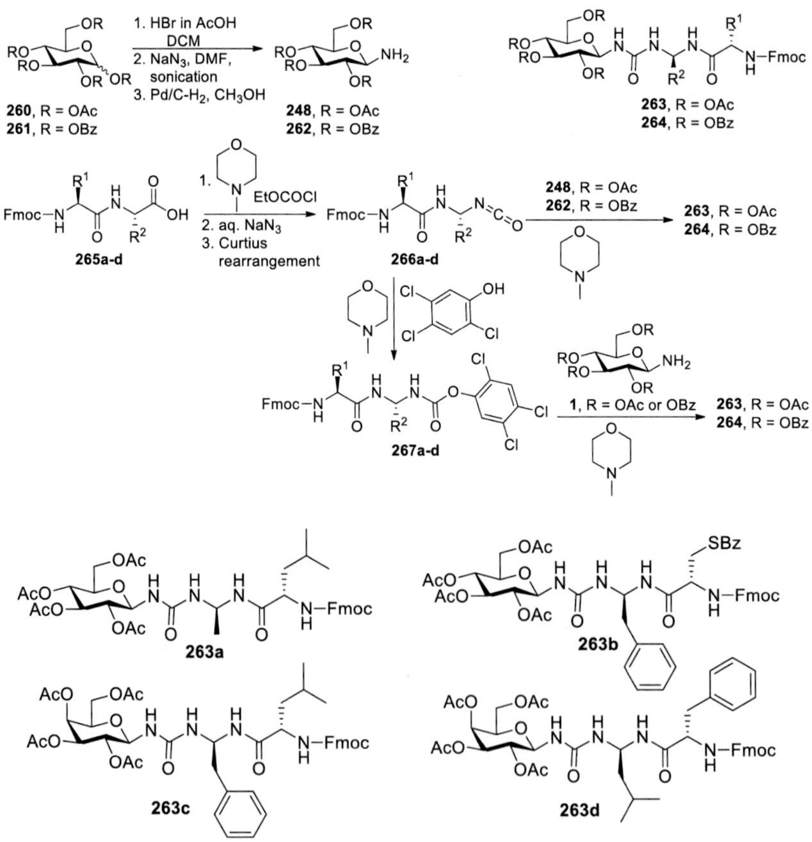

SCHEME 12.37 Synthesis of N-linked glycopeptide derived by incorporation of urea moiety.[78]

azide **269** to obtain **270**, with subsequent removal of the phthaloyl group from **269** using ethylenediamine and acetylation of free amine to yield **271**, which was coupled with Fmoc-Asp(OPfp)-OPac **275** under modified Staudinger conditions. The synthesized GlcNAc-Asn derivative **272** was obtained in 82% yield. Deprotection of the phenacyl group by Zn reduction to give compound **273** did not take place due to the poor solubility of **272** in CH_2Cl_2-CH_3COOH. **270** is converted to TMB-substituted N-acetylglucosaminyl azide **274**, which further condenses with aspartic acid derivative **275** to afford **276**, which on Zn reduction afforded the required N-TMB-substituted GlcNAc-Asn unit **268** (Scheme 12.38).[79]

In another strategy, N-glycopeptide corresponding to erythropoietin (EPO) residues was prepared. Firstly, 2-acetamido-2-deoxy-3,4,6-tri-*O*-acetyl-α-D-glucopyranosyl chloride **155** was synthesized from treatment of N-acetyl glucosamine **277** with acetyl chloride, which was then reacted with tetrabutylammonium hydrogen sulfate and sodium azide to form the corresponding β-glycosyl azide **278**. Further, on subjection to platinum (IV) oxide under an H_2 atmosphere, β-glycosyl amine **279** was prepared, which formed N-Fmoc-N-β-(2-N-acetylamido-2-deoxy-β-D-glucopyranosyl)-2-*tert*-butoxy-L-asparagine **280** on reaction with Fmoc-Asp-O*t*Bu in the presence of 2-ethoxy-1-ethoxycarbonyl-1,2-dihydroquinoline (EEDQ). On subsequent

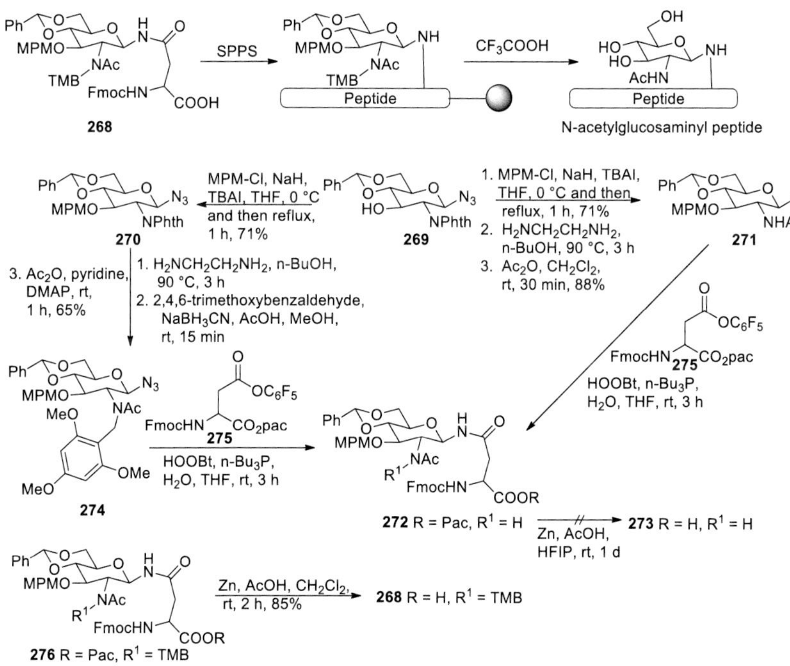

SCHEME 12.38 Synthesis of GlcNAc-Asn derivative using one-step deprotection strategy.[79]

deprotection with TFA, N-Fmoc-N-β-(2-N-acetylamido-2-deoxy-β-D-glucopyranosyl)-L-asparagineglycoconjugate **281** was obtained, which participated in SPPS to form H-AEN((OAc)$_3$(GlcNAc)ITTGC-OH N-linked glycopeptide.[80] Tanaka et al. utilized Boc groups for the protection of hydroxyl groups unlike the conventional amine-protecting group of carbohydrates and synthesized N-linked glycopeptide via SPPS-Fmoc/*t*Bu (Scheme 12.39).[81]

New asparagine-based glycopeptides containing mono- and disaccharides were synthesized using Fmoc/*t*-Bu strategy reported by Sarsan et al. Glycosyl amines **248** and **282a–e** were used as a starting material and transformed into glycosylated aspartic acids **283** and **284a–e**. Fmoc deprotection and coupling with Fmoc-Phe or Fmoc-Trp using HATU/HBTU, DIPEA, and DMF gives dipeptide **285**, **286a–e** and **287**, **288a–e**, while treating the formed dipeptides with Fmoc-Ala using similar reaction conditions results in tripeptide **289**, **290a–e**, and **291**, **292a–e** (Scheme 12.40).[82]

In another study, Rodríguez et al. synthesized the building block of Fmoc-Asn-β-(2,3,4,6-tetra-*O*-Acetyl-D-glucopyranosyl)-OH **295**. This building block contains monosaccharides and is used to synthesize *N*-linked glycopeptides derived from LfcinB using SPPS-Fmoc/*t*Bu strategy. This strategy is very useful for the synthesis of N-linked glycopeptides as it does not affect the integrity of sugar moieties attached to peptides due to the stability of N-glycopeptide during side-chain deprotection and resin cleavage reactions.[83] (Walter A. Szarek) 1-azido-2,3,4,6-tetra-*O*-acety β-D-glucose **137** was used as a precursor synthesized from D-glucose **293** followed by catalytic reduction of azide group to amine **248**. Penta-acetylglycosylamine **248** on treatment with Fmoc-Asp(OH)-O*t*Bu gives **294**, which on treatment with TFA removes the O*t*Bu group and results in building block **295**. Lastly, building block **295** was incorporated into N- and/or C-terminal ends of sequences derived from LfcinB (Scheme 12.41).[84]

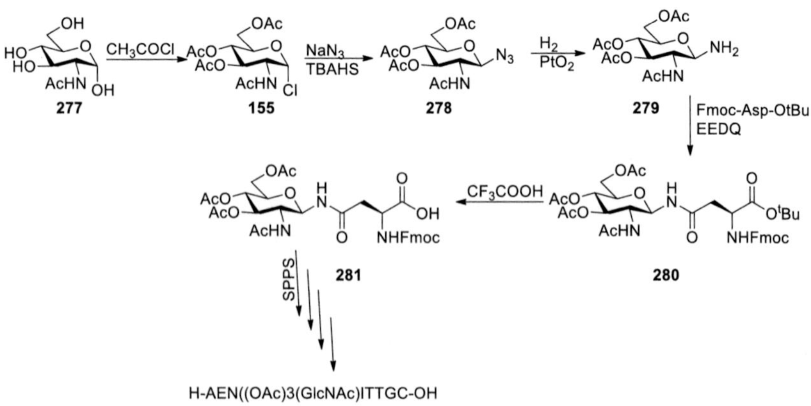

SCHEME 12.39 Synthesis of erythropoietin-related N-glycopeptide.[80]

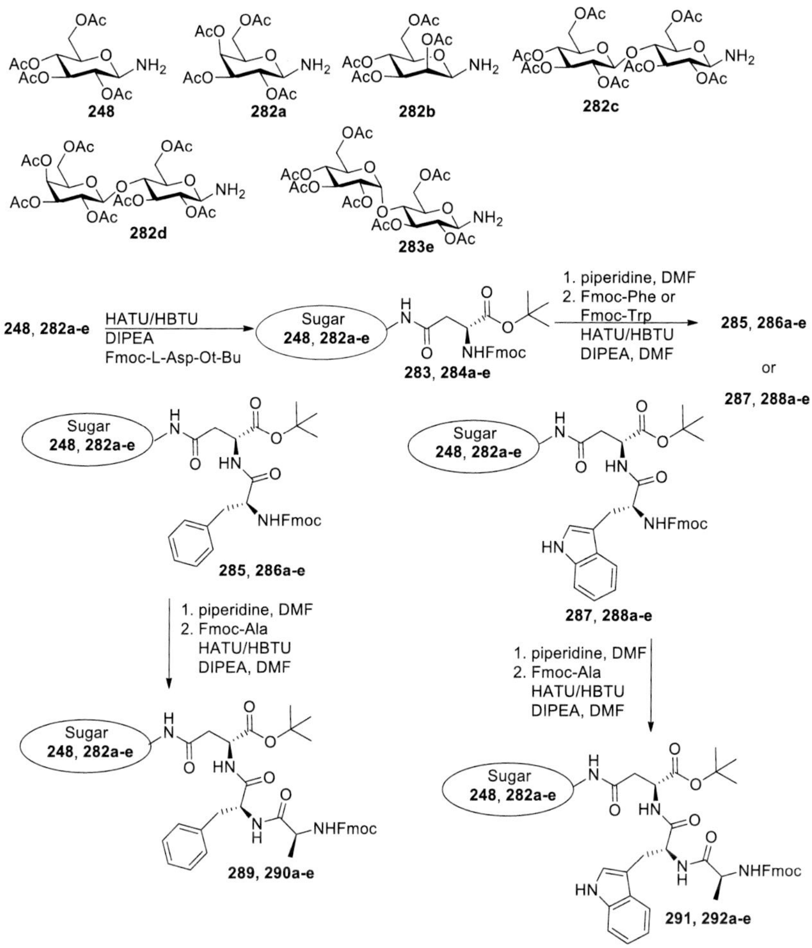

SCHEME 12.40 Synthesis of asparagine-based glycopeptides using Fmoc/*t*-Bu strategy.[82]

Another work demonstrates that auxiliary-based glycopeptide synthesis using Seitz's auxiliary for efficient peptide coupling with bulky Asn-(glycan)-OH moieties was reported by Kajihara et al. Initially, 2-mercapto-2-phenethyl auxiliary **298** from 1-phenylethane-1,2-diol **295** was synthesized and coupled with Asn-(glycan)-OH **299** to get complex-type biantennary sialyloligosaccharide bearing the auxiliary **301**. Treating 1-phenylethane-1,2-diol **295** with trityl chloride followed by mesylation and substitution with thioacetate results in **296**. Acetate groups were removed by treating later with hydrazine in methanol and the trityl group was migrated to the thiol group under acidic conditions resulting in primary alcohol **297**. Dess-Martin

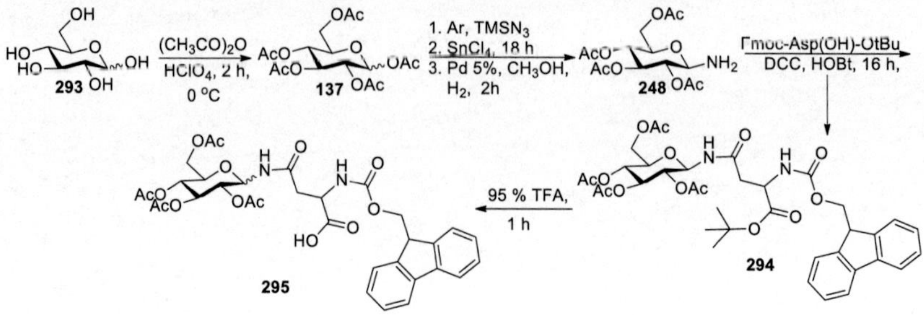

SCHEME 12.41 Synthesis of O-glycosylated Fmoc-Asn-β-(2,3,4,6-tetra-*O*-Acetyl-D-glucopyranosyl)-OH building block.[84]

oxidation of the later compound yielded the desired auxiliary substrate **298**, which reacted with the amine group of compound **299** and yielded compound **300** under reductive amination conditions. Treating later with Cs_2CO_3 and phenacyl bromide in DMF followed by treating it with TFA/TIPS to remove the trityl group results in compound **301**. In the last step, it was coupled with peptide-thioesters **302** using native chemical ligation conditions to afford **303** (Scheme 12.42).[85]

12.5.2 Glycopeptides containing arginine

The sugar moiety is joined to the ω-nitrogen of the guanidine moiety during the N-glycosylation of protein via arginine (Arg). It is found in humans, bacteria, and plants. Wang et al. demonstrated a method for obtaining biologically active glycopeptide by synthesizing GlcNAcylated arginine building blocks. Glycosyl β-isothiocyanate **304** treated with Pbf-NH_2 **305** resulted in nonisolated Pbf-thiourea **306**, which was immediately treated with ethyl iodide (EtI) to result in ethyl isothio-urea **307**. Guanidinylation of compound **307** with Fmoc-Orn-OAll in the presence of mercury salt results in allyl ester **308**, allyl deprotection of compound **308** in $Pd(PPh_3)_4$, and $PhSiH_3$ using THF as a solvent results in desired arginine building block **309** (Scheme 12.43).[86]

Another class of N-linked peptide was found by the researchers containing rhamnosylated arginine moiety. Rhamnosyl acetate **310** was converted into rhamnosyl chloride **311** to get α-isothiocynate **312**, by treating compound **311** with potassium thiocyanate (KSCN) using a catalytic amount of tetra-*n*-butylammonium iodide (TBAI) under the refluxing conditions in CH_3CN. This results in 1,2-*trans* (α) stereoselectivity, and to get β-isomer glycosyl acetate **310** was treated with $SnCl_4$ and $TMSN_3$ forming glycosyl azide **313**, which on reduction gives glycosylamine **314** using H_2, Pd/C. It was then processed to produce the glycosyl isothiocyanate 122, β-isomer, using thiophosgene and calcium carbonate. To get rhamnosylated arginine building blocks, **312** and **315** were converted into Pfb-protected thioureas by treating them directly with Pfb-NH_2 **305** and then alkylated with EtI and K_2CO_3 in acetonitrile gives ethyl isothioureas **316** and **317**. The chemicals that are individually guanidinylated utilizing Fmoc-Orn-OAll in the presence of mercury salt results in allyl ester **318** and **319**; allyl deprotection of each compound in $Pd(PPh_3)_4$ and $PhSiH_3$ using THF as a solvent results in desired rhamnosylated arginine building block products **320** and **321** (Scheme 12.44).[87] The Fmoc-strategy SPPS integrates the amino acid rhamnosylated

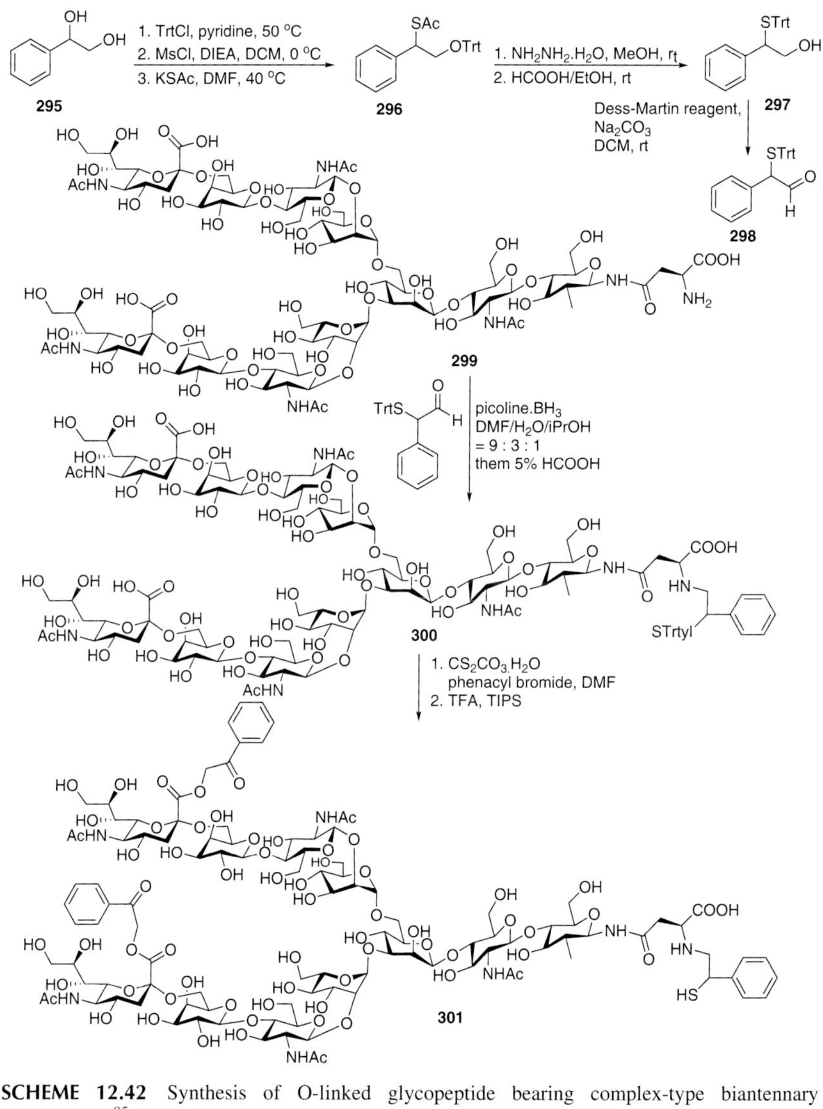

SCHEME 12.42 Synthesis of O-linked glycopeptide bearing complex-type biantennary sialylglycans.[85]

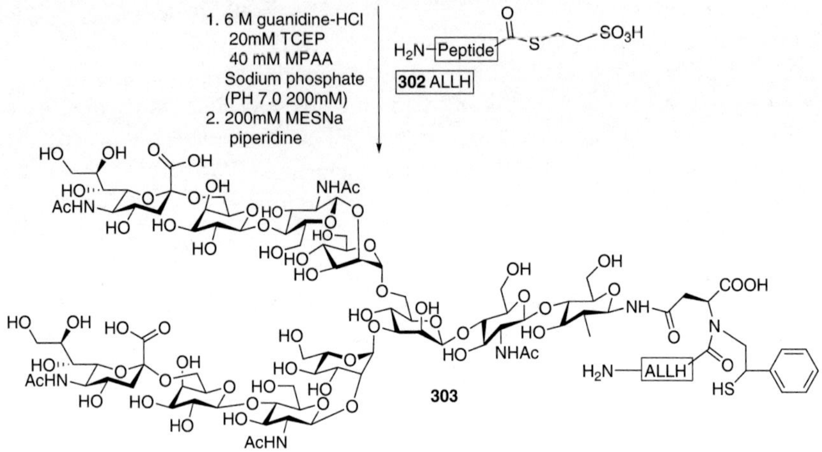

SCHEME 12.42 Continued

arginine **320** and **321**[87] or the arginine building block **309**[86] directly into glycopeptides called "cassettes."

By synthesizing glycopeptides with one or more arginine N-GlcNAcylated moieties, Pan et al. demonstrated how to create antibodies that can discriminate arginine N-GlcNAcylated peptides or proteins in a way that is independent of sequence. N-Glycosyl-S-alkyl-isothiourea **323** was synthesized using reactant 2-(acetylamino)-2-deoxy-D-glucose **322** by converting it firstly into glycosyl chloride **155**, which on treatment with KSCN and TBAI in anhydrous CH_3CN got converted into glycosyl isothiocyanate **304**. It was further treated with NH_3 in tetrahydrofuran and in the next step, treatment with EtI and Boc-anhydride gives **323**.

To synthesize a linear peptide on a solid support using the Fmoc-SPPS method to obtain the desired N-linked glycopeptide, 2-chlorotrityl resin was utilized. On resin, Fmoc-Orn(alloc)-OH was initially employed to obtain a complete peptide after removing the Alloc group using $Pd(PPh3)_4$ resulting in compound **324**. It was then treated with compound **323** in the presence of $AgNO_3$ to afford the glycopeptide **325** on resin. To remove the acetyl group on GlcNAc, it was treated with hydrazine in dimethylformamide, and further on reaction with 5% triisopropylsilane in trifluoroacetic acid results in desired glycopeptide **326** (Scheme 12.45).[88]

Jiang et al. use the silver-promoted solid-phase glycosylation (SPG) process to synthesize arginine-based N-linked glycopeptide. This process has two advantages: (1) synthesis of gram-scale S-alkyl-isothiourea glycosyl donors facilitates the subsequent SSG procedure and (2) the simultaneous introduction of both side-chain sugar motif and arginine residue. Key S-alkyl-isothiourea glycosyl donors were employed as a starting material to

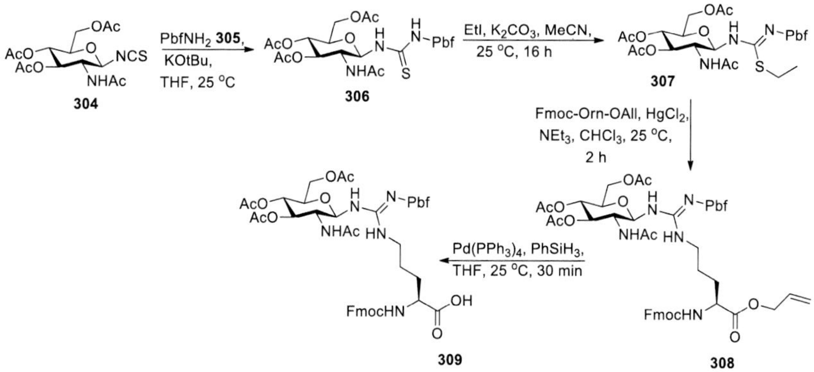

SCHEME 12.43 Synthetic production of the basic building blocks of GlcNAcylated arginine.[86]

synthesize the required N-linked glycopeptide, and several sugar moieties were used in this crucial phase. D-Glucose **327** is used as a starting material to synthesize key S-alkyl-isothiourea glycosyl donors (Scheme 12.46). Compound **327** was treated with acetyl chloride to form glycosyl chloride **328**, which on treatment with potassium thiocyanate and tetrabutylammonium iodide under refluxing conditions using acetonitrile as a solvent resulted in glycosyl isothiocyanate **329**. It was converted into the desired N-glycosyl-S-ethyl-isothiourea **330** by treating later firstly with Pbf-NH_2 **305** using potassium *tert*-butoxide as a base and THF as a solvent and then with ethyl iodide using K_2CO_3 as a base and acetonitrile as a solvent.

2-Chlorotrityl resin as a solid support was used as a solid support and Fmoc-Orn (Dde)-OH as a precursor using the standard Fmoc-based SPPS to synthesize the assembled linear peptide **331**, which on treatment with 2% hydrazine in DMF gives peptide **332** by selectively removing Dde protective group from resin-bound peptide **331**. Coupling of this peptide with N-glycosyl-S-ethyl-isothiourea **330** on resin by $AgNO_3$ in DMF as a solvent and triethyl amine as a base gives resin-bound glycopeptide **333**. Deacetylation of sugar moiety and removal of resin from the glycopeptide **333** results in the synthesis of desired N-linked glycopeptide **334** (Scheme 12.46).[89]

12.5.3 Click chemistry glycosylation

Two molecular building blocks can be bound together resulting in the formation of desired product by using click chemistry. This chemistry is very important as it results in removable side products and stereospecific also provides a very good yield of the desired product in simple reaction conditions. This reaction is distinctive since it occasionally yields no by-products. Azide-alkyne cycloaddition is a process in click chemistry that is catalyzed by Cu(I) (CuAAC). Click chemistry compares a variety of chemical reactions, including the Diels–Alder, Michael addition, strain-promoted azide-

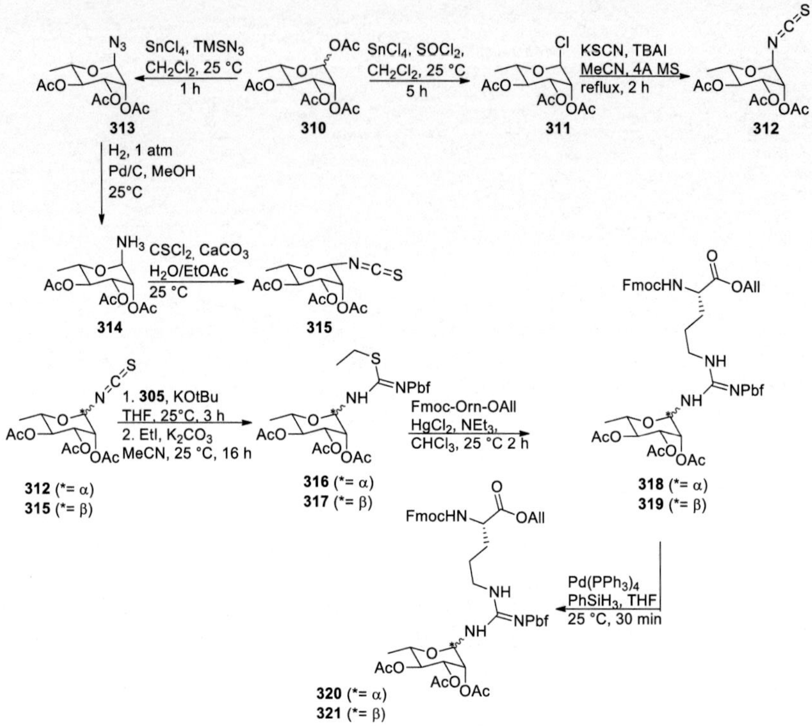

SCHEME 12.44 Synthetic route to synthesize N-linked rhamnosylated arginine building block.[87]

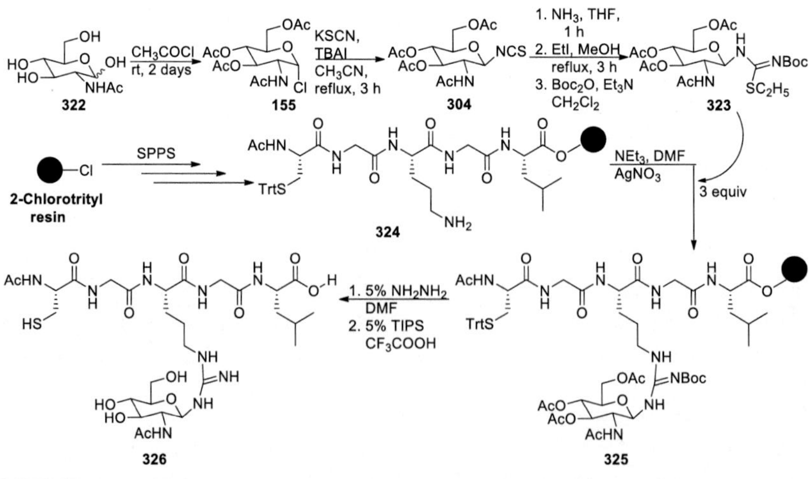

SCHEME 12.45 Route to synthesize N-linked glycopeptide comprising GlcNAcylated arginine unit.[88]

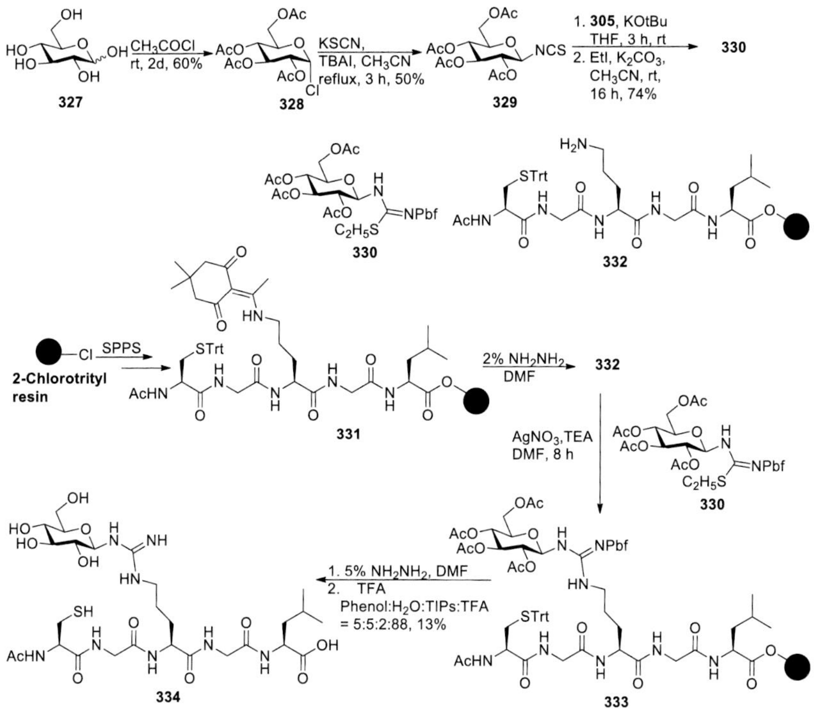

SCHEME 12.46 Synthesis of arginine-based N-linked glycopeptide using silver-promoted solid-phase glycosylation.[89]

alkyne cycloaddition (SPAAC), oxyme, thiolene, and pyridyl sulfide reaction.[90] Synthesis of glycopeptide containing triazole ring via click chemistry possesses biological activities along with structural diversity, which is very beneficial for medicinal chemistry.[91,92] The solubility and conformation of N-linked glycopeptides are determined by the sugar moiety linked to them. The polarity and capacity to establish the H-bond of the triazole ring produced via click chemistry during the synthesis of N-linked glycopeptide and the amide bond are identical. This technique can also be used to synthesize glycopeptide building blocks, which can then be used to make bigger or natural peptides.[93] Marchiori et al. synthesized 1,2,3-triazole-linked glycoconjugate by using Cu(I) catalyzed reaction, that is, click chemistry reaction. Many compounds were formed by the reaction of amino acid-containing azide group (N_3-ThrOBn, N_3-PheOBn, N_3-N-Boc-TrpOBn, N_3-N-Boc-LysOBn, etc.) with corresponding sugar-containing alkyne group such as 3-*O*-propynyl-GalOMe using the click chemistry. The distinguishing characteristic of the aforementioned glycoconjugates is their strong propensity for galectin-3 binding. Galectin-3, which promotes the tumor also helps the tumor to escape from the immune surveillance. Additionally, it is to blame for melanoma and glioblastomas, pancreatic and hepatocellular carcinomas, breast, gastric, lung, and colorectal cancers as well as their development and malignancy.[94] The triazole ring, which is generated by combining azide and alkyne sugar derivatives, is an isostere of the amide bond. Alternative methods for enantioselective triazole ring synthesis include nickel-catalyzed alkyne—azide cycloaddition (NiAAC).[95,96] we can use either sugar of our choice containing azide group with corresponding acetylenic peptide building block or acetylenic sugar with azide containing peptides to giveα- and β-triazole-linked glycopeptides via Cu(I)-[catalyzed 3 + 2] cycloaddition reaction. Kuijpers et al. synthesized triazole-linked glycopeptide via click chemistry. Initially, (R)-N-Boc-propargylglycine methyl ester **335** was coupled with acetyl protected azide group containing derivatives of glucose and galactose, that is, azidoglucoside **237** and azidogalactoside **336** in the presence of Cu(I) salts and bases like NEt_3 and DIPEA to give the product **337** and **338**, respectively. The production of a triazole-linked glycopeptide is achieved by employing the same procedure to couple the acetylenic group located at the anomeric carbon of 2,3,4,6-tetra-*O*-benzyl-D-glucose with the congruent azide group containing peptide (Scheme 12.47).[97]

Another technique involves mixing reducing sugar and peptide in a single pot via click chemistry, with the anomeric hydroxy group being activated by the reagent 2-azido-1,3-dimethylimidazolinium hexafluorophosphate (ADMP). It is also employed as an azide source by using 3 equiv. of ADMP. To synthesize the desired glycopeptide, 3 equiv. of ADMP and NEt_3 was added to the mixture containing D_2O and CH_3CN resulting in glycosyl azide, which further react with propargylic-containing peptide under Cu catalysis resulting in 1,2,3-triazole-linked glycopeptide.[98] In another approach, triazole-linked ^{18}F-glycopeptides were synthesized as imaging agents via click chemistry.[99]

Kim et al. describe a one-pot method to synthesize the desired 1,2,3-triazole-linked glycopeptide via click chemistry directly from reducing sugar in aqueous conditions. ADMP and Cu-catalyzed cycloaddition reaction results in the stereoselective and convergent synthesis of triazole-linked glycopeptide in high yield.[100]

Marchiori et al. also demonstrated the method to synthesize the triazole-linked glycopeptide by using appropriate sugar and peptide using CuAAC reaction. Triazole derivative αGlcNAc-1-*O*-triazol-2Manα-ThrOH **351** and Gal-β1,4-αGlcNAc-1-*O*-triazol-2Manα-ThrOH **352** were synthesized utilizing the SPPS method to produce NHAcThrValαGlcNAc-1-triazol-2Manα] ThrIleArgGlyOH **353** and NHAcThrValGal-β1,4-αGlcNAc-1-triazol-2Manα]ThrIleArgGlyOH **354**, respectively. Newly formed **353** and **354** are analogs of α-dystroglycan, essential components for both structural and functional stability, relaxation, and contraction of muscle fiber. To synthesize these molecules, alkyne and azide derivatives were formed, and coupled with

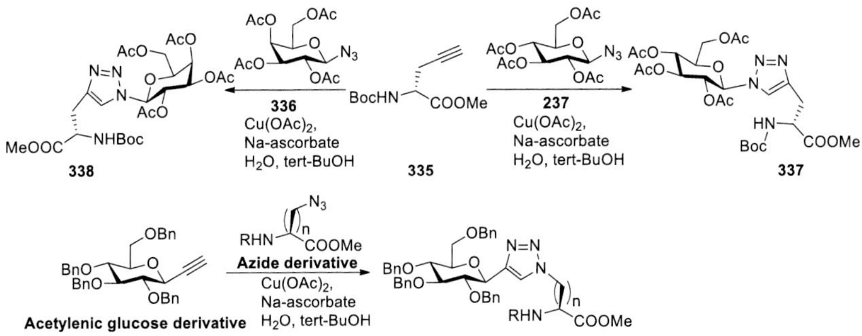

SCHEME 12.47 Synthesis of triazole containing N-linked and C-linked glycopeptides using Cu(I) catalyzed reaction.[97]

each other, they result in glycosyl amino acid, which on incorporation into peptide using SPPS strategy gives the desired glycopeptide.[101]

In case (i), using anhydrous CH_2Cl_2 as a solvent and $BF_3 \cdot OEt_2$ as a catalyst, per-*O*-acetyl-2-azido-2-deoxy-mannosyl acetate **339** was treated with FmocThrOBn **340**, resulting in the azide derivative $ManN_3$-FmocThrOBn **341**. (ii) Alkyne derivative 2′-propynyl-αGlcNAc **342** was synthesized by the glycosylation of 1,3,4,6-tetra-O-acetyl-2-azido-2-deoxy-D-glucopyranose **343** in anhydrous solvent condition using catalyst $BF_3 \cdot OEt_2$ with *p*-thiocresol in $CH_2Cl_2GlcN_3STol$ **344**, which on further treatment with prop-2-yn-1-ol in the presence of anhydrous CH_2Cl_2 using catalytic amount of NIS/TMSOTf affords 2′-propynyl-GlcN3 **345**, and azide group when reduced and acetylated using Zn powder in THF and Ac_2O/AcOH for 1 hour gives **342**. (iii) Another alkyne sugar derivative 2′-propynyl-Gal-β1,4-αGlcNAc **346** was synthesized. In the first step, the anomeric nitro group present in lactose azido-nitrate **347** was replaced with the acetate group by treating it with NaOAc/AcOH followed by treatment with *p*-thiocresol in anhydrous CH_2Cl_2 and catalyst $BF_3 \cdot OEt_2$ gives an anomeric mixture $LacN_3STol$ **348** as a glycosyl donor. This mixture on treatment with propargyl alcohol with catalyst NIS/TMSOTf in anhydrous solvent CH_2Cl_2 results in 2′-propynyl-$LacN_3$, which on treatment with Zn and copper sulfate in THF and Ac_2O/AcOH gives **346**. After the synthesis of these alkyne and azide derivatives, CuAAC reaction in DMF and catalyst $CuSO_4$/sodium ascorbate in microwave heating was used to couple these derivatives and they result in products **349** and **350**. Hydrogenation of these compounds in H_2/Pd-C 10% removes the benzyl ester group and gives the desired product **351** and **352** (Scheme 12.48).[101–103]

Angiogenesis is overexpressed in angiogenic blood vessels and in tumor cells induced by tumor hypoxia required for tumor cell survival. Asparagine–glycine–arginine (NGR) peptide derivatives bind selectively to the aminopeptidase N receptor (APN/CD13), which makes them crucial

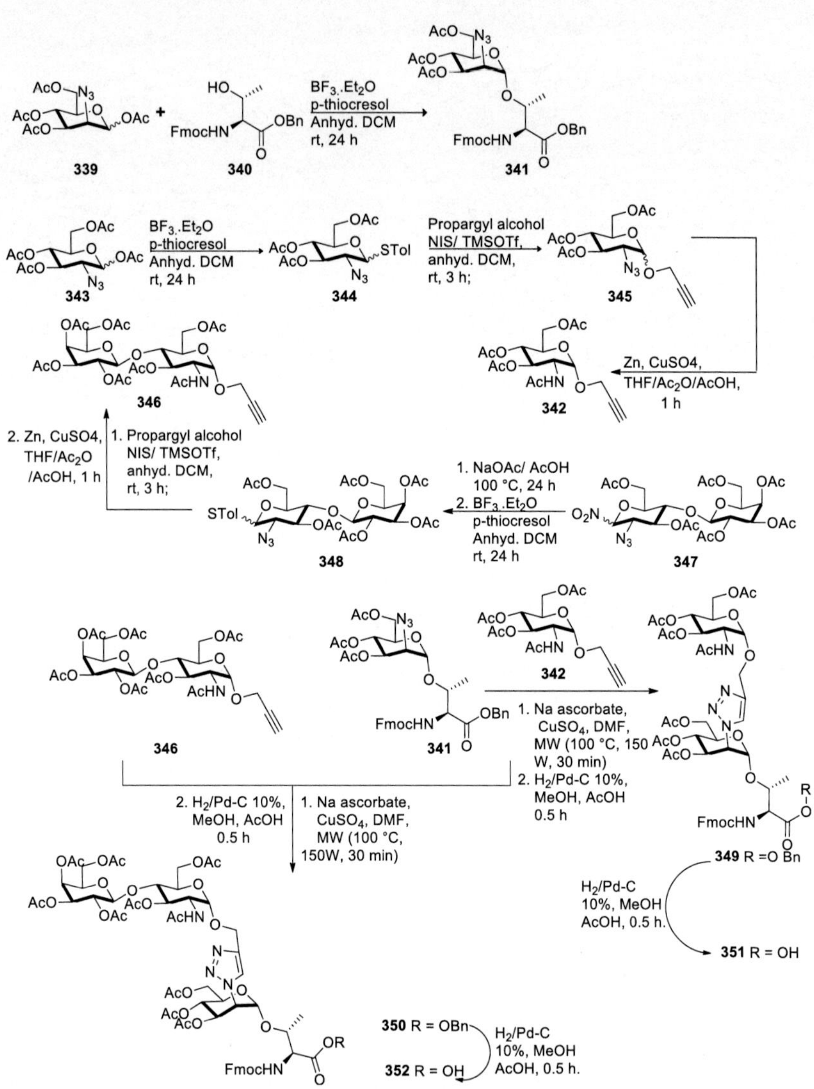

SCHEME 12.48 Synthetic route to synthesize triazole derived αGlcNAc-1-O-triazol-2Manα-ThrOH **351** and Gal-β1,4-αGlcNAc-1-O-triazol-2Manα-ThrOH **352** using Cu(I) catalyzed reaction.[101]

vector molecules in the development of Positron emission tomography radiotracer, which possess the ability to detect APN-rich tumors. Gyuricza et al. reported dual targeting novel 68Ga-NODAGA-LacN-Ec(RGDfK)]$_2$ **364** and ^{68}Ga-labeled cNGR-based N-linked glycopeptides using copper-free click reaction. To synthesize desired novel compound ^{68}Ga-NODAGA-LacN-Ec(RGDfK)]$_2$ **364**, firstly, 3-azido-propyl-β-D-galactopyranosyl-(1→4)-(2-amino-2-deoxy)-β-D-glucopyranoside

360 was prepared by taking 3-azidopropyl 2-phtalimido-2-deoxy-β-D-glucopyranoside **355** as starting material. Treating **355** with *tert*-butyldiphenylsilyl chloride results in the silylation of primary hydroxy groups, which on reacting with 2,3,4,6-tetra-*O*-acetyl-β-D-galactopyranosyl trichloroacetimidate **356** using TMSOTf as a catalyst in CH_2Cl_2 solvent gives disaccharide **357**. Protecting groups were removed by using ethylene diamine in ethanol and trifluoroacetic acid to remove *tert*-butyldiphenylsilyl protecting group to get compound **358**. It was treated with NODAGA-NHS chelator **359** using polar aprotic solvent DMSO and 0.1 M borate buffer (pH 8.4) gives **360**. In the next step, cyclooctyne moiety was introduced in the Ec(RGDfK)]$_2$ peptide **361** by coupling it with DBCO-NHS **362** in dry DMSO using DIPEA as a base to give DBCO-Ec(RGDfK)]$_2$ derivative **363**. In the last step, strain-promoted copper-free click reaction was applied to compounds **360** and **363** in DMSO at room temperature and resulted in **364**. ^{68}Ge/^{68}Ga-generator was used for radiolabeling the synthesized precursors ^{68}Ga-NODAGA-LacN-Ec(RGDfK)]$_2$ **364** (Scheme 12.49).[104] Using a similar approach, ^{68}Ga-labeled cNGR-based N-linked glycopeptides **9** and **10** were also reported by Gyuricza et al. Firstly, chelator-conjugated lactosamine derivative **366** was prepared by coupling 3-azido-propyl-β-D-galactopyranosyl-(1→4)-(2-amino-2-deoxy)-β-D-glucopyranoside **358** with p-NCS-Bn-DOTA-GA **365** in 0.1 M sodium carbonate using dimethyl sulfoxide as a solvent at room temperature. Secondly, cKNGRE **367** was coupled with DBCO-NHS **368** to introduce alkyne moiety to make it suitable for alkyne–azide strain-promoted copper-free click reaction in the presence of DIPEA using DMSO as a solvent to get compound **369**. While treating **367** with DBCO-PEG$_4$-NHS **370** in the presence of DIPEA using DMSO results in **371**. Lastly, compounds **369** and **371** were coupled with **366** using strain-promoted click reaction in DMSO giving desired glycopeptides **372** and **373**, respectively. ^{68}Ge/^{68}Ga-generator was used for radiolabeling the synthesized precursors **372** and **373** producing ^{68}Ga-isotopes with high radiochemical purity (Scheme 12.50).[105]

12.5.4 Hybrid molecule synthesis

Glycopeptides volunteer a highly active class of antibiotics with few of them approved by the FDA. Vancomycin glycopeptide antibiotic is characterized by a tricyclic heptapeptide having a disaccharide consisting of vancosamine and D-glucose.[106] It is widely used to treat bacterial infections including methicillin-resistant *Staphylococcus aureus* (MRSA) and other gram-positive beta-lactam-resistant bacteria. It works by attacking lipid II, which is a specific component for bacteria cell wall synthesis, binds to the C-terminal of pentapeptide and thereby blocks the formation of the cell wall. Being approved by FDA, it soon became the key antibiotic against *S. aureus* with MIC of 0.25–10 μg/mL. However, drug resistance has become a common phenomenon among bacteria, which alarms the need for new antibiotic hybrids that can work efficiently as antibacterial drugs. Many antibiotic

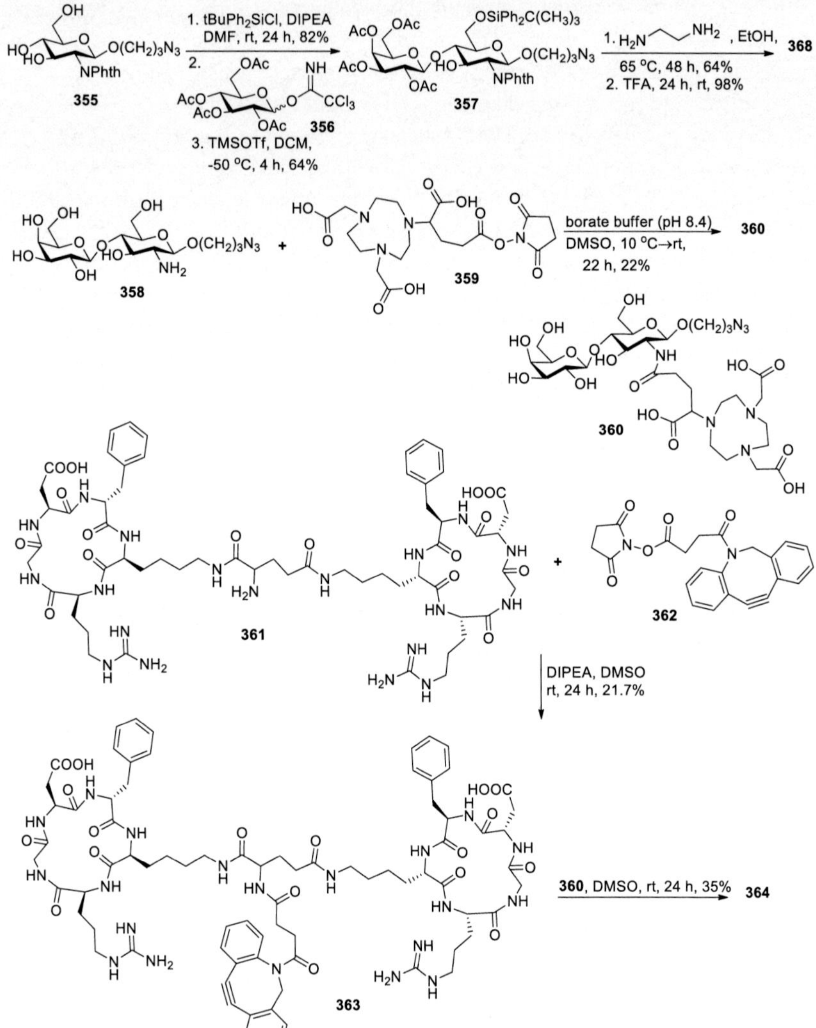

SCHEME 12.49 Route to synthesize ^{68}Ga-labeled ^{68}Ga-NODAGA-LacN-Ec(RGDfK)]$_2$ N-linked glycopeptide.[104]

hybrids have been reported to have many-folded efficiency than their individual counterparts.[107–109] A recently developed hybrid of vancomycin and nisin fragments has shown to be more effective than the individual counterparts. Both the components belong to very different classes of antibiotics; however, both exhibit a common feature by first attacking the lipid II precursor of the cell wall. Vancomycin attacks the C-terminal of the peptide chain and prevents further growth of the cell wall, which results in the death of

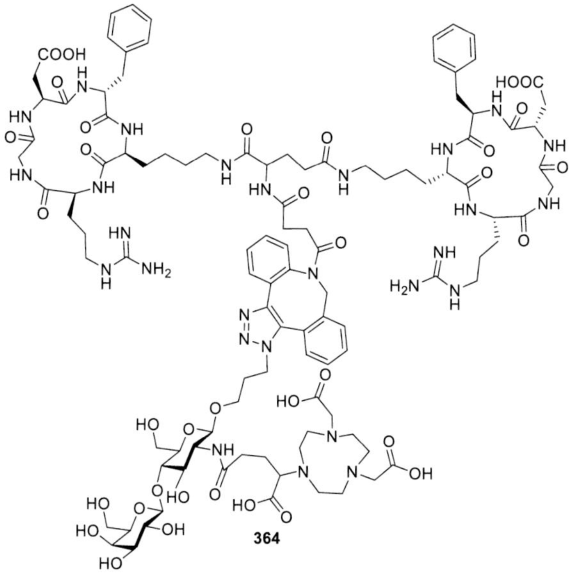
364

SCHEME 12.49 Continued

bacterial cells due to osmotic pressure. In contrast to this, risin targets the pyrophosphate part present on lipid II by a pore-forming mechanism through its N-terminal portion, that is, nisin (1–12), and subsequently forms other complexes with lipid II through its C-terminal and forms a lethal pore. However, both these antibiotics have suffered from bacterial resistance; thus, a combination of these two has been proven to be effective against bacterial infection by their dual mode of action. The simultaneous dual weak interactions of these two with different sites on lipid II allow them to form a strong affinity for lipid II, thus restoring their antibacterial mode of action. The different connection sites and spacer lengths were optimized through computational modeling to bind lipid II efficiently at a suitable distance. The best possible mode chosen was the linkage between the C-terminus of nisin and the N- or C-terminus of vancomycin. The vancomycin-risin conjugate was synthesized via click chemistry by outfitting each component with either alkyne or azide moiety. Standard solution-phase peptide synthesis was used to create the C-terminally functionalized alkyne derivatives **374** and **375** and the N-terminally functionalized vancomycin **376** using PyBroP. The azido-modified nisin (1–12) fragment linked to C-terminus compound **377** was prepared using trypsin digestion followed by treatment with 3-azido-1-propylamine, BOP, and DIPEA in DMF. The subsequent mixing of suitable vancomycin and risin derivative in H_2O containing DMF traces, further subjected to microwave radiations in the presence of $CuSO_4$ resulted in

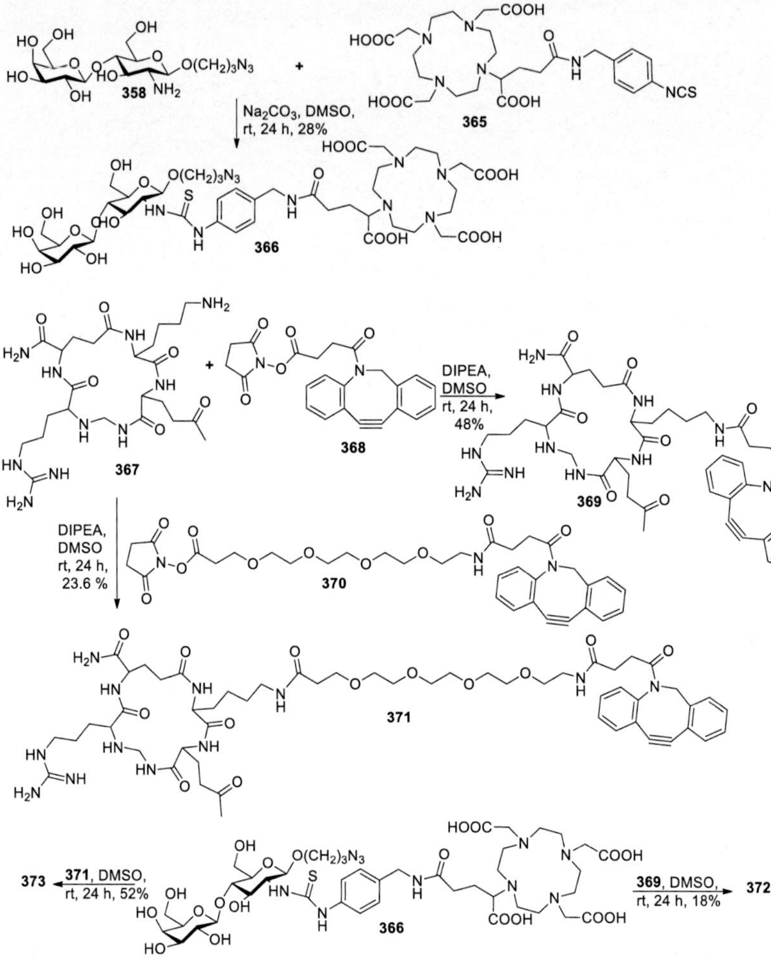

SCHEME 12.50 Route to synthesize ^{68}Ga-labeled cNGR-based N-linked glycopeptide.[105]

the formation of vancomycin-risin conjugate. The hybrid antibiotic **378** was found to be 40 times more effective against vancomycin-resistant enterococci (VRE) bacteria than individual components due to the dual mode of action (Scheme 12.51).[110] This kind of multivalency where a combination of multiple weak interactions leads to strong affinity has also been investigated to develop a vancomycin/cephalosporin hybrid. This hybrid antibiotic blocks the synthesis pathway of the bacterial cell wall and displays high antibacterial activity compared to individual components against gram-positive pathogen MRSA. Cephalosporin, being less complex than other β-lactams was chosen to bind through an amide linkage with vancomycin synthon having less hindered ethyl primary amino unit. Further, the same group optimized

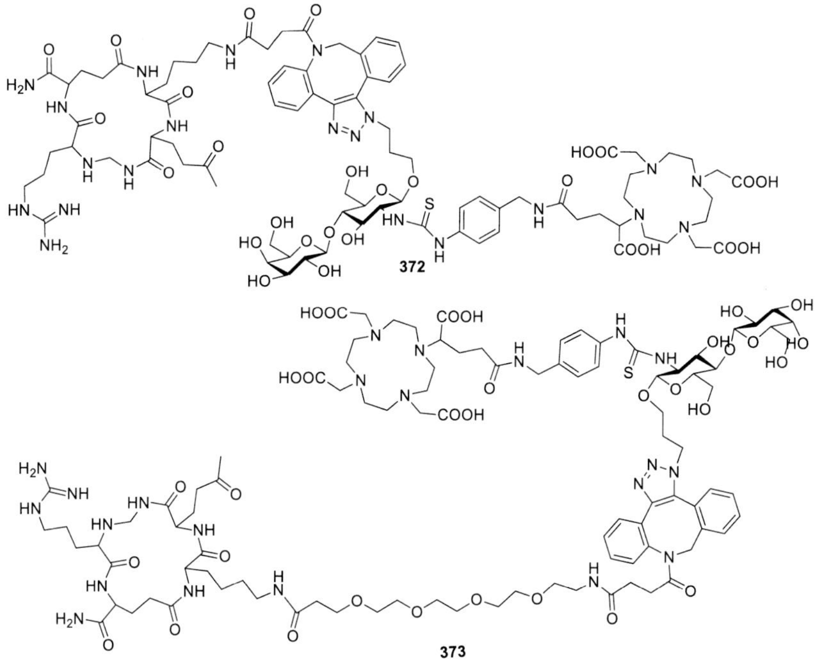

SCHEME 12.50 Continued

the various attachment points for this glycopeptide/β-lactam hybrid and selected compound 35 as the potential clinical candidate, TD-1792, which exhibits >40 fold in vivo efficacy of vancomycin.[111] TD-1792 has entered phase-II clinical trials for the treatment of gram-positive induced skin infections.[112]

Another efficient antibiotic was reported by Tevyashova et al. in which glycopeptide-azithromycin hybrid was synthesized by attachment of glycopeptide to 4″position of cladinose moiety of azithromycin.[106] Azithromycin is a second-generation broad-spectrum antibacterial macrolide, which is readily absorbed by body cells.[106,113] Firstly, the hydroxyl group in azithromycin **380** was acetylated by using acetic anhydride in trimethylamine to yield 2-O-acetylazithromycin **383**, which further reacts with 1,1′-carbonyldiimidazole (CDI) to form 11,12-carbonate 4″-O-acylimidazolide azithromycin **384**. This activated azithromycin combines with vancomycin **379** to give a mixture of products that are difficult to separate. Therefore **384** is treated with ethylenediamine or 1,3-diaminopropane in DBU to give corresponding 4″-O-(aminoalkyl) carbamates **385** or **386**. Then the **385** or **386** was then attached to three glycopeptides namely, vancomycin **379**, eremomycin **381**, and teicoplanin aglycon **382** in the presence of benzotriazol-1-yl-oxytripyrrolidinophosphonium hexafluorophosphate (PyBOP).

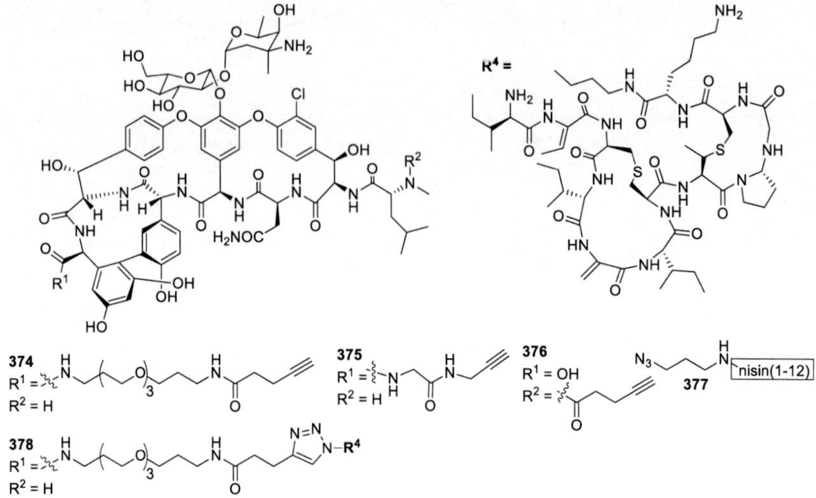

SCHEME 12.51 Structures of vancomycin alkyne hybrids and vancomycin-nisin (1–12) hybrid.[110]

2′-O-acylated derivatives **387–391** were used for early purification and the 2′-O-acetyl group in compounds **387–391** was removed in MeOH to obtain **392–396** hybrid antibiotics (Scheme 12.52).[106] All the synthesized hybrids showed equal or superior activity against gram-positive bacteria when compared to individual components. The high and moderate activity against *Streptococcus pneumoniae* and *Enterococcus faecium* was reported by all five and two hybrids, respectively.[106] Further, efforts have also been made to develop semisynthetic glycopeptides to improve antibacterial activity and reduce the side reactions by amidation of vancomycin and eremomycin with pyrrolidine and piperidine.[114] Thus all these hybrids have proven to be more effective than individual glycopeptides against a variety of bacteria.

Zong et al. developed a novel chemoenzymatic technique for the manufacture of N-linked HIV-1 V1V2 glycopeptides. First, the $Man_5GlcNAc$ oxazoline **399** and GlcNAc-peptide **397** were joined at the N160 location in the presence of the mutant enzyme EndoM-N175Q, resulting in a novel glycopeptide **400** that contains a Man_5 glycan unit. When it is treated with 2.5% aqueous NH_2NH_2, acetylated GlcNAc at the N173 site is no longer protected, producing compound **401**. It is interesting to note that compound **403** was created when SCT-oxazoline **402** and **401** were combined with the enzyme EndoM-N175Q, which also contains sialylated complex-type (SCT) N-sugar. While the enzyme EndoM-N175Q couples, the molecules **399** and **401** to produce the glycosylated product **404**, which includes two Man_5 glycan units. Another approach for its synthesis is to couple GlcNAc-peptide **398** with an excess of **399** using the same enzyme to get the desired product **404** in admirable yield (Scheme 12.53).[115]

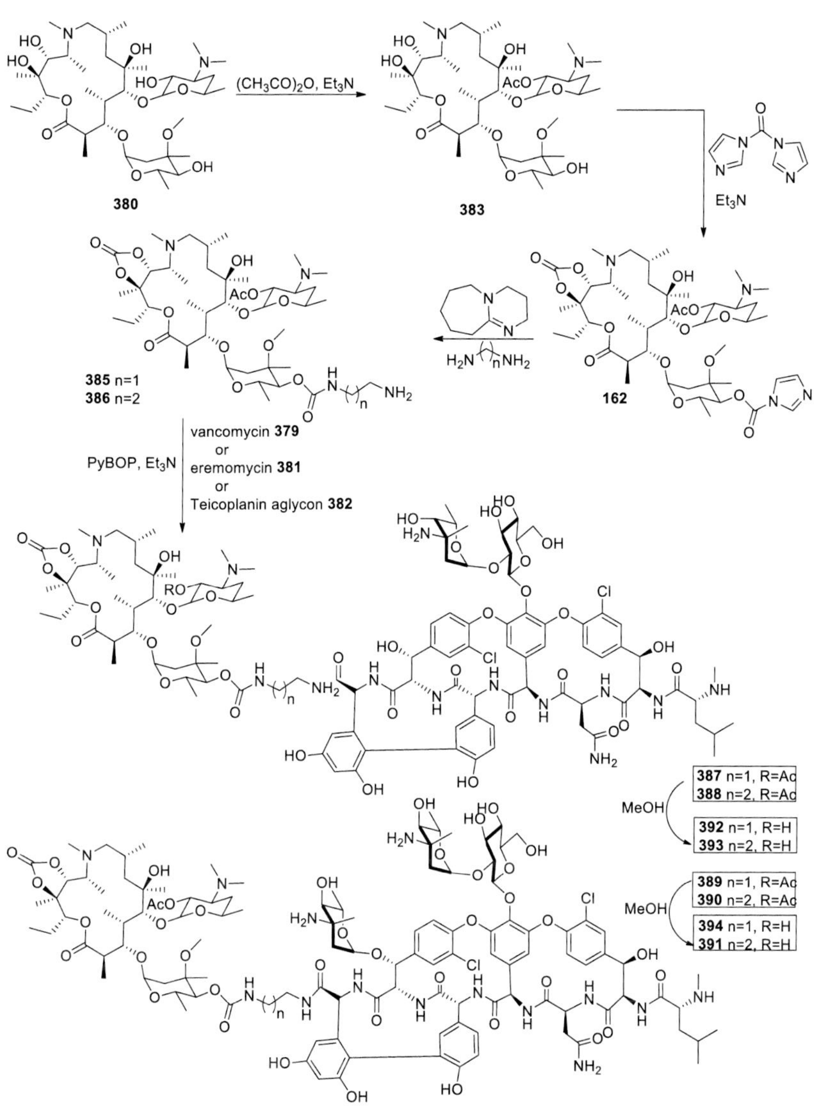

SCHEME 12.52 Synthesis of azithromycin-derived glycopeptides.[106]

A selenoester-mediated synthesis of N-linked glycopeptide has been reported to facilitate rapid amide bond ligation, which is highly favorable for free carboxylic acid and hydroxyl groups of peptides and carbohydrates. The solid-phase synthesis of peptidyl selenoester proceeded via the addition of Fmoc-AA-OH and DIPEA in DMF to pretreated resin for peptide assembly. Then peptidyl selenoester was prepared using diphenyl diselenide (DPDS)

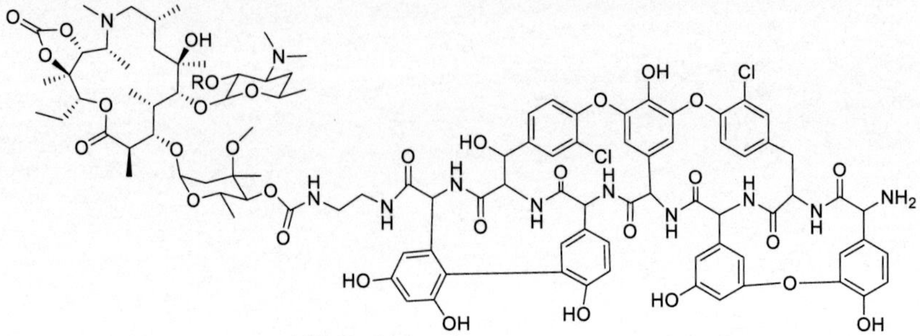

MeOH: **391** R = Ac → **396** R = H

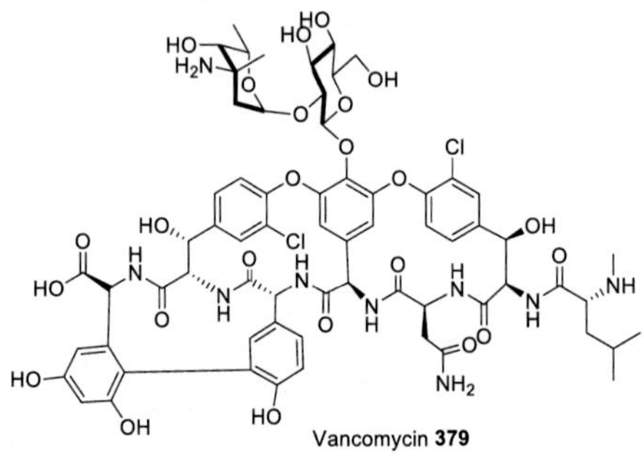

Vancomycin **379**

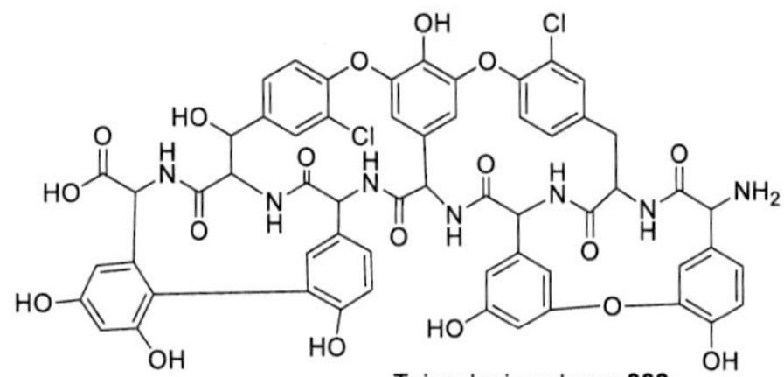

Teicoplanin aglycon **382**

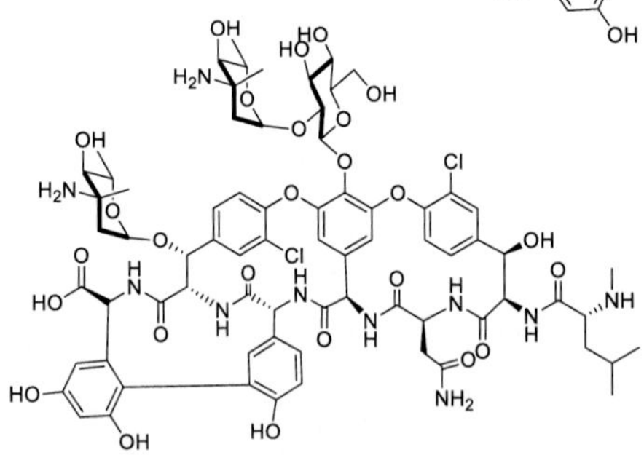

Eremomycin **381**

SCHEME 12.52 Continued

and tributylphosphine, which was employed for the synthesis of N-linked glycopeptide with monosaccharide, chitobiose, and undecasaccharide.[116]

Glycosylation of peptides has proven to be an effective approach toward increased stability in vivo and augmenting blood–brain barrier permeability.

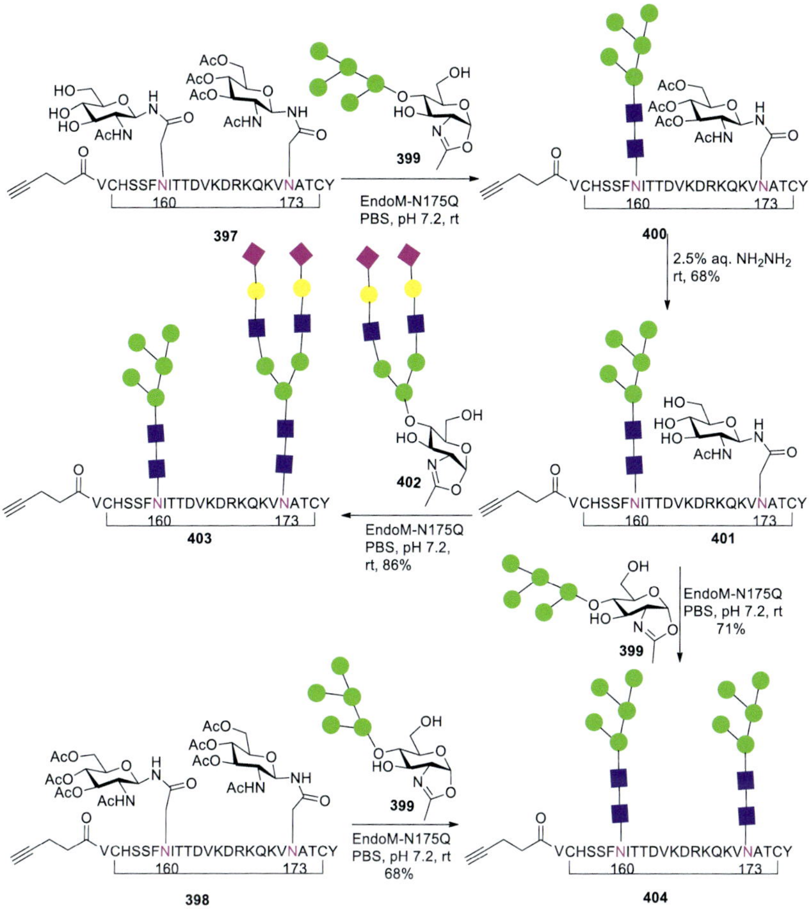

SCHEME 12.53 Synthetic route for the chemoenzymatic synthesis of HIV-1 V1V2 glycopeptides.[115]

Apostol et al. glycosylated pituitary adenylate cyclase activating peptide (PACAP), which is a potential contender as an antineurodegenerative drug. PACAP peptide sequence was prepared using dipeptide building blocks to circumvent the difficulty of aspartimide formation and peptide–peptide aggregation on resin. Next, the glycopeptide was synthesized by loading Fmoc-protected glycosyl amino acid, Fmoc-Ser(O*t*Bu)-OH along with Cl-HOBt and DIC in *N*-methyl-2-pyrrolidone (NMP), further treating with DIPEA/Ac_2O in DCM as capping agents and coupling all the residues using suitable coupling agents depending upon the nature of amino acid. Lastly, $NH_2NH_2 \cdot H_2O$ in NMP was used to facilitate the deprotection of carbohydrate moieties.[117]

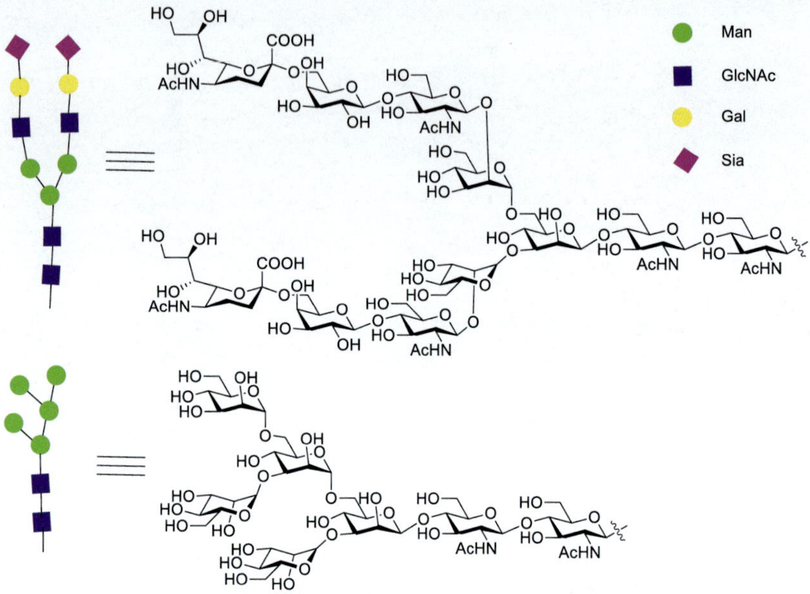

SCHEME 12.53 Continued

Ma et al. demonstrate the synthesis of novel vancomycin analogs containing the polar moieties and lipophilic substitution on seventh amino acid phenyl ring and vancosamine moiety, respectively. Introduction of a lipophilic side chain on the nitrogen of sugar moiety in vancomycin takes place using the method of activated N-hydroxysuccinimide ester (NHS-ester). Initially, NHS-esters **407a–f** were synthesized and in the last step coupling with vancomycin as a starting material takes place, which results in the synthesis of desired analogs. Carboxylic acid precursor **405** reacts with N-hydroxysuccinimide **406** in the presence of DDC using THF as a solvent to afford NHS-esters **407a–f**, which on treatment with vancomycin in the presence of triethyl amine using DMSO/H_2O (v: v = 4: 1) as a solvent gives vancomycin intermediate **408** followed by treatment with (aminomethyl) phosphonic acid or phenylethylamine by Mannich reaction results in desired glycopeptides **409a–k** (Scheme 12.54).[118]

Bereczki et al. reported that the first dimeric derivatives of the glycopeptide antibiotic teicoplanin **410** as its covalent dimers are unknown due to its polar tag. Partial hydrolysis of teicoplanin **410** leads to the removal of N-acetylated D-glucosamine moieties resulting in a single aglycone derivative **411**. Firstly, linker reagent diisothiocyanate **416** and diethylaminopropyl amide **417** derived from teicoplanin **410** were prepared and coupled with aglycone derivative **411** to get desired dimers. Linker reagent **416** was synthesized using *n*-decyclamine **412** by treating it with chloro derivative of triethylene glycol **413** in the presence of potassium iodide using DMF as a solvent at 110°C gives **414** containing two hydroxyl groups that were transformed to primary amino groups by tosylation of free hydroxyl groups followed by the nucleophilic substitution with azide and Staudinger reduction to give compound **415**. Amino groups present in **415** were transformed to isothiocyanate resulting in linker reagent **416**. Teicoplanin **410** was converted to its diethylaminopropyl amide derivative **417**. Coupling of **411** or **417** with linker reagent **416** using triethyl amine and DMF at 25°C gives

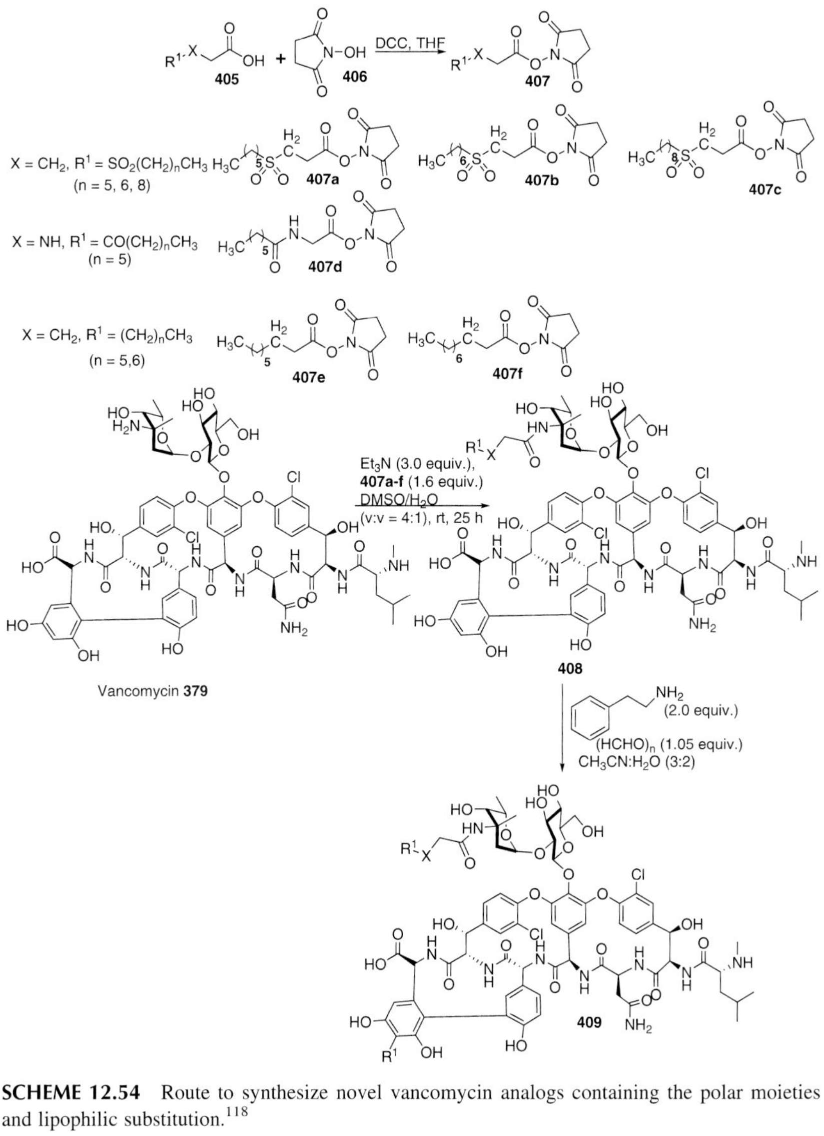

SCHEME 12.54 Route to synthesize novel vancomycin analogs containing the polar moieties and lipophilic substitution.[118]

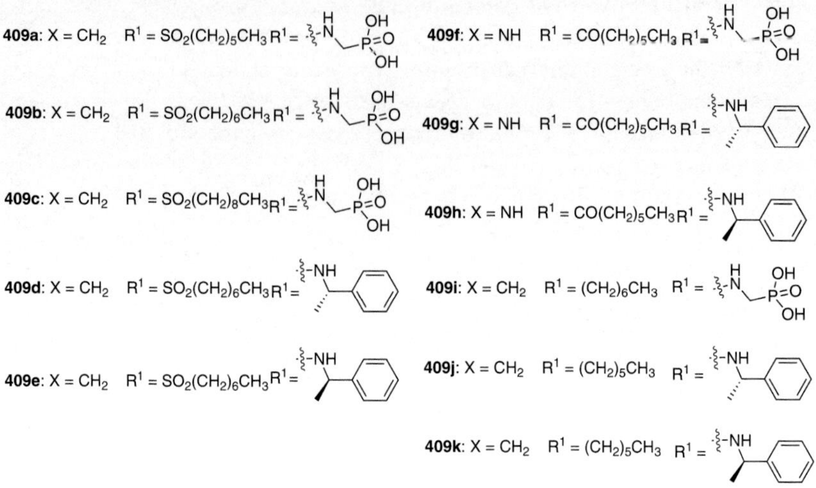

SCHEME 12.54 Continued

monoisothiocyanates **418** and **419**, which on further treatment with **411** or **417** gives desired dimer **420** and **421** of teicoplanin pseudoaglycone **411** with *N–N* orientation (Scheme 12.55).[119]

Moore et al. reported the use of two glycosyltransferases GftE and GftD for enzymatic glycosylation of [ψ [C(=S)NH]Tpg^4] vancomycin aglycon and vancomycin aglycon itself in single-pot, two-step reaction. Firstly, uridine diphosphate (UDP)-vancomycin glycosyl **426** donor was prepared from vancomycin **379** by treating later with imidazole-1-sulfonyl azide using a catalytic amount of $ZnCl_2$ followed by acid-promoted methanolysis to give methyl glycoside **422** with free hydroxyl group at C-4, which on treating with BzCl in DMAP and 80% CH_3COOH/H_2O afforded benzyl-protected compound **423**, which acts as a UV-chromophore. Stable intermediate **424** was formed by treating compound **423** with oxalyl chloride in DMF followed by reacting with bis-tetrabutylammonium salt of UDP (Bu_4N-UDP). Benzoate hydrolysis of **424** gives azide **425**, which on neutralizing with tris-(2-carboxyethyl)phosphine (TCEP · HCl) gives UDP-vancomycin glycosyl **426**. In the next step, [ψ [C(=S)NH]Tpg^4] vancomycin aglycon **428** was synthesized by enzymatic glycosylation of residue 4 thioamide analog **427** of vancomycin using GtfD and GtfE as a catalyst, UDP-glucose, and UDP-vancosamine in the presence of alkaline phosphatase while, using the same reaction conditions, enzymatic glycosylation of vancomycin aglycon **429** was carried out to get vancomycin pseudoaglycon **430** that on further treatment with enzyme GtfD and azide **425** that got in situ reduction to compound **426** results in vancomycin **379** (Scheme 12.56).[120]

12.6 Recent developments for the synthesis of glycohybrids

Glycohybrids are the molecules where a carbohydrate part is linked, merged, or fused with privileged drug-like molecules. This chapter includes some recent developments on the synthesis of such molecules and their associated bioactivity, and these molecules are grouped here based on their bioactivity. The low cost and

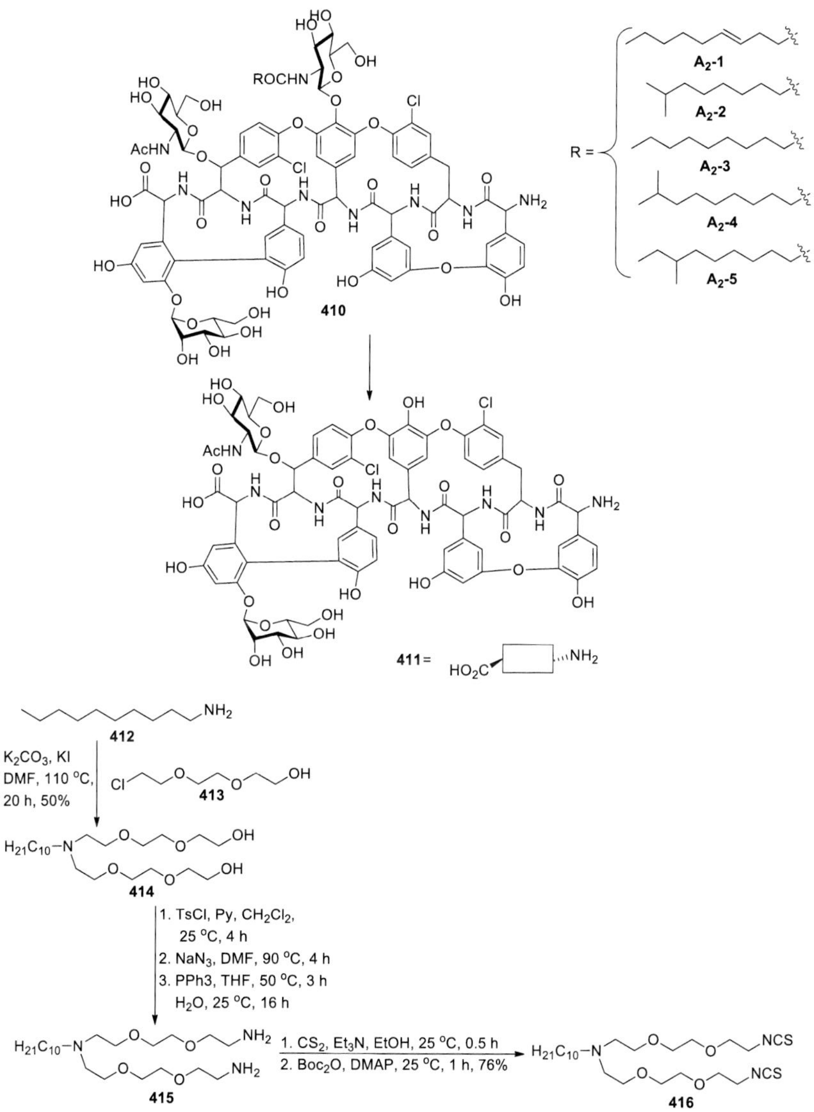

SCHEME 12.55 Synthetic route to synthesize dimeric derivatives of the glycopeptide antibiotic teicoplanin.[119]

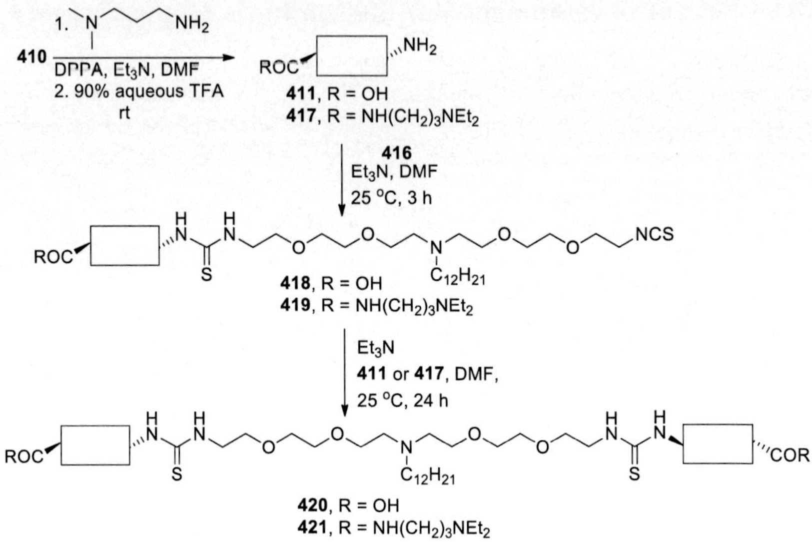

SCHEME 12.55 Continued

simplicity of handling sugar molecules, make them ideal for combining with non-sugar compounds (privileged molecules) and synthesizing natural products or molecules that resemble natural products. These molecules are known as carbohybrids or glycohybrids and are used in the creation of pharmaceuticals due to their biological activity, high diversity, improved pharmacokinetic and pharmacodynamic properties, molecular recognition, intracellular functions, enhanced bioavailability, high water solubility, and low toxicity. Poor bioavailability of carbohydrate molecules due to the presence of free polar −OH groups at many carbon centers, which can be changed by converting those into hydrophobic acyl groups, which may get cleaved through the enzymatic hydrolysis in the blood of human's body.[121] Glycohybrids structures are formed by, with the help of linkers that can be heterocyclic ring or hydrocarbon chain, connecting carbohydrate moiety with biologically relevant molecules without changing much in their initial structure.[122] Their unique property of showing biological activities such as anticancer, antiviral, antimalarial, antifungal, antibacterial, antiinflammatory, immunomodulatory, carbonic anhydrase inhibitors, glycosidase inhibitors, and PTP1B inhibitors made the researchers interest more toward the synthesis of such glycohybrid molecules for their chemical and biological purpose.

12.6.1 Anticancer glycohybrids

Khattab et al. reported the synthesis of a hybrid of thienopyrimidine conjugate linked to different sugar moieties by 1,2,3-triazole core via the CuACC method. By using the 3-(4,5-dimethylthiazol-2-yl)-2,5-diphenyl-2*H*-

tetrazolium bromide (MTT) assay, the glycohybrid produced is cytotoxic to the cell lines MCF-7 and HCT-116. The advantage of glycohybrids with triazole cores over doxorubicin and other well-known compounds is that they have enhanced action against HCT-116. The sugar moiety presented in freshly synthesized thienopyrimidine conjugate hybrids is acetylated, which results in epidermal growth factor receptor (EGFR) inhibitory action, as these glycohybrids were successfully tested against mutant EGFR kinase-type L858R.[123]

Desired glycohybrid derivatives were formed by the reaction of 3-amino-2-thioxothieno[2,3-d]pyrimidin-4-one derivative **431** with 3-bromoprop-1-yne in the presence of weak base K_2CO_3, which results in acetylenic functionalized thienopyrimidine **432**, which couples with different glycosyl azide (2,3,4,6-tetra-*O*-acetyl-β-D-glucopyranosyl azide and 2,3,4-tri-*O*-acetyl-β-D-xylopyranosyl azide) resulting in corresponding glycohybrids **433** and **434** in 65%–60% yield via using click chemistry. Deacetylation of these compounds in the presence of NH_3/MeOH gives the compounds **435** and **436**. These compounds show cytotoxicity to the cell lines HCT-116 and MCF-7,

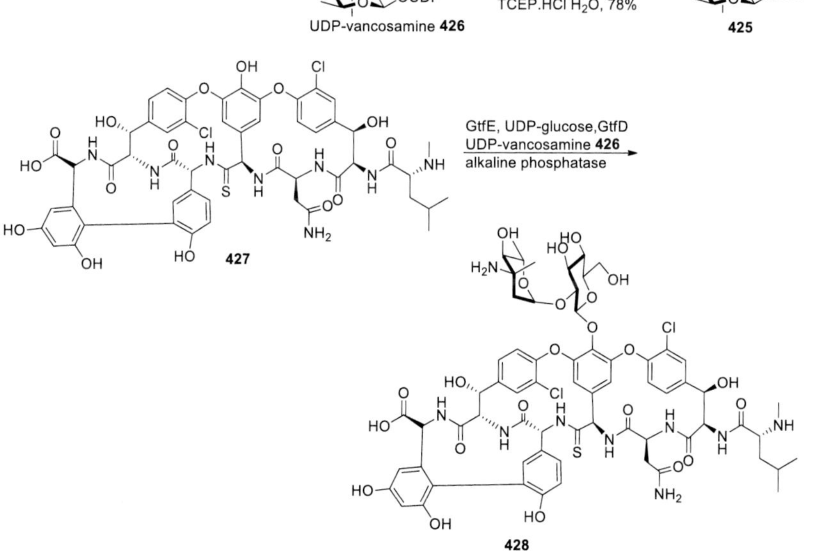

SCHEME 12.56 Route for the enzymatic glycosylation of [ψ [C(=S)NH]Tpg[4]] vancomycin aglycon and vancomycin aglycon.[120]

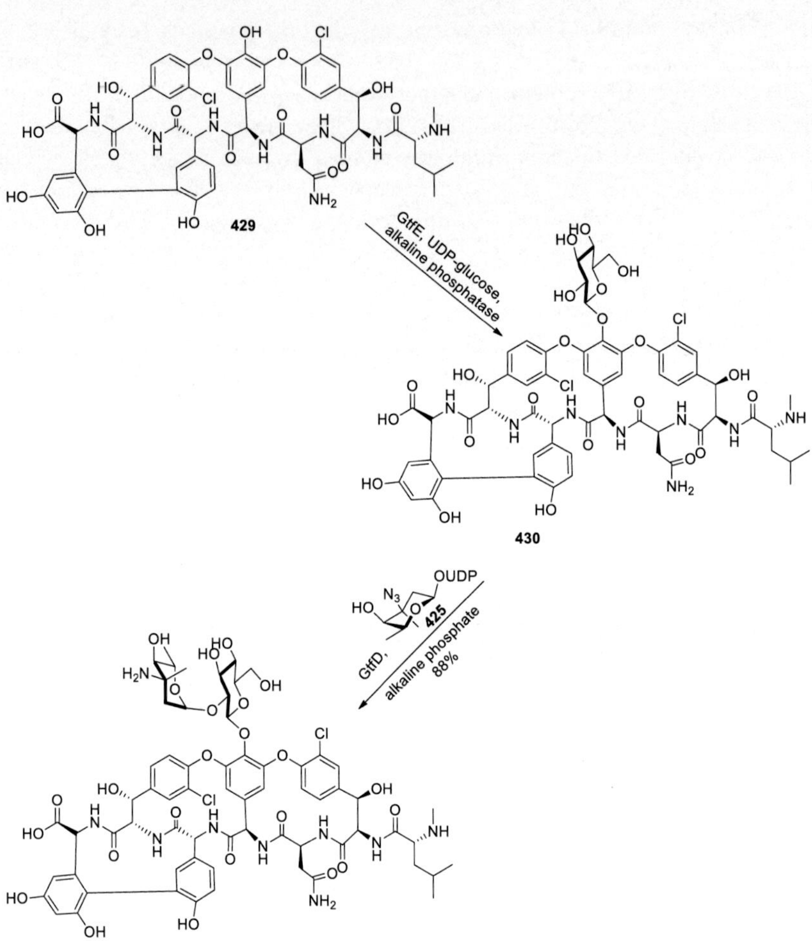

SCHEME 12.56 Continued

while other compounds that show EGFR inhibition were synthesized by reacting **431** with 1,2-dibromoethane or 1,4-dibromobutane in the presence of weak base K_2CO_3 and DMF as a solvent affords bromoalkyl derivatives **437** and **438** and further on treating with sodium azide in DMF gives corresponding azide derivatives **439** and **440**, respectively. These compounds were coupled with acetylated acetylenic glucose, galactose, or xylose via the CuAAC method giving corresponding **441–446** desired glycohybrid molecules (Scheme 12.57).[123]

Kumari et al. synthesized a library of carbohybrids formed by 1,2,3-triazole coupled with N-glycosides derivatives of quinolones **457–464** and coumarins **465–476** using microwave-assisted click chemistry. The reaction employed 1-azido-2,3,4,6-tetra-*O*-acetyl β-D-glucose **237** and 1-azido-

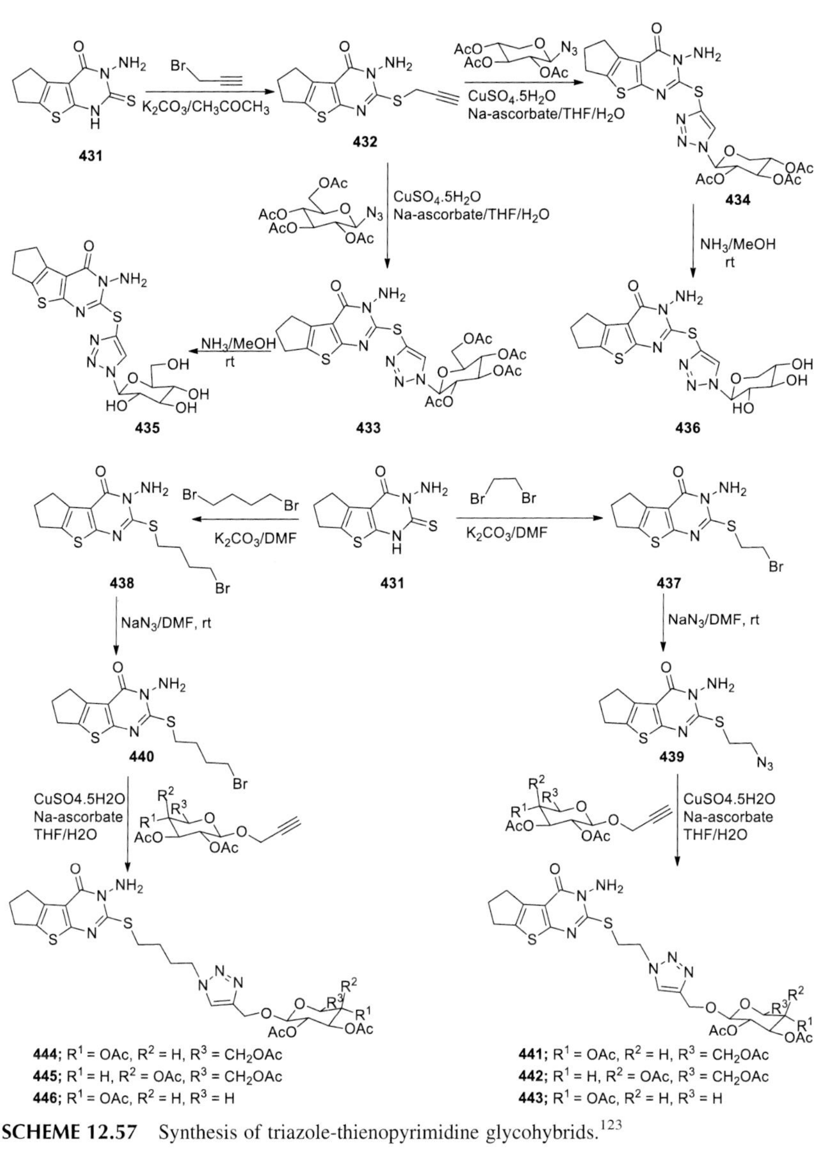

SCHEME 12.57 Synthesis of triazole-thienopyrimidine glycohybrids.[123]

2,3,4,6-tetra-*O*-acetyl β-D-galactose **336** coupling with various 4-*O*-propargyl quinolones **447–450** and 4-O-propargyl coumarins **451–456** (Scheme 12.58).[122] The anticancer activity of carbohybrids was determined against MCF-7, HCT-116, HepG2, and Huh-7.5 cell lines. The origination of reactive oxygen species (ROS) invokes the selective anticancer activity of the selected member in the library against breast cancer cell lines. IC_{50} for compound **468** was found to be 10.97 μM.[122] A significant class of compounds

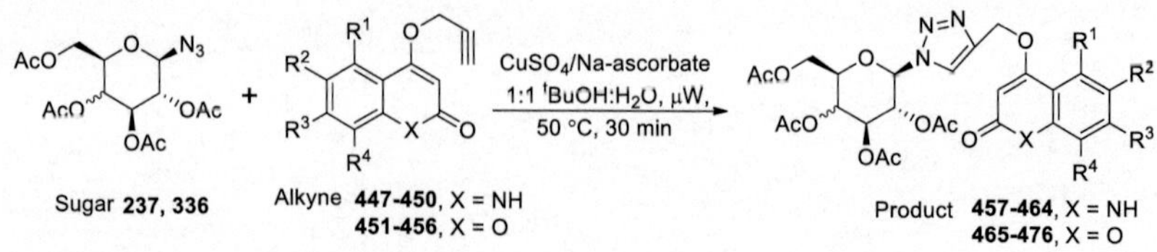

Sugar	R^1	R^2	R^3	R^4	X	Alkyne	Product	Sugar	R^1	R^2	R^3	R^4	X	Alkyne	Product
1	H	H	H	H	NH	447	457	1	H	Cl	H	H	O	453	467
1	H	H	H	F	NH	448	458	1	H	CH_3	H	H	O	454	468
1	H	H	H	NO_2	NH	449	459	1	H	Br	H	H	O	455	469
1	H	H	H	OCF_3	NH	450	460	1	H	H	F	H	O	456	470
2	H	H	H	H	NH	447	461	2	H	H	H	H	O	451	471
2	H	H	H	F	NH	448	462	2	H	OCH_3	H	H	O	452	472
2	H	H	H	NO_2	NH	449	463	2	H	Cl	H	H	O	453	473
2	H	H	H	OCF_3	NH	450	464	2	H	CH_3	H	H	O	454	474
1	H	H	H	H	O	451	465	2	H	Br	H	H	O	455	475
1	H	OCH_3	H	H	O	452	466	2	H	H	F	H	O	456	476

SCHEME 12.58 Synthesis of triazole-linked N-glycopyranosides.[122]

known as C-glycosides has also found a wide range of biological functions that serve as a structural block in many natural chemistry products. Researchers from the same group synthesize glucose, galactose, and mannose-derived C-glycosides **477–479** coupled with indole derivatives **480–483** via aldole condensation reaction followed by treating with hydroxyl amine and hydrazine that resulted in isoxazole and pyrazoline derivatives **484–495** and **496–507**, respectively. The reaction flourished under microwave conditions via condensation of α,β-unsaturated ketone derivative of a sugar moiety with hydrazine hydrate and hydroxyl amine hydrochloride under different pH conditions within 15 minutes (Scheme 12.59).[124] The anticancer activity of synthesized hybrids was investigated using an MTT assay against various cancer cell lines. The two of the selected hybrids showed MCF-7 (breast cancer cell line) selective toxicity at the nanomolar range, whereas no such toxicity was observed against MCF-10A, a normal cancer cell line. The carbohydrate moiety is shown to bind COX-2 via H-bonding, and the active molecules inhibit the COX-2 enzyme, resulting in high cytotoxicity.[125]

Another work reported the synthesis of thioglycosides coupled to substituted pyrimidines in good yield through glycosylation of 1,3,4-thiadiazole thiol derivatives in multistep reaction. The 1,2,3-triazole derivatives associated with sugar moiety were synthesized employing click chemistry. The in vitro cytotoxicity of compounds was studied against liver cancer, breast cancer, and epithelial cell line via MTT assay. The low binding affinity to the investigated biological targets resulted in moderate anticancer activity.[126]

Glycoconjugation reaction at anomeric or C6 positions has been reported between carbohydrates moieties and H_2S-releasing molecules. The study

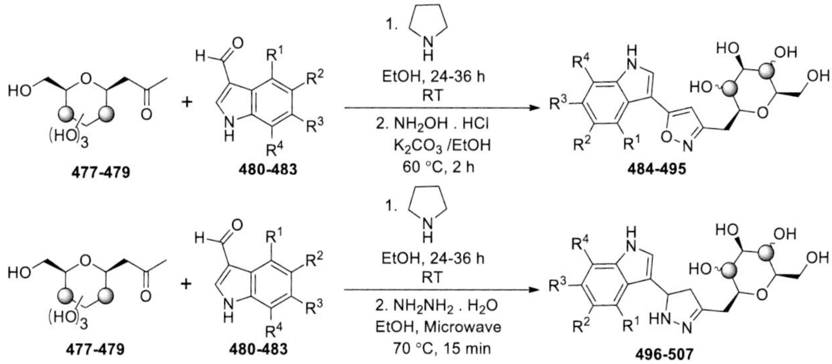

Entry	Sugar	R^1	R^2	R^3	R^4	Product (isoxazole)	Product pyrazoline
1.	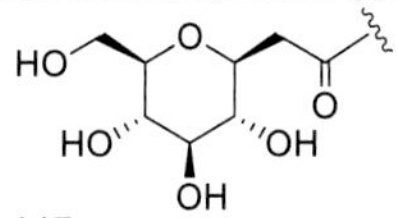	H	H	H	H	484	496
2.		H	NO_2	H	H	485	497
3.		H	Br	H	H	486	498
4.	447	H	Cl	H	H	487	499
5.	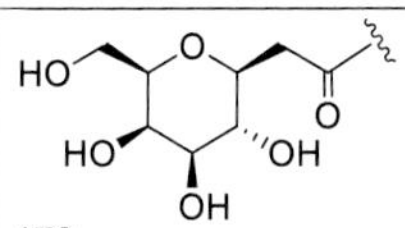	H	H	H	H	488	500
6.		H	NO_2	H	H	489	501
7.		H	Br	H	H	490	502
8.	478	H	Cl	H	H	491	503
9.	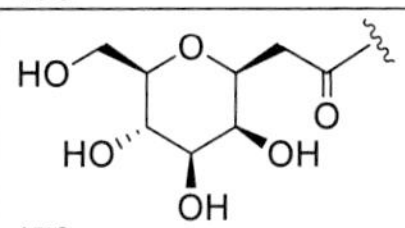	H	H	H	H	492	504
10.		H	NO_2	H	H	493	505
11.		H	Br	H	H	494	506
12.	479	H	Cl	H	H	495	507

SCHEME 12.59 Synthesis of isoxazole and pyrazoline bridged indole C-glycoside hybrids.[124]

demonstrated glucose, galactose, and mannose coupled to H_2S-releasing molecules, that is, dithiolethione- or isothiocyanate owing to their renowned capacity to release hydrogen sulfide in cells and the synthetic accessibility to chimeric molecules containing them. The C1 glycohybrids were stereoselectively synthesized via trichloroacetimidate glycosyl donors, whereas the C6 glycohybrids were obtained by Mitsunobu-based transformation. The anticancer activity was assessed against AsPC$_{-1}$, pancreatic adenocarcinoma tumor cells cell line and the two inhibitors **508** and **509** showed promising results by releasing high amounts of H_2S (Scheme 12.60).[127] Synthesis of two water-soluble glycoconjugates has been reported by Kuzmina et al. using conjugates of zinc-containing chlorin e_6 derivative with β-D-galactose and β-D-glucose. The synthesis proceeded via $CuSO_4$/sodium ascorbate catalyzed 1,3-dipolar cycloaddition of glycosyl azides **510**, **511** with alkyne containing chlorine **512** using tris(1-benzyltriazol-4-ylmethyl) amine (TBTA) ligand in dimethylformamide to yield **513** and **514** conjugates. These conjugates were

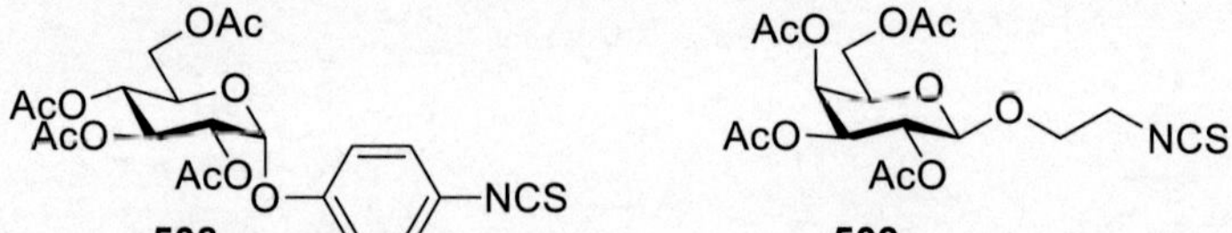

508 509

SCHEME 12.60 O-linked glycohybrids possessing anticancer activity.[127]

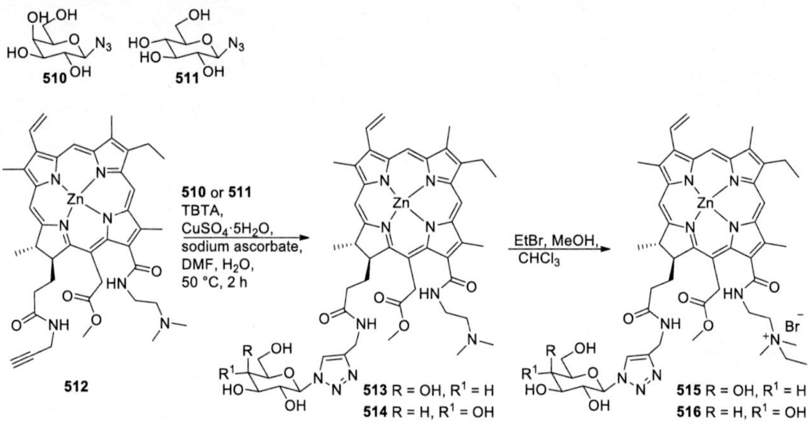

SCHEME 12.61 Synthesis of chlorine e_6 glycohybrids containing triazole using click chemistry.[128]

converted to corresponding quarternary ammonium salts using ethyl bromide to increase their solubility. Hence, the so-obtained conjugates **515**, **516** showed photoinduced cytotoxicity against A431 cell lines through in vitro studies at low micromolar concentrations (Scheme 12.61).[128]

Another work reported copper-catalyzed reaction protocol of glucoglycero and 1,2,3-triazole groups coupled to 1,2,4-oxadiazole through alkyne–azide cycloaddition and N-alkylation. The synthesized glycohybrids were investigated for cytotoxicity against numerous human cancer cell lines. However, the compounds N-propargyl-1,2,4-oxadiazoles **524** and **525** showed excellent antiproliferative activity against lung cancer cell line NCIH-460. Along with this, the report described other molecules that showed potent antituberculosis activity (Scheme 12.62).[131]

Synthesis of 2,3-dihydro-1,3-thiazole-coumarin glycohybrids of D-glucose and D-galactose were reported by Toan et al. using molecular hybridization of 3-acetyl and 6/7-alkoxy-4-formylcoumarin N-(2′,3′,4′,6′-tetra-*O*-acetyl-β-D-glycopyranosyl) thiosemicarbazones with ω-bromoacetophenone. The synthesized semithiocarbazones **517–523** were coupled with ω-bromoacetophenone under basic conditions using sodium acetate (anhydrous) that resulted in the formation of 2,3-dihydro-2(3H)-thiazoles

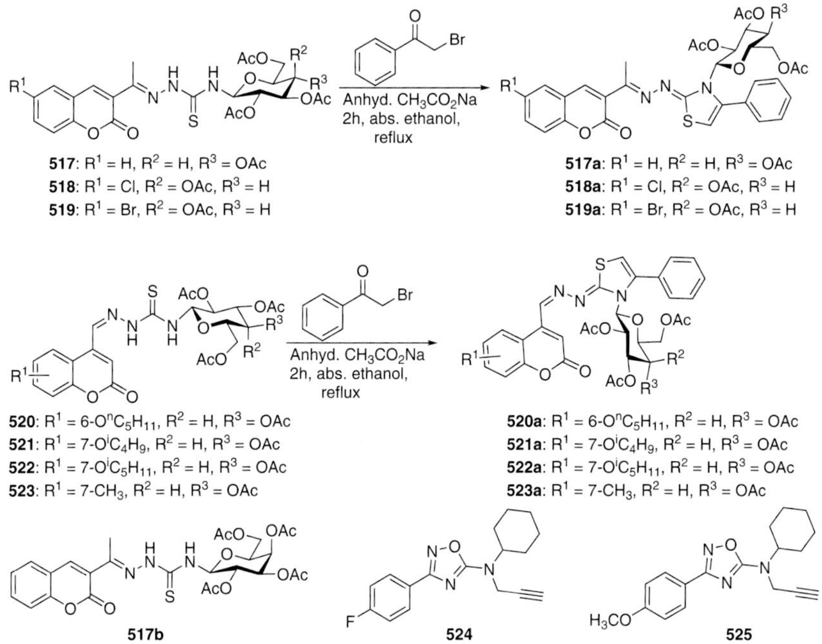

SCHEME 12.62 Synthesis of carbohybrid containing thiazole-coumarin derivative.[129–131]

517a–523a. For this, the performance of different solvents (dichloromethane, chloroform, carbon tetrachloride, 1,2-dichloroethane, and absolute ethanol) was investigated by selecting **517** as model semithiocarbazone and it was concluded that absolute ethanol maximized the yields of desired glycohybrids **517a–523a**. The cytotoxicity of these compounds was investigated against HepG-2, MCF-7, HeLa, SK-Mel-2, and LU-1. The cytotoxicity of normal cell line MRC-5 (fibroblast cell line) was also checked to study the desired selectivity. The compounds showed substantial anticancer activity against cancerous cell lines and high selectivity against MRC-5. The cytotoxicity was also compared with the standard drugs sorafenib, doxorubicin, and 5-fluorouracil. The glycohybrids **521a–523a** being the most active showed potential inhibition against EGFR and HER2 kinases against the standard drug sorafenib.[129]

The same research group reported the microwave-assisted synthesis of thiosemicarbazones glycohybrids of coumarin and D-galactose using suitable 3-acetyl coumarin derivatives and thiosemicarbazide. The antiproliferative activity of synthesized compounds was evaluated against cancerous cell lines (MCF-7, HepG-2, HeLa, SK-Mel-2, Lu-1) using MTT assay, and the selectivity was assessed using WI-38 normal cell line. The thiosemicarbazones **517b**, **518** showed good activity against EGFR-L858R and EGFR-T790M tyrosine kinase.[130]

12.6.2 Antibacterial glycohybrids

Malah et al. described a method to synthesize biologically active compounds possessing antibacterial activity that shows activity against gram-positive *S. aureus* and gram-negative *Pseudomonas aeruginosa* using antibiotic "Ampicillin" as a reference. Synthesis of such compounds takes place by coupling 2-(2-(2-methoxyethoxy)ethoxy)-ethyl 3-azidobenzoate **526** with acetylated acetylenic D-galactose **527**, D-glucose **528**, and D-xylose **529** via click chemistry and affords the products **530–532**, which contains 1,2,3-triazole core as a linker to connect sugar and nonsugar moieties in 75%–81%, respectively. The triethylene glycol unit's function in the resulting glycohybrid molecules is to connect via H-bonding with the D-Ala-D-Ala region of the bacterial cell wall (Scheme 12.63).[132]

In another approach, Salama et al. synthesized different glycohybrids possessing antimicrobial activity; guanidine attached to glycan enhances its antimicrobial property. Four different derivatives of chitosan were synthesized and tested against *S. aureus, P. aeruginosa, Bacillus subtilis*, and *Candida albicans*. Chitosan **533** was treated with acetic acid and cyanamide using Sc$(OTf)_3$ as a Lewis acid catalyst resulting in N-guanidinium chitosan acetate **534** while the reaction of **533** with HCl and cyanamide using $Sc(OTf)_3$ as a Lewis acid catalyst resulted in N-guanidinium chitosan chloride **535**. Other compounds were formed by treating **533** with 1-butyl-3-methylimidazolium chloride (BMIM Cl) at 90°C for 48 hours. Then, it was treated with N,N′-dicyclohexylcarbodiimide at 100°C for 2 days to afford the product N-guanidinium chitosan (N,N′-dicyclohexyl) chloride **536** while treating it with N-(3-dimethylaminopropyl)-N'-ethylcarbodiimide hydrochloride using the same conditions gives N-guanidinium chitosan (N-(3-dimethylaminopropyl)-N'-ethyl hydrochloride) chloride **537**. Chitosan is hydrophobic and insoluble in water but in an acidic medium their hydrophilicity increases and becomes water soluble, which makes them more suitable for medicinal studies. In acidic conditions, chitosan inhibits microbial growth as its cationic charges

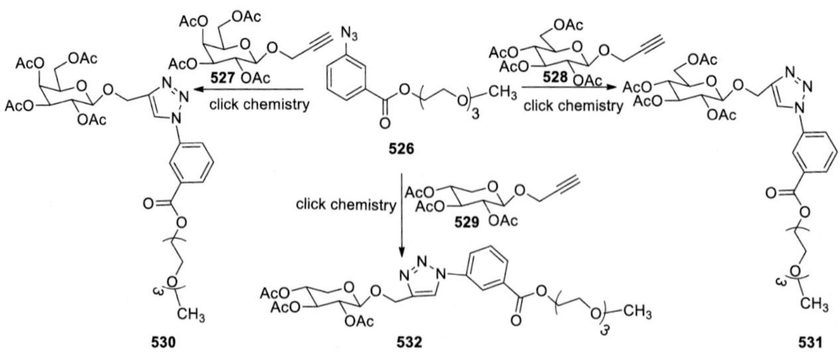

SCHEME 12.63 Synthesis of 1,2,3-triazole glycoside clickamers.[132]

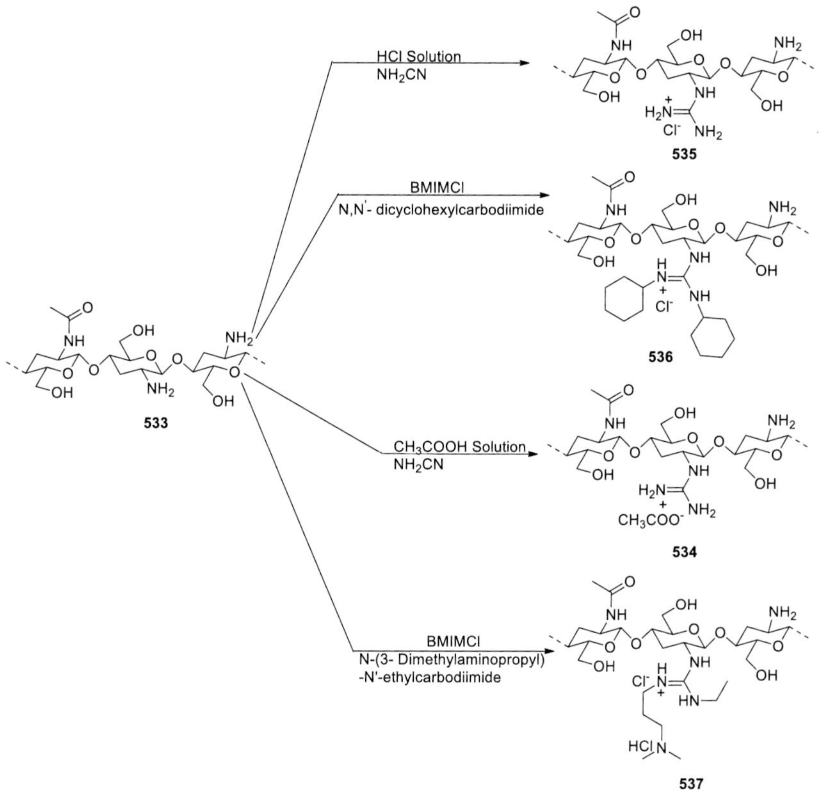

SCHEME 12.64 Synthesis of new chitosan derivatives.[133]

bind with the anionic component in the bacterial membrane. The aforementioned compounds exhibit both antibacterial and antifungal activities (Scheme 12.64).[133]

12.6.3 Antimalarial glycohybrids

The protozoan parasite *Plasmodium* spp. causes the contagious illness malaria. Gupta et al. synthesized glycohybrids derived from natural coumarin linked with sugar moieties. Carbohydrate-fused pyrano[3,2-c]pyranone glycohybrid **540**, which possesses antimalarial activity was synthesized by reacting 4-hydroxycoumarin derivatives with 2-C-formyl galactal. It has action against chloroquine-sensitive (Pf3D7) and chloroquine-resistant (PfRKL9) strains; intriguingly, its lack of cytotoxicity draws researchers' attention more. **540** has a strong level of growth inhibition against Plasmodium, with an IC_{50} value of 5.861 μM. It was synthesized by coupling 4-hydroxycoumarin **539** with 2-C-formyl galactal **538** under microwave heating conditions in the presence of 0.5 equiv. of pyrrolidine in toluene at 80°C affording desired pyrano[3,2-c]pyranone glycohybrid **540** in 93% isolated yield. This reaction follows the intriguing mechanism through C-1,2 nucleophilic addition followed by the 6π-electron electrocyclization reaction (Scheme 12.65).[134,135]

12.6.4 Antiviral glycohybrids

Rahman et al. synthesized uridine-based glycohybrid with strong electrophilicity and antiviral properties. Uridine-derived 4-aminophenyl 1-thioglucosides **541–545** follows the Lipinski rule as these compounds have less than 5 H-bond donors (HBDs), less than 10 H-bond acceptors (HBAs), more than 3 number of rotatable bonds (NRBs), molecular mass less 500 Daltons and octanol–water partition coefficient (logPo/w) is not greater than 5, which makes them more suitable to be used as a drug. These uridine-based glycohybrids possess lower lipophilicity and more hydrophilicity than standard drug sofosbuvir **546**. Docking these glycohybrids with 3su4 and 5e0i protein results shows that they possess a better binding affinity with hepatitis B virus (HBV) protease (5e0i) than hepatitis C virus (HCV) protease (3su4) (Scheme 12.66).[136]

12.6.5 Acetylcholinesterase inhibitor glycohybrids

Gulati et al. synthesized tacrine-linked triazole glycohybrids via coupling of azide-derived sugar moieties with acetylene-derived tacrine and it results in the desired hybrid molecules possessing in vitro acetylcholinesterase (AchE) inhibition activity and no hepatotoxicity. Hybrid molecules show nontoxicity on the HePG2 cell line at 200 μM. To synthesize these hybrids, firstly different sugars **547a–f** were acetylated **548a–f** and then bromination **549a–f** was done at the anomeric position in the presence of HBr in glacial acetic acid; this on further treatment with sodium azide in DMF at 50°C gives sugar azides **550a–f**. Sugar azides **550a–f** on treating with propargylated tacrine **552**, produced by treating tacrine **551** with propargyl bromide in the presence of sodium hydride, gives acetylated glycol-conjugate **553a–f** using

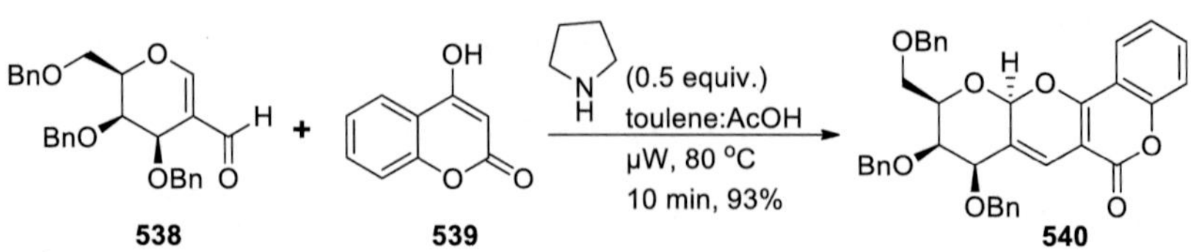

SCHEME 12.65 Synthetic route to synthesize pyrano[3,2-c]pyranone glycohybrid using organocatalyst.[134]

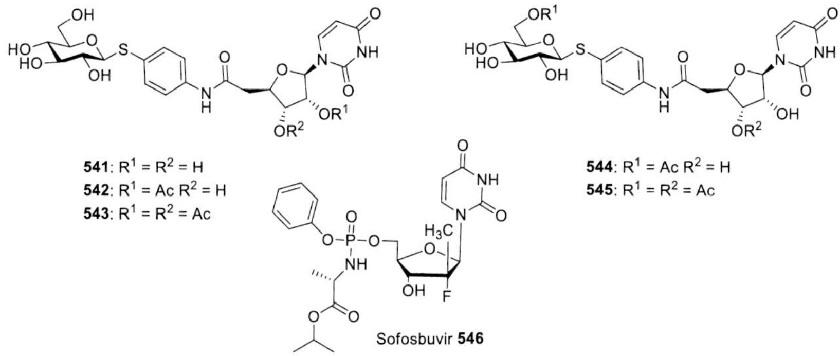

SCHEME 12.66 Structure of uridine-derived 4-aminophenyl 1-thioglucosides and sofosbuvir drug.[136]

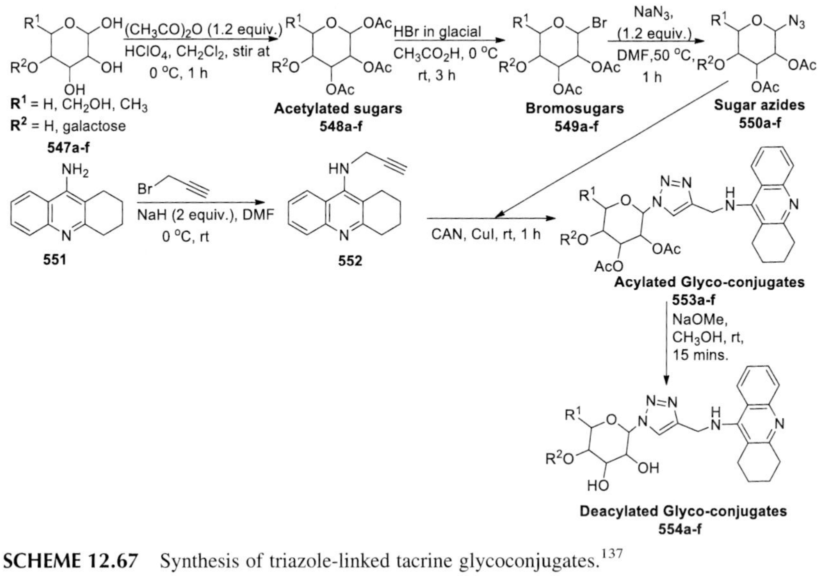

SCHEME 12.67 Synthesis of triazole-linked tacrine glycoconjugates.[137]

CAN and CuI at room temperature. It was further treated with NaOMe in methanol to give desired glycohybrids **554a–f** (Scheme 12.67).[137]

12.6.6 Glycosidase inhibitor glycohybrids

Wang et al. synthesized isofagomine-derived glycohybrid as glycosidase inhibitor as such kind of compound possesses enormous drug potential and also shows pH-dependent fluorescence. Isofagomine **555** was treated with 2-(10-chloroanthracen-9-yl)acetaldehyde **556** and 3-(10-chloroanthracen-9-yl) propanal **557** using sodium cyanoborohydride in $CH_3OH/CHCl_3$ and

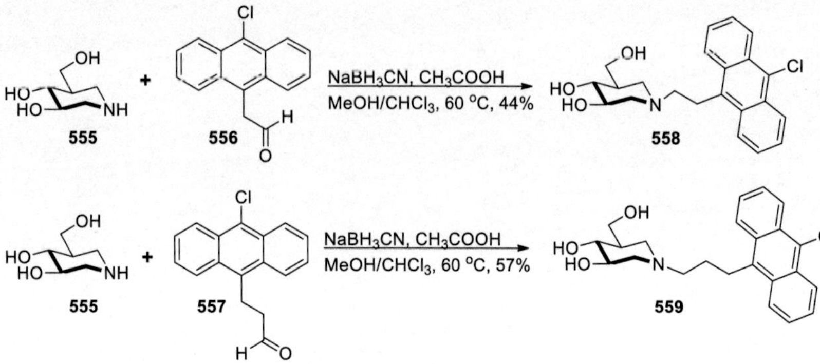

SCHEME 12.68 Synthesis of isofagomine derivatives.[138]

CH_3COOH at 60°C to give the desired glycohybrid products **558** and **559**. Hybrid formed shows poor hydrophilicity, so inhibition constants (K_i) were measured using 10% DMSO added as cosolvent and these compounds were competitive inhibitors of thermotoga maritima β-glucosidase at pH 6.5 with K_i values from 0.37 to 4.6 Mm (Scheme 12.68).[138]

12.6.7 Carbonic anhydrase inhibitor glycohybrids

Human carbonic anhydrases (hCAs) have a crucial role in the regulation of hypertension, epilepsy, hypoxic malignancies, arteriosclerosis, and other conditions in the human body. Pratesi et al. created glycomimetic-based selective inhibitors of hCAs. Two glycohybrids **560** and **561** containing sulfonamide moiety were synthesized in multistep, and hybrid **562** molecule containing aldehyde was synthesized using D-mannose, which on treating with N-benzyl hydroxylamine in dry methylene chloride at room temperature affords nitrone **563**. Reductive amination of this compound with Pd/C, H2, in acetic acid and methanol, which on further reduction with Ambersep 900-OH in MeOH at room temperature affords **564**. Further alkylation in the presence of $Br(CH_2)_6N_3$ and weak base K_2CO_3 using CH_3CN/H_2O as a solvent affords terminal azide-derived sugar derivative, which on treatment with sulfonamide **565** via CuACC strategy affords the desired glycohybrid **560**, while coupling of nitrone **563** with levoglucosenone **566** in dry toluene at 60°C gives region and stereoselective product **567** in 78% yield. The carbonyl group present in levoglucosenone selectively reduced by using L-selectride gives a single stereoisomer, which on catalytic hydrogenation in Pd/C, H2, in acetic acid and methanol gives product **568**. Alkylation of piperidine nitrogen takes place in the presence of $Br(CH_2)_6N_3$ and weak base K_2CO_3 using CH_3CN/H_2O solvent conditions to give **569**. The crude material was coupled with **560** via click chemistry to give the desired product **561**. The extensively produced hCA VII in glycohybrid **560** exhibits substantial suppression of the central nervous system (K_I of 7.4 nM), while **561** shows inhibition against the mitochondrial hCA VA over the VB and the tumor-associated hCA IX (K_Is of 34.1 and 35.9 nM, respectively) (Scheme 12.69).[139]

12.6.8 Antifungal glycohybrids

Goswami et al. synthesized the molecules fused with carbohydrates that show the antifungal activity against *Aspergillus fumigates*.[140] During the pandemic COVID-19, researchers observed that there are high chances of coinfections among the persons suffering from COVID-19, that is, 19.6%–33.3%.[141] Using the Cu(I) metal, molecules derived from eugenol **571** were joined with sugar molecules that contained azide to create the newly synthesized compound **570**. It shows an IC_{50} value of 5.42 μM against *A.*

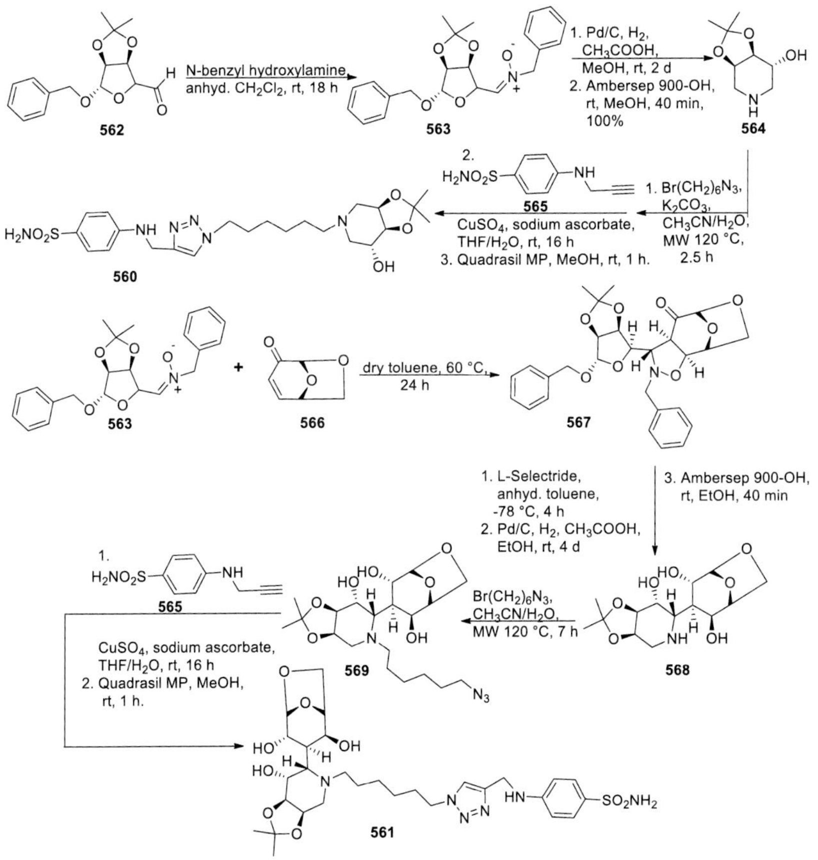

SCHEME 12.69 Synthesis of glycomimetic-based selective inhibitors of hCAs.[139]

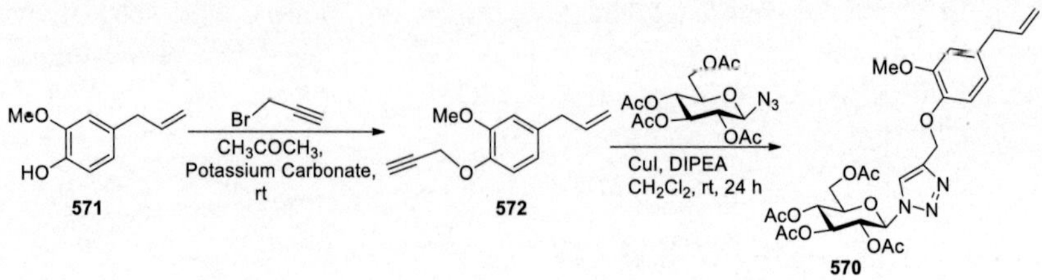

SCHEME 12.70 Synthesis of eugenol-derived 1,2,3-triazole-linked glycohybrid.[140]

fumigates. To create this, molecule **571** propargylated derivative **572** was first synthesized by treating it with 3-bromoprop-1-yne while utilizing base potassium carbonate and solvent CH_3COCH_3 at room temperature. The intended glycohybrid **570** is produced in an 81% yield when compound **572** is coupled with 1-azido-2,3,4,6-tetra-*O*-acetyl β-D-glucose utilizing metal iodide (CuI) as a catalyst and DIPEA as a base in the presence of methylene chloride at room temperature (Scheme 12.70).[140]

12.7 Conclusions and future outlook

From the previous discussion, it becomes very clear that the development of new methods for the synthesis of glycopeptides and glycohybrids becomes a need to time. The methods that could prepare these molecules in an efficient and stereoselective manner with higher yields are very much needed. It is quite clear many methods reported in this chapter could be useful methods for synthesizing diverse glycopeptides (S, N, O, C-linked) and glycohybrids. In this chapter, the conventional and recent methods for the synthesis of glycopeptides such as SPPS, silver-promoted SPG, NCL (Native Chemical Ligation), ball mill, click chemistry, chemoenzymatic, visible-light-mediated, and C−H activation strategy have been discussed. Further, this chapter also gives an elaborated description of the recent synthesis and applications of glycohybrids as, antibacterial, antimalarial, antiviral, anticancer drugs, and many more. These biologically active glycohybrids hold enormous potential to be used as novel drugs for the treatment of many deadly diseases. This chapter includes some green methods; however, efforts should be made to introduce more greener strategies to synthesize glycopeptides that can help in the reduction of uses of harmful chemicals.

References

1. Singhamahapatra, A.; Sahoo, L.; Varghese, B.; Loganathan, D. *RSC Adv.* **2014,** *4* (35), 18038−18043.
2. Herzner, H.; Reipen, T.; Schultz, M.; Kunz, H. *Chem. Rev.* **2000,** *100*, 4495−4537.
3. Payne, R. J.; Wong, C. H. *Chem. Commun.* **2010,** *46*, 21−43.
4. Brik, A.; Wong, C. H. *Chem. Eur. J.* **2007,** *13*, 5670−5675.

5. Spiro, R. G. *Glycobiology* **2002,** *12* (4).. Available from: https://doi.org/10.1093/glycob/12.4.43R.
6. De Leon, C. A.; Levine, P. M.; Craven, T. W.; Pratt, M. R. *Biochemistry* **2017,** *56*, 3507–3517.
7. Koester, D. C.; Holkenbrink, A.; Werz, D. B. *Synthesis.* **2010,** *19*, 3217–3242.
8. Blaskovich, M. A. T.; Hansford, K. A.; Butler, M. S.; Jia, Z.; Mark, A. E.; Cooper, M. A. *ACS Infect. Dis.* **2018,** *4*, 715–735.
9. Moellering, R. C. *Vancomycin: A 50-Year Reassess.* **2006,** *42* (Suppl 1), 3–4.
10. Binda, E.; Cappelletti, P.; Marinelli, F.; Marcone, G. L. *Antibiotics* **2018,** *7*, 1–10.
11. Mühlberg, E.; Umstätter, F.; Kleist, C.; Domhan, C.; Mier, W.; Uhl, P. *Can. J. Microbiol.* **2020,** *66*, 11–16.
12. Sonawane, A.; Mohanty, S.; Jagannathan, L.; Bekolay, A.; Banerjee, S. *Crit. Rev. Microbiol.* **2012,** *38*, 250–266.
13. Michaud, G.; Visini, R.; Bergmann, M.; Salerno, G.; Bosco, R.; Gillon, E.; Richichi, B.; Nativi, C.; Imberty, A.; Stocker, A., et al. *Chem. Sci.* **2016,** *7*, 166–182.
14. Xiao, A.; Zheng, X. J.; Song, C.; Gui, Y.; Huo, C. X.; Ye, X. S. *Org. Biomol. Chem.* **2016,** *14*, 7226–7237.
15. Zhou, N.; Pan, T.; Zhang, J.; Li, Q.; Zhang, X.; Bai, C.; Huang, F.; Peng, T.; Zhang, J.; Liu, C., et al. *J. Biol. Chem.* **2016,** *291*, 9218–9232.
16. Alam, S. M.; Aussedat, B.; Vohra, Y.; Meyerhoff, R. R.; Cale, E. M.; Walkowicz, W. E.; Radakovich, N. A.; Anasti, K.; Armand, L.; Parks, R., et al. *Sci. Transl. Med.* **2017,** 9. Available from: https://doi.org/10.1126/scitranslmed.aai7521.
17. McDonald, D. M.; Byrne, S. N.; Payne, R. J. *Front. Chem.* **2015,** *3*, 1–8. Available from: https://doi.org/10.3389/fchem.2015.00060.
18. Sun, L.; Wu, X.; Xiong, D. C.; Ye, X. S. *Angew. Chem. - Int. Ed.* **2016,** *55*, 8041–8044.
19. Singh, Y.; Demchenko, A. V. *Chem. Eur. J.* **2019,** *25*, 1461–1465.
20. Lefever, M. R.; Szab, L. Z.; Anglin, B.; Ferracane, M.; Hogan, J.; Cooney, L.; Polt, R. *Carbohydr. Res.* **2012,** *351*, 121–125.
21. Dziadek, S.; Griesinger, C.; Kunz, H.; Reinscheid, U. M. *Chem. Eur. J.* **2006,** *12*, 4981–4993.
22. Lindhorst, T. K.; Bruegge, K.; Fuchs, A.; Sperling, O. A. *Beil. J. Org. Chem.* **2010,** *6*, 801–809.
23. Karch, F.; Hoffmann-Röder, A. *Beil. J. Org. Chem.* **2010,** *6*, 1–8.
24. Mo, K. F.; Fang, T.; Stalnaker, S. H.; Kirby, P. S.; Liu, M.; Wells, L.; Pierce, M.; Live, D. H.; Boons, G. J. *J. Am. Chem. Soc.* **2011,** *133*, 14418–14430.
25. Hasegawa, A.; Ogawa, H.; Ishida, H.; Kiso, M. *Carbohydr. Res.* **1992,** *224*, 175–184.
26. Konradsson, P.; Mootoo, D. R.; Mcdevitt, R. E.; Fraser-reid, B. *J. Chem. Soc. Chem. Commun.* **1990,** *2*, 270–272.
27. Sliedregt, L. A. J. M.; van der Marel, G. A.; van Boom, J. H. *Tetrahedron Lett.* **1994,** *35*, 4015–4018.
28. Crich, D.; Smith, M. *J. Am. Chem. Soc.* **2001,** *123*, 9015–9020.
29. Sardar, M. Y. R.; Krishnamurthy, V. R.; Park, S.; Mandhapati, A. R.; Wever, W. J.; Park, D.; Cummings, R. D.; Chaikof, E. L. *Carbohydr. Res.* **2017,** *452*, 47–53.
30. Ren, L.; Liu, Y.; Yu, G. H.; Hao, J. Z.; Cheng, M. S. *Chem. Pap.* **2014,** *68*, 525–530.
31. Zheng, X. J.; Yang, F.; Zheng, M.; Huo, C. X.; Zhang, Y.; Ye, X. S. *Org. Biomol. Chem.* **2015,** *13*, 6399–6406.
32. Yang, F.; Zheng, X. J.; Huo, C. X.; Wang, Y.; Zhang, Y.; Ye, X. S. *ACS Chem. Biol.* **2011,** *6*, 252–259.

33. Keyari, C. M.; Polt, R. *J. Carbohydr. Chem.* **2010,** *29*, 181–206.
34. Mitchell, S. A.; Pratt, M. R.; Hruby, V. J.; Polt, R. *J. Org. Chem.* **2001,** *66*, 2327–2342.
35. Garner, J.; Harding, M. M. *ChemBioChem* **2010,** *11*, 2489–2498.
36. Peltier, R.; Brimble, M. A.; Wojnar, J. M.; Williams, D. E.; Evans, C. W.; Devries, A. L. *Chem. Sci.* **2010,** *1*, 538–551.
37. Corcilius, L.; Santhakumar, G.; Stone, R. S.; Capicciotti, C. J.; Joseph, S.; Matthews, J. M.; Ben, R. N.; Payne, R. J. *Bioorganic Med. Chem.* **2013,** *21*, 3569–3581.
38. Arsequell, G.; Sàrries, N.; Valencia, G. *Tetrahedron Lett.* **1995,** *36*, 7323–7326.
39. Harding, M. M.; Anderberg, P. I.; Haymet, A. D. J. *Eur. J. Biochem.* **2003,** *270*, 1381–1392.
40. Urbańczyk, M.; Jewgiński, M.; Krzciuk-Gula, J.; Góra, J.; Latajka, R.; Sewald, N. *Beil. J. Org. Chem.* **2019,** *15*, 1581–1591.
41. Jensen, K. J.; Meldal, M.; Bock, K. *J. Chem. Soc. Perkin Trans.* **1993,** *1*, 2119–2129.
42. Shinohara, H.; Matsubayashi, Y. *Plant. Cell Physiol.* **2013,** *54*, 369–374.
43. Kumar, V.; Giri, S. K.; Venugopalan, P.; Kartha, K. P. R. *Chempluschem* **2014,** *79*, 1605–1613.
44. Chemistry, O. F. O. *Chem. Asian Journal.* **2010,** *5*, 13–22.
45. Wu, X.; McKay, C.; Pett, C.; Yu, J.; Schorlemer, M.; Ramadan, S.; Lang, S.; Behren, S.; Westerlind, U.; Finn, M. G., et al. *ACS Chem. Biol.* **2019,** *14*, 2176–2184.
46. Wu, X., et al. *Org. Biomol. Chem.* **2021,** *19*, 2448–2455.
47. Zemlyakov, A. E.; Tsikalova, V. N.; Tsikalov, V. V. *Chem. Nat. Compd.* **2020,** *56*, 193–196.
48. Zemlyakov, A. E.; Tsikalov, V. V.; Tsikalova, V. N. *Chem. Nat. Compd.* **2018,** *54*, 1139–1142.
49. Li, T.; Zhang, Y.; Li, T.; Zhuang, H.; Wang, F.; Wang, N.; Schmidt, R. R.; Peng, P. *J. Chem.* **2022,** *40*, 1571–1577.
50. Bi, F.; Zhang, J.; Wei, Z.; Yu, D.; Zheng, S.; Wang, J.; Li, H.; Hua, Z.; Zhang, H.; Yang, G. *Biomacromolecules* **2022,** *23*, 128–139.
51. Zhu, X.; Schmidt, R. R. *Chem. Eur. J.* **2004,** *10*, 875–887.
52. Zhu, X.; Haag, T.; Schmidt, R. R. *Org. Biomol. Chem.* **2004,** *2*, 31–33.
53. Calce, E.; Digilio, G.; Menchise, V.; Saviano, M.; De Luca, S. *Chem. Eur. J.* **2018,** *24*, 6231–6238.
54. Roytman, R.; Adler-Abramovich, L.; Kumar, K. S. A.; Kuan, T. C.; Lin, C. C.; Gazit, E.; Brik, A. *Org. Biomol. Chem.* **2011,** *9*, 5755–5761.
55. Thiem, J.; Laupichler, L. *J. Carbohydr. Chem.* **2018,** *37*, 442–460.
56. Huo, C. X.; Zheng, X. J.; Xiao, A.; Liu, C. C.; Sun, S.; Lv, Z.; Ye, X. S. *Org. Biomol. Chem.* **2015,** *13*, 3677–3690.
57. Hsieh, Y. S. Y.; Wilkinson, B. L.; O'Connell, M. R.; MacKay, J. P.; Matthews, J. M.; Payne, R. J. *Org. Lett.* **2012,** *14*, 1910–1913.
58. Brimble, M. A.; Edwards, P. J.; Harris, P. W. R.; Norris, G. E.; Patchett, M. L.; Wright, T. H.; Yang, S. H.; Carley, S. E. *Chem. Eur. J.* **2015,** *21*, 3556–3561.
59. Zhang, G. L.; Gadi, M. R.; Cui, X.; Liu, D.; Zhang, J.; Saikam, V.; Gibbons, C.; Wang, P. G.; Li, L. *Green. Chem.* **2021,** *23*, 2907–2912.
60. Kaur, S.; Luciano, D. P.; Fan, X.; Zhao, G.; Messier, S.; Walker, M. M.; Zhang, Q.; Wang, T. *Tetrahedron Lett.* **2021,** *86*, 153499.
61. Banisalman, K. A. F.; Polykandritou, A.; Barnieh, F. M.; Ribeiro Morais, G.; Falconer, R. A. *J. Org. Chem.* **2022,** *87*, 14026–14036.
62. Bertozzi, C. R.; Hoeprich, P. D.; Bednarski, M. D. *J. Org. Chem.* **1992,** *57*, 6092–6094.

63. Manabe, S.; Marui, Y.; Ito, Y. *Chem. Eur. J.* **2003,** *9*, 1435–1447.
64. Vangala, M.; Dhokale, S. A.; Gawade, R. L.; Pattuparambil, R. R.; Puranik, V. G.; Dhavale, D. D. *Org. Biomol. Chem.* **2013,** *11*, 6874–6878.
65. Wu, J.; Kaplaneris, N.; Ni, S.; Kaltenhäuser, F.; Ackermann, L. *Chem. Sci.* **2020,** *11*, 6521–6526.
66. Jiang, Y.; Wang, Q.; Zhang, X.; Koh, M. J. *Chem* **2021,** *7*, 3377–3392.
67. Shang, W.; Su, S. N.; Shi, R.; Mou, Z. D.; Yu, G. Q.; Zhang, X.; Niu, D. *Angew. Chem. - Int. Ed.* **2021,** *60*, 385–390.
68. Wang, Q.; Fu, Y.; Zhu, W.; An, S.; Zhou, Q.; Zhu, S. F.; He, G.; Liu, P.; Chen, G. *CCS Chem.* **2021,** *3*, 1729–1736.
69. Mao, R.; Xi, S.; Shah, S.; Roy, M. J.; John, A.; Lingford, J. P.; Gäde, G.; Scott, N. E.; Goddard-Borger, E. D. *J. Am. Chem. Soc.* **2021,** *143*, 12699–12707.
70. Nyandoro, K.; Lamb, C. M. G.; Yu, H.; Shi, J.; Macmillan, D. *Org. Biomol. Chem.* **2022,** *20*, 8506–8514.
71. Pinzón Martín, S. M.; Medina, R. F.; Iregui Castro, C. A.; Rivera Monroy, Z. J.; García Castañeda, J. E. *Int. J. Pept. Res. Ther.* **2018,** *24*, 291–298.
72. Ibatullin, F. M.; Selivanov, S. I. *Tetrahedron Lett.* **2009,** *50*, 6351–6354.
73. Urge, L.; Kollat, E.; Hollosi, M.; Laczko, I.; Wroblewski, K.; Thurin, J.; Otvos, L., Jr. *Tetrahedron Lett.* **1991,** *32*, 3445–3448.
74. Colombo, C.; Bernardi, A. *Eur. J. Org. Chem.* **2011,** *6*, 3911–3919.
75. Xue, J.; Guo, M.; Gu, G.; Guo, Z. *J. Carbohydr. Chem.* **2012,** *31*, 105–113.
76. Brooke, J. S. *Clin. Microbiol. Rev.* **2021,** *34* (3).. Available from: https://doi.org/10.1128/CMR.00030-19.
77. Garcia, J. E.; Fierro, R.; Puentes, A.; Cortés, J.; Bermúdez, A.; Cifuentes, G.; Vanegas, M.; Patarroyo, M. E. *Biochem. Biophys. Res. Commun.* **2007,** *355*, 409–418.
78. Sureshbabu, V. V.; Venkataramanarao, R.; Hemantha, H. P. *Int. J. Pept. Res. Ther.* **2008,** *14*, 34–40.
79. Asahina, Y.; Kanda, M.; Suzuki, A.; Katayama, H.; Nakahara, Y.; Hojo, H. *Org. Biomol. Chem.* **2013,** *11*, 7199–7207.
80. Premdjee, B.; Adams, A. L.; MacMillan, D. *Bioorg. Med. Chem. Lett.* **2011,** *21*, 4973–4975.
81. Tanaka, T.; Shiraishi, M.; Matsuda, A.; Mizuno, M. *Tetrahedron Lett.* **2019,** *60*, 151106.
82. Sršan, L.; Ziegler, T. *Beil. J. Org. Chem.* **2020,** *16*, 888–894.
83. Szarek, W. A.; Grynkiewicz, G.; Doboszewski, B.; Hay, G. W. *Chem. Lett.* **1984,** *13*, 1751–1754.
84. Rodríguez, V.; Román, J.; Insuasty-Cepeda, D.; Fierro-Medina, R.; Rivera-Monroy, Z.; García-Castañeda, J. *ChemistrySelect* **2021,** *6*, 4083–4088.
85. Kajihara, Y.; Nishikawa, R.; Maki, Y.; Okamoto, R. *Arkivoc* **2021,** 230–240.
86. Wang, S.; Corcilius, L.; Sharp, P. P.; Payne, R. J. *Bioorg. Med. Chem.* **2017,** *25*, 2895–2900.
87. Wang, S.; Corcilius, L.; Sharp, P. P.; Rajkovic, A.; Ibba, M.; Parker, B. L.; Payne, R. J. *Chem. Sci.* **2017,** *8*, 2296–2302.
88. Pan, M.; Li, S.; Li, X.; Shao, F.; Liu, L.; Hu, H. G. *Angew. Chem. - Int. Ed.* **2014,** *53*, 14517–14521.
89. Jiang, Y.; Cheng, Z.; Chen, S.; Li, L.; Zhang, W.; Li, X.; Hu, H. *Bioorg. Chem.* **2023,** *130*, 106267.
90. Castro, V.; Rodríguez, H.; Albericio, F. *ACS Comb. Sci.* **2016,** *18*, 1–14.

91. Seifried, B. M.; Qi, W.; Yang, Y. J.; Mai, D. J.; Puryear, W. B.; Runstadler, J. A.; Chen, G.; Olsen, B. D. *Bioconjug. Chem.* **2020,** *31*, 554–566.
92. Aarjane, M.; Slassi, S.; Tazi, B.; Maouloua, M.; Amine, A. *Struct. Chem.* **2020,** *31*, 1523–1531.
93. Singhamahapatra, A.; Sahoo, L.; Loganathan, D. *J. Org. Chem.* **2013,** *78*, 10329–10336.
94. Marchiori, M. F.; Pires Souto, D. E.; Oliveira Bortot, L.; Francisco Pereira, J.; Kubota, L. T.; Cummings, R. D.; Dias-Baruffi, M.; Carvalho, I.; Campo, V. L. *Bioorg. Med. Chem.* **2015,** *23*, 3414–3425.
95. Liu, E. C.; Topczewski, J. J. *J. Am. Chem. Soc.* **2021,** *143*, 5308–5313.
96. Wan, Q.; Chen, J.; Chen, G.; Danishefsky, S. J. *J. Org. Chem.* **2006,** *71*, 8244–8249.
97. Kuijpers, B. H. M.; Groothuys, S.; Keereweer, A. R.; Quaedflieg, P. J. L. M.; Blaauw, R. H.; Van Delft, F. L.; Rutjes, F. P. J. T. *Org. Lett.* **2004,** *6*, 3123–3126.
98. Tomabechi, Y. *Trends Glycosci. Glycotechnol.* **2015,** *27*, 63–65.
99. Maschauer, S.; Einsiedel, J.; Haubner, R.; Hocke, C.; Ocker, M.; Hübner, H.; Kuwert, T.; Gmeiner, P.; Prante, O. *Angew. Chem. - Int. Ed.* **2010,** *49*, 976–979.
100. Lim, D.; Brimble, M. A.; Kowalczyk, R.; Watson, A. J. A.; Fairbanks, A. J. *Angew. Chem. - Int. Ed.* **2014,** *53*, 11907–11911.
101. Marchiori, M. F.; Iossi, G. P.; Bortot, L. O.; Dias-Baruffi, M.; Campo, V. L. *Carbohydr. Res.* **2019,** *472*, 23–32.
102. Ngoje, G.; Li, Z. *Org. Biomol. Chem.* **2013,** *11*, 1879–1886.
103. Kaeothip, S.; Demchenko, A. V. *Carbohydr. Res.* **2011,** *346*, 1371–1388.
104. Gyuricza, B.; Szabó, J. P.; Arató, V.; Szücs, D.; Vágner, A.; Szikra, D.; Fekete, A. *Pharmaceutics* **2021,** *13* (6).. Available from: https://doi.org/10.3390/pharmaceutics13060796.
105. Gyuricza, B.; Szabó, J. P.; Arató, V.; Dénes, N.; Szűcs, Á.; Berta, K.; Kis, A.; Szücs, D.; Forgács, V.; Szikra, D., et al. *Pharmaceutics* **2021,** *13* (12), 1–14.
106. Tevyashova, A. N.; Bychkova, E. N.; Korolev, A. M.; Isakova, E. B.; Mirchink, E. P.; Osterman, I. A.; Erdei, R.; Szücs, Z.; Batta, G. *Bioorg. Med. Chem. Lett.* **2019,** *29*, 276–280.
107. Breukink, E.; de Kruijff, B. *Nat. Rev. Drug. Discov.* **2006,** *5* (4), 321–323.
108. Álvarez, R.; Cortés, L. E. L.; Molina, J.; Cisneros, J. M.; Pachón, J. *Antimicrob. Agents Chemother.* **2016,** *60*, 2601–2609.
109. Ashford, P. A.; Bew, S. P. *Chem. Soc. Rev.* **2012,** *41*, 957–978.
110. Arnusch, C. J.; Bonvin, A. M. J. J.; Verel, A. M.; Jansen, W. T. M.; Liskamp, R. M. J.; De Kruijff, B.; Pieters, R. J.; Breukink, E. *Biochemistry* **2008,** *47*, 12661–12663.
111. Long, D. D.; Aggen, J. B.; Chinn, J.; Choi, S.; Christensen, B. G.; Fatheree, P. R.; Green, D.; Hegde, S. S.; Judice, J. K.; Kaniga, K., et al. *J. Antibiot.* **2008,** *61* (10), 603–614.
112. Stryjewski, M. E.; Potgieter, P. D.; Li, Y. P.; Barriere, S. L.; Churukian, A.; Kingsley, J.; Corey, G. R. *Antimicrob. Agents Chemother.* **2012,** *56* (11), 5476–5483.
113. Parnham, M. J.; Haber, V. E.; Giamarellos-Bourboulis, E. J.; Perletti, G.; Verleden, G. M.; Vos, R. *Pharmacol. Ther.* **2014,** *143*, 225–245.
114. Olsufyeva, E. N.; Shchekotikhin, A. E.; Bychkova, E. N.; Pereverzeva, E. R.; Treshalin, I. D.; Mirchink, E. P.; Isakova, E. B.; Chernobrovkin, M. G.; Kozlov, R. S.; Dekhnich, A. V., et al. *Drug. Des. Devel. Ther.* **2018,** *12*, 2875–2885.
115. Zong, G.; Toonstra, C.; Yang, Q.; Zhang, R.; Wang, L. X. *Int. J. Mol. Sci.* **2021,** *22*, 1–13.
116. Du, J. J.; Zhang, L.; Gao, X. F.; Sun, H.; Guo, J. *Front. Chem.* **2020,** *8*.. Available from: https://doi.org/10.3389/fchem.2020.00396.

117. Apostol, C. R.; Tanguturi, P.; Szabò, L. Z.; Varela, D.; Gilmartin, T.; Streicher, J. M.; Polt, R. *Molecules* **2021,** *26*, 1–13.
118. Ma, C.; He, N.; Ou, Y.; Feng, W. *ChemistrySelect* **2020,** *5*, 6670–6673.
119. Bereczki, I.; Szűcs, Z.; Batta, G.; Nagy, T. M.; Ostorházi, E.; Kövér, K. E.; Borbás, A.; Herczegh, P. *Pharmaceuticals* **2022,** *15* (1), 14–18.
120. Moore, M. J.; Qin, P.; Keith, D. J.; Boger, D. L. *Tetrahedron* **2023,** *131*, 133211.
121. Kumari, P.; Narayana, C.; Tiwari, G.; Sagar, R. Glycohybrid Molecules in Medicinal Chemistry*:* Present Status and Future Prospective*;* Elsevier Inc., 2020.
122. Kumari, P.; Dubey, S.; Venkatachalapathy, S.; Narayan, C.; Gupta, A.; Sagar, R. *N. J. Chem.* **2019,** *43*, 18590–18600.
123. Click Khattab, R. H.; Alshamari, A. K.; Hassan, A. A.; Elganzory, H. H.; El-Sayed, W. A.; Awad, H. M.; Nossier, E. S.; Hassan, N. A. *J. Enzyme Inhib. Med. Chem.* **2021,** *36*, 504–516.
124. Kumari, P.; Mishra, V. S.; Narayana, C.; Khanna, A.; Chakrabarty, A.; Sagar, R. *Sci. Rep.* **2020,** *10* (1), 1–16.
125. Kumari, P.; Mishra, V. S.; Narayana, C.; Khanna, A.; Chakrabarty, A.; Sagar, R. *Sci. Rep.* **2020,** *10* (1), 6660.
126. Khalaf, H. S., et al. *Nucleotides Nucleic Acids* **2020,** *39* (7), 1036–1056.
127. S Fortunato, S., et al. *Bioconjug. Chem.* **2019,** *30* (3), 614–620.
128. Kuzmina, N. S., et al. *Mendeleev Commun.* **2020,** *30*, 159–161.
129. Toan, V. N., et al. *N. J. Chem.* **2021,** *45*, 10636–10653.
130. Toan, V. N.; Nguyen, T. D. *Med. Chem. Res.* **2021,** *30*, 1868–1885.
131. Melo de Oliveira, V. N.; Flávia do Amaral Moura, C.; Peixoto, A. dos S.; Gonçalves Ferreira, V. P.; Araújo, H. M.; Lapa Montenegro Pimentel, L. M.; Pessoa, C. do Ó.; Nicolete, R.; Versiani dos Anjos, J.; Sharma, P. P., et al. *Eur. J. Med. Chem.* **2021,** 220. Available from: https://doi.org/10.1016/j.ejmech.2021.113472.
132. Malah, T. E., et al. *Molecules* **2020,** *25* (4), 790.
133. Salama, A; Hasanin, M; Hesemann, P *Carbohydr. Polym.* **2020,** *241*, 116363. Available from: https://doi.org/10.1016/j.carbpol.2020.116363.
134. Kumari, P.; Gupta, S.; Narayana, C.; Ahmad, S.; Vishnoi, N.; Singh, S.; Sagar, R. *N. J. Chem.* **2018,** *42*, 13985–13997.
135. Gupta, S., et al. *Malar. J.* **2019,** *18*, 1–14.
136. Rahman, M. A.; Chakma, U.; Kumer, A.; Rahman, M. R.; Matin, M. M. *Biointerface Res. Appl. Chem.* **2023,** *13* (1), 1–15.
137. Kaur Gulati, H.; Choudhary, S.; Kumar, N.; Ahmed, A.; Bhagat, K.; Vir Singh, J.; Singh, A.; Kumar, A.; Singh Bedi, P. M.; Singh, H., et al. *Bioorg. Chem.* **2022,** *118*, 105479.
138. Wang, B.; Bogh, S. A.; Poulsen, J. C. N.; Laursen, B. W.; Bols, M. *Eur. J. Org. Chem.* **2020,** *26*, 3989–3996.
139. Pratesi, D.; Matassini, C.; Goti, A., et al. *ACS Med. Chem. Lett.* **2020,** *11*, 727–731.
140. Goswami, L.; Gupta, L.; Paul, S.; Vermani, M.; Vijayaraghavan, P.; Bhattacharya, A. K. *RSC Med. Chem.* **2022,** *13*, 955–962.
141. Lai, C. C.; Yu, W. L. *J. Microbiol. Immunol. Infect.* **2021,** *54*, 46–53.

Chapter 13

Carbohydrate-based chiral organocatalysts

Vinod K. Tiwari[1] **and Hélène Pellissier**[2]

[1]*Department of Chemistry, Institute of Science, Banaras Hindu University, Varanasi, Uttar Pradesh, India,* [2]*Aix-Marseille Univ, CNRS, Centrale Marseille, iSm2, Marseille, France*

13.1 Introduction

Carbohydrates constitute a functionally diverse class of molecules that plays a fundamental role in normal cell functions including energy storage, modulation of protein function, transport, adhesion, signal transduction, and viral and bacterial cell surface recognition.[1] Besides their biological significance, carbohydrates are considered to be an efficient tool for drug discovery[2] and asymmetric synthesis.[3,4] Their natural abundance and occurrence in the form of monomers, oligomers, or polymers possessing several stereogenic centers and differently configured hydroxyl group makes them readily available for structural variations. They have rigid conformation, diverse functionalities, known stereochemistry, and also synthetic fertility.[4] These properties enhance their synthetic utility and facilitate them to act as potential building blocks, auxiliaries, reagents, ligands, or catalysts in organic reactions[4–7] Carbohydrate-derived chiral auxiliaries present a wide range of applications to impose stereocontrol on several important chemical transformations. Stereochemically pure adducts can be achieved with these auxiliaries in [4 + 2] cycloadditions (both as dienes and dienophiles),[6a] [2 + 2] cycloadditions, 1,3-dipolar cycloadditions, cyclopropanations, alkylations, and acylations.[6b] Also, sugar-based chiral molecules are effectively utilized for asymmetric aldol condensations,[5b] other cycloadditions,[6a–c] oxidations and reductions, selective displacement reactions at sulfur and phosphorus atoms,[6d–e] Overman rearrangements, Wittig rearrangements,[6f] photochemical transformations,[7g] and kinetic resolutions. Acting as chiral ligands in asymmetric catalysis, carbohydrate chiral pool has been employed in the form of different phosphorus-, nitrogen-, and sulfur-based ligands, in combination with various metals, such as rhodium, nickel, cobalt, palladium, titanium, zinc, copper, and manganese among others.[7] More recently, the applicability

Synthetic Strategies in Carbohydrate Chemistry. DOI: https://doi.org/10.1016/B978-0-323-91729-2.00021-5

of carbohydrates in asymmetric synthesis has brought the attention of chemists toward their utilization in the emerging field of organocatalysis. Their role in asymmetric organocatalysis has led to an increased demand for carbohydrate-based organocatalysts and their complete synthetic investigations.[8] Indeed, organocatalysts are of enormous use for a wide array of synthetically useful transformations and have proved to be a better choice over transition metals and enzyme catalysts.[8,9] Despite the wide utility and selectivity of organocatalysts, there are only a few selected basic moieties that are very efficient and promising substrates for developing efficient organocatalysts.[8–10] Therefore, the need to develop new organic moieties having promising catalytic utility is challenging. In this context, carbohydrate moiety consisting of chirality, functional diversity, stereogenic centers, and other fascinating structural aspects prompted modern chemists to design and develop new and efficient methodologies for preparing carbohydrate-based organocatalysts in an efficient way. The efforts successfully brought forward some very interesting carbohydrate-derived molecules that can efficiently be used to catalyze different organic reactions[10–12] Due to this reason, the synthesis of carbohydrate-based organocatalysts and the number of reactions that they catalyze has been increasing many folds day by day.[11,12] This chapter highlights the historical background, various basic and structural aspects related to organocatalyts, and recent impressive progress in the area of synthesis and functional utility of carbohydrate-based organocatalysts. Categorization of chiral sugar-based organocatalysts has been done on the basis of their chemical structures and functionalities. To throw light on progressive development in this field, the best efforts have been made to put all this information in chronological order for each category.

13.2 Organocatalysts: historical background and basic aspects

13.2.1 Historical background

An organocatalyst in simplified terms can be defined as a small organic molecule, not bound to a metal atom, which accelerates an organic reaction when used in substoichiometric amounts.[13] The emergence of organocatalysis is linked to the aim of developing such types of simple catalysts that can reduce the shortcomings associated with earlier known metal and enzyme catalysts and is equivalently useful for organic transformations. Metal catalysts have applicability in a wide range of chemical reactions but are toxic in nature. On the other hand, enzymes are environmentally benign but have limited substrate scopes. Removing the shortcomings of metal catalysts and having catalytic efficiency similar to enzymes, this new class of catalysts is moreover nontoxic and ecofriendly in nature, which makes them very useful in pharmaceutical product synthesis. Furthermore, they have wide substrate scope and are inexpensive. Moreover, organocatalysts are inert toward moisture and oxygen and are generally robust, unlike metal and enzyme catalysts. The advantage of organocatalysis over earlier catalysis also includes its applicability in multicomponent, tandem and domino reactions, which can give structurally more complex products with high stereocontrol in single steps. The history of organocatalytic reactions is supposed to be started from the origin of life on earth. This view finds support by the evidence of such type of catalysis playing a determinant role in the formation of prebiotic key building blocks during the process of evolution. In this manner, these reactions have led to the introduction and widespread use of homochirality and enantioselectivity in the living world.[14] From a variety of enantiomerically enriched amino acids available in carbonaceous meteorites, such as L-alanine and L-isovaline, that may be present in up to 15% enantiomeric excess (ee) in these meteorites, have the ability to catalyze the aldol-type dimerization of glycoaldehyde as well as the reaction between glycoaldehyde and formaldehyde producing sugar derivatives. Pizzarello and Weber have successfully demonstrated that L-isovaline, an amino acid identified from carbon-rich Murchison meteorite, generates aldol products (such as L-therose and D-erythrose) by self-aldol reaction of glycoaldehyde in water.[14] The use of simple organic molecules as catalysts for organic synthesis has started from an early age. The earliest example of an organocatalytic reaction was accidentally discovered by Justus von Liebig in 1860, who noticed the conversion of dicyan to oxamide in the presence of an aqueous solution of acetaldehyde (Scheme 13.1A).[15] Moreover, the discovery of enzymes and their functionalities had a great impact on the development of asymmetric organocatalytic reactions. A decarboxylative kinetic resolution discovered by Pasteur was the first asymmetric reaction to be noticed.[16] His investigation confirmed that

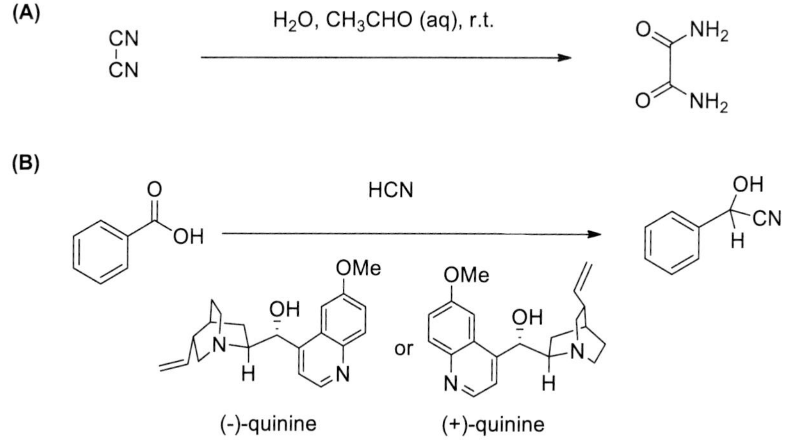

SCHEME 13.1 (A) Von Liebig's oxamide synthesis. (B) Catalysis with quinine and quinidine.

from a racemic solution of ammonium tartrate, the organism *Penicillium Glauca* destroyed more rapidly only one of the two enantiomers, that is, d-form. The first asymmetric C–C bond-forming reaction is attributed to Breding.[13o] The basis of this achievement is related to Rosenthaler's work that has attained success in preparing mandelonitrile by the addition of hydrocyanic acid to benzaldehyde in the presence of an isolated enzyme emulsion.[17] Breding also succeeded in performing an asymmetric reaction in the presence of pseudoenantiomeric quinine and quinidine alkaloids as catalysts (Scheme 13.1B).[18]

In the mid of 1950, Prelog promoted the concept of asymmetric catalysis by reinvestigating Breding's work on asymmetric cyanohydrin synthesis.[19] The progress of enantioselectivity in such types of reactions reached a significant level in late 1950 when Pracejus reported that by using *O*-acetylquinine as a catalyst, methyl phenyl ketene could be converted into (−)-α-phenyl methylpropionate with 74% ee (Scheme 13.2A).[20] This impressive result compelled other researchers to explore such reactions using the cinchona catalyst system. Hence, amino acids and other nitrogen alkaloids, such as strychnine and brucine, were investigated as organocatalysts in several asymmetric versions of organic reactions. In 1970, the impressive discovery of L-proline-catalyzed asymmetric Robinson annulation marked a key event in

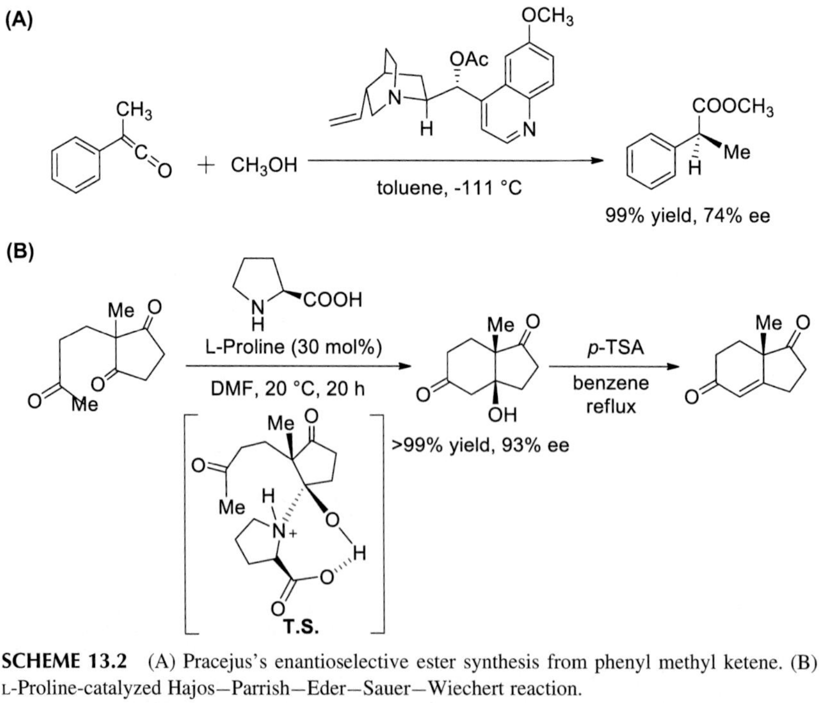

SCHEME 13.2 (A) Pracejus's enantioselective ester synthesis from phenyl methyl ketene. (B) L-Proline-catalyzed Hajos–Parrish–Eder–Sauer–Wiechert reaction.

the history of organocatalytic reaction. This intramolecular aldol reaction also called as Hajos–Parrish–Eder–Sauer–Wiechert reaction led to the formation of some key intermediates for the synthesis of natural products of chemotherapeutic values (Scheme 13.2B).[13c] Actually, this remarkable discovery marked the turning point of organocatalysis in the 1970s when a wide range of reactions proceeded via ion-pair mechanism were well investigated. Later, the reinvestigation of the Hajos–Parrish–Eder–Sauer–Wiechert reaction by the 2021 Nobel prize winners List and MacMillan opened a new arena for many established organic named reactions, such as asymmetric intermolecular cross aldol reactions, Mannich reactions, Michael reactions, Diels–Alder-type reactions, and miscellaneous reactions. Ever since, through the efforts of different research groups, the field of organocatalysis has been enriched with different types of catalysts and reactions catalyzed by them. Now, we have various derivatives of proline, cinchona alkaloids, quinine, and 4-dimethylaminopyridine (DMAP) as well as completely novel catalysts that catalyze a wide number of impressively enantioselective reactions.

13.2.2 Basic aspects of organocatalysts

Organocatalysts have not been classified in an appropriate manner till date because of mechanistic details, especially kinetic details, and of most of the organocatalytic reactions are not well explained. Asymmetric organocatalysis can follow different modes of activation, which can be classified according to the covalent or noncovalent character of the substrate–organocatalyst interaction and to the chemical nature of the catalyst (Lewis bases, Lewis acids, Brønsted bases, Brønsted acids).[13j] According to simplified catalytic cycles, it is shown that the Lewis base catalysts (**B:**) initiate the catalytic cycle by nucleophilic addition to substrates (**S**) and the complex thus formed then releases the product (**P**) and the catalyst for further turnover. Nucleophilic substrates (**S:**) are activated similarly by Lewis acid catalysts (**A**). Initiation of Brønsted bases and acid catalytic cycles are done by partial deprotonation or protonation, respectively. The different catalytic cycles are depicted in Fig. 13.1.[13j]

13.2.2.1 Lewis base catalysts

Most of the Lewis base organocatalysts are molecules containing N, O, S, C, and P atoms that tend to convert the substrate either into activated nucleophiles or electrophiles by acting through diverse mechanisms. The conventional reactive intermediates are iminium ions, acyl ammonium ions, enamines, 1-, 2-, or 3-ammonium enolates, etc. (Fig. 13.2).[13j]

13.2.2.2 Lewis acid catalysts

An important example of applications of Lewis acid catalysts is the asymmetric epoxidation of olefins. Epoxidations of stilbenes,[21] terminal olefins,[22] and

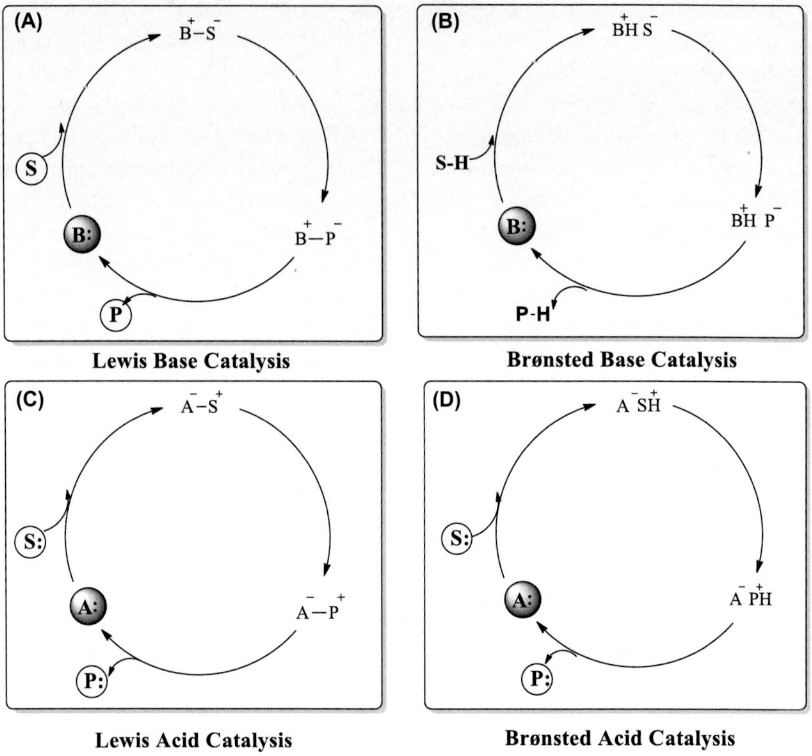

FIGURE 13.1 Four main types of organocatalytic cycles in classical catalysis.

α,β-unsaturated esters[23] were enantioselectively achieved by using carbohydrate-based ketone catalysts derived from D-fructose (Scheme 13.3A). Another example is the asymmetric epoxidation of *trans*-stilbene promoted by *N*-acetylglucosamine-based ketone organocatalyst (Scheme 13.3B).[24]

13.2.2.3 Brønsted base catalysts

Asymmetric hydrocyanation reactions, such as the Strecker reaction and cyanohydrin synthesis, are common examples of applications of Brønsted base catalysts. For example, the use of chiral C_2-symmetric guanidine as a catalyst in the asymmetric Strecker reaction (Scheme 13.4A)[25] and that of a cyclopeptide catalyst for the addition of various *N*-benzhydryl imines to α-aminonitriles (Scheme 13.4B)[26] have been successfully developed. The plausible mechanism by which they work is considered to be the interaction of hydrogen cyanide with nitrogen-containing bases via hydrogen bonding interactions to produce cyanide ions that can then add to carbonyl compounds or with imines coordinated with peptides or with guanidine hydrogen by hydrogen bonding. Additionally, the Michael reaction is done under

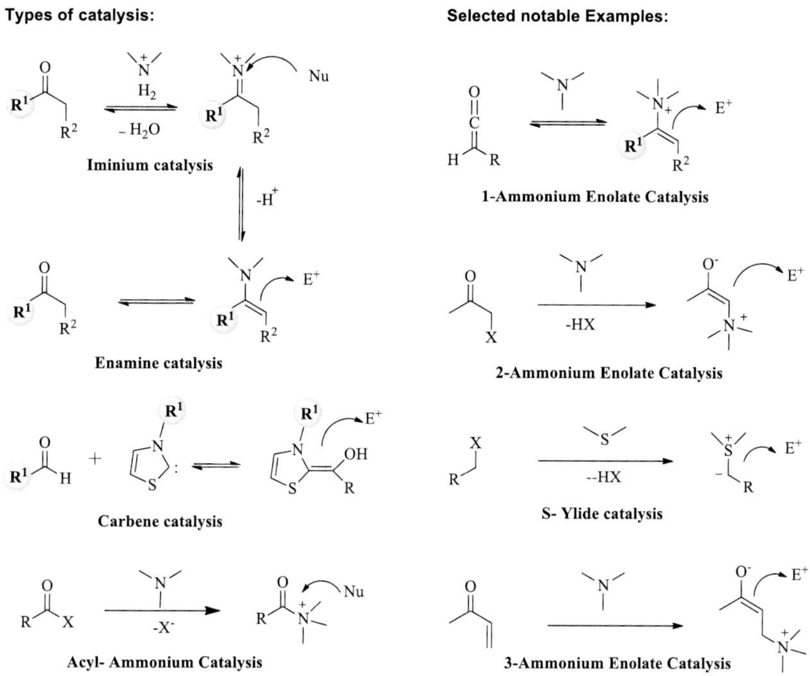

FIGURE 13.2 Examples of Lewis base organocatalysts and their applications.

solvent-free conditions of a prochiral glycine derivative *tert*-butyl diphenyliminoacetate with the help of modified guanidine as a catalyst is another example of applications of Brønsted base catalysts (Scheme 13.4C).

13.2.2.4 Brønsted acid catalysts

In general, Brønsted acid catalysis can be described as the occurrence of hydrogen-bonding interactions in the transition state of the enzyme catalysis type of reactions. For example, the bifunctional organocatalyst, also known as the Takemoto catalyst, developed by using chiral thiourea derivatives with neighboring tertiary amino groups is found to activate the nitro compounds for enantioselective Michael and aza-Henry reactions.[27,28] In the case of such reactions, the nucleophile is being activated by the tertiary amine group, and with the help of hydrogen-bonding activation, the thiourea moiety interacts with the nitro group. This can be well explained by the example of the reaction of nitroolefins with malonates to form the corresponding Michael adducts in good yields and with enantioselectivities of up to 93% ee in the presence of the chiral thiourea-derived catalyst. Similarly, another group of scientists has reported several other imine-based reactions, such as Strecker,[29] Pictet-Spengler,[30] hydrophosphonylation,[31] and Mannich reactions,[32] using chiral urea- and thiourea-derived catalysts. The mode of

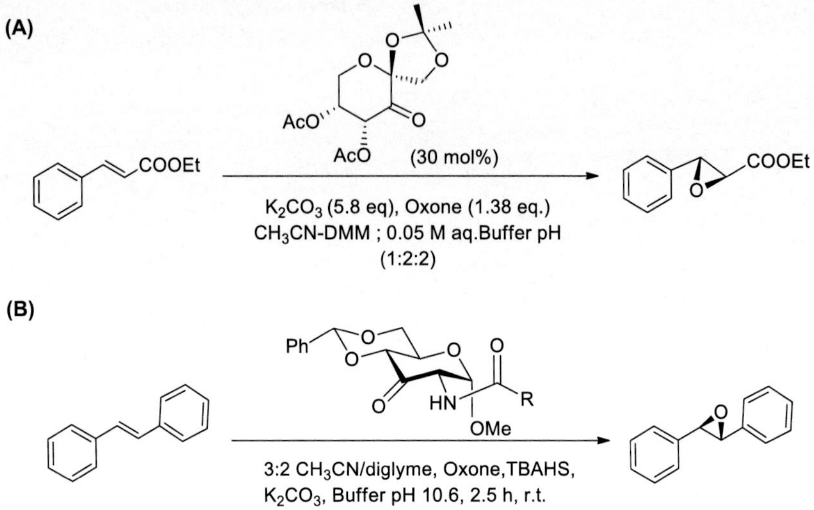

SCHEME 13.3 Examples of some Lewis acid organocatalysts and their use in asymmetric epoxidation.

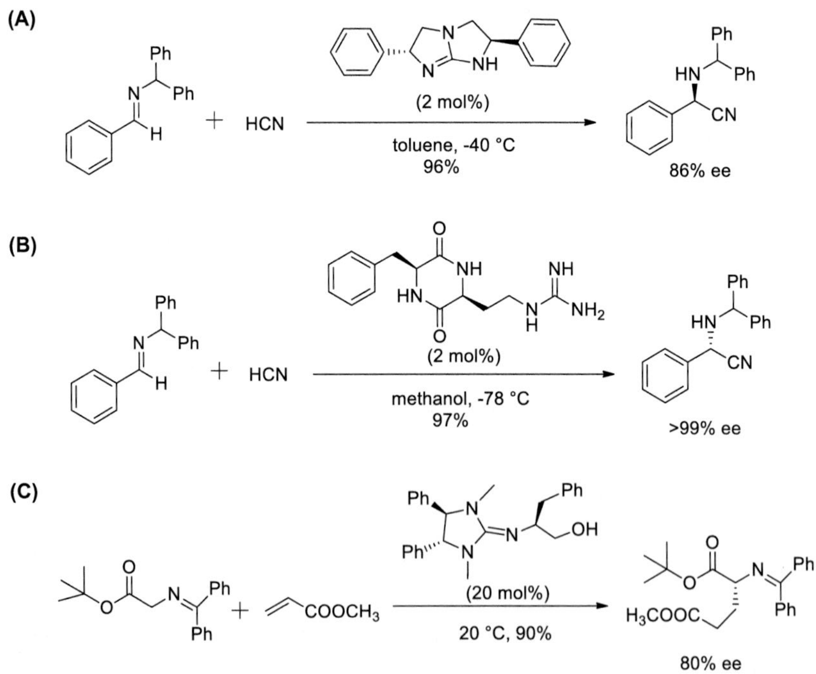

SCHEME 13.4 Examples of Brønsted base catalysts and their applications.

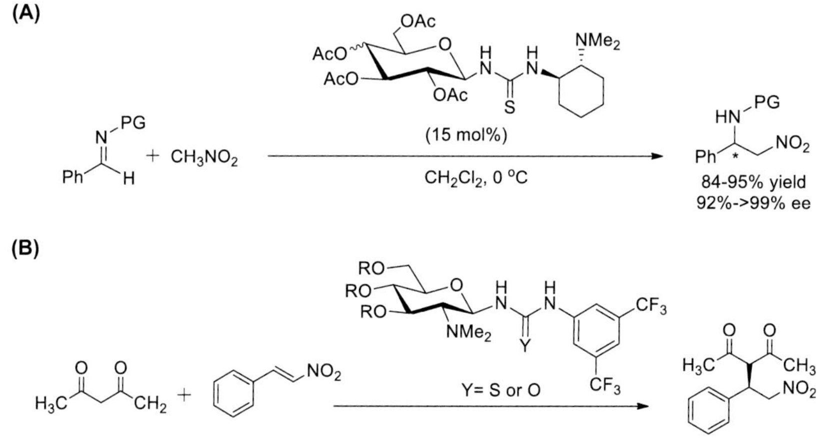

SCHEME 13.5 Examples of Brønsted acid catalysts and their applications.

activation of imine substrate through these catalysts is supposed to be via the hydrogen bonds that are formed in a bridging fashion from the urea hydrogens to the imine nitrogen.[33] Some examples of this type of organocatalysis are given in Scheme 13.5.[34,35]

13.3 Carbohydrates as potent scaffolds for organocatalysis

Along with other organocatalysts, carbohydrates have emerged as very promising catalysts. One of their advantages is that sugars constitute a relatively cheap source of chiral materials. Moreover, both the isomers of some monosaccharides and their derivatives are easily available. Some sugar pairs behave as pseudoenantiomer pairs, for example, L-rhamnose and D-mannose,[36] or D-arabinose and D-galactose (Fig. 13.3A).[37] Also, the stereoisomer of one sugar derivative can often be synthesized from other sugars, and thus the selectivity imposed by the sugar auxiliary can be reversed by using its enantiomer.[38]

The basic structure of carbohydrates also plays an important role in controlling the stereochemistry of the group attached to it. Different conformations of basic pyranose and furanose structures of sugars facilitate variations in geometry. Various hydroxy groups attached to the ring give a choice of the environment for a group to be attached and can be used as the site to attach bulky groups for blocking space around the reactive site.[36,39] The lone pairs of the ring oxygen also influence the coordination of the sugar with metal atoms or other molecules that regulate the selectivity of the asymmetric reaction (Fig. 13.3B).[40]

The conformation of a pyranose ring is governed by equatorial preferences of the substituents on the ring (steric effect) and anomeric effect that impose a stereoelectronic influence on the ring,[36,39,41] whereas bicyclic

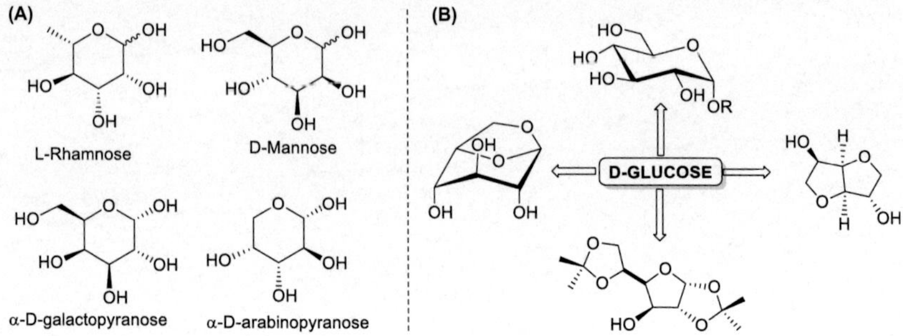

FIGURE 13.3 (A) Sugar-pseudoenantiomer pairs. (B) Some of the possible structural variations with D-glucose.

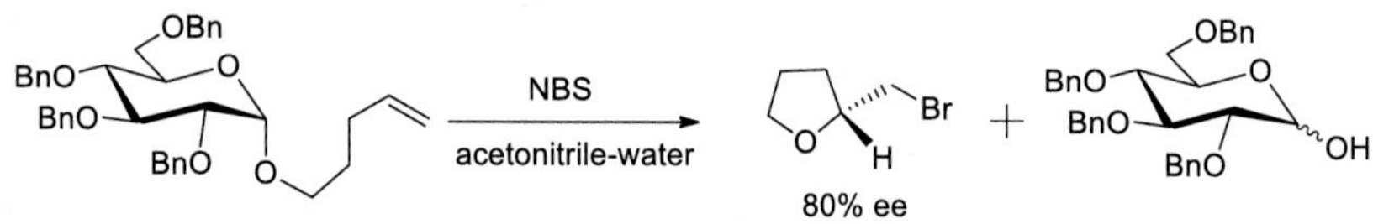

SCHEME 13.6 Stereocontrol through exo-anomeric effect of carbohydrate moiety.

acetal or ketal derivatives of a furanose ring can restrict the flexibility of its complex conformations. Thus, a rigid skeleton for stereoselective reactions can be provided by sugar moieties. The exo-anomeric effect also imposes stereo control over the reactions of glycosides by establishing a *gauche* relationship between C_1-ring oxygen bond and O_1-aglycon bond.[41] A good example of this effect is NBS-promoted enantioselective hydrolysis of *n*-pentyl α-glucosides (Scheme 13.6).[42]

The nuclear magnetic resonance (NMR) signals of protected sugars are easily distinguishable, which enhances the use of this moiety in stereoselective transformations. Protons of a ring with established conformations can be used as fixed points in nuclear overhauser effect (NOE) experiments to determine the stereochemical transformation performed on the attached group.[43]

13.4 Carbohydrate-based ketones: organocatalysts for enantioselective epoxidation

In 1996, Wang and Shi brought a promising solution to the problem of epoxidation of unfunctionalized *trans*-olefins achieved with high enantioselectivity.[21] Indeed, these authors designed a new ketone-based chiral catalyst for dioxirane-mediated asymmetric epoxidation, on the basis of structural features such as: (1) stereogenic center close to reacting center for better stereochemical communication between catalyst and substrates, (2) a quaternary center and a closed ring adjacent to the carbonyl group to minimize the epimerization, (3) other faces of catalyst blocked for competitive approaches.

Ketone **1**, structurally in accordance with above-mentioned criteria, can be easily synthesized from D-fructose in two steps and its enantiomer **1′** can be derived from L-fructose synthesized from L-sorbose (Scheme 13.7).[38] These ketones were used as catalysts with oxone as an oxidant for asymmetric epoxidation of a wide range of *trans*-olefins. The reaction was carried out with an excess of ketone **1** to avoid the loss in enantiomeric excess by decomposition of the catalyst and the reaction time was kept short. Under these conditions, high enantioselectivities of up to 98% ee were achieved in the epoxidation of unfunctionalized *trans*-olefins as well as *trans*-olefins with a variety of functional groups (Fig. 13.4).[44,45]

In 1998, further experiments to optimize the pH, solvent system, and temperature for the reaction revealed that at pH range 10.5–12 and lower temperatures (0°C to −10°C) with CH_3CN or a 1:2 (v/v) mixture of CH_3CN and dimethoxymethane as a solvent, high enantioselectivities and substrate conversions were obtained. At pH range 10.5–12, the decomposition of catalyst **1**, due to the Baeyer–Villiger reaction (proposed by Wang and Shi, Fig. 13.5), was minimized as well as the nucleophilicity of oxone was increased, which resulted in the high efficiency of catalyst **1** for enantioselective epoxidation.[38,46]

Catalyst **1** was also found able to promote the enantio- and chemoselective epoxidation of a variety of enynes exhibiting different substitutions.[39,47] In all cases of substrates, asymmetric epoxidation occurred at the olefinic bond chemoselectively with high enantioselectivity. Afterward, the experiment was extended to enynes with different substitutions, such as alkyl, trimethylsilyl (TMS), and ester groups, on the alkyne end as well as cyclic enynes of different ring sizes, which resulted in the chemoselective epoxidation of the double

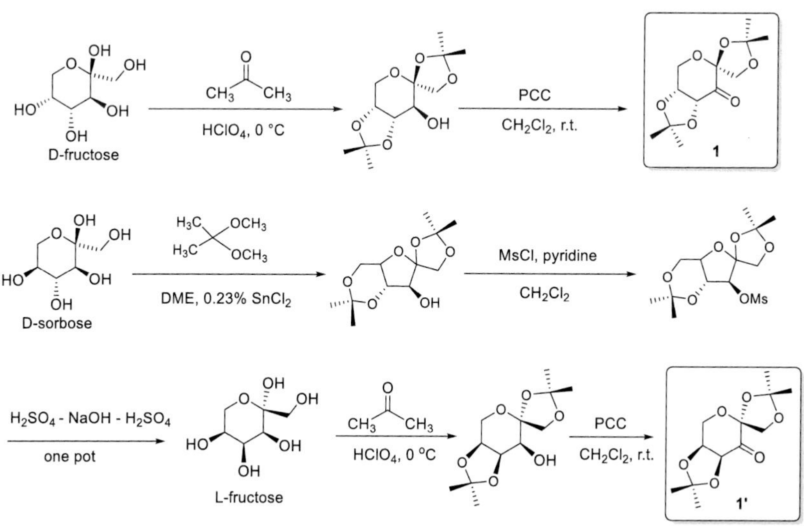

SCHEME 13.7 Synthesis of catalysts **1** and **1′**.

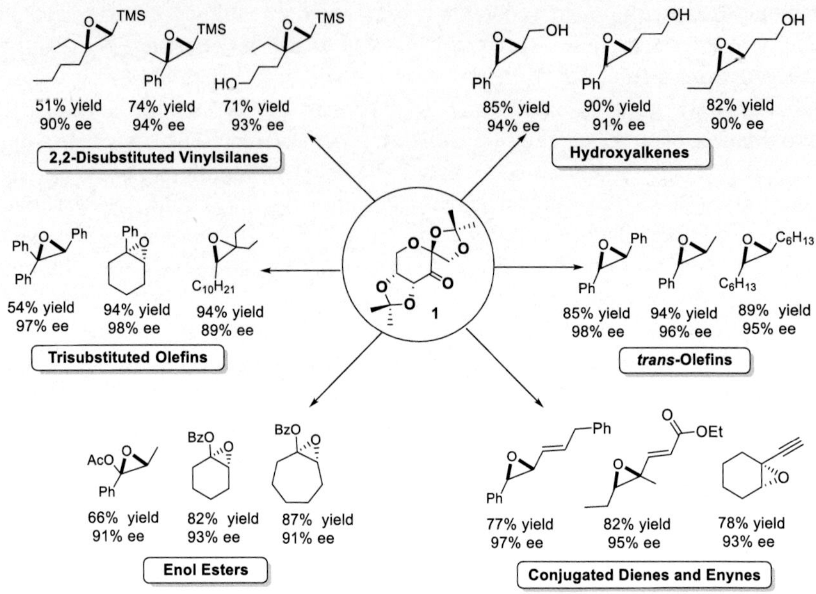

FIGURE 13.4 Enantioselective epoxidations catalyzed by D-fructose-derived ketone **1**.

bond of respective enyne. The enantioselectivity of trisubstituted enynes was found to be more stereoselective than that of isolated trisubstituted olefins, which suggests that the alkyne group was beneficial for enantioselectivity.[47] By controlling the amount of catalyst, efforts were made to apply the same epoxidation methodology to conjugated dienes, for their enantioselective monoepoxidation. High selectivity of monoepoxidation over bisepoxidation (upto 26:1) and appreciable conversion were achieved for both symmetric and asymmetric conjugated dienes when 25 mol% of catalyst **1** was used at 0°C with 1.5 hours reaction time. Results showed that the regioselectivity of monoepoxidation of asymmetric dienes was governed by steric and electronic effects of substituted groups. An electron-withdrawing group deactivated the adjacent olefin either by conjugation (if the substituted group was in conjugation with olefin) or by inductive effect (if the substituted group was not conjugated to olefin). Similarly, making one of the olefin bond sterically hindered resulted in high regioselectivity. In some examples, trisubstituted olefin was epoxidized preferably.[48] The scope of catalyst **1** was expanded with the successful application of this epoxidation method, in the case of epoxidation of enol silyl ethers and enol esters.[49] Starting with 1-*t*-butyldimethylsiloxy-1-phenylpropene, this method was applied to cyclic TMS enol ether of α-tetralone along with other silyl ethers. The resulting product was a mixture of α-hydroxy ketone and α-silyloxy ketone, which on treatment with 5% HCl/MeOH gave α-hydroxy ketone as the sole product. In the first two cases, good yield and enantioselectivity were obtained but the epoxidation products, that

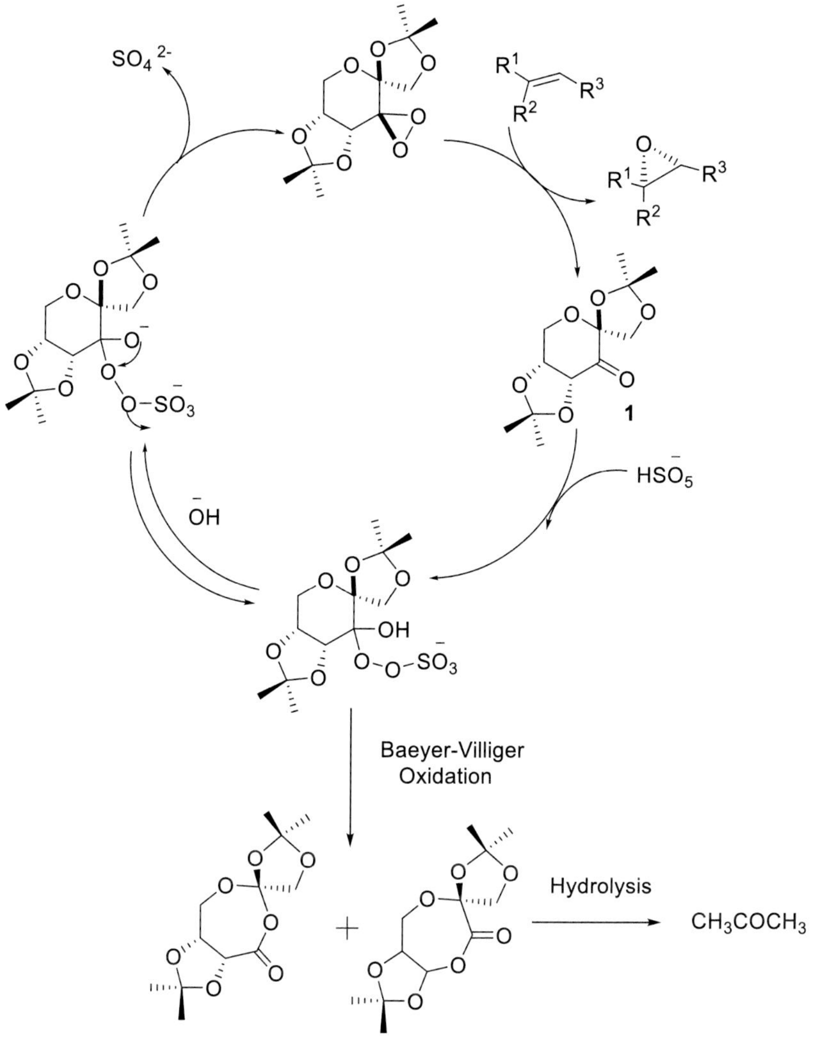

FIGURE 13.5 Possible mechanism for epoxidation catalyzed by **1**.

is, α-hydroxy ketones were prone to dimerization and racemization in other cases, especially in the case of cyclic systems. Enol acetate and enol benzoate of various cyclic and acyclic systems were epoxidized under the same conditions and isolated using triethylamine (TEA)-buffered silica gel column to afford stable epoxides in good yields and high enantiomeric excesses.[49,50] By analyzing the steric effect in the transition state in dioxirane-mediated epoxidation of alkenes, Baumstark and McCloskey proposed a spiro-transition state to be favorable over a planar transition state depicted in Fig. 13.6.[51] Formation of (*R*,*R*)-isomer of stilbene oxide predominantly in this case also supported

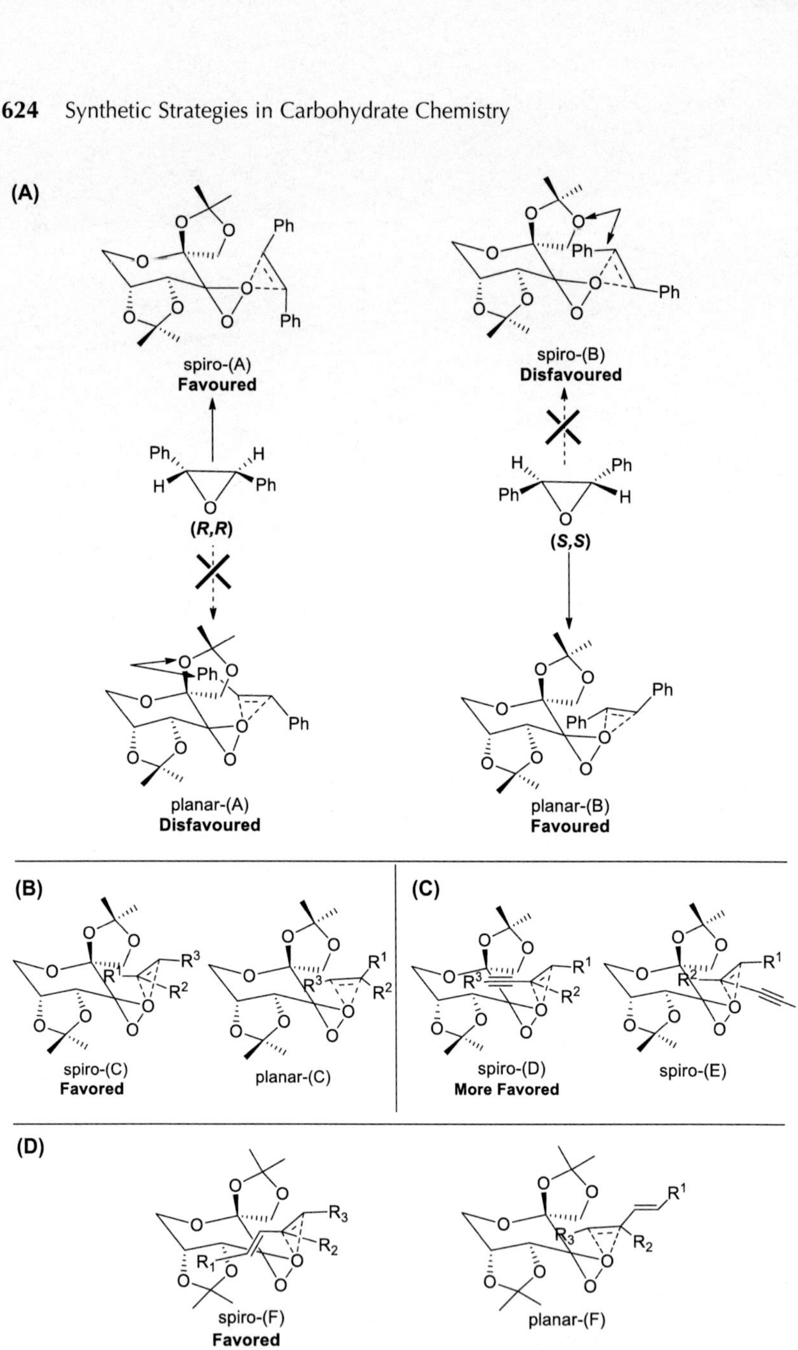

FIGURE 13.6 A) Proposed competitive transition states for epoxidation of stilbene catalyzed by **1**. B) Proposed competitive transition states for epoxidation of trisubstituted olefins catalyzed by **1**. C) Proposed competitive transition states for epoxidation of conjugated enynes catalyzed by **1**. D) Proposed competitive transition states for epoxidation of conjugated dienes catalyzed by **1**.

spiro (**A**) transition state (Fig. 13.6A). The lone pair of dioxirane oxygen is supposed to interact with π^* orbital of olefin, which enhances the stability of the spiro-transition state.[52–54] The enantioselectivity in the case of trisubstituted olefins depends on the two competitive transition states, spiro (C) and planar (C) (Fig. 13.6B). Higher enantioselectivity can be achieved by increasing the size of R^3 and decreasing the size of R^1.[39,47] In the case of conjugated enynes, the lowest unoccupied molecular orbital (LUMO) of olefin is lowered due to conjugation, so the stabilizing interaction between lone pair of dioxirane oxygen and LUMO of alkene increases, which further increases the stability of transition state and results in high enantioselectivity. Spiro state (D) is favored due to the sterically less demanding nature of the acetylene group (Fig. 13.6C).[47] For the same reason, conjugated dienes also provide high enantioselectivity regardless of whether they are disubstituted or trisubstituted (Fig. 13.6D).[48]

Instead of oxone, H_2O_2 was also applied for the asymmetric oxidation of *trans*-olefins in the presence of catalyst **1**. Best results were obtained with 10–30 mol% of catalyst **1**, 2 M K_2CO_3 in 4×10^{-4} M of ethylenediamine tetraacetic acid (EDTA) (for pH maintenance) and 3 equiv. of H_2O_2 in acetonitrile/another solvent (DMM/EtOH–CH_2Cl_2) system at 0°C. In this reaction, acetonitrile not only behaves as a solvent but also as a reagent to produce peroxyimidic acid by reacting with H_2O_2. This in situ generated peroxyimidic acid reacts with ketone catalyst to form dioxirane (Fig. 13.7).[55]

Although catalyst **1** was proven to be efficient to catalyze asymmetric epoxidation of a variety of olefins, the two main drawbacks associated with this catalyst were: (1) decomposition through Baeyer–Villiger reaction under oxidative conditions (Fig. 13.5), and (2) low enantioselectivity in case of *cis*- and terminal alkenes. To overcome these drawbacks, some modifications in the structure of the catalyst were undertaken by Shi et al., in 2001.[56] As C_4 in ketone **1** was found to have more migrating tendency than C_2 in the Baeyer–Villiger reaction, to minimize the decomposition of catalyst, electron-deficiency in C_4 was increased by replacing fused ketal in the structure of ketone **1** by oxazolidinone (**2a**, Fig. 13.8). Similarly, spiroketal of ketone **1** was replaced by nitrogen analog to create an environment in which the substituents on olefin could be electronically or sterically differentiated to enhance selectivity (**2b–2f**, Fig. 13.8).[57] In the case of catalyst **2a** (synthesis in Scheme 13.8), effective catalyst loading could be reduced to 1–5 mol% while the yield and enantioselectivity were similar as those obtained in the case of 20–30 mol% loading of catalyst **1**.

Ketones **2b–f**, the synthesis of which is depicted in Scheme 13.9, successfully provided high enantioselectivities (up to 97% ee) in the epoxidation of *cis*-olefins and terminal olefins. The favored spiro-transition state proposed for epoxidation of *cis*-olefin (Fig. 13.9) reveals the attractive interaction of R groups having π system (R_π) and oxazolidinone moiety, which results in high enantioselectivity.[57] When the nitrogen atom of oxazolidinone

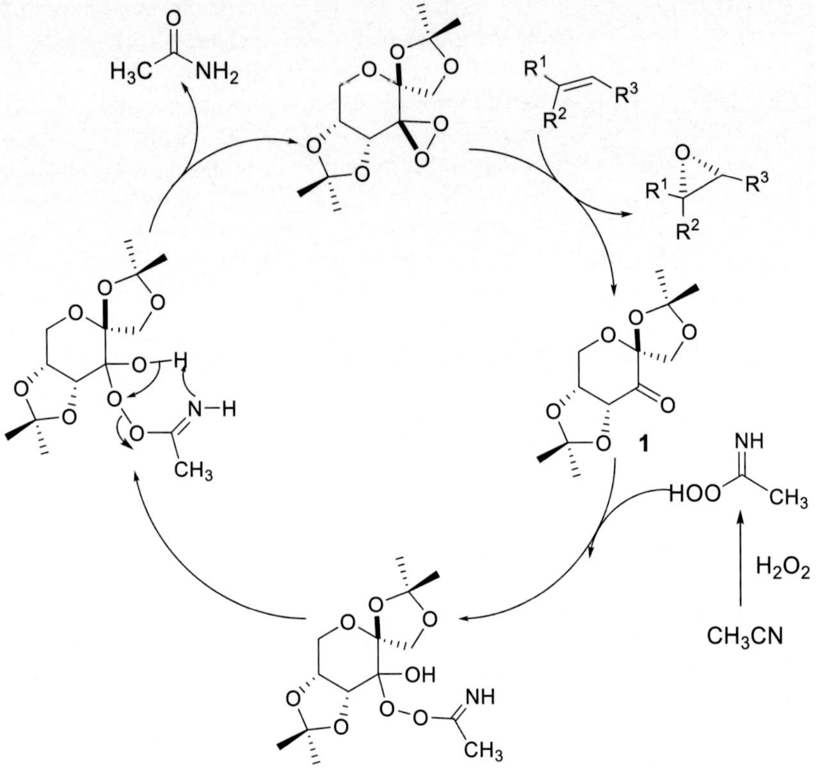

FIGURE 13.7 Possible mechanism for epoxidation catalyzed by **1** in the presence of H_2O_2 and CH_3CN.

was linked with an aryl group (ketones **2g–j**, Fig. 13.8), its efficiency to reach high enantioselectivity was increased. A further increase in selectivity was observed when the *N*-aryl group was substituted with an electron-withdrawing group. This increase in selectivity is a result of a more attractive interaction between R_π of olefin and oxazolidinone ring of the catalyst due to the presence of an electron-withdrawing group on the nitrogen atom of the oxazolidinone ring (Fig. 13.9).[59] Ketone **1** was found ineffective in catalyzing epoxidation of α,β-unsaturated esters. For this purpose, **2k** (Fig. 13.8), an acetate analog of ketone **1**, was found effective enough to give up to 96% yield and up to 97% ee in catalysis of a variety of α,β-unsaturated esters.[23] Another modification in the structure of ketone **1** was achieved in the form of its carbocyclic analogs, but epoxidation of olefins using these analogs provided both lower reactivity and enantioselectivity than reactions performed with ketone **1**, which suggests that the oxygen atom of pyranose ring of ketone **1** has an impact on both enantioselectivity and catalytic activity.[60]

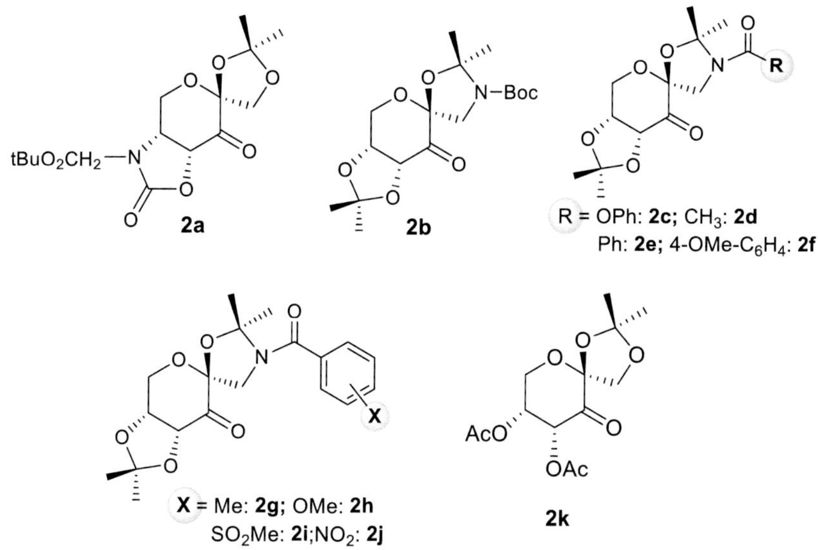

FIGURE 13.8 Structure of carbohydrate-based ketone catalysts **2a–k**.

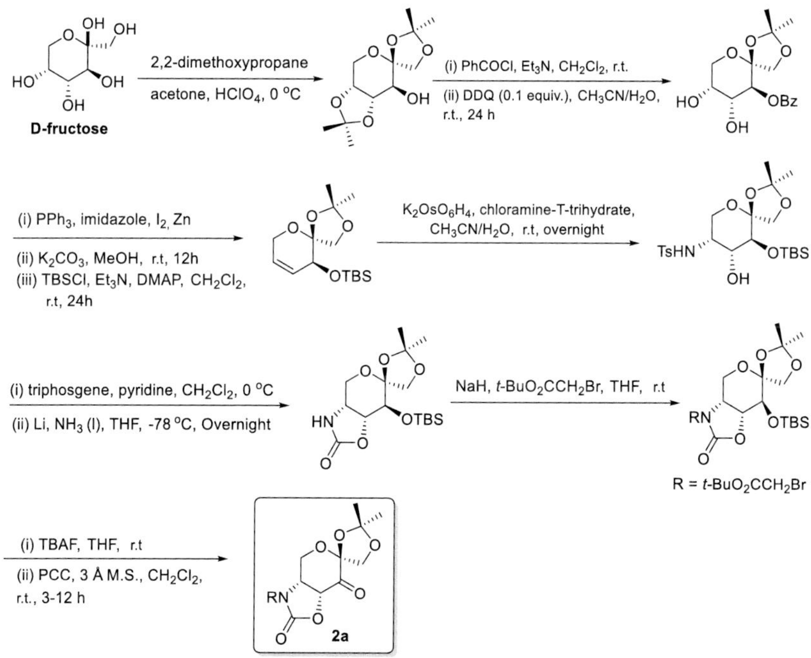

SCHEME 13.8 Synthesis of catalyst **2a**.

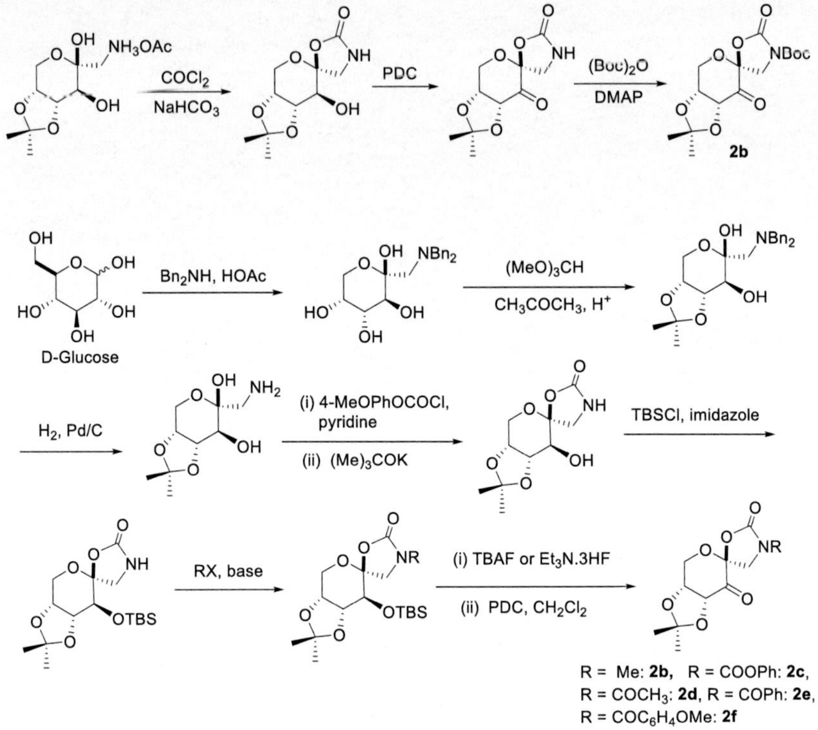

SCHEME 13.9 Synthesis of ketones **2b–2f** oxazolidinone analogs of ketone **1**.[57,58]

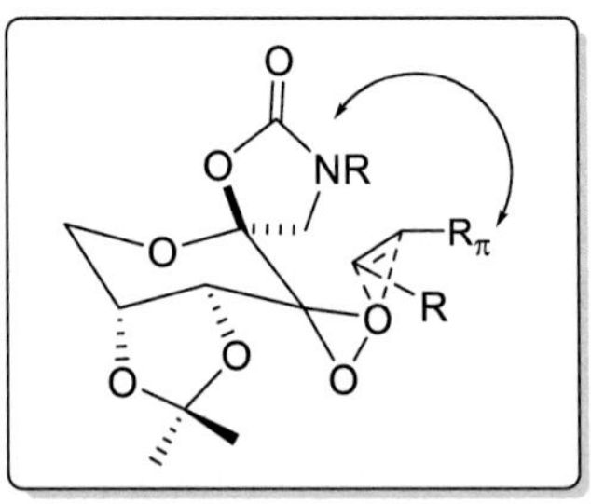

FIGURE 13.9 Favored spiro-transition state for catalysis with ketones **2b–f**.

Some ketones derived from sugars other than fructose were also screened for catalysis of asymmetric epoxidation of a variety of olefins. Among screened ketones, **3a–i** (Fig. 13.10) gave satisfactory results with respect to activity and selectivity. D-Glucose-derived ketone **1a** catalyzed epoxidation of *trans*-trisubstituted alkenes with up to 73% yield and 74% ee.[61] D-Glucosamine-derived ketone **3b** gave high catalytic activity (up to >99%

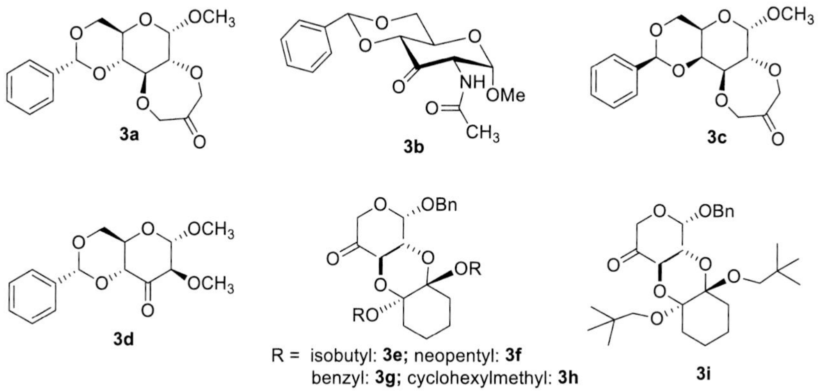

FIGURE 13.10 Ketone catalysts **3a–i** derived from sugars other than fructose.

conversion) and moderate selectivity (up to 81% ee) in epoxidation of *trans*-trisubstituted terminal olefins.[24] Ketone **3c** derived from galactopyranoside showed moderate activity (up to 74% yield) and excellent enantioselectivity (up to >99% ee) when screened for epoxidation of *trans*-trisubstituted alkenes. The (*S,S*)-enantiomer selectively formed with **3c** suggests a favorable spiro-transition state (i) for catalysis (Fig. 13.11).[62] Ketone **3d** derived from mannose also showed efficiency in catalysis of epoxidation of *trans*-disubstituted and trisubstituted alkenes in good yields and high enantioselectivities (upto 90% ee) for (*R,R*)-isomers. The most favored spiro-transition state (ii) supporting (*R,R*)-selectivity is shown in Fig. 13.11.[63] Arabinose-derived ketones **3e–i** exhibited good selectivity toward *trans*-disubstituted and trisubstituted alkenes. Formation of (*S,S*)-enantiomers with these ketones can be supported by a spiro-transition state (iii) (Fig. 13.11), which exerts a minimum steric interaction between olefin substrate and steric blocker of catalyst.[64–66]

13.5 Carbohydrate-based prolinamides, pyrrolidine amides and pyrrolidines: organocatalysts for aldol and Michael reactions

As early demonstrated by Hajos and Parrish in the 1970s,[13c] proline is a key organocatalyst having a vast area of application in asymmetric synthesis. Proline itself as well as its numerous derivatives have proved to be powerful tools for many asymmetric transformations, such as the aldol reaction, Mannich reaction, and Michael addition.[8,67] However, in some cases, proline-catalyzed reactions do not provide satisfactory enantioselectivity when performed in aqueous media. Water interrupts the hydrogen bonding of the carboxylic group present in proline, which in turn affects the transition state and results in a decrease in enantioselectivity.[68] Inspired by aldolase-catalyzed,

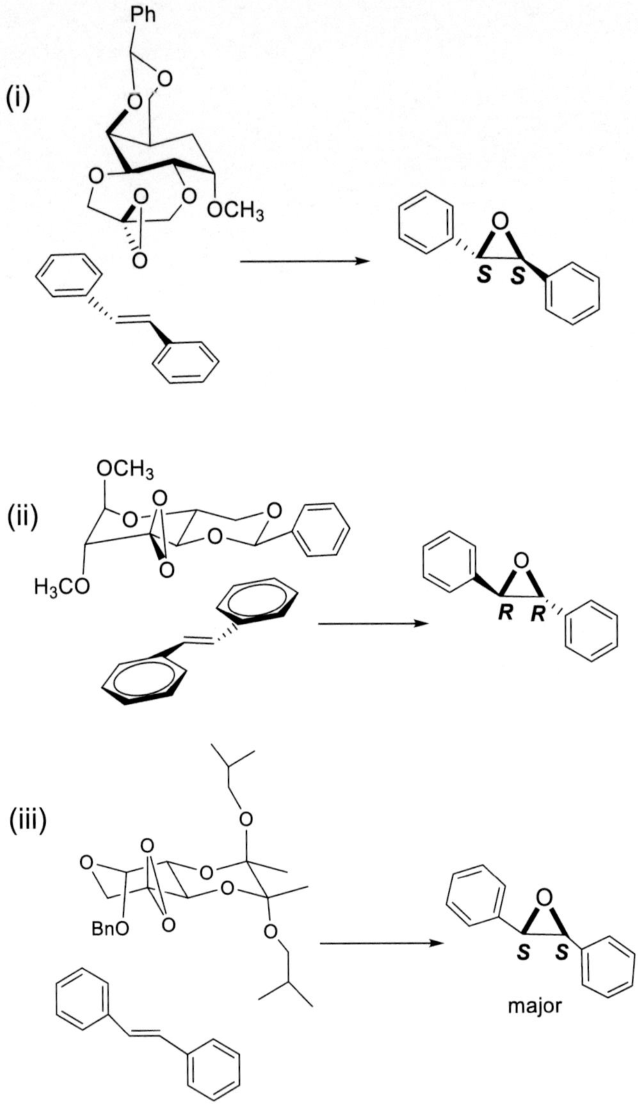

FIGURE 13.11 Favored transition states with **3c**-(i), **3d**-(ii), **3e**-(iii).

aqueous aldol reaction and the importance of the hydroxyl group of the catalyst in the proposed mechanism of aqueous aldol reaction,[69,70] In 2007, Tsutsui et al. successfully developed sugar-based prolinamides to catalyze aldol reactions in aqueous media.[71] In their studies, methyl 2-deoxy-2-(L-prolyl)amido-α-D-glucopyranoside (**4a**) and its isomer 2-deoxy-2-(D-prolyl)amido-α-D-glucopyranoside (**4b**) (Fig. 13.12) were found to be efficient catalysts

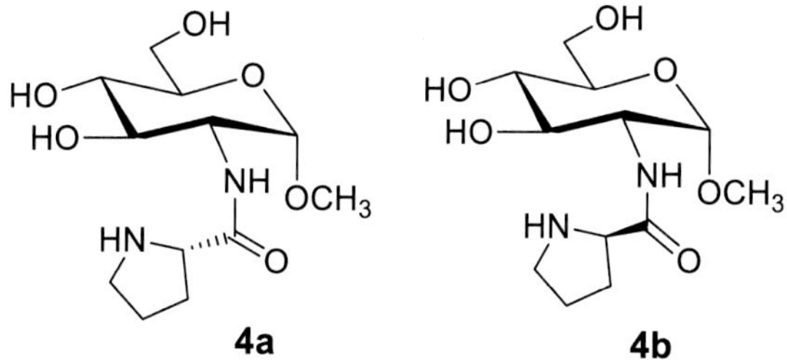

FIGURE 13.12 Sugar-based prolinamides **4a** and **4b**.

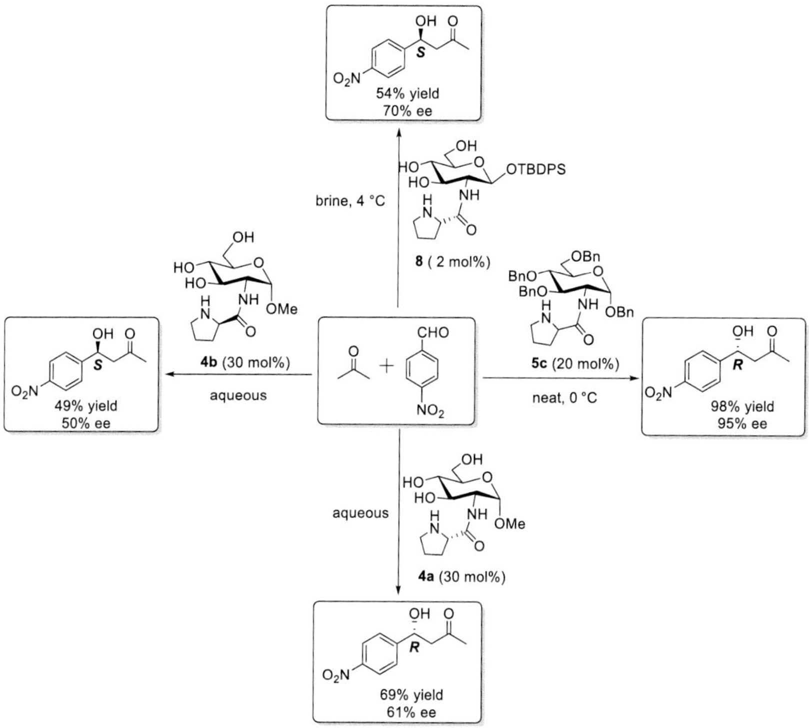

SCHEME 13.10 Aldol reaction of p-nitro-benzaldehyde with acetone catalyzed by carbohydrate-based prolinamides.

with respect to yield and enantioselectivity for model aldol reaction of acetone with 4-nitrobenzaldehyde (Scheme 13.10).[71]

Like the general trend of selectivities shown by amino acid derivatives for aldol reaction, the reaction catalyzed by **4a** provided the *R*-configured product, whereas the use of **4b** led to the formation of the *S*-product. Variations in acetone/water ratio in reaction media showed that lower content of water in reaction media resulted in a low reaction rate and high enantioselectivity. Efficiencies of these catalysts were found to be the best when the acetone/water ratio was of 20:1. With this ratio, **4a** gave aldol product with 69% yield and 61% ee, whereas **4b** led to 49% yield and 50% ee. This effect of the water content ratio in reaction media suggests that hydroxyl groups attached to the sugar moiety of the catalyst compete with water molecules and stabilize the transition state (Fig. 13.13).[55]

Later in 2010, Agarwal and Peddinti investigated sugar-based prolinamide catalysts in solvent-free aldol reactions.[72] Their initial idea consisted of blocking one face of the aldehyde with a substituted sugar moiety having an α-alkyloxy group at the anomeric position, which resulted in a probable hydrogen bonding between the carbonyl group of the aldehyde and the amide proton of the prolinamide catalyst to induce asymmetry. With these aspects, compounds **5a–c** were screened (Fig. 13.14).[72,73] Initial investigations were undertaken for the asymmetric aldol reaction of 3-nitrobenzaldehyde with cyclohexanone. Best results were obtained by using 20 mol% of catalyst **5c** at 0°C under neat conditions, delivering the product in 96% yield, 91:9 (*anti:syn*) diastereomeric ratio, and 93% ee. To check the substrate scope, the reaction was further tested with different aromatic aldehydes with various substituents (Scheme 13.11). In all cases, the yields were found high and the antiproducts were obtained with both high diastereo- and enantioselectivities. However, the rate of the reaction was found to be affected by the type and position of the substituents on the aromatic ring of aldehyde. Nitrobenzaldehydes, irrespective of the position of the nitro group, reacted with cyclohexanone with high yields (96%–98%) and high diastereomeric and enantiomeric excesses (91%–96%) in comparatively lesser reaction time (5 hours). In the case of halo-benzaldehydes, 4-fluorobenzaldehyde took a lesser reaction time (9 hours) in comparison to

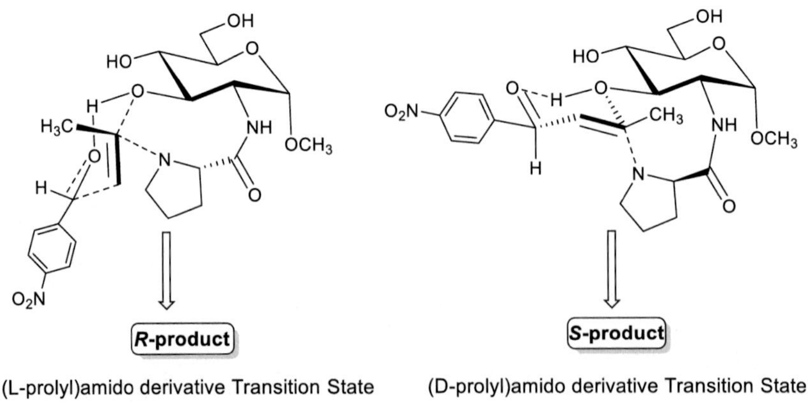

FIGURE 13.13 Proposed transition states for aldol reaction catalyzed by catalysts **4a** and **4b**.

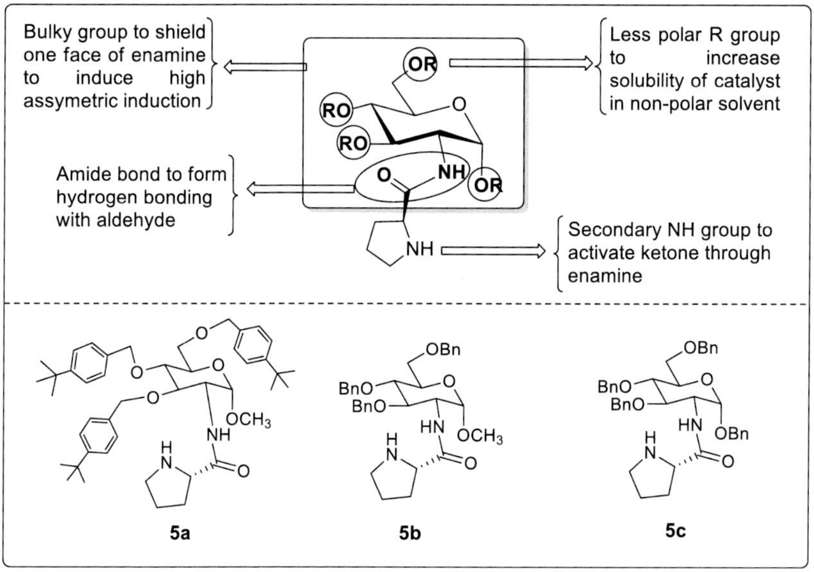

FIGURE 13.14 Structural aspects of prolinamide catalysts **5a–c** screened by Peddinti et al.

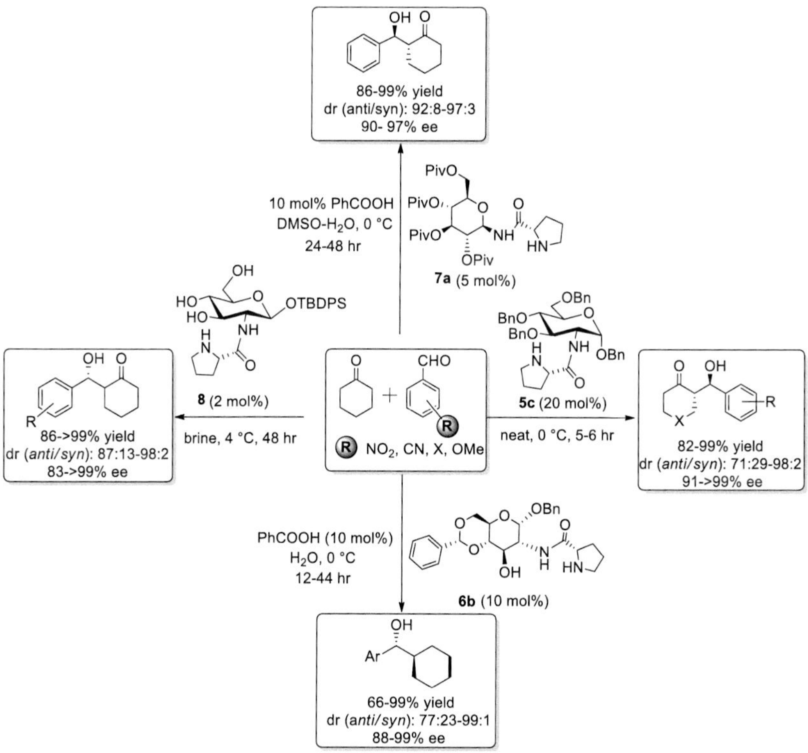

SCHEME 13.11 Aldol reactions between aryl aldehydes and cyclohexanone catalyzed by carbohydrate-based prolinamides.

other halo-benzaldehydes (reaction time 16–21 hours) in the same reaction conditions as earlier. The slowest reaction was observed in the case of 1-naphthylaldehyde and methoxybenzaldehyde but the yields and enantiomeric excesses were high in both cases. Prolinamide **5c** also catalyzed the reaction with aldol donors other than cyclohexanone, such as tetrahydropyran-4-one and tetrahydrothiopyran-4-one, and in each case, results were found excellent. However, when the reaction was performed in solvent for example in the case of solid ketone tetrahydrothiopyran-4-one, the reaction was slowed down significantly. High yield (98%), as well as high enantioselectivity (95% ee), was observed when the reaction was carried out between acetone and aromatic aldehydes (Scheme 13.10); however, in the case of cyclic pentanone, the diastereoselectivity was not found to be appreciable.[72]

Better results obtained by using catalyst **3c** in comparison to other studied sugar-based prolinamides are probably due to the fact that insertion of the benzyl group in sugar moiety increases its solubility and enhances the reaction rate. Among the two plausible transition states (i) and (ii), transition state (i) was supposed to be favorable over transition state (ii) for steric reasons (Fig. 13.15).[72,73]

In 2012, Shen et al. developed other sugar-based prolinamides to be used as promoters in aqueous aldol reactions (Fig. 13.16).[74,75] Among all these prolinamides, **6b** and **7a** were proven to be the most efficient catalysts for asymmetric aldol reaction between cyclohexanone and 4-nitro benzaldehyde performed in water. With prolinamide **6b**, the best results (95% yield and 92% ee) were obtained at 0°C and by using 10 mol% of catalyst loading in the presence of acidic additives. Evaluation of the substrate scope of the reaction revealed that it evolved smoothly with various aromatic aldehydes having different substituents on the aromatic ring and

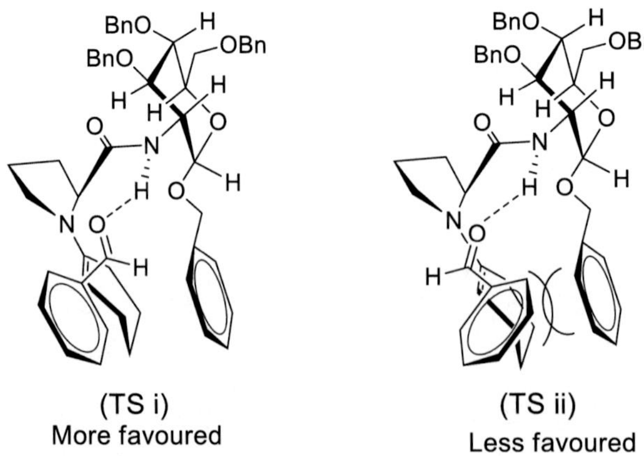

FIGURE 13.15 Two possible transition states in the case of catalyst **5c**.

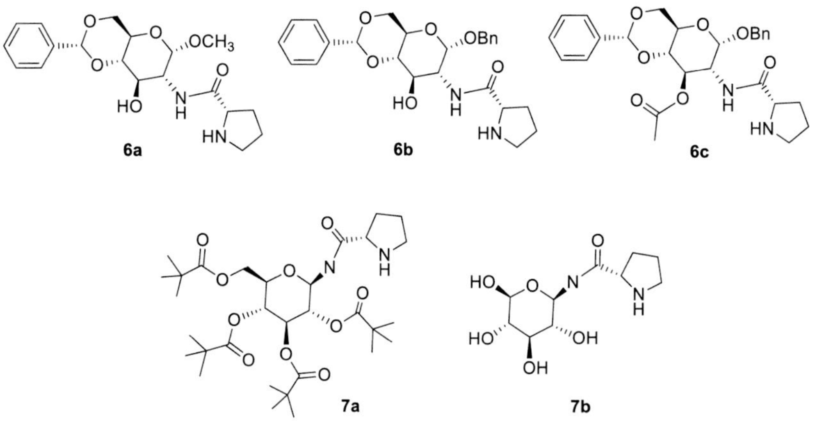

FIGURE 13.16 Prolinamides screened for catalysis by Shen.

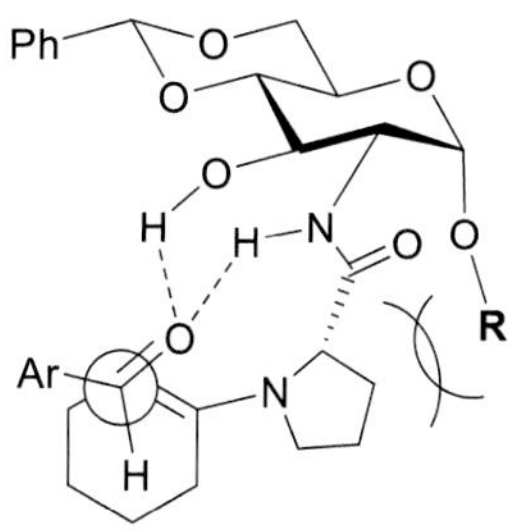

FIGURE 13.17 Plausible mechanism for the catalysis with **6b**.

different types of aldol donors, such as cyclohexanone, cyclopentanone, and butanone (Scheme 13.11). In each case (except with electron-rich aromatic aldehydes), the products were obtained in high yields (66%–99%) with moderate-to-high diastereoselectivities (54%–98% de) and uniformly high enantioselectivities (88%–99% ee). The main feature of catalyst **6b** for aldol reaction is that it can be recovered very easily from the reaction mixture after completion of the reaction. The recovered catalyst could be used for subsequent aldol reactions without the addition of a new amount of catalyst. On the basis of the stereochemistry of the product, the catalyst was supposed to follow an enamine mechanism and plausible transition state depicted in Fig. 13.17.[74]

In continuation of their work, Shen et al. synthesized two new sugar-based prolinamide catalysts **7a** and **7b** and examined the efficacy of these catalysts on various parameters for enantioselective aldol reaction (Fig. 13.16).[75] Catalyst **7a** provided good enantioselectivity and moderate yield, whereas the use of catalyst **7b** resulted in moderate ee value and good yield when applied to an aldol reaction of 4-nitrobenzaldehyde and cyclohexanone in the presence of benzoic acid as an additive. The use of a 1:1 DMSO/water mixture as the solvent at 0°C and that of 5 mol% of catalyst **7a** constituted the most suitable reaction conditions. The reaction of a variety of electron-deficient and electron-rich aromatic aldehydes as well as less reactive benzaldehyde gave good yields (86%–99%) and high enantioselectivities (90%–97% ee) (Scheme 13.11). In all cases, the *anti*-diastereomers were produced selectively with high anti/*syn* ratios (92:8–97:3). In 2012 a novel catalyst was synthesized by Nisco et al. from prolinamide of glucosamine by substitutingan hydrogen atom of the anomeric hydroxyl group by a *tert*-butyldiphenylsilyl (TBDPS) group (Scheme 13.12).[76] This prolinamide **8** was found to be highly efficient in the asymmetric aldol condensation of cyclohexanone with various substituted benzaldehydes (Scheme 13.11). High yields combined with both high diastereo- and enantioselectivities were obtained in each case by using 2 mol% of catalyst loading at 48°C in brine after 48 hours of reaction time. However, in the reaction of 4-nitrobenzaldehyde with acetone, only both moderate yield (54%) and enantioselectivity (70% ee) were obtained (Scheme 13.10). This is due to the extremely low solubility of aromatic aldehydes in a hydrophilic medium (brine), which makes them move toward the hydrophobic site of the catalyst and orient themselves efficiently for enamine nucleophilic attack (Fig. 13.18). The catalyst could be easily recovered for further use. The efficiency of this catalyst can be an account of the aggregation of catalyst molecules in aqueous media to form a hydrophobic microenvironment, which provides a suitable pocket for the reaction to occur. At lower temperatures, as the freedom of molecules is restricted, the aggregation of molecules to form a hydrophobic environment is facilitated, which leads to better reaction results.[76]

A pyrrolidine ring attached to a carbohydrate moiety by either an amide or any other linkage has been proven to be efficient in catalyzing asymmetric

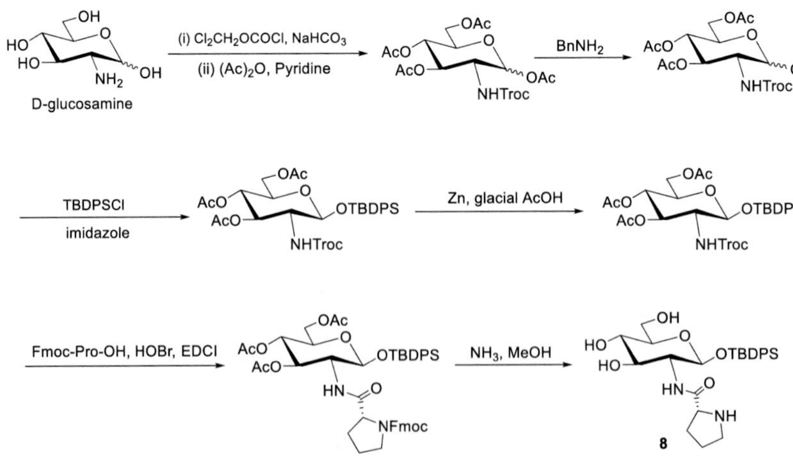

SCHEME 13.12 Synthesis of catalyst **8**.

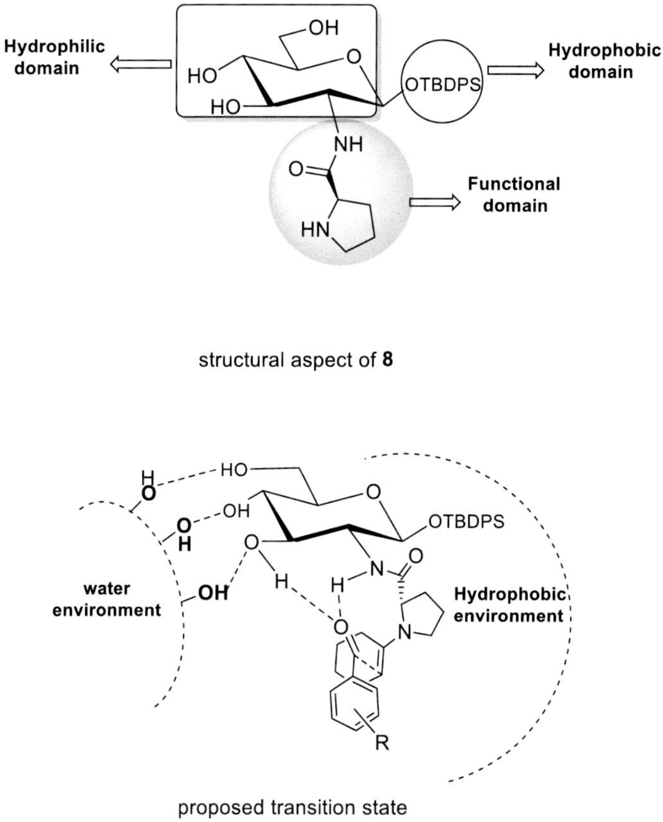

FIGURE 13.18 Structural aspects of catalyst **8** and proposed transition state.

Michael reactions. For example, in 2010 Wang et al. designed a sugar-based pyrrolidine catalyst to be applied to promote Michael's addition of ketones to various nitroolefins.[77] Catalyst **9** was synthesized by click reaction of an L-proline-derived azide with an α-D-glucose-derived alkyne (Scheme 13.13). The Michael addition of different cyclic ketones to various aromatic nitroolefins performed in the presence of 10 mol% of catalyst **9** in the absence of any solvent led to the corresponding products with high yields (98%), high diastereoselectivities (*syn*/anti 94:6–99:1), and high enantioselectivities (91– > 99% ee) at room temperature. However, the reaction between cyclopentanone and acetone as Michael donors afforded excellent yields but moderate-to-good diastereoselectivities and enantioselectivities. Sugar-based prolinamides **5a–c** (Fig. 13.14) were also screened for catalysis of Michael reaction by Agarwal and Peddinti, in 2011.[78] The involvement of 20 mol% of catalyst **5c** in the presence of 20 mol% of benzoic acid as additive provided the best results in the asymmetric Michael reaction of cyclohexanone

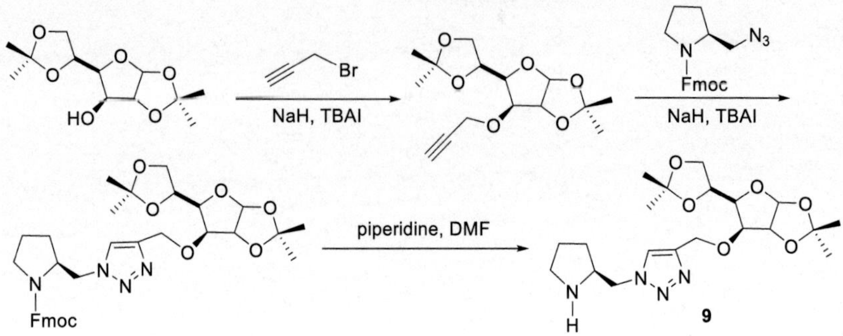

SCHEME 13.13 Synthesis of catalyst **9**.

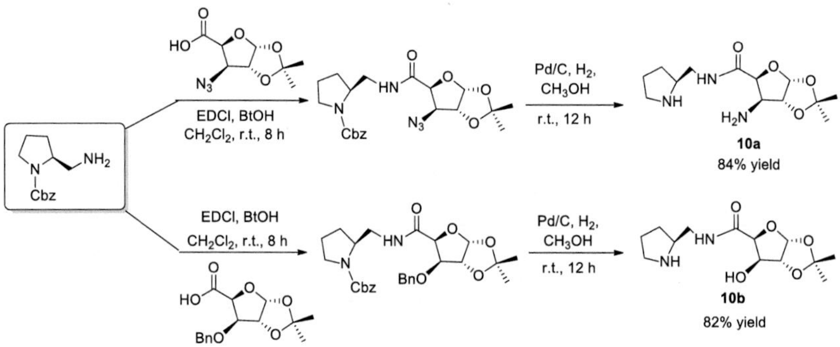

SCHEME 13.14 Synthesis of catalysts **10a–b**.

with various substituted nitrostyrenes under neat conditions at −20°C. In all cases, high yields (88%–96%), good *syn/anti*-diastereoselectivities (86:14–>99:1), and moderate enantioselectivities (68% ee) were obtained (Scheme 13.15).

In 2014, Kumar and Balaji developed sugar-based pyrrolidine amide catalysts **10a** and **10b** derived from pyrrolidine amine and D-glucose-derived sugar acids (Scheme 13.14) to be applied as promotors of asymmetric Michael additions of ketones to nitroolefins.[79] The reaction employed 20 mol% of catalyst **10b** in the absence of any additive at room temperature under neat conditions. In these conditions, the Michael addition of cyclohexanone to differently substituted β-nitrostyrenes resulted in good yields (88%–96%) as well as high diastereomeric ratios (*syn/anti*- = 93:7–99:1) and good enantioselectivities (85%–91% ee) (Scheme 13.15). Although catalyst **10a** also gave good yield and good diastereoselectivity under the same reaction conditions, the observed enantioselectivity was low. A probable reason behind this low enantioselectivity was

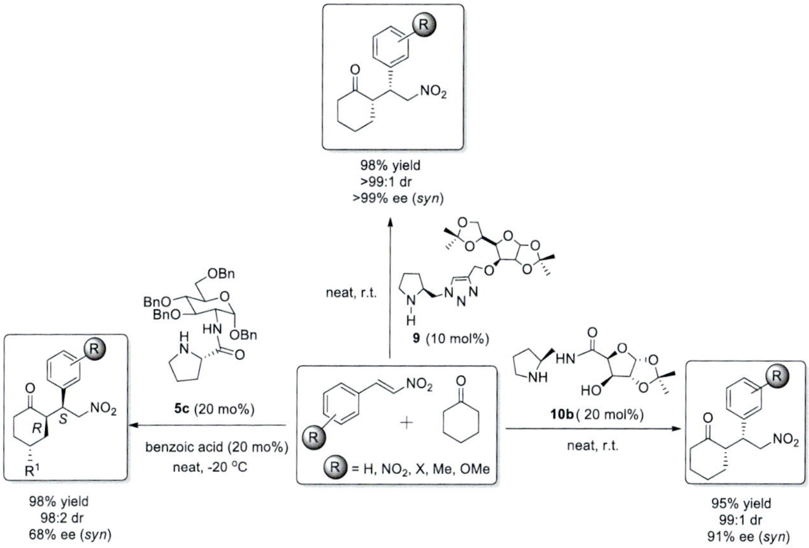

SCHEME 13.15 Michael additions of cyclohexanone to β-nitrostyrenes catalyzed by carbohydrate-based prolinamide **5c** and pyrrolidines **9** and **10b**.

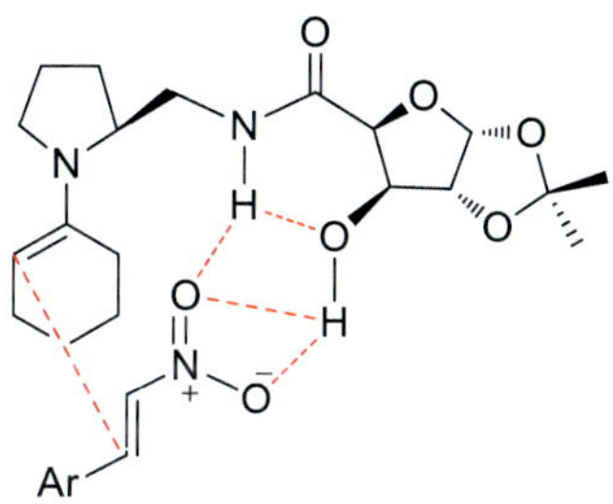

FIGURE 13.19 Proposed transition state for Michael addition catalyzed by **10b**.

given as the enamine formation with the primary amine of sugar moiety in catalyst **10a** may also activate ketone along with the secondary amine of the pyrrolidine ring in the catalyst. In the proposed transition state for the reaction with catalyst **10b**, the secondary amine present in the pyrrolidine ring of the catalyst plays the role of activating ketone through enamine formation, whereas the sugar moiety of the catalyst incorporates stereoselectivity. The free hydroxyl group present in the catalyst enhances the stabilization of the nitroolefin through additional hydrogen-bonding interactions along with the amide group (Fig. 13.19).

13.6 Carbohydrate-derived amino alcohols: organocatalysts for asymmetric aldol and Mannich reactions

To the best of our knowledge, the first example of a carbohydrate-based amino alcohol catalyst was reported in 2008 by Singh et al. when naturally occurring D-glucosamine, a biologically important aminosugar, was found to be able to catalyze an aldol reaction.[80] Indeed, 10 mol% of D-glucosamine efficiently catalyzed the aldol reaction of cyclohexanone with aromatic aldehydes at room temperature in the water as a solvent to afford the corresponding aldol adduct (and no dehydrated product) in moderate-to-good yields (59%–75%). The ratio of *syn*- and *anti*-products (0.8:1–4:1) was found to be dependent on the nature and position of substituents exhibited on the aromatic aldehyde (Scheme 13.16). Unfortunately, the enantioselectivities were found low (10%–54% ee).

Later in 2011, Peddintiand Agarwal investigated the catalytic effect of differently protected glucosamines in the asymmetric aldol reaction of cyclohexanone with aromatic aldehydes.[81] Among the screened molecules, glucosamine derivative **11**, the synthesis of which is detailed in Scheme 13.17, was found to be the most efficient for catalyzing the reaction. For example, the aldol reaction between cyclohexanone and 3-nitrobenzaldehyde led to the corresponding enantiopure (>99% ee) *anti*-product in 96% yield by using 20 mol% of catalyst **11** at −20°C under neat conditions. Further experiments for the substrate scope revealed that the reaction evolved regardless of the nature and position of the substituents on the aromatic aldehyde (Scheme 13.18). In all cases, catalyst **11** was found efficient to produce the *anti*-aldol products in high-to-excellent chemical yields (80%–98%) and moderate-to-excellent enantioselectivities (47– >99% ee). However, the ee values for the *syn*-products were low to good (11%–90% ee) in all cases. The success of catalysis by **11** was considered to be a result of its proline-like functionality. A free hydroxyl group in **11** was supposed to be involved in hydrogen bonding with the carbonyl group of the substrate.

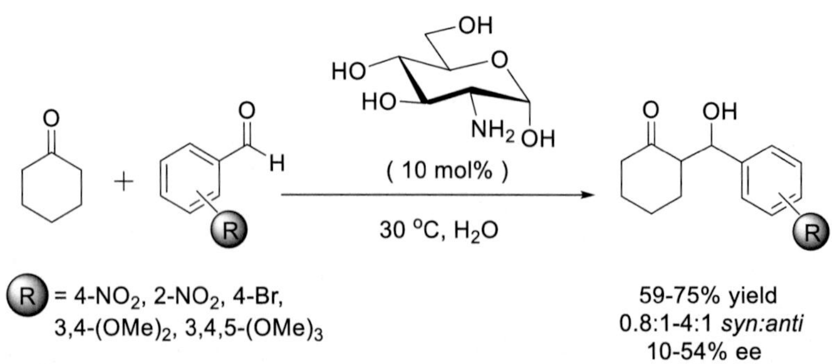

SCHEME 13.16 Aldol reaction catalyzed by D-glucosamine.

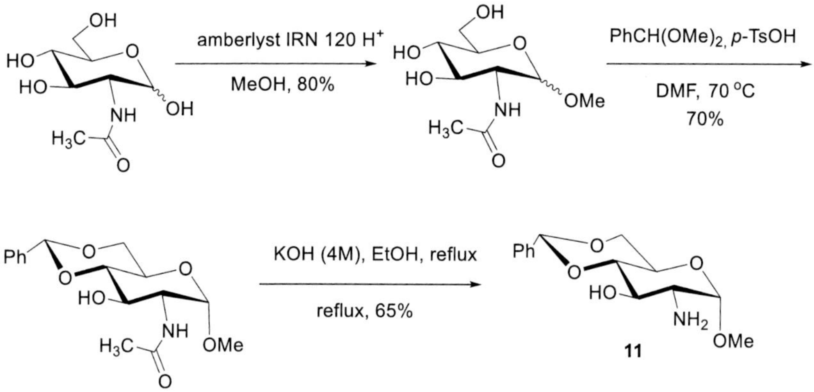

SCHEME 13.17 Synthesis of catalyst **11** from *N*-acetyl D-glucosamine.[82,83]

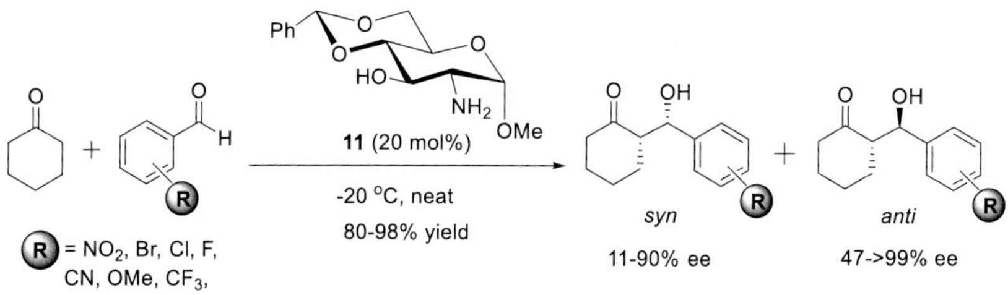

SCHEME 13.18 Aldol reaction catalyzed by catalyst **11**.

A few months later, Shen et al. presented an example of asymmetric induction in an aldol reaction between ketones and isatins catalyzed by carbohydrate-based amino alcohols.[84] The authors developed amino alcohols **12a** and **12b** (Fig. 13.20) derived from D-glucosamine for catalysis of these reactions. Catalyst **12b** was found to be more efficient as compared to catalyst **12a** with respect to yield and enantioselectivity. The involvement of 10 mol% of catalyst **12b** at 0°C in dichloromethane as the solvent provided the best results. Aldol reaction between differently substituted isatins and acetone under these conditions afforded moderate-to-excellent yields (80%–99%) and moderate enantiomeric excesses of *S*-isomers (63%–75%ee) in 1–4 days reaction time (Scheme 13.19). Although the yield of the reaction using catalyst **12a** was good, the observed enantioselectivity was lower. Also, reactions with isatins having electron-withdrawing groups as substituents gave lower enantioselectivities as compared to electron-donating group-substituted isatins. The reaction time for the isatins easily soluble in solvents (such as *N*-methyl- and *N*-benzyl-protected isatins) was found to be much lesser (1 day) in comparison to other isatins (reaction time 4 days). However, the reaction of cyclohexanone with isatin gave better yields and enantioselectivities when water was used as the solvent.

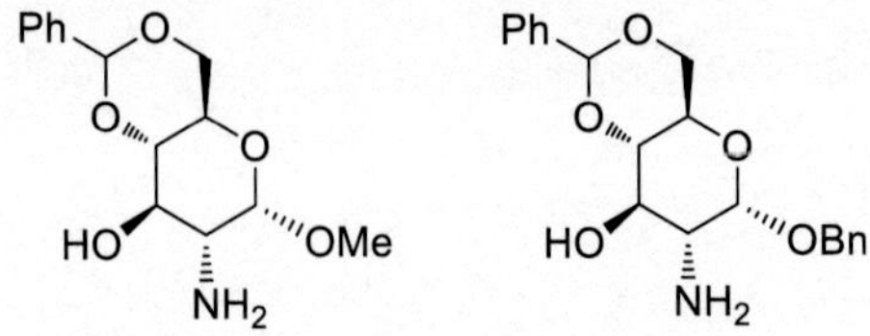

12a **12b**

FIGURE 13.20 Amino alcohols **12a–b** for aldol reaction.

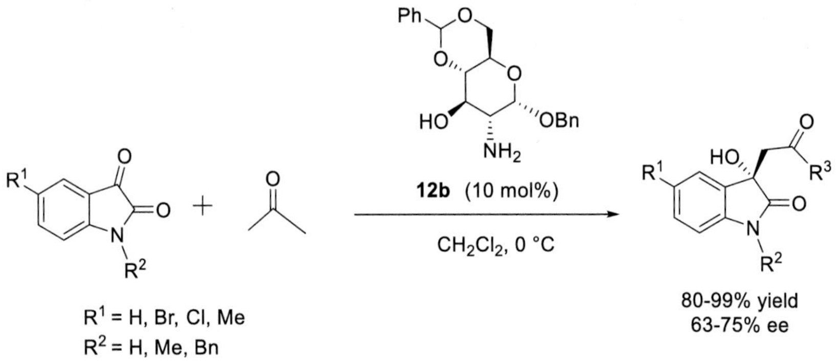

SCHEME 13.19 Aldol reaction between isatins and acetone catalyzed by **12b**.

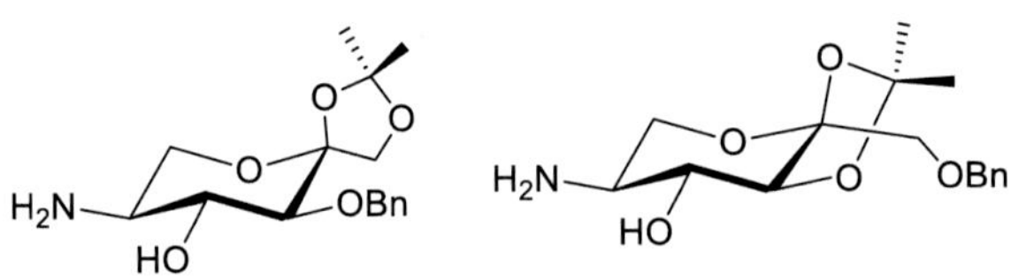

13a **13b**

FIGURE 13.21 Amino alcohols **13a–b** screened for aldol reaction.

In 2013, Li et al. selected D-fructose-derived β-amino alcohols **13a** and **13b** (Fig. 13.21) for catalysis of asymmetric aldol reaction of cyclic ketones with arylaldehydes.[85] After analyzing the effects of additives, temperature, and solvents on the reactions catalyzed by **13a** and **13b**, the optimal conditions for efficient catalysis of aldol reaction between various substituted benzaldehydes and cyclic ketones were found to be 20 mol% of catalyst **13a** combined with 15 mol% of *p*-nitrophenol as an additive at −20°C or −40°C in the absence of solvent (Scheme 13.20). Yields (21%–98%) and enantioselectivities (6%–74% ee for *syn*-products, 16%–87% ee for *anti*-products) of the reaction were found to be dependent on the electronic nature as well as the position

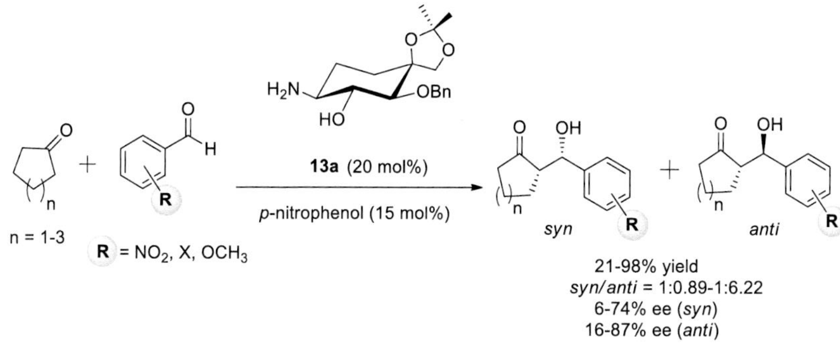

SCHEME 13.20 Aldol reaction of cyclic ketones with aromatic aldehydes catalyzed by **13a**.

of the substituent exhibited on the aromatic ring of the aldehyde. For example, 3-nitrobenzaldehyde reacted with cyclohexanone to give the corresponding product in 94% yield with 59% ee for the *syn*-product and 83% ee for the *anti*-products at −20°C in 23 hours reaction time, whereas the reaction of 2-nitrobenzaldehyde provided 96% yield albeit combined with lower ee values (45% ee for the *syn*-product and 45% ee for the *anti*-products) under the same conditions. Halo-substituted benzaldehydes produced comparatively lower yields and enantioselectivities, whereas methoxy substituent on the aromatic ring enhanced the selectivity. Temperature also affected yield and selectivity as well rate of reaction. In most cases, reaction at −20°C afforded both higher yield and enantioselectivity with shorter reaction time as compared to reaction performed at −40°C. Also, cyclohexanone gave better results regarding yield and enantioselectivity as compared to cyclopentanone and cycloheptanone probably because of lower ring strain.

A possible mechanism for this reaction followed three steps: (1) dehydration of carbinolamine intermediate, (2) nucleophilic attack of the amine to carbonyl group of cyclic ketone, (3) C−C bond formation and hydrolysis of amine-aldol. Activation of the carbonyl group of the aromatic aldehyde was supposed to be done through hydrogen bonding with *p*-nitrophenol and hydroxyl group of fructose derivative at C-4 position (Fig. 13.22). A recovery of the catalyst (77%−84%) constituted an advantage of using this catalyst.

In 2018, Peddinti and Sharma disclosed an enantioselective three-component Mannich reaction catalyzed by 20 mol% of carbohydrate-based primary amine **11**.[86] This catalyst arose from commercially available *N*-acetyl-D-glucosaminehydrochloride. The Mannich reaction was performed at room temperature in the presence of benzoic acid as an additive in dichloromethane as the solvent. It resulted in the formation of the corresponding chiral Mannich products in generally high yields (72%−94%) as major *syn*-diastereomers (20%−90% de) and variable enantioselectivities

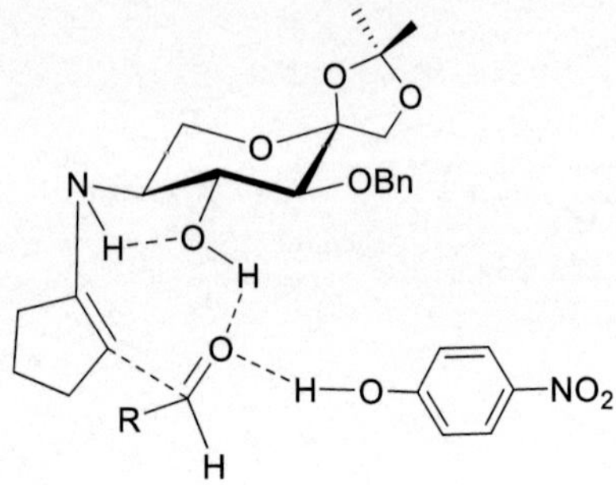

FIGURE 13.22 Transition state for aldol reaction catalyzed by **13a**.

(20%–99% ee), as shown in Scheme 13.21. The catalyst system tolerated the presence of different substituents at the *para* position of the aniline and at the *para*- or *meta*-positions of the benzaldehyde. The stereochemical outcome significantly depended on the electronic nature of these substituents. For example, enantioselectivities were generally high (80%–99% ee) except in cases of *p*-methoxy- and *p*-methylbenzaldehydes which afforded the corresponding chiral β-amino carbonyl compounds with lower enantioselectivities (20%–60% ee). This work constituted the first sugar-derived amine-catalyzed direct asymmetric Mannich reaction using cyclohexanone, an aniline, and a benzaldehyde, through the generation of an imine in situ. The authors proposed the mechanism depicted in Scheme 13.21 in which the Mannich reaction proceeded through an enamine pathway, as the nucleophilic addition of the enamine generated in situ would be faster on an imine than on an aldehyde. Indeed, the sequence begins with the formation of enamine **A** from the reaction between cyclohexanone and the catalyst. The stereochemical outcome can be explained by the plausible transition state **B**. The hydroxy group of the sugar moiety activates the imine through hydrogen bonding and also allows the *Re*-face of the enamine to attack the *Si*-face of the imine, leading to a *syn*-stereochemistry between the two stereogenic centers.

In 2020, an inclusion complex was crafted for the first time by Agarwal and Kaur between a carbohydrate-based molecule and a β-cyclodextrin hydrophobic cavity to be applied as promotors in asymmetric aldol reactions.[87] This novel D-glucosamine-based inclusion catalyst was synthesized in high yield using an innovative and proficient acoustic cavitation technology and well characterized using various techniques, such as infrared spectroscopy, nuclear magnetic resonance spectroscopy, ultraviolet – visible spectroscopy, thermogravimetric analysis, scanning electron microscopy, and other spectroscopic techniques. It was observed that the inclusion of a D-glucosamine derivative into the hydrophobic cavity of β-cyclodextrin increased its surface area and thermal stability. This catalytic system worked well at 10 mol% of catalyst loading in water at 5°C

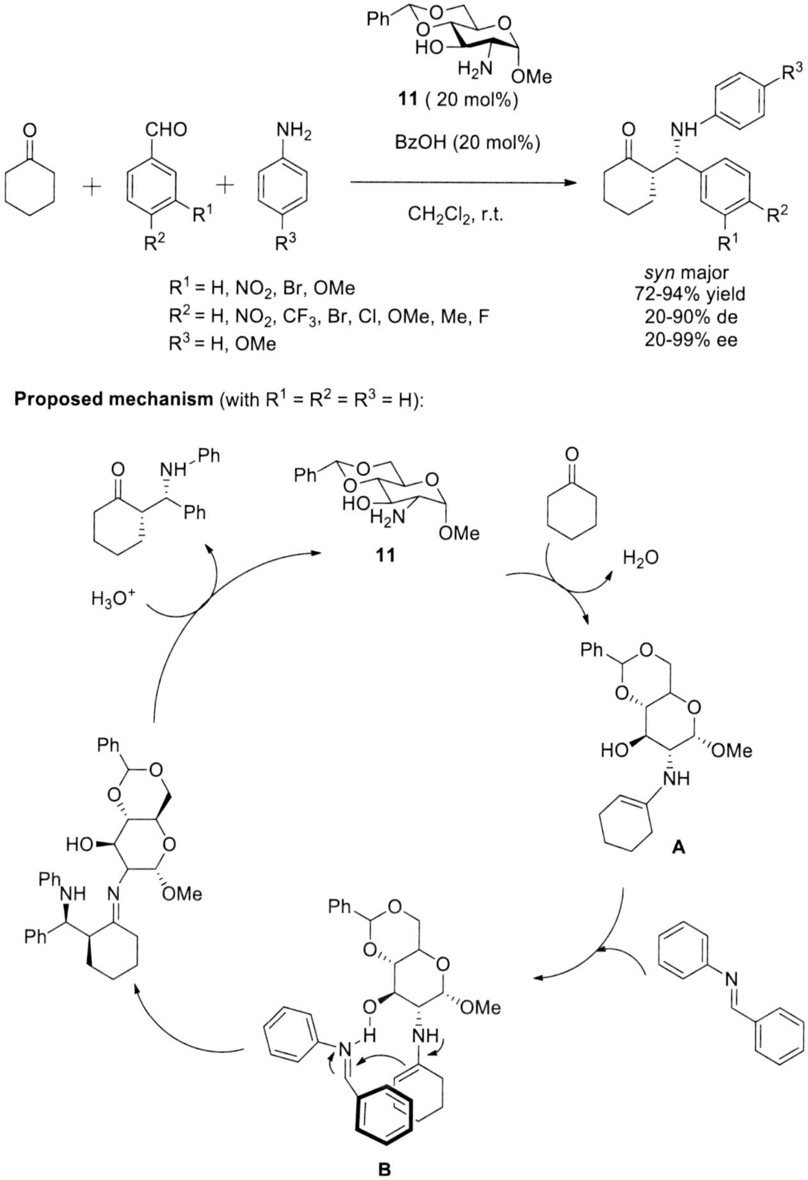

SCHEME 13.21 Mannich reaction catalyzed by **11**.

for the direct aldol reaction between cyclohexanone and variously substituted benzaldehydes to afford the corresponding chiral products in uniformly excellent yields (89%–99%) with low to complete *syn*-diastereoselectivities (8– >98% de) and variable enantioselectivities (10%–99% ee).

13.7 Carbohydrate-based thiourea amines: bifunctional organocatalysts for Michael, Biginelli, aza-Henry, and Mannich reactions

Thiourea catalysts are known to induce asymmetry through hydrogen-bonding interactions with the substrate.[88–90] This class of noncovalent organocatalysts is highly efficient to catalyze reactions with catalyst loadings as low as 0.001 mol%.[91b] Inexpensive synthesis and easy handling are other advantages related to these catalysts. A combination with amine functionality increases the utility and selectivity of these catalysts. Many thiourea-amine bifunctional catalysts have been developed so far to promote many types of asymmetric reactions, such as Morita–Baylis–Hillman reactions and Michael additions.[91] A range of research groups started developing carbohydrate-based thiourea-amine bifunctional catalysts during the first decade of the 21st century and observed highly valuable results.

13.7.1 Carbohydrate-based thiourea-primary amine bifunctional organocatalysts

In 2007, Liu et al. developed a new class of thiourea catalysts which consisted of a sugar moiety and a primary amine.[92] These catalysts were synthesized by the addition of chiral 1,2-cyclohexyldiamines to isothiocyanate of the corresponding carbohydrates. For example, catalysts **14a** and **14b** arose from glycosyl isothiocyanate, and similarly, **14c** and **14d were** derived from isothiocyanates of maltose and lactose, respectively (Scheme 13.22). Efficiency of compounds **14a–d** was tested for catalysis of the asymmetric Michael addition of aromatic ketones to nitroolefins in optimum reaction conditions, that is,

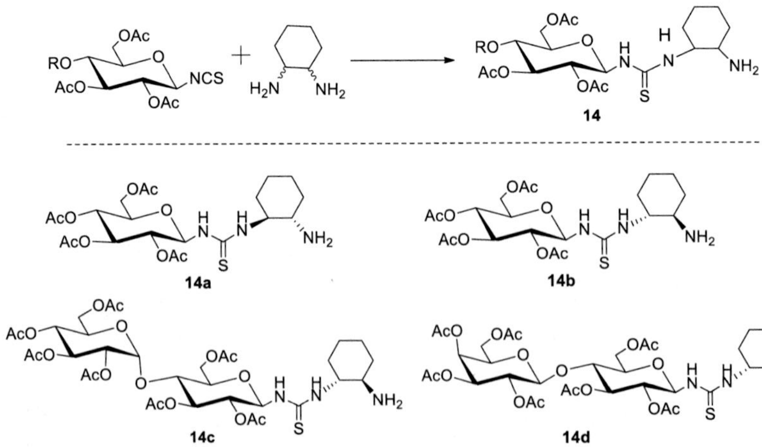

SCHEME 13.22 Synthesis of thiourea bifunctional catalysts **14a–d**.

at room temperature with 15 mol% of catalyst loading in dichloromethane as the solvent. Although all these four catalysts provided high enantioselectivities (87%–97% ee) for the test reaction of acetophenone with nitrostyrene, the efficacy of catalyst **14b** was found optimal for product yield (60%). Notably, catalysts **14a** and **14b** having an opposite configuration of 1,2-cyclohexyldiamines moiety also induced opposite stereochemistry in the product, that is, **14a** with (*S*,*S*)-configuration of the 1,2-diaminocyclohexane moiety gave *R*-selectivity, whereas **14b** having (*R*,*R*)-configuration of the 1,2-diaminocyclohexane moiety induced *S*-selectivity in the product. Reactions with a broad spectrum of nitroolefins and aromatic ketones catalyzed by **14b** took place in low to excellent yields (20%–99%) and uniformly very high enantioselectivities (94%–98% ee), as shown in Scheme 13.23A. A typical thiourea-amine bifunctional mechanism was proposed for the catalyst working in which thiourea moiety enhanced the electrophilicity of the nitro group of the nitroolefin by hydrogen bonding, whereas the amine group of the catalyst activated the ketone through the enamine mechanism. This enamine intermediate approached the *Si*-face of the nitroolefin predominately and generated the (*S*)-configurated adduct. The *Re*-face of the nitroolefin was restricted by the cyclohexyl ring of the catalyst (Fig. 13.23). Later in 2009, Gu et al. successfully promoted the asymmetric Michael addition of acetone with a variety of nitroolefins with catalyst **14b**.[93]

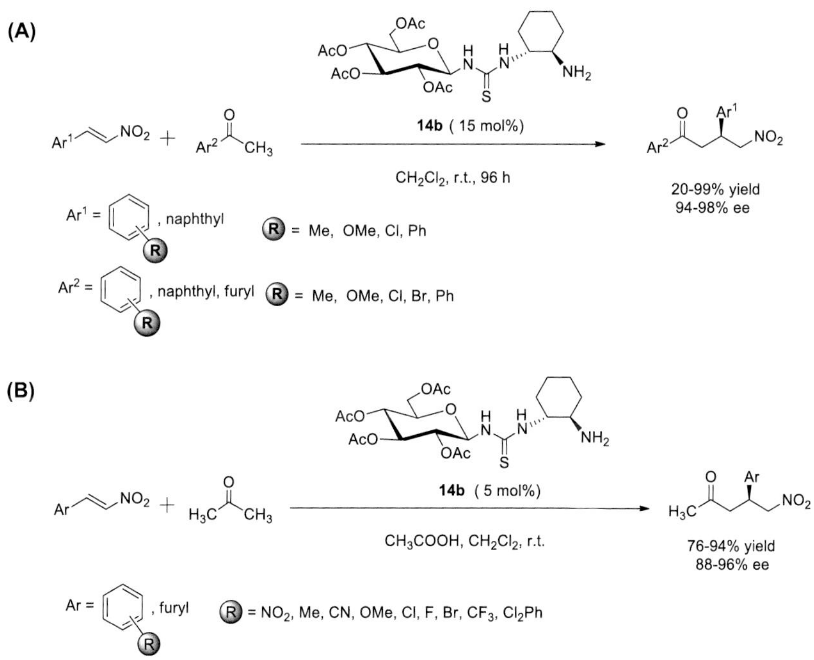

SCHEME 13.23 A) Michael addition between aromatic ketones and aromatic nitroolefins with catalyst **14b**. B) Michael addition between acetone and aromatic nitroolefins with catalyst **14b**.

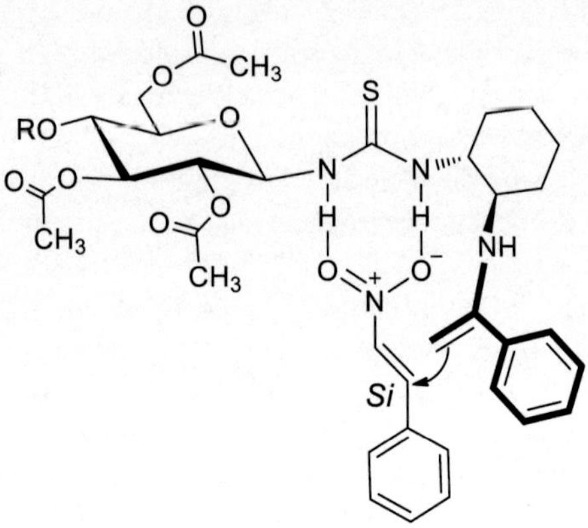

FIGURE 13.23 Transition state for Michael addition catalyzed by **14b**.

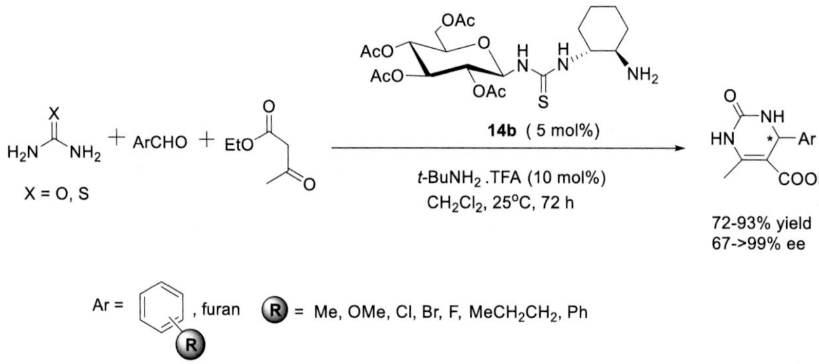

SCHEME 13.24 Biginelli reaction with catalyst **14b**.

The use of 5 mol% of catalyst **14b** in the presence of acetic acid in dichloromethane as solvent allowed the formation of the Michael products to be achieved in both high yields (76%–94%) and enantioselectivities (88%–96% ee) (Scheme 13.23B).

The same year, Wang et al. applied the same catalysts **14a** and **14b** to the Biginelli reaction of urea/thiourea, aromatic aldehydes, and ethyl acetoacetate.[94] The best results were obtained when the three-membered reaction was performed in the presence of 1.5 equiv. of aldehyde, 3 equiv. of ethyl acetoacetate, 5 mol% of catalyst **14b**, and 10 mol% of 2,4,6-trichlorobenzoic acid (TCBA) containing 10 mol% of t-$BuNH_2$.TFA in dichloromethane as the solvent. In these optimal conditions, a variety of aromatic aldehydes with electron-donating and electron-withdrawing substituents, urea/thiourea, and ethyl acetoacetate reacted to give the corresponding chiral products in good yields (72%–93%) and moderate-to-excellent enantioselectivities of *S*-isomers (67- > 99% ee) (Scheme 13.24). In the reaction of furanaldehyde, the

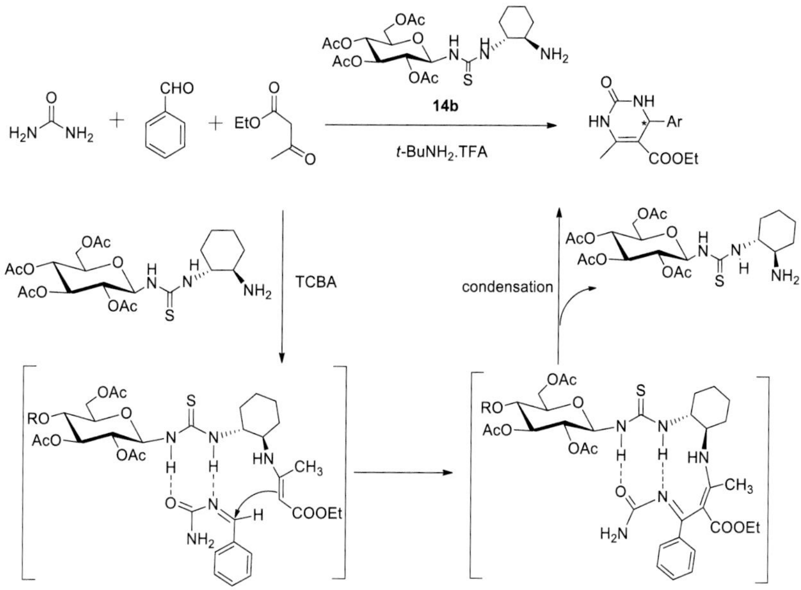

FIGURE 13.24 Proposed mechanism for Biginelli reaction catalyzed by **14b**.

stereochemistry of the product changed to *R* may be due to hydrogen bonding between the furan ring and thiourea-amine catalyst. When catalyst **14a** was employed instead of catalyst **14b**, similar results were obtained with opposite stereochemistry. According to the proposed mechanism, the thiourea moiety of the catalyst interacted with the acyl group of the benzylidene urea, while the primary amine group of the catalyst activated the ethyl acetoacetate through enamine formation. This enamine intermediate approached the *Si*-face of the imine predominantly due to a restricted approach toward the *Re*-face, thus resulting in the formation of the *S*-product (Fig. 13.24).

Another use of catalyst **14b** was reported by Nie et al., in 2011.[95] Asymmetric aldol reaction of trifluoroacetaldehyde hemiacetal with aromatic ketones was achieved at room temperature in the presence of 15 mol% of **14b** and 5 mol% of H_2O in dichloromethane as the solvent. Low-to-moderate yields (13%–42%) and moderate enantioselectivities (50%–62% ee) were obtained for a broad range of aromatic ketones under these conditions (Scheme 13.25). The predominant formation of the *R*-configured products was favored by the transition state proposed by Nie et al. in which the *Re*-face of in situ generated trifluoroacetaldehyde was effectively approached by enamine (formed by amine of catalyst and ketone) due to the restricted approach toward the *Si*-face blocked by the thiourea moiety of the catalyst (Fig. 13.25).

In search of a suitable catalyst for asymmetric Michael addition of aromatic ketones to aromatic nitrodienes, Ma et al. found in 2010 that catalyst **14b** was not sufficiently efficient with respect to yield whereas catalysts **15a**

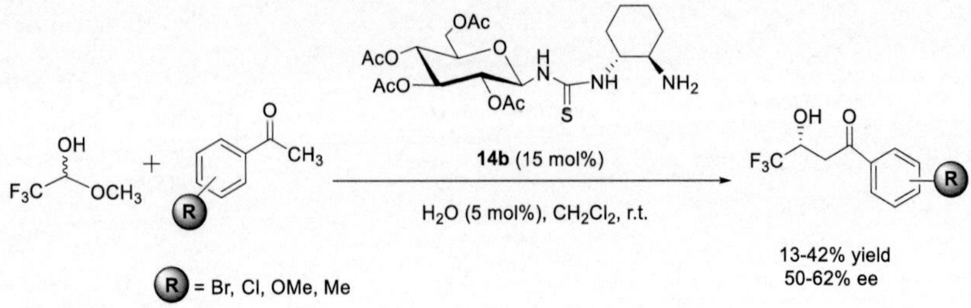

SCHEME 13.25 Aldol reaction of trifluoroacetaldehyde hemiacetal and aromatic ketones catalyzed by **14b**.

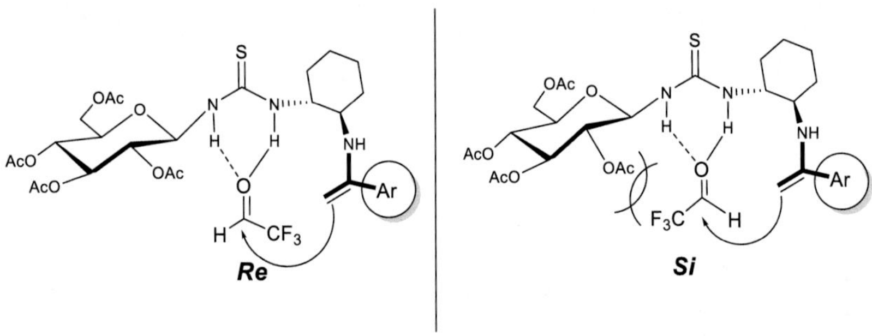

FIGURE 13.25 Transition state for aldol reaction of trifluoroacetaldehyde hemiacetal and aromatic ketones catalyzed by**14b**.

and **15b** (Fig. 13.26) provided moderate-to-high yields and high enantioselectivities.[96] Indeed, the use of 15 mol% of catalyst **15b** efficiently catalyzed reactions of a broad spectrum of (hetero)aromatic ketones with phenylnitrodiene at room temperature in the presence of 5 mol% of benzoic acid as additive in dichloromethane as solvent to afford the corresponding products in good to high yields (72%–90%) and uniformly high enantioselectivities (85%–98% ee), as illustrated in Scheme 13.26. The *S*-enantioselectivity of the reaction could be due to the fact that the *Re*-face of the nitrodiene was sterically hindered by the thiourea moiety of the catalyst (Fig. 13.27). However, this catalyst could not catalyze the Michael addition of cyclic ketones to phenylnitrodiene. To fill this gap, another catalyst **15c** was screened which provided good results with a variety of aliphatic ketones, such as cyclohexanone (94% yield, 95% ee), cyclopentanone (97% yield, 97% ee), and also acyclic acetone (75% yield, 95% ee).

A typical thiourea-amine bifunctional catalysis mechanism was suggested for this reaction in which the electrophilicity of the nitrodiene was increased through hydrogen-bonding interaction with the thiourea moiety of the catalyst whereas the ketone was activated through iminium ion (I) or enamine

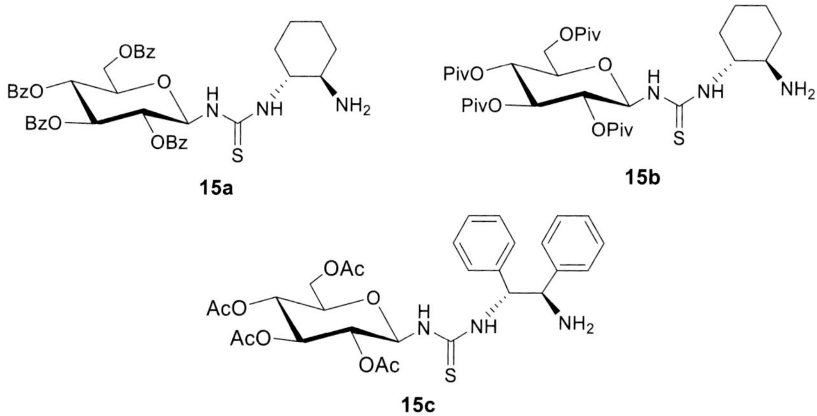

FIGURE 13.26 Catalysts **15a–c** for Michael addition of ketones to nitrodienes.

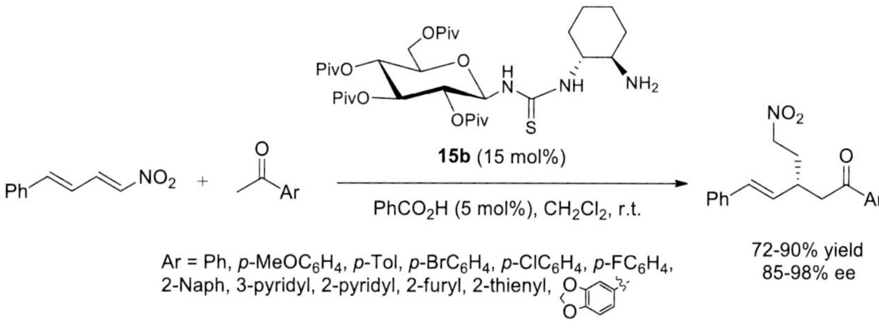

SCHEME 13.26 Michael addition of aromatic ketones to phenylnitrodiene catalyzed by **15b**.

intermediate (II), arising from the reaction between the acid additive and the amine part of the catalyst. This enamine approached the *Si*-face of the β-carbon of nitrodiene, resulting in the formation of imine intermediate (III), which on hydrolysis with water furnished the final product and regenerated catalyst (Fig. 13.27).

In 2012, Tang et al. reported that catalyst **15c** was also able to promote an intramolecular asymmetric Michael addition to synthesize chiral dihydrobenzofuran derivatives with uniformly high yields (91– >99%) and enantioselectivities (94– >99% ee).[97] As shown in Scheme 13.27, when 20 mol % of catalyst **15c** was employed at 20°C with 10 mol% of 4-nitrobenzoic acid as an additive in tetrahydrofuran (THF) as a solvent, the reaction was achieved with high diastereoselectivity (*trans/cis*: 88/12) and high enantioselectivity (96% ee). Diastereoselectivity was further improved to 96/4 (*trans/cis*) by refluxing the reaction mixture for 8 hours after completion of the reaction (Scheme 13.27).

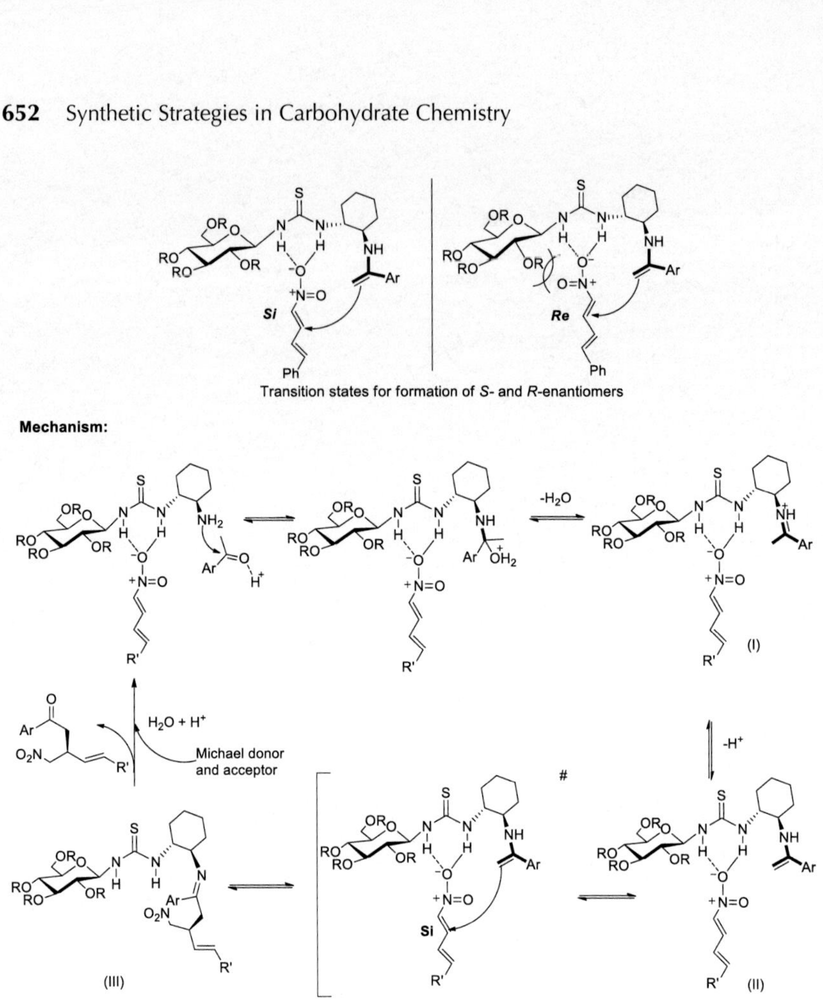

FIGURE 13.27 Proposed mechanism for Michael reaction catalyzed by **15a–b**.

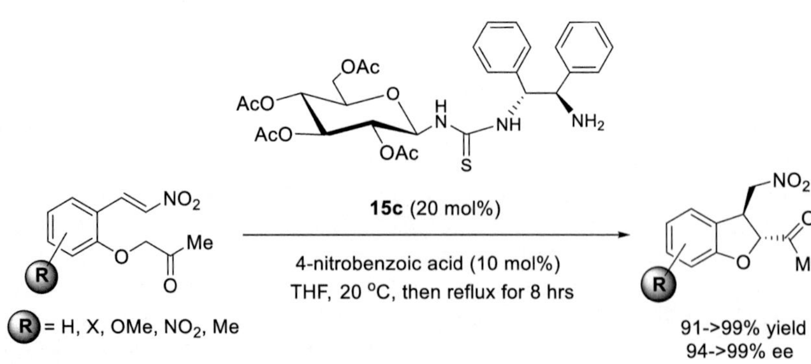

SCHEME 13.27 Intramolecular Michael addition catalyzed by **15c**.

13.7.2 Carbohydrate-based thiourea-secondary amine bifunctional organocatalysts

In 2009, Lu et al. designed new sugar-based pyrrolidine-thioureas **16a** and **16b**, respectively from (*S*)- and (*R*)-*tert*-butyl 2-(aminomethyl) pyrrolidine-1-carboxylate and glucosyl isothiocyanate (Scheme 13.28).[98] Trifluoroacetate derivatives of these catalysts were kept for storage and a catalyst was generated in situ from these derivatives by treatment with triethylamine for reaction purposes. For asymmetric Michael addition of cyclohexanone to nitroolefins, the best results were obtained by using 20 mol% of the catalyst **16a** combined with 20 mol% of TEA and 10 mol% of butyric acid as cocatalyst at −15°C. Although catalyst **16b** produced high enantioselectivity under the same reaction conditions, the reaction became very sluggish. On the other hand, in the presence of catalyst **16a**, good yields (71->99%), high enantioselectivities (92%–97% ee), and high diastereomeric ratios (up to >99/1 *syn/anti*) were obtained for the reaction of cyclohexanone with a broad spectrum of nitroolefins irrespective of the position and electronic nature of substituents (Scheme 13.29). However, the reaction with acetone (75% yield, 40% ee) and cyclopentanone (75/25 dr, 79% ee) was comparatively not efficient, and with

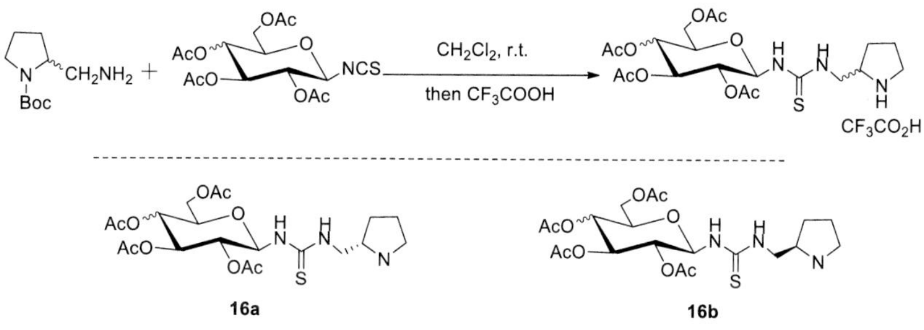

SCHEME 13.28 Synthesis of trifluoroacetate derivatives of catalysts **16a–b**.

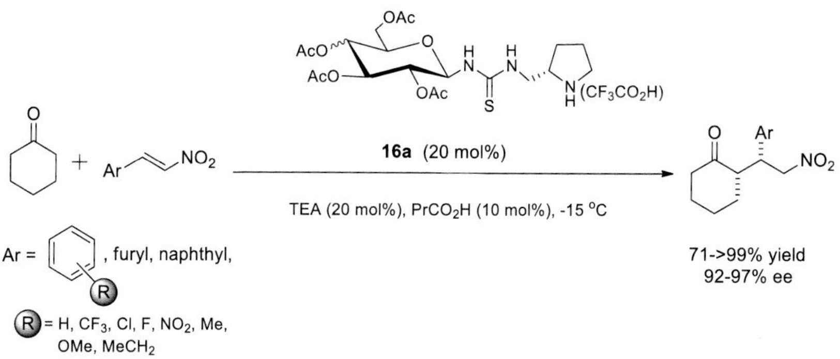

SCHEME 13.29 Michael addition of cyclohexanone to nitroolefins catalyzed by**16a**.

cycloheptanone, no corresponding adduct was obtained. According to the proposed mechanism, the pyrrolidine moiety of the catalyst activated the ketone through enamine formation with the help of an acidic cocatalyst whereas the thiourea part as well as the Brønsted acid additive activated the nitroolefin through hydrogen bonding. The product formation took place through a *Re*-face attack of the enamine to the nitroolefin (Fig. 13.28).

In 2014, a series of novel sugar-derived thiourea-secondary-amine bifunctional catalysts were developed by Ágoston and Fügedi (Fig. 13.29).[99] However, these catalysts showed low-to-moderate yields and uniformly low enantioselectivities for the asymmetric Michael addition of acetylacetone to β-nitrostyrene.

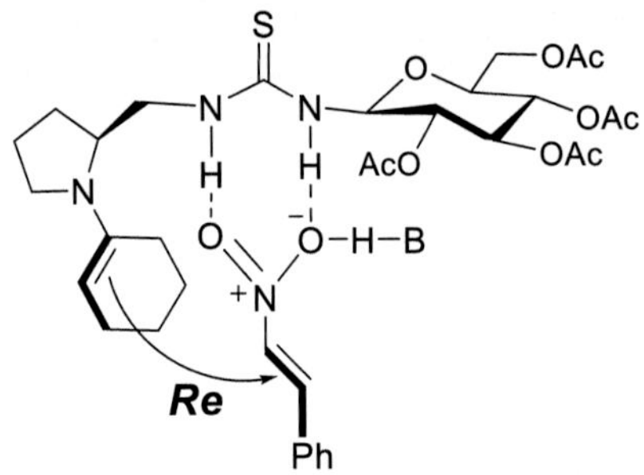

FIGURE 13.28 Proposed transition state for Michael addition catalyzed by **16a**.

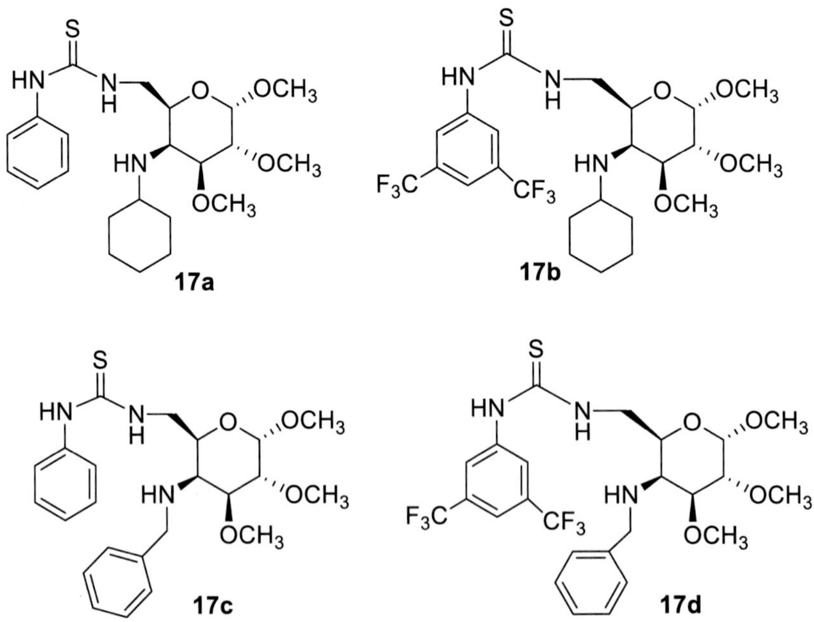

FIGURE 13.29 Catalysts **17a–d** developed by Ágoston and Fügedi.

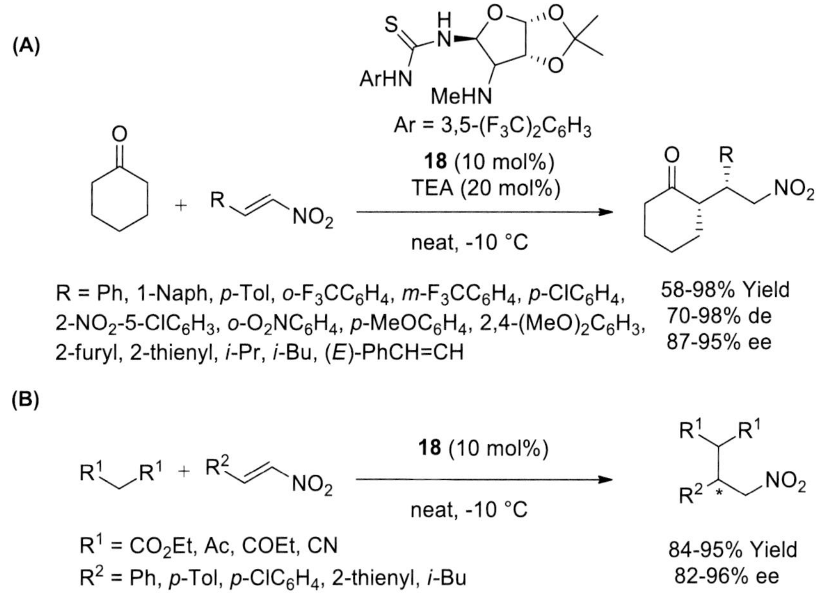

SCHEME 13.30 A) Michael addition of cyclohexanone to nitroolefins catalyzed by **18**. B) Michael addition of other Michael donors to nitroolefins catalyzed by **18**.

In 2016, Khan and Narula described the synthesis of a series of novel chiral thiourea-based bifunctional organocatalysts having D-glucose as a core scaffold to be further investigated as promotors for the asymmetric Michael addition of cyclohexanone to aryl/alkyl *trans*-β-nitrostyrenes.[100] The use of 10 mol% of optimal catalyst **18** in the presence of 20 mol% of TEA as an additive led at −10°C in the absence of solvent to the formation of the desired chiral Michael products with moderate to quantitative yields (58%−98%) and *syn*-diastereoselectivities (70%−98% de) and uniformly high enantioselectivities (87%−95% ee), as shown in Scheme 13.30A. The reaction conditions could be extended to the Michael addition of other Michael donors having active methylene, such as ethyl malonate ($R^1 = CO_2Et$), malononitrile ($R^1 = CN$), biacetyle ($R^1 = Ac$), and 3,5-heptanedione ($R^1 = COEt$), which afforded by reaction with aryl as well as alkyl nitroolefins the corresponding chiral Michael products with both high yields (84%−95%) and ee values (82%−96% ee), as illustrated in Scheme 13.30B.

13.7.3 Carbohydrate-based thiourea-tertiary amines bifunctional organocatalysts

In 2008, Wang et al. developed new carbohydrate-based thiourea-tertiary amine catalysts to be investigated as promotors in asymmetric aza-Henry reactions.[34] Carbohydrate-derived thiourea-tertiary amine bifunctional

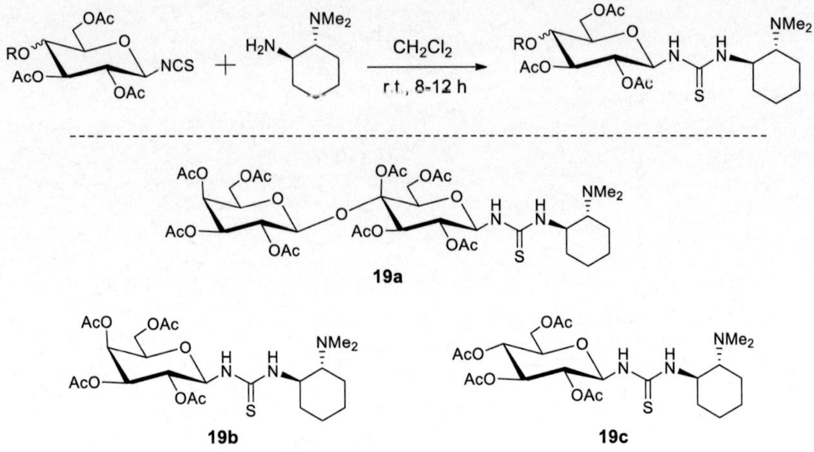

SCHEME 13.31 Synthesis of catalysts **19c**.

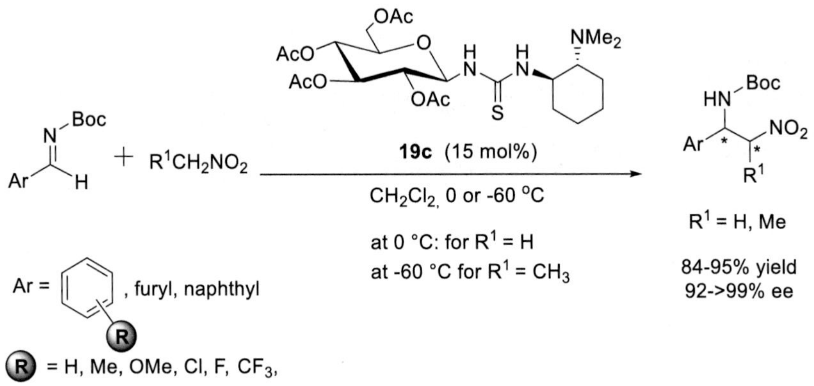

SCHEME 13.32 Aza-Henry reaction of nitroalkanes with *N*-Boc-protected imines catalyzed by **19c**.

catalysts **19a–c** were synthesized according to Scheme 13.31. All these three catalysts were found efficient to produce high yields and moderate-to-high enantioselectivities in the reaction of *tert*-butyl benzylidene carbamate with nitromethane. The best result (86% yield and >99% ee) was obtained at −78°C by using 15 mol% of catalyst **19c** in dichloromethane as solvent. With a broad range of *N*-Boc-protected imines, the reaction resulted in high yields (84%–95%) and excellent enantioselectivities (92–>99% ee). Although nitroethane was found to be less reactive, its reaction proceeded smoothly with good yield and enantioselectivity at −60°C (Scheme 13.32).

Catalysts **19a–c** were further screened for asymmetric Michael addition of acetylacetone to various nitroolefins by Wang et al.[34] The best results were obtained when employing 10 mol% of catalyst **19c** at −40°C in toluene

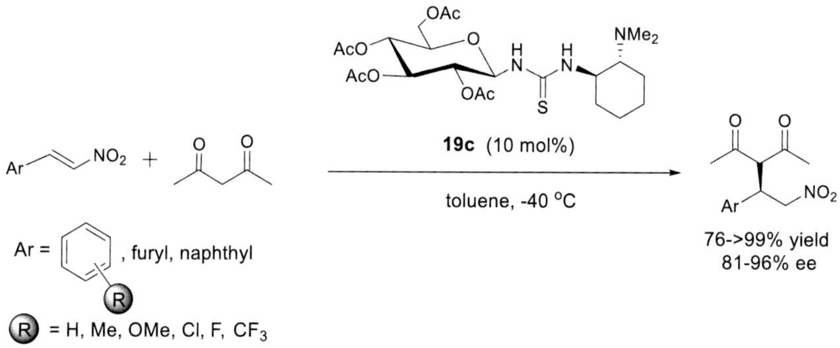

SCHEME 13.33 Michael addition of acetylacetone to nitroolefins catalyzed by **19c**.

as solvent. Under these optimal reaction conditions, **19c** efficiently catalyzed the reaction of acetylacetone with a broad range of nitroolefins bearing aryl and heteroaryl groups. In all cases, except for nitro-substituted aryl nitroolefin, high yields (up to >99%), and excellent enantioselectivities (up to 96% ee) were obtained (Scheme 13.33).

In 2009, a series of new chiral tertiary amine-thiourea catalysts was developed by Pu et al. through the combination of α-amino acids with D-glucopyranose (Scheme 13.34).[101,102] These catalysts were screened for asymmetric conjugate addition of acetylacetone to *trans*-β-nitrostyrene performed at room temperature in the presence of 10 mol% of catalyst loading. L-Valine- and D-glucopyranose-derived catalyst **20a** showed good reactivity and enantioselectivity (88% yield, 85% ee (*S*)) for the reaction performed in THF as solvent. Catalyst **20b**, derived from D-valine and D-glucopyranose, gave opposite enantioselectivity in toluene as a solvent with almost the same enantiomeric excess (90% yield, 86% ee (*R*)). Similarly, catalysts **20c** (derived from L-leucine and D-glucopyranose) and **20d** (derived from D-leucine and D-glucopyranose) showed enantioselectivities opposite to each other. Indeed, catalysis with **20c** performed in THF resulted in the formation of the *S*-configured product (89% yield, 80% ee) whereas the use of **20d** resulted in the formation of the corresponding *R*-product with almost the same enantiomeric excess (89% yield, 79% ee). Catalysts **20e** and **20f**, which are sugar derivatives of L-phenylalanine and D-phenylalanine, respectively, provided high yields (**20e**: 88%, **20f**: 89%) and higher enantioselectivities (**20e**: 90% ee for *S*-isomer, **19f**: 90% ee for *R*-isomer) at 0°C in toluene as solvent. Related catalysts derived from L-phenyl glycine and D-phenyl glycine (**20g** and **20h**, respectively) and L-tryptophan and D-tryptophan (**20i** and **20j**, respectively) also provided good results. Equal access to both isomers of the product is the main feature that makes this class of catalysts important (Scheme 13.35). The efficiency of these catalysts was supposed to be a result of double stereocontrol over the reaction by sugar and amino acid moieties of catalysts. The proposed transition state includes hydrogen-bonding

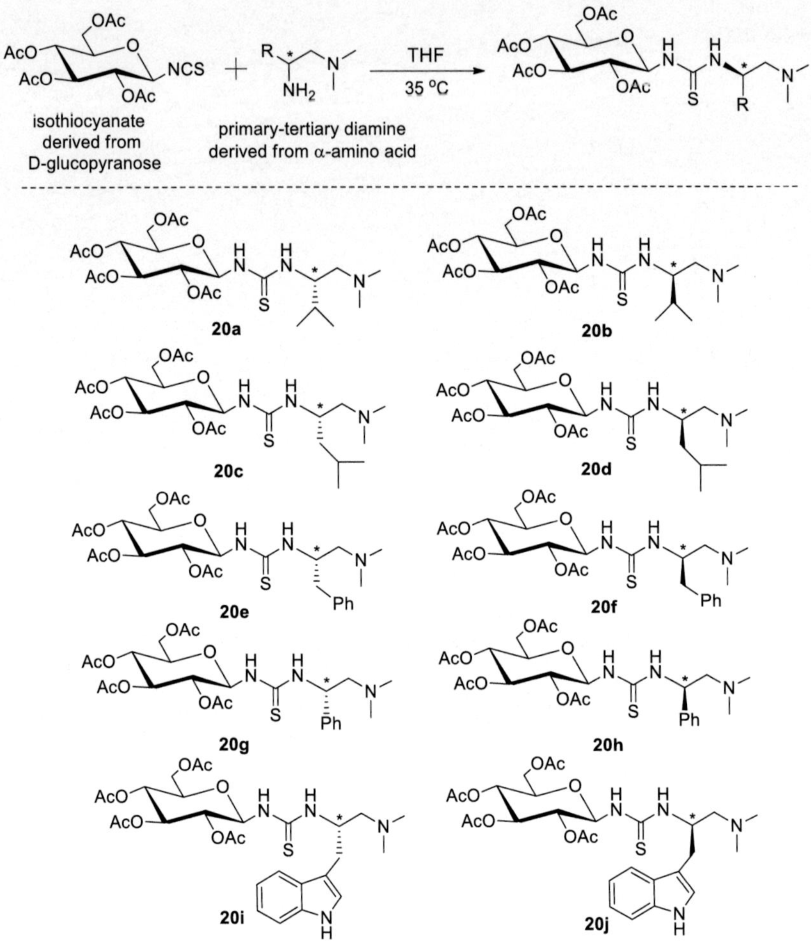

SCHEME 13.34 Synthesis of catalysts **20a–j**.

interaction between the thiourea moiety of the catalyst and the nitro group of the nitroalkene while the tertiary amine group of the catalyst deprotonates the acidic proton of the acetyl acetone (Fig. 13.30).

In 2011, novel thiourea bifunctional chiral catalyst **21** was found the most suitable one among all the catalysts developed by Puglisi et al. (Scheme 13.36) to promote asymmetric nucleophilic additions of acetylacetone to nitroolefins and imines.[103] Indeed, 10 mol% of catalyst **20** efficiently catalyzed the asymmetric Michael addition of acetylacetone to differently substituted β-nitrostyrenes in dichloromethane as solvent at room temperature. The corresponding chiral products were formed with both moderate-to-high yields (67%–93%) and enantioselectivities (55%–85% ee) (Scheme 13.35). A

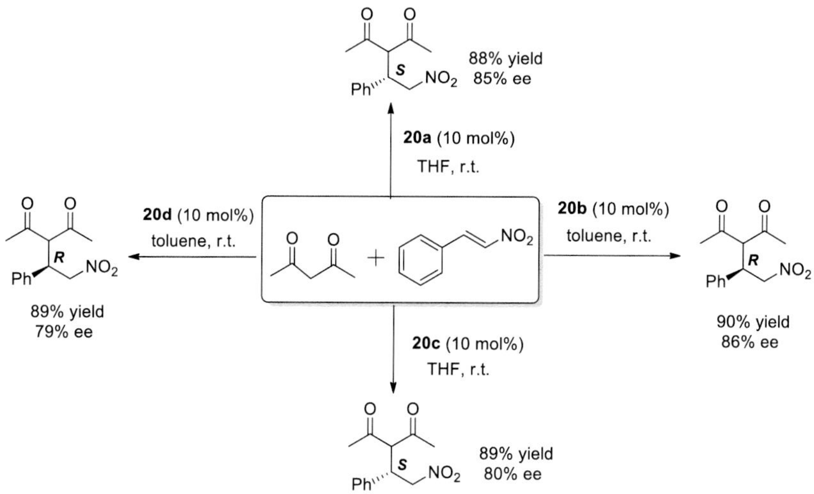

SCHEME 13.35 Michael addition of acetylacetone to β-nitrostyrene catalyzed by **20a–d**.

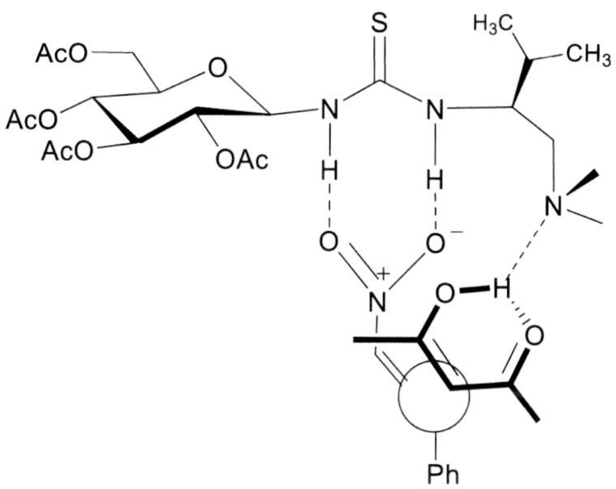

FIGURE 13.30 Proposed transition state for Michael addition of acetylacetone to β-nitrostyrene catalyzed by **20a**.

transition state was proposed by the authors, which is depicted in Fig. 13.31. However, 25% yield and 81% ee were obtained in the addition of diethyl malonate to the *N*-Boc imine of benzaldehyde catalyzed by 10 mol% of **21** (Scheme 13.37).

In 2012, Kong et al. reported the use of chiral thiourea bifunctional catalyst **22**, having cinchona alkaloid and carbohydrate moieties, to promote the asymmetric addition of trimethylsilyl cyanide (TMSCN) to α-ketophosphonates.[104] The best results were obtained when 10 mol% of catalyst **22** combined with 10 mol% of *p*-nitrophenol as an additive was employed at −78°C in toluene as solvent. High yields (80%–90%) and moderate-to-high enantioselectivities

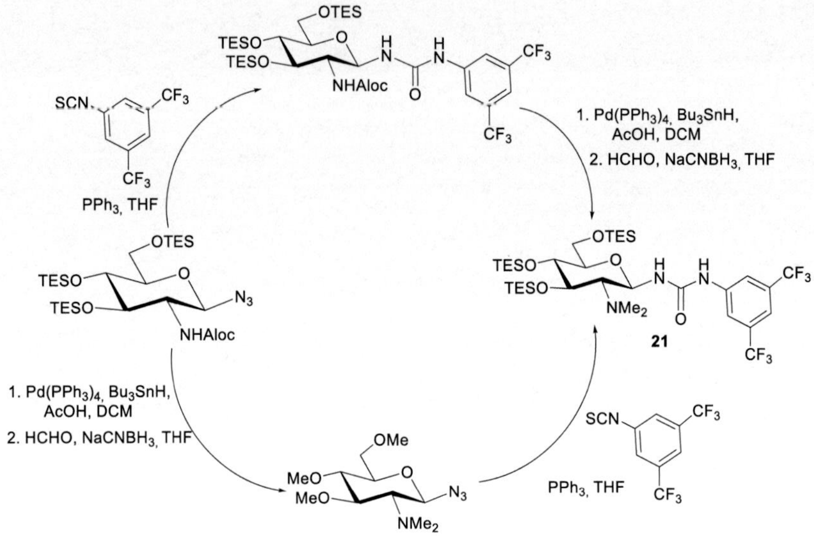

SCHEME 13.36 Synthesis of catalyst **21**.

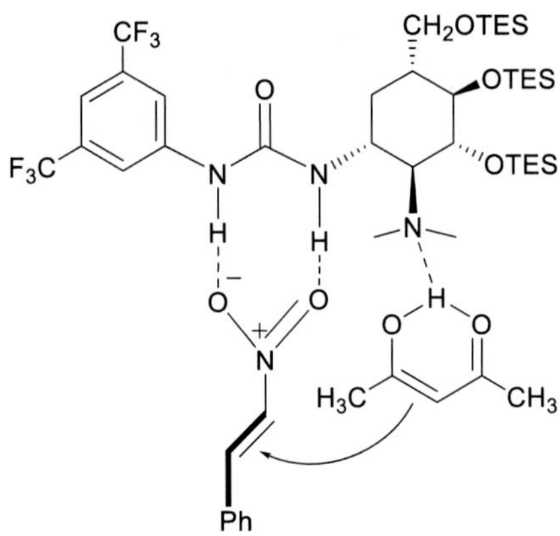

FIGURE 13.31 Transition state for Michael addition of acetylacetone to β-nitrostyrene catalyzed by **21**.

(56%–99% ee) were obtained for a variety of aromatic α-ketophosphonates (Scheme 13.38). Substrates with methylphosphonate moiety gave products with better enantioselectivities than substrates with ethylphosphonate moiety. Low enantioselectivities were obtained with aliphatic α-ketophosphonates. A possible mechanism of the reaction suggested by the authors (Fig. 13.32) shows that the thiourea moiety of the catalyst interacted with the α-ketophosphonate through hydrogen bonding and HCN generated by additive *p*-nitrophenol attacked from the *Si*-face of the α-ketophosphonate. The *Re*-face

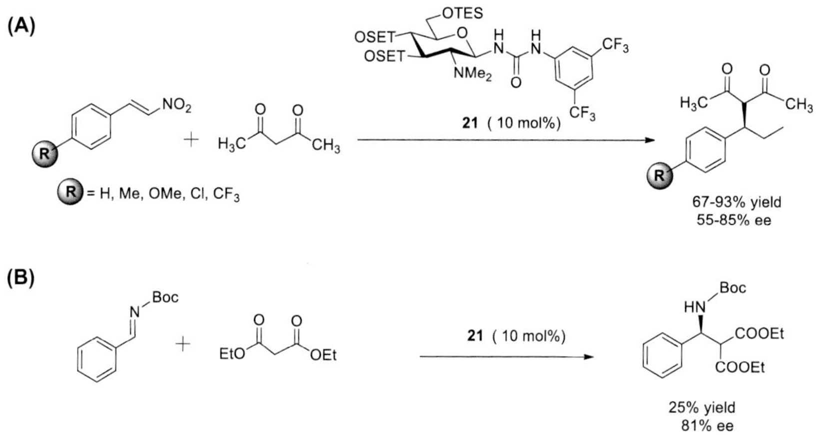

SCHEME 13.37 Nucleophilic additions of acetylacetone to nitroolefins and diethyl malonate to an imine catalyzed by **21**.

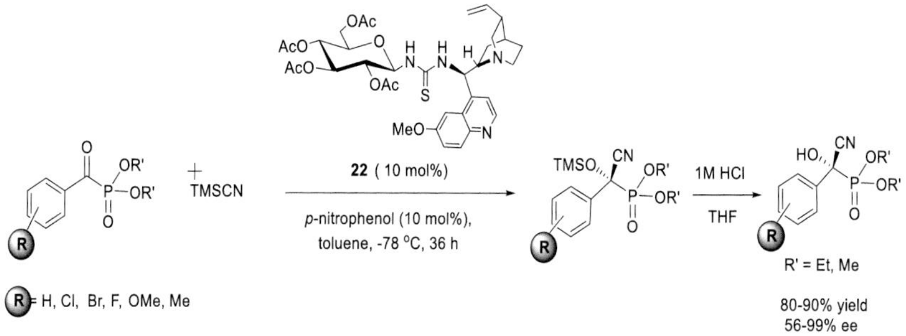

SCHEME 13.38 Addition of TMSCN to α-ketophosphonates catalyzed by **22**.

of the α-ketophosphonate was restricted by the cinchona alkaloid moiety in the catalyst.

In 2013, a successful catalysis of enantioselective decarboxylative Mannich reaction of β-ketoacids with ketimines promoted by catalysts **23a** and **23b** was reported by Yuan et al.[105,106] In the presence of 10 mol% of these catalysts, the reaction of *ortho*-, *meta*-, and *para*-substituted phenyl β-ketoacids with trifluoromethyl ketimine afforded the desired chiral products in both high yields (94%–99%) and enantioselectivities (91%–99% ee) (Scheme 13.39). The reaction catalyzed by **23b** was applied to the synthesis of a precursor for anti-HIV drug DPC 083. To this aim, this precursor was further submitted to successive chemical modifications, such as selective reduction with $NaBH_4$, oxidation with pyridinium chlorochromate (PCC), Grignard reaction with cyclopropyl magnesium bromide, and treatment with

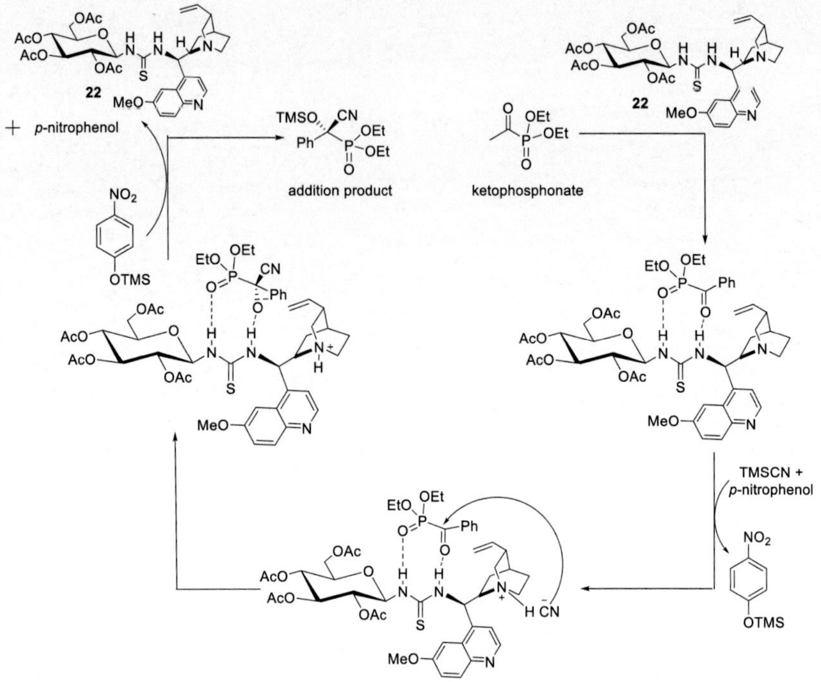

FIGURE 13.32 Possible mechanism of the addition of TMSCN to α-ketophosphonates catalyzed by **22**.

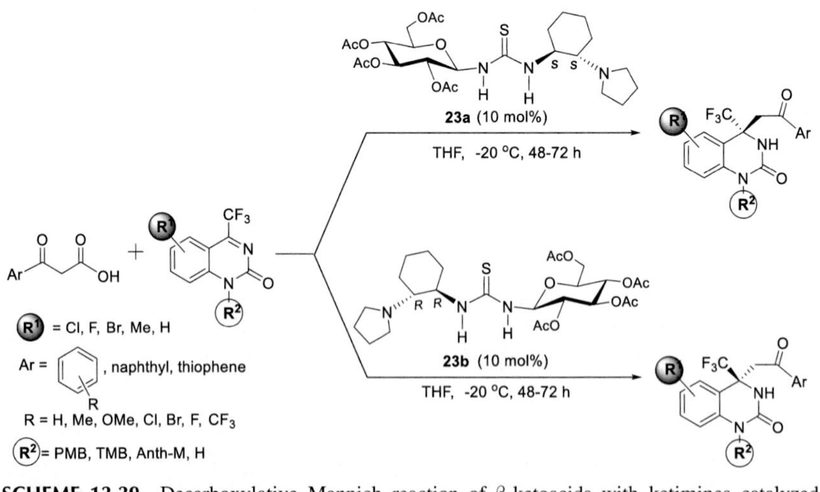

SCHEME 13.39 Decarboxylative Mannich reaction of β-ketoacids with ketimines catalyzed by **23a–b**.

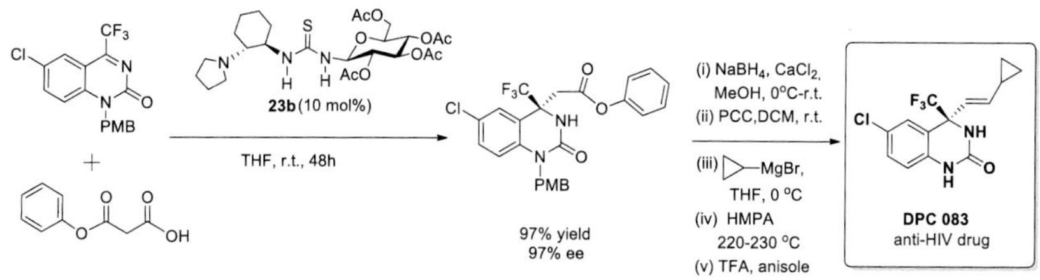

SCHEME 13.40 Application of **23b** to the total synthesis of anti-HIV drug DPC 083.

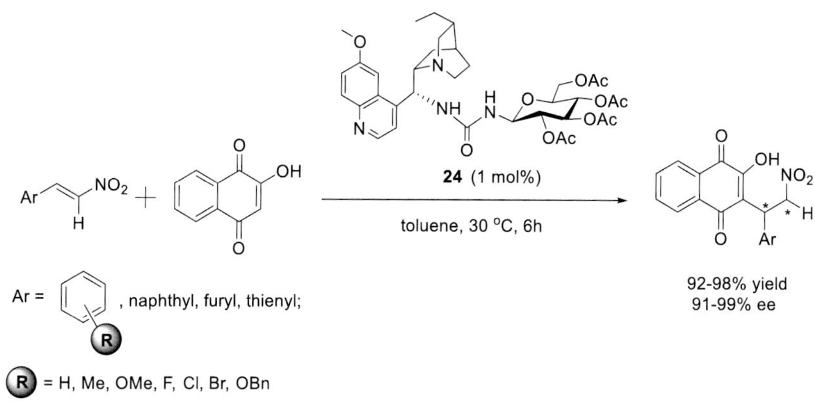

SCHEME 13.41 Michael addition of 2-hydroxy-1,4-naphthoquinone to β-nitroalkenes catalyzed by **24**.

hexamethylphosphoramide (HMPA) and deprotection, which finally afforded anti-HIV drug DPC 083 (Scheme 13.40).

The same year, Reddy et al. screened some cinchona alkaloid-based thiourea catalysts to be used as promotors of asymmetric Michael additions of 2-hydroxy-1,4-naphthoquinone to β-nitroalkenes.[107] Among these catalysts, **24** showed the greatest catalytic efficiency (92%–98% yield) and enantioselectivity (91%–99% ee) when employed at 30°C at 1 mol% of catalyst loading in toluene as solvent (Scheme 13.41). According to the proposed mechanism, the thiourea moiety in the catalyst interacted with the nitro group of β-nitroalkene while the tertiary amine group of cinchona moiety in catalyst deprotonated 2-hydroxy-1,4-naphthoquinone. The *Re*-face of deprotonated 2-hydroxy-1,4-naphthoquinone was then attacked by the nitroalkene to afford the (*S*)-configured product, whereas the other face was blocked by the sugar moiety of catalyst (Fig. 13.33).

In 2014, Ágoston and Fügedi developed a series of novel carbohydrate-based thiourea catalysts **25a–e** (Fig. 13.34). Although all of these catalysts provided only low enantioselectivities for catalysis of the Michael addition

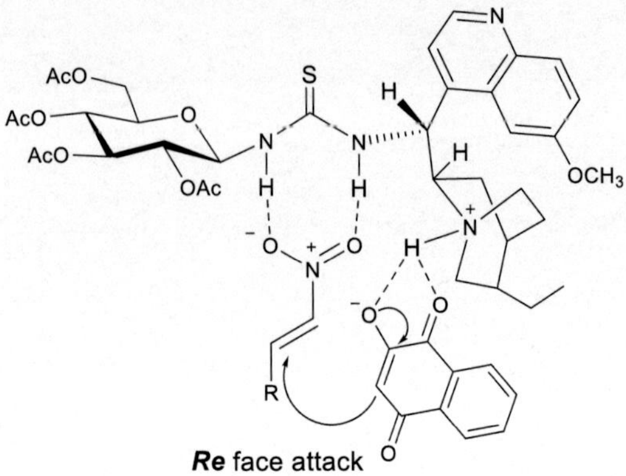

FIGURE 13.33 Proposed mechanism for Michael addition catalyzed by **24**.

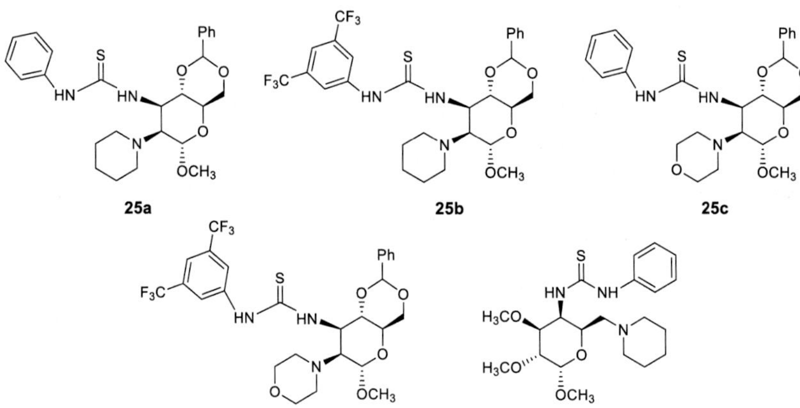

FIGURE 13.34 Catalysts **25a–e** developed by Ágoston and Fügedi.

of acetylacetone to β-nitrostyrene, they could be efficient for other enantioselective transformations.[99]

In 2020, Tvrdonova et al. reported the synthesis of a series of novel chiral carbohydrate-derived bifunctional thioureas to be investigated as catalysts in the asymmetric Michael addition of symmetrical 1,3-dicarbonyl compounds to *trans*-β-nitrostyrenes.[108] The latter were easily prepared in a single-step procedure by treatment of the known 3-deoxy-1,2:5,6-di-O-isopropylidene-3-isothiocyanato-3-C-vinyl-a-D-glucofuranose with the corresponding commercially available amines in very good yields (Scheme 13.42). Among them, catalyst **26** bearing a quinidine-derived cinchona alkaloid scaffold was found optimal to promote the Michael addition of 1,3-diketones and dimethylmalonate to β-nitrostyrenes, providing the corresponding chiral products in

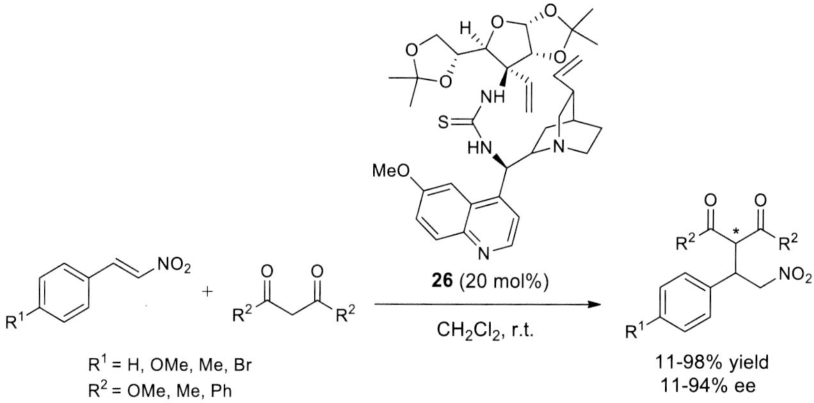

SCHEME 13.42 Michael addition of 1,3-dicarbonyl compounds to β-nitrostyrenes catalyzed by **26**.

variable yields (11%–98%) and enantioselectivities (11%–94% ee), as illustrated in Scheme 13.42. Uniformly very good results (73%–98% yield, 76%–94% ee) were generally obtained in the reaction of 1,3-diketones with β-nitrostyrenes, which could bear different substituents on the phenyl ring. In only the case of a bromo-substituted substrate (R^1 = Br), both lower yield (29%) and ee value (60% ee) were obtained. The scope of the process was extended to the Michael addition of dimethyl malonate (R^2 = OMe), but the corresponding products were obtained with a dramatic loss in enantioselectivity (8%–17% ee) and activity (11%–42% yield). The different reactivity of this malonate toward the Michael addition in comparison with 1,3-diketones could be explained by the lower acidity of the methylene group protons in the case of malonate.

13.8 Other organocatalysts derived from carbohydrates

13.8.1 Organocatalysts with amine functionality

Various amines and amino acids are well-known Lewis base organocatalysts that accelerate different chemical transformations in a stereoselective manner. In 2006, a series of glycosyl-β-amino acids and their derivatives **27a–e** (Fig. 13.35) we developed by Dwivedi et al. for catalysis of aldol reaction of different aldehydes with acetone.[109] Among these amino acids and esters, glycosyl-β-amino acid **27c** showed the highest efficiency in catalyzing the reaction with 20 mol% of catalyst loading at 30°C in DMSO as solvent. Moderate-to-good yields (45%–67%) and low to excellent enantioselectivities (34%–99% ee) were obtained with different aryl aldehydes (Scheme 13.43). In some cases, dehydrated products were also obtained along with aldol products. The authors proposed a mechanism for the reaction similar to that of

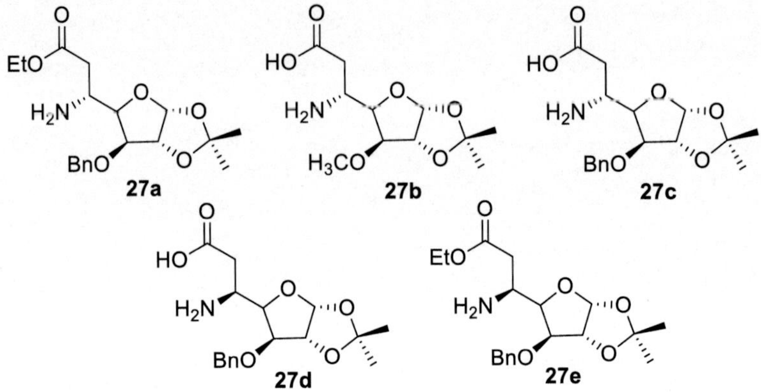

FIGURE 13.35 D-Glucose-derived β-amino acid and ester catalysts **27a–e** for aldol catalysis.

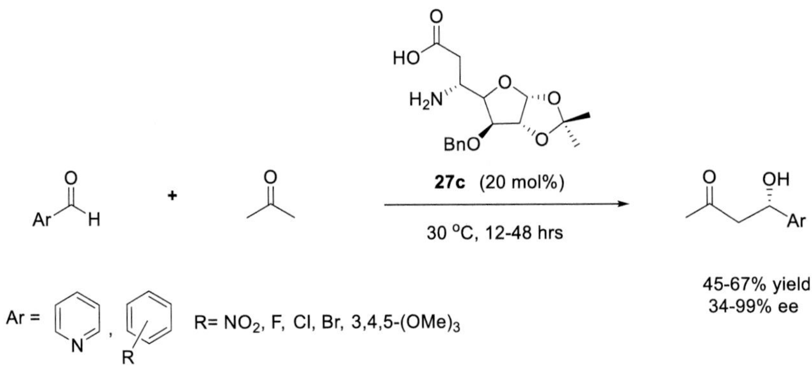

SCHEME 13.43 Aldol reaction of acetone with aryl aldehydes catalyzed by **27c**.

the L-proline-catalyzed aldol reaction. The amine group of a catalyst formed an enamine with acetone via carbonylamine intermediate. This enamine then attacked the carbonyl carbon of aryl aldehyde to afford the aldol product. The hydrogen-bonding interaction of the carboxylic acid group of the catalyst and the carbonyl oxygen of the aryl aldehyde contributed to enantioselectivity. The *Si*-face attack of enamine was preferred over the *Re*-face attack due to the 1,3-synaxial interaction of methyl and phenyl groups in the six-membered cyclic transition state of the *Re*-face attack (Fig. 13.36).

Another class of organocatalysts having a carbohydrate-based amine linked to a tetrazole moiety in its structure was reported by Wynands et al., in 2013.[110] A series of such amines **28a–f** was synthesized (Scheme 13.44) and screened for catalysis of asymmetric aldol reaction between acetone and aryl aldehydes. Catalyst **28c** was found to be the optimal catalyst. Indeed, the use of 10 mol% of this catalyst allowed moderate-to-excellent yields

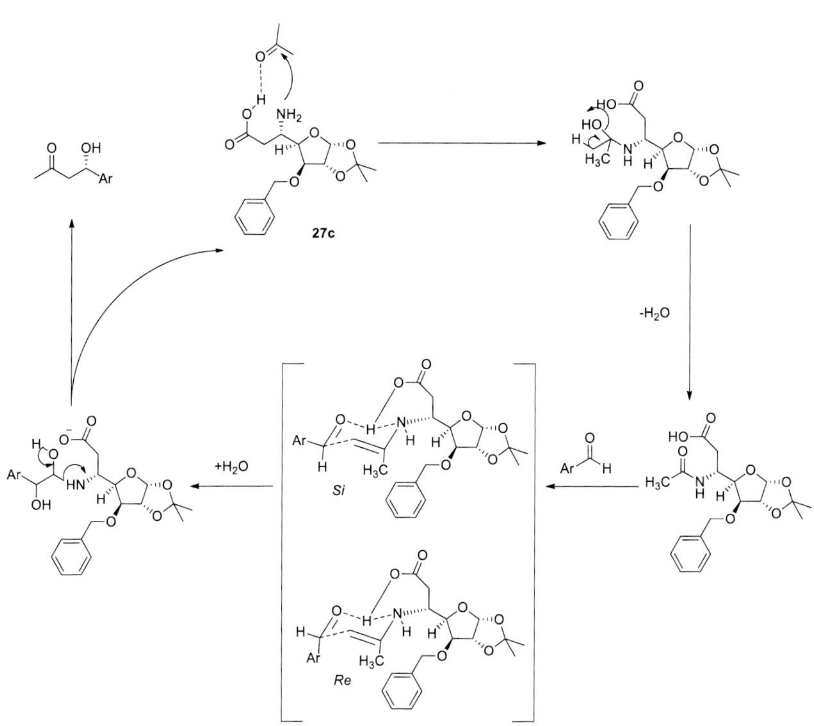

FIGURE 13.36 Proposed mechanism for aldol reaction catalyzed by **27c**.

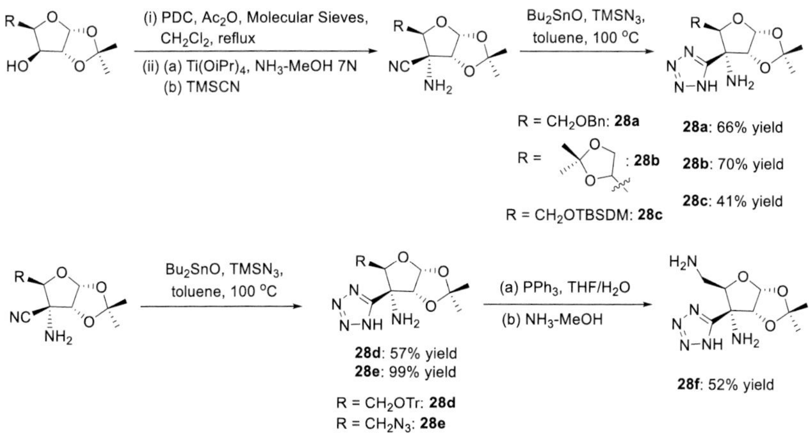

SCHEME 13.44 Synthesis of catalysts **28a–f**.

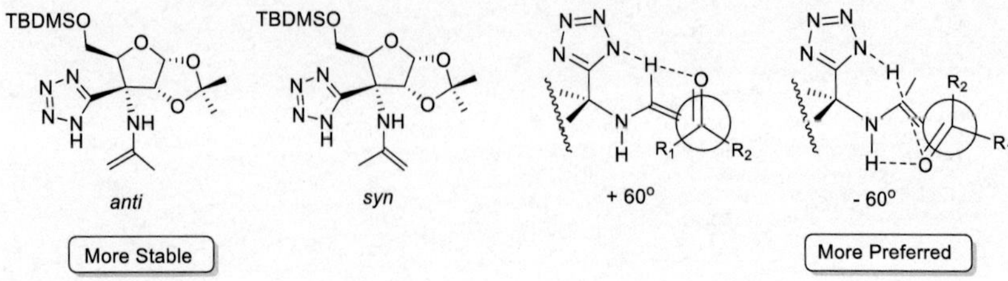

FIGURE 13.37 Enamine intermediate orientations and preferred attacking angle in aldol reaction catalyzed by **28c**.

(34%–99%) and high enantioselectivities (75%–98% ee) to be achieved in the reaction of acetone with a broad range of substituted benzaldehydes. By employing an increased catalyst loading (20 mol%) of related catalysts, good results were also obtained. For example, catalyst **28a** also provided excellent enantioselectivities for different aryl aldehydes but yields were not as good as in the case of using catalyst **28c**. With all the aryl halides (except for 2-chloro-6-fluorobenzaldehyde) the *R*-configured products were formed selectively. The efficiency and selectivity of catalyst **28c** were rationalized by density functional theory (DFT) studies in which the antienamine intermediate formed by the interaction of the catalyst with acetone was found to be more stable than *syn*-enamine intermediate and the −60 degrees was considered as a preferred attacking angle to be more stable *anti*-enamine, which resulted in the formation of the *R*-configured product (Fig. 13.37). However, in the case of 2-chloro-6-fluorobenzaldehyde, because of the electrostatic stabilizing interaction of the proton of the methyl group of enamine and the fluorine substituent on the aromatic aldehyde, the *Si*-face attack on *anti*-enamine was preferred, affording the *S*-configured product.

13.8.2 Bifunctional thiourea/urea organocatalysts

In 2012, Yang et al. developed a series of chiral thiourea-phosphine catalysts (**29a–f**) by condensation of isothiocyanates derived from different sugars (D-glucose, D-mannose and D-galactose) to 2-amino-1-(diphenylphosphino) cyclohexane (Fig. 13.38).[111] D-Glucose-based thiourea-phosphines **29a** and **29d** showed optimal efficiency in catalysis of Morita–Baylis–Hillman reaction of acrylates with aryl aldehydes. As shown in Scheme 13.45, the use of 10 mol% of catalyst **29d** at 25°C in THF as solvent provided the best results (68%–96% yield, 50%–83% ee) for the reaction of a variety of acrylates with nitro benzaldehydes. However, the reaction of *tert*-butyl acrylate and aryl acrylates did not give good results. On the other hand, reactions of *p*-trifluoromethyl benzaldehyde gave satisfactory results (41%–80% yield, 70%–77% ee), while benzaldehyde and halo-benzaldehydes reacted with

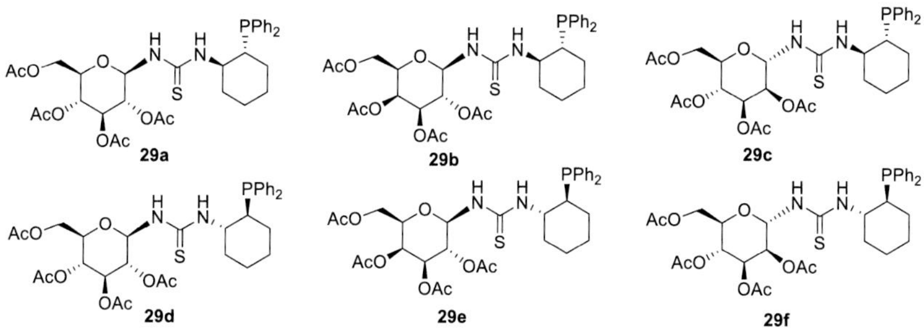

FIGURE 13.38 Catalysts **29a–f** for Morita–Baylis–Hillman reaction.

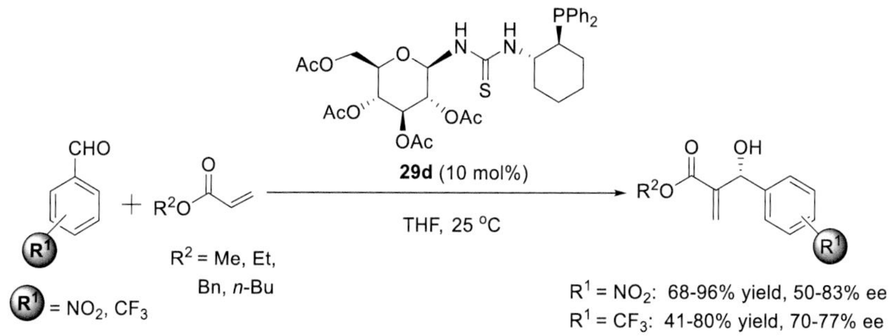

SCHEME 13.45 Morita–Baylis–Hillman reaction catalyzed by **29d**.

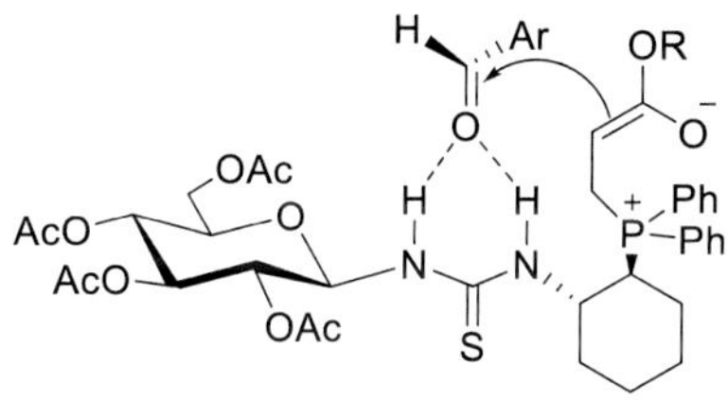

FIGURE 13.39 Proposed transition state for Morita–Baylis–Hillman reaction catalyzed by **29d**.

low yields. According to the transition state proposed by the authors, the phosphine group of the catalyst associated with the acrylate to form an enolate while the carbonyl group of aldehyde was activated through hydrogen-bonding interaction with the thiourea scaffold of the catalyst. Then, the enolate attacked the activated carbonyl group of aldehyde from the *Si*-face to afford the *R*-configured product (Fig. 13.39).

In 2014, Porwanski reported the synthesis of a series of carbohydrate-based organocatalysts **30a–f** including urea-phosphine bifunctionality (Scheme 13.46).[112] When these catalysts were applied to promote the asymmetric Morita–Baylis–Hillman reaction of ethyl acrylate with *p*-nitro

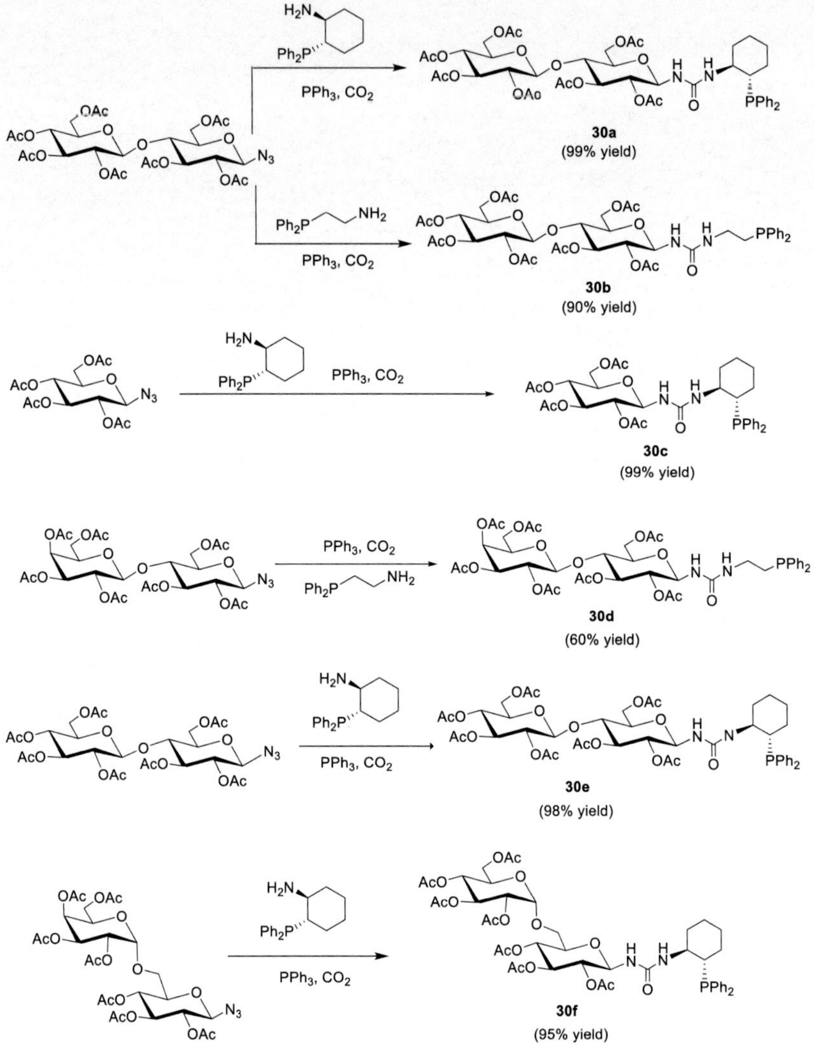

SCHEME 13.46 Synthesis of catalysts **30a–f**.

benzaldehyde, the best results (60% yield, 80% ee) were obtained by using catalyst **30e** whereas catalyst **30f** provided both lower enantioselectivity (72% ee) and yield (36%). The use of catalysts **30a** and **30c** resulted in low yields and moderate enantioselectivities, while catalysts **30b** and **30d** provided low yields and very low enantioselectivities (Scheme 13.47). The authors proposed a transition state for the reaction in which an alkoxy enolate (formed by the association of the phosphino group of the catalyst to acrylate) attacked the *Si*-face of the carbonyl group of benzaldehyde

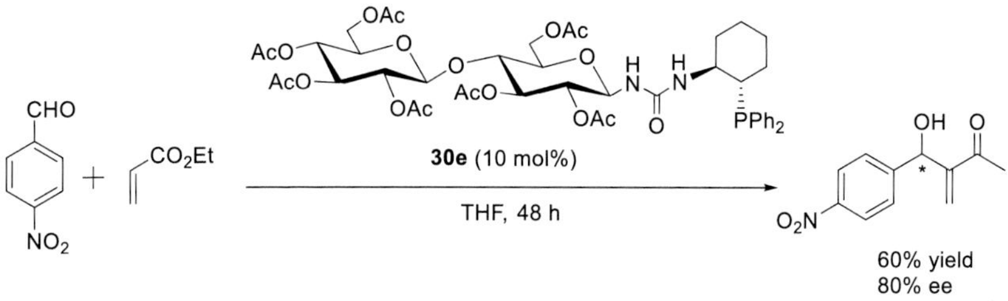

SCHEME 13.47 Morita–Baylis–Hillman reaction catalyzed by **30e**.

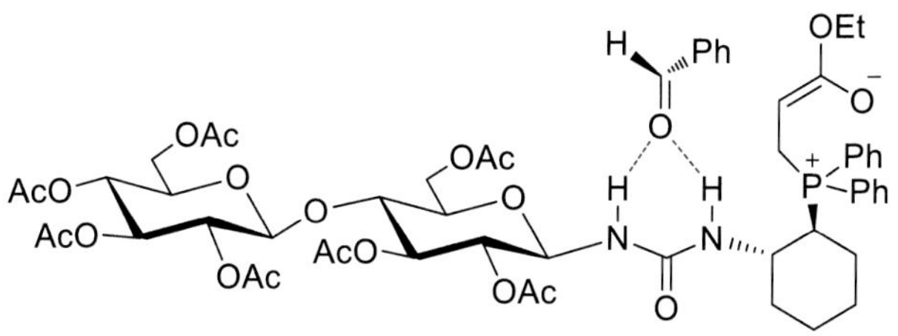

FIGURE 13.40 Proposed transition state for Morita–Baylis–Hillman reaction catalyzed by **30e**.

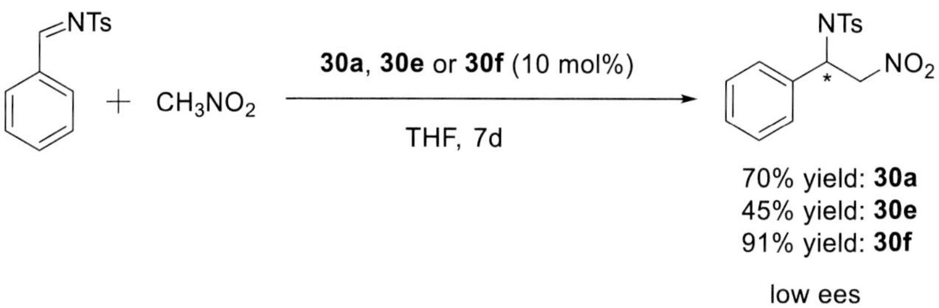

SCHEME 13.48 Aza-Henry reaction catalyzed by **30a**,**e**–**f**.

(activated through hydrogen-bonding interaction with the urea scaffold of catalyst) to afford the *R*-configured product (Fig. 13.40).

Applied to promote the aza-Henry reaction of nitromethane with *N*-tosylimine, catalysts **30a** and **30f** provided good yields, whereas catalyst **30e** gave a low yield after 7 days of reaction time. Enantioselectivities obtained in all cases of catalysts were low (Scheme 13.48).[112]

In 2020, Rawisza and Porwanski described the synthesis of novel chiral bifunctional urea **31** bearing a carbohydrate ring and an optically active base to be investigated as a catalyst in the related asymmetric aza-Henry reaction between a substituted *N*-tosylimine and nitromethane.[113] As shown in Scheme 13.49, the corresponding chiral product **32** was produced with both remarkable yield (98%) and enantioselectivity (99% ee).

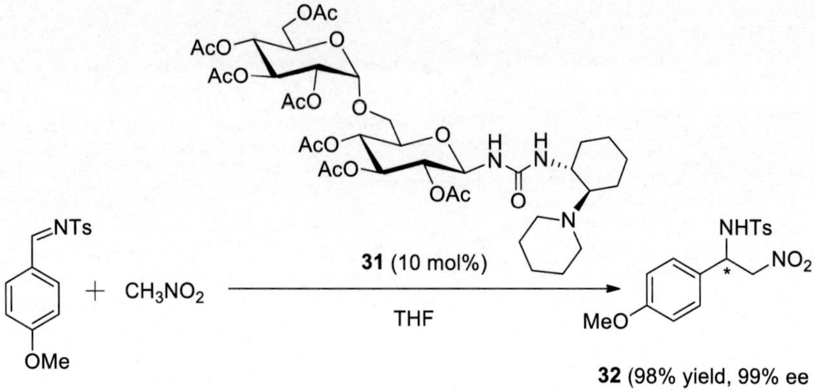

SCHEME 13.49 Aza-Henry reaction catalyzed by **31**.

13.9 Conclusions and future outlook

Besides their biological and synthetic importance, carbohydrates have emerged as promising motifs to incorporate asymmetry in a chemical transformation. Their use in the development of inexpensive and environmentally benign organocatalysts has been proven a fruitful step in the field of synthetic chemistry. Due to the easy accessibility of structural variations in carbohydrate moiety, sugar-based organocatalysts possess a broad spectrum of functionalities. These catalysts have brought a new dimension in asymmetric insertion in a number of useful chemical reactions, including the aldol reaction, Michael addition, Mannich reaction, Morita–Baylis–Hillman reaction, and aza-Henry reaction. These catalysts have played a major role in the development of new facile routes and techniques for the synthesis of drugs as well as medicinally and synthetically important products. Some of these catalysts can efficiently promote more than one type of transformation with high yields and stereoselectivities. With this class of organocatalysts, reactions can be easily catalyzed in organic as well as aqueous media or even without any solvent. Though some of these catalysts have shown low efficiency for the reaction they were screened for, further research will prove undoubtedly their potency in catalyzing other types of transformations. With ongoing research, the development of a number of carbohydrate-derived organocatalysts is increasing day by day; however, there is always a need for some more efficient carbohydrate-based catalysts with new functionalities that can also catalyze those reactions that are still untouched by the catalysts developed so far.

References

1. (a) Varki, A. *Glycobio.* **1993,** *3*, 97–130.
   (b) Bertozzi, C. R.; Kiessling, L. L. *Science* **2001,** *291*, 2357–2364.
   (c) Jelinek, R.; Kolusheva, S. *Chem. Rev.* **2004,** *104*, 5987–6015.

(d) Tiwari, V. K.; Kumar, A.; Schmidt, R. R. *Eur. J. Org. Chem.* **2012,** *15*, 2945–2956.

(e) Tiwari, V. K. *Chem. Rec.* **2021,** *21*, 3029–3048.

(f) Agrahari, A. K.; Bose, P.; Singh, A. S.; Rajkhowa, S.; Jaiswal, M. K.; Hotha, S.; Mishra, N.; Tiwari, V. K. *Chem. Rev.* **2021,** *121*, 7638–7956.

2. (a) Galan, M. C.; Benito-Alifonso, D.; Watt, G. M. *Org. Biomol. Chem.* **2011,** *9*, 3598–3610.

(b) Osborn, H. M. I.; Evans, P. G.; Gemmell, N.; Osborne, S. D. *J. Pharm. Pharmacol.* **2004,** *56*, 691–702.

(c) Gruner, S. A. W.; Locardi, E.; Lohof, E.; Kessler, H. *Chem. Rev.* **2002,** *102*, 491–514.

(d) Wang, D. *Proteomics* **2003,** *3*, 2167–2175.

(e) Tiwari, V. K.; Mishra, R. C.; Sharma, A.; Tripathi, R. P. *Mini Rev. Med. Chem.* **2012,** *12*, 1497–1519.

(f) Tiwari, V. K., Ed. *Carbohydrates in Drug Discovery and Development;* Elsevier Inc: The Netherland, 2020.

(g) Mishra, N.; Tiwari, V. K.; Schmidt, R. R. Recent Trends and Challenges on Carbohydrate-Based Molecular Scaffolding: General Consideration Toward Impact of Carbohydrates in Drug Discovery and Development. In *Carbohydrates in Drug Discovery and Development;* Tiwari, V. K., Ed.; Elsevier Inc., 2020; pp 1–69.

3. (a) Kunz, H.; Sager, W.; Pfrengle, W.; Schanzenbach, D. *Tetrahedron Lett.* **1988,** *29*, 4397–4400.

(b) Kunz, H.; Pfrengle, W. *Tetrahedron* **1988,** *44*, 5487–5494.

4. (a) Blaser, H. U. *Chem. Rev.* **1992,** *92*, 935–952.

(b) Hollingsworth, R. I.; Wang, G. *Chem. Rev.* **2000,** *100*, 4267–4282.

(c) Nicolaou, K. C.; Mitchell, H. J. *Angew. Chem. Int. Ed.* **2001,** *40*, 1576–1624.

(d) Duus, J. O., St.; Hilaire, P. M.; Meldal, M.; Bock, K. *Pure Appl. Chem.* **1999,** *71*, 755–765.

(e) Pellissier, H. *Tetrahedron* **2005,** *61*, 2947–2993.

(f) Boysen, M. M. K. *Chem. Eur. J.* **2007,** *13*, 8648–8659.

(g) Pellissier, H. In *Domino and Intramolecular Rearrangement Reactions as Advanced Synthetic Methods in Glycoscience;* Witczak, Z. J., Bielski, R., Eds.; Wiley: Weinheim, 2016.

5. (a) Hultin, P. G.; Earle, M. A.; Sudharshan, M. *Tetrahedron* **1997,** *53*, 14823–14870.

(b) Lutzen, A.; Koll, P. *Tetrahedron: Asymmetry* **1997,** *8*, 1193–1206.

6. (a) Lubineau, A.; Bienayme, H.; Queneau, Y. *Carbohydr. Res.* **1995,** *270*, 163–179.

(b) Chastanet, J.; Fathallah, H.; Negron, G.; Roussi, G. *Heterocycles* **1992,** *34*, 1565–1572.

(c) Kałża, Z.; Furman, B.; Chmielewski, M. *Tetrahedron: Asymmetry* **1995,** *6*, 1719–1730.

(d) Kolodiazhnyi, O. I.; Grishkun, E. V. *Tetrahedron: Asymmetry* **1996,** *7*, 967–970.

(e) Llera, J. M.; Fernández, I.; Alcudia, F. *Tetrahedron Lett.* **1991,** *32*, 7299–7302.

(f) Kakinuma, K.; Li, H. Y. *Tetrahedron Lett.* **1989,** *30*, 4157–4160.

(g) Piva, O.; Pete, J. P. *Tetrahedron: Asymmetry* **1992,** *3*, 759–768.

7. (a) Dieguez, M.; Pamies, O.; Claver, C. *Chem. Rev.* **2004,** *104*, 3189–3215.

(b) Diéguez, M.; Pàmies, O.; Ruiz, A.; Diaz, Y.; Castillón, S.; Claver, C. *Coord. Chem. Rev.* **2004,** *248*, 2165–2192.

(c) Dieguez, M.; Claver, C.; Pamies, O. *Eur. J. Org. Chem.* **2007,** 4621–4634.

(d) Mishra, A.; Mishra, N.; Tiwari, V. K. *Curr. Org. Synth.* **2016,** *13*, 176–219.

(e) Singh, S. K.; Mishra, N.; Kumar, S.; Jaiswal, M. K.; Tiwari, V. K. *ChemistrySelect* **2022,** *7* e2022013141.

8. Dalko, P. I.; Moisan, L. *Angew. Chem. Int. Ed.* **2001,** *40*, 3726–3748.

9. Dalko, P. I.; Moisan, L. *Angew. Chem. Int. Ed.* **2004,** *43*, 5138–5175.
10. Minuth, T.; Boysen, M. M. K. *Beilstein J. Org. Chem.* **2010,** *6*, 1–7.
11. Phillips, A. M. F. *Eur. J. Org. Chem.* **2014,** 7291–7303.
12. Shen, C.; Zhang, P. F. *Curr. Org. Chem.* **2013,** *17*, 1507–1524.
13. (a) Knoevenagel, E. *Chem. Ber.* **1896,** *29*, 172–174.
(b) Eder, U.; Sauer, G.; Wiechert, R. *Angew. Chem. Int. Ed.* **1971,** *10*, 496–497.
(c) Hajos, Z. G.; Parrish, D. R. *J. Org. Chem.* **1974,** *39*, 1615–1621.
(d) Ahrendt, K. A.; Borths, C. J.; MacMillan, D. W. C. *J. Am. Chem. Soc.* **2000,** *122*, 4243–4244.
(e) List, B.; Lerner, R. A.; Barbas, C. F. *J. Am. Chem. Soc.* **2000,** *122*, 2395–2396.
(f) List, B. *Tetrahedron* **2002,** *58*, 5573–5590.
(g) Jarvo, E. R.; Miller, L. *Angew. Chem. Int. Ed.* **2002,** *58*, 2481–2495.
(h) Westermann, B. *Angew. Chem. Int. Ed.* **2003,** *42*, 151–153.
(i) Berkessel, A.; Gröger, H. Asymmetric Organocatalysis – From Biomimetic Concepts to Powerful Methods for Asymmetric Synthesis; Wiley-VCH: Weinheim, 2005.
(j) Seayad, J.; List, B. *Org. Biomol. Chem.* **2005,** *3*, 719–724.
(k) Taylor, M. S.; Jacobsen, E. N. *Angew. Chem. Int. Ed.* **2006,** *45*, 1520–1543.
(l) List, B. *Chem. Commun.* **2006,** 819–824.
(m) Lelais, G.; MacMillan, D. W. C. *Aldrichimica Acta* **2006,** *39*, 79–87.
(n) Enders, D.; Niemeier, O.; Henseler, A. *Chem. Rev.* **2007,** *107*, 5606–5655.
(o) Dalko, P. I. *Enantioselective Organocatalysis;* Wiley-VCH: Weinheim, 2007.
(p) Dalko, P. I. *Chimia* **2007,** *61*, 213–218.
(q) Pellissier, H. *Tetrahedron* **2007,** *63*, 9267–9331.
(r) Doyle, A. G.; Jacobsen, E. N. *Chem. Rev.* **2007,** *107*, 5713–5743.
(s) Gaunt, M. G.; Johansson, C. C. C.; McNally, A.; Vo, N. C. *Drug Discovery Today* **2007,** *12*, 8–27.
(t) Special Issue on Organocatalysis B., List, Ed. *Chem. Rev.*, 107; 2007, pp 5413–5883.
(u) Akiyama, T. *Chem. Rev.* **2007,** *107*, 5744–5758.
(v) Mukherjee, S.; Yang, J. W.; Hoffmann, S.; List, B. *Chem. Rev.* **2007,** *107*, 5471–5569.
(w) Wurz, R. P. *Chem. Rev.* **2007,** *107*, 5570–5595.
(x) Gaunt, M. J.; Johansson, C. C. C. *Chem. Rev.* **2007,** *107*, 5596–5605.
(y) Colby Davie, E. A.; Mennen, S. M.; Xu, Y.; Miller, S. J. *Chem. Rev.* **2007,** *107*, 5759–5812.
(z) de Figueiredo, R. M.; Christmann, M. *Eur. J. Org. Chem.* **2007,** 2575–2600.
(aa) MacMillan, D. W. C. *Nature* **2008,** *455*, 304–308.
(ab) Yu, X.; Wang, W. *Chem. Asian. J.* **2008,** *3*, 516–532.
(ac) Dondoni, A.; Massi, A. *Angew. Chem. Int. Ed.* **2008,** *47*, 4638–4660.
(ad) Melchiorre, P.; Marigo, M.; Carlone, A.; Bartoli, G. *Angew. Chem. Int. Ed.* **2008,** *47*, 6138–6171.
(ae) Peng, F.; Shao, Z. *J. Mol. Catal. A* **2008,** *285*, 1–13.
(af) Barbas, C. F. *Angew. Chem. Int. Ed.* **2008,** *47*, 42–47.
(ag) Palomo, C.; Oiarbide, M.; Lopez, R. *Chem. Soc. Rev.* **2009,** *38*, 632–653.
(ah) Bertelsen, S.; Jørgensen, K. A. *Chem. Soc. Rev.* **2009,** *38*, 2178–2189.
(ai) Xu, L.-W.; Luo, J.; Lu, Y. *Chem. Commun.* **2009,** 1807–1821.
(aj) Bella, M.; Gasperi, T. *Synthesis* **2009,** *10*, 1583–1614.
(ak) Gong, L. *Chin. Sci. Bull.* **2010,** *55* (17).
(al) List, B. *Top. Curr. Chem.* **2010,** *291*, 1–456.

(am) Pellissier, H. *Recent Developments in Asymmetric Organocatalysis;* Royal Society of Chemistry: Cambridge, 2010.
(an) Marquès-Lopez, E.; Herrera, R. P.; Christmann, M. *Nat. Prod. Rep.* **2010,** *27*, 1138−1167.
(ao) Mahrwald, R. *Enantioselective Organocatalysed Reactions, Vols. I and II;* Springer: Berlin, 2011.
(ap) Scheffler, U.; Mahrwald, R. *Chem. Eur. J.* **2013,** *19*, 14346−14396.
(aq) Volla, C. M. R.; Atodiresei, I.; Rueping, M. *Chem. Rev.* **2014,** *114*, 2390−2431.
(ar) Chauhan, P.; Mahajan, S.; Kaya, U.; Hack, D.; Enders, D. *Adv. Synth. Catal.* **2015,** *357*, 253−281.
(as) Vetica, F.; Marcia de Figueiredo, R.; Orsini, M.; Tofani, D.; Gasperi, V. *Synthesis* **2015,** *47*, 2139−2184.
(at) Tian, L.; Luo, Y.-C.; Hu, X.-Q.; Xu, P.-F. *Asian J. Org. Chem.* **2016,** *5*, 580−607.
(au) Nayak, S.; Panda, P.; Bhakta, S.; Mishra, S. K.; Mohapatra, S. *RSC Adv.* **2016,** *6*, 96154−96175.
(av) Sahoo, B. M.; Banik, B. K. *Curr. Organocatal.* **2019,** *6*, 92−105.
(aw) Kristofikova, D.; Modrocka, V.; Meciarova, M.; Sebesta, R. *ChemSusChem* **2020,** *13*, 2828−2858.
(ax) Xiang, S.-H.; Tan, B. *Nature Commun.* **2020,** *11*, 3786−3791.

14. (a) Gal, J. *Chirality* **2012,** *24*, 959−976.
(b) Lepola, U.; Wade, A.; Andersen, H. F. *Int. Clin. Psychopharmacol.* **2004,** *19*, 149−155.
15. Justus von Liebig. *Annalen der Chem. und Pharmacie* **1860,** *113*, 246−247.
16. Kagan, H. B.; Jacobsen, E. N.; Pfaltz, A.; Yamamoto, H. *Comprehensive Asymmetric Catalysis*, 1. Springer-Verlag: Berlin, 1999.
17. Sharma, M.; Sharma, N. N.; Bhalla, T. C. *Enz. Microb. Tech.* **2005,** *37*, 279−294.
18. List, B. *Asymmetric Organocatalysis*, 2. Springer-Verlag: Berlin, 2009.
19. Prelog, V.; Wilhelm, M. *Helv. Chim. Acta.* **1954,** *37*, 1634−1660.
20. Torres, R. R. *Stereoselective Organocatalysis, Bond Formation Methodologies and Activation Modes;* Wiley & Sons: New Jersey, 2013.
21. Tu, Y.; Wang, Z. X.; Shi, Y. *J. Am. Chem. Soc.* **1996,** *118*, 9806−9807.
22. Tian, H.; She, X.; Xu, J.; Shi, J. *Org. Lett.* **2001,** *3*, 1929−1931.
23. Wu, X. Y.; She, X.; Shi, Y. *J. Am. Chem. Soc.* **2002,** *124*, 8792−8793.
24. Boutureira, O.; McGouran, J. F.; Stafford, R. L.; Emmerson, D. P. G.; Davis, B. G. *Org. Biomol. Chem.* **2009,** *7*, 4285−4288.
25. Corey, E. J.; Grogan, M. J. *Org. Lett.* **1999,** *1*, 157−160.
26. Iyer, M. S.; Gigstad, K. M.; Namdev, N. D.; Lipton, M. *J. Am. Chem. Soc.* **1996,** *118*, 4910−4911.
27. Okino, T.; Hoashi, Y.; Takemoto, Y. *J. Am. Chem. Soc.* **2003,** *125*, 12672−12673.
28. Okino, T.; Nakamura, S.; Furukawa, T.; Takemoto, Y. *Org. Lett.* **2004,** *6*, 625−627.
29. Sigman, M. S.; Jacobsen, E. N. *J. Am. Chem. Soc.* **1998,** *120*, 4901−4902.
30. Taylor, M. S.; Jacobsen, E. N. *J. Am. Chem. Soc.* **2004,** *126*, 10558−10559.
31. Joly, G. D.; Jacobsen, E. N. *J. Am. Chem. Soc.* **2004,** *126*, 4102−4103.
32. Wenzel, A. G.; Jacobsen, E. N. *J. Am. Chem. Soc.* **2002,** *124*, 12964−12965.
33. Nugent, B. M.; Yoder, R. A.; Johnston, J. N. *J. Am. Chem. Soc.* **2004,** *126*, 3418−3419.
34. Wang, C.; Zhou, Z.; Tang, C. *Org. Lett.* **2008,** *10*, 1707−1710.
35. Gao, P.; Wang, C.; Wu, Y.; Zhou, Z.; Tang, C. *Eur. J. Org. Chem.* **2008,** 4563−4566.

36. DiCesare, J. C. *Carbohydrates as chiral auxiliaries in organometallic reactions;* Georgia Institute of Technology, 1992.
37. Kunz, H.; Pfrengle, W.; Rück, K.; Sager, W. *Synthesis.* **1991,** 1039−1042.
38. Wang, Z. X.; Tu, Y.; Frohn, M.; Zhang, J.; Shi, Y. *J. Am. Chem. Soc.* **1997,** *119*, 11224−11235.
39. Cao, G. A.; Wang, Z. X.; Tu, Y.; Shi, Y. *Tetrahedron Lett.* **1998,** *39*, 4425−4428.
40. Capon, B. *Chem. Rev.* **1969,** *69*, 407−498.
41. (a) Juaristi, E.; Cuevas, G. *The Anomeric Effect;* CRC Press: Boca Raton, 1995.
(b) Box, V. G. S. *Heterocycles* **1990,** *31*, 1157−1181.
(c) Box, V. G. S. *Heterocycles* **1991,** *32*, 795−807.
42. Llera, J. M.; Lopez, J. C.; Fraser-Reid, B. *J. Org. Chem.* **1990,** *55*, 2997−2998.
43. Kotowycz, G.; Lemieux, R. U. *Chem. Rev.* **1973,** *73*, 669−698.
44. Wong, O. A.; Shi, Y. *Chem. Rev.* **2008,** *108*, 3958−3987.
45. Shi, Y. *Acc. Chem. Res.* **2004,** *37*, 488−496.
46. Wang, Z. X.; Shi, Y. *J. Org. Chem.* **1998,** *63*, 3099−3104.
47. Wang, Z. X.; Cao, G. A.; Shi, Y. *J. Org. Chem.* **1999,** *64*, 7646−7650.
48. Frohn, M.; Dalkiewicz, M.; Tu, Y.; Wang, Z. X.; Shi, Y. *J. Org. Chem.* **1998,** *63*, 2948−2953.
49. Zhu, Y.; Tu, Y.; Yu, H.; Shi, Y. *Tetrahedron Lett.* **1998,** *39*, 7819−7822.
50. Zhu, Y.; Shu, L.; Tu, Y.; Shi, Y. *J. Org. Chem.* **2001,** *66*, 1818−1826.
51. Baumstark, A. L.; McCloskey, C. J. *Tetrahedron Lett.* **1987,** *28*, 3311−3314.
52. Bach, R. D.; Andres, J. L.; Owensby, A. L.; Schlegel, H. B.; McDouall, J. J. W. *J. Am. Chem. Soc.* **1992,** *114*, 7207−7217.
53. Yang, D. *Acc. Chem. Res.* **2004,** *37*, 497−505.
54. Houk, K. N.; Liu, J.; DeMello, N. C.; Condroski, K. R. *J. Am. Chem. Soc.* **1997,** *119*, 10147−10152.
55. (a) Shu, L.; Shi, Y. *Tetrahedron Lett.* **1999,** *40*, 8721−8724.
(b) Shu, L.; Shi, Y. *Tetrahedron* **2001,** *57*, 5213−5218.
56. Tian, H.; She, X.; Shi, Y. *Org. Lett.* **2001,** *3*, 715−718.
57. Tian, H.; She, X.; Yu, H.; Shu, L.; Shi, Y. *J. Org. Chem.* **2002,** *67*, 2435−2446.
58. Shu, L.; Shen, Y. M.; Burke, C.; Goeddel, D.; Shi, Y. *J. Org. Chem.* **2003,** *68*, 4963−4965.
59. Shu, L.; Wang, P.; Gan, Y.; Shi, Y. *Org. Lett.* **2003,** *5*, 293−296.
60. Wang, Z. X.; Miller, S. M.; Anderson, O. P.; Shi, Y. *J. Org. Chem.* **2001,** *66*, 521−530.
61. Vega-Pérez, J. M.; Holm, M. V.; Martínez, M. L.; Blanco, E.; Iglesias-Guerra, F. *Eur. J. Org. Chem.* **2009,** 6009−6018.
62. Vega-Perez, J. M.; Vega-Holm, M.; Perinan, I.; Palo-Nieto, C.; Iglesias-Guerra, F. *Tetrahedron* **2011,** *67*, 364−372.
63. Vega-Perez, J. M.; Perinan, I.; Vega-Holm, M.; Palo-Nieto, C.; Iglesias-Guerra, F. *Tetrahedron* **2011,** *67*, 7057−7065.
64. Shing, T. K. M.; Leung, G. Y. C.; Luk, T. *J. Org. Chem.* **2005,** *70*, 7279−7289.
65. Shing, T. K. M.; Luk, T. *Tetrahedron: Asymmetry* **2009,** *20*, 883−886.
66. Shing, T. K. M.; Leung, G. Y. C.; Yeung, K. W. *Tetrahedron Lett.* **2003,** *44*, 9225−9228.
67. Wang, L.; Tang, R.; Yang, H. *J. Korean Chem. Soc.* **2013,** *57*, 591−598.
68. Lindstrom, U. M. *Chem. Rev.* **2002,** *102*, 2751−2772.
69. Dickerson, T. J.; Janda, K. D. *J. Am. Chem. Soc.* **2002,** *124*, 3220−3221.
70. Dickerson, T. J.; Lovell, T.; Meijler, M. M.; Noodleman, L.; Janda, K. D. *J. Org. Chem.* **2004,** *69*, 6603−6609.

71. Tsutsui, A.; Takeda, H.; Kimura, M.; Fujimoto, T.; Machinami, T. *Tetrahedron Lett.* **2007,** *48*, 5213–5217.
72. Agarwal, J.; Peddinti, R. K. *Tetrahedron: Asymmetry* **2010,** *21*, 1906–1909.
73. Agarwal, J.; Peddinti, R. K. *Eur. J. Org. Chem.* **2012,** 6390–6406.
74. Shen, C.; Shen, F.; Zhou, G.; Xia, H.; Chen, X.; Liu, X.; Zhang, P. *Cat. Comm.* **2012,** *26*, 6–10.
75. Shen, C.; Liao, H.; Shen, F.; Zhang, P. *Cat. Comm.* **2013,** *41*, 106–109.
76. Nisco, M. D.; Pedatella, S.; Bektas, S.; Nucci, A.; Caputo, R. *Carbohydr. Res.* **2012,** *356*, 273–277.
77. Wang, L.; Liu, J.; Miao, T.; Zhou, W.; Li, P.; Ren, K.; Zhang, X. *Adv. Synth. Catal.* **2010,** *352*, 2571–2578.
78. Agarwal, J.; Peddinti, R. K. *Tetrahedron Lett.* **2011,** *52*, 117–121.
79. Kumar, T. P.; Balaji, S. V. *Tetrahedron: Asymmetry* **2014,** *25*, 473–477.
80. Singh, N.; Pandey, J.; Tripathi, R. P. *Cat. Comm.* **2008,** *9*, 743–746.
81. Agarwal, J.; Peddinti, R. K. *J. Org. Chem.* **2011,** *76*, 3502–3505.
82. Emmerson, D. P. G.; Hems, W. P.; Davis, B. G. *Tetrahedron: Asymmetry* **2005,** *16*, 213–221.
83. Bornaghi, L. F.; Poulsen, S.-A. *Tetrahedron Lett.* **2005,** *46*, 3485–3488.
84. Shen, C.; Shen, F.; Xia, H.; Zhang, P.; Chen, X. *Tetrahedron: Asymmetry* **2011,** *22*, 708–712.
85. Li, L.; Fang, Z.; Fang, J.; Zhou, J.; Xiang, Y. *RSC Adv.* **2013,** *3*, 21084–21091.
86. Sharma, A.; Peddinti, R. K. *Synlett* **2018,** *29*, 630–634.
87. Rani, D.; Sethi, A.; Kaur, K.; Agarwal, J. *J. Org. Chem.* **2020,** *85*, 9548–9557.
88. (a) Schreiner, P. R. *Chem. Soc. Rev.* **2003,** *32*, 289–296.
    (b) Zhang, Z.; Schreiner, P. R. *Synlett.* **2007,** *9*, 1455–1457.
89. (a) Petri, M. P. *Angew. Chem. Int. Ed.* **2004,** *43*, 2062–2064.
    (b) Mike, K.; Schreiner, P. *Synthesis* **2007,** *5*, 779–790.
90. (a) Lukas, W.; Cabrele, C.; Vanejews, M.; Schreiner, P. R. *Eur. J. Org. Chem.* **2007,** *9*, 1474–1490.
    (b) Schreiner, P. R.; Alexander, W. *Org. Lett.* **2002,** *4*, 217–220.
91. (a) Jian, W.; Li, H.; Yu, X.; Zu, L.; Wang, W. *Org. Lett.* **2005,** *7*, 4293–4296.
    (b) Benedek, V.; Varga, S.; Csámpai, A.; Soós, T. *Org. Lett.* **2005,** *7*, 1967–1969.
    (c) McCooey, S. H.; Connon, S. J. *Angew. Chem. Int. Ed.* **2005,** *44*, 6367–6370.
92. Liu, K.; Cui, H. F.; Nie, J.; Dong, K. Y.; Li, X. J.; Ma, J. A. *Org. Lett.* **2007,** *9*, 923–925.
93. Gu, Q.; Guo, X. T.; Wu, X. Y. *Tetrahedron* **2009,** *65*, 5265–5270.
94. Wang, Y.; Yang, H.; Yu, J.; Miao, Z.; Chen, R. *Adv. Synth. Catal.* **2009,** *351*, 3057–3062.
95. Nie, J.; Li, J.; Zheng, X. D. H.; Zhang, F. G.; Cui, S.; Ma, J. A. *J. Fluor. Chem.* **2011,** *132*, 468–473.
96. Ma, H.; Liu, K.; Zhang, F. G.; Zhu, C. L.; Nie, J.; Ma, J. A. *J. Org. Chem.* **2010,** *75*, 1402–1409.
97. Lu, A.; Hu, K.; Wang, Y.; Song, H.; Zhou, Z.; Fang, J.; Tang, C. *J. Org. Chem.* **2012,** *77*, 6208–6214.
98. Lu, A.; Gao, P.; Wu, Y.; Wang, Y.; Zhou, Z.; Tang, C. *Org. Biomol. Chem.* **2009,** *7*, 3141–3147.
99. Ágoston, K.; Fügedi, P. *Carbohydr. Res.* **2014,** *389*, 50–56.
100. Azad, C. S.; Khan, I. A.; Narula, A. K. *Org. Biomol. Chem.* **2016,** *14*, 11454–11461.

101. Pu, X. W.; Li, P.; Peng, F.; Li, X.; Zhang, H.; Shao, Z. *Eur. J. Org. Chem.* **2009,** 4622–4626.
102. Pu, X. W.; Peng, F. Z.; Zhang, H. F.; Shao, Z. H. *Tetrahedron* **2010,** *66*, 3655–3661.
103. Puglisi, A.; Benaglia, M.; Raimondi, L.; Lay, L.; Poletti, L. *Org. Biomol. Chem.* **2011,** *9*, 3295–3302.
104. Kong, S.; Fan, W.; Wu, G.; Miao, Z. *Angew. Chem. Int. Ed.* **2012,** *51*, 8864–8867.
105. Yuan, H. N.; Wang, S.; Nie, J.; Meng, W.; Yao, Q.; Ma, J. A. *Angew. Chem. Int. Ed.* **2013,** *52*, 3869–3873.
106. Yuan, H. N.; Li, S.; Nie, J.; Zheng, Y.; Ma, J. A. *Chem. Eur. J.* **2013,** *19*, 15856–15860.
107. Reddy, B. V. S.; Reddy, S. M.; Swain, M. *RSC Adv.* **2013,** *3*, 930–936.
108. Roncak, R.; Tvrdonova, M.; Gonda, J.; Elecko, J. *Tetrahedron* **2020,** *76*, 131339.
109. Dwivedi, N.; Bisht, S. S.; Tripathi, R. P. *Carbohydr. Res.* **2006,** *341*, 2737–2743.
110. Wynands, L.; Delacroix, S.; Nhien, A. N. V.; Soriano, E.; Contelles, J. M.; Postel, D. *Tetrahedron.* **2013,** *69*, 4899–4907.
111. Yang, W.; Sha, F.; Zhang, X.; Yuan, K.; Wu, X. *Chin. J. Chem.* **2012,** *30*, 2652–2656.
112. Porwanski, S. *Carbohydr. Res.* **2014,** *394*, 7–12.
113. Robak, J.; Koselak, K.; Zawisza, A.; Porwański, S. *Arkivoc* **2020,** *viii*, 150–160.

Chapter 14

Emerging aspects of domino reactions in carbohydrate Chemistry☆

Mangal S. Yadav, Abhishek Gupta, Sunil Kumar and Vinod K. Tiwari
Department of Chemistry, Institute of Science, Banaras Hindu University, Varanasi, Uttar Pradesh, India

14.1 Introduction

One of the primary goals of modern organic synthesis is to achieve synthetic efficiency in terms of minimizing the synthetic steps involved therein while increasing molecular complexity.[1] In light of this, the attention to the development of domino reactions in organic synthesis has grown rapidly in the last two decades.[2–4] Prof. Tietze defined a domino reaction as a chemical process that involves two or more bond-forming modifications that occur within the same reaction conditions even without the addition of additional reagents or catalysts, in which the successive reactions are the outcome of the functionality formed in the preceding step.[5] Domino reactions enable the highly efficient synthesis of complex organic compounds, as these processes do not require any intermediate recovery steps, and they also significantly diminish production costs and times.[6] In the field of carbohydrate chemistry, synthetic efficiency is particularly significant.[7] The synthesis of well-defined sugar derivatives has long been a goal pursued by researchers from various fields due to their remarkable biological importance.[8,9] However, because of their structural complexity, carbohydrates are always not easy to synthesize. As carbohydrates are biocompatible molecules, using them as synthetic building blocks frequently involves various reaction steps and the manipulation of numerous protecting groups.[10–13] In light of this, there has recently been a great interest in the use of domino reactions in carbohydrate chemistry.[14] This chapter summarizes the various domino synthetic approaches explored in carbohydrate chemistry including the use of chiral analogs, chiral catalysts, ring fusion, and ring fragmentation. In addition, some

☆This chapter is dedicated in memory of (Late) Prof. Ivar Karl Ugi, a renowned scientist known for the contributions on multicomponent reactions (MCR).

Synthetic Strategies in Carbohydrate Chemistry. DOI: https://doi.org/10.1016/B978-0-323-91729-2.00013-6

selected total synthesis of biologically relevant scaffolds and molecules and various other conversions are covered herein.

14.2 Asymmetric tandem reactions using chiral carbohydrate derivatives

Despite the rapid evolution of asymmetric domino methods, such as multicomponent reactions, chiral carbohydrate derivative-based domino reactions are still uncommon in the literature. Regarding organocatalysts derived from carbohydrates, this is especially true. A number of significant works that address stereocontrolled domino reactions of chiral carbohydrate scaffolds and enantioselective domino reactions catalyzed by chiral carbohydrate derivatives are discussed herein. A report documented the stereoselective synthesis of 1,4-dihydropyridine C-glycoconjugates **4** through the three-component Hantzsch reaction.[15] This cyclocondensation reaction was catalyzed by L-proline domino reaction with chiral C-glycosyl aldehyde **1**, an enamine **2**, and β-diketone **3** as starting materials. The product was obtained with although 50% yield, but excellent diastereomeric excess (over 98% de) was

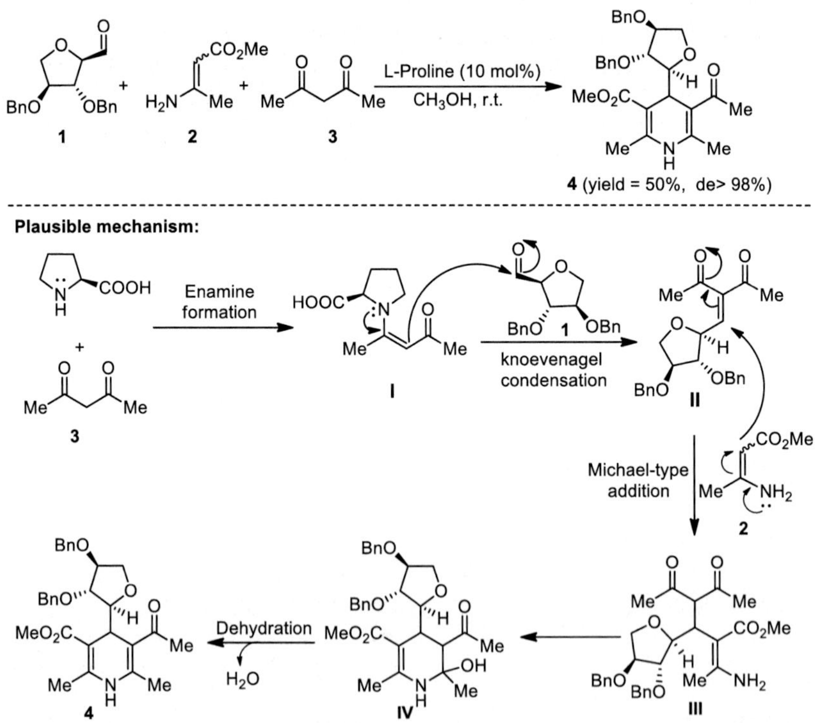

SCHEME 14.1 Proline-catalyzed Hantzsch reaction of C-glycosyl ulose **1** with enamine **2** and diketone **3** to give diverse 1,4-dihydropyridine C-glycoconjugates **4**.

the notable feature (Scheme 14.1). This methodology is based on the principle of double asymmetric induction, originating from the use of C-glycosyl aldehyde as a chiral auxiliary along with L-proline as a chiral organocatalyst. This strategy made the synthesis of biologically relevant C-nucleosides feasible, which were not accessible through uncatalyzed approaches.

The mechanism initiates with the formation of enamine **I** from β-diketone **3** and L-proline through nucleophilic substitution reaction. The enamine attacks the carbonyl carbon of C-glycosyl aldehyde **1** in Knoevenagel fashion to give intermediate **II** by loss of water molecule. The next step involves the Michael-type addition of the enamino ester **2** with intermediate **II** to give intermediate **III**. The reaction then undergoes an intramolecular nucleophilic attack in intermediate **III** resulting in a cyclized intermediate **IV**, which finally on dehydration gives the target product **4** in good yields (Scheme 14.1).[16]

The current idea of multicomponent reaction is closely related to the reactions established with isocyanide reagents. A four-component Ugi reaction is the reaction of carbonyl compounds (preferably, aldehydes), isocyanides, amines, and carboxylic acid to produce α-amino acid derivatives.[17] The reaction proceeds with the in situ formation of imine by the reaction of aldehyde with primary amine and then the addition of isocyanide to imine and carboxylic acid leads to the formation of corresponding α-amino acid derivatives. This reaction was initially studied in 1959 and was even more popular than other multicomponent in that era. In their initial work, Ugi et al. stated that the use of chiral acids or isonitriles did not give any degree of stereoselectivity in the reaction. Another reaction of chiral ferrocenylamine afforded the nonracemic amino acid derivatives with very low or moderate stereoselectivity.[18–20] Later, Kunz et al. established a versatile route in Ugi reaction using a chiral carbohydrate derivative. The reaction of galactosylamine derivatives **5**, aldehyde **6**, and isonitrile **7** afforded 94% de of (*R*)-amino acids **8** with a high degree of diastereoselectivity (Scheme 14.2A).[21] The fact that high levels of stereoselectivity have been reported to only reactions using *tert*-butyl isonitrile was a disadvantage of this asymmetric Ugi reaction. Additionally, an arabinosylamine precursor **9**, *tert*-butyl isonitrile **10**, was used to accomplish the asymmetric synthesis of (*S*)-amino acids **11** with exceptional diasterioselectivities of up to 96% de via Ugi reaction (Scheme 14.2B).[22] Goebel et al. reported a single variant of Kunz's investigation in 1991. The authors demonstrated that 2,3,4,6-tetra-*O*-alkyl-β-D-glucopyranosylamines once used as chiral amine components were even more efficient than the Kunz's *O*-acyl-aldopyranosylamides used to achieving diastereoselectivities of resultant molecules up to >99% de.[23,24]

Another prominent multicomponent reaction in synthetic organic chemistry is the Passerini reaction, which involves the condensation of carbonyl compounds, carboxylic acids, and isocyanides to give corresponding α-acyloxycarboxamides.[25] This reaction needs mild conditions and has a wide scope. Krishna et al. explored the synthetic utility of (*p*-toluenesulfonyl) methylisocyanide (TosMIC) **12** as the isonitrile component in the

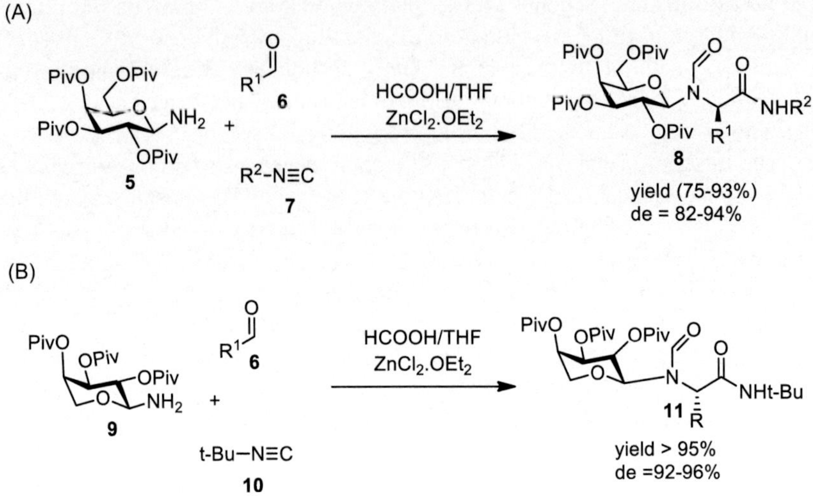

SCHEME 14.2 A tandem Ugi reaction using a chiral carbohydrate amines.

diastereoselective Passerini reaction.[26] In the process, carbohydrate-derived aldehydes **13** were taken as the chiral auxiliaries and allowed to react with TosMIC and carboxylic acids **14**, giving mandelamides **15** as the sole product in moderate yields (35%–45%) and diastereoselectivities (50%–52% de) (Scheme 14.3).[26] In a similar way, several other chiral auxiliaries and substrates have also been employed to explore the stereochemical outcome of the Passerini reaction.[27,28] The mechanism of the reaction starts with the protonation of the carbonyl oxygen of aldehyde **13** with acid **14** to form intermediate **I**, followed by nucleophilic addition of the TosMIC to produce the nitrilium ion **II**, which is then attacked by the carboxylate anion to give the intermediate **III**, which undergoes acyl transfer followed by amide tautomerization to yield the final product **15** (Scheme 14.3).

The synthesis of β-aminocarbonyl derivatives can be carried out very efficiently using the well-known Mannich reaction that employs condensation of an enolizable carbonyl compound, a non enolizable carbonyl compound (e.g., formaldehyde), and an ammonia or amine derivative (primary or secondary) as starting components.[29] This reaction was also explored as an asymmetric multicomponent reaction using various possible chiral substrates.[30–34] A report published by Dondoni et al. emphasized the synthesis of chiral C-glycosyl α,α-dimethyl β-amino esters **19** and **21** from respective chiral C-glycosyl aldehydes **16** and **20**, using the three-component Mannich reaction, along with the addition of 1-methoxy-2-methyl-1-trimethylsilyloxypropene **17** and *p*-methoxybenzylamine **18** in each separate reaction. The reaction was catalyzed by $InCl_3$ and the formation of the products in high yields with excellent diastereoselectivity (Scheme 14.4).[35]

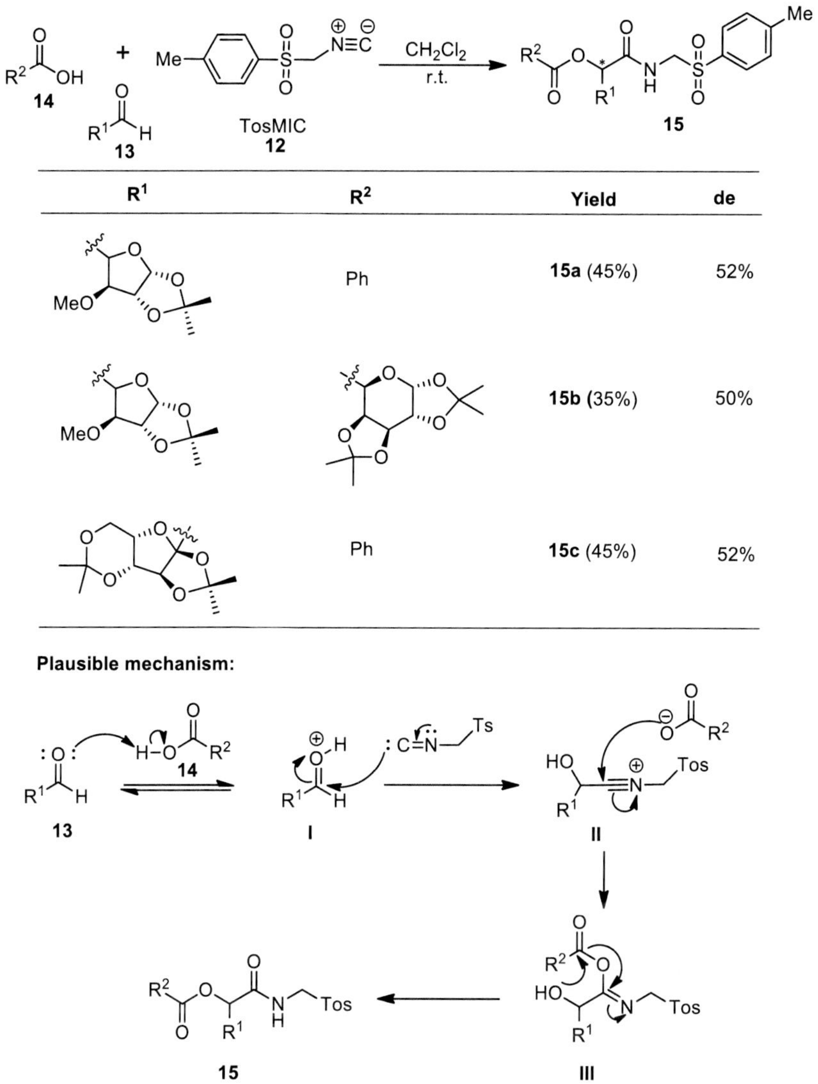

SCHEME 14.3 Passerini reaction of the chiral carbohydrate-derived aldehydes.

Further, Dondoni et al. developed diastereoselective four-component domino Reformatsky-type reactions catalyzed by Ni(II) for the synthesis of chiral C-glycosyl β-amino esters **24** and **25** from corresponding chiral C-glycosyl aldehydes **16** and **20** along with the addition of ethyl bromoacetate **22**, *p*-methoxybenzylamine **23**, and dimethyl zinc as catalyst. The reaction proceeds initially with the formation of imines from in situ coupling of chiral C-glycosyl aldehydes and *p*-methoxybenzylamine. The reactants

(A)

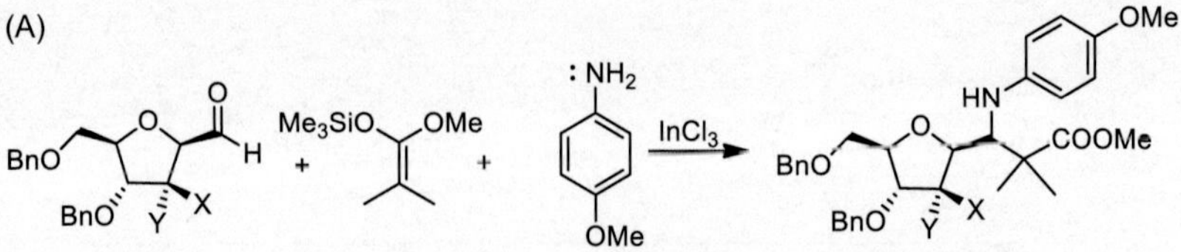

16 17 18 19, de > 99%

X	Y	Yield
OBn	H	**19a** (75%)
H	OBn	**19b** (85%)

(B)

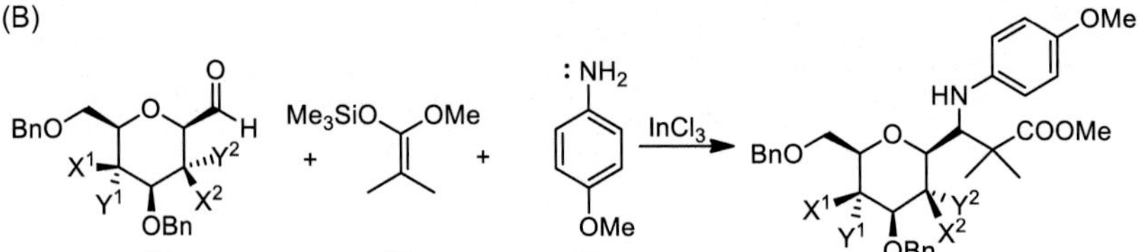

20 17 18 21, de >99%

X^1	Y^1	X^2	Y^2	Yield
OBn	H	H	OBn	**21a** (80%)
H	OBn	OBn	H	**21b** (60%)
H	OBn	H	OBn	**21c** (90%)

SCHEME 14.4 Highly diastereoselective 3CR Mannich reaction of C-glycosyl aldehydes.

dimethyl zinc and ethyl bromoacetate reacted in situ to generate bromozinc enolate, which then reacts with the generated imines to provide the desired product (Scheme 14.5).[35] The most important feature of these reactions was their high stereoselectivity because the chiral Reformatsky products were obtained as single diastereoisomers in moderate-to-good yields.

14.3 Organocatalyzed tandem reaction

Organocatalyzed reactions got more attention in a few decades. A rapid progression of organocatalysis paved the way for C-glycosylation These new techniques, as opposed to the reactions previously mentioned, not only allow access to the desired building blocks from unprotected and unactivated carbohydrates but also facilitate catalytic processes. The enhanced new strategies for C-glycosylation reactions involve the use of organocatalysts such as L-Proline for easy access to the target structural motif. These new methods can also be applied to unprotected and unactivated carbohydrates. There had been ground breaking developments in the field of C-glycosylation in the last few years.[36] A brief description of this development is discussed in Schemes 14.6 and 14.7 with the examples of direct synthesis of C-glycosides

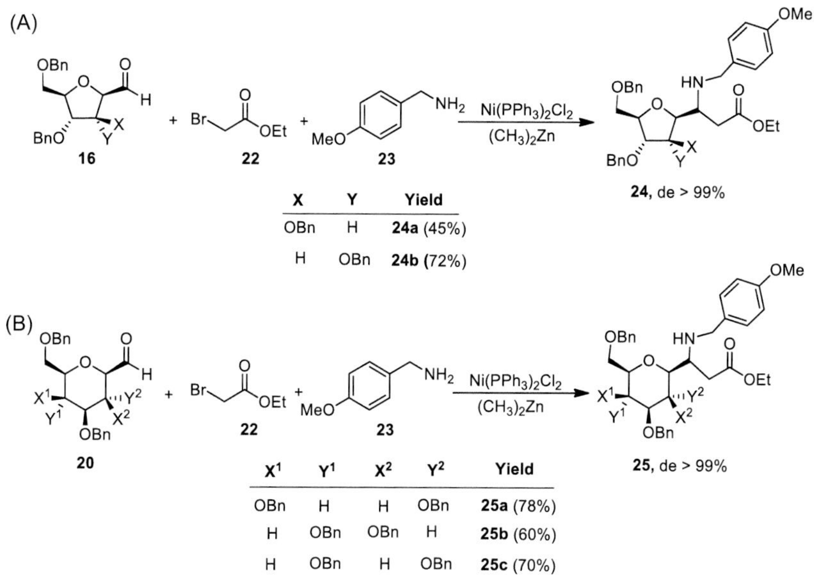

(A)

X	Y	Yield
OBn	H	**24a** (45%)
H	OBn	**24b** (72%)

(B)

X^1	Y^1	X^2	Y^2	Yield
OBn	H	H	OBn	**25a** (78%)
H	OBn	OBn	H	**25b** (60%)
H	OBn	H	OBn	**25c** (70%)

SCHEME 14.5 Domino Reformatsky reaction of chiral C-glycosyl aldehydes with ethyl bromoacetate and amine to give C-glycosyl β-amino esters.

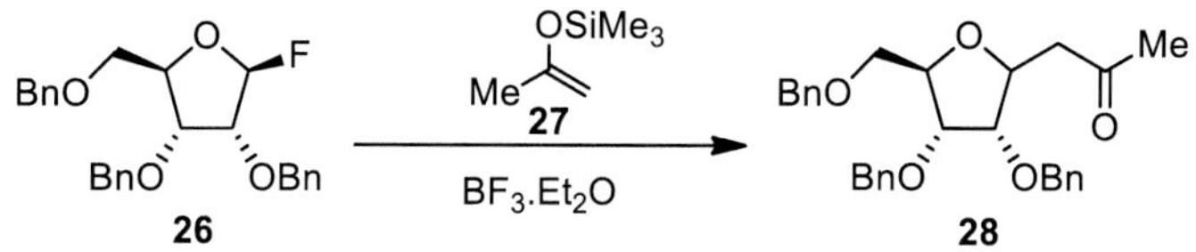

SCHEME 14.6 Synthesis of C-glycosides of acetone.

of unprotected and unactivated pentoses (except Scheme 14.6) like ribose, xylose, and acetone. The first glycoside of acetone **28** was synthesized by the Mukaiyama reaction nearly two decades ago. In this reaction, protected and activated carbohydrates **26** were allowed to react with silyl enol ether **27** of acetone along with the catalytic amount of Lewis acids, for example, $BF_3 \cdot Et_2O$ (Scheme 14.6).[37,38]

A report was published by Lubineau et al. some 10 years later, which depicted the methods to access different mixtures of pyranoid and furanoid structures of α- and β-configured C-glycosides **31**[39,40] via direct reaction of unprotected carbohydrates **29** with acetylacetone **30** in aqueous alkali media (Scheme 14.7A).[41] Early cascade reactions (Knoevenagel/intramolecular oxa-Michael addition) of unprotected and unactivated carbohydrates with barbituric acid derivatives were reported by Perez and Clarkson group.[42,43] The starting 1,3-dicarbonyl component, that is, acetylacetone undergoes loss

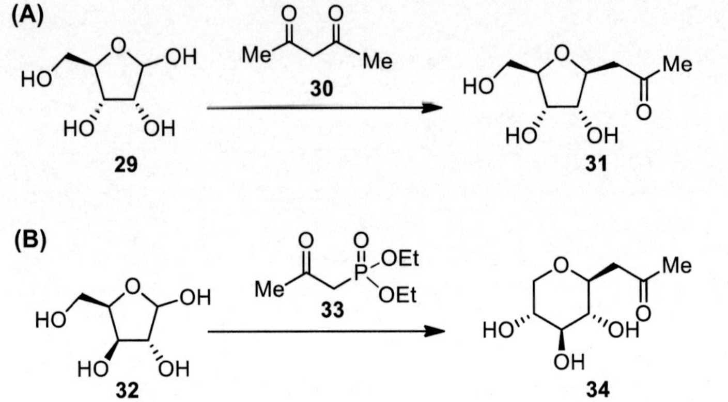

SCHEME 14.7 Synthesis of C-glycosides of acetone.

of a C2 fragment when this Knoevenagel/Michael/retro-Claisen-Aldol cascade reaction is carried out at high temperature (60°C–90°C).

Wittig and Horner–Emmons–Wordsworth olefination methodologies provide an alternate route for the C-glycosylation of unprotected carbohydrates **32** via the use of corresponding phosphonate **33**. The further intramolecular oxa-Michael reaction provides access to the glycoside of acetone **34** (Scheme 14.7B).[44,45]

Easy access to C-glycosides **36** under mild conditions was carried out via the reaction of unprotected carbohydrates **29** with acetone **35** catalyzed by a combination of tertiary amines and proline. For the access of C-glycosides of acetone deployed an equimolar mixture of proline and 1,8-diazabicyclo [5.4.0]undec-7-ene (DBU) as the catalytic system. Unprotected acetone glycosides were obtained in good yields of approximately 70% through this method (Scheme 14.8).[46]

This observation was further explored with various substrates including several unprotected carbohydrates **29** and methyl ketones. Consequently, the optimization for an efficient catalytic system revealed that proline and DBU should be used in equimolar amounts at room temperature for obtaining desired products in good-to-excellent yield or sometimes in quantitative yield. The optimized condition is essential for the transformation of various unsymmetrical methyl ketones **38** into their corresponding C-glycosides (Scheme 14.9).[46]

The substrates were further varied and expanded from methyl ketones to 1,3-dicarbonyl compounds. This gave rise to chain-elongated carbohydrate products instead of C-glycosylated products. The chain elongation takes place by the assimilation of the employed 1,3-dicarbonyl compound into the carbon skeleton of the starting carbohydrate molecule.[47]

The initial experiment was carried out by taking 2-deoxy ribose **39** as a carbohydrate component and ethyl acetoacetate **40** as the 1,3-dicarbonyl

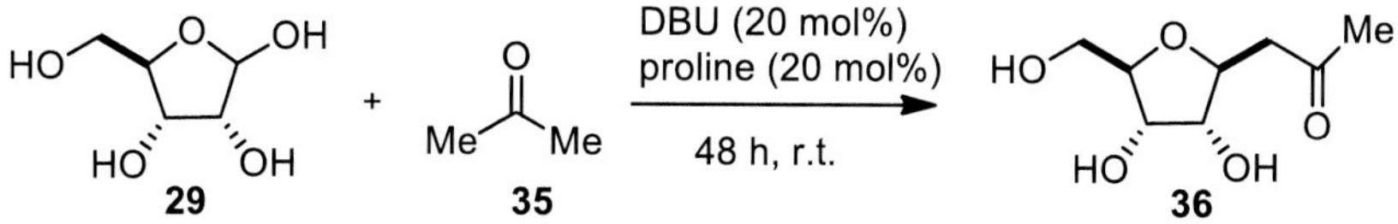

SCHEME 14.8 DBU/L-proline-catalyzed cascade reactions of D-ribose and acetone.

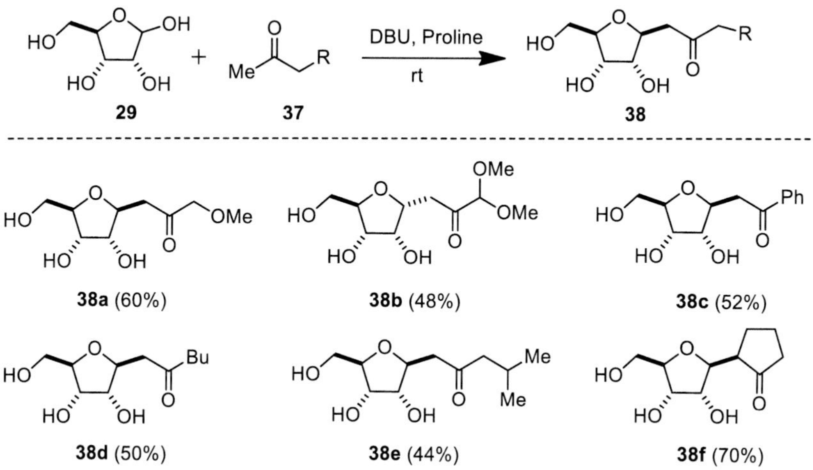

SCHEME 14.9 Synthesis of C-glycosides by cascade reaction of ribose with various methyl ketones.

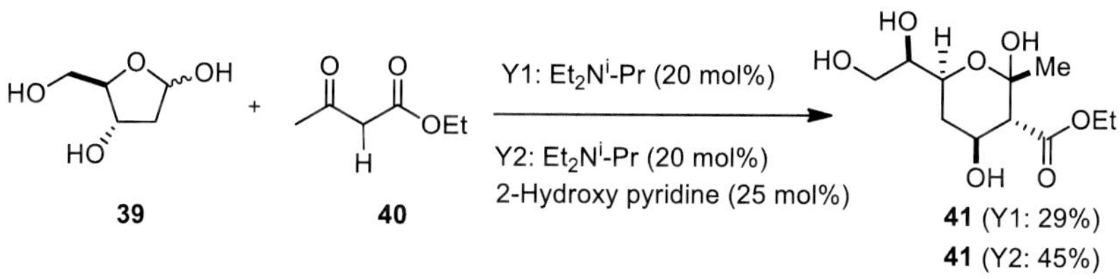

SCHEME 14.10 Amine catalyzed Knoevenagel-addition of 2-deoxyribose with ethyl acetoacetate.

compound in the presence of the catalytic amount of *N*,*N*-diisopropylethylamine (DIPEA). The reaction proceeded through a Knoevenagel-addition/ketalization cascade mechanism and produced a chain-elongated 1-deoxy ketose **41** with an excellent degree of stereoselectivity but in a very moderate yield (29%) (Scheme 14.10).[47] An exceptionally high degree of *syn*-diastereoselectivity was observed in the C–C bond formation in the product with no trace of the corresponding anticonfigured compound. Further extension of this observation to other pentoses produces the corresponding Knoevenagel-addition adducts, although with lower yields. Then 2-hydroxy pyridine was tested as an additional catalyst to minimize the problem of lower yields of the Knoevenagel-addition products. This resulted in the increment of the yield from 29% to 45% in the reaction of 2-deoxy ribose **39** with ethyl acetoacctate **40** (Scheme 14.10). The 2-hydroxy pyridine or 2-pyridone was chosen as an additional catalyst due to its known property of having a large effect on the anomerization equilibration of carbohydrates by the formation of hydrogen bonds.[47–56] A lucid representation of the configurative results of the reaction between D-ribose **29** and ethyl acetoacetate is depicted in Fig. 14.1.

Carbohydrates containing (ethoxycarbonyl) methylene moiety have been efficiently utilized as suitable scaffolds for the synthesis of bicyclic, sugar-fused, five-membered α,β-unsaturated lactones.[57–59] A report was published for the synthesis of carbohydrates possessing (ethoxycarbonyl) methylene group through a tandem approach by the reaction of 2-*O*-acyl protected-D-ribo-3-uloses with [(ethoxycarbonyl)methylene]triphenylphosphorane using the approach of Wittig olefination. This reaction was explored in two ways by utilizing two different solvents and both methods afforded products having different regio- and stereoselectivity. In the first reaction, 2-*O*-acyl protected-D-ribo-3-uloses **42** was allowed to react with [(ethoxycarbonyl) methylene]triphenylphosphorane in acetonitrile and produced regio- and stereoselective 2-(*Z*)-alkenes **43** under microwave irradiation (Scheme 14.11).

The second reaction employed both the same reactants using chloroform as solvent, and it yielded 3-(*E*)-alkenes **44** in regio- and stereospecific

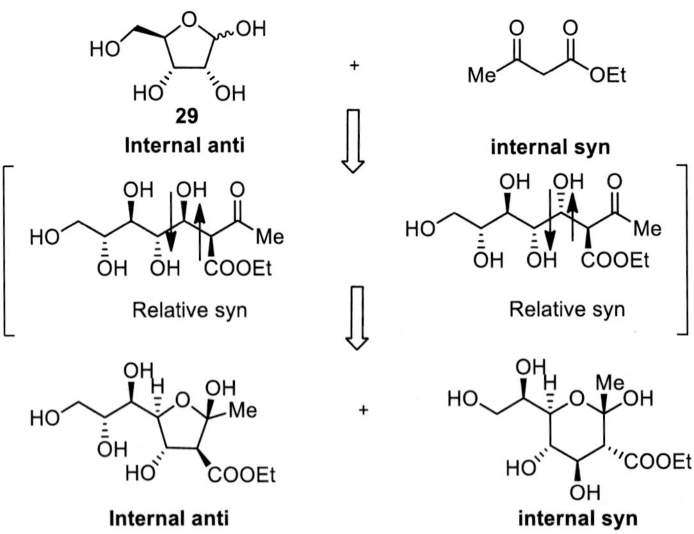

FIGURE 14.1 Configurative course of aldol reaction of D-ribose with ethyl acetoacetate.

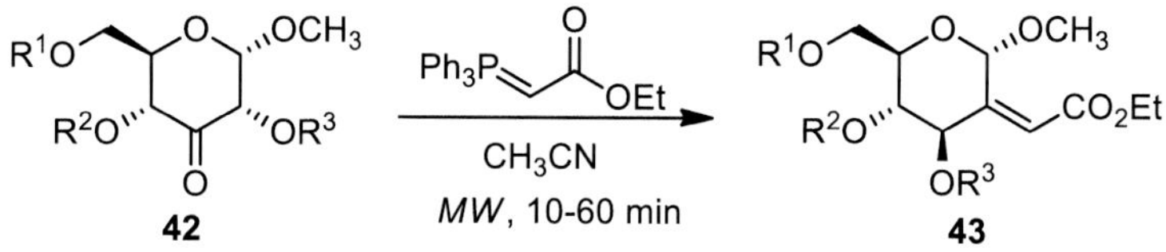

SCHEME 14.11 Synthesis of regio- and stereoselective 2-(*Z*)-alkene.

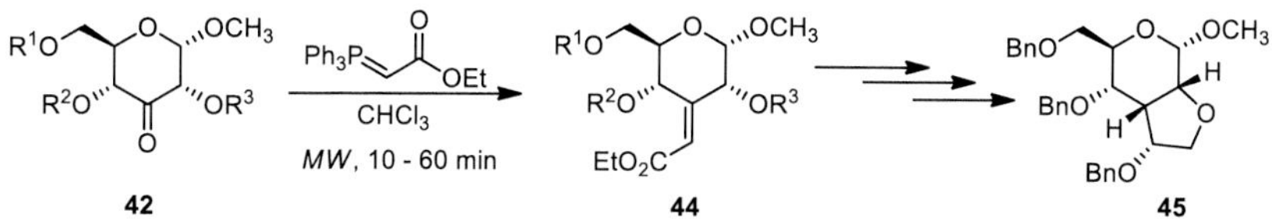

SCHEME 14.12 Synthesis of regio- and stereoselective 3-(*E*)-alkene.

manner. The product of this reaction is also the key precursor for the miharamycin's bicyclic sugar moiety (Scheme 14.12).

The efficacy of Wittig olefination was explored for the synthesis of these compounds. It has been observed that this domino reaction proceeded through the tautomerization of 3-ulose to enol, acyl migration, tautomerization to 3-*O*-acyl-2-ulose, and the Wittig reaction.[60]

A novel protocol for the synthesis of carbohydrates heterocycle derivatives was reported by Mahrwald and coworkers. The group applied the concept of multicomponent reaction by taking L-proline, unprotected maltose **46**, and ethyl isocyanoacetate **47** as starting materials and DIPEA (Huning's base) as catalyst. The construction of glycopeptide mimetic as functionalized lactone **48** was the outcome of this reaction. The product functionalized lactone was isolated in a 38% yield as syn/anti mixture 83/17. This work is illustrated in Scheme 14.13.[61]

Glycosidases are one of the major classes of carbohydrate-processing biocatalysts that catalyze the hydrolysis of glycosyl linkages. Also, the enzyme-catalyzed hydrolysis of carbohydrates is a biologically widespread process; therefore, inhibition of these glycosidases has profound effects on various biological mechanisms in living systems. This property makes glycosidase inhibitors as potential therapeutic candidates for use in several diseases such as viral infections, cancer, and genetic disorders. Natural products such as 1-deoxynojirimycin and its analogs are the best-known examples for the inhibition of glycosidases and glycosyltransferases.[62–65]

Iminosugars derivatives such as *N*-butyl-1-deoxynojirimycin are found to have the desired glycosidase inhibition property for the treatment of various disorders like lysosomal storage disorders (LSDs).[66] Iminosugar-C-glycosides have been found to be having appropriate therapeutic activities in this regard. Thus the synthesis of 1,5-dideoxy-1,5-iminoribitol C-glycosides has been carried out via a new domino strategy.

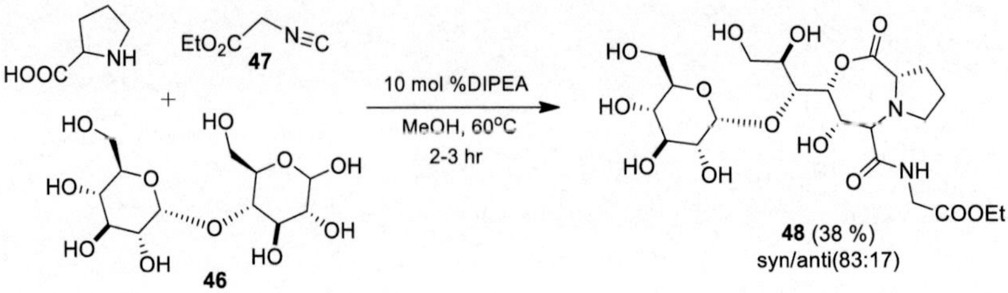

SCHEME 14.13 Mahrwald's synthesis of carbohydrate heterocycles using L-proline, ethyl isocyanoacetate, and maltose.

This synthetic strategy employed 2,3-*O*-isopropylidine-5-*O*-tosyl-α-D-ribofuranose **49** as the starting material, which was synthesized by the previous reported methods.[67,68] This synthetic approach involves six-membered sugar nitrone as a key intermediate. The domino pathway involves first the formation of a sugar oxime (from D-ribose) followed by its intramolecular *N*-alkylation with the departure of the *O*-tosyl group to give the corresponding six-membered cyclic nitrone **51**, which simultaneously undergoes an intermolecular 1,3-dipolar cycloaddition with alcohols **50** present to give the expected isoxazolidine **52** in a one-pot fashion. The corresponding 1,5-dideoxy-1,5-iminoribitol C-glycosides **53** are obtained in 70%–80% yields after treatment of individual isooxazolidines with TFA.H_2O for the hydrolysis of the acetonide group followed by reductive cleavage of the N – O bond using $PdCl_2$ under hydrogen pressure (Scheme 14.14).[69]

Another multicomponent reaction for the sustainable synthesis of various heterocycles was reported, which includes unactivated carbohydrates, oxoacetonitriles, and ammonium acetate as starting components. This reaction was metalfree, catalyzed by triethylamine, and it produced densely functionalized pyrroles in good-to-excellent yield. In the following two examples, D (+) glucose and D(+) maltose were taken as the carbohydrate components along with benzoylacetonitrile, ammonium acetate, and a catalytic amount of triethylamine and heated at 80°C in dimethylformamide (DMF) for 4 hours to yield corresponding polyhydroxyalkyl 2-phenyl pyrrole-3-carbonitrile **54** and **55** in 92% yield. In the same fashion, this tandem reaction was carried out by taking various carbohydrate components such as D (+) galactose, D (+) ribose, D (+) xylose, D (+) cellobiose, and L (+) lactose along with several substituted benzoylacetonitriles under similar reaction conditions (Scheme 14.15).[70]

Judeh et al. reported the high-yielding synthesis of densely functionalized polyhydroxyalkyl 2-amino-3-furanonitriles **56** and polyhydroxyalkyl 2-amino-3-furanocarboxamide **57** in basic aqueous media by condensation of unprotected carbohydrates with malononitrile and malonamide nitrile, respectively, in one pot. Disaccharides enabled the production of new furanoglycosides. Based on the rate at which the furans formed, it was determined

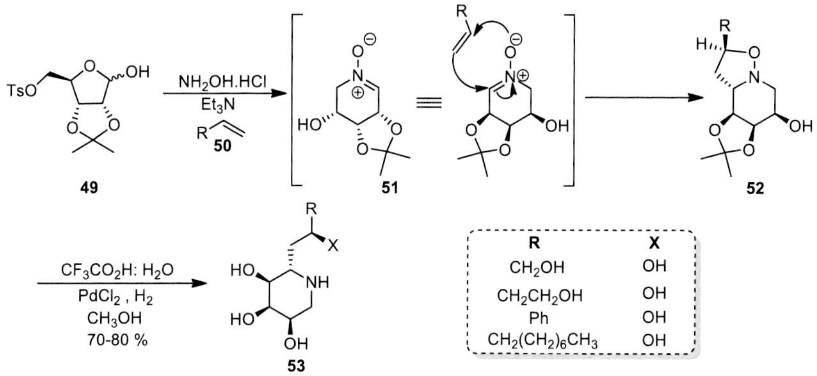

SCHEME 14.14 Synthesis of 1,5-dideoxy-1,5-iminoribitol C-glycosides.

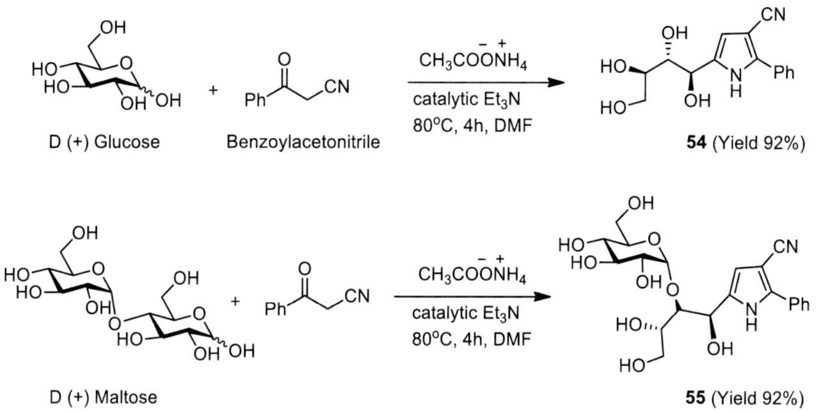

SCHEME 14.15 Synthesis of various heterocycles was reported including unactivated carbohydrates.

that Et_3N was better at catalyzing the reaction than DBU, K_2CO_3, and NaOMe. This tandem reaction was carried out through a series of reactions, including Knoevenagel condensation, oxo-Micheal addition, Thorpe-Ziegler type reaction, and ring-opening-induced aromatization. This straightforward and metal-free reaction progressed with exclusive chemo-, regio-, and stereo-selectivity (Scheme 14.16).[71]

14.4 Glycals in domino reaction

The cyclic enol ether derivatives of sugars that are commonly known as glycals are readily available chiral building blocks for the synthesis of various natural products and their analogs. A substantial development in glycal-based methodologies and their applications in the synthesis of biologically

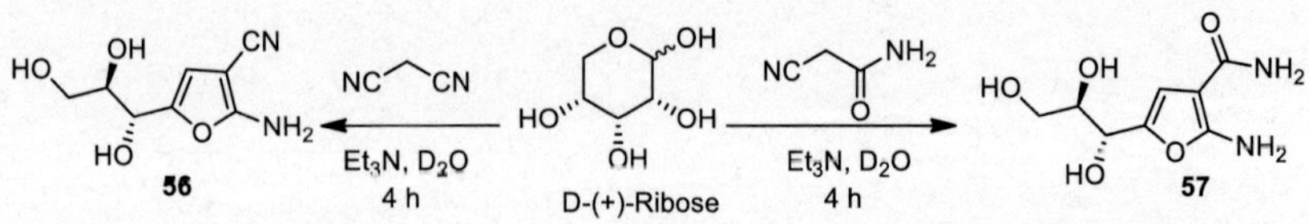

SCHEME 14.16 Synthesis of densely functionalized furans from unprotected carbohydrates.

important molecules has been recorded over the years.[72–79] As glycals contain enol ether double bond, they are challenging substrates for diarylation reactions.

In 2015 Mukherjee and coworkers developed a strategy for efficient vicinal diarylation of glycals and pseudoglycals. The protocol was a kind of Heck–Suzuki arylation that employed aryl boronic acids as aryl coupling partners in acetic acid solvent under Pd catalysis. 2,2,6,6-tetramethylpiperidine-1-oxyl (TEMPO) was also used in this reaction as an oxidizing agent that played a crucial role in glycoside synthesis due to its radical nature. Firstly, tri-*O*-acetyl-D-glucal **58** was allowed to react with phenyl boronic acid **59** in acetic acid solvent using TEMPO as a radical oxidative agent and palladium acetate as a catalyst to afford vicinal diaryl glycal **60**. The yield of this reaction was 65%–80% at 30°C–40°C temperature. Notably, the selectivity was C1 − C2(α,α) in the case of glycals (Scheme 14.17).[80]

Pseudoglycals **61** were also explored under this strategy by utilizing the optimized reaction conditions for glycals. 2,3-diarylated glycosides **62** in 60%–70% yield and with excellent stereoselectivity were obtained as the reaction product at a reduced temperature of 10°C. Two-dimensional nuclear magnetic resonance (2D-NMR) spectroscopic studies revealed that the stereochemistry of the product is *cis*(β,β) at C2 − C3 carbon centers. In nuclear overhauser effect spectroscopy (NOESY) of compound **62**, cross-peaks between H3(δ 3.46)/H5 (δ 3.77) and no cross-peak between H2(δ 3.1)/H4(δ 5.61) were observed, which confirms the stereochemistry of aryl groups as *cis*(β − β) at C2–C3 (Scheme 14.18).[80]

Simply changing the solvent in a triflic acid-catalyzed domino/cascade stereoselective reaction of glycals **63** with styrenes **64** yields complex oxabicyclic frameworks such as *cis*-oxadecalins **65** or *cis*-cyclopentanofurans **66** (Scheme 14.19). In the existence of contributing solvents such as benzene/toluene, cascade Ferrier C-glycosylation, and the double Friedel Crafts reaction gives densely chiral benzofused oxadecalins **65**, whereas in dichloromethane (DCM), an unprecedented ring-opening, ring-closing sequence yielded *cis*-cyclopentanofurans **66**. The above domino sequences have been attempted to explain using intermediate trapping, kinetic experiments, and Baldwin's rules.[81]

Chromans and isochromans are biologically and pharmaceutically active and valuable scaffolds and find widespread applications in the area of medicinal chemistry, natural products chemistry, and synthetic organic chemistry.

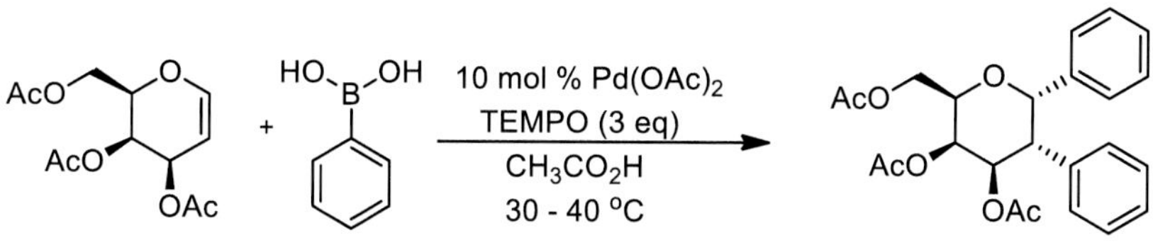

58 59 60

SCHEME 14.17 Vicinal diarylation of tri-*O*-acetyl-D-glucal.

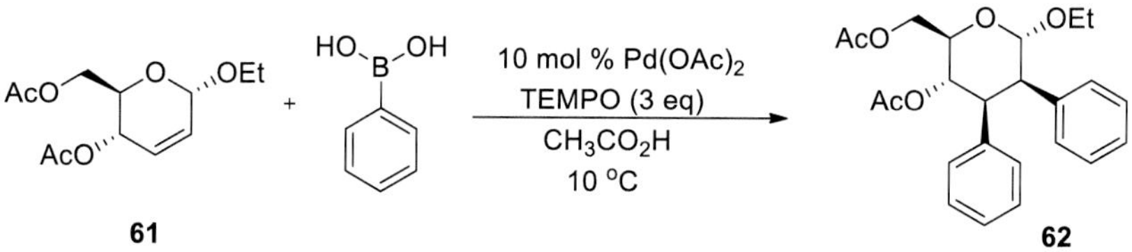

61 62

SCHEME 14.18 Vicinal diarylation of pseudoglycals.

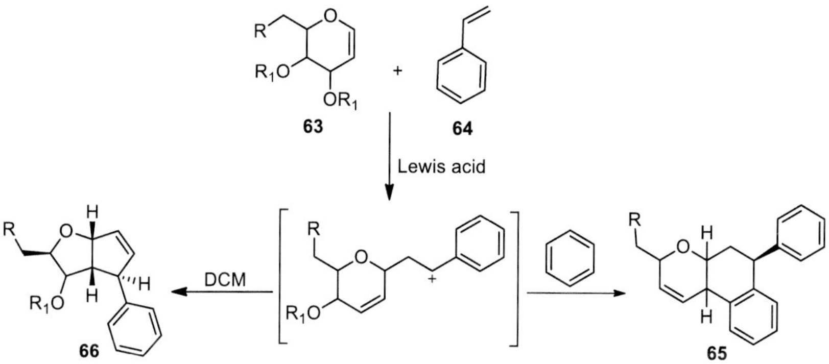

SCHEME 14.19 Triflic acid-catalyzed synthesis of complex oxabicyclic framework by tweaking the solvent in domino fashion.

A broad range of biological activities such as antiviral, antimicrobial, antiproliferative, and antitumor activities are shown by chroman-derived natural products. Chromans and isochromans can be obtained by annulation of a benzene moiety to the pyran system of carbohydrates.[82,83] Werz and coworkers developed an efficient synthetic methodology for the construction of heavily substituted chromans **70** and isochromans **73** through a Pd-catalyzed domino protocol. They utilized 2-bromoglycals **67** as starting materials, which were synthesized from commercially available glycals. First of all, a wide range of different domino precursors **69** were synthesized by attaching various dialkyne chains **68** to 2-bromoglycals. The precursor **69** then undergoes a domino reaction in the presence of a Pd(0) catalyst such as $Pd(PPh_3)_4$ under microwave irradiation at 120°C to furnish corresponding chromans **70** in good yield, which after deprotection afforded compound **71** in very good yield (Scheme 14.20). A broad variety of isochromans **73** were also prepared from appropriate precursors **72** using a similar reaction protocol (Scheme 14.21).[84,85]

The authors also proposed a domino mechanistic pathway for their developed synthetic route. The initial step of this cascade reaction is the oxidative addition of the Pd(0) catalyst into the C–Br bond of the domino precursor **69** to generate intermediate **I**, which undergoes two consecutive intramolecular carbopalladation reactions to generate the intermediate **II**, finally furnish a 1,3,5-triene species **III**.[86,87] The final step of the mechanism is a disrotatory 6π-electrocyclization reaction produced by intermediate **IV** followed by β-hydride elimination leading to the construction of benzene moiety in the final product **70**. The mechanistic picture is provided in Scheme 14.22.

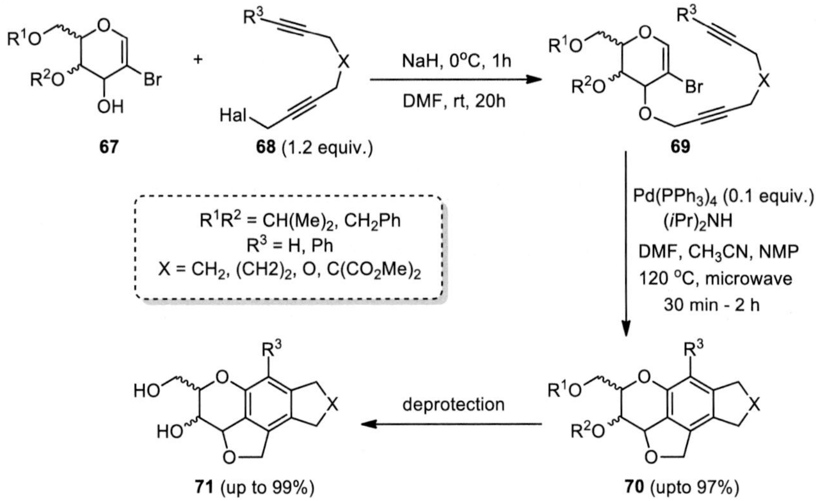

SCHEME 14.20 Synthesis of substituted chromans from 2-bromo glycals through domino approach.

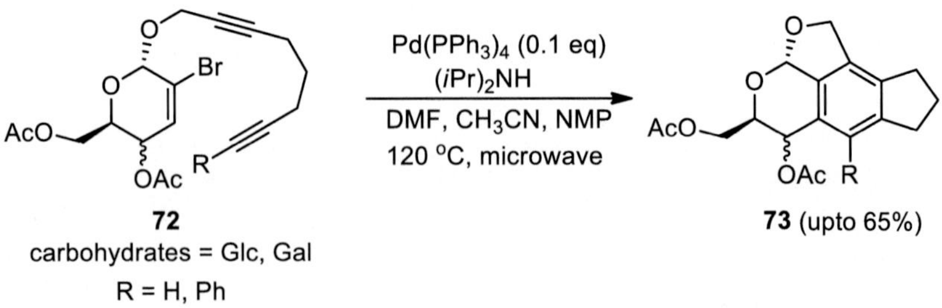

SCHEME 14.21 Synthesis of substituted isochromans **73** through a Pd-catalyzed domino protocol.

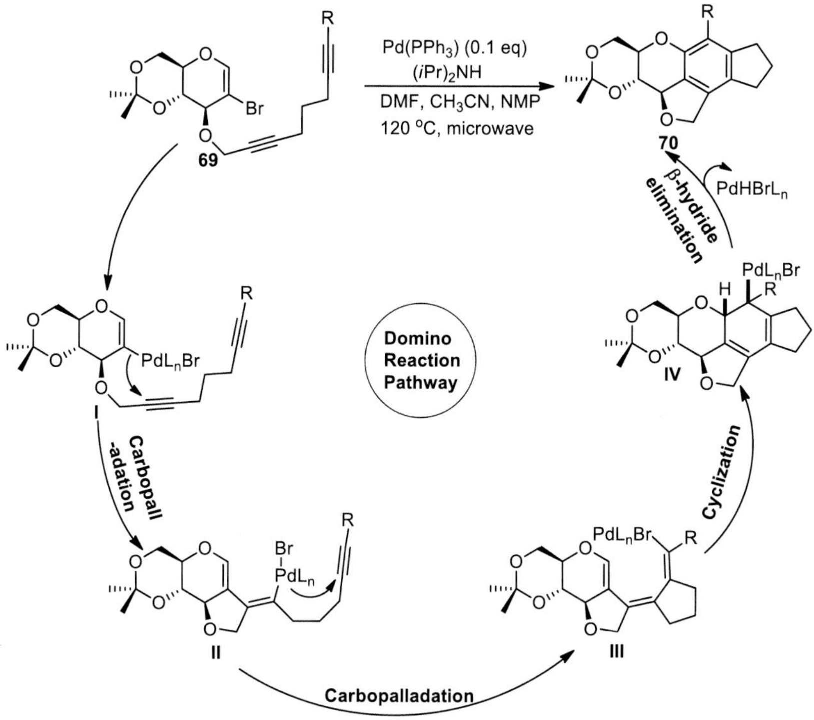

SCHEME 14.22 Domino reaction mechanism for the synthesis of chromans **70**.

This developed methodology was limited to the substitution pattern of the chroman and isochroman framework due to the involvement of an intramolecular domino reaction. The same group overcame this limitation and reported a more flexible intermolecular version of the previous work. The six-membered 2-bromoglycal ring **74** and **75** was employed as the starting material for the construction of chroman **77** and isochroman motifs **78**. In case of chromans, an alkynyl chain is attached to C_3 position of **74** whereas in case of isochromans the alkynyl chain is bonded to C_1 position of **75**. The reaction was carried out in the presence of Fu's salt ($[(t\text{Bu})_3\text{PH}]\text{BF}_4$) and 20 equivalents of the external alkyne **76** for improved yield (Scheme 14.23). Contrary to the previously reported methodology, which was totally in intramolecular fashion, in this approach, the second carbopalladation reaction was performed in an intermolecular manner with external alkynes, hence making this protocol more inclusive and flexible. A broad range of various products were synthesized by varying the carbohydrate part, the attached alkyne as well as the external alkyne. Carbohydrates derived from glucose and galactose in association with silyl-, alkyl-, and aryl substituted alkynes were efficiently employed for the synthetic purpose.[88]

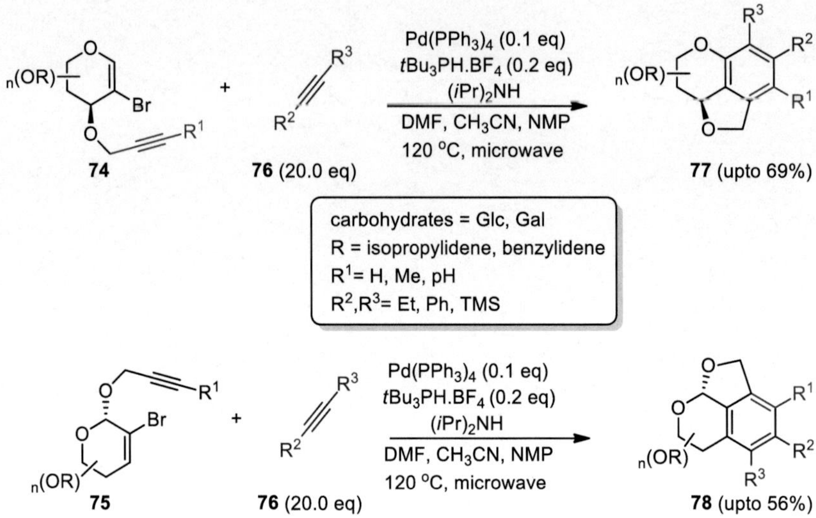

SCHEME 14.23 Domino protocol in intermolecular fashion using external alkynes.

The success of previously developed tandem-domino approaches opens new opportunities for the group to explore this protocol toward the synthesis of other important scaffolds. In this endeavor, Werz et al. extended their work in the field of C–C coupling reactions for the preparation of carbohydrate-derived biphenyl structures **82** and **83** with a chiral axis. Previously described domino sequence was applied to both sides of a linear C_2-symmetric precursor molecule **81** leading to the generation of a chiral axis in the final cyclization step of the reaction. Properly protected 2-bromoglycals **79** with a free hydroxyl group at C_3 and linear tetra alkyne species **80** were employed as starting materials for the synthesis of domino precursor **81**. The reaction was catalyzed by 20 mol% of a palladium catalyst in the presence of 40 mol% of Fu's salt and di-isopropylamine (8.0 equiv) as a base in a mixture comprising DMF, acetonitrile, and *N*-methyl pyrrolidone (NMP) (8:8:1) at 120°C (Scheme 14.24). Glucose and galactose-derived carbohydrates were employed but benzylidene-protected galactose units displayed best yields (61%–66%) and diastereoselectivity.[89]

Fairbanks and colleagues in 2005 described an elegant method for the synthesis of α-(1–6)-C-disaccharide **86** in tandem fashion followed by methylenation using Tebbe reagent and then Claisen rearrangement. The synthesis begins with specially functionalized uronic acids that have been esterified using *allo*glycal. The *allo*glycal **84** underwent Tebbe methylenation to afford the vinyl ether intermediate **85**. The thermal-induced Claisen rearrangement of intermediate vinyl ethers yielded α-(1–6)-C-disaccharides **86** in satisfactory outcome with total control over the anomeric stereochemistry (Scheme 14.25). It is interesting to note that under the same reaction

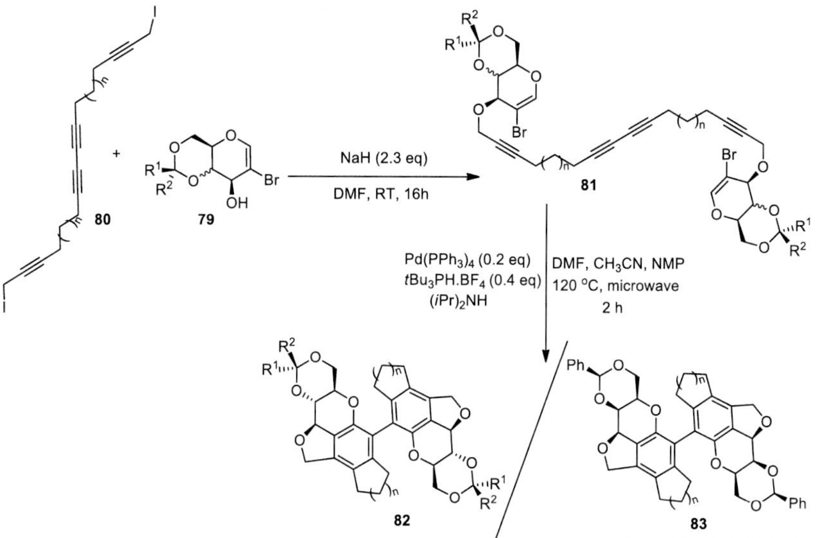

sugar	R^1	R^2	*n*	Yield	d.r.
Glc	Me	Me	1	**82a** (30%)	1:1
Glc	Me	Me	2	**82b** (38%)	4:1
Glc	Ph	H	1	**82c** (41%)	2:1
Glc	Ph	H	2	**82d** (60%)	10:1

sugar	*n*	Yield
Gal	1	**83a** (66%)
Gal	2	**83b** (60%)

Single diastereomer in both cases

SCHEME 14.24 Synthesis of carbohydrate-derived biphenyl structures through the tandem-domino approach.

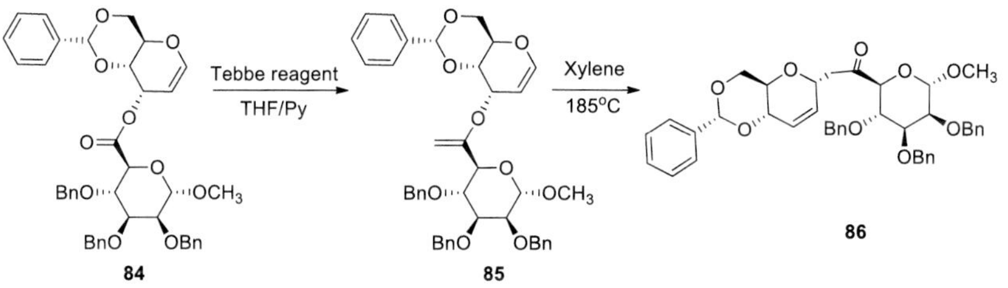

SCHEME 14.25 Synthesis of α-(1−6)-C-disaccharide using esterified uronic acids *allo*glycal **84**.

conditions, the respective *galacto* stereoisomer is completely unreactive. The assertion that stereochemistry plays an important role in the reaction mechanism is further supported by this fact.[90]

Williams and coworkers, recently reported the synthesis of benzopyrans **89** starting from 3,4,6,tri-*O*-acetyl-D-galactal **87**. This interesting methodology involved the formation of a C−C bond via a coupling reaction of various functionalized phenols **88** with 3,4,6,tri-*O*-acetyl-D-galactal system in the presence of aluminum triflate as a catalyst. This novel approach for the

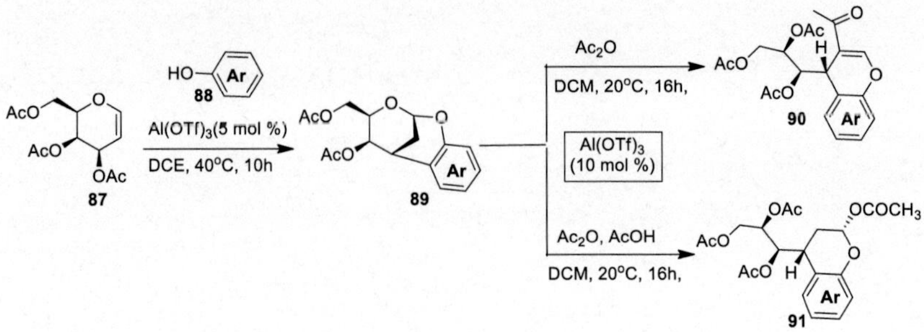

SCHEME 14.26 Alluminium triflate-mediated selective synthesis of chromenes **4** and chromans **5** from benzopyran system **3** (formed by the reaction of 3,4,6,tri-*O*-acetyl-D-galactal **87** and functionalized phenols **88**).

synthesis of the benzopyran system is depicted in Scheme 14.26.[91] The reaction of benzopyran **89** with aluminum triflate (10 mol%) and acetic anhydride afforded chromenes **90**, while the same reagents along with acetic acid furnished chromans **91** in a very good yield (Scheme 14.26).

14.5 Reductive ring opening in tandem reactions of carbohydrates

Low-valent organometallic reagents have recently generated a lot of interest in the field of carbohydrate chemistry because of providing favorable reaction conditions and excellent diastereoselectivity, which can be obtained by stabilizing the process through chelation to nearby keto, hydroxyl, and amino groups.

Pioneer work was proposed by Bernet and Vasella for the synthesis of aldehydes via ring opening of haloglycosides in the presence of activated zinc and acetic acid (Scheme 14.27). In this work, Zn/AcOH was added to the solution of methyl bromoglucoside **92** in ethanol and the resultant was stirred under refluxing conditions to give the corresponding aldehyde **93**.[92] The proposed mechanism shows that, initially, the oxidative insertion of zinc to carbon–halogen bond of methyl bromoglucoside **92** takes place and an intermediate species **A** was formed, which on reductive ring opening followed by elimination furnished the corresponding aldehyde **93** (Scheme 14.27).[93] Later, hex-5-enals and pent-4-enals were synthesized from the ring opening of 6-halopyranosides and 5-halofuranosides, respectively, by the same group under the above-mentioned condition.[94,95]

Bernet–Vasella reductive ring-opening reaction has a wide application in glycochemistry including natural product synthesis.[96,97] The main problem encountered during the reaction is the generation of aldehydes, which frequently construct the number of by-products and decomposition reaction.[98,99] For removing these drawbacks, the aldehyde was trapped in situ by carbon

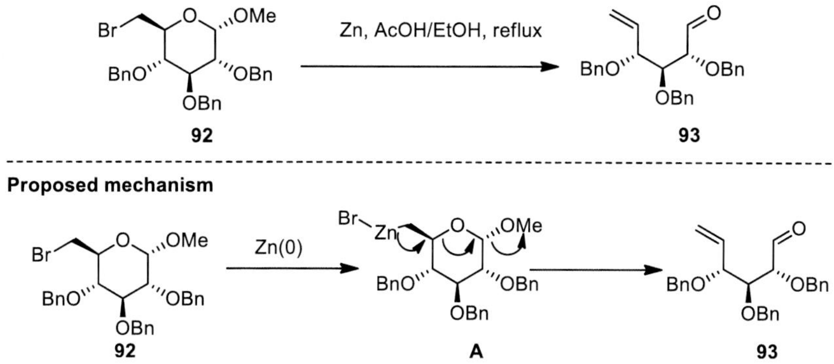

SCHEME 14.27 Zinc/AcOH-mediated ring opening of methyl bromoglucoside 92 (Bernet–Vasella Reaction).

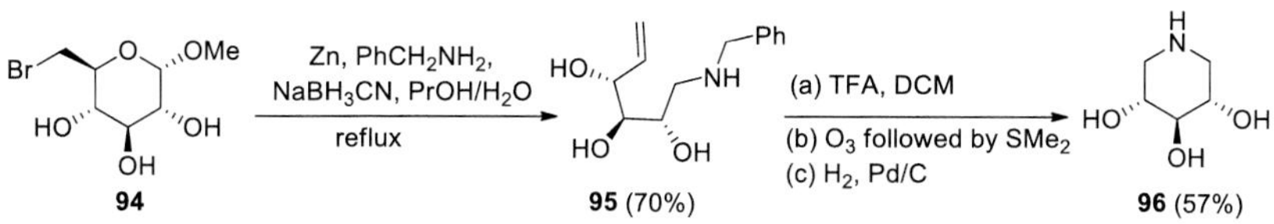

SCHEME 14.28 Synthesis of imino sugar 96 via Bernet–Vasella reductive amination.

and nitrogen-containing nucleophiles. Therefore the total syntheses of a series of biologically relevant compounds were achieved by tandem reactions based on the Bernet–Vasella reaction.

To incorporate an in situ reductive amination of the alkenyl aldehyde, Bernotas and the group proposed a modified reductive fragmentation reaction.[100] The reaction of bromoglycoside **94** with zinc (in excess) in existence with sodium cyanoborohydride and benzylamine and refluxing in PrOH/H_2O (19:1), afforded corresponding amino alkene **95** in 70% yield. This is the first instance of both a reductive opening of an unprotected glycoside and a domino Bernet–Vasella reaction/reductive amination. After forming the trifluoroacetate salt of amino alkene, the process of ozonolysis was followed by hydrogenolysis to produce iminosugar **96** in 57% yield, a powerful inhibitor of β-glucosidase (Scheme 14.28). Later, Dangerfield et al. introduce a modified one-pot approach for the synthesis of primary amines from methyl glycosides through tandem Bernet–Vasella reaction/reductive amination by using ammonia as a nitrogen source.[101]

Combining reductive fragmentation with the reductive formation of a C–C bond greatly increased the synthetic importance of the Bernet–Vasella reaction.[102] For instance, using ring-closing metathesis, the resulting functionalized dienes can then be transformed into carbocycles.[103] Therefore cyclohexene **99** was produced in a 95% yield when the major diene isomer **98**, which was produced by the domino reductive fragmentation/allylation of iodoglycoside **97**, was allowed to treat with commercially available Grubbs' catalyst in DCM (Scheme 14.29).[104,105]

Numerous carbocyclic natural products have been equipped using this approach, which is based on zinc-promoted domino reductive fragmentation/Barbier-type allylation followed by ring-closing metathesis. For instance, Madsen et al. described the synthesis of conduritols, by using a domino zinc-mediated fragmentation of protected methyl 5-deoxy-5-iodo-D-pentofuranosides followed by ring-closing olefin metathesis.[106] Thus the reaction of 3-bromopropenyl acetate and methyl furanoside **100** was performed with zinc in a THF/H_2O mixture and allowed to sonicate at 40°C; compound **101** was obtained as the major product in 72% yield, containing a 4:1 mixture of two isomers. When ring-closing metathesis and deprotection were applied to diene **101**, the compound yielded condurotol D **102** as a final product (Scheme 14.30).[106]

Fused aromatic-carbohydrate hybrid molecules are always of potential interest for both synthetic and medicinal chemists. For instance, *N*-ribosyldimethyl benzimidazole is one of the most promising natural benzimidazole-fused carbohydrate molecules, which plays the role of an axial ligand for cobalt metal in vitamin B_{12}.[107] In 2018 Gonzalez et al. reported an expeditious synthetic route for the synthesis of fused aromatic-carbohydrate hybrids starting from the hemiacetal analogs of carbohydrates via tandem radical fragmentation/cyclization approach. In their report, they described the synthesis of benzimidazole-fused imino sugars **104** through a tandem β-fragmentation-intramolecular cyclization reaction hemiacetal analogs **103** (Scheme 14.31).[108]

In their synthetic strategy, the authors utilized phenyliodosophthalate (PhI (Phth)), a new metal-free and low-toxic hypervalent iodine reagent as a radical promoter and benzimidazole ring present in hemiacetal analogs **103**

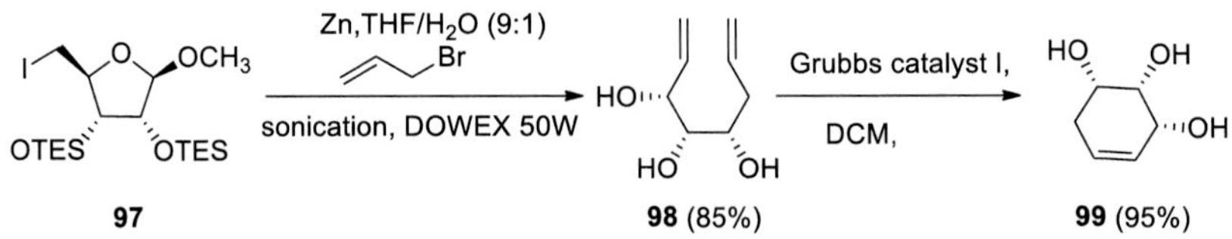

SCHEME 14.29 Synthesis of cyclohexene with iodoglycoside by the application of Bernet–Vasella reaction.

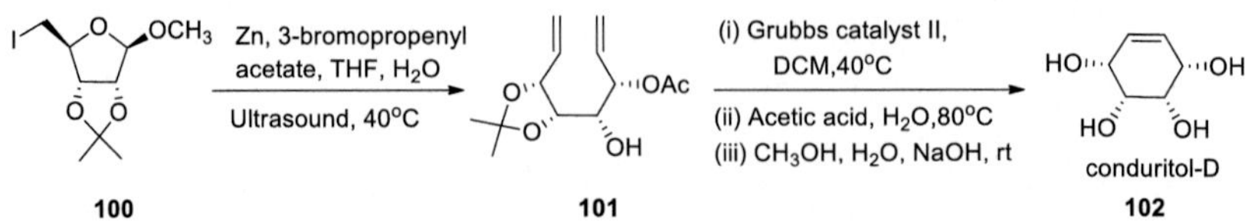

SCHEME 14.30 Synthesis of conduritol-D using domino ring-closing metathesis.

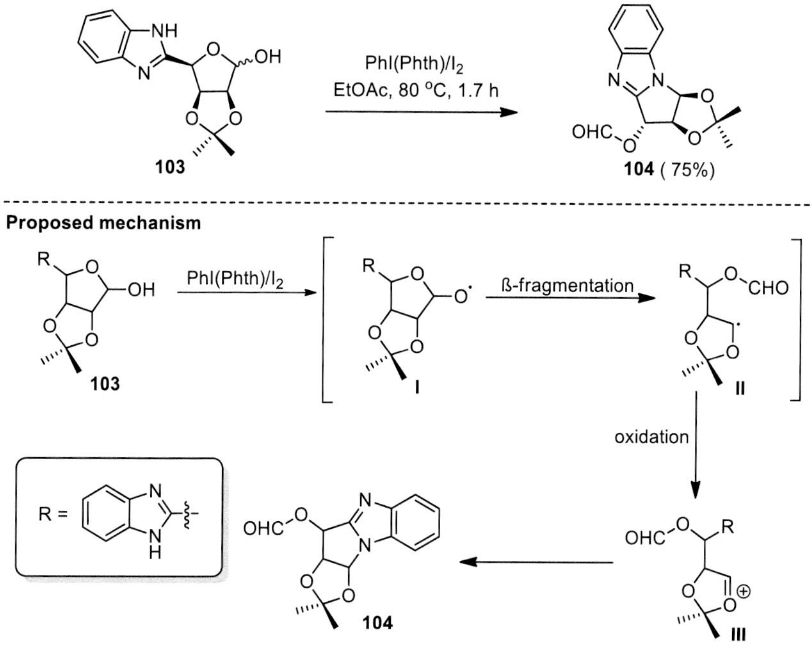

SCHEME 14.31 Synthesis of the fused benzimidazolo carbohydrate via anomeric alkoxyl radical fragmentation (ARF) mechanism.

works as an internal nucleophile in this reaction. The reaction proceeds through the anomeric alkoxyl radical fragmentation (ARF) mechanism and completes in four successive steps in which the first step is the generation of the anomeric alkoxyl radical **I** from the starting hemiacetal form of sugars **103** by the use of PhI(Phth)/I_2 system. The next step is the β-fragmentation of the generated alkoxy radical to furnished C-radical **II**, which is followed by oxidation to produce the corresponding oxocarbenium ion **III** in the third step. In the final step, the trapping of the cationic intermediate takes place to give the desired product **104**.[108]

14.6 Miscellaneous examples

14.6.1 Butenolides and spiro compound

Butenolides are five-membered α,β-unsaturated lactones possessing various therapeutic properties such as antibiotic, antihelmintic, antifungal, insecticidal, antitumor, antiviral, antiinflammatory, cytostatic, antiallergenic, and antihuman immunodeficiency virus (HIV) activities. When sugars are fused to butenolide rings, a bicyclic structure is created with high potential as scaffolds for synthesizing biologically active molecules.[109] But synthesis of butenolide fused sugars is a challenge for glycochemists because stereochemical hindrance problems may alter the reactivity of butenolides, thereby changing the biological properties by enhancing the selectivity or activity. In such situations, the domino reaction approach provides an easy and efficient pathway. Rauter et al. established a pathway for the synthesis of pyranose-fused butenolides **110** and **111**. The desired molecular entities **110** and **111** can be successfully synthesized from the Wittig products **107a** and **107b**, respectively, in a domino fashion. The compounds **107a** and **107b** can be synthesized from diacetonide glucose **105** in two steps. The first step involves the oxidation of **105** with pyridinium chlorochromate (PCC) to give compound **106**, which on olefination with [(ethoxycarbonyl)methylene]triphenylphosphorane afforded the mixture of *(E)* and *(Z)*-isomers **107a** and **107b**, respectively.[58,110] The *(E)*-isomer **107a** undergoes acid-catalyzed hydrolysis in the presence of a strong acid resin and methanol to afford butenolides **108** fused at 3 and 4 positions of pyranose moiety, while *(Z)*-isomer **107b** under similar reaction condition afforded butenolides **109**, fused at 2 and 3 positions of pyranose moiety. Acetate-protected analogs **110** and **111** can easily be obtained in 63% and 78% yield, respectively, by the reaction of compounds **108** and **109** with acetic anhydride in pyridine at room temperature (Scheme 14.32).[110] Thus bicyclic products such as sugar-fused butenolides can be synthesized in a single step and can be isolated as their fully protected acetates in good yields through the application of domino reaction (Scheme 14.32).

Since both the isopropylidene and *tert*-butyldimethylsilyl (TBDMS) protecting groups are acid labile, the earlier synthetic strategy can also be applied to pentofuranoses of type **112** with the primary hydroxy group protected with (TBDMS) group. Under acidic conditions, compound **112** undergoes deprotection followed by a sugar ring-opening reaction to form pentopyranose **115**, which further lactonizes to generate the bicyclic structure **116** in good yield in a single step (Scheme 14.33).[110]

Thio-sugars are very important sugar analogs for the study of cell surface recognition.[111] Bicyclic thio-sugars such as thio-sugar-fused butenolides **121** can also be accessed through the domino approach starting from compound **117**. The acetylated form of butenolides **122** was achieved in good yields by the reaction of **121** with acetic anhydride in pyridine at room temperature as depicted in (Scheme 14.34).[59]

Compounds comprising a butenolide C–C linked to a functionalized chain with an alcohol and an amide are building blocks for pyrrolizidine alkaloids[112] and also analogs of these compounds have been reported to have antilarval settlement activity, an important feature to avoid marine organisms to deposit in marine installations.[113] These compounds can be synthesized by using amino sugar **123** through a tandem reaction approach. Amino sugar **123** in an acidic condition led to hydrolysis product, followed by neutralization with aqueous NaOH gives a sugar-fused butenolide intermediate **124**, which further undergoes deprotonation of the anomeric hydroxy group and cleavage of the C–C bond between positions 1 and 2 to form a resonance

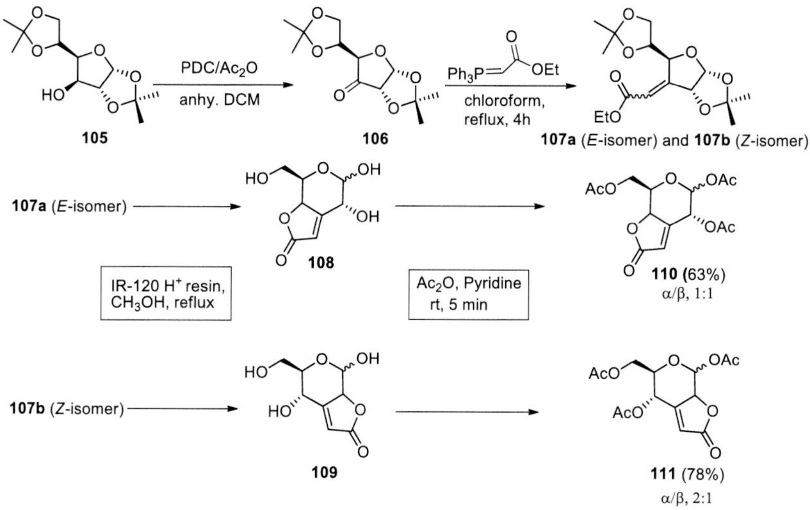

SCHEME 14.32 Synthesis of sugar-fused butenolides **110** and **111** through domino reaction.

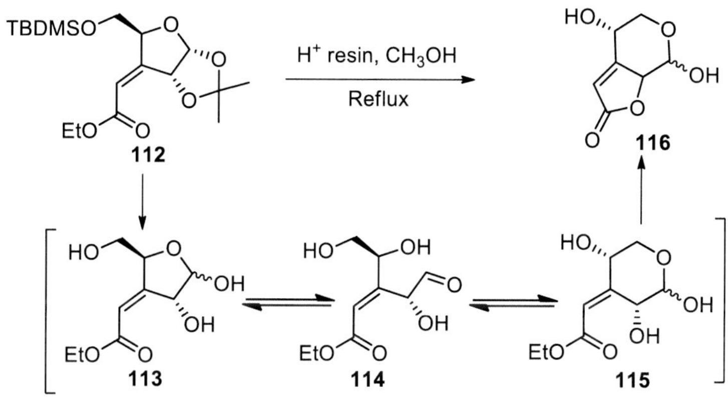

SCHEME 14.33 Domino-mediated synthesis of sugar-fused butenolides.

stabilized carbanions **125** and **126**, which in the presence of the in situ formed acid leads to the compound **127** in 65% yield. The compound **127** on acetylation afforded compound **128** (Scheme 14.35).

Zhang and coworkers put forward an innovative and effective method for creating analogs of α-sialyl spirohydantoin **131** using peracetylated Neu5Ac **129** and a variety of carbodiimides **130** (Scheme 14.36). $BF_3 \cdot Et_2O$ facilitating intramolecular *N*-sialylation with exemplary *R*-selectivity is first depicted. Respective acyl urea intermediate was formed in reasonable yield during the reaction of acid and carbodiimides, which was well characterized by LC-Ms and NMR spectroscopy. Intramolecular *N*-sialylation of acyl urea

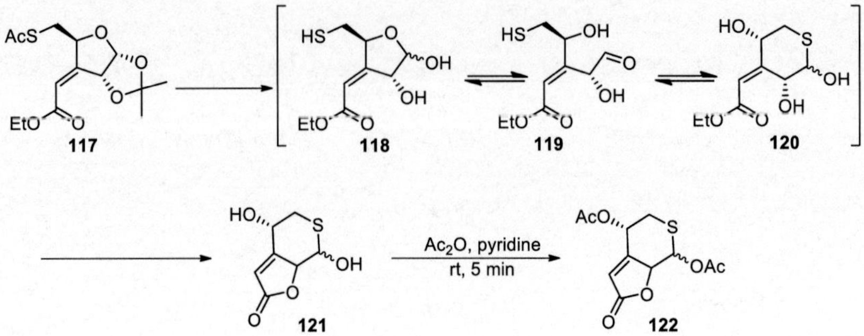

SCHEME 14.34 Synthesis of thio-sugar-fused butenolides through domino approach.

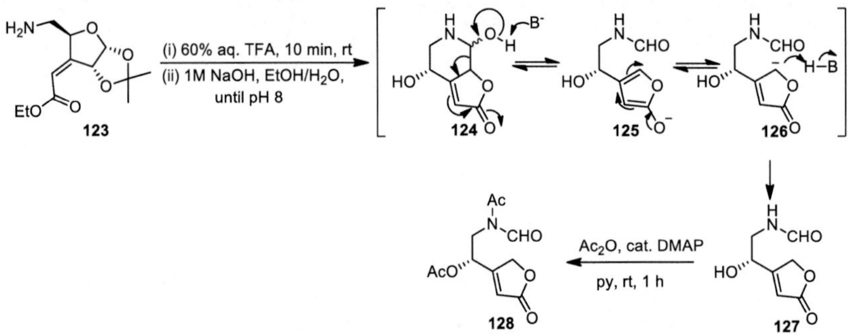

SCHEME 14.35 Synthesis of amide functionality bearing butenolides via domino approach.

was accomplished by using $BF_3 \cdot Et_2O$ to afford analogs of α-sialyl spirohydantoin derivatives **131**. To avoid isolation of acyl urea, the reaction was performed in one pot.[114]

D. B. Werz and group devised a methodology for an efficient synthesis of [*n*,5]-spiroketals ($n = 5$, 6) using exocyclic enol ethers **132** as starting materials and $Yb(OTf)_3$ as a soft Lewis acid catalyst. The developed reaction protocol was based on a domino process involving simultaneous oxidation and rearrangement reactions. The whole reaction sequence involved three synthetic steps, where the first step is the cyclopropanation reaction of exocyclic enol ethers **132** with ethyl diazoacetate **133** in the presence of copper. The resulting ester functionality bearing spiroannelated cyclopropane derivative **134** is further reduced into corresponding alcohols **135** using $LiAlH_4$ reagent in the next step. In the end, the resulting alcohol was oxidized by hypervalent iodine reagent into corresponding cyclopropyl aldehydes **136**, which underwent a ring enlargement process of the three-membered cyclopropyl ring into the desired five-membered [*n*,5]-spiroketals **137** in the key step. The use of Lewis acid catalyst $Yb(OTf)_3$ accelerates the ring enlargement and increased the overall yield of the transformation. Also, during the

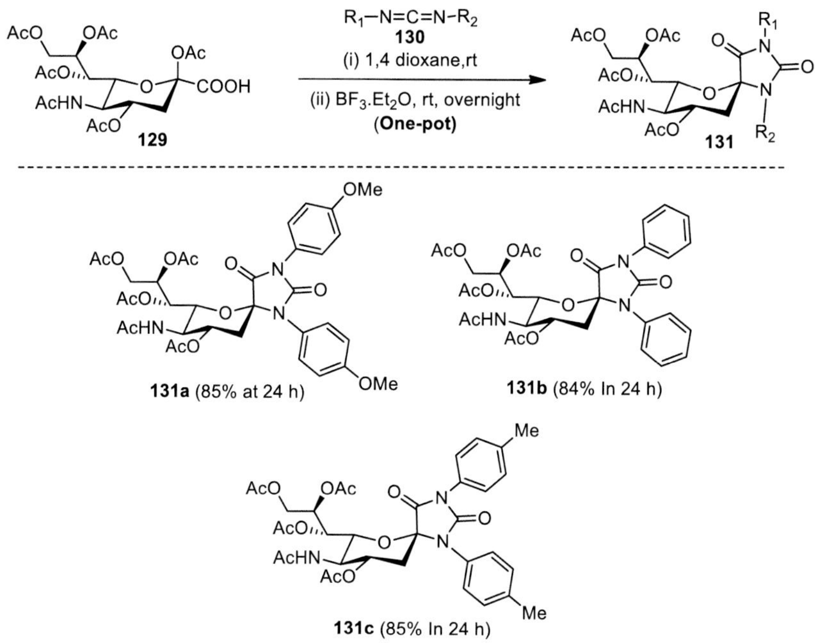

SCHEME 14.36 Highly α-Selective Synthesis of sialyl spirohydantoins by regiospecific domino condensation.

ring enlargement, the stereochemistry of the spirocenter was not conserved because in presence of $Yb(OTf)_3$, the reaction proceeds through an open transition state **A** rather than a concerted mechanism. The major product is the thermodynamically more favored anomeric [*n*,5]-spiroketal **137** (Scheme 14.37).[115]

Compounds embodying an α,β-unsaturated δ-lactam-fused furanose ring structures **141** are of huge importance to medicinal chemists because these structures are precursors for the synthesis of biologically significant alkaloids[116] and pharmacological drugs, namely the δ-amino acid homobaclofen **143**,[117] which is a homolog of the γ-aminobutyric acid B (GABAB) receptor, that demonstrated an outstanding pharmacological activity.[118] The structures **141** can be easily synthesized via the domino strategy by utilizing an analog of compound **123**, having an azido functionality in place of the amino group as starting material **139**. First, the substrate **139** undergoes Staudinger reduction to yield an intermediate **140** having an amino group in place of the azido group and then intramolecular nucleophilic attack of the amino group to the carbonyl carbon of ester functionality followed by elimination of the ethoxy group generates the product **141** in 89% yield. Also, its *N*-acetyl derivative **142** was synthesized in 85% yield by reaction with acetic

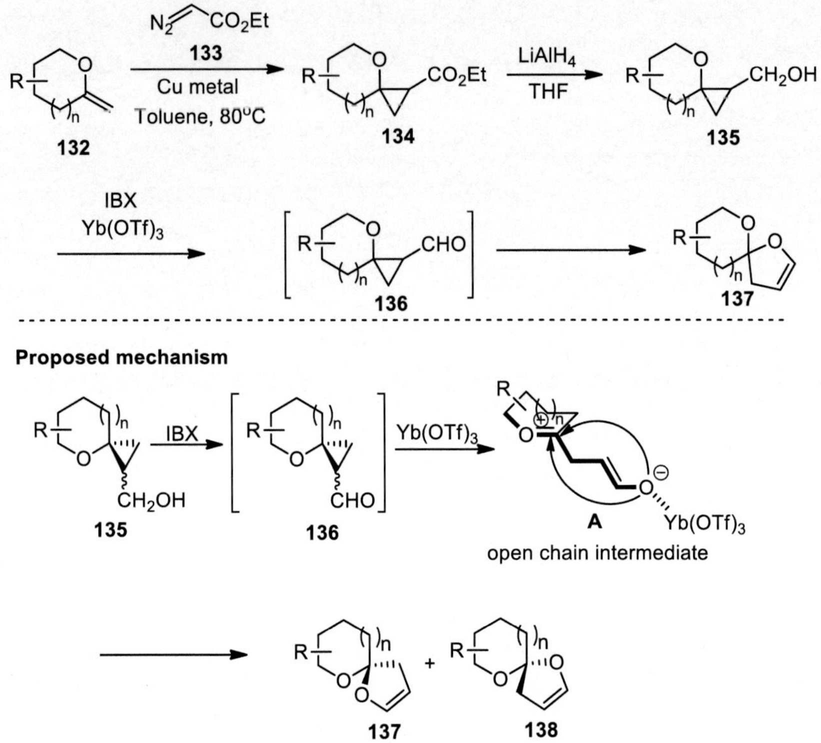

SCHEME 14.37 $Yb(OTf)_3$ catalyzed synthesis of [*n*,5]-spiroketals from exocyclic enol ethers through a domino process.

anhydride in pyridine in the presence of the catalytic amount of DMAP (Scheme 14.38).[119]

Deoxy sugars are a significant class of sugars that are an essential component of various biologically important natural products and synthetic compounds.[120] Among deoxy sugars, 2-deoxy sugars are most studied and can be easily synthesized through electrophilic addition to glycals, whereas several reaction steps are required for the deoxygenation at other positions. A. P. Rauter et al. established a single-step domino approach for effective synthesis of 4-deoxy keto sugars **148** using easily available 4-hydroxy keto sugars **144** as substrates. The reaction was carried out in the presence of triphenylphosphine and iodine/imidazole system. First, the triphenylphosphine activates the hydroxy group of α-hydroxy keto sugar **144**, to give activated species **145**, which then undergoes a nucleophilic substitution reaction with iodide ion to furnish a halogenide intermediate **146**. This halogenide reacts with another iodide ion in situ to generate an enolate intermediate **147** which then tautomerizes to afford the desired 4-deoxy keto sugar **148** in 61% yield (Scheme 14.39).[121,122]

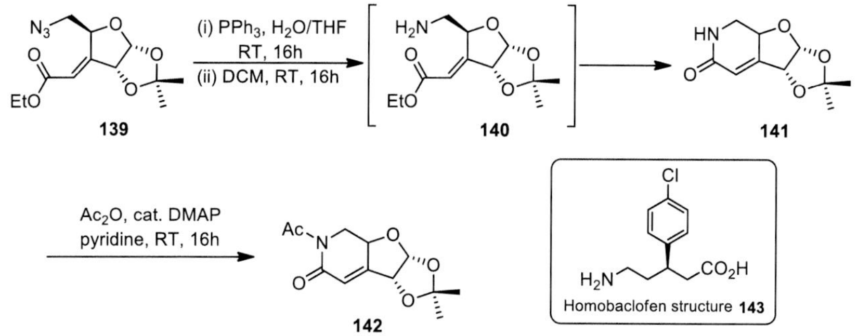

SCHEME 14.38 Synthesis of α,β-unsaturated δ-lactam fused furanose ring structures via domino strategy.

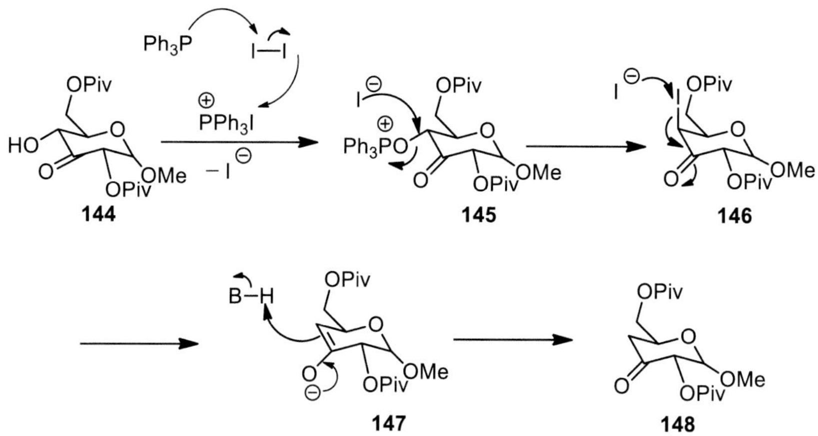

SCHEME 14.39 One-step domino synthesis of 4-deoxy keto sugar from 4-hydroxy keto sugar.

1,2-Dihydropyridin-3-ones are very important structural features of a variety of bioactive molecules[123] and also they are precursors for the synthesis of a diverse range of imino sugar glycosidase inhibitors. The effective γ-secretase inhibitor molecule **160**, which is a target enzyme for the treatment of Alzheimer's disease, comprises the 1,2-dihydropyridin-3-one moiety.[124] Therefore the development of an easy and low-cost synthetic approach for the synthesis of 1,2-dihydropyridin-3-ones **159** becomes very crucial for the effective treatment of various frontline diseases. These compounds can be synthesized very easily through domino protocol starting with 6-azido sugar molecules type **149**. The domino reaction mechanism involves first the hydrolysis of the isopropylidene group of **149** followed by the reduction of its azido group to amine functionality, which leads to the formation of a six-membered ring hemiaminal intermediate **153**. Elimination of a water

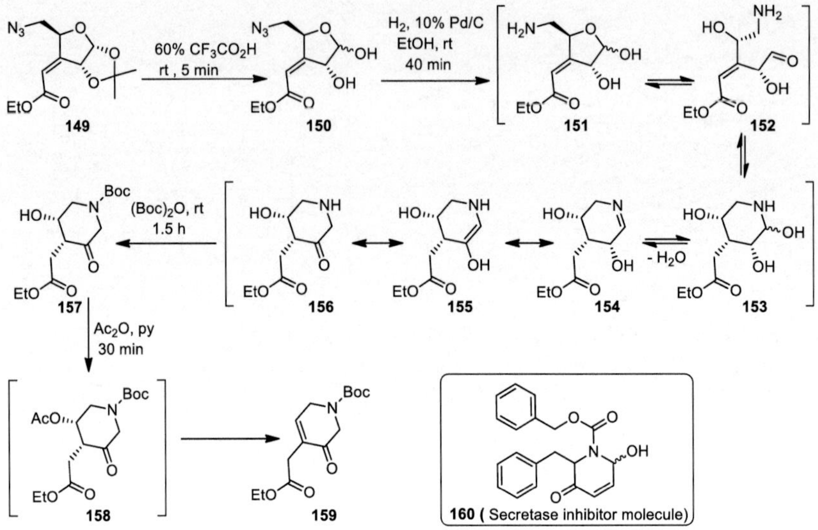

SCHEME 14.40 Domino protocol mediated synthesis of biologically important 1,2-dihydropyridin-3-ones.

molecule from **153** produces an imine intermediate **154**, which isomerizes to enol form **155** and then finally tautomerizes to the keto form **156**. Boc protection and acetylation of the free hydroxy group followed by elimination of an acetic acid molecule afforded the final product **159** in 42% yield (Scheme 14.40).[119]

14.6.2 Sugar ring contraction

Bicyclic structures having two five-membered fused rings are quite difficult to synthesize but they are present in a variety of biologically important molecules, for instance the synthetic nucleosides that comprise bis(tetrahydrofuran)-fused structures have application in the study of structural constriction of sugar component.[125] Also, these fused structures make a part of various bioactive molecules like goniofufurone **163** and anti-HIV drugs such as darunavir **164**. The synthesis of these bicyclic structures can be easily performed through the domino protocol from bicyclic structures comprising a five-membered ring fused to a six-membered ring, which are core structural features of miharamycins, in a Brønsted or Lewis acid medium.

The acetolysis of the structures type **161** with concentrated sulfuric acid in acetic anhydride for 45 minutes results in the replacement of the anomeric methoxy group by the acetoxy group, which is followed by a domino reaction and leads to the formation of product **162** through ring contraction in 85% isolated yield (Fig. 14.2).[126]

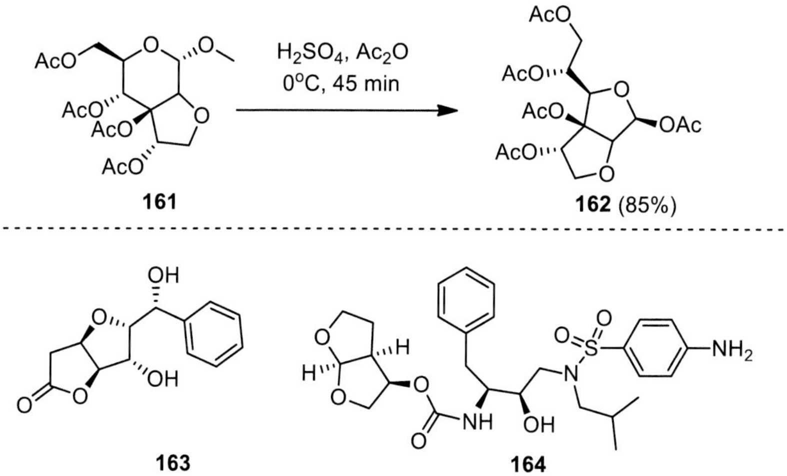

FIGURE 14.2 Synthesis of biologically important fused bicyclic structures via domino approach.

Oxetanes are four-membered cyclic ethers and are a very important class of compounds in medicinal chemistry. They found extensive applications in the area of material science in which they are used as monomers, in synthetic organic chemistry as versatile precursors and also they are found in various biologically active compounds and natural products[127] such as Oxetin **168** and Oxetanocin-A **169**. The synthesis of oxetane derivatives such as oxetane carboxylic esters through ring contraction of 2-hydroxy gamma-lactones bearing a triflyl substituent at position 2 was carried out by Fleet and coworkers in 1990.[128] The group synthesized oxetane carboxylic esters from five-membered aldonolactones, for example, xylonolactone **165** having a good leaving group, such as a triflate at position 2 in a basic medium in methanol via a domino strategy. The methoxy anion generated from the base and methanol undergoes an addition-elimination reaction with lactone leading to the formation of an alkoxy anion **166**, which being a strong nucleophile eliminates the triflate group at position 2 via nucleophilic substitution to afford the desired oxetane analogs **167** (Scheme 14.41).

The above domino ring contraction methodology was also applied for the synthesis of oxetane δ-amino acids by taking aldonolactones such as xylonolactone[129] and ribonolactone as starting materials.[130]

14.6.3 Levoglucosenone

A base-catalyzed tandem stereoselective reaction of levoglucosenone **170**, a chiral α,β-unsaturated ketone, with furfural **171** was developed by Isobe and coworkers. In this newly developed protocol, levoglucosenone forms carbon-

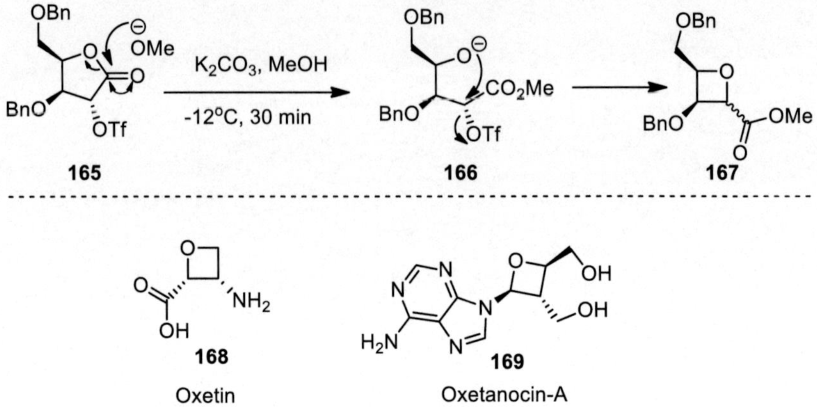

SCHEME 14.41 Ring contraction of lactone triflates to oxetane carboxylic esters in domino fashion.

carbon bond at the C-3 position with furfural. The reaction was carried out at room temperature in a saturated sodium bicarbonate solution for 24 hr to afford the desired product in good yield.[131] Samet and group reported the stereoselective Oxa-Michael-aldol reaction of 2-hydroxybenzaldehyde derivatives **173** with levoglucosenone under the same reaction condition but it was not successful in the formation of similar adduct as in case of furfural. In this devised methodology, the attack of phenolate-anion from the opposite side to the anhydro-ring of levoglucosenone takes place resulting in the inversion of stereochemistry to afford the compound **174**. The reaction sequence is illustrated in Scheme 14.42.[132]

In 2014 Witczak, Z. J. et al. synthesized the heterocyclic thio-chromene systems **176** via a similar domino thio-click Michael-aldol approach using thiohydroxybenzaldehyde **175** and levoglucosenone **170** as starting materials. The reaction involved a base-catalyzed thio-click reaction and proceeds through a classical stereoselective 1,4-addition reaction of thiol to the C-4 position of levoglucosenone, followed by a domino cyclization reaction leading to the C–C bond formation at C-3 of functionalized levoglucosenone. The whole reaction sequence is depicted in Scheme 14.43.[133]

Semenov and group extended the earlier works and investigated the domino reaction with dihydrolevoglucosenone. First, the levoglucosenone **170** was reduced in the presence of dihydrogen and 2% of palladium in charcoal with ethyl acetoacetate as a solvent at 40°C to give the dihydrolevoglucosenone **177**. Stereoselective addition of Grignard reagents for functionalization of C-2 dihydrolevoglucosenone to produce ethynyl derivatives **178**, which react with benzyl azide **179** to furnish the corresponding 1,2,3-triazole derivatives **180** and reaction of **178** with dipolar species **181** to furnish

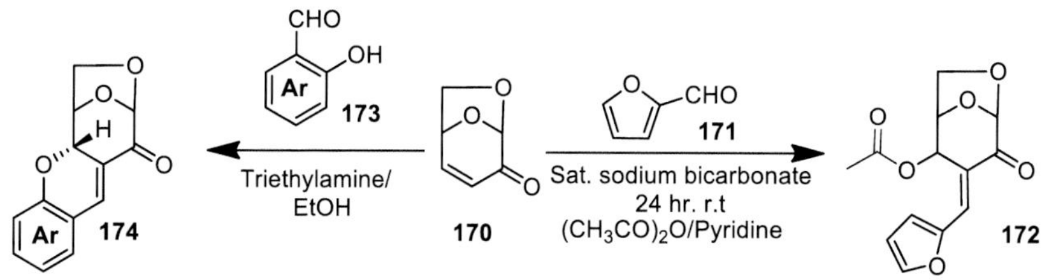

SCHEME 14.42 Stereoselective tandem reaction of levoglucosenone **170** with *o*-hydroxybenzaldehyde **173** and furfural **171**.

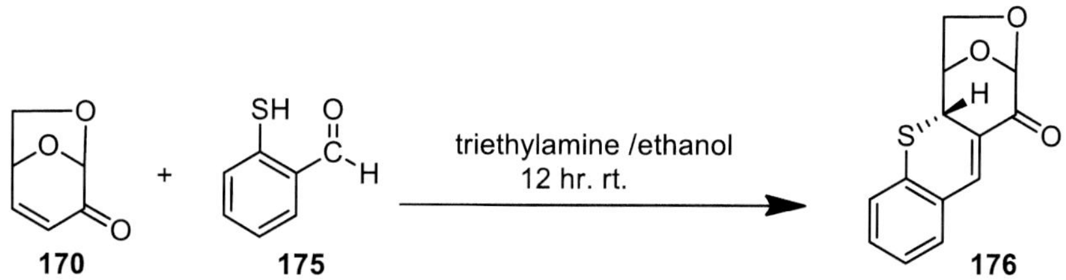

SCHEME 14.43 Stereoselective tandem reaction of levoglucosenone **170** with *o*-thiohydroxybenzaldehyde **175**.

isoxazoles **182** through a [3 + 2] dipolar cycloaddition reaction as illustrated in (Scheme 14.44).[134]

In 2014 Ramasastry et al developed a novel domino approach for the synthesis of bicyclic pyrofuranones **185** and **186** by utilizing acetoxy pyranone **183** and 1,3-dicarbonyls **184** such as acetylacetone as starting materials. The reaction mechanism as proposed by the authors includes a classical base-catalyzed Michael addition followed by cyclization of formed enols with in situ generated oxonium ions, that is, cyclo acetalization reaction to furnish desired bicyclic pyrofuranone derivatives (Scheme 14.45).[135]

Chattopadhyay and coworkers combined the concept of domino reaction with the popular intramolecular azide-olefin [3 + 2] cycloaddition reaction and developed a novel tandem methodology for the synthesis of benzooxazocines **188** and benzodiazocines **189**. The reaction strategy involved functionalized carbohydrate scaffolds **187** as precursors that undergo a Cu-catalyzed aryl azidation followed by stereoselective aziridine formation to produce desired benzooxazocines **188** and benzodiazocines **189**. This novel strategy is explained in (Scheme 14.46).[136]

The application of *N,N*-dimethylethylenediamine (DMEDA) as a base in an aqueous solution of ethanol/toluene made the copper-catalyzed click reaction for stereoselective [3 + 2] cycloaddition reaction is highly efficient. Formation of an unstable intermediate triazoline **A**, which undergoes rapid loss of nitrogen to generate aziridines-fused furobenzoxazocine **188** or furobenzodiazocine **189** as a sole product, is shown in Scheme 14.47.[137]

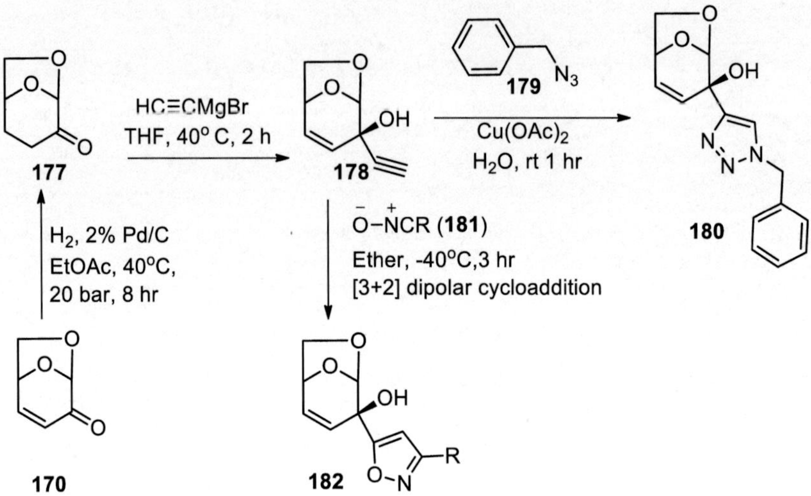

SCHEME 14.44 Synthesis of functionalized isoxazoles **182** and 1,2,3-triazole derivatives **180** from levoglucosenone and Grignard reagent.

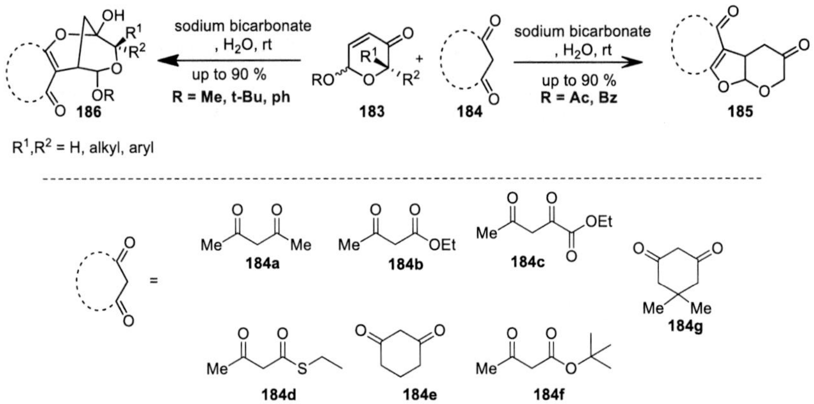

SCHEME 14.45 Bicyclic furopyranones synthesized by using carbohydrates enones.

A new approach to the synthesis of functionalized thioimidazoles **194**, **195**, and **196** via multicomponent reaction using respective starting materials dihydroxy acetone dimer (DHA) **191**, ketoses **192**, and aldoses **193** was reported by Bauman and Axendale. The reaction was performed using precursors DHA or ketoses or hexoses individually along with diverse benzylamine hydrogen chloride salts **190** and potassium isothiocyanate (KSCN) to furnish corresponding thioimidazoles. This approach is based on the Marckwald multicomponent reaction as shown in Scheme 14.48.[138]

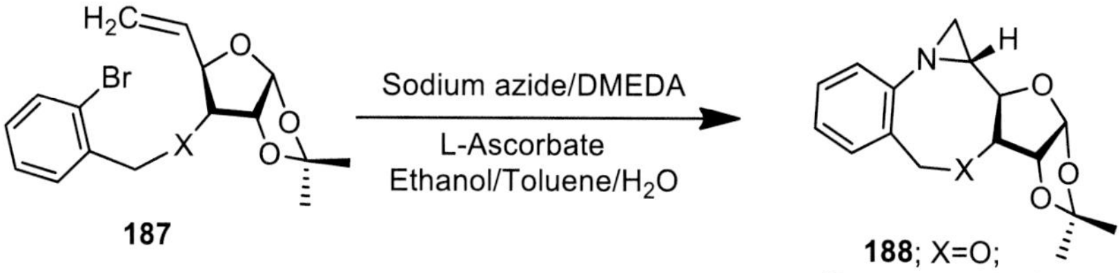

187

X=O, NTs, NCbz

188; X=O;
(Benzoxazocine)
189; X= NTs, NCbz;
(Benzodiazocine)

SCHEME 14.46 Synthesis of benzooxazocines and benzodiazocines through intramolecular azide-olefin [3 + 2] cycloaddition of functionalized carbohydrates scaffolds.

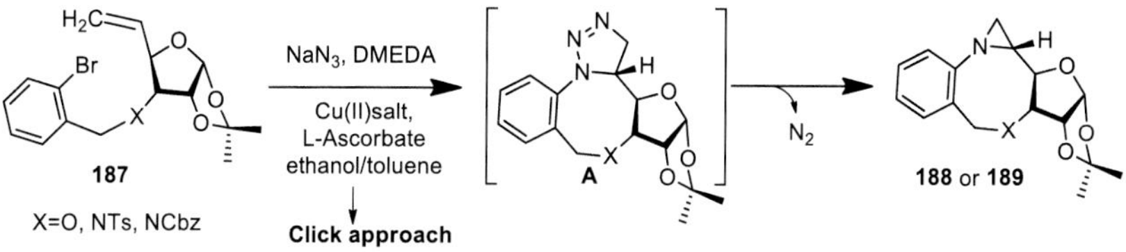

SCHEME 14.47 Development of aziridine-fused benzodiazocine.

14.6.4 Macrocyclic synthesis

The major limitation in the synthesis of macrocyclic bislactone structures is that it involves several steps to complete the synthesis. A. P. Rauter and group overcame this drawback and introduced a novel domino methodology for efficient synthesis of sugar-tethered 10-membered bislactone derivatives. Partially protected carbohydrates having two free vicinal hydroxy groups were utilized as starting materials. For instance, methyl glucopyranoside **197** bearing selective pivaloyl protection at positions 2 and 6 serves as a precursor and undergoes PCC oxidation of both free hydroxy groups to afford diulose intermediate **198**. Subsequently, a nucleophilic addition reaction of the hydroxy groups of monosaccharide precursor **197** with the carbonyl group of diulose moiety **198** produces a bis(hemiacetal) intermediate **199**, which undergoes oxidative cleavage of the C–C bond to furnish desired 10-membered macrocyclic bislactone **200** in moderate yield (Scheme 14.49).[139] The authors also synthesized another 10-membered bislactone derivative through PCC oxidation of methyl-4,6-*O*-benzylidene-α-D-glucopyranoside using similar domino strategy.

Gin et al. reported the synthesis of symmetrical cyclotrimer macrocycles **202** starting from anomeric sugar azide **201** with 4-propargyl ether at the opposite ends. This reaction was catalyzed by copper iodide in the presence of DBU and produced the C3- and C2-symmetric oligosaccharide macrocycles in a highly convergent fashion as illustrated in Scheme 14.50.[137]

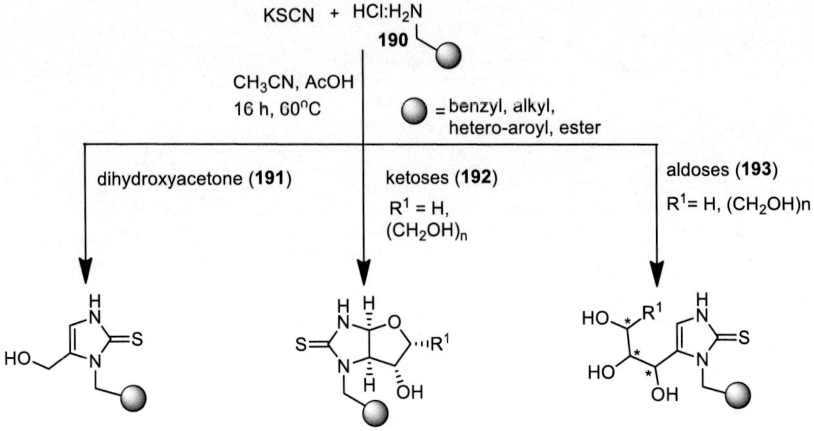

194 195 196

SCHEME 14.48 Functionalized thioimidazoles synthesis via a three-component reaction using unprotected ketoses and aldoses.

14.6.5 Sugar ligands

Sugar-derived ligands attain much advancement in organic synthesis because of proper ligation sites and the tendency to stabilize metals and provide a suitable environment for binding. It covers a wide area of targeted synthesis including cross-coupling, hydrogenation, 1,2- and 1,4-additions of nucleophiles to C = O and C = NR-based substrates, oxidation, hydroformylation, and other many reactions.[140]

Tiwari et al. in 2019 established a protocol for combinatorial synthesis of a library of benzoxazoles **206**, benzothiazole **205** from 2-halobenzamide **203**, and 2-halobenzothioamide **204**, respectively. Glycosyl triazole **207** was established as an effective ligand for Cu-catalyzed intramolecular cross-coupling. The glycosyl triazole ligand provides a suitable binding site through which it ligates with copper and stabilizes the metal oxidation state during the course of the reaction (Scheme 14.51).[141]

D-Glucosamine **213**, a naturally occurring, affordable, and easily accessible sugar, has been investigated as an effective ligand for the tandem Cu(I)-catalyzed regio- and stereoselective synthesis of a variety of (*Z*)-3-methyleneisoindoline-1-ones **210** and (*E*)-*N*-aryl-4*H*-thiochromen-4-imines **211** using 2-halobenzamide **208** and 2-halobenzothioamide **209**, respectively, as starting component along with catalytic amount of copper iodide and terminal alkynes **212** (Scheme 14.52).[142]

In addition, the group developed a new pyridyl glycosyl triazole/CuI catalytic framework via the modular click pathway. This system was investigated in the instant one-pot synthesis of 2-amino-3-substituted quinazolinones **215** and 3-substituted quinazolinones **216** from amides **214**. This catalytic system

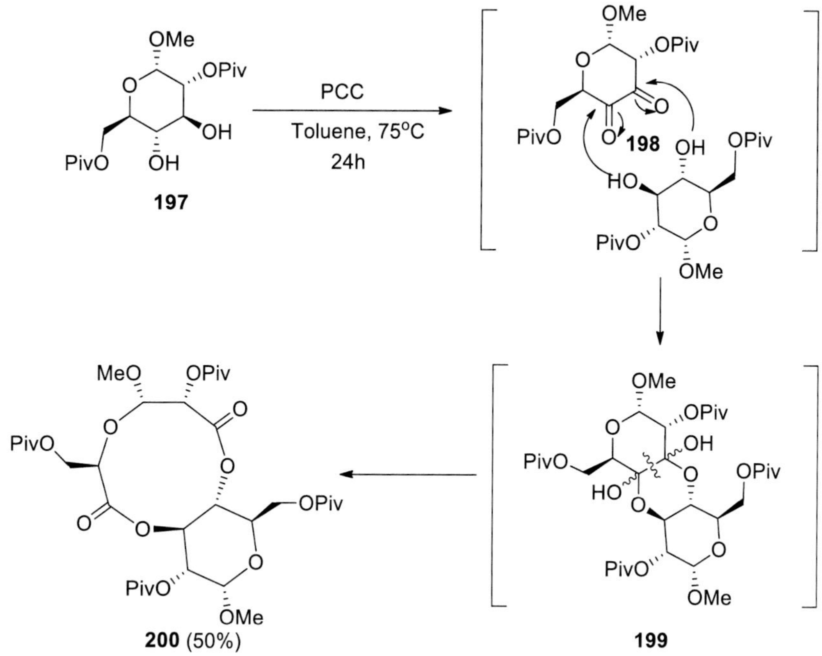

SCHEME 14.49 Synthesis of macrocyclic sugar bislactones via PCC oxidation in a domino fashion.

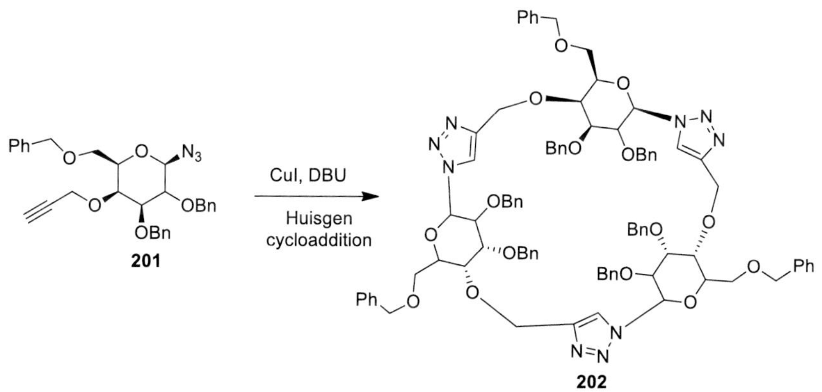

SCHEME 14.50 Synthesis of C3- and C2-symmetrical trimeric oligosaccharide macromolecules.

is applicable for the Ullmann coupling, followed by additive cyclization, to construct a number of novel quinazolinone motifs with good-to-excellent yields (Scheme 14.53).[143]

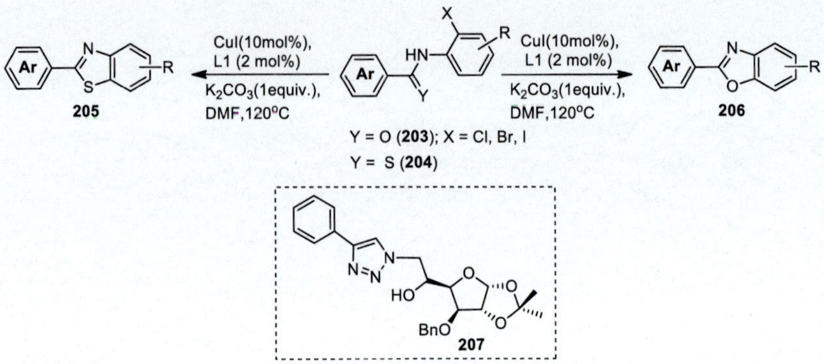

SCHEME 14.51 Cu-catalyzed domino synthesis of 2-substituted benzothiazoles and benzoxazoles.

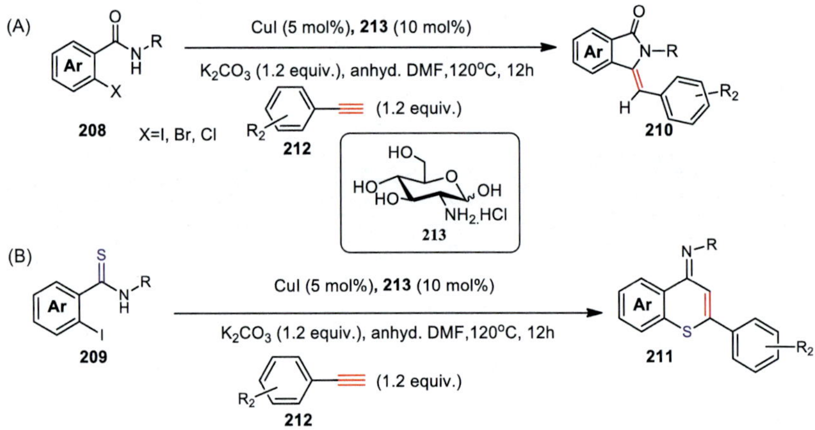

SCHEME 14.52 Synthesis of (*Z*)-3-methyleneisoindoline-1-ones **210**, (*E*)-*N*-aryl-4*H*-thiochromen-4-imines **211** in tandem fashion using D-glucosamine **213**.

14.6.6 Glycosylation

The regio- and stereocontrol during the formation of glycosidic bonds usually require numerous steps, numerous synthetic modifications, significant protecting group adjustment, and time-consuming intermediate purification that slows down the entire synthesis, lowers production efficiency, and takes months to complete. A novel "one-pot sequential chemoselective glycosylation" is now a day's employed to get around these restrictions and hasten the synthesis of biologically relevant oligosaccharides. This process involves a series of reactions between different glycosyl donors and acceptors using appropriate promoters in a one-pot manner.[144]

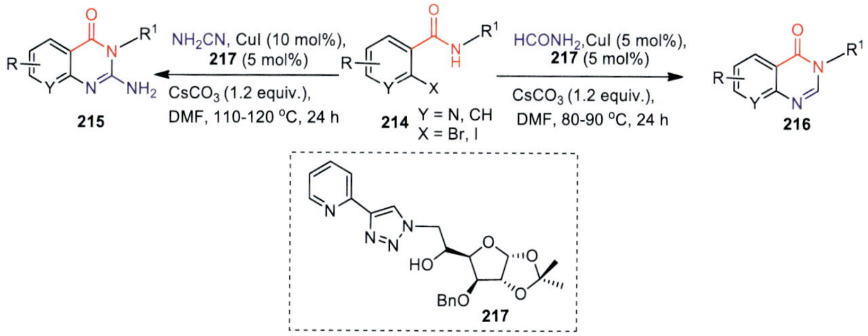

SCHEME 14.53 Domino synthesis of quinazolinones **215** and **216** using pyridyl glycosyl triazole **217** as ligand.

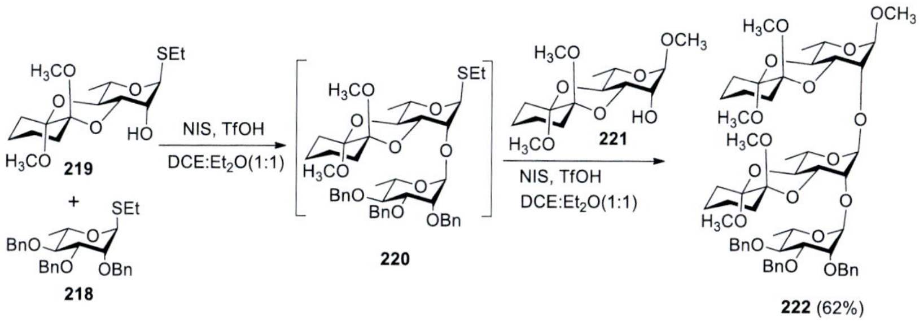

SCHEME 14.54 Ley's one-pot synthesis of trisaccharide **222** using CDA-protected rhamnosides.

Toward this end, a number of one-pot procedures based on reactivity were carried out by considering the reactivity difference of the building blocks in glycosylation methodology. In such protocol, a donor having the greater anomeric reactivity could be activated over bifunctional donor–acceptor with a lesser reactivity containing a similar leaving assembly. Ley et al. in 1994 reported a method for the synthesis of trisaccharide using intermediate levels of reactivity-based one-pot glycosylation. As a result, the reaction of rhamnoside **218** with rhaminoside **219** containing cyclohexane-1,2-diacetyl protecting group, in the presence of *N*-iodosuccinimide and triflic acid as promoter furnished intermediate **220**. Further addition of disaccharide acceptor **221** finally yielded the desired trisaccharide **222** in good yield and desired anomeric selectivity (Scheme 14.54).[145]

Wong and coworkers in 1999 proposed programmable one-pot synthesis of tetrasaccharides, based on the relative anomeric reactivity values (RRVs). This RRV is obtained from programmable software (called as Optimer) that provides data on the reactivity order of glycosyl donors. Greater the RRV, the higher will be the reactivity of the glycosyl donor. The reactivity of various thioglycoside donors **223** (RRV = 17,000), **224** (RRV = 162.9), **225**

(RRV = 13.1), and **226** has been considered for the two glycosylation steps in the synthesis of the target tetrasaccharide **227**. The primary issue with this process was its poor performance, as determined by Optimer analysis (Scheme 14.55).[146] Numerous mono- and disaccharide building blocks were constructed and their RRVs were assigned.[147]

Another most important carbohydrate explored in chemical biology is sialic acids, a nine-carbon negatively charged terminal sugar.[148] Almost all cells in nature are covered with a complex array of these sugars, which displayed a significant role on cell surfaces and are known to mediate and modulate many important cellular interactions. The chemical synthesis of well-defined sialic acid-containing glycans is extremely difficult and time consuming. Thus Prof. Chen Group at UC-Davis established a highly facile one-pot multienzyme (OPME) route for obtaining desired α-2,3-linked sialosides and also α-2,6-linked sialosides.[149–151] The six-carbon monosaccharides, including D-Mannose, ManNAc, ManNGc, or their derivatives (**228**), on sialic acid aldolase-mediated aldol addition with pyruvic acid **229** gave good yield of the respective nine-carbon natural and nonnatural analogs of sialic acid or neuraminic acid analogs **230** (**Step 1**).[152a,b] The resulting sialic acid analogs **230** was then activated with the aid of a second enzyme, that is, CMP-sialic acid synthetase (NmCSS) and afforded a high yield of desired cytidine-5-monophospho-nonulosonate analogs **231** desired donor (**Step 2**). In the end, the resulting donor **231** was subjected to enzymatic glycosylation in the presence of PmST1 or Nm2,3ST sialyltransferase to transfer the donor **231** to the respective acceptors **232** (e.g., galactose, GalNAc, lactose, and their functionalized derivatives) to afford high-to-excellent yield of target α-2,3-linked sialosides **233** (Scheme 14.56).[149–151]

The OPME method is nicely explored in sialylated glycan microarray, an important tool to examine the biological importance of the modified sialic acids in protein–glycan interactions.[153a] Also, the α-2,3 or α-2,6-linked sialosides obtained via chemoenzymatic OPME route is further utilized for an

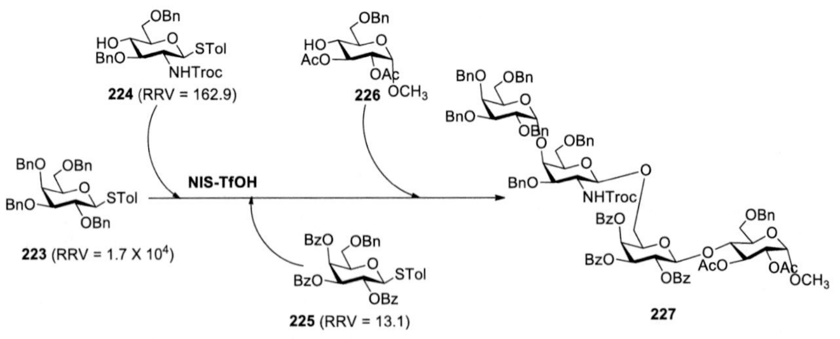

SCHEME 14.55 One-pot glycosylation process for the synthesis of targeted tetrasaccharide **227** using thioglycoside donors.

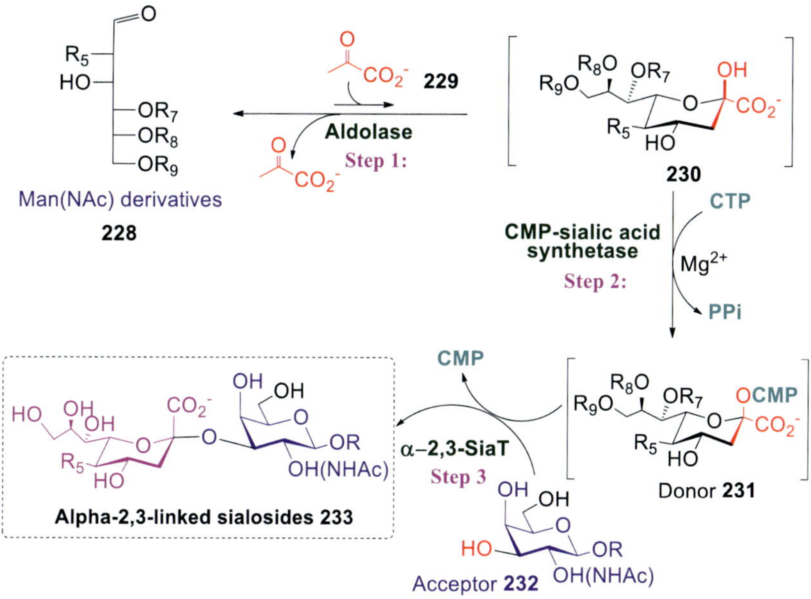

SCHEME 14.56 One-pot multienzyme (OPME) route for easy access to diverse α-2,3-linked sialosides **233**.

easy synthesis of desired GD3 and GT3 oligosaccharides of chemotherapeutic values.[153b] The resulting sialic acid-containing glycans have found diverse applications in the study of carbohydrate–protein interaction.

For the contemporary medicinal chemist, multicomponent reactions and diversity-oriented synthesis are both essential tools. Despite the importance of these compounds in examining multivalency on glycoside inhibition, their use in the synthesis of multivalent glycomimetics has not yet been fully explored. Bellucci et al. introduce a multicomponent cascade process for the combinatorial synthesis of diversity-oriented hetero di- and trivalent glycomimetics.[154] The method is high yielding and very common, functioning properly with readily available sugar precursors, including glycosyl azides **234**, sugar isothiocyanates **235**, which leads to the in situ generation of *N*-sugar, *N*′-sugar carbodiimides **236** via Staudinger reaction. The reaction of **236** with fumaric acid monoester **237** along with 2,4,6-trimethylaminopyridine (TMP) **237**, afforded glycohydantoin conjugate **238**. The glycohydantoin conjugate formed in one pot via domino condensation followed by aza-Michael/N–O migration. The similar reaction condition when performed with *N*-sugar, *N*′-sugar carbodiimides **236**, along nucleophilic sugar amines **239**, furnished glycopeptides **240** (Scheme 14.57).[155b] The authors also established the synthesis of various glycomimetics by varying isocyanates, azides, and other reactive component.

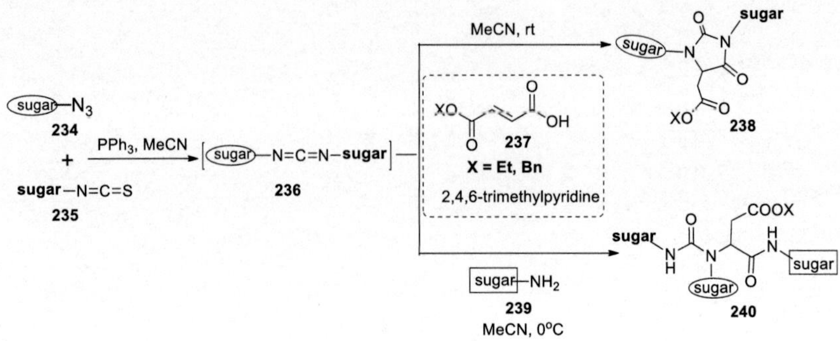

SCHEME 14.57 Domino synthesis of multivalent glycomimetics.

The earlier method has been applied to synthesize aminoglycoside conjugates (neomycin) **243** using glycosylazide **241** and isocyanate **242** (Scheme 14.58).[155b]

Roy and group established a method for the synthesis of 3-C-linked mannopyranosyl coumarins using well-known α-C-allyl mannopyranoside **244** as starting material. Reductive ozonolysis of α-C-allyl mannopyranoside followed by α-methylenation using aqueous formaldehyde and pyrrolidine afforded a 1:1 mixture of conjugated α-aldehyde **245** and its epimeric β-anomer **246** in 69% yield. The oxidation of **245** and **246** to their respective carboxylic acids under benign conditions, followed by the formation of methyl esters. Toward this end, acrylates **247** and **248** were obtained in 78% and 79% yields, from the conjugated aldehydes **245** and **246**, respectively, over the course of two steps. Heck couplings onto methyl acrylates **247** and **248** were then performed with 2-iodophenol under mentioned reaction conditions, the α-anomer **247** give Heck adduct **249** and the coumarin **250** in 77% yield in a 1:5 ratio. Under basic conditions (NaOMe, MeOH), Heck adduct **249** was easily converted into coumarin **250** with an 83% yield. Likewise, conjugated β-methyl ester **248** was subjected to Heck conditions, but only coumarin **251** was isolated (Scheme 14.59).[156]

Vankar et al. reported the unprecedented $BF_3 \cdot OEt_2$-mediated domino synthesis of 1,6-anhydro sugars **254** via the participation of 4-OBn group of sugar alcohol **252**.[157] Although the involvement of a 6-OBn group in the formation of 1,6-anhydro sugars is well known[155,158] and 1,4-anhydro sugars has been also synthesized previously, by the participation of free 4-OH.[159,160] To explain the formation of this intriguing product, the group proposes that $BF_3 \cdot OEt_2$ first activates the aldehyde **253** to form an oxocarbenium ion **B** from **A**. As a result, Prins cyclization may occur, yielding the tetrahydropyranyl cation **C**, which then undergoes a pinacol-type rearrangement to yield the oxocarbenium ion intermediate **D**, followed by C4-OBn participation (via 6-exo trig mode) to yield product **254** (Scheme 14.60).

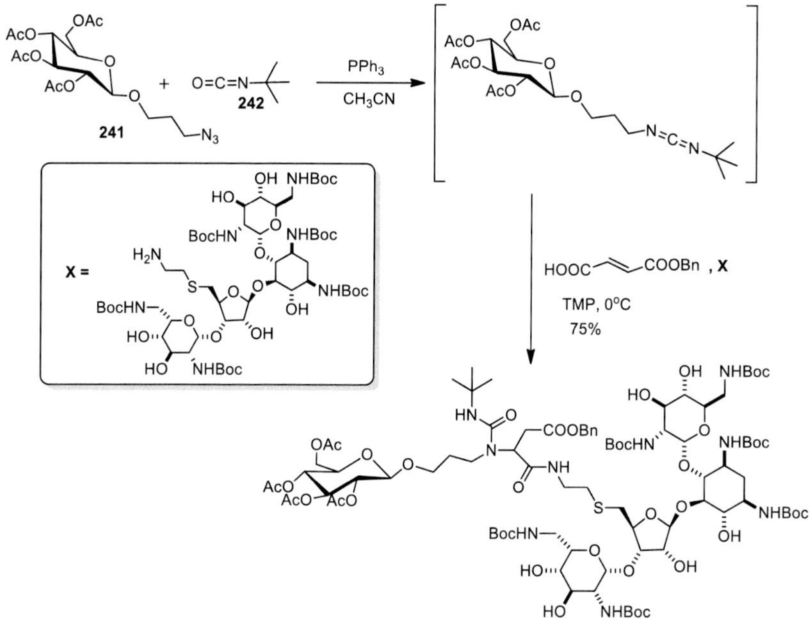

SCHEME 14.58 Multicomponent synthesis of neomycin derivative.

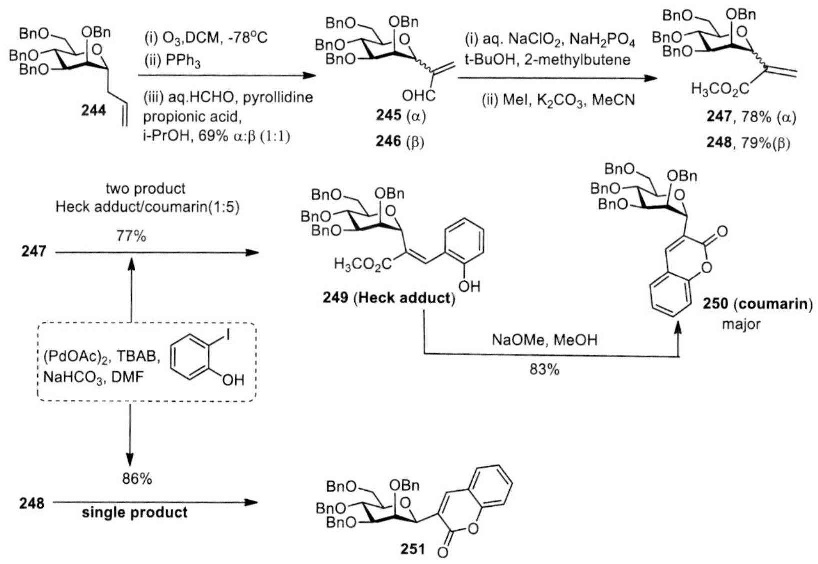

SCHEME 14.59 Synthesis of 3-C-linked mannopyranosyl coumarins via domino Heck/lactonization approach.

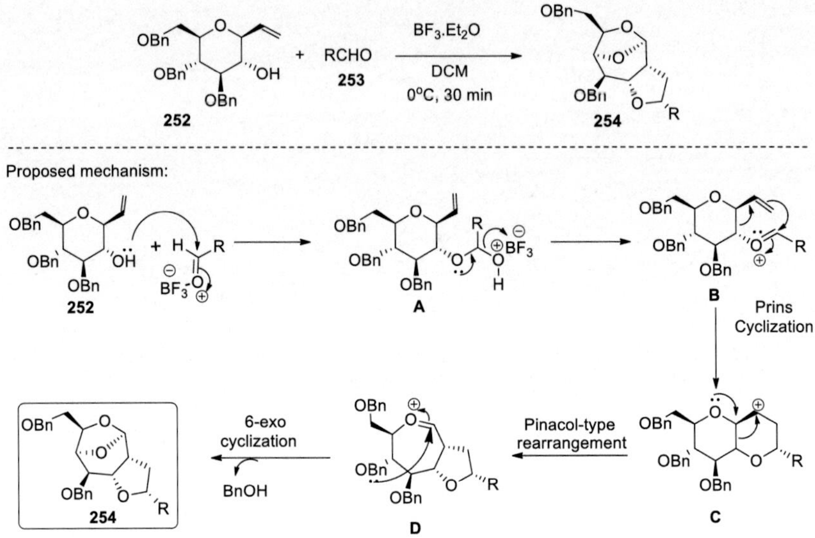

SCHEME 14.60 A domino synthesis of 1,6-anhydro-heptopyranose skeletons.

Anthracyclines belong to the group of aromatic polyketides and are a prevalent class of natural products of utmost biological and medicinal significance.[161] Carbohydrates play a very important role in the biological activity of anthracyclines through binding to the minor groove of double-stranded DNA.[162,163] Therefore the challenging synthesis of various anthracycline carbohydrate derivatives become a major field of attraction for medicinal chemists. D. B. Werz et al. investigated this challenging synthesis and came up with a robust approach to synthesize anthracycline aglycone derivatives **263**. The starting materials employed in their methodology were bromoglycals **255** and a benzene moiety tethered with two alkynyl groups **256**. The same domino carbopalladation technique, which was applied for chromans, was utilized here for the synthesis of a tetracyclic anthracycline scaffold in a single step. First, the building blocks **255** and **256** were linked together through a silyl ether linkage to make an intermediate **257** for a successful intramolecular carbopalladation reaction. In the second key step, the fourfold annulated ring system of anthracycline **258** was obtained through a sequence of carbopalladation reactions. Treatment of **258** with acetyl chloride in anhydrous methanol results in cleavage of the silyl group to give **259**, which is further treated with tetrabutylammonium fluoride (TBAF) to furnish a completely silyl-free intermediate **260**. The TBDMS protection of the hydroxyl functionalities of **260** afforded **261**, which subsequently undergo an iron (III)-catalyzed benzylic oxidation to afford intermediate **262**. Deprotection of **262** followed by hydrolysis with HCl produced the desired anthracycline aglycone derivatives **263** in good yields. Different substituted monosaccharides

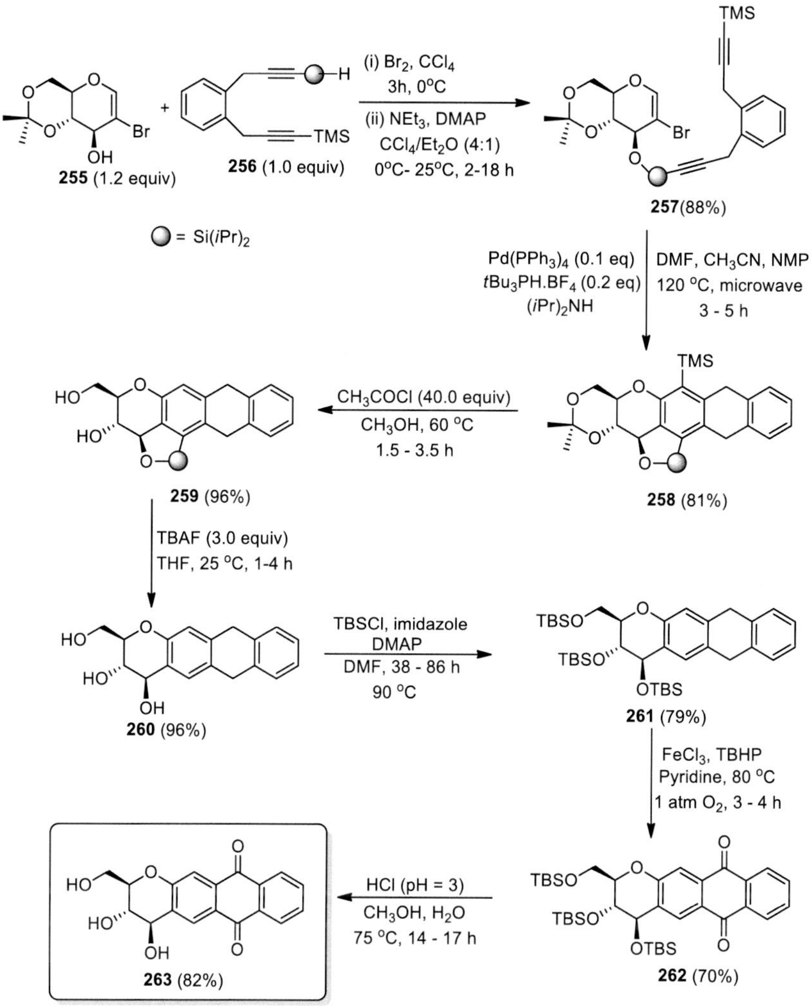

SCHEME 14.61 Synthetic route involving domino carbopalladation sequence for the preparation of anthracycline sugar derivatives.

and dialkyne chains were employed for the synthetic purpose. The whole reaction sequence with bromoglucal and trimethylsilyl (TMS) attached dialkyne is depicted in (Scheme 14.61).[164]

14.7 Conclusions and future outlooks

Conclusively, this chapter summarizes the utility of domino synthesis in carbohydrate chemistry. Although it is not limited here, it explores a distinct area of organic chemistry. In this chapter, we briefly highlighted the involvement of sugar and its analogs in a number of important transformations including the use of chiral carbohydrate derivatives and catalysts. A number of significant works that address stereocontrolled domino reactions of chiral carbohydrate scaffolds and enantioselective domino reactions catalyzed by chiral carbohydrate derivatives were discussed. The role of this technique in the construction of desired building blocks is deliberated herein. The application of low-valent organometallic reagents for the synthesis of various carbohydrate frameworks via tandem route was also discussed. Rather than earlier discussed transformation, the role of the domino reaction in the synthesis of butenolides, spiro derivatives, and macrocycles is also reviewed.

Acknowledgments

The authors sincerely express thanks to the Science and Engineering Research Board (SERB), New Delhi (Grant No. CRG/2021/002776) and Council of Scientific & Industrial Research (CSIR), New Delhi (Scheme No.: 02(0345)/19/EMR-II) for the financial assistance. The authors thankfully acknowledge the Institute of Eminence (IOE), BHU for the incentive grant. MSY, AG, and SK acknowledge the CSIR, New Delhi for the fellowship.

References

1. Trost, B. M. *Angew. Chem. Int. Ed. Engl.* **1995,** *34*, 259–281.
2. Denmark, S. E.; Thorarensen, A. *Chem. Rev.* **1996,** *96*, 137–166.
3. Parsons, P. J.; Penkett, C. S.; Shell, A. J. *Chem. Rev.* **1996,** *96*, 195–206.
4. Eilbracht, P.; Barfacker, L.; Buss, C.; Hollmann, C.; Kitsos-Rzychon, B. E.; Kranemann, C. L.; Rische, T.; Roggenbuck, R.; Schmidt, A. *Chem. Rev.* **1999,** *99*, 3329–3366.
5. Tietze, L. F. *Chem. Rev.* **1996,** *96*, 115–136.
6. Bruggink, A.; Schoevaart, R.; Kieboom, T. *Org. Proc. Res. Dev.* **2003,** *7*, 622–640.
7. Witczak, Z. J.; Bielski, R., Eds. *Domino and Intramolecular Rearranagement Reactions as Advances Synthetic Methods in Glycoscience;* Wiley VCH, 2016.
8. Cipolla, L.; Araújo, A. C.; Bini, D.; Gabrielli, L.; Russo, L.; Shaikh, N. *Expert. Opin. Drug. Discov.* **2010,** *5*, 721–737.
9. Tiwari, V. K.; Mishra, B. B.; Mishra, K. B.; Mishra, N.; Singh, A. S.; Chen, X. *Chem. Rev.* **2016,** *116*, 3086–3240.
10. Agrahari, A. K.; Bose, P.; Jaiswal, M. K.; Rajkhowa, S.; Singh, A. S.; Hotha, S.; Mishra, N.; Tiwari, V. K. *Chem. Rev.* **2021,** *12*, 7638–7956.
11. Tiwari, V. K. *Chem. Rec.* **2021,** *21*, 3029–3048.
12. Hanessian, S. *Preparative Carbohydrate Chemistry;* Marcel Dekker Inc: New York, 1997.
13. Santra, S.; Andreana, T.; Bourgault, J.-P.; Andreana, P. R.; Murphy, P. V. Convertible Isocyanides: Application in Small Molecule Synthesis, Carbohydrate Synthesis, and Drug Discovery, in Domino and Intramolecular Rearranagement Reactions as Advances Synthetic Methods. In *Glycoscience;* Witczak, Z. J., Bielski, R., Eds.; Chapter 7; Wiley VCH, 2016; pp 121–194.
14. Cachatra, V.; Rauter, A. P. Domino Reactions Toward Carbohydrate Frameworks for Applications Across Biology and Medicine, in Domino and Intramolecular Rearranagement

Reactions as Advances Synthetic Methods. In *Glycoscience;* Witczak, Z. J., Bielski, R., Eds.; Chapter 4; Wiley VCH, 2016; pp 76–95.

15. Hantzsch, A. *Justus Liebigs Ann. Chem.* **1882,** *215*, 1–82.
16. Ducatti, D. R. B.; Massi, A.; Noseda, M. D.; Duarte, M. E. R.; Dondoni, A. *Org. Biomol. Chem.* **2009,** *7*, 1980–1986.
17. Ugi, I.; Meyr, R.; Fetzer, U.; Steinbruckner, C. *Angew. Chem.* **1959,** *11*, 386.
18. Domling, A.; Ugi, I. *Angew. Chem., Int. Ed.* **2000,** *39*, 3168–3210.
19. Ugi, I. *J. Prakt. Chem.* **1997,** *339*, 499–516.
20. Sigmuller, F.; Herrmann, R.; Ugi, I. *Tetrahedron* **1986,** *42*, 5931–5941.
21. Kunz, H.; Pfrengle, W. *Tetrahedron* **1988,** *44*, 5487–5494.
22. Kunz, H.; Pfrengle, W.; Sager, W. *Tetrahedron Lett.* **1989,** *30*, 4109–4110.
23. Goebel, M.; Ugi, I. *Synthesis* **1991,** 1095–1098.
24. Lehnhoff, S.; Goebel, M.; Karl, R. M.; Klösel, R.; Ugi, I. *Angew. Chem., Int. Ed. Engl.* **1995,** *34*, 1104–1107.
25. Passerini, M. *Gazz. Chim. Ital.* **1921,** *51*, 126–129.
26. Krishna, P. R.; Dayakar, G.; Reddy, P. V. N. *Tetrahedron Lett.* **2006,** *47*, 5977–5980 A.; Sharma, S.; Gupt, M. P.; Bajpai, V.; Hamidullah; Kumar, B.; Kaushik, M. P.; Konwar, R.; Ampapathi, R. S.; Tripathi, R. P. *Org. Lett.* **2012**, *14*, 4306–4309.
27. Xia, T.-Yu; Li, Y.-B.; Yin, Z.-J.; Meng, X.-B.; Li, S.-C.; Li, Z.-J. *Chin. Chem. Lett.* **2014,** *25*, 1220–1224.
28. Scherrmann, M. C. *Top. Curr. Chem.* **2010,** *295*, 1–18.
29. Mannich, C.; Krosche, W. *Arch. Pharm.* **1912,** *250*, 647–667.
30. Arend, M.; Westermann, B.; Risch, N. *Angew. Chem., Int. Ed.* **1998,** *37*, 1044–1070.
31. Cordova, A. *Acc. Chem. Res.* **2004,** *37*, 102–112.
32. Verkade, J. M. M.; Van Hemert, L. J. C.; Quaedflieg, P. J. L. M.; Rutjes, F. P. J. T. *Chem. Soc. Rev.* **2008,** *37*, 29–41.
33. Arrayas, R. G.; Carretero, J. C. *Chem. Soc. Rev.* **2009,** *38*, 1940–1948.
34. (a) Wong, C.-H., Ed. *Carbohydrate-based Drug Discovery;* Wiley-VCH: Weinheim, 2003. (b) Tiwari, V. K., Ed. *Carbohydrates in drug discovery and development;* Elsevier, 2020.
35. Dondoni, A.; Massi, A. *Acc. Chem. Res.* **2006,** *39*, 451–463.
36. (a) Taillefumier, C.; Chapleur, Y. *Chem. Rev.* **2004,** *104*, 263–292. (b) Lalitha, K.; Muthusamy, K.; Prasad, Y. S.; Vemula, P. K.; Nagargan, S. *Carb. Res* **2015,** *402*, 158–171.
37. Araki, Y.; Kobayashi, N.; Ishido, Y. *Carb. Res.* **1987,** *171*, 125–139.
38. Csuk, R.; Schaade, M. *Tetrahedron* **1994,** *50*, 3333–3348.
39. Feng, W.; Fang, Z.; Yang, J.; Zheng, B.; Jiang, Y. *Carb. Res.* **2011,** *346*, 352–356.
40. Wang, J.; Li, Q.; Ge, Z.; Li, R. *Tetrahedron* **2012,** *68*, 1315–1320.
41. Rodridues, F.; Canac, Y.; Lubineau, A. *Chem. Comm.* **2000,** 2049–2050.
42. Gonzalez, M. A.; Requejo, J. L. J.; Albarran, J. C. P.; Perez, J. A. G. *Carb. Res.* **1986,** *158*, 53–66.
43. Wulff, G.; Clarkson, G. *Carb. Res.* **1994,** *257*, 81–95.
44. Ranoux, A.; Lemiegre, L.; Benoit, M.; Guegan, J. P.; Benvegnu, T. *Eur. J. Org. Chem.* **2010,** 1314–1323.
45. Ranoux, A.; Lemiegre, L.; Benvegnu, T. *Sci. China. Chem.* **2010,** *53*, 1957–1962.
46. Witte, N.; Voigt, B.; Mahrwald, R. *Synthesis* **2015,** *47*, 2249–2255.
47. Voigt, B.; Mahrwald, R. *Chem. Comm.* **2012,** *48*, 5304–5306.
48. Morpurgo, S.; Bossa, M. *Phys. Chem. Chem. Phys.* **2003,** *5*, 1181–1189.
49. Arai, K. *J. Polym. Sci., Pt A: Polym. Chem.* **1993,** *31*, 193–197.

50. Kjaer, A. M.; Nielsen, H.; Soerensen, P. E.; Ulstrup, J. *Acta Chem.. Scand. Ser. A: Phys. Inorg. Chem* **1980,** *A34*, 281–288.
51. Morgan, D. M. L.; Neuberger, A. *Proc. R. Soc. Lond., Ser. A: Math. Phys. Eng. Sci.* **1974,** *337*, 317–332.
52. Smith, W. T.; Hearn, T. L. *Bioorg. Chem.* **1972,** *2*, 39–43.
53. Rony, P. R. *J. Am. Chem. Soc.* **1968,** *90*, 2824–2831.
54. Bentley, R.; Bhate, D. S. *J. Biol. Chem.* **1960,** *235*, 1225–1233.
55. Swain, C. G.; Brown, J. F. *J. Am. Chem. Soc.* **1952,** *74*, 2534–2537.
56. Swain, C. G.; Brown, J. F. *J. Am. Chem. Soc.* **1952,** *74*, 2538–2543.
57. Xavier, N. M.; Rauter, A. P. *Org. Lett.* **2007,** *9*, 3339–3341.
58. Xavier, N. M.; Silva, S.; Madeira, P. J. A.; Florencio, M. H.; Silva, F. V. M.; Justino, J.; Thiem, J.; Rauter, A. P. *Eur. J. Org. Chem.* **2008,** 6134–6143.
59. Xavier, N. M.; Madeira, P. J. A.; Florencio, M. H.; Rauter, A. P. *Eur. J. Org. Chem.* **2009,** 4983–4991.
60. Cachatra, V.; Almeida, A.; Sardinha, J.; Lucas, S. D.; Gomes, A.; Vaz, P. D.; Florencio, M. H.; Nunes, R.; Vila-Vicosa, D.; Calhorda, M. J., et al. *Org. Lett.* **2015,** *17*, 5622–5625.
61. Voigt, B.; Linke, M.; Mahrwald, R. *Org. Lett.* **2015,** *17*, 2606–2609.
62. Stutz, A. E. *Iminosugars as glycosidase inhibitors. Nojirimycin and Beyond;* Wiley-VCH: Weinheim, Germany, 1999.
63. Compain, P.; Martin, O. R. *Iminosugars: From Synthesis to Therapeutic Applications;* Wiley: New York, 2007.
64. Winchester, B.; Fleet, G. W. J. *Glycobiology* **1992,** *2*, 199–210.
65. Jespersen, T. M.; Dong, W.; Skrydstrup, T.; Sierks, M. R.; Lundt, I.; Bols, M. *Angew. Chem., Int. Ed. Engl.* **1994,** *33*, 1778–1779.
66. Boyd, R. E.; Lee, G.; Rybczynski, P.; Benjamin, E. R.; Khanna, R.; Wustman, B. A.; Valenzano, K. J. *J. Med. Chem.* **2013,** *56*, 2705–2725.
67. Moon, H. R.; Choi, W. J.; Kimb, H. O.; Jeonga, L. S. *Tetrahedron: Asymmetry* **2002,** *13*, 1189–1193.
68. Attia, M. I.; Timmermann, M.; Högger, P.; Herdeis, C. *Eur. J. Org. Chem.* **2007,** *22*, 3669–3675.
69. Siriwardena, A.; Sonawane, D. P.; Bande, O. P.; Markad, P. R.; Yonekawa, S.; Tropak, M. B.; Ghosh, S.; Chopade, B. A.; Mahuran, D. J.; Dhavale, D. D. *J. Org. Chem.* **2014,** *79*, 4398–4404.
70. Xia, M.; Lambu, M. R.; Tatina, M. B.; Judeh, Z. M. A. *J. Org. Chem.* **2021,** *86*, 837–849.
71. Lambu, M. R.; Judeh, Z. M. A. *Green. Chem.* **2019,** *21*, 821–829.
72. Nicolaou, K. C.; Mitchell, H. J. *Angew. Chem., Int. Ed.* **2001,** *40*, 1576–1624.
73. Isobe, M.; Nishizawa, R.; Hosokawa, S.; Nishikawa, T. *Chem.Commun.* **1998,** 2665–2676.
74. Ichikawa, Y.; Tsuboi, K.; Jiang, Y.; Naganawa, A.; Isobe, M. *Tetrahedron Lett.* **1995,** *36*, 7101–7104.
75. Jiang, Y.; Isobe, M. *Tetrahedron* **1996,** *52*, 2877–2892.
76. Tsuboi, K.; Ichikawa, Y.; Jiang, Y.; Naganawa, A.; Isobe, M. *Tetrahedron* **1997,** *53*, 5123–5142.
77. Isobe, M.; Yenjai, C.; Tanaka, S. *Synlett* **1994,** *11*, 916–918.
78. Yenjai, C.; Isobe, M. *Tetrahedron* **1998,** *54*, 2509–2520.
79. Rosenthal, A.; Abson, D. J. *J. Am. Chem. Soc.* **1964,** *86*, 5356–5357.
80. Kusunuru, A. K.; Jaladanki, C. K.; Tatina, M. B.; Bharatam, P. V.; Mukherjee, D. *Org. Lett.* **2015,** *17*, 3742–3745.
81. Bhardwaj, M.; Rasool, F.; Tatina, M. B.; Mukherjee, D. *Org. Lett.* **2019,** *21*, 3038–3042.

82. Ellis, G. P.; Lockhart, I. M. *The Chemistry of Heterocyclic Compounds, Chromenes, Chromanones, and Chromones;* Wiley-VCH: New York, 2007.
83. Shen, H. C. *Tetrahedron* **2009,** *65*, 3931–3952.
84. Leibeling, M.; Koester, D. C.; Pawliczek, M.; Schild, S. C.; Werz, D. B. *Nat. Chem. Biol.* **2010,** *6*, 199–201.
85. Leibeling, M.; Koester, D. C.; Pawliczek, M.; Kratzert, D.; Dittrich, B.; Werz, D. B. *Bioorg. Med. Chem.* **2010,** *18*, 3656–3667.
86. Meyer, F. E.; de Meijere, A. *Synlett* **1991,** 777–778.
87. Blond, G.; Bour, C.; Salem, B.; Suffert, J. *Org. Lett.* **2008,** *10*, 1075–1078.
88. Leibeling, M.; Milde, B.; Kratzert, D.; Stalke, D.; Werz, D. B. *Chem. Eur. J.* **2011,** *17*, 9888–9892.
89. Leibeling, M.; Werz, D. B. *Chem. Eur. J.* **2012,** *18*, 6138–6141.
90. Chambers, D. J.; Evans, G. R.; Fairbanks, A. J. *Tetrahedron* **2005,** *61*, 7184–7192.
91. Simelane, S. B.; Kinfe, H. H.; Muller, A.; Williams, B. G. *Org. Lett.* **2014,** *16*, 4543–4545.
92. Bernet, B.; Vasella, A. *Helv. Chim. Acta* **1979,** *62*, 1990 –1919.
93. Cintas, P. *Activated Metals in Organic Synthesis;* CRC Press, Taylor & Francis Group, 1993.
94. Bernet, B.; Vasella, A. *Helv. Chim. Acta* **1979,** *62*, 2400.
95. Bernet, B.; Vasella, A. *Helv. Chim. Acta* **1979,** *62*, 2411.
96. Desire, J.; Prandi, J. *Tetrahedron Lett.* **1997,** *38*, 6189.
97. Hummer, W.; Dubois, E.; Gracza, T.; Jager, V. *Synthesis* **1997,** 634.
98. Ferrier, R. J.; Furneaux, R. H.; Prasit, P.; Tyler, P. C.; Brown, K. L.; Gainsford, G. J.; Diehl, J. W. *J. Chem. Soc. Perkin Trans. 1* **1983,** 1621.
99. Dransfield, P. J.; Moutel, S.; Shipman, M.; Sik, V. *J. Chem. Soc. Perkin Trans. 1* **1999,** 3349.
100. Bernotas, R. C.; Papandreou, G.; Urbach, J.; Ganem, B. *Tetrahedron Lett.* **1990,** *31*, 3393.
101. Dangerfield, E. M.; Plunkett, C. H.; Win-Mason, A. L.; Stocker, B. L.; Timmer, M. S. M. *J. Org. Chem.* **2010,** *75*, 5470.
102. Jorgensen, M.; Hadwiger, P.; Madsen, R.; Stutz, A. E.; Wrodnigg, T. M. *Curr. Org. Chem.* **2000,** *4*, 565.
103. Prunet, J. *Eur. J. Org. Chem.* **2011,** 3634.
104. Hyldtoft, L.; Madsen, R. *J. Am. Chem. Soc.* **2000,** *122*, 8444.
105. Skaanderup, P. R.; Madsen, R. *J. Org. Chem.* **2003,** *68*, 2115.
106. Keinicke, L.; Madsen, R. *Org. Biomol. Chem.* **2005,** *3*, 4124.
107. Bonnett, R. *Chem. Rev.* **1963,** *63*, 573–605.
108. Andre-Joyaux, E.; Santana, A. G.; Gonzalez, C. C. *J. Org. Chem.* **2019,** *84*, 506–515.
109. Patel, R. M.; Puranik, V. G.; Argade, N. P. *Org. Biomol. Chem.* **2011,** *9*, 6312–6322.
110. Xavier, N. M.; Rauter, A. P. *Org. Lett.* **2007,** *9*, 3339–3341.
111. Jayakanthan, K.; Mohan, S.; Pinto, B. M. *J. Am. Chem. Soc.* **2009,** *131*, 5621–5626.
112. Dai, M.; Zhang, X.; Khim, S.-K.; Schultz, A. G. *J. Org. Chem.* **2005,** *70*, 384–387.
113. Li, Y.; Zhang, F.; Xu, Y.; Matsumura, K.; Han, Z.; Liu, L.; W..; Jia, Y.; Qian, P.-Y. *Biofouling* **2012,** *28*, 857–864.
114. Zhang, D.; Ye, D.; Feng, E.; Wang, J.; Shi, J.; Jiang, H.; Liu, H. *J. Org. Chem.* **2010,** *75*, 3552–3557.
115. Brand, C.; Rauch, G.; Zanoni, M.; Dittrich, B.; Werz, D. B. *J. Org. Chem.* **2009,** *74*, 8779–8786.

116. Chen, B.-F.; Tasi, M.-R.; Yang, C.-Y.; Chang, J.-K.; Chang, N.-C. *Tetrahedron* **2004,** *60*, 10223–10231.
117. He, Z.-T.; Wei, Y.-B.; Yu, H.-J.; Sun, C.-Y.; Feng, C.-G.; Tian, P.; Lin, G.-Q. *Tetrahedron* **2012,** *68*, 9186–9191.
118. Jensen, A. A.; Madsen, B. E.; Krogsgaard-Larsen, P.; Brauner-Osborne, H. *Eur. J. Pharmacol.* **2001,** 177–180.
119. Xavier, N. M.; Queneau, Y.; Rauter, A. P. *Eur. J. Org. Chem.* **2011,** *2011*, 713–720.
120. Issa, J. P.; Bennett, C. S. *J. Am. Chem. Soc.* **2014,** *136*, 5740–5744.
121. Rauter, A. P.; Fernandes, A. C.; Czernecki, S.; Valery, J.-M. *J. Org. Chem.* **1996,** *61*, 3594–3598.
122. Rauter, A. P.; Fernandes, A. C.; Figueiredo, J. A. *J. Carbohydr. Chem.* **1998,** *17*, 1037–1045.
123. Jida, M.; Ollivier, J. *Eur. J. Org. Chem.* **2008,** *2008*, 4041 4019.
124. Takahashi, Y.; Fuwa, H.; Kaneko, A.; Sasaki, M.; Yokoshima, S.; Koizumi, H.; Takebe, T.; Kan, T.; Iwatsubo, T.; Tomita, T., et al. *Bioorg. Med. Chem. Lett.* **2006,** *16*, 3813–3816.
125. Mathe, C.; P´erigaud, C. *Eur. J. Org. Chem.* **2008,** *2008*, 1489–1505.
126. Marcelo, F. Total Synthesis and Stereochemical Assignment of Miharamycins. PhD Thesis, University of Lisbon, University Pierre and Marie Curie–Paris VI, 2008.
127. Davis, O. A.; Bull, J. A. *Angew. Chem.* **2014,** *53*, 14230–14234.
128. Witty, D. R.; Fleet, G. W. J.; Vogt, K.; Wilson, F. X.; Wang, Y.; Storer, R.; Myers, P. L.; Wallis, C. J. *Tetrahedron Lett.* **1990,** *31*, 4787–4790.
129. Lucas, S. D.; Iding, H.; Alker, A.; Wessel, H. P.; Rauter, A. P. *J. Carbohydr. Chem.* **2006,** *25*, 187–196.
130. Lucas, S. D.; Rauter, A. P.; Schneider, J.; Wessel, H. P. *J. Carbohydr. Chem.* **2009,** *28*, 431–446.
131. Nishikawa, T.; Araki, H.; Isobe, M. *Biosci. Biotechnol. Biochem.* **1998,** *62*, 190–192.
132. Samet, A. V.; Lutov, D. N.; Firgang, S. I.; Lyssenko, K. A.; Semenov, V. V. *Tetrahedron Lett.* **2011,** *52*, 3026–3028.
133. Witczak, Z.J. Thio-click and domino reaction tandem approach to carbohydrate heterocycles, abstracts of 248th ACS National Meeting, August 12–17, 2014, San Francisco, CA, Abstract CARB-23.
134. Brel, V. K.; Samet, A. V.; Konyushkin, L. D.; Stash, A. I.; Belsky, V. K.; Semenov, V. V. *Mendeleev Commun.* **2015,** *25*, 44–46.
135. Kasare, S.; Bankar, S. K.; Ramasastry, S. V. *Org. Lett.* **2014,** *16*, 4284–4287.
136. Bhattacharya, D.; Ghorai, A. U.; Pal, N.; Maiti, C.; Chattopadhyay, P. *RSC Adv.* **2014,** *4*, 4155–4162.
137. Bodine, K. D.; Gin, D. Y.; Gin, M. S. *Org. Lett.* **2005,** *7*, 4479–4482.
138. Baumann, M.; Baxendale, I. R. *Org. Lett.* **2014,** *16*, 6076–6079.
139. Rauter, A. P.; Piedade, F.; Almeida, T.; Ramalho, R.; Ferreira, M. J.; Resende, R.; Amado, J.; Pereira, H.; Justino, J.; Neves, A., et al. *Carbohydr. Res.* **2004,** *339*, 1889–1897.
140. Woodward, S.; Dieguezb, M.; Pamiesb, O. *Coord. Chem. Rev.* **2010,** *254*, 2007–2030.
141. Mishra, N.; Singh, A. S.; Agrahari, A. K.; Singh, S. K.; Singh, M.; Tiwari, V. K. *ACS Comb. Sci.* **2019,** *21*, 389–399.
142. Singh, S. K.; Yadav, M. S.; Singh, A. S.; Agrahari, A. K.; Mishra, N.; Kumar, S.; Tiwari, V. K. *ACS Omega* **2021,** *6*, 21125–21138.

143. Singh, S. K.; Kumar, S.; Yadav, M. S.; Tiwari, V. K. *J. Org. Chem.* **2022,** *87*, 15389–15402.
144. Sangwan, R.; Mandal, P. K. *Curr. Org. Chem.* **2021,** *25*, 2499–2541.
145. Ley, S. V.; Priepke, W. M. *Angew. Chem. Int. Ed. Engl.* **1994,** *33*, 2292–2294.
146. Zhang, Z.; Ollmann, I. R.; Ye, X.-S.; Wischnat, R.; Baasov, T.; Wong, C.-H. *J. Am. Chem. Soc.* **1999,** *121*, 734–753.
147. Koeller, K. M.; Wong, C. H. *Chem. Rev.* **2000,** *100*, 4465–4494.
148. Bose, P.; Agrahari, A.K.; Singh, A.S.; Jaiswal, M.K.; Tiwari, V.K. (Tiwari, V. K., Ed.); Elsevier Inc., 2020, pp 213 – 266.
149. Yu, H.; Chokhawala, H. A.; Huang, S.; Chen, X. *Nat. Protoc.* **2006,** *1*, 2485–2492.154.
150. Cheng, J.; Yu, H.; Lau, K.; Huang, S.; Chokhawala, H. A.; Li, Y.; Tiwari, V. K.; Chen, X. *Glycobiology* **2008,** *18*, 686–697.
151. Yu, H.; Cao, H.; Tiwari, V. K.; Li, Y.; Chen, X. *Bio. Org. Med. Chem. Lett.* **2011,** *21*, 5037–5040.
152. (a) Li, Y.; Yu, H.; Cao, H.; Lau, K.; Muthana, S.; Tiwari, V. K.; Son, B.; Chen, X. *Appl. Microb. Biotech.* **2008,** *79*, 963–970.
    (b) Huynh, N.; Aye, A.; Li, Y.; Yu, H.; Cao, H.; Tiwari, V. K.; Shin, D.-W.; Chen, X.; Fisher, A. J. *Biochemistry* **2013,** *52*, 8570–8579.
153. (a) Song, X.; Yu, H.; Chen, X.; Lasanajak, Y.; Tappert, M. M.; Air, G. M.; Tiwari, V. K.; Cao, H.; Chokhawala, H. A., et al. *J. Biol. Chem.* **2011,** *286*, 31610–31622.
    (b) Yu, H.; Cheng, J.; Ding, L.; Khedri, Z.; Chen, Y.; Lau, K.; Tiwari, V. K.; Chen, X. *J. Am. Chem. Soc.* **2009,** *131*, 18467–18477.
154. (a) Sganappa, A.; Bellucci, M. C.; Nizet, V.; Tor, Y.; Volonterio, A. *Synthesis* **2016,** *48*, 4443–4450.
    (b) Bellucci, M. C.; Sani, M.; Sganappa, A.; Volonterio, A. *ACS Comb. Sci.* **2014,** *16*, 711–720.
155. Miranda, P. O.; Brouard, I.; Padron, J. I.; Bermejo, J. *Tetrahedron Lett.* **2003,** *44*, 3931.
156. Giguere, D.; Cloutier, P.; Roy, R. *J. Org. Chem.* **2009,** *74*, 8480–8483.
157. Rajasekaran, P.; Singh, G. P.; Hassam, M.; Vankar, Y. D. *Chem. Eur. J.* **2016,** *22*, 18383–18387.
158. Stanetty, C.; Walter, M.; Kosma, P. *J. Org. Chem.* **2014,** *79*, 582.
159. Caravano, A.; Mengin-Lecreulx, D.; Brondello, J.-M.; Vincent, S. P.; Sinaÿ, P. *Chem. Eur. J.* **2003,** *9*, 5888.
160. Nokami, T.; Werz, D. B.; Seeberger, P. H. *Helv. Chim. Acta* **2005,** *88*, 2823.
161. Laatsch, H.; Fotso, S. *Top. Curr. Chem.* **2008,** *282*, 3–74.
162. Leng, F.; Savkur, R.; Fokt, I.; Przewloka, T.; Priebe, W.; Chaires, J. B. *J. Am. Chem. Soc.* **1996,** *118*, 4731–4738.
163. Chaires, J. B.; Satyanarayana, S.; Suh, D.; Fokt, I.; Przewloka, T.; Priebe, W. *Biochemistry* **1996,** *35*, 2047–2053.
164. Leibeling, M.; Werz, D. B. *Beil. J. Org. Chem.* **2013,** *9*, 2194–2201.

Index

Note: Page numbers followed by "*f*" and "*t*" refer to figures and tables, respectively.

A

Acetolysis, 338–339
Acetonitrile solvent, 171
Acetylcholinesterase inhibitor glycohybrids, 600–601, 601*f*
Achmatowicz's approach, 287
Active-site specific pharmacological chaperone (ASSPC), 190–192
α-cyclodextrin synthesis, 58–60
Aerobic glycolysis, 1
AFGPs. *See* Antifreeze glycoproteins
AgOTf/p-ArSCl promoter system
  benzenesulfinyl piperidine (BSP), 126
  chondroitin sulfate glycopeptide, 123–126
  thiogalactoside acceptor, 126
Aldol and Michael reactions, 629–639
  carbohydrate-based prolinamides, 633*f*, 634–635
  plausible mechanism, 634–635, 635*f*
  prolinamide catalysts, 632–634, 633*f*
  proposed transition states, 631–632, 632*f*
  pyrrolidine ring, 636–638
  sugar-based prolinamides, 629–631, 631*f*
Alkoxyl radical fragmentation (ARF), 700–701
Amine group installation
  arginine glycosylation method, 180–181, 182*f*
  azaglycosylation, 177–178, 178*t*
  azide group, 170–173
  N-heterocyclic glycosides, 178–179
  palladium catalyst, 179–180, 180*f*
  plausible mechanism, 181*f*
  sulfonamide, 173–175
  tert-butyl carbamate, 176–177
Ananatoside A synthesis, 413–415
  chemical structure, 413, 414*f*
  glycosylation ring-closing methods, 415
  macrolactonization, 413, 414*f*
  rhamnose containing macrodilatones, 415
1,5-Anhydro-D-fructose, 306–308
Anticancer glycohybrids, 590–597
Antifreeze glycoproteins (AFGPs), 533–535
Antifungal glycohybrids, 603–604, 604*f*
Antimalarial glycohybrids, 599–600, 600*f*
Antiviral glycohybrids, 600, 601*f*
ARF. *See* Alkoxyl radical fragmentation
Arginine glycosylation, 180–181, 182*f*
Arginine glycosylation method, 180–181, 182*f*
Aryl *C*-glycosides synthesis, 367–373
ASSPC. *See* Active-site specific pharmacological chaperone
Asymmetric domino methods, chiral carbohydrate derivatives, 680–684
Automated glycan assembly, 289–291
Auxiliary-based glycopeptide synthesis, 567–568
Azaglycosylation, 177–178, 178*t*
Aza-heterocycles, 308–311
Aza-linked sugar-based macrocycles synthesis, 473–476
Azepanes, 349–350
Azetidines iminosugars, 336–339
Azide group, 170–173

B

Bacillus subtilis, 598–599
Ball-mill method, 536–537, 537*f*
Batatin VI synthesis, 434–437
Benzenesulfinyl piperidine (BSP), 126
Benzopyrans, 697–698
Benzyl groups, 344
Bernet–Vasella reaction, 699
β – fructofuranoside (β – Fruf), 77, 78*f*
Bicyclic iminosugars, 350–361
  indolizidines, 354–357
  nortropanes, 360–361
  pyrrolizidines, 350–353
  quinolizidines, 357–360

Biomass, 320–322
Birch Reduction, 73
Bismuth catalysts, 33–34
Bisphenol A (BPA), 65–66
Bovine serum albumin (BSR), 549
BPA. *See* Bisphenol A
Bromination reaction, 17
Brønsted acid catalysts, 617–619
Brønsted base catalysts, 616–617
BSP. *See* Benzene sulfinyl piperidine
β-stereoselectivity, 31–32
Butenolides and spiro compound, 701–708

C

Calysolin IX synthesis, 425–428, 426*f*, 427*f*
Candida albicans, 598–599
Carbohydrate-based chiral organocatalysts
  Aldol and Michael reactions, 629–639
  amino alcohols, 640–645
  Brønsted acid catalysts, 617–619
  Brønsted base catalysts, 616–617
  enantioselective epoxidation, 620–629
  exo-anomeric effect, 620*f*
  historical background, 612–619
  ketones, 620–629
  Lewis acid catalysts, 615–616
  Lewis base catalysts, 615
  prolinamides, 629–639
  pyrrolidine amides, 629–639
  pyrrolidines, 629–639
  quinine and quinidine catalysis, 612–614, 648*f*
  ring (steric effect) and anomeric effect, 619–620
  structure of, 619
  sugar-pseudoenantiomer pairs, 619, 620*f*
  thiourea-primary amines, 646–652
  thiourea-secondary amines, 653–655
  thiourea-tertiary amines, 655–665
  Von Liebig's oxamide synthesis, 648*f*
Carbohydrate-based ketone catalysts, 625, 627*f*
Carbohydrates
  complexity of, 2–3
  role of, 2
  solar energy, 1
Carbohydrates, protection-deprotection strategies
  acyl migrations, 286
  anomeric position, 277–279
  automated glycan assembly, 289–291
  benzylidene acetals, 279–282
  cyclic protecting group, 279–282
  de novo asymmetric synthesis, 286–287
  glycofuranoses, 288–289
  N-protecting groups, 282–283
  primary position of saccharides, 273–276
  secondary hydroxyl groups, 276–277
  sialic acid derivatives, 288
  sulfated glycosaminoglycans, 289
Carbonic anhydrase inhibitor glycohybrids, 602–603, 603*f*
C-aryl glycosides synthesis, 401–402, 402*f*
Casuarinin synthesis, 444–445, 444*f*
C–C bond forming reactions, 464–473
  acetylene coupling reaction, 470–472, 471*f*
  aldol condensation reactions, 472–473
  macrocyclization, 472–473, 472*f*
  RCM strategy, 464–469, 465*f*
  Sonogashira and Heck reaction, 469–470, 470*f*
C-2 functionalized polyhydroxyazepane, 220–221
C-glycosides for glycosidase inhibition
  piperidine-based iminosugars, 190–194
    active-site-specific pharmacological chaperone (ASSPC), 190–192
    C-glycosides to azepane-sugars, 192–194
    deoxynojirimycin (DNJ), 190
    isofagomine (IFG), 190–192, 191*f*
    pharmacological chaperones (PCs), 190–192
    substrate reduction therapy (SRT), 190–192
  polyhydroxyazepanes, 194–211
C-glycosides synthesis, 373–377
C-glycosides to azepane-sugars, 192–194
Chemical glycosylations, 7–8, 41
Chemoselective glycosylations, 10–11
  one-pot, 108–110
  polymerization, 104–105
  type A, 105–106
  type B, 106–107
  type C, 107–108
Chiral carbohydrate derivatives, 680–684
Chiral ketones
  (+)-coriandrone A synthesis, 306, 307*f*
  epoxidation catalysts, 306, 307*f*
  hydantocidin, 304–305, 306*f*
Chiron approach synthesis, 396
Chloromethylfurfural (CMF), 320–322

Chondroitin sulfate glycopeptide, 123–126
Cistanche tubulosa, 120–121
Clemahexapetoside A synthesis, 408–411
Clemoarmanoside A synthesis, 408–411
Click macrocyclizations, 482–514
  azobenzene containing
    glycomacrocycles synthesis, 507–509, 508*f*, 509*f*
    sugar-based macrocycles, 509–511, 510*f*, 511*f*
  cyclodextrin synthesis, 498–507, 499*f*, 500*f*, 503*f*
  cyclooligosaccharide derivatives, 512–514, 513*f*
  hexose sugar-embedded macrocycles, 489–496, 491*f*, 492*f*, 493*f*
  macrocyclization, 497–498, 498*f*
  multiple component click reactions, 482–483, 483*f*
  pentose sugar-based macrocycles, 484–488, 485*f*
  pyranose-based macrocycle libraries, 512, 512*f*
  sucrose-based macrocycle, 497
C-linked glycopeptides synthesis, 551–558
(+)-Coriandrone A synthesis, 306, 307*f*
Crown ether-like macrocycles synthesis
  carbohydrate-based aza-macrocycles, 453, 454*f*
  chiral sensors, 452–453, 453*f*
  description, 450–451, 451*f*
  glucose-based crown ether, 451, 451*f*
  glucose-based thia-crown ether, 453–454, 455*f*
  glucose-derived aza, 452, 453*f*
  nitrogen/thia linkages, 452, 452*f*
  polymeric thia-crown ether, 454–456, 455*f*
  sugar-based cryptands, 456, 456*f*
  thia macrocycles, 452, 453*f*
Cyclic protecting group, benzylidene acetals, 279–282
Cyclodextrin synthesis, 498–507, 499*f*, 500*f*, 503*f*

D

D-arabinitol (DAB), 309
DAST. *See* Diethyl aminosulfur trifluoride
de novo asymmetric synthesis, 286–287
Deoxynorijimycin (DNJ), 190, 200, 201*f*
Depezay synthesis, 195–196, 196*f*
D-fructopyranoside, 327–328
D-fructose, 307–308, 308*f*
  1,5-anhydro-D-fructose, 306–308
  aza-heterocycles, 308–311
  biomass and fructose, 320–322
  chiral ketones
    (+)-coriandrone A synthesis, 306, 307*f*
    epoxidation catalysts, 306, 307*f*
    hydantocidin, 304–305, 306*f*
  di-D-fructose dianhydrides, 298–299, 299*f*
  epoxides
    dibenzylated hexofuranoside, 303
    Mitsunobu reaction, 301–302, 302*f*
    primary amines, 302
    pyranosyl epoxide, 304, 305*f*
    pyrrolidine and piperidine, 304
  isonucleosides
    furan, 318–319, 319*f*
    nucleobases, 319, 320*f*
    nucleosides, 317–318
  overview, 297–298
  partial protection
    fructofuranoside, 300–301, 301*f*
    pyranoside and furanoside, 299–300, 300*f*
    sucrose, 300, 301*f*
  vinyl selenone-modified
    aryl selenoyl, 313
    aziridines, 314–315, 314*f*
    cyclopropanated, 316, 316*f*
    dihydrofuran, 316–317, 317*f*
    hexopyranoside, 315–316
    oxiraneand thiirane, 317, 318*f*
    phenyl selenide, 313–314
  vinyl sulfone-modified
    amino nucleophiles, 311–312
    cyclopropanated, 312–313, 313*f*
    magnesium monoperoxyphthalate (MMPP), 312
  zeolite catalysts, 297–298, 298*f*
D-glucal, 306, 307*f*
DHA. *See* Dihydroxy acetone dimer
Diastereomers, 3–4
di-*tert*-butylsilylene (DTBS), 241–242
Dichloroethane (DCE), 237–238
Didehydroazepanes synthesis, 197, 197*f*
1,6-dideoxy-1,6-imino-D/L-heptitols
  Aspergillus niger, 201–203
  cyclohex-2-ene-triols, 207
  D-Arabinose-derived alkene, 197
  deoxynorijimycin (DNJ), 200, 201*f*
  depezay synthesis, 195–196, 196*f*
  didehydroazepanes synthesis, 197, 197*f*

1,6-dideoxy-1,6-imino-D/L-heptitols (*Continued*)
dihydroxylation, 197–198
D-Ribose-derived derivative, 209–210, 210*f*
α-galactosidase, 198
Grubb's catalyst, 206–207
hemiacetal carbon, 205
hydroxymethyl unit, 197, 198*f*, 210–211, 211*f*
iminoheptitols, 200, 202*f*
oxazolidinyl azacycles, 203–204
piperidine-based isofagomine, 210–211
polyhydroxylated azepanes, 203
Di-D-fructose dianhydrides, 298–299, 299*f*
Diethyl aminosulfur trifluoride (DAST), 27–28, 275, 325–327
Dihydroxy acetone dimer (DHA), 712
Dihydroxylation, 352–353
Diisobutylaluminium hydride (DIBALH), 338–339
Dimethylimidazolinium chloride (DMC), 167–168, 169*f*
Dimethyl methylthio sulfonium trifluoro methanesulfonate (DMTST), 237
Dinitrosalicylate (DISAL), 86–87, 87*f*
Dipole effect, 4–5
D-lyxose, 344
DMC. *See* Dimethylimidazolinium chloride
DMTST. *See* Dimethyl methylthio sulfonium trifluoro methanesulfonate
Domino reactions in carbohydrate
butenolides and spiro compound, 701–708
chiral carbohydrate derivatives, 680–684
glycals in domino reaction, 691–698
glycosylation, 716–723
levoglucosenone, 709–712
macrocyclic synthesis, 713
organocatalyzed tandem reaction, 684–691
ethoxycarbonyl, 688
Huning's base, 689
lysosomal storage disorders (LSDs), 689
synthesis of *C*-glycosides of acetone, 685*f*
sugar ligands, 714–715
sugar ring contraction, 708–709
tandem reactions in carbohydrates, 698–701
alkoxyl radical fragmentation (ARF), 700–701
Bernet–Vasella reaction, 699
carbocyclic natural products, 700
fragmentation, 699–700
fused aromatic-carbohydrate, 700
Grubbs catalyst, 699–700
phenyliodosophthalate, 700–701

E

EDG. *See* Electron donating group
Edward-Lemieux effect, 4
Electrochemical O-glycosylation, 35
Electron donating group (EDG), 12
Electron-withdrawing group (EWG), 12
Enantioselective epoxidations
carbohydrate-based ketone catalysts, 625, 627*f*
competitive transition states, 621–625, 624*f*
D-fructose-derived ketone, 622*f*
favored transition states, 628–629, 630*f*
H_2O_2 and CH_3CN, 625, 626*f*
mechanism, 621, 623*f*
oxazolidinone ring, 625–626, 628*f*
Enterococcus faecium, 581–582
Epoxidation catalysts, 306, 307*f*
Epoxides derived
dibenzylated hexofuranoside, 303
Mitsunobu reaction, 301–302, 302*f*
primary amines, 302
pyranosyl epoxide, 304, 305*f*
pyrrolidine and piperidine, 304
EWG. *See* Electron-withdrawing group
Exo-anomeric effect, 620*f*

F

Ferrier azaglycosylation, 170–171
Ferrier method, 548–549
Ferrier-type glycosylations, 31–32
Fluoroazetidine iminosugars, 338–339
Forsythenethoside A synthesis, 412, 412*f*, 413*f*
Fructofuranoside, 300–301, 301*f*
Fructose, 320–322
Furanoid glycals, 393–402
C-aryl glycosides, 402*f*
gram-scale synthesis, 395
natural products, 396
Chiron approach synthesis, 396
Heck–Matsuda arylation, 396
nucleosides, 397–401
C-nucleosides, 397–398, 398*f*
Heck coupling, 398, 399*f*
iodo aglycon, 398
palladium-mediated coupling, 397–398, 399*f*

pyranoid glycals, 396, 396f
synthesis of, 393–396
threo- and erythro-furanoid glycals, 393–395
Furanoid glycals synthesis, 393–396
Furanoside, 299–300, 300f
Fused aromatic-carbohydrate, 700

G

GABAB, 705–706
Glycals in domino reaction, 691–698
benzopyrans, 697–698
Heck–Suzuki arylation, 692
N-methyl pyrrolidone (NMP), 696
Glycodiversification through enzymatic glycosylation
aldolase-mediated, 250–251
biocatalyst, 248
chemotherapeutic potential, 256
CMP-sialic acid donors, 256–258
equatorial hydroxyl, 254–255
fluorine atoms, 250
fluoro-sialoside, 252–254
gangliosides and glycoproteins, 256
GD3 and GT3, 256, 257f
hemagglutinin, 251, 253f
mammalian cell surface, 248
monophospho-nonulosonate, 251
monosaccharides, 248, 249f
monosialylated analog, 259
penultimate sugar, 250
PmNanA-catalyzed, 248–250
sialoconjugate decasaccharide, 256
sialosides, 251
structure-activity relationship (SAR), 259
Glycodiversification through modular triazole
azido-alkyne, 260, 261f
biosynthesis, 260–262
chemoenzymatic, 263
macrocyclization, 259–260, 260f
molecular recognition, 262–263
polar immobilization, 263–264
sialoconjugates, 262
Glycofuranoses
D-glucose and D-mannose, 289
ketalization, 289
pentofuranoses and hexofuranoses, 288–289
Glycohybrids synthesis, 589–604
acetylcholinesterase inhibitor glycohybrids, 600–601, 601f
antibacterial glycohybrids, 598–599
chitosan derivatives, 598–599, 599f
1,2,3-triazole glycoside clickamers, 598, 598f
anticancer glycohybrids, 590–597
carbohybrid containing thiazole-coumarin derivative, 596, 597f
chlorine e_6 glycohybrids, 594–596, 596f
isoxazole, 592–594, 595f
O-linked glycohybrids, 594, 596f
pyrazoline bridged indole C-glycoside hybrids, 592–594, 595f
triazolelinked N-glycopyranosides, 592–594, 594f
triazolethienopyrimidine glycohybrids, 591–592, 593f
antifungal glycohybrids, 603–604, 604f
antimalarial glycohybrids, 599–600, 600f
antiviral glycohybrids, 600, 601f
carbonic anhydrase inhibitor glycohybrids, 602–603, 603f
description of, 589–590
glycosidase inhibitor glycohybrids, 601–602, 602f
Glycomes, 2–3
Glycomics, 2–4
Glycopeptides, 523
arginine, 568–571
GlcNAcylated arginine, 568, 571f
N-linked rhamnosylated arginine, 568–570, 572f
silver-promoted solid phase glycosylation, 570–571, 573f
asparagine, 558–568, 559f
anomeric ratio, 560–561
asparagine-based glycopeptides, 566, 567f
auxiliary-based glycopeptide synthesis, 567–568
erythropoietin-related N-glycopeptide, 565–566, 566f
GlcNAc- Asn derivative, 564–565, 565f
glycosyl amine, 558
N-glycosyl Asparagine, 558, 559f, 560–561
N-linked Fmoc-Asn-OH building block, 558–560, 560f
proplylphosphonic acid cyclic anhydride, 561–562, 563f
urea moieties, 562–564, 564f
Glycosaminoglycans (GAGs), 289
Glycosidase inhibitor glycohybrids, 601–602, 602f

Glycosidase inhibitors
piperidine-based iminosugars
active-sitespecific pharmacological chaperone (ASSPC), 190–192
C-glycosides to azepane-sugars, 192–194
deoxynojirimycin (DNJ), 190
isofagomine (IFG), 190–192, 191*f*
pharmacological chaperones (PCs), 190–192
substrate reduction therapy (SRT), 190–192
Glycosyl acetates, 143–144, 158
peracetylated synthesis, 143–144, 145*f*
Vorbruggen classic method, 143, 145*f*
Glycosylation reactions
acceptor role, 40–41
glycosyl chlorides and bromides, 15–23
glycosyl fluorides, 25–30
glycosyl iodides, 23–24
glycosyl trichloroacetimidates, 35–39
thioglycosides, 30–35
Glycosylation ring-closing methods, 415
Glycosyl azides, 158–159, 159*f*
Glycosyl chlorides and bromides, 15–23
acetonitrile, 17–18
benzoylated glycosyl bromides activation, 20, 21*f*
bis (iodoimidazolium) complex, 22, 22*f*
boronic acid, 20–21
bromination reaction, 17
deoxy glycosyl bromides activation, 21*f*
iron (III) ferrocenium salts, 21–22, 22*f*
Koenigs–Knorr method, 15, 16*f*
lower shelf life, 19–20
moisture-lability, 19–20
preparation of, 18
room temperature ionic liquids (RTILs), 22–23
silver oxide (Ag_2O), 16, 16*f*
silver salt activator, 16
stereospecific glycoside product, 20
sugar hemiacetals, 17
Vilsmeier–Haack adduct, 19
Glycosyl donors
glycosyl chlorides and bromides, 15–23
acetonitrile, 17–18
benzoylated glycosyl bromides activation, 20, 21*f*
bis (iodoimidazolium) complex, 22, 22*f*
boronic acid, 20–21
bromination reaction, 17
deoxy glycosyl bromides activation, 21*f*
iron (III) ferrocenium salts, 21–22, 22*f*
Koenigs–Knorr method, 15, 16*f*
lower shelf life, 19–20
moisturc-lability, 19–20
preparation of, 18
room temperature ionic liquids (RTILs), 22–23
silver oxide (Ag_2O), 16, 16*f*
silver salt activator, 16
stereospecific glycoside product, 20
sugar hemiacetals, 17
Vilsmeier–Haack adduct, 19
glycosyl fluorides, 25–30
acid-labile iso-propylidiene, 26
activation of, 29*f*
bromine–fluorine exchange, 26–27
Mitsonobu-type reaction, 25, 26*f*
$SnCl_2$–$AgClO_4$ activator, 28–29
synthesis of, 25, 27*f*, 28*f*
two-step orthogonal activation, 29, 29*f*
glycosyl iodides, 23–24
molecular iodine, 24, 24*f*
polymer-bound imidazole, 24, 24*f*
glycosyl trichloroacetimidates, 35–39
acceptor role, 40*f*
activation of, 38–39
armed trichloroacetimidate, 38, 38*f*
hyaluronic acid formation, 37*f*
reactivity difference, 38–39, 39*f*
selective activation, 38*f*
stereochemistry, 36*f*
versatility of, 36
thioglycosides, 30–35
activation of, 32, 33*t*, 34*f*
bismuth catalysts, 33–34
chemoselective activation, 35, 35*f*
direct preparation, 30
light-induced activation, 34–35
ring-opening mechanism, 30, 31*f*
Glycosyl fluorides, 25–30
acid-labile iso-propylidiene, 26
activation of, 29*f*
bromine–fluorine exchange, 26–27
Mitsonobu-type reaction, 25, 26*f*
$SnCl_2$–$AgClO_4$ activator, 28–29
synthesis of, 25, 27*f*, 28*f*
two-step orthogonal activation, 29, 29*f*
Glycosyl halide, 145, 158–159
Glycosyl iodides, 23–24
molecular iodine, 24, 24*f*
polymer-bound imidazole, 24, 24*f*

Glycosyl *o*-(1-phenylvinyl) benzoate donor, 151–153
Glycosyl *ortho*-hexynylbenzoates, 149–150
Glycosyl trichloroacetimidates, 35–39, 160–165
Glycosyl (Z)-ynenoates, 150–151
Glycoterpenoid synthesis, 445–446, 445*f*, 446*f*
Gram-scale synthesis, 395
Grubb's catalyst, 206–207, 464–465, 465*f*
Grubb's II catalyst, 353

H

Halide glycosyl donors, 111–113
Hantzsch reaction, 680–681
Heck–Matsuda arylation, 396
Heck–Suzuki arylation, 692
Hexamethyldisilane (HMDS), 24
Hexofuranoses, 288–289
High-resolution mass spectrometry (HRMS), 549–550
Hoveyda–Grubbs catalyst, 217–218
HRMS. *See* High-resolution mass spectrometry (HRMS)
Hybrid molecule synthesis, *N*-linked glycopeptides, 577–588
Hydantocidin, 304–305, 306*f*
Hydroxymethylfurfural (HMF), 320
Hydroxyproline (HyP), 63–65
Hyperconjugative model, 5–6, 5*f*

I

IDCP. *See* Iodonium dicollidine perchlorate
Indolizidines, 354–357
Intramolecular aglycon delivery (IAD), 67–68, 68*f*
  anhydromannose donor, 84–85, 86*f*
  β-L-rhamnosides, 80, 80*f*
  β-mannosides, 81, 82*f*
  boronic ester, 84, 85*f*
  glycosyl acceptor, 75–76
  glycosyl donor, 75–76
  lipopolysaccharide (LPS), 84–85
  naphthylidene acetal, 80–81
  NAP-mediated IAD, 78–80, 79*f*
  NIS and DTBMP, stereoselective disaccharides, 76, 76*f*
  PMB-mediated IAD, 77–78, 78*f*
  silyl acetal-based IAD, 81
Iodonium dicollidine perchlorate (IDCP), 105–106
Ipomoeassin F synthesis, 415–425
  chemical structures, 415–416, 416*f*
  fucose-truncated monosaccharide, 420, 421*f*
  glycosyl acceptors, 418–419, 418*f*
  glycosyl donor, 418–419, 418*f*
  5-oxa/aza analogs, 424, 424*f*
  retrosynthetic analysis, 417–418, 417*f*, 422*f*
  side-chain esters, 421, 421*f*
Isofagomine (IFG), 190–192, 191*f*, 344
Isonucleosides
  furan, 318–319, 319*f*
  nucleobases, 319, 320*f*
  nucleosides, 317–318

K

Koenigs–Knorr method, 15, 16*f*

L

L-ascorbic acid (vitamin C), 323–324, 324*f*
Lawesson's reagent, 30
Leaving group approach
  anomeric center, 86, 87*f*
  boronic acid-derived intramolecular glycosylation, 87–88, 88*f*
  decarboxylative glycosylation, 85–86, 86*f*
  decarboxylative technique, 88
  dinitrosalicylate (DISAL), 86–87, 87*f*
  glucosyl carboxylates, 88–89, 89*f*
  pentenyl-type activation, 89, 90*f*
Levoglucosenone, 709–712
  dihydroxy acetone dimer (DHA), 712
  N-dimethylethylenediamine (DMEDA), 711
  Oxa-Michael-aldol, 709–710
  potassium isothiocyanate (KSCN), 712
Lewis acid, 357–358
Lewis acid catalysts, 158, 159*f*, 615–616
Lewis base catalysts, 615
L-fructose, 324, 325*f*
Lipopolysaccharides (LPSs), 230–231
LSDs. *See* Lysosomal storage disorders
L-Sorbose
  aldehyde, 325
  benzoyl chloride, 324–325
  D-fructopyranoside, 327–328
  diethylaminosulfur trifluoride (DAST), 325–327
  L-ascorbic acid (vitamin C), 323–324, 324*f*
  L-fructose, 324, 325*f*
  tautomers, 322–323, 323*f*
Lysosomal storage disorders (LSDs), 689

M

Macaranganin synthesis, 442–443, 443*f*
Macrocycles synthesis, 468–469, 469*f*
  C–C bond forming reactions, 464–473
    acetylene coupling reaction, 470–472
    aldol condensation reactions, 472–473
    macrocyclization, 472–473
    RCM strategy, 464–469, 465*f*
    Sonogashira and Heck reaction, 469–470
  click macrocyclizations, 482–514
    azobenzene containing glycomacrocycles synthesis, 507–509, 508*f*, 509*f*
    azobenzene-containing sugar-based macrocycles, 509–511, 510*f*, 511*f*
    cyclodextrin synthesis, 498–507, 499*f*, 500*f*, 503*f*
    cyclooligosaccharide derivatives, 512–514, 513*f*
    hexose sugar-embedded macrocycles, 489–496, 491*f*, 492*f*, 493*f*
    macrocyclization, 497–498, 498*f*
    multiple component click reactions, 482–483, 483*f*
    pentose sugar-based macrocycles, 484–488, 485*f*
    pyranose-based macrocycle libraries, 512, 512*f*
    sucrose-based macrocycle, 497
  glycosylation, 497
  intermolecular reductive amination, 475–476, 475*f*
  intramolecular reductive amination, 473–475, 473*f*, 474*f*
  macrolactamization, 476–478, 476*f*, 477*f*
  macrolactonization, 479–481, 480*f*, 481*f*
Macrocyclic synthesis, 713
Magnesium monoperoxyphthalate (MMPP), 312
Mallotusinin synthesis, 441–442, 442*f*
Mechanistic studies, oligosaccharide synthesis, 7–15
  active-latent glycosylation, 12, 12*f*
  anomeric selectivity, 8–9
  armed–disarmed approach, 10–11
  chemical glycosylation, 7–8
  glycosylation pathway, 7*f*, 8*f*
  nucleophilic attack, 7–9
  remote participation, 12–15, 13*f*
  step-wise S_N1 mechanism, 8, 8*f*
  two-step S_N1 type reaction protocol, 9
Merremoside D synthesis, 432–433, 434*f*, 435*f*
Methyl glycoside, 159
Mitsonobu-type reaction, 25, 26*f*
MMPP. *See* Magnesium monoperoxyphthalate
Molecular clamping approach, 54*f*, 55
  Birch reduction, 73
  bisphenol A (BPA), 65–66
  α-cyclodextrin synthesis, 58–60
  hydroxyproline (HyP), 63–65
  intramolecular aglycan delivery (IAD), 67–68, 68*f*
  intramolecular glycosylation, 56, 56*f*
  macrocycle, 73–75
  oligosaccharide synthesis, 62–63, 63*f*
  peptide template-based glycosidation, 62–63, 64*t*
  phthaloyl spacer, 58–60
  phthaloyl tether, 56–57
  trisaccharide synthesis, 60–61, 61*f*
  trisaccharide via remote glycosidation, 58, 58*f*
  xylene-tethered intermediate, 60
Monocyclic iminosugars
  four-membered iminosugars (azetidines), 336–339
  five-membered iminosugars (pyrrolidines), 341–342
  six-membered iminosugars (piperidines), 344–348
  seven-membered iminosugars (azepanes), 349–350
Monosaccharide-based crown ether derivatives, 447–450
  altrose based azacrown ethers, 447–448, 448*f*
  altrose-based chiral crown ether, 448–449, 449*f*
  α-D-galactose-based azacrown ether, 447–448, 447*f*
  D-mannitol-based monoaza-15-crown-5 type lariat ethers, 450, 450*f*
  glucose-based crown ether derivatives, 448, 449*f*
  L-threitol-based crown ethers, 448, 449*f*
  sugar-based chiral glucose crown ethers, 448–449, 449*f*
Morita–Baylis–Hillman reaction, 668–669, 669*f*
Morquio disease, 344
Murucoidins IV and V synthesis, 428–431
  disaccharide, 429, 430*f*
  macrolactonization., 430–431, 431*f*
  retrosynthetic analysis, 428–429, 429*f*
  trisaccharide, 429, 430*f*

N

National Research Council of the National Academy of Sciences, 2–3
Natural product-based glycomacrocycles, 408–446
  Ananatoside A synthesis, 413–415
    chemical structure, 413, 414*f*
    glycosylation ring-closing methods, 415
    macrolactonization, 413, 414*f*
    rhamnose containing macrodilatones, 415
  Batatin VI synthesis, 434–437
  Calysolin IX synthesis, 425–428, 426*f*, 427*f*
  casuarinin synthesis, 444–445, 444*f*
  Clemahexapetoside A synthesis, 408–411
  Clemoarmanoside A synthesis, 408–411
  Forsythenethoside A synthesis, 412, 412*f*, 413*f*
  glycoterpenoid synthesis, 445–446, 445*f*, 446*f*
  Ipomoeassin F synthesis, 415–425
    chemical structures, 415–416, 416*f*
    fucose-truncated monosaccharide, 420, 421*f*
    glycosyl acceptors, 418–419, 418*f*
    glycosyl donor, 418–419, 418*f*
    5-oxa/aza analogs, 424, 424*f*
    retrosynthetic analysis, 417–418, 417*f*, 422*f*
    side-chain esters, 421, 421*f*
  macaranganin synthesis, 442–443, 443*f*
  mallotusinin synthesis, 441–442, 442*f*
  Merremoside D synthesis, 432–433, 434*f*, 435*f*
  Murucoidins IV and V synthesis, 428–431
    disaccharide, 429, 430*f*
    macrolactonization., 430–431, 431*f*
    retrosynthetic analysis, 428–429, 429*f*
    trisaccharide, 429, 430*f*
  Neostrictinin synthesis, 439–440, 439*f*
  Phyllanemblinin B synthesis, 440–441, 440*f*
  steroid derived glycomacrocyles, 445–446
  sugar-based macrocycles, 408
    macrocyclic dilactones, 408, 410*f*
    structures of, 408, 409*f*
  tricolorin synthesis, 431–432, 433*f*
N-dimethylethylenediamine (DMEDA), 711
Neostrictinin synthesis, 439–440, 439*f*
N-glycosides
  amine group installation
    arginine glycosylation method, 180–181, 182*f*
    azaglycosylation, 177–178, 178*t*
    azide group, 170–173
    palladium catalyst, 180
    plausible mechanism, 181*f*
    sulfonamide, 173–175
  benzyl alcohol, 379–381
  excluding nucleosides
    direct replacement, 157
    glycosyl acetates, 158
    glycosyl azides, 158–159, 159*f*
    glycosyl halide, 158–159
    glycosyl trichloroacetimidate, 160–165
    methyl glycoside, 159
  Lewis acid, 379, 380*f*
  nucleosides via *N*-glycosylation
    alkynyl glycosyl carbonate, 153
    anhydroses, 146
    3,5-Dimethyl-4-(2-phenylethynylphenyl) phenyl glycoside donors, 153
    glycosyl acetates, 143–144
    glycosyl halide, 145
    glycosyl *o*-(1-phenylvinyl) benzoate donor, 151–153
    glycosyl *ortho*-hexynylbenzoates, 149–150
    glycosyl (Z)-ynenoates, 150–151
    8-(methyltosylaminoethynyl)-1-naphthyl glycosides, 157
    N-phenyl trifluoroacetimidate glycosides, 153–157
    pentenyl glycosides, 146–147
    propargyl-1,2-orthoesters, 147–148
    thioglycosides, 146
  palladium-catalyzed, 379, 380*f*
  pyranoid glycals, 379–381
  sigmatropic rearrangement, 379–381, 380*f*
  synthesis of, 379–381
    aminosugars, 139
    chemical synthesis of nucleosides, 141–181
    nucleosides and derivatives, 140–141
  triphenylphosphine, 379
N-glycosylation
  alkynyl glycosyl carbonate, 153
  anhydroses, 146
  3,5-dimethyl-4-(2-phenylethynylphenyl) phenyl glycoside donors, 153
  glycosyl acetates, 143–144
  glycosyl halide, 145
  glycosyl *o*-(1-phenylvinyl) benzoate donor, 151–153
  glycosyl ortho-hexynylbenzoates, 149–150

N-glycosylation (*Continued*)
glycosyl (Z)-ynenoates, 150–151
8-(methyltosylaminoethynyl)-1-naphthyl glycosides, 157
N-phenyl trifluoroacetimidate glycosides, 153–157
nucleophilic attack, 142, 143*f*
pentenyl glycosides, 146–147
propargyl-1,2-orthoesters, 147–148
synthesis of nucleosides, 142–143, 144*f*
thioglycosides, 146
N-heterocyclic glycosides, 178–179
NiAAC. *See* Nickel-catalyzed alkyne–azide cycloaddition
Nickel-catalyzed alkyne–azide cycloaddition (NiAAC), 571–574
N-iodosuccinimide (NIS), 103, 146–147, 235–236, 239–240
N-linked glycopeptides synthesis, 558–588
click chemistry glycosylation, 571–577
nickel-catalyzed alkyne–azide cycloaddition (NiAAC), 571–574
1,2,3-triazole-linked glycopeptide, 574
triazole ring, 571–574
glycopeptides containing arginine, 568–571
GlcNAcylated arginine, 568, 571*f*
N-linked rhamnosylated arginine, 568–570, 572*f*
silver-promoted solid phase glycosylation, 570–571, 573*f*
glycopeptides containing asparagine, 558–568, 559*f*
anomeric ratio, 560–561
asparagine-based glycopeptides, 566, 567*f*
auxiliary-based glycopeptide synthesis, 567–568
erythropoietin-related *N*-glycopeptide, 565–566, 566*f*
GlcNAc- Asn derivative, 564–565, 565*f*
glycosyl amine, 558
N-glycosyl Asparagine, 558, 559*f*, 560–561
N-linked Fmoc-Asn-OH building block, 558–560, 560*f*
proplylphosphonic acid cyclic anhydride, 561–562, 563*f*
urea moieties, 562–564, 564*f*
hybrid molecule synthesis, 577–588
azithromycin-derived glycopeptides, 581–582, 584*f*
glycopeptide-azithromycin hybrid, 581–582
HIV-1 V1V2 glycopeptides, 582, 586*f*
teicoplanin, 586–588, 590*f*
vancomycin analogs, 586, 588*f*
vancomycin glycopeptide antibiotic, 577–581
N-methyl pyrrolidone (NMP), 696
Nortropanes, 360–361
N-phenyl trifluoroacetimidate donors, 116
N-phenyl trifluoroacetimidate glycosides, 153–157
oxocarbenium ion, 156–157
phenyl trifluoroacetimidate (PTFAI), 153
N-protecting groups, 282–283
N-trifluoroacetyl (*N*-TFA), 239, 240*f*
Nucleosides
chemical synthesis
amine group installation, 168–181
hydroxyl group of sugar, 165–168
N-glycosides excluding nucleosides, 157–165
nucleosides via *N*-glycosylation, 142–157
furanoid glycals, 397–401
C-nucleosides, 397–398, 398*f*
Heck coupling, 398, 399*f*
iodo aglycon, 398
palladium-mediated coupling, 397–398, 399*f*
glycogen phosphorylase, 140–141
pyranoid glycals, 377–379
glycals and unactivated nucleobase, 377–378, 378*f*
natural antibiotics, 377, 378*f*
structural composition, 140, 141*f*
synthesis of, 377–379, 397–401
types of, 140

O

O-glycosides, 383*f*
C2-hydroxyglucopyranosides, 381
2-deoxyglycosides from glycals, 385–386, 386*f*
Ferrier-type O-glycosylation, 383
glycosyl acceptors, 385
2-hydroxy-α-mannopyranoside from glycals, 381, 381*f*
4-(*N*-mesylamino)-2,3-unsaturated α-O-glycoside synthesis, 382–383, 383*f*
synthesis of, 381–386
2-unsaturated α-O-glycosides from vinyl oxirane, 381–382, 382*f*

Oligosaccharides
  anomeric effect, 4–6
  classical approaches, 6–7
  continuum in the mechanistic pathway, 8*f*
  de novo asymmetric synthesis, 286–287
  mechanistic pathway, 7*f*
  mechanistic studies, 7–15
    active-latent glycosylation, 12, 12*f*
    anomeric selectivity, 8–9
    armed–disarmed approach, 10–11
    chemical glycosylation, 7–8
    glycosylation pathway, 7*f*, 8*f*
    nucleophilic attack, 7–9
    remote participation, 12–15, 13*f*
    step-wise S_N1 mechanism, 8, 8*f*
    two-step S_N1 type reaction protocol, 9
  N-protecting groups, 282
O-linked glycopeptides synthesis, 524–544, 525*f*
  antifreeze glycoproteins (AFGPs), 533–535
  ball-mill method, 536–537, 537*f*
  bivalent O-linked glycopeptide, 527, 531*f*
  carbonyl diimidazole, 536–537
  α-dystroglycan, 529–531, 534*f*
  Fmoc-β^3-homo-threonine, 527–529, 532*f*
  glycopeptide amphiphilic dendrimers, 542–544
  hydroxyproline-derived O-glycosylated building block, 533–535, 535*f*
  I_2-promoted O-glycosylated amino acid, 531, 535*f*
  kinetically controlled α-selective method, 537, 538*f*
  Koenigs–Knorr reactions, 524–526
  O-benzyl-protected galactosyl bromide, 526–527
  α-phenyl-D-glucosaminide peracetate, 539–540
  (2,6)-sialyl-T glycosylated MUC1 eicosapeptide, 526–527, 529*f*
  solid-phase glycopeptide synthesis (SPGS), 526–527
One-pot multienzyme (OPME) method, 718–719
Organocatalysts, 629–639
  Aldol and Michael reactions, 629–639
  Brønsted acid catalysts, 617–619
  Brønsted base catalysts, 616–617
  carbohydrates
    amine functionality, 665–668
    amino alcohols, 640–645
    bifunctional thiourea/urea organocatalysts, 668–671
    enantioselective epoxidation, 620–629
    exo-anomeric effect, 620*f*
    ketones, 620–629
    prolinamides, 629–639
    pyrrolidine amides, 629–639
    pyrrolidines, 629–639
    ring (steric effect) and anomeric effect, 619–620
    structure of, 619
    sugar-pseudoenantiomer pairs, 619, 620*f*
    thiourea-primary amines, 646–652
    thiourea-secondary amines, 653–655
    thiourea-tertiary amines, 655–665
  history of, 612–619
  isolated enzyme emulsion, 612–614
  Lewis acid catalysts, 615–616
  Lewis base catalysts, 615
  metal and enzyme catalysts, 612–614
  quinine and quinidine catalysis, 612–614, 648*f*
  Von Liebig's oxamide synthesis, 648*f*
Organocatalyzed tandem reaction, 684–691
  ethoxycarbonyl, 688
  Horner–Emmons–Wordsworth, 686
  Huning's base, 689
  lysosomal storage disorders (LSDs), 689
  synthesis of *C*-glycosides of acetone, 685*f*
Orthogonal glycosylations
  halide glycosyl donors, 111–113
    heptaglucoside, 111–112
    S-dimethyl phenyl (SDMP), 112–113
    type II LacNAc-repeating oligosaccharides, 112, 113*f*
  N-phenyltrifluoroacetimidate donors, 116
  phenylpropiolate glycosyl (PPG), 119
  streptococcus pneumoniae disaccharide, 116
  thioglycoside acceptors, 113–116
  thioglycoside acceptors with glycosyl donors, 116–119
  thioglycoside-based one-pot, 119–122
    cistanche tubulosa, 120–121
    glycosyl phosphate donor, 119, 120*f*
    perbenzoylated, 119
    thioglycoside disaccharide, 122
  trichloroacetimidate donors, 113–116
Oxa-Michael-aldol, 709–710

P

Palladium catalyst, 179–180, 180*f*
Partial protection of D-fructoses
  fructofuranoside, 300–301, 301*f*
  pyranoside and furanoside, 299–300, 300*f*
  sucrose, 300, 301*f*
PCC. *See* Pyridinium chlorochromate
Pentaatomic cyclic nitrones, 350–351
Pentofuranoses, 288–289
Perbenzoylated glycoside, 147–148
Pharmacological chaperones (PCs), 190–192
Phenyliodine (III) bis (trifluoroacetate) (PIFA), 33
Phenyliodosophthalate, 700–701
Phenylpropiolate glycosyl (PPG), 119
Phenyl trifluoroacetimidate (PTFAI), 153
Phosphoenolpyruvate (PEP), 232
Phyllanemblinin B synthesis, 440–441, 440*f*
Piperidine-based iminosugars
  active-sitespecific pharmacological chaperone (ASSPC), 190–192
  C-glycosides to azepane-sugars, 192–194
  deoxynojirimycin (DNJ), 190
  isofagomine (IFG), 190–192, 191*f*
  pharmacological chaperones (PCs), 190–192
  substrate reduction therapy (SRT), 190–192
Piperidines
  anomeric oxygen deprotection, 348
  benzyl groups, 344
  cis-hydroxy group, 348
  D-lyxose, 344
  isofagomine (IFG), 344
  monocyclic iminosugars, 344–348
  one-pot iminium-ion-based strategy, 344–345
  pseudoaxial approach, 344–345
p-methoxybenzylated (PMB), 106
Polyhydroxyazepanes
  C-glycosides
    adenophorine, 216, 216*f*
    (−)- adenophorine, 216–217, 217*f*
    amino cyclization, 213
    azepanes, 216, 216*f*
    azidolactol precursor, 217, 217*f*
    D-gulose, 214–216, 215*f*
    Hoveyda–Grubbs catalyst, 217–218
    N-alkylated azepanes, 218–220, 219*f*
    N-methylated and *N*-butylated, 213
    pyridinium chlorochromate (PCC), 211–212
  1,6-dideoxy-1,6-imino-D/L-heptitols
    aspergillus niger, 201–203
    cyclohex-2-ene-triols, 207
    D-Arabinose-derived alkene, 197
    deoxynorijimycin (DNJ), 200, 201*f*
    didehydroazepanes epoxidation, 198, 199*f*
    dihydroxylation, 197–198
    D-ribose-derived derivative, 209–210, 210*f*
    α-galactosidase, 198
    Grubb's catalyst, 206–207
    hemiacetal carbon, 205
    hydroxymethyl unit, 197, 198*f*, 210–211, 211*f*
    iminoheptitols, 200, 202*f*
    oxazolidinyl azacycles, 203–204
    piperidine-based isofagomine, 210–211
    polyhydroxylated azepanes, 203
    (−)-Shikimic acid–derived, 208–209, 209*f*
    tertiary amide, 205, 207*f*
    tetrahydroxyazepanes, 200, 201*f*
    trans-orientation, 198–200, 200*f*
  glycosidase inhibitors, 194
  nomenclature and numbering, 194–195
  tetrahydroxyazepane, 194–195, 195*f*
Posttranslational method, 546–547
Potassium isothiocyanate (KSCN), 712
PPG. *See* Phenylpropiolate glycosyl
Preactivation-based glycosylation strategies
  AgOTf/p-ArSCl promoter system, 123–127
    benzenesulfinyl piperidine (BSP), 126
    chondroitin sulfate glycopeptide, 123–126
    thiogalactoside acceptor, 126
  one-pot glycosylations, 128–131
  preactivation-based one-pot
    DMF, 128, 129*f*
    hexasaccharide, 128–129, 129*f*
    preactivation-based glycosylation, 129–130, 130*f*
    thiotolyl glycoside donors, 128
Prolinamide catalysts, 632–634, 633*f*
Protection-deprotection strategies, carbohydrates
  acyl migrations, 286
  anomeric position, 277–279
    carbohydrates, 279, 280*f*
    orthoesters, 278–279
    regioselective esterification, 278
    small methyl, 278

thexyl dimethyl silyl (TDS), 279
thioglycosides, 279
trifluoroacetyl (TFA), 278
benzylidene acetals, 279–282
cyclic protecting group, 279–282
de novo asymmetric synthesis, 286–287
glycofuranoses, 288–289
D-glucose and D-mannose, 289
ketalization, 289
pentofuranoses and hexofuranoses, 288–289
N-protecting groups, 282–283
regioselective protection
acyl group, 285–286
catalyzed protocols, 284–285
galactopyranose, 284
O-benzylidene, 284
silyl glycosides, 284
tetra-butyl ammonium fluoride (TBAF), 286
TMS glucosides, 284–285, 285*f*
trityl chloride, 283–284
saccharides, 273–276
secondary hydroxyl groups, 276–277
acetates and benzoates, 276–277
anomeric position, 276
axial and equatorial, 277
protection and deprotection, 276, 277*t*
silyl groups, 276
sialic acid derivatives, 288
sulfated glycosaminoglycans, 289
glycosaminoglycans (GAGs), 289
heparin, 289, 290*f*
proteoglycans (PGs), 289
Protein folding, 2
Proteoglycans (PGs), 289
Pseudoaxial approach, 344–345
PTFAI. *See* Phenyl trifluoroacetimidate
Pummerer reaction-mediated glycosylation, 428–433, 429*f*
Pyranoid glycals
aryl *C*-glycosides, 367–373
C2 functionalization, 386–389
C-glycosides, 373–377
glycals are stable donors, 366–367
N-glycosides, 379–381
benzyl alcohol, 379–381
Lewis acid, 379, 380*f*
palladium-catalyzed, 379, 380*f*
sigmatropic rearrangement, 379–381, 380*f*
triphenylphosphine, 379
nucleosides, 377–379
glycals and unactivated nucleobase, 377–378, 378*f*
natural antibiotics, 377, 378*f*
O-glycosides, 383*f*
C2-hydroxyglucopyranosides, 381
2-deoxyglycosides from glycals, 385–386, 386*f*
Ferrier-type O-glycosylation, 383
glycosyl acceptors, 385
2-hydroxy-α-mannopyranoside from glycals, 381, 381*f*
4-(*N*-mesylamino)-2,3-unsaturated α-O-glycoside synthesis, 382–383, 383*f*
2-unsaturated α-O-glycosides from vinyl oxirane, 381–382, 382*f*
one-pot C1 and C2 functionalization, 389–393
Pyranoid glycals synthesis, 366–367
Pyranoside, 299–300, 300*f*
Pyridinium chlorochromate (PCC), 211–212, 701–702
Pyrrolidines
carbohydrates, organocatalysts, 629–639
1,4-dideoxy-imino-D-arabinitol (DAB), 340–341
metal-free Overman rearrangement, 343–344
N-ethyl hydrazide imide, 340–341
pyrroline-type iminosugars, 340
tetrahydroxylated, 341–342
Pyrrolizidines, 350–353
dihydroxylation, 352–353
Grubb's II catalyst, 353
hydrogen peroxide oxidation, 352–353
hydroxy groups, 350–351
organometal, 353
pentaatomic cyclic nitrones, 350–351
stereodivergent synthesis, 351
steric and stereoelectronic, 351
vicinal benzyloxy, 351

Q

Quinine and quinidine catalysis, 612–614, 648*f*
Quinolizidines, 357–360
diastereoisomeric, 357–358
Lewis acid, 357–358
pyrrolizidines and indolizidines, 357

R

Rabbit ear effect, 4–5
RCM. *See* Ring-closing metathesis
Reduction of the resulting olefin (RCM-Re), 465–466
Regioselective protection of carbohydrates
  acyl group, 285–286
  catalyzed protocols, 284–285
  galactopyranose, 284
  O-benzylidene, 284
  silyl glycosides, 284
  tetra-butyl ammonium fluoride (TBAF), 286
  TMS glucosides, 284–285, 285*f*
  trityl chloride, 283–284
Relative reactivity values (RRVs), 105–106, 717–718
Remote participation
  acyl groups, 13*f*, 14–15
  C6-acyl group, 15*f*
  nonparticipating protecting group, 13
Ring-closing metathesis (RCM), 417–418, 464–469, 465*f*
  cyclization reaction, 467–468
  D-Glucose-derived macrocycles, 467–468, 467*f*
  Grubbs catalysts, 464–465, 465*f*
  macrocycles synthesis, 468–469, 469*f*
  14-membered ring macrolides, 466*f*
  sucrose-based macrocyclic compounds, 468, 468*f*
  sugar-based macrocycles, 465–466, 465*f*
Room temperature ionic liquids (RTILs), 22–23, 308–309, 309*f*
RRVs. *See* Relative reactivity values
Ruthenium (III), 172–173

S

Saccharides
  acetic anhydride, 275–276
  diethylaminosulfur trifluoride (DAST), 275
  hydroxyl group, 273–274
  pivaloate group, 274
  tert-butyldimethylsilyl (TBDMS), 274
  tosylation, 274
  triethylsilyl (TES), 274
  triisopropylsilyl (TIPS), 274
  trimethylsilyl (TMS), 274
S-benzoxazolyl (SBox), 106
S-dimethyl phenyl (SDMP), 112–113
Secondary hydroxyl groups
  acetates and benzoates, 276–277
  anomeric position, 276
  axial and equatorial, 277
  protection and deprotection, 276, 277*t*
  silyl groups, 276
Sequential glycosylation, 10–11
Sialic acids, 288
  carbohydrate motifs, 227–228
  characteristics, 228–234
    enzymatic glycosylation, 232–233
    enzymatic synthesis, 231
    eukaryotic and prokaryotic, 232
    lipooligosaccharides, 228–230
    lipopolysaccharides (LPSs), 230–231
    metabolism, 232
    phosphoenolpyruvate (PEP), 232
    terminal sugar, 230
  chemical synthesis and glycosylation
    acetonitrile intermediate, 238–239
    anomeric selectivity, 234–235
    C3-phenoxythiocarbonyloxy, 245
    C3-SPh substitution, 245, 246*f*
    dichloroethane (DCE), 237–238
    dimethyl methylthio sulfonium trifluoro methanesulfonate (DMTST), 237
    di-tert-butylsilylene (DTBS), 241–242
    methylumbelliferyl glycosides, 246–247
    N-iodosuccinimide (NIS), 235–236, 239–240
    4-nitropicoloyl-protected donor, 240*f*
    N-phenyltrifluoroacetimidate, 245, 245*f*
    N-trifluoroacetyl (*N*-TFA), 239, 240*f*
    N-Troc-protected, 239
    oxacarbenium, 235
    oxazolidinone, 241
    S-adamantyl group in donor, 242*f*
    α-sialosides, 234
    stereoselective sialylation, 242–244
    thioglycoside donor, 244
    2-thiomethyl glycosides of Neu5Ac, 241
    thiosialosides, 238, 238*f*
    xanthate donors, 242
  chiral molecule, 227–228
  glycodiversification through enzymatic glycosylation, 248–259
    aldolase-mediated, 250–251
    biocatalyst, 248
    chemotherapeutic potential, 256
    CMP-sialic acid donors, 256–258
    equatorial hydroxyl, 254–255
    fluorine atoms, 250

fluoro-sialoside, 252–254
gangliosides and glycoproteins, 256
GD3 and GT3, 256, 257*f*
hemagglutinin, 251, 253*f*
mammalian cell surface, 248
monophospho-nonulosonate, 251
monosaccharides, 248, 249*f*
monosialylated analog, 259
penultimate sugar, 250
PmNanA-catalyzed, 248–250
sialic acids-containing glycans, 251–252, 254*f*
sialoconjugate decasaccharide, 256
sialosides, 251
structure-activity relationship (SAR), 259
glycodiversification through modular triazole, 259–264
azido-alkyne, 260, 261*f*
biosynthesis, 260–262
chemoenzymatic, 263
macrocyclization, 259–260, 260*f*
molecular recognition, 262–263
polar immobilization, 263–264
sialoconjugates, 262
monosaccharides, 227–228
sialic acids, 234
S-linked glycopeptides synthesis, 545–551
bovine serum albumin (BSR), 549
diphenylalanine scaffolds, 547–548, 548*f*
Ferrier method, 548–549
high-resolution mass spectrometry (HRMS), 549–550
posttranslational method, 546–547
Tamm-Horsfall glycoprotein, 547
thioglycopeptide, 550, 551*f*
visible-light-mediated S-linked glycopeptide, 550, 551*f*
Solid-phase glycopeptide synthesis (SPGS), 526–527
Solid-phase oligosaccharide synthesis (SPOS), 41–42, 289–291
Solvent effect, 5–6
Sonogashira and Heck reaction, 469–470
SPOS. *See* Solid-phase oligosaccharide synthesis
SRT. *See* Substrate reduction therapy (SRT)
Stereoselective intramolecular *O*-glycosylation
glycosyl acceptor, 53–54
intramolecular aglycon delivery (IAD), 75–85
anhydromannose donor, 84–85, 86*f*
β-L-rhamnosides, 80, 80*f*
β-mannosides, 81, 82*f*
boronic ester, 84, 85*f*
dimethylketal-tethered intermediates, 75*f*
glycosyl acceptor, 75–76
glycosyl donor, 75–76
lipopolysaccharide (LPS), 84–85
naphthylidene acetal, 80–81
NAP-mediated IAD, 78–80, 79*f*
NIS and DTBMP, stereoselective disaccharides, 76, 76*f*
PMB-mediated IAD, 77–78, 78*f*
silyl acetal-based IAD, 81
leaving group approach, 85–90
anomeric center, 86, 87*f*
boronate ester attacked, 87–88
boronic acid-derived intramolecular glycosylation, 87–88, 88*f*
decarboxylative glycosylation, 85–86, 86*f*
decarboxylative technique, 88
dinitrosalicylate (DISAL), 86–87, 87*f*
glucosyl carboxylates, 88–89, 89*f*
pentenyl-type activation, 89, 90*f*
molecular clamping approach, 54*f*, 55–75
Birch Reduction, 73
bisphenol A (BPA), 65–66
α-cyclodextrin synthesis, 58–60
hydroxyproline (HyP), 63–65
intramolecular aglycan delivery (IAD), 67–68, 68*f*
intramolecular glycosylation, 56, 56*f*
macrocycle, 73–75
oligosaccharide synthesis, 62–63, 63*f*
peptide template-based glycosidation, 62–63, 64*t*
phthaloyl spacer, 58–60
phthaloyl tether, 56–57
trisaccharide synthesis, 60–61, 61*f*
trisaccharide via remote glycosidation, 58, 58*f*
xylene-tethered intermediate, 60
Stereoselectivity, 3–4
Steric effect, 619–620
Steroid derived glycomacrocyles, 445–446
Streptococcus pneumoniae, 581–582
Structure-activity relationship (SAR), 259
Substrate reduction therapy (SRT), 190–192
Sucrose, 300, 301*f*
Sucrose-based crown ether, 457–463
cyclotriveratrylene sucrose-based capsules, 462–463, 462*f*
fluorescent sucrose molecular cages, 463, 464*f*

Sucrose-based crown ether (*Continued*)
selective silyation, 457
sucrose based aza compounds, 457–458
sucrose based-macrocyclic compound, 457, 458*f*
sucrose based molecular cages, 463, 463*f*
sucrose containing cryptands, 460–462, 460*f*, 461*f*
sucrose-containing macrocycles, 457
sulfur containing macrocycles, 458–460
visible light responsive sucrose-containing macrocycles, 460, 461*f*
Sucrose-based macrocycles synthesis, 497
Sugar-based macrocycles, 408
Sugar-based prolinamides, 629–631, 631*f*
Sugar-fused macrocyclic lactones synthesis, 469–470
Sugar ligands, 714–715
Sugar ring contraction, 708–709
Sulfated glycosaminoglycans
glycosaminoglycans (GAGs), 289
heparin, 289, 290*f*
proteoglycans (PGs), 289
Sulfonamide
acetylated glucal, 174
aliphatic and aromatic, 174–175
azaglycosylation, 174
enopyranoside II, 175
glycoconjugates, 173, 173*t*
Sulfonamidoglycosides, methyl glycoside, 160*f*
Superarming, 10–11, 11*f*

T

Tandem reactions in carbohydrates, 698–701
alkoxyl radical fragmentation (ARF), 700–701
Bernet–Vasella reaction, 699
carbocyclic natural products, 700
fragmentation, 699–700
fused aromatic-carbohydrate, 700
Grubbs catalyst, 699–700
phenyliodosophthalate, 700–701
TDS. *See* Thexyl dimethyl silyl
tert-butyl carbamate, 176–177
tert-butyldimethylsilyl (TBDMS), 274, 702
tert-butyldimethylsilyl (TBS) ether, 538–539
Tetra-butyl ammonium fluoride (TBAF), 286
Tetrahydroxyazepane, 194–195, 195*f*
Thexyl dimethyl silyl (TDS), 279
Thioglycoside acceptors, 113–116
Thioglycoside-based glycosylations
activation of thioglycoside donors, 100–104
allyl and pentenyl, 103–104
halonium activators, 103
N-iodosuccinimide (NIS), 103
perbenzylated thioglucoside, 100–103
versatile donors, 100
chemoselective glycosylations, 104–110
one-pot, 108–110
polymerization, 104–105
type A, 105–106
type B, 106–107
type C, 107–108
microheterogeneous, 95
one-pot synthesis, 99–100
orthogonal glycosylations, 111–122
halide glycosyl donors, 111–113
N-phenyltrifluoroacetimidate donors, 116
other glycosyl donors, 116–119
thioglycoside-based one-pot, 119–122
trichloroacetimidate donors and thioglycoside acceptors, 113–116
peracetylated sugars, 96–98, 97*t*
preactivation-based glycosylation strategies, 123–131
AgOTf/p-ArSCl promoter system, 123–127
one-pot glycosylations, 128–131
preparation, 96–100
Thioglycoside-based one-pot orthogonal glycosylations
glycosyl phosphate donor, 119, 120*f*
ortho-(methylthio) benzyl thioaglycon, 121–122, 121*f*
perbenzoylated, 119
thioglycoside disaccharide, 122
type I LacNAc-repeating octasaccharide, 122, 123*f*
Thioglycoside donor activation
allyl and pentenyl, 103–104
halonium activators, 103
N-iodosuccinimide (NIS), 103
perbenzylated thioglucoside, 100–103
versatile donors, 100
Thioglycosides, 279
activation of, 32, 33*t*, 34*f*
bismuth catalysts, 33–34
chemoselective activation, 35, 35*f*
direct preparation, 30
light-induced activation, 34–35
ring-opening mechanism, 30, 31*f*
Thiourea-primary amines, 646–652

Thiourea-secondary amines, 653–655
Thiourea-tertiary amines, 655–665
Trichloroacetimidate donors, 113–116
Tricolorin synthesis, 431–432, 433*f*
Triethylsilyl (TES), 274
Trifluoroacetyl (TFA), 278
Triisopropylsilyl (TIPS), 274
Trimethoxyphenyl phosphine (TTMPP), 164
Trimethylsilyl (TMS), 146, 274
Trimethylsilyl iodide (TMSI), 23
Trisaccharide synthesis, 60–61, 61*f*

V

Vancomycin-resistant enterococci (VRE) bacteria, 577–581
Vicinal benzyloxy, 351
Vilsmeier–Haack adduct, 19
Vinyl selenone-modified
 aryl selenoyl, 313
 aziridines, 314–315, 314*f*
 cyclopropanated, 316, 316*f*
 dihydrofuran, 316–317, 317*f*
 hexopyranoside, 315–316
 oxiraneand thiirane, 317, 318*f*
 phenyl selenide, 313–314
Vinyl sugar enol ethers
 furanoid glycals, 393–402
  C-aryl glycosides, 402*f*
  natural products, 396
  nucleosides, 397–401
  pyranoid glycals, 396, 396*f*
  synthesis of, 393–396
  threo- and erythro-furanoid glycals, 393–395
 overview, 365–366
 pyranoid glycals
  aryl *C*-glycosides, 367–373
  C2 functionalization, 386–389
  C-glycosides, 373–377
  glycals are stable donors, 366–367
  N-glycosides, 379–381
  nucleosides, 377–379
  one-pot C1 and C2 functionalization, 389–393
  synthesis of, 366–367
Vinyl sulfone-modified
 amino nucleophiles, 311–312
 cyclopropanated, 312–313, 313*f*
 magnesium monoperoxyphthalate (MMPP), 312
Von Liebig's oxamide synthesis, 648*f*
Vorbruggen classic method, 143, 145*f*

Z

Zeolite catalysts, 297–298, 298*f*
Zinc borohydride, 356–357

Printed in the United States
by Baker & Taylor Publisher Services